Springer Monographs in Mathematics

More information about this series at http://www.springer.com/series/3733

Michel L. Lapidus
Goran Radunović
Darko Žubrinić

Fractal Zeta Functions and Fractal Drums

Higher-Dimensional Theory of Complex Dimensions

With 52 illustrations

Springer

Michel L. Lapidus
Department of Mathematics
University of California
Riverside, California, USA

Goran Radunović
Department of Applied Mathematics
University of Zagreb
Zagreb, Croatia

Darko Žubrinić
Department of Applied Mathematics
University of Zagreb
Zagreb, Croatia

ISSN 1439-7382 ISSN 2196-9922 (electronic)
Springer Monographs in Mathematics
ISBN 978-3-319-44704-9 ISBN 978-3-319-44706-3 (eBook)
DOI 10.1007/978-3-319-44706-3

Library of Congress Control Number: 2016954237

Mathematics Subject Classification (2010): Primary – 11M41, 28A12, 28A75, 28A80, 28B15, 30D10, 35P20, 42B20, 44A05, 58J32; Secondary – 11M06, 30D30, 37C30, 37C45, 40A10, 42B35, 42B37, 44A10, 45Q05, 81Q20

Printed on acid-free paper

This Springer imprint is published by Springer Nature
The registered company is Springer International Publishing AG
The registered company address is: Gewerbestrasse 11, 6330 Cham, Switzerland

To my wife, Odile, with infinite love
To my beautiful and extraordinary children
Julie Anne Myriam and Michael Alex Serge

Michel L. Lapidus

To my parents
Nevenka née Bilić and Ranko

Goran Radunović

To the memory of my parents
Katarina née Suntešić and Velimir

Darko Žubrinić

Overview

Recently, the first author has extended the definition of the zeta function associated with fractal strings to arbitrary bounded subsets A of the N-dimensional Euclidean space $\mathbb{R}^N$, where N is any integer ≥ 1. It is defined by

$$\zeta_A(s) = \int_{A_\delta} d(x,A)^{s-N} \mathrm{d}x,$$

where $d(x,A)$ denotes the distance from x to A and A_δ is a δ-neighborhood of A. In this monograph, we investigate various properties of this "distance zeta function". In particular, we prove that the zeta function is holomorphic in the half-plane $\{\mathrm{Re}\, s > \overline{\dim}_B A\}$, and that under mild hypotheses, the bound $\overline{\dim}_B A$ is optimal. Furthermore, we show that the abscissa of convergence of ζ_A is always equal to $\overline{\dim}_B A$, which generalizes to arbitrary dimensions a well-known result for fractal strings (or equivalently, for arbitrary compact subsets of the real line $\mathbb{R}$). Here, $\overline{\dim}_B A$ denotes the upper box (or Minkowski) dimension of A. Extended to a meromorphic function ζ_A, this "distance zeta function" is shown to be an efficient tool for finding the box dimension of several new classes of subsets of $\mathbb{R}^N$, like fractal nests, geometric chirps and multiple string chirps. It is also used to develop a higher-dimensional theory of complex dimensions of arbitrary fractal sets in Euclidean spaces.

For the sake of simplicity, we pay particular attention in this monograph to the principal complex dimensions of A, defined as the poles of ζ_A located on the "critical line" $\{\mathrm{Re}\, s = \overline{\dim}_B A\}$. We also introduce a new zeta function, denoted by $\tilde{\zeta}_A$ and called a "tube zeta function", and show, in particular, how to calculate the Minkowski content of a suitable (Minkowski measurable) bounded set A in $\mathbb{R}^N$ in terms of the residue of $\tilde{\zeta}_A(s)$ at $s = \dim_B A$, the box dimension of A. More generally, without assuming that A is Minkowski measurable, we obtain analogous results, but now expressed as inequalities involving the upper and lower Minkowski contents of A. In addition, we obtain a new class of harmonic functions generated by fractal sets and represented via singular integrals. Furthermore, a class of sets is constructed with unequal upper and lower box dimensions, possessing alternating

Overview

Recently, the first author has extended the definition of the zeta function associated with fractal strings to arbitrary bounded subsets A of the N-dimensional Euclidean space $\mathbb{R}^N$, where N is any integer ≥ 1. It is defined by

$$\zeta_A(s) = \int_{A_\delta} d(x,A)^{s-N} \mathrm{d}x,$$

where $d(x,A)$ denotes the distance from x to A and A_δ is a δ-neighborhood of A. In this monograph, we investigate various properties of this "distance zeta function". In particular, we prove that the zeta function is holomorphic in the half-plane $\{\mathrm{Re}\, s > \overline{\dim}_B A\}$, and that under mild hypotheses, the bound $\overline{\dim}_B A$ is optimal. Furthermore, we show that the abscissa of convergence of ζ_A is always equal to $\overline{\dim}_B A$, which generalizes to arbitrary dimensions a well-known result for fractal strings (or equivalently, for arbitrary compact subsets of the real line $\mathbb{R}$). Here, $\overline{\dim}_B A$ denotes the upper box (or Minkowski) dimension of A. Extended to a meromorphic function ζ_A, this "distance zeta function" is shown to be an efficient tool for finding the box dimension of several new classes of subsets of $\mathbb{R}^N$, like fractal nests, geometric chirps and multiple string chirps. It is also used to develop a higher-dimensional theory of complex dimensions of arbitrary fractal sets in Euclidean spaces.

For the sake of simplicity, we pay particular attention in this monograph to the principal complex dimensions of A, defined as the poles of ζ_A located on the "critical line" $\{\mathrm{Re}\, s = \overline{\dim}_B A\}$. We also introduce a new zeta function, denoted by $\tilde{\zeta}_A$ and called a "tube zeta function", and show, in particular, how to calculate the Minkowski content of a suitable (Minkowski measurable) bounded set A in $\mathbb{R}^N$ in terms of the residue of $\tilde{\zeta}_A(s)$ at $s = \dim_B A$, the box dimension of A. More generally, without assuming that A is Minkowski measurable, we obtain analogous results, but now expressed as inequalities involving the upper and lower Minkowski contents of A. In addition, we obtain a new class of harmonic functions generated by fractal sets and represented via singular integrals. Furthermore, a class of sets is constructed with unequal upper and lower box dimensions, possessing alternating zeta functions. Moreover, by using a suitable notion of equivalence between zeta functions, we simplify some aspects of the theory of geometric zeta functions attached to fractal strings.

In addition, we study the problem of the existence and the construction of the meromorphic extensions of zeta functions of fractals; in particular, we provide a natural sufficient condition for the existence of such extensions. An analogous problem is studied in the context of spectral zeta functions associated with bounded open subsets in Euclidean spaces with fractal boundary. We introduce transcendentally quasiperiodic sets, and construct a class of such sets, using generalized Cantor sets with two parameters, along with the Gel'fond–Schneider theorem from the theory of transcendental numbers.

With the help of this construction, we obtain an explicit example of a maximally hyperfractal set; namely, a compact set $A \subset \mathbb{R}^N$ such that the associated distance and tube zeta functions have the critical line $\{\mathrm{Re}\, s = \overline{\dim}_B A\}$ as a natural boundary. Actually, for this example, much more is true: every point of the critical line is a nonisolated singularity of the fractal zeta functions ζ_A and $\tilde{\zeta}_A$; so that given any point s on the critical line, ζ_A and $\tilde{\zeta}_A$ cannot be extended meromorphically (and hence, also holomorphically) to some punctured open neighborhood of s.

Furthermore, we introduce the notion of relative fractal drum, which extends the usual notions of fractal string and of fractal drum. The associated definition of relative box dimension is such that it can achieve negative values as well, provided the underlying geometry is sufficiently "flat". Using known results about the spectral asymptotics of fractal drums, and some of our earlier work, we recover known results about the existence of a (nontrivial) meromorphic extension of the spectral zeta function of a fractal drum. We also use some of our new results to establish the optimality of the upper bound obtained for the corresponding abscissa of meromorphic continuation of the spectral zeta function.

Moreover, we develop a higher-dimensional theory of fractal tube formulas, with or without error terms, for relative fractal drums (and, in particular, for bounded sets) in $\mathbb{R}^N$, for any $N \geq 1$. Such formulas, interpreted either pointwise or distributionally, enable us to express the volume of the tubular neighborhoods of the underlying fractal drums in terms of the associated complex dimensions. Therefore, they make apparent the deep connections between the theory of complex dimensions and the intrinsic oscillations of fractals. Accordingly, a geometric object is said to be "fractal" if it has at least one nonreal complex dimension or else, the corresponding fractal zeta function has a partial natural boundary along a suitable curve; so that its associated fractal zeta function cannot be meromorphically continued beyond this curve. We also formulate and establish a Minkowski measurability criterion for relative fractal drums (and, in particular, for bounded sets) in $\mathbb{R}^N$, for any $N \geq 1$. More specifically, under suitable assumptions, a relative fractal drum (and, in particular, a bounded set) in $\mathbb{R}^N$ is shown to be Minkowski measurable if and only if its only complex dimension with real part equal to its Minkowski dimension D is D itself, and it is simple.

We also obtain certain results about fractal tube formulas and Minkowski measurability criteria showing the concrete geometric role played not only by the poles but also by the essential singularities of fractal zeta functions, thereby suggesting a further possible extension of the notion of complex dimensions.

Throughout the book, we illustrate our results by a variety of examples, such as the Cantor set and string, the Cantor dust, a version of the Cantor graph (i.e., the "devil's staircase"), fractal strings (including self-similar strings), fractal sprays (including self-similar sprays), the Sierpiński gasket and carpet as well as their higher-dimensional counterparts, along with non self-similar examples, including fractal nests and geometric chirps.

Finally, we propose a classification of bounded open sets in Euclidean spaces, based on the properties of their tube functions (that is, the volume of their δ-neighborhoods, viewed as a function of the small positive number δ), and propose a number of open problems concerning distance and tube zeta functions, along with their natural extensions in the context of "relative fractal drums". Moreover, we suggest several directions for future research in the higher-dimensional theory of the fractal complex dimensions of arbitrary compact subsets of Euclidean spaces (as well as more generally, of suitable metric measure spaces).

We stress that a significant advantage of the present theory of fractal zeta functions, and therefore, of the corresponding higher-dimensional theory of complex dimensions of fractal sets developed in this book, is that it is applicable to arbitrary bounded (or equivalently, compact) subsets of $\mathbb{R}^N$, for any $N \geq 1$. (At least in principle, it can also be extended to arbitrary compact metric measure spaces, although this is not explicitly done in this work.) In particular, no assumption of self-similarity or, more generally, of "self-alikeness" of any kind, is made about the underlying fractals or, within the broader theory developed here, about the relative fractal drums under consideration.

Preface

The present research monograph is a testimony to the fact that Fractal Analysis is deeply connected to numerous areas of contemporary Mathematics. Here, we have in mind, in particular, Complex Analysis, Geometry (including Fractal Geometry, Spectral Geometry and Geometric Measure Theory), Harmonic Analysis, Number Theory, Oscillation Theory, Mathematical Physics and Partial Differential Equations.

This monograph is a natural consequence of the rapid and exciting development of the theory of geometric zeta functions of fractal strings and their complex dimensions over the past twenty years, the foundations of which have been laid out by the first author and his collaborators, including especially, Machiel van Frankenhuijsen [Lap-vFr1–3], since the early 1990s. An important impetus for the present work came from an interesting and little known result from 1970, due to Harvey and Polking [HarPol, p. 42] in their study of the singularities of the solutions of certain linear partial differential equations, providing some sufficient conditions for the Lebesgue integrability of the (negative) powers of the distance function $x \mapsto d(x,A)$ over any bounded open neighborhood Ω of a given compact subset A of Euclidean space $\mathbb{R}^N$, expressed in terms of its upper box (or Minkowski) dimension $\overline{\dim}_B A$: for any real number γ, we have that

$$\gamma < N - \overline{\dim}_B A \qquad \Longrightarrow \qquad \int_\Omega d(x,A)^{-\gamma} \mathrm{d}x < \infty. \tag{A}$$

Inspired by this result, in 2009, the first author realized the possibility of introducing a new kind of fractal zeta function, expressed as a Dirichlet-type integral and denoted by ζ_A, which we call the *distance zeta function* of A:

$$\zeta_A(s) := \int_\Omega d(x,A)^{s-N} \mathrm{d}x. \tag{B}$$

Here, s is a complex number with a sufficiently large real part. Moreover, it immediately follows from relation (A) that $\zeta_A(s)$ is well defined when $\mathrm{Re}\, s > \overline{\dim}_B A$. It has enabled us to extend the existing theory of geometric zeta functions $\zeta_{\mathscr{L}}$ (of bounded fractal strings $\mathscr{L}$) to zeta functions ζ_A of arbitrary compact subsets A of Euclidean spaces.

A key consequence of this new definition is the possibility to define the *complex dimensions* of any given bounded (fractal) subset A, as the poles of the distance zeta function ζ_A (suitably meromorphically extended), analogously to what was done earlier in the case of bounded fractal strings in [Lap-vFr1–3]. Clearly, the set of complex dimensions of A is at most countable, since the poles of any meromorphic function are isolated points in the complex plane.

This definition has led to a number of fruitful results in Chapter 2 (and elsewhere in the book), where, in particular, we have shown that the abscissa of (absolute) convergence $D(\zeta_A)$ of ζ_A coincides with $\overline{\dim}_B A$, the upper box dimension of A; see Theorem 2.1.11. Also, ζ_A is shown to be holomorphic for $\mathrm{Re}\, s > \overline{\dim}_B A$ and hence, all of the complex dimensions of A lie on or strictly to the left of the "critical line" $\{\mathrm{Re}\, s = \overline{\dim}_B A\}$. Furthermore, the residue of ζ_A computed at $s = D$, where D is the box (or Minkowski) dimension of A (assumed to exist), is very closely related to the lower and upper *Minkowski contents* of A; see Theorem 2.2.3.

Moreover, our study of generalized Cantor sets (with two parameters) has enabled us to construct a class of *quasiperiodic sets* with finitely many (and even infinitely many) quasiperiods; see Theorem 3.1.15. In order to establish this latter result, we have used a deep theorem from Analytic Number Theory, due to Alan Baker and for which he was awarded the Fields medal in 1970. The precise statements of the corresponding results can be found in Chapter 3 and, in a more general context, in Chapter 4; see, in particular, Theorem 3.1.20 and Corollary 4.6.28. These explicit constructions are used both to explore the boundaries of the notion of "fractality" in our context (the so-called maximally hyperfractal sets) and to show that the bounds obtained earlier by the first author for the abscissae of meromorphic continuation of the *spectral zeta functions* of fractal drums are best possible, in general.

During our study of bounded fractal strings and their geometric zeta functions, it became clear that fractal strings could be viewed from a much broader perspective, as *relative fractal drums* (RFDs) in the real line. In fact, the notion of RFDs can even be introduced in arbitrary Euclidean spaces of any dimension, and this is described and investigated in detail in Chapter 4. As was alluded to just above, some of the applications include the study of the meromorphic continuations of the spectral zeta functions associated with elliptic differential operators defined on bounded domains with fractal boundary in Euclidean spaces; see Section 4.3 about the spectral asymptotics of fractal drums.

Why are the complex dimensions $\omega \in \mathbb{C}$ of a given bounded subset A of $\mathbb{R}^N$ important? One of the principal reasons lies in the fact that, under fairly general assumptions, they enable us to reconstruct the *tube function* of the set A, defined by $(0,1) \ni t \mapsto |A_t|$; here, $A_t := \{x \in \mathbb{R}^N : d(x,A) < t\}$ is the t-neighborhood of A and $|A_t|$ is the N-dimensional Lebesgue measure of A_t. More specifically, under suitable hypotheses, the following *fractal tube formula* (without error term) holds:

$$|A_t| = \sum_{\omega} c_\omega t^{N-\omega}, \qquad \text{(C)}$$

for all $t > 0$ sufficiently small, where ω ranges over the set of complex dimensions of A and the complex coefficients c_ω depend only on the set A and on the ambient dimension N. (In fact, in the present case of simple poles, we have that $c_\omega = \operatorname{res}(\zeta_A, \omega)/(N-\omega)$.) Furthermore, the sum on the right-hand side of the identity (C) is, in general, an infinite series, and not just a sum involving finitely many terms (as is the case for convex and smooth geometries).

We point out that the nonreal complex dimensions ω always appear in complex conjugate pairs, and $c_{\overline{\omega}} = \overline{c_\omega}$. Hence, by writing $\omega = (\operatorname{Re}\omega) + \mathrm{i}(\operatorname{Im}\omega)$ and $c_\omega = |c_\omega|\exp(\mathrm{i}\varphi_\omega)$, where $\varphi_\omega \in \mathbb{R}$ and $\mathrm{i} := \sqrt{-1}$, we obtain that

$$
\begin{aligned}
c_\omega t^{N-\omega} + c_{\overline{\omega}} t^{N-\overline{\omega}} &= 2t^N \operatorname{Re}\{c_\omega t^{-\omega}\} = 2t^{N-\operatorname{Re}\omega} \operatorname{Re}\{c_\omega t^{-\mathrm{i}(\operatorname{Im}\omega)}\} \\
&= 2|c_\omega| \cdot t^{N-\operatorname{Re}\omega} \cos\left((\operatorname{Im}\omega)(\log t^{-1}) + \varphi_\omega\right).
\end{aligned}
$$

Therefore, the series in (C) reduces to

$$
\begin{aligned}
|A_t| = 2 &\sum_{\omega,\, \operatorname{Im}\omega>0} |c_\omega| \cdot t^{N-\operatorname{Re}\omega} \cos\left((\operatorname{Im}\omega)(\log t^{-1}) + \varphi_\omega\right) \\
+ &\sum_{\omega,\, \operatorname{Im}\omega=0} c_\omega \cdot t^{N-\omega}.
\end{aligned}
\tag{D}
$$

As we can see from (D), any *nonreal* complex dimension ω of the set A (i.e., such that $\operatorname{Im}\omega \neq 0$) is the source of *oscillations* in the fractal tube formula (C), viewed as a function of sufficiently small values of $t > 0$. The larger the imaginary part of ω, the greater the *oscillation* rate (i.e., the *frequency* of the oscillations) corresponding to the complex dimension ω. Furthermore, still for small enough $t > 0$, the larger the real part of ω, the greater the amplitude $|c_\omega| t^{N-\operatorname{Re}\omega}$ of the oscillations corresponding to ω. The oscillations of the function $t \mapsto |c_\omega| \cdot t^{N-\operatorname{Re}\omega} \cos\left((\operatorname{Im}\omega)(\log \frac{1}{t}) + \varphi_\omega\right)$ for small $t > 0$, corresponding to nonreal complex dimensions ω, are called *intrinsic oscillations* in the geometry of A.

The fractal tube formulas associated with fractal sets (and their generalizations, relative fractal drums) are the focus of Chapter 5. They extend in part the classical tube formulas due to Steiner, Minkowski, Weyl and Federer, dealing with special bounded subsets A of $\mathbb{R}^N$ (namely, convex compact sets, smooth compact submanifolds and, more generally, compact sets of positive reach) in which the tube formula (C) only involves a sum containing finitely many terms, corresponding to the finite set of complex dimensions of A (which happen to all be real numbers and, in fact, integers in $\{0,1,\dots,N\}$). They also fully extend to arbitrary dimensions and to arbitrary bounded sets (or, more generally, RFDs) in $\mathbb{R}^N$ the fractal tube formulas obtained for bounded fractal strings in [Lap-vFr1–3] and for fractal sprays, in [LapPe3, LapPeWi1].

It is often stated that fractals are not well defined mathematical objects. In his celebrated book [Man1], Benoît Mandelbrot, in response to such a criticism, has proposed to define a "fractal" as a geometric object whose Hausdorff dimension is strictly greater than its topological dimension. However, an obvious conterexample to this definition (and of which Mandelbrot was aware of) is the classic Cantor curve (or "devil's staircase"). This paradox has always bothered and intrigued the first author. As a result, it has served as a powerful stimulus for developing the mathematical theory of complex dimensions.

Following the earlier work in [Lap-vFr1–3] but now equipped with a general definition of fractal zeta function and hence, of complex dimensions (valid, in particular, for any bounded subset of $\mathbb{R}^N$, with $N \geq 1$ arbitrary), we say that a geometric object is "fractal" if it admits at least one *nonreal* complex dimension. We also allow in our definition for the possibility of more complicated singularities or the nonexistence of suitable meromorphic extensions. Therefore, according to the above discussion of fractal tube formulas, nonreal complex dimensions are a signature of fractality. We will illustrate this statement by a number of examples, including (variants of) the Cantor curve, several classic self-similar fractals, and new or known non self-similar geometries. Classic Euclidean shapes, such as circles, triangles and squares, or compact convex sets and smooth subvarieties, will also be shown to be non-fractal, in the above sense.

If we were to select just a dozen of the most important results appearing in the present monograph, our choice would be the following:

Chapter 2

- Theorem 2.1.11 (abscissa of convergence of the distance zeta function ζ_A and Minkowski dimension of the bounded set A) on page 57
- Theorem 2.2.3 (residue of ζ_A and Minkowski content of A) on page 114
- Theorem 2.3.37 (meromorphic extension of ζ_A, Minkowski measurable and non-Minkowski measurable cases) on page 166

Chapter 3

- Theorem 3.1.15 (construction of transcendentally n-quasiperiodic sets) on page 198
- Theorem 3.3.6 (construction of complex dimensions of higher order) on page 213

Chapter 4

- Theorem 4.1.7 (abscissa of convergence of the relative distance zeta function $\zeta_{A,\Omega}$ and relative Minkowski dimension of the RFD (A,Ω)) on page 250
- Theorem 4.1.14 (residue of $\zeta_{A,\Omega}$ and Minkowski content) on page 253
- Theorem 4.2.19 (principal complex dimensions of arbitrarily prescribed finite or infinite order) on page 288
- Theorem 4.6.9 (construction of transcendentally ∞-quasiperiodic RFDs) on page 376

Chapter 5

- Theorem 5.1.11 (pointwise fractal tube formula with error term) on page 421
- Theorem 5.3.13 and Corollary 5.3.14 (exact pointwise fractal tube formula, via $\zeta_{A,\Omega}$) on pages 446 and 447, respectively
- Corollary 5.4.26 (characterization of the Minkowski measurability of RFDs in terms of the nonreal principal complex dimensions) on page 472

A more detailed selection of about fifty new results from this book can be found on pages xxix–xxxi.

The bibliography provided at the end of the present book represents just a small portion of the vast literature concerning various aspects of fractal analysis and fractal or spectral geometry. This monograph complements (but is independent of) the one that the first author wrote in collaboration with Machiel van Frankenhuijsen, [Lap-vFr3]. In some sense, it can be considered as its natural continuation, now developing the higher-dimensions theory of fractal zeta functions (and the associated complex dimensions) as well as opening the door to many new possible research directions. Therefore, for further study, we would recommend to the reader to consult [Lap-vFr3], which contains many other references, as well as a systematic survey of numerous results of the first author and his coauthors, obtained during the past twenty five years.

The present book is intended for researchers and graduate students working in Fractal Analysis, Fractal Geometry, Geometric Measure Theory, Nonsmooth Geometry and Analysis, Harmonic Analysis, Complex Analysis, Number Theory, Dynamical Systems, Oscillation Theory, Mathematical Physics, Spectral Geometry, the Spectral Theory of Elliptic Equations, as well as in a number of related areas. Specialists in Number Theory may find it interesting to see an application of the well-known Baker theorem from the Theory of Transcendental Numbers in the construction of transcendentally quasiperiodic sets provided in Sections 3.1 and 4.6. In some sense, this construction, involving a countable collection of two-parameter generalized Cantor sets, can be viewed as a fractal-geometric interpretation of Baker's theorem.

This monograph is accessible to graduate students of Mathematics and Physics. In particular, we only assume that the reader is familiar with the basics of Real and Complex Analysis and with the fundamentals of Lebesgue Integration Theory. Throughout the book, we have illustrated our main results by a variety of detailed examples, in order to facilitate the understanding of the theory. Furthermore, at the end of the monograph, we have provided a relatively long list of open problems (see Subsections 6.2.2 and 6.2.3), as a stimulus for further research.

Comments and suggestions concerning the content of the book are welcome and can be sent directly to the authors.

Riverside, California, USA and Paris, France *Michel L. Lapidus,*
Zagreb, Croatia *Goran Radunović* and *Darko Žubrinić*

Contents

List of Figures

Chapter 1

Chapter 2

Chapter 3

Chapter 4

Chapter 5

Chapter 6

Appendix A

Key Words

Among the key words listed below, we emphasize those notions which seem to appear for the first time, at least in our present context, or are given names which may not be standard.

• zeta function, geometric zeta function, *distance zeta function*, *tube zeta function*, spectral zeta function, perturbed Riemann zeta function, *weighted distance zeta function*, *surface zeta function*, *fractal zeta functions*, tube function,

• fractal string, fractal set, fractal drum, *relative fractal drum*, *relative fractal zeta functions*, *relative distance zeta function*, *relative tube zeta function*, fractal spray, *relative fractal spray*, *hyperfractal* or *maximally hyperfractal set*,

• Minkowski content, upper and lower Minkowski contents, relative Minkowski content, upper and lower relative Minkowski contents, average Minkowski content, Minkowski measurable set, Minkowski nondegenerate set, gauge function,

• box (or Minkowski) dimension, upper and lower box dimensions, *average Minkowski dimension*, complex dimensions, *principal complex dimensions*, relative box (or Minkowski) dimension, upper and lower relative box dimensions, *negative box dimension*, oscillatory period, *oscillatory amplitude*,

• Dirichlet series, Dirichlet integral, holomorphic function, meromorphic extension, abscissa of (absolute) convergence, *abscissa of holomophic continuation*, *abscissa of meromorphic continuation*, critical line (of absolute convergence), *critical line of holomorphy*, *critical line of meromorphy*, half-plane of (absolute) convergence, *half-plane of holomorphic continuation*, *half-plane of meromorphic continuation*, isolated singularity, removable singularity, essential singularity, pole, residue (at a pole), (holomorphic) natural boundary, *domain of holomorphy*, *(meromorphic) natural boundary*, *domain of meromorphy*, *(meromorphic) partial natural boundary*, *partial domain of meromorphy*, Lebesgue integral, *limit Lebesgue space*, Fourier transform, singular integral, harmonic function,

• *constant set* (i.e., Minkowski measurable set), *nonconstant set* (i.e., Minkowski nonmeasurable set), *periodic set*, *nonperiodic set*, generalized Cantor set, *quasiperiodic set*, *transcendentally quasiperiodic set*, *algebraically quasiperiodic set*, *order of quasiperiodicity* (finite or infinite), *hyperfractal, strong hyperfractal, maximal hyperfractal, fractal nest*, chirp, *geometric chirp*, *multiple string*, zigzagging fractal, discrete and continuous spirals,

• cone property of relative fractal drums, *flatness of relative fractal drums*, *tensor product of fractal strings*, Laplace operator, Dirichlet eigenvalue problem,

• *exponent sequence* (or *exponent vector*) of a positive integer, *stalactites, stalagmites and caves* associated with fractal sets,

• *Cantor set of higher order*, *Cantor set of infinite order*, *fractal grill*,

• languidity, strong languidity, tube function, fractal tube formula (pointwise and distributional), *relative shell zeta function*, *relative Mellin zeta function.*

A detailed list of key words, accompanied with the corresponding page numbers, can be found in the Subject Index located at the end of this monograph. See also the Glossary on pages xxxiii–xl for some of these terms and their notation.

Selected Key Results

Below we select some of the most important results appearing in this book.

Chapter 2

- Theorem 2.1.11 (abscissa of convergence of the distance zeta function ζ_A and Minkowski dimension of the bounded set A) on page 57
- Proposition 2.1.59 and Corollary 2.1.61 (geometric and distance zeta functions of fractal strings) on pages 91 and 92, respectively
- Theorem 2.2.3 (residue of ζ_A and Minkowski content of A) on page 114
- Theorem 2.2.11 (abscissa of convergence of the tube zeta function $\tilde{\zeta}_A$) on page 121
- Theorem 2.3.12 (meromophic extensions of geometric zeta functions of Dirichlet strings) on page 150
- Theorem 2.3.18 (meromorphic extension of $\tilde{\zeta}_A$, Minkowski measurable case) on page 154
- Theorem 2.3.25 (meromorphic extension of $\tilde{\zeta}_A$, Minkowski nonmeasurable case) on page 157
- Theorem 2.3.37 (meromorphic extension of ζ_A, Minkowski measurable and non-Minkowski measurable cases) on page 166
- Theorem 2.4.3 (residue of ζ_A and average Minkowski content) on page 178

Chapter 3

- Theorem 3.1.15 (transcendentally n-quasiperiodic sets) on page 198
- Theorem 3.3.3 (geometric zeta function of extended self-similar strings) on page 211
- Theorem 3.3.6 (construction of complex dimensions of higher order) on page 213
- Theorem 3.4.4 (abscissa of convergence of weighted distance zeta functions) on page 218
- Theorem 3.7.2 (dimension of a zigzagging set) on page 241

Chapter 4

- Theorem 4.1.7 (abscissa of convergence of the relative distance zeta function $\zeta_{A,\Omega}$ and Minkowski dimension of the RFD (A,Ω)) on page 250
- Theorem 4.1.14 (residue of $\zeta_{A,\Omega}$ and Minkowski content) on page 253
- Proposition 4.1.35 and Corollary 4.1.38 (construction of RFDs with arbitrarily negative finite or infinite dimensions) on pages 262 and 265, respectively
- Theorem 4.1.40 (scaling property of $\zeta_{A,\Omega}$) on page 267
- Theorem 4.2.17 (distance zeta functions and complex dimensions of self-similar sprays) on page 287
- Theorem 4.2.19 (principal complex dimensions of arbitrarily prescribed finite or infinite order) on page 288
- Proposition 4.2.25 and Example 4.2.26 (distance zeta function and complex dimensions of the Sierpiński N-gasket RFD, $N \geq 2$) on pages 294 ($N = 2$) and 294–303 ($N \geq 2$), respectively, and, especially, on pages 296–297 (N arbitrary) and 300–301 ($N = 3$)
- Theorem 4.3.21 (optimality of the upper bound on the abscissa of meromorphic continuation of the spectral zeta function $\zeta^*_{\Omega_0}$ of a fractal drum, using quasiperiodic RFDs of infinite order) on page 336
- Theorem 4.3.25 (meromorphic extension of the spectral zeta function $\zeta^*_{\Omega_0}$ of a fractal drum) on page 340
- Equation (4.5.2) (functional equation connecting the relative distance and tube zeta functions, $\zeta_{A,\Omega}$ and $\tilde{\zeta}_{A,\Omega}$) on page 351
- Theorem 4.5.1 (meromorphic extension of $\tilde{\zeta}_{A,\Omega}$, Minkowski measurable RFDs) on page 353
- Theorem 4.5.2 (meromorphic extension of $\tilde{\zeta}_{A,\Omega}$, Minkowski nonmeasurable RFDs) on page 355
- Theorem 4.5.6 (tube zeta function $\tilde{\zeta}_A$ and complex dimensions for compact sets of positive reach) on page 359
- Theorem 4.5.8 and Figure 4.16 (precise meromorphic extensions of the fractal zeta functions of disjoint unions of RFDs) on pages 360 and 367, respectively
- Theorem 4.6.9 (construction of transcendentally ∞-quasiperiodic RFDs) on page 376
- Theorem 4.6.13, Corollary 4.6.17 and Corollary 4.6.28 (construction of maximally hyperfractal sets) on pages 382, 384 and 390, respectively
- Theorem 4.7.3 and Theorem 4.7.10 (invariance of the complex dimensions under embeddings into higher-dimensional Euclidean spaces; case of bounded sets and RFDs, respectively) on pages 394 and 398, respectively

Chapter 5

- Theorem 5.1.11 (pointwise fractal tube formula with error term, via $\tilde{\zeta}_{A,\Omega}$) on page 421
- Theorem 5.1.13 (exact pointwise fractal tube formula via $\tilde{\zeta}_{A,\Omega}$) on page 424

- Theorem 5.2.2 (distributional fractal tube formula with error term, via $\tilde{\zeta}_{A,\Omega}$) on page 431
- Theorem 5.2.4 (exact distributional fractal tube formula via $\tilde{\zeta}_{A,\Omega}$) on page 434
- Theorem 5.2.11 (estimate for the distributional error term) on page 438
- Theorem 5.3.11 (pointwise fractal tube formula with error term, via $\zeta_{A,\Omega}$) on page 445
- Theorem 5.3.13 and Corollary 5.3.14 (exact pointwise fractal tube formulas via $\zeta_{A,\Omega}$) on pages 446 and 447, respectively
- Theorem 5.3.16 and Theorem 5.3.17 (pointwise fractal tube formula via $\zeta_{A,\Omega}$ at level $k = 0$; special case of simple poles) on pages 448 and 449, respectively
- Theorem 5.3.19 (distributional fractal tube formula with error term, via $\zeta_{A,\Omega}$) on page 450
- Theorem 5.3.20 (exact distributional fractal tube formula via $\zeta_{A,\Omega}$) on page 450
- Theorem 5.4.2 (sufficient condition for Minkowski measurability) on page 453
- Theorem 5.4.15 (necessary condition for Minkowski measurability) on page 463
- Theorem 5.4.20 (Minkowski measurability criterion in terms of $\zeta_{A,\Omega}$) on page 466
- Corollary 5.4.23 (characterization of the Minkowski measurability of self-similar strings) on page 469
- Theorem 5.4.25 (Minkowski measurability criterion in terms of $\tilde{\zeta}_{A,\Omega}$) on page 471
- Corollary 5.4.26 (characterization of the Minkowski measurability of RFDs in terms of the nonreal principal complex dimensions) on page 472
- Theorem 5.4.27 (generating h-Minkowski measurable RFDs) on page 473
- Subsection 5.5.2 (recovering the tube formulas for fractal strings), especially on pages 484–488
- Example 5.5.14 (fractality of the Cantor graph RFD or 'devil's staircase') on pages 496–502
- Subsection 5.5.6 (self-similar sprays: fractal tube formulas and Minkowski measurability criteria) on pages 511-537 (especially, on pages 511, 514, 529–530 and 532–535)

A less detailed selection of about a dozen of the results from the book can be found towards the end of the preface, on pages xiv–xv.

Glossary

$a_k \sim b_k$ as $k \to \infty$, asymptotically equivalent sequences of complex numbers....41

$a_k \asymp b_k$ as $k \to \infty$, mutually asymptotically bounded sequences................41

$\sum_{k=1}^{\infty} a_k \asymp \sum_{k=1}^{\infty} b_k$, simultaneous convergence or divergence of both series......41

$f \sim g$, equivalence of the DTI f and the meromorphic function g..............98

$f \overset{\text{asymp}}{\sim} g$, asymptotic equivalence of the meromorphic functions f and g......602

$A \setminus B$, the set of elements in A that are not in B..............................45

$\overline{A}$, closure of a subset A in $\mathbb{R}^N$ (or in $\mathbb{C}$)....................................7

$A_{\mathscr{L}} := \{a_j := \sum_{k=j}^{\infty} \ell_k : j \in \mathbb{N}\}$, canonical geometric representation of a bounded fractal string $\mathscr{L} = (\ell_k)_{k \geq 1}$..............................90

$A_t := \{x \in \mathbb{R}^N : d(x,A) < t\}$, t-neighborhood of a subset A of $\mathbb{R}^N$ ($t > 0$).......31

$A_{t,\delta} := A_\delta \setminus \overline{A_t}$, the open shell around the set A ($0 < t < \delta$)..................440

A_N, the inhomogeneous Sierpiński N-gasket..................................296

(A_N, Ω_N), the inhomogeneous Sierpiński N-gasket RFD.....................296

$\mathbf{am}(A)$, oscillatory amplitude of a bounded set A in $\mathbb{R}^N$.......................541

(A, Ω), relative fractal drum (RFD)..247

$(A, \Omega)_M := (A_M, \Omega \times (-1,1)^M)$, embedding of the relative fractal drum (A, Ω) of $\mathbb{R}^N$ into $\mathbb{R}^{N+M}$ ($M \in \mathbb{N}$)..............................396

$\mathfrak{a}$, the spectral operator..548

$\mathscr{A}$, positive elliptic differential operator (of order $2m$)......................338

$\mathscr{A}_{(E,F)}$, a space analogous to the classic Fock space.........................599

$\mathrm{B}(a,b) := \int_0^1 t^{a-1}(1-t)^{b-1}\,\mathrm{d}t$, the Euler beta function 393

$\mathrm{B}_x(a,b) := \int_0^x t^{a-1}(1-t)^{b-1}\,\mathrm{d}t$, the incomplete beta function 404

$B_r(a)$, open ball in $\mathbb{R}^N$ (or in $\mathbb{C}$) of radius r and with center at $a \in \mathbb{R}^N$ (or in $\mathbb{C}$) 254

$B_n^{(\sigma)}(x)$, the n-th generalized Bernoulli polynomial 471

$C^{(a)}$, generalized Cantor set with one parameter a 115

$C^{(m,a)}$, generalized Cantor set with two parameters m and a 186

$\#A$, cardinality of the finite set A 326

$\mathbb{C}$, the field of complex numbers 41

$\widetilde{\mathbb{C}} := \mathbb{C} \cup \{\infty\}$, the Riemann sphere 38

$C^m(\mathbb{R})$, space of m times continuously differentiable functions on $\mathbb{R}$ 158

conv S, convex hull of a subset S of $\mathbb{R}^N$ 234

$\mathrm{cave}(A) = \mathrm{cave}(A, \delta, r)$, the cave associated with a bounded set A 106

$\mathrm{cave}(A, \Omega) = \mathrm{cave}(A, \Omega, r)$, the cave associated with a relative fractal drum ... 272

$\mathscr{C}(s_0, f)$, the cluster set of a holomorphic function $f : U \to \mathbb{C}$ at s_0 37

$\mathscr{D}(0, \delta) := C_c^\infty(0, \delta)$, the space of infinitely differentiable (complex-valued) test functions with compact support contained in $(0, \delta)$ 429

$\mathscr{D}(0, +\infty) := C_c^\infty(0, +\infty)$, the space of infinitely differentiable (complex-valued) test functions with compact support contained in $(0, +\infty)$ 462

$\mathscr{D}'(0, \delta)$, the space of Schwartz distributions (the dual of $\mathscr{D}(0, \delta)$) 429

$\mathscr{D}'(0, +\infty)$, the space of Schwartz distributions on $\mathscr{D}(0, +\infty)$ 462

$\underline{\dim}_B A$, $\overline{\dim}_B A$, $\dim_B A$, box dimensions (Minkowski dimensions) of $A \subset \mathbb{R}^N$ ... 31

$\underline{\dim}_B(A, \Omega)$, $\overline{\dim}_B(A, \Omega)$, $\dim_B(A, \Omega)$, relative box dimensions of (A, Ω) 249

$\underline{\dim}_{av} A$, $\overline{\dim}_{av} A$, $\dim_{av} A$, average Minkowski dimensions of $A \subset \mathbb{R}^N$ 181

$\dim_{PC} A$, the set of principal complex dimensions of a bounded subset A of $\mathbb{R}^N$. 96

$\dim_{PC}(A, \Omega)$, the set of principal complex dimensions of the RFD (A, Ω) 253

$\dim_{PC} \mathscr{L}$, the set of principal complex dimensions of a fractal string $\mathscr{L} = (\ell_j)_{j \geq 1}$ 253

$\dim_{PC}^* \mathscr{L}$, the set of principal spectral complex dimensions of a fractal string .. 554

$\dim_{PC}^* \Omega$, the set of principal spectral complex dimensions of an open set Ω ... 554

$\dim_{PC}^*(A, \Omega)$, the set of principal spectral complex dimensions of a relative fractal drum (RFD) (A, Ω) 554

$d(x, A) := \inf\{|x - a| : a \in A\}$, Euclidean distance from x to A in $\mathbb{R}^N$ vii

$d_H(A, B) := \inf\{\varepsilon > 0 : A \subseteq B_\varepsilon,\ V \subseteq A_\varepsilon\}$, Hausdorff distance between two bounded subsets A and B of $\mathbb{R}^N$ 109

$D(f)$, abscissa of (absolute) convergence of the Dirichlet series or integral 55

$D_{\mathrm{cond}}(f)$, abscissa of conditional convergence of the Dirichlet series f 69

$D_{\mathrm{hol}}(f)$, abscissa of holomorphic continuation of f 64

$D_{\mathrm{mer}}(f)$, abscissa of meromorphic continuation of f 85

$\partial\Omega$, boundary of a subset Ω of $\mathbb{R}^N$ 31

$\mathscr{D}_{\mathrm{qp}}$, the family of quasiperiodic relative fractal drums 375

$\mathscr{D}_{\mathrm{aqp}}$, the family of algebraically quasiperiodic relative fractal drums 375

$\mathscr{D}_{\mathrm{tqp}}$, the family of transcendentally quasiperiodic relative fractal drums 375

DTI, Dirichlet-type integral 579

$\mathfrak{D} = \mathfrak{D}_{\mathfrak{S}}$, the set of scaling complex dimensions 513

$\bigoplus$, vector space direct sum 599

$\mathrm{e} \approx 2.718$, base of the natural logarithm 40

$|E| = |E|_N$, N-dimensional Lebesgue measure of a measurable set $E \subset \mathbb{R}^N$ 30

EDTI, extended Dirichlet-type integral 593

$\mathbf{e}(m)$, exponent sequence of an integer $m \geq 2$ 374

$f[\omega]_m$, $m \in \mathbb{Z}$, the m-th coefficient in the Laurent expansion of f around ω 508

$\mathrm{epi}(f)$, epigraph of a real-valued function f 106

$\mathscr{E}_{(E,\varphi)}$, the class of tamed EDTIs associated with (E,φ) 598

$\mathscr{E}^{\cdot n}_{(E,\varphi)}$, the space of products of n elements of $\mathscr{E}_{(E,\varphi)}$ 599

Fib, the Fibonacci string 488

$\mathrm{fl}(A,\Omega)$, flatness of a relative fractal drum (A,Ω) 266

$\mathscr{F}_{\mathrm{qp}}$, the set of quasiperiodic functions 194

$\mathscr{F}_{\mathrm{tqp}}$, the set of transcendentally quasiperiodic functions 194

$\mathscr{F}_{\mathrm{aqp}}$, the set of algebraically quasiperiodic functions 194

$\varphi^{\otimes n}$, n-fold tensor product of φ by itself 599

$\varphi_a(t) := \frac{1}{a}\varphi\left(\frac{t}{a}\right)$, for a test function φ and $a > 0$ 437

$\Gamma(t) := \int_0^{+\infty} x^{t-1}\mathrm{e}^{-x}\,\mathrm{d}x$, the gamma function 40

h, gauge function 544

$H^D(A)$, D-dimensional Hausdorff measure of the set A 129

$\mathbb{H}_c := L^2(\mathbb{R}, \mathrm{e}^{-2ct}\mathrm{d}t)$, weighted Hilbert space with parameter $c \in \mathbb{R}$ 548

χ_A, the characteristic (or indicator) function of the set A 49

$\mathrm{Hol}(\Omega)$, the space of holomorphic functions on an open set $\Omega \subseteq \mathbb{C}$ 606

$\mathscr{H}(f) := \{\mathrm{Re}\, s > D_{\mathrm{hol}}(f)\}$, the half-plane of holomorphic continuation of f ... 72

$\mathbb{i} = \sqrt{-1}$, the imaginary unit 40

IFS, iterated function system 308

$\inf S := \inf_{\tau \in \mathbb{R}} S(\tau)$, infimum of the screen S 411

$(\mathrm{ISP})_D$, inverse spectral problem for fractal strings of Minkowski dimension D 548

$\mathscr{K}(0,\delta)$, a suitable subspace of $\mathscr{D}(0,\delta)$ 429

$\mathscr{K}(0,+\infty)$, a suitable subspace of $\mathscr{D}(0,+\infty)$ 457

$\mathscr{K}'(0,\delta)$, the corresponding space of Schwartz distributions on the open interval $(0,\delta)$ 429

$\mathscr{K}'(0,+\infty)$ the corresponding space of Schwartz distributions on the open interval $(0,+\infty)$ 462

κ, the languidity exponent 412

κ_d, the d-languidity exponent 445

$\mathscr{L} = (\ell_j)_{j=1}^{\infty}$, fractal string with lengths ℓ_j 87

$(\ell_j)_{j=1}^{\infty}$, sequence of lengths of a fractal string $\mathscr{L}$ written in nonincreasing order 87

$(l_j)_{j=1}^{\infty}$, sequence of distinct lengths of a fractal string $\mathscr{L}$ written in (strictly) decreasing order (and with multiplicities b_j) 68

$\log_a x$, the logarithm of $x > 0$ with base $a > 0$; $y = \log_a x \Leftrightarrow x = a^y$ 40

$\log x := \log_{\mathrm{e}} x$, the natural logarithm of x; $y = \log x \Leftrightarrow x = \mathrm{e}^y$ 40

$L^{\infty)}(\Omega) := \cap_{p>1} L^p(\Omega)$, limit L^∞-space 217

$L^{p)}(\Omega) := \cap_{q<p} L^q(\Omega)$, limit L^p-space 217

$\ell^\infty(\mathbb{R})$, the Banach space of bounded sequences of real numbers $(\tau_j)_{j\geq 1}$ 374

$\mathscr{L}_1 \otimes \mathscr{L}_2$, tensor product of two bounded fractal strings 41

$\mathscr{L}_1 \sqcup \mathscr{L}_2$, union of two bounded fractal strings 41

$\mathscr{L}_{\mathrm{qp}}$, the family of quasiperiodic bounded fractal strings 202

$\mathscr{L}_{\mathrm{aqp}}$ the family of algebraically quasiperiodic bounded fractal strings 202

$\mathscr{L}_{\mathrm{tqp}}$, the family of transcendentally quasiperiodic bounded fractal strings 202

$\{\mathfrak{L}\sigma\}(s) := \int_0^{+\infty} \mathrm{e}^{-st}\sigma(t)\,\mathrm{d}t$, the Laplace transform of the function σ 452

$\mathscr{M}_*^r(A)$ and $\mathscr{M}^{*r}(A)$, lower and upper r-dimensional Minkowski contents of a bounded set $A \subset \mathbb{R}^N$, where $r \geq 0$ 31

$\mathscr{M}^D(A) := \lim_{t\to 0^+} \frac{|A_t|}{t^{N-D}}$, D-dimensional Minkowski content of a Minkowski measurable bounded set $A \subset \mathbb{R}^N$ 32

$\mathscr{M}_*^r(A,\Omega)$ and $\mathscr{M}^{*r}(A,\Omega)$, lower and upper relative r-dimensional Minkowski contents of the relative fractal drum (A,Ω) in $\mathbb{R}^N$, where $r \in \mathbb{R}$ 249

$\mathscr{M}_*^D(A,\Omega,h)$, $\mathscr{M}^{*D}(A,\Omega,h)$, gauge relative lower and upper Minkowski contents of the RFD (A,Ω) in $\mathbb{R}^N$ (with respect to the gauge function h) 352

$\mathscr{M}^D(A,\Omega) := \lim_{t\to 0^+} \frac{|A_t\cap\Omega|}{t^{N-D}}$, D-dimensional Minkowski content of a Minkowski measurable relative fractal drum (A,Ω) in $\mathbb{R}^N$ 249

$\tilde{\mathscr{M}}^D(A)$, $\tilde{\mathscr{M}}^{*s}(A)$, $\tilde{\mathscr{M}}_*^s(A)$, average Minkowski contents of $A\subset\mathbb{R}^N$ 178, 181

$\mathrm{Mer}(f) := \{\mathrm{Re}\, s > D_{\mathrm{mer}}(f)\}$, the half-plane of meromorphic continuation of f . 85

$\{\mathfrak{M}f\}(s) := \int_0^{+\infty} t^{s-1} f(t)\,\mathrm{d}t$, the Mellin transform of the function f 416

$\mu'_{\mathscr{A}}(x)\,\mathrm{d}x$, Browder–Gårding measure 339

$N_b(x)$, box-counting function of a bounded set $A\subset\mathbb{R}^N$ 34

$N_g(x) = N_{g,\mathscr{L}}(x)$, geometric counting function of a fractal string $\mathscr{L}$ 69

$N_v(x)$, spectral (or eigenvalue) counting function of a (relative) fractal drum ... 547

$\mathbb{N} := \{1, 2, 3, \ldots\}$, the set of positive integers 68

$\mathbb{N}_0 := \mathbb{N}\cup\{0\} = \{0, 1, 2, 3, \ldots\}$, the set of nonnegative integers 273

$(\mathbb{N}_0)_c^\infty$, the set of all sequences $\mathbf{e}$ with components in $\mathbb{N}_0 = \mathbb{N}\cup\{0\}$, such that all but at most finitely many components are equal to zero 374

$\binom{n}{k} := \frac{n(n-1)\ldots(n-k+1)}{k!}$ for $k\in\{1,2,\ldots,n\}$, $\binom{n}{0} := 1$, binomial coeffcients 129

$\binom{|\alpha|}{\alpha_1,\alpha_2,\ldots,\alpha_n} := \frac{|\alpha|!}{\alpha_1!\,\alpha_2!\cdots\alpha_n!}$, $|\alpha| := \alpha_1+\cdots+\alpha_n$, multinomial coefficient 283

$\mathrm{osc}_a f$, oscillation of a function f at a point a 160

OSC, the open set condition (for a self-similar set, spray or RFD in $\mathbb{R}^N$) 525

$O(t^{(\beta)}) := \bigcap_{\varepsilon>0} O(t^{\beta+\varepsilon})$, limit big $O(t^\beta)$ as $t\to+\infty$ 146

$O(t^{\alpha)}) := \bigcap_{\varepsilon>0} O(t^{\alpha-\varepsilon})$, limit big $O(t^\alpha)$ as $t\to 0^+$ 155

$O(t^{0)}) := \bigcap_{\varepsilon>0} O(t^{-\varepsilon})$, functions of slow growth tending to $+\infty$ as $t\to 0^+$ 544

ω_N, N-dimensional Lebesgue measure of the unit ball in $\mathbb{R}^N$ 40

$\Omega_{\mathrm{can}} = \Omega_{\mathrm{can},\mathscr{L}}$, canonical geometric realization of a fractal string $\mathscr{L}$ 89

$\Omega_A := \Omega\setminus\overline{A}$, the open set associated with a relative fractal drum (A,Ω) 319

$\mathrm{or}(A,\Omega)$, order of the relative fractal drum (A,Ω) 475

$\otimes$, tensor product of a base relative fractal drum and a fractal string 273

$\mathbf{p}$, oscillatory period of a lattice self-similar set (or string or spray or RFD) 105

$\mathbf{p}(A,\rho)$, oscillatory period of a bounded set A in $\mathbb{R}^N$, with respect to ρ 541

$\mathscr{P}(f)$, the set of poles of a meromorphic function f 97

$\mathscr{P}(f,W)$, the set of poles of a meromorphic function f contained in the interior of the set $W\subseteq\mathbb{C}$ 97

$\mathscr{P}_c(f)$, the set of poles of a (tamed) Dirichlet-type integral (i.e., DTI) f on the critical line 97

$\Pi(f) := \{\mathrm{Re}\, s > D(f)\}$, the half-plane of (absolute) convergence of the Dirichlet series or integral f 69

$\mathbb{Q}$, the field of rational numbers 195

$\overline{\mathbb{Q}}$, the field of algebraic numbers 192

$\mathbb{R}$, the field of real numbers vii

$\mathrm{res}(f,\omega)$, residue of the meromorphic function f at the point ω 37

RFD, relative fractal drum (A,Ω) in $\mathbb{R}^N$ 247

$\mathrm{RFD}_\Lambda(\mathbb{R}^N)$, the familly of all Λ-sprayable RFDs in $\mathbb{R}^N$ 285

S_N, the classic N-dimensional Sierpiński gasket 298

$S = S(\tau)$, screen (viewed as a function) 95

$\boldsymbol{S} := \{S(\tau) + \mathrm{i}\tau : \tau \in \mathbb{R}\}$, screen (viewed as a graph or set) 95

$\boldsymbol{S}_{|n}$, truncated screen 418

$\|S\|_{\mathrm{Lip}}$, the Lipschitz constant of the Lipschitz continuous function S 411

$(s)_k := \frac{\Gamma(s+k)}{\Gamma(s)}$, the Pochhammer symbol for $s \in \mathbb{C}$ and $k \in \mathbb{Z}$ 415

$\mathscr{S}_{\mathrm{qp}}$, the family of quasiperiodic sets 194

$\mathscr{S}_{\mathrm{aqp}}$, the family of algebraically quasiperiodic sets 194

$\mathscr{S}_{\mathrm{tqp}}$, the family of transcendentally quasiperiodic sets 194

$\mathrm{Spray}(\Omega_0, (\lambda_j)_{j\geq 1}, (a_j)_{\geq 1})$, relative fractal spray in $\mathbb{R}^N$ 273

$\sup S := \sup_{\tau\in\mathbb{R}} S(\tau)$, supremum of the screen S 411

supp $\mathbf{e}$, support of a sequence $\mathbf{e} \in (\mathbb{N}_0)^\infty_c$ 374

supp m, support of an integer $m \geq 2$ 374

$\overline{s} := x - \mathrm{i}y$, complex conjugate of a complex number $s = x + \mathrm{i}y$ 60

$\sigma(A,\Omega)$, spectrum of a relative fractal drum (A,Ω) in $\mathbb{R}^N$ 321

$\mathscr{S}^{(I)}_{(E,\varphi),\rho}$, the class of SEDTIs of type I 599

$\mathscr{S}^{(II)}_{(E,\varphi),\rho}$, the class of SEDTIs of type II 599

$\bigsqcup_{j=1}^\infty \mathscr{L}_j$, union of a countable family of fractal strings 368

$\bigsqcup_{j\in J}(A_j,\Omega_j)$, disjoint union of a countable family of relative fractal drums 271

$\bigsqcup_{j\in J}\lambda_j(A_0,\Omega_0)$, countable disjoint union of isometric images (i.e., 'copies') of the relative fractal drum (A_0,Ω_0) scaled by the factors λ_j 271

$V(t) = V_A(t) := |A_t|$, volume of the t-neighborhood of the bounded set $A \subset \mathbb{R}^N$ 415

$V(t) = V_{A,\Omega}(t) := |A_t \cap \Omega|$, volume of the t-neighborhood of the relative fractal drum $(A,\boldsymbol{\Omega})$ 415

$V^{[k]}(t) = V_A^{[k]}(t)$, the k-th primitive function of $V_A(t)$ for $k \geq 0$ 415

$V^{[k]}(t) = V_{A,\Omega}^{[k]}(t)$, the k-th primitive function of $V_{A,\Omega}(t)$ for $k \geq 0$ 415

$\mathscr{V} = \mathscr{V}_{A,\Omega}$, the regular distribution generated by $V_{A,\Omega}(t)$ 430

$\mathscr{V}^{[k]} = \mathscr{V}_{A,\Omega}^{[k]}$, the regular distribution generated by the k-th primitive function of $V_{A,\Omega}(t)$ for $k \geq 0$, or the $|k|$-th distributional derivative of $\mathscr{V}_{A,\Omega}$ for $k < 0$ 430

$\boldsymbol{W}$, window 95

$\boldsymbol{W}_{|n}$, truncated window (associated with the truncated screen $\boldsymbol{S}_{|n}$) 418

$\lfloor x \rfloor = [x]$, the floor (or the integer part) of x; that is, the greatest integer less than or equal to x 163

$\lceil x \rceil$, the ceiling of x; that is, the smallest integer greater than or equal to x 189

$\{x\} := x - \lfloor x \rfloor$, the fractional part of x 163

$\mathbb{Z} := \{\ldots, -2, -1, 0, 1, 2, \ldots\}$, the ring of integers 105

$\zeta(s) = \zeta_R(s) := \sum_{j=1}^{\infty} j^{-s}$, the Riemann zeta function 145

$\zeta_{\mathscr{L}}(s) := \sum_{j=1}^{\infty} (\ell_j)^s$, the geometric zeta function of a fractal string $\mathscr{L}$ 86

$\zeta_A(s) := \int_{A_\delta} d(x,A)^{s-N} \mathrm{d}x$, the distance zeta function of a bounded subset A of $\mathbb{R}^N$ 45

$\tilde{\zeta}_A(s) := \int_0^\delta t^{s-N-1} |A_t| \, \mathrm{d}t$, the tube zeta function of a bounded subset A of $\mathbb{R}^N$ ..118

$\zeta_{A,\Omega}(s) := \int_\Omega d(x,A)^{s-N} \mathrm{d}x$, the relative distance zeta function of a relative fractal drum $(A,\boldsymbol{\Omega})$ 247

$\zeta_{A,\Omega}(s;\delta) := \int_{\Omega \cap A_\delta} d(x,A)^{s-N} \mathrm{d}x$, the relative distance zeta function of a relative fractal drum $(A, \boldsymbol{\Omega} \cap A_\delta)$ 252

$\tilde{\zeta}_{A,\Omega}(s) := \int_0^\delta t^{s-N-1} |A_t \cap \Omega| \, \mathrm{d}t$, the relative tube zeta function of a relative fractal drum $(A,\boldsymbol{\Omega})$ 350

$\check{\zeta}_{A,\Omega}$, the relative shell zeta function of a relative fractal drum $(A,\boldsymbol{\Omega})$ 440

$\zeta_{A,\Omega}^{\mathfrak{M}}$, the relative Mellin zeta function of a relative fractal drum $(A,\boldsymbol{\Omega})$ 458

$\zeta_{A,\Omega}^{*}$, the spectral zeta function of a relative fractal drum $(A,\boldsymbol{\Omega})$ 321

$\zeta_{\Omega_0}^{*}$, the spectral zeta function of a bounded open subset $\boldsymbol{\Omega}_0$ of $\mathbb{R}^N$ 325

$\zeta_{\mathscr{L}}^{*} = \zeta_{v,\mathscr{L}}$, the spectral zeta function of a fractal string $\mathscr{L}$ 554

$\zeta_A(\cdot, \partial)$, the surface zeta function of a bounded subset A of $\mathbb{R}^N$ 142

$\zeta_A(\cdot, w)$, the weighted distance zeta function of a bounded set A, with weight w 216

$\zeta_A(\cdot, \mu)$, the distance zeta function of a bounded set A, with Borel measure μ . 216

$\zeta_{E,\varphi,\mu}(s) := \int_E \varphi(x)^s \mathrm{d}\mu(x)$, Dirichlet-type integral (DTI) 579

$\zeta_{\mathfrak{S}}$, the scaling zeta function (of a fractal spray or of as self-similar RFD) 512

$\zeta_{O,\mathrm{out}}(s) := \zeta_{K,K^c}(s)$, the relative distance zeta function corresponding to the outer neighborhoods of $K := \overline{O}$, where O is an admissible open set for the given IFS 523

Chapter 1
Introduction

Das Wesen der Mathematik liegt in ihrer Freiheit.
[The essence of Mathematics lies in its freedom.]
Georg Cantor (1845–1918)

Abstract This research monograph provides a potentially useful and significant extension of the theory of zeta functions for fractal strings (which can be viewed as objects associated to bounded fractal sets on the real line), to fractal sets and arbitrary compact sets in Euclidean spaces of any dimension. The zeta function on which it is based has been introduced in 2009 by the first author (M. L. Lapidus); see its definition given below in Equation (1). We denote this zeta function by ζ_A and refer to it as a "distance zeta function". Here, by a fractal set, we mean any bounded set A of the N-dimensional Euclidean space $\mathbb{R}^N$, with $N \geq 1$. Fractality refers to the fact that the notion of fractal dimension, more precisely, of the upper box dimension of a bounded set (also called the upper Minkowski dimension), is a basic tool in the study of the properties of the associated zeta functions considered in this book. This new class of zeta functions enables us to extend in a useful manner the definition of the *complex dimensions of fractal strings*, introduced by Lapidus and van Frankenhuijsen, to arbitrary bounded fractal sets and more generally, to arbitrary bounded or compact sets in Euclidean spaces of any dimension. More specifically, given any bounded set $A \subset \mathbb{R}^N$, its *distance zeta function* ζ_A is defined by

$$\zeta_A(s) = \int_{A_\delta} d(x,A)^{s-N} \mathrm{d}x, \tag{1}$$

for all $s \in \mathbb{C}$ with $\operatorname{Re} s$ sufficiently large. Here, $A_\delta = \{x \in \mathbb{R}^N : d(x,A) < \delta\}$ is the δ-neighborhood of A and $d(x,A)$ denotes the Euclidean distance from $x \in \mathbb{R}^N$ to A. The dependence of ζ_A on the choice of δ is inessential. Note that without loss of generality, we could assume that A is an arbitrary compact set in $\mathbb{R}^N$. A similar comment could be made about the *tube zeta function* $\tilde{\zeta}_A$, also studied in this book and defined by

$$\tilde{\zeta}_A(s) = \int_0^\delta t^{s-N-1} |A_t| \mathrm{d}t, \tag{2}$$

for all $s \in \mathbb{C}$ with $\operatorname{Re} s$ sufficiently large. It involves the *tube function* $(0,\delta) \ni t \mapsto |A_t|$ of the set A, where $|A_t|$ is the N-dimensional Lebesgue measure of A_t. The basic

M.L. Lapidus et al., *Fractal Zeta Functions and Fractal Drums*,
Springer Monographs in Mathematics, DOI 10.1007/978-3-319-44706-3_1

property of both of these *fractal zeta functions* (namely, the distance zeta function ζ_A and the tube zeta function $\tilde{\zeta}_A$) is that they are absolutely convergent in the sense of Lebesgue for all $s \in \mathbb{C}$ such that $\operatorname{Re} s > \overline{\dim}_B A$ and define a holomorphic function on the open half-plane $\{\operatorname{Re} s > \overline{\dim}_B A\}$, where $\overline{\dim}_B A$ is the upper Minkowski dimension of the set A. More specifically, $\overline{\dim}_B A$ coincides with the *abscissa of convergence* of both ζ_A and $\tilde{\zeta}_A$; i.e., $\{\operatorname{Re} s > \overline{\dim}_B A\}$ is the largest open right half-plane on which each of the integrals defining ζ_A and $\tilde{\zeta}_A$ in (1) and (2), respectively, is absolutely convergent (and hence, convergent). Further, under mild hypotheses, it is also the largest open right half-plane on which ζ_A and $\tilde{\zeta}_A$ are holomorphic. We also introduce fractal zeta functions in the more general and more flexible context of *relative fractal drums* (or RFDs) (A,Ω), where $A \subseteq \mathbb{R}^N$ is not necessarily bounded and Ω is an open subset of $\mathbb{R}^N$ of finite volume contained in a δ-neighborhood of A for some $\delta > 0$. Then, the distance and tube zeta functions, $\zeta_{A,\Omega}$ and $\tilde{\zeta}_{A,\Omega}$, are defined much as in (1) and (2), respectively, but with A_δ replaced by Ω. (See Chapter 4.) In this general setting, the aim is to study the corresponding *relative tube function* $t \mapsto |A_t \cap \Omega|$ of the RFD (A,Ω), and in particular, to express it as a sum over the underlying complex dimensions (i.e., the poles of $\zeta_{A,\Omega}$, or, equivalently, of $\tilde{\zeta}_{A,\Omega}$); the resulting formula is called a *fractal tube formula*. (See Chapter 5.) New phenomena arise in this setting, including the fact that the *relative Minkowski* (or *box*) *dimension* $\overline{\dim}_B(A,\Omega)$ may be negative, and even take the value $-\infty$, a property related to the "flatness" of the corresponding RFD (A,Ω). The special case of a bounded subset A of $\mathbb{R}^N$ discussed earlier in the text surrounding Equations (1) and (2) then corresponds to the choice of $\Omega = A_\delta$, i.e., to the RFD (A, A_δ). Fractal strings and their higher-dimensional analogs, fractal sprays, are also very special cases of RFDs. As was mentioned above, the complex dimensions of a bounded set (or, more generally, of an RFD), are defined as the poles of the associated zeta function. As such, they form a finite or countable (as well as discrete) subset of the complex plane. The main goal of this book is to develop a comprehensive theory of complex dimensions (and of the associated tube formulas, see Chapter 5), valid for general bounded sets (and RFDs) in $\mathbb{R}^N$, with $N \geq 1$ arbitrary, as well as to illustrate it via a variety of concrete classic and new examples. A number of geometric and spectral applications are also provided throughout the monograph. This book should be accessible and of interest to experts and nonexperts alike, working in a broad range of areas of mathematics (including fractal geometry, dynamical systems, spectral geometry, complex, real and harmonic analysis, number theory, partial differential equations and mathematical physics) and its physical or engineering applications.

Key words: zeta function, distance zeta function, tube zeta function, fractal set, fractal string, intrinsic oscillations of a fractal set, box dimension, complex dimensions, principal complex dimensions, Minkowski content, Minkowski measurable set, singularity, residue, Cantor string, Cantor function.

1.1 Motivations, Goals and Examples

The mathematical concept of *dimension* began to be seriously studied during the 19th century. It has emerged, in particular, out of attempts to define in reasonable generality the notions of a 'line', a 'curve', and of a 'surface', as well as of higher-dimensional algebraic varieties and smooth manifolds. Below, we present a short sketch of the fascinating history of Dimension Theory, by dividing it into the following three parts: the history of integer dimensions, fractal dimensions and then, complex dimensions.

Integer dimensions. Until the beginning of the 20th century, the notion of 'dimension' has been in use exclusively in its usual intuitive meaning, namely, as a *nonnegative integer*. In the 19th century, it was rigorously introduced for linear spaces, appearing in Linear Algebra. More specifically, the dimension of a given linear space was defined as the number of elements of any of its bases, as we still define it today. Soon, several other integer dimensional quantities have been introduced in much more general situations, in General Topology, in order to study various properties of arbitrary subsets of Euclidean spaces and their generalizations. These fundamental topological dimensions are now known as the small inductive dimension (Menger–Urysohn), the large inductive dimension (Brouwer–Čech) and the covering dimension (Čech–Lebesgue). A detailed account of the history of the extremely complex subject of topological dimensions can be found in the survey article [CriJo].

Fractal dimensions. The foundations of the theory of *fractal dimensions*, which may assume arbitrary nonnegative real values instead of just integer values, were laid out in the 1920s, in the works of Minkowski, Hausdorff, Besicovich and Bouligand, in order to better understand geometric properties of very general subsets of Euclidean spaces. These developments resulted in the *Hausdorff dimension* and the *Minkowski dimension* or the *Minkowski–Bouligand dimension* (also called the *box dimension*), which have become basic tools of contemporary *Fractal Geometry* and related fields. There are also numerous other fractal dimensional quantities, which we do not mention here.

Many distinguished researchers have contributed in various ways to spreading and developing this seemingly counterintuitive concept of fractal dimension. See, for example, [Man1, Chapter XI]. The methods of Fractal Geometry are today frequently used in various scientific fields, not only within Mathematics, but in other areas as well, ranging from Physics, Engineering, Computer Science, Biology and Medicine to Economy and Finance, and even to the Visual Arts. It is therefore not surprising that there are now several high-quality professional research journals dedicated exclusively to the study of problems emerging from Fractal Geometry. An overview of the history of Fractal Geometry can be found in [Man1], as well as in [Lap11]. A history of the notions of fractal dimensions appearing in the theory of Dynamical Systems is discussed in [ŽupŽu].

Complex dimensions. The idea of *complex dimension* of bounded fractal strings has been proposed at the beginning of the 1990s by the first author of this book, based in part on earlier work in [Lap1–3, LapPo1–2, LapMa1–2]. A *bounded fractal string* is a bounded open subset Ω of the real line $\mathbb{R}$. In most applications, $\partial\Omega$ is a fractal subset of $\mathbb{R}$.

A well-known example of a fractal string is provided by the *Cantor string* $\mathscr{L}_{CS}$ or Ω_{CS}, whose boundary $\partial\Omega_{CS}$ is the classic ternary Cantor set C. Then, Ω_{CS} is defined as the complement of C in $[0,1]$ or, equivalently, as the disjoint union of the deleted open intervals (the open 'middle thirds') in the usual construction of the Cantor set. Alternatively, $\mathscr{L}_{CS}$, viewed as a sequence of lengths (or 'scales'), namely, the lengths (repeated according to their multiplicities) of the deleted intervals, is given by

$$\mathscr{L}_{CS} = (\ell_j)_{j=1}^{\infty} := \left(\frac{1}{3}, \frac{1}{9}, \frac{1}{9}, \frac{1}{27}, \frac{1}{27}, \frac{1}{27}, \frac{1}{27}, \dots\right), \tag{1.1.1}$$

where the length $1/3^n$ is repeated 2^{n-1} times, for each $n \in \mathbb{N}$.

As we shall see in Equation (2.1.79) of Subsection 2.1.4 below, bounded fractal strings $\mathscr{L}$ can also be identified with certain bounded subsets $A_{\mathscr{L}}$ of the positive real line.[1] In order to define the complex dimensions of a given bounded fractal string $\mathscr{L}$, one has to assign to $\mathscr{L}$ the corresponding (geometric) *zeta function* $\zeta_{\mathscr{L}}$. More specifically,

$$\zeta_{\mathscr{L}}(s) := \sum_{j=1}^{\infty} \ell_j^s, \tag{1.1.2}$$

for all $s \in \mathbb{C}$ with $\operatorname{Re} s$ sufficiently large, where $\mathscr{L} = (\ell_j)_{j=1}^{\infty}$ is the sequence of lengths of the open intervals of which any geometric realization Ω of the fractal string is comprised. Note that $\mathscr{L}$ is independent of the choice of this realization.

The 'complex dimensions' of the bounded fractal string are then defined as the poles of a suitable meromorphic extension of $\zeta_{\mathscr{L}}$, assuming that the meromorphic extension exists. The development of the mathematical theory of complex dimensions of fractal strings and their generalizations can be found in the extensive monograph [Lap-vFr3]. See also the earlier books [Lap-vFr1–2].

In the present research monograph, we define the notion of 'complex dimensions' for *any* nonempty bounded subset A of a given Euclidean space. To this end, we introduce a suitable zeta function ζ_A (see Equation (1.1.6) below), called the *distance zeta function* of A and such that its poles can be considered as the 'complex dimensions' of a given set A (assuming that a suitable meromorphic extension of ζ_A is possible).

For example, for the Cantor string $\mathscr{L}_{CS}$ and in light of Equations (1.1.1) and (1.1.2) above, we have that

$$\zeta_{\mathscr{L}_{CS}}(s) = \sum_{n=1}^{\infty} 2^{n-1}(3^{-n})^s = 3^{-s}\sum_{n=1}^{\infty}(2\cdot 3^{-s})^{n-1} = \frac{3^{-s}}{1-2\cdot 3^{-s}} = \frac{1}{3^s-2},$$

[1] The set $A_{\mathscr{L}} := \{a_k : k \in \mathbb{N}\}$, associated to a given bounded fractal string $\mathscr{L} := (\ell_k)_{k=1}^{\infty}$, is uniquely determined by the following two conditions: (*i*) $a_k \to 0^+$ as $k \to \infty$ and (*ii*) $a_k - a_{k+1} = \ell_k$, for all $k \in \mathbb{N}$. See also Figure 2.7 in Subsection 2.1.4.

for all $s \in \mathbb{C}$ such that $\mathrm{Re}\, s > \log_3 2$; hence, upon meromorphic continuation, we obtain that

$$\zeta_{\mathscr{L}_{CS}}(s) = \frac{1}{3^s - 2}, \quad \text{for all } s \in \mathbb{C}. \tag{1.1.3}$$

It follows that the complex dimensions of $\mathscr{L}_{CS}$ are obtained by solving the equation $3^s = 2$, $s \in \mathbb{C}$, and, hence, are given by[2]

$$\mathscr{D}^*_{CS} = \left\{ \log_3 2 + \frac{2\pi}{\log 3} k\mathbb{i} : k \in \mathbb{Z} \right\}, \tag{1.1.4}$$

where $\mathbb{i} := \sqrt{-1}$ is the imaginary unit and $D := \log_3 2$ is the Minkowski (or box) dimension of the Cantor string $\mathscr{L}_{CS}$ or, equivalently, of the ternary Cantor set, the boundary of Ω_{CS}. In the present case, $D = \log_3 2$ happens to also coincide with the Hausdorff dimension of the Cantor set.

Note, however, that $\zeta_{\mathscr{L}}$ and hence also, D and $\mathscr{D}^*_{CS}$, would remain the same if we were to rearrange the intervals of which the fractal string is comprised so that the boundary would become a sequence of distinct points in $[0,1]$ decreasing to zero (and with zero as its only accumulation point). Then, as was observed just above, the Minkowski dimension would remain invariant under such rearrangements (namely, $D = \log_3 2$), whereas the Hausdorff dimension D_H would change from $\log_3 2$ to 0 (namely, $D_H = 0$), because the boundary would become a countable set.

It is a general fact that the distance zeta function ζ_A, where $A = A_{\mathscr{L}_{CS}}$ or $\partial\Omega_{CS}$, yields the same result as in (1.1.4); that is, the same complex dimensions (except possibly at $s = 0$). See Example 2.1.82 of Chapter 2, where the distance zeta function ζ_A of the ternary Cantor set is computed; see also Example 1.1.2 below. In fact, it turns out that for the Cantor string (or, more generally, for any bounded fractal string $\mathscr{L}$), we have

$$\zeta_{A_{\mathscr{L}}}(s) = \frac{2^{1-s}}{s} \zeta_{\mathscr{L}}(s) + v(s), \tag{1.1.5}$$

where $v(s)$ is a holomorphic function on the open right half-plane $\{\mathrm{Re}\, s > 0\}$; see Equation (2.1.85) in Subsection 2.1.4 below.[3]

More details about the history of the study of the notions of Minkowski content, Minkowski measurability and Minkowski dimension in Euclidean spaces can be found in Subsection 6.1.2 of Chapter 6.

The meaning of the dimension D as a nonnegative integer is intuitively clear for the simplest classes of (piecewise smooth) subsets of Euclidean spaces, such as, for example, balls and cubes (for which $D = 3$), polygons ($D = 2$), lines ($D = 1$), a point ($D = 0$), etc. In that context, the dimension D can roughly be understood as a degree of 'spaciousness' of the set under consideration. However, for general bounded

---

[2] In terms of the distance zeta function, the set of complex dimensions $\mathscr{D}_{CS}$ is given by $\mathscr{D}_{CS} = \mathscr{D}^*_{CS} \cup \{0\}$; see Equation (1.1.5) and footnote 3 on page 5, along with Equation (1.1.15).

[3] More specifically, $v = v(s)$ can be uniquely meromorphically extended to the whole complex plane $\mathbb{C}$, with $s = 0$ as its only pole, which is simple; see footnote 25 in Subsection 2.1.4, on page 90.

subsets of $\mathbb{R}^N$, it is natural to extend the definition of the dimension D to include "fractional" values as well, that is, to take on as possible values all nonnegative real numbers in $[0,N]$. One can think, for example, of subsets of the form of a 'cloud' in the three-dimensional Euclidean space $\mathbb{R}^3$, which are locally highly irregular. In that case, the fractal dimension D can take any noninteger value D in $[0,3]$, depending, roughly, on the local complexity of A. Intuitively, the more irregular (or 'rough') the set A is, the larger the value of its fractal dimension D.

Among the various fractal dimensions that have been introduced in the course of the 20th century, the *Minkowski* (or *box*) *dimension*, to be defined by Equation (1.3.4) below, is of special significance in this monograph.

If A is a given 3-dimensional body in $\mathbb{R}^3$, it is natural to consider its 3-dimensional Lebesgue volume, as a measure of its "3-dimensional content". More generally, if we take *any* (Minkowski measurable) nonempty bounded subset A of $\mathbb{R}^N$, of (possibly noninteger) Minkowski dimension $D \in [0,N]$, one can nevertheless define its *D-dimensional Minkowski content* (denoted by $\mathscr{M}^D(A)$; see Equation (1.3.1) below, for $r := D$ and with the upper limit replaced by a true limit). Intuitively, the "D-dimensional Minkowski content" of A can be thought of as its "D-dimensional fractal volume".[4] For integral values of D, this "content" coincides with the usual volume (more specifically, with the N-dimensional Lebesgue measure), up to a multiplicative constant depending on N; see Remark 1.3.1 below.

We will be especially interested in the *intrinsic ocillations* of a given nonempty bounded subset A of $\mathbb{R}^N$, with $N \geq 1$ arbitrary. Heuristically, these intrinsic oscillations may be thought as being associated with *geometric* (spectral, or dynamical) *waves* whose *amplitudes* (resp., *frequencies*) are directly connected to the *real parts* (resp., the *imaginary parts*) of the underlying complex dimensions.

The existence of intrinsic oscillations is closely related to the existence of nonreal complex dimensions of A. By 'intrinsic (or inner) oscillations' of A, we mean, for example, the oscillations of the *tube function* $t \mapsto |A_t|$ as $t \to 0^+$, where $|A_t|$ is the N-dimensional Lebesgue measure of A_t, the open t-neighborhood of A (i.e., the set of points in the ambient space $\mathbb{R}^N$ which lie within a distance less than t from A). More specifically, we are interested, in particular, in the case when the subset A under consideration is such that its D-dimensional lower and upper Minkowski contents of A (introduced in Equation (1.3.1) of Subsection 1.3.1 below) are (i) not equal and (ii) are respectively positive and finite; such a set is said to be (i) *Minkowski nonmeasurable* and (ii) *Minkowski nondegenerate*.[5]

As an example, consider a bounded subset A of $\mathbb{R}^N$ such that the corresponding tube function has the following asymptotics:

$$|A_t| = t^{N-D}\Big(G\big(\log t^{-1}\big) + O(t^\alpha)\Big) \quad \text{as} \quad t \to 0^+,$$

[4] Here, we should note that, for a noninteger dimension D, the "D-dimensional Minkowski content" does not satisfy the usual countable additivity property, by contrast to the standard N-dimensional Lebesgue measure.

[5] When the lower and upper Minkowski contents of A are equal and are nontrivial (i.e., when $\mathscr{M}^D(A) := \lim_{t\to 0^+} |A_t|/t^{N-D}$ exists in $(0,+\infty)$), then A is said to be *Minkowski measurable*; see Subsection 1.3.1.

where G is a nonconstant periodic function of positive amplitude and $\alpha > 0$.[6] The source of the leading oscillations in the geometry of A in this case is the nonconstant and multiplicatively periodic function $t \mapsto f(t) := G(\log t^{-1})$, which is oscillating faster and faster as $t \to 0^+$. This class of examples includes the Cantor ternary set, the Sierpiński gasket, the Sierpiński carpet, the N-Sierpiński carpet in $\mathbb{R}^N$ for any $N \geq 3$, as well as many other classes of bounded subsets of Euclidean spaces. All of them have *principal* complex dimensions (that is, the complex dimensions s having the largest possible real part, i.e., $\mathrm{Re}\, s = D$) of the form

$$D + \mathbf{p}k\mathrm{i} \quad \text{for} \quad k \in \mathbb{Z},$$

where $\mathbf{p}$ is a positive constant called the *oscillatory period* of the set A. In other words, the principal complex dimensions of A form an *arithmetic set* (or equivalently, a vertical arithmetic progression of length $\mathbf{p}$) contained in the *critical line* $\{\mathrm{Re}\, s = D\}$.

As was already mentioned above, this research monograph provides a potentially very useful and significant extension of the theory of zeta functions for fractal strings (which can be viewed as objects associated to bounded fractal sets on the real line), to fractal sets and arbitrary compact sets in Euclidean spaces of any dimension. The zeta function on which it is based has been introduced in 2009 by the first author (M. L. Lapidus); see its definition given below in Equation (1.1.6). We denote this zeta function by ζ_A and refer to it as a "distance zeta function". Here, by a fractal set, we mean any bounded set A of the Euclidean space $\mathbb{R}^N$, with $N \geq 1$. Fractality refers to the fact that the notion of fractal dimension, more precisely, of the upper box dimension of a bounded set (also called the upper Minkowski dimension, Bouligand dimension, or limit capacity, etc.) is a basic tool in the study of the properties of the associated zeta functions considered in this monograph. As was already mentioned, this new class of zeta functions enables us to extend in a useful manner the definition of the *complex dimensions of fractal strings*, introduced by the authors of [Lap-vFr1], [Lap-vFr2] and [Lap-vFr3], to arbitrary bounded fractal sets and more generally, to arbitrary bounded or compact sets in Euclidean spaces of any dimension.

More specifically, given any bounded set $A \subset \mathbb{R}^N$, its *distance zeta function* ζ_A is defined by

$$\zeta_A(s) = \int_{A_\delta} d(x,A)^{s-N} \mathrm{d}x, \tag{1.1.6}$$

for all $s \in \mathbb{C}$ with $\mathrm{Re}\, s$ sufficiently large. Here, $A_\delta = \{x \in \mathbb{R}^N : d(x,A) < \delta\}$ is the δ-neighborhood of A and $d(x,A)$ denotes the (Euclidean) distance from $x \in \mathbb{R}^N$ to A. As will be shown in Proposition 2.1.76, the dependence of ζ_A on the choice of δ is inessential. Note that without loss of generality, we could assume that A is an arbitrary compact set in $\mathbb{R}^N$. Indeed, replacing A with $\overline{A}$, the closure of A, does not change the δ-neighborhood or the distance zeta function: $A_\delta = (\overline{A})_\delta$ since $d(\cdot,A) = d(\cdot,\overline{A})$, and so $\zeta_A = \zeta_{\overline{A}}$. A similar comment could be made about the tube

[6] Then, a posteriori, the function G must be nonnegative and continuous.

zeta function $\tilde{\zeta}_A$, studied in Subsection 2.2.2 of Chapter 2. The just mentioned *tube zeta function* of A is defined by

$$\tilde{\zeta}_A(s) = \int_0^{\delta} t^{s-N-1}|A_t|\,\mathrm{d}t, \tag{1.1.7}$$

for all $s \in \mathbb{C}$ with $\operatorname{Re} s$ sufficiently large. It involves the *tube function* $(0,\delta) \ni t \mapsto |A_t|$ of the set A, where $|A_t|$ is the N-dimensional Lebesgue measure of A_t.

The basic property of both of these *fractal zeta functions* (namely, the distance zeta function ζ_A and the tube zeta function $\tilde{\zeta}_A$) is that they are absolutely convergent in the sense of Lebesgue for all $s \in \mathbb{C}$ such that $\operatorname{Re} s > \overline{\dim}_B A$ and define a holomorphic function on the open half-plane $\{\operatorname{Re} s > \overline{\dim}_B A\}$, where $\overline{\dim}_B A$ is the upper Minkowski dimension of the set A. More specifically, $\overline{\dim}_B A$ coincides with the *abscissa of convergence* of both ζ_A and $\tilde{\zeta}_A$; i.e., $\{\operatorname{Re} s > \overline{\dim}_B A\}$ is the largest open right half-plane on which each of the integrals defining ζ_A and $\tilde{\zeta}_A$ in (1.1.6) and (1.1.7), respectively, is absolutely convergent (and hence, convergent).

As we will now explain, it is also natural and extremely useful to consider not just bounded subsets A of $\mathbb{R}^N$, per se, but, more generally, suitable ordered pairs (A,Ω), that we call *relative fractal drums* (RFDs; see Definition 4.1.2 in Chapter 4.1), in which A is a (not necessarily bounded) subset of $\mathbb{R}^N$, while Ω is an open subset of $\mathbb{R}^N$ of finite N-dimensional Lebesgue measure. In this more general context, the aim is to study the fractal properties of the set A *relative to* Ω. More specifically, the corresponding *relative tube function* of the RFD (A,Ω) is now $t \mapsto |A_t \cap \Omega|$. The associated *relative Minkowski* (or *box*) *dimension* $\dim_B(A,\Omega)$ of the RFD (A,Ω) is then defined so that its value can be even negative, i.e., $\dim_B(A,\Omega) \in [-\infty, N]$; see Definition 4.1.4 in Section 4.1. The case when the relative Minkowski dimension is negative corresponds to the intuitive idea of *flatness* of A with respect to Ω. A special case of the relative box dimension is the so-called 'one-sided box dimension', introduced independently in 2010 by C. Tricot in [Tri4], where it was noticed that it could sometime be negative. In Subsection 4.1.1 of Chapter 4, we shall introduce the *distance zeta function of a relative fractal drum* (A,Ω), denoted by $\zeta_{A,\Omega}$, and which will enable us to define the complex dimensions of RFDs. It is defined by

$$\zeta_{A,\Omega}(s) := \int_{\Omega} d(x,A)^{s-N}\,\mathrm{d}x, \tag{1.1.8}$$

for all $s \in \mathbb{C}$ with $\operatorname{Re} s$ large enough, where as above, $d(x,A)$ denotes the Euclidean distance from $x \in \mathbb{R}^N$ to A. Similarly, the tube zeta function $\tilde{\zeta}_{A,\Omega}$ of the RFD (A,Ω) is defined exactly as in (1.1.7), except for $|A_t|$ being replaced with $|A_t \cap \Omega|$.

The special case of a bounded subset A of $\mathbb{R}^N$ discussed earlier around Equation (1.1.6) then corresponds to the choice of $\Omega = A_\delta$, i.e., to the RFD (A, A_δ).

We note that the set of complex dimensions of a bounded open set (or, more generally, of a relative fractal drum) is a finite or countable set of complex numbers, with finite multiplicities (as poles of the associated fractal zeta function). Moreover, since it is the set of poles of a meromorphic function, it is a discrete subset of the complex plane.

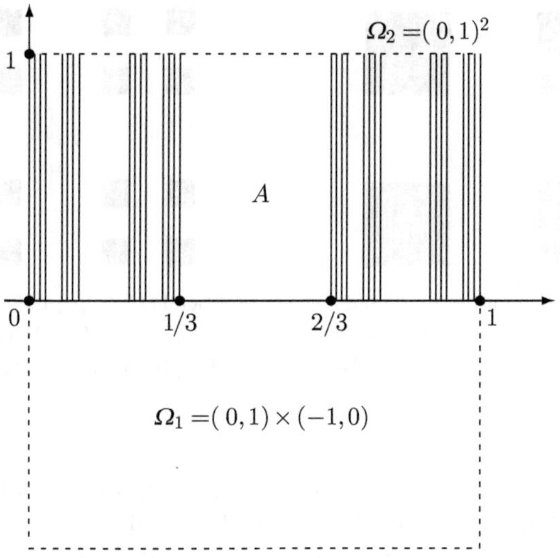

Fig. 1.1 The *Cantor grill* $A := C \times [0,1]$, where C is the ternary Cantor set, viewed from the open squares $\Omega_1 := (0,1) \times (-1,0)$ and $\Omega_2 := (0,1)^2$, has different respective *relative box dimensions*: $\dim_B(A,\Omega_1) = \dim_B C = \log_3 2$, while $\dim_B(A,\Omega_2) = \dim_B C + 1 = \log_3 2 + 1$. See Example 1.1.1.

Example 1.1.1. (*Cantor grill*). As an example illustrating the notion of relative box dimension, let A be a subset of the plane $\mathbb{R}^2$ defined by $A = C \times [0,1]$ (which we call the *Cantor grill*), where C is Cantor's ternary set; see Figure 2.15 in Subsection 2.2.4 of Chapter 2, along with Figure 1.1. Then, for $\Omega_1 := (0,1) \times (-1,0)$, we clearly have $\dim_B(A,\Omega_1) = \log_3 2$, while for $\Omega_2 := (0,1)^2$, we have $\dim_B(A,\Omega_2) = 1 + \log_3 2$. Indeed, viewed from Ω_1, the set Ω_1 is 'seen' just as the usual Cantor ternary set $C \times \{0\}$, contained in the boundary of Ω_1, while viewed from Ω_2, the set A has the form of the Cantor grill.

The above example exhibits a situation where the notion of RFD can be useful and interesting. Moreover, throughout the book, we will see how in many situations the distance zeta function of a complicated fractal set A can be explicitly computed by subdividing the set A into a union of simple 'relative fractal subdrums' and then computing the relative distance zeta function of each subdrum separately. This method can be illustrated most prominently in the process of computing the zeta function of the Cantor dust (Example 4.7.15), where we also use a result about the invariance of the complex dimensions with respect to the dimension of the ambient space.

Even more interestingly, the example of the *Cantor dust* $C \times C$ (i.e., the Cartesian product of two ternary Cantor sets; see Figure 1.2) then shows that not only the expected complex dimensions,[7] $\log_3 4 + \mathbf{p}k\mathrm{i}$, for $k \in \mathbb{Z}$ and with $\mathbf{p} := \frac{2\pi}{\log 3}$, are

[7] These complex dimensions are expected since $\dim_B(C \times C) = 2\log_3 2 = \log_3 4$. More precisely, some but not all of these complex dimensions which are nonreal may (in principle) be canceled by means of zero-pole cancellations.

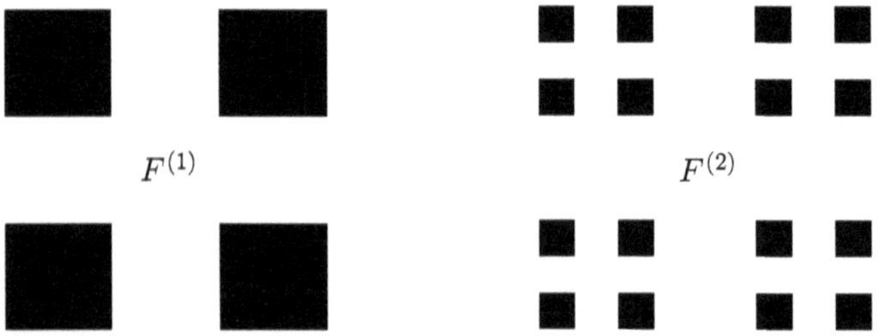

Fig. 1.2 The first two iterations of the sequence of prefractals $(F^{(k)})_{k\geq 1}$ defininig the *Cantor dust*. It is clear that $F^{(k+1)} \subset F^{(k)} \subset [0,1]^2$, for all $k \geq 1$, and $C \times C = \cap_{k\geq 1} F^{(k)}$. Here, $F^{(k)} := E^{(k)} \times E^{(k)}$, with $k \geq 1$, where $E^{(k)} \subset [0,1]$ is the usual k-th prefractal approximation of the ternary Cantor set C.

obtained by means of the corresponding fractal zeta functions but also the complex dimensions $\log_3 2 + \mathbf{p}k\mathrm{i}$, for $k \in \mathbb{Z}$, which actually coincide with the principal complex dimensions of the ternary Cantor set C itself. In other words, this is an example in which we can explicitly see how the complex dimensions also 'encode' the 'lower-dimensional fractality' of the Cantor dust $C \times C$, which evidently coincides with an uncountable union of vertical translates of the ternary Cantor set $C \times \{0\}$ in the plane; namely,

$$C \times C = \bigcup_{c\in C} (C \times \{c\}) = \bigcup_{c\in C} \big(C \times \{0\} + (0,c)\big). \tag{1.1.9}$$

Therefore, this geometric fact is naturally reflected in the set of complex dimensions of the Cantor dust.

Many analogous results are discussed throughout the book, where the geometric properties of the given bounded sets or of the RFDs are reflected in the correreponding sets of complex dimensions. We expect this type of phenomenon to be a general property of complex dimensions which has yet to be precisely formulated as a suitable principle and then properly established.

Also, the notion of RFDs provides us with a unified category under which fractal strings, bounded subsets of Euclidean spaces (of arbitrary dimension) and open subsets of Euclidean spaces with fractal boundary (also known as fractal drums) fall into. By developing the theory in this generality, we can apply it to all of these settings simultaneously, without the need to distinguish them separately; this broad fexibility is one of the powers and great advantages of the present theory.

Example 1.1.2. (*The Cantor set revisited: the Cantor string RFD and its distance zeta function*). We now illustrate the above discussion by revisiting the example of the Cantor string from the new perspective of relative fractal drums and the associated distance zeta functions. (Some readers may wish to note the main results for now and then return to this example later, if and when necessary.) As before, let C be the ternary Cantor set constructed in $[0,1]$ and let $I := (0,1)$, so that our associated

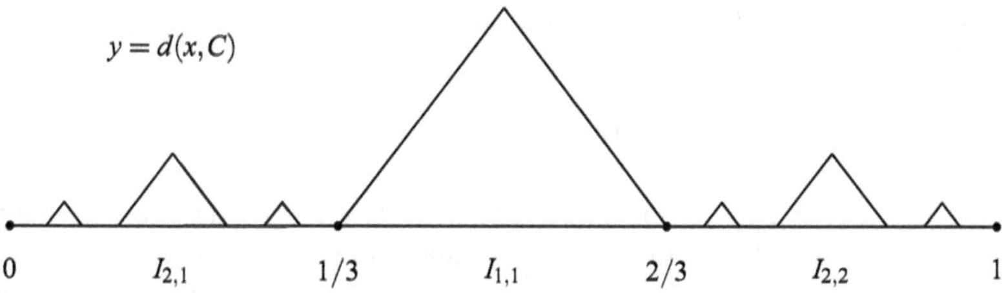

Fig. 1.3 The graph of the distance function $x \mapsto d(x,C)$, where C is the ternary Cantor set. The intervals $I_{n,k}$, where $k = 1,2,\dots,2^{n-1}$, correspond to the n-th generation of deleted open intervals during the construction of the Cantor set. See Equation (1.1.11) in Example 1.1.2.

relative fractal drum is (C,I). It is now straightforward to compute the distance zeta function of (C,I) by integrating over the set I:

$$\zeta_{C,I}(s) = \int_I d(x,C)^{s-1}\,\mathrm{d}x = \sum_{n=1}^{\infty}\sum_{k=1}^{2^{n-1}} \int_{I_{n,k}} d(x,C)^{s-1}\,\mathrm{d}x, \tag{1.1.10}$$

where $I_{n,k}$ denotes the k-th open interval removed from I in the n-th step of the construction of the ternary Cantor set; see Figure 1.3. Furthermore, for obvious reasons of symmetry, for a fixed $n \geq 1$ the integrals over the intervals $I_{n,k}$ are all equal to each other. Moreover, since $I_{n,k}$ is an interval of length 3^{-n} and $d(x,C) = d(x,\partial I_{n,k})$ for all $x \in I_{n,k}$, it is easy to deduce, by using local coordinates (see Remark 1.1.3 on page 13 below), that

$$\begin{aligned}\zeta_{C,I_{n,k}}(s) &= \int_{I_{n,k}} d(x,C)^{s-1}\,\mathrm{d}x \\ &= \int_{I_{n,k}} d(x,\partial I_{n,k})^{s-1}\,\mathrm{d}x = 2\int_0^{\frac{1}{2}3^{-n}} x^{s-1}\,\mathrm{d}x = \frac{2^{1-s}}{3^{ns}s},\end{aligned} \tag{1.1.11}$$

for all $s \in \mathbb{C}$ such that $\operatorname{Re} s > 1$ (and then, upon meromorphic continuation, wee see that $\zeta_{C,I_{n,k}}$ is still given by the right-hand side of Equation (1.1.11) for all $s \in \mathbb{C}$). Combining the above equality with (1.1.10) now leads to

$$\zeta_{C,I}(s) = \frac{2^{-s}}{s}\sum_{n=1}^{\infty}\left(\frac{2}{3^s}\right)^n = \frac{2^{1-s}}{s(3^s-2)}, \tag{1.1.12}$$

valid initially for all $s \in \mathbb{C}$ such that $\operatorname{Re} s > \log_3 2$. Clearly, upon meromorphic continuation, we deduce that

$$\zeta_{C,I}(s) = \frac{2^{1-s}}{s(3^s-2)}, \tag{1.1.13}$$

for all $s \in \mathbb{C}$. As a consequence, the set of all complex dimensions of the RFD (C,I) (that is, the set of poles of $\zeta_{C,I}$) is given by

$$\mathscr{P}(\zeta_{C,I}) = \{0\} \cup (D + \mathbb{i}\mathbf{p}\mathbb{Z}), \tag{1.1.14}$$

where $D+\mathbb{i}\mathbf{p}\mathbb{Z} := \{D+\mathbb{i}k\mathbf{p} : k \in \mathbb{Z}\}$. Observe that

$$\zeta_{C,I}(s) = \frac{2^{1-s}}{s}\zeta_{\mathscr{L}_{CS}}(s), \tag{1.1.15}$$

for all $s \in \mathbb{C}$, where $\zeta_{\mathscr{L}_{CS}}$ is the geometric zeta function of the Cantor string. We stress that a functional equation precisely analogous to (1.1.15) is actually valid for any fractal string $\mathscr{L}$, where the associated relative fractal drum we consider is equal to $(\partial\Omega_{\mathscr{L}}, \Omega_{\mathscr{L}})$ and $\Omega_{\mathscr{L}}$ is any geometric realization of the given fractal string $\mathscr{L}$; i.e., $\Omega_{\mathscr{L}}$ is a disjoint union of open intervals of lengths equal to the lengths (written in nonincreasing order) of the fractal string $\mathscr{L}$ counted with multiplicities; see Equation (1.1.27) below.

Another powerful and very useful way to obtain the expression for $\zeta_{C,I}$ is to use the *scaling property* of the distance zeta function. In short, the scaling property gives us a functional equation which connects the distance zeta function of an RFD (A,Ω) and its scaled version $(\lambda A, \lambda\Omega)$, where $\lambda > 0$ is an arbitrary scaling factor and $\lambda A := \{\lambda x : x \in A\}$. More specifically, the scaling property states that for any RFD (A,Ω), we have

$$\zeta_{\lambda A,\lambda\Omega}(s) = \lambda^s \zeta_{A,\Omega}(s); \tag{1.1.16}$$

this identity is valid on any open connected set $U \subseteq \mathbb{C}$ to which any of the two zeta functions above has a meromorphic continuation. In other words, the complex dimensions of an RFD are invariant under scaling.

Returning to the Cantor set RFD (C,I), we observe that since $I = (0,3^{-1}) \cup [3^{-1}, 2\cdot 3^{-1}] \cup (2\cdot 3^{-1}, 1)$, we can subdivide as follows the integral initially defining $\zeta_{C,I}$ in Equation (1.1.10):

$$\begin{aligned}\zeta_{C,I}(s) &= \int_{(0,3^{-1})} d(x,C)^{s-1}\,\mathrm{d}x + \int_{[3^{-1},2\cdot 3^{-1}]} d(x,C)^{s-1}\,\mathrm{d}x + \int_{(2\cdot 3^{-1},1)} d(x,C)^{s-1}\,\mathrm{d}x \\ &= \int_{(0,3^{-1})} d(x,3^{-1}C)^{s-1}\,\mathrm{d}x + \int_{1/3}^{2/3} d(x,C)^{s-1}\,\mathrm{d}x \\ &\qquad + \int_{(2\cdot 3^{-1},1)} d(x,3^{-1}C+2/3)^{s-1}\,\mathrm{d}x.\end{aligned} \tag{1.1.17}$$

In the first integral appearing in the second line of Equation (1.1.17), we have used the *self-similarity* of the Cantor set C, i.e., the identity

$$C = 3^{-1}C \cup (3^{-1}C + 2\cdot 3^{-1}), \tag{1.1.18}$$

in order to conclude that the part of C contained inside the interval of integration $(0,3^{-1}) = 3^{-1}I$ is equal to $3^{-1}C$ (except for the values of 0 and 1/3, which are inessential). Analogously, in the last integral appearing in Equation (1.1.17), the part of the Cantor set C contained in the interval of integration $(2\cdot 3^{-1},1) = 3^{-1}I + 2/3$ is equal to $3^{-1}C + 2/3$ (except for the values 2/3 and 1, which are inessential as well).

Note that the fourth and sixth integral appearing in Equation (1.1.17) above actually represent the distance zeta functions of the RFDs $\big(3^{-1}C,(0,3^{-1})\big)$ and

$$\big(3^{-1}C+2\cdot 3^{-1},(2\cdot 3^{-1},1)\big)=\big(3^{-1}C+2/3,(0,3^{-1})+2/3\big),$$

respectively, which are copies of the original RFD (C,I) scaled by the factor $1/3$ (with the second one being translated by $2/3$ to the right). In light of this, Equation (1.1.17) reduces to

$$\zeta_{C,I}(s)=\zeta_{3^{-1}C,3^{-1}I}(s)+\int_{1/3}^{2/3}d(x,C)^{s-1}\,\mathrm{d}x+\zeta_{3^{-1}C+2/3,3^{-1}I+2/3}(s). \tag{1.1.19}$$

By the scaling property of the distance zeta function stated in Equation (1.1.16), we now deduce that (see Remark 1.1.4 below)

$$\begin{aligned}\zeta_{C,I}(s)&=\zeta_{3^{-1}C,3^{-1}I}(s)+\int_{1/3}^{2/3}d(x,C)^{s-1}\,\mathrm{d}x+\zeta_{3^{-1}C,3^{-1}I}(s)\\&=2\cdot 3^{-s}\zeta_{C,I}(s)+\int_{I_{1,1}}d(x,C)^{s-1}\,\mathrm{d}x,\end{aligned} \tag{1.1.20}$$

valid for all $s\in\mathbb{C}$ with $\operatorname{Re}s$ sufficiently large. Recalling that the integral over $I_{1,1}:=(1/3,2/3)$ is given by (1.1.11), with $n=k=1$ (see Remark 1.1.3 just below) and solving the above equation for $\zeta_{C,I}$, we recover (1.1.13).

Remark 1.1.3. Note that $d(x,C)$ is in fact equal to the distance from $x\in(1/3,2/3)$ to the boundary of $I_{1,1}$, i.e., to the two-point set $\partial I_{1,1}=\{3^{-1},2\cdot 3^{-1}\}$. Hence, we have that

$$\begin{aligned}\int_{I_{1,1}}d(x,C)^{s-1}\,\mathrm{d}x&=\int_{1/3}^{2/3}d(x,C)^{s-1}\,\mathrm{d}x=\int_{1/3}^{2/3}d(x,\{3^{-1},2\cdot 3^{-1}\})^{s-1}\,\mathrm{d}x\\&=\int_{1/3}^{1/2}(x-3^{-1})^{s-1}\,\mathrm{d}x+\int_{1/2}^{2/3}(2\cdot 3^{-1}-x)^{s-1}\,\mathrm{d}x.\end{aligned}$$

for all $s\in\mathbb{C}$ with $\operatorname{Re}s>0$. Instead of computing the last two easy integrals immediately, we can proceed more elegantly as follows. These two integrals correspond to the distance zeta functions of relative fractal drums $\big(\{3^{-1}\},(3^{-1},2^{-1})\big)$ and $\big(\{2\cdot 3^{-1}\},(2^{-1},2\cdot 3^{-1})\big)$, respectively. Translating both RFDs by -3^{-1} and $-2\cdot 3^{-1}$, respectively, we see that the corresponding distance zeta functions coincide with the one associated with the RFD $\big(\{0\},(0,1/6)\big)$ (whereby in the case of the second RFD, it is convenient to orient the x-axis in the negative direction); so that

$$\int_{I_{1,1}}d(x,C)^{s-1}\,\mathrm{d}x=2\zeta_{\{0\},(0,1/6)}(s)=2\int_0^{1/6}x^{s-1}\,\mathrm{d}x=2\cdot 6^{-s}s^{-1},$$

for all $s\in\mathbb{C}$ with $\operatorname{Re}s>0$. See Figure 1.4.

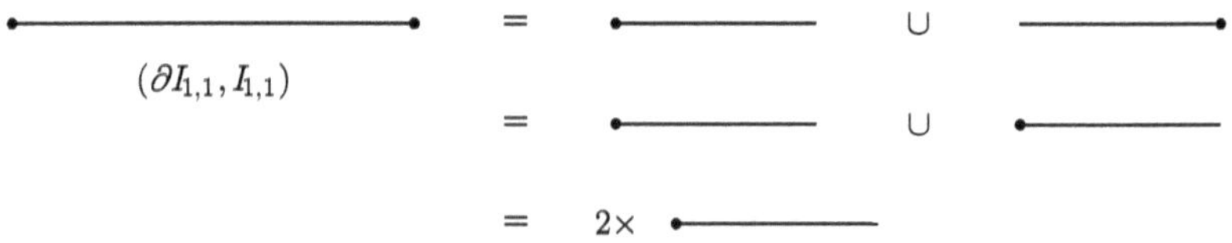

Fig. 1.4 Intuitive explanation of the identity $\zeta_{\partial I_{1,1}, I_{1,1}}(s) = 2 \cdot \zeta_{\{0\},(0,1/6)}(s)$, for all $s \in \mathbb{C}$ such that $\operatorname{Re} s > 0$, appearing in Remark 1.1.3. Here, $I_{1,1} = (1/3, 2/3)$.

Remark 1.1.4. It is easy to verify that the distance zeta function of a relative fractal drum (A, Ω) in $\mathbb{R}^N$ remains unchanged by translating the RFD. More specifically, for any vector $a \in \mathbb{R}^N$, we have that $\zeta_{A,\Omega}(s) = \zeta_{A+a,\Omega+a}(s)$, for all $s \in \mathbb{C}$ with $\operatorname{Re} s$ sufficiently large. A more general statement can be found in Equation (4.2.63) appearing in Lemma 4.2.23 of Subsection 4.2.3. In short, the distance zeta functions (and hence also, the complex dimensions) of RFDs are invariant under the group of displacements of $\mathbb{R}^N$ (the group generated by the rotations, translations and reflections of $\mathbb{R}^N$).

In general, we stress that the procedure described in Example 1.1.2 above can be applied to a wide variety of RFDs exhibiting a self-similar structure, as will be shown in many examples in this monograph. See, e.g., Theorem 4.2.17 in Section 4.2 of Chapter 4 and Equations (4.2.47) and (4.2.48) preceding it (self-similar sprays viewed as RFDs), Theorem 4.2.19 and its proof (self-similar sprays and their higher order counterparts), as well as Examples 4.2.10 (higher-order Cantor sets), 4.2.24 (relative Sierpiński gasket), 4.2.26 (inhomogeneous Sierpiński N-gasket RFD, with $N \geq 2$ arbitrary), 4.2.29 (relative Sierpiński carpet), 4.2.31 (the classic Sierpiński N-carpet, with $N \geq 2$ arbitrary), 4.2.33 (the $1/2$-square fractal), 4.2.34 (the $1/3$-square fractal) and 4.2.35 (a self-similar fractal nest).

For instance, the above procedure, based on using the self-similarity of the RFD and the scaling property of its associated fractal zeta function, can be applied to the well-known Sierpiński gasket A (contained in the unit triangle), in which case the calculation of its zeta function is then reduced to the calculation of the zeta function of the RFD $(\partial\Omega_0, \Omega_0)$, where Ω_0 is the middle open equilateral triangle of side lengths $1/2$ which is removed in the first step of the construction of A. We encourage the interested reader to try to calculate the distance zeta function of A by choosing the parameter δ from Equation (1.1.6) to be greater than $1/4$, as an exercise and motivation for further reading of this monograph or, alternatively, to see Proposition 3.2.3 for details. Here, we only give the closed form for the corresponding distance zeta function of the Sierpiński gasket:

$$\zeta_A(s) = \frac{6(\sqrt{3})^{1-s}2^{-s}}{s(s-1)(2^s-3)} + 2\pi\frac{\delta^s}{s} + 3\frac{\delta^{s-1}}{s-1}, \tag{1.1.21}$$

meromorphic on all of $\mathbb{C}$ and with $\delta > 1/4$. Notice also that the dependence on $\delta > 1/4$ on the right hand side of (1.1.21) is inessential, in the sense that it does not affect the principal part of the above zeta function in any way. Moreover, observe that it follows from Equation (1.1.21) that (independently of δ) the complex dimensions of the Sierpiński gasket A are 0, 1 and $\log_2 3 + \frac{2\pi}{\log 2} k\mathrm{i}$ for every $k \in \mathbb{Z}$, and that all of them are simple.

An even deeper connection between the complex dimensions of a given RFD (or, in particular, of a bounded subset) in $\mathbb{R}^N$ and its inner geometry can be seen in the fractal tube formulas which we obtain in Chapter 5 of this monograph. Roughly speaking, a fractal tube formula for an RFD (A,Ω) is an asymptotic formula for its relative tube function $t \mapsto |A_t \cap \Omega|$ as $t \to 0^+$, expressed as a sum of residues of $\zeta_{A,\Omega}$ (or $\tilde{\zeta}_{A,\Omega}$) taken over the set of (visible) complex dimensions of (A,Ω). These formulas are valid under suitable mild conditions, with or without error term and pointwise or in the sense of Schwartz distributions. Generally, the validity of these formulas for a given RFD (A,Ω) depends on the existence of a meromorphic continuation of the corresponding distance or tube zeta function to a suitable connected open subset $U \subseteq \mathbb{C}$ and on its growth properties on that set. A large class of RFDs satisfy these conditions and, as will be shown in detail in Chapter 5 (and particularly, in Section 5.1), we can apply the theory developed in this book in order to obtain their fractal tube formulas.[8]

For instance, in the case when the zeta function of an RFD (A,Ω) in $\mathbb{R}^N$ can be meromorphically extended to all of $\mathbb{C}$ with only simple poles and when suitable growth conditions are satisfied, we obtain the following exact pointwise fractal tube formula:

$$|A_t \cap \Omega| = \sum_{\omega \in \mathscr{P}(\zeta_{A,\Omega})} \frac{t^{N-\omega}}{N-\omega} \operatorname{res}\left(\zeta_{A,\Omega}, \omega\right), \tag{1.1.22}$$

valid for all $t > 0$ sufficiently small. Here, the set $\mathscr{P}(\zeta_{A,\Omega})$ denotes all of the complex dimensions of (A,Ω).

Example 1.1.5. (*The fractal tube formula for the Cantor string RFD*). Both the Cantor set RFD (C,I) and the Sierpiński gasket A discussed above satisfy the appropriate conditions and we can use Equation (1.1.22) to obtain their well-known fractal tube formulas. In the case of the Cantor set RFD, we have[9]

[8] We refer the interested reader to the introduction of Chapter 5 (on pages 408–411) for a discussion of the history and the geometric interpretation of the classic tube formulas of Steiner, Weyl and Federer (among many others) for "nice" subsets of $\mathbb{R}^N$ (e.g., compact convex sets and smooth, compact submanifolds), prior to the advent of fractal tube formulas in the late 1990s.

[9] The exact computation is given in Example 5.5.3 of Subsection 5.5.2.

$$
\begin{aligned}
|C_t \cap I| &= \sum_{\omega \in \mathscr{P}(\zeta_{C,I})} \frac{t^{1-\omega}}{1-\omega} \operatorname{res}(\zeta_{C,I}, \omega) \\
&= \frac{1}{2\log 3} \sum_{k=-\infty}^{\infty} \frac{(2t)^{1-\omega_k}}{(1-\omega_k)\omega_k} - 2t \\
&= \frac{(2t)^{1-D}}{2\log 3} \sum_{k=-\infty}^{\infty} \frac{(2t)^{-\mathbb{i}k\mathbf{p}}}{(1-\omega_k)\omega_k} - 2t \\
&= t^{1-D} G\left(\log_3 (2t)^{-1}\right) - 2t,
\end{aligned}
\tag{1.1.23}
$$

where $\mathscr{P}(\zeta_{C,I}) := \{0\} \cup (D + \mathbb{i}\mathbf{p}\mathbb{Z})$ is the set of all of the complex dimensions of the RFD (I,C) and $\omega_k := D + \mathbb{i}k\mathbf{p}$ for each $k \in \mathbb{Z}$; furthermore, $D := \dim_B(C,I) = \log_3 2$ and $\mathbf{p} := \frac{2\pi}{\log 3}$ denote, respectively, the relative Minkowski dimension and the 'oscillatory period' of the Cantor string RFD (C,I). Here, G is a positive, nonconstant 1-periodic function, which is bounded away from zero and infinity; specifically, it is given by the following Fourier series expansion (which is absolutely convergent and hence, pointwise convergent for all $x \in \mathbb{R}$):

$$
G(x) := \frac{2^{-D}}{\log 3} \sum_{k \in \mathbb{Z}} \frac{\mathrm{e}^{2\pi \mathbb{i} kx}}{\omega_k(1-\omega_k)}. \tag{1.1.24}
$$

Observe that the presence of the nonconstant periodic function G in (1.1.23) implies that, as is well known (see [LapPo2] and [Lap-vFr3]), the Cantor set RFD (or the Cantor string) is not Minkowski measurable. This fact and the presence of nonreal complex dimensions of the RFD (C,I) on the critical line $\{\operatorname{Re} s = D\}$ are not coincidental but in fact, closely related. More precisely, in this book, we also obtain a criterion for the Minkowski measurability of RFDs which is formulated in terms of the locations of the principal complex dimensions (i.e., the poles of the fractal zeta function with real part D); this criterion generalizes the analogous known result for fractal strings (see [Lap-vFr3, Section 8.3]). More specifically, it states that under suitable assumptions, the Minkowski measurability of an RFD (A,Ω) is equivalent to the absence of nonreal complex dimensions on the critical line $\{\operatorname{Re} s = \dim_B(A,\Omega)\}$, along with the condition that the complex dimension $D := \dim_B(A,\Omega)$ is simple.

We point out that in the case when $N = 1$, the fractal tube formula (1.1.22) becomes the well-known fractal tube formula for fractal strings (see [Lap-vFr3, Sections 8.1 and 8.4]). More precisely, under appropriate conditions and in the case when the geometric zeta function $\zeta_{\mathscr{L}}$ of a given fractal string $\mathscr{L}$ is meromorphic on all of $\mathbb{C}$ and has only simple poles, we have

$$
\begin{aligned}
V_{\mathscr{L}}(t) &:= |(\partial\Omega_{\mathscr{L}})_t \cap \Omega_{\mathscr{L}}| \\
&= \sum_{\omega\in\mathscr{P}(\zeta_{\partial\Omega_{\mathscr{L}},\Omega_{\mathscr{L}}})} \frac{t^{N-\omega}}{N-\omega}\operatorname{res}\left(\zeta_{A,\Omega},\omega\right) \\
&= \sum_{\omega\in\mathscr{P}(\zeta_{\mathscr{L}})} \frac{(2t)^{1-\omega}}{\omega(1-\omega)}\operatorname{res}\left(\zeta_{\mathscr{L}},\omega\right) + \{2t\zeta_{\mathscr{L}}(0)\}.
\end{aligned}
\tag{1.1.25}
$$

Here, the term $\{2t\zeta_{\mathscr{L}}(0)\}$ is equal to $2t\zeta_{\mathscr{L}}(0)$ if 0 is not a pole of $\zeta_{\mathscr{L}}$. If, however, 0 is a simple pole of $\zeta_{\mathscr{L}}$, then we replace $\{2t\zeta_{\mathscr{L}}(0)\}$ on the right-hand side of (1.1.25) with the term

$$
2t(1-\log(2t))\operatorname{res}(\zeta_{\mathscr{L}},0) + 2t\zeta_{\mathscr{L}}[0]_0, \tag{1.1.26}
$$

where $\zeta_{\mathscr{L}}[0]_0$ stands for the constant term in the Laurent series expansion of $\zeta_{\mathscr{L}}$ around $s=0$. This is in agreement with [Lap-vFr3, Corollary 8.10] under the assumption of exactness, i.e., the absence of an error term. The last equality in Equation (1.1.25) follows from the general functional equation connecting the zeta functions $\zeta_{\mathscr{L}}$ and $\zeta_{\partial\Omega_{\mathscr{L}},\Omega_{\mathscr{L}}}$ which, as has already been mentioned, is the generalization to any fractal string of Equation (1.1.15) obtained above; i.e., we have that

$$
\zeta_{\partial\Omega_{\mathscr{L}},\Omega_{\mathscr{L}}}(s) = \frac{2^{1-s}}{s}\zeta_{\mathscr{L}}(s), \tag{1.1.27}
$$

a key identity which is valid for all $s\in\mathbb{C}$ with $\operatorname{Re}s > \overline{\dim}_B(\partial\Omega_{\mathscr{L}},\Omega_{\mathscr{L}})$ and hence, more generally, on every connected open set $U\subseteq\mathbb{C}$ to which any (and thus both) of the two zeta functions has a meromorphic continuation.

Going back to the Sierpński gasket A, since all of its complex dimensions are simple, we can obtain its fractal tube formula by using (1.1.22) and (1.1.21):

$$
\begin{aligned}
|A_t| &= \sum_{\omega\in\mathscr{P}(\zeta_A)} \frac{t^{2-\omega}}{2-\omega}\operatorname{res}\left(\zeta_A,\omega\right) \\
&= t^{2-\log_2 3}\frac{6\sqrt{3}}{\log 2}\sum_{k=-\infty}^{\infty}\frac{(4\sqrt{3})^{-\omega_k}t^{-\mathbb{i}k\mathbf{p}}}{(2-\omega_k)(\omega_k-1)\omega_k} + \left(\frac{3\sqrt{3}}{2}+\pi\right)t^2,
\end{aligned}
\tag{1.1.28}
$$

valid for all $t\in(0,1/2\sqrt{3})$, where $\omega_k := \log_2 3 + \mathbb{i}k\mathbf{p}$ (for each $k\in\mathbb{Z}$) and $\mathbf{p} := 2\pi/\log 2$. The presence of the oscillatory function in (1.1.28) (namely, the nonconstant multiplicatively periodic function represented by its absolutely convergent and hence, convergent Fourier series) shows that the Sierpiński gasket is not Minkowski measurable, a well-known fact which is reflected in the presence of nonreal complex dimensions of the Sierpiński gasket located on the critical line $\{\operatorname{Re}s = \log_2 3\}$.

Many further examples of computations of fractal zeta functions, complex dimensions and fractal tube formulas are provided throughout this monograph, both for classic self-similar and non self-similar fractals, as well as for new classes of bounded sets and relative fractal drums.

Along with several collaborators, the first author has undertaken since the early 1990s a systematic study of zeta functions associated with fractal strings and their counterparts in certain higher-dimensional situations, namely, for fractal sprays; see, in particular, his joint papers with C. Pomerance [LapPo1–3]. In a series of papers and several research monographs (including three books with M. van Frankenhuijsen [Lap-vFr1–3], and the book [Lap6]), it has grown into a well-established theory and is today an active and rapidly growing area of research. For the theory of fractal strings and/or complex dimensions in a variety of situations, beside the aforementioned books, see, for example, [CranMH], [DemDenKoÜ], [DemKoÖÜ], [DenKoÖÜ], [deSLapRRo] [DubSep], [ElLapMacRo], [Es1–2], [EsLi1–2], [Fal2], [Fr], [FreKom], [HamLap], [HeLap], [HerLap1–5], [KeKom], [Kom], [LalLap1–2], [Lap1–3], [Lap7–10], [LapLéRo], [LapLu1–3], [LapLu-vFr1–2], [LapMa1–2], [LapPe1–3], [LapPeWi1–2], [LapPo1–3], [LapRo1–2], [MorSep], [MorSepVi1–2], [Ol1–2], [Pe], [PeWi], [Ra1–2], [RatWi2], [Steinh], [Tep1–2], [Wi], [WiZä], [Zä4–5], along with the relevant references therein. In addition, we point out that Chapter 13 of [Lap-vFr3] contains an exposition of several recent developments in the theory, prior to the present general higher-dimensional theory of complex dimensions.

Other, very different approaches to a higher-dimensional theory of certain fractal sets (namely, fractal sprays and self-similar tilings) were developed in references [LapPe1–3] and [LapPeWi1–2] by the first author, E. Pearse and S. Winter, as well as in the related works [Pe] and [PeWi], via fractal tube formulas and the associated scaling and tubular zeta functions. An earlier approach, based directly on tube formulas but not using any kind of zeta function, was proposed in [LapPe1]. See, respectively, [Lap-vFr3, Sections 13.1 and 12.2.1] of the second revised and enlarged edition of [Lap-vFr2], for an exposition of these approaches.

Since the zeta functions introduced in [LapPe2–3] and [LapPeWi1] are very different in nature from those studied in this monograph, it would be of interest to determine when they give rise to the same or closely related results, as far as the complex dimensions are concerned. This will be done in Subsection 5.5 by using the general higher-dimensional theory of fractal tube formulas developed in Chapter 5. We note that in [LapPe3], an example is provided for which the complex dimensions depend on the choice of the iterated function system giving rise to that self-similar fractal set. It might also be interesting to see whether one can obtain fractal tube formulas in the setting of our monograph, and, in the special case of the Koch snowflake curve, compare the resulting formula with the one obtained in [LapPe1], while in the case of fractal sprays and self-similar tilings, with the tube formulas obtained and used in [LapPe2–3] and [LapPeWi1–2].

We stress that a significant advantage of the present theory of fractal zeta functions—and, therefore, of the corresponding higher-dimensional theory of complex dimensions developed in this book—is that it is applicable to arbitrary bounded (or equivalently, compact) sets in $\mathbb{R}^N$, and can be extended (at least, in principle) to the general setting of arbitrary compact metric measure spaces. In particular, no assumption of self-similarity, or, more generally, of "self-alikeness" of any kind, is made about the underlying fractals (or, within the broader setting of Chapter 4, about the relative fractal drums under consideration). In addition, the fractals in question do not have to be the boundaries of fractal sprays (in the sense of [LapPo3]), that is, of countable disjoint unions of scaled copies of a given bounded set.

We hope that the results obtained in this monograph will provide a new impetus and direction to the higher-dimensional theory of complex dimensions of fractal sets. The zeta functions studied thus far have proved to be an important tool in various fields of mathematics, including complex analysis, number theory, arithmetic geometry, operator algebras, representation theory, fractal geometry, functional analysis, mathematical physics, differential equations, and dynamical systems. Likewise, the new zeta functions introduced and studied in this book should prove to be useful tools in the rapidly expanding theory connecting aspects of fractal geometry, number theory, harmonic analysis, geometric analysis, differential equations and dynamical systems.

1.2 A Short Survey of the Contents

Let us briefly describe the contents of this monograph. In Section 2.1, we establish the holomorphy of the distance zeta function ζ_A on the right half-plane $\{\operatorname{Re} s > \overline{\dim}_B A\}$ of the complex plane $\mathbb{C}$, where $\overline{\dim}_B A$ is the upper box (or Minkowski) dimension of A, and show that the lower bound $\overline{\dim}_B A$ is optimal; see Theorem 2.1.11. In other words, we establish the equality of $\overline{\dim}_B A$ and the abscissa of (absolute) convergence of ζ_A. Therefore, if we know the zeta function of a fractal set A, we can determine the upper box dimension of A; see Corollary 2.1.63. Moreover, under some mild additional hypotheses on A, we show that the half-plane of convergence $\{\operatorname{Re} s > \overline{\dim}_B A\}$ and the half-plane of holomorphic continuation of ζ_A coincide.

In Subsection 2.1.5, we introduce an equivalence relation between zeta functions of fractal sets; see Definition 2.1.69. This enables us to allow for more flexibility in the study and the understanding of the main features of the zeta functions discussed in this monograph. We also show that the distance zeta function has a suitable continuity property with respect to any nonincreasing sequence of compact sets $(A_k)_{k=1}^{\infty}$; see Theorem 2.1.78.

In Section 2.2, we show that the residue of the meromorphic extension of the distance zeta function of a fractal set to a connected open neighborhood of $D := \dim_B A$ (provided the extension exists), computed at the simple pole $s = D$, is closely related to the D-dimensional Minkowski content of A; see Theorem 2.2.3.

A new fractal zeta function, denoted by $\tilde{\zeta}_A$, is introduced in Equation (2.2.20) of Definition 2.2.8; it is referred to as the "tube zeta function" of A and involves the function $(0,\delta) \ni t \mapsto |A_t|$ instead of the function $A_\delta \ni x \mapsto d(x,A)$ in Equation (2.1.1) of Definition 2.1.1, where $|A_t|$ and $d(x,A)$ denote, respectively, the N-dimensional volume of the t-neighborhood of A and the distance from x to A. If a bounded or compact set $A \subset \mathbb{R}^N$ is Minkowski measurable (and under some mild additional hypotheses), we show that the residue of $\tilde{\zeta}_A(s)$ at $s = \dim_B A$, the Minkowski (or box) dimension of A, is equal to the Minkowski content of A. More generally, even if A is not Minkowski measurable, we obtain analogous results, expressed as inequalities involving the upper and lower Minkowski contents of A; see Theorem 2.2.14. Finally, we show that, provided $\overline{\dim}_B A < N$, the half-plane of (absolute) convergence of $\tilde{\zeta}_A$ is exactly the same as for ζ_A (as described above). Hence, the abscissa of convergence of the Dirichlet integral initially defining $\tilde{\zeta}_A$ coincides with $\overline{\dim}_B A$, the upper box dimension of A; see Equation (2.2.52) in Proposition 2.2.19.

Moreover, still assuming that $\overline{\dim}_B A < N$, we show that given any domain $U \subseteq \mathbb{C}$ (containing the critical line $\{\mathrm{Re}\, s = \overline{\dim}_B A\}$), $\tilde{\zeta}_A$ has a meromorphic continuation to U if and only if ζ_A does, and in that case, $\tilde{\zeta}_A$ and ζ_A have the same poles (with the same multiplicities) in U. Hence, the (visible) complex dimensions of A can be defined indifferently via ζ_A or $\tilde{\zeta}_A$. It also follows that, in addition to having the same half-plane of convergence $\{\mathrm{Re}\, s > \overline{\dim}_B A\}$, ζ_A and $\tilde{\zeta}_A$ have the same half-plane of holomorphic continuation. Note that the condition according to which $\overline{\dim}_B A < N$ is satisfied by most fractals of interest and implies that $|A|_N = 0$.

Section 2.3 is devoted to solving (in certain frequently encountered situations) the problem of the existence and the construction of the meromorphic extensions of various zeta functions. We first deal with zeta functions associated with the perturbation of the Riemann strings (Theorem 2.3.2) and of Dirichlet strings (Theorem 2.3.10). We then deal with the distance and tube zeta functions of a class of Minkowski measurable sets in Euclidean spaces (Theorem 2.3.18 and Theorem 2.3.37), as well as with the fractal zeta functions of a class of Minkowski nonmeasurable sets (Theorem 2.3.25, Corollary 2.3.26 and Theorem 2.3.37). In particular, we provide natural sufficient conditions for the existence of the meromorphic continuation of those distance and tube zeta functions. (The case of the distance zeta function is dealt with in Theorem 2.3.37.) This is significant from the point of view of future developments, in light of the fact that the complex dimensions of the given fractal set can be defined as the poles of the meromorphic continuation (in a suitable region) of the associated distance (or tube) zeta function.

In Section 3.1, we introduce a class of quasiperiodic sets. Using generalized Cantor sets with two parameters, we provide a construction of such sets, based on the Gel'fond–Schneider theorem from the theory of transcendental numbers.

In Section 3.2, we study the distance zeta functions of the Sierpiński carpet and the Sierpiński gasket. We also compute the corresponding principal complex dimensions. The method used in the computation of these distance zeta functions will serve as a motivation to introduce the notion of 'relative fractal drums', which will be the central object of study in Chapter 4.

In Section 3.3, we construct a class of bounded fractal strings $\mathscr{L}$ with principal complex dimensions of *any* prescribed order; see Theorem 3.3.6. Furthermore, a class of fractal strings with principal complex dimensions of infinite order (that is, with *essential singularities* on the corresponding critical line) is also constructed in the same theorem. The construction is based on using iterated *tensor products* of suitably chosen bounded fractal strings.

In Section 3.4, we also introduce the notion of weighted zeta function, establish a corresponding holomorphicity result, and show that the derivative of a weighted zeta function is again a weighted zeta function; see Theorem 3.4.4. Furthermore, in Corollary 3.4.7, we obtain new classes of harmonic functions associated with fractal sets.

In Section 3.5, we introduce the notion of fractal nest. Using our zeta functions, we extend to the N-dimensional case Tricot's formula for the box dimension of a discrete spiral of the focus type or of the limit cycle type; see Equations (3.5.14) and (3.5.17).

In Section 3.6, we introduce the notion of geometric chirp and compute the associated distance zeta function. This enables us to extend Tricot's formula for the box dimensions of related chirp curves in $\mathbb{R}^2$ to spherically symmetric chirp-like surfaces in $\mathbb{R}^N$, for any $N \geq 2$; see Proposition 3.6.2. The case when $N = 3$ is of particular interest since it arises naturally in the study of spherically symmetric solutions of p-Laplace boundary value problems. We also introduce the notions of multiple strings and string chirps, which include geometric chirps as a special case, and study their fractal zeta functions. At the end of this section, we introduce the notion of Cartesian product of fractal strings, and in Theorem 3.6.5, we determine the associated distance zeta function as well as the corresponding upper box dimension.

The aim of Section 3.7 is to study the fractal zeta functions of a class of fractal sets (called 'zigzagging fractals') for which the upper and lower box dimensions do not coincide. It is noteworthy that the associated zeta functions are alternating; see Theorem 3.7.2 and Equation (3.7.2).

In Section 4.1, we introduce the notion of *relative zeta function* associated to an ordered pair (A,Ω), where A is a possibly unbounded subset of $\mathbb{R}^N$ and Ω is an open subset of $\mathbb{R}^N$ having finite N-dimensional Lebesgue measure (but being also possibly unbounded). We propose to call (A,Ω) a "relative fractal drum" (or RFD, in short). The associated optimal half-plane of holomorphic continuation involves $\overline{\dim}_B(A,\Omega)$, the relative upper box dimension of A with respect to Ω. In other words, under mild hypotheses, we show that the relative zeta function $\zeta_{A,\Omega}(s)$ is holomorphic on the right half-plane $\{\mathrm{Re}\, s > \overline{\dim}_B(A,\Omega)\}$, and that the lower bound is optimal; see Theorem 4.1.7. More precisely, we prove that the abscissa of (absolute) convergence of $\zeta_{A,\Omega}(s)$ always coincides with $\overline{\dim}_B(A,\Omega)$ (determining the maximal right half-plane of absolute or Lebesgue convergence; see part (b) of Theorem 4.1.7), and that under mild assumptions, it also coincides with the abscissa of holomorphic continuation of $\zeta_{A,\Omega}$ (determining the maximal right half-plane of holomorphic continuation; see part (c) of Theorem 4.1.7). We note that in this generality, the relative box dimension had been introduced earlier in [Žu4]. (Earlier special cases had been used, for example, in [BroCar], [Lap1–3], [LapPo2–3], [LapMa2], [HeLap] and [Lap-vFr1–3].) In light of the above result, we deduce that using zeta functions, it is possible to consider unbounded geometric chirps relative to the associated bounding envelope Ω, and then to show that the natural extension of Tricot's formula also holds in this case; see Example 4.4.1.

Note that any bounded subset $A \subset \mathbb{R}^N$ can be viewed as a relative fractal drum of the form (A,A_δ), for any $\delta > 0$. Consequently, the theory of fractal zeta functions of RFDs and their associated complex dimensions developed in Chapter 4 (and Chapter 5) extends naturally its counterpart for bounded subsets of $\mathbb{R}^N$ developed in Chapters 2 and 3.

Example 4.2.10 provides an explicit construction of a relative fractal drum of $\mathbb{R}$ which possesses an infinite set of poles of arbitrary order or even essential singularities located on the critical line in arithmetic progression. The construction is based on an "iterated Cantor spray" and can be generalized to a large class of relative fractal drums of $\mathbb{R}^N$ (with $N \geq 1$ arbitrary), as is stated in Theorem 4.2.19 and Remark 4.2.21.

In Theorem 4.4.5, we consider the zeta function of the Cartesian product of three strings relative to the related bounding rectangular parallelepiped. We also study the fractal zeta functions of a class of relative fractal drums with logarithmic gauge functions; see Theorem 4.5.1 for the Minkowski measurable case, and Theorem 4.5.2 for the Minkowski nonmeasurable case. Hence, such relative fractal drums scale naturally according to a non power law.

In Section 4.3, we discuss some of the known results about the spectral asymptotics of a (relative) fractal drum, focusing on the leading term (of the asymptotics of the eigenvalues of the Dirichlet Laplacian) and a corresponding (sharp) error term, obtained in [Lap1] and expressed in terms of the (upper) Minkowski (or box) dimension of the boundary. We then apply this remainder estimate, along with some of our earlier techniques, to establish the existence of a (nontrivial) meromorphic extension of the spectral zeta function of a fractal drum, a result already obtained by the first author in [Lap3] (in a slightly different manner). We also use our results in Sections 4.5 and 4.6 below in a key manner in order to establish the optimality of the upper bound obtained for the corresponding abscissa of meromorphic continuation; this latter result is new.

In Section 4.5, we construct a class of relative fractal drums with explicit values of the abscissa of meromorphic continuation of the corresponding relative zeta functions. We interpret these results in terms of the geometric zeta functions of fractal strings, as well as in terms of the distance zeta functions of bounded sets on the real line. The fractal string interpretation is obtained by using a countable union of generalized Cantor strings $C^{(a_j)}$, with suitably chosen parameters $a_j \in (0,1/2)$, involving a sequence of prime numbers; see Theorem 4.5.20. This shows, in particular, that our main results on meromorphic extensions of tube and distance zeta functions, obtained in Section 2.3, are in general optimal.

In Section 4.6, we construct a class of quasiperiodic relative fractal drums possessing infinitely many algebraically independent quasiperiods; see Theorem 4.6.9. These drums are said to be transcendentally ∞-quasiperiodic. Furthermore, we construct a relative fractal drum (A,Ω) such that each of the points on the 'critical line' $\{\mathrm{Re}\, s = \overline{\dim}_B(A,\Omega)\}$ is a nonisolated singularity of the corresponding relative distance or tube zeta function; see Theorem 4.6.13. These drums are also transcendentally quasiperiodic of infinite order, in the sense of Definition 4.6.7, and their explicit construction provided in this section makes an essential use of the celebrated Baker theorem (Theorem 3.1.14 on page 198) about transcendental numbers, itself a generalization of the aforementioned Gel'fond–Schneider theorem (Theorem 3.1.7 on page 192). We also construct fractal strings and bounded sets in the real line with the same property; see Corollary 4.6.17. We call these new geometric objects (maximal) hyperfractals; see Definition 4.6.23. For these latter constructions, we use a suitable family of generalized Cantor sets $C^{(m,a)}$ with two parameters; see Definition 3.1.1. More generally, *strong hyperfractals* are subsets A of $\mathbb{R}^N$ such that the associated fractal zeta function ζ_A (or, equivalently, $\tilde{\zeta}_A$) admits the critical line $\{\mathrm{Re}\, s = \overline{\dim}_B A\}$ as a (meromorphic) *partial natural boundary* (that is, ζ_A cannot be meromorphically continued beyond the critical line); see Definition 4.6.23. Finally, even more generally, *hyperfractals* have the same property, but with respect to some suitable curve, called a screen, and not just the critical line.

In Section 4.7, we show that the complex dimensions of relative fractal drums are preserved under embeddings into higher-dimensional spaces. As a result of the proof, so are the residues of the corresponding fractal zeta functions at any (simple) complex dimension. This provides a significant generalization of Kneser's result [Kne, Satz 7] (see also [Res]), where the independence of the normalized Minkowski content on the dimension of the ambient space was established. Theorem 4.7.9 provides a connection between the relative tube zeta function of the original relative fractal drum and the relative tube zeta function of its embedding into a higher-dimensional space. The results of Section 4.7 can be used in order to determine the possible complex dimensions of special types of higher-dimensional relative fractal drums without explicitly computing the corresponding distance (or tube) zeta functions. This application is nicely illustrated in Example 4.7.15, where the (possible) complex dimensions of the Cantor dust are determined.

The main goal of Chapter 5 is to obtain and establish general pointwise and distributional tube formulas for relative fractal drums in $\mathbb{R}^N$ (with $N \geq 1$ arbitrary), and, in particular, for bounded subsets of $\mathbb{R}^N$. These results extend to arbitrary dimensions the (pointwise and distributional) fractal tube formulas originally obtained for fractal strings in [Lap-vFr1–3](see, especially, [Lap-vFr3, Chapters 5 and 8]) and then extended to suitable fractal sprays and self-similar tilings in [LaPe2–3], and, more generally, in [LapPeWi1]. We also extend to higher dimensions the Minkowski measurability criterion obtained for fractal strings in [Lap-vFr1–3] (see, especially, [Lap-vFr3, Section 8.3]), as well as illustrate those results by means of a variety of examples, including fractal strings, self-similar fractal sprays, the Sierpiński gasket and carpet and their higher-dimensional analogs, along with non self-similar examples such as "fractal nests" and "geometric chirps".

Recall from [Lap-vFr3] that, essentially, "fractal tube formulas" consist in expressing the N-dimensional volume of the t-neighborhoods of relative fractal drums in terms of the underlying complex dimensions, appearing as the (co-)exponents of the resulting generalized Fourier series (in the t-variable). Accordingly, fractal tube formulas enable us to obtain a very precise understanding of the *intrinsic oscillations* of fractals and thereby, to make explicit *the key relationship between complex dimensions and oscillatory phenomena in fractal geometry*. The theory developed in Chapter 5 extends this essential connection to any dimension (i.e., to any Euclidean space $\mathbb{R}^N$, with $N \geq 1$), without making any assumption of self-similarity or of a particular type of underlying fractal geometry.

More specifically, the contents of Chapter 5 can be described in more detail as follows:

In Section 5.1, the main result is Theorem 5.1.14, which provides a pointwise fractal tube formula, with or without an error term, depending on the growth properties of the corresponding relative tube zeta function.

In Section 5.2, in order to weaken the growth conditions imposed in Theorem 5.1.14, we use a distributional approach and derive a fractal tube formula that holds distributionally (on an appropriate space of test functions), also with or without an error term (depending on the hypotheses). Accordingly, the main results of Section 5.2 are Theorem 5.2.6, which is the distributional analog of Theorem 5.1.14, and Theorem 5.2.11, which provides an estimate for the corresponding distributional error term.

In Section 5.3, we "translate" under essentially the same growth assumptions the results of Sections 5.1 and 5.2 (which were expressed in terms of the relative tube zeta function) in terms of the more flexible and practical (as well as geometric) notion of relative distance zeta function. In order to do so, we introduce in Definition 5.3.1 a new type of fractal zeta function, called the *relative shell zeta function*. In the process, we derive the main properties of the relative shell zeta function as well as functional equations that connect it to the relative tube and distance zeta functions; see Theorems 5.3.2, 5.3.3 and 5.3.6. Finally, we note that the main results of Section 5.3 are the pointwise and distributional tube formulas of Theorems 5.3.16 and 5.3.21, respectively.

In Section 5.4, under suitable hypotheses, we obtain a necessary and sufficient criterion for the Minkowski measurability of a large class of relative fractal drums, expressed in terms of their fractal zeta functions and to be described below. The sufficiency part (Theorem 5.4.2) of this criterion is a consequence of the well-known Wiener–Pitt Tauberian theorem. In short, it states that a relative fractal drum (and, in particular, a bounded set) in $\mathbb{R}^N$ is Minkowski measurable if the only pole of its corresponding fractal zeta function located on the critical line is real and simple. Furthermore, this pole is then equal to the relative box dimension of the drum. Moreover, Theorem 5.4.2 then establishes a useful connection between the Minkowski content of the given relative fractal drum and the residue of the corresponding fractal zeta function evaluated at this pole. On the other hand, if, in addition, there are other poles on the critical line, the Wiener–Pitt Tauberian theorem only yields an upper bound for the upper Minkowski content of the relative fractal drum under consideration (see Theorem 5.4.4).

In order to establish the other direction of the characterization of Minkowski measurability, we introduce (in Definition 5.4.6) a new fractal zeta function, called the *relative Mellin zeta function*. Its basic properties are given in Theorems 5.4.7, 5.4.9 and 5.4.10. This new zeta function is needed in order to extend the distributional tube formula of Theorem 5.3.21 to a larger space of test functions, which allows one to use the uniqueness theorem for almost periodic distributions in the proof of Theorem 5.4.15. Finally, by combining Theorem 5.4.2 and 5.4.15, we obtain the sought for Minkowski measurability criterion, in Theorem 5.4.20.

More specifically, under suitable hypotheses, the Minkowski measurability criterion obtained in Theorem 5.4.20 states that an RFD (A,Ω) (and, in particular, a bounded set) in $\mathbb{R}^N$ is Minkowski measurable if and only if it does not admit any nonreal principal complex dimensions (i.e., $D=\dim_B(A,\Omega)$ is its only complex dimension of real part D) and D is simple. According to the fractal tube formulas obtained in Chapter 5, this means that, under appropriate assumptions, an RFD (A,Ω) (or, in particular, a bounded set) in $\mathbb{R}^N$ is Minkowski measurable if and only if its tube function (expressing the volume $|A_t\cap\Omega|_N$ of its t-neighborhoods $A_t\cap\Omega$) does not have any oscillations of leading order (i.e., of order t^{N-D} as $t\to 0^+$) or, equivalently, in the terminology introduced and used in various places in Chapters 4 and 5, if and only if it is not "fractal in dimension D".

Subsection 5.4.4 is dedicated to the notion of h-Minkowski measurability, relative to a suitable (and nontrivial) gauge function (and thereby corresponding physically and geometrically to a scaling behavior which does not obey a pure power law). More specifically, a general result about a class of relative fractal drums having (at most) finitely many complex dimensions located on the critical line $\{\mathrm{Re}\, s = D\}$ and such that the multiplicity m of $s = D$ is strictly greater than the multiplicities of the nonreal poles with real part D, is given in Theorems 5.4.27 and 5.4.32 of Subsection 5.4.4. In short, such a relative fractal drum is then Minkowski degenerate with infinite Minkowski content, but is also h-Minkowski measurable, with respect to an appropriate gauge function. More specifically, $h(t) := (\log t^{-1})^{m-1}$ for all $t \in (0,1)$ and m is the order of the associated complex dimension D. Furthermore, an explicit expression for the h-Minkowski content is also given in terms of the $-m$-th coefficient in the Laurent expansion of the corresponding fractal zeta function around the pole D. Theorem 5.4.29 shows that the optimal tube function asymptotic expansion involves the difference between the abscissa of (absolute) convergence and the abscissa of meromorphic continuation of the fractal zeta function. Moreover, Theorem 5.4.30 can be viewed as the converse of Theorem 4.5.1. In particular, the results obtained in this subsection also show that our results obtained in Chapter 2 and 4 about the existence of meromorphic extensions of fractal zeta functions are, in some sense, optimal.

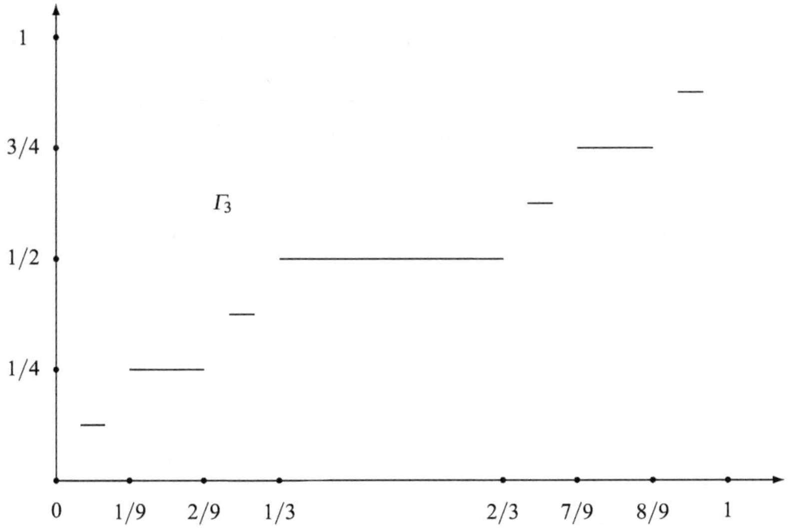

Fig. 1.5 The third approximation Γ_3 of the graph Γ of the Cantor function $f_C : [0,1] \to \mathbb{R}$, defined in the caption of Figure 1.6 just below. The first approximation is defined by $\Gamma_1 := [1/3, 2/3] \times \{1/2\}$, that is, as the longest line-segment (i.e., of length $1/3$), while the second approximation, $\Gamma_2 := \Gamma_1 \cup \big([1/9, 2/9] \times \{1/4\}\big) \cup \big([7/9, 8/9] \times \{3/4\}\big)$, consists of Γ_1 and the next two longest line-segments, of length $1/9$ each. Here, we have that $\Gamma_1 \subset \Gamma_2 \subset \Gamma_3$ and Γ_3 is equal to the union of seven pairwise disjoint line-segments. As we can see, the approximations Γ_j, for $j = 1, 2, 3, \ldots$, of the graph of the Cantor function f_C follow the construction of Cantor's ternary set.

In Section 5.5, we discuss in details several interesting examples and applications of the fractal tube formulas for relative fractal drums developed in Sections 5.1–5.4. Notable among them is the example of a relative fractal drum based on a version of the graph of the *Cantor function* (Example 5.5.14), often called (after [Man1]) the *devil's staircase* in the literature on fractal geometry; see Figures 1.5, 1.6 and 1.7, along with Remark 1.2.1 just below.

Remark 1.2.1. The Cantor function, whose graph is depicted in Figure 1.7 near the point P of its graph Γ (see also Figure 1.6), is α-Hölderian, with $\alpha = \log_3 2$. Here, we can see a striking difference between the scaling rate in the vertical direction (which is equal to 2^{-j} at step $j \in \mathbb{N}$ of the construction) and the scaling rate in the horizontal direction (which is equal to 3^{-j}). Since $2^{-j}/3^{-j} = (3/2)^j \to \infty$ as $j \to \infty$, we have a dramatic elongation in the vertical direction near the points of the graph of f_C, corresponding to the points of the middle-third Cantor set. This is a reflection of the fact that the Cantor graph Γ is *not* self-similar, but is instead an inhomogeneous *self-affine* set; i.e. (see Figure 1.5),

$$\Gamma = S(\Gamma) \cup \Gamma_1, \tag{1.2.1}$$

where

$$\begin{aligned} S(\Gamma) &:= M\Gamma \cup \left(M\Gamma + (2/3, 1/2)^\top\right), \\ M &:= \begin{bmatrix} 1/3 & 0 \\ 0 & 1/2 \end{bmatrix}, \quad M\Gamma := \{Mx : x \in \Gamma\} \end{aligned} \tag{1.2.2}$$

and $\Gamma_1 := [1/3, 2/3] \times \{1/2\}$ is a nonhomogeneous part of Equation (1.2.1). For example, $M\Gamma$ corresponds to the 'left third' of Γ (i.e., to the subset $\{x = (x_1, x_2) \in \Gamma : 0 \le x_1 \le 1/3\}$ of Γ), which is obtained from Γ by scaling it by the factor $1/3$ in the horizontal direction and then by the factor $1/2$ in the vertical direction. The matrix M is called the *affinity matrix*. The 'right third' of Γ (i.e., the subset $\{x = (x_1, x_2) \in \Gamma : 2/3 \le x_1 \le 1\}$ of Γ) is obtained by translating its 'left third' $M\Gamma$ by $(2/3, 1/2)^\top$ (here, by $\top$ we denote the matrix operation of transposition). More information about inhomogeneous self-affine sets can be found in Remark 2.1.87 of Subsection 2.1.6.

Note that the points $(3^{-j}, 2^{-j})$, for $j \in \mathbb{N}$, near the point $(0,0)$ of the graph of the Cantor function, satisfy the equation $y = x^\alpha$ with $\alpha = \log_3 2$. Also note that the part of Γ_7 contained in the rectangle of width 3^{-4}, with vertices at P and Q, is congruent to the part of Γ_7 in the analogous rectangle near the origin, containing the points $(0,0)$ and $(3^{-4}, 2^{-4})$. In fact, the same is true for all subsequent approximations Γ_j of Γ, where $j \ge 7$. It is easy to see that the subsets Γ_j of the plane converge to Γ in the Hausdorff metric, as $j \to \infty$.

As is well known, the box dimension (and hence also, the Hausdorff dimension) of the graph of the Cantor function is trivial; that is, it is equal to one because the graph is rectifiable (i.e., it has finite length). Therefore, according to Mandelbrot's definition of fractality given in [Man1], the Cantor graph is not "fractal". On the

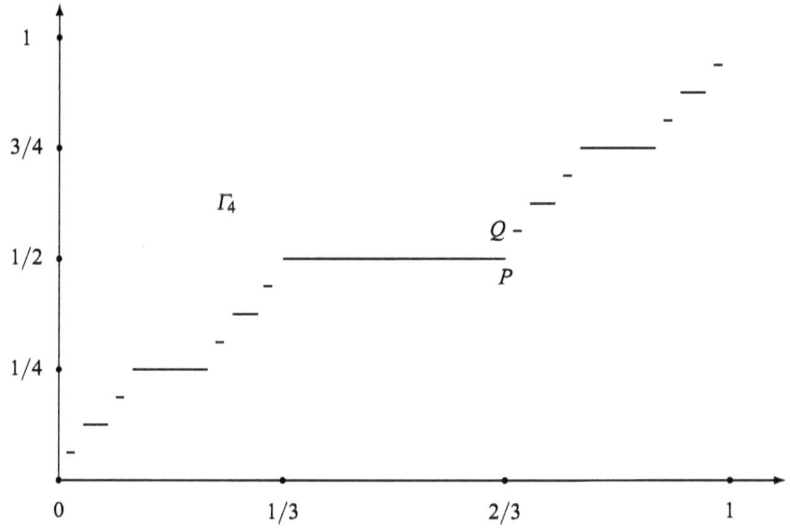

Fig. 1.6 The fourth approximation Γ_4 of the graph Γ of the *Cantor function* $f_C : [0,1] \to \mathbb{R}$. Here, Γ is defined as the closure of the subset $\cup_{j=1}^{\infty} \Gamma_j$ of the plane $\mathbb{R}^2$, where the approximations Γ_j, for $j = 1,2,3,\ldots$, are indicated in this figure and in Figure 1.5 above. For example, Γ_4 is equal to the union of 15 pairwise disjoint line segments. Note that $\Gamma_j \subset \Gamma_{j+1}$ for all $j \in \mathbb{N}$. The Cantor function f_C is continuous and nonconstant, but its pointwise derivative is equal to 0 almost everywhere in $[0,1]$ (more specifically, it vanishes identically on the complement of the Cantor set with respect to $[0,1]$). Hence, as is well known, it is not absolutely continuous, since the Newton–Leibniz formula, $f_C(y) - f_C(x) = \int_x^y f_C'(t)\,dt$, is no longer valid for all $x, y \in [0,1]$. For example, $f_C(1) - f_C(0) = 1$, while $\int_0^1 f_C'(t)\,dt = 0$. A part of the seventh approximation Γ_7 between the points P and Q is shown in Figure 1.7 below.

other hand, intuitively, one would still like to refer to this graph as being "fractal"; the results obtained in Example 5.5.14 provide a partial justification for that. Namely, they show that the zeta function of the relative fractal drum based on the Cantor graph has nonreal poles located to the left of the critical line $\{\mathrm{Re}\, s = 1\}$ and having real part equal to the box dimension of the middle-third Cantor set (that is, to $\log_3 2 \approx 0.63$), as was predicted in [Lap-vFr1–3] on the basis of an approximate tube formula (see, in particular, [Lap-vFr3, Section 12.1]). We then deduce from the theory developed in Chapter 5 that *these poles generate lower-order oscillations* in the asymptotic expansion of the tube function of the relative fractal drum associated to the Cantor graph. According to the definition of fractality introduced in Chapter 4 and further discussed as well as refined in Chapter 5, it follows that the Cantor graph RFD is "fractal". More specifically, it is not fractal in dimension $D = 1$ (the largest possible real dimension), but it is fractal in dimension $d = \log_3 2$, the dimension of the ternary Cantor set. We also conjecture that the same is true for the actual Cantor graph and provide partial evidence towards this conjecture.

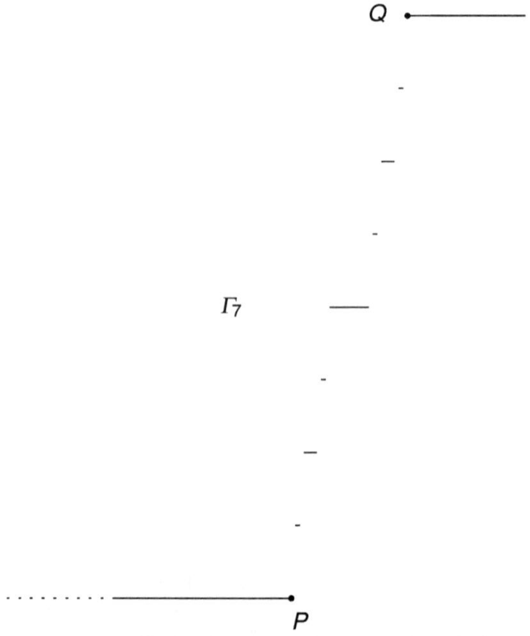

Fig. 1.7 A part of the seventh approximation Γ_7 of the graph Γ of the Cantor function f_C, shown near the point $P(2/3,1/2)$. (The upper right interval near the point Q is of length $3^{-4} \approx 0.01$, and also belongs to Γ_4; see also Figure 1.6 just above.) The slope of the line joining the points P and Q is equal to $2^{-4}/3^{-4} = (3/2)^4 \approx 5$. The slope of the line joining the point P with the left endpoint of the shortest line-segment of Γ_7 just above P, is equal to $2^{-7}/3^{-7} = (3/2)^7 \approx 17$. Moreover, it is easy to see that the right derivative of the Cantor function f_C at the point P is equal to $+\infty$. An analogous property holds for the right endpoint of *any* of the segments appearing in $\cup_{j=1}^{\infty}\Gamma_j$. See also Remark 1.2.1 on page 26. If we denote by $\sigma(\Gamma_j)$ the number of line segments of the j-th approximation Γ_j of the graph of the Cantor function, then it is easy to see that $\sigma(\Gamma_j) = 2^j - 1$ for all positive integers j. In particular, we have that $\sigma(\Gamma_7) = 2^7 - 1 = 127$. This figure shows only 9 of the 127 line-segments of Γ_7.

Alongside other examples in Section 5.5, we also analyze fractal nests (Example 5.5.16) and unbounded geometric chirps (Example 5.5.19), which are not self-similar. The example of a family of fractal nests depending on a real parameter exhibits an interesting new phenomenon; namely, two simple complex dimensions (i.e., simple poles of the associated fractal zeta function) which "merge" for a particular value of the parameter, form a single complex dimension of second order (i.e., a pole of multiplicity two). This second order complex dimension then generates logarithmic terms in the asymptotic expansion of the associated (relative) tube zeta function.

Towards the end of Section 5.5, we show how some of the already established results about the complex dimensions of *self-similar sprays* (see [LaPe2–3, LaPeWi1, DeKÖÜ]) can be recovered from the results of Chapter 5 as well as significantly extended and placed in the much broader context of the new higher-dimensional theory developed in this book.

According to the definition of fractality proposed in [Lap-vFr1–3], a geometric object is said to be "fractal" if it has at least one nonreal complex dimension[10] (see, e.g., [Lap-vFr3, Subsection 12.1.2, pp. 337–342]) or else (according to the refined definition proposed in [Lap-vFr1–3], see e.g., [Lap-vFr3, Subsection 13.4.3, pp. 473–474]) if it has a partial natural boundary (along a suitable curve). Therefore, since (by the results of Chapter 2) distance (and tube) zeta functions are holomorphic on the open right half-plane $\{\mathrm{Re}\, s > \overline{\dim}_B A\}$, maximal hyperfractals are the most fractal objects possible. The construction of maximal hyperfractals discussed above (in the description of Chapter 4, especially of Section 4.6) makes it intuitively clear that such objects are plentiful, especially among random fractals. (See, e.g., the comment following Definition 4.6.23.) However, we expect that many (deterministic) classical fractals (such as self-similar sets, for example) are not of this type. It remains to be determined whether (possibly after having chosen a suitable gauge function) some of the fractals encountered in complex dynamics (such as Julia sets and the Mandelbrot set) or in conformal geometry (such as limit sets of Fuchsian or Kleinian groups) are hyperfractal or even maximally hyperfractal. (See, in particular, Problems 6.2.20–6.2.21, along with Problem 6.2.32.) In Section 6.1, we first propose (in Subsection 6.1.1) a classification of bounded sets in Euclidean spaces, based on their tube functions, while later, in Subsection 6.1.2, we briefly comment on the history of some aspects of this topic, with particular attention to the notions of Minkowski measurability and Minkowski nondegeneracy.

Section 6.2 contains several concluding remarks and many open problems, along with suggestions for further investigation concerning the possible use of distance and tube zeta functions, as well as their weighted or their relative counterparts, in a variety of situations.

Finally, the main text of the book is completed by three appendices. In Appendix A, we introduce a general notion of Dirichlet-type integral (DTI) or function, of which all fractal zeta functions introduced and used in this book are special cases, and develop many aspects of the resulting theory. Furthermore, in Appendix B, we introduce a suitable notion of local distance zeta function. Moreover, in Appendix C, we provide a table of the distance zeta functions and the principal complex dimensions of several basic relative fractal drums used throughout the text.

Some of the many new results presented in this monograph are also discussed in the research articles [LapRaŽu1–6], as well as in the survey articles [LapRaŽu7] and [LapRaŽu8].

Although we have tried to keep this research monograph relatively self-contained, we have preferred to keep the overlap with the research monographs [Lap-vFr1–3] to a minimum. Hence, we refer the interested reader to those monographs, and especially, to [Lap-vFr3], the second edition of [Lap-vFr2], entitled

[10] Complex dimensions are interpreted in this book as the poles (of a meromorphic extension) of the associated zeta function; that is, in the present geometric situation, as the poles of the associated distance (or, equivalently, tube) zeta function. We note that in the theory developed in [Lap-vFr1–3], except in very special (but important) situations (such as fractal strings and fractal sprays) no suitable general geometric definition of a zeta function attached to a (higher-dimensional) fractal is provided.

Fractal Geometry, Complex Dimensions and Zeta Functions (and subtitled *Geometry and Spectra of Fractal Strings*), for a detailed exposition of fractal strings theory and of the associated theory of complex fractal dimensions, as well as for a variety of applications, including, for example, to self-similar fractal strings, generalized explicit formulas, fractal tube formulas, spectral asymptotics, a reformulation of the Riemann hypothesis in terms of inverse spectral problems for fractal strings, along with the distributions of zeros of arithmetic zeta functions and other aspects of number theory. We note that except when needed (for instance, to indicate the chronology of a given result), we will usually refer to [Lap-vFr3], rather than to [Lap-vFr1] or [Lap-vFr2] (or else, to related papers).

We close this section with a few words about the interdependence of some of the chapters and sections. Chapter 2 provides the foundations for the remaining part of the monograph. Sections 3.4–3.7, 4.3, 4.4.2 and 4.6 can be omitted upon a first reading. Indeed, the results therein are of independent interest but are not used in the rest of the book.[11] Also, Sections 3.1 through 3.7 are independent of one another. Finally, Sections 2.3 and 3.1 are prerequisites for Sections 4.3.2, 4.5 and 4.6.

1.3 Basic Notation and Definitions

In Subsection 1.3.1, we first introduce the notions of Minkowski contents and box dimensions, with a special emphasis on their scaling properties. In Subsection1.3.2, we review and introduce several definitions pertaining to the singularities of analytic functions, as they will be needed in several parts of this book. Finally, Subsection 1.3.3 provides a short review of standard mathematical notation which we shall need throughout this monograph. Additional notation and definitions are introduced throughout the text, as well as in the glossary.

1.3.1 Minkowski Contents and Box (or Minkowski) Dimensions of Bounded Sets

By $|E|_N$, we denote the N-dimensional Lebesgue measure of a measurable subset E of $\mathbb{R}^N$. When no ambiguity may arise, we simply write $|E|$ instead of $|E|_N$. The *upper r-dimensional Minkowski content* $\mathscr{M}^{*r}(A)$ of a bounded subset A of $\mathbb{R}^N$, $r \in \mathbb{R}$,[12] is defined by

---

[11] However, the notion of hyperfractal introduced in Subsection 4.6.2 and the construction (in the same subsection) of ∞-quasiperiodic relative fractal drums and strings that are maximally hyperfractal should play an important role in the future developments of the (higher-dimensional) theory of complex dimensions and of fractal zeta functions.

[12] It suffices to consider $r \geq 0$ here, but we want to emphasize that the more general situation when $r \in \mathbb{R}$ will be important in the case of relative box dimensions; see, especially, page 249, along

$$\mathcal{M}^{*r}(A) = \limsup_{t\to 0^+} \frac{|A_t|}{t^{N-r}}, \tag{1.3.1}$$

and the *lower r-dimensional Minkowski content* of A, denoted by $\mathcal{M}_*^r(A)$, is defined analogously (with a lower limit instead of an upper limit in the counterpart of (1.3.1)). Clearly, we always have $0 \le \mathcal{M}_*^r(A) \le \mathcal{M}^{*r}(A) \le \infty$. Here and in the sequel, as in Section 1.1,

$$A_t := \{x \in \mathbb{R}^N : d(x,A) < t\} \tag{1.3.2}$$

denotes the *t-neighborhood* (or *tubular neighborhood of radius t*) of A. If for some real number r we have $\mathcal{M}_*^r(A) = \mathcal{M}^{*r}(A) \in [0,+\infty]$, then the common value is denoted by $\mathcal{M}^r(A)$ and called the *r-dimensional Minkowski content of A*.

All of the conclusions in this monograph remain valid if, instead of using the definition given in Equation (1.3.2), we define $A_t := \{x \in \mathbb{R}^N : d(x,A) \le t\}$. This follows from the fact that the boundary of A_t is negligible in the Lebesgue sense, that is, $|\partial(A_t)| = 0$; see [Sta]. We point out that there exist open subsets U of $\mathbb{R}^N$ such that $|\partial U| > 0$; see Remark 2.1.2 on page 46.

Remark 1.3.1. Note that if A is any bounded subset of $\mathbb{R}^N$, then we can easily conclude from the definition of the Minkowski content that $\mathcal{M}^N(A)$ exists and

$$\mathcal{M}^N(A) = |\overline{A}|_N. \tag{1.3.3}$$

Indeed, since $\inf_{t\in(0,\delta)} |A_t| = \lim_{t\to 0^+} |A_t| = |\overline{A}|$, then by letting $\delta \to 0^+$ we obtain that $\mathcal{M}_*^N(A) = |\overline{A}|$, while from $\sup_{t\in(0,\delta)} |A_t| = |A_\delta|$ and then letting $\delta \to 0^+$, we conclude that $\mathcal{M}^{*N}(A) = |\overline{A}|$. Note that the present remark shows that the claim stated in [Fed2, Theorem 3.2.39] holds without any rectifiability assumption on A, provided $m = N$ in that theorem.

The *upper box* (or *Minkowski*) *dimension* of A is defined by

$$\overline{\dim}_B A = \inf\{r \in \mathbb{R} : \mathcal{M}^{*r}(A) = 0\}; \tag{1.3.4}$$

it is easy to see that we also have

$$\overline{\dim}_B A = \sup\{r \in \mathbb{R} : \mathcal{M}^{*r}(A) = +\infty\} \tag{1.3.5}$$

and

$$\overline{\dim}_B A = \inf\{r \in \mathbb{R} : \mathcal{M}^{*r}(A) < \infty\}. \tag{1.3.6}$$

Furthermore, as was observed in footnote 12 on page 30, in the present case of bounded subsets A of $\mathbb{R}^N$, it would clearly suffice to consider $r \in \mathbb{R}$, $r \ge 0$ in Equations (1.3.4)–(1.3.6).

with Corollary 4.1.38 and Remark 4.1.39 where the associated (relative) Minkowski dimension is equal to $-\infty$.

The *lower box* (or *Minkowski*) *dimension* of A, denoted by $\underline{\dim}_B A$, is defined analogously as in either of Equations (1.3.4)–(1.3.6), with $\mathcal{M}_*^r(A)$ instead of $\mathcal{M}^{*r}(A)$ in the counterpart of (1.3.4)–(1.3.6). Since A is bounded, we always have

$$0 \le \underline{\dim}_B A \le \overline{\dim}_B A \le N. \tag{1.3.7}$$

Moreover, if both dimensions $\overline{\dim}_B A$ and $\underline{\dim}_B A$ coincide, then their common value is denoted by $\dim_B A$, and is called the *box dimension* of A (or Minkowski–Bouligand dimension of A), or else, simply, the *Minkowski dimension* of A. (We then say that $\dim_B A$ exists.) According to (1.3.7), the upper and lower box dimensions of a bounded set $A \subset \mathbb{R}^N$ belong to $[0,N]$. Hence, the same is true of the box (or Minkowski) dimension of A, when it exists.

The values of the upper and lower Minkowski contents of a given bounded subset A of $\mathbb{R}^N$ depend on N, in general, since A can also be viewed as a subset of $\mathbb{R}^{N+1}$. On the other hand, it is easy to see that the values of $\underline{\dim}_B A$ and $\overline{\dim}_B A$ do not depend on N; see [Res, Proposition 1]. (See also the much earlier reference [Kne] where, however, no clear distinction seems to have been made between $\underline{\dim}_B A$ and $\overline{\dim}_B A$.)

If A is a bounded subset of $\mathbb{R}^N$ such that $\underline{\dim}_B A < N$, then $|\overline{A}| = 0$, where $\overline{A}$ denotes the closure of A in $\mathbb{R}^N$. By contraposition, this property can be stated equivalently as follows:

$$\text{If } |\overline{A}| > 0, \text{ then } \dim_B A \text{ exists and is equal to } N.^{13} \tag{1.3.8}$$

To prove this property, note that the condition $\underline{\dim}_B A < N$ implies that $\mathcal{M}_*^N(A) = 0$ (by the counterpart of Equation (1.3.4) for $\underline{\dim}_B A$); i.e., $\liminf_{t\to 0^+} |A_t| = 0$. Since $\cap_{t>0} A_t = \overline{A}$ and $t \mapsto |A_t|$ is nondecreasing, we deduce that $|\overline{A}| = \lim_{t\to 0^+} |A_t| = 0$.

If there exists $D \ge 0$ such that[14]

$$0 < \mathcal{M}_*^D(A) \le \mathcal{M}^{*D}(A) < \infty, \tag{1.3.9}$$

we say that A is *Minkowski nondegenerate*. (By definition, the condition (1.3.9) of Minkowski nondegeneracy is equivalent to $|A_t| \asymp t^{N-D}$ as $t \to 0^+$. The notation $\asymp$ is explained in Subsection 1.3.3 on page 41.) Otherwise, we say that A is *Minkowski degenerate*. In other words, if A is degenerate, then either (*a*) $\underline{\dim}_B A < \overline{\dim}_B A$ (i.e., A is *strongly degenerate*) or (*b*) $D := \dim_B A$ exists and either $\mathcal{M}_*^D(A) = 0$ or $\mathcal{M}^{*D}(A) = +\infty$ (i.e., A is *weakly degenerate*). More details about degenerate sets can be found on pages 544 and 550. Note that if A is nondegenerate, it then follows from (1.3.4)–(1.3.5) and their counterpart for $\mathcal{M}_*^r(A)$ that $\dim_B A$ exists and is equal to D.

Finally, as we have already stated, if $\mathcal{M}_*^D(A) = \mathcal{M}^{*D}(A)$, then their common value is denoted by $\mathcal{M}^D(A)$ and called the *Minkowski content* of A. If, in addition,

$$\mathcal{M}^D(A) \in (0, +\infty), \tag{1.3.10}$$

---

[13] If $\underline{\dim}_B A = N$, then the inequalities in (1.3.7) imply that $\overline{\dim}_B A = N$ as well.

[14] It is easy to verify that, if $D \ge 0$ is such that $0 < \mathcal{M}_*^D(A) \le \mathcal{M}^{*D}(A) < \infty$, then $\dim_B A$ exists and $\dim_B A = D$.

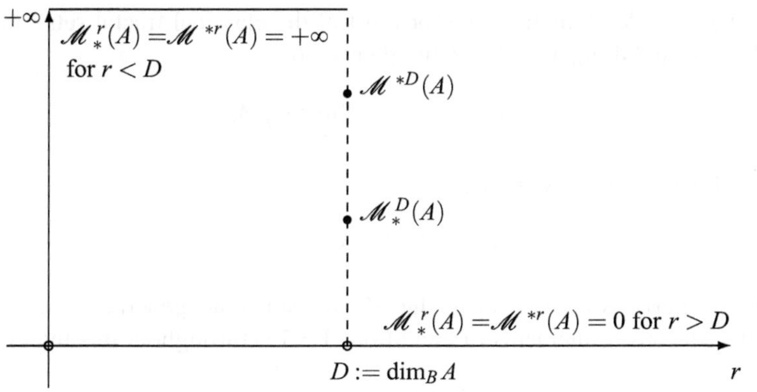

Fig. 1.8 The graphs of the functions $r \mapsto \mathscr{M}_*^r(A)$ and $r \mapsto \mathscr{M}^{*r}(A)$, assuming that A is *Minkowski nondegenerate and nonmeasurable*; i.e., $D := \dim_B A$ exists and $0 < \mathscr{M}_*^D(A) < \mathscr{M}^{*D}(A) < \infty$.

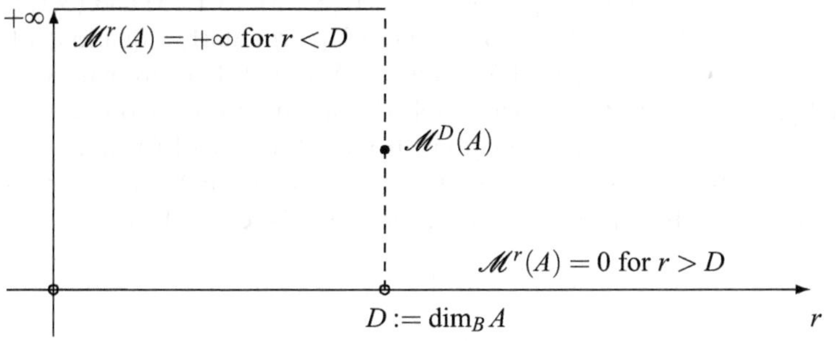

Fig. 1.9 The graph of the function $r \mapsto \mathscr{M}^r(A)$, assuming that A is *Minkowski measurable*; i.e., both $D := \dim_B A$ and $\mathscr{M}^D(A)$ exist and $0 < \mathscr{M}^D(A) < \infty$.

then A is said to be *Minkowski measurable.*[15] (See Figures 1.8 and 1.9.) Hence, a Minkowski measurable set is necessarily Minkowski nondegenerate. If A is not Minkowski measurable, we say that it is *Minkowski nonmeasurable.*[16]

The intuitive meaning of the box dimension of a bounded set A in $\mathbb{R}^N$ can be best understood by considering the asymptotic behavior of the associated *tube function* $t \mapsto |A_t|$ as $t \to 0^+$. If we assume that A is such that

$$|A_t| \asymp t^\gamma \quad \text{as} \quad t \to 0^+, \tag{1.3.11}$$

---

[15] Condition (1.3.10) of Minkowski measurability is easily seen to be equivalent to $|A_t| \sim Ct^{N-D}$ as $t \to 0^+$, where $C \in (0, +\infty)$; then, we must have $\mathscr{M}^D(A) = C$. The notation $\sim$ is explained in Subsection 1.3.3 on page 41.

[16] Note that A is Minkowski nonmeasurable if and only if either (a) $\underline{\dim}_B A < \overline{\dim}_B A$, or (b) $D := \dim_B A$ exists but $\mathscr{M}_*^D(A) = 0$ or else $\mathscr{M}^{*D}(A) = +\infty$ or $0 < \mathscr{M}_*^D(A) < \mathscr{M}^{*D}(A) < \infty$.

for some $\gamma \in (0,N]$ (which is true for most of the classical fractal sets), then $D := \dim_B A$ exists and $\dim_B A = N - \gamma$. In other words,

$$\dim_B A = N - \lim_{t \to 0^+} \log_t |A_t|. \tag{1.3.12}$$

Clearly, (1.3.11) can be written as

$$|A_t| \asymp t^{N-D} \quad \text{as} \quad t \to 0^+, \tag{1.3.13}$$

and this property is equivalent to the Minkowski nondegeneracy of A. We shall encounter various refinements of condition (1.3.13) throughout this monograph, in the form

$$|A_t| = t^{N-D}(F(t) + o(1)) \quad \text{as} \quad t \to 0^+,$$

for various classes of functions F. See, for instance, Section 2.3 (especially, Equations (2.3.26), (2.3.30), (2.3.54)–(2.3.55), (2.3.70) and (2.3.77)–(2.3.78)), Section 2.4, Section 3.1 (especially, Equations (3.1.2)–(3.1.3), (3.1.13) and (3.1.29), Theorems 3.1.12, 3.1.15, and 3.1.20), Section 4.3.2 [particularly Equations (4.5.9) and (4.5.22), along with Theorems 4.5.1, 4.5.2, 4.5.8 (and their proof)], Section 4.6.1 [especially, Equation (4.6.4), Theorem 4.6.9 (and its proof), Corollary 4.6.17 and Example 4.6.21], Section 6.1 (Equations (6.1.1), (6.1.4)–(6.1.6), along with Definitions 6.1.4 and 6.1.7 as well as Example 6.1.5, Problems 6.2.2, 6.2.3 and 6.2.16).

The asymptotic behavior (1.3.13) is expected to be equivalent[17] to

$$N_b(\delta) \asymp \delta^{-D} \quad \text{as} \quad \delta \to 0^+, \tag{1.3.14}$$

where the *box-counting function* $N_b(\delta)$ is the number of δ-mesh cubes in $\mathbb{R}^N$ that intersect A. (It can easily be shown that it suffices to consider diadic meshes of the form $\delta = 2^{-k}$ as $k \to \infty$.) In particular,

$$\dim_B A = \lim_{\delta \to 0^+} \log_{1/\delta} N_b(\delta);$$

see [Fal1, p. 41]. Passing to the general case, that is, assuming that A is *any* bounded set A in $\mathbb{R}^N$, it is not difficult to show that

$$\begin{aligned} \underline{\dim}_B A &= N - \limsup_{t \to 0^+} \log_t |A_t|, \\ \overline{\dim}_B A &= N - \liminf_{t \to 0^+} \log_t |A_t| \end{aligned} \tag{1.3.15}$$

[17] When $N = 1$, this equivalence is established in [LapPo1–2], using the language of fractal strings and hence, is valid for any compact (or, equivalently, bounded) subset A of $\mathbb{R}$.

and

$$\begin{aligned}\underline{\dim}_B A &= \liminf_{\delta\to 0^+} \log_{1/\delta} N_b(\delta),\\ \overline{\dim}_B A &= \limsup_{\delta\to 0^+} \log_{1/\delta} N_b(\delta).\end{aligned} \tag{1.3.16}$$

If we use as an alternative notation $\underline{\lim}$ and $\overline{\lim}$ for lim inf and lim sup, respectively, both equations appearing in (1.3.16) can be written even more succintly as follows:

$$\underline{\dim}_B A = \underline{\lim}_{\delta\to 0^+} \log_{1/\delta} N_b(\delta), \quad \overline{\dim}_B A = \overline{\lim}_{\delta\to 0^+} \log_{1/\delta} N_b(\delta).$$

As was already mentioned, the map $\delta \mapsto N_b(\delta)$, $\delta > 0$, is called the *box-counting function* of A. Equation (1.3.16) therefore provides a natural motivation for the name of the upper and lower *box-counting dimensions*. Other variations of (1.3.16), that is, of the definition of $N_b(\delta)$, can be found in [Fal1, p. 41].

Throughout this book, we will assume, most often implicitly, that the bounded set $A \subset \mathbb{R}^N$ is nonempty, in order to avoid trivial exceptions to the statements of some of our results.

We conclude this subsection with a few words about the *scaling properties of Minkowski contents*. In the sequel, for any given nonempty bounded subset A of $\mathbb{R}^N$ and $\lambda \in \mathbb{R}$, we let $\lambda A := \{\lambda x : x \in A\}$.

Let $\mathscr{P}_b(\mathbb{R}^N)$ be the family of all bounded subsets of $\mathbb{R}^N$ and let $A \in \mathscr{P}_b(\mathbb{R}^N)$ be a given nonempty bounded subset. A function $\mathscr{M} : \mathscr{P}_b(\mathbb{R}^N) \to [0,+\infty]$ is said to be *homogeneous with respect to* A (or $D(A)$-homogeneous) if there exists a real number $D(A)$ such that

$$\mathscr{M}(\lambda A) = \lambda^{D(A)} \mathscr{M}(A), \quad \text{for all} \quad \lambda > 0.$$

If $\mathscr{M}$ is homogeneous with respect to every nonempty bounded subset A in $\mathbb{R}^N$, we say that the function $\mathscr{M}$ *is homogeneous*. For example, the function $\mathscr{M}^* : \mathscr{P}_b(\mathbb{R}^N) \to [0,+\infty]$ defined by $\mathscr{M}^*(A) := \mathscr{M}^{*\overline{D}(A)}(A)$ (i.e., the upper $\overline{D}(A)$-dimensional Minkowski content of A), where $\overline{D}(A) := \overline{\dim}_B A$, is homogeneous; that is,

$$\mathscr{M}^{*\overline{D}(A)}(\lambda A) = \lambda^{\overline{D}(A)} \mathscr{M}^{*\overline{D}(A)}(A), \quad \text{for all} \quad \lambda > 0. \tag{1.3.17}$$

We say for short that the upper Minkowski content is *homogeneous*. See Remark 1.3.2. Much as in Equation (1.3.17), the lower Minkowski content is homogeneous in the following sense:

$$\mathscr{M}_*^{\underline{D}(A)}(\lambda A) = \lambda^{\underline{D}(A)} \mathscr{M}_*^{\underline{D}(A)}(A), \quad \text{for all} \quad \lambda > 0, \tag{1.3.18}$$

where $\underline{D}(A) := \underline{\dim}_B A$. These scaling results are easily obtained by noting that $\lambda(A_t) = (\lambda A)_{\lambda t}$ for any $t > 0$. It is easy to see that the scaling (or homogeneity) properties (1.3.17) and (1.3.18) are equivalent to the following equations:

$$\mathscr{M}^{*s}(\lambda A) = \lambda^s \mathscr{M}^{*s}(A), \quad \mathscr{M}_*^s(\lambda A) = \lambda^s \mathscr{M}_*^s(A) \tag{1.3.19}$$

for all $\lambda > 0$ and $s \in \mathbb{R}$. Indeed, if $s < \overline{D}(A)$, then both sides of the first equation are equal to $+\infty$, whereas for $s > \overline{D}(A)$ both sides are equal to zero. We can argue analogously in the case of the second equation. A more general result, extending the scaling properties (1.3.19) from bounded subsets of $\mathbb{R}^N$ to the setting of relative fractal drums, can be found in Lemma 4.6.10.

Remark 1.3.2. It is clear that $\overline{\dim}_B(\lambda A) = \overline{\dim}_B A$, i.e., $\overline{D}(\lambda A) = \overline{D}(A)$, for every $\lambda > 0$, and similarly for the lower box dimension. This is a very special case of the property of *bi-Lipschitz invariance* of box dimensions: if A is a bounded set in $\mathbb{R}^N$ and $f : A \to \mathbb{R}^N$ is a *bi-Lipschitz function*; that is, there exist positive constants c_1 and c_2 such that $c_1|x-y| \le |f(x)-f(y)| \le c_2|x-y|$ for all $x, y \in A$, then (see [Fal1, p. 44]), we have

$$\overline{\dim}_B f(A) = \overline{\dim}_B A \quad \text{and} \quad \underline{\dim}_B f(A) = \underline{\dim}_B A.$$

Remark 1.3.3. Unless explicitly stated otherwise, the bounded sets $A \subseteq \mathbb{R}^N$ considered in this book are implicitly assumed to be nonempty; furthermore, when working with a relative fractal drum (A, Ω), as in Chapters 4–6, we will also assume that A is nonempty and, in addition, that the open set $\Omega \subseteq \mathbb{R}^N$ is nonempty.

Remark 1.3.4. Since $d(\cdot, A) = d(\cdot, \overline{A})$, where $\overline{A}$ denotes the closure of A in $\mathbb{R}^N$, it follows that the t-neighborhood A_t of a bounded subset A of $\mathbb{R}^N$ is equal to that of its closure; namely, $A_t = (\overline{A})_t$, for every $t \ge 0$. As a result, the same is true for the volume $|A_t|$, as well as for the (upper, lower) Minkowski dimension and for the (upper, lower) Minkowski content of A. For example, $\dim_B A = \dim_B \overline{A}$ and $\mathscr{M}^{*D}(A) = \mathscr{M}^{*D}(\overline{A})$. Since, in a Euclidean space $\mathbb{R}^N$, a subset A is compact if and only if it is closed and bounded, it follows that when we work with the distance zeta function (or later on, the tube zeta function) of a bounded set $A \subset \mathbb{R}^N$, we may as well asume that A is compact.

1.3.2 Singularities of Analytic Functions

Since singularities of holomorphic functions play an important role in this book, we briefly recall their definitions and classification.

Let $f : U \setminus \{s_0\} \to \mathbb{C}$ be a given holomorphic function, where U is a connected open subset of the complex plane $\mathbb{C}$ and $s_0 \in U$. We say that s_0 is an *isolated singularity* of f if it is either a removable singularity or a pole or an essential singularity. Here, we say that:

(*a*) s_0 is a *removable singularity* of f if the limit $\lim_{s \to s_0} f(s)$ exists and is a complex number;[18] equivalently, there exists a holomorphic extension F of the function

[18] All of the limits as $s \to s_0$ are implicitly assumed to hold as $s \to s_0$, $s \in U$.

f from $U \setminus \{s_0\}$ to the whole of U (in other words, $F : U \to \mathbb{C}$ is holomorphic and $f(s) = F(s)$ for all $s \in U \setminus \{s_0\}$, i.e., $f = F|_{U \setminus \{s_0\}}$);

(*b*) s_0 is a *pole of order* k of f, where $k \in \mathbb{N}$, if the limit $\lim_{s \to s_0} (s - s_0)^k f(s)$ exists and is equal to a complex number different from zero. Equivalently, s_0 is a pole of f if $\lim_{s \to s_0} |f(s)| = +\infty$ and in that case, the order of s_0 is the integer k for which the above nonzero limit exists. If $k = 1$, the pole s_0 is said to be *simple* and in that case, the *residue* of f at s_0 is given by $\operatorname{res}(f, s_0) := \lim_{s \to s_0} (s - s_0) f(s)$;

(*c*) s_0 is an *essential singularity* of f if it is neither a removable singularity nor a pole of f; in other words, the limit $\lim_{s \to s_0} (s - s_0)^k f(s)$ does not exist in $\mathbb{C}$, for any $k \in \mathbb{N}$.

Remark 1.3.5. Simple illustrations of isolated singularities of the types (*a*)–(*c*) above, respectively, and in the case when $s_0 := 0$ and $U := \mathbb{C} \setminus \{0\}$, are: (*a*) $f(s) = \sin s / s$, (*b*) $f(s) = 1/s^k$ for some $k \in \mathbb{N}$, and (*c*) $f(s) = \exp(1/s) = \sum_{k=1}^{\infty} 1/(k! s^k)$.

Equivalently, $s_0 \in U$ is an isolated singular point of a holomorphic function $f : U \setminus \{s_0\} \to \mathbb{C}$ if there exists an ε-neighborhood $B_\varepsilon(s_0)$ of s_0, in the complex plane, where $\varepsilon > 0$ and such that f is holomorphic on the punctured ε-neighborhood $B_\varepsilon(s_0) \setminus \{s_0\}$ of s_0. (Here, $B_\varepsilon(s_0)$ denotes the open disk of center s_0 and radius ε.) Recall that it then follows that f can be expanded into a Laurent series arround s_0 in the punctured disk $B_\varepsilon(s_0) \setminus \{s_0\}$:

$$f(s) = \sum_{n=-N}^{\infty} a_n (s - s_0)^n, \tag{1.3.20}$$

where $N \in \mathbb{N}_0 \cup \{\infty\}$ and $a_n \in \mathbb{C}$ for every finite n. Then, s_0 is a removable singularity of f if and only if $N = 0$. Furthermore, it is a pole of order N of f if and only if $N \in \mathbb{N}$ and $a_{-N} \neq 0$; in that case, $\sum_{n=-N}^{-1} a_n (s - s_0)^n$, the *principal part* of f at s_0, is nontrivial but contains only finitely many terms. Finally, s_0 is an essential singularity of f if and only if $N = \infty$; more precisely, if and only if $N = \infty$ and the *principal part* of f at s_0, $\sum_{n=-\infty}^{-1} a_n (s - s_0)^n$, contains infinitely many terms such that $a_n \neq 0$ with $n < 0$.

In several places in this book, particularly in Chapters 4 and 6,[19] we will also work with the nonisolated singularities of a holomorphic function f defined on a given connected open subset U of $\mathbb{C}$. In that case, such singularities will consist of the set of accumulation points s_0 of U such that $B_\varepsilon(s_0) \setminus \{s_0\} \subseteq U$ for sufficiently small $\varepsilon > 0$ and the associated cluster set $\mathscr{C}(s_0, f)$ is not reduced to a single point.[20] Here, the *cluster set* $\mathscr{C}(s_0, f)$ of f at $s_0 \in \widetilde{\mathbb{C}}$ (where $\widetilde{\mathbb{C}} := \mathbb{C} \cup \{\infty\}$ is the Riemann

[19] In Chapter 4, see, e.g., Theorem 4.3.21 in Subsection 4.3.2, Lemma 4.5.10 and Figure 4.16 in Subsection 4.5.2, Theorem 4.6.9 in Subsection 4.6.1, Theorem 4.6.13 and Corollary 4.6.17 in Subsection 4.6.2, Corollary 4.6.28 in Subsection 4.6.4, along with Subsection 4.6.3. In Chapter 6, see, e.g., Problem 6.2.18 and Problems 6.2.21–6.2.26.

[20] If $\mathscr{C}(s_0, f)$ consists of a single point, and we assume that f is holomorphic in $B_\varepsilon(s_0) \setminus \{s_0\}$, then s_0 is an isolated singularity which is either a pole or a removable singularity of f.

sphere equipped with its natural topology) is the set of all $\tau \in \widetilde{\mathbb{C}}$ such that there exists a sequence $(s_n)_{n=1}^{\infty}$ satisfying $s_n \in U$ for all $n \geq 1$, $s_n \to s_0$ and $f(s_n) \to \tau$ in $\widetilde{\mathbb{C}}$, as $n \to \infty$. Such a nonisolated singularity is also sometimes called an "essential singularity point of f". (See, e.g., [Haz].) This is, of course, a more general notion than the usual type of essential singularity encountered in elementary complex analysis and discussed earlier in this subsection.

Assume that $f : U \to \mathbb{C}$ is a holomorphic function, where U is a connected open subset of the complex plane $\mathbb{C}$ and s_0 belongs to the boundary of U, i.e., $s_0 \in \partial U$. If there is no $\varepsilon > 0$ such that f can be holomorphically extended to a punctured ε-neighborhood $B_\varepsilon(s_0) \setminus \{s_0\}$ of s_0, we say that s_0 is a *nonisolated singularity* of f.

For example, if there exists a sequence of isolated singularities $(s_k)_{k\geq 1}$ of f converging to s_0, then s_0 is a nonisolated singularity of f.

Definition 1.3.6. Given a holomorphic function $f : U \to \mathbb{C}$, where U is a connected open subset of $\mathbb{C}$, as above, we say that $K := \partial U$ is a (*holomorphic*) *natural boundary* of f if there is no $s \in K$ and $\varepsilon > 0$ for which an analytic (i.e., holomorphic) continuation of f is possible to $B_\varepsilon(s)$. (Here, as before in this subsection, $B_\varepsilon(s)$ denotes the open disk of center s and radius ε in $\mathbb{C}$.) Then, U is called a *domain of holomorphy* for f.

Remark 1.3.7. The only nonisolated singularity of the function $f(s) := 1/\sin(1/s)$ is $s = 0$. Indeed, $s_k = 1/(k\pi)$, with $k \in \mathbb{N}$, are simple poles of f converging to 0 as $k \to \infty$; here, $U := \mathbb{C} \setminus (\{1/(k\pi) : k \in \mathbb{N}\} \cup \{0\})$. Another well-known example is provided by the lacunary series defined by $g(s) := \sum_{k=1}^{\infty} s^{2^k} = 1 + s^2 + s^4 + s^8 + \dots$, where $s \in U := B_1(0)$, for which it can be shown that the unit circle $K = \{s \in \mathbb{C} : |s| = 1\}$ is a corresponding (holomorphic) natural boundary of g; in particular, each point $s_0 \in K$ is a nonisolated singularity of the function g. Equivalently, the open unit disk $U := \{s \in \mathbb{C} : |s| < 1\}$ is a domain of holomorphy for g, in the sense of Definition 1.3.6.

Let U be a connected open subset of the complex plane $\mathbb{C}$ and let $\widetilde{\mathbb{C}} := \mathbb{C} \cup \{\infty\}$ be the one-point compactification of $\mathbb{C}$, also referred to as the *Riemann sphere*. We say that a function $f : U \to \widetilde{\mathbb{C}}$ is *meromorphic* if there exists a subset S of U consisting of isolated points in U, such that f is holomorphic on $U \setminus S$ and each $s_0 \in S$ is a pole of f. Here, since it is discrete, the set S is necessarily at most countable, possibly empty. As is well known, if f is extended to become a $\widetilde{\mathbb{C}}$-valued function $\tilde{f}$ (i.e., $\tilde{f} : U \to \widetilde{\mathbb{C}}$, with $\tilde{f}(s) := \infty$ for all $s \in S$ and $\tilde{f}(s) := f(s)$ for all $s \in U \setminus S$) and $\widetilde{\mathbb{C}}$ is viewed as a compact Riemann surface, then the meromorphic function $f : U \to \mathbb{C}$ becomes a holomorphic function $\tilde{f} : U \to \widetilde{\mathbb{C}}$. The converse of this statement is also true (and in that case, $S := f^{-1}(\{\infty\})$ is the set of poles of f); see, e.g., [Ebe].

Finally, let us still assume that U is a connected open subset (i.e., a *domain*) of $\mathbb{C}$. Then, recall that a function $f : U \to \mathbb{C}$ is meromorphic if and only if it can be written as the ratio of two holomorphic functions on U: $f = \varphi/\psi$, where φ and ψ are holomorphic on U. In that case, provided φ and ψ do not have any common zeros, the set S of poles of f coincides with the zeros of ψ: $S := \{s \in U : \psi(s) = 0\}$.

(In general, it is clearly contained in S.) The above factorization property implies that, as is well known, the principle of analytic continuation extends to meromorphic functions. More specifically, if two meromorphic functions f and g on a given domain U of $\mathbb{C}$ coincide on a subset of U having a limit point in U, then they coincide everywhere in U. This 'principle of meromorphic continutation' will be used throughout this book, and often referred to as the principle of analytic continuation. Furthermore, recall that as a simple consequence of the principle of analytic continuation, a meromorphic function admits at most one meromorphic extension to a given connected open set; in other words, on any domain V of $\mathbb{C}$, the meromorphic continuation of f, if it exists, is unique.

In the remainder of this book, we will need the following nonstandard definitions, especially when discussing the notions of hypefractality (and strong hyperfractality), as well as maximal hyperfractality, in Subsection 4.6.3 of Chapter 4. For now, the reader may wish to omit these definitions upon a first reading, and return to them later, when necessary.

Definition 1.3.8. Let $f : U \to \mathbb{C}$ be a meromorphic function, where U is a connected open set in $\mathbb{C}$ with boundary $K := \partial U$.

(*i*) We say that f admits K as a (*meromorphic*) *partial natural boundary* (or that K is a (meromorphic) *partial natural boundary* of f) if f cannot be meromorphically continued beyond K or, more precisely, if there exist $s_0 \in K$ and $\varepsilon > 0$ such that f cannot be meromorphically extended to $B_\varepsilon(s_0)$. Equivalently, given any open set V of $\mathbb{C}$ containing $U \cup \{s_0\}$, f cannot be meromorphically continued to V. (See also Remark 1.3.9 below.) We then say that U is a *partial domain of meromorphy* for f.

(*ii*) Moreover, K is called a (*meromorphic*) *natural boundary* of f if there does not exist $s_0 \in K$ and $\varepsilon > 0$ for which a meromorphic extension of f is possible to $B_\varepsilon(s_0)$. Equivalently, given any $s_0 \in K$ and any open set V of $\mathbb{C}$ containing $U \cup \{s_0\}$, f cannot be meromorphically extended to V. We then say that U is a *domain of meromorphy* for f.[21]

Note that unlike the traditional notion of (holomorphic) natural boundary recalled earlier (on page 38), the above definition of (meromorphic) partial natural boundary or of (meromorphic) natural boundary refers to the meromorphic continuation rather than to the holomorphic continuation of f.

We will only need to use the notion of (meromorphic) partial natural boundary, not its obvious holomorphic counterpart (which is *not* an equivalent notion). Therefore, when no ambiguity may arise, we will sometimes omit the adjective "meromorphic" when referring to a partial natural boundary.

---

[21] Clearly, a meromorphic natural boundary is a holomorphic natural boundary (in the sense recalled in Definition 1.3.6 on page 38), but the converse is not true, in general. Indeed, if a fractal zeta function cannot be extended meromorphically to some connected open subset in $\mathbb{C}$, then a fortiori, it cannot be extended holomorphically. Equivalently, if U is a domain of meromorphy for f, then it is also a domain of holomorphy for f (still in the sense of Definition 1.3.6).

Remark 1.3.9. (*a*) We will only use Definition 1.3.8 when K is a screen $\boldsymbol{S}$ (with associated window $\boldsymbol{W}$), in the sense of [Lap-vFr3] recalled in Definition 5.1.1, and f is one of the fractal zeta functions discussed in this book. In that case, U is the interior of the window and its boundary $K = \boldsymbol{S} = \partial U$ is a suitable curve in $\mathbb{C}$ (extending vertically in both directions). Consequently, f is assumed to be meromorphic to the right of the screen $\boldsymbol{S}$. We point out that in the important special case when $K = \boldsymbol{S}$ is the critical line $\{\mathrm{Re}\, s = D\}$, this hypothesis is automatically satisfied because by part (*a*) of Theorem 2.1.11 below, the fractal zeta function is then holomorphic for $\mathrm{Re}\, s > D$.

(*b*) It would be reasonable to strenghten part (*i*) of Definition 1.3.8 by requiring that there exists an infinite sequence of distinct points $s_n \in K$ and of positive numbers ε_n such that for each $n \geq 1$, f cannot be meromorphically extended to $B_{\varepsilon_n}(s_n)$. The examples of strongly hyperfractal RFDs (that are not maximally hyperfractal) given in this book would still satisfy this stronger property, with f being the associated fractal zeta function and K coinciding with the critical line $\{\mathrm{Re}\, s = D\}$.

1.3.3 Standard Mathematical Symbols and Conventions

Throughout this book, we will use the special symbol $\mathbb{i}$ for the imaginary unit:

$$\mathbb{i} := \sqrt{-1}. \tag{1.3.21}$$

This will enable us, in particular, to use the ordinary symbol i as a running index, either as a subscript or superscript. We use the upright ('Roman') e and d to denote, respectively, the base of the natural logarithm and the differentiation sign.

The logarithm of $x > 0$ in base $a > 0$ is denoted by $\log_a x$. Recall that, by definition, $y = \log_a x$ is equivalent to $x = a^y$. The *natural logarithm*, that is, the logarithm in base $\mathrm{e} \approx 2.718$, is denoted by $\log x$. It is easy to see that for any $c > 0$, the following useful property holds: $\log_a x = \log_c x \,/\, \log_c a$. In particular, $\log_a x = \log x \,/\, \log a$.

We shall also need to use the notation

$$\omega_m := \frac{2\pi^{m/2}}{m\Gamma(m/2)}, \tag{1.3.22}$$

where Γ is the gamma function and m is any positive real number. If m is a positive integer, then ω_m is equal to the m-dimensional Lebesgue measure of the unit ball in $\mathbb{R}^m$. Recall that $\Gamma(x+1) = x\Gamma(x)$ for any positive real number x and $\Gamma(1) = 1$, so that $\Gamma(n+1) = n!$ for any $n \in \mathbb{N}$. Also, $\Gamma(1/2) = \sqrt{\pi}$, so that $\Gamma(m/2)$ can be easily calculated when m is odd.

Recall that given a function $f : (0,a) \to \mathbb{R}$, with fixed $a > 0$, its *lower and upper limits as* $t \to 0^+$ are defined, respectively, by

$$\liminf_{t\to 0^+} f(t) := \lim_{\delta\to 0^+} \inf_{t\in(0,\delta)} f(t), \quad \limsup_{t\to 0^+} f(t) := \lim_{\delta\to 0^+} \sup_{t\in(0,\delta)} f(t).$$

For any two sequences of positive real numbers $(a_k)_{k=1}^{\infty}$ and $(b_k)_{k=1}^{\infty}$, we write

$$a_k \asymp b_k \quad \text{as } k \to \infty$$

if there exists a positive constant c such that $c \leq a_k/b_k \leq c^{-1}$ for all $k \geq 1$. If $a_k/b_k \to 1$ (or more generally, if $b_k = a_k(1+o(1))$) as $k \to \infty$, we write $a_k \sim b_k$ as $k \to \infty$. [Here, $o(1)$ ("little o" of one) means that the implied error term tends to zero as $k \to \infty$.] We adopt entirely analogous notation for functions of a real variable and for the asymptotic behavior of such functions at finite or infinite values. Furthermore, the same symbol $\sim$ will be used for the equivalence of zeta functions (see Definition 2.1.69), and the different meanings of the symbol $\sim$ used in the text should be clear from the context. Also, we write $\sum_{k=1}^{\infty} a_k \asymp \sum_{k=1}^{\infty} b_k$ if the series are simultaneously convergent or divergent. This is the case if $a_k \asymp b_k$ as $k \to \infty$.

We say that a sequence $(l_j)_{j\geq 1}$ of real numbers is: *increasing* if $l_j < l_{j+1}$ for all $j \geq 1$; *decreasing* if $l_j > l_{j+1}$ for all $j \geq 1$; *nondecreasing* if $l_j \leq l_{j+1}$ for all $j \geq 1$: *nonincreasing* if $l_j \geq l_{j+1}$ for all $j \geq 1$. We adopt an analogous terminology for real functions of a real variable.

We also recall the notion of a *bounded fractal string* $\mathscr{L}$ (see [Lap-vFr1–3] and the beginning of Subsection 2.1.4). It is defined as a nonincreasing sequence $\mathscr{L} = (\ell_k)_{k\in\mathbb{N}}$ of positive real numbers with finite *total length* $|\mathscr{L}|_1$; i.e., such that $|\mathscr{L}|_1 := \sum_{k=1}^{\infty} \ell_k < \infty$. Alternatively, we will also consider and use the geometric definition of a bounded fractal string, namely, as an open subset Ω of $\mathbb{R}$ such that $|\Omega|_1 < \infty$; see Subsection 2.1.4 along with [Lap-vFr3, LapPo2, Lap3]. Furthermore, we define the *tensor product* $\mathscr{L}_1 \otimes \mathscr{L}_2$ of two bounded fractal strings $\mathscr{L}_1$ and $\mathscr{L}_2$ as the bounded fractal string consisting of all possible products $\lambda \cdot \mu$ with $\lambda \in \mathscr{L}_1$ and $\mu \in \mathscr{L}_2$, counted with their corresponding multiplicities. See Definition 4.2.2 for more details. It is clear that $|\mathscr{L}_1 \otimes \mathscr{L}_2|_1 = |\mathscr{L}_1|_1 \cdot |\mathscr{L}_2|_1$. Similarly, we can define the *union* of two bounded fractal strings, $\mathscr{L}_1 \sqcup \mathscr{L}_2$, as the union of the corresponding multisets; that is, as the usual union but also with the corresponding multiplicities taken into account (see also Definition 4.5.11).

One can easily check that $|\mathscr{L}_1 \sqcup \mathscr{L}_2|_1 = |\mathscr{L}_1|_1 + |\mathscr{L}_2|_1$. Furthermore, if we denote the collection of all bounded fractal strings by $\mathscr{L}_b$, it is easy to see that both $(\mathscr{L}_b, \otimes)$ and $(\mathscr{L}_b, \sqcup)$ are commutative semigroups, while $\mathscr{L}_b$ is a convex cone in the standard Banach space $(\ell_1(\mathbb{R}), +)$ of absolutely summable sequences $(x_k)_{k\in\mathbb{N}}$ of real numbers. The union $\sqcup$ can be extended to include an infinite sequence of bounded fractal strings $(\mathscr{L}_k)_{k\in\mathbb{N}}$. More specifically, the infinite union $\sqcup_{k=1}^{\infty} \mathscr{L}_k$ is a bounded fractal string, provided $\sum_{k=1}^{\infty} |\mathscr{L}_k|_1 < \infty$. Finally, for any bounded fractal string $\mathscr{L}$ and $c > 0$, we define the *c-scaled string* $c\mathscr{L}$ as $c\mathscr{L} := (c\lambda)_{\lambda\in\mathscr{L}}$.

Given $\alpha \in \mathbb{R}$, we denote by $\{\mathrm{Re}\, s > \alpha\}$ the open right half-plane

$$\{s \in \mathbb{C} : \mathrm{Re}\, s > \alpha\}.$$

We use a similar convention for the left half-plane $\{\operatorname{Re} s < \alpha\}$, for example. Analogously, we denote by $\{\operatorname{Re} s = \alpha\}$ the vertical line

$$\{s \in \mathbb{C} : \operatorname{Re} s = \alpha\}.$$

Finally, by a *countable* set A, we mean an infinite set which is in bijection with the set of natural numbers $\mathbb{N} = \{1, 2, 3, \cdots\}$. Equivalently, a countable set can be represented as an infinite sequence, $A = (a_k)_{k\in\mathbb{N}}$, such that $a_i \neq a_j$ for any pair of distinct indices i and $j \in \mathbb{N}$.

Chapter 2
Distance and Tube Zeta Functions

Le plus court chemin entre deux vérités dans le domaine réel passe par le domaine complexe.

[The shortest path between two truths in the real domain passes through the complex domain.]

Jacques Hadamard (1865–1963)

Abstract Distance and tube zeta functions of fractals in Euclidean spaces can be considered as a bridge between the geometry of fractal sets and the theory of holomorphic functions. This is first seen from their fundamental property: the upper box dimension of any bounded fractal is equal to the abscissa of convergence of its distance and tube zeta functions. Furthermore, under some natural conditions, the residue of the tube zeta function of a fractal, evaluated at its abscissa of convergence, is equal to its Minkowski content, a fractal analog of its volume. It is possible to obtain very general results dealing with the problem of meromorophic continuation of these two fractal zeta functions. We show, in particular, that the distance zeta function and the tube zeta function contain essentially the same information, both from the point of view of their meromorphic continuation (when it exists) to a given domain of the complex plane, of their poles (called visible *complex dimensions*) and their residues (or, more generally, their principal parts), which are related in a simple manner. Consequently, the higher-dimensional theory of complex dimensions can be developed by using either of these two fractal zeta functions, and much preferably, both of them since one of these zeta functions is often more natural or simpler to use in a given situation or example. A variety of examples are studied from this point of view throughout the book (including in this chapter, the $(N-1)$-dimensional sphere, generalized Cantor sets and the a-string, and in later chapters, the N-dimensional analogs of the Sierpiński carpet and the Sierpiński gasket, as well as fractal nests, self-similar fractal sprays, two-parameter generalized Cantor sets, discrete and continuous spirals, geometric chirps, etc.). In the one-dimensional case (that is, in the case of fractal strings), we show that the geometric zeta function of a fractal string and the corresponding distance zeta function are equivalent (in a suitable sense), and, in fact, define the same complex dimensions (except possibly at $s=0$); in particular, they have the same abscissa of convergence, equal to the upper Minkowski dimension of the fractal string (or, equivalently, of the associated fractal subset of the real line). As we shall see in later chapters, distance and tube zeta functions can also be viewed as a bridge to the transcendental theory of numbers. For these reasons, these new fractal zeta functions deserve to be seriously studied.

M.L. Lapidus et al., *Fractal Zeta Functions and Fractal Drums*,
Springer Monographs in Mathematics, DOI 10.1007/978-3-319-44706-3_2

In fact, as is suggested by the title of this research monograph, they are the central object of investigation in this chapter and, along with their poles (or 'complex dimensions'), throughout the entire book.

Key words: distance zeta function, tube zeta function, geometric zeta function, fractal set, fractal string, box dimension, complex dimensions, principal complex dimensions, Minkowski content, Minkowski measurable set, residue, Dirichlet series, Dirichlet-type integral, meromorphic extension, abscissae of meromorphic and absolute convergence, generalized Cantor set, Sierpiński carpet, average Minkowski content, average Minkowski dimension.

Distance and tube zeta functions of fractals in Euclidean spaces can be considered as a bridge between the geometry of fractal sets and the theory of holomorphic functions. This is first seen from their fundamental property: the upper box dimension of any bounded fractal is equal to the abscissa of convergence of its distance and tube zeta functions (see Theorem 2.1.11 and Theorem 2.2.11). Furthermore, under some natural conditions, the residue of the tube zeta function of a fractal, evaluated at its abscissa of convergence, is equal to its Minkowski content, a fractal analog of its volume; see Subsection 2.2.2, along with its counterpart for distance zeta functions, Subsection 2.2.1. It is possible to obtain very general results dealing with the problem of meromorophic continuation of these two fractal zeta functions; see, especially, Theorems 2.3.18, 2.3.25 and 2.3.37.

We show, in particular, that the distance zeta function and the tube zeta function contain essentially the same information, both from the point of view of their meromorphic continuation (when it exists) to a given domain of the complex plane, of their poles (called visible *complex dimensions*) and their residues (or, more generally, their principal parts), which are related in a simple manner; see, especially, Corollary 2.2.20. Consequently, the higher-dimensional theory of complex dimensions can be developed by using either of these two fractal zeta functions, and much preferably, both of them since one of these zeta functions is often more natural or simpler to use in a given situation or example. A variety of examples are studied from this point of view throughout the book (including in this chapter, the $(N-1)$-dimensional sphere, generalized Cantor sets and the a-string, and in later chapters, the N-dimensional analogs of the Sierpiński carpet and the Sierpiński gasket, as well as fractal nests, fractal sprays, two-parameter generalized Cantor sets, discrete and continuous spirals, geometric chirps, etc.).

In the one-dimensional case (that is, in the case of fractal strings), we show that the geometric zeta function of a fractal string and a corresponding distance zeta function are equivalent (in a suitable sense), and, in fact, define the same complex dimensions (except possibly at $s = 0$); in particular, they have the same abscissa of convergence, equal to the upper Minkowski dimension of the fractal string (or, equivalently, of the associated fractal subset of the real line). See Subsection 2.1.4, especially, Proposition 2.1.59.

As we shall see in later chapters (Sections 3.1 and 4.6), distance and tube zeta functions can also be viewed as a bridge to the transcendental theory of numbers.

For these reasons, we believe that these new fractal zeta functions deserve to be seriously studied. In fact, as is suggested by the title of this research monograph, they are the central object of investigation in this chapter and, along with their poles (or 'complex dimensions'), throughout the entire book.

2.1 Basic Properties of the Zeta Functions of Fractal Sets

In this section, we introduce a new class of zeta functions, which we call distance zeta functions; see Definition 2.1.1. They represent a natural extension of the geometric zeta functions of bounded fractal strings, introduced by the first author in the early 1990s in [Lap1–3] (see also [LapPo1–3], [LapMa1–2] and [HeLap]) and studied extensively in [Lap-vFr1–3]. Especially important is the notion of equivalence of zeta functions, as well as the definition of principal complex dimensions of fractals; see Section 2.1.5.

2.1.1 Definition of the Distance Zeta Functions of Fractal Sets

We study some of the basic properties of the distance zeta function $\zeta_A = \zeta_A(s)$ associated with an arbitrary bounded set A in $\mathbb{R}^N$. Here, s is a complex variable. The definition of this new fractal zeta function, introduced by the first author in 2009 (see Definition 2.1.1), involves the Euclidean distance from x to A, denoted by $d(x,A)$, and the δ-neighborhood (or tubular neighborhood) of A, that is,

$$A_\delta = \{x \in \mathbb{R}^N : d(x,A) < \delta\}.$$

Definition 2.1.1. Let δ be a fixed positive number. Then, the zeta function ζ_A of a bounded set A in $\mathbb{R}^N$, or *distance zeta function* of A, is defined by

$$\zeta_A(s) = \int_{A_\delta} d(x,A)^{s-N} \mathrm{d}x, \tag{2.1.1}$$

for all $s \in \mathbb{C}$ with $\operatorname{Re} s$ sufficiently large. The integral is taken in the Lebesgue sense (and hence, is absolutely convergent).

We shall see in Theorem 2.1.11 below that ζ_A is holomorphic in the half-plane $\{\operatorname{Re} s > \overline{\dim}_B A\}$, with an expression still given by (2.1.1), and that the lower bound $\overline{\dim}_B A$ is the best possible. The integral occurring in Equation (2.1.1) above can be taken over $A_\delta \setminus \overline{A}$ instead of A_δ; see Proposition 2.1.22. Also, we shall extend the definition of the distance zeta function so that the value of δ will become inessential; see Definition 2.1.79. Furthermore, we will simplify (or rather, supplement and extend) the original definition of the complex dimensions of fractal strings introduced by the first author and M. van Frankenhuijsen in [Lap-vFr1–3]; see, in particular, Subsections 2.1.4 and 2.1.5. Here, we deal with the *principal complex dimensions* (i.e., the poles of ζ_A located on the *critical line* $\{\mathrm{Re}\, s = \overline{\dim}_B A\}$), as well as with the *visible complex dimensions* (i.e., the poles of ζ_A within a suitable subset of $\mathbb{C}$ containing the closed half-plane $\{\mathrm{Re}\, s \geq \overline{\dim}_B A\}$), in the higher-dimensional case; see, respectively, Definition 2.1.67 and Definition 2.1.68.

We will place some emphasis on the principal complex dimensions of A. Indeed, as is the case in [Lap-vFr1–3] where is developed a general theory of *explicit tube formulas* and of *fractal tube formulas*, these principal complex dimensions should also provide in the higher-dimensional case the leading asymptotic behavior in the geometry (for example, the volume of the tubular neighborhoods or t-tubes of fractals as $t \to 0^+$), the spectrum (of associated fractal drums), and the dynamics (of underlying dynamical systems, when applicable). In other words, they should give rise to the *oscillatory terms* with the largest amplitudes; see, especially, Chapters 5–11 of [Lap-vFr3]. Indeed, this is the case in the geometric situation, as is amply demonstrated in Chapter 5 on fractal tube formulas.

We will do this mostly for technical reasons, one of the main goals of the present monograph being to illustrate the power of this method in the effective computation of the box dimension for some classes of fractal sets, in particular, for the fractal nests introduced in Definition 3.5.3, the geometric chirps and string chirps; see Sections 3.5 and 3.6. The situation with general complex dimensions is already quite nontrivial in the one-dimensional case; see [Lap-vFr1–3]. As we shall see later in this book, all of the visible complex dimensions play a role in the fractal tube formulas obtained in Chapter 5.

The value of the distance zeta function in (2.1.1) remains unchanged if we replace the domain of integration A_δ with its closure $\overline{A_\delta}$. This follows from the fact that for any $\delta > 0$, the boundary $\partial(A_\delta)$ of A_δ is $(N-1)$-Minkowski measurable; see [Sta, Theorem 2], and hence, its N-dimensional Lebesgue measure is equal to zero, that is, $|\partial(A_\delta)| = 0$.

Remark 2.1.2. The analogous claim is not true for an arbitrary bounded open set U in $\mathbb{R}^N$. Indeed, in this more general situation, one may have $|\partial U| > 0$. For example, let $N = 1$ and let U be the open subset of the unit interval $I = (0,1)$, obtained as the union of the deleted intervals during a slightly modified Cantor ternary procedure, in which instead of deleting the usual sequence of the 'middle' open intervals of lengths 3^{-k}, with multiplicities 2^{k-1}, $k \in \mathbb{N}$, we delete the halves of the indicated lengths. It is then easy to see that if U is the union of the deleted intervals, then $|U|_1 = 1/2$. Note that the set $I \setminus U$ is totally disconnected, but has positive Lebesgue measure.

By a slight change of the argument, one can construct an open subset U of $[0,1]$, the boundary of which has Lebesgue measure arbitrarily close to 1. The idea is to follow the Cantor construction, but by deleting very small open intervals (instead of the middle thirds), such that the (one-dimensional) Lebesgue measure of their union is equal to an arbitrarily small prescribed positive real number ε. The boundary of the union U of these deleted open intervals is equal to $[0,1] \setminus \Omega$, and its Lebesgue measure is equal to $1 - \varepsilon$.

2.1.2 Analyticity of the Distance Zeta Functions

The main result of this section is stated in Theorem 2.1.11. It shows that the zeta function ζ_A is analytic (i.e., holomorphic) in the half-plane $\{\mathrm{Re}\, s > \overline{\dim}_B A\}$, and that under mild assumptions, the lower bound $\overline{\dim}_B A$ cannot be improved; see parts (*a*) and (*b*) of Theorem 2.1.11, along with Corollary 2.1.20. In addition, we show that this bound is always best possible from the point of view of the convergence of the Lebesgue integral defining ζ_A; see part (*a*) of Theorem 2.1.11 along with Corollary 2.1.20 In other words, the *abscissa of convergence* $D(\zeta_A)$ of the 'Dirichlet-type integral' on the right-hand side of (2.1.1) is equal to $\overline{\dim}_B A$, the upper box dimension of A; see Definition 2.1.28.

In the proof, we shall need an interesting result, due to Harvey and Polking, and stated without proof in [HarPol, p. 42]. We formulate it in a different, but equivalent, way than in [HarPol]. Following [Žu3, Lemma 1], we use the *dyadic decomposition* of the set $A_\delta \setminus \overline{A}$, i.e., of the *deleted* δ*-neighborhood of* A, in order to establish this result. We note that the goal of [HarPol] was to study the singularities of the solutions of certain linear partial differential equations.

Lemma 2.1.3 (Harvey–Polking, [HarPol, p. 42]). *Assume that A is an arbitrary bounded subset of $\mathbb{R}^N$ and let δ be an arbitrary positive number.*

$$\text{If } \gamma \in (-\infty, N - \overline{\dim}_B A), \text{ then } \int_{A_\delta} d(x,A)^{-\gamma} \mathrm{d}x < \infty, \text{ where } \gamma \text{ is real.} \tag{2.1.2}$$

Note that, in light of Definition 2.1.1, this statement precisely means that for any real number $s > \overline{\dim}_B A$, we have that $\zeta_A(s) < \infty$.

Proof. Observe that the claim of the lemma is obviously true when $\gamma \in (-\infty, 0]$, since then the function $x \mapsto d(x,A)^{-\gamma}$ is continuous and thus bounded on the set A_δ (since it must be bounded on the compact set $\overline{A_\delta}$). Therefore, it suffices to assume that $\gamma > 0$.

Let us choose any real number $s \in (\overline{\dim}_B A, N - \gamma)$. Note that the latter interval is nonempty, since by assumption, $\gamma < N - \overline{\dim}_B A$. The function $(0,\delta] \ni t \mapsto |A_t|/t^{N-s}$ is continuous; hence, since $\mathscr{M}^{*s}(A) = 0$, the supremum of this function is finite. If we denote the supremum by $C(\delta)$, then $|A_t| \le C(\delta) t^{N-s}$, for all $t \in (0,\delta]$.

We now use the following dyadic decomposition of the set $A_\delta \setminus \overline{A}$:

$$A_\delta = \overline{A} \cup \Big(\bigcup_{i=1}^{\infty} B_i \Big), \quad B_i := A_{2^{-i}\delta} \setminus A_{2^{-i-1}\delta}. \tag{2.1.3}$$

Let us first show that

$$I(A) := \int_{\overline{A}} d(x,A)^{-\gamma} \mathrm{d}x < \infty.$$

If $|\overline{A}| = 0$, then the integral is equal to zero. In the case where $|\overline{A}| > 0$, we have $\overline{\dim}_B A = N$. (See Equation (1.3.8) and the discussion following it, on page 32).

Hence, $\gamma \in (-\infty,0]$, and in this case, as we have noted, the claim of the lemma is clear.

We may assume without loss of generality that $\delta \in (0,1]$. (Indeed, if $\delta > 1$, then we choose any $\delta_1 \in (0,1]$ and write

$$\int_{A_\delta} d(x,A)^{-\gamma}\mathrm{d}x = \int_{A_{\delta_1}} d(x,A)^{-\gamma}\mathrm{d}x + \int_{A_\delta \setminus A_{\delta_1}} d(x,A)^{-\gamma}\mathrm{d}x.$$

The last integral is finite since $d(x,A)^{-\gamma}$ is bounded on $A_\delta \setminus A_{\delta_1}$.) Using (2.1.3) and the assumption $0<\gamma<N-s$ (with $s\in\mathbb{R}$) from the beginning of the proof, we have successively:

$$\begin{aligned}\int_{A_\delta} d(x,A)^{-\gamma}\mathrm{d}x &= I(A) + \sum_{i=0}^{\infty}\int_{B_i} d(x,A)^{-\gamma}\mathrm{d}x \le I(A) + \sum_{i=0}^{\infty}\int_{A_{2^{-i}\delta}} d(x,A)^{-\gamma}\mathrm{d}x \\ &\le I(A) + \sum_{i=0}^{\infty}(2^{-i-1}\delta)^{-\gamma}|A_{2^{-i}\delta}| \\ &\le I(A) + C(\delta)\sum_{i=0}^{\infty}(2^{-i-1}\delta)^{-\gamma}(2^{-i}\delta)^{N-s} \\ &\le I(A) + \frac{2^\gamma C(\delta)}{1-2^{\gamma-N+s}}\delta^{N-s-\gamma} < \infty.\end{aligned}$$

Here, the first inequality follows from the fact that for all $i \ge 0$,

$$B_i := A_{2^{-i}\delta} \setminus A_{2^{-i-1}\delta} \subseteq A_{2^{-i}\delta}.$$

This completes the proof of the lemma. □

Lemma 2.1.3 can be viewed as a far-reaching extension of the following simple fact from the basic theory of integration in $\mathbb{R}^N$: for any fixed $\delta > 0$, if $\gamma < N$ then $\int_{B_\delta(0)} |x|^{-\gamma}\mathrm{d}x < \infty$, where $B_\delta(0)$ is the open unit ball centered at 0 (i.e., $B_\delta(0)$ is the δ-neighborhood of $\{0\}$). Note that in this case, $A = \{0\}$ and $\dim_B A = 0$.

For a discussion of Lemma 2.1.3 and its various extensions, see [Žu3, Theorem 2], [Žu4, Sections 3 and 4] and [Žu5, Theorem 4.1]. (If we assume that $D = \dim_B A$ exists and $\mathscr{M}_*^D(A) > 0$, then the converse of Lemma 2.1.3 holds as well, i.e., the condition $\gamma \in (-\infty, N - \dim_B A)$ is equivalent to $\int_{A_\delta} d(x,A)^{-\gamma}\mathrm{d}x < \infty$; see [Žu5, Theorem 4.1].) Here, we state and prove a more general fact than in Lemma 2.1.3, because we shall need it later on.

Lemma 2.1.4. *Let A be any bounded set in $\mathbb{R}^N$. Then, for every value of the parameter γ in the open interval $(-\infty, N - \overline{\dim}_B A)$, the following identity holds:*

$$\int_{A_\delta} d(x,A)^{-\gamma}\,\mathrm{d}x = \delta^{-\gamma}|A_\delta| + \gamma\int_0^\delta t^{-\gamma-1}|A_t|\,\mathrm{d}t. \tag{2.1.4}$$

Furthermore, both of the integrals appearing in (2.1.4) *are finite; hence, they are convergent as Lebesgue integrals.*

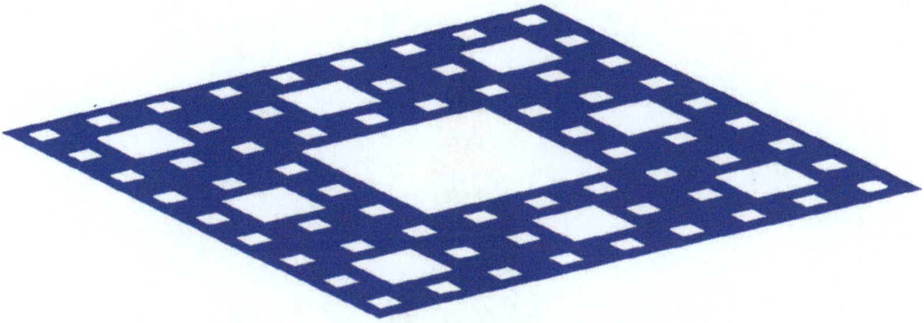

Fig. 2.1 The Sierpiński carpet A is obtained by consecutive deletion of open squares from the closed unit square. Only the first three generations of deleted squares are indicated. Figure 2.2 below shows the graph of the distance function associated with A.

To prove Lemma 2.1.4, we shall need the following well-known technical result. For completeness, we provide a proof of this elementary result. An alternative and independent proof of Lemma 2.1.4 is provided on pages 53 and 55 below.

Lemma 2.1.5 (See, e.g., [Foll, p. 198]). *Let $f : \mathbb{R}^N \to [0,+\infty]$ be any nonnegative Lebesgue measurable function and let $\alpha \in (0,+\infty)$. Then*

$$\int_{\mathbb{R}^N} f(x)^\alpha \mathrm{d}x = \alpha \int_0^{+\infty} t^{\alpha-1} |\{f > t\}| \, \mathrm{d}t, \tag{2.1.5}$$

where $\{f > t\} := \{x \in \mathbb{R}^N : f(x) > t\}$.

Proof. First, observe that if $|\{f > t\}| = +\infty$ for some $t > 0$, then both of the integrals in (2.1.5) are infinite. If this is not the case we first consider $f : \mathbb{R}^N \to \mathbb{R}$ to be a simple function and check that the identity holds. Let $A_1, \dots, A_n$ be a finite family of pairwise disjoint Lebesgue measurable subsets of $\mathbb{R}^N$ and define

$$f(x) := \sum_{i=1}^n a_i \chi_{A_i}(x), \tag{2.1.6}$$

where $0 < a_1 < a_2 < \cdots < a_n$ and for each $i = 1, \dots, n$, χ_{A_i} denotes the *characteristic function* of the set A_i; that is, $\chi_{A_i}(x) := 1$ for $x \in A_i$ and $\chi_{A_i}(x) := 0$ for $x \in \mathbb{R}^N \setminus A_i$. Note that for any $t \in (a_{i-1}, a_i)$ we have $\{f > t\} = A_i \cup A_{i+1} \cup \cdots \cup A_n$, where we let $a_0 := 0$, and hence, $|\{f > t\}| = \sum_{j=i}^n |A_j|$. Starting from the right-hand side of (2.1.5), we have

$$\begin{aligned} \alpha \int_0^{+\infty} t^{\alpha-1} |\{f > t\}| \, \mathrm{d}t &= \alpha \sum_{i=1}^n \int_{a_{i-1}}^{a_i} t^{\alpha-1} |\{f > t\}| \, \mathrm{d}t = \alpha \sum_{i=1}^n \int_{a_{i-1}}^{a_i} t^{\alpha-1} \sum_{j=i}^n |A_j| \, \mathrm{d}t \\ &= \sum_{j=1}^n |A_j| \sum_{i=1}^j \int_{a_{i-1}}^{a_i} \alpha t^{\alpha-1} \, \mathrm{d}t = \sum_{j=1}^n |A_j| a_j^\alpha = \int_{\mathbb{R}^N} f(x)^\alpha \, \mathrm{d}x. \end{aligned}$$

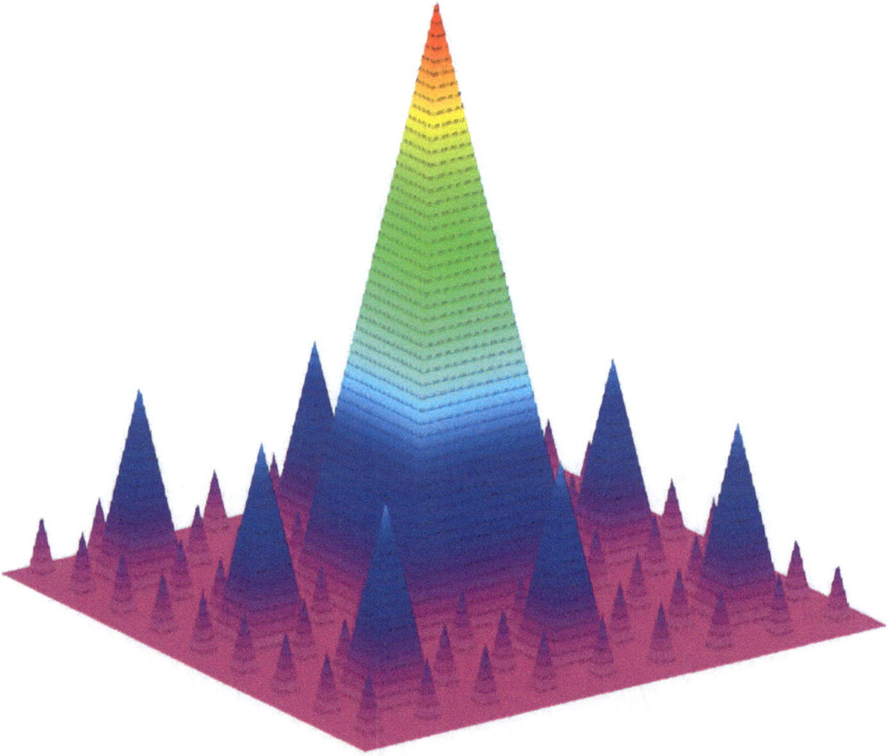

Fig. 2.2 Fractal stalagmites associated with the Sierpiński carpet. The graph of the distance function $y = d(x,A)$, defined on the unit square, where A is the Sierpiński carpet. Only the first three generations of the countable family of pyramidal tents (called 'stalagmites', see page 107) are shown. The figure is scaled vertically by the factor 3; i.e., it represents in fact the graph of $y = 3\,d(x,A)$.

Finally, in order to establish the lemma in the general case, we take a nondecreasing sequence of simple functions $(g_n)_{n=1}^{\infty}$ that monotonically converges (i.e., increases) to f. We then observe that $\{f > t\}$ is equal to the increasing union of the measurable sets $\{g_n > t\}$, where $n = 1,2,\ldots$, from which we conclude that $(\lambda_{g_n})_{n=1}^{\infty}$ pointwise increases and converges to λ_f, where $\lambda_g(t) := t^{\alpha-1}|\{g > t\}|$ for a given measurable function g on $[0,+\infty)$. It now suffices to apply the monotone convergence theorem in order to complete the proof of the lemma. □

Proof of Lemma 2.1.4. We consider the following three cases:

(a) Case when $\gamma > 0$: Since $0 < \gamma < N - \overline{\dim}_B A$, we proceed much in the same way as in [Žu5, Lemma 4.1 and Theorem 4.1(a)]. We shall use Lemma 2.1.5, with

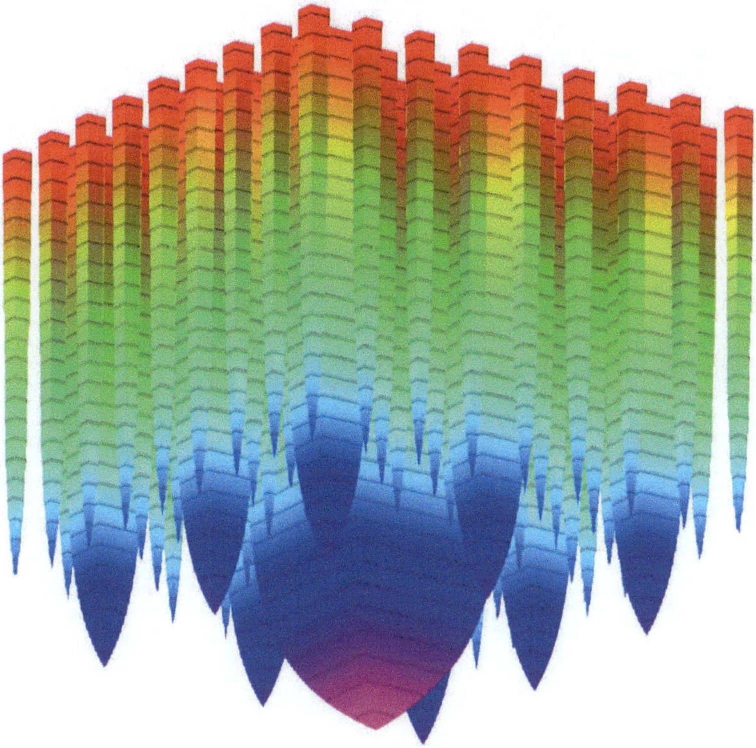

Fig. 2.3 Fractal stalactites associated with the Sierpiński carpet depicted in Figure 2.1 on page 49. The graph of the function $y = d(x,A)^{-\gamma}$, defined on the unit square, where A is the Sierpiński carpet. Since A is known to be Minkowski nondegenerate (see [Lap3], [Lap-vFr3] or [HorŽu]), the function is Lebesgueintegrable if and only if $\gamma \in (-\infty, 2-D)$, $D = \dim_B A = \log_3 8$ (see Lemma 2.1.3). For $\gamma > 0$, the graph consists of countably many connected components, called 'stalactites' (see Definition 2.1.83 on page 106), all of which are unbounded. In the present figure, $\gamma := 0.1$.

$\alpha := \gamma$ and the Borel (and hence, Lebsesgue) measurable function $f : \mathbb{R}^N \to [0,+\infty]$ given by

$$f(x) := \begin{cases} d(x,A)^{-1}, & \text{for } x \in A_\delta, \\ 0, & \text{for } x \in \mathbb{R}^N \setminus A_\delta. \end{cases}$$

Here, by definition, $f(x) = +\infty$ for $x \in \overline{A}$; furthermore, note that since $\overline{\dim}_B A < N$, then $|\overline{A}| = 0$ (see the discussion preceding Equation (1.3.9), on page 32). It is easy to see that the set $\{x \in \mathbb{R}^N : f(x) > t\}$ is equal to $A_{1/t}$ for $t > \delta^{-1}$ and to a constant set A_δ for $t \in (0, \delta^{-1})$. Therefore,

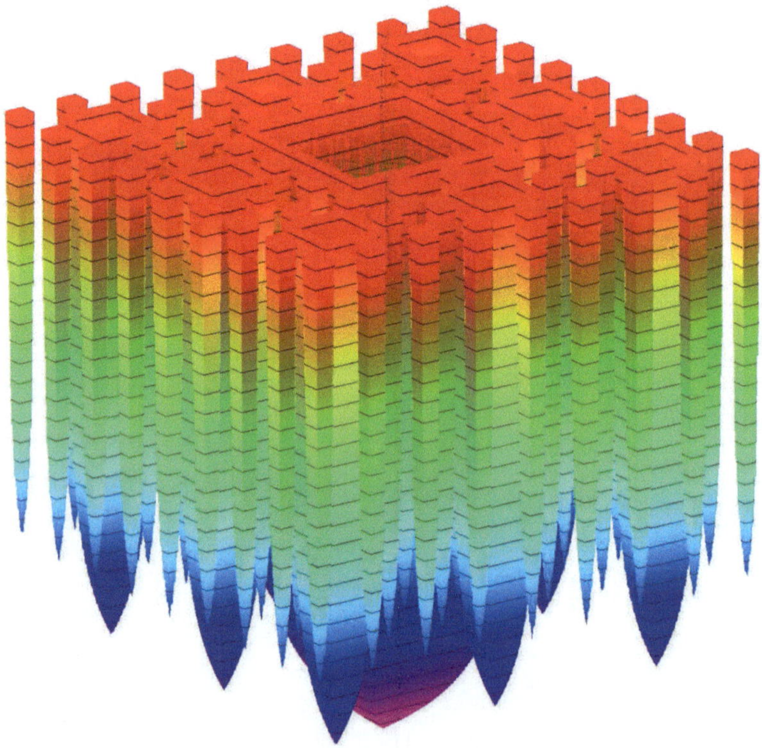

Fig. 2.4 Fractal stalactites associated with the Sierpiński carpet, revisited. Another view of the graph of the function $y = d(x,A)^{-\gamma}$, where A is the Sierpiński carpet, similar to the one provided in Figure 2.3. The level set of this function tends to the Sierpiński carpet in the Hausdorff metric, when the level tends to $+\infty$. This is a special case of Proposition 2.1.89 on page 109. Here, we have let $\gamma := 0.1$, as in Figure 2.3.

$$\begin{aligned}\int_{A_\delta} d(x,A)^{-\gamma}\mathrm{d}x &= \gamma\Big(\int_0^{1/\delta} + \int_{1/\delta}^{+\infty}\Big)t^{\gamma-1}|\{f > t\}|\,\mathrm{d}t \\ &= \gamma|A_\delta|\int_0^{1/\delta} t^{\gamma-1}\mathrm{d}t + \gamma\int_{1/\delta}^{+\infty} t^{\gamma-1}|A_{1/t}|\,\mathrm{d}t.\end{aligned}$$

Equation (2.1.4) now follows by using the change of variable $\tau = 1/t$ in the last integral. In order to show that the last integral in (2.1.4) is finite, let $\varepsilon > 0$ be small enough so that $\gamma \in (0, N - D - \varepsilon)$, where $D := \overline{\dim}_B A$. Then $\mathscr{M}^{*(D+\varepsilon)}(A) = 0$, and thus, there exists a positive constant $C = C(\delta)$ such that $|A_t| \le Ct^{N-D-\varepsilon}$ for all $t \in (0, \delta]$. Hence,

$$\int_0^\delta t^{-\gamma-1}|A_t|\,\mathrm{d}t \le C\int_0^\delta t^{N-D-\varepsilon-\gamma-1}\mathrm{d}t < \infty.$$

(b) Case when $\gamma = 0$: If we assume that $\gamma = 0 < N - \overline{\dim}_B A$ (which implies that $|A| = 0$), then it suffices to show that $I := \int_0^\delta t^{-1}|A_t|\,dt < \infty$. Letting $D = \overline{\dim}_B A$, we then have $D + \varepsilon < N$ for $\varepsilon > 0$ small enough; hence, since $\mathscr{M}^{*(D+\varepsilon)}(A) = 0$, there exists a positive constant C such that $|A_t| \le C t^{N-D-\varepsilon}$ for all $t \in (0, \delta)$. This immediately implies that $I \le C \int_0^\delta t^{N-D-\varepsilon-1}\mathrm{d}t < \infty$.

(c) Case when $\gamma < 0$: It is clear that in this case, the left-hand side of (2.1.4) is finite. We shall use Lemma 2.1.5, where $\alpha = -\gamma$ and the Borel (and hence, Lebesgue) measurable function $f : \mathbb{R}^N \to [0, +\infty]$ is given by

$$f(x) := \begin{cases} d(x,A), & \text{for } x \in A_\delta, \\ 0, & \text{for } x \in \mathbb{R}^N \setminus A_\delta. \end{cases}$$

First, note that $\{f > t\} = \emptyset$ for $t \ge \delta$, and $\{f > t\} = A_\delta \setminus \overline{A_t}$ for $0 < t < \delta$, where $\overline{A_t}$ is the closure of the set A_t. We now show that for any $t > 0$, $|A_t| = |\overline{A_t}|$. This is an immediate consequence of the fact that the boundary of A_t is of Lebesgue measure zero, that is, $|\partial(A_t)| = 0$. Indeed, as was noted earlier, according to [Sta, Theorem 2], the set $\partial(A_t)$ is Minkowski measurable and $\dim_B \partial(A_t) = N - 1$ for any $t > 0$; so that, by Equation (1.3.3), $|\partial(A_t)| = \mathscr{M}^N(\partial(A_t)) = 0$. (Since $N > N - 1$, the second equality follows from the definition of the Minkowski dimension; see Figures 1.8 and 1.9 on page 33, along with Equation (1.3.4) on page 31.) Therefore, for $0 < t < \delta$ we have

$$|\{f > t\}| = |A_\delta| - |\overline{A_t}| = |A_\delta| - |A_t|.$$

Using (2.1.5), we then obtain

$$\int_{A_\delta} d(x,A)^\alpha \mathrm{d}x = \alpha \int_0^\delta t^{\alpha-1}(|A_\delta| - |A_t|)\,\mathrm{d}t = \delta^\alpha |A_\delta| - \alpha \int_0^\delta t^{\alpha-1}|A_t|\,\mathrm{d}t,$$

which proves (2.1.4). □

For the sake of completeness, we provide an alternative proof of (2.1.4), and thus of Lemma 2.1.4, based on the generalized change of variables formula and on an integration by parts. It is essentially the same as in [Žu2, Theorem 2.9(a)].

An alternative proof of Lemma 2.1.4. We shall need the assumption that $\gamma < N - \overline{\dim}_B A$ from Lemma 2.1.4 only in case (c) below. Introducing the new variable $t = d(x,A)$ and the function $V(t) = |A_t|$, $t > 0$, we can arrive to the desired result by using the following formal computation (with $\nabla f(x)$ denoting the pointwise almost everywhere defined gradient of f):

$$\begin{aligned}\int_{A_\delta} d(x,A)^{-\gamma}\mathrm{d}x &\overset{(a)}{=} \int_0^\delta t^{-\gamma}\mathrm{d}V(t)\\ &\overset{(b)}{=} t^{-\gamma}V(t)\Big|_{t=0}^{\delta} - \int_0^\delta V(t)(-\gamma)t^{-\gamma-1}\mathrm{d}t \\ &\overset{(c)}{=} \delta^{-\gamma}|A_\delta| + \gamma\int_0^\delta t^{-\gamma-1}|A_t|\,\mathrm{d}t.\end{aligned} \tag{2.1.7}$$

Let us justify this computation in the following three steps:

(a) In order to prove the first equality in (2.1.7), let us set $f(x) = d(x,A)$ and $g(x) = d(x,A)^{-\gamma}$ for all x in A_δ, and $f(x) = g(x) = 0$ for all x in $\mathbb{R}^N \setminus A_\delta$. We then have

$$\begin{aligned}\int_{A_\delta} d(x,A)^{-\gamma}\mathrm{d}x &= \int_{A_\delta} g(x)\,|\nabla f(x)|\,\mathrm{d}x = \int_0^\delta \left[\int_{f^{-1}(t)} g(x)\,\mathrm{d}H^{N-1}\right]\mathrm{d}t \\ &= \int_0^\delta t^{-\gamma}H^{N-1}(\partial(A_t))\,\mathrm{d}t = \int_0^\delta t^{-\gamma}\mathrm{d}V(t),\end{aligned} \tag{2.1.8}$$

where H^{N-1} is the $(N-1)$-dimensional Hausdorff measure on $f^{-1}(t) = \partial(A_t)$. The first equality in (2.1.8) follows from the fact that $|\nabla f| = 1$ a.e. in A_δ, which is a consequence of Rademacher's theorem, according to which a Lipschitz function on $\mathbb{R}^N$ is (Lebesgue) almost everywhere differentiable (see, e.g., [EvGa, Theorem 2 in Section 3.1.2]), and the proof of [Fed2, Lemma 3.2.34]. The second equality in (2.1.8) follows from the generalized change of variables formula (see, e.g., [EvGa, Theorem 2 in Section 3.4.3] or [JohLap, Theorem 3.3.2]). Now, the first equality in (2.1.7) follows from the fact that $V'(t) = H^{N-1}(\partial(A_t))$ for (Lebesgue) a.e. $t > 0$, where $V(t) = |A_t|$ as above and V' is the Lebesgue almost everywhere defined derivative of V; see Stachó's result given in [Sta, Theorem 2 and Lemma 2(ii)]. (Moreover, the identity $V'(t) = H^{N-1}(\partial(A_t))$ holds for all $t > 0$ outside a countable set; see [Sta, Theorem 2 and Lemma 2(ii)].) Note that the Lebesgue integrability of the function $d(\cdot,A)^{-\gamma}$, defined on A_δ, is ensured by Lemma 2.1.3.

(b) The second equality in (2.1.7) is due to the integration by parts formula for Lebesgue–Stieltjes integrals; see [Foll, Theorem 3.36]. Indeed, it suffices to use this result on intervals of the form $(\varepsilon,\delta]$ for $\varepsilon \in (0,\delta)$ (note that both $t^{-\gamma}$ and $V(t)$ are of bounded variation and continuous on these intervals), and then pass to the limit as $\varepsilon \to 0^+$.

(c) In order to justify the last equality in (2.1.7), we must show that

$$\lim_{t\to 0^+} t^{-\gamma}V(t) = 0.$$

The assumption $\gamma < N - \overline{\dim}_B A$ implies that the open interval $(\overline{\dim}_B A, N-\gamma)$ in $\mathbb{R}$ is nonempty. Let us take any $d \in (\overline{\dim}_B A, N-\gamma)$. Since $d > \overline{\dim}_B A$, there exists $C_d > 0$

such that $V(t) \le C_d t^{N-d}$ for all $t \in (0,\delta]$. We conclude that $0 < t^{-\gamma}V(t) \le C_d t^{N-d-\gamma}$, and the claim follows by passing to the limit as $t \to 0^+$. □

In the sequel, we shall also need the following result, which complements Lemma 2.1.3.

Lemma 2.1.6. *Let A be a bounded set in $\mathbb{R}^N$ and $\delta > 0$. If $\gamma > N - \overline{\dim}_B A$, then $\int_{A_\delta} d(x,A)^{-\gamma} \mathrm{d}x = +\infty$.*

Proof. Note that for all $\gamma > 0$, we have

$$I_\delta := \int_{A_\delta} d(x,A)^{-\gamma} \mathrm{d}x = \delta^{-\gamma}|A_\delta| + \gamma \int_0^\delta s^{-\gamma-1}|A_s|\,\mathrm{d}s \ge \delta^{-\gamma}|A_\delta|, \tag{2.1.9}$$

where the second equality of (2.1.9) is precisely the content of [Žu5, Lemma 4.1].

For the sake of completeness, we next reproduce the proof of the second equality in (2.1.9). The argument is similar to the proof of Lemma 2.1.4. Let us define $f(x) = d(x,A)^{-1}$ for $x \in A_\delta$, and $f(x) = 0$ for $x \in \mathbb{R}^N \setminus A_\delta$. Since $\{x : f(x) > t\} = A_{1/t}$ for $t > 1/\delta$, and $\{x : f(x) > t\} = A_\delta$ for $t < 1/\delta$, we deduce from Lemma 2.1.5 that

$$\int_{A_\delta} d(x,A)^{-\gamma} \mathrm{d}x = \gamma |A_\delta| \int_0^{1/\delta} t^{\gamma-1} \mathrm{d}t + \gamma \int_{1/\delta}^{\infty} t^{\gamma-1}|A_{1/t}|\,\mathrm{d}t.$$

The desired equality follows by using the change of variable $s = 1/t$.

We now continue the proof of Lemma 2.1.6. Let us write $d = \overline{\dim}_B A$ and choose $\sigma < d$ sufficiently close to d so that $\gamma > N - \sigma$. Then $\mathscr{M}^{*\sigma}(A) = +\infty$ (see (1.3.5)), which implies that there exists a sequence of positive numbers s_k converging to zero and such that

$$C_k := \frac{|A_{s_k}|}{s_k^{N-\sigma}} \to +\infty \quad \text{as} \quad k \to \infty.$$

Since $\delta \mapsto I_\delta$ is nondecreasing (see also (2.1.9)), we have for all k large enough

$$I_\delta \ge I_{s_k} \ge (s_k)^{-\gamma}|A_{s_k}| = C_k \cdot s_k^{N-\sigma-\gamma} \to +\infty$$

as $k \to \infty$. Hence, $I_\delta = +\infty$, as desired. □

Remark 2.1.7. If $\gamma := N - \overline{\dim}_B A$, then the conclusion of Lemma 2.1.6 does not hold, in general. A class of counterexamples is provided in [Žu4, Theorem 4.3].

Definition 2.1.8. Following and extending the commonly used terminology for Dirichlet series and integrals (see, e.g., [Ser], [Pos] and Subsection 2.1.3 below), we define the *abscissa of convergence* $D(\zeta_A)$ of ζ_A by the following equality:[1]

---

[1] This is really the abscissa of *absolute* convergence of the (generalized) Dirichlet integral ζ_A, but we will not stress this point in the sequel. In fact, since the integral defining ζ_A in Equation (2.1.1) is interpreted here as a Lebesgue integral (and since the corresponding integrand, $x \mapsto d(x,A)^{s-N}$, is continuous and hence, Borel measurable on $\mathbb{R}^N$), there is no difference between absolute convergence and convergence of the integral.

$$D(\zeta_A) := \inf\left\{\alpha \in \mathbb{R} : \int_{A_\delta} d(x,A)^{\alpha-N}\mathrm{d}x < \infty\right\}. \tag{2.1.10}$$

Note that $D(\zeta_A) \in \mathbb{R} \cup \{\pm\infty\}$. Alternatively, in light of Lemma 2.1.9 below, $D(\zeta_A)$ can be uniquely defined by the property according to which the open right half-plane $\{\mathrm{Re}\, s > D(\zeta_A)\}$, called the *half-plane of convergence* of ζ_A and denoted by $\Pi(\zeta_A)$, is the *maximal* (i.e., the largest) open right half-plane (of the form $\{\mathrm{Re}\, s > \alpha\}$, with $\alpha \in \mathbb{R} \cup \{\pm\infty\}$), on which the integral defining ζ_A is absolutely convergent (i.e., $\int_{A_\delta} |d(x,A)^{s-N}|\,\mathrm{d}x = \int_{A_\delta} d(x,A)^{\mathrm{Re}\, s-N}\,\mathrm{d}x < \infty$), or equivalently, on which the Lebesgue integral $\int_{A_\delta} d(x,A)^{s-N}\mathrm{d}x$ is convergent (see footnote 1 on page 55).[2]

Finally, when $D(\zeta_A)$ is real, the vertical line $\{\mathrm{Re}\, s = D(\zeta_A)\}$ is called the *critical line of convergence* of ζ_A (or simply, the *critical line*, when no ambiguity may arise).

Lemma 2.1.9. *If the Lebesgue integral $\zeta_A(s) := \int_{A_\delta} d(x,A)^{s-N}\mathrm{d}x$ converges for some $s = s_0 \in \mathbb{C}$, then it also converges for any $s \in \mathbb{C}$ such that* $\mathrm{Re}\, s > \mathrm{Re}\, s_0$.

Proof. Assume the hypothesis of the lemma and fix $s \in \mathbb{C}$ with $\mathrm{Re}\, s > \mathrm{Re}\, s_0$. Without loss of generality, we may assume that $\delta \le 1$. Indeed, if $\delta > 1$, we write the disjoint union $A_\delta = A_1 \cup (A_\delta \setminus A_1)$, and hence,

$$\int_{A_\delta} |d(x,A)^{s-N}|\,\mathrm{d}x = \int_{A_1} d(x,A)^{\mathrm{Re}\, s-N}\mathrm{d}x + \int_{A_\delta \setminus A_1} d(x,A)^{\mathrm{Re}\, s-N}\mathrm{d}x =: I_1 + I_2.$$

Since the function $x \mapsto |d(x,A)^{s-N}| = d(x,A)^{\mathrm{Re}\, s-N}$ is continuous and nowhere vanishing on the compact set $\overline{A_\delta} \setminus A_1$ (indeed, note that this set is contained in the complement of the 1-neighborhood of $\overline{A}$, which is the set of zeros of the function), the function $x \mapsto |d(x,A)|^{\mathrm{Re}\, s-N}$ is continuous as well on $\overline{A_\delta} \setminus A_1$ (regardless of the sign of the exponent $\mathrm{Re}\, s - N$), and therefore, $I_2 < \infty$.

More precisely, if $\mathrm{Re}\, s \ge N$, then $|d(x,A)|^{\mathrm{Re}\, s-N} \le \delta^{\mathrm{Re}\, s-N}$ for all $x \in A_\delta \setminus A_1$, while for $\mathrm{Re}\, s < N$ we have that $|d(x,A)|^{\mathrm{Re}\, s-N} \le 1^{\mathrm{Re}\, s-N} = 1$ for the same values of x. Therefore,

$$I_2 \le |A_\delta \setminus A_1|_N \cdot \max\{\delta^{\mathrm{Re}\, s-N}, 1\} < \infty.$$

Next, let us assume that $0 < \delta \le 1$. Then[3]

$$\begin{aligned}\int_{A_\delta} |d(x,A)^{s-N}|\,\mathrm{d}x &= \int_{A_\delta} d(x,A)^{\mathrm{Re}\, s-N}\mathrm{d}x \\ &\le \int_{A_\delta} d(x,A)^{\mathrm{Re}\, s_0-N}\mathrm{d}x = \int_{A_\delta} |d(x,A)^{s_0-N}|\,\mathrm{d}x < \infty,\end{aligned}$$

---

[2] By convention, when $D(\zeta_A) = +\infty$, we have $\Pi(\zeta_A) = \emptyset$ while when $D(\zeta_A) = -\infty$, we have $\Pi(\zeta_A) = \mathbb{C}$. A similar comment could be made later on about the half-plane of holomorphic continuation $\mathscr{H}(\zeta_A)$ when $D_{\mathrm{hol}}(\zeta_A) = \pm\infty$, as well as for the half-plane of meromorphic continuation $\mathrm{Mer}\,(\zeta_A)$ when $D_{\mathrm{mer}}(\zeta_A) = \pm\infty$.

[3] If $\mathrm{Re}\, s_0 \le N$, then the assumed convergence of the Lebesgue integral $\zeta_A(s_0)$ implies that $|\overline{A}| = 0$. On the other hand, if $\mathrm{Re}\, s_0 > N$, then $|\overline{A}|$ may be positive.

by hypothesis. Note that in the inequality just above, we have used the fact that $d(x,A) \le 1$ for all $x \in A_\delta$ since $\delta \le 1$. This concludes the proof of the lemma. □

Remark 2.1.10. Here and in the sequel, given $w \in \mathbb{C}$, we use the following convention:

$$0^w := \begin{cases} 0, & \text{for} \quad \mathrm{Re}\, w > 0, \\ +\infty, & \text{for} \quad \mathrm{Re}\, w < 0. \end{cases} \tag{2.1.11}$$

Furthermore, note that if the zero-set

$$Z := \{x \in \mathbb{R}^N : d(x,A) = 0\} \tag{2.1.12}$$

is of positive N-dimensional Lebesgue measure (i.e., $|Z| > 0$), then for all $s \in \mathbb{C}$ with $\mathrm{Re}\, s < N$, the integral $\int_{A_\delta} |d(x,A)^{s-N}|\,\mathrm{d}x$ is equal to $+\infty$, and hence, the Lebesgue integral $\int_{A_\delta} d(x,A)^{s-N}\mathrm{d}x$ does not converge (i.e., does not exist). Therefore, if the latter integral is convergent, we must have $\mathrm{Re}\, s \ge N$ and so $D(\zeta_A) \ge N$. But since by part (*b*) of Theorem 2.1.11 below, $D(\zeta_A) = \overline{\dim}_B A \in [0,N]$, it then follows that $\overline{\dim}_B A = N$.

In light of (2.1.12), $Z = \overline{A}$ and $Z \cap A_\delta = \overline{A}$ for any $\delta > 0$. (Indeed, given $\delta > 0$, we have $\overline{A} \subset A_\delta$ because if we take any $\delta' \in (0,\delta)$, then the closed set $\{x \in A : d(x,A) \le \delta'\}$ is contained in A_δ and so, A_δ must contain $\overline{A}$.) It follows that $Z \cap A_\delta$ (i.e., Z) is of positive measure is equivalent to $|\overline{A}| > 0$, and in light of (1.3.8), this implies that $\dim_B A$ exists and $\dim_B A = N$, which is consistent with the above claim.

It is now easy to deduce from Lemma 2.1.9 that the half-plane of convergence of ζ_A, defined as above by $\Pi(\zeta_A) := \{\mathrm{Re}\, s > D(\zeta_A)\}$, where $D(\zeta_A)$ is the abscissa of convergence of ζ_A defined by (2.1.10), is indeed the maximal open right half-plane of convergence of the Lebesgue integral defining ζ_A in (2.1.1), as stated above. We leave the verification as an exercise for the reader.

We are now ready to state the main result of this subsection.

Theorem 2.1.11. *Let A be an arbitrary bounded subset of $\mathbb{R}^N$ and let $\delta > 0$. Then:*

(*a*) *The distance zeta function ζ_A defined by* (2.1.1) *is holomorphic* (*i.e., analytic*) *in the open right half-plane $\{\mathrm{Re}\, s > \overline{\dim}_B A\}$, and for all complex numbers s in that region, its complex derivative is given as follows*:

$$\zeta_A'(s) = \int_{A_\delta} d(x,A)^{s-N} \log d(x,A)\,\mathrm{d}x. \tag{2.1.13}$$

(*b*) *The lower bound in the open right half-plane $\{\mathrm{Re}\, s > \overline{\dim}_B A\}$ is optimal, from the point of view of the* (*absolute*) *convergence of the Dirichlet-type integral defining ζ_A. In other words,*

$$\overline{\dim}_B A = D(\zeta_A), \tag{2.1.14}$$

where $D(\zeta_A)$ is the abscissa of convergence of ζ_A, as defined in Equation (2.1.10).[4] *It follows that $D(\zeta_A) \in [0,N]$. (See also Corollary 2.1.20 below for more detailed information.) Furthermore, the identity* (2.1.1) *continues to hold in the half-plane of (absolute) convergence $\{\operatorname{Re} s > \overline{\dim}_B A\}$ of ζ_A. Moreover, we have*[5]

$$D(\zeta_A) = \inf\left\{\alpha \in [0,N] : \int_{A_\delta} d(x,A)^{\alpha-N} \mathrm{d}x < \infty\right\}. \tag{2.1.15}$$

(c) If the box (or Minkowski) dimension $D := \dim_B A$ exists, $D < N$, and $\mathcal{M}_*^D(A) > 0$, then $\zeta_A(s) \to +\infty$ as $s \in \mathbb{R}$ converges to D from the right.[6] *According to part (ii) of Corollary 2.1.20 below, it then follows that (under the additional hypotheses of part (c) of the theorem), we have*

$$\dim_B A = D(\zeta_A) = D_{\mathrm{hol}}(\zeta_A), \tag{2.1.16}$$

where $D_{\mathrm{hol}}(\zeta_A)$, the abscissa of holomorphic continuation of ζ_A (as given by (2.1.27) *below), is defined so that $\{\operatorname{Re} s > D_{\mathrm{hol}}(\zeta_A)\}$ be the maximal right half-plane of the form $\{\operatorname{Re} s > \alpha\}$, for some $\alpha \in \mathbb{R} \cup \{\pm\infty\}$, to which ζ_A can be holomorphically continued. (For more details, see Corollary 2.1.20 and the text preceding it.)*

Proof. (a) We give here a direct and elementary proof of the holomorphicity of ζ_A, not requiring any additional assumption about A. (See also Remark 2.1.51, based on Theorem 2.1.45 in Subsection 2.1.3 below, for a different approach.) Let us denote the right-hand side of (2.1.13) by $I(s)$. To prove the holomorphicity of ζ_A, let us fix any s such that $\operatorname{Re} s > \overline{\dim}_B A$. We then have to show that

$$\begin{aligned} R(h) &:= \frac{\zeta_A(s+h) - \zeta_A(s)}{h} - I(s) \\ &= \int_{A_\delta} \left(\frac{d(x,A)^h - 1}{h} - \log d(x,A) \right) d(x,A)^{s-N} \mathrm{d}x \end{aligned} \tag{2.1.17}$$

converges to zero as $h \to 0$ in $\mathbb{C}$, with $h \neq 0$.

---

[4] See Subsection 2.1.3.2 below, along with Appendix A, for the more general setting of Dirichlet-type integrals.

[5] A priori, the infimum should be taken over all real numbers $\alpha \in \mathbb{R}$, but since $D(\zeta_A) = \overline{\dim}_B A \in [0,N]$) (in light of (2.1.14)) and for $\alpha > N$ the function $x \mapsto d(x,A)^{\alpha-N}$ is bounded on A_δ, it can be taken over all $\alpha \in [0,N]$.

[6] Hence, D is a singularity (which may or may not be a pole) of ζ_A. Naturally, if ζ_A possesses a meromorphic continuation to a connected open neighborhood of D, then it follows that D is a pole of ζ_A. In Section 2.3 and Section 4.5 will be provided several sufficient conditions under which ζ_A can be meromorphically continued beyond the critical line $\operatorname{Re} s = D$, and hence, in particular, to a connected open neighborhood of D.

Letting $d := d(x,A) \in (0,\delta)$, we first consider

$$f(h) := \frac{d^h - 1}{h} - \log d = \frac{1}{h}(\mathrm{e}^{(\log d)h} - 1) - \log d. \tag{2.1.18}$$

Using the MacLaurin series $\mathrm{e}^z = \sum_{j\geq 0} \frac{z^j}{j!}$, which converges for all $z \in \mathbb{C}$, we obtain that

$$f(h) = h(\log d)^2 \sum_{k=0}^{\infty} \frac{1}{(k+2)(k+1)} \cdot \frac{(\log d)^k h^k}{k!}, \tag{2.1.19}$$

for all $h \in \mathbb{C}$. Furthermore, assuming without loss of generality that $0 < \delta \leq 1$ (if $\delta > 1$, see Lemma 2.1.15 below applied to A and $U := A_\delta \setminus \overline{A}_1$) and hence, $\log d \leq 0$, we have

$$\begin{aligned}|f(h)| &\leq \frac{1}{2}|h|\,(\log d)^2 \sum_{k=0}^{\infty} \frac{(|\log d|\,|h|)^k}{k!} \\ &= \frac{1}{2}|h|\,(\log d)^2 \mathrm{e}^{-(\log d)|h|} = \frac{1}{2}|h|\,(\log d)^2 d^{-|h|}.\end{aligned}$$

Therefore,

$$|R(h)| \leq \frac{1}{2}|h| \int_{A_\delta} |\log d(x,A)|^2 d(x,A)^{\mathrm{Re}\, s - N - |h|} \mathrm{d}x. \tag{2.1.20}$$

Let $\varepsilon > 0$ be a sufficiently small number, to be specified below. Taking $h \in \mathbb{C}$ such that $|h| < \varepsilon$, since $\delta \leq 1$ and hence, $d(x,A) \leq 1$ for all $x \in A_\delta$, we have

$$|R(h)| \leq \frac{1}{2}|h| \int_{A_\delta} |\log d(x,A)|^2 d(x,A)^{\varepsilon} d(x,A)^{\mathrm{Re}\, s - N - 2\varepsilon} \mathrm{d}x.$$

Clearly, there exists a positive constant $C = C(\delta,\varepsilon)$ such that $|\log \rho|^2 \rho^\varepsilon \leq C$ for all $\rho \in (0,\delta)$. This implies that

$$|R(h)| \leq \frac{1}{2} C|h| \int_{A_\delta} d(x,A)^{\mathrm{Re}\, s - N - 2\varepsilon} \mathrm{d}x. \tag{2.1.21}$$

Letting $\gamma := 2\varepsilon + N - \mathrm{Re}\, s$, we see that the integrability condition $\gamma < N - \overline{\dim}_B A$ stated in Equation (2.1.2) of Lemma 2.1.3 is equivalent to $\mathrm{Re}\, s > \overline{\dim}_B A + 2\varepsilon$. Observe that this latter inequality holds for all positive ε small enough, due to the assumption $\mathrm{Re}\, s > \overline{\dim}_B A$. Hence, $R(h) \to 0$ as $h \to 0$ in $\mathbb{C}$, with $h \neq 0$. Therefore, we conclude that $\zeta_A(s)$ is holomorphic for $\mathrm{Re}\, s > \overline{\dim}_B A$, with (complex) derivative $\zeta_A'(s)$ given by (2.1.13) as desired. This establishes part (a) of the theorem.

(b) In light of part (a), this follows immediately from Lemma 2.1.6. Indeed, the latter result implies that for any real number $\alpha < D := \overline{\dim}_B A$ (and hence, $\gamma := N - \alpha > N - D$, as required in Lemma 2.1.6), we have

$$\int_{A_\delta} d(x,A)^{\alpha - N} \mathrm{d}x = +\infty.$$

The claim now follows since in light of Lemma 2.1.3, we know that

$$\zeta_A(\alpha) = \int_{A_\delta} d(x,A)^{\alpha-N}\,\mathrm{d}x < \infty$$

for any real number $\alpha > D$, as desired. We therefore deduce from the definition (2.1.10) of $D(\zeta_A)$ that $D(\zeta_A) = \overline{\dim}_B A$, as desired.

Finally, the fact that the identity (2.1.1) continues to hold in the half-plane $\{\mathrm{Re}\, s > D\}$ follows from the principle of analytic continuation and the holomorphicity of ζ_A on the domain (i.e., connected open set) of $\mathbb{C}$ given by $\{\mathrm{Re}\, s > D\}$. Recall that the latter property of ζ_A has been established in part (a) of the proof or else, alternatively, follows from the well-known properties of a (generalized) Dirichlet integral (see Subsection 2.1.3.2, including, especially, Theorem 2.1.47 and Remark 2.1.51 below; see also Appendix A). This completes the proof of part (b).

(c) Note that since $\mathscr{M}_*^D(A) > 0$, then for any fixed $\delta > 0$ there exists $C > 0$ such that for all $t \in (0,\delta)$, we have $|A_t| \geq Ct^{N-D}$. Using Lemmas 2.1.3 and 2.1.4, we see that for any $\gamma \in (0, N-D)$,

$$\begin{aligned}\infty > I(\gamma) &= \int_{A_\delta} d(x,A)^{-\gamma}\mathrm{d}x = \delta^{-\gamma}|A_\delta| + \gamma\int_0^\delta t^{-\gamma-1}|A_t|\,\mathrm{d}t \\ &\geq \gamma C\int_0^\delta t^{N-D-\gamma-1}\mathrm{d}t = \gamma C\frac{\delta^{N-D-\gamma}}{N-D-\gamma}.\end{aligned}$$

Therefore, if $\gamma \in \mathbb{R}$ is such that $\gamma \to N-D$ from the left, then $I(\gamma) \to +\infty$. Equivalently, if $s \in \mathbb{R}$ is such that $s \to D$ from the right, then $\zeta_A(s) \to +\infty$. For the proof of the identity stated in Equation (2.1.16), that is, for the proof of the additional equality $D(\zeta_A) = D_{\mathrm{hol}}(\zeta_A)$ under the hypotheses of part (c) of Theorem 2.1.11, we refer to the corresponding part of the proof of part (ii) of Corollary 2.1.20 below (as well as to the text preceding it for the necessary definitions).

This concludes the proof of the theorem. □

Remark 2.1.12. It is clear that for real s, the values of $\zeta_A(s)$ are also real. Furthermore, using the principle of reflection (see, e.g., [Tit1, p. 155]), we deduce that for all complex numbers s such that $\mathrm{Re}\, s > \overline{\dim}_B A$, we have $\overline{\zeta_A(s)} = \zeta_A(\overline{s})$. Naturally, this identity remains valid upon meromorphic continuation (in any region $U \subseteq \mathbb{C}$ to which the distance zeta function ζ_A can be meromorphically extended). It follows from the above observation about the symmetry of ζ_A that provided the given domain $U \subseteq \mathbb{C}$ is symmetric with respect to the real axis, the *nonreal (visible) complex dimensions of* ζ_A in U (i.e., the poles of ζ_A in U) *come in complex conjugate pairs*. The same is true for the complex dimensions of (ordinary) fractal strings; see [Lap-vFr3, Remarks 1.6 and 1.16].

Proposition 2.1.13. *Assuming that* $|\overline{A}| = 0$ *(which is always the case if* $\overline{\dim}_B A < N$, *see Equation* (1.3.8) *on page 32), and given any* $\delta > 0$, *we can compute the distance zeta function* ζ_A *in Equation* (2.1.1) *as follows, for every* $s \in \mathbb{C}$ *with* $\mathrm{Re}\, s > \overline{\dim}_B A$:

$$\zeta_A(s) = \lim_{\varepsilon\to 0^+} \int_{A_\delta \setminus \overline{A_\varepsilon}} d(x,A)^{s-N} \mathrm{d}x. \tag{2.1.22}$$

Proof. Fix $\delta > 0$ and assume that $0 < \varepsilon < \delta$, in what follows. Then, the characteristic function $\chi_{A_\delta \setminus A_\varepsilon}$ converges to χ_{A_δ} (pointwise) a.e. in A_δ as $\varepsilon \to 0^+$. We now claim that (2.1.22) holds and that the convergence on the right-hand side of (2.1.22) is uniform as $\varepsilon \to 0^+$, with respect to all s such that $\mathrm{Re}\, s > \xi_0$, where $\xi_0 > \overline{\dim}_B A$. To see this, for such a complex number s and assuming without loss of generality that $\varepsilon \in (0,1)$, we note that

$$\left|\int_{A_\varepsilon} d(x,A)^{s-N}\mathrm{d}x\right| \le \int_{A_\varepsilon} d(x,A)^{\mathrm{Re}\, s-N}\mathrm{d}x \le \int_{A_\varepsilon} d(x,A)^{\xi_0-N}\mathrm{d}x.$$

Let us choose any $d \in (\overline{\dim}_B A, \xi_0)$. Since $d > \overline{\dim}_B A$, then $\mathscr{M}^{*d}(A) = 0$, and therefore, there exists a positive constant $C = C(d,N,A)$ such that $|A_t| \le Ct^{N-d}$ for all $t \in (0,\varepsilon]$. Using Lemma 2.1.4 with $\gamma := N - \xi_0 < N - \overline{\dim}_B A$, it follows that

$$\begin{aligned}\int_{A_\varepsilon} d(x,A)^{\xi_0-N}\mathrm{d}x &= \varepsilon^{-\gamma}|A_\varepsilon| + \gamma\int_0^\varepsilon t^{\gamma-1}|A_t|\,\mathrm{d}t \\ &\le \varepsilon^{-\gamma}C\varepsilon^{N-d} + \gamma\int_0^\varepsilon t^{\gamma-1}Ct^{N-d}\mathrm{d}t = C_1\cdot\varepsilon^{\xi_0-d},\end{aligned}$$

where $C_1 := C(N-d)/(\xi_0-d)$. Hence, using $d < \xi_0$, we conclude that

$$\sup_{\mathrm{Re}\, s>\xi_0}\left|\int_{A_\varepsilon} d(x,A)^{s-N}\mathrm{d}x\right| \le C_1\cdot\varepsilon^{\xi_0-d} \to 0^+ \quad \text{as } \varepsilon\to 0^+. \tag{2.1.23}$$

Equation (2.1.22) now follows from Definition 2.1.1. □

We deduce from the proof of Proposition 2.1.13 that the distance zeta function satisfies the following asymptotic property; see Equation (2.1.23) just above.

Proposition 2.1.14. *Assume that A is a bounded subset of $\mathbb{R}^N$, $\varepsilon \in (0,1)$, and define the corresponding distance zeta function ζ_{A,A_ε} by*

$$\zeta_{A,A_\varepsilon}(s) := \int_{A_\varepsilon} d(x,A)^{s-N}\,\mathrm{d}x,$$

for $s \in \mathbb{C}$ with $\mathrm{Re}\, s > \overline{\dim}_B A$. Then, for any $\xi_0 > \overline{\dim}_B A$ and $d \in (\overline{\dim}_B A, \xi_0)$, there exists a positive constant $C_1 = C_1(\xi_0,d,N,A)$ such that

$$\sup_{\mathrm{Re}\, s>\xi_0} |\zeta_{A,A_\varepsilon}(s)| \le C_1\varepsilon^{\xi_0-d}, \quad \text{for all} \quad \varepsilon\in(0,1).$$

In other words, $\sup_{\mathrm{Re}\, s>\xi_0}|\zeta_{A,A_\varepsilon}(s)| = O(\varepsilon^{\xi_0-d})$ as $\varepsilon\to 0^+$. Moreover, assuming that $\mathscr{M}^{*D}(A) < \infty$, where $D = \overline{\dim}_B A$, then the same conclusion holds with d replaced by D.

The following lemma deals with the distance zeta function associated with an ordered pair (A,U) of suitable subsets of $\mathbb{R}^N$; see (2.1.24). Such distance zeta functions will be studied in a significantly more general setting (that of 'relative fractal drums') in Chapter 4 (as well as in Appendix A, in even greater generality). The lemma is a special case of Theorem 2.1.45(c) below.

Lemma 2.1.15. *Let A and U be bounded sets in $\mathbb{R}^N$ which have disjoint closures, that is, such that $\overline{A}\cap\overline{U}=\emptyset$. Further assume that U is Lebesgue measurable. Then*

$$F:\mathbb{C}\to\mathbb{C},\quad F(s):=\int_U d(x,A)^{s-N}\mathrm{d}x, \tag{2.1.24}$$

is an entire function and we have

$$F'(s)=\int_U d(x,A)^{s-N}\log d(x,A)\,\mathrm{d}x \tag{2.1.25}$$

for all $s\in\mathbb{C}$.

Proof. Let s be a fixed complex number and set $R(h)=\frac{1}{h}(F(s+h)-F(s))-I_1(s)$, for $h\in\mathbb{C}$, $h\neq 0$, where $I_1(s)$ is defined by the right-hand side of (2.1.25). By using the same procedure as in the proof of Theorem 2.1.11, we deduce that the identity (2.1.19) involving $f(d)$ defined by (2.1.18) yields

$$|f(d)|\leq\frac{1}{2}|h|\,|\log d|^2\exp(|\log d|\,|h|),$$

and from this it follows that

$$|R(h)|\leq\frac{1}{2}|h|\int_U|\log d(x,A)|^2\exp(|\log d(x,A)|\,|h|)\,d(x,A)^{\mathrm{Re}\,s-N}\mathrm{d}x. \tag{2.1.26}$$

The conditions on A and U imply the existence of positive and finite constants d_1 and d_2 such that $d_1\leq d(x,A)\leq d_2$ for all $x\in U$. Therefore, the function under the integral sign in (2.1.26), when restricted to U, is bounded from above by a positive and finite constant C, uniformly for all $h\in\mathbb{C}$ such that $|h|\leq\varepsilon$, where $\varepsilon>0$ is fixed:

$$\begin{aligned}C:=&\max\{(\log d_1)^2,(\log d_2)^2\}\exp(\max\{|\log d_1|,|\log d_2|\}\,\varepsilon)\\&\times\max\{d_1^{\mathrm{Re}\,s-N},d_2^{\mathrm{Re}\,s-N}\}.\end{aligned}$$

Hence, $|R(h)|\leq\frac{1}{2}|h|\,C\,|\Omega|$, and therefore $R(h)\to 0$ as $h\to 0$ in $\mathbb{C}$, with $h\neq 0$. It follows that $F(s)$ is holomorphic in s, with complex derivative $F'(s)$ equal to $I_1(s)$; that is, $F'(s)$ is given by the right-hand side of (2.1.25). Since $s\in\mathbb{C}$ is arbitrary, we deduce that $F(s)$ is entire and that (2.1.25) holds for all $s\in\mathbb{C}$, as desired. ☐

Lemma 2.1.15 can also be obtained as a consequence of Theorem 2.1.45(c) below, in which we let $\varphi(x):=d(x,A)$ and $\mathrm{d}\mu(x):=d(x,A)^{-N}\mathrm{d}x$.

We next comment on the hypotheses of part (c) of Theorem 2.1.11 (further comments about Theorem 2.1.11 and its corollary, Corollary 2.1.20 below, will be provided in Remark 2.1.21):

(i) The condition $\mathcal{M}_*^D(A) > 0$ in Theorem 2.1.11(c) cannot be omitted. Indeed, for $N = 1$, there is a class of subsets $A \subset [0,1]$ such that $D = \dim_B A$ exists and $\mathcal{M}_*^D(A) = 0$, while $\zeta_A(D) = \int_{A_\delta} d(x,A)^{D-N} \mathrm{d}x < \infty$; see [Žu4, Theorem 4.3]. Using the Lebesgue dominated convergence theorem, it is then easy to see that for $s \in \mathbb{R}$, we have $\lim_{s\to D^+} \zeta_A(s) = \zeta_A(D)$. Indeed, in order to verify this, it suffices to assume without loss of generality that $\delta < 1$, and observe that for all $s \in \mathbb{R}$ with $s > D$ we have $d(x,A)^{s-N} \le d(x,A)^{D-N} \in L^1(A_\delta)$. We note that this class of subsets of $\mathbb{R}$ can be easily extended to $\mathbb{R}^N$ for any $N \ge 2$ by letting $B := A \times [0,1]^{N-1} \subset \mathbb{R}^N$ and using the results of Subsection 2.2.4 about the fractal zeta functions of fractal grills (see, especially, Theorem 2.2.32 and Example 2.2.34).

(ii) The assumptions of Theorem 2.1.11(c) according to which $D = \dim_B A$ exists, $D < N$ and $\mathcal{M}_*^D(A) > 0$ are fulfilled by practically all of the standard examples of fractal sets.[7] However, it is possible to construct fractal sets A for which $D = \dim_B A$ either does not exist (that is, $\underline{\dim}_B A < \overline{\dim}_B A$; see [Fal1, p. 53] or [Žu4, Theorem 1.2]), or, as we have mentioned, for which D exists and $\mathcal{M}_*^D(A) = 0$. See also Section 3.7 of this monograph, dealing with zigzagging fractals.

Example 2.1.16. The present simple example shows that the condition $D < N$ in Theorem 2.1.11(c) cannot be omitted. To see this, take $A := [0,1] \subset \mathbb{R}$, so that $D = N = 1$. It is easy to see that $\zeta_A(s) = 2\delta^s s^{-1}$ for $s > 1$ (and not for $s \in (0,1)$, since in this case $\zeta_A(s) = 2\delta^s s^{-1} + \int_0^1 0^{s-1} \mathrm{d}x = +\infty$). Indeed, for $s > 1$ we have that

$$\zeta_A(s) = \int_{-\delta}^{1+\delta} d(x,A)^{s-1} \mathrm{d}x = \int_{-\delta}^{0} |x|^{s-1} \mathrm{d}x + \int_0^1 0\,\mathrm{d}x + \int_1^{1+\delta} (x-1)^{s-1} \mathrm{d}x = 2\delta^s s^{-1}.$$

It follows that the largest open right half-plane to which the distance zeta function of A can be holomorphically extended is equal to $\{\mathrm{Re}\, s > 0\}$; i.e., $\mathcal{H}(\zeta_A) = \{\mathrm{Re}\, s > 0\}$, in the notation of the second part of Definition 2.1.17 just below. (Note that ζ_A can be meromorphically extended in a unique way from $\{\mathrm{Re}\, s > 0\}$ to the whole complex plane by letting $\zeta_A(s) := 2\delta^s s^{-1}$ for all $s \in \mathbb{C}$; hence, $\mathrm{Mer}(\zeta_A) = \mathbb{C}$, in the notation introduced in Equation (2.1.70) of Definition 2.1.53 below.) This example, along with more complicated ones to be discussed further on in this book, motivates us to introduce the following definition.[8]

---

[7] One notable exception is the boundary A of the Mandelbrot set (viewed as a subset of $\mathbb{R}^2 \simeq \mathbb{C}$), for which $\dim_H A = 2$ (and hence, in particular, $\dim_B A$ exists and $\dim_B A = 2$, since $2 = \dim_H A \le \underline{\dim}_B A \le \overline{\dim}_B A \le 2$; see the third displayed equation on page 77 of [Mat]), according to Shishikura's well-known theorem [Shi]. For the Mandelbrot set, one can try to use $\tilde{\zeta}_A$, the tube zeta function of A, for which the condition $D < N$ is no longer needed for the counterpart of Theorem 2.1.11(c). However, it does not seem to be known whether $\mathcal{M}_*^2(A) > 0$ or whether a different gauge function (other than one based on a mere power law, see [HeLap]) should be used in this case in order to define the lower and upper Minkowski contents of A.

[8] We are grateful to Erin Pearse for having provided us with this example.

Definition 2.1.17. In part (c) of Theorem 2.1.11 above as well as in the sequel, we denote by $D_{\rm hol}(\zeta_A)$ the extended real number (i.e., $D_{\rm hol}(\zeta_A)\in\mathbb{R}\cup\{\pm\infty\}$) defined by

$$D_{\rm hol}(\zeta_A):=\inf\big\{\alpha\in\mathbb{R}:\zeta_A \text{ is holomorphic on } \{{\rm Re}\, s>\alpha\}\big\} \tag{2.1.27}$$

and called the *abscissa of holomorphic continuation* of ζ_A. We stress that in Equation (2.1.27), when we write that f is holomorphic on $\{{\rm Re}\, s>\alpha\}$, we mean that f has a holomorphic continuation (necessarily unique) to the open right half-plane (and hence, connected open set) $\{{\rm Re}\, s>\alpha\}$. We will use a similar convention, most often implicitly, in Subsections 2.1.3 and 2.1.5, as well as elsewhere in this book.

Much as in Definition 2.1.8, we let

$$\mathscr{H}(\zeta_A):=\{{\rm Re}\, s>D_{\rm hol}(\zeta_A)\} \tag{2.1.28}$$

and call it the *half-plane of holomorphic continuation* of ζ_A. See also Definition 2.1.62 in Subsection 2.1.5 below for a more general setting.

Alternatively, in light of (2.1.27) and according to the principle of analytic continuation, $\mathscr{H}(\zeta_A)$ is also the *maximal* (i.e., the largest) open right half-plane (of the form $\{{\rm Re}\, s>\alpha\}$, for some $\alpha\in\mathbb{R}\cup\{\pm\infty\}$) to which ζ_A can be holomorphically extended.

Definition 2.1.18. Finally, when $D_{\rm hol}(\zeta_A)\in\mathbb{R}$, the vertical line $\{{\rm Re}\, s=D_{\rm hol}(\zeta_A)\}$ is called the *critical line of holomorphic continutation* (or the *holomorphy critical line*) of ζ_A.

Remark 2.1.19. In the case of Example 2.1.16 above, in which $A=[0,1]$, we have that $D(\zeta_A)=1$, while $D_{\rm hol}(\zeta_A)=0$.

We can now state the following corollary of Theorem 2.1.11.

Corollary 2.1.20. (i) *Let A be an arbitrary bounded subset of $\mathbb{R}^N$. Then we have the following inequality*:

$$D_{\rm hol}(\zeta_A)\le D(\zeta_A)=\overline{\dim}_B A, \tag{2.1.29}$$

and hence, $\Pi(\zeta_A)\subseteq\mathscr{H}(\zeta_A)$.

(ii) *Furthermore, if, in addition, we assume (as in part (c) of Theorem 2.1.11) that $D:=\dim_B A$ exists, $D<N$ and $\mathscr{M}_*^D(A)>0$, then we actually have the following identity*:

$$D_{\rm hol}(\zeta_A)=D(\zeta_A)=\overline{\dim}_B A, \tag{2.1.30}$$

and hence, $\Pi(\zeta_A)=\mathscr{H}(\zeta_A)$. In particular, $D_{\rm hol}(\zeta_A)\in[0,N]$, so that we also have

$$D_{\rm hol}(\zeta_A)=\inf\big\{\alpha\in[0,N]:\zeta_A \text{ is holomorphic on } \{{\rm Re}\, s>\alpha\}\big\}, \tag{2.1.31}$$

where

$$\Pi(\zeta_A):=\{{\rm Re}\, s>D(\zeta_A)\}=\{{\rm Re}\, s>\overline{\dim}_B A\} \tag{2.1.32}$$

is the half-plane of convergence of ζ_A *introduced in Definition 2.1.8 (where* ζ_A *is viewed as a Dirichlet-type integral, in the sense of Subsection 2.1.3.2 and Appendix A below) and* $\mathscr{H} := \{\mathrm{Re}\, s > D_{\mathrm{hol}}(\zeta_A)\}$ *is the half-plane of holomorphic continuation* of ζ_A*, introduced in Definition 2.1.17.*

Proof. (*i*) The first part of the corollary (Equation (2.1.29) of the corollary) follows readily from parts (*a*) and (*b*) of Theorem 2.1.11.

(*ii*) The second part of the corollary (Equation (2.1.30) and the equality following it) follows from part (*c*) of Theorem 2.1.11 (along with the definitions of $D(\zeta_A)$ and $D_{\mathrm{hol}}(\zeta_A)$ given in Equations (2.1.10) and (2.1.27), respectively, and the corresponding definitions of $\Pi(\zeta_A)$ and $\mathscr{H}(\zeta_A)$). Indeed, according to Theorem 2.1.11(*c*), if we assume that ζ_A admits a holomorphic continuation to $\{\mathrm{Re}\, s > \alpha\}$ for some $\alpha < D(\zeta_A)$, we obtain a contradiction, since then, D is a pole of ζ_A, because $|\zeta_A(s)|$ blows-up (i.e., tends to $+\infty$) as $s \to D^+$ along the real axis. Therefore, $D_{\mathrm{hol}}(\zeta_A) \geq D(\zeta_A)$. But then, we must have $D_{\mathrm{hol}}(\zeta_A) = D(\zeta_A)$, since we always have $D_{\mathrm{hol}}(\zeta_A) \leq D(\zeta_A)$, according to Equation (2.1.29).

This concludes the proof of the corollary. □

Remark 2.1.21. (*a*) In Remark 2.1.19, we have given a very simple example of a subset A of the real line (namely, $A := [0,1]$) for which $D_{\mathrm{hol}}(\zeta_A) < D(\zeta_A)$. It would be interesting to find (if possible) a class of bounded subsets A of $\mathbb{R}^N$ for which the corresponding distance zeta function ζ_A (meromorphically extended to a connected open neighborhood of $\{\mathrm{Re}\, s \geq D(\zeta_A)\}$) possesses nonreal poles with positive real parts, and such that $D_{\mathrm{hol}}(\zeta_A) < D(\zeta_A)$. A natural candidate could be the bounded sets $A \subset [0,1]$ studied in [Žu4, Theorem 4.3] and for which $D := \dim_B A$ exists and $\mathscr{M}_*^D(A) = 0$ but $\zeta_A(D) = \int_{A_\delta} d(x,A)^{D-N} \mathrm{d}x < \infty$. Observe that by letting $B := A \times [0,1]^{N-1}$, one would then obtain a bounded subset of $\mathbb{R}^N$ (for $N \geq 2$ arbitrary) having the exact same properties as A.

(*b*) The inequality (2.1.29) in Corollary 2.1.20 is sharp, i.e., is best possible, in general. Indeed, we will construct in Corollary 4.6.17 to Theorem 4.6.9 an example of a maximally hyperfractal bounded subset A of $\mathbb{R}^N$.[9] In particular (in the terminology of Subsection 4.6.2 below; see parts (*ii*) and (*iii*) of Definition 4.6.23), ζ_A has a partial natural boundary along the vertical line $\{\mathrm{Re}\, s = D\}$, where $D = D(\zeta_A) = \overline{\dim}_B A$; this means that ζ_A cannot be meromorphically (and let alone, holomorphically) extended to a connected open neighborhood of this vertical line. (In fact, for this example, all of the points of this line are singularities of ζ_A; hence the name "maximal hyperfractal". Therefore, $\{\mathrm{Re}\, s = D\}$ is a holomorphic natural boundary of ζ_A, in the sense of Definition 1.3.6 of Subsection 1.3.2.) It follows that $D_{\mathrm{hol}}(\zeta_A) \geq D$. Since we also have $D_{\mathrm{hol}}(\zeta_A) \leq D$, in light of (2.1.29), it follows

[9] Of course, in the terminology of part (*ii*) of Definition 4.6.23 it would suffice for A to be strongly hyperfractal but Corollary 4.6.17 provides an even more singular geometric object, namely, a maximal hyperfractal. Moreover, in Corollary 4.6.17, A is a bounded subset of $\mathbb{R}$, but as is noted in Remark 4.6.19, for any fixed $N \geq 2$, by considering the Cartesian product $B := A \times [0,1]^{N-1}$, one can readily obtain a corresponding bounded subset of $\mathbb{R}^N$ having the exact same properties.

that for this example of maximal hyperfractal obtained in Corollary 4.6.17 (and in Remark 4.6.19), we have

$$D := \overline{\dim}_B A = D(\zeta_A) = D_{\rm hol}(\zeta_A).$$

We note in passing that with $D_{\rm mer}(\zeta_A)$ denoting the abscissa of meromorphic continuation of ζ_A (defined like $D_{\rm hol}(\zeta_A)$, except for "holomorphic" replaced by "meromorphic"), we also have

$$D := \overline{\dim}_B A = D(\zeta_A) = D_{\rm hol}(\zeta_A) = D_{\rm mer}(\zeta_A),$$

while in general, for any bounded subset A of $\mathbb{R}^N$, we have

$$D_{\rm mer}(\zeta_A) \leq D_{\rm hol}(\zeta_A) \leq D(\zeta_A) = \overline{\dim}_B A. \tag{2.1.33}$$

Therefore, this new string of inequalities in Equation (2.1.33) is also sharp.

The following simple result is important for the development of the theory. It will be used throughout the book, most often implicitly. In hindsight, it shows that we do not have to worry about the set $\overline{A}$ on which the distance function $x \mapsto d(x,A)$ (precisely) vanishes identically and hence, on which the integrand $x \mapsto d(x,A)^{s-N}$ (in the definition of $\zeta_A(s)$ given in Equation (2.1.1) above) is singular, i.e., is identically equal to $+\infty$, when $\mathrm{Re}\, s < N$. See Remark 2.1.10 above for more details.

Proposition 2.1.22. *Let A be an arbitrary bounded subset of $\mathbb{R}^N$. Then, we can change the domain of integration in Equations (2.1.1) and (2.1.13) from A_δ to the set $A_\delta \setminus \overline{A}$ without modifying ζ_A and hence, without changing its abscissa of convergence. In particular, for any $\delta > 0$ and every $s \in \mathbb{C}$ such that* $\mathrm{Re}\, s > \overline{\dim}_B A$, *we have*:

$$\begin{aligned} \zeta_A(s) &= \int_{A_\delta \setminus \overline{A}} d(x,A)^{s-N} \mathrm{d}x, \\ \zeta_A'(s) &= \int_{A_\delta \setminus \overline{A}} d(x,A)^{s-N} \log d(x,A)\, \mathrm{d}x. \end{aligned} \tag{2.1.34}$$

Proof. Indeed, if $\overline{\dim}_B A < N$, then (according to the comment preceding Equation (1.3.9) on page 32) the set $\overline{A}$ is necessarily of Lebesgue measure zero in $\mathbb{R}^N$, so that $d(x,A)^{\mathrm{Re}\, s-N} = +\infty$ on a (Lebesgue) negligible set only, provided $\mathrm{Re}\, s \in (\overline{\dim}_B A, N)$. If $\overline{\dim}_B A = N$, then we have $\mathrm{Re}\, s - N > 0$ (see the comment preceding Equation (1.3.9) on page 32), so that $d(x,A)^{s-N} = 0$ for $x \in \overline{A}$, in light of Theorem 2.1.11(b); see also Remark 2.1.10 above. □

Remark 2.1.23. We can easily conclude that if A is any bounded, Lebesgue nonmeasurable set in $\mathbb{R}^N$, then $\dim_B A$ exists and $\dim_B A = N$. Indeed, assume the contrary, i.e., that $\underline{\dim}_B A < N$. We then deduce that $|\overline{A}| = 0$ (see Equation (1.3.8) on page 32 above); so that A is of Lebesgue measure zero, and hence, is Lebesgue measurable (by the completeness of the N-dimensional Lebesgue measure). However, this is a contradiction. (See [LapRoŽu, Example 2.21].)

Remark 2.1.24. The distance $d(x,A)$ in Theorem 2.1.11 has been defined via the Euclidean norm. Nevertheless, Theorem 2.1.11 remains valid, with the same proof, for a distance $d_*(x,A)$ associated to any other norm $\|\cdot\|_*$ in $\mathbb{R}^N$. Indeed, as is well known (see, e.g., [Foll]), all such norms are equivalent, from which one easily deduces that the corresponding distance zeta functions have the same abscissa of convergence. However, the corresponding zeta functions do not necessarily have the same (visible) poles; in other words, the choice of the norm (and hence, of metric) on $\mathbb{R}^N$ may change the set of (visible) complex dimensions of A. More generally, transforming A via a bi-Lipschitz homeomorphism of $\mathbb{R}^N$ does not necessarily preserve the complex dimensions of A.

Remark 2.1.25. In the proof of Theorem 2.1.11, we have shown that

$$\zeta_A(s+h)-\zeta_A(s)=\zeta_A'(s)h+o(h) \quad \text{as } h\to 0,$$

where (in view of (2.1.21)) the remainder term $o(h)=hR(h)$, with $R(h)$ given by (2.1.17), can be estimated by

$$|o(h)|\leq |h|^2C(\delta,\varepsilon)\int_{A_\delta} d(x,A)^{\mathrm{Re}\,s-N-2\varepsilon}\mathrm{d}x, \tag{2.1.35}$$

provided $\varepsilon\in(0,\frac{1}{2}(\mathrm{Re}\,s-\overline{\dim}_BA))$ and $\delta\in(0,1]$ are fixed, and $|h|\leq\varepsilon$. The constant $C(\delta,\varepsilon)=\max\{|\log d|^2d^\varepsilon : d\in(0,\delta]\}$ can be explicitly computed:

$$C(\delta,\varepsilon)=\begin{cases}\frac{4}{\mathrm{e}^2}\varepsilon^{-2} & \text{if } \mathrm{e}^{-2/\varepsilon}<\delta,\\ |\log\delta|^2\delta^\varepsilon & \text{if } \mathrm{e}^{-2/\varepsilon}\geq\delta,\end{cases}$$

while the integral appearing on the right-hand side of (2.1.35), can be estimated as follows, assuming that $\mathscr{M}^{*D}(A)<\infty$:

$$\int_{A_\delta} d(x,A)^{\mathrm{Re}\,s-N-2\varepsilon}\mathrm{d}x\leq\frac{N-D}{\mathrm{Re}\,s-D-2\varepsilon}\left(\sup_{t\in(0,\delta)}\frac{|A_t|}{t^{N-D}}\right)\delta^{\mathrm{Re}\,s-D-2\varepsilon},$$

where $D=\overline{\dim}_BA$; see [Žu2, Theorem 3.1(a)]. Using (2.1.35), and minimizing with respect to $\varepsilon\in I:=(0,\frac{1}{2}(\mathrm{Re}\,s-D))$, we obtain:

$$|o(h)|\leq|h|^2(N-D)\left(\sup_{t\in(0,\delta)}\frac{|A_t|}{t^{N-D}}\right)\left(\inf_{\varepsilon\in I}\frac{C(\delta,\varepsilon)}{\mathrm{Re}\,s-D-2\varepsilon}\delta^{\mathrm{Re}\,s-D-2\varepsilon}\right).$$

It is easy to see that the infimum is achieved for some $\varepsilon_0\in I$, since the corresponding function is continuous and tends to infinity as ε tends to either of the endpoints of the interval I, while remaining within I. The above inequality holds for all $h\in\mathbb{C}$ such that $|h|\leq\varepsilon_0$.

The following result shows that the distance zeta function has a natural additivity property.

Proposition 2.1.26. *Let A and B be two bounded subsets of $\mathbb{R}^N$ which are a positive distance apart; that is, $d(A,B) > 0$, where $d(A,B) := \inf\{|x-y| : x \in A,\ y \in B\}$. Then, for any $\delta \in (0, \frac{1}{2}d(A,B))$, where δ is the positive number used in Equation (2.1.1) for defining ζ_A, ζ_B and $\zeta_{A\cup B}$, we have*

$$\zeta_{A\cup B}(s) = \zeta_A(s) + \zeta_B(s), \tag{2.1.36}$$

for $\operatorname{Re} s > \max\{\overline{\dim}_B A, \overline{\dim}_B B\}$.

Proof. This follows easily from Equation (2.1.1). Fix $\delta \in (0, \frac{1}{2}d(A,B))$ and $s \in \mathbb{C}$ with $\operatorname{Re} s > \max\{\overline{\dim}_B A, \overline{\dim}_B B\}$. Then, since $(A\cup B)_\delta = A_\delta \cup B_\delta$, we have $\zeta_{A\cup B}(s) = \int_{A_\delta} d(x, A\cup B)^{s-N}\mathrm{d}x + \int_{B_\delta} d(x, A\cup B)^{s-N}\mathrm{d}x$. Now, for any $x \in A_\delta$, we obviously have $d(x, A\cup B) = d(x,A)$, and analogously for $x \in B_\delta$. Hence, the desired conclusion follows from Definition 2.1.1. □

2.1.3 Dirichlet Series and Dirichlet Integrals

The goal of this subsection is to review the notion of abscissa of convergence of Dirichlet series and integrals, and to describe some of its basic properties. A concise introduction to the theory of (generalized) Dirichlet series can be found in [Ser, Section V.2.2]. We also refer to [HardWr] for a thorough exploration of the theory of classical Dirichlet series, which are of the form $\sum_{j=1}^{\infty} b_j/j^s$, with $b_j \in \mathbb{C}$; that is, for which $l_j := 1/j$, for all $j \in \mathbb{N}$. Furthermore, a discussion of Dirichlet integrals is provided in [Pos, esp., Section 2.3].

2.1.3.1 Dirichlet Series

It is interesting (and well known) that it is possible to give an explicit expression for the abscissa of convergence (see Definition 2.1.28 below) of the (generalized) Dirichlet series

$$f(s) := \sum_{j=1}^{\infty} b_j l_j^s; \tag{2.1.37}$$

see Theorem 2.1.33 below. Here, we assume that $l_j > l_{j+1} > 0$ for all $j \in \mathbb{N}$, and $l_j \to 0^+$ as $j \to \infty$. For simplicity, we assume that the 'multiplicities' b_j are positive real numbers. In the applications to fractal strings, they are natural numbers, interpreted as the actual multiplicities of the distinct 'lengths' (or 'scales') l_j. In order to formulate the corresponding result (stated in Theorem 2.1.33 below), we need to introduce the following counting function b, defined on $(0, +\infty)$ by

$$b(x) := \sum_{\{j : \log l_j^{-1} \le x\}} b_j. \tag{2.1.38}$$

If $b_j \in \mathbb{N}$ for each j, then clearly, $b(x) \to +\infty$ as $x \to +\infty$, and $b(x)$ is called the (*geometric*) *counting function* of the associated fractal string (defined as the *multiset* consisting of all the distinct numbers l_j with multiplicities b_j). More precisely, in order to be consistent with the terminology introduced in [Lap-vFr1–3], let us recall that the *geometric counting function* $N_{g,\mathscr{L}}$ of a fractal string $\mathscr{L} = (\ell_j)_{j=1}^{\infty}$, with $\ell_j \geq \ell_{j+1}$ for all $j \in \mathbb{N}$ and $\ell_j \to 0^+$ as $j \to \infty$, is defined by

$$N_{g,\mathscr{L}}(x) = \sum_{\{j:l_j^{-1} \leq x\}} b_j,$$

for every $x > 0$. Hence, the counting function b can be viewed as the counterpart of $N_{g,\mathscr{L}}$ when the 'reciprocal scales' are measured on a logarithmic scale.

In 1894, Cahen [Cah] proved the following important and classical result concerning the convergence of Dirichlet series.

Theorem 2.1.27 (Cahen, [Cah]). *If a Dirichlet series $f(s) = \sum_{j=0}^{\infty} b_j l_j^s$ converges absolutely* (*and hence, converges*) *for some $s_1 \in \mathbb{C}$, then it converges absolutely* (*and hence, converges*) *on the open right half-plane* $\{\mathrm{Re}\, s > \mathrm{Re}\, s_1\}$. *Consequently, the Dirichlet series is either convergent absolutely* (*and hence, convergent*) *for all complex numbers s, or divergent absolutely for all $s \in \mathbb{C}$, or else there exists a* (*necessarily unique*) *real number D such that the Dirichlet series converges absolutely* (*and hence, converges*) *on* $\{\mathrm{Re}\, s > D\}$ *and diverges absolutely on* $\{\mathrm{Re}\, s < D\}$ (*i.e., is not absolutely convergent at any point $s \in \mathbb{C}$ with* $\mathrm{Re}\, s < D$).

Note that in this generality, nothing can be said about the convergence of the Dirichlet series for complex numbers s on the vertical line $\{\mathrm{Re}\, s = D\}$. Furthermore, since the coefficients b_j are assumed to be positive, the Dirichlet series converges absolutely for $\mathrm{Re}\, s > D$.

Definition 2.1.28. The unique value of $D \in \mathbb{R}$ appearing in Theorem 2.1.27 is called the *abscissa of* (*absolute*) *convergence* of the Dirichlet series, and is denoted by $D = D(f)$. Following the usual conventions, we extend this definition to the case where $D \in \mathbb{R} \cup \{\pm\infty\}$. Accordingly, the case where $D = -\infty$ or $D = +\infty$ corresponds, respectively, to the first or second situation described in the statement of the theorem.

Furthermore, much as in Subsection 2.1.2 above (see Equation (2.1.10) and the text surrounding it), we let $\Pi(f) := \{\mathrm{Re}\, s > D(f)\}$ denote the *half-plane of* (*absolute*) *convergence* of f. Then, according to Cahen's theorem (Theorem 2.1.27), $\Pi(f)$ is the maximal open right half-plane on which the Dirichlet series f converges absolutely (and hence, is convergent).

Finally, when $D \in \mathbb{R}$, the vertical line $\{\mathrm{Re}\, s = D\}$ is called the *critical line* of f (or sometimes in this book, the '*critical line of convergence*').

Remark 2.1.29. For a Dirichlet series with complex (rather than positive) coefficients, there is an analogous theorem for the notion of ordinary (rather than absolute) convergence of a Dirichlet series, allowing one to define the *abscissa of conditional* (rather than absolute) *convergence* of a Dirichlet series, denoted by $D_{\rm cond}(f)$ and which (thanks to Theorem 2.1.34 and Corollary 2.1.35) does not exceed $D_{\rm hol}(f)$; see, e.g., [Ser], *loc. cit.* Furthermore, for a Dirichlet series with positive coefficients, as is assumed in the present subsection (i.e., in Subsection 2.1.3.1), the abscissae of absolute and conditional convergence are equal and also coincide with the abscissa of holomorphic continuation: $D(f) = D_{\rm cond}(f) = D_{\rm hol}(f)$; see Corollary 2.1.36 below.

In contrast, a classic example of Dirichlet series with real coefficients for which $D(f)$, $D_{\rm cond}(f)$ and $D_{\rm hol}(f)$ are all distinct is provided by $f(s) := \sum_{n=1}^{\infty}(-1)^{n-1}/n^s$. Then, a simple (but clever) computation shows that $\zeta(s) = f(s) + 2^{1-s}\zeta(s)$, where $\zeta = \zeta(s)$ denotes the classic Riemann zeta function; i.e., $f(s) = (1-2^{1-s})\zeta(s)$. This last identity shows that the simple pole of $\zeta(s)$ at $s = 1$ is precisely cancelled by the simple zero of $1-2^{1-s}$ at $s = 1$; so that f is an entire function (i.e., has a holomorphic extension to all of $\mathbb{C}$) and hence,

$$D(f) = 1 > D_{\rm cond}(f) = 0 > D_{\rm hol}(f) = -\infty, \qquad (2.1.39)$$

where $D_{\rm cond}(f)$ is the abscissa of conditional convergence of f. It is easy to check that $D_{\rm cond}(f) = 0$ by using Abel's partial summation theorem and noting that clearly, the series initially defining $f(s)$ diverges at $s = 0$.

Example 2.1.30. A well-known and interesting example of Dirichet series f for which the abscissa of absolute convergence $D(f)$ and of conditional convergence $D_{\rm cond}(f)$ are expected to be different is given by the *Möbius Dirichlet series*

$$f(s) := \sum_{n=1}^{\infty} \frac{\mu(n)}{n^s}, \qquad (2.1.40)$$

where μ is the *Möbius function* defined by $\mu(1) = 1$, $\mu(n) = (-1)^k$ if n is the product of k distinct primes (with $k \in \mathbb{N}$), and $\mu(n) = 0$ otherwise (i.e., if n is not square-free). Then, we always have that $D(f) = 1$ and $1/2 \le D_{\rm cond}(f) \le 1$. Furthermore, $D_{\rm cond}(f) = 1/2$ if and ony if the *Riemann hypothesis* is true (i.e., if and only if $\zeta(s) = 0$ for some $s \in \mathbb{C}$ with $0 < \mathrm{Re}\, s < 1$ implies that $\mathrm{Re}\, s = 1/2$).[10] More specifically, we always have that, independently of the truth of the Riemann hypothesis (i.e., unconditionally),

$$D_{\rm cond}(f) = \sup\{\alpha \in [1/2, 1) : \zeta(s) = 0 \text{ for some } s \in \mathbb{C} \text{ with } \mathrm{Re}\, s = \alpha\}, \qquad (2.1.41)$$

where ζ denotes the meromorphic continuation to all of $\mathbb{C}$ of the classic Riemann zeta function.

The statement concerning $D_{\rm cond}(f)$ follows from the fact that

$$\sum_{n=1}^{\infty} \frac{\mu(n)}{n^s} = \frac{1}{\zeta(s)}, \quad \text{for } \mathrm{Re}\, s > 1 \qquad (2.1.42)$$

[10] It is known from Hadamard's theorem and the functional equation satisfied by ζ that $\zeta(s) \neq 0$ for all $s \in \mathbb{C}$ with $\mathrm{Re}\, s = 0$ or $\mathrm{Re}\, s = 1$; see, e.g., [Edw] or [Tit3].

(itself a consequence of Möbius' inversion formula, see [Edw, Tit3]). Indeed, in light of Equation (2.1.42) and since ζ can be meromorphically continued to all of $\mathbb{C}$, we deduce that f can be meromorphically continued to all of $\mathbb{C}$ and that

$$f(s) = \frac{1}{\zeta(s)}, \quad \text{for all } s \in \mathbb{C}. \tag{2.1.43}$$

Consequently, $f(s)$ has a pole at every zero of $\zeta(s)$ and so $D_{\text{cond}}(f)$ is given by the right-hand side of Equation (2.1.41) since $\zeta(s)$ has zeros along the critical line $\{\text{Re}\, s = 1/2\}$.

Now, to prove that $D(f) = 1$, we proceed as follows. First, since $|\mu(n)| \leq 1$ for all $n \in \mathbb{N}$, it is clear that $\sum_{n=1}^{\infty} \frac{|\mu(n)|}{n^{\text{Re}\, s}} < \infty$ for $\text{Re}\, s > 1$. Hence, $D \leq 1$. Furthermore, for $s = 1$, we have successively (with $p_1, \ldots, p_k, p$ running through the set of all prime numbers, and such that in the first equality, the product $p_1 \ldots p_k$ is square-free, for any $k \in \mathbb{N}$):[11]

$$\begin{aligned} \sum_{n=1}^{\infty} \frac{|\mu(n)|}{n} &= 1 + \sum_{k=1}^{\infty} \sum_{p_1,\ldots,p_k} \frac{1}{p_1 \cdots p_k} \\ &\geq \sum_{p_1} \frac{1}{p_1} = \sum_{p} \frac{1}{p} = +\infty. \end{aligned} \tag{2.1.44}$$

Hence, $\sum_{n=1}^{\infty} \frac{|\mu(n)|}{n} = +\infty$ and thus, $D(f) \geq 1$. This shows that $D(f) = 1$, as claimed.

Furthermore, we note that since $\zeta(s)$ does not have any zeros for $\text{Re}\, s > 1$ (because it is given by a convergent infinite product, the Euler product, in the right half-plane $\{\text{Re}\, s > 1\}$), we must have (in light of Equation (2.1.43)) that $D_{\text{hol}}(f) \leq 1$. However, since $\zeta(s)$ has zeros on the critical line $\{\text{Re}\, s = 1/2\}$ and since (also in light of Equation (2.1.43)) $f(s)$ has a pole at $s \in \mathbb{C}$ if and only if $\zeta(s) = 0$, it follows that $1/2 \leq D_{\text{hol}}(f) \leq 1$ and that, in fact, unconditionally, $D_{\text{hol}}(f) = D_{\text{cond}}(f)$. We deduce, in particular, that $D_{\text{hol}}(f)$ is also given by the right-hand side of Equation (2.1.41). Moreover, we observe that it follows from the last equality and from Equation (2.1.41) that unconditionally

$$1 = D(f) > D_{\text{hol}}(f) = D_{\text{cond}}(f) \tag{2.1.45}$$

due to existence of zero-free regions which are asymptotic to the vertical line $\{\text{Re}\, s = 1\}$ (see, e.g., [Edw] or [Tit3]). Indeed, in light of Equation (2.1.41) above, the hypothesis that $D_{\text{cond}}(f) = 1$ is equivalent to the existence of a sequence $(s_j)_{j \geq 1}$ of critical zeros of $\zeta = \zeta(s)$ tending to the vertical line $\{\text{Re}\, s = 1\}$ (i.e., such that $\text{Re}\, s_j \to 1$ as $j \to \infty$, with $\text{Re}\, s_j < 1$) and this, in turn, contradicts the existence of a zero-free region.

[11] In the last equality of Equation (2.1.44), we are using the well-known fact according to which the series of reciprocal primes, $\sum_p \frac{1}{p}$, is divergent; see, e.g., [Edw], [Tit3] or [Ser, Section VI.3.1, Corollary 2].

Finally, combining (2.1.41) and (2.1.45), we see that still unconditionally, $D_{\rm cond}(f)$
$= (D_{\rm hol}(f)) \in [1/2, 1)$ and that (as was already noted above) $D_{\rm cond}(f) = 1/2$ if and only if the Riemann hypothesis is true.

This concludes the discussion of this example.

The following definition will be useful in the sequel. See also Definition 2.1.62 in Subsection 2.1.5 below for a more general setting.

Definition 2.1.31. As before in Subsection 2.1.2, we denote by $D_{\rm hol}(f)$ the *abscissa of holomorphic continuation* of f and we call $\mathscr{H}(f) := \{{\rm Re}\, s > D_{\rm hol}(f)\}$ the *half-plane of holomorphic continuation* of f. More specifically, $D_{\rm hol}(f) \in \mathbb{R} \cup \{\pm\infty\}$ is given by (2.1.27), with ζ_A replaced by f, and it follows that $\mathscr{H}(f) := \{{\rm Re}\, s > D_{\rm hol}(f)\}$ is the maximal open right half-plane (of the form $\{{\rm Re}\, s > \alpha\}$, for some $\alpha \in \mathbb{R} \cup \{\pm\infty\}$) to which the Dirichlet series f can be holomorphically continued.

Finally, the vertical line $\{{\rm Re}\, s = D_{\rm hol}(f)\}$ is called the *holomorphy critical line* of $f(s) = \sum_{j=1}^{\infty} b_j l_j^s$.

Remark 2.1.32. The Dirichlet series $\sum_{j=1}^{\infty} b_j l_j^s$ can be viewed as a special case of Stieltjes–Dirichlet integral, corresponding to the Laplace transform of a discrete measure $\nu_1 := \sum_{j=1}^{\infty} b_j \delta_{\log l_j^{-1}}$ (or equivalently, to the Mellin transform of the discrete measure $\nu_2 := \sum_{j=1}^{\infty} b_j \delta_{l_j^{-1}}$). Here, for $x > 0$, δ_x denotes the (unit) Dirac mass (or measure) concentrated at $\{x\}$.

Next, we formulate a result, also due to Cahen [Cah], containing an explicit formula for the computation of the abscissa of convergence of a general Dirichlet series.

Theorem 2.1.33 (Cahen, [Cah]). *Let b be the counting function defined by (2.1.38). Assume that $b(x) \to +\infty$ as $x \to +\infty$. Then the abscissa of convergence D of the Dirichlet series $f(s) = \sum_{j=1}^{\infty} b_j l_j^s$ is nonnegative and given by*

$$D = \limsup_{x \to +\infty} \frac{\log b(x)}{x}. \tag{2.1.46}$$

Moreover, in terms of the sequence $(b_j)_{j=1}^{\infty}$ of 'multiplicities', this value is also given by

$$D = \limsup_{n \to \infty} \frac{1}{\log l_n^{-1}} \log \Big(\sum_{j=1}^{n} b_j \Big). \tag{2.1.47}$$

The following classic result, due to Perron, is a well-known extension of Cahen's theorem (Theorem 2.1.27). It will be used, in particular, in the proof of Theorem 2.1.39 below.

Theorem 2.1.34 (Perron, [Per]). *If a (generalized) Dirichlet series converges for some $s_0 \in \mathbb{C}$, then it converges uniformly in any sector of the form* ${\rm Re}(s - s_0) > 0$, $|\arg(s - s_0)| \le \Theta$, *with $\Theta \in (0, \pi/2)$.*

The proof of Perron's theorem can be found, for example, in [Ser, Proposition 6, Section VI.2.2]. It relies, in particular, on Abel's partial summation formula (that is, on a discrete analog of integration by parts).

Corollary 2.1.35. *If a (generalized) Dirichlet series converges for some* $s_0 \in \mathbb{C}$, *then it converges in the open right half-plane* $\{\mathrm{Re}\, s > \mathrm{Re}\, s_0\}$ *and the associated function so defined is holomorphic in that region.*

A priori, in light of Corollary 2.1.35, it follows from the definition of $D(f)$ and $D_{\mathrm{hol}}(f)$ given in Definition 2.1.28 and Definition 2.1.31, respectively, that we only have the following inequality: $D_{\mathrm{hol}}(f) \le D(f)$. However, the next result (Corollary 2.1.36) will show that in our present situation, we actually have the following equality (due to the positivity of the coefficients of the Dirichlet series): $D_{\mathrm{hol}}(f) = D(f)$.

We note that, on the other hand, there are elementary examples of Dirichlet series with real or complex coefficients for which $D_{\mathrm{hol}}(f) < D(f)$. For instance, as was explained in Remark 2.1.29 (see, especially, Equation (2.1.39) in that remark), for the Dirichlet L-function $f(s) := \sum_{n=1}^{\infty}(-1)^{n-1}n^{-s}$, which will be revisited in the text following Remark 2.1.37 below, we have $D_{\mathrm{hol}}(f) = -\infty$ whereas $D(f) = 1$; see, e.g., [Ser, Section VI.3] for other examples of such L-functions (namely, Dirichlet L-functions with nontrivial primitive characters). Also, for such Dirichlet L-functions, we have $D_{\mathrm{cond}}(f) = 0$ and $D(f) = 1$.

The next corollary of Perron's theorem (Theorem 2.1.34 above) is well known and relies in a crucial way on the fact that the Dirichlet series has *positive coefficients*. See, e.g., Proposition 7 in Section 2.3 of [Ser], where this result is stated in a different, but equivalent manner.

Corollary 2.1.36. *Assume that the coefficients* b_j *are positive real numbers for all* $j \in \mathbb{N}$. *Then, the Dirichlet series* $f(s) := \sum_{j=1}^{\infty} b_j l_j^s$ *tends to* $+\infty$ *as* s *tends to* $D(f)$ *from the right, along the real axis.*[12] *Consequently, we have the following equality:*

$$D(f) = D_{\mathrm{hol}}(f). \tag{2.1.48}$$

Remark 2.1.37. It follows from Corollary 2.1.36 and the definitions of $D(f)$ and $D_{\mathrm{hol}}(f)$ that we have the following equality, which is a restatement of Equation (2.1.48):

$$\begin{aligned} D(f) &:= \inf\left\{\alpha \in \mathbb{R} : \sum_{j=1}^{\infty} b_j l_j^{\alpha} < \infty\right\} \\ &= \inf\{\alpha \in \mathbb{R} : f \text{ is holomorphic on } \{\mathrm{Re}\, s > \alpha\}\} =: D_{\mathrm{hol}}(f). \end{aligned} \tag{2.1.49}$$

In the case when the sequence defining the Dirichlet series is (infinite and) associated with an ordinary fractal string $\mathscr{L} = (\ell_j)_{j\ge1}$, represented by a bounded open

[12] Hence, if $D := D(f) \in \mathbb{R}$, then D is a singularity of f (located on the real axis). Moreover, if, in addition, f can be meromorphically extended to a connected open neighborhood of the critical line $\{\mathrm{Re}\, s = D\}$, then D is a pole of f.

subset Ω of $\mathbb{R}$, we can even assume that the infimum is taken over $\alpha \in [0,1]$ in the first and second equalities of (2.1.49), and we also have several additional equalities:

$$D_{\text{hol}}(\zeta_{\mathscr{L}}) = D(\zeta_{\mathscr{L}}) = \overline{\dim}_B \partial\Omega = \overline{\dim}_B A, \tag{2.1.50}$$

where $\partial\Omega$ is the boundary of Ω, $\zeta_{\mathscr{L}}$ is the geometric zeta function of $\mathscr{L}$, and $A = A_{\mathscr{L}}$ denotes the bounded subset of $\mathbb{R}$ associated with $\mathscr{L}$ (as explained in Subsection 2.1.4 below). See Subsection 2.1.4, especially Theorem 2.1.55 and Corollary 2.1.57.

Assume that we are in the setting of the first part of the previous remark (Remark 2.1.37); that is, $f(s)$ is a generalized Dirichlet series with positive coefficients, as in the rest of the present subsection. In particular, Equation (2.1.48) holds. Let $D \in \mathbb{R} \cup \{\pm\infty\}$ denote the common value of $D_{\text{hol}}(f)$ and $D(f)$. It then follows that the Dirichlet series $f(s) := \sum_{j=1}^{\infty} b_j l_j^s$, where the coefficients b_j are positive, is absolutely convergent (and hence, convergent) for $\operatorname{Re} s > D$ and diverges for $\operatorname{Re} s < D$ (as stated in Theorem 2.1.27 above).

By contrast, as was noted earlier, for a Dirichlet series with real or complex (but not positive) coefficients b_j, we may have $D_{\text{hol}}(f) < D(f)$. This is the case of all Dirichlet L-functions with nontrivial primitive characters (see, e.g., [Ser, Tit2, ParsSh1–2]), for which $D_{\text{hol}}(f) = -\infty$ whereas $D_{\text{cond}}(f) = 0$ and $D(f) = 1$. For instance, for the (classic) Dirichlet series $f(s) := \sum_{n=1}^{\infty} (-1)^{n-1} n^{-s}$ (which is an example of such a Dirichlet L-function), we have $D_{\text{hol}}(f) = -\infty$, $D_{\text{cond}}(f) = 0$ and $D(f) = 1$, where $D_{\text{cond}}(f)$ is the abscissa of conditional convergence of f. (See Equation (2.1.39) in Remark 2.1.29.) This implies that the series converges absolutely (and hence, converges) for $\operatorname{Re} s > 1$, ceases to converge absolutely for $\operatorname{Re} s \leq 1$, whereas it converges for $\operatorname{Re} s > 0$ and diverges for $\operatorname{Re} s < 0$. In particular, the Dirichlet series converges conditionally (but hence, not absolutely) for $0 < \operatorname{Re} s \leq 1$, i.e., for all $s \in \mathbb{C}$ such that $0 < \operatorname{Re} s \leq D(f)$.

In the classic terminology (see, e.g., [Ser] and [Pos]), $D(f) = 1$ is the *abscissa of absolute convergence* of f (in the precise sense of Theorem 2.1.27 and Definition 2.1.28 above), whereas (as was observed in Remark 2.1.29 on page 69) the *abscissa of conditional convergence* of f (in the sense of Theorem 2.1.27 and Definition 2.1.31, but with "convergence" replaced with "conditional convergence") is equal to 0. This situation is in sharp contrast with that encountered for (generalized) Dirichlet series with *positive* coefficients, which is the situation of interest in this book.

By contrast, recall from part (a) of Remark 2.1.21 that we do not know whether there exists generalized (Dirichlet) integrals $f = f(s)$ of the type of the distance zeta function (or, equivalently, of the tube zeta function) of a bounded subset A of $\mathbb{R}^N$ for which $D < N$ and $D_{\text{hol}}(f) < D(f)$, with $f := \zeta_A$ (or, equivalently, $f := \tilde{\zeta}_A$).[13] We do not know of such an example even for a relative fractal drum (in the sense of Section 4.1). On the other hand, recall from part (b) of Remark 2.1.21 above that we have explicit examples of bounded subsets of $\mathbb{R}^N$ for which $D < N$ and $D_{\text{hol}}(f) = D(f)$, for both $f = \zeta_A$ and $f = \tilde{\zeta}_A$.

[13] In the case of the tube zeta function $\tilde{\zeta}_A$, we do not need to assume that $D < N$.

It is noteworthy (although elementary) that any *fractal string* $\mathcal{L} = (\ell_j)_{j\geq 1}$ (that is, here, a nonincreasing sequence of positive real numbers $(\ell_j)_{j\geq 1}$ converging to zero, see the beginning of Subsection 2.1.4 below) is uniquely determined by its *geometric (or scaling) zeta function* $\zeta_{\mathcal{L}}(s) = \sum_{j=1}^{\infty} \ell_j^s$.

More precisely, the following result holds, and is a consequence of a well-known uniqueness result about (generalized) Dirichlet series (see, e.g., [Ser, Corollary 4, Section VI.2.2]). For completeness, we will nevertheless include a proof of this theorem.

Remark 2.1.38. Here and in the sequel, we adopt the convention of [Lap-vFr2–3] according to which $(l_j)_{j\geq 1}$ refers to a strictly decreasing sequence of positive numbers whereas $(\ell_j)_{j\geq 1}$ refers to a nonincreasing sequence of positive numbers (repeated according to their multiplicities). Hence, for all $s \in \mathbb{C}$ with $\operatorname{Re} s > D(\zeta_{\mathcal{L}})$, we have

$$\zeta_{\mathcal{L}}(s) = \sum_{j=1}^{\infty} \ell_j^s = \sum_{j=1}^{\infty} b_j l_j^s, \tag{2.1.51}$$

where for each $j \geq 1$, b_j is the multiplicity of l_j.

Theorem 2.1.39. *Assume that $\mathcal{L} = (\ell_j)_{j\geq 1}$ and $\mathcal{L}' = (\ell'_j)_{j\geq 1}$ are two fractal strings such that their geometric zeta functions are holomorphic in an open right half-plane $G = \{\operatorname{Re} s > \sigma\}$ (or, equivalently, are such that their abscissae of convergence do not exceed σ), for some $\sigma \in \mathbb{R} \cup \{-\infty\}$. If $\zeta_{\mathcal{L}}(s_k) = \zeta_{\mathcal{L}'}(s_k)$ for a sequence $(s_k)_{k\geq 1}$ of elements of G possessing an accumulation point in G, then $\mathcal{L} = \mathcal{L}'$; i.e., $\ell_j = \ell'_j$ for all $j \geq 1$. Equivalently, in the notation of Remark 2.1.38, we have $l_j = l'_j$ and $b_j = b'_j$ for all $j \geq 1$.*

Proof. Without loss of generality, we may assume that the sequences $(\ell_j)_{j\geq 1}$ and $(\ell'_j)_{j\geq 1}$ are nonincreasing; see Subsection 2.1.4. By hypothesis, we must have $\sigma \geq \max\{D(\zeta_{\mathcal{L}}), D(\zeta_{\mathcal{L}'})\}$, where $D(\zeta_{\mathcal{L}})$ and $D(\zeta_{\mathcal{L}'})$ denote the abscissae of convergence of $\mathcal{L}$ and $\mathcal{L}'$, respectively. Hence, according to the principle of analytic continuation (see, e.g., [Con, Corollary 3.8]), the condition $\zeta_{\mathcal{L}}(s_k) = \zeta_{\mathcal{L}'}(s_k)$ for all $k \geq 1$ implies that $\zeta_{\mathcal{L}}(s) = \zeta_{\mathcal{L}'}(s)$ in G. Assume by contradiction that $\ell_{j_0} > \ell'_{j_0}$, with the smallest possible $j_0 \in \mathbb{N}$. Canceling the first $j_0 - 1$ terms in the sums defining the two geometric zeta functions, we may assume without loss of generality that $j_0 = 1$. Dividing $\zeta_{\mathcal{L}}(s) = \zeta_{\mathcal{L}'}(s)$ by ℓ_1^s, we obtain the equality

$$1 + (\ell_2\ell_1^{-1})^s + \cdots + (\ell_j\ell_1^{-1})^s + \cdots = (\ell'_1\ell_1^{-1})^s + (\ell'_2\ell_1^{-1})^s + \cdots + (\ell'_j\ell_1^{-1})^s + \cdots. \tag{2.1.52}$$

Passing to the limit as $s \to +\infty$ in $\mathbb{R}$, we deduce that $1 = 0$, which is a contradiction. Actually, we deduce that $n = 0$, where the integer $n \geq 1$ is equal to the multiplicity of ℓ_1 (that is, to the number of times the value ℓ_1 is repeated in the sequence representing $\mathcal{L}$); i.e., $n = b_1 > 0$. (Alternatively, in the notation of Remark 2.1.38, we have $\ell_1 = \cdots = \ell_n = l_1$.) Note that in order to justify the interchange of limit and sum in (2.1.52), we use Theorem 2.1.34, where we have taken $s_0 > \sigma$. More specifically, Theorem 2.1.34 guarantees the uniform convergence in a sector containing the half-line $\{s \in \mathbb{R} : s > s_0\}$ (and hence, in a neighborhood of $+\infty$ in the extended real line) of each of the Dirichlet series appearing in Equation (2.1.52). In turn, this enables us to justify the interchange of the limits as $s \to +\infty$ and of the infinite sums. □

2.1.3.2 Dirichlet Integrals

Consider the *Dirichlet-type integral* (DTI, for short)

$$F(s) := \int_E \varphi(x)^s \mathrm{d}\mu(x), \tag{2.1.53}$$

where E is a measurable space, φ is a suitable positive (or, more generally, under a suitable assumption, nonnegative, see Remark 2.1.50) measurable function on E, and μ is a local (or locally bounded) positive or complex measure on E.[14] (Recall that if μ is positive, then its total variation measure $|\mu|$ satisfies $|\mu| = \mu$, while if μ is a local complex measure, then $|\mu|$ is a positive and locally bounded measure; for more detailed information, see Definition A.1.1 in Appendix A and, e.g., [Coh], [Foll] or [Ru], for classic measure theory.) The interested reader can find in Appendix A a thorough discussion of Dirichlet-type integrals (DTIs) and of extended DTIs, along with some of their main properties.

For the purposes of the more general theory of Dirichlet-type integrals (DTIs) developed in Appendix A, we will assume E to be a locally compact, Hausdorff (or metrizable) topological space and that μ is a *local* positive (or complex) measure. Roughly speaking, a local measure on E is a set-function on $\mathscr{B}(E)$, the Borel σ-algebra of E, its total variation measure whose restriction to every compact subset of E is bounded. Hence, if μ is a local complex measure, then $|\mu|$ is only locally bounded. See Definition A.1.1 for the precise definition of a local measure.

In the sequel, we shall need the following definitions, which have already been introduced in a less general, but closely related context in Subsections 2.1.2 and 2.1.3.1. (See also Subsection 2.1.5.) For the definitions of $D(F)$ and $\Pi(F)$, introduced just below, to be meaningful, we have to assume that the function φ is nonnegative and bounded from above in the following sense:

$$\text{There exists a constant } C = C(F) > 0 \text{ such that } 0 \le \varphi(x) \le C \ |\mu|\text{-a.e. on } E. \tag{2.1.54}$$

We assume throughout this chapter that this condition is satisfied. (In Appendix A such a DTI is said to be *tamed*.) See also Theorem 2.1.45 which discusses some other possibilities. Note that in the case of the distance zeta function $F := \zeta_A$, we have $E := A_\delta$ and $\varphi(x) := d(x,A) \in [0,\delta]$ for all $x \in A_\delta$, so that condition (2.1.54) is clearly satisfied.

Define the abscissa of convergence $D(F)$ of the generalized Dirichlet integral $F(s) := \int_E \varphi(t)^s \mathrm{d}\mu(t)$ in exactly the same manner as for (the special case of) the distance zeta function ζ_A in Equation (2.1.10) of Subsection 2.1.2, except for the

[14] When $\varphi(x) = 0$, we let $\varphi(x)^s := 0$. (This is quite reasonable, at least for $\operatorname{Re} s > 0$.)

integral $\int_{A_\delta} d(x,A)^{s-N} dx$ replaced by the Dirichlet-type integral $\int_E \varphi(t)^s d\mu(t)$ (and with μ replaced by the total variation measure $|\mu|$ if μ is not a positive measure). Namely,[15]

$$D(F) := \inf\left\{\alpha \in \mathbb{R} : \int_E \varphi(t)^\alpha d|\mu|(t) < \infty\right\}. \tag{2.1.55}$$

Furthermore, call

$$\Pi(F) := \{\mathrm{Re}\, s > D(F)\} \tag{2.1.56}$$

the *half-plane of (absolute) convergence* of F. The value of $D(F)$ is well defined due to the assumption (2.1.54), and for the same reason, the open right half-plane $\Pi(F)$ is also well defined; see Theorem 2.1.45 below.[16] As before, we have $D(F) \in \mathbb{R} \cup \{\pm\infty\}$ and we use the standard convention according to which $\Pi(F) = \emptyset$ or $\mathbb{C}$ if $D(F) = +\infty$ or $-\infty$, respectively.

Moreover, if $D(F) \in \mathbb{R}$, the vertical line $\{\mathrm{Re}\, s = D(F)\}$ is called the *critical line* of F, when no ambiguity may arise (or, less briefly, the *critical line of convergence* of F).

As was the case in Subsection 2.1.2, one then deduces from the counterpart of Lemma 2.1.9 in the present context that the half-plane of convergence $\Pi(F) := \{\mathrm{Re}\, s > D(F)\}$ is the *maximal* right open half-plane of convergence (of the form $\{\mathrm{Re}\, s > \alpha\}$, for some $\alpha \in \mathbb{R} \cup \{\pm\infty\}$) of the Lebesgue integral defining $F(s)$ in (2.1.53); see Theorem A.1.4 in Appendix A. Note that for $s \in \mathbb{C}$, $\int_E |\varphi(t)^s|\, d|\mu|(t) = \int_E \varphi(t)^{\mathrm{Re}\, s}\, d|\mu|(t) < \infty$ implies that $\mathrm{Re}\, s > \alpha$, and conversely.

Finally, we define $D_{\mathrm{hol}}(F)$, the *abscissa of holomorphic continuation* of F, exactly as for ζ_A in Equation (2.1.27) of Subsection 2.1.2, except for the fact that ζ_A is now replaced by F. Then, $\mathscr{H}(F) := \{\mathrm{Re}\, s > D_{\mathrm{hol}}(F)\}$ is called the *half-plane of holomorphic continuation* of F and the vertical line $\{\mathrm{Re}\, s = D_{\mathrm{hol}}(F)\}$ is referred to as the *critical line of continuation* (or simply, the *holomorphy critical line*) of F. See, also, Definition 2.1.62 at the beginning of Subsection 2.1.5 below.

The following three examples should be helpful to the reader, as they play a central role in the book.

Example 2.1.40. (Distance zeta functions). Let A be a bounded subset of $\mathbb{R}^N$ and ζ_A be the associated distance zeta function (as in Subsection 2.1.1), initially defined by (2.1.1) for some $\delta > 0$. Let $E := A_\delta$, $\varphi(x) := d(x,A)$ for all $x \in A_\delta$ (so that $\varphi \equiv 0$ on $\overline{A}$, the closure of A) and consider the positive measure defined by $\mu(dx) := \rho(x)\, dx$, where $\rho : E \to \mathbb{R}$ is defined by

$$\rho(x) := \begin{cases} d(x,A)^{-N}, & \text{for } x \in E \setminus \overline{A}, \\ 0, & \text{for } x \in \overline{A}, \end{cases} \tag{2.1.57}$$

[15] Recall that all of the integrals are taken in the Lebesgue sense.

[16] If we replace the condition appearing in (2.1.54) by $\varphi(x) \geq C$ $|\mu|$-a.e. on E for a positive constant C, then in the definition of $D(F)$ in (2.1.55), we have to replace inf by sup and the corresponding Dirichlet-type integral $F(s)$ is then (absolutely) convergent on the open *left* half-plane $\{\mathrm{Re}\, s < D(F)\}$. See case (b) of Theorem 2.1.45 below.

and dx denotes the Lebesgue measure on $\mathbb{R}^N$. Note that $\mu = |\mu|$ (since μ is positive). Then, in light of Theorem 2.1.45(a) below (or, simply, by definition of φ and μ) and for Res large enough, using the common convention $0 \cdot (+\infty) = 0$ in Lebesgue's integration theory, it follows that ζ_A coincides with the following Dirichlet-type integral given by (2.1.53):

$$\zeta_A(s) := \int_{A_\delta} d(x,A)^{s-N} \mathrm{d}x = \int_E \varphi(x)^s \mathrm{d}\mu(x) =: F(s).$$

Furthermore, according to parts (a) and (b) of Theorem 2.1.11, we have $D(\zeta_A) = \overline{\dim}_B A$, the upper box dimension of A, and $D_{\text{hol}}(\zeta_A) \le D(\zeta_A)$. Moreover, if in addition, we assume that $D := \dim_B A$ exists, $D < N$ and $\mathscr{M}_*^D(A) > 0$, then it follows from part (c) of Theorem 2.1.11 that we also have the following string of equalities:

$$D_{\text{hol}}(\zeta_A) = D(\zeta_A) = \dim_B A \; (= D).$$

In closing this example, we note that the tameness condition (2.1.54) is clearly satisfied since $0 \le \varphi(x) = d(x,A) \le \delta$ for all $x \in A_\delta$ and hence, for all $x \in E$.

Example 2.1.41. (*Relative distance zeta functions*). Looking ahead to Chapter 4, we mention that a discussion entirely analogous to the one given in Example 2.1.40 can be provided (with $E := \Omega$ instead of $E := A_\delta$, $\varphi(x) := d(x,A)$ for all $x \in \Omega$ and $\mu(\mathrm{d}x) := \rho(x)\mathrm{d}x$, where ρ is still given by (2.1.57)) for the relative distance zeta function $\zeta_{A,\Omega}$ of a 'relative fractal drum' (A,Ω) (introduced in Definition 4.1.2 of Chapter 4), where A is any subset of $\mathbb{R}^N$, Ω is an open subset of $\mathbb{R}^N$ such that $|\Omega| < \infty$, and there exists $\delta > 0$ such that $\Omega \subseteq A_\delta$:

$$\zeta_{A,\Omega}(s) := \int_\Omega d(x,A)^{s-N} \mathrm{d}x, \tag{2.1.58}$$

for all $s \in \mathbb{C}$ with Res sufficiently large. (Neither A nor Ω is assumed to be bounded, in that case. However, since $\Omega \subseteq A_\delta$, it follows that $d(x,A) < \delta$ for all $x \in \Omega$. Hence, in light of (2.1.58), the DTI $\zeta_{A,\Omega}$ is clearly tamed; more precisely, condition (2.1.54) is satisfied with $C := \delta$.) There too, we have (in light of parts (a) and (b) of Theorem 4.1.7 along with Corollary 4.1.10 in Section 4.1.1 below) that $D(\zeta_{A,\Omega}) = \overline{\dim}_B(A,\Omega)$, the relative upper box dimension of (A,Ω), and $D_{\text{hol}}(\zeta_{A,\Omega}) \le D(\zeta_{A,\Omega})$; see Remark 2.1.42 below.

As will be stressed in Section 4.1, one important difference between ζ_A and $\zeta_{A,\Omega}$ is that $\overline{\dim}_B A \in [0,N]$ whereas $\overline{\dim}_B(A,\Omega) \in [-\infty,N]$. Also, in light of part (c) of Theorem 4.1.7 below, and provided $D := \dim_B(A,\Omega)$ exists, $D < N$ and $\mathscr{M}_*^D(A,\Omega) > 0$, we then have

$$D_{\text{hol}}(\zeta_{A,\Omega}) = D(\zeta_{A,\Omega}) = \dim_B(A,\Omega) \; (=: D).$$

We note that in this latter situation, the infimum implicit in the definitions of $D_{\text{hol}}(\zeta_{A,\Omega})$ and $D(\zeta_{A,\Omega})$ must be taken over all $\alpha \in \mathbb{R}$ (in fact, since $\overline{\dim}_B(A,\Omega) \in$

$[-\infty,N]$, it suffices to take the infimum over all $\alpha \in (-\infty,N])$. The reason is that in the definition of upper and lower Minkowski contents of a relative fractal drum (A,Ω), we take the infimum over all $r \in \mathbb{R}$; see Equation (4.1.4) below.

Remark 2.1.42. The analog of Proposition 2.1.22 also holds in the more general setting of Example 2.1.41 because the function $\varphi(x) := d(x,A)$, defined on $E := \Omega$, vanishes identically on $\overline{A} \cap \Omega$, while (much as in Example 2.1.40, and using the standard convention in Lebesgue's theory, according to which $0 \cdot (+\infty) = 0$), $\mu(dx) := \rho(x)\,\mathrm{d}x$, where as was explained above, the function $\rho : E \to \mathbb{R}$, with $E := \Omega$, is defined by

$$\rho(x) := \begin{cases} d(x,A)^{-N}, & \text{for } x \in E \setminus \overline{A}, \\ 0, & \text{for } x \in \overline{A}. \end{cases} \tag{2.1.59}$$

Example 2.1.43. (*Tube zeta functions and their relative counterparts*). Also looking ahead towards the rest of the book, we mention that the tube zeta function $\tilde{\zeta}_A$ of a bounded subset A of $\mathbb{R}^N$ (see Definition 2.2.8 of Subsection 2.2.2 below) is a tamed DTI. This statement is explained in detail in Lemma 2.2.9 (and its proof). More generally, the tube zeta function $\tilde{\zeta}_{A,\Omega}$ of a relative fractal drum (A,Ω) in $\mathbb{R}^N$ (see Equation (4.5.1) of Subsection 4.5.1 below) is a tamed DTI. This statement is proved much as for its counterpart for the relative distance function in Example 2.1.41 above and is explained in the proof of part (1) of Proposition A.2.4 of Appendix A.

Example 2.1.44. (*Generalized Dirichlet series and geometric zeta functions*). The generalized Dirichlet series studied in Subsection 2.1.3.1 can be easily viewed as (tamed) Dirichlet-type integrals (DTIs, in short) of the form (2.1.53), as we now explain. Indeed, let $E := [\log(1/\ell_1),+\infty)$ and with the notation of Subsection 2.1.3.1,[17]

$$\varphi(t) := \mathrm{e}^{-t} \quad \text{and} \quad \mu(dt) := \sum_{j=1}^{\infty} b_j \delta_{\log(1/l_j)} \tag{2.1.60}$$

(or, alternatively, $E := (0,\ell_1]$, $\varphi(t) := t$ for all $t \in E$ and $\mu(dt) := \sum_{j=1}^{\infty} b_j \delta_{l_j} = \sum_{j=1}^{\infty} \delta_{\ell_j}$; see, especially, Equation (2.1.51) in Remark 2.1.38 above). Then, the generalized Dirichlet series $f(s) := \sum_{j=1}^{\infty} b_j l_j^s = \sum_{j=1}^{\infty} \ell_j^s$, as given by (2.1.37) and (2.1.51), coincides with the Dirichlet-type integral $F(s) := \int_E \varphi(t)^s \mathrm{d}\mu(t)$, as given by (2.1.53). Moreover, since

$$0 < \varphi(t) \leq C := \varphi(\log(1/\ell_1)) = \ell_1$$

[17] Recall from Subsection 2.1.3.1 (and, especially, Remark 2.1.38) that $(\ell_j)_{j=1}^{\infty}$ is a nonincreasing sequence of positive numbers such that $\ell_j \downarrow 0$ as $j \to \infty$, and that $(l_j)_{j=1}^{\infty}$ is the sequence composed of the *distinct* values of the ℓ_k's. Furthermore, for each $j \geq 1$, l_j has multiplicity b_j, when it appears in the sequence $(\ell_k)_{k=1}^{\infty}$. In particular, we have $\ell_1 = \cdots = \ell_{b_1} = l_1$.

for all $t \in E = [\log(1/\ell_1), +\infty)$ (alternatively, $0 < \varphi(t) \leq C := \varphi(\ell_1) = \ell_1$ for all $t \in E = (0, \ell_1]$), condition (2.1.54) is satisfied; i.e., f is a tamed DTI, in the sense of Definition A.1.3 of Appendix A. Here, if φ is given by (2.1.60) and the sequence $(\ell_j)_{j=1}^{\infty}$ is nonincreasing, while the optimal choice of C is given by $C = \ell_1 = l_1$ (still with the notation of Subsection 2.1.3.1).

As a result, for the geometric zeta function $\zeta_{\mathscr{L}}(s) := \sum_{j=1}^{\infty} \ell_j^s$ of a bounded[18] fractal string $\mathscr{L} := (\ell_j)_{j=1}^{\infty}$ (in the sense of Subsection 2.1.4 below), we have

$$D_{\text{hol}}(\zeta_{\mathscr{L}}) = D(\zeta_{\mathscr{L}}) = \inf\left\{\alpha \in [0,1] : \sum_{j=1}^{\infty} \ell_j^{\alpha} < \infty\right\}. \tag{2.1.61}$$

More generally, the geometric zeta function ζ_η of a local (positive or complex) measure η on $(0, +\infty)$ (as defined in [Lap-vFr3, Chapter 4] by $\zeta_\eta(s) := \int_0^{+\infty} x^{-s} \eta(dx)$ for all $s \in \mathbb{C}$ with $\operatorname{Re} s$ sufficiently large) is a tamed DTI; see the second half of part (2) of Proposition A.2.4 (and its proof). In particular, all arithmetic zeta functions occurring in number theory (see, e.g., [ParsSh1–2, Edw, Tit3], [Lap-vFr3, Appendix A], as well as [Lap6, Appendices B, C and E]) are tamed DTIs. As a special case appearing in various parts of Subsection 2.1.3, let us mention the Dirichlet L-functions and, in particular, the classic Riemann zeta function.

Finally, we close this example by pointing out that the spectral zeta functions of relative fractal drums (RFDs) studied in Section 4.3 and defined in Definition 4.3.4 below (and, in particular, the spectral zeta functions of fractal strings) are tamed DTIs since they can clearly be viewed as geometric zeta functions of generalized fractal strings (and really, as generalized Dirichlet series, in the sense of Subsection 2.1.3.1).

We are now ready to state a theorem concerning the holomorphicity of the Dirichlet-like integral F. It will imply, in particular, part (a) of Theorem 2.1.11 in Subsection 2.1.2 above, as will be explained later.

We should point out that part (a) of Theorem 2.1.45 just below is a very special case of a significantly more general result, provided in Appendix A (see Theorem A.2.6 and Corollary A.2.7) and proved in a similar manner, by using Theorem 2.1.47 below. In particular, we should note that in part (a) of Theorem 2.1.45 below, it suffices to assume that μ is a (positive or complex) *local* measure on E (in the sense of Definition A.1.1, roughly speaking, a locally bounded set-function on $\mathscr{B}(E)$, the Borel σ-algebra of E) and that φ is a positive measurable function satisfying condition (2.1.54); i.e., $F = \zeta_{(E,\varphi,\mu)}$ (as given by (2.1.62) below and in the notation of Appendix A, see Equation (A.1.2) of Appendix A) is a tamed Dirichlet-type integral (DTI, in short); see Definition A.1.3. Moreover, for the fractal zeta functions of interest considered in this book, φ is easily seen to be bounded from above (see the hypothesis made in Theorem 2.1.45(a) below); for example, as was noted in Examples 2.1.40 and 2.1.41, $\varphi(x) := d(x,A) < \delta$ for all $x \in A_\delta$. As a result, all of

[18] We say that a fractal $\mathscr{L} := (\ell_j)_{j=1}^{\infty}$ string is *bounded* if $\sum_{j=1}^{\infty} \ell_j < \infty$.

these fractal zeta functions are tamed. See also Remark 2.1.46 below, along with Proposition A.2.4 and Corollary A.2.7 of Appendix A.

Theorem 2.1.45. *Let $(E,\mathscr{B}(E),\mu)$ be a measure space, where E is a locally compact metrizable space, $\mathscr{B}(E)$ is the Borel σ-algebra of E, and μ is a positive or complex (local) measure, with total variation (local) measure denoted by $|\mu|$. Furthermore, let $\varphi : E \to (0,+\infty)$ be a measurable function satisfying condition (2.1.54).*[19] *Then:*

(a) If φ is essentially bounded (that is, if there exists $C > 0$ such that $\varphi(t) \le C$ for $|\mu|$-a.e. $t \in E$), and if there exists $\sigma \in \mathbb{R}$ such that $\int_E \varphi(t)^\sigma \mathrm{d}|\mu|(t) < \infty$, then

$$F(s) := \int_E \varphi(t)^s \mathrm{d}\mu(t) \tag{2.1.62}$$

is holomorphic on $\{\mathrm{Re}\, s > \sigma\}$, and $F'(s) = \int_E \varphi(t)^s \log \varphi(t)\, \mathrm{d}\mu(t)$ in that region. Furthermore, the abscissa of convergence $D(F)$ of F (defined as in Equation (2.1.55) above) and the abscissa of holomorphic continuation $D_{\mathrm{hol}}(F)$ of F (defined either as above or as in Definition 2.1.62 of Subsection 2.1.5 below) satisfy the following inequality:

$$D_{\mathrm{hol}}(F) \le D(F). \tag{2.1.63}$$

(b) If there exists $C > 0$ such that $\varphi(t) \ge C$ for $|\mu|$-a.e. $t \in E$, and if there exists $\sigma \in \mathbb{R}$ such that $\int_E \varphi(t)^{-\sigma} \mathrm{d}|\mu|(t) < \infty$, then

$$G(s) := \int_E \varphi(t)^{-s} \mathrm{d}\mu(t) \tag{2.1.64}$$

is holomorphic on $\{\mathrm{Re}\, s > \sigma\}$, and $G'(s) = -\int_E \varphi(t)^{-s} \log \varphi(t)\, \mathrm{d}\mu(t)$ in that region. Here, $D(G)$, the abscissa of convergence of G, satisfies $D_{\mathrm{hol}}(G) \le D(G)$.

(c) Finally, if there exist positive constants C_1 and C_2 such that $C_1 \le \varphi(t) \le C_2$ for $|\mu|$-a.e. $t \in E$, and there exists $\sigma \in \mathbb{R}$ such that $\int_E \varphi(t)^\sigma \mathrm{d}|\mu|(t) < \infty$, then the Dirichlet-type integrals F and G in (a) and (b), respectively, are entire functions (and hence, $D(F) = D_{\mathrm{hol}}(F) = -\infty$, where $D = D(F)$ is the abscissa of convergence of F and $D_{\mathrm{hol}}(F)$ is the abscissa of holomorphic continuation of F, and analogously for G).

In many applications of interest to us in this book, the local measure μ is positive and hence, $|\mu| = \mu$. Furthermore, the inequality (2.1.63) is actually an equality. This is the case, for example, for the (generalized) Dirichlet series studied in Subsection 2.1.3.1 above (in light of Example 2.1.44, this is a consequence of Theorem 2.1.34 and Corollary 2.1.35) as well as, similarly, for the classic Dirichlet integrals (of the form $F(s) := \int_0^{+\infty} \mathrm{e}^{-st} \mathrm{d}\mu(t)$, for some suitable positive Borel measure μ on $[0,+\infty)$, supported away from 0; see, e.g., [Pos]). Under mild assumptions (namely, the hypotheses of part (c) of Theorem 2.1.11 and Corollary 2.1.20), this is also

[19] See Remark 2.1.50 (and the text following it) for the case when $\varphi > 0$ $|\mu|$-almost everywhere.

the case for the distance zeta functions of bounded sets. More generally, under the hypotheses of part (c) of Theorem 4.1.7, it is also the case for the distance zeta functions of relative fractal drums.

Remark 2.1.46. It is easy to check that all of the fractal zeta functions $f = \zeta_{(E,\varphi,\mu)}$ discussed in Examples 2.1.40–2.1.44 above (and, more generally, all those encountered in this book) are such that φ is bounded from above (pointwise everywhere) on E (i.e., not only $|\mu|$-bounded from above). Indeed, either $\varphi(x) := d(x,A) < \delta$ for all $x \in A_\delta$, where A is any (bounded or unbounded) subset of $\mathbb{R}^N$, or $\varphi(t) := t < \delta$ for all $t \in (0,\delta)$. The latter choice of φ will correspond to tube zeta functions (and their relative counterparts) to be introduced in Subsection 2.2.2 (and Subsection 4.5.1).

Theorem 2.1.45 will follow from a more general and well-known result (see, e.g., [Carl, pp. 295–296] or [CarMi, pp. 152–153], dealing with the holomorphicity of integrals depending on a parameter, of the form

$$H(s) = \int_E f(s,t)\,\mathrm{d}\mu(t). \tag{2.1.65}$$

We state it without proof. When $n = 1$, the interested reader can easily establish it by combining the Lebesgue dominated convergence theorem, Cauchy's integral formula, and/or Fubini's theorem along with Morera's characterization of holomorphic functions (as having zero contour integrals along closed loops contained in the open set V). Note, however, that the latter characterization of holomorphicity is not really needed if one shows directly that H is holomorphic with complex derivative given by (2.1.66) below, with $k := 1$.

Theorem 2.1.47. *Let V be an open set in $\mathbb{C}$ (or even in $\mathbb{C}^n$, for any $n \in \mathbb{N}$). Furthermore, let $(E,\mathscr{B}(E),\mu)$ be a measurable space, as in Theorem 2.1.45, with a positive or complex (local) measure μ and the corresponding total variation (local) measure denoted by $|\mu|$; see Equation* (A.1.1) *of Definition A.1.1 in Appendix A. Assume that a function $f : V \times E \to \mathbb{C}$ is given, satisfying the following three conditions:*

(1) *$f(\cdot,t)$ is holomorphic for $|\mu|$-a.e. $t \in E$,*

(2) *$f(s,\cdot)$ is μ-measurable for all $s \in V$, and*

(3) *a suitable growth property on f is fulfilled: for every compact set K contained in V, there exists $g_K \in L^1(|\mu|)$ such that $|f(s,t)| \le g_K(t)$ for all $s \in V$ and $|\mu|$-a.e. $t \in K$.*

Then, the function H defined by (2.1.65) *is holomorphic on V. Moreover, one can interchange the derivative and the integral. More precisely, for every $s \in V$ and every $k \in \mathbb{N}$, we have*

$$H^{(k)}(s) = \int_E \frac{\partial^k}{\partial s^k} f(s,t)\,\mathrm{d}\mu(t). \tag{2.1.66}$$

Note that conditions (1) and (2), appearing in Theorem 2.1.47, imply that the complex-valued function $f(s,t)$ satisfies the well-known *Carathéodory conditions*; that is, $f(s,t)$ is continuous with respect to $s \in V$ for $|\mu|$-a.e. $t \in E$, and is μ-measurable with respect to $t \in E$ for all $s \in V$.

Remark 2.1.48. According to [Mattn] and as is well known, if conditions (1) and (2) from Theorem 2.1.47 are satisfied, then condition (3) is equivalent to the following condition, which is generally slightly more practical to verify:

(3′) $\int_E |f(\cdot,t)|\,\mathrm{d}|\mu|(t)$ is locally bounded; that is, for each fixed $s_0 \in V$, there exists $\delta > 0$ such that

$$\sup_{s\in V,\, |s-s_0|<\delta} \int_E |f(s,t)|\,\mathrm{d}|\mu|(t) < \infty. \tag{2.1.67}$$

(See also Remark 2.1.49.) In other words, we can replace condition (3) with condition (3′) in the statement of Theorem 2.1.47. This is the case because the notion of holomorphicity is *local*. For the same reason, we can verify the holomorphicity of F on a relatively compact neighborhood of any given $s_0 \in V$, and therefore, work with a local measure under the hypotheses indicated in the statement of Theorem 2.1.47.

Remark 2.1.49. We note that if μ is assumed to be a standard positive or complex measure, then the counterpart of Theorem 2.1.47 is valid on an arbitrary measure space; see, e.g., [JohLap, Lemma 15.2.9], where it is stated much more generally.

Proof of Theorem 2.1.45. We use Theorem 2.1.47. In our case, $f(s,t) := \varphi(t)^s$, $V := \{\mathrm{Re}\, s > \sigma\}$. Note that for any $\sigma_1 > \sigma$, we have $\varphi(t)^{\sigma_1} \leq \|\varphi\|_\infty^{\sigma_1-\sigma}\varphi(t)^\sigma$, so that $\varphi^\sigma \in L^1(|\mu|)$ implies that $\varphi^{\sigma_1} \in L^1(|\mu|)$. In particular, since $|f(s,t)| = \varphi(t)^{\mathrm{Re}\, s}$, it follows that $f(s,t) = \varphi(t)^s \in L^1(|\mu|)$ for all $s \in \mathbb{C}$ such that $\mathrm{Re}\, s > \sigma$.

Let K be a compact subset of $V = \{\mathrm{Re}\, s > \sigma\}$. Since

$$|f(s,t)| = \varphi(t)^{\mathrm{Re}\, s} \leq \|\varphi\|_\infty^{\mathrm{Re}\, s-\sigma}\varphi(t)^\sigma, \tag{2.1.68}$$

we have that $|f(s,t)| \leq g_K(t) := C_K\varphi(t)^\sigma$ for all $s \in K$ and $|\mu|$-a.e. $t \in E$, where $C_K = \max_{s\in K}\|\varphi\|_\infty^{\mathrm{Re}\, s-\sigma}$. This proves part (a) of the theorem.

Part (b) follows from part (a) applied to $\varphi(t)^{-1}$.

Finally, part (c) follows similarly as in (a), by noting that

$$|f(s,t)| = \varphi(t)^{\mathrm{Re}\, s} \leq \max\{C_1^{\mathrm{Re}\, s-\sigma}, C_2^{\mathrm{Re}\, s-\sigma}\}\varphi(t)^\sigma, \tag{2.1.69}$$

for every complex number s. ☐

Remark 2.1.50. Theorem 2.1.45 extends without any difficulty if $\varphi(t) \geq 0$ for $|\mu|$-a.e. $t \in E$ and $|\mu|(\{t \in E : \varphi(t) = 0\}) = 0$.

In our present case, corresponding to the setting of Theorem 2.1.11, we have $E := A_\delta$, $\varphi(x) := d(x,A)$ for $x \in \mathbb{R}^N$ (note that φ restricted to E is bounded, with values in $[0,\delta)$) and $\mathrm{d}\mu(x) := d(x,A)^{-N}\mathrm{d}x$, where $\mathrm{d}x$ is the Lebesgue measure on $\mathbb{R}^N$. More precisely, we let $E := A_\delta \setminus \{\varphi = 0\}$ and observe that the set $\{\varphi = 0\} = \{x \in \mathbb{R}^N : d(x,A) = 0\}$ does not contribute to the integral defining $\zeta_A(s)$ in Equation (2.1.1). Note that we know from Lemma 2.1.3 that $\sigma \geq \overline{\dim}_B A$ and hence, $\{\mathrm{Re}\, s > \sigma\} \subseteq \{\mathrm{Re}\, s > 0\}$.

Since $\varphi(x) = d(x,A) = 0$ for $x \in \overline{A} \subset A_\delta$, the function φ vanishes on a set which may not be of zero measure. If $|A| > 0$, then $\overline{\dim}_B A = N$, and by using Theorem 2.1.11(b) we conclude that $D(\zeta_A) = N$. On the other hand, if $|A| = 0$, then $\overline{\dim}_B A \leq N$, and we conclude that $D(\zeta_A) \leq N$.

Remark 2.1.51. It is also worth pointing out that Theorem 2.1.11(a) can be derived by using Theorem 2.1.45(a). Indeed, in light of this latter result, it suffices to show that for any real number σ such that $\sigma > \overline{\dim}_B A$ we have that

$$\int_{A_\delta} \varphi(x)^\sigma \mathrm{d}\mu(x) = \int_{A_\delta} d(x,A)^{\sigma - N} \mathrm{d}x < \infty,$$

where $\varphi(x) := d(x,A)$ and $\mathrm{d}\mu(x) := d(x,A)^{-N}\mathrm{d}x$. But this follows immediately from Lemma 2.1.3, using $\gamma := N - \sigma < N - \overline{\dim}_B A$:

$$\int_{A_\delta} d(x,A)^{\sigma - N} \mathrm{d}x = \int_{A_\delta} d(x,A)^{-\gamma} \mathrm{d}x < \infty.$$

We therefore conclude from Theorem 2.1.45(a) that ζ_A is holomorphic for $\sigma = \mathrm{Re}\, s > \overline{\dim}_B A$, and that its (complex) derivative $\zeta_A'(s)$ is given by (2.1.13), which is precisely the statement of Theorem 2.1.11(a).

Since for any bounded set A in $\mathbb{R}^N$ the abscissa of convergence $D(\zeta_A)$ of the distance zeta function ζ_A of A is equal to the upper box dimension of A, that is, $D(\zeta_A) = \overline{\dim}_B A$ (see Theorem 2.1.11(b)), and $\overline{\dim}_B A \in [0,N]$, it is clear that $D(\zeta_A) \geq 0$. The following lemma provides a direct proof of the fact that $D(\zeta_A)$ cannot be negative. On the other hand, for relative fractal drums, which we will introduce in Section 4.1, the abscissa of convergence of the distance zeta function can be negative, and even equal to $-\infty$; see Subsection 4.1.2.

Lemma 2.1.52. *For any bounded subset A of $\mathbb{R}^N$, we have $D(\zeta_A) \geq 0$.*

Proof. Assume the contrary, namely, that $D(\zeta_A) < 0$. Then $\zeta_A(s)$ is well defined and continuous for $s \in (D(\zeta_A), +\infty)$, and in particular, it is continuous at $s = 0$.

Let us take any $a \in A$. Since $A_\delta \supseteq B_\delta(a)$, and $d(x,A) \leq |x - a|$, we have that for every real number $s \in (0,N)$,

$$\begin{aligned}\zeta_A(s) &= \int_{A_\delta} d(x,A)^{s-N} \mathrm{d}x \geq \int_{B_\delta(s)} d(x,A)^{s-N} \mathrm{d}x \\ &\geq \int_{B_\delta(a)} |x-a|^{s-N} \mathrm{d}x = N\omega_N \int_0^\delta r^{s-N} r^{N-1} \mathrm{d}r = N\omega_N \frac{\delta^s}{s},\end{aligned}$$

where we have passed to polar coordinates with the point a as the origin (and where ω_N is the volume of the unit ball in $\mathbb{R}^N$). Therefore, $\zeta_A(s) \to +\infty$ as $s \to 0^+$, $s \in \mathbb{R}$. This clearly contradicts the continuity of ζ_A at $s = 0$. □

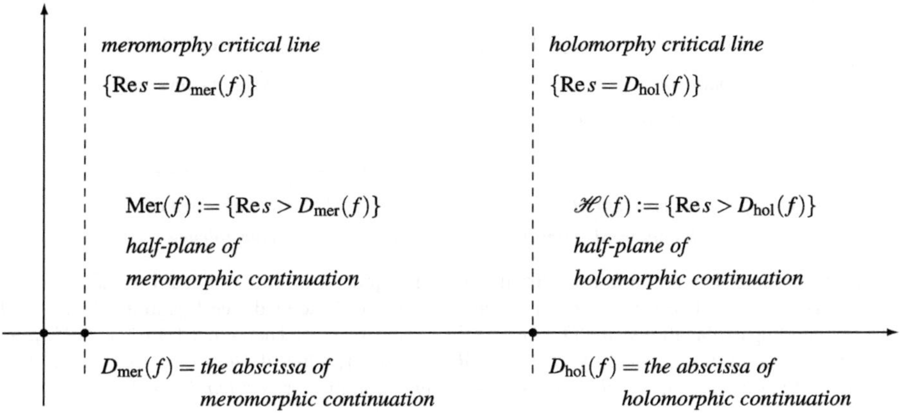

Fig. 2.5 The *holomorphy critical line* $\{\mathrm{Re}\, s = D_{\mathrm{hol}}(f)\}$, on the right, and the *meromorphy critical line* $\{\mathrm{Re}\, s = D_{\mathrm{mer}}(f)\}$, on the left, associated with a given function f. Here, $D_{\mathrm{hol}}(f)$ is the *abscissa of holomorphic continuation* of f and $D_{\mathrm{mer}}(f)$ is the *abscissa of meromorphic continuation* of f. The *half-plane of meromorphic continuation* of f is defined by $\mathrm{Mer}(f) := \{\mathrm{Re}\, s > D_{\mathrm{mer}}(f)\}$, while the *half-plane of holomorphic continuation* of f is defined by $\mathcal{H}(f) := \{\mathrm{Re}\, s > D_{\mathrm{hol}}(f)\}$; see Definition 2.1.53 and Definition 2.1.62, respectively.

In this monograph, we shall pay particular attention to meromorphic functions f. It is therefore natural to introduce the following definition, which is analogous to the definition of the abscissa of convergence $D(f)$. Compare with Definition 2.1.28 and Equation (2.1.49) above.

Definition 2.1.53. Let $f : U \to \mathbb{C}$ be a meromorphic function on a domain $U \subseteq \mathbb{C}$. We define the *abscissa of meromorphic continuation* $D_{\mathrm{mer}}(f)$ of f as the infimum of all real numbers α such that f possesses a meromorphic extension to the open right half-plane $\{\mathrm{Re}\, s > \alpha\}$. Equivalently,

$$\mathrm{Mer}(f) := \{\mathrm{Re}\, s > D_{\mathrm{mer}}(f)\} \tag{2.1.70}$$

is the largest open half-plane (of the form $\{\mathrm{Re}\, s > \alpha\}$, for some $\alpha \in \mathbb{R} \cup \{\pm\infty\}$) to which f can be meromorphically extended. We then call $\mathrm{Mer}(f)$ the *half-plane of meromorphic continuation* of f, while, if $D_{\mathrm{mer}}(f) \in \mathbb{R}$, $\{\mathrm{Re}\, s = D_{\mathrm{mer}}(f)\}$ is referred to as the *critical line of meromorphic continuation*, or, more briefly, the *meromorphy critical line* of f; see Figure 2.5.

We invite the reader to consult Remarks 2.3.39 and 2.3.40 on page 169 (in Subsection 2.3.3 below), which provide additional information concerning Definition 2.1.53 (and Definition 2.1.62 below). For now, we limit ourselves to the following comment.

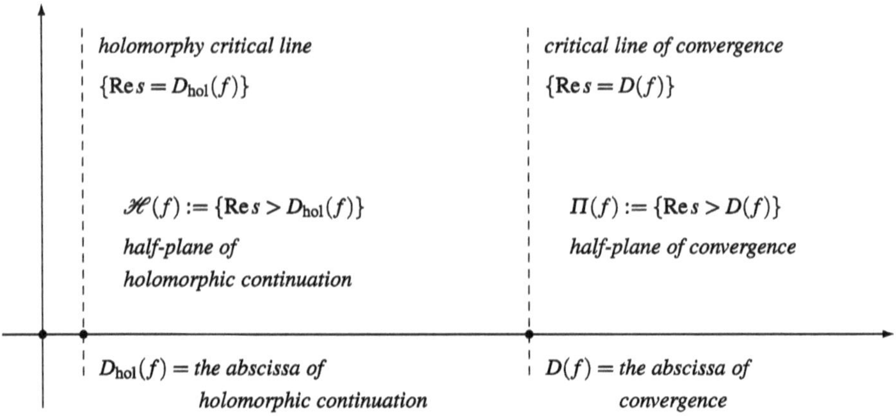

Fig. 2.6 Assume that f is a tamed Dirichlet-type integral, as in Subsection 2.1.3.2 above or in Appendix A. Then, its *abscissa of convergence* $D(f)$ is well defined (see Equation (2.1.55) and the surrounding explanations) and $D_{\mathrm{hol}}(f) \le D(f)$; see part (a) of Theorem 2.1.45. The *half-plane of convergence* of f is defined by $\Pi(f) := \{\mathrm{Re}\, s > D(f)\}$, while if $D(f) \in \mathbb{R}$, the vertical line $\{\mathrm{Re}\, s = D(f)\}$ is called the *critical line of convergence* (or, in short, *critical line*) of f.

Remark 2.1.54. It is clear that $\mathscr{H}(f) \subseteq \mathrm{Mer}(f)$, and therefore, that $D_{\mathrm{mer}}(f) \le D_{\mathrm{hol}}(f)$, where $D_{\mathrm{hol}}(f)$ is the abscissa of holomorphic continuation of f (as given in Definition 2.1.62 below). If in addition, we assume that f is given by a tamed Dirichlet-type integral (in the sense of Subsection 2.1.3.2 above, then we also have $D_{\mathrm{hol}}(f) \le D(f)$); see Theorem 2.1.45$(a)$ and Figure 2.6. In Theorem 4.5.20, we will construct a class of bounded fractal strings $\mathscr{L} = (\ell_j)_{j\ge1}$ such that $D_{\mathrm{mer}}(\zeta_{\mathscr{L}}) < D_{\mathrm{hol}}(\zeta_{\mathscr{L}})$, with a prescribed value of $D_{\mathrm{mer}}(\zeta_{\mathscr{L}})$. Furthermore, in Corollary 4.6.17(a), we will construct a class of bounded fractal strings such that $D_{\mathrm{mer}}(\zeta_{\mathscr{L}}) = D_{\mathrm{hol}}(\zeta_{\mathscr{L}}) = D(\zeta_{\mathscr{L}}) =: D$ and all the points of the holomorphy critical line $\{\mathrm{Re}\, s = D\}$ are nonisolated singularities of $\zeta_{\mathscr{L}}$. (Such fractal strings will be said to be *maximally hyperfractal*; see Definition 4.6.23 and the comments surrounding it.) In both cases, the fractal string $\mathscr{L}$ will be constructed as a union of a suitable sequence of generalized Cantor strings.

2.1.4 Zeta Functions of Fractal Strings and of Associated Fractal Sets

The following example shows that Definition 2.1.1 provides a natural extension of the zeta function associated with a fractal string $\mathscr{L} = (\ell_j)_{j\ge1}$, where $(\ell_j)_{j\ge1}$ is a nonincreasing sequence of positive numbers such that $\sum_{j=1}^{\infty} \ell_j < \infty$:

$$\zeta_{\mathscr{L}}(s) = \sum_{j=1}^{\infty} \ell_j^s, \tag{2.1.71}$$

for all $s \in \mathbb{C}$ with $\mathrm{Re}\, s$ sufficiently large.

This zeta function arose naturally in the early 1990s in joint work of the first author [Lap1–3] with Carl Pomerance [LapPo1–3] and with Helmut Maier [LapMa1–2] (see also, e.g., [Lap1–2] and [HeLap]) while investigating direct and inverse spectral problems associated with the vibrations of a fractal string. Such a zeta function, $\zeta_{\mathscr{L}}$, called the *geometric zeta function* of $\mathscr{L}$, has been studied in a number of references, including several monographs [Lap-vFr1–3].

Recall that geometrically, a *fractal string* is a bounded open set $\Omega \subseteq \mathbb{R}$. It can be uniquely written as a disjoint union of (bounded) open intervals I_j ($\Omega = \cup_{j=1}^{\infty} I_j$) with lengths ℓ_j. Without loss of generality, one may assume that $(\ell_j)_{j\geq 1}$ is written in nonincreasing order and that $\ell_j \to 0^+$ as $j \to \infty$:

$$\ell_1 \geq \ell_2 \geq \cdots \geq \ell_j \geq \cdots > 0. \tag{2.1.72}$$

From the point of view of fractal string theory, one may identify a fractal string with the sequence $\mathscr{L}$ of its *lengths (or scales)*: $\mathscr{L} = (\ell_j)_{j\geq 1}$. For example, the volume (i.e., length) of the (inner) tubular neighborhoods of a fractal string of Ω depends only on $\mathscr{L} = (\ell_j)_{j\geq 1}$. Consequently, the (inner) Minkowski content[20] of the boundary $\partial\mathscr{L} = \partial\Omega$ of $\mathscr{L}$, as well as the complex dimensions of $\mathscr{L}$ (to be discussed just below), depend solely on $\mathscr{L} = (\ell_j)_{j\geq 1}$; see [LapPo2], [Lap-vFr3, Section 8.1]. The same is true for the spectrum (i.e., the eigenvalues) of the Dirichlet Laplacian on Ω.

In the sequel, any bounded open set $\Omega = \cup_{j=1}^{\infty} I_j \subset \mathbb{R}$ of the above type (i.e., such that $\mathscr{L} = (\ell_j)_{j=1}^{\infty}$ is the sequence of lengths of its connected components $(I_j)_{j=1}^{\infty}$) will be called a *geometric realization* of the fractal string $\mathscr{L}$. Furthermore, $\mathscr{L}$ will often be called a *bounded* fractal string in order to emphasize that its *total length* $|\mathscr{L}|_1 = \sum_{j=1}^{\infty} \ell_j$ is finite. (Of course, it would suffice to suppose that $|\Omega|_1 < \infty$ instead of assuming that Ω is bounded.)

We now recall a basic property of $\zeta_{\mathscr{L}}$, first observed in [Lap2], using a key result of Besicovich and Taylor [BesTay]. For a direct proof, see [Lap-vFr3, Theorem 1.10 or Theorem 13.111]; see also [LapLu-vFr2].

Theorem 2.1.55. *If $\mathscr{L}$ is nontrivial (i.e., if $\mathscr{L} = (\ell_j)_{j\geq 1}$ is an infinite sequence), then the abscissa of convergence $D(\zeta_{\mathscr{L}})$ of $\zeta_{\mathscr{L}}$ coincides with the (inner) upper Minkowski dimension[21] $\delta_{\partial\Omega}$ of $\partial\mathscr{L} = \partial\Omega$:[22]*

$$D(\zeta_{\mathscr{L}}) = D_{\text{hol}}(\zeta_{\mathscr{L}}) = \delta_{\partial\Omega}. \tag{2.1.73}$$

---

[20] In the terminology introduced in Section 4.1 below, the inner Minkowski content is called the Minkowski content of the relative fractal drum $(\partial\Omega, \Omega)$, or the Minkowski content of $\partial\Omega$ relative to Ω.

[21] The inner Minkowski dimension $\delta_{\partial\Omega}$ is a special case of the notion of a relative box dimension, defined in Section 4.1. In this case it is denoted by $\overline{\dim}_B(\partial\Omega, \Omega)$.

[22] In [Lap-vFr3], the abscissa of convergence of $\zeta_{\mathscr{L}}$ is denoted by $\sigma_{\mathscr{L}}$. Furthermore, if $\mathscr{L} = (\ell_j)_{j\geq 1}$ is a finite sequence, then $D(\zeta_{\mathscr{L}}) = -\infty$ whereas $\delta_{\partial\Omega} = 0$; so that in general, we have that $\delta_{\partial\Omega} = \max\{D(\zeta_{\mathscr{L}}), 0\}$.

Recall that

$$D(\zeta_{\mathscr{L}}) := \inf\Big\{\alpha \in \mathbb{R} : \sum_{j=1}^{\infty} \ell_j^{\alpha} < \infty\Big\}, \tag{2.1.74}$$

while $\delta_{\partial\Omega}$ is defined in terms of the volume of the inner epsilon neighborhoods of $\partial\Omega$, namely, $(\partial\Omega)_\varepsilon \cap \Omega = \{x \in \Omega : d(x,\partial\Omega) < \varepsilon\}$; see [Lap-vFr3, Definition 1.2, p. 11].

More specifically, we must let $N = 1$ and replace $|A_t|$ by $|(\partial\Omega)_t \cap \Omega|_1$ in order to define the (inner) upper Minkowski content of $A := \partial\Omega$ and then define $\delta_{\partial\Omega}$ by the counterpart of Equation (1.3.4). Alternatively, in light of the counterpart of Equation (1.3.15), we have that

$$\delta_{\partial\Omega} = 1 - \liminf_{t\to 0^+} \log_t |(\partial\Omega)_t \cap \Omega|_1. \tag{2.1.75}$$

Finally, in the notation to be introduced in Section 4.1, for the relative fractal drum $(A,\Omega) := (\partial\Omega,\Omega)$, we have

$$\delta_{\partial\Omega} = \overline{\dim}_B(\partial\Omega,\Omega), \tag{2.1.76}$$

the relative upper Minkowski dimension of $(\partial\Omega,\Omega)$; see Equations (4.1.3) and (4.1.4).

Remark 2.1.56. We note that within the framework of the fractal zeta functions developed in this book, yet another (albeit indirect) proof of Theorem 2.1.55 can be provided; see Proposition 2.1.72 and Remark 2.1.73 in Subsection 2.1.5, which themselves make use of Example 2.1.58 below and part (b) of Theorem 2.1.11 above.

It follows from Theorem 2.1.55 that $\zeta_{\mathscr{L}}$ is holomorphic for $\mathrm{Re}\, s > \delta_{\partial\Omega}$ and that $\{s \in \mathbb{C} : \mathrm{Re}\, s > \delta_{\partial\Omega}\}$ is the largest open right half-plane having this property. See Subsection 2.1.3 above.

In fractal string theory, one is particularly interested in the meromorphic continuation of $\zeta_{\mathscr{L}}$ to a suitable region (when it exists), along with its poles, which are called the *complex dimensions* of $\mathscr{L}$. In particular, in the theory of complex dimensions developed in [Lap-vFr1–3], are obtained explicit tube formulas applicable to various counting functions associated with the geometry and the spectra of fractal strings. These explicit formulas are expressed in terms of the complex dimensions (i.e., the poles of $\zeta_{\mathscr{L}}$) and the associated residues. Furthermore, they enable one to obtain a very precise understanding of the oscillations underlying the geometry and spectra of fractal strings (as well as of more general fractal-like objects); see [Lap-vFr3].

From the perspective of the theory developed in the present monograph, a convenient choice of geometric realization of the (nontrivial) fractal string $\mathscr{L} = (\ell_j)_{j=1}^{\infty}$ and therefore, of the set $A = A_{\mathscr{L}}$ corresponding to $\mathscr{L}$, is the following *canonical geometric realization* of $\mathscr{L}$:

$$\Omega_{\text{can}} = \Omega_{\text{can},\mathscr{L}} := \bigcup_{k=1}^{\infty} (a_{k+1}, a_k), \tag{2.1.77}$$

where

$$a_k := \sum_{j \geq k} \ell_j \quad \text{for each } k \geq 1; \tag{2.1.78}$$

so that, assuming that $\mathscr{L}$ is *nontrivial* (i.e., $\mathscr{L}$ is an infinite sequence), we have $a_k \downarrow 0$ as $k \to \infty$; see Figure 2.7 below. (Note that, by construction, the length of each connected component (a_{k+1}, a_k) of Ω_{can} is given by $\ell_k = a_k - a_{k+1}$, for $k \geq 1$.) Then,

$$A := A_{\mathscr{L}} = \{a_k : k \geq 1\} = \partial\Omega_{\text{can}} \setminus \{0\} \tag{2.1.79}$$

and we call this set the *canonical geometric representation of the fractal string* $\mathscr{L}$.

As follows easily from Theorem 2.1.55 and the definition of $A_{\mathscr{L}}$, the function $\zeta_{\mathscr{L}}$ in (2.1.71) is holomorphic for $\mathrm{Re}\, s > \overline{\dim}_B A_{\mathscr{L}}$. Moreover, still assuming that the fractal string $\mathscr{L}$ is nontrivial, then this bound is optimal.[23] In other words, $\overline{\dim}_B A$ coincides with the abscissa of convergence of $\mathscr{L}$. Furthermore, $\zeta_{\mathscr{L}}(s) \to +\infty$ as $s \in \mathbb{R}$ converges to $\overline{\dim}_B A$ from the right; see [Lap-vFr3, p. 15] or [Ser, Section VI.2.3]. Compare with Theorem 2.1.11.

As a result, $\{\mathrm{Re}\, s > \overline{\dim}_B A\}$ is the maximal right open half-plane to which $\zeta_{\mathscr{L}}$ can be holomorphically continued (i.e., $D_{\text{hol}}(\zeta_{\mathscr{L}}) = \overline{\dim}_B A$) and hence, in light of (2.1.73) of Theorem 2.1.55 above, combined with parts (a) and (b) of Theorem 2.1.11, we have

$$D(\zeta_{\mathscr{L}}) = D_{\text{hol}}(\zeta_{\mathscr{L}}) = D(\zeta_{A_{\mathscr{L}}}) = D_{\text{hol}}(\zeta_{A_{\mathscr{L}}}) = \overline{\dim}_B A = \delta_{\partial\Omega}, \tag{2.1.80}$$

where $\delta_{\partial\Omega}$ is given as in Theorem 2.1.55 and *is therefore independent of the geometric realization Ω of $\mathscr{L}$* (in particular, we have $\delta_{\partial\Omega} = \delta_{\partial\Omega_{\text{can}}}$).

In summary, we can state the following result:

Corollary 2.1.57. *In light of Theorem 2.1.55 and of Theorem 2.1.11, assuming that $\mathscr{L}$ is nontrivial, we have the following equalities*:

$$\overline{\dim}_B A_{\mathscr{L}} = D(\zeta_{\mathscr{L}}) = \delta_{\partial\Omega}. \tag{2.1.81}$$

Example 2.1.58. With the above notation, let $I_k = (a_{k+1}, a_k)$, $k \geq 1$, and let s be a complex variable. Using (2.1.1), we see that the distance zeta function of A is given by[24]

$$\zeta_A(s) = 2\int_0^{\delta} x^{s-1} \mathrm{d}x + \sum_{k=1}^{\infty} \int_{I_k} d(x, \partial I_k)^{s-1} \mathrm{d}x = 2s^{-1}\delta^s + \sum_{k=1}^{\infty} J_k(s), \tag{2.1.82}$$

[23] If a fractal string $\mathscr{L}$ is trivial, i.e., a finite sequence, then the bound is not optimal since in this case $D(\zeta_{\mathscr{L}}) = -\infty$, while $\overline{\dim}_B A_{\mathscr{L}} = 0$ (and hence, $\dim_B A_{\mathscr{L}}$ exists and $\dim_B A_{\mathscr{L}} = 0$). Therefore, for any bounded fractal string $\mathscr{L}$, we have that $D(\zeta_{\mathscr{L}}) \in [0,1] \cup \{-\infty\}$.

[24] The second integral appearing in Equation (2.1.82), that is, $\zeta_{\partial I_k, I_k}(s) := \int_{I_k} d(x, \partial I_k)^{s-1} \mathrm{d}x$, is the distance zeta function of the so-called 'relative fractal drum' $(\partial I_k, I_k)$, a notion which will be introduced and studied in detail in Chapter 4.

for all $s \in \mathbb{C}$ such that $\operatorname{Re} s > D(\zeta_A)$ (ensuring the convergence of the series), where the first term corresponds to boundary points of the interval $(0, a_1)$ and for all $k \geq 1$, $J_k(s)$ is defined in (2.1.83) just below. Assuming that $\delta \geq \ell_1/2$, we have that for all k,

$$J_k(s) := 2\int_0^{\ell_k/2} x^{s-1}\mathrm{d}x = s^{-1}2^{1-s}\ell_k^s. \tag{2.1.83}$$

Here, we have integrated with respect to the local coordinate system placed at the left endpoint of the interval I_k. In light of (2.1.71) and (2.1.82)–(2.1.83), we obtain the following relation:

$$\zeta_A(s) = s^{-1}2^{1-s}\zeta_{\mathscr{L}}(s) + 2s^{-1}\delta^s. \tag{2.1.84}$$

The case when $0 < \delta < \ell_1/2$ yields an analogous relation:

$$\zeta_A(s) = u(s)\zeta_{\mathscr{L}}(s) + v(s), \tag{2.1.85}$$

where again $u(s) := s^{-1}2^{1-s}$, with a single, simple pole at $s = 0$. Note that here, $u(s)$ and $v(s) = v(s, \delta)$ are holomorphic functions in the open right half-plane $\{\operatorname{Re} s > 0\}$; (initially, for $\operatorname{Re} s > D(\zeta_{\mathscr{L}})$ and then, after analytic continuation, for $\operatorname{Re} s > 0$).[25] Hence, since $\zeta_{\mathscr{L}}$ is holomorphic for $\operatorname{Re} s > \overline{\dim}_B A$, the same relation still holds for the meromorphic extension of ζ_A (when it exists) to any subdomain U of the right half-plane $\{\operatorname{Re} s > 0\}$. An immediate consequence of (2.1.84) is that

$$D(\zeta_{A_{\mathscr{L}}}) = \max\{0, D_{\mathrm{hol}}(\zeta_{\mathscr{L}})\}, \tag{2.1.86}$$

which is in accordance with Lemma 2.1.52. This example will serve as a basic motivation and guide for introducing and studying fractal nests, geometric chirps and string chirps in higher dimensions; see, especially, Sections 3.5 and 3.6.

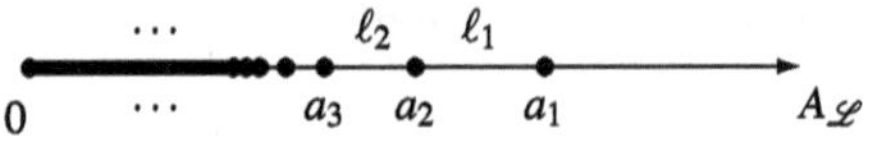

Fig. 2.7 Any nontrivial bounded fractal string $\mathscr{L} = (\ell_j)_{j\geq 1}$ generates the sequence $(a_k)_{k\geq 1}$, where $a_k := \sum_{j\geq k}\ell_j$, converging to the origin as $k \to \infty$. We denote by $A = A_{\mathscr{L}}$ the corresponding subset of the real line: $A_{\mathscr{L}} := \{a_k : k \geq 1\}$. Note that $\ell_k = a_k - a_{k+1}$ for all $k \geq 1$.

[25] It can be easily shown that the function $v = v(s)$, appearing in Equation (2.1.85), has the form

$$v(s) := \frac{2\delta^s}{s} + e(s),$$

where $e = e(s)$ is an entire function and δ is the positive constant from Equation (2.1.1) defining the distance zeta function ζ_A. For $\delta \geq \ell_1/2$, as we have already seen in Equation (2.1.84), we have that $v(s) \equiv 0$. For $\delta \in (0, \ell_1/2)$, the claim follows from the fact that the difference $e(s) := \zeta_A(s;\delta) - \zeta_A(s;\ell_1/2)$ is an entire function; see Proposition 2.1.76 below.

The following result is in accordance with Lemma 2.1.52 and Theorem 2.1.55; see Figure 2.7.

Proposition 2.1.59. *Let $\mathscr{L} = (\ell_j)_{j\geq 1}$ be a nontrivial, bounded fractal string, and let $A_{\mathscr{L}} = \{a_k = \sum_{j\geq k}\ell_j : k \geq 1\}$. Then*

$$D(\zeta_{A_{\mathscr{L}}}) = D_{\mathrm{hol}}(\zeta_{A_{\mathscr{L}}}) = D(\zeta_{\mathscr{L}}) = D_{\mathrm{hol}}(\zeta_{\mathscr{L}}) = \overline{\dim}_B A_{\mathscr{L}}. \tag{2.1.87}$$

Furthermore, the sets of poles of the meromorphic extensions of $\zeta_{A_{\mathscr{L}}}$ and $\zeta_{\mathscr{L}}$ to any open right half-plane $\{\mathrm{Re}\, s > c\}$, with $c \geq 0$ (if the extensions exist), coincide. Moreover, the poles of $\zeta_{A_{\mathscr{L}}}$ and $\zeta_{\mathscr{L}}$ (in such a half-plane) have the same multiplicities.

In particular, if either the geometric zeta function $\zeta_{\mathscr{L}}$ or the corresponding distance zeta function $\zeta_{A_{\mathscr{L}}}$ admits a meromorphic continuation (necessarily unique) to a subdomain of the open right half-plane $\{\mathrm{Re}\, s > 0\}$ containing the critical line $\{\mathrm{Re}\, s = D(\zeta_{\mathscr{L}})\}$, then so does the other one. Furthermore, in that case, these two fractal zeta functions have exactly the same poles (or 'visible complex dimensions') within $\{\mathrm{Re}\, s > 0\}$, with the same orders (or multiplicities). See also Remark 2.1.60 below.

Proof. The first claim, except for the first equality in Equation (2.1.87), follows from Theorem 2.1.55 combined with parts (a) and (b) of Theorem 2.1.11.

To prove the first equality in (2.1.85), assume, by contradiction, that $D(\zeta_{A_{\mathscr{L}}}) > D_{\mathrm{hol}}(\zeta_{A_{\mathscr{L}}})$. We consider the following two cases:

(a) If $D_{\mathrm{hol}}(\zeta_{A_{\mathscr{L}}}) \geq 0$, then from Equation (2.1.85) we obtain that $\zeta_{\mathscr{L}}$ can be *holomorphically* extended at least to the open right half-plane $\{\mathrm{Re}\, s > D_{\mathrm{hol}}(\zeta_{A_{\mathscr{L}}})\}$. Therefore, $D_{\mathrm{hol}}(\zeta_{\mathscr{L}}) \leq D_{\mathrm{hol}}(\zeta_{A_{\mathscr{L}}}) < D(\zeta_{A_{\mathscr{L}}}) = D(\zeta_{\mathscr{L}})$. However, this contradicts the equality $D_{\mathrm{hol}}(\zeta_{\mathscr{L}}) = D(\zeta_{\mathscr{L}})$.

(b) The case when $D_{\mathrm{hol}}(\zeta_{A_{\mathscr{L}}}) < 0$ is dealt with analogously, by taking into account the fact that the functions u and v, appearing in (2.1.85), have $s = 0$ as their unique pole, and furthermore, this pole is simple. Indeed, the function $\zeta_{A_{\mathscr{L}}}(s) - v(s)$ has $s = 0$ as the only pole located in the corresponding (meromorphy) half-plane $\{\mathrm{Re}\, s > D_{\mathrm{mer}}(\zeta_{A_{\mathscr{L}}})\}$. Therefore, the function $s(\zeta_{A_{\mathscr{L}}}(s) - v(s))$ is holomorphic in this half-plane, which, in light of Equation (2.1.85), implies that $\zeta_{\mathscr{L}}$ is holomophic on the same half-plane as well. We deduce that $D(\zeta_{\mathscr{L}}) < 0$. However, this contradicts the fact that for any bounded fractal string $\mathscr{L}$, we always have $D(\zeta_{\mathscr{L}}) \in [0,1]$.

The second and the third claims are an immediate consequence of (2.1.85). This completes the proof of the proposition. □

Remark 2.1.60. (a) In the first part of Proposition 2.1.59, one can replace the half-plane $\{\mathrm{Re}\, s > c\}$ by an arbitrary connected open subset U of $\mathbb{C} \setminus \{0\}$. A similar comment can be made about the second part of Proposition 2.1.59, with $\{\mathrm{Re}\, s > 0\}$ replaced by a domain $U \subseteq \mathbb{C} \setminus \{0\}$ containing the critical line $\{\mathrm{Re}\, s = D(\zeta_{\mathscr{L}})\}$.

(b) This fact is significant from the point of view of the theory of complex dimensions of fractal strings developed in [Lap-vFr3]. Indeed, recall that according to [Lap-vFr3], a fractal string $\mathscr{L}$ (or the associated fractal set $A_{\mathscr{L}}$) is 'fractal' if it has at least one complex dimension with positive real part. Hence, by the present proposition, the notion of fractality is independent of the choice of the fractal zeta function used to define it. See, in particular, part (c) of the present remark.

(c) According to the results obtained further on in this chapter (see, especially, Theorem 2.2.11 and Proposition 2.2.19 of Subsection 2.2.2), the exact same results (as in Proposition 2.1.59 above and Corollary 2.1.61 below) hold if the distance zeta function $\zeta_{A_{\mathscr{L}}}$ is replaced by the corresponding tube zeta function $\tilde{\zeta}_{A_{\mathscr{L}}}$ (in the sense of Definition 2.2.8), throughout Proposition 2.1.59 above and Corollary 2.1.61 below.

One may naturally wonder whether in the above results, the geometric zeta function $\zeta_{\mathscr{L}}$ can only be compared to the distance zeta function $\zeta_{A_{\mathscr{L}}}$, where $A = A_{\mathscr{L}}$ is defined by (2.1.79). In particular, $A_{\mathscr{L}} \cup \{0\} = \partial\Omega_{\rm can}$ is the boundary of the *canonical* geometric representation of $\mathscr{L}$. In other words, do the statements of these results (including Theorem 2.1.55, Corollary 2.1.57, Example 2.1.58 and Proposition 2.1.59) depend on the choice of the geometric realization Ω of $\mathscr{L}$? The answer is that they do *not* depend on the geometric realization of $\mathscr{L}$ by a bounded open set $\Omega \subset \mathbb{R}$, which is a very useful fact, indeed. The reason for the simplicity of this answer is that in fractal string theory (see [Lap1–3], [LapPo2], [LapMa2], [HeLap], [Lap-vFr1–3], [Lap6–10], ...), all of the quantities involved, $V(\varepsilon) = V_{inn}(\varepsilon) := |(\partial\Omega)_\varepsilon \cap \Omega|_1$, D (the Minkowski dimension or the upper box dimension), etc., are defined in terms of the *inner* ε-neighborhood of $\partial\Omega$ (*relative* to Ω), and therefore (as was noted earlier) depend only on $\mathscr{L} = (\ell_j)_{j=1}^\infty$ and not on the geometric realization Ω of $\mathscr{L}$.

Corollary 2.1.61. (i) *The exact same results as in Theorem 2.1.55, Corollary 2.1.57, Example 2.1.58 and Proposition 2.1.59* (*along with Remark 2.1.60*) *hold if given a bounded fractal string $\mathscr{L}$, one replaces $A_{\mathscr{L}}$ by $A_{\partial\Omega}$* (*the boundary of Ω*), *where Ω is any geometric realization of $\mathscr{L}$ by a bounded open subset of $\mathbb{R}$.*

(ii) *Moreover, still for an arbitrary geometric realization Ω of $\mathscr{L}$, one can replace $\zeta_{A_{\mathscr{L}}}$, the distance zeta function of $A_{\mathscr{L}}$, by $\zeta_{\partial\Omega,\Omega}$, the distance zeta function of the relative fractal drum* ($\partial\Omega,\Omega$), *in the sense of Definition 4.1.1 of Subsection 4.1.1 below. In that case, the functional equation connecting the relative distance zeta function $\zeta_{\partial\Omega,\Omega}$ and the geometric zeta function $\zeta_{\mathscr{L}}$ is even simpler than in the counterpart of Equation* (2.1.84) *or* (2.1.85). *More specifically, for any geometric realization Ω of the bounded fractal string $\mathscr{L}$, we have*

$$\zeta_{\partial\Omega,\Omega}(s) = \frac{2^{1-s}}{s}\zeta_{\mathscr{L}}(s), \tag{2.1.88}$$

for all $s \in \mathbb{C}$ such that $\mathrm{Re}\, s > D(\zeta_{\mathscr{L}})$, *where*

$$D(\zeta_{\mathscr{L}}) = D(\zeta_{\partial\Omega,\Omega}) = D_{\rm hol}(\zeta_{\mathscr{L}}) = D_{\rm hol}(\zeta_{\partial\Omega,\Omega}) = \overline{\dim}_B(\partial\Omega,\Omega). \tag{2.1.89}$$

It follows that given any domain U of $\mathbb{C}$ containing $\{\mathrm{Re}\, s > D\}$ (or equivalently, containing the critical line $\{\mathrm{Re}\, s = D\}$),[26] where $D := D(\zeta_{\mathscr{L}})$, $\zeta_{\mathscr{L}}$ has a meromorphic continuation to U if and only if $\zeta_{\partial\Omega,\Omega}$ does, and in that case, $\zeta_{\mathscr{L}}$ and $\zeta_{\partial\Omega,\Omega}$ have the same poles in $U \setminus \{0\}$ and with the same multiplicities. Furthermore, for any simple pole $\omega \in U \setminus \{0\}$ of $\zeta_{\partial\Omega,\Omega}$ (and hence, also of $\zeta_{\mathscr{L}}$), we have that

$$\mathrm{res}(\zeta_{\partial\Omega,\Omega}, \omega) = \frac{2^{1-\omega}}{\omega} \mathrm{res}(\zeta_{\mathscr{L}}, \omega). \tag{2.1.90}$$

Proof. Part (i) is established exactly in the same way as for the specific canonical geometric realization $\Omega_{\mathrm{can},\mathscr{L}}$ of $\mathscr{L}$ in the rest of this subsection (i.e., Subsection 2.1.4).

Part (ii) relies on a computation analogous to that in Equations (2.1.82) and (2.1.83) above, as we now explain:

$$\begin{aligned} \zeta_{\partial\Omega,\Omega}(s) &:= \int_{\Omega} d(x, \partial\Omega)^{s-1} \mathrm{d}x = \sum_{k=1}^{\infty} \int_{I_k} d(x, \partial I_k)^{s-1} \mathrm{d}x \\ &= \sum_{k=1}^{\infty} s^{-1} 2^{1-s} \ell_k^s = s^{-1} 2^{1-s} \zeta_{\mathscr{L}}(s), \end{aligned} \tag{2.1.91}$$

for all $s \in \mathbb{C}$ such that $\mathrm{Re}\, s > D(\zeta_{\mathscr{L}})$, which proves Equation (2.1.88). The rest of (ii) follows immediately from this functional equation. (In fact, Equation (2.1.88) can be deduced immediately from Equation (2.1.84), by noticing that $\zeta_{A_{\mathscr{L}}}(s) = \zeta_{\partial\Omega,\Omega}(s) + 2s^{-1}\delta^{-s}$, provided $\delta \geq \ell_1/2$.[27]) $\square$

Finally, we note that for the exact same reason as explained above, all of the results concerning bounded fractal strings discussed in this book could be expressed in terms of an arbitrary geometric realization of $\mathscr{L}$ via a bounded open subset of $\mathbb{R}$; that is, the distance zeta function $\zeta_{A_{\mathscr{L}}}$ (or its counterpart, the tube zeta function $\tilde{\zeta}_{A_{\mathscr{L}}}$) could be replaced by $\zeta_{\partial\Omega,\Omega}$, the distance zeta function of the relative fractal drum $(\partial\Omega, \Omega)$, in the sense of Definition 4.1.1 of Subsection 4.1.1 below (or respectively, by $\tilde{\zeta}_{\partial\Omega,\Omega}$, the tube zeta function of the relative fractal drum $(\partial\Omega, \Omega)$). Therefore, the choice of the canonical geometric realization Ω_{can} and $A \cup \{0\} = A_{\mathscr{L}} \cup \{0\} = \partial\Omega_{\mathrm{can}}$ given by (2.1.77)–(2.1.79) is merely convenient, but not necessary. We will not always recall this important fact in the sequel. See, however, Remark 2.1.73 in Subsection 2.1.5 below.

[26] Recall from Corollary 2.1.35 and Corollary 2.1.36 that $\zeta_{\mathscr{L}}$ is holomorphic on $\{\mathrm{Re}\, s > D\}$.

[27] To see this, it suffices to note that $\zeta_{\partial\Omega,\Omega} = \zeta_{\partial\Omega_{\mathrm{can},\mathscr{L}},\Omega_{\mathrm{can},\mathscr{L}}}$ and $(A_{\mathscr{L}})_\delta = (-\delta, 0) \cup \Omega_{\mathrm{can},\mathscr{L}} \cup (\ell, \ell + \delta)$, provided $\delta \geq \ell_1/2$. Here, $\ell := |\Omega_{\mathrm{can},\mathscr{L}}|_1 = \sum_{k=1}^{\infty} \ell_k$. We mention in passing that Equation (2.1.90) is equivalent to Equation (5.5.15) appearing in Proposition 5.5.4 of Subsection 5.5.2 below.

2.1.5 Equivalent Fractal Zeta Functions

In this section, we shall introduce an equivalence relation $\sim$ on the set of zeta functions (see Definition 2.1.69) aimed, in particular, at identifying the zeta function ζ_A with its simpler form $\zeta_{\mathscr{L}}$, by removing the inessential functions $a(s)$ and $b(s)$ appearing in Equation (2.1.85) above. As a result, the distance zeta function ζ_A will enable us to recover some of the main features of the geometric zeta function $\zeta_{\mathscr{L}}$.

Definition 2.1.62. Given a meromorphic function f, it will be convenient to define the *abscissa of holomorphic continuation* of f (if it exists), $D_{\text{hol}}(f)$, as the infimum of the real numbers such that f admits a holomorphic continuation to the open right half-plane $\{\mathrm{Re}\, s > \alpha\}$:[28]

$$D_{\text{hol}}(f) = \inf\{\alpha \in \mathbb{R} : f \text{ is holomorphic for } \mathrm{Re}\, s > \alpha\}. \tag{2.1.92}$$

In general, either $D_{\text{hol}}(f)$ does not exist (i.e., $D_{\text{hol}}(f) = +\infty$), or $D_{\text{hol}}(f) \in [-\infty, +\infty)$. In the sequel, and as was the case in Subsections 2.1.3.1 and 2.1.3.2, we follow the usual convention and say that $D_{\text{hol}}(f)$ always exists as an extended real number: $D_{\text{hol}}(f) \in \mathbb{R} \cup \{\pm\infty\}$.

The largest open right half-plane (of the form $\{\mathrm{Re}\, s > \alpha\}$, for some $\alpha \in \mathbb{R} \cup \{\pm\infty\}$) on which f is holomorphic, $\mathscr{H}(f) := \{\mathrm{Re}\, s > D_{\text{hol}}(f)\}$, is called *the half-plane of holomorphic continuation* of f. Here or in Equation (2.1.92), when we write that f is holomorphic for $\mathrm{Re}\, s > \alpha$, we mean that f admits a holomorphic continuation to the open right half-plane $\{\mathrm{Re}\, s > \alpha\}$.

Finally, the vertical line $\{\mathrm{Re}\, s = D_{\text{hol}}(f)\}$ is called the *critical line of holomorphic continuation* of f (or, more briefly, the *holomorphy critical line*); see Figures 2.5 and 2.6 on page 85 and page 86.

From Theorem 2.1.11(a) and (b), we immediately deduce the following result, which will be useful for the computation of the upper box dimension of some classes of fractal sets. Note that according to Section 2.1.3.2, we have $D_{\text{hol}}(\zeta_A) \leq D(\zeta_A)$, the abscissa of convergence of the Dirichlet-type integral defining the distance zeta function ζ_A in (2.1.1).

Corollary 2.1.63. *If A is any bounded subset of $\mathbb{R}^N$, then*

$$\overline{\dim}_B A = D(\zeta_A). \tag{2.1.93}$$

The next very useful result is just a restatement of part (ii) of Corollary 2.1.20.

Corollary 2.1.64. *If, in addition, we assume that $D := \dim_B A$ exists and $\mathscr{M}_*^D(A) > 0$, then*

$$\dim_B A = D(\zeta_A) = D_{\text{hol}}(A). \tag{2.1.94}$$

---

[28] Initially, f is defined on some domain $U \subseteq \mathbb{C}$. It is clear, then, that while f could a priori be any complex-valued function on U, it must a posteriori be holomorphic on U.

Following and adapting [Lap-vFr3, Sections 1.2.1 and 5.1] to our present more general situation,[29] in order to be able to define the key notions of complex dimensions and of principal complex dimensions (see Definition 2.1.68 and Definition 2.1.67, respectively), we assume that the set A has the property that ζ_A can be extended to a meromorphic function defined on $G \subseteq \mathbb{C}$, where G is an open and connected neighborhood of the *window* $\boldsymbol{W}$ defined by

$$\boldsymbol{W} = \{s \in \mathbb{C} : \operatorname{Re} s \geq S(\operatorname{Im} s)\}.$$

Here, the function $S : \mathbb{R} \to (-\infty, D(\zeta_A)]$, called the *screen*, is assumed to be Lipschitz continuous. The graph

$$\boldsymbol{S} := \{S(\tau) + \mathrm{i}\tau : \tau \in \mathbb{R}\} \tag{2.1.95}$$

of the function S, with the horizontal and vertical axes interchanged, is also called the *screen* and the precise meaning will always be clear from the context. Note that the closed set $\boldsymbol{W}$ has for boundary $\boldsymbol{S}$ (i.e., $\partial \boldsymbol{W} = \boldsymbol{S}$) and contains the *critical line* $\{\operatorname{Re} s = D(\zeta_A)\}$; in fact, $\boldsymbol{W}$ also contains the closed half-plane $\{\operatorname{Re} s \geq D(\zeta_A)\}$. In other words, we assume that A is such that its distance zeta function can be extended meromorphically to an open domain G containing the closed half-plane $\{\operatorname{Re} s \geq D(\zeta_A)\}$. (Following the usual conventions, we still denote by ζ_A the meromorphic continuation of ζ_A to G, which is necessarily unique due to the principle of analytic continuation. Furthermore, as in [Lap-vFr3], we assume that the screen does not contain any poles of ζ_A.) A set A satisfying this property and for which ζ_A is 'languid' (in the sense of [Lap-vFr3, Definition 5.2], that is, roughly speaking, grows at most polynomially along the screen and a suitable sequence of horizontal lines avoiding the poles of ζ_A) is said to be *admissible*. (There exist nonadmissible fractal sets; see Corollary 4.6.17 and [Lap-vFr3, Example 5.32].) In the present monograph, we will need to consider, in particular, the *set of poles of* ζ_A *located on the critical line* $\{\operatorname{Re} s = D(\zeta_A)\}$:[30]

$$\mathscr{P}_c(\zeta_A) = \{\omega \in \boldsymbol{W} : \omega \text{ is a pole of } \zeta_A \text{ and } \operatorname{Re} \omega = D(\zeta_A)\}, \tag{2.1.96}$$

called the *set of principal complex dimensions of* A (see Definition 2.1.67). Since, as was noted before, the window $\boldsymbol{W}$ contains the critical line $\{\operatorname{Re} s = D(\zeta_A)\}$, $\mathscr{P}_c(\zeta_A)$ is a subset of the *set of all poles of* ζ_A in $\boldsymbol{W}$ (i.e., the set of all *visible complex dimensions* of A, which we denote by $\mathscr{P}(\zeta_A)$ or $\mathscr{P}(\zeta_A, \boldsymbol{W})$ (see Definition 2.1.68).

Remark 2.1.65. We stress that because, in most of this monograph (and with the exception of Chapter 5), we will not use or extend the pointwise and distributional explicit tube formulas obtained in [Lap-vFr1–3] (and for the validity of which the above polynomial growth conditions are essential, see [Lap-vFr3, Chapter 5]), we

[29] We also take into account the fact that according to Theorem 2.1.11 and Corollary 2.1.20, $D_{\text{hol}}(\zeta_A) \leq D(\zeta_A)$ (with equality if the hypotheses of part (*c*) of Theorem 2.1.11 or, equivalently, of Corollary 2.1.20, are satisfied). See also Figure 2.6 on page 86 for the important special case where f is given by a tamed Dirichlet-type integral, in the sense of Subsection 2.1.3.2.

[30] Note that clearly (and in contrast to $\mathscr{P}(\zeta_A) = \mathscr{P}(\zeta_A, \boldsymbol{W})$ to be introduced in Definition 2.1.68), $\mathscr{P}_c(\zeta_A)$ is independent of the choice of the window $\boldsymbol{W}$.

do not need to include these polynomial growth conditions in our definition of admissibility. Therefore, throughout most of this monograph, an admissible set A is one for which a meromorphic continuation of ζ_A exists in a suitable open neighborhood of the given window $\boldsymbol{W}$ (but without requiring any growth conditions on ζ_A) and not having any pole along the screen $\boldsymbol{S}$. The one exception to this general statement will be provided by Chapter 5 in which we will establish and apply fractal tube formulas in the general context of the higher-dimensional theory of complex dimensions developed in this book. In that chapter, we will then make the appropriate hypotheses on the screen $\boldsymbol{S}$ and, especially, assume the growth (languidity) conditions satisfied by the fractal zeta functions under consideration, much as in [Lap-vFr3, Chapters 5 and 8].

Remark 2.1.66. In general, it seems to be difficult to check whether a set A is admissible. For example, we do not know if the Mandelbrot set or if 'typical' Julia sets are admissible (even in the weaker form of Remark 2.1.65). Further on, however, we will provide natural sufficient conditions on a given bounded subset A of $\mathbb{R}^N$ guaranteeing that ζ_A has an appropriate meromorphic continuation to a nontrivial open right half-plane (strictly containing the half-plane $\mathscr{H}(\zeta_A) = \{\mathrm{Re}\, s > D(\zeta_A)\}$ of holomorphic continuation) and hence, that A is admissible (in the sense of Remark 2.1.65). See Section 2.3; see also, more generally, Section 4.5.1, the main results of which guarantee the admissibility of a large class of relative fractal drums (A,Ω), in the sense of Section 4.1.1.

A class of Minkowski measurable admissible sets is described in Theorem 2.3.18, and a class of Minkowski nonmeasurable admissible sets in Theorem 2.3.25. Furthermore, a class of Minkowski measurable admissible sets on the real line, generated by perturbed Dirichlet strings, is described in Theorem 2.3.17.

The following definition is a slight modification of the notion of complex dimension for fractal strings introduced by the first author and Machiel van Frankenhuijsen in [Lap-vFr1–3], which depends not only on the string, but also on the window $\boldsymbol{W}$; see [Lap-vFr3, Subsection 1.2.1]. The simplification is only introduced here for technical reasons, and is useful especially when one of the main goals is the computation of the box dimension of some new classes of fractal sets in $\mathbb{R}^N$; see Section 3.5.

Another situation where such a notion is potentially very useful is when one wants to understand the leading oscillatory behavior (that is, the oscillations of largest amplitudes) in the geometry of fractal sets. This latter theme should be significantly developed in later extensions of the theory (and of the present monograph), especially when investigating 'fractal tube formulas' in higher dimensions; see, e.g., Problem 6.2.38 in Subsection 6.2.3. See also many of the examples provided in Section 5.5 and illustrating the general fractal tube formulas obtained in Chapter 5 (Sections 5.1–5.3).

Definition 2.1.67. Let A be an admissible subset of $\mathbb{R}^N$. The *set of principal complex dimensions* of A, denoted by $\dim_{PC} A$, is defined as the set of poles of ζ_A which are located on the critical line $\{\mathrm{Re}\, s = D(\zeta_A)\}$:

$$\dim_{PC} A := \mathscr{P}_c(\zeta_A), \tag{2.1.97}$$

where $\mathscr{P}_c(\zeta_A)$ is given by (2.1.96).

As we see, in Definition 2.1.67, if $A \subset \mathbb{R}^N$ is bounded, the singularities of ζ_A we are interested in are located on the vertical line $\{\operatorname{Re} s = \overline{\dim}_B A\}$. This observation follows from part (a) of Theorem 2.1.11 above.

Following and extending the definition of complex dimensions of fractal strings (and other fractals) provided in [Lap-vFr1–3], we also introduce the following natural higher-dimensional generalization in our context.

Definition 2.1.68. Let A be an admissible subset of $\mathbb{R}^N$. Then, the *set of visible complex dimensions* of A *with respect to a given window* $\boldsymbol{W}$ (often called, in short, the *set of complex dimensions of A relative to* $\boldsymbol{W}$, or simply the *set of (visible) complex dimensions* of A if no ambiguity may arise or if $\boldsymbol{W} = \mathbb{C}$), is defined as the set of all the poles of ζ_A which are located in the window $\boldsymbol{W}$:

$$\mathscr{P}(\zeta_A) = \{\omega \in \boldsymbol{W} : \omega \text{ is a pole of } \zeta_A\}. \tag{2.1.98}$$

When necessary, one may also denote it by $\mathscr{P}(\zeta_A, \boldsymbol{W})$, for clarity, and reserve the notation $\mathscr{P}(\zeta_A)$ to the cases where $\boldsymbol{W} = \mathbb{C}$ or to where no ambiguity may arise.

Next, we would like to extend the class of zeta functions to which a slight modification of Definition 2.1.67 and Definition 2.1.68 can be applied. This definition will be convenient, in particular, for the study of the zeta functions encountered in the examples discussed in Section 3.5. Given a tamed Dirichlet-type integral f (i.e., a DTI given by (2.1.53) and satisfying condition (2.1.54)) which has a meromorphic extension to a domain $G \subseteq \mathbb{C}$ containing the vertical line $\{\operatorname{Re} s = D(f)\}$, we define the set $\mathscr{P}_c(f)$ in much the same way as in (2.1.96):

$$\mathscr{P}_c(f) = \{\omega \in G : \omega \text{ is a pole of } f \text{ and } \operatorname{Re}\omega = D(f)\}. \tag{2.1.99}$$

It is a subset of the set $\mathscr{P}(f)$ of all the poles of the meromorphic function f belonging to G. In other words,

$$\mathscr{P}(f) = \{\omega \in G : \omega \text{ is a pole of } f\}. \tag{2.1.100}$$

When necessary, one may write $\mathscr{P}(f, G)$ instead of $\mathscr{P}(f)$, for more precision.

If $f = \zeta_A$, where A is an admissible set for a given window $\boldsymbol{W}$, then (with $G := \mathring{\boldsymbol{W}}$, the interior of the window) $\mathscr{P}_c(f) = \mathscr{P}_c(\zeta_A)$, the set of principal complex dimensions of A, while $\mathscr{P}(f, \mathring{\boldsymbol{W}}) = \mathscr{P}(f) = \mathscr{P}(\zeta_A) = \mathscr{P}(\zeta_A, \boldsymbol{W})$, the set of (visible) complex dimensions of A (relative to $\boldsymbol{W}$). This follows from the fact that since A is admissible, ζ_A does not have any poles along the screen $\boldsymbol{S}$; see the discussion following Corollary 2.1.63.

Note that $\mathscr{P}_c(f)$ is independent of the choice of the domain G containing the critical line $\{\operatorname{Re} s = D(f)\}$. Moreover, since by Theorem 2.1.45(a) above (or, more generally, by CorollaryA.2.7 in Appendix A below), the function f is holomorphic for $\mathrm{Re}\, s > D(f)$, there are no poles of f located in the open right half-plane $\{\mathrm{Re}\, s > D(f)\}$; this is why we could equivalently require that the domain $G \subseteq \mathbb{C}$ contains the closed half-plane $\{\mathrm{Re}\, s \geq D(f)\}$ in order to define $\mathscr{P}_c(f)$ and $\mathscr{P}(f)$. An entirely analogous comment can be made about $\mathscr{P}_c(\zeta_A)$ and $\mathscr{P}(\zeta_A)$ in Definition 2.1.67 and Definition 2.1.68.

Finally, recall from Definition 2.1.62 that we call $\{\mathrm{Re}\, s > D_{\mathrm{hol}}(f)\}$ the half-plane of holomorphic continuation of f. Indeed, it is the largest open right half-plane of the form $\{\mathrm{Re}\, s > \alpha\}$ (for some $\alpha \in \mathbb{R} \cup \{\pm\infty\}$) to which f can be holomorphically extended. See Figure 2.5 and also, in the important special case where f is initially given by a tamed Dirichlet-type integral, Figure 2.6.

We next define the equivalence of two meromororphic functions having the form of tamed Dirichlet-type integrals, a notion which will be very useful to us in the sequel; see Remark 2.1.70 below.

Definition 2.1.69. Let f and g be two tamed Dirichlet-type functions (i.e., integrals) f and g, having a (necessarily unique) meromorphic extension to a connected open neighborhood $U \subseteq \mathbb{C}$ of the closed right half-plane $\{\mathrm{Re}\, s \geq D(f)\}$.[31] Then, they are said to be *equivalent* if $D(f) = D(g)$ (and this common value is a real number), see Equation (2.1.55), and their sets of poles located on their common critical line (of convergence) $\{\mathrm{Re}\, s = D(f) = D(g)\}$ coincide (see Equation (2.1.96)). More succinctly, we have

$$f \sim g \quad \overset{\text{def.}}{\Longleftrightarrow} \quad D(f) = D(g)\ (\in \mathbb{R}) \quad \text{and} \quad \mathscr{P}_c(f) = \mathscr{P}_c(g). \tag{2.1.101}$$

Remark 2.1.70. We also refer to Definition A.5.1 in Appendix A for a slightly more general definition of equivalence relation, within the context of (tamed) DTIs of type I (in the sense of Definition A.4.5), as well as to Definition A.6.6 in Appendix A (and the text surrounding it, including Remark A.6.7) for a somewhat different (but analogous) definition of "asymptotic equivalence" which is no longer a true equivalence relation but presents the advantage of allowing the function g to be merely assumed to be meromorphic in a suitable domain of $\mathbb{C}$. The latter notion of asymptotic equivalence should also be useful as a practical tool in many applications of the theory developed in this book.

Remark 2.1.71. In Definition 2.1.69 above, the multiplicities of the poles should be taken into account when writing $\mathscr{P}_c(f) = \mathscr{P}_c(g)$. (The only known exception to this general statement is provided by Example 2.1.80 below, which is really corresponding to a "borderline case" of equivalence.) In other words, we must view here the set of principal poles $\mathscr{P}_c(f)$ as a multiset (i.e., as a set with multiplicities). Indeed, to our knowledge, all the other examples encountered in this book (apart from Example 2.1.80) and for which we use the symbol $\sim$ (in the present sense of the equivalence of zeta functions), the multiplicities of the principal poles are preserved.

[31] As follows from the complete definition, this closed half-plane is actually the closure of the common half-plane of convergence of f and g, given by $\Pi := \Pi(f) = \Pi(g)$.

If a tamed Dirichlet-type function f is given, the aim in this context is to find an equivalent function g, defined by a simpler expression. Examples of tamed Dirichlet-type functions g encountered in this context include functions of the form $g(s) = 1/P(s)$, where $P(s)$ is a given polynomial with complex coefficients, as well as $g(s) = 1/(e^{\omega s} - \rho)$, where ω and ρ are nonzero real numbers. (See Theorem A.3.2 and Corollary A.3.3 in Appendix A.) Satisfactory results can already be obtained with functions g of the form $g(s) = u(s)f(s) + v(s)$, for a suitable choice of the holomorphic functions u and v, with u nowhere vanishing in the given domain.

Proposition 2.1.72. *Assume that $u(s)$, $v(s)$ and $f(s)$ are meromorphic functions on the right half-plane $\{\mathrm{Re}\, s > 0\}$, with f a tamed Dirichlet-type integral. Furthermore, assume that $u(s)$ and $v(s)$ have no poles on the closed right half-plane $\{\mathrm{Re}\, s \geq D(f)\}$, and that $u(s)$ has no zeros on the critical line $\{\mathrm{Re}\, s = D(f)\}$. If $D(f) > 0$, then*

$$f(s) \sim u(s)f(s) + v(s).$$

In particular, the geometric zeta function $\zeta_{\mathscr{L}}$ of a bounded fractal string $\mathscr{L} = (\ell_j)_{j\geq 1}$ such that $D(\zeta_{\mathscr{L}}) > 0$ is equivalent to the distance zeta function ζ_A of the corresponding set $A = A_{\mathscr{L}} = \{a_k = \sum_{j\geq k} \ell_j : k \in \mathbb{N}\} \subseteq \mathbb{R}$; that is,

$$\zeta_{\mathscr{L}} \sim \zeta_A.$$

Hence, according to Definition 2.1.69, we have that

$$D(\zeta_{\mathscr{L}}) = D(\zeta_A) \quad \text{and} \quad \mathscr{P}_c(\zeta_{\mathscr{L}}) = \mathscr{P}_c(\zeta_A).$$

Proof. The first claim follows immediately from the definition. The second claim follows from the first one, combined with Equation (2.1.85), obtained in Example 2.1.58. □

Remark 2.1.73. In keeping with the discussion concluding Subsection 2.1.4 above, the exact analog of Proposition 2.1.72 holds if A is replaced by the relative fractal drum $(\partial\Omega, \Omega)$ in the sense of Definition in Chapter 4, where Ω is any geometric realization of the fractal string $\mathscr{L}$ as a bounded open subset of $\mathbb{R}$ and $\partial\Omega$ is the topological boundary of Ω, while $\zeta_{\partial\Omega,\Omega}$ is the corresponding relative zeta function defined in Subsection 4.1.1 below. It is clear that $\zeta_{A_{\mathscr{L}}}(s) = \zeta_{\partial\Omega,\Omega}(s) + 2s^{-1}\delta^s$, provided δ, appearing in (2.1.1), is such that $\delta \geq \ell_1/2$, so that $\overline{\dim}_B \mathscr{L} = \overline{\dim}_B(\partial\Omega, \Omega)$. Furthermore, the principal complex dimensions (and their corresponding multiplicities) of the functions $\zeta_{\mathscr{L}}$, $\zeta_{A_{\mathscr{L}}}$ and $\zeta_{\partial\Omega,\Omega}$ coincide.

Moreover, under the analog of the hypotheses (c) of Theorem 2.1.11, we can also state that $D_{\mathrm{hol}}(\zeta_{A,\Omega}) = D(\zeta_{A,\Omega})$ and hence, that the lower bound $\overline{\dim}_B A$ is also optimal from the point of view of holomorphic continuation: $\mathscr{H}(\zeta_{A,\Omega}) = \Pi(\zeta_{A,\Omega}) = \{\mathrm{Re}\, s > \overline{\dim}_B A\}$.

Example 2.1.74. It is easy to construct two bounded sets A and B which are not bi-Lipschitz equivalent, but are such that $\zeta_A \sim \zeta_B$. It suffices, for example, to consider the ternary Cantor set B on the real line. Let $\mathscr{L} = (\ell_j)_{j\geq 1}$ be the associated

Cantor string. Specifically, $B = \partial\mathscr{L}$ and $\mathscr{L}$ is the sequence of lengths 3^{-n} with multiplicities 2^{n-1}, for $n = 1,2,3,\ldots$; see [Lap-vFr3, Section 2.3.1]. Now define the set $A = \{a_k\}_{k\geq 1}$ by $a_k = \sum_{j\geq k} l_j$, for each $k \geq 1$. Then $\zeta_A(s) = \zeta_B(s)$. In particular, $\dim_B A = \dim_B B$. Note that $\dim_H B = \log_3 2$, while $\dim_H A = 0$, so that A and B are not bi-Lipschitz equivalent. (Here, we denote by $\dim_H A$ the Hausdorff dimension of A.) Similar examples of sets A and B generating the same fractal strings, but having different Hausdorff dimensions have been discussed in [Lap1, Example 5.1 and 5.1'], [Lap-vFr1–3], as well as revisited in [LapRo1], [LapLéRo] and [ElLapMacRo], where certain "multifractal zeta functions" are introduced in order to account, in particular, for such phenomena.

Definition 2.1.75. Let A be a bounded set in $\mathbb{R}^N$, and let Ω be a bounded open set such that $\overline{A} \subset \Omega$. We define $\zeta_{A,\Omega}$, the *distance zeta function of A relative to Ω* (or *relative distance zeta function*,[32] in short), by

$$\zeta_{A,\Omega}(s) := \int_\Omega d(x,A)^{s-N}\mathrm{d}x, \tag{2.1.102}$$

for all $s \in \mathbb{C}$ with $\operatorname{Re} s$ sufficiently large.

This zeta function is holomorphic in the half-plane $\{\operatorname{Re} s > \overline{\dim}_B A\}$ and the lower bound $\overline{\dim}_B A$ is optimal from the point of view of the convergence of the Lebesgue integral on the right-hand side of (2.1.102); that is, the abscissa of (absolute) convergence of $\zeta_{A,\Omega}$ coincides with $\overline{\dim}_B A$, and hence, $\Pi(\zeta_{A,\Omega}) = \{\operatorname{Re} s > \overline{\dim}_B A\}$ is the half-plane of (absolute) convergence of $\zeta_{A,\Omega}$; see Equations (2.1.55) and (2.1.56) and Figure 2.6. (We will discuss this issue in much greater generality in Subsection 4.1.1; see, especially, Theorem 4.1.7.) Indeed, the condition $\overline{A} \subseteq \Omega$ implies that there exists $\delta > 0$ such that $A_\delta \subseteq \Omega$. Now, we apply Lemma 2.1.15 to the part of the integral in (2.1.102) over the set $\Omega \setminus A_\delta$ and deduce that it is an entire function of s. Applying Theorem 2.1.11 to the part of the integral in (2.1.102) over the set A_δ, we reach the desired conclusion.

We still denote by $\zeta_{A,\Omega}$ its meromorphic extension (if it exists, that is, if A is admissible) to some open domain G containing a window $\boldsymbol{W}$.

If $\overline{A}$ and $\overline{\Omega}$ are disjoint, then $\zeta_{A,\Omega}$, defined by (2.1.102), is an entire function; see Lemma 2.1.15. The usual distance zeta function of A (see Definition 2.1.1) corresponds to the choice $\Omega = A_\delta$, for some $\delta > 0$: $\zeta_A(s) = \zeta_{A,A_\delta}(s)$.

Proposition 2.1.76. *Let A be a bounded set of $\mathbb{R}^N$, and let Ω_1, Ω_2 be bounded open sets in $\mathbb{R}^N$ containing $\overline{A}$. Then:*

(a) The difference $\zeta_{A,\Omega_1} - \zeta_{A,\Omega_2}$ can be extended to an entire function, and in particular, $\zeta_{A,\Omega_1} \sim \zeta_{A,\Omega_2}$. As a special case, if δ_1 and δ_2 are any two positive real numbers, then $\zeta_{A,A_{\delta_1}} - \zeta_{A,A_{\delta_2}}$ can be identified with an entire function, and in particular, $\zeta_{A,A_{\delta_1}} \sim \zeta_{A,A_{\delta_2}}$.

[32] The notion of relative distance zeta function will be further extended in Section 4.1, where we drop the condition $\overline{A} \subset \Omega$ and no longer assume that A and Ω are bounded; see Definition 4.1.1.

(*b*) *Let* δ_1 *and* δ_2 *be any two positive real nymbers. If a complex number* s_0 *is a simple pole of* $\zeta_{A,A_{\delta_1}}$, *then it is a simple pole of* $\zeta_{A,A_{\delta_2}}$ *as well and the residues of these two functions computed at* s_0 *coincide*:

$$\operatorname{res}(\zeta_{A,A_{\delta_1}}, s_0) = \operatorname{res}(\zeta_{A,A_{\delta_2}}, s_0). \tag{2.1.103}$$

Proof. (*a*) For $\operatorname{Re} s > \overline{\dim}_B A$, the difference of the functions $\zeta_{A,\Omega_1}(s)$ and $\zeta_{A,\Omega_2}(s)$ is equal to

$$\int_{\Omega_1 \setminus (\Omega_1 \cap \Omega_2)} d(x,A)^{s-N} \mathrm{d}x - \int_{\Omega_2 \setminus (\Omega_1 \cap \Omega_2)} d(x,A)^{s-N} \mathrm{d}x. \tag{2.1.104}$$

In light of the inclusion $\overline{A} \subset \Omega_1 \cap \Omega_2$, the set $\overline{A}$ is disjoint from $\Omega_j \setminus (\Omega_1 \cap \Omega_2)$, for $j = 1, 2$. Therefore, both integrals in (2.1.104) are entire functions; see Lemma 2.1.15.

(*b*) The claim follows immediately from part (*a*):

$$\begin{aligned}
\operatorname{res}(\zeta_{A,A_{\delta_1}}, s_0) &= \operatorname{res}(\zeta_{A,A_{\delta_2}}, s_0) + \operatorname{res}\left(\zeta_{A,A_{\delta_1}} - \zeta_{A,A_{\delta_2}}, s_0\right) \\
&= \lim_{s \to s_0} (s - s_0)\zeta_{A,A_{\delta_2}}(s) + \lim_{s \to s_0} (s - s_0)\left(\zeta_{A,A_{\delta_2}}(s) - \zeta_{A,A_{\delta_2}}(s)\right) \\
&= \operatorname{res}(\zeta_{A,A_{\delta_2}}, s_0).
\end{aligned}$$

This concludes the proof of the proposition. □

The following result deals with the scaling property of the distance zeta function. Here, as usual, we write $\zeta_{A,A_\delta}(s) := \int_{A_\delta} d(x,A)^{s-N} \mathrm{d}x$, for $\operatorname{Re} s > \overline{\dim}_B A$.

Proposition 2.1.77 (Scaling property of distance zeta functions)**.** *For any bounded subset A of* $\mathbb{R}^N$, $\delta > 0$ *and* $\lambda > 0$, *we have* $D(\zeta_{\lambda A, \lambda(A_\delta)}) = D(\zeta_{A,A_\delta}) = \overline{\dim}_B A$ *and*

$$\zeta_{\lambda A, \lambda(A_\delta)}(s) = \lambda^s \zeta_{A,A_\delta}(s), \tag{2.1.105}$$

for all $s \in \mathbb{C}$ *with* $\operatorname{Re} s > \overline{\dim}_B A$. *Furthermore, if* $\omega \in \mathbb{C}$ *is a simple pole of the meromorphic extension of* $\zeta_{A,A_\delta}(s)$ *to some connected open neighborhood of the critical line* $\{\operatorname{Re} s = \overline{\dim}_B A\}$ (*as usual, we use the same notation for the meromophically extended function*), *then*

$$\operatorname{res}(\zeta_{\lambda A, \lambda(A_\delta)}, \omega) = \lambda^\omega \operatorname{res}(\zeta_{A,A_\delta}, \omega). \tag{2.1.106}$$

Proof. Equation (2.1.105) follows easily by noting that $\lambda(A_\delta) = (\lambda A)_{\lambda\delta}$; we leave the details to the interested reader. To prove Equation (2.1.106), note that

$$\begin{aligned}
\operatorname{res}(\zeta_{\lambda A, \lambda(A_\delta)}, \omega) &= \lim_{s \to \omega} (s - \omega)\zeta_{\lambda A, \lambda(A_\delta)}(s) \\
&= \lim_{s \to \omega} (s - \omega)\lambda^s \zeta_{A,A_\delta}(s) = \lambda^\omega \operatorname{res}(\zeta_A, \omega),
\end{aligned}$$

as desired. □

An extension of Proposition 2.1.77, formulated in the context of relative fractal drums, can be found in Theorem 4.1.40 and Corollary 4.1.42.

The relative distance zeta function $\zeta_{A,\Omega}$ has a suitable continuity property with respect to any nonincreasing sequence of compact sets, as we now explain.

Theorem 2.1.78. *Let Ω be a bounded open set, $(A_k)_{k\geq 1}$ a nonincreasing sequence of compact subsets of Ω such that $A_k \downarrow A$ as $k \to \infty$; that is, $A_k \supseteq A_{k+1}$ for all k and $A = \cap_{k\geq 1} A_k$. Then*

$$\zeta_{A_k,\Omega}(s) \to \zeta_{A,\Omega}(s) \quad \text{as} \quad k \to \infty, \tag{2.1.107}$$

pointwise for $\mathrm{Re}\, s > N$.

Proof. For the sake of brevity and only in this proof, we write $\zeta_A(s) = \zeta_{A,\Omega}(s)$. Let us fix any $s \in \mathbb{C}$ such that $\mathrm{Re}\, s > N$. It suffices to show that for any subsequence (k') of (k), there exists a further subsequence (k'') of (k') such that $\zeta_{A_{k''}}(s) \to \zeta_A(s)$ as $k'' \to \infty$.

In order to prove this, we use the Lebesgue dominated convergence theorem. Let us fix $x \in \Omega$. Consider a subsequence (k') of (k). By the compactness assumption, for each k', there exists $a_{k'} = a_{k'}(x) \in A_{k'}$ such that $d(x, A_{k'}) = d(x, a_{k'})$. The sequence $(a_{k'})$ is bounded; therefore, there exists a convergent subsequence $(a_{k''})$ converging to some $a \in A_1$. It is easy to see that, in fact, $a \in A$. Indeed, defining $B_k = \{a_j : j \geq k\}$, we have $B_k \subset A_k$ and from this, it follows that $a \in \overline{B}_k \subseteq A_k$ for each k. Hence, $a \in \cap_k \overline{B}_k \subseteq \cap_k A_k = A$ (clearly, $\cap_k \overline{B}_k = \{a\}$). Therefore,

$$|d(x, A_{k''}) - d(x, A)| = |d(x, a_{k''}) - d(x, a)| \leq |a_{k''} - a| \to 0, \tag{2.1.108}$$

as $k \to \infty$, and hence $d(x, A_{k''})^{s-N} \to d(x, A)^{s-N}$ for $\mathrm{Re}\, s > N$ pointwise on Ω. The sequence of complex-valued functions $d(x, A_{k''})^{s-N}$ is dominated as follows:

$$|d(x, A_{k''})^{s-N}| = d(x, A_{k''})^{\mathrm{Re}\, s-N} \leq d(x, A)^{\mathrm{Re}\, s-N} =: F(x), \tag{2.1.109}$$

where the last inequality results from the inclusion $A \subseteq A_{k''}$. Here, $F \in L^1(\Omega)$ because F is continuous on $\overline{\Omega}$ since $\mathrm{Re}\, s > N$. Therefore, $\zeta_{A_{k''}}(s) \to \zeta_A(s)$, as desired. □

In Subsection 4.1.1, in connection with so-called "relative fractal drums" (A, Ω), we will study the relative distance zeta function $\zeta_{A,\Omega}$ under more general assumptions on A and Ω than those of Definition 2.1.75. In particular, Ω (and A) will be allowed to be unbounded. As will be mentioned in Remark 4.1.9, Theorem 2.1.78 still holds (with essentially the same proof) in the general context of Section 4.1.1.

We now extend the notion of the zeta function of a fractal set A, as follows.

Definition 2.1.79. Let A be a bounded subset of $\mathbb{R}^N$. Assume that A is admissible with associated window $\boldsymbol{W}$. The *zeta function* ζ_A is defined as the family of meromorphic and tamed Dirichlet-type functions $f: G \to \mathbb{C}$ (where $G \subseteq \mathbb{C}$ is a connected open set containing $\boldsymbol{W}$), such that $D(f) = D(\zeta_{A,A_\delta})$ and $\mathscr{P}(f) = \mathscr{P}(\zeta_{A,A_\delta})$ for some $\delta > 0$. Each member of the family will be denoted by ζ_A, in short. In other words, we identify the distance zeta function ζ_A (initially defined by (2.1.1)) with f if $\zeta_A \sim f$ in the sense of Definition 2.1.69, and the zeta function can be viewed as the equivalence class $\zeta_A = [\zeta_A] = \{f : f \sim \zeta_A\}$ with respect to the relation $\sim$ defined on the set of all such meromorphic and tamed Dirichlet-type functions.

The aim is to find a representative $f \in [\zeta_A]$ of the zeta function of a given set A which is as simple as possible.

In Definition 2.1.79, one could replace the notion of equivalence introduced in Definition 2.1.69 by the more general and flexible notion of asymptotic equivalence given in Definition A.6.6 of Appendix A.

The definitions and results about equivalent zeta functions given here are in the spirit of (but not identical to) the corresponding ones obtained towards the beginning of Chapter 5 of [Lap6] (see, especially, [Lap6, Sections 5.2 and 5.3]), where various notions of equivalences of fractal strings (and of associated zeta functions) are introduced in order to define and study the corresponding moduli spaces of fractal strings (and fractal membranes).

Example 2.1.80. It is easy to check that the derivative ζ_A' is holomorphic precisely where the function ζ_A is. If we assume that a meromorphic extension of ζ_A to a domain $G \subseteq \mathbb{C}$ containing the abscissa of convergence $D(\zeta_A)$ exists,[33] then the meromorphic extensions of ζ_A and ζ_A' to G have identical sets of poles (i.e., $\mathscr{P}(\zeta_a) = \mathscr{P}(\zeta_A')$): if ζ_A has a pole at $\omega \in G$, then ζ_A' has a pole at the same point, and furthermore, if ζ_A is holomorphic at $s \in G$, so is ζ_A'. Hence, for the fractal string $\mathscr{L} = (\ell_j)_{j=1}^\infty$ in Example 2.1.58, we have $\zeta_\mathscr{L} \sim \zeta_\mathscr{L}^{(m)}$, where $\zeta_\mathscr{L}^{(m)}$ is the m-th derivative of $\zeta_\mathscr{L}$ for any fixed positive integer m, and we conclude that $\sum_{j\geq 1} \ell_j^s \sim \sum_{j\geq 1} \ell_j^s (\log \ell_j)^m$.

Note, however, that the orders of the poles are not the same: indeed, if ω is a pole of ζ_A of order n, then ω is a pole of ζ_A' of order $n+1$. Therefore, but only for this "borderline" example of use of Definition 2.1.69 (see Remark 2.1.71), we do not view here the set of poles $\mathscr{P}(\zeta_A)$ as a multiset, that is, as a set with multiplicities.

More generally, for any positive integer m, we have that

$$\int_{A_\delta} d(x,A)^{s-N} \mathrm{d}x \sim \int_{A_\delta} d(x,A)^{s-N} (\log d(x,A))^m \mathrm{d}x, \tag{2.1.110}$$

where the integrals are viewed as functions of a complex variable s, meromorphically extendable to a domain $G \subseteq \mathbb{C}$ containing their abscissae of convergence. The last integral is a special case of a weighted zeta function, which we shall study in Section 3.4.

[33] Note that this is essential in order to avoid situations in which ζ_A could have a singularity that is not a pole. A simple example is $f(s) = \log s$ and $f'(s) = 1/s$.

The following simple lemma will often be used without explicit mention. Its proof is easy, and we omit it.

Lemma 2.1.81. *Assume that* $h(s) = f(s)g(s)$, *where* g *and* h *are meromorphic functions defined on a domain containing* $\{\mathrm{Re}\, s > \alpha\}$ *for some real number* α, *and* f *has no zeros in that domain. Then* $D(h) = \max\{D(f), D(g)\}$. *If, in addition,* f *is holomorphic, then* $D(h) = D(g)$ *and* $\mathscr{P}(h) = \mathscr{P}(g)$.[34] *In particular, if* $h(s) = (s-c)^{-1}g(s)$ *for some real number* c, *then*

$$D(f) = \max\{c, D(g)\}. \tag{2.1.111}$$

Example 2.1.82. (*Ternary Cantor set*). Let $C^{(1/3)}$ be the standard ternary Cantor set in $[0,1]$. In [Žu3, Example 7], it is shown that for any $\gamma < 1 - \log_3 2$, the following Lebesgue integral can be explicitly computed, using a simple summation of the corresponding integrals over the open intervals I_k, $k \geq 1$, defined by $[0,1] \setminus C^{(1/3)} = \bigcup_{k \geq 1} I_k$:

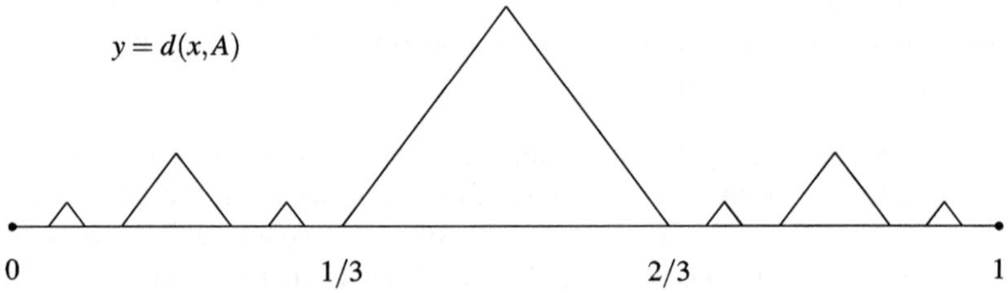

Fig. 2.8 The graph of the distance function $x \mapsto d(x,A)$, where A is the ternary Cantor set. Only the first three generations of the countable family of tents are shown here.

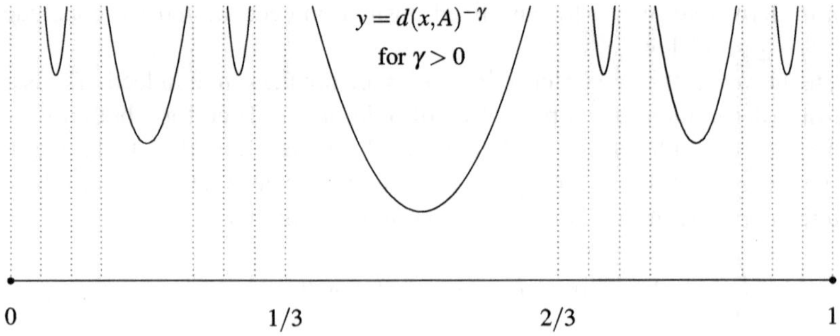

Fig. 2.9 For the ternary Cantor set $A = C^{(1/3)}$, the function $y = d(x,A)^{-\gamma}$, $x \in \Omega := (0,1)$, is Lebesgue integrable if and only if $\gamma < 1 - \log_3 2$. For $\gamma > 0$, its graph has countably many connected components (all of which are unbounded) and uncountably many vertical asymptotes. For any $\gamma < 1 - \log_3 2$, the area of the set $\{(x,y) \in \Omega \times \mathbb{R} : 0 < y < d(x,A)^{-\gamma}\}$ is equal to $\zeta_A(1-\gamma, \Omega)$.

---

[34] Observe that in light of Definition 2.1.69, this implies that $h \sim g$, but is also a significantly stronger statement.

$$\int_0^1 d(x, C^{(1/3)})^{-\gamma} \mathrm{d}x = \frac{2 \cdot 6^{\gamma-1}}{(1-\gamma)(1-2\cdot 3^{\gamma-1})}. \tag{2.1.112}$$

See the graphs of the functions $x \mapsto d(x, C^{(1/3)})$ and $y = d(x, C^{(1/3)})^{-\gamma}$ in Figures 2.8 and 2.9 on page 104. Using the same procedure and the same formula with $\gamma = 1 - s \in \mathbb{C}$, where $\operatorname{Re} s > \log_3 2$, we obtain successively (see (2.1.1)):

$$\begin{aligned}
\zeta_{C^{(1/3)}}(s) &= \int_0^1 d(x, C^{(1/3)})^{s-1} \mathrm{d}x + \int_{-\delta}^0 |x|^{s-1} \mathrm{d}x + \int_1^{1+\delta} (x-1)^{s-1}\, \mathrm{d}x \\
&= \int_0^1 d(x, C^{(1/3)})^{s-1} \mathrm{d}x + 2\int_0^{\delta} x^{s-1}\, \mathrm{d}x \qquad (2.1.113)\\
&\sim \int_0^1 d(x, C^{(1/3)})^{s-1} \mathrm{d}x = 2\sum_{k=1}^{\infty} 2^{k-1} \int_0^{\frac{1}{2}3^{-k}} x^{s-1} \mathrm{d}x \\
&= \frac{2\cdot 6^{-s}}{s(1-2\cdot 3^{-s})} \sim \frac{1}{1-2\cdot 3^{-s}},
\end{aligned}$$

which is equivalent to the geometric zeta function of the Cantor string ζ_{CS} obtained in [Lap-vFr3, p. 22]. In particular, we recover the well-known fact (see [Lap-vFr3, Section 2.3.1]) according to which the set of principal complex dimensions of the Cantor set is given by

$$\dim_{PC} C^{(1/3)} = \{\log_3 2 + k\mathbf{p}\mathbb{i} : k \in \mathbb{Z}\} =: \log_3 2 + \mathbf{p}\mathbb{i}\mathbb{Z}, \tag{2.1.114}$$

where $\mathbf{p} = 2\pi/\log 3$ is the *oscillatory period* of the Cantor set, using the notation and terminology from [Lap-vFr3, pp. 22-23]. Recall our convention according to which $\mathbb{i} := \sqrt{-1}$ denotes the imaginary unit; see Equation (1.3.21) on page 40 above. From Corollary 2.1.63, we deduce another well-known fact: $\dim_B A = \log_3 2$.

For the generalized Cantor set $C^{(a)}$ with $a \in (0, 1/2)$, defined in Example 2.2.6,[35] we obtain in a similar way

$$\zeta_{C^{(a)}}(s) \sim \frac{1}{1-2a^s}. \tag{2.1.115}$$

(See [Žu3, Example 7].) Hence, we recover the fact that (see [Lap-vFr3, p. 284]):

$$\dim_{PC} C^{(a)} = \log_{1/a} 2 + \mathbf{p}\mathbb{i}\mathbb{Z}, \tag{2.1.116}$$

where $\mathbf{p} = 2\pi/\log(1/a)$ is the oscillatory period of $C^{(a)}$, and $d = \dim_B C^{(a)} = \log_{1/a} 2$. An extension of this fact to a two-parameter family of Cantor sets is obtained in Proposition 3.1.2.

[35] Generalized Cantor sets and their oscillations are studied in detail in [Lap-vFr3, Chapter 10].

2.1.6 Stalactites, Stalagmites and Caves Associated with Fractal Sets and Fractal Strings

Let A be a given bounded subset of $\mathbb{R}^N$, and let δ be a fixed positive real number. The set $A_\delta \setminus \overline{A}$ can be represented as a disjoint union of its connected components $U_k = U_k(A,\delta)$, $k \in J$; that is,

$$A_\delta \setminus \overline{A} = \bigcup_{k\in J} U_k, \tag{2.1.117}$$

where the index set J is at most countable. Note that the topological structure of the set $A_\delta \setminus \overline{A}$ can vary from being very simple to extremely complex, depending on the choice of the fractal set A.[36] On the other hand, if A is the von Koch curve, then $A_\delta \setminus \overline{A}$ has a very simple topological structure (it is connected), while its metric properties near A are complicated.

For any given nonzero real number r, we consider the function

$$f : A_\delta \to [0,+\infty], \quad f(x) := d(x,A)^r. \tag{2.1.118}$$

(If $r < 0$, we let $0^r := +\infty$.) For each $k \in J$, let f_k be the restriction of f to the connected component U_k of the open set $A_\delta \setminus \overline{A}$; i.e., $f_k := f|_{U_k}$. Note that $f(x) \in (0,+\infty)$ for each $x \in U_k$ and $k \in \mathbb{N}$.

Definition 2.1.83. Assume that $r \neq 0$. Then, for each $k \in J$, the graph of the function f_k, viewed as a subset of $U_k \times (0,+\infty) \subset \mathbb{R}^{N+1}$,[37] is called the k-th *stalactite* associated with the fractal set A (and with parameters r and δ).

Furthermore, the set $\operatorname{cave}(A) = \operatorname{cave}(A,\delta,r)$, defined by

$$\operatorname{cave}(A) := \{(x,u) \in A_\delta \times (0,+\infty) : 0 < u < f(x)\} \tag{2.1.119}$$

and viewed as a subset of $A_\delta \times (0,+\infty) \subset \mathbb{R}^{N+1}$, is called the A-*cave* (corresponding to the choice of r and $\delta > 0$). In other words, the A-cave is the open subset of $A_\delta \times (0,+\infty)$ located (strictly) between the graph of f and the x-hyperplane.

It is also convenient to define the *epigraph of* f as the closed[38] subset of the Cartesian product $A_\delta \times [0,+\infty]$ lying either above the graph of f or on it. More precisely,

$$\operatorname{epi}(f) := \{(x,u) \in A_\delta \times [0,+\infty] : f(x) \le u\}. \tag{2.1.120}$$

Note that for $r < 0$, each stalactite is unbounded since $f(x) \to +\infty$ as $x \in U_k$ and $x \to \partial U_k$.

---

[36] If $A = \{a\}$, that is, if A is a one-point set, then the set $A_\delta \setminus \overline{A} = B_\delta(a) \setminus \{a\}$ is connected for $N \ge 2$, while if A is the Cantor set contained in the real line, then $A_\delta \setminus \overline{A}$ has countably many connected components.

[37] Note that $U_k \subseteq A_\delta \setminus \overline{A}$, so that $x \mapsto d(x,A)$ does not vanish on U_k; hence, $f_k(x) < \infty$ for all $x \in U_k$, even if $r < 0$.

[38] The epigraph of f is a *closed* subset of $A_\delta \times [0,+\infty]$ with respect to the relative topology in $A_\delta \times [0,+\infty]$.

If $r > 0$, then the union of all of the stalactites is a bounded subset of $\mathbb{R}^{N+1}$. Furthermore, note that $f(x) \to 0$ as $x \to \partial U_k$. If we view the graph of f from any point in its epigraph, then the stalactites look like stalagmites. Therefore, in the case when $r > 0$, we shall call them *stalagmites* instead. Note that, according to our terminology, stalagmites are a special case of stalactites; namely, stalagmites are bounded stalactites.

If A is the Sierpiński carpet, then Figures 2.10 and 2.11 illustrate the corresponding stalagmites, i.e., the graphs of $y = d(x,A)^r$ for $r > 0$. For $r = 1$, see Figure 2.3 on page 51.

In the following proposition, we collect some basic properties of stalactites, stalagmites and caves associated with fractal sets. Its proof is easy, and we leave it as an exercise for the interested reader. It is instructive to illustrate its content in the case of the Sierpiński carpet; see Figures 2.3 and 2.4 (on pages 51 and 52) corresponding to the case when r is negative, along with Figures 2.2, 2.10 and 2.11 (respectively on pages 50, 107 and 108) corresponding to the case when r is positive.[39]

Proposition 2.1.84. *Let A be a bounded set in $\mathbb{R}^N$. Assume that δ is a fixed positive real number and let $r \neq 0$. Then:*

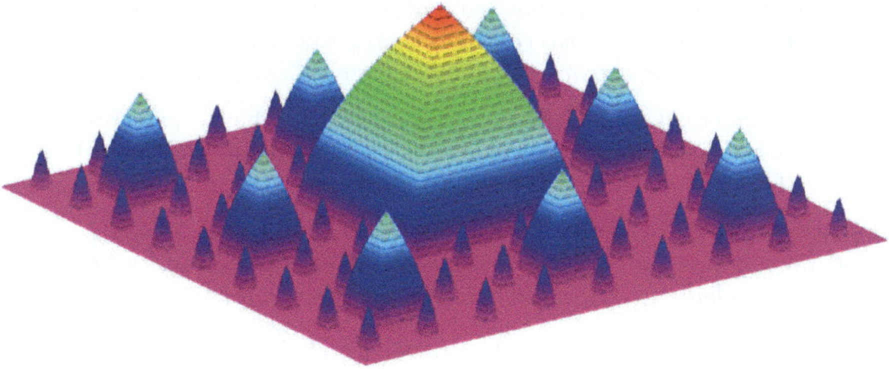

Fig. 2.10 Sierpiński stalagmites in the *Hölder case*, i.e., when $r \in (0,1)$. Here, $r = 0.5$.

(a) If $r < 0$, then there is a natural bijection between the family of connected components of A_δ and the family of connected components of the A-cave. In particular, if the set A_δ is connected, then the same property holds for the A-cave.

Furthermore, there is a natural bijection between the family of connected components of the set $A_\delta \setminus \overline{A}$ and the family of stalactites (corresponding to A, δ and r).

[39] We plan to create a virtual gallery of fractal caves generated by various planar fractal sets A and for different values of $r \neq 0$. The notion of a 'fractal cave' already exists in mathematical geology, at least since 1986, motivated in part by the earlier work of Mandelbrot [Man1]; see its definition in [Cur] on page 168. That definition, based on the idea of the so-called 'cave modulus', is completely different from ours.

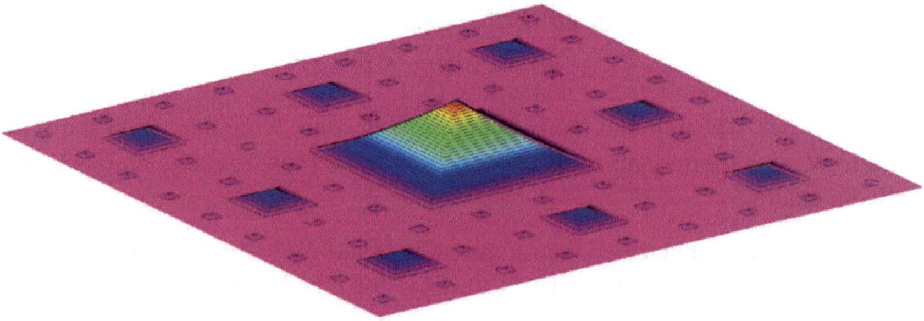

Fig. 2.11 Sierpiński stalagmites in the *Lipschitz case*, i.e., when $r \geq 1$. Here, $r = 1.3$.

In particular, if the set $A_\delta \setminus \overline{A}$ *has countably many connected components, then there are countably many stalactites. All of the stalactites are unbounded.*

(*b*) *If* $r > 0$, *then there is a natural bijection between the family of connected components of* A_δ *and the family of connected components of the interior of the epigraph of the function* f *defined by* (2.1.120).

Furthermore, there is a natural bijection between the family of connected components of the set $A_\delta \setminus \overline{A}$ *and the family of stalagmites* (*corresponding to* A, δ *and* r). *In particular, if the set* $A_\delta \setminus \overline{A}$ *has countably many connected components, then there are countably many stalagmites. The union of these stalagmites is bounded.*

Next, let us formulate an interesting geometric result, which is just a restatement of the corresponding one appearing in Lemma 2.1.3, and due to Harvey and Polking. Note that for the exponents $r \geq 0$, the claim is trivial.

Proposition 2.1.85. *Assume that* A *is any bounded fractal set in* $\mathbb{R}^N$. *Let us fix* $\delta > 0$. *If* $r > \overline{\dim}_B A - N$, *then the corresponding* A*-cave in* $A_\delta \times (0, +\infty)$ *is of finite volume (i.e., of finite* $(N+1)$*-dimensional Lebesgue measure).*[40]

Example 2.1.86. If A is the Sierpiński carpet in the plane, then for $r = 1$ (and $\delta > 1/6$) the corresponding countably infinite family of stalagmites can be found in Figure 2.2 on page 50. On the other hand, for $r < 0$, the corresponding countably infinite family of stalactites, along with the associated *Sierpiński cave*, can be found in Figure 2.3 on page 51, and its total volume is finite if and only if $r > \log_3 8 - 2$.

If A is the ternary Cantor set, then for $r = 1$ (and $\delta > 1/6$) the corresponding countably infinite family of stalagmites can be found in Figure 2.8 on page 104, while for $r < 0$ the corresponding stalactites, along with the associated *Cantor cave*, can be found in Figure 2.9 on page 104. The Cantor cave is of finite volume if and only if $r > \log_3 2 - 1$.

[40] If A is such that there exists $D := \dim_B A$ and $\mathscr{M}_*^D(A) > 0$, then the converse implication holds as well; see [Žu5, Theorem 4.1].

Remark 2.1.87. It is easy to check that if A is the ternary Cantor set, then the *graph* of the function $y = d(x,A)^r$, $x \in [0,1]$, where r is a nonzero real number, is an *inhomogeneous self-affine set* (for the precise definition, see Fraser [Fra2], along with the references cited therein, including [BarDemk]), with the affinity matrix equal to $\mathrm{diag}(1/3, 1/3^r)$; see Figures 2.8 and 2.9. Let us provide a very brief description of the notion of inhomogeneous self-affine set in this case. Recall that the ternary Cantor set A is a *homogeneous self-similar* set, in the following sense:

$$A = S(A),$$

where $S(A) := S_1(A) \cup S_2(A)$ and the similarity transformations $\{S_i\}_{i=1}^2$ of $\mathbb{R}$ are defined by $S_1(x) = \frac{1}{3}x$ and $S_2(x) = \frac{1}{3}x + \frac{2}{3}$, $x \in [0,1]$. If we denote by G the graph of the function $y = d(x,A)^r$, $x \in [0,1]$, and by G_0 its subset corresponding to $x \in [1/3, 2/3]$ (that is, to the middle interval), then G satisfies the following *inhomogeneous* fixed point equation:

$$G = \tilde{S}(G) \cup G_0. \tag{2.1.121}$$

Here, $\tilde{S}(G) := \tilde{S}_1(G) \cup \tilde{S}_2(G)$ and $\tilde{S}_i(x,y) := (S_i(x), \frac{1}{3^r}y)$, for $i = 1,2$.

In the case when A is the Sierpiński carpet, the function $y = d(x,A)^r$, $x \in [0,1]^2$, where r is a nonzero real number, also has an inhomogeneous self-affine graph (see Figures 2.2, 2.3 and 2.10), with the corresponding affinity matrix equal to $\mathrm{diag}(1/3, 1/3, 1/3^r)$. More precisely, the Sierpiński carpet A satisfies the homogeneous fixed point equation $A = S(A)$, where $S(A) = S_1(A) \cup \cdots \cup S_8(A)$ and for $i = 1, \ldots, 8$, S_i is the similarity transformation of $\mathbb{R}^2$ defined by $S_i(x) = \frac{1}{3}x + a_i$, for $x \in [0,1]^2$, while $a_i \in \mathbb{R}^2$ is a suitable translation vector. Let $\tilde{S} := \tilde{S}_1 \cup \cdots \cup \tilde{S}_8$, where $\tilde{S}_i(x,y) := (S_i(x), \frac{1}{3^r}y)$, and let G be the graph of $y = d(x,A)^r$, $x \in [0,1]^2$; also, denote by G_0 the subset of the graph corresponding to $x \in [1/3, 2/3]^2$ (that is, to the middle square). Then G satisfies the *inhomogeneous* fixed point equation $G = \tilde{S}(G) \cup G_0$.

For $r = 1$, in both cases we obtain functions which have *inhomogeneous self-similar graphs. Inhomogeneous self-similar sets* have been introduced by Barnsley and Demko in [BarDemk]; see also [Bar] and [Fra1–2, BakFraMa]. Each inhomogeneous self-similar set is obviously an inhomogeneous self-affine set.

Example 2.1.88. Let A be the von Koch curve in the plane. Let δ be a fixed positive number. For $r < 0$, the corresponding *von Koch cave* is an unbounded connected set. According to Proposition 2.1.85, since $\dim_B A = \log_3 4$, we conclude that von Koch's cave is of finite volume if and only if $r > 1 - \log_3 4$.

Note that the topology of the stalactites and the stalagmites depends on the choice of the parameter δ. This can be already seen in the case when A is just a two-point set. If A is the Sierpiński carpet, then the content of the following proposition is illustrated in Figure 2.2 (on page 50) in case (a), and in Figure 2.4 (on page 52) in case (b). Recall that the *Hausdorff distance* between two (possibly unbounded) subsets A and B of $\mathbb{R}^N$ is defined as the infimum of all $\varepsilon > 0$ such that $A \subseteq B_\varepsilon$ and $B \subseteq A_\varepsilon$. We denote it by $d_H(A,B)$.

Proposition 2.1.89. *Let A be a bounded set in $\mathbb{R}^N$. Given $M \geq 0$, let $f^{-1}(M)$ be the M-level set of the function $f : A_\delta \to [0,+\infty]$, defined by $f(x) := d(x,A)^r$, where r is a nonzero real number. Then the following properties hold:*

(a) Assume that $r > 0$. Then, as $M \to 0^+$, the M-level set $f^{-1}(M)$ tends to the boundary ∂A in the Hausdorff metric; that is,

$$\lim_{M\to 0^+} d_H(f^{-1}(M), \partial A) = 0.$$

(b) Assume that $r < 0$. Then, as $M \to +\infty$, the M-level set $f^{-1}(M)$ tends to the boundary ∂A in the Hausdorff metric; that is,

$$\lim_{M\to +\infty} d_H(f^{-1}(M), \partial A) = 0.$$

Proof. Let first us consider the case when $r = 1$. Let ε be a given (arbitrarily small) positive real number. If we choose M such that $M \in (0,\varepsilon)$, then we have that $f^{-1}(M) \subset (\partial A)_\varepsilon$ and $\partial A \subset (f^{-1}(M))_\varepsilon$. Hence, $d_H(f^{-1}(M), \partial A) < \varepsilon$; that is, $\lim_{M\to 0^+} d_H(f^{-1}(M), \partial A) = 0$.

The case when $r \neq 0$ (either for $r > 0$ or $r < 0$, respectively) is easily reduced to the case when $r = 1$. □

It is clear that, for any nontrivial bounded fractal string $\mathscr{L} = (\ell_j)_{j\in\mathbb{N}}$, it is possible to introduce analogous definitions. More specifically, let $A := A_{\mathscr{L}} = \{a_k\}_{k\in\mathbb{N}}$ be the bounded subset of $\mathbb{R}$ defined by $a_k := \sum_{j\geq k} \ell_j$, and let $U_k := (a_{k+1}, a_k)$ for any $k \in \mathbb{N}$; see Figure 2.7 on page 90. Much as was done above, for a given nonzero real number r, we can introduce the function $f : (0,a_1) \to [0,+\infty]$ defined by $f(x) = d(x, A_{\mathscr{L}})^r$. (As before, we let $0^r = +\infty$ if $r < 0$.) For any $k \geq 1$, the *k-th stalactite* associated with the bounded fractal string $\mathscr{L}$ is the graph of the function $f_k := f|_{U_k}$, the restriction of f to the k-th interval $U_k = (a_{k+1}, a_k)$. Furthermore, the *fractal cave* associated with the fractal string $\mathscr{L}$ (and with r) is defined as the subset of $(0,a_1) \times (0,+\infty) \subset \mathbb{R}^2$ given by

$$\operatorname{cave}\mathscr{L} := \{(x,u) \in (0,a_1) \times (0,+\infty) : 0 < u < f(x)\}. \tag{2.1.122}$$

For $r < 0$, the corresponding fractal cave is always simply connected, while for $r > 0$ it has countably many connected components.

Since (for $\operatorname{Re} s > \overline{\dim}_B A$) the distance zeta function $\zeta_A(s) := \int_{A_\delta} d(x,A)^{s-N} \mathrm{d}x$ is defined via the function $g : A_\delta \to \mathbb{C}$ given by $g(x) := d(x,A)^{s-N}$, it is of interest to know the geometry of the graph of g for various values of the complex number s.

We first consider the case when s is a real number. We then have the following two possibilities:

(a) If $s \in (\overline{\dim}_B A, N)$ (here, we assume that $\overline{\dim}_B A < N$), then the corresponding value $r := s - N$ is negative, and the graph of the function g consists of at most countably many stalactites. Furthermore, according to Proposition 2.1.85, the volume of the associated A-cave is finite for the indicated values of s.

(*b*) If $s > N$, then the graph of the function g consists of at most countably many stalagmites.

The case when $s \in \mathbb{C} \setminus \mathbb{R}$ is considered in the following subsection.

2.1.7 Oscillatory Nature of the Function $x \mapsto d(x,A)^{s-N}$

Let A be a given subset of $\mathbb{R}^N$ and let $\delta > 0$. Let s be a fixed nonreal complex number; that is, $s := \xi + \eta \mathrm{i}$, with $\xi, \eta \in \mathbb{R}$ and $\eta \neq 0$. Since the distance zeta function ζ_A is defined via the function $g : A_\delta \setminus \overline{A} \to \mathbb{C}$, where $g(x) := d(x,A)^{s-N}$, it is of interest to consider some basic properties of this complex valued-function. To this end, let us fix $t \in (0, \delta)$. Note that since $\partial(A_t) = \{x \in \mathbb{R}^N : d(x,A) = t\}$, then

$$A_\delta \setminus \overline{A} = \{x \in \mathbb{R}^N : d(x,A) \in (0,\delta)\} = \bigcup_{t \in (0,\delta)} \partial(A_t),$$

where the union is disjoint. The set $\partial(A_t)$ can be viewed as a *t-shell around* A. Since the function g is constant on $\partial(A_t)$, and equal to t^{s-N}, it is natural to study the behavior of the function

$$h : (0,\delta) \to \mathbb{C}, \quad h(t) := t^{s-N} \tag{2.1.123}$$

as $t \to 0^+$. In other words, we are interested in the behavior of the function g when the shells 'tend' to the set A. Since

$$h(t) = t^{\xi - N} \mathrm{e}^{\mathrm{i}\eta \log t}$$

and $\eta \neq 0$, we see that the function h is *oscillatory*, in the sense that its range in $\mathbb{C}$ is the curve defined in polar coordinates (r, θ) by $r = t^{\xi - N}$, $\theta = \eta \log t$, for $t \in (0, \delta)$. It is easy to see (by eliminating the parameter t) that the corresponding curve in the complex plane is of the form $r = \exp(\frac{\xi - N}{\eta}\theta)$; that is,

$$r = \mathrm{e}^{\frac{\mathrm{Re}\, s - N}{\mathrm{Im}\, s}\theta}, \quad \theta \in (-\infty, (\mathrm{Im}\, s) \log \delta). \tag{2.1.124}$$

Let us assume that $\eta = \mathrm{Im}\, s > 0$. We have the following three possibilites:

(*a*) if $\mathrm{Re}\, s < N$, then the curve described by (2.1.124) is the exponential spiral tending to $+\infty$ as $t \to 0^+$ (that is, $\theta \to -\infty$);[41]

(*b*) if $\mathrm{Re}\, s = N$, then the curve described by (2.1.124) is the circle of radius 1;

---

[41] Recall that, since $t = d(x,A)$, then the condition $t \to 0^+$ is equivalent to $d(x,A) \to 0^+$.

(c) if $\mathrm{Re}\, s > N$, then the curve described by (2.1.124) is the exponential spiral converging to zero as $t \to 0^+$.

In particular, when $\mathrm{Re}\, s < N$, then the function

$$(0,\delta) \ni t \mapsto \mathrm{Re}\, h(t) = t^{\mathrm{Re}\, s - N} \cos((\mathrm{Im}\, s)\log t)$$

behaves like an *unbounded chirp* when $t \to 0^+$. Here, by a *chirp* we mean a function $g : (0,\delta) \to \mathbb{R}$ of the form $f(t) = a(t)\cos b(t)$, where $b(t) \to +\infty$ or $b(t) \to -\infty$ as $t \to 0^+$, while either $a(t) \to 0^+$ (for *bounded chirps*) or $a(t) \to +\infty$ (for *unbounded chirps*) as $t \to 0^+$.

An analogous discussion can be carried out when $\mathrm{Im}\, s < 0$.

2.2 Residues of Zeta Functions and Minkowski Contents

In this section, we show that some important information concerning the geometry of fractal sets in $\mathbb{R}^N$ is encoded in their associated fractal zeta functions. Therefore, the distance zeta functions, as well as the tube zeta functions which we introduce below (see Definition 2.2.8), can be considered as a useful tool in the study of the geometric properties of fractals.

2.2.1 Distance Zeta Functions of Fractal Sets and Their Residues

We show here that the residue of any meromorphic extension of the distance zeta function of a fractal set A in $\mathbb{R}^N$ is closely related to the Minkowski content of the set; see Theorem 2.2.3 and Theorem 2.2.14. We use the notation $\zeta_{A,A_\delta}(s)$ for the zeta function instead of $\zeta_A(s)$, since we want to stress the dependence of the zeta function on δ. We start with a result which is interesting in itself, and which leads to a new class of zeta functions called tube zeta functions and described by Definition 2.2.8 below. We shall see in Equation (2.2.23) that *this result can be interpreted as a key functional equation connecting the distance and tube zeta functions.*

Theorem 2.2.1. *Let A be a bounded set in $\mathbb{R}^N$, and let δ be a fixed positive number. Then, for all $s \in \mathbb{C}$ such that* $\mathrm{Re}\, s > \overline{\dim}_B A$, *the following identity holds*:

$$\int_{A_\delta} d(x,A)^{s-N}\mathrm{d}x = \delta^{s-N}|A_\delta| + (N-s)\int_0^\delta t^{s-N-1}|A_t|\,\mathrm{d}t. \tag{2.2.1}$$

Furthermore, the function $\tilde{\zeta}_A$ defined by $\tilde{\zeta}_A(s) := \int_0^\delta t^{s-N-1}|A_t|\,\mathrm{d}t$ is absolutely convergent (and hence, in particular, holomorphic) on the open right half-plane $\{\mathrm{Re}\, s > \overline{\dim}_B A\}$.

Proof. Letting $s := N - \gamma$ (with $\gamma < N - \overline{\dim}_B A$) and using Lemma 2.1.4, we deduce that equality (2.2.1) holds for all real numbers $s \in (\overline{D}, +\infty)$. Let us denote the left-hand side of (2.2.1) by $f(s)$, and the right-hand side by $g(s)$. Since $f(s) = g(s)$ on the subset $(\overline{D}, +\infty) \subset \mathbb{C}$, to prove the theorem, it suffices to show that $f(s)$ and $g(s)$ are both holomorphic in the region $\{\mathrm{Re}\, s > \overline{D}\}$. Indeed, the fact that (2.2.1) then holds for all $s \in \mathbb{C}$ with $\mathrm{Re}\, s > \overline{D}$ follows from the principle of analytic continuation; see, e.g., [Con, Corollary 3.8]. The holomorphicity of $f(s)$ in that region is precisely the content of Theorem 2.1.11(a).

To prove the holomorphicity of $g(s)$ on $\{\mathrm{Re}\, s > \overline{D}\}$, it suffices to consider $\tilde{\zeta}_A(s) := \int_0^\delta t^{s-N-1} |A_t|\, \mathrm{d}t$. Note that $\tilde{\zeta}_A(s)$ has the form of the tamed Dirichlet-type integral, $\tilde{\zeta}_A(s) = \int_E \varphi(t)^s \mathrm{d}\mu(x)$, where $E := (0, \delta)$, $\varphi(t) := t$,[42] $\mathrm{d}\mu(x) := t^{-N-1}|A_t|\, \mathrm{d}t$, and the latter measure is positive; see Section 2.1.3.2 and Theorem 2.1.45(a). Therefore, it suffices to show that for any s such that $\mathrm{Re}\, s > \overline{D}$, the Dirichlet-type integral $\tilde{\zeta}_A(s)$ is well defined. To see this, let $\varepsilon > 0$ be small enough, so that $\mathrm{Re}\, s > \overline{D} + \varepsilon$. Since $\mathscr{M}^{*(\overline{D}+\varepsilon)}(A) = 0$, there exists $C_\delta > 0$ such that $|A_t| \le C_\delta t^{N - \overline{D} - \varepsilon}$ for all $t \in (0, \delta]$. Then

$$
\begin{aligned}
|\tilde{\zeta}_A(s)| &\le \int_0^\delta t^{\mathrm{Re}\, s - N - 1} |A_t|\, \mathrm{d}t \\
&\le C_\delta \int_0^\delta t^{\mathrm{Re}\, s - \overline{D} - \varepsilon - 1} \mathrm{d}t = C_\delta \frac{\delta^{\mathrm{Re}\, s - \overline{D} - \varepsilon}}{\mathrm{Re}\, s - \overline{D} - \varepsilon} < \infty,
\end{aligned}
$$

which completes the proof of the theorem. □

Theorem 2.2.1 also extends the identity (2.1.4) from the case of real numbers $\gamma \in (-\infty, N - \overline{D})$ to all complex numbers γ such that $\mathrm{Re}\, \gamma < N - \overline{\dim}_B A$. Furthermore, observe that the identity (13.129) for fractal strings appearing in [Lap-vFr3, Lemma 13.110, p. 442] is a special case of our identity (2.2.1), obtained for $N = 1$ and $\delta = l_1$.

We can formulate Theorem 2.2.1 in a more condensed form, as follows.

Corollary 2.2.2. *Let A be a bounded set in $\mathbb{R}^N$, and let $\delta > 0$ be fixed. If* $\mathrm{Re}\, s > \overline{\dim}_B A - N$, *then*

$$
\int_{A_\delta} d(x, A)^s \mathrm{d}x = \delta^s |A_\delta| - s \int_0^\delta t^{s-1} |A_t|\, \mathrm{d}t. \tag{2.2.2}
$$

The following theorem is a higher-dimensional generalization of Equation (8.25) [Lap-vFr3, Theorem 8.15], and provides more information than the latter result when $N = 1$, in the case when the underlying set is not Minkowski measurable. (Here and in the sequel, given a meromorphic function $\varphi = \varphi(s)$ in a connected open neighborhood of $s = \omega \in \mathbb{C}$, we denote by $\mathrm{res}(\varphi, \omega)$ its residue at $s = \omega$.)

---

[42] Note that the Dirichlet-type integral $\tilde{\zeta}_A$ is tamed, in the sense of condition (2.1.54) (or of Definition A.1.3 of Appendix A), since $\varphi(t) \in (0, \delta)$ for all $t \in (0, \delta)$. See the proof of Lemma 2.2.9 in Subsection 2.2.2 below for more details.

Theorem 2.2.3. *Assume that the bounded set $A \subset \mathbb{R}^N$ is Minkowski nondegenerate, that is, $0 < \mathscr{M}_*^D(A) \le \mathscr{M}^{*D}(A) < \infty$ (in particular, $\dim_B A = D$), and $D < N$. If, in addition, $\zeta_{A,A_\delta}(s)$ can be extended meromorphically to a connected open neighborhood of $s = D$, then D is necessarily a simple pole of $\zeta_{A,A_\delta}(s)$, and*

$$(N-D)\mathscr{M}_*^D(A) \le \operatorname{res}(\zeta_{A,A_\delta}, D) \le (N-D)\mathscr{M}^{*D}(A). \tag{2.2.3}$$

Furthermore, the value of $\operatorname{res}(\zeta_{A,A_\delta}, D)$ does not depend on $\delta > 0$. In particular, if A is Minkowski measurable, then

$$\operatorname{res}(\zeta_{A,A_\delta}, D) = (N-D)\mathscr{M}^D(A). \tag{2.2.4}$$

Proof. Since $\mathscr{M}_*^D(A) > 0$, by using Theorem 2.1.11(c) we conclude that $s = D$ is a pole. Therefore, it suffices to show that the order of the pole at $s = D$ is not larger than 1. Let us take any fixed $\delta > 0$, and let

$$C_\delta := \sup_{t \in (0,\delta]} \frac{|A_t|}{t^{N-D}}. \tag{2.2.5}$$

Note that $C_\delta < \infty$ because $\mathscr{M}^{*D}(A) < \infty$. Then, in light of (2.2.1), for all $s \in \mathbb{R}$ with $D < s < N$, we have

$$\begin{aligned} \zeta_{A,A_\delta}(s) &= \int_{A_\delta} d(x,A)^{s-N} \mathrm{d}x = \delta^{s-N}|A_\delta| + (N-s)\int_0^\delta t^{s-N-1}|A_t|\,\mathrm{d}t \\ &\le C_\delta \delta^{s-D} + C_\delta (N-s)\frac{\delta^{s-D}}{s-D} = C_\delta (N-D)\delta^{s-D}\frac{1}{s-D}. \end{aligned} \tag{2.2.6}$$

Therefore, $0 < \zeta_{A,A_\delta}(s) \le C_1(s-D)^{-1}$ for all $s \in (D,N)$. This shows that $s = D$ is a pole of $\zeta_{A,A_\delta}(s)$ which is at most of order 1, and the first claim is established. Namely, D is a simple pole of $\zeta_{A,A_\delta}(s)$.

It is easy to see that for any positive real numbers δ and δ_1, with $\delta < \delta_1$, the difference

$$\zeta_{A,A_{\delta_1}}(s) - \zeta_{A,A_\delta}(s) = \int_{A_{\delta_1}\setminus A_\delta} d(x,A)^{s-N}\mathrm{d}x$$

is an entire function of s, since $\delta \le d(x,A) \le \delta_1$ for any $x \in A_{\delta_1} \setminus A_\delta$; see Lemma 2.1.15 or Theorem 2.1.45(c). Therefore, the residue of $\zeta_{A,A_\delta}(s)$ at D does not depend on δ.

In order to prove the second inequality in (2.2.3), is suffices to multiply (2.2.6) by $s-D$, with s real and $s > D$, and then take the limit as $s \to D^+$:

$$\operatorname{res}(\zeta_{A,A_\delta}, D) \le (N-D)\lim_{s\to D^+} C_\delta \delta^{s-D} = (N-D)C_\delta. \tag{2.2.7}$$

Since the residue of $\zeta_{A,A_\delta}(s)$ at D does not depend on δ, (2.2.3) follows from (2.2.7) by recalling the definition of C_δ given in (2.2.5) and passing to the limit as $\delta \to 0^+$

in the right-hand side of (2.2.7); note that the function $\delta \mapsto C_\delta$ is nondecreasing and $C_\delta \to \mathscr{M}^{*D}(A)$ as $\delta \to 0^+$. The first inequality in (2.2.3) is proved analogously. □

Example 2.2.4. There is a class of fractal sets A for which $D := \dim_B A$ exists with $\mathscr{M}^D(A) = 0$ and

$$\int_{A_\delta} d(x,A)^{D-N} \mathrm{d}x < \infty, \tag{2.2.8}$$

that is, $\zeta_{A,A_\delta}(D) < \infty$; see [Žu4, Theorem 4.2]. In particular, it follows that

$$\lim_{s \to D^+} \zeta_{A,A_\delta}(s) = \zeta_A(D).$$

This shows that there is no meromorphic continuation of ζ_{A,A_δ} to a connected open neighborhood of $s = D$, since in that case we would have $\lim_{s \to D^+} \zeta_{A,A_\delta}(s) = +\infty$. Here, the limit as $s \to D^+$ is taken along the real axis (i.e., for $s > D$) or, more generally, within a sector ($\operatorname{Re} s > D$, $|\arg(s - D)| < \Theta$) with half-angle Θ satisfying $0 \le \Theta < \pi/2$; see, e.g., [Ser, Section VI.2] or [HardWr].

Example 2.2.5. Let $A = \{0\}$ in $\mathbb{R}$. Then $\zeta_{A,A_\delta}(s) = 2\delta^s s^{-1}$. Hence, $s = 0$ is a simple pole of ζ_{A,A_δ}, and for each $\delta > 0$,

$$\operatorname{res}(\zeta_{A,A_\delta}, 0) = \lim_{s \to 0} 2\delta^s = 2.$$

Example 2.2.6. (*Residues of the zeta function of the generalized Cantor set*). Let $A = C^{(a)}$ be the generalized Cantor set[43] defined by the parameter $a \in (0, 1/2)$. Recall that $C^{(a)}$ is obtained by deleting the middle interval of length $1 - 2a$ from the interval $[0, 1]$, and then continuing in the usual way, scaling by the factor a at each step (for $a = 1/3$, we obtain the middle third Cantor set). By a direct computation, or using [Žu3, Equation (15) with $\gamma := N - s$], we obtain the corresponding zeta function:

$$\zeta_{A,A_\delta}(s) = \frac{2^{1-s}(1-2a)^s}{s(1-2a^s)} + 2\delta^s s^{-1}, \tag{2.2.9}$$

where δ is a fixed positive real number. In particular,

$$\zeta_{A,A_\delta}(s) \sim \frac{1}{1-2a^s}. \tag{2.2.10}$$

Its residue at $D = D(a) := \dim_B A = \log_{1/a} 2$ (see [Žu3]) is independent of δ (in accordance with Theorem 2.2.3) and given by

$$\operatorname{res}(\zeta_{A,A_\delta}, D) = \frac{2}{\log 2} \left(\frac{1}{2} - a \right)^D. \tag{2.2.11}$$

On the other hand, the values of the lower and upper D-dimensional Minkowski contents are respectively equal to (see [Žu2, Equations (3.12) and (3.13) for $m = 2$]):

[43] An even more general class of Cantor sets will be introduced in Definition 3.1.1.

$$\mathcal{M}_*^D(A) = \frac{1}{D}\left(\frac{2D}{1-D}\right)^{1-D}, \quad \mathcal{M}^{*D}(A) = 2(1-a)\left(\frac{1}{2}-a\right)^{D-1}, \tag{2.2.12}$$

and thus $\mathcal{M}_*^D(A) < \mathcal{M}^{*D}(A)$. It follows that $C^{(a)}$ is not Minkowski measurable (for a much more general result, see [Lap-vFr3, Theorem 2.16]; see also [Lap-vFr3, Chapter 10]). (We note that in the case of the classical Cantor set, where $a = 1/3$ and $D = \log_3 2$, the values in (2.2.12) have been first obtained in [LapPo2, Theorem 2.4].) Therefore, for any generalized Cantor set $A = C^{(a)}$, with $a \in (0,1/2)$, we have that

$$(1-D)\mathcal{M}_*^D(A) < \operatorname{res}(\zeta_{A,A_\delta}, D) < (1-D)\mathcal{M}^{*D}(A); \tag{2.2.13}$$

see Figure 2.12. This is in agreement with (2.2.3) in Theorem 2.2.3, and also with the inequalities in the first line of (2.3.62) in Theorem 2.3.37 below. In particular, since the functions $(0,1/2) \ni a \mapsto \mathcal{M}_*^D(A)$ and $a \mapsto \mathcal{M}^{*D}(A)$ are bounded, and $D = \log_{1/a} 2 \to 1^-$ as $a \to 1/2^-$, we have that for any positive δ,

$$\lim_{a \to 1/2^-} \operatorname{res}(\zeta_{A,A_\delta}, D) = 0.$$

The residues of $\zeta_{A,A_\delta}(s)$ at the poles $s_k := D + k\mathbf{p}\mathrm{i}$, $k \in \mathbb{Z}$, on the critical line $\{\operatorname{Re} s = D\}$, expressed in terms of the residue at D and of the oscillatory period $\mathbf{p} := 2\pi/\log(1/a)$ of A, are the following:

$$\operatorname{res}(\zeta_{A,A_\delta}, s_k) = \frac{D 2^{-k\mathbf{p}\mathrm{i}}(1-2a)^{k\mathbf{p}\mathrm{i}}}{s_k a^{k\mathbf{p}\mathrm{i}}} \operatorname{res}(\zeta_{A,A_\delta}, D), \quad k \in \mathbb{Z}. \tag{2.2.14}$$

It is noteworthy that these residues tend to zero as $k \to \pm\infty$; more precisely,

$$|\operatorname{res}(\zeta_{A,A_\delta}, s_k)| = \frac{D}{|s_k|} \operatorname{res}(\zeta_{A,A_\delta}, D) \asymp \frac{1}{k} \quad \text{as} \quad k \to \pm\infty. \tag{2.2.15}$$

This situation is different from that of the zeta function of the Cantor string $\mathcal{L}$ for which $a = 1/3$ (see [Lap-vFr3, Subsection 2.3.1, p. 41–43]), where we have that the residues of the meromorphic continuation of the geometric zeta function $\zeta_{\mathcal{L}}(s) = \sum_j \ell_j^s$ at s_k are all equal to $1/\log 3 = 2 \cdot 3^{-s_k}/\log 3$, with $D = \log_3 2$, $\mathbf{p} = 2\pi/\log 3$. (See [Lap-vFr3, Subsection 2.3.1] for the Cantor set with basic length equal to 3; a general result is stated in Theorem 2.16 and in Remark 2.18 of [Lap-vFr3].) This is easily explained by the presence of the factor s^{-1} in (2.1.84).

Note that for $a \in (0,1/2)$ the corresponding oscillatory period

$$\mathbf{p}(a) := \frac{2\pi}{\log(1/a)}$$

of the generalized Cantor set set $A = C^{(a)}$ tends to zero as $a \to 0^+$, which means that the set of poles

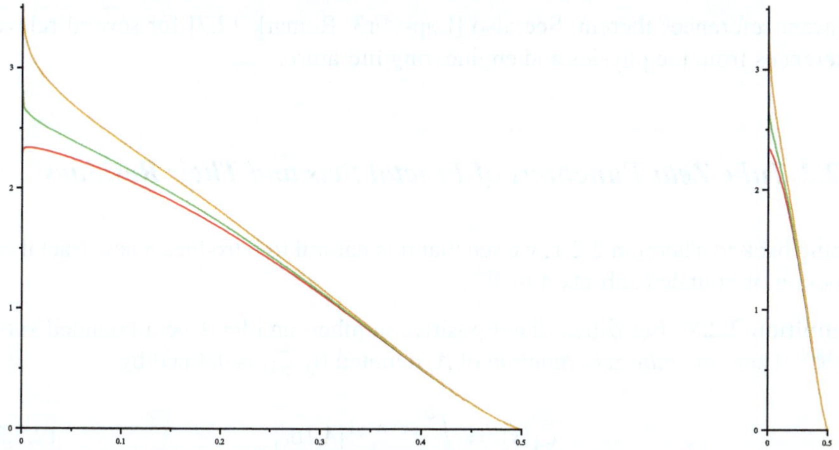

Fig. 2.12 On the left, the graphs of (respectively, from top to bottom) $(1-D)\mathscr{M}^{*D}(A)$, $\operatorname{res}(\zeta_{A,A_\delta},D)$ and $(1-D)\mathscr{M}_*^D(A)$, viewed as functions of $a\in(0,1/2)$, are depicted in the case of the generalized Cantor set $A=C^{(a)}$ studied in Example 2.2.6. Here, $D=\log_{1/a}2$. The horizontal a-axis is expanded ten times with respect to the vertical axis. On the right, the same figure is represented, but with the natural scale on the horizontal axis. This illustrates the inequality (2.2.13); see also (2.2.3).

$$\mathscr{P}(a)=\{D(a)+k\mathbf{p}\mathbb{i}:k\in\mathbb{Z}\}=D(a)+\mathbf{p}\mathbb{i}\mathbb{Z} \tag{2.2.16}$$

of the zeta function $\zeta_{A,A_\delta}(s)$ converges to the imaginary axis in the Hausdorff metric, as $a\to0^+$. Furthermore,

$$D(a)=\frac{\log 2}{2\pi}\,\mathbf{p}(a)=\log_{1/a}2. \tag{2.2.17}$$

Also, $D(a)=\dim_B A\to0^+$ and $\frac{\mathrm{d}}{\mathrm{d}a}D(a)\to+\infty$ as $a\to0^+$, while $D(a)\to1$ and $\frac{\mathrm{d}}{\mathrm{d}a}D(a)\to2/\log2$ as $a\to1/2^-$. The behavior of the residue for a near 0 is the following:

$$\lim_{a\to0^+}\operatorname{res}(\zeta_{A,A_\delta},D)=\frac{2}{\log2},\quad \lim_{a\to0^+}\frac{\mathrm{d}}{\mathrm{d}a}\operatorname{res}(\zeta_{A,A_\delta},D)=-\infty. \tag{2.2.18}$$

Also,

$$\lim_{a\to0^+}\frac{\mathrm{d}}{\mathrm{d}a}(1-D)\mathscr{M}_*^D(A)=+\infty,\quad \lim_{a\to0^+}(1-D)\frac{\mathrm{d}}{\mathrm{d}a}\mathscr{M}^{*D}(A)=-\infty. \tag{2.2.19}$$

Remark 2.2.7. Much of the discussion in Example 2.2.6 parallels [Lap-vFr3, Subsection 12.1.3] about the lacunarity of a family of generalized Cantor sets. The (semi-heuristic) notion of *lacunarity* has been introduced by Mandelbrot in [Man1, Chapter X]; see also [Man2]. Mathematically, it has been further explored from different points of view in [BedFi], [Man2], [Lap-vFr3, Subsection 12.1.3], and in the relevant references therein. See also [Lap-vFr3, Remark 12.7] for several relevant references from the physics and engineering literature.

2.2.2 *Tube Zeta Functions of Fractal Sets and Their Residues*

Going back to Theorem 2.2.1, we see that it is natural to introduce a new fractal zeta function of bounded subsets A of $\mathbb{R}^N$.

Definition 2.2.8. Let δ be a fixed positive number, and let A be a bounded subset of $\mathbb{R}^N$. Then, the *tube zeta function* of A, denoted by $\tilde{\zeta}_A$, is defined by

$$\tilde{\zeta}_A(s) := \int_0^\delta t^{s-N-1}|A_t|\,\mathrm{d}t, \tag{2.2.20}$$

for all $s \in \mathbb{C}$ with $\operatorname{Re} s > \overline{\dim}_B A$; see Theorem 2.2.1. As we know from the proof of Theorem 2.2.1 (see also footnote 42 on page 113), the tube zeta function is a tamed Dirichlet-type integral.

An immediate consequence of Theorem 2.2.1 is that for any bounded subset A of $\mathbb{R}^N$, we have

$$D_{\mathrm{hol}}(\tilde{\zeta}_A) \le D(\tilde{\zeta}_A) \le \overline{\dim}_B A. \tag{2.2.21}$$

Furthermore, assuming that $\overline{\dim}_B A < N$, then $D(\tilde{\zeta}_A) = \overline{\dim}_B A$, as we shall see in Corollary 2.2.10 below.

Note that the underlying space of scales can be viewed as the multiplicative group $(0,+\infty)$ equipped with its natural *Haar measure* $\mathrm{d}t/t$. The measure $\mathrm{d}\mu(t) := \mathrm{d}t/t$ is the standard 'Haar measure' on the group $G := (0,+\infty)$, in the sense that it is invariant under multiplication. More precisely, for any μ-measurable subset A of G and for any $g \in G$, we have $\mu(gA) = \mu(A)$. In our case, this follows easily from the fact that $\mu(gA) = \int_{gA} \mathrm{d}t/t = \int_A \mathrm{d}\sigma/\sigma = \mu(A)$, where we have used the change of variables $\sigma = g^{-1}t$. Hence, Equation (2.2.20) can be rewritten as follows (for $\operatorname{Re} s > \overline{\dim}_B A$):

$$\tilde{\zeta}_A(s) = \int_0^\delta t^{s-N}|A_t|\,\frac{\mathrm{d}t}{t}. \tag{2.2.22}$$

A similar comment can be made about the integral appearing on the right-hand side of (2.1.4) in Lemma 2.1.4 and of (2.2.1) in Theorem 2.2.1.

We first note that $\tilde{\zeta}_A$ is a tamed Dirichlet-type integral (in the sense that it is of the form (2.1.53) and satisfies condition (2.1.54)). Hence, it follows from Theorem A.1.4 of Appendix A that $D(\tilde{\zeta}_A)$, the abscissa of convergence of $\tilde{\zeta}_A$ (as given by (2.1.55)), is well defined and $\Pi(\tilde{\zeta}_A)$, the half-plane of convergence of $\tilde{\zeta}_A$ (as given by (2.1.56)) is the largest open right half-plane on which the Lebesgue integral appearing on the right-hand side of (2.2.20) is convergent (i.e., is absolutely convergent).

The following lemma and its proof supplement Example 2.1.43 above. See also part (1) of Proposition A.2.4 in Appendix A.

Lemma 2.2.9. *The tube zeta function $\tilde{\zeta}_A$ is a tamed DTI.*

Proof. With the notation of (2.1.53), we can let $E := (0,\delta)$, $\varphi(t) := t$ for all $t \in E$, and $\mu(dt) := t^{-N-1}|A_t|\,dt = t^{-N}|A_t|\,\frac{dt}{t}$, viewed as a local positive measure on E. Clearly, we then have

$$\tilde{\zeta}_A(s) = \int_0^\delta t^{s-N-1}|A_t|\,dt = \int_E \varphi(t)^s \mu(dt),$$

for all $s \in \mathbb{C}$ such that $\operatorname{Re} s > \overline{\dim}_B A$. Furthermore, we obviously have that $0 < \varphi(t) = t < \delta$ for all $t \in E$ (so that condition (2.1.54) holds with $C := \delta$). Hence, $\tilde{\zeta}_A$ is a tamed DTI.

Alternatively, if we want to view dt/t as the scale invariant measure on $(0,+\infty)$, we can let $E := (0,+\infty)$,

$$\varphi(t) := \begin{cases} t, & \text{for } t \in (0,\delta), \\ \delta, & \text{for } t \in E \setminus (0,\delta), \end{cases}$$

and $\mu(dt) := \tilde{\rho}(t)\,dt/t$, where

$$\tilde{\rho}(t) := \begin{cases} t^{-N}|A_t|, & \text{for } t \in (0,\delta), \\ 0, & \text{for } t \in E \setminus (0,\delta). \end{cases}$$

Again, it is clear that $\tilde{\zeta}_A$ is a tamed DTI, also with the choice of $C := \delta$ in condition (2.1.54), since $0 < \varphi(t) \leq \delta$ for all $t \in E$. □

We call $\tilde{\zeta}_A$ the tube zeta function of A since its definition involves the *tube function* $(0,\delta) \ni t \mapsto |A_t|$. Relation (2.2.1) can be written as the following *functional equation*:

$$\zeta_{A,A_\delta}(s) = \delta^{s-N}|A_\delta| + (N-s)\tilde{\zeta}_A(s), \tag{2.2.23}$$

for any $\delta > 0$ and for all s such that $\operatorname{Re} s > \overline{\dim}_B A$. From Theorem 2.2.1, we see that $\tilde{\zeta}_A(s)$ is holomorphic on $\{\operatorname{Re} s > \overline{\dim}_B A\}$ and that the lower bound is optimal, from the point of view of the convergence (i.e., absolute convergence) of the Lebesgue integral defining $\tilde{\zeta}_A(s)$ in Equation (2.2.20) or, equivalently, Equation (2.2.22).

Note that in light of the functional equation (2.2.23) connecting the tube and distance zeta functions, one can have the impression that, provided $\overline{\dim}_B A < N$, the tube zeta function has a simple pole at $s = N$. However, as we have seen in Theorem 2.2.1, the distance zeta function ζ_A is holomorphic in the open right half-plane$\{\operatorname{Re} s > \overline{\dim}_B A\}$. It follows, in particular, that $\tilde{\zeta}_A$ is regular (i.e., holomorphic) at $s = N$ provided $\overline{\dim}_B A < N$.

Assuming that there exists a meromorphic extension of $\tilde{\zeta}_A(s)$ (or, equivalently, of $\zeta_A(s)$) to a connected open neighborhood of $\overline{D} := \overline{\dim}_B A$, and $\overline{D}$ is a simple pole, with $\overline{D} < N$,[44] then it follows from (2.2.23) that

$$\operatorname{res}(\tilde{\zeta}_A, \overline{D}) = \frac{1}{N - \overline{D}} \operatorname{res}(\zeta_{A,A_\delta}, \overline{D}). \tag{2.2.24}$$

Indeed,

$$\begin{aligned}\operatorname{res}(\zeta_{A,A_\delta}, \overline{D}) &= \lim_{s \to \overline{D}} (s - \overline{D})[\delta^{s-N}|A_\delta| + (N-s)\tilde{\zeta}_A(s)] \\ &= (N - \overline{D}) \lim_{s \to \overline{D}} (s - \overline{D})\tilde{\zeta}_A(s) \\ &= (N - \overline{D}) \operatorname{res}(\tilde{\zeta}_A, \overline{D}).\end{aligned}$$

The following corollary is an immediate consequence of Proposition 2.2.19 below and of the relevant definitions given earlier in Section 2.1.

Corollary 2.2.10. *Let us assume that* $\overline{\dim}_B A < N$. *Then, not only* (2.2.52) *in Proposition 2.2.19 below holds (i.e.,* $D(\zeta_A) = D(\tilde{\zeta}_A) = \overline{\dim}_B A$*), but* ζ_A *and* $\tilde{\zeta}_A$ *also have the same abscissae of holomorphic continuation and meromorphic continuation, respectively*:

$$D_{\rm hol}(\zeta_A) = D_{\rm hol}(\tilde{\zeta}_A), \quad D_{\rm mer}(\zeta_A) = D_{\rm mer}(\tilde{\zeta}_A) \tag{2.2.25}$$

with

$$D_{\rm mer}(\zeta_A) \le D_{\rm hol}(\zeta_A) \le D(\zeta_A) = \overline{\dim}_B A. \tag{2.2.26}$$

Therefore, ζ_A *and* $\tilde{\zeta}_A$ *have the same half-planes of* (*absolute*) *convergence, holomorphic continuation and meromorphic continuation; that is,*

$$\begin{aligned}\{\operatorname{Re} s > \overline{\dim}_B A\} = \Pi(\zeta_A) = \Pi(\tilde{\zeta}_A), \\ \mathcal{H}(\zeta_A) = \mathcal{H}(\tilde{\zeta}_A) \quad \text{and} \quad \operatorname{Mer}(\zeta_A) = \operatorname{Mer}(\tilde{\zeta}_A).\end{aligned} \tag{2.2.27}$$

In particular, ζ_A *and* $\tilde{\zeta}_A$ *have the same critical line* $\{\operatorname{Re} s = \overline{\dim}_B A\}$, *as well as the same holomorphy critical line* $\{\operatorname{Re} s = D_{\rm hol}(\zeta_A)\}$ *and the same meromorphy critical line* $\{\operatorname{Re} s = D_{\rm mer}(\zeta_A)\}$.

Recall from part (b) of Remark 2.1.21 (which relies on key results from Section 4.6 below, namely, Theorem 4.6.9 and Corollary 4.6.17 along with Remark 4.6.19) that the inequalities in (2.2.26) are sharp. More specifically, given any $N \ge 1$, there exists an explicitly constructible bounded subset A of $\mathbb{R}^N$ for which $\overline{D} := \overline{\dim}_B A < N$ and the following string of inequalities holds:

$$\begin{aligned}D_{\rm mer}(\zeta_A) = D_{\rm mer}(\tilde{\zeta}_A) = D_{\rm hol}(\zeta_A) = D_{\rm hol}(\tilde{\zeta}_A) \\ = D(\zeta_A) = D(\tilde{\zeta}_A) = \overline{\dim}_B A.\end{aligned} \tag{2.2.28}$$

Moreover, the set A can be constructed to be maximally hyperfractal (in the sense of part (*iii*) of Definition 4.6.23) and transcendentally ∞-quasiperiodic (in the sense of Subsection 4.6.1).

[44] Since $A \subseteq \mathbb{R}^N$, we always have that $\overline{D} \le N$.

We summarize part of this discussion in the following theorem, which follows form Equation (2.2.23) and Theorem 2.1.11. It is the exact counterpart for tube zeta functions of Theorem 2.1.11 for distance zeta functions. A detailed comparison between ζ_A and $\tilde{\zeta}_A$ is provided in Proposition 2.2.19 and Corollary 2.2.20 below, along with Corollary 2.2.10 above. We note that in light of the presence of the factor $(s-N)$ on the right-hand side of Equation (2.2.23), the case when $\overline{\dim}_B A = N$ is discussed separately in the proof of Theorem 2.2.11 below.

Theorem 2.2.11. *Let A be an arbitrary bounded subset of $\mathbb{R}^N$ and let $\delta > 0$. Then:*

(a) The tube zeta function $\tilde{\zeta}_A$ defined by (2.2.20) *is holomorphic (i.e., analytic) in the open right half-plane $\{\mathrm{Re}\, s > \overline{\dim}_B A\}$, and for all complex numbers s in that region,*

$$\tilde{\zeta}_A'(s) = \int_0^\delta t^{s-N-1} \log t \, |A_t| \, \mathrm{d}t. \tag{2.2.29}$$

(b) The lower bound in the open right half-plane $\{\mathrm{Re}\, s > \overline{\dim}_B A\}$ is optimal, from the point of view of the (absolute) convergence of the tamed Dirichlet-type integral defining $\tilde{\zeta}_A$. In other words,

$$\overline{\dim}_B A = D(\tilde{\zeta}_A), \tag{2.2.30}$$

where $D(\tilde{\zeta}_A)$ is the abscissa of convergence of $\tilde{\zeta}_A$, as defined in Equation (2.1.55).[45] *It follows that $D(\tilde{\zeta}_A) \in [0,N]$. (See also Corollary 2.2.10 above for more detailed information.) Furthermore, the identity* (2.2.20) *continues to hold in the half-plane of (absolute) convergence $\{\mathrm{Re}\, s > \overline{\dim}_B A\}$ of $\tilde{\zeta}_A$. Moreover, we have (see part (a) of Remark 2.2.12 below)*

$$D(\tilde{\zeta}_A) = \inf\left\{ \alpha \in [0, N+\tfrac{1}{10}] : \int_0^\delta t^{\alpha-N-1} |A_t| \, \mathrm{d}t < \infty \right\}. \tag{2.2.31}$$

(c) If the box (or Minkowski) dimension $D := \dim_B A$ exists and $\mathscr{M}_*^D(A) > 0$, then $\tilde{\zeta}_A(s) \to +\infty$ as $s \in \mathbb{R}$ converges to D from the right. According to (2.2.21), *it then follows that (under the additional hypotheses of the present part (c) of the theorem), we have*

$$\dim_B A = D(\tilde{\zeta}_A) = D_{\mathrm{hol}}(\tilde{\zeta}_A), \tag{2.2.32}$$

where $D_{\mathrm{hol}}(\tilde{\zeta}_A)$, the abscissa of holomorphic continuation of $\tilde{\zeta}_A$ (as given by (2.1.27) *above), is defined so that $\{\mathrm{Re}\, s > D_{\mathrm{hol}}(\tilde{\zeta}_A)\}$ be the maximal right half-plane of the form $\{\mathrm{Re}\, s > \alpha\}$, for some $\alpha \in \mathbb{R} \cup \{\pm\infty\}$, to which $\tilde{\zeta}_A$ can be holomorphically continued. For more details, see Corollary 2.2.10 and Proposition 2.2.19 along with part (b) of Remark 2.2.12 below.*

---

[45] See Subsection 2.1.3.2 above and Appendix A below for the more general setting of tamed Dirichlet-type integrals.

Proof. (b) If $\overline{\dim}_B A < N$, the claim follows from the functional equation (2.2.23). Therefore, let us consider the case when $\overline{\dim}_B A = N$. Recall from Equation (1.3.7) that we always have $0 \le \overline{\dim}_B A \le N$. Let $\varepsilon > 0$ and define $\sigma := N - \varepsilon$. We then have that $\mathscr{M}^{*\sigma}(A) = +\infty$, which implies that there exists a sequence of positive numbers $s_k \in (0, \delta)$ converging to zero and such that

$$C_k := \frac{|A_{s_k}|}{s_k^{N-\sigma}} = \frac{|A_{s_k}|}{s_k^{\varepsilon}} \to +\infty, \quad \text{as} \quad k \to \infty.$$

Since $t \mapsto |A_t|$ is nondecreasing, we now have

$$\begin{aligned}
\int_0^\delta t^{\sigma-N-1}|A_t|\,\mathrm{d}t &= \int_0^\delta t^{-\varepsilon-1}|A_t|\,\mathrm{d}t \ge \int_{s_k}^\delta t^{-\varepsilon-1}|A_t|\,\mathrm{d}t \\
&\ge |A_{s_k}|\int_{s_k}^\delta t^{-\varepsilon-1}\,\mathrm{d}t = \frac{|A_{s_k}|}{\varepsilon s_k^{\varepsilon}} - \frac{|A_{s_k}|}{\varepsilon\delta^{\varepsilon}} \\
&\ge \frac{|A_{s_k}|}{\varepsilon s_k^{\varepsilon}} - \frac{|A_{\delta}|}{\varepsilon\delta^{\varepsilon}}.
\end{aligned}$$

By letting $k \to \infty$, we conclude that $\int_0^\delta t^{\sigma-N-1}|A_t|\,\mathrm{d}t = +\infty$, and since this is true for every $\varepsilon > 0$, we have that $D(\tilde{\zeta}_A) = N = \overline{\dim}_B A$.

(c) If $\dim_B A < N$, the claim follows from Equation (2.2.23). Let us therefore assume that $\dim_B A = N$. Since $\mathscr{M}_*^N(A) > 0$, for a fixed $\delta > 0$, there exists a constant $C > 0$ such that for all $t \in (0, \delta)$, we have $|A_t| \ge C t^{N-N} = C$. By choosing $\sigma > N$, we then have

$$\tilde{\zeta}_A(\sigma) = \int_0^\delta t^{\sigma-N-1}|A_t|\,\mathrm{d}t \ge C\int_0^\delta t^{\sigma-N-1}\,\mathrm{d}t = \frac{C\delta^{\sigma-N}}{\sigma-N}$$

and hence, $\tilde{\zeta}_A(\sigma) \to +\infty$ when $\sigma \to N^+$. □

Remark 2.2.12. (a) A priori, the infimum in Equation (2.2.31) of part (b) of Theorem 2.2.11 should be taken over all real numbers $\alpha \in \mathbb{R}$, but since $D(\tilde{\zeta}_A) = \overline{\dim}_B A \in [0, N]$ (in light of (2.2.30)) and for $\alpha > N$ the function $t \mapsto t^{\alpha-N-1}|A_t|$ is integrable on $(0, \delta)$ (namely, it can be dominated by the function $t \mapsto C \cdot t^{\alpha-N-1}$, where $C := |A_\delta|$, which is clearly integrable on $(0, \delta)$), it can be taken over all $\alpha \in [0, N + \frac{1}{10}]$. In fact, instead of $\frac{1}{10}$, we may take any positive real number.

(b) It follows from part (c) of Theorem 2.2.11 that D is a singularity (which may or may not be a pole) of $\tilde{\zeta}_A$. Naturally, if $\tilde{\zeta}_A$ possesses a meromorphic continuation to an open connected neighborhood of D, then it follows that D is a pole of $\tilde{\zeta}_A$. In Section 2.3 and Section 4.5 will be provided several sufficient conditions under which $\tilde{\zeta}_A$ can be meromorphically continued beyond the critical line $\operatorname{Re} s = D$, and hence, in particular, to a connected open neighborhood of D.

The following proposition is the counterpart for tube zeta functions of Proposition 2.1.76. It will be useful in the proof of Theorem 2.2.14 below.

Proposition 2.2.13. *Let A be a bounded subset of $\mathbb{R}^N$, and let δ_1, δ_2 be two positive real numbers. Let us denote the corresponding two tube zeta functions by $\tilde{\zeta}_A(\,\cdot\,;\delta_j)$, for $j = 1,2$. Then:*

(a) The difference $\tilde{\zeta}_A(\,\cdot\,;\delta_1) - \tilde{\zeta}_A(\,\cdot\,;\delta_2)$ can be extended to an entire function, and in particular, $\tilde{\zeta}_A(\,\cdot\,;\delta_1) \sim \tilde{\zeta}_A(\,\cdot\,;\delta_2)$. As a special case, if δ_1 and δ_2 are any two positive real numbers, then $\tilde{\zeta}_A(\,\cdot\,;\delta_1) - \tilde{\zeta}_A(\,\cdot\,;\delta_2)$ can be identified with an entire function, and in particular, $\tilde{\zeta}_A(\,\cdot\,;\delta_1) \sim \tilde{\zeta}_A(\,\cdot\,;\delta_2)$.

(b) If a complex number s_0 is a simple pole of $\tilde{\zeta}_A(\,\cdot\,;\delta_1)$, then it is also a simple pole of $\tilde{\zeta}_A(\,\cdot\,;\delta_2)$ and we have that

$$\operatorname{res}(\tilde{\zeta}_A(\,\cdot\,;\delta_1), s_0) = \operatorname{res}(\tilde{\zeta}_A(\,\cdot\,;\delta_2), s_0). \tag{2.2.33}$$

Proof. (a) Assuming without loss of generality that $\delta_1 < \delta_2$, we have that

$$\tilde{\zeta}_A(s;\delta_2) - \tilde{\zeta}_A(s;\delta_1) = \int_{\delta_1}^{\delta_2} t^{s-N-1}\mathrm{d}s. \tag{2.2.34}$$

In the notation of (2.1.53), the last integral can be viewed as a tamed Dirichlet-type integral (DTI) with $E := (\delta_1, \delta_2)$, $\varphi(t) := t$ for all $t \in E$ and $\mathrm{d}\mu(t) := t^{-N-1}\mathrm{d}t$, by noting that $\varphi(t) \in (\delta_1, \delta_2)$ for all $t \in E$. This DTI defines an entire function, by case (c) of Theorem 2.1.45. This completes the proof of case (a) of the proposition.

The proof of case (b) follows much in the same way as the proof of part (b) of Proposition 2.1.76 above. □

In light of the discussion surrounding Equation (2.2.24) and preceding the statement of Theorem 2.2.11, the following result is an immediate consequence of Theorem 2.2.3 (and relation (2.2.1)). More specifically, in light of those earlier results, only the case when $\dim_B A = N$ will be discussed in the proof.

Theorem 2.2.14. *Assume that A is a Minkowski nondegenerate bounded subset of $\mathbb{R}^N$ and there exists a (necessarily unique) meromorphic extension of $\tilde{\zeta}_A$ to a connected open neighborhood of $D := \dim_B A$.*[46] *Then D is a simple pole of $\tilde{\zeta}_A$, and for any positive δ, $\operatorname{res}(\tilde{\zeta}_A, D)$ is independent of δ. Furthermore, we have*

$$\mathscr{M}_*^D(A) \leq \operatorname{res}(\tilde{\zeta}_A, D) \leq \mathscr{M}^{*D}(A). \tag{2.2.35}$$

In particular, if A is Minkowski measurable, then the residue of the tube zeta function of A at $s = D$ is equal to the D-dimensional Minkowski content of A; that is,

$$\operatorname{res}(\tilde{\zeta}_A, D) = \mathscr{M}^D(A). \tag{2.2.36}$$

---

[46] Recall from Subsection 1.3.1 that $\dim_B A$ exists since A is Minkowski nondegenerate.

Proof. Let us only consider the case when $\dim_B A = N$. In the proof of Theorem 2.2.11, it is shown that $s = N$ is at least a simple pole of $\tilde{\zeta}_A$ since $\mathcal{M}_*^N(A) > 0$. We will show that its order cannot be greater than 1. Fix $\delta > 0$ and let

$$C_\delta := \sup_{t\in(0,\delta]} \frac{|A_t|}{t^{N-N}} = \sup_{t\in(0,\delta]} |A_t| = |A_\delta| < \infty. \tag{2.2.37}$$

Then, for all $s \in \mathbb{R}$ such that $s > N$, we have

$$\tilde{\zeta}_A(s) = \int_0^\delta t^{s-N-1}|A_t|\,dt \le \frac{C_\delta \delta^{s-N}}{s-N}. \tag{2.2.38}$$

Therefore, we have that $0 < \tilde{\zeta}_A(s) \le C_1(s-N)^{-1}$ for all real numbers $s > N$; hence, the order of the pole $s = N$ is at most 1. It follows that $\tilde{\zeta}_A$ has a simple pole at $s = N$. By case (b) of Proposition 2.2.13 just above, we know that the residue of $\tilde{\zeta}_A$ at $s = N$ does not depend on δ. We now multiply (2.2.38) by $(s-N)$ and take the limit as $s \to N^+$ of the resulting inequality in order to conclude that

$$\operatorname{res}(\tilde{\zeta}_A, N) \le \lim_{s\to N^+} C_\delta \delta^{s-N} = C_\delta. \tag{2.2.39}$$

By recalling the definition of C_δ and letting $\delta \to 0^+$ (note that $\delta \mapsto C_\delta$ is nondecreasing) we get that $\operatorname{res}(\tilde{\zeta}_A, N) \le \mathcal{M}^{*N}(A)$. The inequality $\mathcal{M}_*^N(A) \le \operatorname{res}(\tilde{\zeta}_A, N)$ is proved analogously. □

Refinements and extensions of this result can be found in Theorems 2.3.18 and 2.3.25 below.

Remark 2.2.15. According to Remark 1.3.1 on page 31, we conclude that if A is such that $D := \dim_B A$ exists and $D = N$, then $\mathcal{M}^N(A)$ exists as well and $\mathcal{M}^N(A) = |\overline{A}|$. Moreover, in this case, it follows from Theorem 2.2.14 that we also have that $\operatorname{res}(\tilde{\zeta}_A, N) = |\overline{A}|$.

Remark 2.2.16. Returning to the case of the unit interval $I := [0,1]$ in $\mathbb{R}$, already considered in Example 2.1.16, we will demonstrate how the distance zeta function fails to provide useful information about the Minkowski content of I in this example (since $\dim_B I = 1$), but the tube zeta function still does. Recall that, by the aforementioned example, the distance zeta function of I is given by

$$\zeta_I(s) = \frac{2\delta^s}{s} \tag{2.2.40}$$

and is meromorphic on $\mathbb{C}$ with a simple pole at $s = 0$. Also recall that $D_{\text{hol}}(\zeta_I) = 0 < 1 = D(\zeta_I)$.

One the other hand, one has that $|I_t| = 1 + 2t$ and by an easy calculation,

$$\tilde{\zeta}_I(s) = \int_0^\delta t^{s-2}|A_t|\,dt = \int_0^\delta t^{s-2}(1+2t)\,dt = \frac{2\delta^s}{s} + \frac{\delta^{s-1}}{s-1}, \tag{2.2.41}$$

for all $s \in \mathbb{C}$ with $\mathrm{Re}\, s > 1$. Upon analytic continuation, we deduce that $\tilde{\zeta}_I$ has a meromorphic extension to all of $\mathbb{C}$ given by

$$\tilde{\zeta}_I(s) = \frac{2\delta^s}{s} + \frac{\delta^{s-1}}{s-1}, \quad \text{for all } s \in \mathbb{C}. \tag{2.2.42}$$

Therefore, $D_{\mathrm{hol}}(\tilde{\zeta}_I) = D(\tilde{\zeta}_I) = 1$ and

$$\mathrm{res}(\tilde{\zeta}_I, 1) = 1 = |I| = \mathscr{M}^1(I). \tag{2.2.43}$$

We will encounter this type of bad behavior of the distance zeta function whenever $s = N$ is a pole of the tube zeta function. This statement follows from the general functional equation

$$\zeta_A(s) = \delta^{s-N}|A_\delta| - (s-N)\tilde{\zeta}_A(s) \tag{2.2.44}$$

obtained in (2.2.23) for bounded subsets A of $\mathbb{R}^N$, since clearly the potential pole of the tube zeta function at $s = N$ (which is simple if it exists) is canceled by the factor $(N-s)$ and, hence the distance zeta function is then holomorphic at $s = N$.

Also note that if $s = N$ is a simple pole of the tube zeta function $\tilde{\zeta}_A$ (viewed via the meromophic extension), then

$$\zeta_A(N) = |A_\delta| - \mathrm{res}(\tilde{\zeta}_A, N). \tag{2.2.45}$$

In particular, if $s = N$ is a simple pole of $\tilde{\zeta}_A$, then $\zeta_A(N) < |A_\delta|$.[47] We point this out because one could wrongly conclude that the equality holds by substituting $s = N$ in the integral $\int_{A_\delta} d(x,A)^{s-N}\,\mathrm{d}x$ defining the distance zeta function ζ_A and evaluating it. See also Equation (2.1.34) in Proposition (2.1.22). If $|\overline{A}| > 0$ (and, hence, $\dim_B A = N$ according to (1.3.8)), the apparent contradiction can be explained by the fact that we are integrating the indeterminate form 0^0 over a set $\overline{A}$ of positive Lebesgue measure. In the case when $\underline{\dim}_B A < \overline{\dim}_B A = D$, it would be interesting to prove or disprove that the tube zeta function cannot have a meromorphic continuation to a connected open neighborhood of $s = D$.

Proposition 2.2.17. *Assume that A is a bounded subset of $\mathbb{R}^N$ such that its tube zeta function $\tilde{\zeta}_A$ can be meromorphically extended to a connected open neighborhood of $s = N$. Then*

$$\zeta_A(N) = |A_\delta| - \mathrm{res}(\tilde{\zeta}_A, N). \tag{2.2.46}$$

Moreover, if A is Minkowski measurable, then

$$\zeta_A(N) = |A_\delta \setminus \overline{A}|_N. \tag{2.2.47}$$

In other words, the value of $\zeta_A(N)$ is equal to the Lebesgue measure of the deleted δ-neighborhood of $\overline{A}$, i.e., of the set $A_\delta \setminus \overline{A}$.

Proof. Equation (2.2.46) was already proved in Equation (2.2.45) in the case when $s = D$ is a nonremovable singularity (i.e., in the present case, a simple pole) of the

[47] By contrast, we have the equality $\zeta_A(N) = |A_\delta|$ here, whenever $\overline{\dim}_B A < N$.

tube zeta function $\tilde{\zeta}_A$. Since, in the case when $s=N$ is a removable singularity, the claim is obvious (by passing to the limit as $s\to N$ in Equation (2.2.44)), we have thus proved the first result.

Equation (2.2.47) follows from Equation (2.2.36) of Theorem 2.2.14 and from a result stated in Remark 1.3.1 on page 31:

$$\begin{aligned}\zeta_A(N) &= |A_\delta| - \operatorname{res}(\tilde{\zeta}_A, N) = |A_\delta| - \mathcal{M}^N(A)\\ &= |A_\delta| - |\overline{A}| = |A_\delta \setminus \overline{A}|.\end{aligned} \tag{2.2.48}$$

This concludes the proof of the proposition. □

The next remark follows essentially from the discussion preceding Theorem 2.2.14 and will be used in the sequel, most of the time implicitly.

Remark 2.2.18. Assume that $\overline{D} := \overline{\dim}_B A < N$. In light of Equation (2.2.23), for any $\delta > 0$, the two zeta functions $\zeta_A(s) := \zeta_{A,A_\delta}(s)$ and $\tilde{\zeta}_A(s)$ differ by a function $\delta^{s-N} + (N-s-1)\tilde{\zeta}_A(s)$, which according to (2.2.21), is holomorphic at least on the open right half-plane $\{\operatorname{Re} s > \overline{D}\}$. Therefore, if one of them has a meromorphic continuation to some connected open set $U \supseteq \{\operatorname{Re} s > \overline{D}\}$, so does the other, and then they have exactly the same poles in U (with the same multiplicities): $\mathcal{P}(\tilde{\zeta}_A) = \mathcal{P}(\zeta_{A,A_\delta})$ or, more precisely, $\mathcal{P}(\tilde{\zeta}_A, U) = \mathcal{P}(\zeta_{A,A_\delta}, U)$. In particular,

$$\mathcal{P}_c(\tilde{\zeta}_A) = \mathcal{P}_c(\zeta_{A,A_\delta}). \tag{2.2.49}$$

Furthermore, in that case, if $\omega \in U$ is a simple pole of $\tilde{\zeta}_A(s)$, then it is also a simple pole of $\zeta_{A,A_\delta}(s)$ and the corresponding residues are related in much the same way as in Equation (2.2.24) above (note that we must have $\omega \neq N$ since $\operatorname{Re}\omega \leq \overline{D}$ and $\overline{D} < N$):

$$\operatorname{res}(\tilde{\zeta}_A, \omega) = \frac{1}{N-\omega}\operatorname{res}(\zeta_{A,A_\delta}, \omega). \tag{2.2.50}$$

More generally, if $\omega \in U$ is a pole of higher order, then the corresponding principal parts (at $s=\omega$) of the two zeta functions $\tilde{\zeta}_A$ and ζ_A are related in a simple way, which can also be deduced from the identity (2.2.23).

In light of Equation (2.2.23), it is not surprising that, under some natural conditions, the distance and tube zeta functions of a given bounded subset of $\mathbb{R}^N$ are equivalent, in the sense of Definition 2.1.69. Further information will be provided in Corollary 2.2.20 below.

Proposition 2.2.19. *Assume that A is a bounded subset of $\mathbb{R}^N$ such that $\overline{\dim}_B A < N$. Assume that either ζ_A or $\tilde{\zeta}_A$ possesses a meromorphic extension to an open, connected open neighborhood U of a window $\boldsymbol{W}$, containing the closed half-plane $\{\operatorname{Re} s \geq \overline{\dim}_B A\}$. Then, both ζ_A and $\tilde{\zeta}_A$ can be (uniquely) meromophically extended to U. Furthermore, $\mathcal{P}(\zeta_A, \boldsymbol{W}) = \mathcal{P}(\tilde{\zeta}_A, \boldsymbol{W})$.*

More specifically, the tube and distance zeta functions ζ_A and $\tilde{\zeta}_A$, meromorphically extended to U,[48] *have exactly the same set of poles in* $\boldsymbol{W}$*, with the same multiplicities. Hence, the fractal set A has the same set of visible complex dimensions, whether they are defined via ζ_A or via $\tilde{\zeta}_A$. In particular,*

$$\zeta_A \sim \tilde{\zeta}_A, \quad \text{that is,} \quad \dim_{PC} A := \mathscr{P}_c(\zeta_A) = \mathscr{P}_c(\tilde{\zeta}_A). \tag{2.2.51}$$

Also, the upper box dimension of A coincides with the abscissae of (absolute) convergence of ζ_A and $\tilde{\zeta}_A$:

$$\overline{\dim}_B A = D(\zeta_A) = D(\tilde{\zeta}_A). \tag{2.2.52}$$

Proof. In light of Equation (2.2.23), and since $s \mapsto \delta^{s-N}|A_\delta|$ is an entire function, we have that $\zeta_A(s) \sim (N-s)\tilde{\zeta}_A(s)$. Using the inequality $\overline{\dim}_B A < N$ and the same equation, we conclude from Theorem 2.2.11(a) (or from Theorem 2.2.1) that $s = N$ cannot be a singularity of $\tilde{\zeta}_A$, since, otherwise, it would also be a singularity of ζ_A. Therefore, $(N-s)\tilde{\zeta}_A(s) \sim \tilde{\zeta}_A(s)$, and the claim follows immediately by transitivity of the equivalence relation $\sim$.

Finally, the last part of the proposition (namely, Equations (2.2.51) and (2.2.52)) follows from the earlier part combined with Theorem 2.1.11(b). □

The next corollary follows at once from Theorems 2.1.11, 2.2.3, 2.2.11 and 2.2.14, combined with Proposition 2.2.19, which it complements.

Corollary 2.2.20. *If $D := \dim_B A$ exists and satisfies $D < N$, then D is a simple pole of both ζ_A and $\tilde{\zeta}_A$. Furthermore, we have*

$$\operatorname{res}(\zeta_A, D) = (N-D)\operatorname{res}(\tilde{\zeta}_A, D), \tag{2.2.53}$$

while if, in addition, A is Minkowski measurable, we have

$$\operatorname{res}(\tilde{\zeta}_A, D) = \mathscr{M}^D(A), \quad \operatorname{res}(\zeta_A, D) = (N-D)\mathscr{M}^D(A). \tag{2.2.54}$$

Moreover, if ω is a pole of the meromorphic continuation (assumed to exist) of $\tilde{\zeta}_A$ to a domain $U \subseteq \mathbb{C}$ containing $\{\operatorname{Re} s > D\}$ (or equivalently, containing the critical line $\{\operatorname{Re} s = D\}$),[49] *then it is a pole (of the same order) of the meromorphic continuation of ζ_A to U, while if, in addition, ω is a simple pole of $\tilde{\zeta}_A$ (and hence, also, of ζ_A),*[50] *then we also have*

$$\operatorname{res}(\zeta_A, \omega) = (N-\omega)\operatorname{res}(\tilde{\zeta}_A, \omega), \tag{2.2.55}$$

which is independent of δ.[51]

---

[48] As usual, the notation ζ_A and $\tilde{\zeta}_A$ for the meromorphic extensions is left unchanged.

[49] Recall from Theorem 2.1.11(a) and Theorem 2.2.11(a) that ζ_A and $\tilde{\zeta}_A$ are holomorphic on $\{\operatorname{Re} s > D\}$.

[50] Naturally, if ω is a multiple pole, and in light of Equation (2.2.20), the principal parts of $\tilde{\zeta}_A$ and ζ_A are related similarly.

[51] In light of Proposition 2.2.19, the exact same results hold if the roles of $\tilde{\zeta}_A$ and ζ_A are interchanged in the statement of Corollary 2.2.20.

In the following example, we compute the complex dimensions of the $(N-1)$-dimensional sphere in $\mathbb{R}^N$.[52]

Example 2.2.21. Let $B_R(0)$ be the ball in $\mathbb{R}^N$ centered at the origin and of radius $R>0$, and let $A := \partial B_R(0)$ be its boundary; i.e., the $(N-1)$-dimensional sphere of radius R. We would like to compute its complex dimensions. To this end, we first compute the corresponding tube zeta function $\tilde{\zeta}_A$. Let us fix any $\delta \in (0,R)$, and let $c_k := 1-(-1)^k$; i.e., $c_k := 0$ for even k and $c_k := 2$ for odd k, where $k \in \{0,1,\dots,N\}$. Since $|A_t| = \omega_N(R+t)^N - \omega_N(R-t)^N$, where $t \in (0,R)$ and ω_N is the N-dimensional Lebesgue measure of the unit ball in $\mathbb{R}^N$ (see Equation (1.3.22)), we have that for any fixed $\delta \in (0,R)$,

$$\begin{aligned}
\tilde{\zeta}_A(s) &= \int_0^\delta t^{s-N-1}|A_t|\,\mathrm{d}t = \omega_N \int_0^\delta t^{s-N-1}\left((R+t)^N-(R-t)^N\right)\mathrm{d}t \\
&= \omega_N \int_0^\delta t^{s-N-1}\left(\sum_{k=1}^N \binom{N}{k} R^{N-k}(1-(-1)^k)t^k\right)\mathrm{d}t \\
&= \omega_N \sum_{k=0}^N c_k R^{N-k}\binom{N}{k}\int_0^\delta t^{s-N+k-1}dt \\
&= \omega_N \sum_{k=0}^N c_k R^{N-k}\binom{N}{k}\frac{\delta^{s-N+k}}{s-(N-k)},
\end{aligned}$$

for all $s \in \mathbb{C}$ with $\operatorname{Re} s > N-1$. (Here, the numbers $\binom{N}{k}$ stand for the usual binomial coefficients.) The last expression can be meromorphically extended to the whole complex plane, and we still denote it by $\tilde{\zeta}_A(s)$. Therefore, we have

$$\tilde{\zeta}_A(s) = \omega_N \sum_{k=0}^N c_k R^{N-k}\binom{N}{k}\frac{\delta^{s-N+k}}{s-(N-k)}, \tag{2.2.56}$$

for all $s \in \mathbb{C}$, where the constants c_k are defined as above. It follows that

$$\dim_B A = D(\tilde{\zeta}_A) = N-1 \tag{2.2.57}$$

and moreover, the set of complex dimensions of A (i.e., the set of poles of $\tilde{\zeta}_A$ in all of $\mathbb{C}$, see Proposition 2.2.19), is given by (with $\lfloor x \rfloor$ denoting the integer part of $x \in \mathbb{R}$)

$$\begin{aligned}
\mathscr{P}(\tilde{\zeta}_A) &= \left\{N-(2j+1) : j = 0,1,2,\dots,\left\lfloor\frac{N-1}{2}\right\rfloor\right\} \\
&= \left\{N-1, N-3, \dots, N-\left(2\left\lfloor\frac{N-1}{2}\right\rfloor+1\right)\right\}.
\end{aligned} \tag{2.2.58}$$

[52] In Appendix B, we will propose a suitable notion of local distance and tube zeta functions. In particular, in Example B.0.6, we will see that the set of complex dimensions generated by the local tube zeta function is given by $\dim_{loc}\mathbb{R}^N = \{0,1,\dots,N\}$; see Equation (B.0.12).

For odd N, the last number in this set is equal to 0, while for even N, it is equal to 1.[53] Furthermore, the residue of the tube zeta function $\tilde{\zeta}_A$ at any of its poles $N-k \in \mathscr{P}(\tilde{\zeta}_A)$ is given by

$$\operatorname{res}(\tilde{\zeta}_A, N-k) = 2\omega_N \binom{N}{k} R^{N-k}. \tag{2.2.59}$$

Since $\binom{N}{k} = \binom{N}{N-k}$, we can write this result in an even more 'symmetric' form:

$$\operatorname{res}(\tilde{\zeta}_A, d) = 2\omega_N \binom{N}{d} R^d, \quad \text{for all} \quad d \in \mathscr{P}(\tilde{\zeta}_A). \tag{2.2.60}$$

Note that in the case when $d = D := N-1$, we obtain

$$\operatorname{res}(\tilde{\zeta}_A, D) = 2N\omega_N R^{N-1} = \mathscr{M}^D(A), \tag{2.2.61}$$

where the last equality is easily obtained from the definition of the Minkowski content:

$$\mathscr{M}^D(A) = \lim_{t\to 0+} \frac{|A_t|}{t^{N-D}} = \lim_{t\to 0+} \frac{\omega_N (R+t)^N - \omega_N (R-t)^N}{t} = 2N\omega_N R^{N-1}.$$

In other words, A is Minkowski measurable and

$$\mathscr{M}^D(A) = 2H^D(A), \tag{2.2.62}$$

where H^D denotes the D-dimensional Hausdorff measure.[54] Equation (2.2.61) is in accordance with Equation (2.2.36) in Theorem 2.2.14. See also the corresponding Example 4.1.19 in the context of relative fractal drums, studied in Chapter 4.

Let $A := \partial B_R(0)$, as in Example 2.2.21. Since $\dim_B A = N-1 < N$, it follows from Proposition 2.2.19 that the sphere A has the same complex dimensions, whether they are computed via the distance or the tube zeta function. Namely, $\mathscr{P}(\zeta_A) = \mathscr{P}(\tilde{\zeta}_A)$, as given by (2.2.58). Moreover, since $D := D(\zeta_A) = D(\tilde{\zeta}_A) = N-1$, we deduce from Equation (2.2.50) that for each $m \in \mathscr{P}(\zeta_A)$,

$$\operatorname{res}(\zeta_A, m) = (N-m)\operatorname{res}(\tilde{\zeta}_A, m), \tag{2.2.63}$$

as given by the right-hand side of (2.2.60).

---

[53] As we can see, the 0-dimensional sphere in $\mathbb{R}$ (which is just the pair of points $\{-1,1\}$) has 0 as its only complex dimension. Similarly, the only complex dimension of the 1-dimensional sphere in $\mathbb{R}^2$ (i.e., the unit circle S^1) is equal to 1, while the 2-dimensional sphere in $\mathbb{R}^3$ has exactly two complex dimensions (0 and 2), as well as the 3-dimensional sphere in $\mathbb{R}^4$ (namely, 1 and 3). All of these complex dimensions are simple.

[54] Equation (2.2.62) is a special case of a much more general result proved by Federer in [Fed2, Theorem 3.2.39].

Tube zeta functions of fractal sets have a scaling property analogous to that of distance zeta functions in Propositions 2.1.77; see Proposition 2.2.22 just below. We use the following notation:

$$\tilde{\zeta}_A(s;\delta) := \int_0^\delta t^{s-N-1}|A_t|\,\mathrm{d}t \quad \text{for} \quad \operatorname{Re}s > \overline{\dim}_B A.$$

Proposition 2.2.22 (Scaling property of tube zeta functions). *For any bounded subset A of $\mathbb{R}^N$, $\delta > 0$ and $\lambda > 0$, we have $D(\tilde{\zeta}_{\lambda A}(\,\cdot\,;\lambda\delta)) = D(\tilde{\zeta}_A(\,\cdot\,;\delta)) = \overline{\dim}_B A$ and*

$$\tilde{\zeta}_{\lambda A}(s;\lambda\delta) = \lambda^s \tilde{\zeta}_A(s;\delta), \tag{2.2.64}$$

for all $s \in \mathbb{C}$ with $\operatorname{Re}s > \overline{\dim}_B A$. Furthermore, if ω is a simple pole of a meromorphic extension of $\tilde{\zeta}_A(s;\delta)$ to an open connected neighborhood of the critical line (as always, we use the same notation for the meromophically extended function), then

$$\operatorname{res}(\tilde{\zeta}_{\lambda A}(\,\cdot\,;\delta),\omega) = \lambda^\omega \operatorname{res}(\tilde{\zeta}_A(\,\cdot\,;\delta),\omega). \tag{2.2.65}$$

Proof. Since $|(\lambda A)_t| = |(\lambda A)_t| = \lambda^N |A_{t/\lambda}|$, passing to the new variable $\tau = t/\lambda$ we obtain successively:

$$\begin{aligned}\tilde{\zeta}_{\lambda A}(s;\lambda\delta) &= \int_0^{\lambda\delta} t^{s-N-1}|(\lambda A)_t|\,\mathrm{d}t = \int_0^{\lambda\delta} t^{s-N-1}\lambda^N|A_{t/\lambda}|\,\mathrm{d}t \\ &= \int_0^\delta (\lambda t)^{s-N-1}\lambda^N|A_\tau|\,\lambda\mathrm{d}\tau = \lambda^s\int_0^\delta \tau^{s-N-1}|A_\tau|\,\mathrm{d}\tau = \lambda^s\tilde{\zeta}_A(s;\delta).\end{aligned}$$

Equation (2.2.65) is then obtained in much the same manner as in the proof of Proposition 2.1.77. □

Proposition 2.2.22 will be further extended in Proposition 4.6.11 to tube zeta functions of relative fractal drums.

2.2.3 Zeta Functions of Generalized Cantor Sets and a-Strings

We provide here two examples illustrating some of the main results of this section, as well as of Theorems 2.3.18, 2.3.25 and 2.3.37 below.

Example 2.2.23. (Generalized Cantor sets, Example 2.2.6 continued). Note that the Minkowski contents appearing in (2.2.35) and (2.2.36) depend on N as well; see (1.3.1). An illustration of inequality (2.2.35) in the case of generalized Cantors sets, $A = C^{(a)}$, $a \in (0,1/2)$, is provided in Figure 2.13 on page 131. It is worth observing (see Figure 2.13) that $C^{(a)}$ becomes almost like a Minkowski measurable set for a close to $1/2$, since both $\mathcal{M}^{*D}(A)$ and $\mathcal{M}_*^D(A)$ (where $D = D(a) = \log_{1/a}2)$) tend to the common limit 1 as $a \to (1/2)^-$.

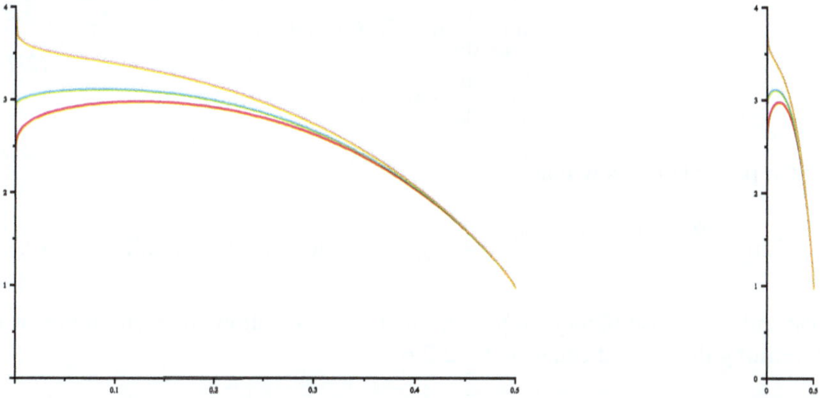

Fig. 2.13 On the left, the graphs of $\mathcal{M}^{*D}(A)$, $\operatorname{res}(\tilde{\zeta}_A, D)$ and $\mathcal{M}_*^D(A)$, viewed as functions of $a \in (0,1/2)$, are respectively depicted from top to bottom in the case of the generalized Cantor set $A = C^{(a)}$. Here, $D = \log_{1/a} 2$. The horizontal a-axis is expanded ten times with respect to the vertical axis. The same graphs are exhibited on the right with adjusted scales on both axes. This illustrates the inequality (2.2.35), as well as (2.3.36). For $a = 1/2$ we have that $D = 1$ and $\mathcal{M}^1(A)$ exists, which is in accordance with our discussion in Remark 1.3.1 on page 31.

Moreover,

$$\lim_{a\to 0^+} \mathcal{M}^{*D}(A) = 4, \qquad \lim_{a\to 0^+} \mathcal{M}_*^D(A) = 2, \tag{2.2.66}$$

and the function $a \mapsto \mathcal{M}^{*D}(A) - \mathcal{M}_*^D(A)$ is decreasing on $(0,1/2)$ from 2 to 0.

If we define the *oscillatory amplitude* of the Cantor set A by

$$\mathbf{am}(A) := \mathcal{M}^{*D}(A) - \mathcal{M}_*^D(A) = 2(1-a)\left(\frac{1}{2}-a\right)^{D-1} - \frac{1}{D}\left(\frac{2D}{1-D}\right)^{1-D},$$

then $\mathbf{am}(A)$ is monotonically decreasing as a function of $a \in (0,1/2)$; see Figure 2.13. (A general definition of the oscillatory amplitude valid for a wide class of bounded sets in $\mathbb{R}^N$ is provided in Subsection 6.1.1.1 on page 541.) Furthermore,

$$\lim_{a\to 0^+} \mathbf{am}(A) = 2, \qquad \lim_{a\to 1/2^-} \mathbf{am}(A) = 0. \tag{2.2.67}$$

The following limits describe the behavior of the D-dimensional Minkowski contents for a close to 0 and $1/2$ (see the right side of Figure 2.13):

$$\begin{aligned} &\lim_{a\to 0^+} \frac{\mathrm{d}}{\mathrm{d}a}\mathcal{M}^{*D}(A) = -\infty, && \lim_{a\to 0^+} \frac{\mathrm{d}}{\mathrm{d}a}\mathcal{M}_*^D(A) = +\infty, \\ &&\text{and}&& \\ &\lim_{a\to 1/2^-} \frac{\mathrm{d}}{\mathrm{d}a}\mathcal{M}^{*D}(A) = -\infty, && \lim_{a\to 1/2^-} \frac{\mathrm{d}}{\mathrm{d}a}\mathcal{M}_*^D(A) = -\infty, \end{aligned} \tag{2.2.68}$$

where we have used (2.2.12). The above limits at $1/2$ help explain the spikes observed in Figure 2.13. Using (2.2.11), we deduce that

$$\lim_{a\to 0^+} \frac{\mathrm{d}}{\mathrm{d}a} \operatorname{res}(\tilde{\zeta}_A, D) = +\infty,$$
$$\lim_{a\to 1/2^-} \frac{\mathrm{d}}{\mathrm{d}a} \operatorname{res}(\tilde{\zeta}_A, D) = -\infty. \tag{2.2.69}$$

It is also possible to show that

$$\lim_{a\to 1/2^-} \frac{\mathscr{M}^{*D}(A) - \mathscr{M}_*^D(A)}{1/2 - a} = 4(\log 2 - \log(\log 2) - 1) \approx 0.2388. \tag{2.2.70}$$

In particular, the oscillatory amplitude of A has the following asymptotics at $a = 1/2$, refining the second equation in (2.2.67):

$$\mathbf{am}(A) \sim c \cdot \Big(\frac{1}{2} - a\Big) \quad \text{as} \quad a \to 1/2^-, \tag{2.2.71}$$

where c is the value of the limit in (2.2.70). Indeed, we deduce from (2.2.12) that

$$\lim_{a\to 1/2^-} \frac{\mathscr{M}^{*D}(A) - \mathscr{M}_*^D(A)}{1 - D} = \log 2 - \log(\log 2) - 1 \approx 0.0597,$$

and (2.2.70) follows since $\lim_{a\to 1/2^-} \frac{1-D}{1/2-a} = 4$.

The lower D-dimensional Minkowski content $\mathscr{M}_*^D(A)$, viewed as a function of $a \in (0, 1/2)$, attains its maximum at $a = 1/8$, and the maximum value is $\mathscr{M}_*^D(C^{(1/8)}) = 3$. Furthermore, the residue of $\tilde{\zeta}_A(s, A_\delta)$ at $s = D$ attains its maximum at $a \approx 0.08649194033$, and its maximum value is approximately given by $\operatorname{res}(\tilde{\zeta}_A(\,\cdot\,, A_\delta), D) \approx 3.134663524$.

Example 2.2.24. (*a-strings*). Given $a > 0$, let $A := \{j^{-a} : j \in \mathbb{N}\}$. This set is Minkowski measurable,

$$\mathscr{M}^D(A) = \frac{2^{1-D}}{1-D} a^D, \quad D = D(a) = \frac{1}{1+a}, \tag{2.2.72}$$

and the related string $\mathscr{L} = (\ell_j)_{j\geq 1}$ defined by $\ell_j = j^{-a} - (j+1)^{-a}$ is called the *a-string*; see [Lap1], [LapPo2] and [Lap-vFr3, Section 6.5.1] for the study of its various properties. (Geometrically, the a-string is realized as the complement of A in $[0, 1]$; therefore, its boundary is equal to $A \cup \{0\}$.) Due to (2.2.4) and (2.2.36), we know that

$$\operatorname{res}(\zeta_{A,A_\delta}, D) = (1-D)\mathscr{M}^D(A), \quad \operatorname{res}(\tilde{\zeta}_A, D) = \mathscr{M}^D(A). \tag{2.2.73}$$

The graphs of these two residues, viewed as functions of $a > 0$, are shown in Figure 2.14 on page 133. Using (2.2.72), we see that for any fixed positive number δ, we have

$$\lim_{a\to 0^+} \operatorname{res}(\zeta_{A,A_\delta}, D) = 0, \quad \lim_{a\to 0^+} \operatorname{res}(\tilde{\zeta}_A, D) = 1,$$

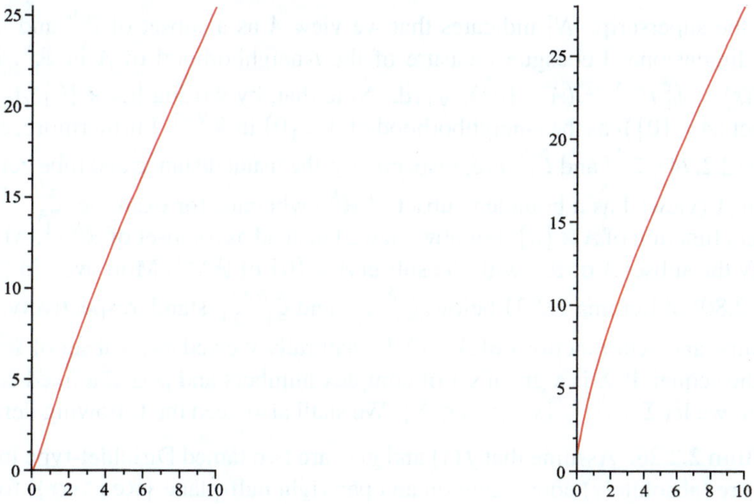

Fig. 2.14 The graphs of the functions $a \mapsto \operatorname{res}(\zeta_{A,A_\delta}, D)$ and $a \mapsto \operatorname{res}(\tilde{\zeta}_A, D)$, where $a > 0$, $A = \{j^{-a} : j \in \mathbb{N}\}$ and $D = 1/(1+a)$. (See Example 2.2.24.) These two graphs coincide with the graphs of $a \mapsto (1-D)\mathscr{M}^D(A)$ and $a \mapsto \mathscr{M}^D(A)$; see Equation (2.2.73).

$$\lim_{a\to 0^+} \frac{\mathrm{d}}{\mathrm{d}a} \operatorname{res}(\zeta_{A,A_\delta}, D) = 1, \quad \lim_{a\to 0^+} \frac{\mathrm{d}}{\mathrm{d}a} \operatorname{res}(\tilde{\zeta}_A, D) = +\infty,$$

and

$$\lim_{a\to +\infty} \frac{\mathrm{d}}{\mathrm{d}a} \operatorname{res}(\zeta_{A,A_\delta}, D) = 2, \quad \lim_{a\to +\infty} \frac{\mathrm{d}}{\mathrm{d}a} \operatorname{res}(\tilde{\zeta}_A, D) = 2.$$

2.2.4 Distance and Tube Zeta Functions of Fractal Grills

It is of interest to understand the behavior of the distance and tube zeta functions with respect to the Cartesian products of sets. We consider a very special type of Cartesian products, called fractal grills, and we shall study their distance and tube zeta functions.

Definition 2.2.25. Let A be a bounded subset of $\mathbb{R}^N$ and let k be a positive integer. Then, the subset of $\mathbb{R}^{N+k}$ of the form $A \times [0,1]^k$ is called the *fractal grill* (generated by A). More generally, we can consider fractal grills of the form $A \times [a,b]^k \subset \mathbb{R}^{N+k}$, where a and b are positive real numbers with $a < b$.

Since a given bounded subset A of $\mathbb{R}^N$ can be naturally identified with $A \times \{0\} \subset \mathbb{R}^{N+1}$, it will be convenient to introduce the following notation (for all $s \in \mathbb{C}$ with $\operatorname{Re} s$ sufficiently large):

$$\zeta_A^{[N]}(s) := \int_{A_\delta} d(x,A)^{s-N}\,\mathrm{d}x, \quad \tilde{\zeta}_A^{[N]}(s) := \int_0^\delta t^{s-N-1}|A_t|_N\,\mathrm{d}t, \tag{2.2.74}$$

where the superscript $[N]$ indicates that we view A as a subset of $\mathbb{R}^N$ and $|A_t|_N$ is the N-dimensional Lebesgue measure of the t-neighborhood of A in $\mathbb{R}^N$. Hence, $\tilde{\zeta}_A^{[N+1]}(s) = \int_0^\delta t^{s-N-2}|(A\times\{0\})_t|_{N+1}\mathrm{d}t$. Note that, by writing $|(A\times\{0\})_t|_{N+1}$, we interpret $(A\times\{0\})_t$ as the t-neighborhood of $A\times\{0\}$ in $\mathbb{R}^{N+1}$. Furthermore, observe that, in (2.2.74), $\zeta_A^{[N]}$ and $\tilde{\zeta}_A^{[N]}$ are, respectively, the usual distance and tube zeta functions of A (viewed as a bounded subset of $\mathbb{R}^N$) whereas, for example, $\tilde{\zeta}_A^{[N+1]}$ is the tube zeta function of $A\times\{0\}$, but now viewed instead as a subset of $\mathbb{R}^{N+1}$, where we identify the subset A of $\mathbb{R}^N$ with the subset $A\times\{0\}$ of $\mathbb{R}^{N+1}$. Moreover, in (2.2.79) and (2.2.80) of Lemma 2.2.31 below, $\zeta_{A\times[0,1]}^{[N+1]}$ and $\tilde{\zeta}_{A\times[0,1]}^{[N+1]}$ stand, respectively, for the usual tube and zeta functions of $A\times[0,1]$ (naturally viewed as a subset of $\mathbb{R}^{N+1}$).

In the sequel, if Σ is a given set of complex numbers and $\rho\in\mathbb{C}$ a fixed complex number, we let $\Sigma+\rho := \{s+\rho : s\in\Sigma\}$. We shall also need the following definition.

Definition 2.2.26. Assume that $f(s)$ and $g(s)$ are two tamed Dirichlet-type integrals which are (absolutely) convergent on an open right half-plane $\{\mathrm{Re}\, s>\alpha\}$, for some $\alpha\in\mathbb{R}$. Let their difference $h(s) := f(s)-g(s)$ be such that $D(h)<D(g)$.[55] Then we say that f and g are *weakly equivalent* and write $f\simeq g$.

Remark 2.2.27. The difference $h := f-g$, appearing in Definition 2.2.26, is a tamed DTI. To see this, it suffices to apply Theorem A.2.3 of Appendix A with $\alpha=1$ and $\beta=-1$. It then follows from Theorem A.1.4 of Appendix A that both $D(h)$ and $\Pi(h)$ are well defined.

Note that in Definition 2.2.26, we do *not* assume that g possesses a meromorphic continuation to a connected open neighborhood of any point on its critical line $\{\mathrm{Re}\, s = D(g)\}$. This is in contrast to the definition of equivalence $\sim$ introduced in Definition 2.1.69 of Subsection 2.1.5 and extended in Section A.5 of Appendix A.

Case (c) of Lemma 2.2.28 below provides a simple and useful condition for the implication $f\simeq g \implies f\sim g$ to hold, where the equivalence $\sim$ is described in Definition 2.1.69 above.

Lemma 2.2.28. *Assume that f and g are two Dirichlet-type integrals such that $f\simeq g$, in the sense of Definition 2.2.26 above. Then, the following properties hold:*

(a) *We have $D(f)=D(g)$.*

(b) *The relation $\simeq$ is reflexive and symmetric. If in Definition 2.2.26 we consider the class of tamed DTIs in the $($complex$)$ function space $\mathscr{E}_{(E,\varphi)}$ introduced in Definition A.6.1 of Appendix A $($see also Definitions A.1.1 and A.1.2$)$, with a given pair (E,φ), then $\simeq$ is a relation of equivalence on this vector space.*

(c) *If there exists a connected open set $U\subseteq\{\mathrm{Re}\, s>D(f-g)\}$ containing the critical line $\{\mathrm{Re}\, s=D(g)\}$ and such that g can be meromorphically continued to U, then f has the same property and $\mathscr{P}_c(f)=\mathscr{P}_c(g)$. In particular, $f\sim g$ in the sense of Definition 2.1.69.*

[55] Alternatively, but equivalently, assume that there exists a real number β, $\beta<D(g)$, such that the integral defining h is absolutely convergent (and hence, holomorphic) on $\{\mathrm{Re}\, s>\beta\}$.

Proof. (a) Since, by Definition 2.2.26, $f(s) = g(s) + h(s)$ and $D(h) < D(g)$, we conclude that $D(f) \leq D(g)$. If $D(f) < D(g)$, we would then have

$$\max\{D(f), D(h)\} < D(g). \tag{2.2.75}$$

On the other hand, the integral defining the function (i.e., the DTI) $g(s) = f(s) - h(s)$ is absolutely convergent on $\{\operatorname{Re} s > \max\{D(f), D(h)\}\}$, which is impossible in light of (2.2.75). This contradiction proves that $D(f) = D(g)$.

Property (b) follows at once from (a) and Definition 2.2.26.

Property (c) follows easily from the relation $f(s) = g(s) + h(s)$. Indeed, the reflexivity $f \simeq f$ follows by taking $h = 0$. If $f \simeq g$ in the sense of Definition 2.2.26 (with a function $h := f - g$), then by noting that $D(h) = D(-h)$, we conclude that $g \simeq f$ (with the function $-h = g - f$), which proves the symmetry of the relation $\simeq$. Finally, in order to prove the transitivity of the relation (under the stated additional conditions), assume that $f, g, h \in \mathscr{E}_{(E,\varphi)}$ are such that $f \simeq g$ (with the corresponding function h_1) and $g \simeq h$ (with respect to h_2). It then follows that $f \simeq h$ with respect to $h_1 + h_2$, by observing that $D(h_1 + h_2) \leq D(h_1) + D(h_2)$.

Note that by Theorem A.2.3 of Appendix A, the functions h_1, h_2 and $h_1 + h_2$ are also tamed DTIs contained in the vector space $\mathscr{E}_{(E,\varphi)}$ of all tamed DTIs of the form $\zeta_{E,\varphi,\rho}$ (with the pair (E, φ) fixed and the local measure ρ arbitrary) introduced in Definition A.6.1 (see also Definition A.1.2). $\square$

The following simple lemma is crucial, since it shows that the tube function of the fractal grill $A \times [0,1]$ in $\mathbb{R}^{N+1}$ is equal to the sum of the tube function of the subset A of $\mathbb{R}^N$ and the tube function of the subset $A \times \{0\}$ of $\mathbb{R}^{N+1}$.

Lemma 2.2.29 ([Res, Remark 1]). *Let A be a bounded subset of $\mathbb{R}^N$. Then*

$$|(A \times [0,1])_t|_{N+1} = |A_t|_N + |(A \times \{0\})_t|_{N+1}. \tag{2.2.76}$$

Proof. We can represent the subset $(A \times [0,1])_t$ of $\mathbb{R}^{N+1}$ as the union of three pairwise disjoint subsets:

$$\begin{aligned}(A \times [0,1])_t = A_t &\times [0,1] \\ &\cup (A \times \{0\})_t \cap \{x_{n+1} < 0\}) \\ &\cup (A \times \{1\})_t \cap \{x_{n+1} > 1\}),\end{aligned} \tag{2.2.77}$$

where $\{x_{N+1} < 0\} := \{x = (x_1, \ldots x_N, x_{N+1}) \in \mathbb{R}^{N+1} : x_{N+1} < 0\}$ and similarly for $\{x_{n+1} > 1\}$. If we translate the subset $(A \times \{1\})_t \cap \{x_{n+1} > 0\}$ by the vector $(0,, \ldots, 0, -1) \in \mathbb{R}^{N+1}$, the resulting subset $(A \times \{0\})_t \cap \{x_{n+1} > 0\}$ still remains disjoint with respect to the second set appearing on the right-hand side of Equation (2.2.77). Therefore,

$$
\begin{aligned}
|(A\times[0,1])_t|_{N+1} &= |A_t\times[0,1]|_{N+1}\\
&\quad +|(A\times\{0\})_t\cap\{x_{n+1}<0\}\cup(A\times\{1\})_t\cap\{x_{n+1}>1\}|_{N+1}\\
&= |A_t|_N\cdot 1\\
&\quad +|(A\times\{0\})_t\cap\{x_{n+1}<0\}\cup(A\times\{0\})_t\cap\{x_{n+1}>0\}|_{N+1}\\
&= |A_t|_N+|(A\times\{0\})_t\setminus\{x_{N+1}=0\}|_{N+1}.
\end{aligned}
$$

However, since the hyperplane $\{x_{N+1}=0\}$ has $(N+1)$-dimensional Lebesuge measure equal to zero, we conclude that $|(A\times\{0\})_t\setminus\{x_{N+1}=0\}|_{N+1}=|(A\times\{0\})_t|_{N+1}$. Hence, $|(A\times[0,1])_t|_{N+1}=|A_t|_N+|(A\times\{0\})_t|_{N+1}$ and this completes the proof of the lemma. □

Remark 2.2.30. By slightly modifying the proof of Lemma 2.2.29, we conclude that for any bounded subset A of $\mathbb{R}^N$ and for any two real numbers a and b such that $a<b$, we have

$$|(A\times[a,b])_t|_{N+1}=|A_t|_N(b-a)+|(A\times\{0\})_t|_{N+1}. \tag{2.2.78}$$

Lemma 2.2.31. *Let A be a bounded subset of $\mathbb{R}^N$. Then*

$$\zeta^{[N+1]}_{A\times[0,1]}(s)=\zeta^{[N]}_A(s-1)+\zeta^{[N+1]}_A(s) \tag{2.2.79}$$

and

$$\tilde{\zeta}^{[N+1]}_{A\times[0,1]}(s)=\tilde{\zeta}^{[N]}_A(s-1)+\tilde{\zeta}^{[N+1]}_A(s), \tag{2.2.80}$$

for all $s\in\mathbb{C}$ with $\operatorname{Re}s>\overline{\dim}_BA+1$. In particular, if A is such that ζ_A or (equivalently, provided $\overline{\dim}_BA<N$) $\tilde{\zeta}_A$ admits a (necessarily unique) meromorphic continuation to a connected open neighborhood of the critical line of Lebesgue (absolute) convergence $\{\operatorname{Re}s=D(\zeta_A)\}$,[56] *then*

$$\zeta^{[N+1]}_{A\times[0,1]}(s)\simeq\zeta^{[N]}_A(s-1)\quad\text{and}\quad\tilde{\zeta}^{[N+1]}_{A\times[0,1]}(s)\simeq\tilde{\zeta}^{[N]}_A(s-1). \tag{2.2.81}$$

Hence, if ζ_A can be meromorphically continued to a connected, open set U containing the critical line $\{\operatorname{Re}s=D(\zeta_A)\}$, then $\mathscr{P}_c(\zeta_{A\times[0,1]})=\mathscr{P}_c(\zeta_A)+1$; that is,

$$\dim_{PC}(A\times[0,1])=\dim_{PC}A+1. \tag{2.2.82}$$

In particular, if $\overline{\dim}_BA<N$, then

$$
\begin{aligned}
D(\zeta^{[N+1]}_{A\times[0,1]})&=D(\zeta^{[N]}_A)+1=D(\tilde{\zeta}^{[N]}_A)+1=D(\tilde{\zeta}^{[N+1]}_{A\times[0,1]})\\
&=\overline{\dim}_B(A\times[0,1])=\overline{\dim}_BA+1.
\end{aligned} \tag{2.2.83}
$$

Proof. Let us first prove Equation (2.2.80). Substituting Equation (2.2.76) from Lemma 2.2.29 into the second equality of (2.2.74), we conclude that

[56] Recall from part (a) of Theorem 2.1.11 that $D(\zeta_A)=\overline{\dim}_BA$.

$$
\begin{aligned}
\tilde{\zeta}^{[N+1]}_{A\times[0,1]}(s) &= \int_0^\delta t^{s-N-2}(|A_t|_N + |(A\times\{0\})_t|_{N+1})\,\mathrm{d}t \\
&= \int_0^\delta t^{(s-1)-N-1}|A_t|_N\mathrm{d}t + \int_0^\delta t^{s-(N+1)-1}|(A\times\{0\})_t|_{N+1}\mathrm{d}t \\
&= \tilde{\zeta}^{[N]}_A(s-1) + \tilde{\zeta}^{[N+1]}_A(s),
\end{aligned}
\tag{2.2.84}
$$

for all $s\in\mathbb{C}$ with $\mathrm{Re}\,s > \overline{\dim}_B A + 1$. Here, we also use the easily verified fact that $\overline{\dim}_B A$ is the same in the case when $A\subset\mathbb{R}^{N+1}$, as in the case when $A\subset\mathbb{R}^N$; that is, the upper box dimension of a bounded set, as well as the lower box dimension, does not depend on N; see [Had] or, e.g., [Res, Proposition 1]. This completes the proof of Equation (2.2.80).

Moreover, let us note that all tube zeta functions can be viewed as tamed DTIs based on the same underlying pair (E,φ), with $E := (0,\delta)$ and $\varphi(t) := t$ for all $t\in E$. (See Definition 2.2.8 and the proof of Lemma 2.2.9 above.) It then easily follows that $\tilde{\zeta}^{[N+1]}_{A\times[0,1]}$ and $\tilde{\zeta}^{[N]}_A(\cdot-1)$ are tamed DTIs based on the same pair (E,φ), so that in light of Theorem A.2.3 of Appendix A below, the second weak equivalence in (2.2.81) holds (see Definition 2.2.26).

Let us next establish Equation (2.2.79). To this end, we use (2.2.23), which we write in the following form:

$$
\tilde{\zeta}^{[N]}_A(s) = \frac{\zeta^{[N]}_A(s) - \delta^{s-N}|A_\delta|_N}{N-s},
\tag{2.2.85}
$$

for all $s\in\mathbb{C}$ with $\mathrm{Re}\,s > \overline{\dim}_B A$ and $s\neq N$. Making use of Equation (2.2.84), we deduce that

$$
\begin{aligned}
\frac{\zeta^{[N+1]}_{A\times[0,1]}(s) - \delta^{s-N-1}|(A\times[0,1])_\delta|_{N+1}}{(N+1)-s} &= \frac{\zeta^{[N]}_A(s-1) - \delta^{(s-1)-N}|A_\delta|_N}{N-(s-1)} \\
&\quad + \frac{\zeta^{[N+1]}_A(s) - \delta^{s-(N+1)}|(A\times\{0\})_\delta|_{N+1}}{(N+1)-s},
\end{aligned}
\tag{2.2.86}
$$

for all $s\in\mathbb{C}$ with $\mathrm{Re}\,s > \overline{\dim}_B A$ and $s\neq N+1$. Since, in light of (2.2.76), we have $|(A\times[0,1])_\delta|_{N+1} = |A_\delta|_N + |(A\times\{0\})_\delta|_{N+1}$, we conclude from (2.2.86) after a short computation that

$$
\zeta^{[N+1]}_{A\times[0,1]}(s) = \zeta^{[N]}_A(s-1) + \zeta^{[N+1]}_A(s),
\tag{2.2.87}
$$

for all $s\in\mathbb{C}$ with $\mathrm{Re}\,s > \overline{\dim}_B A + 1$, where we have also used the principle of analytic continuation. This completes the proof of Equation (2.2.79).

Note that, according to Theorem 2.1.11, both $\zeta^{[N]}_A(s-1)$ and $\zeta^{[N+1]}_{A\times[0,1]}(s)$ are holomorphic on $\{\mathrm{Re}\,s > \overline{\dim}_B A + 1\}$ (recall that $\overline{\dim}_B(A\times[0,1]) = \overline{\dim}_B A + 1$, see [Fal1]), while, according to the same theorem, the function

$$\zeta_{A\times[0,1]}^{[N+1]}(s)-\zeta_A^{[N]}(s-1)=\zeta_A^{[N+1]}(s) \tag{2.2.88}$$

is holomorphic on $\{\operatorname{Re} s>\overline{\dim}_B A\}$. Therefore, since

$$D(\zeta_A^{[N+1]})=\overline{\dim}_B A<\overline{\dim}_B A+1=D(\zeta_A^{[N]}(\cdot-1)), \tag{2.2.89}$$

it follows from Definition 2.2.26 that $\zeta_{A\times[0,1]}^{[N+1]}(s)\simeq\zeta_A^{[N]}(s-1)$. This completes the proof of the first weak equivalence in (2.2.81).

The remaining part of Lemma 2.2.31 can be deduced from part (c) of Lemma 2.2.28 by noting that since $\zeta_A(s)$ can be meromorphically continued to the set U, then $\zeta_A(s-1)$ can be meromorphically continued to the set $U+1$. Hence, by Lemma 2.2.28(c), we have $\zeta_{A\times[0,1]}^{[N+1]}(s)\sim\zeta_A^{[N]}(s-1)$ in the sense of Definition 2.1.69, and therefore,

$$\mathscr{P}_c(\zeta_{A\times[0,1]}^{[N+1]})=\mathscr{P}_c(\zeta_A^{[N]}(\cdot-1))=\mathscr{P}_c(\zeta_A^{[N]})+1,$$

or, equivalently, $\dim_{PC}(A\times[0,1])=\dim_{PC}A+1$. This completes the proof of the lemma. □

Theorem 2.2.32. *Let A be a bounded subset of $\mathbb{R}^N$ and let d be a positive integer. Then the following properties hold:*

(a) The distance and tube zeta functions of $A\times[0,1]^d\subset\mathbb{R}^{N+d}$ are given, respectively, by

$$\zeta_{A\times[0,1]^d}^{[N+d]}(s)=\sum_{k=0}^{d}\binom{d}{k}\zeta_A^{[N+k]}(s-d+k) \tag{2.2.90}$$

and

$$\tilde{\zeta}_{A\times[0,1]^d}^{[N+d]}(s)=\sum_{k=0}^{d}\binom{d}{k}\tilde{\zeta}_A^{[N+k]}(s-d+k), \tag{2.2.91}$$

for all $s\in\mathbb{C}$ with $\operatorname{Re} s>\overline{\dim}_B A+d$.

(b) If the distance zeta function ζ_A can be meromophically extended to a connected open set containing the critical line $\{\operatorname{Re} s=\overline{\dim}_B A\}$, then

$$\zeta_{A\times[0,1]^d}^{[N+d]}(s)\sim\zeta_A^{[N]}(s-d),\quad \tilde{\zeta}_{A\times[0,1]^d}^{[N+d]}(s)\sim\tilde{\zeta}_A^{[N]}(s-d) \tag{2.2.92}$$

and $\mathscr{P}_c(\zeta_{A\times[0,1]^d})=\mathscr{P}_c(\zeta_A)+d$; that is,

$$\dim_{PC}(A\times[0,1]^d)=\dim_{PC}A+d. \tag{2.2.93}$$

In particular, if $\overline{\dim}_B A<N$, then

$$\begin{aligned}D(\zeta_{A\times[0,1]^d}^{[N+d]})&=D(\zeta_A^{[N]})+d=D(\tilde{\zeta}_A^{[N]})+d=D(\tilde{\zeta}_{A\times[0,1]^d}^{[N+d]})\\&=\overline{\dim}_B(A\times[0,1]^d)=\overline{\dim}_B A+d.\end{aligned} \tag{2.2.94}$$

Proof. (a) Let us first establish Equation (2.2.90). We do so by using mathematical induction on d. The case when $d=1$ has already been established in Lemma 2.2.31.

Now, let us assume that the claim holds for some fixed positive integer $d \geq 1$. We deduce from (2.2.79) that

$$\zeta^{[N+d+1]}_{A\times[0,1]^{d+1}}(s) = \zeta^{[N+d]}_{A\times[0,1]^{d}}(s-1) + \zeta^{[(N+1)+d]}_{A\times[0,1]^{d}}(s).$$

Therefore,

$$\begin{aligned}\zeta^{[N+d+1]}_{A\times[0,1]^{d+1}}(s) &= \sum_{k=0}^{d}\binom{d}{k}\zeta_A^{[N+k]}(s-1-d+k) + \sum_{k=0}^{d}\binom{d}{k}\tilde{\zeta}_A^{[N+1+k]}(s-d+k)\\ &= \zeta_A^{[N]}(s-d-1) + \sum_{k=0}^{d-1}\binom{d}{k+1}\zeta_A^{[N+k+1]}(s-d+k)\\ &\quad + \sum_{k=0}^{d-1}\binom{d}{k}\zeta_A^{[N+1+k]}(s-d+k) + \zeta_A^{[N+1+d]}(s)\\ &= \sum_{k=0}^{d+1}\binom{d+1}{k}\zeta_A^{[N+k]}(s-(d+1)+k),\end{aligned}$$

where in the last equality we have used the fact that $\binom{d}{k} + \binom{d}{k+1} = \binom{d+1}{k+1}$. This completes the proof of Equation (2.2.90).

Equation (2.2.91) can be proved by mathematical induction in much the same way as in the case of the distance zeta function. This completes the proof of part (a) of the theorem.

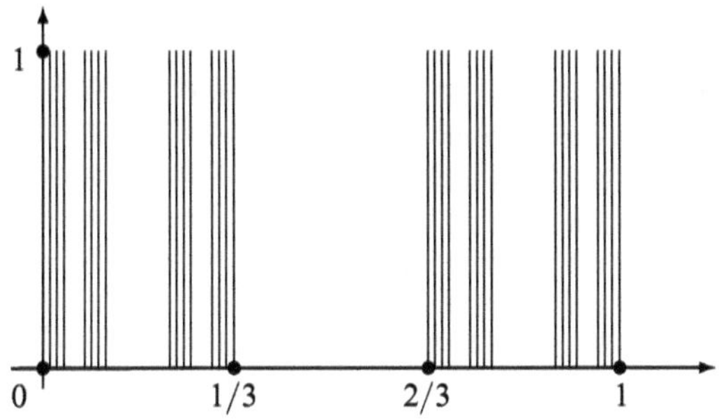

Fig. 2.15 The distance zeta function associated with the *Cantor grill* $A := C^{(1/3)} \times [0,1]$ satisfies $\zeta_A(s) = \zeta_{C^{(1/3)}}(s-1) \sim (1-2\cdot(1/3)^{s-1})^{-1} \sim (3^{s-1}-2)^{-1}$, and the corresponding set of principal complex dimensions of A is given by $\dim_{PC} A = (\log_3 2+1) + \frac{2\pi}{\log 3}\mathrm{i}\mathbb{Z}$; see Example 2.2.34.

(b) To prove that $\zeta^{[N+d]}_{A\times[0,1]^{d}}(s) \sim \zeta_A^{[N]}(s-d)$, it suffices to note that, by Equation (2.2.90), the function

$$h(s) := \zeta^{[N+d]}_{A\times[0,1]^{d}}(s) - \zeta_A^{[N]}(s-d) = \sum_{k=1}^{d}\binom{d}{k}\zeta_A^{[N+k]}(s-d+k) \tag{2.2.95}$$

has for abscissa of convergence

$$D(h) = \overline{\dim}_B A + (d-1)\} < \overline{\dim}_B A + d = D(\zeta_A^{[N]}(\cdot - d)); \tag{2.2.96}$$

so that $\zeta_{A\times[0,1]^d}^{[N+d]}(s) \simeq \zeta_A^{[N]}(s-d)$. Using part (c) of Lemma 2.2.28, we deduce that $\zeta_{A\times[0,1]^d}^{[N+d]}(s) \sim \zeta_A^{[N]}(s-d)$ in the sense of Definition 2.1.69, which proves the first relation in (2.2.92). The second relation in (2.2.92) can be proved along the same lines. This completes the proof of claim (b), as well as of the entire theorem. □

Remark 2.2.33. The relations appearing in (2.2.92) can be written in a less precise form as follows:

$$\zeta_{A\times[0,1]^d}(s) \sim \zeta_A(s-d) \quad \text{and} \quad \tilde{\zeta}_{A\times[0,1]^d}(s) \sim \tilde{\zeta}_A(s-d). \tag{2.2.97}$$

We propose to call these two properties the *shift properties* of the distance and tube zeta functions, respectively.

Example 2.2.34. (*Generalized Cantor sets and Cantor grills*). Let $A = C^{(a)}$ is the generalized Cantor set introduced above in Example 2.2.6 and let d be a positive integer. Then, using (2.2.92) and (2.2.10), we obtain that

$$\zeta_{C^{(a)}\times[0,1]^d}(s) \sim \frac{1}{1-2a^{s-d}}.$$

Furthermore, we conclude from (2.2.93) that (with $\mathrm{i} := \sqrt{-1}$, as usual)

$$\dim_{PC}(C^{(a)} \times [0,1]^d) = (\log_{1/a} 2 + d) + \frac{2\pi}{\log(1/a)}\mathrm{i}\mathbb{Z}. \tag{2.2.98}$$

Moreover, by noticing that $\zeta_{C^{(a)}\times[0,1]^d}$ can be meromorphically extended to the whole complex plane, we conclude from Equation (2.2.90) above and from the first part of Equation (3.1.6) below that the set of all complex dimensions of $C^{(a)} \times [0,1]^d \subset \mathbb{R}^{1+d}$ is well defined in $\mathbb{C}$ and given by

$$\mathscr{P}(\zeta_{C^{(a)}\times[0,1]^d}) = \{0,1,\dots,d\} \cup \bigcup_{k=0}^{d} \left((\log_{1/a} 2 + k) + \frac{2\pi}{\log(1/a)}\mathrm{i}\mathbb{Z}\right). \tag{2.2.99}$$

The sets of the form $C^{(a)} \times [0,1]^d$ appear, for example, in the study of the Smale horseshoe map; see, e.g., [Sma]. They also appear naturally in the study of the singularities of Sobolev functions and of weak solutions of elliptic equations; see, e.g., [Žu1] and [HorŽu], where they are called 'Cantor grills'.

Equations (2.2.98) and (2.2.99) also hold for a more general class of Cantor grills $C^{(m,a)} \times [0,1]^d$, involving a class of Cantor sets $C^{(m,a)}$ depending on two paramters a and m, which we introduce in Definition 3.1.1 of Chapter 3 below.

Example 2.2.35. (*Fractal combs*). Similarly as in Example 2.2.34, sets of the form $\partial\Omega \times [0,1]^{N-1}$, where $\Omega = \Omega_a$ is a geometric realization of a fractal string (for example, the so-called a-string, $\Omega = \cup_{j=1}^{\infty}((j+1)^{-a}, j^{-a})$, where $a > 0$ and for which $\partial\Omega = \{j^{-a} : j \geq 1\} \cup \{0\}$ satisfies $\overline{\dim}_B\partial\Omega = 1/(a+1)$; see Example 2.2.24), are used in the study of fractal drums to extend certain results from one to higher dimensions $N \geq 2$; see [Lap1, Examples 5.1 and 5.1']. See also Subsection 4.3.2 below for further use of the same technique as in [Lap1–3], in a closely related context.

The open set $\Omega \times (0,1)^{N-1}$, whose boundary is

$$\partial(\Omega \times (0,1)^{N-1}) = (\partial\Omega \times [0,1]^{N-1}) \cup \big([0,1] \times \partial((0,1)^{N-1})\big), \tag{2.2.100}$$

and where $\partial\big(([0,1]^{N-1}\big)$ is taken in the space $\mathbb{R}^{N-1}$, is called a 'fractal comb' in [Lap1–3]. (See also [LapRaŽu7].) Note that the subset $\partial\big((0,1)^{N-1}\big)$ of $\mathbb{R}^{N-1}$ is an $(N-2)$-dimensional Lipschitz submanifold (which for $N = 2$ degenerates to a pair of points); hence, the box dimension of $[0,1] \times \partial((0,1)^{N-1})$ is equal to $N - 1$. Therefore, by the property of 'finite stability' of the upper box dimension (see [Fal1]), we have

$$\begin{aligned}\overline{\dim}_B\partial(\Omega \times (0,1)^{N-1}) &= \max\{\overline{\dim}_B(\partial\Omega \times [0,1]^{N-1}), N-1\} \\ &= \overline{\dim}_B(\partial\Omega \times [0,1]^{N-1}) = \overline{\dim}_B\partial\Omega + N - 1.\end{aligned}$$

Since, according to [Lap-vFr3, Theorem 6.21] (along with Example (2.1.58) and Proposition (2.1.59) above),

$$\dim_{PC}\partial(\Omega_a) = \{\rho, -\rho, -2\rho, -3\rho, \dots\}, \tag{2.2.101}$$

where $\rho := 1/(a+1)$, we deduce from Theorem 2.2.32 that

$$\begin{aligned}&\dim_{PC}\partial(\Omega_a \times (0,1)^{N-1}) = \dim_{PC}(\partial\Omega_a \times [0,1]^{N-1}) \\ &= \{N-1+\rho, N-1-\rho, N-1-2\rho, N-1-3\rho, \dots\},\end{aligned} \tag{2.2.102}$$

still with $\rho = 1/(a+1)$. Furthermore, all of these complex dimensions are simple.

More precisely, it could be that in Equation (2.2.101), beside ρ, which is always a (simple) pole of $\zeta_{\partial\Omega}$, some of the numbers $-n\rho$ $(n \geq 1)$ are not poles of $\zeta_{\partial\Omega}$ (because the coresponding residue of $\zeta_{\partial\Omega}$ happens to vanish, for some arithmetic reason connected with the value of a). And hence, similarly, in Equation (2.2.102).

Finally, we point out that if, instead, $\Omega = \Omega_{CS}$ is the Cantor string (i.e., the complement of the classic ternary Cantor set in $[0,1]$), then according to [Lap-vFr3, Subsection 1.2.2, Equation (1.30)] (or else Equation (2.2.16) on page 117 above, specialized to $a = 1/3$) and Theorem 2.2.32, we have

$$\dim_{PC}\partial(\Omega_{CS} \times (0,1)^{N-1}) = \big((N-1) + \log_3 2\big) + \frac{2\pi}{\log 3}\mathrm{i}\mathbb{Z}, \tag{2.2.103}$$

which is the special case of (2.2.98) corresponding to $m := 2$, $a := 1/3$ and $d := N - 1$.

Remark 2.2.36. Note that, as is expected, the a-string Ω_a (or equivalently, its boundary $\partial\Omega_a$) is not "fractal" in the sense of the theory of fractal dimensions developed in [Lap-vFr1–3] (see, especially, [Lap-vFr3, Section 12.1 and 12.2]). Accordingly, "fractality" is associated with the presence of a *nonreal* complex dimension (with positive real part). This definition of fractality will be extended to higher dimensions in Chapter 4 by using the theory of fractal zeta functions and the associated complex dimensions developed in this book. (See, especially, Subsection 4.6.2, including Remark 4.6.24; see also [LapRaŽu1–8].) Here, in light of Equation (2.2.101), all of the complex dimensions of the a-string (or, equivalently, of the compact set $\partial(\Omega_a) \subset \mathbb{R}$) are real. Hence, as expected, $\partial(\Omega_a) = \{j^{-a} : j \geq 1\} \cup \{0\}$ (or, equivalently, the a-string) is not fractal. Furthermore, in light of Equation (2.2.102), the same is true for the boundary of the Cartesian product of the a-string Ω_a by $[0,1]^{N-1}$ (for any fixed $N \geq 1$)

In contrast, in light of Equation (2.2.16) on page 117 (specialized to $a = 1/3$), the Cantor string Ω_{CS} (or, equivalently, its boundary $\partial(\Omega_{CS}) \subset \mathbb{R}$, namely, the ternary Cantor set) is fractal because it has nonreal complex dimensions with positive real part. Moreover, in light of Equation (2.2.103), the same is true of $\partial(\Omega_{CS} \times [0,1]^{N-1}) \subset \mathbb{R}^N$ for any $N \geq 2$. Namely, in every dimension N, the compact set $A := \partial(\Omega_{CS} \times [0,1]^{N-1}) \subset \mathbb{R}^N$ admits nonreal (in fact, infinitely many) complex dimensions with (positive) real part $D(A) = N - 1 + D(\partial(\Omega_{CS}))$, where $D(\partial(\Omega_{CS})) = \log_3 2$ and $\partial(\Omega_{CS})$ is the classic ternary Cantor set.

2.2.5 Surface Zeta Functions

To any bounded set A in $\mathbb{R}^N$, we can associate its *surface zeta function*, defined by

$$\zeta_A(s, \partial) = \int_0^\delta t^{s-N} H^{N-1}(\partial(A_t))\,\mathrm{d}t \tag{2.2.104}$$

for all $s \in \mathbb{C}$ with $\operatorname{Re} s$ sufficiently large, where δ is a fixed positive number and H^{N-1} denotes the $(N-1)$-dimensional Hausdorff measure. In Subsection 2.1.7, we have already discussed the oscillatory nature of the function $(0,\delta) \ni t \mapsto t^{s-N}$, for any fixed nonreal complex number s.

Proposition 2.2.37. *If* $\operatorname{Re} s > \overline{\dim}_B A$, *then the distance zeta function and surface zeta function are well-defined holomorphic functions and coincide (see Corollary 2.2.38 below for further related results):*

$$\zeta_A(s) = \zeta_A(s, \partial), \tag{2.2.105}$$

for all $s \in \mathbb{C}$ *with* $\operatorname{Re} s > \overline{\dim}_B A$.

Proof. It suffices to use the identity

$$\int_{A_\delta} d(x,A)^{-\gamma}\,\mathrm{d}x = \int_0^\delta t^{-\gamma} H^{N-1}(\partial(A_t))\,\mathrm{d}t,$$

where γ is a real number; see (2.1.8) or [Žu2, Equation (2.19)]. Note that for any $\gamma < N - D$, where $D := \overline{\dim}_B A$, the integral on the left-hand side is finite; see Lemma 2.1.4. This proves (2.2.105) for $s \in (D, +\infty) \subset \mathbb{R}$. The claim then follows much as in the proof of Theorem 2.2.1. □

We say that a bounded set A in $\mathbb{R}^N$ is *surface nondegenerate* if there exists $d \geq 0$ such that $H^{N-1}(\partial(A_t)) \asymp t^{N-d-1}$ as $t \to 0^+$. It can be shown that the set A is Minkowski nondegenerate if and only if it is surface nondegenerate, and in this case, we necessarily have that $\dim_B A$ exists and $d = \dim_B A$. This and other properties of the mapping $t \mapsto H^{N-1}(\partial(A_t))$ have been established by Rataj and Winter [RatWi1]; see also [RatWi2].

In light of the results of Section 2.2.2 about the relationship between the distance zeta function $\zeta_A = \zeta_A(\cdot, A_\delta)$ and the tube zeta function $\tilde{\zeta}_A$ (see, especially, Remark 2.2.18 and Proposition 2.2.19), the next result follows at once from Proposition 2.2.37 just above (and from its proof).

Corollary 2.2.38. *Let A be a bounded subset of $\mathbb{R}^N$ such that $\overline{\dim}_B A < N$. If either of the fractal zeta functions ζ_A, $\tilde{\zeta}_A$ or $\zeta_A(\cdot,\partial)$ possesses a meromorphic continuation (necessarily unique) to a connected open neighborhood U of a window $\boldsymbol{W}$, then so do the other two zeta functions and the resulting meromorphic extensions of the distance zeta function and of the surface zeta function coincide in U:*

$$\zeta_A(s) = \zeta_A(s,\partial), \quad \text{for all} \quad s \in U.$$

Furthermore, the corresponding sets of (visible) complex dimensions coincide (see Definition 2.1.68 and Equation (2.1.98) *above):*

$$\mathscr{P}(\zeta_A) = \mathscr{P}(\tilde{\zeta}_A) = \mathscr{P}(\zeta_A(\cdot,\partial)).$$

The residues of the distance and surface zeta functions coincide at each (visible) complex dimension. Similarly, the corresponding sets of principal complex dimensions coincide (see Definition 2.1.67 and Equation (2.1.99)*):*

$$\mathscr{P}_c(\zeta_A) = \mathscr{P}_c(\tilde{\zeta}_A) = \mathscr{P}_c(\zeta_A(\cdot,\partial)).$$

2.3 Meromorphic Extensions of Fractal Zeta Functions

The goal of this section is to describe a construction of meromorphic extensions of zeta functions associated with some fractal strings that are in some sense close to classical strings. As a rule, these classical zeta functions possess meromorphic extensions to the entire complex plane. We also study meromorphic extensions of distance and tube zeta functions (henceforth also referred to as 'fractal zeta functions'), and obtain several refinements of Theorem 2.2.3 and Theorem 2.2.14. The main results of this section are stated in Theorems 2.3.2 and 2.3.10, dealing with the fractal zeta functions of the Riemann and Dirichlet strings, respectively, and in Theorems 2.3.18 and 2.3.25, dealing with fractal zeta functions of Minkowski measurable and Minkowski nonmeasurable sets, respectively.

As we have already indicated in Example 2.1.58 and in Theorem 2.1.59, the study of the geometric zeta function $\zeta_{\mathscr{L}}(s) := \sum_{j=1}^{\infty} \ell_j^s$ of any bounded fractal string $\mathscr{L} = (\ell_j)_{j\geq 1}$ is equivalent to the study of the corresponding distance zeta function

$$\zeta_{A_{\mathscr{L}}}(s) := \int_{-\delta}^{a_1+\delta} d(x, A_{\mathscr{L}})^{s-1} \mathrm{d}x \tag{2.3.1}$$

of the subset $A_{\mathscr{L}} = \{a_k := \sum_{j\geq k} \ell_j : k \in \mathbb{N}\}$ of the real line, associated to $\mathscr{L}$, where δ is an arbitrary fixed positive real number. Note that here, $a_1 = \sum_{j\geq 1} \ell_j$ is the total length of $\mathscr{L}$ and $(a_k)_{k\geq 1}$ is a nonincreasing sequence of positive real numbers converging to zero as $k \to \infty$.

More generally, the geometric zeta function $\zeta_{\mathscr{L}}$ can be identified with the *relative zeta function*

$$\zeta_{\partial\Omega,\Omega}(s) := \int_{\Omega} d(x, \partial\Omega)^{s-1} \mathrm{d}x, \tag{2.3.2}$$

where $\Omega := \sqcup_{j=1}^{\infty} I_j$ is a disjoint union of open intervals I_j of length ℓ_j for each $j \geq 1$. Analogously as in Example 2.1.58 of Subsection 2.1.4, it is easy to see that

$$\zeta_{\partial\Omega,\Omega}(s) = \sum_{j=1}^{\infty} \int_{I_j} d(x, \partial I_j)^{s-1} \mathrm{d}x = s^{-1} 2^{1-s} \zeta_{\mathscr{L}}(s) \tag{2.3.3}$$

for all $s \in \mathbb{C}$ with $\operatorname{Re} s > \overline{\dim}_B \mathscr{L}$.

Recall (see the beginning of Subsection 2.1.4) that the disjoint family[57] of open intervals $(I_j)_{j\geq 1}$ is called the *canonical geometric realization* of the fractal string $\mathscr{L}$. Much more general relative zeta functions are studied Chapters 4 and 5.

A *geometric realization* of the fractal string $\mathscr{L} := (\ell_j)_{j\geq 1}$ is an open set $\Omega \subset \mathbb{R}$ or an RFD $(\partial\Omega, \Omega)$ in $\mathbb{R}$, where Ω is any bounded open subset of $\mathbb{R}$ (or, more generally, any open set $\Omega \subset \mathbb{R}$ of finite length $|\Omega|_1 < \infty$) such that $\Omega = \sqcup_{j=1}^{\infty} J_j$, where $(J_j)_{j\geq 1}$ is a disjoint sequence of open intervals of $\mathbb{R}$ such that $|J_j| = \ell_j$ for every $j \geq 1$. Then, it follows from the discussion in Subsection 2.1.4 that the key identity (2.3.3), namely,

$$\zeta_{\partial\Omega,\Omega}(s) = s^{-1} 2^{1-s} \zeta_{\mathscr{L}}(s), \tag{2.3.4}$$

is independent of the choice of the geometric realization of the fractal string $\mathscr{L}$ by an RFD $(\partial\Omega, \Omega)$ in $\mathbb{R}$.

[57] Throughout this book, by a *disjoint family of sets*, we mean a family of pairwise disjoint sets.

2.3.1 Zeta Functions of Perturbed Riemann Strings

The *Riemann string* is defined as $\mathscr{L} = (j^{-1})_{j\geq 1}$. Note that it is unbounded, in the sense that $\sum_{j=1}^{\infty} j^{-1} = +\infty$. Its geometric zeta function $\zeta_{\mathscr{L}}(s) = \zeta_R(s) = \sum_{j=1}^{\infty} j^{-s}$ is the classical Riemann zeta function; see (2.3.5) below. The Riemann string, also called the *harmonic string*, was introduced by the first author in [Lap2, Example 5.4(ii), pp. 171–172] and further discussed in [Lap3, pp. 144–145]; see also [Lap-vFr3, pp. 123–129] and the later work [HerLap1–5] where it plays an important role.

If $(c_j)_{j\geq 1}$ is a given sequence of real numbers, we can consider the *perturbed Riemann fractal string* $\mathscr{L}' = ((j+c_j)^{-1})_{j\geq 1}$ and the corresponding zeta function, see (2.3.6) below. Throughout this subsection and the next one (i.e., Subsections 2.3.1 and 2.3.2), we assume that the perturbation is such that $j+c_j > 0$ for all j, and analogously for other strings.

Our aim in this subsection is to show that if we perturb the classical Riemann zeta function

$$\zeta_R(s) = \sum_{j=1}^{\infty} j^{-s} \tag{2.3.5}$$

by a sufficiently small sequence of real numbers $(c_j)_{j\geq 1}$, in the sense that $c_j = O(j^{\beta})$ as $j \to \infty$, where $\beta < 1$, then the resulting perturbed Riemann zeta function

$$\zeta_{R,pert}(s) = \sum_{j=1}^{\infty} (j+c_j)^{-s} \tag{2.3.6}$$

possesses a (necessarily unique) meromorphic extension to $\{\operatorname{Re} s > \beta\}$. As in Definition 2.1.28, we denote by $D(\zeta_{R,pert})$ the abscissa of convergence of $\zeta_{R,pert}(s)$; see Section 2.1.3.

Remark 2.3.1. Recall that ζ_R is meromorphic in all of $\mathbb{C}$, with a single, simple pole located at $s=1$; furthermore, $\operatorname{res}(\zeta_R, 1) = 1$. See, e.g., [Tit3] or [Edw].

We first state the main result of this subsection:

Theorem 2.3.2. *Let $\beta \in (-\infty, 1)$ be fixed, and assume that the sequence $(c_j)_{j\geq 1}$ satisifes $c_j = O(j^{\beta})$ as $j \to \infty$. Then, for the perturbed Riemann zeta function defined by (2.3.6), we have $D(\zeta_{R,pert}) = 1$, and $\zeta_{R,pert}(s)$ has a (necessarily unique) meromorphic extension (at least) to the open right half-plane*

$$\{\operatorname{Re} s > \beta\}. \tag{2.3.7}$$

Furthermore, $s=1$ is a pole of the meromorphic continuation in this half-plane; it is simple, and $\operatorname{res}(\zeta_{R,pert}, 1) = 1$. The sets of poles of the classical Riemann zeta function and of $\zeta_{R,pert}$, located in $\{\operatorname{Re} s > \beta\}$, coincide, which means in the present case that $s=1$ is the only pole of $\zeta_{R,pert}$ in $\{\operatorname{Re} s > \beta\}$.

In particular, if the sequence $(c_j)_{j\geq 1}$ *is bounded, then there exists a unique meromorphic extension of* $\zeta_{R,pert}(s)$ *(at least) to the open right half-plane* $\{\mathrm{Re}\, s > 0\}$*; its only pole is located at* $s = 1$ *and it is simple.*

Remark 2.3.3. It would be interesting to know whether or not the bound β in (2.3.7) is optimal; see Problem 6.2.10 in Chapter 6.

Remark 2.3.4. It is easy to see that the condition $c_j = O(j^\beta)$ as $j \to \infty$ in Theorem 2.3.2 can be relaxed to $c_j = O(j^{(\beta})$ as $j \to \infty$, by which we mean that $c_j = O(j^{\beta_0})$ as $j \to \infty$ for *all* $\beta_0 > \beta$; that is,

$$O(t^{(\beta}) := \bigcap_{\beta_0 > \beta} O(t^{\beta_0}) \quad \text{as} \quad t \to +\infty. \tag{2.3.8}$$

(Compare with Definition 2.3.20 below.) Equivalently, $O(t^{(\beta}) := \bigcap_{\varepsilon>0} O(t^{\beta+\varepsilon})$ as $t \to +\infty$. An example of such a (weaker) perturbation is $c_j := j^\beta \log j$, for all $j \geq 1$. Indeed, it follows from Theorem 2.3.2 that $\zeta_{R,pert}(s)$ has a unique meromorphic extension to each half-plane $\{\mathrm{Re}\, s > \beta_0\}$, with $\beta_0 > \beta$, and therefore, by the principle of analytic continuation, it has a unique meromorphic extension to the union

$$\bigcup_{\beta_0 > \beta} \{\mathrm{Re}\, s > \beta_0\} = \{\mathrm{Re}\, s > \beta\}.$$

In the proof of Theorem 2.3.2, we shall use the following simple fact.

Lemma 2.3.5. *Assume that* $\zeta_1(s)$ *is a generalized Dirichlet integral with abscissa of convergence equal to* $D(\zeta_1)$*, such that it possesses a meromorphic extension to the open right half-plane* $\{\mathrm{Re}\, s > a_1\}$*, where* $a_1 \in [-\infty, D(\zeta_1))$*. Assume that* $\zeta_2(s)$ *is a holomorphic function with abscissa of convergence* $D(\zeta_2)$ *such that* $a_1 \leq D(\zeta_2) < D(\zeta_1)$*. Then, for* $\zeta_{pert}(s) := \zeta_1(s) + \zeta_2(s)$*, we have* $D(\zeta_{pert}) = D(\zeta_1)$*, and* $\zeta_{pert}(s)$ *possesses a (necessarily unique) meromorphic extension (at least) to the open right half-plane* $\{\mathrm{Re}\, s > D(\zeta_2)\}$*. Furthermore, the poles of* $\zeta_{pert}(s)$ *and* $\zeta_1(s)$ *coincide in this half-plane, as well as their corresponding multiplicities (or orders).*

Proof. The function $\zeta_1(s)$ is meromorphic in $\{\mathrm{Re}\, s > D(\zeta_2)\}$, while $\zeta_2(s)$ is holomorphic in this same half-plane. Hence, their sum, $\zeta_{pert}(s)$, is meromorphic in this half-plane. The equality $D(\zeta_{pert}) = D(\zeta_1)$ is obvious, since $D(\zeta_2) < D(\zeta_1)$. As is well known, the uniqueness of the meromorphic extension of $\zeta_{pert}(s)$ follows from the principle of analytic continuation since any two meromorphic extensions must coincide on $\{\mathrm{Re}\, s > D(\zeta_1)\}$. The poles of $\zeta_1(s)$ in the half-plane $\{\mathrm{Re}\, s > D(\zeta_2)\}$, as well as their corresponding multiplicities (or orders), do not change after adding the holomorphic function $\zeta_2(s)$. □

In the applications of Lemma 2.3.5 which will be considered in this book, we will most often have $a_1 = -\infty$; that is, ζ_1 will be assumed to have a meromorphic extension to all of $\mathbb{C}$.

Proof of Theorem 2.3.2. We shall use Lemma 2.3.5 with $\zeta_1(s) = \zeta_R(s)$ and

$$\zeta_2(s) = \zeta_{R,pert}(s) - \zeta_R(s) = \sum_{j=1}^{\infty} \left((j+c_j)^{-s} - j^{-s}\right). \tag{2.3.9}$$

Since $D(\zeta_1) = 1$ and ζ_1 possesses a meromorphic extension to all of $\mathbb{C}$, it suffices, in light of Lemma 2.3.5, to show that $D(\zeta_2) \le \beta$.

Let us first fix $s = x + y\mathrm{i}$, where $x, y \in \mathbb{R}$, and let us also fix $j \in \mathbb{N}$. Setting $e(t) = \mathrm{e}^{-(y\log t)\mathrm{i}} \in S^1$, for any $t > 0$, we can write $t^{-s} = t^{-x}e(t)$. We have

$$\begin{aligned}(j+c_j)^{-s} - j^{-s} &= (j+c_j)^{-x}e(j+c_j) - j^{-x}e(j)\\ &= \left((j+c_j)^{-x} - j^{-x}\right)e(j+c_j) + j^{-x}\left(e(j+c_j) - e(j)\right),\end{aligned}$$

and hence,

$$|(j+c_j)^{-s} - j^{-s}| \le |(j+c_j)^{-x} - j^{-x}| + j^{-x}|e(j+c_j) - e(j)|. \tag{2.3.10}$$

By the Lagrange mean value theorem applied to $g(t) = t^{-x}$, $t \in [j, j+c_j]$, we have

$$|(j+c_j)^{-x} - j^{-x}| \le |x|\, j^{-x-1}\, |c_j| \le K|x|\, j^{-x-1+\beta}$$

for all $j \in \mathbb{N}$, where the positive constant K is independent of j.

On the other hand, let us set $\varphi(t) = -y\log t$. Due to the geometrically obvious inequality, $|e(t) - e(\tau)| \le |\varphi(t) - \varphi(\tau)| = |y|\,|\log(t/\tau)|$ for any $t, \tau \in \mathbb{R}$, there exists a positive constant K_1 such that

$$\begin{aligned}|e(j+c_j) - e(j)| &\le |y|\log\frac{j+c_j}{j}\\ &= |y|\log\left(1+\frac{c_j}{j}\right) \le |y|\frac{c_j}{j} \le K_1|y|\, j^{\beta-1}\end{aligned}$$

for all $j \in \mathbb{N}$.

Using (2.3.10), we conclude that there is a positive constant K_2 such that

$$|(j+c_j)^{-x} - j^{-x}| \le K_2(|x| + |y|)\, j^{-(x-\beta+1)} \tag{2.3.11}$$

for all $j \in \mathbb{N}$. This proves that the Dirichlet series initially defining $\zeta_2(s)$ in (2.3.9) converges absolutely and uniformly if $|x| + |y| \le C$ for the positive constant C fixed and arbitrarily large, and $x - \beta + 1 > 1 + \varepsilon$ for some positive ε; that is, for $x > \beta + \varepsilon$. Since $x = \operatorname{Re} s$, for any given s satisfying $\operatorname{Re} s > \beta$, we can find $\varepsilon > 0$ small enough so that $\operatorname{Re} s > \beta + \varepsilon$. By letting $\varepsilon \to 0^+$ and $C \to +\infty$, we conclude that $D(\zeta_2) \le \beta$, as desired. In light of Lemma 2.3.5, this proves that $\zeta_{R,pert}(s)$ can be meromorphically extended to the open right half-plane $\{\operatorname{Re} s > \beta\}$, as desired. □

Remark 2.3.6. It is interesting to note that the function $g(t) := t^{-s}$, defined for $t \in (j, j+c_j)$, with $s = x + iy$ fixed, appearing in the proof of Theorem 2.3.2, has the property that for large j, its range is 'almost' equal to the ray $\{\varphi = -y\log j\}$. More specifically, the range of g is contained in the sector between the rays $\{\varphi = \varphi(j)\}$ and $\{\varphi = \varphi(j+c_j)\}$, where $\varphi(t) := -y\log t$, and the opening angle of the sector tends to zero for large j: $|\varphi(j+c_j) - \varphi(j)| = |y|\log\left(1+\frac{c_j}{j}\right) \to 0^+$ as $j \to \infty$.

In the following corollary, we provide some sufficient conditions under which the perturbed Riemann zeta function possesses a unique meromorphic extension to the entire complex plane. The perturbing sequence $(c_j)_{j\geq 1}$ has to converge to zero very fast as $j \to \infty$.

Corollary 2.3.7. *Assume that the sequence $(c_j)_{j\geq 1}$ is such that there exists a sequence $(\beta_k)_{k\geq 1}$ tending to $-\infty$ as $k \to \infty$, and having the following property. For any given $k \geq 1$, there exists a constant $M_k > 0$ such that $|c_j| \leq M_k j^{\beta_k}$ for all $j \geq 1$. Then, the perturbed Riemann zeta function $\zeta_{R,pert}(s)$ defined by (2.3.6) has for abscissa of convergence $D(\zeta_{R,pert}) = 1$ and possesses a (necessarily unique) meromorphic extension to the entire complex plane. The set of poles of $\zeta_{R,pert}$ coincides with the set of poles of the classical Riemann zeta function ζ: $\mathscr{P}(\zeta_{R,pert}) = \mathscr{P}(\zeta) = \{1\}$. More precisely, $\zeta_{R,pert}(s)$ has a single, simple pole, located at $s = 1$ and with residue* $\operatorname{res}(\zeta_{R,pert}, 1) = 1$.

Proof. By Theorem 2.3.2, $\zeta(s)$ is meromorphic on $\{\operatorname{Re} s > \beta_k\}$ for any k, and therefore also on $\bigcup_{k=1}^{\infty}\{\operatorname{Re} s > \beta_k\} = \mathbb{C}$. □

Example 2.3.8. The sequence $(c_j)_{j\geq 1}$ defined by $c_j := (j!)^{-1}$ for every $j \geq 1$ satisfies the condition of Corollary 2.3.7 with $\beta_k := -k$ for every $k \in \mathbb{N}$, since for any fixed $k \in \mathbb{N}$ we have that $\frac{c_j}{j^{-k}} = \frac{j^k}{j!} \to 0^+$ as $j \to \infty$. Therefore, by Corollary 2.3.7, the corresponding perturbed Riemann zeta function

$$\zeta_{R,pert}(s) = \sum_{j=1}^{\infty}\left(j + \frac{1}{j!}\right)^{-s}$$

is meromorphic on $\mathbb{C}$. The set of poles of $\zeta_{R,pert}$ in $\mathbb{C}$ coincides with the set of poles of the classical Riemann zeta function, namely, $s = 1$. Hence, $\zeta_{R,pert}$ has a unique pole in $\mathbb{C}$, located at $s = 1$, and this pole is simple.

Theorem 2.3.2 is easily seen to be equivalent to the following result.

Theorem 2.3.9. *Let $\gamma > 1$, and assume that $(d_j)_{j\geq 1}$ is a sequence of real numbers satisfying $d_j = O(j^{-\gamma})$ as $j \to \infty$. Then, for the zeta function of the perturbed Riemann string $\mathscr{L} = (j^{-1} + d_j)_{j\geq 1}$, defined by*

$$\zeta_{\mathscr{L}}(s) = \sum_{j=1}^{\infty}(j^{-1} + d_j)^s, \tag{2.3.12}$$

we have $D(\zeta_{\mathscr{L}}) = 1$, and $\zeta_{\mathscr{L}}$ possesses a (necessarily unique) meromorphic extension (at least) to the open right half-plane $\{\operatorname{Re} s > 2 - \gamma\}$. Furthermore, $s = 1$ is a simple pole and is the only pole of this meromorphic continuation.

2.3.2 Zeta Functions of Perturbed Dirichlet Strings

The following result shows that the zeta function of a sufficiently small perturbation of the *Dirichlet string*, defined by $\mathscr{L} = (j^{-a})_{j\geq 1}$ where $a > 0$ is fixed, possesses a nontrivial meromorphic extension. If $a > 1$, the Dirichlet string is bounded. The perturbed string $\mathscr{L} = ((j+c_j)^{-a})_{j\geq 1}$ is obtained from the Dirichlet string by adding the sequence $(c_j)_{j\geq 1}$. The claim of Theorem 2.3.10 just below is easily seen to be equivalent to Theorem 2.3.2 by introducing the new variable $s_1 = as$.

Theorem 2.3.10. *Let $a > 0$, $\beta \in (-\infty, 1)$, and $c_j = O(j^\beta)$ as $j \to \infty$. Then, for the zeta function of the perturbed Dirichlet string $\mathscr{L} = ((j+c_j)^{-a})_{j\geq 1}$, defined by*

$$\zeta_{\mathscr{L}}(s) = \zeta_{pert}(s) = \sum_{j=1}^{\infty} (j+c_j)^{-as}, \tag{2.3.13}$$

we have $D := D(\zeta_{\mathscr{L}}) = 1/a$, and $\zeta_{\mathscr{L}}$ has a unique meromorphic extension (at least) to the open right half-plane $\{\mathrm{Re}\, s > \beta/a\}$. Furthermore, $s = 1/a$ is a pole of the meromorphic continuation in this half-plane; it is simple and $\mathrm{res}(\zeta_{\mathscr{L}}, D) = D$.

An analog of Corollary 2.3.7 can easily be formulated and proved in the context of Dirichlet strings. Furthermore, in Theorem 2.3.10 and Corollary 2.3.7, we can relax the condition $c_j = O(j^\beta)$ to $c_j = O(j^{(\beta})$ as $j \to \infty$; see Remark 2.3.4.

If $|\beta|$ is sufficiently large, the meromorphic continuation of $\zeta_{\mathscr{L}}$ in the half-plane $\{\mathrm{Re}\, s > \beta/a\}$ will have additional poles, beside the pole with the largest possible abscissa $1/a$. In fact, we expect that the techniques used in proving [Lap-vFr3, Theorem 6.21] can be useful in obtaining more precise results in this context. See the comments preceding the statement of Corollary 2.3.13 below.

Theorem 2.3.10 is easily seen to be equivalent to the following result (compare with Theorem 2.3.9). As indicated in the introduction to Subsection 2.3.1, we assume that $j^{-a} + d_j > 0$ for all $j \in \mathbb{N}$; that is, the numbers d_j may have negative values as well.

Theorem 2.3.11. *Let $0 < a < \gamma$, and let $(d_j)_{j\geq 1}$ be a sequence of real numbers such that $d_j = O(j^{-\gamma})$ as $j \to \infty$. Then, for the zeta function associated with the perturbed Dirichlet string $\mathscr{L} = (j^{-a} + d_j)_{j\geq 1}$, defined by*

$$\zeta_{\mathscr{L}}(s) = \sum_{j=1}^{\infty} (j^{-a} + d_j)^s, \tag{2.3.14}$$

we have $D := D(\zeta_{\mathscr{L}}) = 1/a$, and $\zeta_{\mathscr{L}}$ possesses a (necessarily unique) meromorphic extension (at least) to the open right half-plane

$$\left\{\mathrm{Re}\, s > \frac{1}{a} - \left(\frac{\gamma}{a} - 1\right)\right\}.$$

Furthermore, $s = 1/a$ is the only pole in the half-plane; it is simple, and the associated residue is given by $\mathrm{res}(\zeta_{\mathscr{L}}, D) = D$.

Proof. Let the sequence $(c_j)_{j\geq 1}$ be defined by

$$j^{-a}+d_j=(j+c_j)^{-a}. \tag{2.3.15}$$

In light of Theorem 2.3.10, it suffices to prove that $c_j=O(j^\beta)$, where $\beta=1+a-\gamma<1$. Indeed, using (2.3.15) we have

$$\begin{aligned}c_j&=(j^{-a}+d_j)^{-1/a}-j=j[(1+j^a d_j)^{-1/a}-1]\\&=j[1+O(j^a d_j)-1]=O(j^{1+a}d_j)=O(j^{1+a-\gamma})\end{aligned}$$

as $j\to\infty$. □

The following result will be useful in the study of spectral zeta functions of relative fractal drums, which we introduce in Section 4.3.1; see the proof of Proposition 4.3.10.

Theorem 2.3.12. *Let $a>0$, $C>0$ and let $(d_j)_{j\geq 1}$ be a sequence of real numbers such that $d_j=O(j^\gamma)$ as $j\to\infty$, where $\gamma<a$ (here, γ may be negative as well). Then, for the zeta function $\zeta_{\mathscr{L}}$, associated with the fractal string $\mathscr{L}=\big((C\cdot j^a+d_j)^{-1}\big)_{j\geq 1}$, defined by*

$$\zeta_{\mathscr{L}}(s)=\sum_{j=1}^{\infty}(C\cdot j^a+d_j)^{-s}, \tag{2.3.16}$$

we have $D:=D(\zeta_{\mathscr{L}})=1/a$, and $\zeta_{\mathscr{L}}$ possesses a unique meromorphic extension (at least) to the open right half-plane

$$\left\{\operatorname{Re}s>\frac{1}{a}-\left(1-\frac{\gamma}{a}\right)\right\}. \tag{2.3.17}$$

Furthermore, $s=1/a$ is the only pole of $\zeta_{\mathscr{L}}$ in the half-plane; it is simple and

$$\operatorname{res}(\zeta_{\mathscr{L}},1/a)=\frac{1}{a}C^{-1/a}. \tag{2.3.18}$$

Proof. Let us define the sequence $(e_j)_{j\geq 1}$ by $j^a+d'_j=(j+e_j)^a$, where $d'_j:=C^{-1}d_j$. Then

$$\begin{aligned}e_j&=(j^a+d'_j)^{1/a}-j=j\left((1+j^{-a}d'_j)^{1/a}-1\right)\\&=jO(j^{-a}d'_j)=O(j^{1-a+\gamma})\quad\text{as}\quad j\to\infty.\end{aligned}$$

Note that

$$\zeta_{\mathscr{L}}(s)=C^{-s}\zeta_{\mathscr{L}'}(s), \tag{2.3.19}$$

where $\mathscr{L}':=\big((j^a+d'_j)^{-1}\big)_{j\geq 1}$. The claim now follows from Theorem 2.3.10 applied to $\mathscr{L}'$, by taking $\beta=1-a+\gamma$. Using (2.3.19) we conclude that $\operatorname{res}(\zeta_{\mathscr{L}},D)=C^{-D}\operatorname{res}(\zeta_{\mathscr{L}'},D)=C^{-D}D$. □

Let $\mathscr{L} = (\ell_j)_{j\geq 1}$ be the *a-string*, that is, $\ell_j = j^{-a} - (j+1)^{-a}$ for each $j \geq 1$, where $a > 0$ is fixed; see Example 2.2.24.[58] The following result shows that the geometric zeta function $\zeta_{\mathscr{L}}$ of any a-string possesses a unique meromorphic extension to $\{\operatorname{Re} s > 0\}$. We point out that this is a special case of a more general result proved in [Lap-vFr3, Theorem 6.21], according to which the geometric zeta function $\zeta_{\mathscr{L}}$ associated to the a-string possesses a unique meromorphic extension (still denoted by $\zeta_{\mathscr{L}}$) to the entire complex plane. Furthermore, the poles of $\zeta_{\mathscr{L}}$ are all simple, and are located at $D = 1/(a+1)$ (the dimension of the boundary of the string), and at (a subset of) $-D, -2D, -3D, \cdots$. The proof of the corollary below is surprisingly simple. (Although, in some sense, it parallels the beginning of the proof of [Lap-vFr3, Theorem 6.21], it also places it in a broader context.)

Corollary 2.3.13. *Let $\zeta_a(s)$ be the zeta function associated with an a-string; that is, $\zeta_a(s) = \sum_{j=1}^{\infty} \left(j^{-a} - (j+1)^{-a}\right)^s$, for all $s \in \mathbb{C}$ with $\operatorname{Re} s > 1$, where $a > 0$ is fixed.*

(a) Then, $D := D(\zeta_a) = 1/(a+1)$ and the zeta function can be meromorphically extended (at least) to the open right half-plane $\{\operatorname{Re} s > 0\}$.

(b) Furthermore, $D = 1/(a+1)$ is the only pole in the half-plane $\{\operatorname{Re} s > 0\}$; it is simple, and $\operatorname{res}(\zeta_a, D) = Da^D$.

Proof. (a) If we show that $\ell_j = j^{-a} - (j+1)^{-a}$ has the form $\ell_j = aj^{-(a+1)} + d_j$, and $d_j = O(j^{-(a+2)})$ as $j \to \infty$, the claim will follow immediately from Theorem 2.3.11, with $a_1 := a+1$ and $\gamma_1 := a+2$.

Since

$$\left(1+\frac{1}{j}\right)^{-a} = 1 + \binom{-a}{1}\frac{1}{j} + O(j^{-2}) = 1 - \frac{a}{j} + O(j^{-2})$$

as $j \to \infty$, we have that

$$\begin{aligned} d_j = \ell_j - aj^{-a-1} &= j^{-a}\left(1 - (1+\frac{1}{j})^{-a}\right) - aj^{-a-1} \\ &= j^{-a}\left(\frac{a}{j} + O(j^{-2})\right) - aj^{-a-1} = O(j^{-a-2}) \end{aligned}$$

as $j \to \infty$, and the claim is proved.

(b) Since $\ell_j = aj^{-(a+1)} + d_j$, then ζ_a has the form $\zeta_a = \zeta_1 + \zeta_2$, where

$$\zeta_1(s) := \sum_{j=1}^{\infty} (aj^{-(a+1)})^s = a^s \zeta_R((a+1)s) = a^s \zeta_R\left(\frac{s}{D}\right)$$

[58] The a-string was introduced in [Lap1, Example 5.1] and used in [Lap1–3], [LapPo1–3], [HeLap] and [Lap-vFr1–3], in particular, in order to illustrate various results and test or motivate several conjectures.

and ζ_R is the Riemann zeta function, while ζ_2 is holomorphic in $\{\mathrm{Re}\, s > 0\}$. Hence,

$$\mathrm{res}(\zeta_a, D) = \mathrm{res}(\zeta_1, D) = a^D \lim_{s \to D} (s - D)\zeta_R\left(\frac{s}{D}\right).$$

Introducing a new variable $z = s/D$, we obtain

$$\mathrm{res}(\zeta_a, D) = Da^D \lim_{z \to 1} (z-1)\zeta_R(z) = Da^D \,\mathrm{res}(\zeta_R, 1) = Da^D,$$

where we have used the fact that the residue of the Riemann zeta function at its simple pole $z = 1$ is equal to 1; see Remark 2.3.1 above. □

Remark 2.3.14. It was first proved in [Lap1, Theorem C, p. 523] that the a-string (introduced in [Lap1, Example 5.1]) has Minkowski dimension $D = 1/(a+1)$ and is Minkowski measurable with Minkowski content

$$\mathscr{M} = \mathscr{M}^D = \frac{2^{1-D}}{1-D} a^D. \tag{2.3.20}$$

This result helped formulate and illustrate the characterization of Minkowski measurability (of fractal strings, or equivalently, of compact subsets of $\mathbb{R}$) obtained in [LapPo1–2]. Moreover, it was revisited in [Lap-vFr1–3] from the point of view of the theory of complex fractal dimensions and significantly expanded since a fractal tube formula was also obtained for the volume of the (inner) ε-neighborhoods of the a-string; see [Lap-vFr3, Subsection 8.1.2].

Remark 2.3.15. According to [Lap-vFr3, Theorem 8.15], the Minkowski measurability of the a-string $\mathscr{L}$ and the value of its Minkowski content $\mathscr{M}$ (as obtained in [Lap1, Theorem C of Appendix C], see Remark 2.3.14) can be recovered from the fact that $D = 1/(a+1)$ is the only complex dimension of $\mathscr{L}$ located on the critical line $\{\mathrm{Re}\, s = D = D(\zeta_{\mathscr{M}})\}$. On the other hand, according to the Minkowski measurability criterion obtained in [LapPo2, Theorem 2.2], the fact that $\mathscr{L}$ is Minkowski measurable (with Minkowski content given by (2.3.20)) follows from the asymptotic relation $\ell_j = j^{-a} - (j+1)^{-a} \sim a j^{-1/D}$ as $j \to \infty$ (i.e., $\ell_j = a j^{-1/D}(1 + o(1))$ as $j \to \infty$).

Remark 2.3.16. According to [Lap-vFr3, Equation (8.25)], we have (still for the a-string and with $D = 1/(a+1)$)

$$\mathscr{M} = \mathscr{M}^D = \frac{2^{1-D}}{D(1-D)} \mathrm{res}(\zeta_a, D) \tag{2.3.21}$$

and therefore, comparing (2.3.20) and (2.3.21), we deduce that

$$\mathrm{res}(\zeta_a, D) = Da^D, \tag{2.3.22}$$

as claimed in part (b) of Corollary 2.3.13, and not a^D as stated in [Lap-vFr3, Theorem 6.21] and [Lap-vFr3, Equation (8.22)] where there seems to be a misprint.

In the following result, we deal with bounded Dirichlet strings, so that we need the condition $a > 1$.

Theorem 2.3.17. *Let $a > 1$, $\beta \in (-\infty, 1)$, and $\delta > 0$. Assume that $(c_j)_{j\geq 1}$ is a sequence of real numbers such that $c_j = O(j^\beta)$ as $j \to \infty$. Let the set A be defined by*

$$A = \Big\{ a_k = L \sum_{j=k}^{\infty} (j+c_j)^{-a} : k \in \mathbb{N} \Big\},$$

where $L > 0$. In other words, A is the set associated with the perturbed Dirichlet string $\mathscr{L} = (L(j+c_j)^{-a})_{j\geq 1}$. Then:

(a) For the distance zeta function $\zeta_A(s) = \int_{A_\delta} d(x,A)^{s-1} \mathrm{d}x$, with $A_\delta \subset \mathbb{R}$, we have $D := D(\zeta_A) = 1/a$. Furthermore, $\zeta_A(s)$ possess a (necessarily unique) meromorphic extension (at least) to the open right half-plane $\{\mathrm{Re}\, s > \beta/a\}$, and

$$\mathrm{res}(\zeta_A, D) = 2^{1-D} L^D. \tag{2.3.23}$$

(b) For the tube zeta function $\tilde{\zeta}_A(s) = \int_0^\delta t^{s-2} |A_t| \, \mathrm{d}t$, we have $D = D(\tilde{\zeta}_A) = 1/a$. Furthermore, $\tilde{\zeta}_A(s)$ possess a (necessarily unique) meromorphic extension (at least) to the open right half-plane $\{\mathrm{Re}\, s > \beta/a\}$; that is,

$$\mathrm{res}(\tilde{\zeta}_A, D) = L^D \frac{2^{1-D}}{1-D}. \tag{2.3.24}$$

Proof. In both cases (a) and (b), it suffices to consider the geometric zeta function $\zeta_{\mathscr{L}}(s) = \sum_{j=1}^{\infty} (\ell_j)^s$ of the string $\mathscr{L} = (\ell_j)_{j\geq 1}$, where $\ell_j := L(j+c_j)^{-a}$; see (2.1.84) and (2.2.23).

Since $c_j \sim j^\beta$ as $j \to \infty$, we deduce from Theorem 2.3.10 that $\zeta_{\mathscr{L}}(s)$ can be meromorphically extended to $\{\mathrm{Re}\, s > \beta/a\}$.

It is clear that $\ell_j \sim L j^{-a}$ as $j \to \infty$. Using [LapPo2, Theorem 2.2] or the counterpart of [Lap1, Theorem C of Appendix C] (compare Remark 2.3.15), we conclude that the set A is Minkowski measurable, its box dimension is equal to $D = 1/a$, and the D-dimensional Minkowski content of A is given by

$$\mathscr{M}^D(A) = L^D \frac{2^{1-D}}{1-D}. \tag{2.3.25}$$

(It is worth noting that the presence of the factor L^D on the right-hand side of (2.3.25) is due to the scaling property of the Minkowski content; see Equation (1.3.19).) The values of the residues in (a) and (b) are then obtained from Theorem 2.2.3 and Theorem 2.2.14, respectively. (Compare Remark 2.3.16 above.) □

It is easy to see that the conclusions of Theorem 2.3.17(a) hold for the distance zeta function $\zeta_1(s) = \int_0^{a_1} d(x,A)^{s-1} \mathrm{d}x$ as well. Indeed, it suffices to take δ large enough, so that $A_\delta = (-\delta, a_1 + \delta)$, and then drop the integrals corresponding to the intervals $(-\delta_1, 0)$ and $(a_1, a_1 + \delta)$, since they are both equal to $\delta^s s^{-1}$, and therefore, are meromorphic functions on $\mathbb{C}$, with $s = 0$ as their only pole.

2.3.3 Meromorphic Extensions of Tube and Distance Zeta Functions

The following theorem (Theorem 2.3.18) shows that the tube zeta function of a class of Minkowski measurable sets possess a nontrivial meromorphic extension, assuming a mild technical condition on the growth rate of the tube function $t \mapsto |A_t|$. As we see from Theorems 2.3.18 and 2.3.25, the second term in the asymptotic expansion of the tube function plays a crucial role in order to reach such a conclusion. For a counterpart of this result for the distance zeta functions of bounded sets and for the geometric zeta functions of fractal strings, see Theorem 2.3.37 and Theorem 2.3.38, respectively.

We note that conjecturally, and in light of the results obtained in [Lap-vFr3, Chapters 2–3 and Subsection 8.3.3] (including [Lap-vFr3, Theorem 8.3]), as well as of the results of [LapPe2–3, LapPeWi1–2] (as described in part in [Lap-vFr3, Section 13.1]), the class of compact sets to which Theorem 2.3.18 applies should include (under some mild additional assumptions, yet to be specifically determined) all nonlattice self-similar sets (satisfying the open set condition, see [Hut, Fal1]). Provided one assumes in addition that $D < N$, this comment also applies to Theorem 2.3.37 (in the Minkowski measurable case), the counterpart of Theorem 2.3.18 for distance zeta functions. We refer to Problems 6.2.36 and 6.2.38 for a detailed discussion of closely related issues.

Theorem 2.3.18 (Minkowski measurable case). *Let A be a bounded subset of $\mathbb{R}^N$ such that there exist $\alpha > 0$, $\mathscr{M} \in (0,+\infty)$ and $D \geq 0$ satisfying*

$$|A_t| = t^{N-D}(\mathscr{M} + O(t^{\alpha})) \quad \text{as} \quad t \to 0^+. \tag{2.3.26}$$

Then, $\dim_B A$ exists and $\dim_B A = D$. Furthermore, A is Minkowski measurable with Minkowski content $\mathscr{M}^D(A) = \mathscr{M}$. Moreover, the tube zeta function $\tilde{\zeta}_A$ has for abscissa of convergence $D(\tilde{\zeta}_A) = \dim_B A = D$ and possesses a unique meromorphic continuation (still denoted by $\tilde{\zeta}_A$) to (at least) the open right half-plane $\{\operatorname{Re} s > D - \alpha\}$; that is,

$$D_{\text{mer}}(\tilde{\zeta}_A) \leq D - \alpha.$$

The only pole of $\tilde{\zeta}_A$ in this half-plane is $s = D$; it is simple, and $\operatorname{res}(\tilde{\zeta}_A, D) = \mathscr{M}$.

Proof. We have

$$\begin{aligned}\tilde{\zeta}_A(s) &= \int_0^{\delta} t^{s-N-1}|A_t|\,dt = \int_0^{\delta} t^{s-N-1}t^{N-D}(\mathscr{M} + O(t^{\alpha}))\,dt \\ &= \underbrace{\mathscr{M}\frac{\delta^{s-D}}{s-D}}_{\zeta_1(s)} + \underbrace{\int_0^{\delta} t^s O(t^{-D+\alpha-1})\,dt}_{\zeta_2(s)},\end{aligned}$$

provided $\operatorname{Re} s > D$. The function $\zeta_1(s)$ is meromorphic in the entire complex plane and $D(\zeta_1) = D$, while for $\zeta_2(s)$ we have

$$|\zeta_2(s)| \le K \int_0^\delta t^{\mathrm{Re}\,s-D+\alpha-1}\,dt < \infty$$

for $\mathrm{Re}\,s > D-\alpha$, where K is a positive constant. Therefore, $D(\zeta_2) \le D-\alpha < D = D(\zeta_1)$, and the claim now follows from Lemma 2.3.5, since (in the notation of Lemma 2.3.5) we have $a_1 = -\infty$ here. □

Remark 2.3.19. Much as in Remark 2.3.4, a function of order $O(t^\alpha)$ as $t \to 0^+$, appearing in Theorem 2.3.18, can be replaced by a function of order $O(t^{\alpha)})$ as $t \to 0^+$, in the precise sense of Definition 2.3.20 just below. In the statement of Theorem 2.3.18, this enables us to replace functions of order $O(t^\alpha)$ as $t \to 0^+$ with more general functions, for example of the form $t^\alpha \log(1/t)$ or

$$t^\alpha \underbrace{\log\ldots\log}_{q}(1/t),$$

near $t = 0^+$, where the last factor is the q-th iterated logarithm for an arbitrary integer $q \ge 1$.

Definition 2.3.20. Let f be defined on an interval $(0,\delta)$, for some $\delta > 0$. Then, given $\alpha \in \mathbb{R}$, f is said to be *of order* $O(t^{\alpha)})$ *as* $t \to 0^+$ (which we write $f(t) = O(t^{\alpha)})$ as $t \to 0^+$) if for every $\alpha_0 < \alpha$, it is of order $O(t^{\alpha_0})$ as $t \to 0^+$; that is, symbolically,

$$O(t^{\alpha)}) := \bigcap_{\alpha_0<\alpha} O(t^{\alpha_0}) \quad \text{as} \quad t \to 0^+. \tag{2.3.27}$$

Equivalently, $O(t^{\alpha)}) := \bigcap_{\varepsilon>0} O(t^{\alpha-\varepsilon})$ as $t \to 0^+$.

Example 2.3.21. Given an integer $k \ge 0$, let $A(a,k) = \{j^{-a} : j \in \mathbb{N}\} \times [0,1]^k \subset \mathbb{R}^{1+k}$, as in [Lap1, Example 5.1'], where $a > 0$. For the set $A(a) = \{j^{-a} : j \in \mathbb{N}\}$ (associated with the a-string; see Example 2.2.24 and Remark 2.3.15), we have

$$|A(a)_t|_{\mathbb{R}} = t^{1-D}\left(\frac{2^{1-D}a^D}{1-D} + O(t^{1+\frac{D}{2})})\right) \quad \text{as} \quad t \to 0^+,$$

where $D = 1/(a+1)$, the (inner) t-neighborhood of the a-string is taken in $\mathbb{R}$ and t is any positive number; see [Lap-vFr3, Equation (8.21) with $J = 0$] and the comment following it. We therefore obtain that

$$|A(a,k)_t|_{\mathbb{R}^{1+k}} = t^{(1+k)-(k+D)}\left(\frac{2^{1-D}a^D}{1-D} + O(t^{1+\frac{D}{2})})\right)$$

as $t \to 0^+$, where the (inner) t-neighborhood and the Lebesgue measure are now taken in $\mathbb{R}^{1+k}$, and the notation $O(t^{1+\frac{D}{2})})$ is explained in (2.3.27). By using Theorem 2.3.18, we obtain that $D(\zeta_{A(a,k)}) = k+D$, and since $\alpha = 1+\frac{D}{2}$ (see also Remark 2.3.19), it follows that $\zeta_{A(a,k)}(s)$ possesses a unique meromorphic extension (at least) to the open right half-plane $\{\mathrm{Re}\,s > (k+D)-(1+\frac{D}{2}) = k-1+\frac{D}{2}\}$.

Remark 2.3.22. Actually, in the present case, using the much more precise asymptotic expansion for $|A(a)_t|_{\mathbb{R}}$ (and hence, also for $|A(a,k)_t|_{\mathbb{R}^{1+k}} = |A(a)_t|_{\mathbb{R}}$) given by [Lap-vFr3, Equation (8.21)], namely, for every integer $J \geq 1$, we have

$$|A(a)_t|_{\mathbb{R}} = \frac{(2t)^{1-D}}{1-D}a^D - \sum_{j=1}^{J} \frac{(2t)^{1+jD}}{jD(1+jD)} \operatorname{res}(\zeta_{\mathscr{L}_a}, -jD) + O(t^{1+(J+1)D})$$

as $t \to 0^+$, one can show (much as in the proof of Theorem 2.3.18) that for every integer $k \geq 0$, the tube zeta function $\tilde{\zeta}_{A(a,k)}$ (and hence, in particular, $\tilde{\zeta}_{A(a)}$, by letting $k = 0$) has a (unique) meromorphic continuation to all of $\mathbb{C}$, with simple poles located at $k+D$ (and at subset of) $k-D$, $k-2D$, $k-3D, \cdots$. In light of Remark 2.2.18 and since $\dim_B A = k+D < k+1$, the exact same statement holds for the distance zeta function $\zeta_{A(a,k),A(a,k)_\delta}$ for any fixed $\delta > 0$.

Assume that $A \subset \mathbb{R}^N$ is a bounded and Minkowski measurable set, with its D-dimensional Minkowski content denoted by $\mathscr{M} = \mathscr{M}(A)$, where $D := \dim_B A$. Then, it is clear that in general, $\mathscr{M} = \mathscr{M}_N$ depends on N. However, it was shown in the 1950s by Martin Kneser [Kne, Satz 7] that the *normalized Minkowski content*, defined (much as was later done in [Fed2]) by $\mathscr{M}_N/\omega_{N-D}$ is independent of N.[59] (Here, given any $m \in \mathbb{N}$, $\omega_m := 2\pi^{m/2}/m\Gamma(m/2)$ denotes the m-dimensional Lebesgue measure of the unit ball in $\mathbb{R}^m$.) An application of this observation is provided in the following result, which follows from Theorem 2.3.18 and from Theorem 2.2.3 (Equation (2.2.4)), according to which we have, respectively,

$$\operatorname{res}(\tilde{\zeta}_A, D) = \mathscr{M}_N \quad \text{and} \quad (\text{provided } D < N) \quad \operatorname{res}(\zeta_A/(N-D), D) = \mathscr{M}_N.$$

Corollary 2.3.23. *Assume that A is a bounded subset of $\mathbb{R}^N$ satisfying the hypotheses of Theorem 2.3.18. Then, the residue at $D = \dim_B A$ of the normalized tube zeta function $\tilde{\zeta}_A/\omega_{N-D}$ (as well as of the normalized distance zeta function $\zeta_A/(N-D)\omega_{N-D}$, provided $D < N$) is given by the normalized Minkowski content $\mathscr{M}_N/\omega_{N-D}$, and is therefore independent of the embedding dimension N.*

We also refer to the related open Problem 6.2.16 in Subsection 6.2.2.

The following theorem (Theorem 2.3.25) deals with an important class of bounded sets in $\mathbb{R}^N$ that are not Minkowski measurable. More specifically, we deal with the sets A such that $0 \leq \mathscr{M}_*^D(A) < \mathscr{M}^{*D}(A) < \infty$, where $D = \dim_B A$. The case when $\mathscr{M}^{*D}(A) = +\infty$ is more difficult, and is treated in Subsection 4.5.1 in the context of relative fractal drums, using suitable gauge functions.

We note that conjecturally, and in light of the results obtained in [Lap-vFr3, Chapters 2–3 and Section 8.4], as well as of results of [LapPe2–3, LapPeWi1–2] (described in part in [Lap-vFr3, Section 13.1]), the class of compact sets to which Theorem 2.3.25 applies should include (under some mild additional assumptions, yet to be specifically determined; see footnote 62 on page 158) all lattice self-similar sets A (satisfying the open set condition, see [Hut, Fal1]). Provided one assumes in

[59] Unaware of the reference [Kne], Maja Resman has rediscovered this result in [Res, Theorem 4].

addition that $D < N$, this comment also applies to Theorem 2.3.37 (in the Minkowski nonmeasurable case), the counterpart of Theorem 2.3.18 for distance zeta functions. We refer to Problem 6.2.36 as well as to Problem 6.2.38 (and the comments following it) for a detailed discussion of closely related issues.

Before stating Theorem 2.3.25, we must first introduce some notation. Given a locally integrable T-periodic function $G : \mathbb{R} \to \mathbb{R}$, we denote by G_0 its truncation to $[0,T]$, while the Fourier transform of G_0 is denoted by $\hat{G}_0$:

$$G_0(\tau) = \begin{cases} G(\tau), & \text{if } \tau \in [0,T], \\ 0, & \text{if } \tau \notin [0,T], \end{cases} \tag{2.3.28}$$

and

$$\hat{G}_0(t) = \int_{-\infty}^{+\infty} \mathrm{e}^{-2\pi \mathrm{i} t\tau} G_0(\tau)\,\mathrm{d}\tau = \int_0^T \mathrm{e}^{-2\pi \mathrm{i} t\tau} G(\tau)\,\mathrm{d}\tau, \tag{2.3.29}$$

for all $t \in \mathbb{R}$.

Definition 2.3.24. The (additive) T-periodicity of G implies that the function $G_1(t) = G(\log t^{-1})$ is *multiplicatively periodic*, with *multiplicative period* $P = \mathrm{e}^T > 1$; that is, $G_1(Pt) = G_1(t)$, for all $t \in \mathbb{R}$; see Figure 2.16. In particular, this means that for any fixed $t > 0$, the value of $G_1(P^k t)$ is independent of $k \in \mathbb{Z}$.

Conversely, if a function $G_1 : \mathbb{R} \to \mathbb{R}$ is multiplicatively periodic with multiplicative period $P > 1$, then the function $G(\tau) := G_1(e^{-\tau})$ is (additively) T-periodic with $T = \log P$. Note that if P is a multiplicative period of G_1, then $G_1(P^k t) = G_1(\tau)$ for all $k \in \mathbb{Z}$. For example, $G(P^{-1}t) = G(PP^{-1}t) = G(t)$, for all $t \in \mathbb{R}$.

Theorem 2.3.25 (Minkowski nonmeasurable case). *Let A be a bounded subset of $\mathbb{R}^N$ such that there exist $D \geq 0$, $\alpha > 0$, and let $G : \mathbb{R} \to (0,+\infty)$ be a nonconstant periodic function with minimal period $T > 0$, satisfying*

$$|A_t| = t^{N-D}\left(G(\log t^{-1}) + O(t^\alpha)\right) \quad \text{as} \quad t \to 0^+. \tag{2.3.30}$$

Then G is continuous, $\dim_B A$ exists and $\dim_B A = D$. Furthermore, A is Minkowski nondegenerate with upper and lower Minkowski contents respectively given by

$$\mathscr{M}_*^D(A) = \min G, \quad \mathscr{M}^{*D}(A) = \max G. \tag{2.3.31}$$

(Hence, the range of $G|_{[0,T]}$ is equal to the compact interval $[\mathscr{M}_*^D(A), \mathscr{M}^{*D}(A)]$.) Moreover, the tube zeta function $\tilde{\zeta}_A$ has for abscissa of convergence $D(\tilde{\zeta}_A) = D$ and possesses a (necessarily unique) meromorphic extension (still denoted by $\tilde{\zeta}_A$) to (at least) the open right half-plane $\{\mathrm{Re}\, s > D - \alpha\}$; that is,

$$D_{\mathrm{mer}}(\tilde{\zeta}_A) \leq D - \alpha.$$

In addition, the set of all the poles of $\tilde{\zeta}_A$ located in this half-plane (i.e., the set of visible complex dimensions of A) is given by[60]

$$\mathscr{P}(\tilde{\zeta}_A) = \left\{ s_k = D + \frac{2\pi}{T}\mathbb{i}k : \hat{G}_0\left(\frac{k}{T}\right) \neq 0,\, k \in \mathbb{Z} \right\} \tag{2.3.32}$$

(see (2.3.29)*); they are all simple, and the residue at each $s_k \in \mathscr{P}(\tilde{\zeta}_A)$, $k \in \mathbb{Z}$, is given by*

$$\operatorname{res}(\tilde{\zeta}_A, s_k) = \frac{1}{T}\hat{G}_0\left(\frac{k}{T}\right). \tag{2.3.33}$$

If $s_k \in \mathscr{P}(\tilde{\zeta}_A)$, then $s_{-k} \in \mathscr{P}(\tilde{\zeta}_A)$ (reality principle, see Remark 2.3.28), and

$$|\operatorname{res}(\tilde{\zeta}_A, s_k)| \leq \frac{1}{T}\int_0^T G(\tau)\,\mathrm{d}\tau, \quad \lim_{k\to\pm\infty} \operatorname{res}(\tilde{\zeta}_A, s_k) = 0. \tag{2.3.34}$$

Moreover, the set of poles $\mathscr{P}(\tilde{\zeta}_A)$ (i.e., of complex dimensions of A) contains $s_0 = D$, and

$$\operatorname{res}(\tilde{\zeta}_A, D) = \frac{1}{T}\int_0^T G(\tau)\,\mathrm{d}\tau. \tag{2.3.35}$$

In particular, A is not *Minkowski measurable and*

$$\mathscr{M}_*^D(A) < \operatorname{res}(\tilde{\zeta}_A, D) < \mathscr{M}^{*D}(A) < \infty. \tag{2.3.36}$$

If, in addition, $G \in C^m(\mathbb{R})$ (i.e., G is m times continuously differentiable on $\mathbb{R}$)[61] *for some integer $m \geq 1$, and G has an extremum t_0 such that*

$$G'(t_0) = G''(t_0) = \cdots = G^{(m)}(t_0) = 0, \tag{2.3.37}$$

then there exits $C_m > 0$ such that for all $k \in \mathbb{Z}$ and $s_k \in \mathscr{P}(\tilde{\zeta}_A)$ we have

$$|\operatorname{res}(\tilde{\zeta}_A, s_k)| \leq C_m |k|^{-m}. \tag{2.3.38}$$

Before proving Theorem 2.3.25, we state and establish a useful corollary. We note that, conjecturally, the class of compact sets to which this corollary can be applied includes all lattice self-similar sets (see the geometric part of [Lap3, Conjecture 3] and the comment preceding the statement of Theorem 2.3.25).[62]

---

[60] Note that the set defined by (2.3.32) coincides with the set of principal complex dimensions of A; that is, with $\dim_{PC} A := \mathscr{P}_c(\tilde{\zeta}_A)$, in the notation of Definition 2.1.67 and Equation (2.1.99).

[61] We do not know examples of periodic sets A such that the corresponding nonconstant periodic functions G appearing in (2.3.30) are C^1-regular; see Remark 2.3.32 and Problem 6.2.5.

[62] Actually, in light of the main result of [KomPeWi] proving the geometric part of [Lap3, Conjecture 3] for a nonintegral value of D, it is reasonable to expect that Corollary 2.3.26 can be applied to a large class of lattice self-similar sets such that $D \notin \mathbb{N}_0$. The remaining issue to be dealt with, however, is to find appropriate hypotheses on A enabling us to obtain (as is assumed in condition (2.3.30)) a sufficiently good error term, of the form $O(t^\alpha)$ for some $\alpha > 0$ rather than merely $o(1)$ as $t \to 0^+$.

Corollary 2.3.26. *Assume that the hypotheses of the first part of Theorem 2.3.25 hold (i.e., assume that $A \subset \mathbb{R}^N$ is bounded and satisfies (2.3.30)). Then, A is not Minkowski measurable but is Minkowski nondegenerate (provided G takes its values in $(0,+\infty)$ rather than in $[0,+\infty)$) and possesses an average Minkowski content (defined as in Equation (2.4.4) in Definition 2.4.1 below) given by*

$$\tilde{\mathscr{M}}^D(A) = \operatorname{res}(\tilde{\zeta}_A, D) = \frac{1}{T}\int_0^T G(\tau)\,\mathrm{d}\tau. \tag{2.3.39}$$

In particular,

$$\mathscr{M}_*^D(A) < \tilde{\mathscr{M}}^D(A) < \mathscr{M}^{*D}(A). \tag{2.3.40}$$

Proof. Clearly, in light of (2.3.31), A is not Minkowski measurable because G is nonconstant (hence, $\inf G = \mathscr{M}_*^D(A) < \mathscr{M}^{*D}(A) = \sup G$) and continuous and takes its values in $(0,+\infty)$ (hence, $\mathscr{M}_*^D(A) = \inf G > 0$) while $\mathscr{M}^{*D}(A) < \infty$ (because G is periodic and continuous).

Finally, the existence of $\tilde{\mathscr{M}}^D(A)$ follows much as in the proof of [Lap-vFr3, Theorem 8.30] and so, we will omit it here. We then obtain that

$$\tilde{\mathscr{M}}^D(A) = \frac{1}{T}\int_0^T G(\tau)\,\mathrm{d}\tau. \tag{2.3.41}$$

Equation (2.3.39) now follows by combining Equations (2.3.41) and (2.3.35).

We note in closing this proof that the existence and the value of $\tilde{\mathscr{M}}^D(A)$ (as given by Equation (2.3.41)) also follows from part (b) of Theorem 2.4.3 in Section 2.4.1 below; see Equation (2.4.7). (See also the special case of Proposition 3.1.2 and of Corollary 3.1.6 where we have set $m = 2$ and $a = 1/3$.) □

Further postponing the proof of Theorem 2.3.25 for a while, we first provide several remarks.

Remark 2.3.27. The analog of Remark 2.3.19 also applies to Theorem 2.3.25.

Remark 2.3.28. All of the fractal zeta functions appearing in this monograph satisfy the so-called *reality principle*: the nonreal poles of zeta functions (defined on domains which are symmetric under complex conjugation) come in complex conjugate pairs. This property has been mentioned and discussed in [Lap-vFr3, Remark 1.6] in the case of the geometric zeta functions of fractal strings. In particular, if in Theorem 2.3.25 we have that $s_k = D + \frac{2\pi}{T}k\mathrm{i}$ is a pole of the tube zeta function $\tilde{\zeta}_A$ for some $k \in \mathbb{Z}\setminus\{0\}$ (that is, $\hat{G}_0(\frac{k}{T}) \neq 0$), then its complex conjugate $\overline{s}_k = D - \frac{2\pi}{T}k\mathrm{i}$ is also a pole (since $\hat{G}_0(-\frac{k}{T}) = \overline{\hat{G}_0(\frac{k}{T})} \neq 0$). The reality principle for a zeta function (or more generally, a meromorphic function) $f(s)$ follows from the identity $\overline{f(s)} = f(\overline{s})$, which is satisfied if $f(r) \in \mathbb{R}$ for any $r \in (D(f),+\infty)$. It follows from the principle of analytic continuation that this identity, and hence also the 'reality principle', continues to hold in any domain to which f can be meromorphically continued and, in particular, to $\operatorname{Mer}(f)$, the half-plane of meromorphic continuation of f; see Definition 2.1.53. The reality principle is also called *the principle of reflection*; see [Tit1, p. 155] and Remark 2.1.12.

Remark 2.3.29. It would be interesting to find some reasonably general conditions on A (that is, on the function G) under which the set $\mathscr{P}(\tilde{\zeta}_A)$ in Theorem 2.3.25 is arithmetic (i.e., is a full arithmetic progression, namely, the sequence $D+\frac{2\pi}{T}\mathrm{i}\mathbb{Z}$). Equivalently, this amounts to asking under what conditions on G is $\hat{G}_0(k/T)\neq 0$ for all $k\in\mathbb{Z}$ (or equivalently, for all $k\in\mathbb{N}$, $k\neq 0$). See also Problem 6.2.8 on page 556.

In the proof of Theorem 2.3.25, we shall need the following simple lemma.

Lemma 2.3.30. *Let $F:(0,\delta)\to\mathbb{R}$ be continuous, and assume that $G:\mathbb{R}\to\mathbb{R}$ is a T-periodic function, for some $T>0$. If $F(t)=G(\log t^{-1})+o(1)$ as $t\to 0^+$, then G is continuous.*

Proof. In light of the periodicity of G, it suffices to show that G is continuous on $(0,+\infty)$. We reason by contradiction. Hence, we assume that G is not continuous at some $\tau_0>0$. Then, by periodicity, for every $k\geq 1$, we have that G is not continuous at $\tau_k=kT+\tau_0$. Recall that *the oscillation of a function G at a point $x\in\mathbb{R}$ is given by*

$$\operatorname*{osc}_x G:=\lim_{\varepsilon\to 0^+}\Big(\sup_{(x-\varepsilon,x+\varepsilon)}G-\inf_{(x-\varepsilon,x+\varepsilon)}G\Big).$$

Defining $t_k=\mathrm{e}^{-\tau_k}$, we have $\operatorname{osc}_{t_k}G(\log t^{-1})=\operatorname{osc}_{\tau_k}G=c>0$, where (in light of the T-periodicity of G) c does not depend on k. Here and in the sequel, we choose k sufficiently large so that $t_k\in(0,\delta)$. Since $t_k\to 0^+$ as $k\to\infty$, we may take k large enough and fixed, such that $|o(1)|\leq c/2$ for $t=t_k$. Here, $o(1)$ is the function of t given in the statement of the lemma. In particular, $o(1)\to 0$ as $t\to 0^+$. Therefore,

$$\begin{aligned}\operatorname*{osc}_{t_k}F&=\operatorname*{osc}_{t_k}(G(\log t^{-1})+o(1))\\&\geq\operatorname*{osc}_{t_k}(G(\log t^{-1})-\operatorname*{osc}_{t_k}|o(1)|\geq c-\frac{1}{2}c=\frac{1}{2}c>0.\end{aligned}$$

On the other hand, since F is continuous on $(0,\delta)$, we must have $\operatorname{osc}_{t_k}F=0$, which is a contradiction. Hence, G must be continuous everywhere. □

We are now ready to establish Theorem 2.3.25.

Proof of Theorem 2.3.25. To show that G is continuous, it suffices to apply Lemma 2.3.30 to $F(t):=|A_t|t^{D-N}$, which is defined and continuous for $t>0$. We can write $\tilde{\zeta}_A(s)=\zeta_1(s)+\zeta_2(s)$, where

$$\zeta_1(s)=\int_0^\delta t^{s-D-1}G(\log t^{-1})\,\mathrm{d}t,\quad \zeta_2(s)=\int_0^\delta t^sO(t^{-D+\alpha-1})\,\mathrm{d}t,\tag{2.3.42}$$

for some $\delta>0$ fixed. As in the proof of Theorem 2.3.18, we have $D(\zeta_2)=D-\alpha$. Therefore, it suffices to prove that $\zeta_1(s)$ can be meromorphically extended to the whole complex plane. We will show this by computing $\zeta_1(s)$ in a closed form. Since G is T-periodic, we have

$$\zeta_1(s) = \int_0^\delta t^{s-D-1} G(\log t^{-1} + T)\,\mathrm{d}t.$$

Introducing a new variable u defined by $\log u^{-1} = \log t^{-1} + T$, that is, $u = \mathrm{e}^{-T} t$, we obtain

$$\begin{aligned}
\zeta_1(s) &= \mathrm{e}^{T(s-D)} \int_0^{\delta \mathrm{e}^{-T}} u^{s-D-1} G(\log u^{-1})\mathrm{d}u = \mathrm{e}^{T(s-D)} \left(\int_0^\delta + \int_\delta^{\mathrm{e}^{-T}\delta} \right) \\
&= \mathrm{e}^{T(s-D)} \left(\zeta_1(s) + \int_\delta^{\mathrm{e}^{-T}\delta} t^{s-D-1} G(\log t^{-1})\,\mathrm{d}t \right).
\end{aligned}$$

From this, we immediately obtain $\zeta_1(s)$ in closed form:

$$\begin{aligned}
\zeta_1(s) &= \frac{\mathrm{e}^{T(s-D)}}{\mathrm{e}^{T(s-D)} - 1} \int_{\mathrm{e}^{-T}\delta}^{\delta} t^{s-D-1} G(\log t^{-1})\,\mathrm{d}t \\
&= \frac{\mathrm{e}^{T(s-D)}}{\mathrm{e}^{T(s-D)} - 1} \underbrace{\int_{\log \delta^{-1}}^{\log \delta^{-1} + T} \mathrm{e}^{-\tau(s-D)} G(\tau)\,\mathrm{d}\tau}_{I(s)},
\end{aligned} \tag{2.3.43}$$

where in the last equality we have passed to the new variable $\tau := \log t^{-1}$. The last integral $I(s)$ is obviously an entire function of s, since δ is different from 0 and $+\infty$. Here, we have used Theorem 2.1.45(c) with $\varphi(\tau) = \mathrm{e}^{\tau}$. This shows that the function $\zeta_1(s)$ is meromorphic on $\mathbb{C}$, and the set of its poles is equal to the set of complex solutions s_k of $\exp(T(s-D)) = 1$ for which $I(s_k) \neq 0$. If $I(s_k) = 0$, it is easy to see that s_k is a removable singularity of $\zeta_1(s)$:

$$\lim_{s \to s_k} \zeta_1(s) = \lim_{s \to s_k} \frac{s - s_k}{\mathrm{e}^{T(s-D)} - 1} \mathrm{e}^{T(s-s_k)} \frac{I(s)}{s - s_k} = \frac{1}{T} I'(s_k),$$

where I' denotes the derivative of I. Since

$$\begin{aligned}
I(s_k) &= \int_{\log \delta^{-1}}^{\log \delta^{-1} + T} \mathrm{e}^{-2\pi \mathrm{i} \frac{k}{T} \cdot \tau} G(\tau)\,\mathrm{d}\tau \\
&= \int_0^T \mathrm{e}^{-2\pi \mathrm{i} \frac{k}{T} \cdot \tau} G(\tau)\,\mathrm{d}\tau = \hat{G}_0\left(\frac{k}{T}\right),
\end{aligned} \tag{2.3.44}$$

where we have used the fact that both $\tau \mapsto \mathrm{e}^{2\pi \mathrm{i} \frac{k}{T} \cdot \tau}$ and $\tau \mapsto G(\tau)$ are T-periodic functions, we conclude that the set of poles of the tube zeta function $\tilde{\zeta}_A$ is described by (2.3.32). Note that it contains D, since for $k = 0$ we have

$$I(D) = I(s_0) = \hat{G}_0(0) = \int_0^T G(\tau)\,\mathrm{d}\tau > 0. \tag{2.3.45}$$

Indeed, the range of the function $G|_{[0,T]}$ is equal to the interval $[\mathscr{M}_*, \mathscr{M}^*]$, where $\mathscr{M}_* = \mathscr{M}_*^D(A)$ and $\mathscr{M}^* = \mathscr{M}^{*D}(A)$. Since G is assumed to be nonconstant, we deduce from (2.3.30) that $0 \le \mathscr{M}_* < \mathscr{M}^* < \infty$.

Therefore, we have $D(\zeta_1) = D > D - \alpha = D(\zeta_2)$, and by Lemma 2.3.5, since $a_1 = -\infty$ (in the notation of that lemma), we also know that $\tilde{\zeta}_A$ possesses a (necessarily unique) meromorphic extension to the open right half-plane $\{\mathrm{Re}\, s > D - \alpha\}$.

Next, we compute the residue of $\tilde{\zeta}_A$ at $s_k = D + \frac{2\pi}{T} k\mathbb{i} \in \mathscr{P}(\tilde{\zeta}_A)$ for an arbitrary $k \in \mathbb{Z}$, using l'Hospital's rule and (2.3.44):

$$\begin{aligned}\operatorname{res}(\tilde{\zeta}_A, s_k) &= \operatorname{res}(\zeta_1, s_k)\\ &= \lim_{s\to s_k} \frac{s - s_k}{\mathrm{e}^{T(s-D)} - 1}\, \mathrm{e}^{T(s_k - D)} I(s_k) = \frac{1}{T}\hat{G}_0\Big(\frac{k}{T}\Big).\end{aligned} \tag{2.3.46}$$

Substituting $k = 0$, we obtain (2.3.35). The inequalities in (2.3.36) follow from (2.3.35).

As is well known, since $G_0 \in L^1(\mathbb{R})$, we have $|\hat{G}_0(\tau)| \le \|G_0\|_{L^1(\mathbb{R})} = \|G\|_{L^1(0,T)}$ and $\lim_{|t|\to+\infty} \hat{G}_0(t) = 0$ (by the Riemann–Lebesgue lemma; see, e.g., [Ru] or [MitŽu, p. 101]), so that (2.3.34) follows immediately from (2.3.46).

If the function G is of class C^m, it does not mean that G_0 is of the same class. However, we can define $G_1 : \mathbb{R} \to \mathbb{R}$ by

$$G_1(\tau) = \begin{cases} G(t) - \mathscr{M}_*, & \text{if } \tau \in [0,T],\\ 0, & \text{if } \tau \notin [0,T]. \end{cases} \tag{2.3.47}$$

Since the value of $\mathscr{M}_*$ is in the range of G, we may assume without loss of generality that $t_0 = 0$ is a minimum of G; namely, $G(0) = G(T) = \mathscr{M}_*$. Otherwise, we can shift the graph of G in the horizontal direction in order to achieve this. Furthermore, $\mathscr{M}_*$ is equal to the minimal value of G; hence, $G_1(0) = G_1(T) = 0$. This means that G_1 is continuous on $\mathbb{R}$, and moreover, due to (2.3.37), that G_1 has the same regularity as G; that is, $G_1 \in C^m(\mathbb{R})$. A direct computation shows that for each $t \in \mathbb{R}$,

$$\hat{G}_1(t) = \hat{G}_0(t) - \mathscr{M}_* \frac{1 - \mathrm{e}^{-2\pi \mathbb{i} t\cdot T}}{2\pi \mathbb{i} t}, \tag{2.3.48}$$

from which it follows that

$$\operatorname{res}(\tilde{\zeta}_A, s_k) = \frac{1}{T}\hat{G}_0\Big(\frac{k}{T}\Big) = \frac{1}{T}\hat{G}_1\Big(\frac{k}{T}\Big). \tag{2.3.49}$$

Since $G_1 \in C^m(\mathbb{R})$, by a standard result from Fourier analysis obtained by repeated integration by parts (see, e.g. [MitŽu, p. 103]), we know that there exists $C_m > 0$ such that $|\hat{G}_1(t)| \le C_m t^{-m}$ for all $t \in \mathbb{R}$. This proves (2.3.38). Of course, the same conclusion can be achieved by defining $G_1(\tau) = G(\tau) - \mathscr{M}^*$. □

Example 2.3.31. (*Complex dimensions of the ternary Cantor set, revisited*). Let A be the classic ternary Cantor set in $[0,1]$. According to [Lap-vFr3, Equation (1.11)] (or to Theorem 5.3.13, see Example 5.5.3 in Section 5.5.2 below),[63] we have

$$|A_t| = t^{1-D} 2^{1-D} \left(2^{-\{\log_3 (2t)^{-1}\}} + (3/2)^{\{\log_3 (2t)^{-1}\}} \right) \tag{2.3.50}$$

for all $t \in (0,1/2)$. Here, for $x \in \mathbb{R}$, $\{x\} = x - \lfloor x \rfloor \in [0,1)$ denotes the fractional part of x, where $\lfloor x \rfloor$ (the integer part or the 'floor' of x) is defined as the largest integer which is less than or equal to x. Then the condition (2.3.30) in Theorem 2.3.25 is satisfied for $N = 1$, $D = \log_3 2$, $\alpha = D$, and

$$G(\tau) := 2^{1-D} \left(2^{-\left\{\frac{\tau - \log 2}{\log 3}\right\}} + (3/2)^{\left\{\frac{\tau - \log 2}{\log 3}\right\}} \right). \tag{2.3.51}$$

It is easy to see that the function G is periodic, with minimal period $T = \log 3$, and is continuous. However, it is not of class C^1 since it is nondifferentiable at the points $\tau_k = \log 2 + Tk$, $k \in \mathbb{Z}$; see Figure 2.16 or [Lap-vFr3, Figure 1.5]. It is therefore convenient to consider the restriction $G|_I$ of G to the interval $I = [\log 2, \log 2 + T]$, since the value of $\mathscr{M}^{*D}(A)$ is achieved at the endpoints of I, and $G|_I$ is convex. An easy geometric analysis shows that

$$\mathscr{M}^{*D}(A) = 2^{2-D} \approx 2.583,$$

while the minimum value of G is

$$\mathscr{M}_*^D(A) = 2^{1-D} \frac{D^D}{(1-D)^{1-D}} \approx 2.495,$$

achieved at the minimum of $G|_I$, which is easy to compute. (See also [LapPo2, Theorem 4.6] or [Lap-vFr3, Section 1.1.2]) In particular, the oscillatory period of the ternary Cantor set is given by

$$\mathbf{am}(A) := \mathscr{M}_*^D(A) - \mathscr{M}^{*D}(A) \approx 0.08.$$

According to Theorem 2.3.25, the corresponding tube zeta function $\tilde{\zeta}_A$ has for abscissa of convergence $D(\tilde{\zeta}_A) = \log_3 2$; therefore, it can be meromorphically extended to the open right half-plane $\{\mathrm{Re}\, s > \alpha\}$ for any $\alpha > 0$, and hence, to the entire complex plane. The set of poles of the tube zeta function is given by

$$\mathscr{P}(\tilde{\zeta}_A) = \left\{ s_k = D + \frac{2\pi}{\log 3} k\mathrm{i} : k \in \mathbb{Z} \right\} = D + \mathbf{p}\mathrm{i}\mathbb{Z},$$

where $\mathbf{p} := 2\pi / \log 3$ is the oscillatory period of the ternary Cantor set, in agreement with [Lap-vFr3, Equation (1.30)]. Computing the Fourier transform of G_0 directly,

---

[63] In our situation we have $A_t \setminus [0,1] = (-t,0) \cup (1,1+t)$, so that in (2.3.50) we do not have the term $-2t$, unlike in [Lap-vFr3, Equation (1.11)]. Equation (2.3.50) can also be recovered from the general fractal tube formulas obtained in Chapter 5; see the discussion of the Cantor string (viewed as an RFD) in Subsection 5.5.2 below. (See also Equation (1.1.23) of Example 1.1.5 on page 15.)

with its support shifted to $[\log 2, \log 2 + T]$ (since, on this interval, formula (2.3.51) holds without curly brackets), we obtain that for each pole $s_k \in \mathscr{P}(\tilde{\zeta}_A)$,

$$\begin{aligned} \operatorname{res}(\tilde{\zeta}_A, s_k) &= \frac{1}{T}\hat{G}_0\Big(\frac{k}{T}\Big) = \frac{1}{T}\int_{\log 2}^{\log 6} \mathrm{e}^{-2\pi \mathrm{i} \frac{k}{T}\tau} G(\tau)\,\mathrm{d}\tau \\ &= \frac{2^{1-D}}{T}\Big(\frac{2^D}{s_k}(2^{-s_k} - 6^{-s_k}) + \frac{(1.5)^{-D}}{1-s_k}(6^{1-s_k} - 2^{1-s_k})\Big) \\ &= \frac{2^{-s_k}}{T s_k(1-s_k)}, \end{aligned} \tag{2.3.52}$$

where in the last equality we have used the fact that $3^{s_k} = 2$. Since $|2^{-s_k}| = 2^D$, we conclude that $\operatorname{res}(\tilde{\zeta}_A, s_k) \asymp k^{-2}$ as $|k| \to \infty$, which is in agreement with the limit in (2.3.34). It is interesting to note that inequality (2.3.38) is satisfied for $m = 2$, even though G is not of class C^2 and not even of class C^1.

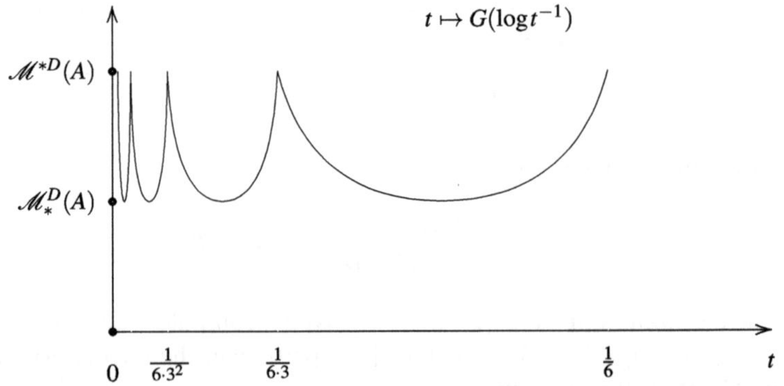

Fig. 2.16 Oscillatory nature of the function $G(\log t^{-1})$ appearing in the tube function $t \mapsto |A_t| = t^{1-D}G(\log t^{-1})$ near $t = 0$ for the ternary Cantor set $A = C^{(1/3)}$, where $D = \dim_B A = \log_3 2$; see Example 2.3.31. Here, $G(\tau)$ is $\log 3$-periodic, or equivalently, $G(\log t^{-1})$ is multiplicatively periodic, with multiplicative period $P = 3$; see Remark 2.3.24 on page 157. The ternary Cantor set is Minkowski nondegenerate, but is not Minkowski measurable; see [LapPo2]. (After [Lap-vFr1–3].)

The residues of the distance zeta function ζ_A and the zeta function of the Cantor string $\zeta_{\mathscr{L}}$ are obtained by using (2.2.23) (with $N = 1$) and (2.1.85), respectively:

$$\operatorname{res}(\zeta_A, s_k) = (1 - s_k)\operatorname{res}(\tilde{\zeta}_A, s_k) = \frac{2^{-s_k}}{T s_k},$$
$$\operatorname{res}(\zeta_{\mathscr{L}}, s_k) = s_k 2^{s_k - 1}\operatorname{res}(\zeta_A, s_k) = \frac{1}{2T}.$$

For the Cantor string $\mathscr{L} = (\ell_j)_{j \geq 1}$ corresponding to the set A, we have

$$\ell_1 = \frac{1}{3}, \quad \ell_2 = \ell_3 = \frac{1}{9}, \quad \ell_4 = \ell_5 = \ell_6 = \ell_7 = \frac{1}{27}, \quad \ldots.$$

Alternatively, we view $\mathscr{L}$ as a decreasing sequence $(\ell_j)_{j\geq 1}$, where for each $j \geq 1$, $l_j = 3^{-j}$ has multiplicity 2^{j-1}. Therefore,

$$\zeta_{\mathscr{L}}(s) = \sum_{j=1}^{\infty} 2^{j-1} 3^{-js} = \frac{3^{-s}}{1 - 2 \cdot 3^{-s}} \tag{2.3.53}$$

(see [Lap-vFr3, Equation (1.29)] or the discussion surrounding Equation (1.1.3) on page 5 of Chapter 1), where the first equality holds for all $s \in \mathbb{C}$ such that $\mathrm{Re}\, s > \log_3 2$. Hence, $\zeta_{\mathscr{L}}$ has a meromorphic extension to all of $\mathbb{C}$, given by the last expression in (2.3.53). It now follows from (2.1.85) and (2.2.23) that both the distance and the tube zeta functions of A possess a meromorphic extension to the entire complex plane, with one additional simple pole at $s = 0$.

Remark 2.3.32. It would be of interest to find examples of bounded sets A in $\mathbb{R}^N$ such that (2.3.30) holds with $G \in C^m(\mathbb{R})$, for a prescribed value of $m \in \mathbb{N}$ (see Problem 6.2.5). Recall that we use the convention according to which $\mathbb{N} = \{1, 2, 3, \ldots\}$. They will typically not be fractal sets since for such sets, one would expect G (when it exists) not to be differentiable on all of $\mathbb{R}$.

Example 2.3.33. The asymptotics of the tube function $t \mapsto |A_t|$ as in (2.3.30) occur naturally in the study of self-similar lattice strings (in the sense of [Lap-vFr3, Chapter 2]); see [Lap-vFr3, Subsection 8.4.4 and, in particular Equation (8.44)]. If we consider an arbitrary (nontrivial) lattice self-similar string $\mathscr{L}$ (also in the sense of [Lap-vFr3, Chapter 2 and Section 8.4]), then it follows from [Lap-vFr3, Corollary 8.27] that there exists $\eta \in \mathbb{R}$, $\eta \leq D$, such that for the associated fractal boundary $A = \partial\mathscr{L}$ we have

$$|A_t| = t^{1-D}\big(G(\log t^{-1}) + O(t^{\eta+\delta})\big) \tag{2.3.54}$$

as $t \to 0^+$, for all sufficiently small $\delta > 0$,[64] where $D = \dim_B A \in (0,1)$ and G is a nonconstant periodic function. (See also Remark 2.3.35 below for a more precise statement of estimate (2.3.54), and [Lap-vFr3, Corollary 8.27] for an even more refined, but more technical, version of (2.3.54).) Since here $\alpha = \eta$, it follows from Theorem 2.3.25 that the corresponding distance and tube zeta functions of A (as well as the geometric zeta function $\zeta_{\mathscr{L}}$ of the self-similar string $\mathscr{L}$) can be meromorphically extended (at least) to the open right half-plane $\{\mathrm{Re}\, s > D - \eta\}$. We note that in [Lap-vFr3, Chapter 2], $\zeta_{\mathscr{L}}$ is given in a closed form and admits a meromorphic extension to all of $\mathbb{C}$, not just to the half-plane $\{\mathrm{Re}\, s > D - \eta\}$.

Remark 2.3.34. We caution the reader that in Example 2.3.33, the set A stands for the boundary of the fractal string $\mathscr{L}$ (i.e., $A = \partial\mathscr{L} = \partial\Omega$, where $\Omega \subseteq \mathbb{R}$ is the open bounded set defining the self-similar string $\mathscr{L}$). This is in contrast with much of the rest of the present monograph where by the set $A = A_{\mathscr{L}}$ associated with a fractal string $\mathscr{L} = (\ell_j)_{j\geq 1}$, we mean $A := \big\{\sum_{j\geq k} \ell_j : k \geq 1\big\} \subset (0, +\infty) \subset \mathbb{R}$; see, especially, Section 2.1.4 above.

[64] In other words, $|A_t| = t^{1-D}\big(G(\log t^{-1}) + O(t^{(\eta})\big)$ as $t \to 0^+$, where the notation $O(t^{(\eta})$ as $t \to 0^+$ is explained by (2.3.8).

Remark 2.3.35. A much more precise statement of Equation (2.3.54) can be found in [Lap-vFr3, Theorem 8.25 and Corollary 8.27]. In particular, if $\mathrm{Re}\, s = \Theta$ is the rightmost vertical line of complex dimensions to the left of $\{\mathrm{Re}\, s = D\}$, then it follows from [Lap-vFr3, Corollary 8.27] that if $\Theta < 0$, then estimate (2.3.54) holds with the optimal values $\eta = D$ and $\delta = 0$, while if $\Theta \geq 0$, then (2.3.54) holds with $\eta = D - \Theta$ and for all $\delta > 0$. Note that if there are no complex dimensions to the left of $\{\mathrm{Re}\, s = D\}$ (as is the case, for example, for the classic ternary Cantor string $\mathscr{L}$ and the associated ternary Cantor set $A = \partial\mathscr{L}$), we have $\Theta = -\infty < 0$ and hence, estimate (2.3.54) then holds with $\eta = D$ and $\delta = 0$.

Example 2.3.36. (*Complex dimensions of the Sierpiński carpet*). Let A be the classic Sierpiński carpet in the plane, as depicted in Figure 2.1 on page 49. Then, using [HorŽu, p. 537] (see also [Lap3] and [Lap-vFr3]), we obtain that

$$|A_t| = t^{2-D}\left(G(\log t^{-1}) + O(t^{D-1})\right) \tag{2.3.55}$$

as $t \to 0^+$, where $D = \log_3 8$, and G is a nonconstant periodic function with period $T = \log 3$.[65] Since here $\alpha = D - 1$ and $T = \log 3$, by using Theorem 2.3.25, we deduce that the distance and tube zeta functions of the Sierpiński carpet possess a unique meromorphic extension (at least) to the open right half-plane $\{\mathrm{Re}\, s > 1\}$, and that the set of complex dimensions of A (in that half-plane) consists of simple poles and is given by

$$\dim_{PC} A = \left\{D + \frac{2\pi}{\log 3}k\mathrm{i} : k \in \mathbb{Z}\right\} = D + \frac{2\pi}{\log 3}\mathrm{i}\mathbb{Z}. \tag{2.3.56}$$

A direct computation shows that both fractal zeta functions are, in fact, meromorphic on all of $\mathbb{C}$.

Theorems 2.3.18 and 2.3.25 are stated for the tube zeta functions of bounded sets $A \subset \mathbb{R}^N$. Here, we first formulate the corresponding results for the distance zeta functions.

Theorem 2.3.37 (Distance zeta functions of bounded sets: Minkowski measurable and nonmeasurable cases). *Let A be a bounded subset of $\mathbb{R}^N$, with $N \geq 1$. In the Minkowski measurable case, we assume that hypothesis* (2.3.26) *of Theorem 2.3.18 holds, while in the Minkowski nonmeasurable case, we assume that hypothesis* (2.3.30) *of Theorem 2.3.25 holds. Furthermore, let $D \geq 0$ be the real number occurring in* (2.3.26) *or in* (2.3.30), *respectively. Then, if $D < N$,*[66] *the conclusions of Theorem 2.3.18* (*respectively, of Theorem 2.3.25*), *concerning the tube zeta function* ($\tilde{\zeta}_A = \tilde{\zeta}_{A,A_\delta}$ *for any fixed* $\delta > 0$), *also hold for the distance zeta*

[65] Alternatively, Equation (2.3.55) also follows by a direct application of the general fractal tube formulas obtained in Chapter 5 below. Conversely, the latter derivation requires the computation of ζ_A (or of $\tilde{\zeta}_A$) which is given in Subsection 3.2.1 below.

[66] Recall that the (upper) box dimension of a bounded subset of $\mathbb{R}^N$ always lies in the closed interval $[0, N]$.

function ζ_A, except for the values of the residues. In the case of the counterpart of Theorem 2.3.18, these values are given by

$$\operatorname{res}(\zeta_A, D) = (N-D)\mathscr{M}^D(A) \tag{2.3.57}$$

and, more generally, in the case of the counterpart of Theorem 2.3.25, by

$$\operatorname{res}(\zeta_A, s_k) = \frac{N-s_k}{T}\hat{G}_0\left(\frac{k}{T}\right), \tag{2.3.58}$$

for each pole $s_k \in \mathscr{P}(\tilde{\zeta}_A)$, with $k \in \mathbb{Z}$; see Equation (2.3.32). We also have the following asymptotics:

$$|\operatorname{res}(\zeta_A, s_k)| = o(|k|), \tag{2.3.59}$$

as $k \to \pm\infty$, and in the case when $G \in C^m(\mathbb{R})$, for some integer $m \geq 1$, we have

$$|\operatorname{res}(\zeta_A, s_k)| = O(|k|^{1-m}), \tag{2.3.60}$$

as $k \to \pm\infty$. Furthermore, in the Minkowski nonmeasurable case and independently of the smoothness of G, we always have that $\tilde{\mathscr{M}}^D(A)$ (the average Minkowski content of A, as given by Equation (2.4.4) *in Definition 2.4.1 below) exists and*

$$\operatorname{res}(\zeta_A, D) = (N-D)\frac{1}{T}\int_0^T G(\tau)\,\mathrm{d}\tau = (N-D)\tilde{\mathscr{M}}^D(A), \tag{2.3.61}$$

and

$$(N-D)\mathscr{M}_*^D(A) < \operatorname{res}(\zeta_A, D) < (N-D)\mathscr{M}^{*D}(A). \tag{2.3.62}$$

Proof. It suffices to use Theorems 2.3.18 and 2.3.25. The claim about the distance zeta function $\zeta_A(s)$ follows immediately from identity (2.2.23). The asymptotics in (2.3.59) and (2.3.60) follow from (2.3.58) and (2.3.38), since $|2^{s_k}| = 2^D$, $|s_k| \sim \frac{2\pi}{T}k$ and $|N - s_k| \sim \frac{2\pi}{T}k$ as $|k| \to \infty$.

The value in (2.3.61) follows from (2.3.58) for $k = 0$. Finally, the inequality in (2.3.62) is a consequence of (2.3.35), (2.3.36) and (2.3.61). □

Note that in Theorem 2.3.37 just above and in agreement with Remark 2.2.18, the distance and tube zeta functions extend meromorphically to the same open right half-plane and have exactly the same poles (with the same multiplicities, equal to 1 here): $\mathscr{P}(\zeta_A) = \mathscr{P}(\tilde{\zeta}_A)$ and, in particular, $\mathscr{P}_c(\zeta_A) = \mathscr{P}_c(\tilde{\zeta}_A)$ (each of these sets being independent of δ). Also, $\dim_B A$ exists and coincides with D: $D = \dim_B A$.

Next, we state the corresponding result for geometric zeta functions associated with bounded fractal strings. Recall that a fractal string $\mathscr{L} = (\ell_j)_{j\geq 1}$ is said to be *bounded* if $\sum_{j=1}^{\infty} \ell_j < \infty$.

Theorem 2.3.38 (Geometric zeta functions of fractal strings). *Let $\mathscr{L} = (\ell_j)_{j\geq 1}$ be a bounded fractal string and let $A = A_{\mathscr{L}} \subseteq (0, +\infty)$ be the corresponding bounded subset of $\mathbb{R}$ defined by $A = \{a_k = \sum_{j\geq k} \ell_j : k \in \mathbb{N}\}$. Much as in*

Theorem 2.3.37, assume that the tube function of A satisfies hypothesis (2.3.26) *of Theorem 2.3.18* (*in the Minkowski measurable case*) *or hypothesis* (2.3.30) *of Theorem 2.3.25* (*in the Minkowski nonmeasurable case*), *respectively. Then, if $D < 1$* (*since $A \subset \mathbb{R}$ and hence, $N = 1$ here*), *and $\alpha < D$,*[67] *the conclusions of Theorem 2.3.18* (*respectively, Theorem 2.3.25*), *concerning the tube zeta function $\tilde{\zeta}_A$, hold for the distance zeta function ζ_A as well as* (*except for the values of the residues*) *for the geometric zeta function $\zeta_{\mathscr{L}}$ of $\mathscr{L}$. For $\zeta_{\mathscr{L}}$, in the case of the counterpart of Theorem 2.3.18, these values are given by*

$$\operatorname{res}(\zeta_{\mathscr{L}}, D) = D2^{D-1}(1-D)\mathscr{M}^D(A), \tag{2.3.63}$$

and (*still for $\zeta_{\mathscr{L}}$*) *in the case of the counterpart of Theorem 2.3.25, by*

$$\operatorname{res}(\zeta_{\mathscr{L}}, s_k) = s_k 2^{s_k-1}\frac{1-s_k}{T}\hat{G}_0\left(\frac{k}{T}\right), \tag{2.3.64}$$

for each pole $s_k \in \mathscr{P}(\tilde{\zeta}_A)$, with $k \in \mathbb{Z}$; see (2.3.32). We also have the following asymptotics:

$$|\operatorname{res}(\zeta_{\mathscr{L}}, s_k)| = o(k^2), \tag{2.3.65}$$

as $k \to \pm\infty$, and in the case when $G \in C^m(\mathbb{R})$, for some integer $m \geq 1$, we have

$$|\operatorname{res}(\zeta_{\mathscr{L}}, s_k)| = O(|k|^{2-m}), \tag{2.3.66}$$

as $k \to \pm\infty$. Furthermore, in the Minkowski nonmeasurable case and independently of the smoothness of G, we always have that $\tilde{\mathscr{M}}^D(A)$ (*the average Minkowski content of A*) *exists and*

$$\operatorname{res}(\zeta_{\mathscr{L}}, D) = D2^{D-1}(1-D)\frac{1}{T}\int_0^T G(\tau)\,\mathrm{d}\tau = D2^{D-1}(1-D)\tilde{\mathscr{M}}^D(A), \tag{2.3.67}$$

and

$$D2^{D-1}(1-D)\mathscr{M}_*^D(A) < \operatorname{res}(\zeta_{\mathscr{L}}, D) < D2^{D-1}(1-D)\mathscr{M}^{*D}(A). \tag{2.3.68}$$

Proof. The claim concerning the distance zeta function ζ_A follows at once from Theorem 2.3.37 applied to the bounded set $A = A_{\mathscr{L}} \subseteq \mathbb{R}$ (and hence, with $N = 1$). Moreover, the claim concerning the geometric zeta function $\zeta_{\mathscr{L}}$ follows from Proposition 2.1.59 since Equation (2.1.84) can be read as follows:

$$\zeta_{\mathscr{L}}(s) = s2^{s-1}(\zeta_A(s) - v(s)),$$

where $v(s)$ is holomorphic for $\operatorname{Re} s > 0$. □

If $\mathscr{L} = (\ell_j)_{j \geq 1}$ is a bounded fractal string and $A = \{a_k = \sum_{j \geq k} \ell_j : k \in \mathbb{N}\}$, then it is easy to see that

---

[67] Indeed, in this case, we have $D - \alpha > 0$; see Proposition 2.1.59 and the condition $c > 0$ assumed there.

$$|A_t| = \sum_{j=1}^{\infty} \tilde{\ell}_j(t) + 2t, \tag{2.3.69}$$

where for each $j \geq 1$, the function $\tilde{\ell}_j : (0, +\infty) \to \mathbb{R}$ is defined by

$$\tilde{\ell}_j(t) = \begin{cases} t, & \text{for } 0 < t < \frac{1}{2}\ell_j, \\ \frac{1}{2}\ell_j, & \text{for } t \geq \frac{1}{2}\ell_j. \end{cases}$$

The formula (2.3.63) appearing in Theorem 2.3.38 is precisely [Lap-vFr3, Equation (8.65)]; see also [Lap-vFr3, Equation (8.25) in Theorem 8.15]. In a later work, we plan to study other applications of some of the results obtained in this section.

An interesting question arises concerning the possible optimality of the domains of the meromorphic extensions appearing in Theorems 2.3.2, 2.3.18, 2.3.25, 2.3.51 and 2.3.52. We illustrate this problem with a result in the context of Theorem 2.3.18, dealing with the Minkowski measurable case; see Theorem 2.3.41 below. Before discussing this result, we make two remarks which complement Definition 2.1.53, the definition of $\mathrm{Mer}(f)$, the half-plane of meromorphic continuation of a given meromorphic function f, initially defined on some domain $U \subseteq \mathbb{C}$.

Remark 2.3.39. In Definition 2.1.53, much as in the standard theory of (generalized) Dirichlet series (see, e.g., [Ser, Section V.2.2]), we let $D_{\mathrm{mer}}(f) := -\infty$ if f admits a meromorphic extension to all of $\mathbb{C}$ (i.e., if $\mathrm{Mer}(f) = \mathbb{C}$), and $D_{\mathrm{mer}}(f) := +\infty$ if f does not admit a meromorphic extension to any open right half-plane (i.e., if $\mathrm{Mer}(f) = \emptyset$). Hence, in every case, $\mathrm{Mer}(f)$ is the union of all the half-planes to which f admits a meromorphic extension. Note that since in light of the principle of analytic continuation, a meromorphic continuation to a given domain, if it exists, is unique (and since the union of an arbitrary family of open right half-planes is itself an open right half-plane),[68] the meromorphic extension of f to $\mathrm{Mer}(f)$ is well defined.

Remark 2.3.40. If $f : U \to \mathbb{C}$ is a meromorphic function given initially (for $\mathrm{Re}\, s$ sufficiently large) by a (generalized) Dirichlet series or more generally, by a Dirichlet-type integral [as is the case for the geometric zeta function of a fractal string (see (2.1.71)) or the distance zeta function of a bounded set $A \subseteq \mathbb{R}^N$ (see (2.1.1))], then its abscissa of (absolute) convergence, $D(f)$, is well defined. Clearly, since f is then holomorphic (and therefore, meromorphic) for $\mathrm{Re}\, s > D(f)$, we must have $D_{\mathrm{mer}}(f) \leq D(f)$. In general, however, we may have $D_{\mathrm{mer}}(f) < D(f)$. For example, for the Riemann zeta function $\zeta = \zeta(s)$ (given initially by the Dirichlet series $\zeta(s) = \sum_{n=1}^{\infty} n^{-s}$, for $\mathrm{Re}\, s > 1$), we have $D_{\mathrm{mer}}(\zeta) = -\infty$ (see Remark 2.3.39) since ζ has a meromorphic continuation to all of $\mathbb{C}$, but it is well known that $D(\zeta) = 1$ since ζ has a pole at $s = 1$ (see Remark 2.3.1).

[68] More precisely, if $\{\alpha_j\}_{j \in J}$ is an arbitrary family of (extended) real numbers and $\alpha := \inf_{j \in J} \alpha_j$, then $\bigcup_{j \in J} \{\mathrm{Re}\, s > \alpha_j\} = \{\mathrm{Re}\, s > \alpha\}$.

Returning to the tube zeta functions $\tilde{\zeta}_A$, we note that the following result provides the precise value of $D_{\rm mer}(\tilde{\zeta}_A)$ for a class of sets A satisfying the asymptotic condition (2.3.70) below.

Theorem 2.3.41 (Minkowski measurable case). *Assume that A is a bounded set in $\mathbb{R}^N$ such that there exist $D \geq 0$, $\mathscr{M} \in (0,+\infty)$, $t_0 \in (0,1]$, such that*

$$|A_t| = t^{N-D}(\mathscr{M} + f(t)), \text{ for all } t \in (0,t_0), \tag{2.3.70}$$

where $f : (0,t_0) \to \mathbb{R}$ has the form $f(t) = \sum_{k=1}^{\infty} c_k t^{\alpha_k}$, with $c_k \neq 0$ for all $k \geq 1$, $\sum_{k=1}^{\infty} |c_k| < \infty$, and the sequence of positive real numbers $(\alpha_k)_{k\geq 1}$ is strictly decreasing and converges to $\alpha > 0$. Then, the largest open right half-plane to which $\tilde{\zeta}_A$ can be meromorphically extended is given by $\mathrm{Mer}(\tilde{\zeta}_A) = \{\mathrm{Re}\, s > D - \alpha\}$;[69] *that is,*

$$D_{\rm mer}(\tilde{\zeta}_A) = D - \alpha. \tag{2.3.71}$$

More specifically, the set of all singularities of $\tilde{\zeta}_A$ is equal to

$$\{D\} \cup \{D - \alpha_k : k \in \mathbb{N}\} \cup \{D - \alpha\}, \tag{2.3.72}$$

where D is a simple pole, each $D - \alpha_k$ is also a simple pole, and $D - \alpha$ is an accumulation point of poles.

Finally, the box dimension $\dim_B A = D$ exists and A is Minkowski measurable, with Minkowski content given by $\mathscr{M}^D(A) = \mathscr{M}$.

Proof. Since $t_0 \leq 1$ and by hypothesis, $\alpha < \alpha_k$ for all $k \geq 1$, we have that

$$|f(t)| \leq t^{\alpha} \sum_{k=1}^{\infty} |c_k| t^{\alpha_k - \alpha} \leq t^{\alpha} \sum_{k=1}^{\infty} |c_k|.$$

Therefore, $f(t)$ is well defined for all $t \in (0,1]$ and $f(t) = O(t^{\alpha})$ as $t \to 0^+$. By using Theorem 2.3.18, we deduce that $D_{\rm mer}(\tilde{\zeta}_A) \leq D - \alpha$. It suffices to prove the reverse inequality. We establish this by proving, in particular, that $(D - \alpha_k)_{k\geq 1}$ is a sequence of poles of $\tilde{\zeta}_A$ (and that each of these poles is simple). Since it converges to $D - \alpha$, then $D - \alpha$ is a singularity of $\tilde{\zeta}_A$, but not a pole, which will show that $D_{\rm mer}(\tilde{\zeta}_A) \geq D - \alpha$.

Let us define ζ_1 and ζ_2 as in the proof of Theorem 2.3.18. In particular, $\tilde{\zeta}_A = \zeta_1 + \zeta_2$, with $\tilde{\zeta}_A$ initially defined (for $\mathrm{Re}\, s$ sufficiently large), by

$$\begin{aligned}
\tilde{\zeta}_A(s) &= \int_0^{\delta} t^{s-N-1} |A_t| \,\mathrm{d}t = \int_0^{\delta} t^{s-N-1} t^{N-D} (\mathscr{M} + f(t)) \,\mathrm{d}t \\
&= \mathscr{M} \int_0^{\delta} t^{s-D-1} \,\mathrm{d}t + \int_0^{\delta} t^{s-D-1} f(t) \,\mathrm{d}t \\
&= \zeta_1(s) + \zeta_2(s).
\end{aligned}$$

[69] See Definition 2.1.53.

(Here, we can choose any fixed positive $\delta < t_0$; in particular, we have $\delta < 1$.) If we assume that $\mathrm{Re}\, s > \alpha_1$, then $\mathrm{Re}\, s > \alpha_k$ for all $k \geq 1$, and in this case we have

$$\begin{aligned}\zeta_2(s) &:= \int_0^\delta t^{s-D-1} f(t)\,\mathrm{d}t = \sum_{k=1}^\infty c_k \int_0^\delta t^{s-D+\alpha_k-1}\,\mathrm{d}t \\ &= \sum_{k=1}^\infty c_k \frac{\delta^{s-D+\alpha_k}}{s-(D-\alpha_k)} =: F(s).\end{aligned} \tag{2.3.73}$$

The function $\zeta_1(s) := \mathscr{M} \int_0^\delta t^{s-D-1}\,\mathrm{d}t = \frac{\mathscr{M}\delta^{s-D}}{s-D}$ is meromorphic in all of $\mathbb{C}$, with a single, simple pole, at $s = D$. Let S be the set of all singularities of F; that is, $S = \{D-\alpha_k : k \in \mathbb{N}\} \cup \{D-\alpha\}$. Since $\sum_{k=1}^\infty |c_k| < \infty$, and since $\delta < 1$, we have that for a given $s_0 \in \mathbb{C} \setminus S$ and for all s in some suitable open disk V centered at s_0 and with sufficiently small radius (so that $d(s,S) \geq d(s_0,S)/2 > 0$ for all $s \in V$),

$$\begin{aligned}\sum_{k=1}^\infty \left| c_k \frac{\delta^{s-D+\alpha_k}}{s-(D-\alpha_k)} \right| &\leq \frac{\delta^{\min_{k\geq 1}|\mathrm{Re}\, s-(D-\alpha_k)|}}{\min_{k\geq 1}|s-(D-\alpha_k)|} \sum_{k=1}^\infty |c_k| \\ &= \frac{\delta^{d(\mathrm{Re}\, s, S)}}{d(s,S)} \sum_{k=1}^\infty |c_k| \\ &\leq \frac{2}{d(s_0,S)} \sum_{k=1}^\infty |c_k| < \infty.\end{aligned}$$

Note that since S is compact and $s_0 \notin S$, then clearly $d(s_0,S) > 0$. Therefore, by the Weierstrass M-test, the last series $F(s) = \sum_{k=1}^\infty c_k \frac{\delta^{s-D+\alpha_k}}{s-(D-\alpha_k)}$ appearing in (2.3.73) is well defined, and it is holomorphic in the open neighborhood V of s_0 and, in particular, at s_0 itself. Since $s_0 \in \mathbb{C} \setminus S$ is arbitrary, we conclude that $\zeta_2(s) = F(s)$ can be meromorphically extended to the open set $\mathbb{C} \setminus S$ and that the largest open right half-plane to which it can be meromorphically extended is $\{\mathrm{Re}\, s > D - \alpha\}$; i.e., $D_{\mathrm{mer}}(\zeta_2) = D - \alpha$. (As usual, we still denote this extension by ζ_2; that is, $\zeta_2 := F$.) Indeed, for each $k \geq 1$, $\zeta_2 = F$ clearly has a simple pole at each $D - \alpha_k$ and furthermore, $D - \alpha = \lim_{k\to\infty}(D - \alpha_k)$ must be a singularity of ζ_2 which is not a pole (because the set of poles of a meromorphic function must be discrete).

Since, according to the above discussion, ζ_1 has a single, simple pole at $s = D$ and is meromorphic in all of $\mathbb{C}$, we deduce that $\tilde{\zeta}_A = \zeta_1 + \zeta_2$ can be meromorphically extended to $\{\mathrm{Re}\, s > D - \alpha\}$, and the set of all of its singularities is equal to $S \cup \{D\} = \{D\} \cup \{D - \alpha_k : k \in \mathbb{N}\} \cup \{D - \alpha\}$. Furthermore, D and each $D - \alpha_k$ $(k \geq 1)$ are simple poles of $\tilde{\zeta}_A$. Since (just as in the case of ζ_2) $D - \alpha$ is a singularity of $\tilde{\zeta}_A$ which is not a pole, we also conclude that $D_{\mathrm{mer}}(\tilde{\zeta}_A) = D - \alpha$.

Finally, the statement concerning the Minkowski dimension and the Minkowski measurability of A follows immediately from hypothesis (2.3.70) and the standard definitions. □

Note that in contrast to Theorem 2.3.18 (where $f(t) = t^\alpha$), we did *not* assume in Theorem 2.3.41 that

$$|A_t| = t^{N-D}\big(\mathscr{M} + O(f(t))\big) \text{ as } t \to 0^+, \tag{2.3.74}$$

a clearly weaker and more flexible hypothesis than (2.3.70). An entirely analogous comment can be made in the context of Remark 2.3.42 just below.

Remark 2.3.42. Much as in Theorem 2.3.41, it is possible to prove an analogous result dealing with Minkowski nonmeasurable sets A, such that

$$|A_t| = t^{N-D}(G(\log t^{-1}) + f(t)),$$

for all $t \in (0, t_0)$, where f is defined exactly as in Theorem 2.3.41, G is a nonconstant, periodic function with values in $(0, +\infty)$ and with minimal period equal to T; see Theorem 2.3.25. The largest open right half-plane to which $\tilde{\zeta}_A$ can be meromorphically continued (i.e., the half-plane of meromorphic continuation of $\tilde{\zeta}_A$, see Definition 2.1.53) is given by $\mathrm{Mer}(\tilde{\zeta}_A) = \{\mathrm{Re}\, s > D - \alpha\}$, and the corresponding set of poles is given by

$$\mathscr{P}(\tilde{\zeta}_A) = \{D\} \cup \{D - \alpha_k : k \in \mathbb{N}\} \cup \left\{D + \frac{2\pi}{T}k\mathrm{i} : \hat{G}_0\left(\frac{k}{T}\right) \neq 0,\ k \in \mathbb{Z}\right\}. \quad (2.3.75)$$

Furthermore, much as in Theorem 2.3.41 above, one can show that $\tilde{\zeta}_A$ has a singularity at $D - \alpha$, which is not a pole. Moreover, the box dimension of A exists and is given by $\dim_B A = D$. In addition to this, A is not Minkowski measurable since $\inf G = \mathscr{M}_*^D < \mathscr{M}^{*D} = \sup G$, but it admits an *average Minkowski content*, denoted by $\tilde{\mathscr{M}}^D(A)$, which will be defined in Definition 2.4.1 of Section 2.4 below, and is given by $\tilde{\mathscr{M}} = \frac{1}{T}\int_0^T G(\tau)\,\mathrm{d}\tau$. The proof of this latter statement can be easily adapted from the proof of [Lap-vFr3, Theorem 8.30]. See also Theorem 2.4.3 below.

In Theorem 4.5.20, we will construct a class of subsets A of the real line such that, for any prescribed values of $D \in (0,1)$ and $\alpha \in (0, D)$, we have $D(\zeta_A) = D$ and $D_{\mathrm{mer}}(\zeta_A) = D - \alpha$. We do this by using an appropriate sequence of relative fractal drums (see Section 4.1), generated by a sequence of generalized Cantor sets. See, in particular, (4.5.56), which parallels condition (2.3.70) in Theorem 2.3.41.

Theorem 2.3.41 and its counterpart discussed in Remark 2.3.42 imply the optimality of the results concerning the existence of a meromorphic extension of the distance zeta function ζ_A in the statement of Theorem 2.3.37 (both in the Minkowski measurable and in the Minkowski nonmeasurable cases). It follows that the same optimality result holds true for the existence of the meromorphic extension of the tube zeta function $\tilde{\zeta}_A$ in the statements of Theorems 2.3.18 and 2.3.25. Indeed, in light of Equation (2.2.23), and since $N \geq D(\zeta_A) = D(\tilde{\zeta}_A)$, we have that

$$\mathrm{Mer}(\zeta_A) = \mathrm{Mer}(\tilde{\zeta}_A) \quad (2.3.76)$$

for any bounded set $A \subset \mathbb{R}^N$.

The following result deals with a class of fractal sets that are Minkowski nonmeasurable, but do not satisfy the 'periodicity' assumption of Theorem 2.3.25 and Remark 2.3.42.

Theorem 2.3.43. *Let A and B be two disjoint bounded sets in $\mathbb{R}^N$ such that their Euclidean distance is positive; that is, $d(A,B) := \inf\{d(x,y) : x \in A,\ y \in B\} > 0$. Assume that there exist $D \geq 0$, $\alpha > 0$, $\beta > 0$, such that*

$$
\begin{aligned}
|A_t| &= t^{N-D}(G(\log t^{-1}) + O(t^{\alpha})) \quad \text{as} \quad t \to 0^+, \\
|B_t| &= t^{N-D}(H(\log t^{-1}) + O(t^{\beta})) \quad \text{as} \quad t \to 0^+,
\end{aligned}
\tag{2.3.77}
$$

where G and H are nonconstant periodic functions on $\mathbb{R}$ with values in $[0,+\infty)$ and minimal periods T and S, respectively, such that the number T/S is irrational, and $t \in (0, \frac{1}{2}d(A,B))$. Then

$$
|(A\cup B)_t| = t^{N-D}\big((G+H)(\log t^{-1}) + O(t^{\min\{\alpha,\beta\}})\big) \quad \text{as} \quad t \to 0^+, \tag{2.3.78}
$$

where the function $G+H$ is nonperiodic,

$$
\inf(G+H) \leq \mathscr{M}_*^D(A\cup B) \leq \mathscr{M}^{*D}(A\cup B) \leq \sup(G+H),
$$

and $\tilde{\zeta}_{A\cup B}$ has a meromorphic extension to the open right half plane

$$
\{\operatorname{Re} s > D - \min\{\alpha,\beta\}\}; \tag{2.3.79}
$$

so that

$$
D_{\text{mer}}(\tilde{\zeta}_{A\cup B}) \leq D - \min\{\alpha,\beta\}. \tag{2.3.80}
$$

Furthermore, the corresponding set of poles of $\tilde{\zeta}_{A\cup B}$ is given by

$$
\mathscr{P}(\tilde{\zeta}_{A\cup B}) = \mathscr{P}(\tilde{\zeta}_A) \cup \mathscr{P}(\tilde{\zeta}_B), \tag{2.3.81}
$$

where

$$
\mathscr{P}(\tilde{\zeta}_A) = \left\{D + \frac{2\pi}{T}k\mathrm{i} : \hat{G}_0\left(\frac{k}{T}\right) \neq 0,\, k \in \mathbb{Z}\right\}, \tag{2.3.82}
$$

$$
\mathscr{P}(\tilde{\zeta}_B) = \left\{D + \frac{2\pi}{S}l\mathrm{i} : \hat{H}_0\left(\frac{l}{S}\right) \neq 0,\, l \in \mathbb{Z}\right\}, \tag{2.3.83}
$$

and so $\mathscr{P}(\tilde{\zeta}_A) \cap \mathscr{P}(\tilde{\zeta}_B) = \emptyset$. Moreover, each of these poles is simple and

$$
\operatorname{res}\left(\tilde{\zeta}_{A\cup B}, D + \frac{2\pi}{T}k\mathrm{i}\right) = \operatorname{res}\left(\tilde{\zeta}_A, D + \frac{2\pi}{T}k\mathrm{i}\right) = \frac{1}{T}\hat{G}_0\left(\frac{k}{T}\right), \tag{2.3.84}
$$

$$
\operatorname{res}\left(\tilde{\zeta}_{A\cup B}, D + \frac{2\pi}{T}l\mathrm{i}\right) = \operatorname{res}\left(\tilde{\zeta}_B, D + \frac{2\pi}{T}l\mathrm{i}\right) = \frac{1}{S}\hat{H}_0\left(\frac{l}{S}\right), \tag{2.3.85}
$$

for $k \neq 0$ ($k \in \mathbb{Z}$) and $l \neq 0$ ($l \in \mathbb{Z}$), with $\hat{G}_0$ defined as in (2.3.28), while

$$
\operatorname{res}(\tilde{\zeta}_{A\cup B}, D) = \frac{1}{T}\int_0^T G(\tau)\,d\tau + \frac{1}{S}\int_0^S H(\tau)\,\mathrm{d}\tau. \tag{2.3.86}
$$

Finally, A, B and $A \cup B$ have the same box dimension D:

$$\dim_B A = \dim_B B = \dim_B(A \cup B) = D. \tag{2.3.87}$$

Furthermore, if, in addition, $D < N$, we have (with $\operatorname{res}(\tilde{\zeta}_{A\cup B}, D)$ *given by* (2.3.86))

$$\begin{aligned} \mathscr{M}_*^D(A) + \mathscr{M}_*^D(B) &\le \mathscr{M}_*^D(A \cup B) \le \operatorname{res}(\tilde{\zeta}_{A\cup B}, D) \\ &\le \mathscr{M}^{*D}(A \cup B) \le \mathscr{M}^{*D}(A) + \mathscr{M}^{*D}(B). \end{aligned} \tag{2.3.88}$$

Proof. Since $t < \frac{1}{2}d(A,B)$, it is easy to see that

$$|(A \cup B)_t| = |A_t \cup B_t| = |A_t| + |B_t|, \tag{2.3.89}$$

which implies that $\tilde{\zeta}_{A\cup B} = \tilde{\zeta}_A + \tilde{\zeta}_B$ and hence that (2.3.78) holds. Much as in the proof of Theorem 2.3.25, we have $\tilde{\zeta}_A = \zeta_1^A + \zeta_2^A$ and $\tilde{\zeta}_B = \zeta_1^B + \zeta_2^B$, from which it follows that

$$\tilde{\zeta}_{A\cup B} = (\zeta_1^A + \zeta_1^B) + (\zeta_2^A + \zeta_2^B). \tag{2.3.90}$$

Since $\zeta_1 := \zeta_1^A + \zeta_1^B$ is meromorphic in the entire complex plane, and $\zeta_2 := \zeta_2^A + \zeta_2^B$ has for abscissa of convergence $D(\zeta_2) \le \max\{D-\alpha, D-\beta\} = D - \min\{\alpha,\beta\}$, it follows from Lemma 2.3.5 that $\tilde{\zeta}_{A\cup B}$ can be meromorphically extended at least to the open right half-plane $\{\operatorname{Re} s > D - \min\{\alpha,\beta\}\}$; i.e., $D_{\rm mer}(\tilde{\zeta}_{A\cup B}) \le D - \min\{\alpha,\beta\}$. The remaining claims about the residues follow from Theorem 2.3.25.

More specifically, according to Theorem 2.3.25 applied to both A and B, we know that the second equality holds in each of Equations (2.3.84) and (2.3.85). Furthermore, we know that

$$\operatorname{res}(\tilde{\zeta}_A, D) = \frac{1}{T}\int_0^T G(\tau)\,\mathrm{d}\tau \tag{2.3.91}$$

and

$$\operatorname{res}(\tilde{\zeta}_B, D) = \frac{1}{S}\int_0^S H(\tau)\,\mathrm{d}\tau. \tag{2.3.92}$$

Moreover, since $\tilde{\zeta}_{A\cup B} = \tilde{\zeta}_A + \tilde{\zeta}_B$, we have

$$\operatorname{res}(\tilde{\zeta}_{A\cup B}, \omega) = \operatorname{res}(\tilde{\zeta}_A, \omega) + \operatorname{res}(\tilde{\zeta}_B, \omega), \tag{2.3.93}$$

for every $\omega \in \mathbb{C}$ such that $\operatorname{Re}\omega > D - \min\{\alpha,\beta\}$. The first equality in (2.3.84) then follows because $\tilde{\zeta}_B$ is holomorphic at the poles of $\tilde{\zeta}_A$ (which, by Theorem 2.3.25, are simple and occur precisely at $D + \frac{2\pi}{T}k\mathrm{i}$, for every $k \in \mathbb{Z}$, provided $\hat{G}_0(k/T) \neq 0$). The fact that the first equality in (2.3.85) holds follows by interchanging the roles of A and B. Moreover, (2.3.86) follows by combining (2.3.93) (applied to $\omega = D$), (2.3.91) and (2.3.92).

Finally, it also follows from Theorem 2.3.25 applied to both A and B that $\dim_B A = \dim_B B = \dim_B(A \cup B)$. It then easily follows from the definition (see, in particular, Equations (1.3.4) and (1.3.5)) that $\dim_B(A \cup B)$ exists and is also equal to D. Furthermore, by taking separately the lower limit and the upper limit in (2.3.89), we deduce that

$$\mathcal{M}_*^D(A) + \mathcal{M}_*^D(B) \leq \mathcal{M}_*^D(A \cup B) \tag{2.3.94}$$

and

$$\mathcal{M}^{*D}(A \cup B) \leq \mathcal{M}^{*D}(A) + \mathcal{M}^{*D}(B). \tag{2.3.95}$$

Next, according to Theorem 2.2.14 (and when $D < N$), applied to $A \cup B$, A and B, we must have that

$$\mathcal{M}_*^D(A \cup B) \leq \operatorname{res}(\tilde{\zeta}_{A\cup B}, D) \leq \mathcal{M}^{*D}(A \cup B). \tag{2.3.96}$$

Equation (2.3.88) now follows by combining the inequalities (2.3.94), (2.3.95) and (2.3.96). This concludes the proof of Theorem 2.3.43. □

Corollary 2.3.44. *Assume that the hypotheses of Theorem 2.3.43 are satisfied. (Here, the assumption according to which $D < N$ is not needed for this corollary to be true.) Then, $A \cup B$ is not Minkowski measurable in general, but it possesses an average Minkowski content (defined as in Equation* (2.4.4) *of Definition 2.4.1 below) given by*

$$\tilde{\mathcal{M}}^D(A \cup B) = \operatorname{res}(\tilde{\zeta}_{A\cup B}, D) = \frac{1}{T}\int_0^T G(\tau)\,\mathrm{d}\tau + \frac{1}{S}\int_0^S H(\tau)\,\mathrm{d}\tau. \tag{2.3.97}$$

Proof. Indeed, Equation (2.3.97) follows by applying Theorem 2.3.25 (or rather, Corollary 2.3.26) to both A and B, in order to deduce that $\tilde{\mathcal{M}}^D(A)$ and $\tilde{\mathcal{M}}^D(B)$ exist and are given as follows:

$$\tilde{\mathcal{M}}^D(A) = \frac{1}{T}\int_0^T G(\tau)\,\mathrm{d}\tau, \quad \tilde{\mathcal{M}}^D(B) = \frac{1}{S}\int_0^S H(\tau)\,\mathrm{d}\tau.$$

One then uses (2.3.89) combined with Equation (2.4.4) (in Definition 2.4.1 below) to deduce that $\tilde{\mathcal{M}}^D(A \cup B)$ exists and is equal to the sum of $\tilde{\mathcal{M}}^D(A)$ and $\tilde{\mathcal{M}}^D(B)$. Hence, in light of (2.3.86), (2.3.97) holds, as desired. □

2.3.4 Landau's Theorem About Meromorphic Extensions

Let $(\mathrm{l_j})_{\mathrm{j}\geq 1}$ be a strictly decreasing sequence of positive numbers and $(b_j)_{j\geq 1}$ be a sequence of positive numbers. For each j, b_j can be thought of as being the 'multiplicity' of $\mathrm{l_j}$. As in Section 2.1.3, let us consider the (generalized) Dirichlet series

$$f(s) = \sum_{j=1}^{\infty} b_j \mathrm{l}_\mathrm{j}^\mathrm{s}, \tag{2.3.98}$$

and define the counting function

$$A(x) = \sum_{\{j : \mathrm{l}_\mathrm{j}^{-1} \leq \mathrm{x}\}} b_j. \tag{2.3.99}$$

We can now formulate a result due to Landau [Lan], which generalized previous results due to Dirichlet and Phragmén.

Theorem 2.3.45 (Landau, [Lan]). *Assume that the Dirichlet series (2.3.98) is such that there exist a positive constant* $\mathscr{M}$ *and* $\gamma \in (0,1)$ *satisfying the following condition*:

$$A(x) = \mathscr{M}x + O(x^{\gamma}) \quad \text{as} \quad x \to \infty. \tag{2.3.100}$$

Then, the function

$$f(s) - \frac{\mathscr{M}}{s-1} \tag{2.3.101}$$

is holomorphic on the half-plane $\{\mathrm{Re}\, s > \gamma\}$.

The conclusion of Theorem 2.3.45 can be formulated equivalently as follows: Then, the function $f(s)$ possesses a meromorphic continuation to the open right half-plane $\{\mathrm{Re}\, s > \gamma\}$, with a unique (and simple) pole located at $s = 1$, and $\mathrm{res}(f,1) = \mathscr{M}$. Furthermore, Landau showed by example that the bound γ, appearing in the open right half-plane $\{\mathrm{Re}\, s > \gamma\}$, cannot be improved.

As we see, Landau's result is of the same nature as parts of Theorems 2.3.2 and 2.3.18 above. A concise introduction to the study of singularities of Dirichlet series can be found in [BerGay, Section 5.2].

Remark 2.3.46. In this section (Section 2.3), we certainly have not exhausted the problem of finding the meromorphic continuation of a given fractal zeta function. In fact, it is far from being the case, as the reader can easily realize. As is well known, the difficult problem of showing the existence of a meromorphic extension (and then studying this extension on a suitable domain) of a given zeta function is one of the fundamental problems in analytic number theory. See, e.g., [Edw, ParsSh1–2, Tit3, WaMLItz, Lap6, Lap-vFr3, Es1–2, EsLapRRo] and the relevant references therein.

Along similar lines, in the study of dynamical (or Ruelle) zeta functions attached to hyperbolic dynamical systems (including subshifts of finite type), powerful techniques were developed by Parry and Pollicott in [ParrPol1–2], as well as by Ruelle [Rue1–4], his collaborators and many other researchers in a variety of dynamical settings. It would be interesting to investigate whether they could be suitably adapted to our geometric situation in order to obtain further meromorphic extension results, applicable to a broader class of bounded sets and later, in Chapter 4, of relative fractal drums (beyond, for example, the self-similar case). We note that in the case of ordinary fractal drums in dimension $N \geq 1$, we shall obtain new results concerning the existence of meromorphic extensions of the spectral zeta functions of such fractal drums, as well as the optimality of certain bounds for the corresponding abscissae of meromorphic continuation of those zeta functions; see Section 4.3, especially Subsection 4.3.2. Moreover, throughout various parts of this book, we will obtain further results about the existence or the non-existence of meromorphic continuations of fractal zeta functions; see, for example, Section 4.5 and Section 4.6, for the case of relative fractal drums.

Some of these results (along with a variety of examples and results in Chapters 4 and 5, see, e.g., Subsection 5.4.4) point to the importance of not only searching for meromorphic extensions (although this is essential in our current theory), but also determining the location and the nature of the nonremovable singularities (in suitable domains of $\mathbb{C}$) of fractal zeta functions. This will lead us, in particular, to extend the definition of fractality proposed in Subsection 4.6.3 (see, especially, Remark 4.6.24) by replacing "poles" with "nonremovable singularities" (including essential singularities) in that definition. The examples of fractal tube formulas obained in several places in Chapter 5 in connection with nonstandard gauge functions and hence, nonremovable singularities that are not poles of the corresponding fractal zeta functions (see also Example 4.2.10) show that the notion of "complex dimensions" itself should be extended accordingly, while still keeping a proper geometric meaning.

2.4 Average Minkowski Contents and Dimensions

From our point of view, introducing average Minkowski contents of a fractal set A can be motivated by the need to understand the behavior of the corresponding tube zeta function near the abscissa of convergence. Its construction leads us in a natural way to a new type of fractal dimensions, that we call (upper and lower) average Minkowski dimensions of A.

2.4.1 Average Minkowski Contents of Bounded Sets in $\mathbb{R}^N$

The following definition is an immediate extension (to bounded subsets of $\mathbb{R}^N$) of the corresponding one in [Lap-vFr3, Definition 8.29], introduced for bounded fractal strings on the real line. See also an analogous expression in Gatzouras' paper [Gat, Theorem 2.3(ii)] and in a different (but related) context in the paper by Bedford and Fisher [BedFi]. This extended definition can be found in the article by Freiberg and Kombrink [FreKom, Definition 1.4(i)], as well as in [Lap-vFr3, Section 13.1] (based on the work by Lapidus and Pearse in [LapPe2–3] and by those same authors and Winter in [LapPeWi1–2]). In order to motivate introducing the average Minkowski content, note that by Theorem 2.2.3, the tube zeta function $\tilde{\zeta}_A(s)$ of any Minkowski nondegenerate fractal set A in $\mathbb{R}^N$ has a simple pole at $s = D := \dim_B A$ (provided there is a meromorphic extension of $\tilde{\zeta}_A$ to a neighborhood of D), and therefore,

$$\tilde{\zeta}_A(D^+) := \lim_{s \to D^+} \int_0^\delta t^{s-N-1} |A_t| \, \mathrm{d}t = +\infty. \tag{2.4.1}$$

It is interesting to know how fast the expression $\int_{1/r}^{\delta} t^{D-N-1} \mathrm{d}t$ tends to infinity as $r \to +\infty$. We shall see in Theorem 2.4.3 below that, under some additional conditions, the growth rate is logarithmic, and moreover,

$$\int_{1/r}^{\delta} t^{D-N-1}|A_t|\,\mathrm{d}t \sim \operatorname{res}(\tilde{\zeta}_A, D)\log r \quad \text{as } r \to +\infty. \tag{2.4.2}$$

Definition 2.4.1. Assume that A is a bounded subset of $\mathbb{R}^N$ such that $D := \dim_B A$ exists. Let $\delta > 0$ be fixed. Then the *average upper D-dimensional Minkowski content* (or, for short, the *average upper Minkowski content*) of A is defined by

$$\tilde{\mathscr{M}}^{*D}(A) := \limsup_{r\to+\infty} \frac{1}{\log r}\int_{1/r}^{\delta} t^{D-N-1}|A_t|\,\mathrm{d}t. \tag{2.4.3}$$

We can analogously define the *average lower Minkowski content* $\tilde{\mathscr{M}}_*^D(A)$, by taking the lower (instead of the upper) limit as $r \to +\infty$ in the counterpart of (2.4.3). If both of these values coincide, that is, if the limit

$$\tilde{\mathscr{M}}^{D}(A) := \lim_{r\to+\infty} \frac{1}{\log r}\int_{1/r}^{\delta} t^{D-N-1}|A_t|\,\mathrm{d}t \tag{2.4.4}$$

exists, then the common value of $\tilde{\mathscr{M}}_*^D(A)$ and $\tilde{\mathscr{M}}^{*D}(A)$ is denoted by $\tilde{\mathscr{M}}^D(A)$ and is called the *average Minkowski content of A*. It is easy to see that the value of the average Minkowski content does not depend on the choice of $\delta > 0$, and therefore, we can assume without loss of generality that $\delta = 1$.

Remark 2.4.2. Note that the integral occurring in (2.4.2), (2.4.3) and (2.4.4) can be rewritten as follows:

$$\int_{1/r}^{\delta} t^{D-N}|A_t|\frac{\mathrm{d}t}{t},$$

where $\mathrm{d}t/t$ is the Haar measure on the multiplicative group $(0,+\infty)$, viewed as the space of scales; see the discussion immediately preceding Equation (2.2.22) on page 118. A similar comment applies to the integral on the right-hand side of (2.4.1), but now with D replaced by s.

Clearly, the integral appearing in the definition of the average Minkowski content is analogous to the one occurring in the definition of the tube zeta function. The following result, Theorem 2.4.3, shows that the average Minkowski content $\tilde{\mathscr{M}}^D(A)$ exists; moreover, $\tilde{\mathscr{M}}^D(A)$ is equal to the residue of the tube zeta function at $D = \dim_B A$, provided the box dimension exists. Part (b) of Theorem 2.4.3 extends part of [Lap-vFr3, Theorem 8.30] to the N-dimensional case. It is easy to check that the technical condition assumed about $f(t)$ in Theorem 2.4.3 is satisfied when $f(t) = O(t^\alpha)$ as $t \to 0^+$, for some $\alpha > 0$; see Proposition 2.4.4.

Theorem 2.4.3. *Assume that A is a bounded set in* $\mathbb{R}^N$ *such that* $D := \dim_B A$ *exists. Then*:

(a) If A is such that $\mathscr{M}^D(A)$ *exists and* $\mathscr{M}^D(A) > 0$ *(and in particular, if A is Minkowski measurable), then* $\tilde{\mathscr{M}}^D(A) = \mathscr{M}^D(A)$*. Furthermore, if there exist positive real numbers* α *and* $\mathscr{M}$ *such that* $|A_t| = t^{N-D}(\mathscr{M} + O(t^\alpha))$ *as* $t \to 0^+$*, then*

$$\tilde{\mathscr{M}}^D(A) = \mathscr{M}^D(A) = \operatorname{res}(\tilde{\zeta}_A, D) = \mathscr{M}. \tag{2.4.5}$$

(b) Let $f : (0,1) \to \mathbb{R}$ be a function such that $f(t) \to 0$ as $t \to 0^+$ and $|f(t)| \le g(\log t^{-1})$, where the function $g : (0,+\infty) \to \mathbb{R}$ satisfies the condition

$$\lim_{L\to+\infty} \frac{1}{L}\int_0^L g(\tau)\,\mathrm{d}\tau = 0. \tag{2.4.6}$$

If $|A_t| = t^{N-D}\left(G(\log t^{-1}) + f(t)\right)$ for all $t \in (0,1)$, where G is a T-periodic function on $\mathbb{R}$,[70] *then*

$$\tilde{\mathscr{M}}^D(A) = \operatorname{res}(\tilde{\zeta}_A, D) = \frac{1}{T}\int_0^T G(\tau)\,\mathrm{d}\tau. \tag{2.4.7}$$

Furthermore, $\mathscr{M}_*^D(A) = \inf G$, $\mathscr{M}^{*D}(A) = \sup G$, and

$$\mathscr{M}_*^D(A) < \tilde{\mathscr{M}}^D(A) < \mathscr{M}^{*D}(A).$$

Proof. (a) The condition $\mathscr{M}^D(A) > 0$ implies that $\tilde{\zeta}_A(D) = +\infty$. Indeed, let $s \in \mathbb{R}$, $s > D$. Then, for any $\delta \in (0,1]$, we have

$$\tilde{\zeta}_A(s) = \int_0^\delta t^{s-N-1}|A_t|\,\mathrm{d}t \le \int_0^\delta t^{D-N-1}|A_t|\,\mathrm{d}t = \tilde{\zeta}_A(D).$$

Letting $s \to D^+$ and by using Theorem 2.1.11(c), we obtain that $\tilde{\zeta}_A(D) = +\infty$. Therefore, since the function $t \mapsto |A_t|$ is clearly continuous on $(0,+\infty)$, we can apply l'Hospital's rule to deduce that $\tilde{\mathscr{M}}^D(A)$ exists and

$$\begin{aligned}\tilde{\mathscr{M}}^D(A) &= \lim_{r\to+\infty} \frac{\int_{1/r}^\delta t^{D-N-1}|A_t|\,\mathrm{d}t}{\log r} = \lim_{r\to+\infty} r^{N-D}|A_{1/r}| \\ &= \lim_{\tau\to 0^+} \frac{|A_\tau|}{\tau^{N-D}} = \mathscr{M}^D(A),\end{aligned}$$

where we have introduced a new variable $\tau := 1/r$. To prove the second equality in (2.4.5), it suffices to also use Theorem 2.3.18.

(b) We have that

$$\begin{aligned}\int_{1/r}^1 t^{D-N-1}|A_t|\,\mathrm{d}t &= \int_{1/r}^1 t^{-1}\left(G(\log t^{-1}) + f(t)\right)\mathrm{d}t \\ &= \int_0^{\log r} G(\tau)\,d\tau + \int_{1/r}^1 t^{-1} f(t)\,\mathrm{d}t,\end{aligned}$$

where we have introduced a new variable $\tau := \log t^{-1}$.

[70] It is clear that G must be nonnegative since $\inf G = \mathscr{M}_*^D(A) \ge 0$.

Step 1: If we write $\log r = kT + \sigma_k$ for r large enough, where $k \in \mathbb{N}$ and $\sigma_k \in [0,T)$ (note that k is uniquely determined by r), it suffices to consider

$$\frac{1}{\log r}\int_0^{\log r} G(\tau)\,\mathrm{d}\tau = \frac{1}{kT+\sigma_k}\left(\int_0^{kT} + \int_{kT}^{kT+\sigma_k}\right) G(\tau)\,\mathrm{d}\tau.$$

Since $r \to +\infty$ implies that $k \to +\infty$,

$$0 \le \int_{kT}^{kT+\sigma_k} G(\tau)\,d\tau \le \int_0^T G(\tau)\,\mathrm{d}\tau, \quad \text{and} \quad \int_0^{kT} G(\tau)\,\mathrm{d}\tau = k\int_0^T G(\tau)\,\mathrm{d}\tau,$$

we have that

$$\lim_{r\to+\infty}\frac{1}{\log r}\int_0^{\log r} G(\tau)\,\mathrm{d}\tau = \lim_{k\to+\infty}\frac{1}{T+\sigma_k/k}\int_0^T G(\tau)\,d\tau = \frac{1}{T}\int_0^T G(\tau)\,\mathrm{d}\tau.$$

Step 2: Using

$$I(r) := \left|\frac{1}{\log r}\int_{1/r}^1 t^{-1} f(t)\,\mathrm{d}t\right| \le \frac{1}{\log r}\int_{1/r}^1 t^{-1} g(\log t^{-1})\,\mathrm{d}t < \infty,$$

where $r > 1$, letting $L := \log r$, and then introducing a new variable $\tau := \log t^{-1}$, we deduce that

$$0 \le \lim_{r\to+\infty} I(r) \le \lim_{L\to+\infty}\frac{1}{L}\int_0^L g(\tau)\,\mathrm{d}\tau = 0.$$

Hence, $\lim_{r\to+\infty} I(r) = 0$.

By combining Steps 1 and 2, we conclude that the average Minkowski content $\tilde{\mathscr{M}}^D(A)$ exists and $\tilde{\mathscr{M}}^D(A) = \frac{1}{T}\int_0^T G(\tau)\,\mathrm{d}\tau$. The remaining equality in (2.4.5) follows from Theorem 2.3.25. □

Proposition 2.4.4. *The technical condition* (2.4.6) *on the function $g(t)$ appearing in part* (b) *of Theorem 2.4.3 is fulfilled if g is nonnegative and satisfies one of the following properties*:

(a) $\displaystyle\int_0^{+\infty} g(\tau)\,\mathrm{d}\tau < \infty$;

(b) $\displaystyle\int_0^{+\infty} g(\tau)\,\mathrm{d}\tau = +\infty$, $g(\tau) \to 0$ as $\tau \to +\infty$, and g is continuous.[71]

Proof. While the sufficiency of (a) is obvious, the sufficiency of (b) follows immediately from l'Hospital's rule:

$$\lim_{L\to+\infty}\frac{1}{L}\int_0^L g(\tau)\,\mathrm{d}\tau = \lim_{L\to+\infty} g(L) = 0.$$

This concludes the proof. □

[71] It suffices to assume that g is continuous in a neighborhood of $+\infty$.

Example 2.4.5. Condition (a) in Proposition 2.4.4 is satisfied if $f(t) = O(t^\alpha)$ as $t \to 0^+$, where $\alpha > 0$ is fixed, since $|f(t)| \le Ct^\alpha = C\exp(-\alpha \log t^{-1})$, and then $g(\tau) := Ce^{-\alpha\tau}$, where C is a positive constant.

On the other hand, condition (b) in Proposition 2.4.4 is satisfied if

$$f(t) = O\big((\log t^{-1})^{-\gamma}\big) \quad \text{as} \quad t \to 0^+,$$

where $\gamma > 0$ is fixed, since then $g(\tau) := C\tau^{-\gamma} \to 0$ as $\tau \to +\infty$, and $\int_0^\infty g(\tau)\, d\tau = +\infty$. Indeed, note that for $\gamma \in (0,1)$ the function g is not integrable near $\tau = +\infty$, while for $\gamma \ge 1$ it is not integrable near $\tau = 0^+$.

This example can be further extended to the case when

$$f(t) = O\big((\log^m t^{-1})^{-\gamma}\big) \quad \text{as} \quad t \to 0^+,$$

where $\gamma > 0$, m is an integer ≥ 2, and $\log^m$ denotes the m-fold composition of log's. In this case, $g(\tau) := C(\log^{m-1} \tau)^{-\gamma} \to 0$ as $\tau \to +\infty$. For notational simplicity, we only check the nonintegrability of G for $m = 2$:

$$\int_e^{+\infty} g(\tau)\, d\tau = C \int_e^{+\infty} (\log \tau)^{-\gamma} d\tau = C \int_1^{+\infty} u^{-\gamma} e^u du = +\infty,$$

since $u^{-\gamma} e^u \to +\infty$ as $u \to +\infty$.

2.4.2 Average Minkowski Dimensions of Bounded Sets in $\mathbb{R}^N$

Much as in the case of Minkowski contents, it is natural to introduce the notion of upper and lower average Minkowski contents, depending on a nonnegative real parameter s. These definitions extend, refine and complement [Lap-vFr3, Definition 8.29].

Definition 2.4.6. Let A be a bounded subset of $\mathbb{R}^N$, and $s \ge 0$. Then, the *upper s-dimensional average Minkowski content of A* is defined by

$$\tilde{\mathscr{M}}^{*s}(A) = \limsup_{r \to +\infty} \frac{1}{\log r} \int_{1/r}^{\delta} t^{s-N-1} |A_t|\, dt, \tag{2.4.8}$$

where $\delta > 0$ is fixed. It is easy to check that the value of $\tilde{\mathscr{M}}^{*s}(A) \in [0, +\infty]$ does not depend on the choice of δ, since

$$\lim_{r \to +\infty} \frac{1}{\log r} \int_{\delta}^{\delta_1} t^{s-N-1} |A_t|\, dt = 0$$

for any two positive real numbers δ and δ_1.

The *lower s-dimensional average Minkowski content* $\tilde{\mathscr{M}}_*^s(A)$ *of* A is defined exactly as in (2.4.8), except for the fact that the upper limit is now replaced with a lower limit.

The following lemma shows that the average Minkowski contents have properties analogous to those of the usual Minkowski contents.

Lemma 2.4.7. *Let A be a bounded set in* $\mathbb{R}^N$. *Then:*

(a) If $s_0 > 0$ *is such that* $\tilde{\mathscr{M}}^{*s_0}(A) > 0$, *then for any positive* $s < s_0$ *we have that* $\tilde{\mathscr{M}}^{*s}(A) = +\infty$. *An analogous claim holds for the lower average Minkowski content.*

(b) If $s_0 > 0$ *is such that* $\tilde{\mathscr{M}}^{*s_0}(A) < \infty$, *then for any positive* $s > s_0$ *we have that* $\tilde{\mathscr{M}}^{*s}(A) = 0$. *An analogous claim holds for the lower average Minkowski content.*

(c) There exists a unique nonnegative real number $\overline{D}_{av}$ *such that*

$$\tilde{\mathscr{M}}^{*s}(A) = \begin{cases} 0, & \text{for } s > \overline{D}_{av}, \\ +\infty, & \text{for } 0 \le s < \overline{D}_{av}. \end{cases} \tag{2.4.9}$$

(d) There exists a unique nonnegative real number $\underline{D}_{av} \ge 0$ *such that*

$$\tilde{\mathscr{M}}_*^{s}(A) = \begin{cases} 0, & \text{for } s > \underline{D}_{av}, \\ +\infty, & \text{for } 0 \le s < \underline{D}_{av}. \end{cases} \tag{2.4.10}$$

Proof. We only provide the proofs for the upper average Minkowski contents. Indeed, for the lower average Minkowski contents, the proof is entirely analogous.

(a) Assume that $\tilde{\mathscr{M}}^{*s_0}(A) > 0$, and let $s < s_0$. Then for any $r > 0$,

$$\int_{1/r}^{\delta} t^{s-N-1}|A_t|\,dt = \int_{1/r}^{\delta} t^{s-s_0}t^{s_0-N-1}|A_t|\,dt \ge \delta^{s-s_0}\int_{1/r}^{\delta} t^{s_0-N-1}|A_t|\,dt.$$

Dividing by $\log r$ and then taking the upper limit as $r \to +\infty$, we obtain that

$$\tilde{\mathscr{M}}^{*s}(A) \ge \delta^{s-s_0}\tilde{\mathscr{M}}^{*s_0}(A).$$

Since average Minkowski contents do not depend on the choice of $\delta > 0$, we can pass to the limit as $\delta \to 0^+$. This proves that $\tilde{\mathscr{M}}^{*s}(A) = +\infty$.

(b) Assume that $\tilde{\mathscr{M}}^{*s_0}(A) < 0$, and let $s > s_0$. Then, reasoning much as in case (a), we obtain that

$$\tilde{\mathscr{M}}^{*s}(A) \le \delta^{s-s_0}\tilde{\mathscr{M}}^{*s_0}(A).$$

Passing to the limit as $\delta \to 0^+$, we conclude that $\tilde{\mathscr{M}}^{*s}(A) = 0$.

(c) follows at once from the first parts of (a) and (b). Analogously, (d) follows from the second parts of (a) and (b). □

An immediate consequence of Lemma 2.4.7 is that the following definition is meaningful.

Definition 2.4.8. Let A be a bounded set in $\mathbb{R}^N$. We define the *upper and lower average Minkowski dimensions of* A as the respective values of $\overline{D}_{av}$ and $\underline{D}_{av}$ in Lemma 2.4.7. In other words, the upper and lower average Minkowski dimensions of A are defined by

$$\begin{aligned}\overline{\dim}_{av}A &:= \inf\{s>0 : \tilde{\mathscr{M}}^{*s}(A)=0\} = \sup\{s>0 : \tilde{\mathscr{M}}^{*s}(A)=+\infty\},\\ \underline{\dim}_{av}A &:= \inf\{s>0 : \tilde{\mathscr{M}}^{s}_{*}(A)=0\} = \sup\{s>0 : \tilde{\mathscr{M}}^{s}_{*}(A)=+\infty\}.\end{aligned} \tag{2.4.11}$$

We use the convention according to which $\sup\emptyset = 0$. If $\overline{\dim}_{av}A = \underline{\dim}_{av}A$, then this common value is denoted by $\dim_{av}A$, and we call it the *average Minkowski dimension of* A.

Proposition 2.4.9. *For any bounded set A in* $\mathbb{R}^N$, *we have*

$$\underline{\dim}_B A \le \underline{\dim}_{av}A \le \overline{\dim}_{av}A \le \overline{\dim}_B A.$$

In particular, if $\dim_B A$ *exists (that is,* $\underline{\dim}_B A = \overline{\dim}_B A$*), then* $\dim_{av} A$ *exists as well and*

$$\dim_{av} A = \dim_B A. \tag{2.4.12}$$

Proof. We first prove the inequality $\overline{\dim}_{av}A \le \overline{\dim}_B A$. Let $\overline{D} := \overline{\dim}_B A$. In view of (2.4.11), it suffices to show that for any $s > \overline{D}$ we have $\tilde{\mathscr{M}}^{*s}(A) = 0$.

Let us choose $s_0 \in (\overline{D}, s)$. Since $\mathscr{M}^{*s_0}(A) = 0$, then there exists a positive constant $C = C(\delta)$ such that $|A_t| \le Ct^{N-s_0}$ for all $t \in (0,\delta)$. Therefore,

$$\int_{1/r}^{\delta} t^{s-N-1}|A_t|\,\mathrm{d}t \le C\int_{1/r}^{\delta} t^{s-s_0-1}\,\mathrm{d}t = C\frac{\delta^{s-s_0}-(1/r)^{s-s_0}}{s-s_0}.$$

Dividing by $\log r$ and letting $r \to +\infty$, we obtain $\tilde{\mathscr{M}}^{*s}(A) \le 0$; that is, $\tilde{\mathscr{M}}^{*s}(A) = 0$.

To prove the inequality $\underline{\dim}_B A \le \underline{\dim}_{av}A$, let us write $\underline{D} := \underline{\dim}_B A$. In view of (2.4.11), it suffices to show that for any $s < \underline{D}$, we have $\tilde{\mathscr{M}}^{*s}(A) = +\infty$.

Let us choose $s_0 \in (s, \underline{D})$. Since $\mathscr{M}^{s_0}_{*}(A) = +\infty$, then there exists a positive constant $C = C(\delta)$ such that $|A_t| \ge Ct^{N-s_0}$ for all $t \in (0,\delta)$. Therefore, much as above, we obtain that

$$\int_{1/r}^{\delta} t^{s-N-1}|A_t|\,\mathrm{d}t \ge C\int_{1/r}^{\delta} t^{s-s_0-1}\,\mathrm{d}t = C\frac{\delta^{s-s_0}-(1/r)^{s-s_0}}{s-s_0} = C\frac{r^{s_0-s}-\delta^{s-s_0}}{s_0-s}.$$

Dividing by $\log r$ and letting $r \to +\infty$, we deduce that $\tilde{\mathscr{M}}^{*s}(A) \ge +\infty$; i.e., $\tilde{\mathscr{M}}^{*s}(A) = +\infty$. □

Regarding the relations between average Minkowski dimensions and standard box dimensions obtained in Proposition 2.4.9, we refer the interested reader to Problems 6.2.28 and 6.2.30.

The notion of average Minkowski content can be extended from bounded sets A in Euclidean spaces to relative fractal drums (A,Ω), which we introduce in Chapter 4. In this way, for any real number s (and not just for any $s \geq 0$), we can define the *average Minkowski contents of relative fractal drums*, $\mathscr{M}^{*s}(A,\Omega)$ and $\mathscr{M}_*^s(A,\Omega)$, and the associated *average Minkowski dimensions of relative fractal drums*, $\overline{\dim}_{av}(A,\Omega)$ and $\underline{\dim}_{av}(A,\Omega)$, which may assume negative values as well, including $-\infty$.

Chapter 3
Applications of Distance and Tube Zeta Functions

Aus dem Paradies, das Cantor uns geschaffen, soll uns niemand vertreiben können.

[No one shall expel us from the paradise that Cantor has created for us.]

David Hilbert (1862–1943)

Abstract In this chapter, we show that some fundamental geometric and number-theoretic properties of fractals can be studied by using their distance and tube zeta functions. This will motivate us, in particular, to introduce several new classes of fractals. Especially interesting among them are the transcendentally quasiperiodic sets, since they can be placed at the crossroad between geometry and number theory. We shall need two deep results from transcendental number theory; namely, the theorem of Gel'fond–Schneider, and its extension due to Baker. In this context, the connecting link between the number theory and the geometry of fractals will be their tube zeta functions. A natural extension of the notion of distance zeta function leads us to introducing a general class of *weighted zeta functions*. Here, we introduce the space $L^{\infty)}(\Omega) := \cap_{p>1} L^p(\Omega)$, called the limit L^∞-space, from which the weight functions are taken. Intuitively, a given weight function w from the space $L^{\infty)}(\Omega)$ may only have very mild singularities, say, of logarithmic type. However, the set of singularities may be large, in the sense that its Hausdorff dimension can be arbitrarily close (and even equal) to N. A typical example is the function $w(x) = \log d(x,A)$ which appears under the integral sign when we differentiate the distance zeta function. We illustrate the efficiency of the use of distance zeta functions by computing the upper box dimension of several new classes of geometric objects, including *geometric chirps*, *fractal nests* and *string chirps*. These sets are closely related to bounded spirals and chirps in the plane. We also recall the construction of a class of fractals, called *zigzagging fractals*, for which the upper and lower box dimensions do not coincide, and show that the associated fractal zeta functions are alternating, in a suitable sense.

Key words: fractal zeta functions, meromorphic extensions, generalized Cantor set, transcendentally n-quasiperiodic set, Sierpiński carpet, Sierpiński gasket, weighted zeta function, tensor product of bounded fractal strings, multiple complex dimensions, essential singularity, fractal nest, geometric chirp, multiple string chirp, zigzagging fractal set.

M.L. Lapidus et al., *Fractal Zeta Functions and Fractal Drums*,
Springer Monographs in Mathematics, DOI 10.1007/978-3-319-44706-3_3

In this chapter, we show that some fundamental geometric and number-theoretic properties of fractals can be studied by using their distance and tube zeta functions. This will motivate us, in particular, to introduce several new classes of fractals. Especially interesting among them are the transcendentally quasiperiodic sets, since they can be placed at the crossroad between geometry and number theory; see Section 3.1. We shall need two deep results from number theory: namely, the theorem of Gel'fond–Schneider, and its extension due to Baker. In this context, the connecting link between the number theory and the geometry of fractals will be their tube zeta functions.

A natural extension of the notion of distance zeta function leads us to introducing a general class of *weighted zeta functions*; see Section 3.4. Here, we introduce the space $L^{\infty)}(\Omega) := \bigcap_{p>1} L^p(\Omega)$, called the limit L^∞-space, from which the weight functions are taken. Intuitively, a given weight function w from the space $L^{\infty)}(\Omega)$ may contain only very mild singularities, say, of logarithmic type. However, the set of singularities may be large, in the sense that its Hausdorff dimension can be arbitrarily close (and even equal) to N. A typical example is the function $w(x) = \log d(x,A)$ which appears under the integral sign when we differentiate the distance zeta function; see Equation (3.4.9) in Theorem 3.4.4(a).

We illustrate the efficiency of the use of distance zeta functions by computing the upper box dimension of several new classes of geometric objects, including *geometric chirps*, *fractal nests* and *string chirps*; see Sections 3.5 and 3.6. These sets are closely related to bounded spirals and chirps in the plane. We also recall the construction of a class of fractals, called *zigzagging fractals*, introduced by the third author in [Žu4, Remark 1.7], for which the upper and lower box dimensions do not coincide, and show that the associated fractal zeta functions are alternating; see Section 3.7.

3.1 Transcendentally Quasiperiodic Sets and Their Zeta Functions

The goal of this section is to define generalized Cantor sets by means of two auxiliary parameters, and to describe a construction of some of the simplest classes of quasiperiodic sets, which we introduce in Definition 3.1.11. The main result is stated in Theorem 3.1.15.

3.1.1 Generalized Cantor Sets Defined by Two Parameters

Here, we introduce a class of generalized Cantor sets $C^{(m,a)}$, depending on two parameters. As a special case we obtain Cantor sets of the form $C^{(a)} := C^{(2,a)}$, which we introduced in Example 2.2.6.

Definition 3.1.1. The generalized Cantor sets $C^{(m,a)}$ are determined by an integer $m \geq 2$ and a positive real number a such that $ma < 1$. In the first step of the analog of Cantor's construction, we start with m equidistant, closed intervals in $[0,1]$ of length a, with $m-1$ 'holes', each of length $(1-ma)/(m-1)$. In the second step, we continue by scaling by the factor a each of the m intervals of length a; and so on, ad infinitum. The (two-parameter) *generalized Cantor set* $C^{(m,a)}$ is defined as the intersection of the decreasing sequence of compact sets constructed in this way. It is easy to check that $C^{(m,a)}$ is a perfect, uncountable compact subset of $\mathbb{R}$. (Recall that a *perfect set* is a closed set without any isolated points.) Furthermore, $C^{(m,a)}$ is also self-similar. For $m := 2$, the sets $C^{(m,a)}$ reduce to the (one-parameter) generalized Cantor sets $C^{(a)}$, defined in Example 2.2.6. In order to avoid any possible confusion, we note that the generalized Cantor sets introduced here are different from the generalized Cantor strings introduced and studied in [Lap-vFr3, Chapter 10]. With our present notation, the classic Cantor set is obtained as $C^{(2,1/3)}$.

We note that the box dimension of $C^{(m,a)}$, as given by Equation (3.1.1) below, is equal to its Hausdorff dimension. The proof of this fact in the case of the classic Cantor set can be found in [Fal1] and is due to Moran [Mora] (in the present case when $N = 1$). This also follows from a general higher-dimensional result in [Hut] (described in [Fal1, Theorem 9.3]) because $C^{(m,a)}$ is a self-similar set satisfying the open set condition.

It can be shown that the generalized Cantor sets $C^{(m,a)}$ have the following properties, which extend the ones established for the sets $C^{(a)}$. Apart from the proof of (3.1.5), which is provided below, the proof of the proposition is similar to that for the standard Cantor set (see [Lap-vFr3, Equation (1.11)]).

Proposition 3.1.2. *If $A := C^{(m,a)} \subset \mathbb{R}$ is the generalized Cantor set introduced in Definition 3.1.1, where m is an integer, $m \geq 2$, and $a \in (0,1/m)$, then*

$$D := \dim_B C^{(m,a)} = D(\zeta_A) = D_{\text{hol}}(\zeta_A) = \log_{1/a} m. \tag{3.1.1}$$

Furthermore, the tube formula associated with $C^{(m,a)}$ is given by

$$|C_t^{(m,a)}| = t^{1-D} G(\log t^{-1}) \tag{3.1.2}$$

for all $t \in (0, \frac{1-ma}{2(m-1)})$, where $G = G(\tau)$ is the following nonconstant periodic function, with minimal period $T = \log(1/a)$, and is defined by

$$G(\tau) := c^{D-1}(ma)^{g(\frac{\tau-c}{T})} + 2c^D m^{g(\frac{\tau-c}{T})}. \tag{3.1.3}$$

Here, $c = \frac{1-ma}{2(m-1)}$, and $g : \mathbb{R} \to [0,+\infty)$ is the 1-periodic function defined by $g(x) := 1-x$ for $x \in (0,1]$.

Moreover,

$$\begin{aligned}\mathscr{M}_*^D(C^{(m,a)}) &= \min G = \frac{1}{D}\left(\frac{2D}{1-D}\right)^{1-D},\\ \mathscr{M}^{*D}(C^{(m,a)}) &= \max G = \left(\frac{1-ma}{2(m-1)}\right)^{D-1}\frac{m(1-a)}{m-1}.\end{aligned} \tag{3.1.4}$$

Finally, if we assume that $\delta \geq \frac{1-ma}{2(m-1)}$, *then the distance zeta function of* $A = C^{(m,a)}$ *is given by*

$$\zeta_A(s) := \int_{-\delta}^{1+\delta} d(x,A)^{s-1}\mathrm{d}x = \left(\frac{1-ma}{2(m-1)}\right)^{s-1}\frac{1-ma}{s(1-ma^s)} + \frac{2\delta^s}{s}. \tag{3.1.5}$$

As a result, $\zeta_A(s)$ *admits a meromorphic continuation to all of* $\mathbb{C}$, *given by the last expression in (3.1.5). In particular, the set of poles of* $\zeta_A(s)$ *(in* $\mathbb{C}$*) and the residue of* $\zeta_A(s)$ *at* $s = D$ *are respectively given by*

$$\begin{aligned}\mathscr{P}(\zeta_A) &= (D + \mathbf{p}\mathrm{i}\mathbb{Z}) \cup \{0\} \quad \text{and}\\ \operatorname{res}(\zeta_A, D) &= \frac{1-ma}{DT}\left(\frac{1-ma}{2(m-1)}\right)^{D-1},\end{aligned} \tag{3.1.6}$$

where $\mathbf{p} = 2\pi/T$ *is the oscillatory period (in the sense of* [Lap-vFr3]*), while the oscillatory amplitude*[1] *of* A *is equal to*

$$\mathbf{am}(A) := \mathscr{M}^{*D}(A) - \mathscr{M}_*^D(A) = \frac{1}{D}\left(\frac{2D}{1-D}\right)^{1-D} - \left(\frac{1-ma}{2(m-1)}\right)^{D-1}\frac{m(1-a)}{m-1}.$$

Furthermore,

$$D = \frac{\log m}{2\pi}\mathbf{p},$$

and both $\mathbf{p} \to 0^+$ *and* $D \to 0^+$ *as* $a \to 0^+$. *In particular,* $\mathscr{P}(\zeta_A)$ *converges to the imaginary axis in the Hausdorff metric, as* $a \to 0^+$. *Finally, each pole in* $\mathscr{P}(\zeta_A)$ *is simple.*

Proof. Part (a): Let us first prove (3.1.5). If we consider an open interval I of length ρ, then it is easy to see that for any $\gamma < 1$,

$$\int_I d(x,\partial I)^{-\gamma}\mathrm{d}x = 2\int_0^{\delta/2} x^{-\gamma}\mathrm{d}x = \frac{2}{1-\gamma}\left(\frac{\rho}{2}\right)^{1-\gamma}. \tag{3.1.7}$$

The generalized Cantor set $A = C^{(m,a)}$, as described in Definition 3.1.1, is an intersection of a decreasing sequence of compact sets $(C_n)_{n\geq 0}$. The largest 'holes' are of

[1] The general definition of the oscillatory amplitude for a class a fractal sets in $\mathbb{R}^N$ can be found on page 541.

length $\rho_0 = \frac{1-ma}{m-1}$ (these 'holes' are connected components of $[0,1]\setminus A$), and there are $m-1$ holes of this size. Each C_n has $m^n(m-1)$ holes of minimal length, equal to $\rho_n = a^n\frac{1-ma}{m-1}$. Using (3.1.7) and the fact that A is negligeable (i.e., of Lebesgue measure zero), we see that the integral below can be computed as an infinite series of integrals over the holes (and hence, bounded open intervals), defined as the connected components of the open subset of $\mathbb{R}$ given by $(0,1)\setminus A$ (see Remark 3.1.3 following this proof):

$$\begin{aligned}\int_0^1 d(x,A)^{-\gamma}\mathrm{d}x &= \sum_{n=0}^{\infty} m^n(m-1)\frac{2}{1-\gamma}\left(\frac{\rho_n}{2}\right)^{1-\gamma} \\ &= \frac{2^\gamma(m-1)^\gamma}{1-\gamma}(1-ma)^{1-\gamma}\sum_{n=0}^{\infty}(ma^{1-\gamma})^n \\ &= \left(\frac{\rho_0}{2}\right)^{-\gamma}\frac{1-ma}{(1-\gamma)(1-m\cdot a^{1-\gamma})}.\end{aligned} \tag{3.1.8}$$

The series converges provided $m\cdot a^{1-\gamma}<1$, that is, if $\gamma<1-D$, where $D=\log_{1/a}m$. Letting $s:=1-\gamma\in(D,+\infty)$, and in light of Equation (2.1.1), the above identity implies that

$$\zeta_A(s) = \left(\frac{\rho_0}{2}\right)^{s-1}\frac{1-ma}{s(1-ma^s)}+\frac{2\delta^s}{s}. \tag{3.1.9}$$

If we now view s as a complex variable, both the left and right-hand sides of (3.1.9) define holomorphic functions on their respective domains. (Note that we know from Theorem 2.1.11 that ζ_A is holomorphic for $\mathrm{Re}\,s > D(\zeta_A) = \dim_B A$, where $A := C^{(m,a)}$.) Since they coincide on the real interval $(D,+\infty)$, it follows that in (3.1.9), we have equality for all complex numbers s in the half-plane $\{\mathrm{Re}\,s > D\}$. The claim then follows by meromorphically extending $\zeta_A(s)$, using the expression on the right-hand side of (3.1.9).

The above reasoning clearly shows that $\{\mathrm{Re}\,s > D\}$ is the largest half-plane on which ζ_A is holomorphic (as well as absolutely convergent); that is, $D(\zeta_A) = D_{\mathrm{hol}}(\zeta_A) = D$ and hence, the identity (3.1.1) holds. It is easy to check that $\dim_B A$ exists in the present case (moreover, as we shall see in part (b) of the proof, we have that $0 < \mathscr{M}_*^D(A) < \mathscr{M}^{*D}(A) < \infty$; see also footnote 14 on page 32), and therefore equals D: $D = \dim_B A = D(\zeta_A) = D_{\mathrm{hol}}(\zeta_A)$, where $D = \log_{1/a} m$.

Part (b): We next compute the Minkowski contents of the generalized Cantor set $A = C^{(m,a)}$. Let $t>0$ be fixed. We first determine the nonnegative integer $n = n(t)$ such that $\rho_{n(t)} \le 2t < \rho_{n(t)-1}$. (Recall from part (a) of the proof that ρ_n was defined as the minimal length of the holes of C_n.) Solving the inequality $\rho_n \le 2t$, we obtain

$$n \ge \frac{\log t^{-1} + \log c}{\log(1/a)} =: h(t),$$

and hence,

$$n(t) = \lceil h(t)\rceil = h(t) + g(h(t)),$$

where by $\lceil x\rceil$ (the 'ceiling' of x) we denote the smallest integer which is not less than x. Note that $g(x) = \lceil x\rceil - x$.

Recall from part (a) that $A = \cap_{n\geq 0} C_n$, where $(C_n)_{n\geq 0}$ is a decreasing family of compact sets defined at the beginning of the proof, and each C_n consists of m^{n+1} disjoint intervals of length a^{n+1}. By our choice of $n = n(t)$, since $2t \geq \rho_n$, we see that all the (smallest) holes corresponding to C_n are filled in by its t-neighborhood. Therefore, we have the disjoint union

$$A_t = (C_{n-1})_t = C_{n-1} \cup ((C_{n-1})_t \setminus C_{n-1}),$$

and C_{n-1} consists of m^n intervals of length a^n, while $(C_{n-1}))_t \setminus C_{n-1}$ consists of $2m^n$ intervals of length t. We conclude that

$$\begin{aligned}|A_t| &= |(C_{n-1})_t| = m^n(a^n + 2t)\\ &= (ma)^{h(t)}(ma)^{g(h(t))} + 2t\, m^{h(t)} m^{g(h(t))}\\ &= m^{h(t)}(a^{h(t)}(am)^{g(h(t))} + 2tm^{g(h(t))}) = t^{1-D} G(\log t^{-1}),\end{aligned}$$

where $G = G(\tau)$ is defined by (3.1.3).

Since the function $G = G(\tau)$ is periodic, and its maximum is achieved at the points $\tau = c + kT$, $k \in \mathbb{Z}$, we deduce that

$$\begin{aligned}\mathscr{M}^{*D}(A) &= G((\log c)^-) = c^{D-1}(ma)^0 + 2c^D m^0\\ &= c^{D-1}(1+2c) = c^{D-1}\frac{m(1-a)}{m-1},\end{aligned}$$

where we have let $G((\tau_0)^-) := \lim_{\tau\to(\tau_0)^-} G(\tau)$ and used the fact that $g(0^-) = 0$.

In order to compute $\mathscr{M}_*^D(A)$, let us define

$$f(x) = c^{D-1}(ma)^x + 2c^D m^x.$$

Since $ma < 1$ and $m > 1$, it is clear that f is a convex function which is bounded from below and that it has a unique minimum value, achieved at a single point, denoted by x_0. Clearly, $\mathscr{M}_*^D(A) = \min f = f(x_0)$. Since we must have $f'(x_0) = 0$, we conclude that

$$x_0 = D\log_m\left((2c)^{-1}\log_m(ma)^{-1}\right).$$

After an elementary computation, we obtain that

$$\mathscr{M}_*^D(A) = f(x_0) = c^{D-1}(ma)^{x_0} + 2c^D m^{x_0} = \frac{1}{D}\left(\frac{2D}{1-D}\right)^{1-D}. \tag{3.1.10}$$

This completes the proof of Proposition 3.1.2. □

The values of the upper and lower Minkowski contents of $C^{(m,a)}$ have been obtained earlier in [Žu2, Equations (3.12) and (3.13)]. It is rather time-consuming to verify the last equality appearing in Equation (3.1.10) by hand. The reader may find it easier to verify this last equality by using a symbolic manipulation computer package.

In the case of the classic Cantor set, that is, for $m := 2$ and $a := 1/3$, we recover the values first obtained in [LapPo2, Theorem 2.4]. Finally, the tube formula (3.1.2) extends the one obtained in [Lap-vFr3, Equation (1.11)]. We also refer to [Lap-vFr3, Chapter 10] where these computations have been significantly extended and refined (both in the geometric and, crucially, in the spectral case) for a certain one-parameter family of generalized (and possibly virtual) Cantor sets and strings in order to prove new results about the distribution and the asymptotic density of the 'critical zeros' of the Riemann zeta function, as well as of a large class of arithmetic zeta functions and generalized Dirichlet series or integrals; see [Lap-vFr3, Chapter 11].

Remark 3.1.3. Note that essentially by definition, in the above proof of Proposition 3.1.2, the 'holes' are deleted (open) intervals in the construction of the two-parameter Cantor set $A = C^{(m,a)}$. Hence, they form an infinite sequence of pairwise disjoint bounded open intervals whose union is equal to $(0,1)\setminus A$, the complement of A in $(0,1)$. In other words, $A \subset \mathbb{R}$ is viewed as the boundary of an ordinary fractal string, realized geometrically as the open bounded set $\Omega_{m,a} := (0,1)\setminus A$ (with the holes as associated intervals) and which can be naturally called the *generalized two-parameter Cantor string* (by analogy with the classic ternary Cantor string $\Omega_{2,1/3}$; see [Lap-vFr3, Subsection 1.1.2]).

Definition 3.1.4. According to the terminology introduced in [Lap-vFr3], the value of $\mathbf{p} = 2\pi/T$, appearing in Proposition 3.1.2, is called the *oscillatory period* of the generalized Cantor set $A = C^{(m,a)}$. As we see from the proof of Proposition 3.1.2, the set of principal complex dimensions of A is then given by

$$\dim_{PC} A := \mathscr{P}_c(\zeta_A) = D + \mathbf{p}\mathrm{i}\mathbb{Z}.$$

Remark 3.1.5. We leave it as an easy exercise for the reader to write down the exact counterpart of Proposition 3.1.2 for the tube zeta function $\tilde{\zeta}_A$. We only mention here that the poles of $\tilde{\zeta}_A$ are the same as those of ζ_A (i.e., $\mathscr{P}(\tilde{\zeta}_A) = \mathscr{P}(\zeta_A)$), as given by (3.1.6) and that they are simple (but that the residues of $\tilde{\zeta}_A$ differ by a multiplicative factor from the corresponding residues of ζ_A). In particular, $D = D(\tilde{\zeta}_A) = D(\zeta_A)$, as given by (3.1.1), and D is a simple pole of $\tilde{\zeta}_A$. Hence, according to (2.2.24), we have that $\operatorname{res}(\tilde{\zeta}_A, D) = \frac{1}{1-D}\operatorname{res}(\zeta_A, D)$, where $\operatorname{res}(\zeta_A, D)$ is given by (3.1.6).

In light of Remark 3.1.5 and part (*b*) of Theorem 2.4.3 (see, especially, Equation (2.4.7)), we can now state the following corollary of Proposition 3.1.2.

Corollary 3.1.6. *Let $A := C^{(m,a)}$ be the above generalized Cantor set and let D and G be given, respectively, as in Equation* (3.1.1) *and Equation* (3.1.3). *Then the average Minkowski content of A exists* (*in* $(0,+\infty)$) *and is given by*

$$\tilde{\mathscr{M}}^D(A) = \operatorname{res}(\tilde{\zeta}_A, D) = \frac{1}{T}\int_0^T G(\tau)\,\mathrm{d}\tau = \frac{1}{1-D}\operatorname{res}(\zeta_A, D), \qquad (3.1.11)$$

with the value of $\operatorname{res}(\zeta_A, D)$ *explicitly given by the last line of Equation* (3.1.6). *In particular, A is Minkowski nondegenerate and*

$$0 < \mathscr{M}_*^D(A) < \tilde{\mathscr{M}}^D(A) < \mathscr{M}^{*D}(A) < \infty.$$

3.1.2 Construction of Transcendentally 2*-Quasiperiodic Sets*

The following example, Example 3.1.8, provides the basic ideas for further definitions, results and constructions. The main result of this subsection is contained in Theorem 3.1.12.

Recall that the *field of algebraic numbers* (often denoted by $\overline{\mathbb{Q}}$ in the literature) can be viewed (up to isomorphism) as the algebraic closure of $\mathbb{Q}$ (the field of rational numbers) and is obtained by adjoining to $\mathbb{Q}$ the roots of the monic polynomial equations with coefficients in $\mathbb{Q}$ (or, equivalently, in $\mathbb{Z}$). Note that, as a result, it is a countable set.

We shall need a classic result due to Gel'fond and Schneider (see [Gel]), proved independently by these two authors in 1934. We state it in a form that will be convenient for our purposes.

Theorem 3.1.7 (Gel'fond–Schneider, [Gel]). *Let ρ be a positive algebraic number different from* 1, *and let x be an irrational algebraic number. Then ρ^x is transcendental.*

Example 3.1.8. We construct a bounded subset of the real line of a box dimension $D \in (0,1)$, possessing two *incommensurable* quasiperiods T and S (and moreover, such that the number T/S is *transcendental*); this set is *transcendentally quasiperiodic* (more precisely, *transcendentally* 2*-quasiperiodic*) in the sense of part (a) of Definition 3.1.11 below. The set will be of the form $A \cup B$, where A and B are two bounded subsets of the real line, satisfying the assumptions of Theorem 2.3.43 and defined as the generalized Cantor sets $A = C^{(a)} := C^{(2,a)} \subset [0,1]$, where $a \in (0,1/2)$, and $B = C^{(3,b)} \subset [2,3]$, where $b \in (0,1/3)$. We choose b so that $D := \log_{1/a} 2 = \log_{1/b} 3$. See Definition 3.1.1 and Proposition 3.1.2 above.

We may take, for example $a := 1/3$ and $b := 3^{-\log_2 3}$. Note that we then have $3b = 3^{1-\log 3_2} < 1$. The functions G_1 and G_2 corresponding to A and B are T and S-periodic, respectively, with $T := \log(1/a) = \log 3$ and $S := \log(1/b)$. Since

$$\frac{T}{S} = \frac{\log 3}{\log(1/b)} = \frac{\log 2}{\log 3} = \log_3 2$$

is irrational (and even transcendental), we see that the conditions of Theorem 2.3.43 (on page 173) are satisfied with $N = 1$ and $O(t^\alpha) \equiv 0$ (that is, with no error term).

Since it solves the equation $3^x = 2$, the positive real number $x = T/S$ is transcendental, in light of the Gel'fond–Schneider theorem which is recalled in Theorem 3.1.7 just above. Indeed, if x were an irrational algebraic number, then according to this result, the number 3^x would be transcendental, which is a contradiction.

It is easy to see that x cannot be a rational number p/q, with $p,q \in \mathbb{N}$, since otherwise, it would follow from the equality $3^{p/q} = 2$ that $3^p = 2^q$, which is also impossible, according to the fundamental theorem of arithmetic. Therefore, x is transcendental. Here, we mention that $x = \log\alpha$ is transcendental for all algebraic numbers $\alpha \neq 0$ not equal to 0 or 1, a result going back to F. von Lindemann and K. Weierstrass; see [Ba, p. 4].

For our later needs, it will be convenient to introduce the following definition of quasiperiodic functions.

Definition 3.1.9. We say that a function $G = G(\tau) : \mathbb{R} \to \mathbb{R}$ is *n-quasiperiodic* (or *quasiperiodic, of order of quasisperiodicity equal to n*) if it is of the form

$$G(\tau) := H(\tau,\ldots,\tau), \tag{3.1.12}$$

where for some $n \geq 2$, $H : \mathbb{R}^n \to \mathbb{R}$ is a function which is nonconstant and T_k-periodic in its k-th component, for each $k = 1,\ldots,n$, and the corresponding periods $T_1,\ldots,T_n$ are rationally independent (i.e., linearly independent over the field of rational numbers). The values of T_i are called the *quasiperiods of G*.

In addition, we say that a function $G = G(\tau)$ is

(a) transcendentally n-quasiperiodic if the periods $T_1,\ldots,T_n$ are algebraically independent. In particular, all of the quotients T_i/T_j, for $i \neq j$, are then transcendental (and hence, irrational) numbers;

(b) algebraically n-quasiperiodic if the corresponding periods $T_1,\ldots,T_n$ are rationally independent and algebraically dependent. More precisely, we assume here that the set of quasiperiods $\{T_1,\ldots,T_n\}$ is algebraically dependent; in other words, there exist algebraic numbers $\lambda_1,\ldots,\lambda_n$, not all of them zero, such that $\lambda_1 T_1 + \cdots + \lambda_n T_n = 0$.

In the existing literature on dynamical systems, mathematical physics and harmonic analysis, there is a wide variety of different definitions of quasiperiodic and almost periodic functions (and sets). See, for example, [WaMLItz], [Sen], [Boh], [Kat], [Lap-vFr3], [Lap6, Appendix F], along with the relevant references therein. Definition 3.1.9 (along with Definition 3.1.11) suits ideally our purposes. Definitions 3.1.9 and 3.1.11 will be further refined and extended to the $n = \infty$ case in Definitions 4.6.6 and 4.6.7, respectively.

The notion of quasiperiodic function provided in Definition 3.1.9 above has been motivated by [Vin]. However, while in [Vin], it is assumed that the *reciprocals* of the quasiperiods $T_1,\ldots,T_n$ are rationally independent, we assume in Definition 3.1.9 that the quasiperiods $T_1,\ldots,T_n$ themselves are rationally independent. The distinction between algebraically n-quasiperiodic and transcendentally n-quasiperiodic functions seems to be new. Furthermore, in Definition 4.6.6 on page 374 below, we shall introduce ∞-quasiperiodic functions, that is, quasiperiodic functions with infinitely many rationally independent quasiperiods.

It is clear from Definition 3.1.9 that each quasiperiodic function is either transcendentally quasiperiodic or algebraically quasiperiodic. In other words, the set

$\mathscr{F}_{\rm qp}$ of quasiperiodic functions is equal to the disjoint union of the set $\mathscr{F}_{\rm tqp}$ of transcendentally quasiperiodic functions and the set $\mathscr{F}_{\rm aqp}$ of algebraically quasiperiodic functions:

$$\mathscr{F}_{\rm qp} = \mathscr{F}_{\rm tqp} \cup \mathscr{F}_{\rm aqp}.$$

Example 3.1.10. If $G(\tau) = G_1(\tau) + G_2(\tau)$, where the functions G_i are nonconstant and T_i-periodic, $i = 1,2$, such that T_1/T_2 is an irrational algebraic number, then G is algebraically 2-quasiperiodic. In this case and in the notation of Definition 3.1.9, we have $H(\tau_1,\tau_2) := G_1(\tau_1) + G_2(\tau_2)$. If T_1/T_2 is transcendental, then G is transcendentally 2-quasiperiodic (in the sense of Definition 3.1.9).

Definition 3.1.11. Assume that a bounded set $A \subset \mathbb{R}^N$ has the following tube formula:

$$|A_t| = t^{N-D}(G(\log(1/t)) + o(1)) \quad \text{as} \quad t \to 0^+, \tag{3.1.13}$$

such that G is nonnegative, $0 < \liminf_{\tau\to+\infty} G(\tau) \le \limsup_{\tau\to+\infty} G(\tau) < +\infty$ and $D \in [0,N]$ is a given constant. Note that it then follows that $\dim_B A$ exists and is equal to D. Moreover, $\mathscr{M}_*^D(A) = \liminf_{\tau\to+\infty} G(\tau)$ and $\mathscr{M}^{*D}(A) = \limsup_{\tau\to+\infty} G(\tau)$.

We say that A is an *n-quasiperiodic set* (of *order of quasiperiodicity equal to n*). if the corresponding function $G = G(\tau)$ is n-quasiperiodic.

In addition, the set A is said to be

(*a*) *transcendentally n-quasiperiodic* if the corresponding function G is transcendentally n-quasiperiodic;

(*b*) *algebraically n-quasiperiodic* if the corresponding function G is algebraically n-quasiperiodic.

In light of Definition 3.1.11 and the comment following Definition 3.1.9, we see that each n-quasiperiodic set is either transcendentally n-quasiperiodic or n-algebraically quasiperiodic. In other words, the family $\mathscr{S}_{\rm qp}(n)$ of n-quasiperiodic sets is equal to the disjoint union of the family $\mathscr{S}_{\rm tqp}(n)$ of transcendentally n-quasiperiodic sets and the family $\mathscr{S}_{\rm aqp}(n)$ of algebraically n-quasiperiodic sets:

$$\mathscr{S}_{\rm qp}(n) = \mathscr{S}_{\rm tqp}(n) \cup \mathscr{S}_{\rm aqp}(n).$$

Note that the family $(\mathscr{S}_{\rm qp}(n))_{n\ge2}$ is *disjoint*, as well as the family $(\mathscr{S}_{\rm tqp}(n))_{n\ge2}$ and the family $(\mathscr{S}_{\rm aqp}(n))_{n\ge2}$. Letting

$$\mathscr{S}_{\rm qp} := \bigcup_{n\ge2} \mathscr{S}_{\rm qp}(n), \quad \mathscr{S}_{\rm tqp} := \bigcup_{n\ge2} \mathscr{S}_{\rm tqp}(n), \quad \mathscr{S}_{\rm aqp} := \bigcup_{n\ge2} \mathscr{S}_{\rm aqp}(n),$$

we have

$$\mathscr{S}_{\rm qp} = \mathscr{S}_{\rm tqp} \cup \mathscr{S}_{\rm aqp}.$$

Example 3.1.8 shows that the family $\mathscr{S}_{\rm tqp}(2)$ is nonempty. In Theorem 3.1.12 we shall see that the family $\mathscr{S}_{\rm tqp}(2)$ is infinite. Moreover, the family $\mathscr{S}_{\rm tqp}(n)$ is infinite for any $n \ge 2$; see Theorem 3.1.15. We do not know if the family $\mathscr{S}_{\rm aqp}$ is nonempty.

A more general definition of quasiperiodic sets can be found on page 542 of Subsection 6.1.1.1. This will lead us to consider the family $\mathscr{S}_{\rm qp}(\infty)$ of ∞-quasiperiodic sets, i.e., the family of quasiperiodic sets with infinitely many quasiperiods; see Definition 4.6.7 on pages 375 and 376.

Generalizing the idea of the above example, we obtain the following result. It is noteworthy that Theorem 3.1.12 involves a simple *geometric condition* (of linear independence over the rationals) on exponent vectors associated with two positive integers. A similar comment can be made about Theorem 3.1.15 in Subsection 3.1.3 below.

Theorem 3.1.12. *Let $A_1 = C^{(m_1,a_1)} \subset [0,1]$ and $A_2 = C^{(m_2,a_2)} \subset [2,3]$ be two generalized Cantor sets (see Definition 3.1.1 and Proposition 3.1.2) such that their box dimensions coincide;*[2] *call D this common value of $\dim_B A_i$ for $i = 1,2$. Let $\{p_1, p_2, \dots, p_k\}$ be the set of all distinct prime factors of m_1 and m_2, and write*

$$m_1 = p_1^{\alpha_1} p_2^{\alpha_2} \dots p_k^{\alpha_k}, \quad m_2 = p_1^{\beta_1} p_2^{\beta_2} \dots p_k^{\beta_k}, \tag{3.1.14}$$

where $\alpha_i, \beta_i \in \mathbb{N} \cup \{0\}$ for $i = 1, \dots, k$. If the exponent vectors

$$(\alpha_1, \alpha_2, \dots, \alpha_k) \quad \text{and} \quad (\beta_1, \beta_2, \dots, \beta_k), \tag{3.1.15}$$

corresponding to m_1 and m_2, are linearly independent over the rationals (in other words, there is no $q \in \mathbb{Q}$ such that $(\alpha_1, \alpha_2, \dots, \alpha_k) = q(\beta_1, \beta_2, \dots, \beta_k)$), then the function $G = G_1 + G_2$, associated with $A = A_1 \cup A_2$, is transcendentally 2-quasiperiodic; that is, the quotient T_1/T_2 of the quasiperiods of G (i.e., of the periods of G_1 and G_2) is transcendental. In other words, the set A itself is transcendentally 2-quasiperiodic.

Moreover, ζ_A can be meromorphically extended to all of $\mathbb{C}$ and we have that

$$\zeta_A(s) \sim \frac{1}{1 - m_1 a_1^s} + \frac{1}{1 - m_2 a_2^s}, \quad D(\zeta_A) = \dim_B A = D \quad \text{and} \quad D_{\rm mer}(\zeta_A) = -\infty.$$

(The exact expression of ζ_A is given in Equation (3.1.18) below.) Hence, the set $\dim_{PC} A = \mathscr{P}_c(\zeta_A)$ of principal complex dimensions of A consists of simple poles of ζ_A and coincides with the following nonarithmetic set (see Remark 3.1.13 below):

$$\dim_{PC} A = \left(D + \frac{2\pi}{T_1}\mathrm{i}\mathbb{Z}\right) \cup \left(D + \frac{2\pi}{T_2}\mathrm{i}\mathbb{Z}\right). \tag{3.1.16}$$

Besides $(\dim_{PC} A) \cup \{0\}$, there are no other poles of the distance zeta function ζ_A in $\mathbb{C}$. Furthermore, all of the complex dimensions of A are simple.

Finally, exactly the same results hold for the tube zeta function $\tilde{\zeta}_A$ of A (in place of the distance zeta function ζ_A).

---

[2] More generally, it suffices to assume that the Cantor sets A_1 and A_2 are contained in two compact unit intervals with disjoint interiors, respectively.

Proof. First of all, using (3.1.2), applied to both A_1 and A_2, we see that for all $t \in (0,1/2)$,

$$|(A_1 \cup A_2)_t| = |(A_1)_t| + |(A_2)_t| = t^{1-D}\left(G_1(\log t^{-1}) + G_2(\log t^{-1})\right).$$

It suffices to show that the quotient T_1/T_2 of the quasiperiods T_1 and T_2 of $G(\tau) := G_1(\tau) + G_2(\tau)$ is transcendental.

From the fact that $D = \log_{1/a_1} m_1 = \log_{1/a_2} m_2$ and $T_i = \log_{1/a_i}$ for $i = 1,2$, we deduce that $x := T_1/T_2$ satisfies the equation $(m_2)^x = m_1$. The exponent x cannot be an irrational algebraic number, since otherwise, by the Gel'fond–Schneider theorem (Theorem 3.1.7), $(m_2)^x$ would be transcendental. If x were rational, say, $x = b/a$, with $a,b \in \mathbb{N}$ (note that $x > 0$, since $m_1 \geq 2$), this would then imply that $(m_1)^a = (m_2)^b$; that is,

$$p_1^{a\alpha_1} p_2^{a\alpha_2} \cdots p_k^{a\alpha_k} = p_1^{b\beta_1} p_2^{b\beta_2} \cdots p_k^{b\beta_k}.$$

Therefore, using the fundamental theorem of arithmetic, we would have

$$a(\alpha_1, \alpha_2, \dots, \alpha_k) = b(\beta_1, \beta_2, \dots, \beta_k).$$

However, this is impossible due to the assumption of linear independence over the rationals of the above exponent vectors. Therefore, x is transcendental.

The claims about the zeta function $\zeta_{A_1 \cup A_2}$ follow from Proposition 3.1.2 applied both to A_1 and A_2. Indeed, since A_1 and A_2 are subsets of two disjoint compact intervals $[0,1]$ and $[2,3]$, we can choose without loss of generality $\delta = 1/2$ (see Proposition 2.1.76), so that $\overline{A_\delta} = [-1/2, 3+1/2]$ and therefore,

$$\zeta_A(s) = \int_0^1 d(x,A_1)^{s-1}\mathrm{d}x + \int_2^3 d(x,A_1)^{s-1}\mathrm{d}x + 4\int_0^{1/2} x^{s-1}\mathrm{d}x. \tag{3.1.17}$$

Note that the last term on the right-hand side of (3.1.17) corresponds to the union of the following four intervals: $(-1/2,0)$, $(1,3/2)$, $(3/2,2)$ and $(3,7/2)$. By using Equation (3.1.5) in Proposition 3.1.2, applied separately to A_1 and A_2, we obtain that for all $s \in \mathbb{C}$ such that $\mathrm{Re}\, s > D$,

$$\begin{aligned} \zeta_A(s) = &\left(\frac{1-m_1a_1}{2(m_1-1)}\right)^{s-1} \frac{1-m_1a_1}{s(1-m_1a_1^s)} \\ &+ \left(\frac{1-m_2a_2}{2(m_2-1)}\right)^{s-1} \frac{1-m_2a_2}{s(1-m_2a_2^s)} + \frac{4\delta^s}{s}. \end{aligned} \tag{3.1.18}$$

Equation (3.1.18) shows that $\zeta_A(s)$ can be meromophically extended to the whole complex plane, and that besides $(\dim_{PC} A) \cup \{0\}$ there are no other poles of ζ_A in $\mathbb{C}$. The same equation shows that $\zeta_A(s) \sim (1-m_1a_1^s)^{-1} + (1-m_2a_2^s)^{-1}$.

Finally, since $D = \dim_B A < 1$, the fact that $\tilde{\zeta}_A$ satisfies the same properties as those of ζ_A stated in Theorem 3.1.12 follows from Remark 2.2.18 on page 126 and from Proposition 2.2.19. □

Remark 3.1.13. Note that in Equation (3.1.16) of Theorem 3.1.12, we can write that

$$\dim_{PC} A = D + \left(\frac{2\pi}{T_1}\mathbb{Z} \cup \frac{2\pi}{T_2}\mathbb{Z}\right)\mathrm{i}, \tag{3.1.19}$$

and (since T_1/T_2 is irrational) the set $\frac{2\pi}{T_1}\mathbb{Z} \cup \frac{2\pi}{T_2}\mathbb{Z}$ is not of the form $c\mathbb{Z}$, for any $c > 0$. This is precisely what we mean here by stating in Theorem 3.1.12 that $\dim_{PC} A$ is a *nonarithmetic set*.

Theorem 3.1.12 provides a construction of the set $A = A_1 \cup A_2$ such that the set $\dim_{PC} A$ of principal complex dimensions of A is equal to the union of two (discrete) sets of complex dimensions, each of which is composed of poles in infinite vertical arithmetic progressions, but with algebraically incommensurable *oscillatory quasiperiods* $\mathbf{p}_1 = 2\pi/T_1$ and $\mathbf{p}_2 = 2\pi/T_2$ of A_1 and A_2, respectively; that is, $\mathbf{p}_1/\mathbf{p}_2$ is transcendental. These oscillatory quasiperiods of A are equal to the oscillatory periods of A_1 and A_2. In Theorem 3.1.15, we will construct a class of bounded sets on the real line possessing an arbitrary prescribed finite number of algebraically incommensurable quasiperiods. This result will be further extended in Section 4.6.1 (see Theorems 4.6.9, 4.6.13, and Corollary 4.6.17), where we will construct a bounded set A_0 on the real line such that all points on the critical line of the corresponding distance zeta function are nonisolated singularities.

The function $G = G_1 + G_2$ appearing in Theorem 3.1.12 is *transcendentally* 2-*quasiperiodic*; that is, for the periods T_j of G_j, $j = 1, 2$, we have that $k_1 T_1 + k_2 T_j \neq 0$ for any nonzero pair of algebraic numbers (k_1, k_2), or equivalently, T_1 and T_2 are algebraically independent. We say that a finite set of real numbers is *algebraically independent* if it is linearly independent over the field of algebraic numbers.

Under the conditions of Theorem 3.1.12, by using Proposition 2.1.26, we see that

$$\zeta_{A_1 \cup A_2}(s) \sim \zeta_{A_1}(s) + \zeta_{A_2}(s),$$

in the sense of Definition 2.1.69. Furthermore, we have that

$$\mathscr{P}(\zeta_{A_1 \cup A_2}) = \mathscr{P}(\zeta_{A_1}) \cup \mathscr{P}(\zeta_{A_2}).$$

3.1.3 Transcendentally n-Quasiperiodic Sets and Baker's Theorem

It is possible to further extend Theorem 3.1.12. The main result of this subsection is contained in Theorem 3.1.15 below.

In the sequel, we shall need the following important theorem from transcendental number theory, due to Baker [Ba, Theorem 2.1]. It represents a nontrivial extension of Theorem 3.1.7, due to Gel'fond and Schneider [Gel].

Theorem 3.1.14 (Baker, [Ba, Theorem 2.1]). *Let $n \in \mathbb{N}$ with $n \geq 2$. If $m_1, \dots, m_n$ are positive algebraic numbers such that* $\log m_1, \dots, \log m_n$ *are linearly independent over the rationals, then*

$$1, \log m_1, \dots, \log m_n$$

are linearly independent over the field of all algebraic numbers (or algebraically independent, in short). In particular, the numbers $\log m_1, \dots, \log m_n$ *are transcendental, as well as all of their (nontrivial) pairwise quotients.*

We now state the main result of this section, which can be considered as a fractal set-theoretic interpretation of Baker's theorem. It extends Theorem 3.1.12 even in the case when $n = 2$.

Theorem 3.1.15. *Let $n \in \mathbb{N}$ with $n \geq 2$. Assume that $A_j = C^{(m_j, a_j)} \subset I_j$, $j = 1, \dots, n$, are generalized Cantor sets (in the sense of Definition 3.1.1) such that their box dimensions are equal to a fixed number $D \in (0,1)$,*[3] *and assume that they are contained in a disjoint family*[4] *of closed unit intervals I_j.*[5] *Let $T_i := \log(1/a_i)$ be the associated periods, and G_i be the corresponding T_i-periodic functions, for $i = 1, \dots, n$. Furthermore, let $\{p_j : j = 1, \dots, k\}$ be the union of all distinct prime factors which appear in the integers m_i, for $i = 1, \dots, n$; that is, $m_i = p_1^{\alpha_{i1}} \dots p_k^{\alpha_{ik}}$, where $\alpha_{ij} \in \mathbb{N} \cup \{0\}$. If the exponent vectors of the numbers m_i,*

$$e_i = (\alpha_{i1}, \dots, \alpha_{ik}), \quad i = 1, \dots, n, \tag{3.1.20}$$

are linearly independent over the rationals, then the real numbers

$$\frac{1}{D}, T_1, \dots, T_n \tag{3.1.21}$$

are linearly independent over the field of all algebraic numbers. It follows that the set $A := A_1 \cup \dots \cup A_n \subset \mathbb{R}$ is transcendentally n-quasiperiodic; see Definition 3.1.11. Furthermore, in the terminology and the notation of Definition 3.1.9 and Definition 3.1.11, an associated transcendentally n-quasiperiodic function G is given by $G := G_1 + \dots + G_n$.

Moreover, ζ_A can be meromorphically extended to all of $\mathbb{C}$ and we have that

$$\zeta_A(s) \sim \sum_{i=1}^{n} \frac{1}{1 - m_i a_i^s}, \quad D(\zeta_A) = D \quad \text{and} \quad D_{\mathrm{mer}}(\zeta_A) = -\infty.$$

Hence, the set $\dim_{PC} A = \mathscr{P}_c(\zeta_A)$ *of principal complex dimensions of A consists of simple poles of ζ_A and coincides with the following nonarithmetic set (see Remark 3.1.13 on page 196 above):*

---

[3] According to Proposition 3.1.2, this can be easily arranged.

[4] See footnote 57 on page 144.

[5] More generally, it suffices to assume that the interiors of the closed unit intervals I_j are pairwise disjoint for $j = 1, \dots, n$.

$$\dim_{PC} A = \bigcup_{i=1}^{n} \left(D + \frac{2\pi}{T_i} \mathrm{i}\mathbb{Z} \right) = D + \left(\bigcup_{i=1}^{n} \frac{2\pi}{T_i} \mathbb{Z} \right) \mathrm{i}. \tag{3.1.22}$$

Besides $(\dim_{PC} A) \cup \{0\}$, *there are no other poles of the distance zeta function* ζ_A *in* $\mathbb{C}$. *Furthermore, all of the complex dimensions of* A *are simple.*

Finally, exactly the same results hold for the tube zeta function $\tilde{\zeta}_A$ *of* A (*in place of the distance zeta function* ζ_A).

Proof. As in the proof of Theorem 3.1.12, using (3.1.2), applied to each A_i, for $i = 1, \ldots, n$, we see that for all $t > 0$ small enough,

$$|A_t| = t^{1-D} \sum_{i=1}^{n} G_i(\log t^{-1}),$$

and for each $i = 1, \ldots, n$, $G_i = G_i(\tau)$ is nonconstant and T_i-periodic, where $T_i := \log(1/a_i)$. We next proceed in three steps:

Step 1: It is easy to see that the numbers $\log p_j$ are rationally independent. Indeed, if we had $\sum_{j=1}^{k} \lambda_j \log p_j = 0$ for some integers λ_j, then $\prod_{j=1}^{k} p_j^{\lambda_j} = 1$. This implies that $\lambda_j = 0$ for all j, since otherwise it would contradict the fundamental theorem of arithmetic. (A moment's reflection shows that this argument is valid even if the numbers λ_j are not all of the same sign.)

Step 2: Let us show that $\log m_1, \ldots, \log m_n$ are linearly independent over the rationals. Indeed, assume that for $i = 1, \ldots, n$, $\mu_i \in \mathbb{Q}$ are such that $\sum_{i=1}^{n} \mu_i \log m_i = 0$. Then

$$\sum_{i=1}^{n} \mu_i \sum_{j=1}^{k} \alpha_{ij} \log p_j = 0. \tag{3.1.23}$$

Changing the order of summation, we have

$$\sum_{j=1}^{k} \left(\sum_{i=1}^{n} \mu_i \alpha_{ij} \right) \log p_j = 0. \tag{3.1.24}$$

Since, by Step 1, the numbers $\log p_j$ are rationally independent, we have that for all $j = 1, \ldots, k$,

$$\sum_{i=1}^{n} \mu_i \alpha_{ij} = 0;$$

that is, $\sum_{i=1}^{n} \mu_i e_i = 0$, where the e_i's are the exponent vectors given by (3.1.20). According to the hypotheses of the theorem, the exponent vectors e_i are rationally independent, and we therefore conclude that $\mu_i = 0$ for all $i = 1, \ldots, n$, as desired.

Step 3: Using [Ba, Theorem 2.1], that is, Theorem 3.1.14 above, we conclude that

$$1, \log m_1, \ldots, \log m_n$$

are linearly independent over the field of algebraic numbers. Since $T_i = \frac{1}{D}\log m_i$, for $i = 1,\dots,n$, it then follows that the numbers listed in (3.1.21) are also linearly independent over the field of algebraic numbers. The function

$$G := G_1 + \cdots + G_n, \quad G(\tau) = G_1(\tau) + \cdots + G_n(\tau),$$

associated with A, is transcendentally n-quasiperiodic; that is, the set A is transcendentally n-quasiperiodic, in the sense of Definition 3.1.11. (Note that here, in the notation of Definition 3.1.9, we have $H(\tau_1,\dots,\tau_n) := G_1(\tau_1) + \cdots + G_n(\tau_n)$.) The last claim, about the distance zeta function ζ_A and its complex dimensions, now follows from Proposition 3.1.2 applied to each of the bounded sets A_i ($i = 1,\dots,n$), much as in the proof of Theorem 3.1.12. We omit the details.

Finally, as was also noted in the proof of Theorem 3.1.12, since $D = \dim_B A < 1$, the fact that $\tilde{\zeta}_A$ satisfies the same properties as those of ζ_A stated in Theorem 3.1.12 follows from Remark 2.2.18 on page 126. □

Defining the *frequencies* f_i of A, for $i = 0,1,\dots,n$, by

$$f_0 := D,\ f_1 := 1/T_1, \dots, f_n := 1/T_n$$

then, under the conditions of Theorem 3.1.15, we conclude that the quotient of any two frequencies f_i/f_j, for $i \neq j$, is a transcendental number.

We leave it as a simple exercise for the interested reader to state (and prove) the counterpart for the tube zeta function $\tilde{\zeta}_A$ of Theorem 3.1.12 (in Subsection 3.1.2) and of Theorem 3.1.15 (in Subsection 3.1.3). Actually, those results hold without change for $\tilde{\zeta}_A$ (instead of for ζ_A) since the values of the residues are not given in Theorems 3.1.12 and 3.1.15.

The following proposition shows that if A is a quasiperiodic subset of $\mathbb{R}^N$, then the subset $A \times [0,1]^d$ of $\mathbb{R}^{N+d}$ is also quasiperiodic.

Proposition 3.1.16. *Assume that A is a quasiperiodic subset of $\mathbb{R}^N$ of a given type, with associated quasiperiodic function $G = G(\tau)$. If d a positive integer and $L > 0$, then the subset $A \times [0,L]^d$ of $\mathbb{R}^{N+d}$ is also quasiperiodic of the same type, with associated quasiperiodic function equal to $L^d \cdot G$. In particular, if $n \geq 2$ is an integer and A is one of the n-quasiperiodic subsets of $\mathbb{R}$ constructed in Theorem 3.1.15, then the subset $A \times [0,L]^d$ of $\mathbb{R}^{1+d}$ is also n-quasiperiodic.*

Proof. Let us first prove the claim for $d = 1$. By hypothesis, we have that

$$|A_t|_N = t^{N-D}\big(G(\log t^{-1}) + o(1)\big) \quad \text{as } t \to 0^+, \tag{3.1.25}$$

where $G = G(\tau)$ is a quasiperiodic function; see Definition 3.1.9. Much as in Equation (2.2.76), we can write

$$\begin{aligned} |(A \times [0,L])_t|_{N+1} &= |A_t|_N \cdot L + |A_t|_{N+1} \\ &= t^{(N+1)-(D+1)}\big(L \cdot G(\log t^{-1}) + o(1)\big) + |A_t|_{N+1} \end{aligned} \tag{3.1.26}$$

as $t \to 0^+$. Since, obviously, $|A_t|_{N+1} \leq |A_t|_N \cdot 2t$, we have that

$$
\begin{aligned}
|A_t|_{N+1} &\leq t^{N+1-D}\big(G(\log t^{-1}) + o(1)\big) = t^{(N+1)-(D+1)} \cdot t\big(G(\log t^{-1}) + o(1)\big) \\
&= t^{(N+1)-(D+1)} \cdot O(t) \quad \text{as } t \to 0^+.
\end{aligned}
\tag{3.1.27}
$$

Therefore,

$$
\begin{aligned}
|(A \times [0,L])_t|_{N+1} &= t^{(N+1)-(D+1)}\big(L \cdot G(\log t^{-1}) + o(1) + O(t)\big) \\
&= t^{(N+1)-(D+1)}\big(L \cdot G(\log t^{-1}) + o(1)\big) \quad \text{as } t \to 0^+.
\end{aligned}
\tag{3.1.28}
$$

Hence, according to Definition 3.1.11, the set $A \times [0,L]$ is quasiperiodic, with associated quasiperiodic function $L \cdot G$. This completes the proof of the proposition for $d = 1$. The general case is easily obtained by induction on d. □

3.1.4 Transcendentally n-Quasiperiodic Fractal Strings

In this subsection, we introduce the notion of n-quasiperiodic fractal strings and describe their construction, based on generalized Cantor strings, that we define below. We first recall the definition of the tube function corresponding to a given bounded fractal string $\mathscr{L} := (\ell_j)_{j\geq 1}$. To this end, we shall need to use the set $A = A_{\mathscr{L}} = \{a_k := \sum_{j\geq k} \ell_j : k \in \mathbb{N}\}$, already introduced in Figure 2.7 on page 90.

Definition 3.1.17. The *tube function corresponding to a given bounded fractal string* $\mathscr{L}$ is defined as the function $t \mapsto |A_t \cap (0, a_1)|$, where $t > 0$, $a_1 := \sum_{j\geq 1} \ell_j$ is the total length of the string, and $A = A_{\mathscr{L}}$.[6]

Definition 3.1.18. Assume that $\mathscr{L}$ is a bounded fractal string such that

$$|A_t \cap (0,a_1)| = t^{1-D}(G(\log(1/t)) + o(1)) \quad \text{as} \quad t \to 0^+, \tag{3.1.29}$$

where $D \in [0,1]$ is a constant, G is nonnegative and $0 < \liminf_{\tau\to+\infty} G(\tau) \leq \limsup_{\tau\to+\infty} G(\tau) < \infty$. Then, clearly, $D = \dim_B \mathscr{L}$, and

$$\mathscr{M}_*^D(\mathscr{L}) = \liminf_{\tau\to+\infty} G(\tau) \quad \text{and} \quad \mathscr{M}^{*D}(\mathscr{L}) = \limsup_{\tau\to+\infty} G(\tau).$$

(*a*) We say that the *fractal string* $\mathscr{L}$ *is periodic* if the corresponding function G appearing in (3.1.29) is nonconstant and periodic.

(*b*) We say that the *fractal string* $\mathscr{L}$ *is n-quasiperiodic* if the function G is n-quasiperiodic; see Definition 3.1.9. In particular, if G is transcendentally n-quasiperiodic, we say that the *fractal string* $\mathscr{L}$ *is transcendentally n-quasiperiodic*. If G is algebraically n-quasiperiodic, we say that $\mathscr{L}$ *is algebraically n-quasiperiodic*.

[6] Since $|A_t| = |A_t \cap (0,a_1)| + 2t$, we consider the tube function $t \mapsto |A_t \cap (0,a_1)|$ for convenience only, instead of $t \mapsto |A_t|$.

As we see, for any fixed integer $n \geq 2$, the family $\mathscr{L}_{qp}(n)$ of all n-quasiperiodic fractal strings is equal to the disjoint union of the family $\mathscr{L}_{tqp}(n)$ of transcendentally quasiperiodic strings of order n and of the family $\mathscr{L}_{aqp}(n)$ of algebraically quasiperiodic strings of order n:

$$\mathscr{L}_{qp}(n) = \mathscr{L}_{tqp}(n) \cup \mathscr{L}_{aqp}(n).$$

It is natural to define

$$\mathscr{L}_{qp} := \bigcup_{n\geq 2} \mathscr{L}_{qp}(n), \quad \mathscr{L}_{tqp} := \bigcup_{n\geq 2} \mathscr{L}_{tqp}(n), \quad \mathscr{L}_{aqp} := \bigcup_{n\geq 2} \mathscr{L}_{aqp}(n). \tag{3.1.30}$$

We then have that

$$\mathscr{L}_{qp} = \mathscr{L}_{tqp} \cup \mathscr{L}_{aqp}.$$

We shall see below that the family $\mathscr{L}_{tqp}(n)$ is nonempty (and, moreover, it is infinite) for any $n \geq 2$. We do not know if the family $\mathscr{L}_{aqp}$ is nonempty.

An example of a fractal string generated by a fractal set is the *Cantor string* $\mathscr{L}_{C^{(1/3)}}$, defined as the sequence of lengths of the open intervals deleted during the construction of the ternary Cantor set $C^{(1/3)}$:

$$\mathscr{L}_{C^{(1/3)}} = (3^{-1}, 3^{-2}, 3^{-2}, \underbrace{3^{-3}, \ldots, 3^{-3}}_{4 \text{ times}}, \underbrace{3^{-4}, \ldots, 3^{-4}}_{8 \text{ times}}, \ldots).$$

In other words, for each $k = 1, 2, \ldots$, the element 3^{-k} appears in the fractal string with multiplicity 2^{k-1}.

We can analogously define the *generalized Cantor string* $\mathscr{L}_{C^{(m,a)}}$, generated by the generalized Cantor set $C^{(m,a)}$, where m is a positive integer and $0 < a < 1/m$; see Definition 3.1.1.

Definition 3.1.19. Let $\mathscr{L}_1$ and $\mathscr{L}_2$ be two bounded fractal strings. We define a new bounded fractal string $\mathscr{L}_1 \sqcup \mathscr{L}_2$, called the *union of the fractal strings* $\mathscr{L}_1$ *and* $\mathscr{L}_2$. It consists of all the elements from the union of these two fractal strings, with the multiplicity of each common element of $\mathscr{L}_1$ and $\mathscr{L}_2$ equal to the sum of its respective multiplicities in $\mathscr{L}_1$ and $\mathscr{L}_2$.[7]

For example, if

$$\mathscr{L}_1 = (1, 1, 1/2, 1/3, 1/4, \ldots) \quad \text{and} \quad \mathscr{L}_2 = (2, 1/2, 1/2, 1/3, 1/5, \ldots),$$

then

$$\mathscr{L}_1 \sqcup \mathscr{L}_2 = (2, 1, 1, 1/2, 1/2, 1/2, 1/3, 1/3, \ldots).$$

We now state the analog of Theorems 3.1.12 in the context of fractal strings.

[7] The union of a countable family of fractal strings is introduced in Definition 4.5.11.

Theorem 3.1.20. *Let $\mathscr{L}_{C(m_1,a_1)}$ and $\mathscr{L}_{C(m_2,a_2)}$ be two generalized Cantor strings such that their box dimensions coincide; denote by D this common value. Let $\{p_1, p_2, \dots, p_k\}$ be the set of all distinct prime factors of m_1 and m_2, and write*

$$m_1 = p_1^{\alpha_1} p_2^{\alpha_2} \dots p_k^{\alpha_k}, \quad m_2 = p_1^{\beta_1} p_2^{\beta_2} \dots p_k^{\beta_k}, \tag{3.1.31}$$

where $\alpha_i, \beta_i \in \mathbb{N} \cup \{0\}$ for $i = 1, \dots, k$. If the exponent vectors

$$(\alpha_1, \alpha_2, \dots, \alpha_k) \quad \text{and} \quad (\beta_1, \beta_2, \dots, \beta_k) \tag{3.1.32}$$

corresponding to m_1 and m_2, are linearly independent over the rationals, then the fractal string $\mathscr{L} = \mathscr{L}_1 \sqcup \mathscr{L}_2$ is transcendentally 2-quasiperiodic; that is, the quotient T_1/T_2 of the quasiperiods of G (i.e., of the periods of G_1 and G_2) is transcendental. In other words, the fractal string $\mathscr{L}$ is transcendentally 2-quasiperiodic in the sense of Definition 3.1.11(a).

Moreover, since $\zeta_{\mathscr{L}} = \zeta_{\mathscr{L}_1} + \zeta_{\mathscr{L}_2}$, $\zeta_{\mathscr{L}}$ can be meromophically extended to all of $\mathbb{C}$ and we have that

$$\zeta_{\mathscr{L}}(s) \sim \frac{1}{1 - m_1 a_1^s} + \frac{1}{1 - m_2 a_2^s}, \quad D(\zeta_{\mathscr{L}}) = D \quad \text{and} \quad D_{\text{mer}}(\zeta_{\mathscr{L}}) = -\infty.$$

Hence, the set $\dim_{PC} \mathscr{L} = \mathscr{P}_c(\zeta_{\mathscr{L}})$ *of principal complex dimensionsof $\mathscr{L}$ consists of simple poles of $\zeta_{\mathscr{L}}$ and is given by*

$$\dim_{PC} A = D + \left(\frac{2\pi}{T_1}\mathbb{Z} \cup \frac{2\pi}{T_2}\mathbb{Z}\right)\mathbb{i}.$$

Besides $(\dim_{PC} \mathscr{L}) \cup \{0\}$, *there are no other poles of the geometric zeta function $\zeta_{\mathscr{L}}$ in $\mathbb{C}$.*

The relationship between the geometric zeta function $\zeta_{\mathscr{L}}$ of a bounded fractal string $\mathscr{L} = (\ell_j)_{j \geq 1}$ and the distance zeta function ζ_A of the set $A := A_{\mathscr{L}} := \{a_k = \sum_{j \geq k} \ell_j : k \in \mathbb{N}\}$ associated with $\mathscr{L}$ can be found in Equation (2.1.85).

For every integer $n \geq 2$, the analog of Theorem 3.1.15 can also be formulated in terms of generalized Cantor strings, and we leave it as an easy exercise for the interested reader.

3.2 Distance Zeta Functions of the Sierpiński Carpet and Gasket

In this section, we shall study the distance zeta functions of two classic fractal sets in the plane; namely, the Sierpiński carpet and the Sierpiński gasket. We also compute the corresponding principal complex dimensions. The method of computation of distance zeta functions anticipates some of the ideas to be developed in Chapter 4. More precisely, this computation will serve as a motivation to introduce the notion of 'relative fractal drums', which will be the central object of study in Chapter 4.

We point out that another point of view (closely related, however, to the one presented in the present section) concerning the Sierpiński carpet and the Sierpiński gasket will be taken in Subsection 4.2.3. More specifically, we shall study these two fractals from the point of view of the so-called 'relative fractal sprays'; see Section 4.2.

3.2.1 Distance Zeta Function of the Sierpiński Carpet

The construction of the Sierpiński carpet is indicated in Figure 2.1 on page 49. In this subsection, we compute its principal complex dimensions. In order to do this, we must first describe the computation of the distance zeta function of the Sierpiński carpet. Here is the main result of Subsection 3.2.1.

Proposition 3.2.1 (Distance zeta function of the Sierpiński carpet). *Let A be the Sierpiński carpet in $\mathbb{R}^2$, constructed in the usual way inside the unit square; see Figure 2.1. Let δ be a fixed positive real number. We assume without loss of generality that $\delta > 1/6$, so that the set A_δ is connected.*[8] *Then, the distance zeta function ζ_A of the Sierpiński carpet is given for all $s \in \mathbb{C}$ by*

$$\zeta_A(s) = \frac{8}{2^s s(s-1)(3^s-8)} + 2\pi\frac{\delta^s}{s} + 4\frac{\delta^{s-1}}{s-1}, \tag{3.2.1}$$

which is meromorphic on the whole complex plane and equivalent to $(3^s-8)^{-1}$, in the sense of Definition 2.1.69. In particular, the set of principal complex dimensions of the Sierpiński carpet is given by

$$\dim_{PC} A = \log_3 8 + \frac{2\pi}{\log 3}\mathrm{i}\mathbb{Z} \tag{3.2.2}$$

and consists only of simple poles of ζ_A. The residues of the distance zeta function ζ_A computed at the principal poles s_k, $k \in \mathbb{Z}$, are given by

$$\operatorname{res}(\zeta_A, s_k) = \frac{2^{-s_k}}{(\log 3)s_k(s_k-1)},$$

where $s_k := \log_3 8 + \frac{2\pi}{\log 3}k\mathrm{i} \in \dim_{PC} A$, for any integer $k \in \mathbb{Z}$.

As we see from Equation (3.2.1), the set of complex dimensions (i.e., the set of poles of ζ_A in all of $\mathbb{C}$) of the Sierpiński carpet is given by

$$\mathscr{P}(\zeta_A) = \{0, 1\} \cup \left(\log_3 8 + \frac{2\pi}{\log 3}\mathrm{i}\mathbb{Z}\right), \tag{3.2.3}$$

and consists only of simple poles of ζ_A.

[8] More precisely, for this choice of δ, A_δ is equal to the δ-neighborhood of the unit square $[0,1]^2$.

For the needs of the proof of Proposition 3.2.1, it will be very convenient here to introduce some auxilliary notation. Let A be a compact subset of $\mathbb{R}^2$ and assume that Ω is a bounded open (or more generally, a bounded and Lebesgue measurable) subset of $\mathbb{R}^2$. Then we define

$$\zeta_{A,\Omega}(s) := \int_{\Omega} d((x,y),A)^{s-2} \mathrm{d}x\,\mathrm{d}y, \tag{3.2.4}$$

for all complex numbers s such that $\mathrm{Re}\,s$ is sufficiently large. We shall call it the *distance zeta functions of A with respect to Ω*, or else, the *relative distance zeta function* of A with respect to Ω. Such distance zeta functions and their generalizations, associated with a suitable ordered pair (A,Ω) of subsets of $\mathbb{R}^N$ (called *relative fractal drums* in $\mathbb{R}^N$; see Definition 4.1.2 on page 247), will be studied in detail in Chapter 4.

Proof of Proposition 3.2.1. In order to evaluate

$$\zeta_A(s) = \int_{A_\delta} d((x,y),A)^{s-2} \mathrm{d}x\,\mathrm{d}y,$$

we integrate (i) first over the set $A_\delta \setminus [0,1]^2$, and then (ii) over the unit square $[0,1]^2$.

Step (i): The integration over the set $A_\delta \setminus (0,1)^2$ leads us to the following result:

$$\begin{aligned} \zeta_{A,A_\delta\setminus[0,1]^2}(s) &= \int_0^{2\pi} \mathrm{d}\varphi \int_0^{\delta} r^{s-2} r\,\mathrm{d}r + 4\int_0^1 \mathrm{d}x \int_0^{\delta} y^{s-2} \mathrm{d}y \\ &= 2\pi\frac{\delta^s}{s} + 4\frac{\delta^{s-1}}{s-1}, \end{aligned} \tag{3.2.5}$$

for all complex numbers s such that $\mathrm{Re}\,s > 1$. Indeed, it suffices to note that the (connected) set $A_\delta \setminus [0,1]^2$ can be viewed as the disjoint union of

• four quarters of the corresponding disks of radius δ, with centers at the vertices of the unit square $[0,1]^2$ (and since all of the corresponding integrals over these four sets are equal, it suffices to consider the integral of $|(x,y)|^{s-2} = r^{s-2}$ over the disk $B_\delta((0,0))$ of radius δ, centered at the origin), and

• of the remaining four rectangles, that are all isometrically isomorphic to $[0,1] \times (-\delta,\delta)$.

See also Table C.1 of Appendix C, on pages 613 and 614.

Step (ii): Now, let us consider $\zeta_{A,[0,1]^2}(s)$. Since the boundary of the unit square $[0,1]^2$ is of 2-dimensional Lebesgue measure zero, it suffices to consider $\zeta_{A,(0,1)^2}(s)$. Furthermore, since A is of 2-dimensional Lebesgue measure zero as well,[9] then it suffices to consider

[9] Indeed, the union of the deleted open squares in $[0,1]^2$, obtained during the construction of the Sierpiński carpet, is of 2-dimensional Lebesgue measure 1, since

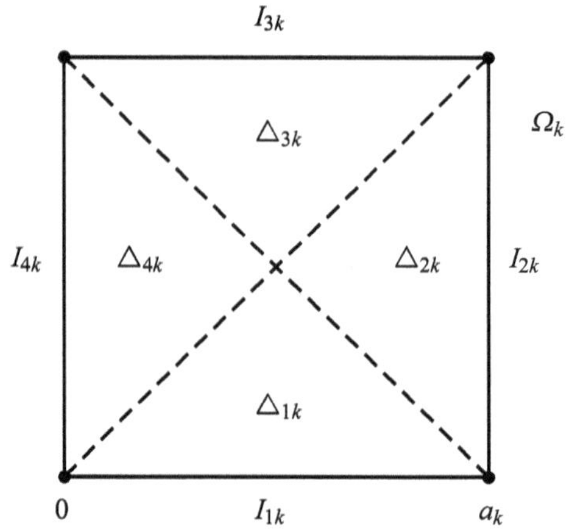

Fig. 3.1 The square Ω_k corresponds to any of the 8^{k-1} deleted open square in the k-th generation during the construction of the Sierpiński carpet. It can be viewed as the union of four triangles, determined by its diagonals. This figure explains a part of Step (*ii*) in the proof of Proposition 3.2.1.

$$\zeta_{A,[0,1]^2}(s) = \zeta_{A,(0,1)^2 \setminus A}(s) = \sum_{k=1}^{\infty} 8^{k-1} \zeta_{A_k,\Omega_k}(s), \tag{3.2.6}$$

where for any positive integer k, Ω_k is a fixed deleted square of sides of length $a_k := 3^{-k}$ in the k-th generation and $A_k = \partial\Omega_k$ is the boundary of Ω_k.[10] Recall that the k-th generation of deleted squares contains precisely 8^{k-1} deleted squares that are all isometric to Ω_k. Furthermore, due to the stated isometry, it is easy to see that all of the distance zeta functions corresponding to the deleted open squares Ω_k in the k-th generation coincide. (This is a special case of Lemma 4.2.23 formulated in terms of general relative fractal drums, which the reader can find on page 291.) Now, if we denote the side length of Ω_k by $a_k := 3^{-k}$, it is an easy exercise to check that

$$\zeta_{A_k,\Omega_k}(s) = \frac{8 \cdot 2^{-s} a_k^s}{s(s-1)}, \tag{3.2.7}$$

for all complex numbers s such that $\operatorname{Re} s > 1$. By the principle of analytic continuation, this same equation as in (3.2.7) then continues to hold for all $s \in \mathbb{C}$. Indeed, for any of the four sides I_{1k}, I_{2k}, I_{3k} and I_{4k} of the square Ω_k, it is natural to consider the set of points $(x,y) \in \Omega_k$ such that

$$d\big((x,y), \partial\Omega_k\big) = d\big((x,y), I_{ik}\big), \quad \text{for } i = 1,2,3,4.$$

$$\sum_{k=1}^{\infty} 8^{k-1} (3^{-k})^2 = \frac{1}{9} \sum_{k=1}^{\infty} \left(\frac{8}{9}\right)^{k-1} = \frac{1}{9} \cdot \frac{1}{1-\frac{8}{9}} = 1.$$

[10] The complement $(0,1)^2 \setminus A$ of the Sierpiński carpet A in $(0,1)^2$ can be thought of as the 'dual Sierpiński carpet' corresponding to the usual Sierpiński carpet A in the plane.

It is easy to see that this set is a triangle, and therefore we can decompose Ω_k into the union of four isosceles right triangles $\triangle_{ik}$, $i = 1,2,3,4$ (each of them corresponding to one of the four sides of the square Ω_k), as indicated in Figure 3.1. Note that the triangles are determined by the two diagonals of Ω_k. Since obviously,

$$\zeta_{A_k,\Omega_k}(s) = 4\zeta_{I_{1k},\triangle_{1k}}(s), \tag{3.2.8}$$

we can proceed as indicated in Figure 4.9 on page 304, corresponding to the case when $a_1 = 1/3$, $k \geq 0$, and with the corresponding computation analogous to that in Equation (4.2.97):

$$\begin{aligned}\zeta_{I_{1k},\triangle_{1k}}(s) &= \int_0^{a_k/2} \mathrm{d}x \int_0^x d((x,y),I_{1,k})^{s-2}\mathrm{d}y + \int_{a_k/2}^{a_k} \mathrm{d}x \int_0^{a_k-x} d((x,y),I_{1,k})^{s-2}\mathrm{d}y \\ &= 2\int_0^{a_k/2} \mathrm{d}x \int_0^x y^{s-2}\mathrm{d}y = \frac{2}{s-1}\int_0^{a_k/2} x^{s-1}\mathrm{d}x \\ &= \frac{2(a_k/2)^s}{s(s-1)},\end{aligned}$$

for all complex numbers s such that $\mathrm{Re}\, s > 1$. See also Table C.1 on pages 613 and 614. Therefore, using Equation (3.2.8) we obtain (3.2.7). Substituting Equation (3.2.7) into (3.2.6), we conclude that

$$\zeta_{A,[0,1]^2}(s) = \frac{2^{-s}}{s(s-1)}\sum_{k=1}^{\infty} 8^k 3^{-ks} = \frac{8}{2^s s(s-1)(3^s-8)},$$

for all complex numbers s such that $\mathrm{Re}\, s > \log_2 8$. Naturally, by analytic continuation, this same equation continues to hold for all $s \in \mathbb{C}$.

Step (*iii*): The resulting expression for the distance zeta function ζ_A stated in Equation (3.2.1) follows from Steps (*i*) and (*ii*). By the principle of analytic continuation, ζ_A can be meromorphically extended to the whole complex plane and is given by the same formula. Clearly, the principal complex dimensions $s_k \in \dim_{PC} A$, $k \in \mathbb{Z}$, coincide with the zeros of $3^s - 8$. Finally, the computation of the corresponding residues $\operatorname{res}(\zeta_A, s_k)$ is left as an easy exercise for the interested reader.

This concludes the proof of Proposition 3.2.1. □

Concerning part (*i*) of the proof of Proposition 3.2.1, it is worth noticing that the relative distance zeta function of the set $A_0 := \partial([0,1]^2)$ (that is, of the boundary of the unit square) with respect to the open connected set $\Omega_0 := A_\delta \setminus [0,1]^2$ is equal

- to the sum of the relative distance zeta functions of the point $\{(0,0)\}$ in $\mathbb{R}^2$ with respect to the open disk $B_\delta((0,0))$ and of the relative distance zeta function of the open interval $I = (0,2) \times \{0\}$ of length 2 with respect to the open set $(0,2) \times (-\delta,\delta)$; or
- to the sum of the relative distance zeta function of the interval $(0,1) \times \{0\}$ with respect to its open δ-neighborhood I_δ in the plane and of the relative distance zeta function of the interval $(0,1) \times \{0\}$ with respect to $(0,1) \times (-\delta,\delta)$.

We leave it as a simple exercise for the interested reader to check the above statements. See Table C.1 of Appendix C, on pages 613 and 614.

Remark 3.2.2. Consistent with the conjecture formulated in the geometric part of [Lap3, Conjecture 3, p. 163], according to which a lattice self-similar set is not Minkowski measurable, the Sierpiński carpet A is not Minkowski measurable. According to [HorŽu, Theorem 4(*a*)], the precise values of the lower and upper Minkowski contents of A are respectively given by

$$\mathscr{M}_*^D(A) = \frac{(D_-)^{D-2}}{D}\Big(\frac{4D_-}{5}+2\Big) \tag{3.2.9}$$

and

$$\mathscr{M}^{*D}(A) = \frac{(D_+)^{D-2}}{D}\Big(\frac{4D_+}{5}+2\Big), \tag{3.2.10}$$

where $D := \log_3 8$ and

$$D_\pm := \frac{7}{2D}\left(\frac{D-1}{5} \pm \sqrt{\frac{(D-1)^2}{25}+\frac{D(D-2)}{7}}\right).$$

It is interesting to note that the lower and upper Minkowski contents of the Sierpiński carpet are rather close to each other, and coincide up to the second decimal. More precisely, $\mathscr{M}_*^D(A) \approx 1.350670$ and $\mathscr{M}^{*D}(A) \approx 1.355617$.

3.2.2 Distance Zeta Function of the Sierpiński Gasket

In order to compute the distance zeta function of the Sierpiński gasket, we can proceed much as in the proof of Proposition 3.2.1. Therefore, we limit ourselves to stating the corresponding result. We leave the details of the proof to the interested reader; see an analogous computation given in Example 4.2.24 of Subsection 4.2.3 below, on pages 292–294, for the case of the corresponding relative Sierpiński gasket.

Proposition 3.2.3 (Distance zeta function of the Sierpiński gasket). *Let A be the Sierpiński gasket in $\mathbb{R}^2$, constructed in the usual way inside the unit equilateral triangle; see Figure 4.5 on page 275. Let δ be a fixed positive real number. We assume without loss of generality that $\delta > \sqrt{3}/12$, so that the set A_δ is connected.*[11] *Then, for all $s \in \mathbb{C}$, the distance zeta function ζ_A of the Sierpiński gasket is given by*

$$\zeta_A(s) = \frac{6(\sqrt{3})^{1-s}2^{-s}}{s(s-1)(2^s-3)} + 2\pi\frac{\delta^s}{s} + 3\frac{\delta^{s-1}}{s-1}, \tag{3.2.11}$$

[11] More precisely, for this choice of δ, A_δ is equal to the δ-neighborhood of the unit triangle ▲.

which is meromorphic on the whole complex plane and equivalent to $(2^s-3)^{-1}$. *In particular, the set of principal complex dimensions of the Sierpiński gasket is given by*

$$\dim_{PC} A = \log_2 3 + \frac{2\pi}{\log 2}\mathrm{i}\mathbb{Z} \tag{3.2.12}$$

and consists only of simple poles of ζ_A. *In particular,*

$$\operatorname{res}(\zeta_A, s_k) = \frac{6(\sqrt{3})^{1-s_k}}{4^{s_k}(\log 2)s_k(s_k-1)},$$

where $s_k := \log_2 3 + \frac{2\pi}{\log 2}k\mathrm{i} \in \dim_{PC} A$, *for every integer* $k \in \mathbb{Z}$.

We deduce from Equation (3.2.12) that the set of complex dimensions (i.e., the set of poles of ζ_A in all of $\mathbb{C}$) of the Sierpiński gasket is given by

$$\mathscr{P}(\zeta_A) = \{0, 1\} \cup \left(\log_2 3 + \frac{2\pi}{\log 3}\mathrm{i}\mathbb{Z}\right), \tag{3.2.13}$$

and consists only of simple poles of ζ_A.

3.3 Tensor Products of Bounded Fractal Strings and Multiple Complex Dimensions of Arbitrary Orders

In this section, we construct a class of bounded fractal strings $\overline{\mathscr{L}}$ with principal complex dimensions of *any* prescribed order (i.e., multiplicity); see Theorem 3.3.6 below. Furthermore, the same theorem provides a construction of a class of fractal strings with principal complex dimensions of infinite order; that is, with *essential singularities* on the corresponding critical line. The idea of the construction is to use iterated *tensor products* of suitably chosen bounded fractal strings.

Let us first recall the definition of a self-similar fractal string (see [Lap3], [LapPe2], [Lap-vFr1–2], [LapPeWi1], [Lap-vFr3, Section 2.1]). In fact, we introduce a more general notion.

Definition 3.3.1. Let $\mathscr{L}_0$ be a bounded fractal string and $\{r_1,\dots,r_J\}$ a multiset of positive numbers ("ratio list") such that

$$r_1 + \dots + r_J < 1. \tag{3.3.1}$$

An *extended self-similar fractal string* $\overline{\mathscr{L}} = \overline{\mathscr{L}}(\mathscr{L}_0; r_1,\dots,r_J)$, generated by $\mathscr{L}_0$ and $\{r_1,r_2,\dots,r_J\}$, is the bounded fractal string defined by

$$\overline{\mathscr{L}} := \bigsqcup_{\alpha\in(\mathbb{N}_0)^J} (r_1^{\alpha_1}\dots r_J^{\alpha_J})\mathscr{L}_0, \tag{3.3.2}$$

where $\alpha := (\alpha_1, \dots, \alpha_J)$ and the notation $\sqcup$ is described towards the end of Section 1.3. Therefore, $\overline{\mathscr{L}}$ can be written as the following tensor product of fractal strings:

$$\overline{\mathscr{L}} = \mathscr{L}_0 \otimes \mathscr{L}(r_1, \dots, r_J), \tag{3.3.3}$$

where the tensor product is defined at the end of Section 1.3 and the fractal string $\mathscr{L}(r_1, \dots, r_J)$ is defined by

$$\mathscr{L}(r_1, \dots, r_J) := \{r_1^{\alpha_1} \dots r_J^{\alpha_J} : \alpha \in (\mathbb{N}_0)^J\}, \tag{3.3.4}$$

viewed as a multiset.

The following lemma provides some basic properties of the geometric zeta function of tensor products of fractal strings.

Lemma 3.3.2. *Let $\mathscr{L}_1$ and $\mathscr{L}_2$ be two bounded fractal strings. Then, their tensor product is also a bounded fractal string and the corresponding geometric zeta function is given by*

$$\zeta_{\mathscr{L}_1 \otimes \mathscr{L}_2}(s) = \zeta_{\mathscr{L}_1}(s) \cdot \zeta_{\mathscr{L}_2}(s) \tag{3.3.5}$$

for all $s \in \mathbb{C}$ such that $\operatorname{Re} s > \max\{\overline{\dim}_B \mathscr{L}_1, \overline{\dim}_B \mathscr{L}_2\}$. *Furthermore,*

$$\overline{\dim}_B(\mathscr{L}_1 \otimes \mathscr{L}_2) = \max\{\overline{\dim}_B \mathscr{L}_1, \overline{\dim}_B \mathscr{L}_2\}, \tag{3.3.6}$$

where $\overline{\dim}_B \mathscr{L}_j$ denotes the upper box (or Minkowski) dimension of $\mathscr{L}_j$. Equivalently,

$$D(\mathscr{L}_1 \otimes \mathscr{L}_2) = \max\{D(\mathscr{L}_1), D(\mathscr{L}_2)\}. \tag{3.3.7}$$

Proof. If $s \in \mathbb{C}$ is such that $\operatorname{Re} s > \max\{\overline{\dim}_B \mathscr{L}_1, \overline{\dim}_B \mathscr{L}_2\}$, then (3.3.5) holds since the series defining $\zeta_{\mathscr{L}_1 \otimes \mathscr{L}_2}(s)$ on the left and $\zeta_{\mathscr{L}_1}(s)$ and $\zeta_{\mathscr{L}_2}(s)$ on the right are absolutely convergent, which allows us to use the corresponding special case of Fubini's theorem.

For any fractal string $\mathscr{L}$ and $\alpha > 0$, we let $\mathscr{L}^\alpha := (\lambda^\alpha)_{\lambda \in \mathscr{L}}$ and $|\mathscr{L}|_1 := \sum_{\lambda \in \mathscr{L}} \lambda$. Since $|\mathscr{L}_1^\alpha|_1$ and $|\mathscr{L}_2^\alpha|_1$ are series with positive entries, we then have that

$$|(\mathscr{L}_1 \otimes \mathscr{L}_2)^\alpha|_1 = |\mathscr{L}_1^\alpha|_1 \cdot |\mathscr{L}_2^\alpha|_1. \tag{3.3.8}$$

If $\alpha < \max\{\overline{\dim}_B \mathscr{L}_1, \overline{\dim}_B \mathscr{L}_2\}$, then either $|\mathscr{L}_1^\alpha|_1 = +\infty$ or $|\mathscr{L}_2^\alpha|_1 = +\infty$, and by (3.3.8), we thus deduce that $|(\mathscr{L}_1 \otimes \mathscr{L}_2)^\alpha|_1 = +\infty$. On the other hand, if $\alpha \in \mathbb{R}$ is such that $\alpha > \max\{\overline{\dim}_B \mathscr{L}_1, \overline{\dim}_B \mathscr{L}_2\}$, then both $|\mathscr{L}_1^\alpha|_1 < \infty$ and $|\mathscr{L}_2^\alpha|_1 < \infty$, and by (3.3.8), we must have that $|(\mathscr{L}_1 \otimes \mathscr{L}_2)^\alpha|_1 < \infty$. This proves Equation (3.3.6). Finally, Equation (3.3.7) follows from Theorem 2.1.55. This completes the proof of the lemma. □

Part (*b*) of the following theorem extends [Lap-vFr3, Theorem 2.3] to the present more general context of extended self-similar strings, while part (*a*) corresponds to the special case of the just mentioned theorem when the geometric self-similar string under consideration has a single gap (see Remark 3.3.4 below).

Theorem 3.3.3. *Let the assumptions of Definition 3.3.1 be satisfied and let* $D \in (0,1)$ *be the (necessarily unique)* real *solution of the Moran equation* $\sum_{j=1}^{J} r_j^D = 1$. *Then:*

(a) The fractal string $\mathscr{L}(r_1,\dots,r_J)$ *generated by the scaling ratios* $r_1,\dots,r_j$ *is bounded and has total length given by*

$$|\mathscr{L}(r_1,\dots,r_J)|_1 = \frac{1}{1-\sum_{j=1}^{J} r_j}. \tag{3.3.9}$$

Furthermore, its geometric zeta function has a meromorphic continuation to the entire complex plane and is given by

$$\zeta_{\mathscr{L}(r_1,\dots,r_J)}(s) = \frac{1}{1-\sum_{j=1}^{J} r_j^s} \tag{3.3.10}$$

for all $s \in \mathbb{C}$. *Moreover, its abscissa of convergence is given by*

$$D(\zeta_{\mathscr{L}(r_1,\dots,r_J)}) = D. \tag{3.3.11}$$

(b) The extended self-similar fractal string $\overline{\mathscr{L}} := \mathscr{L}_0 \otimes \mathscr{L}(r_1,\dots,r_J)$ *is bounded and has total length given by*

$$|\overline{\mathscr{L}}|_1 = \frac{|\mathscr{L}_0|_1}{1-\sum_{j=1}^{J} r_j}. \tag{3.3.12}$$

Furthermore, its geometric zeta function has for abscissa of meromorphic continuation $D_{\mathrm{mer}}(\zeta_{\overline{\mathscr{L}}}) = D_{\mathrm{mer}}(\zeta_{\mathscr{L}_0})$ *and its meromorphic extension is given by*

$$\zeta_{\overline{\mathscr{L}}}(s) = \frac{\zeta_{\mathscr{L}_0}(s)}{1-\sum_{j=1}^{J} r_j^s}, \tag{3.3.13}$$

for all $s \in \mathbb{C}$ *such that* $\operatorname{Re} s > D_{\mathrm{mer}}(\zeta_{\mathscr{L}_0})$. *Moreover, its abscissa of convergence is given by*

$$D(\zeta_{\overline{\mathscr{L}}}) = \max\{D(\zeta_{\mathscr{L}_0}), D\}. \tag{3.3.14}$$

Finally, for a given window $\boldsymbol{W}$ *of* $\zeta_{\mathscr{L}_0}$, *the visible complex dimensions in* $\boldsymbol{W}$ *of* $\overline{\mathscr{L}}$ *satisfy*

$$\mathscr{P}(\zeta_{\overline{\mathscr{L}}}, \boldsymbol{W}) \subseteq \mathfrak{D} \cup \mathscr{P}(\zeta_{\mathscr{L}_0}, \boldsymbol{W}), \tag{3.3.15}$$

where $\mathfrak{D}$ *is the set of complex solutions in* $\boldsymbol{W}$ *of the Moran equation* $\sum_{j=1}^{J} r_j^s = 1$. *Furthermore, if there are no zero-pole cancellations in* (3.3.13), *then we have an equality in* (3.3.15).

Proof. (*a*) It is clear that Equation (3.3.2) defines a fractal string, since $r_j \in (0,1)$ for all $j = 1,\dots,J$, while each element $r_1^{\alpha_1}\dots r_J^{\alpha_J}$ has finite multiplicity $\frac{(\alpha_1+\dots+\alpha_J)!}{\alpha_1!\dots\alpha_J!}$, and $r_1^{\alpha_1}\dots r_J^{\alpha_J} \to 0$ as $|\alpha|_1 := \alpha_1+\dots+\alpha_J \to \infty$. The multiset $\{r_1^{\alpha_1}\dots r_J^{\alpha_J} : \alpha \in (\mathbb{N}_0)^J\}$

is itself a bounded fractal string $\mathscr{L}(r_1,\dots,r_J)$, since of all of its members are listed as monomials in the expansion of the sum $\sum_{k=0}^{\infty}(r_1+\cdots+r_J)^k$ and therefore, we have

$$|\mathscr{L}(r_1,\dots,r_J)|_1 = \sum_{k=0}^{\infty}\Big(\sum_{j=1}^{J} r_j\Big)^k = \frac{1}{1-\sum_{j=1}^{J} r_j}, \tag{3.3.16}$$

due to condition (3.3.1).

Exactly as in the proof of [Lap-vFr3, Theorem 2.3], Equation (3.3.10) follows by direct computation:

$$\begin{aligned}\zeta_{\mathscr{L}(r_1,\dots,r_J)}(s) &= \sum_{\alpha\in(\mathbb{N}_0)^J} (r_1^{\alpha_1}\dots r_J^{\alpha_J})^s \\ &= \sum_{k=0}^{\infty}\Big(\sum_{j=1}^{J} r_j^s\Big)^k = \frac{1}{1-\sum_{j=1}^{J} r_j^s}.\end{aligned} \tag{3.3.17}$$

Note that, a priori, the above computation is valid for $\operatorname{Re} s > D$ (since then, $|\sum_{j=1}^{J} r_j^s| \le \sum_{j=1}^{J} r_j^{\operatorname{Re} s} < 1$), but in fact, the endresult (i.e., Equation (3.3.10) above) remains valid for all $s\in\mathbb{C}$, upon meromorphic continuation.

(b) In light of Equation (3.3.2), the extended self-similar fractal string $\overline{\mathscr{L}}$ can be written as follows:

$$\overline{\mathscr{L}} = \bigsqcup_{\lambda\in\mathscr{L}(r_1,\dots,r_J)} \lambda\mathscr{L}_0. \tag{3.3.18}$$

We then have

$$\begin{aligned}|\overline{\mathscr{L}}|_1 &= \sum_{\lambda\in\mathscr{L}(r_1,\dots,r_J)} |\lambda\mathscr{L}_0|_1 = |\mathscr{L}_0|_1 \sum_{\lambda\in\mathscr{L}(r_1,\dots,r_J)} \lambda \\ &= |\mathscr{L}_0|_1\cdot|\mathscr{L}(r_1,\dots,r_J)|_1 = \frac{|\mathscr{L}_0|_1}{1-\sum_{j=1}^{J} r_j} < \infty,\end{aligned} \tag{3.3.19}$$

where in the last equality we have used Equation (3.3.16). This completes the proof of Equation (3.3.12). Equations (3.3.13) and (3.3.14) are a consequence of part (a) and of Lemma 3.3.2.

Finally, the remaining part of the theorem (i.e., the inclusion (3.3.15)) now easily follows from Equation (3.3.13). □

Note that in order to deduce (3.3.10) from (3.3.13), it suffices to let $\mathscr{L}_0 := \{1\}$, so that $\zeta_{\mathscr{L}_0}(s) = 1$, for all $s\in\mathbb{C}$.

The following comment provides a direct and alternative proof of Equation (3.3.13) and, in fact, of all of Theorem 3.3.3, based on a useful scaling argument.

Remark 3.3.4. The above Definition 3.3.1 coincides with the usual definition of self-similar strings of *total length* equal to 1 given in [Lap-vFr3, Section 2.1] when $\mathscr{L}_0$ is taken to be a finite string with lengths $g_1,\dots,g_K$ (corresponding to the *gaps*) such that

$$\sum_{j=1}^{J} r_j + \sum_{k=1}^{K} g_k = 1. \tag{3.3.20}$$

This situation corresponds to the geometric construction of a self-similar string. For this reason, the resulting fractal string $\overline{\mathscr{L}}$ is then referred to as a geometric self-similar string.

Remark 3.3.5 (Alternative proof of Theorem 3.3.3). Let us consider the extended self-similar fractal string $\overline{\mathscr{L}}$ introduced in Definition 3.3.1 and studied in part (b) of Theorem 3.3.3. Then,

$$\overline{\mathscr{L}} = \mathscr{L}_0 \sqcup \bigsqcup_{j=1}^{J} r_j \overline{\mathscr{L}}. \tag{3.3.21}$$

It follows that the geometric zeta function of $\overline{\mathscr{L}}$ satisfies the following functional equation:

$$\zeta_{\overline{\mathscr{L}}}(s) = \zeta_{\mathscr{L}_0}(s) + \sum_{j=1}^{J} \zeta_{r_j \overline{\mathscr{L}}}(s). \tag{3.3.22}$$

Furthermore, by using the scaling property of the geometric zeta function, the above equation becomes $\zeta_{\overline{\mathscr{L}}}(s) = \zeta_{\mathscr{L}_0}(s) + \sum_{j=1}^{J} r_j^s \zeta_{\overline{\mathscr{L}}}(s)$; that is,

$$\zeta_{\overline{\mathscr{L}}}(s) = \zeta_{\mathscr{L}_0}(s) + \zeta_{\overline{\mathscr{L}}}(s) \sum_{j=1}^{J} r_j^s. \tag{3.3.23}$$

Since the series defining the geometric zeta function $\zeta_{\overline{\mathscr{L}}}(s)$ of the bounded fractal string $\overline{\mathscr{L}}$ introduced in Equation (3.3.2) is (absolutely) convergent for all $s \in \mathbb{C}$ such that $\operatorname{Re} s > 1$ (see Equation (3.3.9) in Theorem 3.3.3(a)), this functional equation yields Equation (3.3.13) directly for all $s \in \mathbb{C}$ such that $\operatorname{Re} s > D_{\mathrm{mer}}(\zeta_{\mathscr{L}_0})$. Indeed, upon meromorphic continuation, each of the meromophic functions involved in the above scaling argument can be interpreted as the meromorphic continuation of the corresponding zeta functions. This completes the proof of part (b) of Theorem 3.3.3.

Note that the special case when $\mathscr{L}_0 = \{1\}$ and hence $\zeta_{\mathscr{L}_0} = 1$, also yields part (a) of Theorem 3.3.3. In particular, Equation (3.3.10) holds for all $s \in \mathbb{C}$ and Equation (3.3.9) holds since $|\mathscr{L}|_1 = \zeta_{\mathscr{L}}(1)$, where $\mathscr{L} := \mathscr{L}(r_1, \dots, r_J)$.

The next theorem gives a general construction of complex dimensions of higher order generated by means of extended self-similar strings.

Theorem 3.3.6. *Let $\overline{\mathscr{L}} := \mathscr{L}_0 \otimes \mathscr{L}(r_1, \dots, r_J)$ be an extended self-similar fractal string in $\mathbb{R}$ generated by a bounded fractal string $\mathscr{L}_0$ and the set of scaling ratios $\{r_1, r_2, \dots, r_J\}$ with $0 < r_j < 1$, such that $\sum_{j=1}^{J} r_j < 1$. Furthermore, assume that $\zeta_{\mathscr{L}_0}$ is meromorphic on $\mathbb{C}$ and that there are no zero-pole cancellations in* (3.3.13)*. Let $\mathfrak{D}$ be the set of complex solutions of the Moran equation $\sum_{j=1}^{J} r_j^s = 1$ and let m be an arbitrary positive integer. Then, one can explicitly construct an extended self-similar fractal string $\overline{\mathscr{L}}_m$ which has exactly the same complex dimensions as $\overline{\mathscr{L}}$ but with the orders (i.e., the multiplicities) of the complex dimensions located in $\mathfrak{D}$ multiplied by m.*

Moreover, if we let $\mathfrak{D}^+ := \mathfrak{D} \cap \{\mathrm{Re}\, s > 0\}$, then one can explicitly construct an extended self-similar fractal string $\mathscr{L}_\infty$ such that all of its complex dimensions contained in $\mathfrak{D}^+$ are of infinite order; that is, they are essential singularities of its geometric zeta function $\zeta_{\mathscr{L}_\infty}$. In particular, we have that $D_{\mathrm{mer}}(\zeta_{\mathscr{L}_\infty}) = D(\zeta_{\mathscr{L}_\infty})$.

Proof. Let $\mathscr{L}_0$ be the generator and let $\{r_1, r_2, \ldots, r_J\}$ with $0 < r_j < 1$ be the associated scaling ratios. Furthermore, we define $\overline{\mathscr{L}} := \mathscr{L}_0 \otimes \mathscr{L}(r_1, \ldots, r_J)$ as in part (b) of Theorem 3.3.3, and we now let this be our new generator; that is, we define a new extended self-similar fractal string $\overline{\mathscr{L}}_2 := \overline{\mathscr{L}} \otimes \mathscr{L}(r_1, \ldots, r_J)$ as the disjoint union of scaled copies of $\overline{\mathscr{L}}$ by scaling factors built by all possible words of multiples of the ratios r_j. This construction implies that

$$\overline{\mathscr{L}}_2 = \overline{\mathscr{L}} \sqcup \bigsqcup_{j=1}^{J} r_j \overline{\mathscr{L}}_2 \tag{3.3.24}$$

and, similarly as before, by the scaling property of the geometric zeta function (see Remark 3.3.5 above) and, in light of part (b) of Theorem 3.3.3, we then have

$$\zeta_{\overline{\mathscr{L}}_2}(s) = \frac{\zeta_{\overline{\mathscr{L}}}(s)}{1 - \sum_{j=1}^{J} r_j^s} = \frac{\zeta_{\mathscr{L}_0}(s)}{\left(1 - \sum_{j=1}^{J} r_j^s\right)^2}. \tag{3.3.25}$$

As is apparent in Equation (3.3.25), the fractal string $\overline{\mathscr{L}}_2$ has exactly the same complex dimensions as $\overline{\mathscr{L}}$, except for the fact that the orders of the ones contained in $\mathfrak{D}$ are multiplied by 2.

We can next proceed inductively by using $\overline{\mathscr{L}}_2$ as our new base fractal string and, for each $n \in \mathbb{N}$, we thus obtain a fractal string $\overline{\mathscr{L}}_n$ such that

$$\zeta_{\overline{\mathscr{L}}_n}(s) = \frac{\zeta_{\mathscr{L}_0}(s)}{\left(1 - \sum_{j=1}^{J} r_j^s\right)^n}; \tag{3.3.26}$$

$\overline{\mathscr{L}}_n$ has exactly the same complex dimensions as $\overline{\mathscr{L}}$, except for the fact that the ones contained in $\mathfrak{D}$ have their orders multiplied by n.

In order to generate essential singularities, we take a disjoint union of the fractal strings $\overline{\mathscr{L}}_n$ scaled by $(n!)^{-1}$. More specifically, we define $\overline{\mathscr{L}}_\infty$ as

$$\overline{\mathscr{L}}_\infty := \bigsqcup_{n=1}^{\infty} (n!)^{-1} \overline{\mathscr{L}}_n. \tag{3.3.27}$$

The construction of $\overline{\mathscr{L}}_\infty$ (see Definition 4.5.7 and Lemma 4.5.10 in Subsection 4.5.2 below) and the scaling property of the geometric zeta function (see Remark 3.3.5) then imply that

$$\zeta_{\overline{\mathscr{L}}_\infty}(s) = \zeta_{\mathscr{L}_0}(s) \sum_{n=1}^{\infty} \frac{1}{(n!)^s \left(1 - \sum_{j=1}^{J} r_j^s\right)^n}. \tag{3.3.28}$$

By the Weierstrass M-test, the above sum defines a holomorphic function on $\{\mathrm{Re}\, s > 0\} \setminus \mathfrak{D}^+$ and $\mathfrak{D}^+$ is the set of essential singularities of the function defined by this sum. □

Example 3.3.7. (The n-th order Cantor string). In this example we introduce the notion of an n-th order Cantor string, where $n \in \mathbb{N}$ is arbitrary. Namely, in the notation of Theorem 3.3.6 define $\mathscr{L}_1$ to be the Cantor string of total length 1 (see Example 2.3.31). Furthermore, define now the 2nd order Cantor string as the extended self-similar string $\mathscr{L}_2 := \overline{\mathscr{L}} = \mathscr{L}_1 \otimes \mathscr{L}(3^{-1}, 3^{-1})$. By using Theorem 3.3.6 and since $\zeta_{\mathscr{L}_1}$ is given by Equation (2.3.53), one now concludes that

$$\zeta_{\mathscr{L}_2}(s) = \frac{\zeta_{\mathscr{L}_1}(s)}{1 - 2 \cdot 3^{-s}} = \frac{3^s}{(3^s - 2)^2}, \tag{3.3.29}$$

that is, $\zeta_{\mathscr{L}_2}$ is meromorphic on all of $\mathbb{C}$. Moreover,

$$\mathscr{P}(\zeta_{\mathscr{L}_2}) = \log_3 2 + \frac{2\pi}{\log 3}\mathrm{i}\mathbb{Z} \tag{3.3.30}$$

and all of the above (principal) complex dimensions are of second order.

To define the n-th order Cantor string for $n \geq 2$, we proceed inductively by defining $\mathscr{L}_n := \mathscr{L}_{n-1} \otimes \mathscr{L}(3^{-1}, 3^{-1})$. The associated geometric zeta function is then given by

$$\zeta_{\mathscr{L}_n}(s) = \frac{3^{s(n-1)}}{(3^s - 2)^n}, \tag{3.3.31}$$

is meromorphic on all of $\mathbb{C}$ and the set of its poles coincides with (3.3.30) but all of them are of n-th order.

Finally, we define $\mathscr{L}_\infty$, the *Cantor string of infinite order*, as the union of the resulting fractal strings (viewed as multisets; that is, taking the multiplicities into account):

$$\mathscr{L}_\infty := \bigsqcup_{n=1}^{\infty} (n!)^{-1} \mathscr{L}_n. \tag{3.3.32}$$

Its geometric zeta function is then given by

$$\zeta_{\mathscr{L}_\infty}(s) = \sum_{n=1}^{\infty} \frac{3^{s(n-1)}}{(n!)^s (3^s - 2)^n}, \tag{3.3.33}$$

and, by the Weierstrass M-test, is holomorphic on the open set

$$\{\mathrm{Re}\, s > 0\} \setminus (\log_3 2 + \frac{2\pi}{\log 3}\mathrm{i}\mathbb{Z}).$$

The line $\{\mathrm{Re}\, s = 0\}$ is a (meromorphic) partial natural boundary for $\zeta_{\mathscr{L}_\infty}$ (in the sense of part (*i*) of Definition 1.3.8 of Subsection 1.3.2, and strengthened as in Remark 1.3.9) and the set $\log_3 2 + \frac{2\pi}{\log 3}\mathrm{i}\mathbb{Z}$ consists of essential singularities of $\zeta_{\mathscr{L}_\infty}$. Anticipating on the future developments to be discussed in Chapter 4, we point out that the fractal string $\mathscr{L}_\infty$ is strongly hyperfractal, in the sense of part (*ii*) of Definition 4.6.23 of Subsection 4.6.3.

We note that this example will be revisited in Example 4.2.10 of Subsection 4.2.2 below in the more general context of relative fractal drums.

Remark 3.3.8. Taking any bounded fractal string $\mathscr{L}_0$ in $\mathbb{R}$ as a base fractal string, we can spray it using two (or more) sets of scaling ratios $\{r_1,\dots,r_J\}$ and $\{\rho_1,\dots,\rho_K\}$ in $(0,1)$. More specifically, we first generate $\overline{\mathscr{L}}$ as the extended fractal string obtained in Definition 3.3.1. Next, taking $\overline{\mathscr{L}}$ as a new base fractal string, we define the corresponding extended fractal string $\mathscr{L}_2$ by using $\{\rho_1,\dots,\rho_K\}$ as the set of scaling ratios. Then, having used twice part (b) of Theorem 3.3.13, we conclude that the geometric zeta function of $\mathscr{L}_2$ is given by

$$\zeta_{\mathscr{L}_2}(s) = \frac{\zeta_{\mathscr{L}_0}(s)}{\left(1-\sum_{j=1}^J r_j^s\right)\cdot\left(1-\sum_{j=1}^K \rho_j^s\right)}. \tag{3.3.34}$$

Finally, if $\zeta_{\mathscr{L}_0}$ can be meromorphically extended to a window $\boldsymbol{W}$ such that it has no common zeros in $\boldsymbol{W}$ with the functions $1-\sum_{j=1}^J r_j^s$ and $1-\sum_{j=1}^K \rho_j^s$, then (taking into account the multiplicities), we have

$$\begin{aligned}\mathscr{P}(\mathscr{L}_2,\boldsymbol{W}) = &\left\{s\in\boldsymbol{W} : \sum_{j=1}^J r_j^s = 1\right\}\cup\left\{s\in\boldsymbol{W} : \sum_{j=1}^K \rho_j^s = 1\right\}\\ &\cup\mathscr{P}(\mathscr{L}_0,\boldsymbol{W}).\end{aligned} \tag{3.3.35}$$

In closing, we mention that the results of this section can be generalized from the case of bounded fractal strings to that of 'relative fractal drums', which also include bounded fractal subsets of Euclidean spaces as a special case, as well as bounded fractal strings; see Subsection 4.2.2 of Chapter 4 below.

3.4 Weighted Zeta Functions

The notion of weighted distance zeta function, associated with a fractal A, is an inevitable consequence of the definition of the usual distance zeta function ζ_A. Namely, upon repeated differentiation of ζ_A, the logarithmic weights emerge immediately. This section can be considered as providing a natural path towards an even more general situation, in which we may consider a fractal A jointly with a Borel measure μ defined on a fixed δ-neighborhood of A, for some (small) $\delta > 0$. This yields the notion of a *distance zeta function with measure*:

$$\zeta_A(s,\mu) = \int_{A_\delta} d(x,A)^s \mathrm{d}\mu(x), \tag{3.4.1}$$

for all $s\in\mathbb{C}$ with $\mathrm{Re}\, s$ sufficiently large.

All of the fractal zeta functions appearing in this section (i.e., Section 3.4) belong to this class.

3.4.1 Definition and Properties of Weighted Zeta Functions

We can also consider the *weighted distance zeta function* $\zeta_A(\cdot, w)$ of a bounded set A in $\mathbb{R}^N$, associated with a given complex-valued *weight function* w defined on A_δ, for a fixed $\delta > 0$:

$$\zeta_A(s,w) := \int_{A_\delta} w(x)\, d(x,A)^{s-N} \mathrm{d}x, \tag{3.4.2}$$

for all $s \in \mathbb{C}$ with $\mathrm{Re}\, s$ sufficiently large.

Here, we shall need the notion of a *limit* L^∞*-space*, or the space of (at most) *weakly singular functions*, defined as the intersection of L^p-spaces (with respect to the Lebesgue measure on $\mathbb{R}^N$) of complex-valued functions, for $1 < p < \infty$,

$$L^{\infty)}(A_\delta) = \bigcap_{1<p<\infty} L^p(A_\delta). \tag{3.4.3}$$

Example 3.4.1. Note that since A_δ is bounded, then $L^\infty(A_\delta) \subset L^{\infty)}(A_\delta)$, and the inclusion is strict. Indeed, the function $f(x) = \log d(x,A)$ is in the space $L^{\infty)}(A_\delta)$ for any positive δ, provided $\overline{\dim}_B A < N$; see [Žu1, proof of Theorem 1(d)]. It suffices to fix $p > 1$ and to note that $|\log d(x,A)|^p \le C d(x,A)^{-\gamma}$ for all $x \in A_\delta \setminus \overline{A}$, where C is a positive constant. Taking $0 < \gamma < N - \overline{\dim}_B A$, we deduce that $f \in L^p(A_\delta)$ in light of Lemma 2.1.3.

Naturally, the standard distance zeta function $\zeta_A = \zeta_{A,A_\delta}$, introduced in Definition 2.1.1 (see Equation (2.1.1)), corresponds to the case when the weight function is constant: $w \equiv 1$.

Proposition 3.4.2. *Let Ω be an open set in $\mathbb{R}^N$. Then, the following properties hold:*

(a) The vector space $L^{\infty)}(\Omega)$ is an algebra of functions with respect to pointwise multiplication.

(b) Assume, in addition, that Ω is bounded. If $f \in L^p(\Omega)$ and $g \in L^{\infty)}(\Omega)$, then

$$fg \in L^{p)}(\Omega) := \bigcap_{1<q<p} L^q(\Omega). \tag{3.4.4}$$

In particular, the product of $L^{p)}$ and $L^{\infty)}$-functions is an $L^{p)}$-function.

Proof. (a) If we take $f, g \in L^{\infty)}(\Omega)$ and a fixed $p > 1$, then we have $fg \in L^p(\Omega)$ since for any $r > 1$, we have by Hölder's inequality, $\| |fg|^p \|_{L^1} \le \| |f|^p \|_{L^r} \| |g|^p \|_{L^{r'}} = (\|f\|_{L^{pr}} \|g\|_{L^{pr'}})^p$ (here, $r' := r/(r-1)$ is the conjugate exponent of r); that is,

$$\|fg\|_{L^p} \le \|f\|_{L^{pr}} \|g\|_{L^{pr'}} < \infty, \tag{3.4.5}$$

and therefore $fg \in L^{\infty)}(A_\delta)$.

(b) Let $q \in (1,p)$ be fixed. Much as in (a), we have $\|fg\|_{L^q} \le \|f\|_{L^{qr}}\|g\|_{L^{qr'}} < \infty$, since $f \in L^p(\Omega)$ and $L^p(\Omega) \subset L^{qr}(\Omega)$ (due to the boundedness of Ω) for $r > 1$ sufficiently close to 1, or more precisely, for $qr < p$. To prove the second part of (b), let $f \in L^{p)}(\Omega)$. Then $f \in L^q(\Omega)$ for any $q \in (1,p)$, and by the first part of (b), we have that $fg \in L^{q)}(\Omega)$. Hence, $fg \in \cap_{1<q<p} L^{q)}(\Omega) = L^{p)}(\Omega)$. □

Remark 3.4.3. We point out that the notion of *limit L^p-space* (denoted here by $L^{p)}(\Omega)$, see case (b) of Proposition 3.4.2 above) is close to the already existing notion of *grand L^p-space*, which we denote by $L^{p)}_{\varphi}(\Omega)$. The latter notion has been introduced in 1992 by T. Iwaniec and C. Sbordone in [IwSb], and the corresponding space $L^{p)}_{\varphi}(\Omega)$ is defined as the set of measurable functions $f : \Omega \to \mathbb{C}$ satisfying

$$\sup_{q\in(1,p)} \varphi(q) \left(\int_\Omega |\varphi(x)|^q \,\mathrm{d}x \right)^{1/q} < \infty, \tag{3.4.6}$$

where $\varphi(q) := \left(\frac{p-q}{|\Omega|_N}\right)^{1/q}$. The grand L^p-space is always contained in the limit L^p-space. More precisely, for any given bounded Lebesgue measurable subset Ω such that $|\Omega|_N > 0$ and for each $p \in (1,+\infty)$, we have that

$$L^p(\Omega) \subset L^{p)}_{\varphi}(\Omega) \subset L^{p)}(\Omega) := \bigcap_{1<q<p} L^q(\Omega). \tag{3.4.7}$$

Also, we caution the reader that the grand Lebesgue space $L^{p)}_{\varphi}(\Omega)$ is often denoted by $L^{p)}(\Omega)$ in the literature.

The following result extends Theorem 2.1.11 to weighted distance zeta functions of fractal sets. It shows, in particular, that the derivative of a weighted distance zeta function of A is again a weighted distance zeta function of A but, of course, for a different weight function.

Theorem 3.4.4. *Let A be a bounded set in $\mathbb{R}^N$, $\delta > 0$, and assume that $1 < p < \infty$. Then:*

(a) If $w \in L^p(A_\delta)$, then the weighted distance zeta function $\zeta_A(s,w)$ of A (see (3.4.2)) is holomorphic in the open half-plane

$$\mathrm{Re}\, s > \frac{1}{p'}\overline{\dim}_B A + \frac{N}{p}, \tag{3.4.8}$$

where $p' := p/(p-1)$ is the conjugate exponent of p, defined by $1/p + 1/p' = 1$. Furthermore, in that same half-plane, we have

$$\zeta_A'(s,w) = \int_{A_\delta} w(x)\, d(x,A)^{s-N} \log d(x,A)\,\mathrm{d}x; \tag{3.4.9}$$

that is, the derivative of the weighted distance zeta function is again a weighted distance zeta function:

$$\zeta_A'(s,w) = \zeta_A(s,w_1),$$

where $w_1(x) := w(x)\log d(x,A) \in L^{p)}(A_\delta)$. *Moreover, if* $w \in L^{\infty)}(A_\delta)$, *then* $w_1 \in L^{\infty)}(A_\delta)$.

(*b*) *Assume, as in part* (*a*), *that* $w \in L^p(A_\delta)$. *If* $\dim_B A =: D$ *exists and* $\mathscr{M}_*^D(A) > 0$, *then the lower bound on the right-hand side of (3.4.8) is optimal for the collection of all weight functions* $w \in L^p(A_\delta)$.

(*c*) *If* $w \in L^{\infty)}(A_\delta)$, *then* $\zeta_A(s,w)$ *is holomorphic in the half-plane* $\mathrm{Re}\, s > \overline{\dim}_B A$. *If, in addition,* $\dim_B A =: D$ *exists and* $\mathscr{M}_*^D(A) > 0$, *then the lower bound for* $\overline{\dim}_B A$ *on the right hand side of (3.4.8) is optimal for the collection of all weight functions* $w \in L^{\infty)}(A_\delta)$.

Proof. (*a*) Let $s \in \mathbb{C}$ with real part satisfying the inequality (3.4.8). Repeating the proof of Theorem 2.1.1 in this more general situation, we obtain the following analog of (2.1.21):

$$|R(h)| \le C|h| \int_{A_\delta} |w(x)|\, d(x,A)^{\mathrm{Re}\, s - N - 2\varepsilon} \mathrm{d}x. \tag{3.4.10}$$

By using Hölder's inequality, we deduce that

$$|R(h)| \le C|h|\, \|w\|_{L^p} \|g\|_{L^{p'}}, \tag{3.4.11}$$

where $g(x) := d(x,A)^{\mathrm{Re}\, s - N - 2\varepsilon}$. Letting $\gamma := 2\varepsilon + N - \mathrm{Re}\, s$, with $\varepsilon > 0$ sufficiently small, and by using Lemma 2.1.3, we see that $g \in L^{p'}(A_\delta)$ provided $p'\gamma < N - \overline{\dim}_B A$. This inequality follows from (3.4.8) (which is assumed here) provided ε is small enough. This shows that $\zeta_A(\cdot, w)$ is holomorphic in the open half-plane defined by (3.4.8), and that $\zeta_A'(\cdot, w)$ exists and is given by (3.4.9) in that same half-plane. (See the corresponding part of Theorem 2.1.11; see also Theorem 2.1.45, in the statement of which the weight function w can be can be absorbed into the measure: $\mathrm{d}\mu(t) = w(t)\,\mathrm{d}t$, in the notation of that theorem.) The $L^{p)}$ and $L^{\infty)}$ claims (more precisely, the fact that for $q = p, \infty$, with $1 < p < \infty$, $w_1(x) = w(x)\log d(x,A)$ belongs to $L^{q)}(A_\delta)$ if $w(x)$ does) follow from Proposition 3.4.2.

(*b*) If $\dim_B A = N$, then the right-hand side of (3.4.8) is equal to N, and the optimality of the bound follows from Theorem 2.1.11(*c*), as we now explain. Here, we assume that $w \in L^p(A_\delta)$ with $1 < p < \infty$. If A is such that $\overline{\dim}_B A < N$, let us define w by $w(x) = d(x,A)^{-\gamma}$ for some γ satisfying $0 < \gamma < \frac{1}{p}(N - \overline{\dim}_B A)$. Therefore, $w \in L^p(A_\delta)$; see Lemma 2.1.3. We have

$$\zeta_A(s,w) = \int_{A_\delta} d(x,A)^{-\gamma + s - N} \mathrm{d}x = \zeta_A(s - \gamma); \tag{3.4.12}$$

so that, by Theorem 2.1.11, the function $\zeta_A(s,w)$ is holomorphic for $\mathrm{Re}(s-\gamma) > \overline{\dim}_B A$, that is, for $\mathrm{Re}\, s > \gamma + \overline{\dim}_B A$. We have $\gamma + \overline{\dim}_B A < \frac{1}{p'}\overline{\dim}_B A + \frac{N}{p} =: r$. Using Theorem 2.1.11(*c*), we see that $\zeta_A(s,w) \to +\infty$ as $\mathbb{R} \ni s \to \gamma + D$ from the right. Since $\gamma + D$ can be made arbitrarily close to r for γ sufficiently close to $\frac{1}{p}(N-D)$, we deduce that the estimate (3.4.8) cannot be improved.

(*c*) If $w \in L^{\infty)}(A_\delta)$, then the weighted zeta function $\zeta_A(s,w)$ is holomorphic in the union of the half-planes defined by (3.4.8), corresponding to all $p > 1$; note that the right-hand side of (3.4.8) is decreasing as p grows and so the half-planes are increasing. The claim now follows from (3.4.8) by letting $p \to +\infty$, since then $p' \to 1$. ☐

Note that the number r defined by the right-hand side of (3.4.8) is a convex combination of $\overline{\dim}_B A$ and N, and therefore, $r \in [\overline{\dim}_B A, N]$. Furthermore, if $p \to +\infty$, then $r \to \overline{\dim}_B A$, while if $p \to 1$, then $r \to N$. It is also worth noticing that the derivative of the standard distance zeta function of A defined by (2.1.1) in Theorem 2.1.11 is equal to the weighted zeta function of A with weight $w(x) = \log d(x,A)$, see (2.1.13):

$$\zeta_A'(s) = \zeta_A(s,w). \tag{3.4.13}$$

Remark 3.4.5. Motivated by the notion of generalized fractal string discussed in [Lap-vFr3, Chapter 4] (and introduced in [Lap-vFr1]), one can generalize the weighted zeta functions (3.4.2) to distance zeta functions associated with positive or complex Borel measures μ on $\mathbb{R}^N$ (or more generally, on A_δ, for some small positive δ):

$$\zeta_A(s,\mu) = \int_{A_\delta} d(x,A)^{s-N} d\mu(x), \tag{3.4.14}$$

for all $s \in \mathbb{C}$ with $\mathrm{Re}\, s$ large enough.

If μ is such that $|d\mu(x)| \le |w(x)|\, dx$, with w as in Theorem 3.4.4, then the counterpart of (3.4.9) in Theorem 3.4.4 reads as follows (see Theorem 2.1.45):

$$\zeta_A'(s,\mu) = \int_{A_\delta} d(x,A)^{s-N} \log d(x,A)\, d\mu(x),$$

for all $s \in \mathbb{C}$ such that $\mathrm{Re}\, s > \overline{\dim}_B A$.

Remark 3.4.6. The singular dimension of $L^{\infty)}(\Omega)$ is equal to 0, while its upper singular dimension is equal to N (see [Žu1, Theorem 1(d)]):

$$\text{s-}\dim L^{\infty)}(\Omega) = 0, \quad \text{s-}\overline{\dim}\, L^{\infty)}(\Omega) = N. \tag{3.4.15}$$

The *singular dimension* (respectively, the *essential singular dimension*) of a vector space X (or just a set) of measurable complex-valued functions $f : \Omega \to \overline{\mathbb{C}}$ is defined by

$$\begin{aligned} \text{s-}\dim X &= \sup\{\dim_H(\mathrm{Sing}\, f) : f \in X\}, \\ \text{s-}\overline{\dim}\, X &= \sup\{\dim_H(\text{e-Sing}\, f) : f \in X\}. \end{aligned}$$

Here, $\dim_H E$ denotes the Hausdorff dimension of $E \subseteq \mathbb{R}^N$, $\mathrm{Sing}\, f$ is the *singular set* of f consisting of all $a \in \Omega$ for which there exists $\gamma > 0$ such that $|f(x)| \geq C|x-a|^{-\gamma}$ Lebesgue a.e. in a neighborhood of a, while e-$\mathrm{Sing}\, f$ is the *extended singular set* of f defined by

$$\text{e-Sing}\, f = \Big\{a \in \Omega : \limsup_{r\to 0^+} \frac{1}{r^N} \int_{B_r(a)} |f(x)|\, \mathrm{d}x = +\infty\Big\}.$$

The extended singular set of f contains, for example, all points $a \in \Omega$ of logarithmic and iterated logarithmic growth of f. For any $f \in L^{\infty)}(\Omega)$, we have $\mathrm{Sing}\, f = \emptyset$, while e-$\mathrm{Sing}\, f$ may be nontrivial; see Example 3.4.1. A detailed analysis of pointwise regularity and local oscillations of functions can be found in the memoir [JaffMey] by Jaffard and Meyer.

The space $L^{\infty)}(\Omega)$ appears naturally in the theory of Sobolev spaces, that is central to the study of partial differential equations. If Ω is a bounded open set in $\mathbb{R}^N$ and $kp = N$, where k is a positive integer and $1 \leq p < \infty$, then by the Sobolev embedding theorem (see, e.g., [Bre]),

$$W^{k,p}(\Omega) \subset L^{\infty)}(\Omega).$$

If Ω is not necessarily bounded, then under the same conditions on k, p and N, we have a more general result:

$$W^{k,p}(\Omega) \subset L^{[p,\infty)}(\Omega) := \bigcap_{p\leq q<\infty} L^q(\Omega).$$

3.4.2 Harmonic Functions Generated by Fractal Sets

Let A be a bounded subset of $\mathbb{R}^N$. Writing $s = \xi + \eta\mathrm{i} \in \mathbb{C}$, where $\xi, \eta \in \mathbb{R}$, and separating the real and imaginary parts u and v of the distance zeta function $\zeta_A(s) = u(\xi,\eta) + v(\xi,\eta)\mathrm{i}$ defined by (2.1.1), we obtain the following functions, defined by means of singular integrals (see also Proposition 2.1.22) for $\xi = \mathrm{Re}\, s$ sufficiently large,

$$\begin{aligned} u(\xi,\eta) &= \int_{A_\delta\setminus\overline{A}} d(x,A)^{\xi-N} \cos(\eta \log d(x,A))\, \mathrm{d}x, \\ v(\xi,\eta) &= \int_{A_\delta\setminus\overline{A}} d(x,A)^{\xi-N} \sin(\eta \log d(x,A))\, \mathrm{d}x. \end{aligned} \tag{3.4.16}$$

Note that in the integrands, the singularities appear on the set $\overline{A}$ both in the function $x \mapsto d(x,A)^{\xi-N}$, provided $\xi < N$, and in the function $x \mapsto \log d(x,A)$. The oscillatory nature of the corresponding chirp-like function $(0,\delta) \ni t \mapsto t^{\xi-N}\cos(\eta \log t)$ has been discussed in Subsection 2.1.7.

Since ζ_A is holomorphic for $\mathrm{Re}\,s > \overline{\dim}_B A$, we can immediately deduce the following consequence, in which we obtain a new class of harmonic functions generated by fractal sets in $\mathbb{R}^N$. It is noteworthy that the natural common domain of definition of these harmonic functions depends on the upper box dimension of A.

Corollary 3.4.7. *Let A be a bounded set in $\mathbb{R}^N$, and δ be a fixed positive number. Define $G = (\overline{\dim}_B A, +\infty) \times \mathbb{R} \subset \mathbb{R}^2$. Then the functions u and v defined by (3.4.16) are of class C^∞ and harmonic on G, that is, $\Delta u = 0$ and $\Delta v = 0$ on G. Furthermore, if $D = \dim_B A$ exists and $\mathscr{M}_*^D(A) > 0$, then $u(\xi, 0) \to +\infty$ as $\xi \to D^+$, with ξ real.*

More general harmonic functions than those obtained in Corollary 3.4.7 can be easily generated by using distance zeta functions with weights, or even measures; see Theorem 3.4.4 and Remark 3.4.5.

Example 3.4.8. Extremely complicated harmonic functions are those generated by the boundary A of the Mandelbrot set; see (3.4.16). Thanks to Shishikura's well-known theorem [Shi], we have that $\dim_B A = \dim_H A = 2$, where $\dim_H A$ denotes the Hausdorff dimension of A;[12] so that the associated distance zeta function $\zeta_A(s)$ is holomorphic for $\mathrm{Re}\,s > 2$, $\mathscr{H}(\zeta_A) = \{\mathrm{Re}\,s > 2\}$ and $D(\zeta_A) = 2$. We do not know if $\mathscr{M}_*^2(A) > 0$; if this is true, then it follows from Corollary 3.4.7 that $\zeta_A(s) \to +\infty$ as $\mathbb{R} \ni s \to 2^+$.

In closing this example, we note that the question of whether or not ζ_A or $\tilde{\zeta}_A$ (the distance or the tube zeta function of the Mandelbrot set, respectively) admits the critical line $\{\mathrm{Re}\,s = 2\}$ as a (meromorphic) partial natural boundary, or even as a (meromorphic) natural boundary (in the sense of Definition 1.3.8 and possibly when ζ_A and $\tilde{\zeta}_A$ are defined via an appropriate gauge function, see footnote 12 on page 222), will be addressed in Problem 6.2.21 of Subsection 6.2.2.

3.5 Zeta Functions of Fractal Nests

In this section, we provide several examples of fractal sets illustrating the use of zeta functions for the computation of their box dimensions. Note that the sets appearing in Example 3.5.1 below are not the boundary of any fractal spray; see [LapPo3], [Lap-vFr3, Section 1.4] for the definition of fractal sprays. In short, a fractal spray (as introduced in [LapPo3]) is a disjoint union of countably many scaled copies of a single bounded and open subset of $\mathbb{R}^N$ (called the 'basic shape' in [Lap-vFr3] or the 'generator' of the spray in [LapPe2–3, LapPeWi1–2]). The scaling is done via a

[12] However, it does not seem to be known whether A is either Hausdorff or Minkowski nondegenerate, and in case it is degenerate, what is a corresponding gauge function h with respect to which it is h-Minkowski (or h-Hausdorff) nondegenerate. See Definition 6.1.4 and the discussion following it.

fractal string $\mathscr{L} = (\ell_j)_{j\geq 1}$, as in Example 3.5.1. As a variant, one can also allow (as in [LapPe2–3, LapPeWi1–2]) finitely many generators for the fractal spray. At first glance, the notion of a fractal nest may seem to be close to that of a fractal spray, but it is in fact essentially different; see Remark 3.5.2 following Example 3.5.1.

Example 3.5.1. (*Fractal nests of concentric circles and spheres*). Let $\mathscr{L} = (\ell_j)_{j\geq 1}$ be a *bounded fractal string*, i.e., a nonincreasing sequence of positive numbers tending to 0, such that $\sum_{j\geq 1} \ell_j < \infty$. Let $A = A_{\mathscr{L}} = \{a_k : k \in \mathbb{N}\}$ be the corresponding bounded subset of $\mathbb{R}$, defined by $a_k := \sum_{j\geq k} \ell_j$ for each $k \geq 1$. Clearly, $A \subset [0, a_1]$.

We stress that whereas up to this point in this monograph, $A = A_{\mathscr{L}}$ was viewed as a subset of the real line $\mathbb{R}$, we will view A in this example as being embedded in a higher-dimensional Euclidean space; namely, we will consider it as a subset of $\mathbb{R}^2$ (in parts (a) and (b)) and as a subset of $\mathbb{R}^N$ (in parts (c) and (d)), where $N \geq 2$ is arbitrary.

(a) We view A as a subset of the x_1-axis in $\mathbb{R}^2$. Let A_1 be a planar set obtained by rotation of A around the origin, that is, as the union of the sequence of concentric circles of radii a_k; see Figure 3.2. By a method similar to the one used in Example 2.1.58, we obtain that A_1 has the following zeta function:

$$\zeta_{A_1}(s) = \frac{2\pi\delta^s}{s} + \frac{2\pi a_1 \delta^{s-1}}{s-1} + \frac{2^{2-s}\pi}{s-1} \sum_{k=1}^{\infty} \ell_k^{s-1}(a_k + a_{k+1}), \tag{3.5.1}$$

where we assume that $\delta \geq l_1/2$. This assumption is inessential due to Proposition 2.1.76. The first two terms on the right-hand side of (3.5.1) correspond to the annulus $a_1 < r < a_1 + \delta$ in $\mathbb{R}^2$, and they are also inessential since it is clear that the box dimension of A is at least 1. Since $a_{k+1} = a_k - l_k$, we deduce from (3.5.1) that

$$f(s) := \sum_{k=1}^{\infty} \ell_k^{s-1}(a_k + a_{k+1}) = \sum_{k=1}^{\infty} \ell_k^{s-1}(2a_k - \ell_k) = \zeta_1(s) + \zeta_2(s) \tag{3.5.2}$$

where $\zeta_1(s) := 2\sum_{k=1}^{\infty} \ell_k^{s-1} a_k$ and $\zeta_2(s) := -\sum_{k=1}^{\infty} \ell_k^s$. Assuming that $D(\zeta_1) > D(\zeta_2)$, from Lemma 2.3.5 we conclude that $f(s) \sim \zeta_1(s)$. Therefore, in light of Lemma 2.1.81, we have

$$\overline{\dim}_B A_1 = \max\{1, D(\zeta_1)\}. \tag{3.5.3}$$

Moreover, this value is equal to $\dim_{PC} A_1$.

Note that if $D(f) = 1$, then $\zeta_A(s) \sim \frac{1}{s-1} + f(s)$ and if $D(f) < 1$, then $\zeta_A(s) \sim \frac{1}{s-1}$; see Equation (3.5.1).

(b) As a special case of (a), let us consider a standard example of a fractal string, namely, the α-string, where $a_k = k^{-\alpha}$ for each $k \geq 1$, and $\alpha > 0$. (See [Lap1, Example 5.1] and [Lap-vFr3, Subsection 6.5.1]. Recall, however, that $A = \{a_k : k \geq 1\}$ is now viewed as a subset of $\mathbb{R}^2$.) Let us check the condition $D(\zeta_1) > D(\zeta_2)$ in this

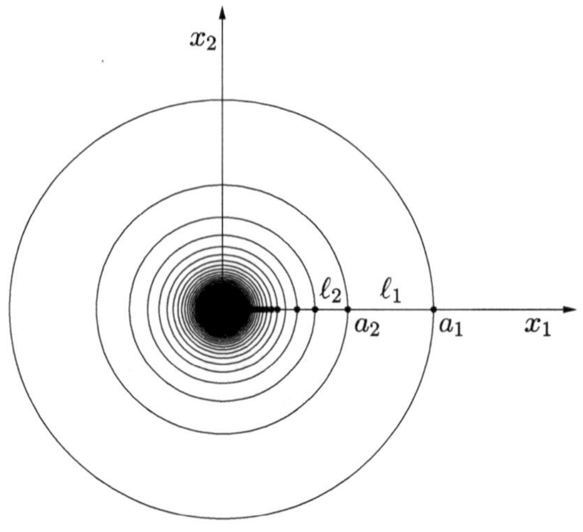

Fig. 3.2 The fractal nest of center type in the plane $\mathbb{R}^2$ generated by the fractal string $\mathscr{L} = (\ell_j)_{j\geq 1}$. Note that for every $k \geq 1$ we have $\ell_k = a_k - a_{k+1}$, where $a_k := \sum_{j\geq k} \ell_j$. Furthermore, we have $A_{\mathscr{L}} := \{a_k : k \geq 1\}$. Compare with Figure 2.7.

situation. Since the case when $\alpha \geq 1$ is easy to deal with,[13] we only consider the case when $\alpha \in (0,1)$.

It is easy to see that $D(\zeta_1) = 2/(1+\alpha) > 1$. Indeed, assuming that $s \in \mathbb{R}$, and since by the Lagrange mean value theorem, $\ell_k = k^{-\alpha} - (k+1)^{-\alpha} \asymp k^{-\alpha-1}$ as $k \to \infty$, we conclude that

$$\zeta_1(s) = 2\sum_{k=1}^{\infty} \ell_k^{s-1} a_k \asymp \sum_{k=1}^{\infty} k^{-(\alpha+1)(s-1)-\alpha} = \sum_{k=1}^{\infty} k^{-((\alpha+1)s-1)}.$$

(The notation $\asymp$ is explained in Subsection 1.3.3, on page 41.) The last Dirichlet series converges if and only if $(\alpha+1)s - 1 > 1$. Hence, $D(\zeta_1) > 2/(1+\alpha)$. In particular, $D(\zeta_1) > 1$ since $\alpha \in (0,1)$.

Similarly, we have $D(\zeta_2) = 1/(1+\alpha)$ since for any $s \in \mathbb{R}$,

$$-\zeta_2(s) = \sum_{k=1}^{\infty} \ell_k^s \asymp \sum_{k=1}^{\infty} k^{-(\alpha+1)s},$$

and the last Dirichlet series converges if and only if $(\alpha+1)s > 1$. Since $D(\zeta_1) > 1$, we deduce from (3.5.1) that

$$\zeta_{A_1}(s) \sim \zeta_1(s) + \zeta_2(s).$$

---

[13] For $\alpha > 1$, the set A_1 is rectifiable (in other words, the sum of circumferences of all circles contained in A_1 is finite). We note that in the case when $\alpha = 1$ we have $\dim_B A_1 = 1$, but the set is Minkowski degenerate. More precisely, its 1-dimensional Minkowski content exists and is equal to $+\infty$. However, it can be shown that $h(t) = \log(1/t)$, for $0 < t < 1$, is the corresponding gauge function of A_1, in the sense of Definition 6.1.4.

Furthermore, since $D(\zeta_1) > D(\zeta_2)$, Lemma 2.3.5 implies that

$$\zeta_1(s) + \zeta_2(s) \sim \zeta_1(s).$$

Therefore, $\zeta_{A_1}(s) \sim \zeta_1(s)$, and by using Theorem 2.1.11, we obtain that for any $\alpha \in (0,1)$,

$$\dim_B A_1 = \frac{2}{1+\alpha}. \tag{3.5.4}$$

We can summarize the above discussion by stating that for any $\alpha > 0$,

$$\dim_B A_1 = \max\left\{1, \frac{2}{1+\alpha}\right\}. \tag{3.5.5}$$

This result was obtained earlier in [ŽuŽup1, Remarks 2 and 8], where it was also noted that the set A_1 is Minkowski measurable if and only if $\alpha \neq 1$. For $\alpha \in (0,1)$ the Minkowski content of A_1 can be explicitly computed directly by analyzing the function $\delta \mapsto |(A_1)_\delta|$ and is given by

$$\mathscr{M}^D(A_1) = \pi(2/\alpha)^{2\alpha/(1+\alpha)}\frac{1+\alpha}{1-\alpha}, \tag{3.5.6}$$

where $D := 2/(1+\alpha)$, while for $\alpha > 1$ we have that $D = 1$ and the corresponding value of $\mathscr{M}^1(A_1)$ is finite and equal to the length of the curve. (Compare with [Lap1, Example 5.1 and Appendix C] and [Lap-vFr3, Subsections 6.5.1 and 8.1.2]; note also that $\mathscr{M}^1(A_1) = \infty$ for $\alpha = 1$.) Furthermore, the value in (3.5.5) is equal to $\dim_{PC} A_1$. Finally, as a consequence of Equation (2.2.4) from Theorem 2.2.3 and Equation (3.5.6), still assuming that $\alpha < 1$, we deduce that the residue of the distance zeta function of A_1 computed at $s = D$ is given by

$$\operatorname{res}(\zeta_{A_1}, d) = (2-D)\mathscr{M}^D(A_1) = \pi(2/\alpha)^{2\alpha/(1+\alpha)}\frac{2\alpha}{1-\alpha}. \tag{3.5.7}$$

See also Example 5.5.16 where the results of Chapter 5 were used to obtain the value (3.5.6) of the Minkowski content of A_1 directly from its distance zeta function. More generally, the results of Chapter 5 give an asymptotic formula for the tube function $t \mapsto |(A_1)_t \cap B_1(0)|$ as $t \to 0^+$ in terms of the complex dimensions of the set A_1.

(c) Next, we view the set $A = \{a_k\}_{k\geq 1}$ from (a), generated by a fractal string $\mathscr{L} = (\ell_j)_{j\geq 1}$, as a subset of $\mathbb{R}^N$ placed on the x_1-axis, where $N \geq 2$. Let A_{N-1} be the subset of $\mathbb{R}^N$ defined as the union of the concentric $(N-1)$-dimensional spheres of radii a_k and with common center at the origin. It is well known that the $(N-1)$-dimensional Hausdorff measure of the unit sphere of $\mathbb{R}^N$ is equal to $N\omega_N$, where ω_N is the N-dimensional Lebesgue measure of the unit ball in $\mathbb{R}^N$ (with explicit values recalled on page 40 in Section 1.3). Passing to spherical coordinates (and by dropping the part depending on δ, corresponding to the shell $a_1 < r < a_1 + \delta$, because it is inessential), we deduce that

$$\zeta_{A_{N-1}}(s) \sim N\omega_N \int_0^{a_1} d(r,A)^{s-N} r^{n-1} \mathrm{d}r = N\omega_N \sum_{k=1}^{\infty} J_k, \tag{3.5.8}$$

where

$$\begin{aligned} J_k &= \int_{a_{k+1}}^{a_k} \left| r - \frac{a_k + a_{k+1}}{2} \right|^{s-N} r^{N-1} \mathrm{d}r \\ &= 2\int_0^{l_k/2} \rho^{s-N} [(\rho + a_{k+1})^{N-1} + (a_k - \rho)^{N-1}] \mathrm{d}\rho. \end{aligned} \tag{3.5.9}$$

It is clear that only the constant terms in square brackets yield (under suitable restrictions, see below) the largest real pole, namely, the box dimension, because the other terms yield the singularities $s = 0, 1, \ldots, N-1$ after integration. Therefore, $\overline{\dim}_B A_{N-1} \geq N-1$, that is, $D(\zeta_{A_{N-1}}) \geq N-1$, which is intuitively clear since A_{N-1} consists of $(N-1)$-dimensional spheres. Let us therefore assume that $D(\zeta_{A_{N-1}}) > N-1$. Dropping all the unnecessary terms from (3.5.9) except for the one involving a_k^{N-1}, we obtain that $J_k \sim a_k^{N-1} \int_0^{\ell_k/2} \rho^{s-N} \mathrm{d}\rho \sim \ell_k^{s-N+1} a_k^{N-1}$, and therefore

$$\zeta_{A_{N-1}}(s) \sim \sum_{k=1}^{\infty} \ell_k^{s-N+1} a_k^{N-1} =: f(s). \tag{3.5.10}$$

We thus deduce from Corollary 2.1.63 that

$$\overline{\dim}_B A_{N-1} = \max\{N-1, D(f)\}. \tag{3.5.11}$$

If we consider the special case of the α-string and of the associated fractal set A_{N-1} in $\mathbb{R}^N$, where $a_k = k^{-\alpha}$ and $l_k = a_k - a_{k+1}$, then using (3.5.10) we obtain the following result generalizing the one obtained in case (a) above:

$$\zeta_{A_{N-1}}(s) \sim \sum_{k=1}^{\infty} \frac{1}{k^{(\alpha+1)(s-N+1)+\alpha(N-1)}} = \sum_{k=1}^{\infty} \frac{1}{k^{(\alpha+1)s-N+1}} \tag{3.5.12}$$

and hence,

$$\overline{\dim}_B A_{N-1} = \frac{N}{\alpha+1}, \tag{3.5.13}$$

provided $0 < \alpha < 1/(N-1)$. Indeed, for $\alpha > 1/(N-1)$ we have $\frac{N}{\alpha+1} < N-1$, so that the half-plane $\{\mathrm{Re}\, s > \frac{N}{\alpha+1}\}$ contains the singularity $s = N-1$, which we have dropped (since in this case, $\zeta_{A_{N-1}}(s) \sim 1/(N-1)$). Therefore, if α is any positive number, we have

$$\overline{\dim}_B A_{N-1} = \max\left\{N-1, \frac{N}{\alpha+1}\right\}. \tag{3.5.14}$$

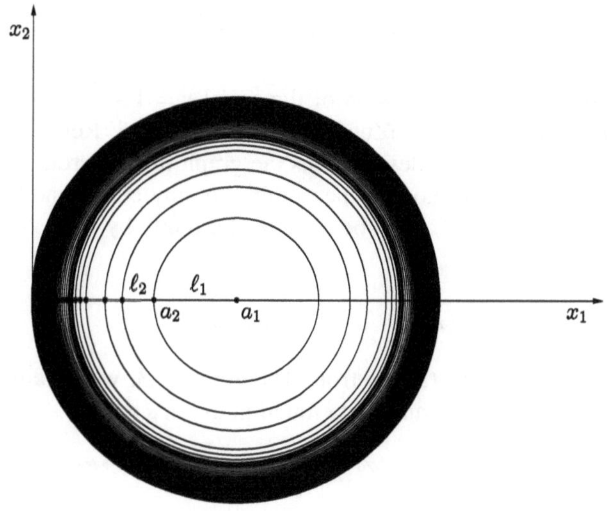

Fig. 3.3 The fractal nest B_1 of outer type in the plane $\mathbb{R}^2$ generated by the fractal string $\mathcal{L} = (\ell_j)_{j\geq 1}$. Note that $\ell_j = a_j - a_{j+1}$ for all $j \geq 1$.

It can be shown that $\dim_B A_{N-1}$ exists in this case. Formula (3.5.14) can be viewed as a generalization to the N-dimensional case of Tricot's formula (3.5.5) (see [Tri3, p. 121] or [DupMenTri]).

(d) Let us now define the subset B_{N-1} of $\mathbb{R}^N$ exactly in the same manner as the set $A_{N-1} \subseteq \mathbb{R}$ in (c), but with center at a_1 (instead of at the origin). In other words, $B_{N-1} \subseteq \mathbb{R}^N$ is the union of the concentric $(N-1)$-dimensional spheres of radii $b_k = \sum_{j\leq k} \ell_k$ and with center at the point a_1 on the x_1-axis; see Figure 3.3 for $N = 2$. Then, using much the same methods as in (c), we obtain that

$$\zeta_{B_{N-1}}(s) \sim \frac{1}{s-N+1} \sum_{k=1}^{\infty} \ell_k^{s-N+1}. \tag{3.5.15}$$

Taking again $a_k = k^{-\alpha}$ with positive α, we have that

$$\zeta_{B_{N-1}}(s) \sim \sum_{k=1}^{\infty} \frac{1}{k^{(\alpha+1)(s-N+1)}}, \tag{3.5.16}$$

which converges for $\operatorname{Re} s > N - 1 + (\alpha+1)^{-1}$. Therefore, using Corollary 2.1.63, we deduce that

$$\overline{\dim}_B B_{N-1} = N - 1 + \frac{1}{\alpha+1}. \tag{3.5.17}$$

This dimension result is intuitively clear since the set B_{N-1} looks locally like a Cartesian product $(0,1)^{N-1} \times \{b_k : k \geq 1\}$ (see [ŽuŽup1, Remark 6]) and $\dim_B(\{b_k : k \geq 1\}) = 1/(\alpha+1)$ (see [Lap1, Examples 5.1 and 5.1']). For $N = 2$, we obtain that

$$d := \dim_B B_1 = \frac{2+\alpha}{1+\alpha},$$

which is precisely the box dimension of the spiral $r = 1 - \theta^{-\alpha}$ of the limit cycle type, where $\theta > \theta_0 > 0$; see [ŽuŽup1, Theorems 2 and 5, Remarks 2 and 8]. Of course, the set B_1 can also be described as the graph of a discrete spiral $r = f(\theta)$, much as in part (c). It is Minkowski measurable for any $\alpha > 0$ and the value of the Minkowski content of $B_1 \subset \mathbb{R}^2$ can be computed directly by analyzing the function $\delta \mapsto |(B_1)_\delta|$ and is given by

$$\mathscr{M}^d(B_1) = 2\pi(1+\alpha)(2/\alpha)^{\alpha/(1+\alpha)}. \tag{3.5.18}$$

Hence, in view of Equation (2.2.4) from Theorem 2.2.3, we obtain the following value of the residue of the distance zeta function ζ_{B_1} at $s = d$:

$$\operatorname{res}(\zeta_{B_1}, d) = (2-d)\mathscr{M}^d(B_1) = 2\pi\alpha(2/\alpha)^{\alpha/(1+\alpha)}. \tag{3.5.19}$$

Remark 3.5.2. In Example 3.5.1 just above, it may seem at first glance that the fractal nest is a 'fractal spray' generated by the unit circle as the 'basic shape' (note that its interior is empty), using the sequence $(a_j)_{j\geq 1}$ as the corresponding fractal string. This is not the case, however. Indeed, in the case of fractal sprays, only the 'inner geometry' of scaled copies is important, whereas for fractal nests, their 'outer geometry' is also essential. More specifically, a fractal spray is *any* disjoint collection of scaled copies of the basic shape; see for, example, Figure 4.5 illustrating the case of the Sierpiński gasket. The corresponding *inner* box dimension of the boundary of the union Ω of scaled copies (more precisely, $\overline{\dim}_B(\partial\Omega, \Omega)$; see Chapter 4 and, in particular, Section 4.2.1) does not depend on the 'arrangement' of the scaled copies in $\mathbb{R}^N$. In contrast, in the case of fractal nests, the arrangement of the scaled copies of the basic shape is essential.

It is convenient to introduce the notion of *fractal nest*, which generalizes the construction of the fractal sets considered in Example 3.5.1. We use the notion of a *basic shape* introduced in the context of fractal sprays in [Lap-vFr3, p. 28] (see also [LapPo3]) for a set which generates the nest.[14]

Definition 3.5.3. Let Ω_0 be a given basic shape in $\mathbb{R}^N$, which we assume here to be a bounded open subset of $\mathbb{R}^N$ that is starshaped with respect to the origin. Recall that Ω_0 is said to be *starshaped* if for any $x \in \Omega$, the open interval $\{tx : t \in (0,1)\}$ joining x and the origin is contained in Ω_0. In particular, the origin belongs to $\overline{\Omega_0}$. Note that for such a set Ω_0, the condition $a_1 > a_2 > 0$ implies that $a_1\Omega_0 \supseteq a_2\Omega_0$.

[14] Note that in [LapPo3, Lap-vFr3], the 'basic shape' (of a fractal spray) is allowed to be an arbitrary bounded open subset of $\mathbb{R}^N$. Recall that the 'basic shape' is also referred to as a 'generator' in [LapPe2–3, LapPeWi1–2, Pe, PeWi].

(a) Let $(a_k)_{k\geq 1}$ be a nonincreasing sequence of positive real numbers, converging to zero. Then define the open set $\Omega = a_1\Omega_0 \setminus \bigcup_{k=1}^{\infty} \partial(a_k\Omega_0)$, viewed as a subset of $\mathbb{R}^N$. Its boundary

$$\partial\Omega = \{0\} \cup \bigcup_{k=1}^{\infty} \partial(a_k\Omega_0). \tag{3.5.20}$$

is called a *fractal nest of center type*. Note that if Ω_0 is the unit ball, then the corresponding fractal nest coincides with the set A_{N-1} in Example 3.5.1(c).

(b) Let $(b_k)_{k\geq 1}$ be a nondecreasing sequence of positive real numbers, converging to b_0. Then define the open set $\Omega = b_0\Omega_0 \setminus \bigcup_{k=1}^{\infty} \partial(b_k\Omega_0)$, viewed as a subset of $\mathbb{R}^N$. Its boundary

$$\partial\Omega = \bigcup_{k=0}^{\infty} \partial(b_k\Omega_0) \tag{3.5.21}$$

is called a *fractal nest of outer type*. Observe that if Ω_0 is the unit ball, then the corresponding fractal nest is congruent to the set B_{N-1} in Example 3.5.1(d).

If we take as a basic shape an open pyramid Ω_0 in $\mathbb{R}^N$ with vertex at the origin, and let $a_k = k^{-\alpha}$ for each $k \geq 1$ and some $\alpha > 0$, then the corresponding set $\partial\Omega$ has box dimension equal to $\max\{N-1, N/(\alpha+1)\}$. This can be easily proved using Example 3.5.1(c) and the property of *finite stability* of the upper box dimension (namely, for any finite family of bounded subsets $C_1,\dots,C_n$ of $\mathbb{R}^N$, we have $\overline{\dim}_B(\bigcup_{k=1}^{n} C_k) = \max\{\overline{\dim}_B C_k : k = 1,\dots,n\}$; see, e.g., Equation (6.1.8) in Subsection 6.1.2 of Chapter 6, [Fal1], [Mat] and [Tri3]).

3.6 Zeta Functions of Geometric Chirps and Multiple String Chirps

In the present section, we introduce geometric chirps and study their fractal zeta functions (modulo equivalence); see Section 3.6.1. We also investigate (still modulo equivalence) the fractal zeta functions of so-called string chirps and multiple strings (in Section 3.6.2) as well as of the Cartesian products of fractal strings (in Section 3.6.3).

3.6.1 Geometric Chirps

By a *geometric* (α,β)*-chirp*, defined by the positive parameters α and β, we mean the planar set A defined by

$$A = \bigcup_{k\geq 1} A_k, \quad A_k = \{k^{-1/\beta}\} \times (0, k^{-\alpha/\beta}). \tag{3.6.1}$$

Here, A_k is the Cartesian product of the one-point set $\{k^{-1/\beta}\}$ and the open interval $(0,k^{-\alpha/\beta})$. The set A is a simplified geometric imitation of the standard *chirp* defined as the graph of the function $y = x^\alpha \sin x^{-\beta}$ for $x \in (0,1]$. See Figures 3.4 and 3.5. Note that the zero points x_k of $y(x)$ have the asymptotics $x_k \asymp a_k := k^{-1/\beta}$ as $k \to \infty$, and that $b_k := y(a_k) \asymp k^{-\alpha/\beta}$. Let $\ell_k := a_k - a_{k+1} \asymp k^{-1-1/\beta}$ as $k \to \infty$. Then

$$\zeta_A(s) = \sum_{k=1}^{\infty} \zeta_k(s) + R(s), \tag{3.6.2}$$

where $\zeta_k(s) := \zeta_{A_k,\Omega_k}(s)$, and Ω_k is the rectangle containing A_k defined by $\Omega_k = (a_k - \frac{\ell_k}{2}, a_k + \frac{\ell_{k-1}}{2})$ for each $k \geq 1$, and $R(s)$ is the remainder term equal to the zeta function of A corresponding to $A_\delta \setminus (\cup_{k\geq 1}\Omega_k)$. The critical line of the function $R(s)$ is located to the left of the critical line corresponding to the sum $\sum_{k\geq 1} J_k$ in (3.6.2), since $R(s)$ corresponds to the boundary of A (a discrete set); we omit the details. As usual, we take δ large enough, so that A_δ contains $\cup_{k\geq 1}\Omega_k$; see Proposition 2.1.76. The corresponding zeta function of the 'needle' A_k with respect to Ω_k is equivalent to $\frac{1}{s-1} b_k \ell_k^{s-1}$, uniformly with respect to k, since

$$\begin{aligned}\zeta_{A_k,\Omega_k}(s) &= \int_0^{\ell_k/2} x^{s-2}\mathrm{d}x \int_0^{b_k} \mathrm{d}y + \int_0^{\ell_{k-1}/2} x^{s-2}\mathrm{d}x \int_0^{b_k} \mathrm{d}y \\ &= \frac{2^{1-s} b_k}{s-1}(\ell_k^{s-1} + \ell_{k-1}^{s-1}) \sim \frac{b_k}{s-1}\ell_k^{s-1}.\end{aligned}$$

Note that we placed the origin of the local coordinate system at $(a_k, 0)$, oriented to the left on the part of the rectangle Ω_k left of A_k, and to the right on the part of Ω_k to the right of A_k. Therefore, the zeta function of the geometric (α,β)-chirp satisfies

$$\begin{aligned}\zeta_A(s) &\sim \sum_{k=1}^{\infty} \zeta_{A_k,\Omega_k}(s) \\ &\sim \frac{1}{s-1}\sum_{k=1}^{\infty} b_k \ell_k^{s-1} \sim \frac{1}{s-1}\sum_{k=1}^{\infty} k^{-\frac{\alpha}{\beta}-(1+\frac{1}{\beta})(s-1)}.\end{aligned}$$

The latter series converges if and only if $\frac{\alpha}{\beta} + (1+\frac{1}{\beta})(\mathrm{Re}\, s - 1) > 1$, and from this we see that $D(\zeta_A) = 1 + \frac{\beta-\alpha}{1+\beta} = 2 - \frac{1+\alpha}{1+\beta}$, provided $D(\zeta_A) > 1$. Therefore, using Corollary 2.1.63, we deduce that the upper box dimension of the geometric (α,β)-chirp is given by

$$\overline{\dim}_B A = \max\left\{1, 2 - \frac{1+\alpha}{1+\beta}\right\}. \tag{3.6.3}$$

Remark 3.6.1. This result can be used to prove Tricot's formula for the box dimension of the graph of the chirp $y = x^\alpha \sin x^{-\beta}$ near the origin (see [Tri3, p. 121]):

$$\dim_B \mathrm{Gr}(y) = \max\left\{1, 2 - \frac{1+\alpha}{1+\beta}\right\}. \tag{3.6.4}$$

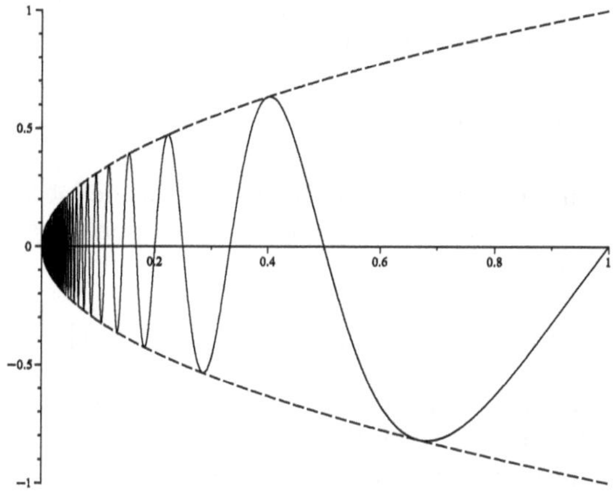

Fig. 3.4 The bounded $(1/2,1)$-chirp defined by $f(x) = x^{1/2}\sin(\pi x^{-1})$, $0 < x < 1$; its graph has box-dimension equal to $5/4$.

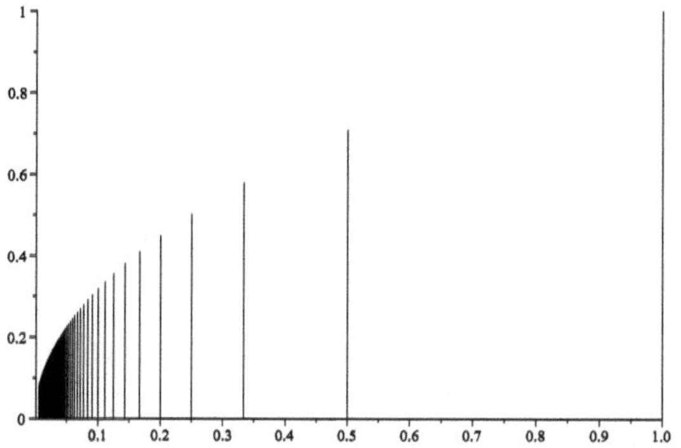

Fig. 3.5 The geometric $(1/2,1)$-chirp; its box-dimension is the same as the box-dimension of the graph of the function in Figure 3.4, and hence is equal to $5/4$.

Observe that the box dimension is nontrivial (i.e., larger than 1) if and only if $\alpha < \beta$. Furthermore, we note that in Section 3.6.2, the notion of a geometric chirp will be generalized to 2-strings.

Let $R_3(A)$ be the subset of $\mathbb{R}^3$ obtained by rotating the geometric (α,β)-chirp A with respect to the vertical axis. We have

$$R_3(A) = \{(x,z) \in \mathbb{R}^2 \times \mathbb{R} : (|x|,z) \in A\},$$

where $|x|$ denotes the Euclidean norm of $x \in \mathbb{R}^2$. This set is a simplified geometric imitation of the (α,β)-*chirp wave* in $\mathbb{R}^3$ defined as the graph of the spherically symmetric function $z(r,\theta) = r^\alpha \sin r^{-1/\beta}$, where we have used polar coordinates (r,θ) in the plane. Using a procedure similar to the above one, we obtain the value of the box dimension of the geometric (α,β)-chirp-like surface in $\mathbb{R}^3$:

$$\dim_B R_3(A) = \max\left\{2, 3 - \frac{2+\alpha}{1+\beta}\right\}. \tag{3.6.5}$$

This box dimension of the surface is nontrivial (i.e., larger than 2) if and only if $\beta - \alpha > 1$.

More generally, if we view $\mathbb{R}^{N+1}$ as $\mathbb{R}^N \times \mathbb{R}$, let us define $R_{N+1}(A)$ as the N-dimensional surface (where N is the topological dimension), obtained by rotating A around the vertical axis:

$$R_{N+1}(A) = \{(x,z) \in \mathbb{R}^{N+1} : (|x|, z) \in A\}, \tag{3.6.6}$$

where $|x|$ is the Euclidean norm of $x \in \mathbb{R}^N$. Note that this surface has countably many connected components. The following result extends Tricot's formula (3.6.4) to N-dimensional chirp-like spherically symmetric surfaces.

Proposition 3.6.2. *The spherically symmetric geometric* (α,β)*-chirp surface in* $\mathbb{R}^{N+1}$ *defined by (3.6.6) and (3.6.1) has box dimension*

$$\overline{\dim}_B R_{N+1}(A) = \max\left\{N, N+1 - \frac{N+\alpha}{1+\beta}\right\}. \tag{3.6.7}$$

Proof. Note that $R_{N+1}(A) = \cup_{k\geq 1} R_{N+1}(A_k)$. We introduce the spherical coordinate system of $\mathbb{R}^N$, i.e., $(r, \theta_1, \dots, \theta_{N-1})$, with respect to the origin. Let

$$\Omega_k := R_{N+1}\left(\left(a_k - \frac{\ell_k}{2}, a_k + \frac{\ell_{k-1}}{2}\right) \times (0, b_k)\right).$$

Passing to the variable $\rho = a_k - r$ in the inner part of Ω_k with respect to $R_{N+1}(A_k)$, and to $\rho = r - a_k$ in the outer part, the zeta function of the surface $R_{N+1}(A_k)$ with respect to Ω_k satisfies

$$\begin{aligned} \zeta_{R_{N+1}(A_k),\Omega_k}(s) &\sim \int_0^{b_k} \mathrm{d}z \int_0^{\ell_k/2} \rho^{s-(N+1)} N\omega_N (a_k - \rho)^{N-1} \mathrm{d}\rho \\ &\quad + \int_0^{b_k} \mathrm{d}z \int_0^{\ell_{k-1}/2} \rho^{s-(N+1)} N\omega_N (\rho - a_k)^{N-1} \mathrm{d}\rho \\ &\sim \frac{1}{s-N} a_k^{N-1} b_k \ell_k^{s-N}, \end{aligned} \tag{3.6.8}$$

where $r^{N-1} = |a_k - \rho|^{N-1}$ is the Jacobian, ω_N is the N-dimensional Lebesgue measure of the unit ball in $\mathbb{R}^N$, and so $N\omega_N$ is the $(N-1)$-dimensional Lebesgue measure of the unit sphere in $\mathbb{R}^N$. In order to justify the last equivalence in (3.6.8), we first note that a_k^{N-1} corresponds to the obvious term in the Jacobian. Upon integration, every other term $a_k^{N-1-j} r^j$, for fixed $j \geq 1$, generates the series

$$\zeta_j(s) \sim \frac{1}{s-N+j} \sum_{k=1}^{\infty} \ell_k^{s-N+j} a_k^{N-1} b_k \sim \frac{1}{s-N+j} \sum_k k^{-[\frac{1+\beta}{\beta}(s-N+j) + \frac{N-1}{\beta} + \frac{\alpha}{\beta}]}.$$

We have $D(\zeta_j) = \max\{N - j, N+1-j-\frac{N-j+\alpha}{1+\beta}\} < N$, and this number is clearly smaller than $D(\zeta_0) = \max\{N, N+1-\frac{N+\alpha}{1+\beta}\} \geq N$. We reason analogously for the term in (3.6.8) containing ℓ_{k-1} instead of ℓ_k. Therefore, the zeta function of the whole surface satisfies

$$\zeta_{R_{N+1}(A)}(s) \sim \zeta_0(s) \sim \frac{1}{s-N} \sum_{k=1}^{\infty} a_k^{N-1} b_k \ell_k^{s-N}. \tag{3.6.9}$$

Using the fact that

$$\zeta_{R_{N+1}(A)}(s) \sim \frac{1}{s-N} \sum_{k=1}^{\infty} k^{-\frac{N-1+\alpha+(1+\beta)(s-N)}{\beta}}, \tag{3.6.10}$$

we deduce that $D(\zeta_{R_{N+1}(A)}) = \max\{N, N+1-\frac{N+\alpha}{1+\beta}\}$. The claim now follows from Corollary 2.1.63. □

It can be shown that in Proposition 3.6.2, the box dimension $\dim_B R_{N+1}(A)$ exists.

Remark 3.6.3. Let the spherically symmetric function $z : \mathbb{R}^N \to \mathbb{R}$ be defined by $z(x) = |x|^\alpha \sin |x|^{-1/\beta}$, $0 < |x| \leq 1$. We expect that the box dimension of its graph in $\mathbb{R}^{N+1}$ is also given by the right-hand side of (3.6.7):

$$\dim_B \mathrm{Gr}(z) = \max\left\{N, N+1-\frac{N+\alpha}{1+\beta}\right\}, \tag{3.6.11}$$

which would extend Tricot's formula (3.6.4). This will be the subject of a further investigation. Note that the box dimension of the surface is nontrivial (i.e., larger than N) if and only if $\beta - \alpha > N-1$.

We can modify the set A in (3.6.1) as follows:

$$A = \bigcup_{k \geq 1} A_k, \quad A_k = \{1 - k^{-1/\beta}\} \times (0, k^{-\alpha/\beta}). \tag{3.6.12}$$

The box dimension of A is clearly the same as in (3.6.3) since the set A is obtained by reflection of an (α, β)-chirp with respect to the vertical line $x = 1/2$. Let us consider the corresponding set $R_{N+1}(A)$, defined by (3.6.6). Using the method of zeta functions, as in the proof of Proposition 3.6.2, we obtain that

$$\overline{\dim}_B R_{N+1}(A) = \max\left\{N, N+1-\frac{1+\alpha}{1+\beta}\right\}. \tag{3.6.13}$$

It can be shown that the graph $\mathrm{Gr}(z)$ of the function

$$z(x) = (1-|x|)^\alpha \sin(1-|x|)^{-1/\beta}, \quad x \in \mathbb{R}^N, \quad 0 < |x| < 1,$$

has the same box dimension:

$$\dim_B \mathrm{Gr}(z) = \max\left\{N, N+1-\frac{1+\alpha}{1+\beta}\right\}. \tag{3.6.14}$$

We note that this formula was also proved by Naito, Pašić, Tanaka and the third author in [NaPaTaŽu, Proposition 1.1 and Example 1.1], by using different methods.

3.6.2 Multiple Strings and String Chirps

In the present section, we first show that the geometric chirp in the plane can be generated by two (fractal) strings, that is, by a so-called 2-string. Assume that two bounded fractal strings $\mathscr{L} = (\ell_j)_{j\geq 1}$ and $\mathscr{M} = (m_j)_{j\geq 1}$ are given; that is, $\mathscr{L}$ and $\mathscr{M}$ are two nonincreasing sequences of positive numbers (ℓ_j) and (m_j) with finite sums. For each $k \geq 1$, let $a_k := \sum_{j\geq k} \ell_j$ and $b_k := \sum_{j\geq k} m_j$. Here, instead, it is more natural to think of two monotone sequences of positive numbers (a_k) and (b_k) given in advance, both converging to zero, so that the sequences $\ell_k = a_k - a_{k+1}$ and $m_k = b_k - b_{k+1}$ are monotone.

(a) Given an arbitrary set $\mathscr{L} \subset \mathbb{R}^N$, we let $\partial_0 \mathscr{L} := \partial \mathscr{L} \cap \mathrm{Int}\,\overline{\mathscr{L}}$ and call it the *inner boundary* of $\mathscr{L}$. We define a 2-*string* $\mathscr{L}_2 = \mathscr{L}_2(\mathscr{L}, \mathscr{M})$ as the union of the interiors of the convex hulls of $A_k \cup A_{k+1}$, for all $k \geq 1$:

$$\mathscr{L} = \bigcup_{k=1}^{\infty} \mathrm{Int}\,(\mathrm{conv}\,(A_k \cup A_{k+1})), \tag{3.6.15}$$

where for each $k \geq 1$, A_k is the vertical interval in $\mathbb{R}^2$ defined by $A_k = \{a_k\} \times (0, b_k)$. Each set $\mathrm{Int}\,(\mathrm{conv}\,(A_k \cup A_{k+1}))$ is a connected component of $\mathscr{L}$, which we call the *k-th slice* of $\mathscr{L}$. It is clear that the set

$$A = \bigcup_{k=1}^{\infty} A_k \tag{3.6.16}$$

is the inner boundary of the 2-string $\mathscr{L}_2$, which we can view as the geometric chirp associated with the 2-string. We call it the 2-*string chirp*. Similarly as above, we can show that its zeta function satisfies

$$\zeta_A(s) \sim \frac{1}{s-1} \sum_{k=1}^{\infty} \ell_k^{s-1} b_k, \tag{3.6.17}$$

and the corresponding upper box dimension is $\overline{\dim}_B A = \max\{1, D(\zeta_A)\}$. Special cases of this situation are the (α, β)-geometric chirps from the beginning of Subsection 3.6.1, where $a_k = k^{-1/\beta}$ and $b_k = k^{-\alpha/\beta}$.

(b) Analogously, if we have three given strings $\mathscr{L} = (\ell_j)$, $\mathscr{M} = (m_j)$ and $\mathscr{N} = (n_j)$ which generate monotone sequences $(a_k)_k$, $(b_k)_k$ and $(c_k)_k$ converging to zero as $k \to \infty$, where $\ell_k = a_k - a_{k+1}$, $m_k = b_k - b_{k+1}$ and $n_k = c_k - c_{k+1}$, then we can define the sequence of rectangles $A_k = \{a_k\} \times (0, b_k) \times (0, c_k)$ in $\mathbb{R}^3$, and the corresponding 3-string (3.6.15), with the set A defined by (3.6.16). Then

$$\zeta_A(s) \sim \frac{1}{s-2} \sum_{k=1}^{\infty} \ell_k^{s-2} b_k c_k. \tag{3.6.18}$$

We can think of the 3-string $\mathscr{L}_3 = \mathscr{L}_3(\mathscr{L}, \mathscr{M}, \mathscr{N})$ as a loaf of bread cut into thinner and thinner slices.

(c) Let N be any integer ≥ 2. Assume that N strings $(\ell_j^{(1)})_j, \ldots, (\ell_j^{(1)})_j$ are given. They generate the sequences $(a_k^{(1)})_k, \ldots, (a_k^{(N)})_k$, where $a_k^{(i)} := \sum_{j \geq k} \ell_j^{(i)}$ for each $k \geq 1$. Then, we can define the sequence of $(N-1)$-dimensional slices as quadrilaterals:

$$A_k = \{a_k\} \times (0, a_k^{(2)}) \times \cdots \times (0, a_k^{(N)}),$$

contained in the hyperplane $\{a_k\} \times \mathbb{R}^{N-1}$ of $\mathbb{R}^N$, and the corresponding N-string $\mathscr{L}$ defined by (3.6.15). Then, for the associated N-string chirp A defined by (3.6.16), we have:

$$\zeta_A(s) \sim \frac{1}{s-N+1} \sum_{k=1}^{\infty} (\ell_k^{(1)})^{s-N+1} a_k^{(2)} \ldots a_k^{(N)}. \tag{3.6.19}$$

In light of Corollary 2.1.63, the corresponding box dimension is given by

$$\dim_B A = \max\{N-1, D(\zeta_A)\}. \tag{3.6.20}$$

Example 3.6.4. If we take α_i-strings $\mathscr{L}_i = (\ell_k^{(i)})$, with $\ell_k^{(i)} = a_k^{(i)} - a_{k+1}^{(i)}$ for each $k \geq 1$, generated by $a_k^{(i)} = k^{-\alpha_i}$, $i = 1, \ldots, N$, where α_i are positive numbers, then for the associated N-string chirp defined by (3.6.16), we deduce from (3.6.19) that

$$\zeta_A(s) \sim \frac{1}{s-N+1} \sum_{k=1}^{\infty} k^{-[(s-N+1)(\alpha_1+1)+\alpha_2+\cdots+\alpha_N]}.$$

The series converges for $\operatorname{Re} s > D$, where $(D-N+1)(\alpha_1+1)+\alpha_2+\cdots+\alpha_N = 1$. Using (3.6.20), we obtain that

$$\begin{aligned} \overline{\dim}_B A &= \max\left\{N-1, N-1+\frac{1-\alpha_2-\cdots-\alpha_N}{1+\alpha_1}\right\} \\ &= \max\left\{N-1, N-\frac{\alpha_1+\alpha_2+\cdots+\alpha_N}{1+\alpha_1}\right\}. \end{aligned} \tag{3.6.21}$$

The value of the upper box dimension lies in the interval $(N-1, N)$ if and only if $\alpha_2 + \cdots + \alpha_N < 1$. Note that it can be shown that the box dimension $\dim_B A$ exists.

3.6.3 Zeta Functions and Cartesian Products of Fractal Strings

Let $\mathscr{L} = (\ell_j)_{j\ge1}$ and $\mathscr{M} = (m_k)_{k\ge1}$ be two bounded fractal strings such that both sequences are nonincreasing, $\ell_j > 0$, $m_k > 0$, and $a_1 = \sum_{j=1}^{\infty} \ell_j < \infty$ and $b_1 = \sum_{k=1}^{\infty} m_k < \infty$. We identify the sequence $\mathscr{L} = (\ell_j)_{j\ge1}$ with the family $\mathscr{L} = (I_j)_{j\ge1}$ of subsets of $[0,a_1]$ consisting of bounded open intervals I_j of length ℓ_j, written in nonincreasing order starting from the right endpoint a_1 and ending at the left endpoint 0 of the interval $[0,a_1]$. We proceed analogously for $\mathscr{M} = (J_k)_{k\ge1}$. Let $a_j := \sum_{i\ge j} \ell_i$, $b_k := \sum_{i\ge k} m_i$, and $A := (a_j)_{j\ge1}$, $B := (b_k)_{k\ge1}$. We define $\overline{\dim}_B \mathscr{L} = \overline{\dim}_B A$ and $\overline{\dim}_B \mathscr{M} = \overline{\dim}_B B$.

The *Cartesian product of the fractal strings* $\mathscr{L}$ and $\mathscr{M}$, defined by

$$\mathscr{L} \times \mathscr{M} = \{I_j \times J_k\}_{j,k\ge1}, \tag{3.6.22}$$

consists of the countable disjoint family of rectangles $R_{jk} = I_j \times I_k$ densely covering the rectangle $[0,a_1] \times [0,b_1]$. Now, denote by $\partial(\mathscr{L} \times \mathscr{M})$ the boundary of the union of all these rectangles: $\partial(\mathscr{L} \times \mathscr{M}) = \partial(\cup_{j,k} R_{jk})$. It is clear that

$$\partial(\mathscr{L} \times \mathscr{M}) = (\overline{A} \times [0,b_1]) \cup ([0,a_1] \times \overline{B}), \tag{3.6.23}$$

and $\overline{A} = A \cup \{0\}$, $\overline{B} = B \cup \{0\}$. See Figure 3.6. In the following theorem, we compute the zeta function of $\partial(\mathscr{L} \times \mathscr{M})$.

Theorem 3.6.5. *Let $\mathscr{L} = (I_j)_{j\ge1}$ and $\mathscr{M} = (J_k)_{k\ge1}$ be two fractal strings, $|I_j| = \ell_j$, $|J_k| = m_k$, where $(\ell_j)_{j\ge1}$ and $(m_k)_{k\ge1}$ are nonincreasing sequences of positive real numbers. Then, for $E := \partial(\mathscr{L} \times \mathscr{M})$, we have*

$$\zeta_E(s) \sim \sum_{j,k=1}^{\infty} \left[|\ell_j - m_k| \min\{\ell_j, m_k\}^{s-1} + \frac{2}{s} \min\{\ell_j, m_k\}^s \right]. \tag{3.6.24}$$

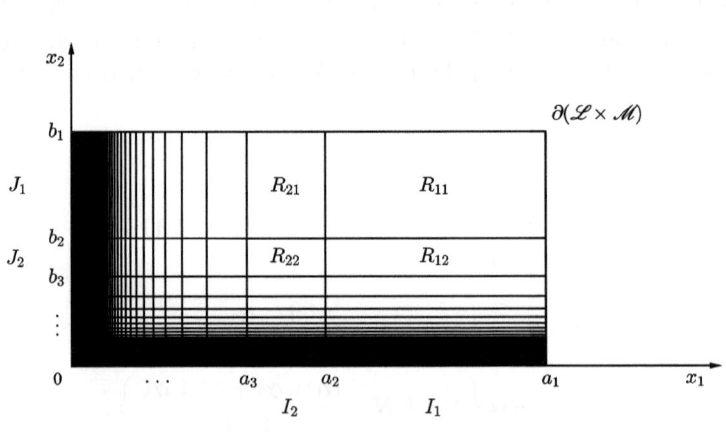

Fig. 3.6 The boundary $\partial(\mathscr{L} \times \mathscr{M})$ of the Cartesian product $\mathscr{L} \times \mathscr{M}$ of two fractal strings $\mathscr{L} = (\ell_j)_{j\ge1}$ and $\mathscr{M} = (m_k)_{k\ge1}$.

Furthermore,

$$\overline{\dim}_B E = 1 + \max\{\overline{\dim}_B \mathscr{L}, \overline{\dim}_B \mathscr{M}\}, \tag{3.6.25}$$

and this value is equal to the abscissa of convergence of the zeta function ζ_E.

Proof. Let us consider a typical rectangle translated at the origin, $R = (0,\ell) \times (0,m)$ with $m > \ell$. If we choose a point $T \in R$, then we consider the distance function $d(T,\partial R)$. We split $\overline{R}$ into the union of eight right-angle triangles with sides $\ell/2$ and $m/2$, and two smaller rectangles with sides $\ell/2$ and $m-\ell$, placed in the middle of R. Denoting a typical triangle by $\triangle$, and placing it in an (x,y)-coordinate system so that $\triangle = \{(x,y) : 0 \le x \le \ell/2;\ 0 \le y \le x\}$, we find that the zeta function of its side $S = [0,\ell/2] \times \{0\}$ relative to $\triangle$ (more information about relative zeta functions can be found in Section 4.1 below) is given by

$$\zeta_{S,\triangle}(s) = \iint_{\triangle} y^{s-2} \mathrm{d}x\mathrm{d}y = \frac{1}{s(s-1)} \left(\frac{\ell}{2}\right)^{s}. \tag{3.6.26}$$

Similarly, the zeta function of the vertical side $V = \{0\} \times [0, m-\ell]$ with respect to the rectangle $R' = [0,\ell/2] \times [0, m-\ell]$ is given by

$$\zeta_{V,R'}(s) = \iint_{R'} x^{s-2} \mathrm{d}x\mathrm{d}y = \frac{m-\ell}{s-1} \left(\frac{\ell}{2}\right)^{s-1}. \tag{3.6.27}$$

Therefore, we can compute

$$\begin{aligned} \zeta_{\partial R,R}(s) &= 2\zeta_{V,R'}(s) + 8\zeta_{S,\triangle}(s) \\ &= \frac{|m-\ell|}{(s-1)2^{s-2}} \min\{\ell,m\}^{s-1} + \frac{1}{s(s-1)2^{s-3}} \min\{\ell,m\}^{s}. \end{aligned} \tag{3.6.28}$$

Using the definition of the distance zeta function in (2.1.1), we can then write

$$\begin{aligned} \zeta_E(s) &= \sum_{j,k=1}^{\infty} \zeta_{\partial R_{jk},R_{jk}}(s) + g(s) \\ &= \sum_{j,k=1}^{\infty} \left[\frac{|\ell_j - m_k|}{(s-1)2^{s-2}} \min\{\ell_j, m_k\}^{s-1} \right. \\ &\quad \left. + \frac{1}{s(s-1)2^{s-3}} \min\{\ell_j, m_k\}^{s} \right] + g(s). \end{aligned} \tag{3.6.29}$$

Here, the term $g(s)$ is unimportant, and for $\delta \ge \frac{1}{2}\max\{\ell_1, m_1\}$, its value is given by

$$\begin{aligned} g(s) &= \iint_{E_\delta \setminus ([0,a_1]\times[0,b_1])} d((x,y), \partial([0,a_1]\times[0,b_1]))^{s-2} \mathrm{d}x\mathrm{d}y \\ &= \frac{2\delta^{s-1}}{s-1}(a_1 + b_1 + \pi). \end{aligned}$$

This follows easily by splitting the domain of integration into four rectangles and four right-angle triangles. Since $g(s)$ is holomorphic, this proves (3.6.24); see Proposition 2.1.76. The second claim follows easily using (3.6.23) and the property of finite stability of the upper box dimension:

$$\begin{aligned}\overline{\dim}_B\partial(\mathscr{L}\times\mathscr{M}) &= \overline{\dim}_B((\overline{A}\times[0,b_1])\cup([0,a_1]\times\overline{B}))\\ &= \max\{\overline{\dim}_B(\overline{A}\times[0,b_1]),\overline{\dim}_B([0,a_1]\times\overline{B})\}\\ &= \max\{1+\overline{\dim}_BA,1+\overline{\dim}_BB\}.\end{aligned}$$

This concludes the proof of Theorem 3.6.5. □

Remark 3.6.6. Any fractal string $\mathscr{L}=(\ell_j)_{j\geq 1}$ (viewed here as a sequence of 'lengths' or 'scales') can be identified with the measure $\eta=\sum_{j\geq 1}\delta_{\ell_j^{-1}}$, and its natural generalization is the weighted string $\eta=\sum_{j\geq 1}w_j\delta_{l_j^{-1}}$, where $(w_j)_{j\geq 1}$ is a sequence of positive real numbers and $(l_j)_{j\geq 1}$ is a decreasing sequence of positive real numbers (corresponding, when the numbers w_j are integers, to the distinct values of the lengths ℓ_j); see [Lap-vFr3, Section 4.1, page 121, and Assumption (P) on page 307], along with Remark 4.1.4. Using the Dirichlet integral $\zeta_\eta(s)=\int_0^{+\infty}x^{-s}\eta(\mathrm{d}x)$ (see [Lap-vFr3, Equation (4.4)]) and viewing η as a positive local measure (as in [Lap-vFr3, Chapter 4] and Definition A.1.1 of Appendix A below), we obtain the following weighted (generalized) Dirichlet series:[15]

$$\zeta_\eta(s)=\sum_{j=1}^{\infty}w_jl_j^s.$$

The distance zeta function of $E=\partial(\mathscr{L}\times\mathscr{M})$ in Theorem 3.6.5 is equivalent to a weighted zeta function of the string $(\ell_j)_{j\geq 1}\cup(m_k)_{k\geq 1}$. Indeed, it follows at once from (3.6.24) that, up to equivalence (in the sense of Definition 2.1.69), $\zeta_E(s)$ can be written as the following weighted Dirichlet series:[16]

$$\zeta_E(s)\sim\sum_j w_j^{(l)}\ell_j^s+\sum_{\{k:m_k\neq\ell_j,\forall j\}}w_k^{(m)}m_k^s,$$

where the weights are respectively given by

$$w_j^{(l)}:=\ell_j^{-1}\sum_{\{k:m_k>l_j\}}(m_k-\ell_j)+\frac{2}{s}(\#\{k:m_k\geq\ell_j\})$$

---

15 In light of Example 2.1.44 above, ζ_η is a tamed DTI, in the sense of Subsection 2.1.3.2 or of Definitions A.1.2 and A.1.3 of Appendix A.

16 Through the end of this discussion, $(m_k)_{k=1}^\infty$ is viewed as a decreasing sequence of positive real numbers with an associated sequence of generalized multiplicities denoted by $(w_k^{(m)})_{k=1}^\infty$.

and

$$w_k^{(m)} := m_k^{-1} \sum_{\{j:\ell_j>m_k\}} (\ell_j - m_k) + \frac{2}{s}(\#\{j : \ell_j > m_k\}),$$

with $\#D$ denoting the cardinality of the finite set D.

Corollary 3.6.7. *Assume that a and b are positive real numbers. Let $\mathscr{L} = (I_j)_{j\geq 1}$ and $\mathscr{M} = (J_k)_{k\geq 1}$ be a and b-strings, respectively, so that $|I_j| \sim a\, j^{-a-1}$, $|J_k| \sim b\, k^{-b-1}$, where the sequences of lengths $(|I_j|)_{j\geq 1}$ and $(|J_k|)_{k\geq 1}$ are nonincreasing. Then, for $E = \partial(\mathscr{L} \times \mathscr{M})$, we have*

$$\zeta_E(s) \sim \sum_{j,k=1}^{\infty} \Big[a\, j^{-a-1} - b\, k^{-b-1} | \min\{a\, j^{-a-1}, b\, k^{-b-1}\}^{s-1} + \frac{2}{s} \min\{a\, j^{-a-1}, b\, k^{-b-1}\}^{s} \Big], \tag{3.6.30}$$

and the abscissa of convergence of the zeta function ζ_E is given by

$$\dim_B E = 1 + \frac{1}{1+\min\{a,b\}}. \tag{3.6.31}$$

(Note that $\dim_B E$ exists in this case.) Furthermore, the set E is Minkowski nondegenerate, and $\zeta_E(s) \to +\infty$ as $\mathbb{R} \ni s \to \dim_B E$ from the right.

Proof. The claim follows from Theorem 3.6.5 and the fact that $\dim_B(\{k^{-a} : k \geq 1\}) = 1/(1+a)$; see [Lap2, Example 5.1 and Appendix C], [Lap-vFr3, Subsection 6.5.1], [Tri3, p. 25] or [LapPo2, Theorem 2.4] for a more general statement. Minkowski nondegeneracy of E follows from the fact that this property is preserved under Cartesian products; see [KraPa, Theorem 3.3.6] or [Žu4, Proposition 4.3] for a more general statement involving gauge functions. For the remaining part, see Theorem 2.1.11(c). □

At this stage, we do not have any information about the possible complex dimensions of $\partial(\mathscr{L} \times \mathscr{M})$ in Theorem 3.6.5.

Remark 3.6.8. Assume that given bounded sets A and B in (possibly different) Euclidean spaces, we know the corresponding distance zeta functions ζ_A and ζ_B. We do not know how the zeta function of the Cartesian product $\zeta_{A\times B}$ is related to them. We do not know this even in the case of the a-string for which $A = B = \{k^{-a} : k \geq 1\}$, where $a > 0$. What does $\zeta_{A\times A}$ look like in this case? This zeta function converges for $\operatorname{Re} s > 2/(1+a)$ and $\zeta_{A\times A}(s) \to +\infty$ as $\mathbb{R} \ni s \to 2/(1+a)$ from the right (see Theorem 2.1.11), since $A \times A$ is Minkowski nondegenerate and $\dim_B(A \times A) = 2/(1+a)$ (see [KraPa, Theorem 3.3.6]). For a related open problem, see Problem 6.2.7 on page 556 below.

3.7 Zigzagging Fractal Sets and Alternating Zeta Functions

In this section, we construct a class of sets A in $[0,1]$ with different upper and lower box dimensions, and we compute their distance zeta functions. Such sets have a *nonuniform oscillating nature*; we call them *zigzagging fractals*. It is a result of intermittent spraying and swarming during their construction; see below. We stress that this kind of oscillations of fractals is different from the type of oscillations discussed and analyzed, for example, in [Lap-vFr1-3]. For an even more general definition of zigzagging fractals, see [Žu4, Remark 1.7].

Definition 3.7.1. Let $(n_j)_{j\geq 1}$ be a given sequence of positive integers. We construct a family of disjoint open intervals I_j in $[0,1]$, and define the corresponding set A to be the boundary of the union. All the intervals will have endpoints on the binary grid, that is, the intervals will be of the form $(k2^{-j},(k+1)2^{-j})$, where $k\in\mathbb{N}_0$, $j\in\mathbb{N}$. Starting from $x=0$ to the right, we take the first n_1 consecutive intervals I_j, $j=1,\dots,n_1$, of lengths 2^{-j}. We then say that we have *sprayed* n_1 intervals in $[0,1]$. In the next n_2 steps, we consecutively halve the remaining portion of the unit interval of length 2^{-n_1}, so that in the last step, we obtain 2^{n_2} new subintervals (we call this n_2*-swarming*). In each of these subintervals of length $2^{-(n_1+n_2)}$, we then spray n_3 new open intervals having lengths $2^{-(n_1+n_2+1)},\dots,2^{-(n_1+n_2+n_3)}$. In the remaining parts of each of the subintervals, we do the n_4-swarming, and so on, by intermittently spraying and swarming. The family of open intervals $\mathscr{L}$ obtained in this way is a fractal string contained in $[0,1]$, and the boundary of the union is the *zigzagging set* A.

If we let $|I_j|=\ell_j$ and use (2.1.82) and (2.1.83), we can compute the distance zeta function of A as follows (we drop two inessential terms corresponding to $x=0$ and 1):

$$\begin{aligned}\zeta_A(s)&\sim\sum_{j=1}^{\infty}\int_{I_j}d(x,\partial I_j)^{s-1}\mathrm{d}x=s^{-1}2^{1-s}\sum_{j=1}^{\infty}\ell_j^s\\&=\frac{2^{1-s}}{s}\Bigg(\sum_{k=1}^{n_1}2^{-ks}+2^{n_2}\sum_{k=1}^{n_3}2^{-(k+n_1+n_2)s}+\\&\qquad 2^{n_2+n_4}\sum_{k=1}^{n_3}2^{-(k+n_1+n_2+n_3+n_4+n_5)s}+\cdots\Bigg);\end{aligned}\tag{3.7.1}$$

that is,

$$\begin{aligned}\zeta_A(s)\sim\frac{2^{1-s}}{s(2^s-1)}\Big(&1-2^{-sn_1}+2^{-s(n_1+n_2)+n_2}-2^{-s(n_1+n_2+n_3)+n_2}\\&+2^{-s(n_1+n_2+n_3+n_4)+n_2+n_4}-2^{-s(n_1+n_2+n_3+n_4+n_5)+n_2+n_4}+\cdots\Big).\end{aligned}$$

Substituting

$$m_k = n_1 + n_2 + \cdots + n_k, \quad e_k = n_2 + n_4 + \cdots + n_{2k},$$

we deduce that the distance zeta function of A can be represented as the following alternating series of complex numbers, which we call an *alternating zeta function*:

$$\begin{aligned}\zeta_A(s) \sim 1 - 2^{-sm_1} + 2^{-sm_2+e_1} - 2^{-sm_3+e_1} + \\ +2^{-sm_4+e_2} - 2^{-sm_5+e_2} + \cdots + 2^{-sm_{2k}+e_k} - 2^{-sm_{2k+1}+e_k} + \cdots . \quad (3.7.2)\end{aligned}$$

More precisely, this is an alternating zeta function generated by the fractal string $\mathscr{L}' = (2^{-m_k})_{k\geq 0}$ (we define $m_0 = 0$) with weights (or multiplicities) $(w_k)_{k\geq 0}$, since (3.7.2) can be written as

$$\zeta_A(s) \sim \sum_{k=0}^{\infty} (-1)^k w_k 2^{-sm_k}, \tag{3.7.3}$$

where $w_k := 2^{e_{\lfloor k/2 \rfloor}}$ and the index $\lfloor k/2 \rfloor$ is the integer part of $k/2$.

Let $\theta_k = \theta_k(A)$ be the number of intervals of the 2^{-k}-grid in $[0,1]$ which have nonempty intersection with A. It is well known that for $s_k := \log_2 \theta_k$ (the *swarming sequence* associated to A, according to the terminology of [Žu4, Section 1]), we have

$$\underline{\dim}_B A = \liminf_{k\to\infty} \frac{s_k}{k}, \quad \overline{\dim}_B A = \limsup_{k\to\infty} \frac{s_k}{k}; \tag{3.7.4}$$

see [Fal1, p. 41] or [Tri3, 24]. Now we compute the sequence θ_k. It is easy to see that during the first spraying, we have $\theta_1 = 2$, $\theta_2 = 4$, $\theta_3 = 6, \ldots, \theta_{n_1} = 2n_1$. Furthermore, during the ensuing swarming, we have $\theta_{n_1+1} = 2n_1 + 2$, $\theta_{n_1+2} = 2n_1 + 2^2, \ldots$, $\theta_{n_1+n_2} = 2n_1 + 2^{n_2}$. The second spraying yields $\theta_{n_1+n_2+n_3} = 2n_1 + 2^{n_2}(2n_3) = 2(n_1 + 2^{n_2}n_3)$. The second swarming results in $\theta_{n_1+n_2+n_3+n_4} = 2(n_1 + 2^{n_2}n_3) + 2^{n_2+n_4}$, etc. We can now recognize the general pattern: for each $k \geq 0$, we have

$$\theta_{m_{2k}} = 2(n_1 + 2^{e_1}n_3 + \cdots + 2^{e_k}n_{2k-1}) + 2^{e_k}, \tag{3.7.5}$$

$$\theta_{m_{2k+1}} = 2(n_1 + 2^{e_1}n_3 + \cdots + 2^{e_k}n_{2k-1} + 2^{e_k}n_{2k+1}). \tag{3.7.6}$$

We note that the mapping $j \mapsto \theta_j$ has intermittent exponential and linear growth rate. More precisely, it is of exponential growth rate for $j \in \{m_{2k-1}, \ldots, m_{2k}\}$ and of linear growth for $j \in \{m_{2k}, \ldots, m_{2k+1}\}$. In other words, odd indices in m_j correspond to switching from linear to exponential growth of the sequence $(\theta_j)_{j\geq 1}$, while even indices correspond to switching from exponential to linear growth.

We are now ready to state and prove the main result of this section.

Theorem 3.7.2. *Let $(n_j)_{j\geq 1}$ be an increasing sequence of positive integers. Let A be the corresponding zigzagging set, as given in Definition 3.7.1. Then*

$$\overline{\dim}_B A = \limsup_{k\to\infty} \frac{n_2 + n_4 + \cdots + n_{2k}}{n_1 + n_2 + \cdots + n_{2k}} \tag{3.7.7}$$

and

$$\underline{\dim}_B A = \liminf_{k\to\infty} \frac{n_2+n_4+\cdots+n_{2k}}{n_1+n_2+\cdots+n_{2k}+n_{2k+1}}. \tag{3.7.8}$$

Furthermore, the corresponding distance zeta function ζ_A is equivalent to the expression given by (3.7.2). Moreover, we have that $\zeta_A(s) \to +\infty$ as $s \to \overline{\dim}_B A$ from the right, with $s \in \mathbb{R}$.

Proof. Since the sequence n_k is nondecreasing, then using (3.7.5) and (3.7.6) we have

$$2^{e_k} \le \theta_{m_{2k}} \le 2(k+1)2^{e_k} n_{2k-1},$$
$$2^{e_k} \le \theta_{m_{2k+1}} \le 2(k+1)2^{e_k} n_{2k+1}.$$

From this we easily deduce that

$$s_{m_{2k}} \sim e_k, \quad s_{m_{2k+1}} \sim e_k \tag{3.7.9}$$

as $k \to \infty$, since

$$e_k \le s_{m_{2k}} \le e_k + \log n_{2k-1} + \log 2(k+1),$$
$$e_k \le s_{m_{2k+1}} \le e_k + \log n_{2k+1} + \log 2(k+1),$$

and (assuming that $k \ge 2$),

$$0 < \frac{\log n_{2k-1}}{m_{2k}} \le \frac{\log m_{2k}}{m_{2k}} \to 0, \quad 0 < \frac{\log(k+1)}{m_{2k}} \le \frac{\log(k+1)}{2k} \to 0,$$

$$0 < \frac{\log n_{2k+1}}{m_{2k+1}} \le \frac{\log m_{2k+1}}{m_{2k+1}} \to 0, \quad 0 < \frac{\log(k+1)}{m_{2k+1}} \le \frac{\log(k+1)}{2k+1} \to 0$$

as $k \to \infty$, due to the lower bound $m_j \ge j$. Let C and D be the right-hand sides in (3.7.8) and (3.7.7), respectively; that is,

$$C = \liminf_{k\to\infty} \frac{e_k}{m_{2k+1}}, \quad D = \limsup_{k\to\infty} \frac{e_k}{m_{2k}}. \tag{3.7.10}$$

In order to prove the theorem, we have to show that for any sequence (j_k) of positive integers such that $j_k \to +\infty$,

$$C \le \liminf_{k\to\infty} \frac{s_{j_k}}{j_k} \le \limsup \frac{s_{j_k}}{j_k} \le D. \tag{3.7.11}$$

It suffices to consider the following two cases:

(a) The case when $j_k \in [m_{2k}, m_{2k+1}]$ for all k. Then $s_{j_k} \in [s_{m_{2k}}, s_{m_{2k+1}}]$, and therefore,

$$\frac{s_{m_{2k}}}{m_{2k+1}} \le \frac{s_{j_k}}{j_k} \le \frac{s_{m_{2k+1}}}{m_{2k}}.$$

Using the right-hand side inequality and (3.7.9), we see that

$$\limsup_{k\to\infty}\frac{s_{j_k}}{j_k}\leq\limsup_{k\to\infty}\frac{s_{m_{2k+1}}}{m_{2k}}=\limsup_{k\to\infty}\frac{e_k}{m_{2k}}=D,$$

and similarly, we show that $\liminf_{k\to\infty}\frac{s_{j_k}}{j_k}\geq C$.

(b) The case when $j_k\in[m_{2k-1},m_{2k}]$ for all k. We have $s_{j_k}\in[s_{m_{2k-1}},s_{m_{2k}}]$, therefore,

$$\frac{s_{m_{2k-1}}}{m_{2k}}\leq\frac{s_{j_k}}{j_k}\leq\frac{s_{m_{2k}}}{m_{2k-1}}.$$

Using the right most inequality and (3.7.9), we obtain that

$$\limsup_{k\to\infty}\frac{s_{j_k}}{j_k}\leq\limsup_{k\to+\infty}\frac{s_{m_{2k}}}{m_{2k-1}}\leq\limsup_{k\to\infty}\frac{e_k}{m_{2k}}=D,$$

and similarly, $\liminf_{k\to\infty}\frac{s_{j_k}}{j_k}\geq D$. The claim follows from (a), (b) and (3.7.4).

The last claim of the theorem follows from the fact that the zeta function in (3.7.1) has nonnegative coefficients; see [Ser, Proposition 7, p. 67]. □

We do not have any information about the possible complex dimensions of A in Theorem 3.7.2. It is easy to see that for real s, the absolute values of members of the alternating series $\zeta_A(s)$ in (3.7.2) are nonincreasing if and only if $s\geq 1$. Indeed, since the exponents must be nonincreasing starting from some k_0, we must have $-sm_{2k+1}+e_k\geq -sm_{2k+2}+e_{k+1}$ for all $k\geq k_0$, which is satisfied if and only if $s\geq 1$. Furthermore, for $s\geq 1$, we have $-sm_{2k}+e_k\leq -m_{2k}+e_k=-n_1-n_3-\cdots-n_{2k+1}\to -\infty$, so that $2^{-sm_{2k}+e_k}\to 0$ as $k\to\infty$. Therefore, still for real s, the Leibniz criterion of convergence for the alternating series defining $\zeta_A(s)$ is applicable only for $s\geq 1$.

In the following corollary, as before, we denote by $\lfloor x\rfloor$ the integer part of a real number x. We obtain a class of fractal sets with unequal values for the upper and lower box dimensions and with explicit expressions for the associated zeta functions.

Corollary 3.7.3. *Let $a>1$ be a fixed real number. Let $(n_k)_{k\geq 1}$ be a sequence of positive integers such that $n_k\sim a^k$ as $k\to\infty$ (for example, $n_k=\lfloor a^k\rfloor$,[17] for all $k\geq 1$). Then, for the corresponding zigzagging set A in $[0,1]$ (as given in Definition 3.7.1), we have*

$$\overline{\dim}_BA=\frac{a}{a+1},\quad \underline{\dim}_BA=\frac{1}{a+1},\tag{3.7.12}$$

and the associated distance zeta function $\zeta_A(s)$ is equivalent to the expression given by the alternating series (3.7.2), at least for $\mathrm{Re}\,s>\overline{\dim}_BA$. *(Note that, in light of (3.7.12), $\underline{\dim}_BA<\overline{\dim}_BA$ if and only if $a>1$.)*

Proof. Since $n_k\sim a^k$ as $k\to\infty$, there exist two monotone sequences, (c_k) increasing and (d_k) decreasing, both converging to 1, such that

$$c_ka^k\leq n_k\leq d_ka^k,\tag{3.7.13}$$

[17] Here, for $x\in\mathbb{R}$, we have $\lfloor x\rfloor=[x]$ (the integer part of x, also called the 'floor' of x).

for all $k \geq 1$. Let j be a fixed even positive integer. Then for $k > j$, writing $m_k := m_{j-1} + n_j + \cdots + n_k$ and using (3.7.13), we obtain that

$$m_{2k} \leq m_{j-1} + d_j a^j (1 + a + \cdots + a^{2k-j}) = m_{j-1} + d_j a^j \frac{a^{2k-j+1} - 1}{a - 1}, \tag{3.7.14}$$

and similarly,

$$m_{2k} \geq m_{j-1} + c_j a^j \frac{a^{2k-j+1} - 1}{a - 1}. \tag{3.7.15}$$

Analogously, writing $e_k := e_{j-2} + n_j + n_{j+2} + \cdots + n_{2k}$, we show that

$$e_{j-2} + c_j a^j \frac{a^{2k-j+2} - 1}{a^2 - 1} \leq e_k \leq e_{j-2} + d_j a^j \frac{a^{2k-j+2} - 1}{a^2 - 1}. \tag{3.7.16}$$

Using (3.7.14), (3.7.15) and (3.7.16), we obtain the inequality

$$\frac{c_j a}{d_j (1 + a)} \leq \liminf_{k \to \infty} \frac{e_k}{m_{2k}} \leq \limsup_{k \to \infty} \frac{e_k}{m_{2k}} \leq \frac{d_j a}{c_j (1 + a)}, \tag{3.7.17}$$

for all $k > j$. Letting $j \to \infty$, we deduce that

$$\lim_{k \to \infty} \frac{e_k}{m_{2k}} = \frac{a}{1 + a}. \tag{3.7.18}$$

The first equality in (3.7.12) then follows by using Theorem 3.7.2; see (3.7.7). The second equality in (3.7.12) follows from the first one since $m_{2k+1} \sim a m_{2k}$; see also (3.7.10). □

Assume that the hypotheses of Corollary 3.7.3 are satisfied. It is noteworthy that if $a \to +\infty$, then $\underline{\dim}_B A \to 0^+$ while $\overline{\dim}_B A \to 1^-$. Furthermore, it is possible to construct a set $A \subset [0,1]$ such that we even have $\underline{\dim}_B A = 0$ and $\overline{\dim}_B A = 1$; see [Žu4, Theorem 1.2], as well as [Fra1] and [RoSha].

Chapter 4
Relative Fractal Drums and Their Complex Dimensions

Only yesterday the practical things of today were decried as impractical, and the theories which will be practical tomorrow will always be branded as valueless games by the practical man of today.

William Feller (1906–1970)

Abstract In this chapter, we introduce the notion of relative fractal drums (or RFDs, in short). They represent a simple and natural extension of two fundamental objects of fractal analysis, simultaneously: that of bounded sets in $\mathbb{R}^N$ (i.e., of fractals) and that of bounded fractal strings (introduced by the first author and Carl Pomerance in the early 1990s). Furthermore, there is a natural way to define their associated Minkowski contents and relative distance as well as tube zeta functions. We stress a new phenomenon exhibited by relative fractal drums: namely, their box dimensions can be negative as well (and even equal to $-\infty$). This can be viewed as a property of their 'flatness', since it is related to the loss of the cone property. In short, a relative fractal drum (RFD) consists of an ordered pair (A,Ω), where A is an arbitrary (possibly unbounded) subset of $\mathbb{R}^N$ and Ω is an open subset of $\mathbb{R}^N$ of finite volume and such that $\Omega \subseteq A_\delta$, for some $\delta > 0$. The corresponding zeta function, either a distance or tube zeta function, is denoted by $\zeta_{A,\Omega}$ or $\tilde{\zeta}_{A,\Omega}$, respectively. We show that $\zeta_{A,\Omega}$ and $\tilde{\zeta}_{A,\Omega}$ are connected via a key functional equation, which implies that their poles (i.e., the *complex dimensions* of the RFD (A,Ω)) are the same. We also extend to this general setting the main results of Chapters 2 and 3 concerning the holomorphicity and meromorphicity of the fractal zeta functions. We introduce the notion of transcendentally quasiperiodic relative fractal drums, using their tube functions. One way of constructing such drums is based on a carefully chosen sequence of generalized Cantor sets, as well as on the use of a classic result by Alan Baker from transcendental number theory. This construction and result extend the corresponding ones obtained in Chapter 3, in which we studied transcendentally quasiperiodic fractal sets. Furthermore, some explicit constructions of RFDs lead us naturally to introduce a new class of fractals, which we call *hyperfractals*. Particulary noteworthy among them are the maximal hyperfractals, for which the critical line $\{\mathrm{Re}\, s = \overline{\dim}_B(A,\Omega)\}$, where $\overline{\dim}_B(A,\Omega)$ is the relative upper box dimension of (A,Ω) and coincides with the abscissa of convergence of $\zeta_{A,\Omega}$ or $\tilde{\zeta}_{A,\Omega}$, consists solely of nonisolated singularities of the corresponding fractal zeta function (i.e., of the relative distance or tube zeta function), $\zeta_{A,\Omega}$ or $\tilde{\zeta}_{A,\Omega}$.

M.L. Lapidus et al., *Fractal Zeta Functions and Fractal Drums*,
Springer Monographs in Mathematics, DOI 10.1007/978-3-319-44706-3_4

Key words: relative fractal drum (RFD), relative Minkowski dimension, relative Minkowski content, relative distance and tube zeta functions, relative fractal zeta functions, scaling, relative fractal spray, flatness of RFDs, compact sets of positive reach, spectral zeta function, modified Weyl–Berry conjecture, meromorphic extension, abscissae of meromorphic and absolute convergence, residue, inhomogeneous Sierpiński N-gasket, relative Sierpiński N-carpet, spectral zeta functions of fractal drums, transcendentally ∞-quasiperiodic RFD, hyperfractals, embeddings of RFDs into higher-dimensional spaces.

In this chapter, we introduce the notion of relative fractal drums. They represent a simple and natural extension of two fundamental objects of fractal analysis, simultaneously: that of bounded sets in $\mathbb{R}^N$ (i.e., of fractals) and that of bounded fractal strings (introduced by the first author and Carl Pomerance in the early 1990s). Furthermore, there is a natural way to define their associated Minkowski contents and relative distance zeta functions. We stress a new phenomenon exhibited by relative fractal drums: namely, their box dimensions can be negative as well (and even equal to $-\infty$). This can be viewed as a property of their 'flatness', since it is related to the loss of the cone property; see Proposition 4.1.33.

In short, a relative fractal drum (RFD) consists of an ordered pair (A,Ω), where A is an arbitrary (possibly unbounded) subset of $\mathbb{R}^N$ and Ω is an open subset of $\mathbb{R}^N$ of finite volume and such that $\Omega \subseteq A_\delta$ for some $\delta > 0$; see Definition 4.1.2.

In Section 4.6, we introduce the notion of transcendentally quasiperiodic relative fractal drums, using their tube functions. One way of constructing such drums, described in Theorem 4.6.9, is based on a carefully chosen sequence of generalized Cantor sets, as well as on a classic result by Alan Baker from transcendental number theory; see Theorem 3.1.14. This construction and result extend the corresponding ones obtained in Section 3.1, in which we studied transcendentally quasiperiodic fractal sets.

Furthermore, some explicit constructions of RFDs lead us naturally to introduce a new class of fractals, which we call *hyperfractals*; see Definition 4.6.23. Particulary noteworthy among them are the maximal hyperfractals, for which the critical line $\{\mathrm{Re}\, s = \overline{\dim}_B(A,\Omega)\}$, where $\overline{\dim}_B(A,\Omega)$ is the relative upper box dimension of (A,Ω) and coincides with the abscissa of convergence of the corresponding zeta function, consists solely of nonisolated singularities; see Corollary 4.6.17. Therefore, for such a (relative) maximally hyperfractal drum, the critical line is a (meromorphic) natural boundary (in the sense of part *(ii)* of Definition 1.3.8) for each of the associated fractal zeta functions $\zeta_{A,\Omega}$ and $\tilde{\zeta}_{A,\Omega}$.

4.1 Zeta Functions of Relative Fractal Drums

We discuss here several natural generalizations of various notions which are central to this and related works, including notably relative distance zeta functions (in which the region of integration need not be bounded but is of finite volume), the associated relative box (and complex) dimensions, and RFDs. As is illustrated in a number of examples, this additional flexibility enables us to account for a broad range of situations and phenomena, including the case of fractal strings (Example 4.1.3) and of unbounded geometric chirps (Example 4.4.1). We also provide sufficient conditions ensuring the existence of a (necessarily unique) meromorphic continuation of the relative distance zeta function.

4.1.1 Relative Minkowski Content, Relative Box Dimension, and Relative Zeta Functions

In this subsection, we introduce the notion of a relative zeta function, associated to an appropriate ordered pair (A,Ω) of two suitable subsets of $\mathbb{R}^N$, which may be unbounded. The relative distance zeta function (see (4.1.1)), is a natural generalization of the standard distance zeta function defined by (2.1.1). We have already briefly encountered it in Section 2.1.5 in a less general context (see especially, Definition 2.1.75, Proposition 2.1.76 and Theorem 2.1.78, where Ω was assumed to be bounded), but we will now significantly relax our earlier assumptions.

Definition 4.1.1. Let Ω be an open subset of $\mathbb{R}^N$, not necessarily bounded, but of finite N-dimensional Lebesgue measure. Let $A\subseteq\mathbb{R}^N$, also possibly unbounded, such that Ω is contained in A_δ for some $\delta>0$.[1] The *distance zeta function* $\zeta_{A,\Omega}$ *of A relative to* Ω (or the *relative distance zeta function*) is defined by

$$\zeta_{A,\Omega}(s):=\int_\Omega d(x,A)^{s-N}\mathrm{d}x, \tag{4.1.1}$$

for all $s\in\mathbb{C}$ with $\operatorname{Re}s$ sufficiently large.

Unlike in (2.1.102), the closure of A is allowed here to intersect the boundary of Ω. (The closures of A and Ω may even be disjoint; see Example 4.1.22.) For this reason, the abscissa of convergence of this new zeta function will depend not only on the set A, but on Ω as well; see Theorem 4.1.7 and Example 4.1.25 below.

Definition 4.1.2. We propose to call the ordered pair (A,Ω), appearing in Definition 4.1.1, a *relative fractal drum* (RFD). Therefore, we shall also use the phrase *zeta functions of relative fractal drums* instead of relative zeta functions.

Example 4.1.3. Any bounded fractal string $\mathscr{L}=(\ell_j)_{j\geq1}$ (initially defined, as usual, as an infinite nonincreasing sequence of positive numbers $(\ell_j)_{j=1}^\infty$ such that $\sum_{j=1}^\infty\ell_j<\infty$) can also be viewed as a relative fractal drum $(A_\mathscr{L},\Omega_\mathscr{L})$. Indeed, the associated sets $A_\mathscr{L}$ and $\Omega_\mathscr{L}$ are

$$A_\mathscr{L}=\Big\{a_k=\sum_{j=k}^\infty\ell_j:k\in\mathbb{N}\Big\},\quad \Omega_\mathscr{L}=\bigcup_{k=1}^\infty(a_{k+1},a_k); \tag{4.1.2}$$

[1] We need this technical condition on A and Ω in order to ensure that the integral defined by Equation (4.1.1) is well defined for all $s\in\mathbb{C}$ with $\operatorname{Re}s$ large enough.

see Subsection 2.1.4. Therefore, the notion of relative fractal drum (A,Ω) is a natural extension of the notion of bounded fractal string $\mathscr{L}=(\ell_j)_{j\geq 1}$. Here, we point out that the notion of "generalized fractal string" already exists, and has a different meaning; see [Lap-vFr3, Chapter 4]. See also Remark 4.1.4 just below.

Remark 4.1.4. In short, a *generalized fractal string* (in the sense of [Lap-vFr3]) is a local positive or complex measure on $(0,+\infty)$ which does not have mass near 0. A local positive measure is simply a locally bounded positive Borel measure on $(0,+\infty)$, while a local complex measure is a locally bounded set-function on (the Borel σ-algebra of) $(0,+\infty)$ whose restriction to every compact subinterval J of $(0,+\infty)$ is a (necessarily bounded) Borel complex measure on J. (See Definition A.1.1 in Appendix A.) For example, an ordinary fractal string is represented by the positive measure $\eta_{\mathscr{L}}:=\sum_{j=1}^{\infty}\delta_{\ell_j^{-1}}$, where for $x>0$, δ_x denotes the unit Dirac mass (or measure) concentrated at x. Note that clearly, since $\ell_j\downarrow 0$ as $j\to\infty$, $\eta_{\mathscr{L}}$ is a generalized fractal string because it does not have any mass near 0. More generally, one could consider *generalized fractal strings* which are discrete but with noninteger multiplicities, say, $\eta=\eta_{\mathscr{L}}:=\sum_{l\in\mathscr{L}}b_l\delta_{l^{-1}}$, where $\mathscr{L}=\{l\}$ is an ordinary fractal string (now consisting of distinct lengths l) and 'multiplicities' or 'weights' b_l (with $b_l\geq 0$ or $b_l\in\mathbb{C}$ for each $l\in\mathscr{L}$); so that its 'geometric zeta function'

$$\zeta_\eta(s):=\int_0^{+\infty}x^{-s}\eta(\mathrm{d}x)$$

is the *Mellin transform* of the generalized Dirichlet series $\sum_{l\in\mathscr{L}}b_l l^s$. Of course, one can also consider continuous analogs, say, $\eta(\mathrm{d}x)=\varphi(x)\mathrm{d}x$, with φ a suitable real-valued function on $(0,+\infty)$. See, especially, [Lap-vFr3, Chapters 4, 5, 9, 10] and [Lap-vFr3, Section 6.3 and 11.1] for a variety of examples and applications of the theory of generalized fractal strings.

Remark 4.1.5. For results and conjectures concerning the spectra and the vibrations of (ordinary) fractal drums (or 'drums with fractal boundary'), we refer, e.g., to [Berr1, Berr2], [BroCar], [SapGoMar], [Lap1–3], [Ger], [GerSc], [FlVa], [Cae], [LapNeuReGr], [LapPa], [MolVai], [HeLap], [vBGilk], [Lap-vFr1–2], [HamLap], as well as [Lap-vFr3, Section 12.5] and the relevant references therein. We note that in the present monograph, however, we study mainly the geometry (rather than the eigenvalue spectrum) of (relative) fractal drums. A short discussion of the spectral zeta functions of a simple class of RFDs in $\mathbb{R}^N$ can be found in Section 4.3.1.

We can define the *relative complex dimensions* of A with respect to Ω (and with respect to a given window $\boldsymbol{W}$) as the set of poles (in $\boldsymbol{W}$) of the meromorphic extension of the relative distance zeta function $\zeta_{A,\Omega}$.

In particular, when $\zeta_{A,\Omega}$ has a meromorphic continuation to an open (connected) neighborhood of $\{\mathrm{Re}\,s=\overline{\dim}_B(A,\Omega)\}$, one can define (much as in Definition 2.1.67) the *set of relative principal complex dimensions* of (A,Ω), which is denoted by $\mathscr{P}(\zeta_{A,\Omega})$ or by $\dim_{PC}(A,\Omega)$, and consists of the poles of $\zeta_{A,\Omega}$ which lie on the critical line $\{\mathrm{Re}\,s=\overline{\dim}_B(A,\Omega)\}$. (We will see in Theorem 4.1.7 and

Remark 4.1.8 that the abscissa of convergence of $\zeta_{A,\Omega}$ coincides with $\overline{\dim}_B(A,\Omega)$ and hence, $\zeta_{A,\Omega}$ is holomorphic in the open half plane $\{\mathrm{Re}\, s > \overline{\dim}_B(A,\Omega)\}$.) See Definition 4.1.13.

In the previous paragraph and in the sequel, we are using the relative upper box dimension $\overline{\dim}_B(A,\Omega)$ (introduced in [Žu4] and generalizing that of [BroCar] and [Lap1–3], where $A := \partial\Omega$), instead of $\overline{\dim}_B(A)$. Its definition is analogous to that of the usual upper box dimension.

First, for any $r \in \mathbb{R}$, we define the *upper r-dimensional*[2] *Minkowski content of A relative to* Ω (or *the upper relative Minkowski content*, or *the upper Minkowski content of the RFD* (A,Ω)) by

$$\mathcal{M}^{*r}(A,\Omega) = \limsup_{t\to 0^+} \frac{|A_t \cap \Omega|}{t^{N-r}}, \tag{4.1.3}$$

and then proceed exactly as in (1.3.4) or in (1.3.5) in order to define $\overline{\dim}_B(A,\Omega)$:

$$\begin{aligned} \overline{\dim}_B(A,\Omega) &= \inf\{r \in \mathbb{R} : \mathcal{M}^{*r}(A,\Omega) = 0\} \\ &= \inf\{r \in \mathbb{R} : \mathcal{M}^{*r}(A,\Omega) < \infty\} \\ &= \sup\{r \in \mathbb{R} : \mathcal{M}^{*r}(A,\Omega) = +\infty\}. \end{aligned} \tag{4.1.4}$$

We call it the *relative upper box dimension* (or *relative upper Minkowski dimension*) of A with respect to Ω (or else the *relative upper box dimension of* (A,Ω)). Note that

$$\overline{\dim}_B(A,\Omega) \in [-\infty, N], \tag{4.1.5}$$

and the values can indeed be negative, and even equal to $-\infty$; see Proposition 4.1.35 and Corollary 4.1.38.

Naturally, $\mathcal{M}_*^r(A,\Omega)$, the *lower r-dimensional Minkowski content* of (A,Ω), is defined as in (4.1.3), except for a lower instead of an upper limit.

We define analogously the *relative lower box* (or *Minkowski*) *dimension* of (A,Ω):

$$\begin{aligned} \underline{\dim}_B(A,\Omega) &= \inf\{r \in \mathbb{R} : \mathcal{M}_*^r(A,\Omega) = 0\} \\ &= \inf\{r \in \mathbb{R} : \mathcal{M}_*^r(A,\Omega) < \infty\} \\ &= \sup\{r \in \mathbb{R} : \mathcal{M}_*^r(A,\Omega) = +\infty\}. \end{aligned} \tag{4.1.6}$$

Furthermore, when $\underline{\dim}_B(A,\Omega) = \overline{\dim}_B(A,\Omega)$, we denote by $\dim_B(A,\Omega)$ this common value and then say that the *relative box* (or *Minkowski*) *dimension* $\dim_B(A,\Omega)$ *exists*. See Remark 4.1.6 below.

If $0 < \mathcal{M}_*^D(A,\Omega) \le \mathcal{M}^{*D}(A,\Omega) < \infty$, we say that the relative fractal drum (A,Ω) is *Minkowski nondegenerate*. It then follows that $\dim_B(A,\Omega)$ exists and is equal to D.

If $\mathcal{M}_*^D(A,\Omega) = \mathcal{M}^{*D}(A,\Omega)$, this common value is denoted by $\mathcal{M}^D(A,\Omega)$ and called the *relative Minkowski content* of (A,Ω). If $\mathcal{M}^D(A,\Omega)$ exists and is

[2] An important novelty here is that we allow *negative values* of r as well.

different from 0 and $+\infty$ (in which case $\dim_B(A,\Omega)$ exists and we necessarily have $D = \dim_B(A,\Omega)$), we say that the relative fractal drum (A,Ω) is *Minkowski measurable*. For relative box (or rather, Minkowski) dimensions and their properties, see [Lap1], [HeLap] and more generally, [Žu4].

For example, if we assume that $A = \partial\Omega$, then $\overline{\dim}_B(\partial\Omega,\Omega)$ is also called the *one-sided box dimension of the boundary* (i.e., with respect to Ω, see [HeLap]) or the *inner Minkowski dimension* of $\partial\Omega$ (see, e.g., [BroCar], [Lap1–3], [LapPo1–3], [LapMa1–2], [FlVa] and [Lap-vFr1–3]). It may be different from $\overline{\dim}_B\partial\Omega$.

Remark 4.1.6. Here and in the sequel, we use interchangeably the terms "relative box dimension" and "relative Minkowski dimension". However, strictly speaking, only the latter term is correct in this general context because we do not have a proper independent (and geometric) definition of relative box dimension. See Problem 6.2.6 on page 556.

If A is a bounded subset of $\mathbb{R}^N$ and Ω is an open subset of $\mathbb{R}^N$ of finite N-dimensional Lebesgue measure, it is clear that

$$\overline{\dim}_B(A,\Omega) \in [-\infty, \overline{\dim}_B A], \tag{4.1.7}$$

and similarly for the lower box dimension. The inequality $\overline{\dim}_B(A,\Omega) \le \overline{\dim}_B A$ may be strict; see Examples 4.1.23 and 4.1.25. An obvious example is when the distance between A to Ω is positive, in which case $\overline{\dim}_B(A,\Omega) = \dim_B(A,\Omega) = -\infty$, no matter what value is taken by $\overline{\dim}_B A$. Furthermore, there are simple examples of disjoint sets $\overline{A}$ and $\overline{\Omega}$ for which $\overline{\dim}_B(A,\Omega)$ is nonzero; see Example 4.1.22. It is interesting that the value of $\overline{\dim}_B(A,\Omega)$ may be negative, whereas $\overline{\dim}_B A$ (as well as $\underline{\dim}_B A$) is always nonnegative. See, especially, Proposition 4.1.35, Corollary 4.1.38 and Remark 4.1.39.

The following result extends Theorem 2.1.11 to the present, more general, setting. To see this, it suffices to take $\Omega = A_\delta$ for any fixed $\delta > 0$.

Theorem 4.1.7. *Let Ω be an open subset of $\mathbb{R}^N$ of finite N-dimensional Lebesgue measure, and let $A \subseteq \mathbb{R}^N$ be such that $\Omega \subseteq A_\delta$ for some $\delta > 0$. Then the following properties hold:*

(a) The relative distance zeta function $\zeta_{A,\Omega}(s)$ is holomorphic in the half-plane $\{\mathrm{Re}\, s > \overline{\dim}_B(A,\Omega)\}$, and for those same values of s, we have

$$\zeta'_{A,\Omega}(s) = \int_\Omega d(x,A)^{s-N} \log d(x,A)\,\mathrm{d}x.$$

(b) The lower bound in the open right half-plane $\{\mathrm{Re}(s) > \overline{\dim}_B(A,\Omega)\}$ is optimal, from the point of view of the (absolute) convergence of the Dirichlet-type integral initially defining $\zeta_{A,\Omega}$ in (4.1.1). In other words,

$$D(\zeta_{A,\Omega}) = \overline{\dim}_B(A,\Omega), \tag{4.1.8}$$

where $D(\zeta_{A,\Omega})$ is the abscissa of convergence of $\zeta_{A,\Omega}$. (See also Remark 4.1.8 and part (i) of Corollary 4.1.10 below.)

(c) If $D = \dim_B(A,\Omega)$ exists, $D < N$, and $\mathscr{M}_*^D(A,\Omega) > 0$, then $\zeta_{A,\Omega}(s) \to +\infty$ as $s \in \mathbb{R}$ converges to D from the right. See also part (ii) of Corollary 4.1.10 below.

Proof. The proof is similar to that of Theorem 2.1.11. Instead of Lemma 2.1.3, we have to use a more general result (see [Žu4, Theorem 3.3]):

$$\text{If } \gamma < N - \overline{\dim}_B(A,\Omega), \text{ then } \int_{A_\delta \cap \Omega} d(x,A)^{-\gamma} \mathrm{d}x < \infty, \tag{4.1.9}$$

where δ is any fixed positive number. Lemma 2.1.6 can be easily adapted to the case of the relative box dimension; see [Žu2]. We omit the details. We simply note that in [Žu4, Theorem 3.3], the result is proven under the assumption that we deal with relative Minkowski contents for $r \geq 0$. Here, we allow $r < 0$ as well and it is easy to see that, nevertheless, this result still holds in this more general context. □

Remark 4.1.8. The claim in Theorem 4.1.7(b) follows easily from (a) and the fact that if $s \in \mathbb{R}$ and $s < \overline{\dim}_B(A,\Omega)$, then the defining integral in (4.1.1) is equal to infinity. Note that it follows from Theorem 4.1.7(b) that the relative upper box dimension, $\overline{\dim}_B(A,\Omega)$, coincides with the abscissa of convergence of the Dirichlet-type integral defining $\zeta_{A,\Omega}$ in (4.1.1). Equivalently, as was stated in Theorem 4.1.7(b), we have

$$\overline{\dim}_B(A,\Omega) = D(\zeta_{A,\Omega}), \tag{4.1.10}$$

where the latter notation is defined in Equation (2.1.92).

Remark 4.1.9. The continuity property stated in Theorem 2.1.78 also holds in the more general case of the relative zeta functions studied in the present subsection (that is, under the general assumptions of Definition 4.1.1). The proof of this fact is completely analogous to that of Theorem 2.1.78.

Since, as we have noted in Example 2.1.41, the relative distance zeta function $\zeta_{A,\Omega}$ is a Dirichlet-type integral satisfying condition (2.1.54) specified in Subsection 2.1.3.2 (i.e., it is a tamed DTI, in the sense of Definition A.1.3 of Appendix A), its abscissa of convergence $D(\zeta_{A,\Omega})$ is well defined. (For more details, see also the proof of part (1) of Proposition A.2.4 in Appendix A.) Exactly the same comment can be made about the relative tube zeta function $\tilde{\zeta}_{A,\Omega}$, to be introduced later in Subsection 4.5.1, Equation (4.5.1).

The following result is the exact analog for RFDs of Corollary 2.1.20 and Corollary 2.2.10 combined with Remark 4.1.11 and Remark 4.1.8. Note, however, that we no longer conclude that $D(\zeta_{A,\Omega}) \geq 0$, as will be further discussed in Subsection 4.1.2. Moreover, the analog of this result holds for the relative tube zeta function $\tilde{\zeta}_{A,\Omega}$.

Corollary 4.1.10. *(i) Let* (A,Ω) *be a relative fractal drum of* $\mathbb{R}^N$*. Then*:

$$D_{\mathrm{mer}}(\zeta_{A,\Omega}) \le D_{\mathrm{hol}}(\zeta_{A,\Omega}) \le D(\zeta_{A,\Omega}) = \overline{\dim}_B(A,\Omega), \tag{4.1.11}$$

and each of these inequalities is sharp, in general.

(ii) If, in addition, we assume that the hypotheses of part (c) *of Theorem 4.1.7 are satisfied, we then have the following equalities*:

$$D_{\mathrm{hol}}(\zeta_{A,\Omega}) = D(\zeta_{A,\Omega}) = \overline{\dim}_B(A,\Omega), \tag{4.1.12}$$

and hence, $\Pi(\zeta_{A,\Omega}) = \mathscr{H}(\zeta_{A,\Omega})$*, whereas under the assumptions of part* (i) *just above, we only have* $\Pi(\zeta_{A,\Omega}) \subseteq \mathscr{H}(\zeta_{A,\Omega})$.

Remark 4.1.11. Much as was noted in part (a) of Remark 2.1.21, we do not know whether there exist RFDs (A,Ω) for which the second inequality in (4.1.11) is strict; namely, $D_{\mathrm{hol}}(\zeta_{A,\Omega}) < D(\zeta_{A,\Omega})$. Such RFDs could not be ordinary fractal strings since we always have an equality in the latter case.

It is easy to find a relative fractal drum (A,Ω) for which $D_{\mathrm{mer}}(\zeta_{A,\Omega}) < D_{\mathrm{hol}}(\zeta_{A,\Omega})$. In fact, for every (nontrivial) fractal string, the equalities in Equation (4.1.12) always hold (without assuming the hypotheses of part (c) of Theorem 4.1.7), and with $\overline{\dim}_B(A,\Omega) \ge 0$, but, for example, for the Cantor string, we have $D_{\mathrm{mer}}(\zeta_{A,\Omega}) = -\infty$.

It is easy to see that, given any subset A and an open set Ω in $\mathbb{R}^N$ with finite N-dimensional Lebesgue measure, the relative zeta function of (A,Ω) can also be defined in the following way:

$$\zeta_{A,\Omega}(s;\delta) := \int_{\Omega\cap A_\delta} d(x,A)^{s-N}\mathrm{d}x, \tag{4.1.13}$$

where δ is a fixed positive number. Namely, for $\Omega' := \Omega \cap A_\delta$, the condition in Theorem 4.1.7 according to which $\Omega' \subseteq A_\delta$ is clearly satisfied.

In our definition of RFDs (A,Ω), we assume that Ω is an open subset of $\mathbb{R}^N$. Actually, Theorem 4.1.7 holds even in the case when Ω is a Borel set in $\mathbb{R}^N$. For example, Ω may have an empty interior and a positive Lebesgue measure. Therefore, it is natural to consider more general fractal drums (A,Ω), for which Ω is just an arbitrary Borel subset of $\mathbb{R}^N$. This issue is pursued in Appendix B, where the notion of 'local zeta function' is discussed.

In the following result, we obtain a simple sufficient condition for two RFDs to be equivalent. Its proof is similar to the proof of Proposition 2.1.76, and therefore we omit it.

Proposition 4.1.12. *Assume that* (A,Ω_1) *and* (A,Ω_2) *are RFDs in* $\mathbb{R}^N$ *such that* $f_j(s) := \int_{\Omega_j\setminus(\Omega_1\cap\Omega_2)} d(x,A)^{s-N}\mathrm{d}x$ *are entire functions, for* $j=1,2$*. Then the corresponding distance zeta functions are equivalent, that is,*

$$\zeta_{A,\Omega_1} \sim \zeta_{A,\Omega_2}.$$

Definition 4.1.13. Assume that (A,Ω) is a relative fractal drum in $\mathbb{R}^N$ such that its distance zeta function possesses a meromorphic extension to a domain which contains the critical line $\{\operatorname{Re} s = D(\zeta_{A,\Omega})\}$ in its interior. The set of poles of $\zeta_{A,\Omega}$ located on the critical line is called *the set of principal complex dimensions of the relative fractal drum* (A,Ω), or *the set of relative principal complex dimensions of* (A,Ω), and is denoted by $\dim_{PC}(A,\Omega)$ or equivalently, $\mathscr{P}_c(\zeta_{A,\Omega})$. (This extends the definition of $\dim_{PC} A = \mathscr{P}_c(\zeta_A)$ given in Definition 2.1.67.) We can analogously define the set $\dim_{PC}\mathscr{L}$ of *principal complex dimensions of any bounded (or unbounded) fractal string* $\mathscr{L} = (\ell_j)_{j\geq 1}$, as the set of poles of $\zeta_{\mathscr{L}}$ contained on the critical line $\{\operatorname{Re} s = D(\zeta_{\mathscr{L}})\}$.

In light of Theorem 2.2.3, we have the following result.

Theorem 4.1.14. *Assume that (A,Ω) is a Minkowski nondegenerate RFD in $\mathbb{R}^N$, that is, $0 < \mathscr{M}_*^D(A,\Omega) \leq \mathscr{M}^{*D}(A,\Omega) < \infty$ (in particular, $\dim_B(A,\Omega) = D$), and $D < N$. If $\zeta_{A,\Omega}(s)$ can be extended meromorphically to a connected open neighborhood of $s = D$, then D is necessarily a simple pole of $\zeta_{A,\Omega}$, the residue $\operatorname{res}(\zeta_{A,\Omega}, D)$ is independent of δ and*

$$(N-D)\mathscr{M}_*^D(A,\Omega) \leq \operatorname{res}(\zeta_{A,\Omega}, D) \leq (N-D)\mathscr{M}^{*D}(A,\Omega). \tag{4.1.14}$$

Furthermore, if (A,Ω) is Minkowski measurable, then

$$\operatorname{res}(\zeta_{A,\Omega}, D) = (N-D)\mathscr{M}^D(A,\Omega). \tag{4.1.15}$$

The next lemma follows immediately from the definition of the relative upper and lower box dimensions.

Lemma 4.1.15. *Assume that we have two RFDs (A_j,Ω_j) in $\mathbb{R}^N$ $(j = 1,2)$, where each Ω_j is of finite Lebesgue measure. If $A_1 \subseteq A_2$ and $\Omega_1 \subseteq \Omega_2$, then $\overline{\dim}_B(A_1,\Omega_1) \leq \overline{\dim}_B(A_2,\Omega_2)$. This is also true for the lower relative box dimensions.*

An immediate consequence is the following simple and useful result.

Lemma 4.1.16. *Assume that $\Omega_1 \subseteq \Omega \subseteq \Omega_2$ are open sets of finite Lebesgue measure in $\mathbb{R}^N$. If*

$$\overline{\dim}_B(A,\Omega_1) = \overline{\dim}_B(A,\Omega_2),$$

then this common value is equal to $\overline{\dim}_B(A,\Omega)$.

The following countable additivity property of zeta functions is a simple consequence of the σ-additivity property of the Lebesgue integral.

Proposition 4.1.17. *Assume that $\Omega = \cup_{j=1}^{\infty} B_j$ is an open subset of $\mathbb{R}^N$ of finite N-dimensional Lebesgue measure, where $(B_j)_{j=1}^{\infty}$ is a sequence of pairwise disjoint* *open subsets of $\mathbb{R}^N$. Furthermore, assume that $A \subseteq \mathbb{R}^N$ and there exists $\delta > 0$ such that $\Omega \subseteq A_\delta$. Then, for all $s \in \mathbb{C}$ such that* $\operatorname{Re} s > \overline{\dim}_B(A,\Omega)$, *we have*

$$\zeta_{A,\Omega}(s) = \sum_{j=1}^{\infty} \zeta_{A,B_j}(s). \tag{4.1.16}$$

Example 4.1.18. Let $\Omega \subset \mathbb{R}$ be a disjoint union of open intervals I_k in the real line, of lengths $1/k^2$ for each $k \geq 1$. Here, Ω may be unbounded. Let $A = \partial\Omega$. Then

$$\zeta_{A,\Omega}(s) = \sum_{k=1}^{\infty} \zeta_{A,I_k}(s) = \frac{2^{1-2s}}{s} \sum_{k=1}^{\infty} k^{-2s} \sim \sum_{k=1}^{\infty} k^{-2s} = \zeta(2s), \tag{4.1.17}$$

where $\zeta(s) = \sum_{j\geq 1} k^{-s}$ is the classical Riemann zeta function (or its meromorphic continuation). The abscissa of convergence of $\zeta_{A,\Omega}(s)$ is therefore equal to $s = 1/2$, and by using Theorem 4.1.7(b) we conclude that $\overline{\dim}_B(A,\Omega) = 1/2$. Note that by analytic continuation, $\zeta_{A,\Omega}$ has a meromorphic extension to all of $\mathbb{C}$, and that $\zeta_{A,\Omega}(s) = \frac{2^{1-2s}}{s}\zeta(2s)$ for all $s \in \mathbb{C}$.

The following example is a relative analog of Example 2.2.21.

Example 4.1.19. Let $\Omega = B_R(0)$ be the open ball in $\mathbb{R}^N$ of radius R and let $A = \partial\Omega$ be the boundary of Ω, i.e, the $N-1$-dimensional sphere of radius R. Then, introducing the new variable $\rho = R - r$, we have

$$\begin{aligned}
\zeta_{A,\Omega}(s) &= N\omega_N \int_0^R (R-r)^{s-N} r^{N-1}\,\mathrm{d}r = N\omega_N \int_0^R \rho^{s-N}(R-\rho)^{N-1}\,\mathrm{d}\rho \\
&= N\omega_N \int_0^R \rho^{s-N} \sum_{k=0}^{N-1} (-1)^k \binom{N-1}{k} R^{N-1-k}\rho^k\,\mathrm{d}\rho \\
&= N\omega_N R^s \sum_{k=0}^{N-1} \binom{N-1}{k} \frac{(-1)^k}{s-(N-k-1)} \\
&= N\omega_N R^s \sum_{j=0}^{N-1} \binom{N-1}{j} \frac{(-1)^{N-j-1}}{s-j}
\end{aligned}$$

for all $s \in \mathbb{C}$ with $\operatorname{Re} s > N-1$, where ω_N is the N-dimensional Lebesgue measure of the unit ball in $\mathbb{R}^N$; see Equation (1.3.22) on page 40. (Note that we have also used the well-known symmetry of the binomial coefficients, $\binom{N-1}{N-1-j} = \binom{N-1}{j}$.) In particular, $\zeta_{A,\Omega}$ can be meromorphically extended to the whole complex plane and is given by

$$\zeta_{A,\Omega}(s) = N\omega_N R^s \sum_{j=0}^{N-1} \binom{N-1}{j} \frac{(-1)^{N-j-1}}{s-j}, \tag{4.1.18}$$

for all $s \in \mathbb{C}$.

Therefore, we have

$$\begin{gathered}\dim_B(A,\Omega) = D(\zeta_{A,\Omega}) = N-1 \\ \mathscr{P}(\zeta_{A,\Omega}) = \{0,1,\ldots,N-1\} \quad \text{and} \quad \dim_{PC}(A,\Omega) = \{N-1\}.\end{gathered} \tag{4.1.19}$$

Furthermore,

$$\operatorname{res}(\zeta_{A,\Omega}, j) = (-1)^{N-j-1} N\omega_N \binom{N-1}{j} R^j \tag{4.1.20}$$

for $j = 0,1,\ldots,N-1$. It is noteworthy that the set $\mathscr{P}(\zeta_{A,\Omega})$ of complex dimensions of (A,Ω) is not the same as the set $\mathscr{P}(\zeta_A)$ of complex dimensions of A; compare Equation (4.1.19) with Equation (2.2.58) of Example 2.2.21 on page 128. As a special case of (4.1.20), for $j = D := N-1$ we obtain that

$$\operatorname{res}(\zeta_{A,\Omega}, D) = N\omega_N R^{N-1} = \mathscr{M}^D(A,\Omega). \tag{4.1.21}$$

The last equality follows from a direct computation:

$$\mathscr{M}^D(A,\Omega) = \lim_{t\to 0^+} \frac{|A_t \cap \Omega|}{t^{N-D}} = \lim_{t\to 0^+} \frac{\omega_N R^N - \omega_N (R-t)^N}{t} = N\omega_N R^{N-1}. \tag{4.1.22}$$

Furthermore, recall that $H^D(A) = H^{N-1}(\partial B_R(0)) = N\omega_N R^{N-1}$, where H^{N-1} is the $(N-1)$-dimensional Hausdorff measure. In particular, the relative fractal drum (A,Ω) is Minkowski measurable and

$$\mathscr{M}^D(A,\Omega) = H^D(A). \tag{4.1.23}$$

Equation (4.1.21) is a special case of Equation (4.5.13) in the case when $m := 0$ in Theorem 4.5.1 on page 353; see also Equation (4.5.1).

Proposition 4.1.20. *(a) For any relative fractal drum (A,Ω), with $|\Omega| < \infty$, we have*

$$\overline{\dim}_B(A,\Omega) = \overline{\dim}_B(\overline{A},\Omega),$$

and similarly for the relative lower box dimension.

(b) The Cartesian product $(A_1 \times A_2, \Omega_1 \times \Omega_2)$ of two Minkowski nondegenerate RFDs (A_1,Ω_1) and (A_2,Ω_2), is also Minkowski nondegenerate. Furthermore,

$$\dim_B(A_1 \times A_2, \Omega_1 \times \Omega_2) = \dim_B(A_1,\Omega_1) + \dim_B(A_2,\Omega_2).$$

Proof. Part (a) follows easily from the fact that $A_t = (\overline{A})_t$ for all $t > 0$, where $\overline{A}$ denotes the closure of A in $\mathbb{R}^N$. Part (b) follows from [Žu2, Proposition 4.3]. □

Some basic open questions about the relative upper box dimension can be found in Problem 6.2.31 of Section 6.2.2.

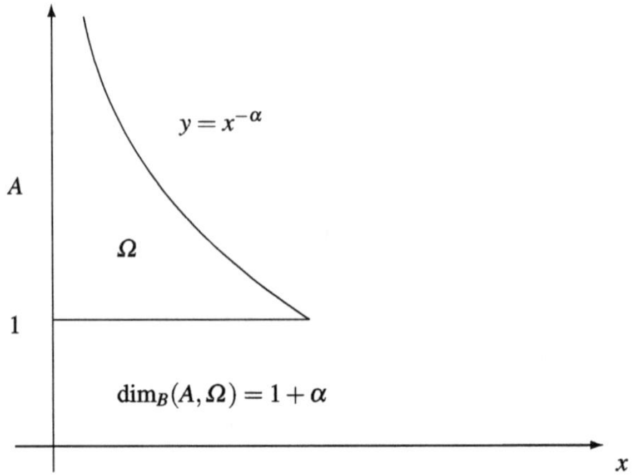

Fig. 4.1 A relative fractal drum (A,Ω) in the plane with relative box dimension $\dim_B(A,\Omega) = 1+\alpha \in (1,2)$, for $\alpha \in (0,1)$; see Example 4.1.21.

Example 4.1.21. Here, we deal with a situation where both of the sets A and Ω are unbounded. This example is based on [Žu2, Example 2.1]. Let $A = \{0\} \times (1,+\infty) \subseteq \mathbb{R}^2$ and $\Omega = \{(x,y) \in \mathbb{R}^2 : x \in (0,1),\ 1 < y < x^{-\alpha}\}$, for some fixed $\alpha \in (0,1)$; see Figure 4.1. Note that Ω is unbounded, but has finite two-dimensional Lebesgue measure. The relative distance zeta function is then given by

$$\begin{aligned}\zeta_{A,\Omega}(s) &= \iint_\Omega d((x,y),A)^{s-2}\mathrm{d}x\,\mathrm{d}y \\ &= \int_0^1 x^{s-2}\mathrm{d}x \int_1^{x^{-\alpha}} \mathrm{d}y = \int_0^1 (x^{s-2-\alpha} - x^{s-2})\,\mathrm{d}x \\ &= \frac{1}{s-1-\alpha} - \frac{1}{s-1} \sim \frac{1}{s-1-\alpha},\end{aligned} \tag{4.1.24}$$

where in the computation of the double integral, we have assumed that $\operatorname{Re} s > 1+\alpha$. It follows that $\sigma = 1+\alpha$ is the abscissa of convergence of the relative zeta function $\zeta_{A,\Omega}$: $D(\zeta_{A,\Omega}) = 1+\alpha$. Therefore (see (4.1.10)), the half-line A has a nontrivial relative box-dimension with respect to Ω, given by

$$\overline{\dim}_B(A,\Omega) = D(\zeta_{A,\Omega}) = 1+\alpha.$$

It is not difficult to show that a stronger result holds; namely, $\dim_B(A,\Omega)$ exists, $\dim_B(A,\Omega) = 1+\alpha$, and the relative fractal drum (A,Ω) is Minkowski measurable.

It follows from the above discussion that the set $\mathscr{P}_c(\zeta_{A,\Omega})$ of relative principal complex dimensions of the half-line A (with respect to the open unit square Ω) is given by

$$\mathscr{P}_c(\zeta_{A,\Omega}) = \{1+\alpha\}.$$

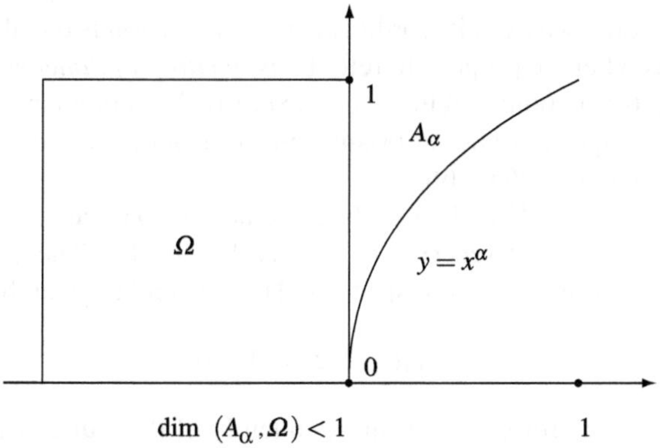

Fig. 4.2 A relative fractal drum (A_α,Ω) such that $\dim_B(A_\alpha,\Omega)=1-\alpha<1$ (here, $0<\alpha<1$), whereas $\dim_B A_\alpha=1$; see Example 4.1.23. This illustrates the *drop of dimension phenomenon* for relative Minkowski dimensions.

Actually, a more precise result holds. Indeed, note that according to the last equality of (4.1.24), $\zeta_{A,\Omega}$ has a meromorphic continuation to all of $\mathbb{C}$ (given by the right-hand side of the last equality of (4.1.24)), and furthermore, the set $\mathscr{P}(\zeta_{A,\Omega})$ of all relative complex dimensions of (A,Ω) is given by

$$\mathscr{P}(\zeta_{A,\Omega})=\{1,1+\alpha\}=\{1+\alpha\}\cup\{1\},$$

the union of $\{1+\alpha\}$, the set of scaling complex dimensions, and $\{1\}$, the set of positive integer dimensions (in the sense of [LapPe2–3] and [LapPeWi1], see also [Lap-vFr3, Section 13.1]). We point out, however, that the theory of [LapPe2, LapPeWi1] cannot be applied to the present example in order to also yield this result. Hence, the relative 'fractal drum' (A,Ω) is not fractal (in the sense of [Lap-vFr1–3]) since it does not have any nonreal principal complex dimensions, which is, of course, natural since both A and Ω are standard Euclidean geometric shapes.

Example 4.1.22. Let Ω be the same as in the preceding example, and define $A=\{(x,y)\in(-1,0)\times\mathbb{R}: y=|x|^{-\alpha}\}$, where $\alpha\in(0,1)$ is fixed. Here, we also have that $\dim_B(A,\Omega)$ exists and

$$\dim_B(A,\Omega)=D(\zeta_{A,\Omega})=1+\alpha.$$

Note that now, the sets $\overline{A}$ and $\overline{\Omega}$ are disjoint.

It is clear that in the case of a bounded set A, we have $\overline{\dim}_B(A,\Omega)\le\overline{\dim}_B A$, and analogously for the lower box dimension. The following example shows that the inequality may be strict.

Example 4.1.23. We provide here an example showing that a smooth rectifiable curve (see part (a) of Remark 4.1.24 below) may have a relative box dimension strictly less than one, whereas its (ordinary) box dimension is equal to one. This example illustrates what we propose to refer to as *the drop of dimension phenomenon*, which is frequently encountered in the context of RFDs. For an even more dramatic example of this important and surprising phenomenon, see Corollary 4.1.38 and Remark 4.1.39 on pages 265–266.

Let $\Omega = (-1,0)\times(0,1)$ and let A_α be the graph of a Hölder continuous function $y = x^\alpha$, $0 < x < 1$, for a fixed $\alpha \in (0,1)$; see Figure 4.2. Then, the relative box dimension of the curve A_α (with respect to Ω) exists and is given by

$$\dim_B(A_\alpha,\Omega) = 1-\alpha.$$

Note that, in contrast, $\dim_B A_\alpha = 1$, independently of the value of $\alpha \in (0,1)$, since A_α is clearly rectifiable, i.e., of finite length. (See part (b) of Remark 4.1.24 below.) Also, it is worth noting that A_α and Ω are disjoint. The relative zeta function $\zeta_{A_\alpha,\Omega}(s)$ is holomorphic on the half-plane $\mathrm{Re}\,s > 1-\alpha$, and the bound is optimal:

$$D(\zeta_{A_\alpha,\Omega}) = \dim_B(A_\alpha,\Omega) = 1-\alpha.$$

Remark 4.1.24. (a) Note that A_α is a C^∞-curve, since it does not contain the origin. Furthermore, the curve $\overline{A}_\alpha$ is at least of class C^1 (more precisely, of class C^k with $k = \lfloor 1/\alpha \rfloor$), since it can be viewed as the graph of the function $x = y^{1/\alpha}$ for $y \in [0,1]$, where the exponent $1/\alpha$ is larger than 1 (and in particular, the function is Lipschitz continuous).

(b) The length of A_α is bounded by the sum of its projections onto the vertical and horizontal axes, that is, by 2. If a curve is rectifiable (i.e. of finite length), then its graph has box (i.e., Minkowski) dimension equal to 1; see, e.g., Federer [Fed2, Theorem 3.2.39] for a more general statement concerning k-rectifiable sets. Namely, the Minkowski (or box) dimension of a closed and *k-rectifiable set* (i.e., of the image in $\mathbb{R}^N$ under a Lipschitz map of a bounded set in $\mathbb{R}^k$) exists and does not exceed k, and, moreover, its k-dimensional Minkowski content exists and is finite. Here, we have $k=1$, $N=2$ and, clearly, the Minkowski dimension of a smooth curve is not smaller than 1.

Example 4.1.25. Slightly modifying the above example, let us set $A' = \{0\}\times(0,1)$ and consider the family of open sets $\Omega'_\alpha = \{(x,y)\in(0,1)^2 : y < x^\alpha\}$, where $\alpha \in (0,1)$. Then

$$D(\zeta_{A',\Omega'_\alpha}) = \dim_B(A',\Omega'_\alpha) = 1-\alpha.$$

This shows that the relative box dimension depends on the domain Ω'_α.

In the following proposition, we extend the well-known property of *finite stability* of the usual upper box dimension $\overline{\dim}_B A$ (see, e.g., [Fal1]) to the more general case of the relative upper box dimension $\overline{\dim}_B(A,\Omega)$; see Equation (4.1.25) in Proposition 4.1.26 below. The claim is not true for the relative lower box dimension $\underline{\dim}_B(A,\Omega)$; see the discussion immediately following Equation (6.1.8) in Subsection 6.1.2 of Chapter 6 below.

Proposition 4.1.26 (Finite stability of the relative upper box dimension). *Let (A,Ω) and (B,Ω) be two relative fractal drums in $\mathbb{R}^N$. Then $(A\cup B,\Omega)$ is an RFD as well, and the following property of* finite stability of the relative upper box dimension *holds*:

$$\overline{\dim}_B(A\cup B,\Omega)=\max\{\overline{\dim}_B(A,\Omega),\overline{\dim}_B(B,\Omega)\}. \tag{4.1.25}$$

Moreover, for any real number $s\in\mathbb{R}$, we have that

$$\max\{\mathscr{M}^{*s}(A,\Omega),\mathscr{M}^{*s}(B,\Omega)\}\le\mathscr{M}^{*s}(A\cup B,\Omega)\le\mathscr{M}^{*s}(A,\Omega)+\mathscr{M}^{*s}(B,\Omega). \tag{4.1.26}$$

Proof. Since (A,Ω) and (B,Ω) are RFDs, then $\Omega\subseteq A_\delta$ and $\Omega\subseteq B_\delta$ for some $\delta>0$; hence, $\Omega\subseteq A_\delta\cup B_\delta=(A\cup B)_\delta$. Therefore, $(A\cup B,\Omega)$ is an RFD as well.

Let us first prove the two inequalities appearing in (4.1.26). The first one follows immediately from the two inclusions $A\subseteq A\cup B$ and $B\subseteq A\cup B$, while the second one is an easy consequence of the fact that $(A\cup B)_t=A_t\cup B_t$, for all $t>0$:

$$\begin{aligned}\mathscr{M}^{*s}(A\cup B,\Omega)&=\limsup_{t\to0^+}\frac{|(A\cup B)_t\cap\Omega|}{t^{N-s}}=\limsup_{t\to0^+}\frac{|(A_t\cup B_t)\cap\Omega|}{t^{N-s}}\\&\le\limsup_{t\to0^+}\left(\frac{|A_t\cap\Omega|}{t^{N-s}}+\frac{|B_t\cap\Omega|}{t^{N-s}}\right)\\&\le\limsup_{t\to0^+}\frac{|A_t\cap\Omega|}{t^{N-s}}+\limsup_{t\to0^+}\frac{|B_t\cap\Omega|}{t^{N-s}}\\&=\mathscr{M}^{*s}(A,\Omega)+\mathscr{M}^{*s}(B,\Omega).\end{aligned} \tag{4.1.27}$$

Now, Equation (4.1.25), which we write as $L=R$, follows easily from Equation (4.1.26). Indeed, assume that (4.1.25) does not hold, i.e., that $L\neq R$. Let us consider the following two cases:

(a) If $L<R$ in (4.1.25), then by taking any real number $s\in(L,R)$, we have that $\mathscr{M}^{*s}(A\cup B,\Omega)=0$ and either $\mathscr{M}^{*s}(A,\Omega)=+\infty$ or $\mathscr{M}^{*s}(B,\Omega)=+\infty$. However, this is impossible, due to the first inequality in (4.1.26).

(b) If $L>R$, then by taking any real number $s\in(R,L)$, we obtain that $\mathscr{M}^{*s}(A\cup B,\Omega)=+\infty$ and $\mathscr{M}^{*s}(A,\Omega)=\mathscr{M}^{*s}(A,\Omega)=0$. This is also impossible, due to the second inequality in (4.1.26).

This completes the proof of Equation (4.1.25), as well as of the proposition. □

Remark 4.1.27. If (A,Ω_1) and (B,Ω_2) are two relative fractal drums in $\mathbb{R}^N$ such that for some $\varepsilon>0$, $A_\varepsilon\cap\Omega_2=\emptyset$ and $B_\varepsilon\cap\Omega_1=\emptyset$, then the property of *finite stability* holds in the following sense:

$$\overline{\dim}_B(A\cup B,\Omega_1\cup\Omega_2)=\max\{\overline{\dim}_B(A,\Omega_1),\overline{\dim}_B(B,\Omega_2)\}. \tag{4.1.28}$$

Moreover, for any real number $s \in \mathbb{R}$, we have that

$$\begin{aligned}\max\{\mathscr{M}^{*s}(A,\Omega_1),\mathscr{M}^{*s}(B,\Omega_2)\} &\le \mathscr{M}^{*s}(A\cup B,\Omega_1\cup\Omega_2)\\ \mathscr{M}^{*s}(A\cup B,\Omega_1\cup\Omega_2) &\le \mathscr{M}^{*s}(A,\Omega_1)+\mathscr{M}^{*s}(B,\Omega_2).\end{aligned} \tag{4.1.29}$$

Note, however, that Equations (4.1.28) and (4.1.29), jointly with the indicated assumptions, do not contain Proposition 4.1.25 as a special case.

In order to prove (4.1.29), it suffices to observe that for any $t \in (0,\varepsilon)$, we have that $(A\cup B)_t \cap (\Omega_1\cap\Omega_2) = (A_t\cap\Omega_1)\cup(B_t\cap\Omega_2)$, and then to proceed analogously as in the first part of the proof of Proposition 4.1.26. Equation (4.1.28) follows from (4.1.29) and the arguments from the second part of the proof of the proposition.

4.1.2 Cone Property and Flatness of Relative Fractal Drums

We introduce the cone property of a relative fractal drum (A,Ω) at a point, in order to ensure that the abscissa of convergence of the associated relative zeta function $\zeta_{A,\Omega}$ be nonnegative. The main result of this subsection is stated in Proposition 4.1.33. We also construct a simple class of RFDs for which the relative box dimension is negative; see Proposition 4.1.35.

Definition 4.1.28. Let $B_r(a)$ be a given ball in $\mathbb{R}^N$ of radius r. Let ∂B be the boundary of the ball, which is an $(N-1)$-dimensional sphere, and assume that G is a closed connected subset contained in a hemisphere of ∂B. [Intuitively, G is a disk-like subset ('calotte') of a hemisphere contained in the sphere ∂B.] We assume that G is open with respect to the relative topology of ∂B. The *cone* $K = K_r(a,G)$ *with vertex at* a, and of radius r, is defined as the interior of the convex hull of the union of $\{a\}$ and G.

Definition 4.1.29. Let (A,Ω) be a relative fractal drum in $\mathbb{R}^N$. We say that a *relative fractal drum* (A,Ω) *has the cone property at a point* $a \in \overline{A}\cap\overline{\Omega}$ if there exists $r>0$ such that Ω contains a cone $K_r(a,G)$ with vertex at a (and of radius r).

Remark 4.1.30. If $a \in \overline{A}\cap\Omega$ (hence, a is an inner point of Ω), then the cone property of the relative fractal drum (A,Ω) is obviously satisfied at this point. So, the cone property is actually interesting only on the boundary of Ω, that is, at $a\in\overline{A}\cap\partial\Omega$.

Example 4.1.31. Given $\alpha>0$, let (A,Ω_α) be the relative fractal drum in $\mathbb{R}^2$ defined by $A=\{(0,0)\}$ and $\Omega_\alpha = \{(x,y)\in\mathbb{R}^2 : 0<y<x^\alpha,\ x\in(0,1)\}$. If $0<\alpha\le 1$, then the cone property of (A,Ω) is fulfilled at $a=(0,0)$, while it is not satisfied (at $a=(0,0)$), for $\alpha>1$. Using these domains, we can construct a one-parameter family of RFDs with negative relative box dimension; see Proposition 4.1.35 below.

Proposition 4.1.33 below is an extension of Lemma 2.1.52, which states that $D(\zeta_A)\ge 0$ for any bounded set A. We first need an auxiliary result.

Lemma 4.1.32. *Assume that $K = K_r(a,G)$ is an open cone in $\mathbb{R}^N$ with vertex at a (and of radius $r > 0$), and $f \in L^1(0,r)$ is a nonnegative function. Then there exists a positive integer m, depending only on N and on the opening angle of the cone, such that*

$$\int_{B_r(a)} f(|x-a|)\,\mathrm{d}x \le m \int_K f(|x-a|)\,\mathrm{d}x. \tag{4.1.30}$$

Proof. Since the sphere ∂B is compact, there exist finitely many calottes $G_1,\dots,G_m$ contained in the sphere, that are all congruent to G (that is, each G_i can be obtained from G by a rigid motion), and which cover ∂B. Let $K_i = K_r(a,G_i)$, $i = 1,\dots,m$, be the corresponding cones with vertex at a. It is clear that the value of

$$\int_{K_i} f(|x-a|)\,\mathrm{d}x \tag{4.1.31}$$

does not depend on i. Since $B_r(a) = \cup_{i=1}^m K_i$, we have

$$\int_{B_r(a)} f(|x-a|)\,\mathrm{d}x \le \sum_{i=1}^m \int_{K_i} f(|x-a|)\,\mathrm{d}x = m \int_K f(|x-a|)\,\mathrm{d}x. \tag{4.1.32}$$

□

Proposition 4.1.33. *Let (A,Ω) be a relative fractal drum in $\mathbb{R}^N$. Then:*

(a) If the sets A and Ω are a positive distance apart (i.e., if $d(A,\Omega) > 0$), then $D(\zeta_{A,\Omega}) = -\infty$; that is, $\zeta_{A,\Omega}$ is an entire function. Furthermore, $\dim_B(A,\Omega) = -\infty$.

(b) Assume that there exists at least one point $a \in \overline{A} \cap \overline{\Omega}$ at which the relative fractal drum (A,Ω) satisfies the cone property. Then $D(\zeta_{A,\Omega}) \ge 0$.

Proof. (a) For $r > 0$ small enough such that $r < d(A,\Omega)$, where $d(A,\Omega)$ is the distance between A and Ω, we have $A_r \cap \Omega = \emptyset$; so that $\zeta_{A,A_r\cap\Omega}(s) \equiv 0$ for all $s \in \mathbb{C}$. Therefore, $D(\zeta_{A,A_r\cap\Omega}) = -\infty$. Since $\zeta_{A,\Omega}(s) - \zeta_{A,A_r\cap\Omega}(s)$ is an entire function, we conclude that we also have that $D(\zeta_{A,\Omega}) = -\infty$. Since $|A_\varepsilon \cap \Omega| = 0$ for all sufficiently small $\varepsilon > 0$, we have $\mathscr{M}^r(A,\Omega) = 0$ for all $r \in \mathbb{R}$, and therefore, $\dim_B(A,\Omega) = -\infty$.

(b) Assume, by reasoning by contradiction, that $D(\zeta_{A,\Omega}) < 0$. In particular, $\zeta_{A,\Omega}(s)$ is continuous at $s = 0$ (because it must then be holomorphic at $s = 0$). By hypothesis, there exists an open cone $K = K_r(a,G)$, such that $K \subset \Omega$. Using the inequality $d(x,A) \le |x-a|$ (valid for all $x \in \mathbb{R}^N$ since $a \in \Omega$) and Lemma 4.1.32, we deduce that for any real number $s \in (0,N)$,

$$\begin{aligned}\zeta_{A,\Omega}(s) \ge \zeta_{A,K}(s) &= \int_K d(x,A)^{s-N}\mathrm{d}x \ge \int_K |x-a|^{s-N}\mathrm{d}x \\ &\ge \frac{1}{m}\int_{B_r(a)} |x-a|^{s-N}\mathrm{d}x = \frac{N\omega_N}{m} r^s s^{-1},\end{aligned}$$

where m is the positive constant appearing in Equation (4.1.30) of Lemma 4.1.32. This implies that $\zeta_{A,\Omega}(s) \to +\infty$ as $s \to 0^+$, $s \in \mathbb{R}$, which contradicts the holomorphicity (or simply, the continuity) of $\zeta_{A,\Omega}(s)$ at $s = 0$. □

The cone condition can be replaced by a much weaker condition, as we will now explain in the following proposition.

Proposition 4.1.34. *Let $(r_k)_{k\geq 0}$ be a decreasing sequence of positive real numbers, converging to zero. We define a subset of the cone $K_r(a,G)$ as follows:*

$$K_r(a,G,(r_k)_{k\geq 0}) = \Big\{x \in K_r(a,G) : |x-a| \in \bigcup_{k=0}^{\infty}(r_{2k}, r_{2k+1})\Big\}. \tag{4.1.33}$$

If we assume that the sequence $(r_k)_{k\geq 1}$ is such that

$$\sum_{k=0}^{\infty}(-1)^k r_k^s \to L > 0 \quad \text{as} \quad s \to 0^+, s \in \mathbb{R}, \tag{4.1.34}$$

then the conclusion of Proposition 4.1.33(b) still holds, with the cone condition involving $K := K(a,G)$ replaced by the above modified cone condition, involving the set $K' := K_r(a,G,(r_k)_{k\geq 0})$ contained in K.

Proof. In order to establish this claim, it suffices to use a procedure analogous to the one used in the proof of Proposition 4.1.33:

$$\begin{aligned}\zeta_{A,\Omega}(s) &\geq \int_{K'}|x-a|^{s-N}\mathrm{d}x \geq \frac{1}{m}\sum_{k=0}^{\infty}\int_{B_{r_{2k}}(a)\setminus B_{r_{2k+1}}(s)}|x-a|^{s-N}\mathrm{d}x\\ &= \frac{N\omega_N}{m}s^{-1}\sum_{k=0}^{\infty}(r_{2k}^s - r_{2k+1}^s) = \frac{N\omega_N}{m}s^{-1}\sum_{k=0}^{\infty}(-1)^k r_k^s.\end{aligned}$$

For example, if $r_k = 2^{-k}$, then condition (4.1.34) is fulfilled since

$$\sum_{k=0}^{\infty}(-1)^k r_k^s = \sum_{k=0}^{\infty}(-1)^k 2^{-ks} = \frac{1}{1+2^{-s}} \to \frac{1}{2} \quad \text{as} \quad s \to 0^+, s \in \mathbb{R}.$$

This concludes the proof of the proposition. □

The following proposition (building on Example 4.1.31 above) shows that the box dimension of a relative fractal drum can be negative. It also shows that the analog of Lemma 2.1.52 does not hold for arbitrary RFDs.

Proposition 4.1.35. *Let $A = \{(0,0)\}$ and*

$$\Omega = \{(x,y) \in \mathbb{R}^2 : 0 < y < x^\alpha,\ x \in (0,1)\}, \tag{4.1.35}$$

where $\alpha > 1$; see Figure 4.3. Then the relative fractal drum (A,Ω) has a negative box dimension. More specifically, $\dim_B(A,\Omega)$ *exists, the relative fractal drum (A,Ω) is Minkowski measurable and*

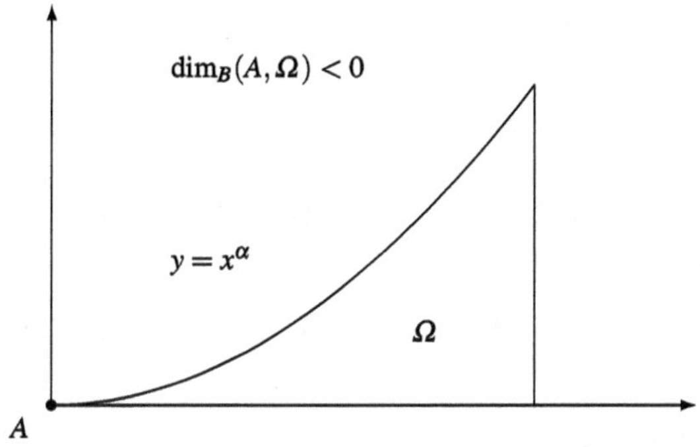

Fig. 4.3 A relative fractal drum (A,Ω) with negative box dimension $\dim_B(A,\Omega) = 1-\alpha < 0$ (here $\alpha > 1$), due to the 'flatness' of the open set Ω at A; see Proposition 4.1.35. This provides a further illustration of *the drop in dimension phenomenon* (for relative box dimensions).

$$\begin{aligned} \dim_B(A,\Omega) &= D(\zeta_{A,\Omega}) = 1-\alpha < 0, \\ \mathscr{M}^{1-\alpha}(A,\Omega) &= \frac{1}{1+\alpha}, \\ D_{\rm mer}(\zeta_{A,\Omega}) &\le 3(1-\alpha). \end{aligned} \tag{4.1.36}$$

Furthermore, $s = 1-\alpha$ is a simple pole of $\zeta_{A,\Omega}$.

Proof. First note that $A_\varepsilon = B_\varepsilon((0,0))$. Therefore, for every $\varepsilon > 0$, we have

$$|A_\varepsilon \cap \Omega| \le \int_0^\varepsilon x^\alpha \mathrm{d}x = \frac{\varepsilon^{\alpha+1}}{\alpha+1}.$$

If we choose a point $(x(\varepsilon), y(\varepsilon))$ such that

$$(x(\varepsilon), y(\varepsilon)) \in \partial(A_\varepsilon) \cap \{(x,y) : y = x^\alpha,\ x \in (0,1)\},$$

then the following equation holds:

$$x(\varepsilon)^2 + x(\varepsilon)^{2\alpha} = \varepsilon^2. \tag{4.1.37}$$

It is clear that

$$|A_\varepsilon \cap \Omega| \ge \int_0^{x(\varepsilon)} x^\alpha \mathrm{d}x = \frac{x(\varepsilon)^{\alpha+1}}{\alpha+1}.$$

Letting $D := 1-\alpha$, we conclude that

$$\frac{1}{\alpha+1}\left(\frac{x(\varepsilon)}{\varepsilon}\right)^{\alpha+1} \le \frac{|A_\varepsilon \cap \Omega|}{\varepsilon^{2-D}} \le \frac{1}{\alpha+1}, \quad \text{for all} \quad \varepsilon > 0. \tag{4.1.38}$$

We deduce from (4.1.37) that $x(\varepsilon) \sim \varepsilon$ as $\varepsilon \to 0^+$, since

$$\frac{x(\varepsilon)}{\varepsilon} = (1 + x(\varepsilon)^{2(\alpha-1)})^{-1/2} \to 1 \quad \text{as} \quad \varepsilon \to 0^+, \tag{4.1.39}$$

and therefore, (4.1.38) implies that $\dim_B(A,\Omega) = D$ and $\mathscr{M}^D(A,\Omega) = 1/(\alpha+1)$.
Using (4.1.38) again, we have that

$$0 \le f(\varepsilon) := \frac{1}{\alpha+1} - \frac{|A_\varepsilon \cap \Omega|}{\varepsilon^{2-D}} \le \frac{1}{\alpha+1}\Big(1 - \Big(\frac{x(\varepsilon)}{\varepsilon}\Big)^{\alpha+1}\Big). \tag{4.1.40}$$

Using (4.1.39) and the binomial expansion, we conclude that

$$\Big(\frac{x(\varepsilon)}{\varepsilon}\Big)^{\alpha+1} = 1 - \frac{\alpha+1}{2} x(\varepsilon)^{2\alpha-2} + o(x(\varepsilon)^{2\alpha-2}) \quad \text{as} \quad \varepsilon \to 0^+.$$

Therefore, we deduce from (4.1.40) that

$$f(\varepsilon) = O(x(\varepsilon^{2\alpha-2})) = O(\varepsilon^{2\alpha-2}) \quad \text{as} \quad \varepsilon \to 0^+.$$

Since $|A_\varepsilon \cap \Omega| = \varepsilon^{2-D}((\alpha+1)^{-1} + f(\varepsilon))$, by using Theorem 2.3.18 (adjusted to the case of RFDs, see Theorem 4.5.1), we then conclude that

$$D_{\mathrm{mer}}(\zeta_{A,\Omega}) \le D - (2\alpha - 2) = 3(1-\alpha).$$

Furthermore, $s = D$ is a simple pole.
Finally, we note that the equality $D(\zeta_{A,\Omega}) = D$ follows from (4.1.10). □

Example 4.1.36. Let (A,Ω) be the relative fractal drum in $\mathbb{R}^2$ defined by $A = \{(0,0)\}$ and $\Omega = \{(x,y) \in \mathbb{R}^2 : 0 < y < x^2,\ x \in (0,1)\}$; see Figure 4.2, for $\alpha = 2$. This relative fractal drum does not satisfy the cone property (at any point). (Note that since $\overline{A} \cap \partial\Omega = \{(0,0)\}$, it suffices to check that (A,Ω) does not have the cone property at $a = (0,0)$, which is the case since $2 > 1$; see Remark 4.1.30 and Example 4.1.31.) According to Proposition 4.1.35, its relative box dimension is equal to -1. We will show directly that the relative distance zeta function $\zeta_{A,\Omega}(s)$ is well defined at $s = 0$, and equal to Catalan's constant. First, using polar coordinates (r,θ), we obtain that for every $s > 0$,

$$\begin{aligned}\zeta_{A,\Omega}(s) &= \int_\Omega d((x,y),A)^{s-2}\,\mathrm{d}x\,\mathrm{d}y = \int_0^1 \mathrm{d}x \int_0^{x^2} (\sqrt{x^2+y^2})^{s-2}\,\mathrm{d}y \\ &= \int_0^{\pi/4} \mathrm{d}\theta \int_{\tan\theta/\cos\theta}^{1/\cos\theta} r^{s-1}\,\mathrm{d}r = \frac{1}{s}\int_0^{\pi/4} \frac{1-\tan^s\theta}{\cos^s\theta}\,\mathrm{d}\theta.\end{aligned}$$

The function under the integral sign is dominated by a constant (independent of s), so we conclude from the Lebesgue dominated convergence theorem that the integral in the last expression above converges to zero. We can now apply l'Hospital's rule and differentiate under the integral sign in order to compute the limit at $s = 0$:

$$\begin{aligned}\lim_{s\to 0^+}\zeta_{A,\Omega}(s) &= \lim_{s\to 0^+}\int_0^{\pi/4}\frac{\partial}{\partial s}\Big(\frac{1-\tan^s\theta}{\cos^s\theta}\Big)\,\mathrm{d}\theta\\ &= \lim_{s\to 0^+}\int_0^{\pi/4}\left[\Big(\frac{\tan\theta}{\cos\theta}\Big)^s\log(\cot\theta)+\frac{\log(\cos\theta)}{\cos^s\theta}(\tan^s\theta-1)\right]\mathrm{d}\theta\\ &= \int_0^{\pi/4}\log(\cot\theta)\,\mathrm{d}\theta=\sum_{n=0}^{\infty}\frac{(-1)^n}{(2n+1)^2}.\end{aligned}$$

The next-to-last equality again follows from an application of Lebesgue's dominated convergence theorem, while the last sum is *Catalan's constant*, which is approximately equal to 0.915.

In the following lemma, we show that for any $\delta>0$, the respective sets of principal complex dimensions corresponding to RFDs (A,Ω) and $(A,A_\delta\cap\Omega)$ coincide.

Lemma 4.1.37. *Assume that (A,Ω) is a relative fractal drum in $\mathbb{R}^N$. Then for any $\delta>0$ we have*

$$\zeta_{A,\Omega}(s)\sim\zeta_{A,A_\delta\cap\Omega}(s). \tag{4.1.41}$$

In particular,

$$\dim_{PC}(A,\Omega)=\dim_{PC}(A,A_\delta\cap\Omega) \tag{4.1.42}$$

and therefore,

$$\overline{\dim}_B(A,\Omega)=\overline{\dim}_B(A,A_\delta\cap\Omega). \tag{4.1.43}$$

Here, the δ-neighborhood of A can be taken with respect to any norm on $\mathbb{R}^N$.[3]

Proof. Recall that according to the definition of a relative fractal drum (A,Ω), there exists $\delta_1>0$ such that $d(x,A)<\delta_1$ for all $x\in\Omega$; see Definition 4.1.2. On the other hand, we have that $d(x,A)>\delta$ for all $x\in\Omega\setminus A_\delta$. Therefore, by using Theorem 2.1.45 with $\varphi(x):=d(x,A)$ and $\mathrm{d}\mu(x):=d(x,A)^{-N}\mathrm{d}x$, we conclude that the difference

$$\zeta_{A,\Omega}(s)-\zeta_{A,A_\delta\cap\Omega}(s)=\int_{\Omega\setminus A_\delta}d(x,A)^{s-N}\mathrm{d}x$$

defines an entire function. This proves the desired equivalence in (4.1.41). The remaining claims of the lemma follow immediately from this equivalence. Finally, the fact that any norm on $\mathbb{R}^N$ can be chosen to define A_δ follows from the equivalence of all the norms on $\mathbb{R}^N$. □

The following result provides an example of a nontrivial relative fractal drum (A,Ω) such that $\dim_B(A,\Omega)=-\infty$. It suffices to construct a domain Ω in $\mathbb{R}^2$ which is *flat* in a connected open neighborhood of one of its boundary points.

Corollary 4.1.38 (A maximally flat RFD). *Let $A=\{(0,0)\}$ and*[4]

$$\Omega'=\{(x,y)\in\mathbb{R}^2:0<y<\mathrm{e}^{-1/x},\ 0<x<1\}. \tag{4.1.44}$$

---

[3] This fact will be used in an essential manner in the proof of Corollary 4.1.38.

[4] The corresponding RFD (A,Ω') is very similar to the RFD (A,Ω) exhibited in Figure 4.2, but now with an extremely sharp spike at the origin.

Then $\dim_B(A,\Omega')$ *exists and*

$$\dim_B(A,\Omega') = D(\zeta_{A,\Omega'}) = -\infty. \tag{4.1.45}$$

Proof. Let us fix $\alpha > 1$. Then, by l'Hospital's rule,

$$\lim_{x\to 0^+} \frac{e^{-1/x}}{x^\alpha} = \lim_{t\to+\infty} \frac{t^\alpha}{e^t} = 0.$$

Hence, there exists $\delta = \delta(\alpha) > 0$ such that $0 < e^{-1/x} < x^\alpha$ for all $x \in (0,\delta)$; that is,

$$\Omega'_{\delta(\alpha)} \subset \Omega_{\delta(\alpha)},$$

where

$$\Omega'_{\delta(\alpha)} := \{(x,y) \in \mathbb{R}^2 : 0 < y < e^{-1/x},\ 0 < x < \delta(\alpha)\}$$

and

$$\Omega_{\delta(\alpha)} := \{(x,y) \in \mathbb{R}^2 : 0 < y < x^\alpha,\ 0 < x < \delta(\alpha)\}.$$

Using Lemma 4.1.37 (with Ω' instead of Ω and with the ∞-norm on $\mathbb{R}^2$ instead of the usual Euclidean norm)[5] and Proposition 4.1.35, we see that

$$\overline{\dim}_B(A,\Omega') = \overline{\dim}_B(A,\Omega'_{\delta(\alpha)}) \le \dim_B(A,\Omega_{\delta(\alpha)}) = 1-\alpha.$$

The claim follows by letting $\alpha \to +\infty$, since then, we have that

$$-\infty \le \underline{\dim}_B(A,\Omega') \le \overline{\dim}_B(A,\Omega') = -\infty.$$

We conclude, as desired, that $\dim_B(A,\Omega)$ exists and is equal to $-\infty$. □

Remark 4.1.39. (*Flatness and 'infinitely sharp blade'*). It is easy to see that Corollary 4.1.38 can be significantly generalized. For example, it suffices to assume that A is a point on the boundary of Ω such that Ω has the *flatness property of* A *relative to* Ω. This can even be formulated in terms of subsets A. We can imagine a bounded open set $\Omega \subset \mathbb{R}^3$ with a Lipschitz boundary $\partial\Omega$, except on a subset $A \subset \partial\Omega$, which may be a line segment, near which Ω is flat. A simple construction of such a set is $\Omega = \Omega' \times (0,1)$, where Ω' is given as in Corollary 4.1.38, and $A = \{(0,0)\} \times (0,1)$; see Equation (4.1.44). Note that this domain is not Lipschitz near the points of A, and not even Hölderian; see Figure 4.4. The *flatness of a relative fractal drum* (A,Ω) can be defined by

$$\mathrm{fl}(A,\Omega) = \left(\overline{\dim}_B(A,\Omega)\right)^-,$$

where $(r)^- := \max\{0,-r\}$ is the negative part of a real number r. *We say that the flatness of* (A,Ω) *is nontrivial if* $\mathrm{fl}(A,\Omega) > 0$*, that is, if* $\overline{\dim}_B(A,\Omega) < 0$. In the example just mentioned above, we have a relative fractal drum (A,Ω) with infinite

[5] Note that $\Omega'_{\delta(\alpha)} = \Omega' \cap B_{\delta(\alpha)}(0)$, where $B_\delta(0) := \{(x,y) \in \mathbb{R}^2 : |(x,y)|_\infty < \delta\}$ and $|(x,y)|_\infty := \max\{|x|,|y|\}$.

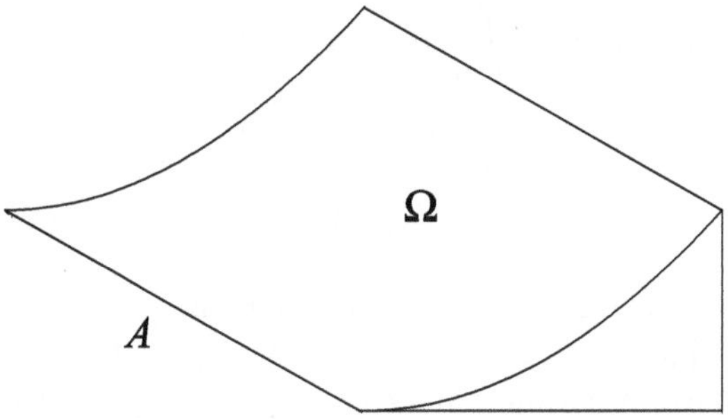

Fig. 4.4 A relative fractal drum (A,Ω) with infinite flatness, as described in Remark 4.1.39. In other words, Ω has infinite flatness near A; equivalently, $\dim_B(A,\Omega) = -\infty$, which provides an even more dramatic illustration of *the drop in dimension phenomenon* (for relative box dimensions).

flatness, i.e., with $\mathrm{fl}(A,\Omega) = +\infty$. Intuitively, it can be viewed as an 'ax' with an 'infinitely sharp' blade.

4.1.3 Scaling Property of Relative Zeta Functions

We start with the following result, which shows that if (A,Ω) is a given relative fractal drum, then for any $\lambda > 0$, the zeta function $\zeta_{\lambda A,\lambda\Omega}(s)$ of the scaled relative fractal drum $(\lambda A,\lambda\Omega)$ is equal to the zeta function $\zeta_{A,\Omega}(s)$ of (A,Ω) multiplied by λ^s. This result extends Proposition 2.1.77.

Theorem 4.1.40 (Scaling property of relative distance zeta functions). *Let $\zeta_{A,\Omega}(s)$ be the relative distance zeta function. Then, for any positive real number λ, we have that $D(\zeta_{\lambda A,\lambda\Omega}) = D(\zeta_{A,\Omega}) = \overline{\dim}_B(A,\Omega)$ and*

$$\zeta_{\lambda A,\lambda\Omega}(s) = \lambda^s \zeta_{A,\Omega}(s), \tag{4.1.46}$$

for $\mathrm{Re}\, s > \overline{\dim}_B(A,\Omega)$ *and any* $\lambda > 0$. (*See also Corollary 4.1.42 below for a more general statement.*)

Proof. The claim is established by introducing a new variable $y = x/\lambda$, and by noting that $d(\lambda y,\lambda A) = \lambda\, d(y,A)$, for any $y \in \mathbb{R}^N$ (which is an easy consequence of the homogeneity of the Euclidean norm). Indeed, in light of Remark 4.1.8 or part (*b*) of Theorem 4.1.7, for $s \in \mathbb{C}$ with $\mathrm{Re}\, s > \overline{\dim}_B(A,\Omega) = D(\zeta_{A,\Omega})$, we have successively:

$$\begin{aligned}\zeta_{\lambda A,\lambda\Omega}(s) &= \int_{\lambda\Omega} d(x,\lambda A)^{s-N}\mathrm{d}x \\ &= \int_{\Omega} d(\lambda y,\lambda A)^{s-N}\lambda^N \mathrm{d}y \\ &= \lambda^s \int_{\Omega} d(y,A)^{s-N}\mathrm{d}y = \lambda^s \zeta_{A,\Omega}(s).\end{aligned}$$

It follows that (4.1.46) holds and $\zeta_{\lambda A,\lambda\Omega}(s)$ is holomorphic for $\operatorname{Re}s > \overline{\dim}_B(A,\Omega)$. Since $D(\zeta_{A,\Omega}) = \overline{\dim}_B(A,\Omega)$ (by part (*b*) of Theorem 4.1.7 and by Remark 4.1.8, as was recalled above), we deduce that $D(\zeta_{\lambda A,\lambda\Omega}) \leq D(\zeta_{A,\Omega})$, for every $\lambda > 0$. But then, replacing λ with its reciprocal λ^{-1} in this last inequality, we obtain the reverse inequality,[6] and hence, we conclude that

$$\overline{\dim}_B(A,\Omega) = D(\zeta_{A,\Omega}) = D(\zeta_{\lambda A,\lambda\Omega}),$$

for all $\lambda > 0$, as desired. □

If $\mathscr{L} = (\ell_j)_{j\geq 1}$ is a fractal string and λ is a positive constant, then for the scaled string $\lambda\mathscr{L} := (\lambda\ell_j)_{j\geq 1}$, the corresponding claim in Theorem 4.1.40 is trivial: $\zeta_{\lambda\mathscr{L}}(s) = \lambda^s\zeta_{\mathscr{L}}(s)$, for every $\lambda > 0$. Indeed, by definition of the geometric zeta function of a fractal string (see Equation (2.1.71) in Section 2.1.4), we have

$$\zeta_{\lambda\mathscr{L}}(s) = \sum_{j=1}^{\infty}(\lambda\ell_j)^s = \lambda^s\sum_{j=1}^{\infty}\ell_j^s = \lambda^s\zeta_{\mathscr{L}}(s),$$

for $\operatorname{Re}s > D(\zeta_{\mathscr{L}})$. (The exact same argument as above then shows that $D(\zeta_{\mathscr{L}}) = D(\zeta_{\lambda\mathscr{L}})$.) Then, by analytic (i.e., meromorphic) continuation, the same identity continues to hold in any domain to which $\zeta_{\mathscr{L}}$ can be meromorphically extended to the left of the critical line $\{\operatorname{Re}s = D(\zeta_{\mathscr{L}})\}$.

Remark 4.1.41. Let $\mathscr{L} := (A,\Omega)$ be a relative fractal drum in $\mathbb{R}^N$ and let $\lambda\mathscr{L} := (\lambda A,\lambda\Omega)$, where $\lambda > 0$. If we define $\zeta_{\mathscr{L}}(s) := \zeta_{A,\Omega}(s) = \int_\Omega d(x,A)^{s-N}\mathrm{d}x$, then we can reformulate Theorem 4.1.40 as follows: $D(\zeta_{\lambda\mathscr{L}}) = D(\zeta_{\mathscr{L}}) = \overline{\dim}_B\mathscr{L}$ and

$$\zeta_{\lambda\mathscr{L}}(s) = \lambda^s\zeta_{\mathscr{L}}(s), \quad \text{for} \quad \operatorname{Re}s > \overline{\dim}_B\mathscr{L} \quad \text{and} \quad \lambda > 0. \tag{4.1.47}$$

More explicitly,

$$\zeta_{\lambda A,\lambda\Omega}(s) = \lambda^s\zeta_{A,\Omega}(s), \quad \text{for} \quad \operatorname{Re}s > \overline{\dim}_B(A,\Omega) \quad \text{and} \quad \lambda > 0. \tag{4.1.48}$$

Clearly, in light of the principle of analytic continuation, the identities (4.1.47) and (4.1.48) continue to hold for all $s \in U$, where U is any domain of $\mathbb{C}$ to which $\zeta_{\mathscr{L}}$ can be meromorphically continued.

[6] More specifically, we replace (A,Ω) with $(\lambda^{-1}A,\lambda^{-1}\Omega)$ to deduce that for every $\lambda > 0$, $D(\zeta_{A,\Omega}) \leq D(\zeta_{\lambda^{-1}A,\lambda^{-1}\Omega})$. We then substitute λ^{-1} for λ in this last inequality in order to obtain the desired reversed inequality: for every $\lambda > 0$, $D(\zeta_{A,\Omega}) \leq D(\zeta_{\lambda A,\lambda\Omega})$.

The following result supplements Theorem 4.1.40 in several different and significant ways.

Corollary 4.1.42. *Fix $\lambda > 0$. Assume that $\zeta_{A,\Omega}$ admits a meromorphic continuation to some open connected neighborhood U of the open half-plane $\{\mathrm{Re}\, s > \overline{\dim}_B(A,\Omega)\}$. Then, so is the case for $\zeta_{\lambda A,\lambda\Omega}$ and the identity (4.1.46) continues to hold for every $s \in U$ which is not a pole of $\zeta_{A,\Omega}$ (and hence, of $\zeta_{\lambda A,\lambda\Omega}$ as well).*

Moreover, if we assume, for simplicity,[7] that ω is a simple pole of $\zeta_{A,\Omega}$ (and hence also, of $\zeta_{\lambda A,\lambda\Omega}$), then the following identity holds:[8]

$$\mathrm{res}(\zeta_{\lambda A,\lambda\Omega},\omega) = \lambda^{\omega}\,\mathrm{res}(\zeta_{A,\Omega},\omega). \tag{4.1.49}$$

Proof. The fact that $\zeta_{\lambda A,\lambda\Omega}$ is holomorphic at a given point $s \in U$ if and only if $\zeta_{A,\Omega}$ is holomorphic at s (i.e., if and only if $\mathrm{Re}\, s > \overline{\dim}_B(A,\Omega)$), follows from (4.1.46) and the equality $D(\zeta_{\lambda A,\lambda\Omega}) = D(\zeta_{A,\Omega}) = \overline{\dim}_B(A,\Omega)$. An analogous statement is true if "holomorphic" is replaced with "meromorphic". More specifically, by analytic continuation of (4.1.46), $\zeta_{\lambda A,\lambda\Omega}$ is meromorphic in the domain U (containing the critical line $\{\mathrm{Re}\, s = \overline{\dim}_B(A,\Omega)\}$) if and only if $\zeta_{A,\Omega}$ is meromorphic in U, and then, clearly, identity (4.1.46) continues to hold for every $s \in U$ which is not a pole of $\zeta_{A,\Omega}$ (and hence also, of $\zeta_{\lambda A,\lambda\Omega}$). Therefore, the first part of the corollary is established.

Next, assume that ω is a simple pole of $\zeta_{A,\Omega}$. Then, in light of (4.1.46) and the discussion in the previous paragraph, we have that for all s in a punctured neighborhood of ω (contained in U but not containing any other pole of $\zeta_{A,\Omega}$),

$$(s-\omega)\zeta_{\lambda A,\lambda\Omega}(s) = \lambda^{s}\left((s-\omega)\zeta_{A,\Omega}(s)\right). \tag{4.1.50}$$

The fact that (4.1.49) holds now follows by letting $s \to \omega$, $s \neq \omega$ in (4.1.50). Indeed, we then have

$$\mathrm{res}(\zeta_{A,\Omega},\omega) = \lim_{s\to\omega}(s-\omega)\zeta_{A,\Omega}(s),$$

and similarly for $\mathrm{res}(\zeta_{\lambda A,\lambda\Omega},\omega)$. □

This important scaling property of distance zeta functions of RFDs, established in Theorem 4.1.40 and Corollary 4.1.42, is analogous to the well-known scaling property of Hausdorff measure in Euclidean space (see, e.g., [Fal2]), but note that in the spirit of the theory of complex fractal dimensions, it now holds for all *complex* values of s (rather than just for the Hausdorff fractal dimension in the case of Hausdorff measure). See, in addition, identity (4.1.49) of Corollary 4.1.42 where a corresponding scaling property also holds for the complex fractal dimensions themselves, at the level of the residues.

---

[7] If s is a multiple pole, then an analogous scaling property holds for the principal parts (instead of the residues) of the zeta functions involved, as the reader can easily verify.

[8] If we use the notation $\mathscr{L} := (A,\Omega)$ and $\lambda\mathscr{L} := (\lambda A,\lambda\Omega)$ from Remark 4.1.41, Equation (4.1.49) can be written more compactly as $\mathrm{res}(\zeta_{\lambda\mathscr{L}},\omega) = \lambda^{\omega}\,\mathrm{res}(\zeta_{\mathscr{L}},\omega)$.

The scaling property of relative zeta functions (established in Theorem 4.1.40 and Corollary 4.1.42) motivates us to introduce the notion of relative fractal spray (Definition 4.2.1), which is very close to (but also subtly different from) the usual notion of fractal spray introduced by the first author and Carl Pomerance in [LapPo3] (see [Lap-vFr3] and the references therein). First, we define the operation of union of (disjoint) families of RFDs (Definition 4.1.43).

Definition 4.1.43. (*Union of relative fractal drums*). Let $(A_j,\Omega_j)_{j\geq 1}$ be a countable family of relative fractal drums in $\mathbb{R}^N$, such that the corresponding family of open sets $(\Omega_j)_{j\geq 1}$ is disjoint (i.e., $\Omega_j\cap\Omega_k=\emptyset$ for $j\neq k$), and the set $\Omega:=\cup_{j=1}^{\infty}\Omega_j$ is of finite N-dimensional Lebesgue measure (but may be unbounded). Then, the *union of the* (finite or countable) *family of relative fractal drums* (A_j,Ω_j) $(j\geq 1)$ is the relative fractal drum (A,Ω), where $A:=\cup_{j=1}^{\infty}A_j$ and $\Omega:=\cup_{j=1}^{\infty}\Omega_j$. We write

$$(A,\Omega)=\bigcup_{j=1}^{\infty}(A_j,\Omega_j). \tag{4.1.51}$$

It is easy to derive the following countable additivity property of the distance zeta functions.

Theorem 4.1.44. *Assume that $(A_j,\Omega_j)_{j\geq 1}$ is a finite or countable family of RFDs satisfying the conditions of Definition 4.1.43, and let (A,Ω) be its union (in the sense of Definition 4.1.43). Furthermore, assume that the following condition is fulfilled:*

$$\text{For any } j\in\mathbb{N} \text{ and } x\in\Omega_j, \text{ we have that } d(x,A)=d(x,A_j). \tag{4.1.52}$$

Then, for all $s\in\mathbb{C}$ such that $\operatorname{Re} s>\overline{\dim}_B(A,\Omega)$, *we have*

$$\zeta_{A,\Omega}(s)=\sum_{j=1}^{\infty}\zeta_{A_j,\Omega_j}(s). \tag{4.1.53}$$

Condition (4.1.52) *is satisfied, for example, if for every $j\in\mathbb{N}$, A_j is equal to the boundary of Ω_j in $\mathbb{R}^N$; that is, $A_j=\partial\Omega_j$.*

Proof. The claim follows from the following computation, which is valid for $\operatorname{Re} s>\overline{\dim}_B(A,\Omega)$:

$$\begin{aligned}\zeta_{A,\Omega}(s)&=\int_{\Omega}d(x,A)^{s-N}\mathrm{d}x=\sum_{j=1}^{\infty}\int_{\Omega_j}d(x,A)^{s-N}\mathrm{d}x\\&=\sum_{j=1}^{\infty}\int_{\Omega_j}d(x,A_j)^{s-N}\mathrm{d}x=\sum_{j=1}^{\infty}\zeta_{A_j,\Omega_j}(s).\end{aligned} \tag{4.1.54}$$

More specifically, clearly, (4.1.54) holds for s real such that $s>\overline{\dim}_B(A,\Omega)\geq D(\zeta_{A,\Omega})$. Therefore, for such a value of s,

$$\zeta_{A,\Omega_j}(s)=\int_{\Omega_j}d(x,A)^{s-N}\mathrm{d}x\leq\int_{\Omega}d(x,A)^{s-N}\mathrm{d}x=\zeta_{A,\Omega}(s)<\infty,$$

for every $j \geq 1$. Hence,

$$\sup_{j\geq 1}\{D(\zeta_{A,\Omega_j})\} \leq D(\zeta_{A,\Omega}) \leq \overline{\dim}_B(A,\Omega), \tag{4.1.55}$$

from which (4.1.54) now follows for all $s \in \mathbb{C}$ with $\operatorname{Re} s > \overline{\dim}_B(A,\Omega)$, in light of the countable additivity of the complex Borel measure (and hence, bounded measure) on Ω, given by $\mathrm{d}\gamma(x) := d(x,A)^{s-N}\mathrm{d}x$. Note that according to the hypothesis of Definition 4.1.43, we have $|\Omega| < \infty$, so that $d\gamma$ is indeed a complex Borel measure; see, e.g., [Foll] or [Ru]. □

Remark 4.1.45. In the statement of Theorem 4.1.44, the numerical series on the right-hand side of (4.1.53) converges absolutely (and hence, converges also in $\mathbb{C}$) for $\operatorname{Re} s > \overline{\dim}_B(A,\Omega)$. In particular, for s real such that $s > \overline{\dim}_B(A,\Omega)$, it is a convergent series of positive terms (i.e., it has a finite sum). It remains to be investigated whether (and under which hypotheses) Equation (4.1.53) continues to hold for all $s \in \mathbb{C}$ in a common domain of meromorphicity of the zeta functions $\zeta_{A,\Omega}$ and ζ_{A_j,Ω_j} for $j \geq 1$ (away from the poles). At the poles, an analogous question could be raised for the corresponding residues (assuming, for simplicity, that the poles are simple). We will encounter a similar issue when discussing 'local distance zeta functions' in Appendix B.

Since, among other things, Theorem 4.1.44 gives a way to compute the distance zeta function of a given relative fractal drum if it can be appropriately subdivided into a disjoint union of relative fractal 'subdrums', we introduce the following important definition.

Definition 4.1.46. (*Disjoint union of relative fractal drums*). Let the conditions of Definition 4.1.43 be satisfied and also assume that condition (4.1.52) is satisfied (so that the conclusion of Theorem 4.1.44 holds). Then, we call the union given in (4.1.51) a *disjoint union of relative fractal drums* and write

$$(A,\Omega) = \bigsqcup_{j=1}^{\infty}(A_j,\Omega_j). \tag{4.1.56}$$

Furthermore, in the special case when for every $j \in \mathbb{N}$, we have that $(A_j,\Omega_j) = \lambda_j(A_0,\Omega_0)$ for some sequence of positive numbers $(\lambda_j)_{j\geq 1}$ and some given relative fractal drum (A_0,Ω_0), we will slightly abuse the notation and write

$$(A,\Omega) = \bigsqcup_{j=1}^{\infty}\lambda_j(A_0,\Omega_0), \tag{4.1.57}$$

in the sense that the scaled RFDs appearing in (4.1.57) are actually isometric images of $\lambda_j(A_0,\Omega_0)$ arranged in such a way that the union (4.1.57) is indeed a disjoint union of relative fractal drums.

4.1.4 Stalactites, Stalagmites and Caves Associated With Relative Fractal Drums

In this subsection, we extend the notions of stalactites, stalagmites and caves, introduced in Subsection 2.1.6, associated with fractal sets. Let (A,Ω) be a given relative fractal drum in $\mathbb{R}^N$. Assume that

$$\Omega\setminus\overline{A}=\bigcup_{k\in J}U_k,$$

where $\{U_k\}_{k\in J}$ is the disjoint family of connected components of the open set $\Omega\setminus\overline{A}$. It is clear that the index set is at most countable. Let r be a given nonzero real number, and let us define the following function:

$$f:\Omega\to[0,+\infty],\quad f(x):=d(x,A)^r.$$

(If $r<0$, we let $0^r=+\infty$.) For each $k\in J$, we also introduce the function $f_k:=f|_{U_k}$.

Definition 4.1.47. For each $k\in J$, the graph of the function f_k is called the k-th *stalactite corresponding to the relative fractal drum* (A,Ω) (and to r). The set $\mathrm{cave}(A,\Omega)=\mathrm{cave}(A,\Omega,r)$ defined by

$$\mathrm{cave}(A,\Omega):=\{(x,u)\in\Omega\times(0,+\infty):0<u<f(x)\}$$

and contained in $\mathbb{R}^{N+1}$, is called the (A,Ω)-*cave associated with the relative fractal drum* (A,Ω) (and corresponding to r).

Note that a connected component U_k of an unbounded open set $\Omega\setminus\overline{A}$ may be unbounded. However, when $r>0$, the corresponding function f_k is bounded, due to the assumption according to which there exists $\delta>0$ such that $\Omega\subset A_\delta$; see Definitions 4.1.1 and 4.1.2.

We could now proceed with further discussion and illustrative examples, in the spirit of Subsection 2.1.6. Instead, we will limit ourselves to stating the analog of Proposition 2.1.84.

Proposition 4.1.48. *If s is a real number and $s>\overline{\dim}_B(A,\Omega)$, then the volume of the (A,Ω)-cave, corresponding to the parameter $r=s-N$, is finite.*

Proof. This follows at once from Theorem 4.1.7. □

4.2 Relative Fractal Sprays With Principal Complex Dimensions of Arbitrary Orders

In this section, we consider a special type of RFDs, called *relative fractal sprays*, and consider their distance zeta functions. We then illustrate the results obtained by computing the complex dimensions of relative Sierpiński sprays. More specifically, we determine the complex dimensions of the relative Sierpiński gasket and of the relative Sierpiński carpet; we also calculate the associated residues.

4.2.1 Relative Fractal Sprays in $\mathbb{R}^N$

We now introduce the definition of relative fractal spray, which is very similar to (but more general than) the notion of fractal spray (see [LapPo3], [Lap-vFr3, Definition 13.2], [LapPe2–3] and [LapPeWi1–2]), itself a generalization of the notion of (ordinary) fractal string [LapPo1–2, Lap1–3, Lap-vFr3].

Definition 4.2.1. Let $(\partial\Omega_0,\Omega_0)$ be a fixed relative fractal drum in $\mathbb{R}^N$ (which we call the *base relative fractal drum*, or *generating relative fractal drum* or else, simply, the *generator*), $(\lambda_j)_{j\geq 0}$ a decreasing sequence of positive numbers (scaling factors), converging to zero, and $(b_j)_{j\geq 0}$ a given sequence of positive integers (multiplicities). The associated *relative fractal spray* is a relative fractal drum (A,Ω) obtained as the disjoint union of a sequence of RFDs $\mathscr{F}:=\{(\partial\Omega_i,\Omega_i): i\in\mathbb{N}_0\}$, where $\mathbb{N}_0:=\mathbb{N}\cup\{0\}$, such that each Ω_i can be obtained from $\lambda_j\Omega_0$ by a rigid motion in $\mathbb{R}^N$, and for each $j\in\mathbb{N}_0$ there are precisely b_j RFDs in the family $\mathscr{F}$ that can be obtained from $\lambda_j\Omega_0$ by a rigid motion. Any relative fractal spray (A,Ω), generated by the base relative fractal drum (or 'basic shape') Ω_0 and the sequences of 'scales' $(\lambda_j)_{j\geq 0}$ with associated 'multiplicities' $(b_j)_{\geq 0}$, is denoted by

$$(A,\Omega):=\mathrm{Spray}(\Omega_0,(\lambda_j)_{j\geq 0},(b_j)_{j\geq 0}). \tag{4.2.1}$$

The family $\mathscr{F}$ is called the *skeleton of the spray*. The distance zeta function $\zeta_{A,\Omega}$ of the relative fractal spray is computed in Theorem 4.2.5 below.

If there exist $\lambda\in(0,1)$ and an integer $b\geq 2$ such that $\lambda_j=\lambda^j$ and $b_j=b^j$, for all $j\in\mathbb{N}_0$, then we simply write

$$(A,\Omega)=\mathrm{Spray}(\Omega_0,\lambda,b).$$

Here, it should be noted that there exist *nonsprayable* RFDs $(\partial\Omega_0,\Omega_0)$ in $\mathbb{R}^N$; see Example 4.2.13 below.

Definition 4.2.2. The relative fractal spray $(A,\Omega)=\mathrm{Spray}(\Omega_0,(\lambda_j)_{j\geq 0},(b_j)_{\geq 0})$ can be viewed as a relative fractal drum generated by $(\partial\Omega_0,\Omega_0)$ and a fractal string $\mathscr{L}=(\ell_j)_{j\geq 1}$, consisting of the decreasing sequence $(\lambda_j)_{j\geq 0}$ of positive real numbers, in which each λ_j has multiplicity b_j. Thus, we can write $(A,\Omega)=\mathrm{Spray}(\Omega_0,\mathscr{L})$. It is also convenient to view the construction of (A,Ω) in Definition 4.2.1 as the *tensor product* of the base relative fractal drum (A_0,Ω_0) and the fractal string $\mathscr{L}$:

$$(A,\Omega)=(\partial\Omega_0,\Omega_0)\otimes\mathscr{L}. \tag{4.2.2}$$

We can also define the *tensor product of two (possibly unbounded) fractal strings* $\mathscr{L}_1 = (\ell_{1j})_{j\geq 1}$ and $\mathscr{L}_2 = (\ell_{2k})_{k\geq 1}$ as the following fractal string (note that here, $\mathscr{L}_1$ and $\mathscr{L}_2$ are viewed as nonincreasing sequences of positive numbers tending to zero, but that we may have $\sum_{j=1}^{\infty} \ell_{1j} = +\infty$ or $\sum_{k=1}^{\infty} \ell_{2k} = +\infty$):

$$\mathscr{L}_1 \otimes \mathscr{L}_2 := (\ell_{1j}\ell_{2k})_{j,k\geq 1}. \tag{4.2.3}$$

The multiplicity of any $l \in \mathscr{L}_1 \otimes \mathscr{L}_2$ is equal to the number of ordered pairs of (ℓ_{1j}, ℓ_{2k}) in the Cartesian product $\mathscr{L}_1 \times \mathscr{L}_2$ of multisets such that $l = \ell_{1j}\ell_{2k}$.

We can easily modify the notion of relative fractal spray in Definition 4.2.1 in order to deal with a finite collection of K basic RFDs (or generating RFDs) $(\partial\Omega_{01}, \Omega_{01}), \dots, (\partial\Omega_{0K}, \Omega_{0K})$, similarly as in [LapPo3], [Lap-vFr3, Definition 13.2] (and [LapPe2–3, LapPeWi1–2]). A slightly more general notion would consist in replacing $(\partial\Omega_0, \Omega_0)$ with any relative fractal drum (A_0, Ω_0); see Definition 4.2.9.

It is important to stress that, from our point of view, the sets Ω_i in the definition of a relative fractal spray (Definition 4.2.1) do not have to be 'densely packed'. In fact, in general, they cannot be 'densely packed', as indicated by Example 4.2.4(*c*) below. They can just be viewed as a union of the *disjoint family* $\{(\partial\Omega_i, \Omega_i)\}_{i\geq 0}$ *of RFDs* in $\mathbb{R}^N$, where the corresponding family of open sets $\{\Omega_i\}_{i\geq 1}$ is disjoint. Its union, $\cup_{i=0}^{\infty}\Omega_i$, can even be unbounded in $\mathbb{R}^N$, although it has to be of finite N-dimensional Lebesgue measure. As an example, we can consider the family of balls $\{\Omega_i := B_{r_i}(a_i)\}_{i\geq 0}$ in $\mathbb{R}^N$, such that $|a_i| \to +\infty$ as $i \to \infty$ and $\sum_{i=0}^{\infty} r_i^N < \infty$.

The following simple lemma provides necessary and sufficient conditions for a relative fractal spray (A, Ω) to be such that $|\Omega| < \infty$.

Lemma 4.2.3. *Assume that* $(A, \Omega) := \mathrm{Spray}(\Omega_0, (\lambda_j)_{j\geq 0}, (b_j)_{\geq 0})$ *in* $\mathbb{R}^N$ *is a relative fractal spray. Then* $|\Omega| < \infty$ *if and only if* $|\Omega_0| < \infty$ *and*

$$\sum_{j=0}^{\infty} b_j \lambda_j^N < \infty. \tag{4.2.4}$$

In that case, we have

$$|\Omega| = |\Omega_0| \sum_{j=0}^{\infty} b_j \lambda_j^N. \tag{4.2.5}$$

In particular, the relative fractal drum (A, Ω) *is well defined and* $\overline{\dim}_B(A, \Omega) \leq N$.

Proof. Let us prove the sufficiency part. For $\Omega_j = \lambda_j \Omega_0$ we have $|\Omega_j| = |\lambda_j \Omega_0| = \lambda_j^N |\Omega_0|$, and therefore,

$$|\Omega| = \sum_{j=0}^{\infty} |\Omega_j| = \sum_{j=0}^{\infty} b_j |\lambda_j \Omega_0| = |\Omega_0| \sum_{j=0}^{\infty} b_j \lambda_j^N.$$

The proof of the necessity part is also easy and is therefore omitted. □

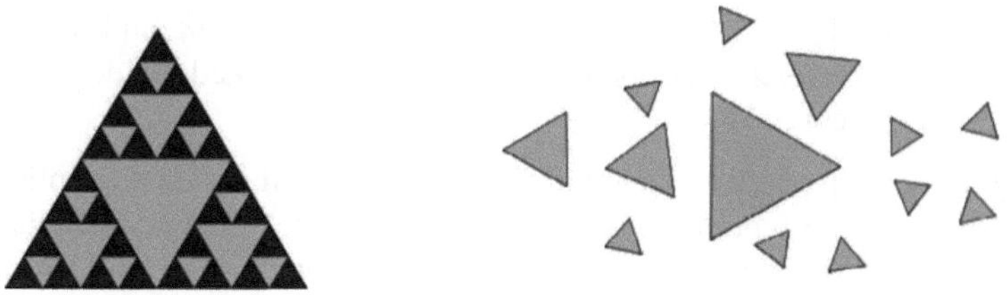

Fig. 4.5 *Left*: The Sierpiński gasket A, viewed as a relative fractal drum (A,Ω), with Ω being the countable disjoint union of open triangles contained in the unit triangle Ω_0. *Right*: An equivalent interpretation of the Sierpiński gasket drum (A,Ω). Here, Ω is a countable disjoint union of open equilateral triangles, and $A=\partial\Omega$. (There are 3^{j-1} triangles with sides 2^{-j} in the union, with $j\in\mathbb{N}$.) Both pictures depict the first three iterations of the construction. We can also view the standard Sierpiński gasket A as a relative fractal drum (A,Ω), in which Ω is just the open unit triangle in the left picture.

Example 4.2.4. Here, we provide a few simple examples of relative fractal sprays:

(a) The ternary Cantor set can be viewed as a relative fractal drum

$$(A,\Omega)=\mathrm{Spray}(\Omega_0,1/3,2)$$

(or the *Cantor relative fractal drum*, or the *relative Cantor fractal spray*), generated by

$$(\partial\Omega_0,\Omega_0)=(\{1/3,2/3\},(1/3,2/3))$$

as the base relative fractal drum, $\lambda=1/3$ and $b=2$. Its relative box dimension is given by $D=\log_3 2$. Of course, this is just an example of ordinary fractal string, namely, the well-known Cantor string.

(b) The Sierpiński gasket can be viewed as a relative fractal drum (or the *Sierpiński relative fractal drum*, or *Sierpiński relative fractal spray*), generated by $(\partial\Omega_0,\Omega_0)$ as the basic relative fractal drum, where Ω_0 is an open equilateral triangle of sides of length $1/2$, $\lambda=1/2$ and $b=3$. Its relative box dimension is given by $D=\log_2 3$.

(c) If Ω_0 is any bounded open set in $\mathbb{R}^2$ (say, an open disk), $\lambda=1/2$ and $b=3$, we obtain a fractal spray $(A,\Omega)=\mathrm{Spray}(\Omega_0,1/2,3)$, in the sense of Definition 4.2.1. In Theorem 4.2.5, we shall see that if Ω_0 has a Lipschitz boundary, then the set of poles of the relative zeta function of this fractal spray (which is a relative fractal drum), as well as the multiplicities of the poles, do not depend on the choice of Ω_0. In this sense, examples (b) and (c) are equivalent. In particular, the box dimension of the *generalized Sierpiński relative fractal drum* is constant, and equal to $D=\log_2 3$.

In other words, the Sierpiński gasket $(A,\Omega)=\mathrm{Spray}(\Omega_0,1/2,3)$, appearing in Example 4.2.4(b), can be viewed as *any* countable disjoint collection of open triangles in the plane (which can be even an unbounded collection) and their bounding triangles, of sizes $\lambda_j = 2^{-j}$ and multiplicities $b_j = 3^j$, $j \in \mathbb{N}_0$, and not just as the standard disjoint collection of open triangles, densely packed inside the unit open triangle. See Figure 4.5.

Using the scaling property stated in Theorem 4.1.40, it is easy to explicitly compute the distance zeta function of relative fractal sprays. Note that the zeta function involves the Dirichlet series $f(s) := \sum_{j=0}^{\infty} b_j \lambda_j^s$. Theorem 4.2.5 just below can be considered as an extension of Theorem 4.1.40.

Theorem 4.2.5 (Distance zeta function of relative fractal sprays). *Let*

$$(A,\Omega) = \mathrm{Spray}(\Omega_0,(\lambda_j)_{j\geq 0},(b_j)_{j\geq 0})$$

be a relative fractal spray in $\mathbb{R}^N$, in the sense of Definition 4.2.1, and such that $|\Omega_0| < \infty$. Assume that condition (4.2.4) of Lemma 4.2.3 is satisfied; that is, $|\Omega| < \infty$. Let Ω be the (countable, disjoint) union of all the open sets appearing in the skeleton, corresponding to the fractal spray. In other words, Ω is the disjoint union of the open sets Ω_j, each repeated with the multiplicity b_j for $j \in \mathbb{N}_0$. Let $f(s) := \sum_{j=0}^{\infty} b_j \lambda_j^s$.[9] *Then, for* $\mathrm{Re}\, s > \max\{\overline{\dim}_B(A,\Omega), D(f)\}$, *the distance zeta function of the relative fractal spray (A,Ω) is given by the* factorization formula

$$\zeta_{A,\Omega}(s) = \zeta_{\partial\Omega_0,\Omega_0}(s) \cdot f(s), \tag{4.2.6}$$

and

$$\overline{\dim}_B(A,\Omega) = \max\{\overline{\dim}_B(\partial\Omega_0,\Omega_0), D(f)\}. \tag{4.2.7}$$

Proof. Clearly, it follows from (4.2.4) that $f(N) < \infty$. Hence, $D(f) \leq N$; so that $\overline{\dim}_B(A,\Omega) \leq N$. Each open set of the skeleton of the relative fractal spray is obtained by a rigid motion of sets of the form $\lambda_j \Omega_0$, and for any fixed $j \in \mathbb{N}_0$, there are precisely b_j such sets. Identity (4.2.6) then follows immediately from Theorems 4.1.40 and 4.1.44. The remaining claims are easily derived by using this identity. □

Note that it follows from Definition 4.2.2 and relation (4.2.6) that the distance zeta function of the tensor product is equal to the product of the zeta functions of its components:

$$\zeta_{(\partial\Omega_0,\Omega_0)\otimes\mathscr{L}}(s) = \zeta_{\partial\Omega_0,\Omega_0}(s) \cdot \zeta_{\mathscr{L}}(s). \tag{4.2.8}$$

Equation (4.2.7) can therefore be written as follows:

$$\overline{\dim}_B((\partial\Omega_0,\Omega_0)\otimes\mathscr{L}) = \max\{\overline{\dim}(\partial\Omega_0,\Omega_0), \overline{\dim}_B\mathscr{L}\}. \tag{4.2.9}$$

Theorem 4.2.6. *Assume that a relative fractal spray* $(A,\Omega) = \mathrm{Spray}(\Omega_0,\lambda,b)$, *introduced at the end of Definition 4.2.1, is such that $|\Omega_0| < \infty$, $\lambda \in (0,1)$, $b \geq 2$ is an integer, and $b\lambda^N < 1$. Then, for* $\mathrm{Re}\, s > \max\{\overline{\dim}_B(\partial\Omega_0,\Omega_0), \log_{1/\lambda} b\})$, *we have*

[9] Note that according to (4.2.4), this Dirichlet series converges absolutely for $\mathrm{Re}\, s \geq N$; hence, $D(f) \leq N$.

$$\zeta_{A,\Omega}(s) = \frac{\zeta_{\partial\Omega_0,\Omega_0}(s)}{1-b\lambda^s}, \tag{4.2.10}$$

and the lower bound for Re s *is optimal. In particular, it is equal to* $D(\zeta_{A,\Omega})$, *and*

$$\overline{\dim}_B(A,\Omega) = D(\zeta_{A,\Omega}) = \max\{\overline{\dim}_B(\partial\Omega_0,\Omega_0), \log_{1/\lambda} b\}.$$

If, in addition, Ω_0 *is bounded and has a Lipschitz boundary* $\partial\Omega_0$ *which can be described by finitely many Lipschitz charts, then* $\dim_B(A,\Omega)$ *exists and*

$$\dim_B(A,\Omega) = \max\{N-1, \log_{1/\lambda} b\}. \tag{4.2.11}$$

If we assume that $\log_{1/\lambda} b \in (N-1,N)$, *then the set* $\dim_{PC}(A,\Omega) = \mathscr{P}_c(\zeta_{A,\Omega})$ *of principal complex dimensions of the relative fractal spray* (A,Ω) *is given by*

$$\dim_{PC}(A,\Omega) = \log_{1/\lambda} b + \frac{2\pi}{\log(1/\lambda)}\mathrm{i}\mathbb{Z}. \tag{4.2.12}$$

Proof. If $\lambda_j = \lambda$ and $b_j = b^j$ for all $j \in \mathbb{N}$, with $b\lambda^N < 1$, then $\sum_{j=0}^{\infty} b^j \lambda^{jN} = \frac{1}{1-b\lambda^N} < \infty$; so that $|\Omega| < \infty$. Identity (4.2.10) follows immediately from (4.2.6), by using the fact that for Ω_0 with a Lipschitz boundary satisfying the stated assumption, we have $\dim_B(\partial\Omega_0,\Omega_0) = \dim_B \partial\Omega_0 = N-1$ (this follows, for example, from [ŽuŽup2, Lemma 3]; see also [Lap1]), together with the property of finite stability of the upper box dimension; see, e.g., [Fal1, p. 44]. □

Example 4.2.7. Here, we construct a relative fractal spray

$$(A,\Omega) = \mathrm{Spray}(\Omega_0, (\lambda_j)_{j\geq 1}, (b_j)_{\geq 1})$$

in $\mathbb{R}^2$ such that $|\Omega_0| < \infty$, $b_j \equiv 1$, $\sum_{j=1}^{\infty} \lambda_j^2 < \infty$ (hence, $|\Omega| < \infty$ by Lemma 4.2.4), and such that the base set Ω_0 is *unbounded*, as well as its boundary $\partial\Omega_0$. Let Ω_0 be any unbounded Borel set of finite 2-dimensional Lebesgue measure, such that both Ω_0 and $\partial\Omega_0$ are unbounded, and Ω_0 is contained in a horizontal strip

$$V_1 := \{(x,y) \in \mathbb{R}^2 : 0 < y < 1\}.$$

We can construct such a set explicitly as

$$\Omega_0 = \{(x,y) \in \mathbb{R}^2 : 0 < y < x^{-\alpha},\ x > 1\},$$

where $\alpha > 1$, so that $|\Omega_0| < \infty$.

Let $(V_j)_{j\geq 1}$ be a countable, disjoint sequence of horizontal strips in the plane, defined by $V_j = V_1 + (0,j)$ for each $j \in \mathbb{N}$. Let $(\lambda_j)_{j\geq 1}$ be a sequence of real numbers in $(0,1)$ such that $\sum_{j=1}^{\infty} \lambda_j^2 < \infty$. It is clear that for any λ_j, $j \geq 2$, the set $\lambda_j\Omega_0$ is congruent (up to a rigid motion) to the subset $\Omega_j := \lambda_j\Omega_0 + (0,j)$ of V_j. Then, the fractal spray

$$(A,\Omega) = \bigcup_{j=1}^{\infty} (\partial\Omega_j, \Omega_j)$$

has the desired properties.

It is clear that the tensor product introduced in Definition 4.2.2 is associative, in the following sense:[10]

$$((A_0,\Omega_0)\otimes\mathscr{L}_1)\otimes\mathscr{L}_2=(A_0,\Omega_0)\otimes(\mathscr{L}_1\otimes\mathscr{L}_2). \tag{4.2.13}$$

This equation shows that the tensor product defines the *action* of bounded fractal strings $\mathscr{L}$ on the set of relative fractal drums. [Here, we consider the set of bounded fractal strings (with the trivial strings included and equipped with the tensor product $\otimes$) as a monoid, where the identity element is the trivial fractal string consisting of only one length $\ell=1$.] We can therefore extend Theorem 4.2.5 as follows.

Theorem 4.2.8. *Assume that (A_0,Ω_0) is a base relative fractal drum in $\mathbb{R}^N$, and let $(\mathscr{L}_k)_{k\geq 0}$ be a sequence of fractal strings. Let (A_k,Ω_k), $k\geq 1$, be a sequence of relative fractal sprays defined by*

$$(A_k,\Omega_k)=(A_{k-1},\Omega_{k-1})\otimes\mathscr{L}_{k-1}. \tag{4.2.14}$$

Then

$$(A_k,\Omega_k)=(A_0,\Omega_0)\otimes\left(\bigotimes_{j=0}^{k-1}\mathscr{L}_j\right). \tag{4.2.15}$$

Furthermore, for each $k\geq 1$, we have

$$\zeta_{A_k,\Omega_k}(s)=\zeta_{A_0,\Omega_0}(s)\cdot\prod_{j=0}^{k-1}\zeta_{\mathscr{L}_j}(s), \tag{4.2.16}$$

for all $s\in\mathbb{C}$ with $\operatorname{Re}s>\max\{\overline{\dim}_B(A_0,\Omega_0),\overline{\dim}_B\mathscr{L}_0,\dots,\overline{\dim}_B\mathscr{L}_{k-1}\}$, and

$$\overline{\dim}_B(A_k,\Omega_k)=\max\{\overline{\dim}_B(A_0,\Omega_0),\overline{\dim}_B\mathscr{L}_0,\dots,\overline{\dim}_B\mathscr{L}_{k-1}\}. \tag{4.2.17}$$

Proof. Relation (4.2.15) follows easily by induction, using the associativity of the tensor product. The remaining claims then follow much as in the proof of Theorem 4.2.5. □

We close this subsection by providing the following generalization of the notion of fractal spray, which is quite natural in our context.

Definition 4.2.9. A *relative fractal spray* is defined exactly as a fractal spray in Definition 4.2.1, except that the generator of the spray is now allowed to be an arbitrary relative fractal drum (A_0,Ω_0), where $A_0\subseteq\mathbb{R}^N$ is arbitrary and $\Omega_0\subseteq\mathbb{R}^N$ is open, but not necessarily bounded; see Definition 4.1.2. (We assume that $\Omega_0\subseteq(A_0)_\delta$, for some $\delta>0$. In addition, we may also require that the total volume of the spray be finite: $|\Omega|<\infty$.) The corresponding relative fractal spray (A,Ω) is denoted by

$$(A,\Omega):=\operatorname{Spray}((A_0,\Omega_0),(\lambda_j)_{j\geq 0},(b_j)_{\geq 0}). \tag{4.2.18}$$

[10] This equality should be understood modulo isometric displacements of scaled copies of (A_0,Ω_0).

In the special case when $\lambda_j = \lambda^j$ and $b_j = b^j$, $j \geq 0$, where $\lambda \in (0,1)$ and an integer $b \geq 2$ are fixed, the corresponding relative fractal spray is denoted by $(A,\Omega) := \mathrm{Spray}((A_0,\Omega_0),\lambda,b)$.

For example, 'spraying' a given relative fractal spray $\mathrm{Spray}((A_0,\Omega_0),\lambda_0,b_0)$ is also possible:

$$(A_1,\Omega_1) = \mathrm{Spray}(\mathrm{Spray}((A_0,\Omega_0),\lambda_0,b_0),\lambda_1,b_1). \tag{4.2.19}$$

By continuing to spray as in Equation (4.2.19), we can define *iterated relative fractal sprays* (A_n,Ω_n) inductively by

$$(A_n,\Omega_n) = \mathrm{Spray}((A_{n-1},\Omega_{n-1}),\lambda_n,b_n), \quad \text{for each} \quad n \geq 1. \tag{4.2.20}$$

The notion of a relative fractal spray will be used in several places in the remainder of this chapter as well as in Chapters 5–6, most often without explicit mention. We leave it to the reader (or to future work) to further explore some of the additional properties of relative fractal sprays and their relative (distance or tube) zeta functions, defined as in Definition 4.1.1 and by using Theorem 4.1.40.

4.2.2 Principal Complex Dimensions of Arbitrary Multiplicities

The goal of this subsection is to show how one can effectively construct fractal sets (as well as fractal strings and even RFDs) which have poles along the critical line (i.e., principal complex dimensions) of any given order (i.e., multiplicity), and even infinitely many essential singularities (see Theorem 4.2.19 and Remark 4.2.21 below). Such fractal strings and more general RFDs are interesting examples of strongly hyperfractal RFDs, in the sense of part (*ii*) of Definition 4.6.23 in Subsection 4.6.3 below. The corresponding method for constructing these RFDs is explained and illustrated in the following example.

Example 4.2.10. (*Cantor sets of higher order*). We will provide here an example of a relative fractal drum of $\mathbb{R}$ such that, for any given $m \in \mathbb{N}$, its distance zeta function has an infinite set of poles of order m in arithmetic progression and located on the critical line. The construction is based on an 'iterated Cantor set', as we now explain.

Let C be the standard middle-third Cantor set contained in $[0,1]$ and let $\Omega := (0,1)$. Then, let (C,Ω) be our base relative fractal drum and let $\mathscr{L} := \mathscr{L}_{CS}$ be the Cantor string with total length 3; that is,

$$\mathscr{L} = (1,3^{-1},3^{-1},\underbrace{3^{-2},\dots,3^{-2}}_{4 \text{ times}},\underbrace{3^{-3},\dots,3^{-3}}_{8 \text{ times}},\dots).$$

We now define the relative fractal drum (C_2,Ω_2) as the tensor product $(C,\Omega) \otimes \mathscr{L}$; see Definitions 4.2.1, 4.2.2 and Figure 4.6. Furthermore, one can see clearly that

$$(C_2,\Omega_2) = (C,\Omega) \sqcup 3^{-1}(C_2,\Omega_2) \sqcup 3^{-1}(C_2,\Omega_2), \tag{4.2.21}$$

where $\sqcup$ denotes a disjoint union of isometric images of scaled copies of (C_2,Ω_2); see Definition 4.1.46. Then, by the scaling property of the relative distance zeta function (see Theorem 4.1.40), we have

$$\zeta_{C_2,\Omega_2}(s) = \zeta_{C,\Omega}(s) + 2\,\zeta_{3^{-1}C_2,3^{-1}\Omega_2}(s) = \zeta_{C,\Omega}(s) + 2\cdot 3^{-s}\,\zeta_{C_2,\Omega_2}(s),$$

for all $s \in \mathbb{C}$ such that $\operatorname{Re} s$ is sufficiently large, or, in other words,

$$\zeta_{C_2,\Omega_2}(s) = \frac{3^s}{3^s-2}\zeta_{C,\Omega}(s) = \frac{2\cdot 3^s}{2^s s(3^s-2)^2}, \tag{4.2.22}$$

where, in the second equality, we have used the expression for $\zeta_C = \zeta_{C,\Omega}$ obtained in Example 2.1.82. In light of the principle of analytic continuation, it is clear that Equation (4.2.22) continues to hold for all $s \in \mathbb{C}$, and hence, ζ_{C_2,Ω_2} is meromorphic on all of $\mathbb{C}$ and

$$\mathscr{P}(\zeta_{C_2,\Omega_2}) = \{0\} \cup \left(\log_3 2 + \frac{2\pi}{\log 3}\mathrm{i}\mathbb{Z}\right). \tag{4.2.23}$$

Furthermore, the poles $\omega_k := \log_3 2 + \frac{2\pi \mathrm{i} k}{\log 3}$ for $k \in \mathbb{Z}$ are all of second order (i.e., of multiplicity two). We conclude that $\overline{\dim}_B(C_2,\Omega_2) = \log_3 2$. More specifically, by Theorem 5.3.16 in Chapter 5 below, and in light of expression (4.2.22) for ζ_{C_2,Ω_2}, we obtain the following exact tube formula for the second order Cantor set, valid pointwise for all $t \in (0,1)$:

$$|(C_2)_t \cap \Omega_2| = t^{1-\log_3 2}\Big(\log t^{-1} G(\log t^{-1}) + H(\log t^{-1})\Big) + 2t, \tag{4.2.24}$$

where $G,H\colon \mathbb{R} \to \mathbb{R}$ are nonconstant, bounded periodic functions with minimal period $T := \log 3$. These functions can be computed explicitly in terms of their Fourier series but the algebraic expressions for their Fourier coefficients are too complicated to be given here in a concise manner. Furthermore, we conclude from the tube formula (4.2.24) that $\dim_B(C_2,\Omega_2)$ exists and $\dim_B(C_2,\Omega_2) = \log_3 2$, and moreover, that $\mathscr{M}^D(C_2,\Omega_2) = +\infty$.

We can now repeat the above process inductively; that is, for each integer $n \geq 2$, we define the relative fractal drum (C_n,Ω_n) as a relative fractal spray generated by (C_{n-1},Ω_{n-1}) and $\mathscr{L}$; that is, $(C_n,\Omega_n) := (C_{n-1},\Omega_{n-1}) \otimes \mathscr{L}$, for each integer $n \geq 2$. Much as before, we obtain that

$$\zeta_{C_n,\Omega_n}(s) = \frac{2\cdot 3^{(n-1)s}}{2^s s(3^s-2)^n}, \quad \text{for all} \quad s \in \mathbb{C}. \tag{4.2.25}$$

The set of complex dimensions of the RFD (C_n,Ω_n) is the same as in the case when $n = 2$ (see Equation (4.2.23) above), but except at $s := 0$ (which is simple), the corresponding multiplicities are not the same and depend on n. (Hence, the *multisets* $\mathscr{P}(\zeta_{C_n,\Omega_n})$ are different for each $n \in \mathbb{N}$.) More specifically, the poles of ζ_{C_n,Ω_n} at

Fig. 4.6 The *second order Cantor set* from Example 4.2.10. Only the first four iterations are shown here. More precisely, from left to right, we have the middle-third Cantor set C in $[0,1]$, then two copies of C scaled by $1/3$, and then four copies of C scaled by $1/9$; and so on, ad infinitum.

$s := \omega_k = \log_3 2 + \frac{2\pi \mathrm{i} k}{\log 3}$ for each $k \in \mathbb{Z}$ are of order n and $D := \dim_B(C_2, \Omega_2) = \log_3 2$. Furthermore, again by Theorem 5.3.16, we have the following exact tube formula, valid pointwise for all $t \in (0,1)$:

$$|(C_n)_t \cap \Omega_n| = t^{1-\log_3 2} \sum_{i=1}^{n} (\log t^{-1})^{i-1} G_i(\log t^{-1}) + 2t, \tag{4.2.26}$$

where for $i = 1, \dots, n$, $G_i \colon \mathbb{R} \to \mathbb{R}$ is a nonconstant, bounded periodic function with minimal period $T := \log 3$. As in the case of the second order Cantor set, each of these functions can be computed explicitly in terms of its Fourier series.

Finally, we can now use the sequence of relative fractal drums (C_n, Ω_n), for $n \in \mathbb{N}$, in order to construct an RFD (A, Ω) which will have an infinite set of essential singularities on the critical line $\{\mathrm{Re}\, s = \overline{\dim}_B(A, \Omega)\}$. The construction is analogous to the one in the proof of Theorem 3.3.6 and in Example 3.3.7 in Section 3.3 above, dealing with Cantor strings of higher order. We let $(C_1, \Omega_1) := (C, \Omega)$, scale down every RFD (C_n, Ω_n) by the factor $3^{-n}/n!$ and define (A, Ω) as the disjoint union of copies of the resulting RFDs; that is,

$$(A, \Omega) := \bigsqcup_{n=1}^{\infty} \frac{3^{-n}}{n!} (C_n, \Omega_n). \tag{4.2.27}$$

(Here, we have used Definition 4.5.7 and Lemma 4.5.10 in Subsection 4.5.2 below.) We then have

$$\begin{aligned} \zeta_{A,\Omega}(s) &= \sum_{n=1}^{\infty} \zeta_{3^{-n}(n!)^{-1}(C_n,\Omega_n)}(s) = \sum_{n=1}^{\infty} \frac{3^{-ns}}{(n!)^s} \zeta_{C_n,\Omega_n}(s) \\ &= \frac{2}{6^s s} \sum_{n=1}^{\infty} \frac{1}{(n!)^s (3^s - 2)^n}. \end{aligned} \tag{4.2.28}$$

By the Weierstrass M-test, $\zeta_{A,\Omega}(s)$ is holomorphic on $\{\operatorname{Re} s > 0\} \setminus \left(\log_3 2 + \frac{2\pi}{\log 3}\mathrm{i}\mathbb{Z}\right)$. More precisely, it has essential singularities at each point of the set $\log_3 2 + \frac{2\pi}{\log 3}\mathrm{i}\mathbb{Z}$. Note that the critical line $\{\operatorname{Re} s = \log_3 2\}$ is clearly not a natural boundary for $\zeta_{A,\Omega}$ since $\zeta_{A,\Omega}$ given by Equation (4.2.28) can be holomorphically continued to the connected open set $\{\operatorname{Re} s > 0\} \setminus \dim_{PC}(A,\Omega)$.

In light of Theorem 5.3.16 of Chapter 5 below, we deduce that the tube formula of (A,Ω) has the following asymptotic expansion:

$$|A_t \cap \Omega| = t^{1-\log_3 2} \sum_{i=1}^{\infty} (\log t^{-1})^{i-1} G_i(\log t^{-1}) + O(t^{1-\alpha}) \quad \text{as} \quad t \to 0^+, \quad (4.2.29)$$

for any $\alpha > 0$, and where, similarly as before, the functions G_i for $i \in \mathbb{N}$ are nonconstant, bounded periodic functions with minimal period $T := \log 3$. Although in Chapter 5, we always assume that the corresponding fractal zeta function has a meromorphic extension to a suitable connected open neighborhood of the critical line, the results of Chapter 5 actually extend to functions having only isolated singularities in a suitable neighborhood of the critical line; that is, the corresponding fractal zeta functions may also have essential singularities. It is now easy to check that $\zeta_{A,\Omega}$ given by (4.2.28) satisfies the conditions of Theorem 5.3.16, with $\kappa_d := -1$; (see Definitions 5.1.3 and 5.3.9), and with the screen $\boldsymbol{S}$ taken as the vertical line $\{\operatorname{Re} s = \alpha\}$. The tube formula (4.2.29) now follows by calculating the residues $\operatorname{res}(t^{1-s}(1-s)^{-1}\zeta_{A,\Omega}(s), \omega_k)$, where $\omega_k \in \log_3 2 + \frac{2\pi}{\log 3}\mathrm{i}\mathbb{Z}$.

We close this discussion by observing that, as was alluded to earlier, the RFD (A,Ω) is a strongly hyperfractal RFD (in the sense of part *(ii)* of Definition 4.6.23 in Subsection 4.6.3 below and as strengthened in both parts of Remark 1.3.9), which is not maximally hyprefractal (in the sense of part *(iii)* of that same definition).

The above construction can be generalized verbatim for any (nontrivial) bounded fractal string $\mathscr{L}$ instead of the Cantor string $\mathscr{L}_{CS}$. *This suggests that the definition of complex dimensions should be extended to also include potential essential singularities (as well as algebraic and transcendental singularities) of the fractal zeta functions*, in the spirit of [Lap-vFr3, Subsection 13.4.3].

Let us now recall the definition of a self-similar spray or tiling (see [LapPe2–3], [LapPeWi1–2], [Lap-vFr3, Section 13.1]). More precisely, let us state this definition slightly more generally and in the context of relative fractal drums.

Definition 4.2.11. (*Self-similar spray or tiling*). Let G be a given open subset (*base set* or *generator*) of $\mathbb{R}^N$ of finite N-dimensional Lebesgue measure and let $\{r_1, r_2, \dots, r_J\}$ be a finite multiset (also called a *ratio list*) of positive real numbers (in $(0,1)$) such that $J \in \mathbb{N}$, $J \geq 2$ and

$$\sum_{j=1}^{J} r_j^N < 1. \quad (4.2.30)$$

Furthermore, let Λ be the multiset consisting of all the possible 'words' of multiples of the scaling factors $r_1, \dots, r_J$; that is, let

$$\Lambda := \{1, r_1, \dots, r_J, r_1 r_1, \dots, r_1 r_J, r_2 r_1, \dots, r_2 r_J, \dots, r_J r_1, \dots, r_J r_J, \\ r_1 r_1 r_1, \dots, r_1 r_1 r_J, \dots\} \tag{4.2.31}$$

and arrange all of the elements of the multiset Λ into a *scaling sequence* $(\lambda_i)_{i\geq 0}$, where $\lambda_0 := 1$. Note that $0 < \lambda_i < 1$, for every $i \geq 1$.

A *self-similar spray* (or *tiling*), generated by the base set G and the ratio list $\{r_1, r_2, \dots, r_J\}$ is an RFD $(\partial\Omega, \Omega)$ in $\mathbb{R}^N$, where Ω is a disjoint union of open sets G_i; i.e.,

$$\Omega := \bigsqcup_{i=0}^{\infty} G_i, \tag{4.2.32}$$

such that each G_i is congruent to $\lambda_i G$, for each $i \geq 0$. Here, the disjoint union $\sqcup$ can be understood as the disjoint union of RFDs given in Definition 4.1.46, with $(A_i, \Omega_i) := (\partial G_i, G_i)$ for every $i \geq 0$, in the notation of that definition.

Remark 4.2.12. Note that in the above definition, the scaling sequence $(\lambda_i)_{i\geq 0}$ consists of all the products of ratios $r_1, \dots, r_J$ appearing in the infinite sum

$$\sum_{n=0}^{\infty} \left(\sum_{j=1}^{J} r_j \right)^n, \tag{4.2.33}$$

after expanding the powers and counted with their multiplicities. More precisely, we have that for every multi-index $\alpha = (\alpha_1, \dots, \alpha_J) \in \mathbb{N}_0^J$, the multiplicity of $r_1^{\alpha_1} r_2^{\alpha_2} \dots r_J^{\alpha_J}$ in the multiset Λ is equal to the multinomial coefficient

$$\binom{|\alpha|}{\alpha_1, \alpha_2, \dots, \alpha_J} = \frac{|\alpha|!}{\alpha_1! \alpha_2! \cdots \alpha_J!}, \tag{4.2.34}$$

where $|\alpha| := \sum_{j=1}^{J} \alpha_j$. Of course, depending on the specific values of the ratios $r_1, \dots, r_J$, some of the numbers $r_1^{\alpha_1} r_2^{\alpha_2} \dots r_J^{\alpha_J}$ may be equal for different multi-indices $\alpha \in \mathbb{N}_0^J$.

Furthermore, the condition (4.2.30) ensures that the set $\Omega = \sqcup_{i\geq 0} G_i$ has finite N-dimensional Lebesgue measure. Indeed, we have

$$\begin{aligned} |\Omega| &= \sum_{i=0}^{\infty} |G_i| = \sum_{i=0}^{\infty} |\lambda_i G| = |G| \sum_{i=0}^{\infty} \lambda_i^N \\ &= |G| \sum_{n=0}^{\infty} \left(\sum_{j=1}^{J} r_j^N \right)^n = \frac{|G|}{1 - \sum_{j=1}^{J} r_j^N} < \infty, \end{aligned} \tag{4.2.35}$$

since (4.2.30) is satisfied. Note that the second to last equality above follows from the construction of the scaling sequence $(\lambda_i)_{i\geq 0}$.

In Definition 4.2.11, it is implicitly assumed that the generator G is such that it is indeed possible to construct the *disjoint* union appearing in (4.2.32), as given in Definition 4.1.46. This can always be achieved when G is bounded, which is the usual assumption made when dealing with self-similar sprays as, for instance, in [LapPe2–3], [LapPeWi1–2] and [Lap-vFr3, Section 13.1]. However, contrary to intuition, this does not have to be the case for a general open set G of finite N-dimensional Lebsgue measure, as is shown by the following example.

Example 4.2.13. Here, we construct an open set G in $\mathbb{R}^N$ of finite N-dimensional Lebesgue measure, and which is dense in $\mathbb{R}^N$. Therefore, *any* isometric image of a scaled copy of G has an intersection with G of positive N-dimensional Lebesgue measure. Let $A = \{a_k \in \mathbb{R}^N : k \in \mathbb{N}\}$ be a countable dense subset of $\mathbb{R}^N$ (for example, the set of points in $\mathbb{R}^N$ with rational coordinates). Let $(\rho_k)_{k\geq 1}$ be a sequence of positive real numbers such that $\sum_{k=1}^{\infty} \rho_k^N < \infty$, and consider the open set G defined as the (not necessarily disjoint) union of the open balls $B_{\rho_k}(a_k)$ of radius ρ_k and with centers at a_k, for $k \geq 1$:

$$G := \bigcup_{k=1}^{\infty} B_{\rho_k}(a_k). \tag{4.2.36}$$

Then, its N-dimensional volume is positive and finite since

$$0 < |G|_N \leq \sum_{k=1}^{\infty} |B_{\rho_k}(a_k)|_N = \omega_N \sum_{k=1}^{\infty} \rho_k^N < \infty, \tag{4.2.37}$$

where ω_N is the volume of the unit ball of $\mathbb{R}^N$. Since $\overline{A} = \mathbb{R}^N$, it follows that A (and hence, G as well) has a nonempty intersection with *any* nonempty open subset of $\mathbb{R}^N$.

We proceed by discussing some interesting properties of the RFD (A,G). Since $\overline{A} = \mathbb{R}^N$ and since $d(x,A) = d(x,\overline{A}) = 0$ for any $x \in \mathbb{R}^N$, we have that $A_t = \mathbb{R}^N$ for any $t > 0$; so that $A_t \cap G = G$, and therefore, $|A_t \cap G| = |G|$ for all $t > 0$. Hence, for any fixed real number s, we have

$$\frac{|A_t \cap G|}{t^{N-s}} = |G| t^{s-N} \sim t^{s-N} \quad \text{as} \quad t \to 0^+; \tag{4.2.38}$$

it follows that

$$\dim_B(A,G) = N. \tag{4.2.39}$$

Let us now compute the tube zeta function $\tilde{\zeta}_{A,G}$ of the RFD (A,G):

$$\tilde{\zeta}_{A,G}(s) := \int_0^{\delta} t^{s-N-1} |A_t \cap G| \,\mathrm{d}t = |G| \int_0^{\delta} t^{s-N-1} \mathrm{d}t = |G| \frac{\delta^{s-N}}{s-N}, \tag{4.2.40}$$

for all $s \in \mathbb{C}$ such that $\mathrm{Re}\, s > N$. Therefore, $\tilde{\zeta}_{A,G}$ can be (uniquely) meromorphically extended to the whole complex plane by letting $\tilde{\zeta}_{A,G}(s) := |G| \frac{\delta^{s-N}}{s-N}$ for all $s \in \mathbb{C}$.

In order to compute the distance zeta function of the RFD (A,G), note (much as before) that for any $x \in \mathbb{R}^N$, we have

$$d(x,A) = d(x,\overline{A}) = d(x,\mathbb{R}^N) = 0. \tag{4.2.41}$$

Therefore, the distance zeta function $\zeta_{A,G}$ satisfies $\zeta_{A,G}(s) := \int_G d(x,A)^{s-N}\mathrm{d}x = 0$ for all $s \in \mathbb{C}$ such that $\operatorname{Re} s > N$. This function can be holomorphically extended to the whole complex plane by letting $\zeta_{A,G} \equiv 0$ on all of $\mathbb{C}$. We have thus constructed an RFD (A,G) in $\mathbb{R}^N$ such that

$$\begin{aligned} D(\zeta_{A,G}) = N, \quad D_{\mathrm{hol}}(\zeta_{A,G}) = D_{\mathrm{mer}}(\zeta_{A,G}) = -\infty, \\ D(\tilde{\zeta}_{A,G}) = D_{\mathrm{hol}}(\tilde{\zeta}_{A,G}) = N, \quad D_{\mathrm{mer}}(\tilde{\zeta}_{A,G}) = -\infty. \end{aligned} \tag{4.2.42}$$

Note that in the case of the relative distance zeta function $\zeta_{A,G}$, we have achieved the maximal possible gap between its abscissa of (absolute) convergence and its abscissa of holomorphic continuation, since for any RFD (A,G) in $\mathbb{R}^N$, we have $D(\zeta_{A,G}), D_{\mathrm{hol}}(\zeta_{A,G}) \in [-\infty, N]$.

It is also worth noting that the open set G has finite N-dimensional Lebesgue measure, while the N-dimensional Lebesgue measure of its boundary ∂G is infinite. Indeed, we have that

$$|\partial G|_N = |\overline{G} \setminus \Omega|_N = |\mathbb{R}^N \setminus G|_N = |\mathbb{R}^N|_N - |G|_N = +\infty - |G|_N = +\infty. \tag{4.2.43}$$

In light of the above example, it is natural to introduce the following definition.

Definition 4.2.14. We let $\mathrm{RFD}_\Lambda(\mathbb{R}^N)$ be the family of all relative fractal drums (A,Ω) in $\mathbb{R}^N$ such that for a given multiset $\Lambda = \Lambda(r_1, \dots, r_J)$ of scaling factors $\lambda \in (0,1)$ defined by (4.2.31), one can construct the *disjoint* union $\sqcup_{\lambda \in \Lambda} \lambda(A,\Omega)$, in the sense of Definition 4.1.46. We then say that (A,Ω) is Λ-*sprayable* in $\mathbb{R}^N$. Furthermore, we say that (A,Ω) is *universally sprayable* if it is sprayable with respect to any finite multiset of scaling factors Λ.

Example 4.2.15. We can provide two simple classes of RFDs $(\partial G, G)$ which are universally sprayable:

(a) Any $(\partial G, G)$, where G is a bounded subset of $\mathbb{R}^N$.

(b) Any $(\partial G, G)$, where G is a *strip-like* subset of $\mathbb{R}^N$; i.e., such that the set G is contained between two parallel hyperplanes in $\mathbb{R}^N$ (more precisely, there exists two real constants a and b and a nonzero vector $c \in \mathbb{R}^N$ such that $a \le c \cdot x \le b$ for all $x \in G$, where $\cdot$ denotes the inner product in $\mathbb{R}^N$).

Note that, according to this definition, each bounded set G is a strip-like set.

Consider now a self-similar spray as a relative fractal drum (A,Ω), which we refer to in the sequel as a *self-similar RFD* or as the *self-similar RFD associated with the self-similar spray* (A,Ω) (see Definition 4.2.20 on page 290 below, along with the corresponding footnote 13); that is, let $A := \partial\Omega$ and $\Omega := \sqcup_{i \ge 0} G_i$ (see Definition 4.2.11). The 'self-similarity' of (A,Ω) is nicely exhibited by the scaling relation (4.2.44) given in the following lemma.

Lemma 4.2.16. *Let (A,Ω) be a self-similar spray in $\mathbb{R}^N$, as in Definition 4.2.11. Then, the relative fractal drum (A,Ω) satisfies the following* 'self-similar identity':

$$(A,\Omega) = (\partial G, G) \sqcup (r_1 A, r_1 \Omega) \sqcup \cdots \sqcup (r_J A, r_J \Omega), \tag{4.2.44}$$

where (with the exception of the first term on the right-hand side of (4.2.44)) *the symbol* $\sqcup$ *indicates that this represents a disjoint union of copies of* (A,Ω) *scaled by factors* $r_1,\dots,r_J$ *and displaced by isometries of* $\mathbb{R}^N$ *(see Definition 4.1.46).*

Proof. Let us reindex the scaling sequence $(\lambda_i)_{i\geq 0}$ in a way that keeps track of the actual construction of the numbers λ_i out of the scaling ratios $r_1,\dots,r_J$; see Equation (4.2.31) above. We let

$$I := \{\emptyset\} \cup \bigcup_{m=1}^{\infty} \{1,\dots,J\}^m \tag{4.2.45}$$

be the set of all finite sequences consisting of numbers $1,\dots,J$ (or, equivalently, of all finite words based on the alphabet $\{1,\dots,J\}$). Furthermore, for every $\alpha \in I$, define

$$\lambda_\alpha := \begin{cases} 1, & \alpha = \emptyset, \\ r_{\alpha_1} r_{\alpha_2} \cdots r_{\alpha_m}, & \alpha \neq \emptyset. \end{cases} \tag{4.2.46}$$

We then deduce from the construction of (A,Ω) that

$$\begin{aligned}(A,\Omega) &= \bigsqcup_{i=0}^{\infty} (\partial G_i, G_i) = \bigsqcup_{i=0}^{\infty} \lambda_i (\partial G, G) \\ &= \bigsqcup_{\alpha \in I} \lambda_\alpha (\partial G, G) = (\partial G, G) \sqcup \bigsqcup_{\alpha \in I \setminus \{\emptyset\}} \lambda_\alpha (\partial G, G).\end{aligned}$$

Observe now that in the last disjoint union above, every $\alpha \in \{1,\dots,J\}^m$ can be written as $\{j\} \times \{1,\dots,J\}^{m-1}$, for some $j \in \{1,\dots,J\}$, provided we identify $\{j\}$ with $\{j\} \times \{\emptyset\}$ when $m = 1$. Note that this identification is consistent with the definition of λ_α, in the sense that $\lambda_{\{j\}\times\beta} = r_j \lambda_\beta$ for all $j \in \{1\dots,J\}$ and $\beta \in I$. In light of this, we can next partition the last union above with respect to which number $j \in \{1,\dots,J\}$ the sequence α begins with:

$$\begin{aligned}(A,\Omega) &= (\partial G, G) \sqcup \bigsqcup_{j=1}^{J} \bigsqcup_{\alpha \in \{j\}\times I} \lambda_\alpha (\partial G, G) = (\partial G, G) \sqcup \bigsqcup_{j=1}^{J} \bigsqcup_{\beta \in I} r_j \lambda_\beta (\partial G, G) \\ &= (\partial G, G) \sqcup \bigsqcup_{j=1}^{J} r_j \Big(\bigsqcup_{\beta \in I} \lambda_\beta (\partial G, G) \Big) = (\partial G, G) \sqcup \bigsqcup_{j=1}^{J} r_j (A,\Omega).\end{aligned}$$

This completes the proof of the lemma. $\square$

In light of (4.2.44) and the additivity of the distance zeta function, it is now clear that the distance zeta function of (A,Ω) satisfies the following functional equation:

$$\zeta_{A,\Omega}(s) = \zeta_{\partial G, G}(s) + \zeta_{r_1 A, r_1 \Omega}(s) + \cdots + \zeta_{r_J A, r_J \Omega}(s), \tag{4.2.47}$$

for all $s \in \mathbb{C}$ with $\operatorname{Re} s$ sufficiently large.[11] Furthermore, for such s, by using the scaling property of the relative distance zeta function (Theorem 4.1.40), we deduce that the above equation then becomes

$$\zeta_{A,\Omega}(s) = \zeta_{\partial G,G}(s) + r_1^s \zeta_{A,\Omega}(s) + \cdots + r_J^s \zeta_{A,\Omega}(s). \tag{4.2.48}$$

Finally, this last identity together with an application of the principle of analytic continuation now yields the following theorem. We note that the second equality in Equation (4.2.50) of Theorem 4.2.17 follows from Equation (4.2.17).

Theorem 4.2.17. *Let G be the generator of a self-similar spray in $\mathbb{R}^N$, and let $\{r_1, r_2, \ldots, r_J\}$, with $r_j > 0$ (for $j = 1, \ldots, J$, $J \geq 2$) and such that $\sum_{j=1}^J r_j^N < 1$, be its scaling ratios counted according to their multiplicities. Furthermore, let $(A, \Omega) := (\partial\Omega, \Omega)$ be the self-similar spray generated by G, as in Definition 4.2.11. Then, the distance zeta function of (A, Ω) is given by*

$$\zeta_{A,\Omega}(s) = \frac{\zeta_{\partial G,G}(s)}{1 - \sum_{j=1}^J r_j^s}, \tag{4.2.49}$$

for all $s \in \mathbb{C}$ with $\operatorname{Re} s$ *sufficiently large. Furthermore,*

$$D(\zeta_{A,\Omega}) = \overline{\dim}_B(A, \Omega) = \max\{\overline{\dim}_B(\partial G, G), \sigma_0\}, \tag{4.2.50}$$

where $\sigma_0 > 0$ is the unique real solution s of the Moran equation $\sum_{j=1}^J r_j^s = 1$ (i.e., σ_0 is the similarity dimension *of the self-similar spray (A, Ω)).*[12]

More specifically, given a connected open neighborhood U of the critical line $\{\operatorname{Re} s = D\}$, where $D := \overline{\dim}_B(A, \Omega)$, $\zeta_{A,\Omega}$ has a meromorphic continuation to U if and only if $\zeta_{\partial G,G}$ does, and in that case, $\zeta_{A,\Omega}(s)$ is given by (4.2.49) *for all $s \in U$. Consequently, the visible complex dimensions of (A, Ω) satisfy*

$$\mathscr{P}(\zeta_{A,\Omega}, U) \subseteq (\mathfrak{D} \cap U) \cup \mathscr{P}(\zeta_{\partial G,G}, U), \tag{4.2.51}$$

where $\mathfrak{D}$ is the set of all the complex solutions s of the Moran equation $\sum_{j=1}^J r_j^s = 1$. Finally, if there are no zero-pole cancellations in (4.2.49), *then we have an equality in* (4.2.51).

We refer to [Lap-vFr3, Chapter 3, esp. Theorem 3.6] for detailed information about the structure of $\mathfrak{D}$; see also the brief discussion given before Corollary 5.4.23 and Problem 6.2.36 below.

Remark 4.2.18. There are two particularly interesting situations in which Theorem 4.2.17 can be applied:

[11] For instance, it suffices to assume that $\operatorname{Re} s > N$ since, by Theorem 4.1.7, all of the zeta functions appearing in (4.2.47) are holomorphic on the right half-plane $\{\operatorname{Re} s > N\}$.

[12] Clearly, $\sigma_0 > 0$ since $J \geq 2 > 1$; furthermore, $\sigma_0 < N$ since $\sum_{j=1}^J r_j^N < 1$.

(i) The case when $U := \{\mathrm{Re}\, s > D_{\mathrm{mer}}(\zeta_{\partial G,G})\}$, the largest open right half-plane to which $\zeta_{\partial G,G}$ can be meromorphically extended.

(ii) The case when $U := \mathring{\boldsymbol{W}}$, where $\boldsymbol{W}$ is an arbitrary window for $\zeta_{\partial G,G}$ and hence also for $\zeta_{A,\Omega}$, either in the sense of Chapter 2 (see Subsection 2.1.5, page 95) or in the sense of Chapter 5 (see Definition 5.1.1 in Subsection 5.1.1). In that case, since the screen $\boldsymbol{S} = \partial \boldsymbol{W}$ associated with the window $\boldsymbol{W}$ does not contain any poles, the inclusion (4.2.51) can be equivalently written as follows:

$$\mathscr{P}(\zeta_{A,\Omega}, \boldsymbol{W}) \subseteq (\mathfrak{D} \cap \boldsymbol{W}) \cup \mathscr{P}(\zeta_{\partial G,G}, \boldsymbol{W}). \tag{4.2.52}$$

Furthermore, if there are no zero-pole cancellations for any $s \in \boldsymbol{W}$ in the right-hand side of (4.2.49), then we have an equality in (4.2.52).

The next theorem gives a general construction of complex dimensions of higher order generated by self-similar sprays. It is stated for RFDs in $\mathbb{R}^N$. For notational simplicity, in that theorem, we assume that $\zeta_{\partial G,G}$ admits a meromorphic continuation to all of $\mathbb{C}$ (which is very often the case, in practice), but the reader will easily be able to extend it to a more general situation, in the spirit of Theorem 4.2.17 and Remark 4.2.18 above.

Theorem 4.2.19. *Let $(A,\Omega) := (\partial\Omega,\Omega)$ be a self-similar spray in $\mathbb{R}^N$ (with $N \geq 1$) generated by an open set G and the set of scaling ratios $\{r_1, r_2, \dots, r_J\}$, with $r_j > 0$ (for $j = 1, \dots, J$, $J \geq 2$) and such that $\sum_{j=1}^{J} r_j^N < 1$; see Definition 4.2.11 above. Furthermore, assume that $\zeta_{\partial G,G}$ has a meromorphic continuation to all of $\mathbb{C}$ and that there are no zero-pole cancellations in (4.2.49); i.e., that $\mathfrak{D} \cap \mathscr{P}(\zeta_{\partial G,G}) = \emptyset$, where $\mathfrak{D}$ is the set of all the complex solutions s of the Moran equation $\sum_{j=1}^{J} r_j^s = 1$ (also called the* scaling complex dimensions *of the self-similar spray (A,Ω) in the sequel); see, e.g., Subsection 5.5.6 or Section 6.2.*

Then, given an arbitrary integer $n \in \mathbb{N}$, there is an explicitly constructible relative fractal drum (A_n, Ω_n) (in fact, a fractal spray also generated by G or, more precisely, with base RFD $(\partial G, G)$) which has exactly the same complex dimensions as (A,Ω), provided the corresponding multiplicities are not *taken into account, but with the orders (i.e., multiplicities) of the complex dimensions belonging to $\mathfrak{D}$ now being multiplied by n.*

Moreover, if we let $U := \{\mathrm{Re}\, s > 0\}$ and $\mathfrak{D}^+ := \mathfrak{D} \cap U$, then there is an explicitly constructible RFD $(A_\infty, \Omega_\infty)$ such that its complex dimensions visible through U are the same as the complex dimensions of (A,Ω) visible through U, provided the multiplicities are not *taken into account, but with the complex dimensions belonging to $\mathfrak{D}^+$ now being of infinite order, that is, being essential singularities of its distance zeta function $\zeta_{A_\infty,\Omega_\infty}$. In particular, we have that*

$$D_{\mathrm{mer}}(A_\infty, \Omega_\infty) = D(\zeta_{A_\infty,\Omega_\infty}) = \overline{\dim}_B(A_\infty, \Omega_\infty). \tag{4.2.53}$$

Proof. We use the RFD (A,Ω) as our new 'generator'; that is, we define a new relative fractal drum (A_2, Ω_2) as a disjoint union of scaled copies of (A,Ω) by scaling factors λ_i, where $(\lambda_i)_{i\geq 0}$ is the scaling sequence of the self-similar spray (A,Ω). Much as in the proof of Lemma 4.2.16, this construction then implies that

$$(A_2,\Omega_2) = (A,\Omega) \sqcup \bigsqcup_{j=1}^{J} (r_j A_2, r_j \Omega_2). \tag{4.2.54}$$

Furthermore, much as before, by the scaling property of the relative distance zeta function (see Theorem 4.1.40) and in light of Theorem 4.2.17, we then obtain (after an application of the principle of analytic continuation) that

$$\zeta_{A_2,\Omega_2}(s) = \frac{\zeta_{A,\Omega}(s)}{1-\sum_{j=1}^{J} r_j^s} = \frac{\zeta_{\partial G,G}(s)}{\left(1-\sum_{j=1}^{J} r_j^s\right)^2}, \tag{4.2.55}$$

for all $s \in \mathbb{C}$, since, by hypothesis, $\zeta_{\partial G,G}$ admits a meromorphic continuation to all of $\mathbb{C}$. From the above identity (4.2.55), we now conclude that the relative fractal drum (A_2,Ω_2) has the same complex dimensions as (A,Ω), but with the orders of those belonging to $\mathfrak{D}$ being multiplied by two.

We can now proceed inductively by using (A_2,Ω_2) as a new 'generator'. Therefore, for each $n \in \mathbb{N}$, we obtain a relative fractal drum (A_n,Ω_n) (in fact, a fractal spray also generated by G or, more precisely, with base RFD $(\partial G, G)$) such that

$$\zeta_{A_n,\Omega_n}(s) = \frac{\zeta_{\partial G,G}(s)}{\left(1-\sum_{j=1}^{J} r_j^s\right)^n}, \tag{4.2.56}$$

for all $s \in \mathbb{C}$; that is, (A_n,Ω_n) has the same complex dimensions as (A,Ω), but the complex dimensions belonging to $\mathfrak{D}$ (i.e., the scaling complex dimensions of the fractal spray (A_n,Ω_n)) have their orders multiplied by n. By convention, we let $(A_1,\Omega_1) := (\partial G, G)$.

In order to generate essential singularities, we take a disjoint union of copies of the relative fractal drums (A_n,Ω_n) scaled by $(n!)^{-1}$. More specifically, we define (A_∞,Ω_∞) as follows:

$$(A_\infty,\Omega_\infty) := (A,\Omega) \sqcup \bigsqcup_{n=2}^{\infty} (n!)^{-1}(A_n,\Omega_n). \tag{4.2.57}$$

The construction of (A_∞,Ω_∞) and the scaling property of the relative distance zeta function (see Theorem 4.1.40) then imply that

$$\zeta_{A_\infty,\Omega_\infty}(s) = \zeta_{\partial G,G}(s) \sum_{n=1}^{\infty} \frac{1}{(n!)^s\left(1-\sum_{j=1}^{J} r_j^s\right)^n}, \tag{4.2.58}$$

for all $s \in \mathbb{C}$ with $\operatorname{Re} s$ sufficiently large. By the Weierstrass M-test, the sum in the above equation (4.2.58) defines a holomorphic function on $\{\operatorname{Re} s > 0\} \setminus \mathfrak{D}^+$ and, furthermore, it is easy to show that $\mathfrak{D}^+$ is the set of essential singularities of the function defined by this sum. This completes the proof of the theorem. □

Clearly, the relative fractal drums constructed in the proof of Theorem 4.2.19 also exhibit some kind of self-similarity. Indeed, we can introduce the notion of a *self-similar RFD*, which generalizes the notion of a self-similar spray used, in particular, in [Lap2–3], [LapPo3], [Lap-vFr1–3], [LapPe2–3] and [LapPeWi1–2]. Namely, we can give the following formal definition.

Definition 4.2.20. Take (A_0,Ω_0) to be a *base* or *generating* relative fractal drum and define the self-similar relative fractal drum (A,Ω) analogously as in Definition 4.2.11; that is, let

$$(A,\Omega) := \bigsqcup_{i=0}^{\infty} \lambda_i(A_0,\Omega_0), \tag{4.2.59}$$

where $(\lambda_i)_{i\geq 0}$ is the scaling sequence corresponding to the multiset Λ defined by (4.2.31). Of course, in this case, we implicitly assume that the base relative fractal drum (A_0,Ω_0) is such that the above *disjoint* union can be constructed (see Examples 4.2.13 and 4.2.15).

When this is the case (for example, when Ω_0 is bounded), (A,Ω) is called a *self-similar RFD*. More specifically, (A,Ω) is called "the" *self-similar RFD with base RFD* (*or generated by the RFD*) (A_0,Ω_0) and with the scaling ratios $r_1,\dots,r_J$.[13] Its *self-similarity dimension* σ_0 is the unique real number s such that $\sum_{j=1}^{J} r_j^s = 1$. Necessarily, we have that $0<\sigma_0<N$.

Note that in the terminology introduced in Definition 4.2.20, the self-similar spray $(\partial G, G)$ of Theorem 4.2.17 is also a self-similar RFD with base RFD (or generated by the RFD) $(\partial G, G)$.

Remark 4.2.21. The iterative construction given in the proof of Theorem 4.2.19 can also be applied in the more general situation where the relative fractal drum (A,Ω) is actually a relative fractal spray. Namely, we fix a fractal string $\mathscr{L}$ and define the sequence of fractal strings $(\mathscr{L}_k)_{k\geq 0}$ from Theorem 4.2.8 by $\mathscr{L}_k := \mathscr{L}$ for every $k \geq 0$. Then, under the assumption that the base relative fractal drum (A_0,Ω_0) is 'nice enough', for each given $k\in\mathbb{N}_0$, the set of complex dimensions of the relative fractal drum (A_k,Ω_k) from Theorem 4.2.19 will contain the set of complex dimensions of $\mathscr{L}$, but with their orders (i.e., multiplicities) now multiplied by k.

4.2.3 Relative Sierpiński Sprays and Their Complex Dimensions

In this subsection, we provide two examples (Example 4.2.24 and 4.2.29) of relative fractal sprays, dealing with the *inhomogeneous relative Sierpiński gasket RFD* and the *relative Sierpiński carpet*, respectively, viewed as RFDs. We also discuss higher-dimensional analogs of these classic examples of self-similar fractals, namely, the *inhomogeneous Sierpiński N-gasket RFD*, also called the *inhomogeneous N-gasket*

[13] Clearly, such an RFD is unique only up to multiple choices of isometries (or displacements) of $\mathbb{R}^N$, corresponding to the countably many copies of the base RFD (A_0,Ω_0) it is composed of.

RFD (Example 4.2.26) and associated with the so-called *inhomogeneous N-gasket*, along with the (relative) Sierpiński N-carpet, (Example 4.2.31), with $N \geq 3$. (In this notation, the $N = 2$ case corresponds to the above standard Sierpiński gasket and carpet RFDs, respectively.) In fact, far from being trivial generalizations to higher-dimensions, these families of examples reveal several interesting new phenomena, which will be discussed especially in Example 4.2.26 (the inhomogeneous N-gasket RFD, with $N \geq 3$) and whose consequences will be considered in several parts of Chapter 5, including Subsection 5.5.6 (particularly, part (c) of Remark 5.5.26), as well as in some of the open problems of Chapter 6 (especially, Problems 6.2.32, 6.2.35 and 6.2.36).

In order to avoid any possible confusion, we stress from the outset that for $N \geq 3$, the inhomogeneous Sierpiński N-gasket can be viewed as a 'self-similar RFD' (or a self-similar spray, called a relative Sierpiński spray in the title of this subsection) but is *not* associated with a self-similar set, in the usual sense of the term; see, e.g., [Hut] or [Fal1, Chapter 9]. Indeed it is not associatd with a 'homogeneous self-similar set', as in the aforementioned references and the classic literature on fractal geometry, but (still for $N \geq 3$) it is instead associated with (in a sense to be specified in Example 4.2.26 below) an 'inhomogeneous self-similar set', in the sense of Barnsley and Demko [BarDemk], a notion already briefly described in a specific example in Remark 2.1.87 above. (See also [Fra2], along with the relevant references therein, for more detailed information about this topic.) The same comment also applies to Examples 4.2.33, 4.2.34 and 4.2.35, except for the fact that the self-similar fractal nest from Example 4.2.35 is a self-similar set in an even more general sense, which will be described below.

We note that aspects of this subsection are closely related to Section 3.2. In the sequel, it will be useful to use the following definition.

Definition 4.2.22. We say that *two given relative fractal drums* (A_1, Ω_1) *and* (A_2, Ω_2) *in* $\mathbb{R}^N$ *are congruent* if there exists an isometry[14] $f : \mathbb{R}^N \to \mathbb{R}^N$ such that $A_2 = f(A_1)$ and $\Omega_2 = f(\Omega_1)$.

It is easy to see that the congruence of RFDs is an equivalence relation.

The following lemma states, in particular, that any two congruent RFDs have the same distance zeta functions. We leave its proof as a simple exercise for the interested reader.

Lemma 4.2.23. *Let* (A_1, Ω_1) *and* (A_2, Ω_2) *be two congruent RFDs in* $\mathbb{R}^N$. *Then, for any* $r \in \mathbb{R}$, *we have*

$$\mathcal{M}_*^r(A_1, \Omega_1) = \mathcal{M}_*^r(A_2, \Omega_2), \quad \mathcal{M}^{*r}(A_1, \Omega_1) = \mathcal{M}^{*r}(A_2, \Omega_2) \tag{4.2.60}$$

and

$$\underline{\dim}_B(A_1, \Omega_1) = \underline{\dim}_B(A_2, \Omega_2), \quad \overline{\dim}_B(A_1, \Omega_1) = \overline{\dim}_B(A_2, \Omega_2) =: \overline{D}. \tag{4.2.61}$$

[14] Recall that, up to a translation, an *isometry* (or *displacement*) of $\mathbb{R}^N$ is necessarily linear, with determinant ± 1.

As a result, $\dim_B(A_1,\Omega_1)$ *exists if and only if* $\dim_B(A_2,\Omega_2)$ *exists and in that case, we have*

$$\dim_B(A_1,\Omega_1) = \dim_B(A_2,\Omega_2). \tag{4.2.62}$$

Furthermore,

$$\zeta_{A_1,\Omega_1}(s) = \zeta_{A_2,\Omega_2}(s), \tag{4.2.63}$$

for any $s \in \mathbb{C}$ *with* $\operatorname{Re} s > \overline{\dim}_B(A_1,\Omega_1)$.

More generally, the identity (4.2.63) *holds for all* $s \in U$*, where* U *is any connected open neighborhood of the common critical line* $\{\operatorname{Re} s = \overline{\dim}_B(A_1,\Omega_1)\}$ *to which* ζ_{A_1,Ω_1} *(or, equivalently,* ζ_{A_1,Ω_1}*) can be meromorphically continued.*

It follows from (4.2.63) that under the hypotheses of Lemma 4.2.23 and given a connected open set $U \subseteq \mathbb{C}$ chosen as in the last part of the theorem (containing the common critical line $\{\operatorname{Re} s = \overline{D}\}$ of the RFDs (A_1,Ω_1) and (A_2,Ω_2); see Equation (4.2.61)), ζ_{A_1,Ω_1} and ζ_{A_2,Ω_2} have the exact same mermomorphic continuation to U, and therefore the same poles in U and associated residues (or more generally, principal parts in the case of multiple poles). In particular, two congruent RFDs have the same (visible) complex dimensions.

Example 4.2.24. (*Relative Sierpiński gasket*). Let A be the Sierpiński gasket in $\mathbb{R}^2$, the outer boundary of which is an equilateral triangle with unit sides. Consider the countable family of all open triangles in the standard construction of the gasket. Namely, these are the open triangles which are deleted at each stage of the construction. If Ω is the largest open triangle (with unit sides), then the *relative Sierpiński gasket* is defined as the ordered pair (A,Ω). The distance zeta function $\zeta_{A,\Omega}$ of the relative Sierpiński gasket (A,Ω) can be computed as the distance zeta function of the following relative fractal spray (see Definition 4.2.1):

$$\operatorname{Spray}(\Omega_0, \lambda = 1/2, b = 3),$$

where Ω_0 is the first deleted open triangle with sides $1/2$. It suffices to apply Equation (4.2.10) from Theorem 4.2.6. Decomposing Ω_0 into the union of six congruent right triangles (determined by the heights of the triangle Ω_0, see Figure 4.7) with disjoint interiors, we have that

$$\begin{aligned}\zeta_{\partial\Omega_0,\Omega_0}(s) &= 6\,\zeta_{A',\Omega'}(s) = 6\int_{\Omega'} d((x,y),A')^{s-2}\mathrm{d}x\mathrm{d}y \\ &= 6\int_0^{1/4}\mathrm{d}x\int_0^{x/\sqrt{3}} y^{s-2}\mathrm{d}y = 6\frac{(\sqrt{3})^{1-s}2^{-s}}{s(s-1)},\end{aligned} \tag{4.2.64}$$

for all $s \in \mathbb{C}$ with $\operatorname{Re} s > 1$. Using Equation (4.2.10) and appealing to Lemma 4.2.23, we deduce that the distance zeta function of the relative Sierpiński gasket (A,Ω) satisfies

$$\zeta_{A,\Omega}(s) = \frac{6(\sqrt{3})^{1-s}2^{-s}}{s(s-1)(1-3\cdot 2^{-s})} \sim \frac{1}{1-3\cdot 2^{-s}}, \tag{4.2.65}$$

where the equality holds for all $s \in \mathbb{C}$ with $\operatorname{Re} s > \log_2 3$ and as usual, the equivalence $\sim$ holds in the sense of Definition 2.1.69. Therefore, by the principle of analytic

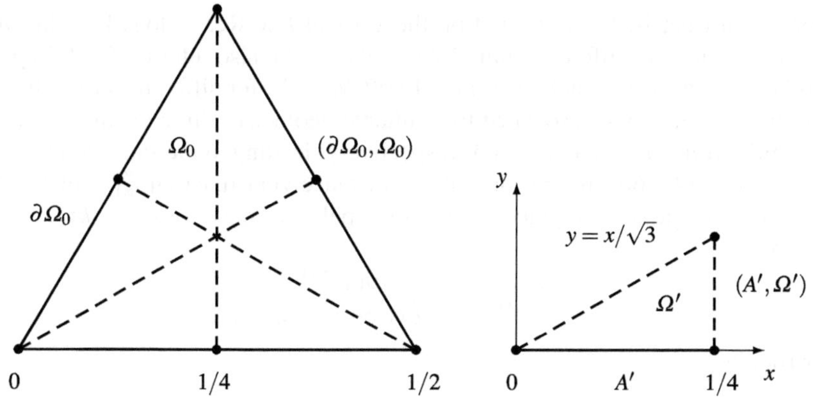

Fig. 4.7 On the left is depicted the base relative fractal drum $(\partial\Omega_0, \Omega_0)$ of the relative Sierpiński gasket, where Ω_0 is the associated (open) equilateral triangle with sides $1/2$. It can be viewed as the (disjoint) union of six RFDs, all of which are congruent to the relative fractal drum (A', Ω') on the right. This figure explains Equation (4.2.64) appearing in Example 4.2.24; see Lemma 4.2.23.

continuation, it follows that $\zeta_{A,\Omega}$ has a meromorphic extension to the entire complex plane, given by the same closed form as in Equation (4.2.65). More specifically,

$$\zeta_{A,\Omega}(s) = \frac{6(\sqrt{3})^{1-s}2^{-s}}{s(s-1)(1-3\cdot 2^{-s})}, \quad \text{for all } s \in \mathbb{C}. \tag{4.2.66}$$

Hence, the set of all of the complex dimensions (in $\mathbb{C}$) of the relative Sierpiński gasket is given by

$$\mathscr{P}(\zeta_{A,\Omega}) = \left(\log_2 3 + \frac{2\pi}{\log 2}\mathrm{i}\mathbb{Z}\right) \cup \{0,1\}. \tag{4.2.67}$$

Each of these complex dimensions is simple (i.e., is a simple pole of $\zeta_{A,\Omega}$). Note that here, $\{0,1\}$ can be interpreted as the set of *integer dimensions* of A, in the sense of [LapPe2–3] and [LapPeWi1] (see also [Lap-vFr3, Section 13.1]). In particular, we deduce from (4.2.67) that $D(\zeta_{A,\Omega}) = \log_2 3$, and we thus recover a well-known result. Namely, the set $\dim_{PC}(A,\Omega) := \mathscr{P}_c(\zeta_{A,\Omega})$ of principal complex dimensions of the relative Sierpiński gasket (A,Ω) is given by

$$\dim_{PC}(A,\Omega) = \log_2 3 + \mathbf{p}\mathrm{i}\mathbb{Z},$$

where $\mathbf{p} = 2\pi/\log 2$ is the oscillatory period of the Sierpiński gasket; see [Lap-vFr3, Subsection 6.6.1].

Note, however, that in [Lap-vFr1–3], the complex dimensions are obtained in a completely different manner (via an associated spectral zeta function corresponding to the Dirichlet Laplacian on the bounded open set Ω) and not geometrically. In addition, all of the complex dimensions of the Sierpiński gasket A are shown to be principal (that is, to be located on the vertical line $\mathrm{Re}\, s = \log_2 3 = \dim_B A$), a conclusion which is different from (4.2.67) above. We also refer to [ChrIvLap] and [LapSar], as well as to [LapPe2–3] and [LapPeWi1–2], for different approaches (via a spectral zeta function associated to a suitable geometric Dirac operator and via a self-similar tiling associated with A, respectively) leading to the same conclusion.

In light of (4.2.66), the residue of the distance zeta function $\zeta_{A,\Omega}$ of the relative Sierpiński gasket computed at any principal pole $s_k := \log_2 3 + \mathbf{p}k\mathrm{i}$, $k \in \mathbb{Z}$, is given by

$$\operatorname{res}(\zeta_{A,\Omega}, s_k) = \frac{6(\sqrt{3})^{1-s_k}}{2^{s_k}(\log 2)s_k(s_k-1)}.$$

In particular,

$$|\operatorname{res}(\zeta_{A,\Omega}, s_k)| \sim \frac{6(\sqrt{3})^{1-D}}{D\,2^D \log 2} k^{-2} \quad \text{as} \quad k \to \pm\infty,$$

where $D := \log_2 3$.

The following proposition shows that the relative Sierpiński gasket can be viewed as the relative fractal spray generated by the relative fractal drum (A', Ω') appearing in Figure 4.7 on the right.

Proposition 4.2.25 (Relative Sierpiński gasket). *Let (A', Ω') be the relative fractal drum defined in Figure 4.7 on the right. Let (A, Ω) be the relative fractal spray generated by the base relative fractal drum (A', Ω'), with scaling ratio $\lambda = 1/2$ and with multiplicities $m_k = 6 \cdot 3^{k-1}$, for any positive integer k:*

$$(A, \Omega) = \operatorname{Spray}((A', \Omega'), \lambda = 1/2,\, m_k = 6 \cdot 3^{k-1} \textit{ for } k \in \mathbb{N}), \tag{4.2.68}$$

in the notation of Definition 4.2.1. (*Observe that we assume here that the base relative fractal drum (A', Ω') has a multiplicity equal to* 8.) *Then, the relative distance zeta function of the relative fractal spray (A, Ω) coincides with the relative distance zeta function of the relative Sierpiński gasket; see Equation* (4.2.66).

Example 4.2.26. (*Inhomogeneous Sierpiński N-gasket RFD*). The usual Sierpiński gasket is contained in the unit triangle in the plane. Its (inhomogeneous) analog in $\mathbb{R}^3$, which we call the *inhomogeneous Sierpiński* 3*-gasket* or *inhomogeneous tetrahedral gasket*, is obtained by deleting the middle open octahedron (from the initial compact, convex unit tetrahedron), defined as the interior of the convex hull of the midpoints of each of the six edges of the initial tetrahedron (thus, four sub-tetrahedra are left after the first step), and so on.

More generally, for $N \geq 2$, the *inhomogeneous Sierpiński N-gasket* A_N, contained in $\mathbb{R}^N$, can be defined as follows. (More briefly, A_N is also referred to as the *inhomogeneous N-gasket*.) Let $V_N := \{P_1, P_2, \dots, P_{N+1}\}$ be a set of N points in $\mathbb{R}^N$ such that the mutual distance of any two points from the set is equal to 1.

The set V_N, where $N \geq 2$, with the indicated property, can be constructed inductively as follows. For $N = 2$, we take V_2 to be the set of vertices of any unit triangle in $\mathbb{R}^2$. We then reason by induction. Given $N \geq 2$, we assume that the set V_N of $N+1$

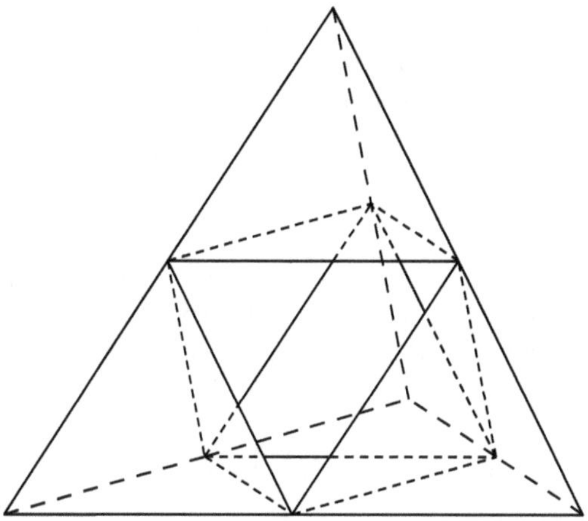

Fig. 4.8 The open octahedron $\Omega_{3,0}$ inscribed into the largest (compact) tetrahedron Ω_3, surrounded with 4 smaller (compact) tetrahedra scaled by the factor $1/2$. Each of them contains analogous scaled open octahedra, etc. The countable family of all open octahedra (viewed jointly with their boundaries) constitutes the tetrahedral gasket RFD or the Sierpiński 3-gasket RFD (A_3, Ω_3). The complement of the union of all open octahedra, with respect to the initial tetrahedron Ω_3, is called the *inhomogeneous Sierpiński* 3*-gasket RFD* (or the *relative Sierpiński* 3*-gasket*).

Unlike the classic Sierpiński 3-gasket (also known as the Sierpiński pyramid or tetrahedron) $S := S_3$, which is a (homogeneous or standard) self-similar set in $\mathbb{R}^3$ and satisfies the usual fixed point equation, $S = \cup_{j=1}^4 \Phi_j(S)$, where $\{\Phi_j\}_{j=1}^4$ are suitable contractive similitudes of $\mathbb{R}^3$ with respective fixed points $\{P_j\}_{j=1}^4$ and scaling ratios $\{r_j\}_{j=1}^4$ of common value $1/2$, the inhomogeneous Sierpiński 3-gasket RFD A_3 is *not* a self-similar set. Instead, it is an *inhomogeneous self-similar set* (in the sense of [BarDemk], see also [Fra2] and Remark 2.1.87 above). More specifically, $A := A_3$ satisfies the following *inhomogeneous* fixed point equation (of which it is the unique solution in the class of all nonempty compact subsets of $\mathbb{R}^3$), $A = \bigcup_{j=1}^4 \Phi_j(A) \cup B$, where B is the boundary $\partial\Omega_{3,0}$ of the first octahedron $\Omega_{3,0}$. In fact, B can simply be taken as the union of four middle triangles on the boundary of the outer tetrahedron Ω_3.

points in R^N has been constructed. Note that the set V_N is contained in a sphere, whose center is denoted by O. Let us consider the line of $\mathbb{R}^{N+1} = \mathbb{R}^N \times \mathbb{R}$ through the point O and perpendicular to the hyperplane $\mathbb{R}^N = \mathbb{R}^N \times \{0\}$ in $\mathbb{R}^{N+1}$. There exists a unique point P_{N+2} in the half-plane $\{x_{N+1} > 0\}$ of $\mathbb{R}^{N+1}$, which is a unit distance from all of the N vertices of V_N. (Here, we identify V_N with $V_N \times \{0\} \subset \mathbb{R}^{N+1}$.) We then define V_{N+1} by $V_{N+1} := V_N \cup \{P_{N+2}\}$.

Let us define Ω_N as the convex hull of the set V_N. As usual, we call it the N-*simplex*. Let $\Omega_{N,0}$, called the N-*plex*, be the open set defined as the interior of the convex hull of the set of midpoints of all of the $\binom{N+1}{2}$ edges of the N-simplex Ω_N.[15] The set $\overline{\Omega_N} \setminus \Omega_{N,0}$ is equal to the union of $N+1$ congruent N-simplices with disjoint interiors, having all of their side lengths equal to $1/2$. This is the first step of the

[15] For example, for $N = 2$, the set $\Omega_{2,0}$ (that is, the 2-plex) is an open equilateral triangle in $\mathbb{R}^2$ of side lengths equal to $1/2$, while for $N = 3$, the set $\Omega_{3,0}$ (that is, the 3-plex) is an open octahedron in $\mathbb{R}^3$ of side lengths equal to $1/2$.

construction. We proceed analogously with each of the $N+1$ compact N-simplices. The compact set A_N obtained in this way is called the *inhomogeneous Sierpiński N-gasket* (or, more briefly, the *inhomogeneous N-gasket*). The corresponding relative fractal spray in $\mathbb{R}^N$, defined by

$$(A_N,\Omega_N) = \mathrm{Spray}\,((\partial\Omega_{N,0},\Omega_{N,0}),\lambda = 1/2, b = N+1), \tag{4.2.69}$$

is called the *inhomogeneous Sierpiński N-gasket RFD* (or, more briefly, the *inhomogeneous N-gasket RFD*). It is a self-similar spray RFD; see the end of Definition 4.2.1 or of Definition 4.2.20. According to Theorem 4.2.6, we have the following factorization formula:

$$\zeta_{A_N,\Omega_N}(s) = f(s)\cdot\zeta_{\partial\Omega_{N,0},\Omega_{N,0}}(s), \tag{4.2.70}$$

where

$$f(s) = \sum_{k=0}^{\infty}(N+1)^k(2^{-k})^s = \frac{1}{1-(N+1)2^{-s}}.$$

Upon analytic continuation, we deduce that $f(s)$ has a meromorphic continuation to all of $\mathbb{C}$ given by

$$f(s) := \frac{1}{1-(N+1)2^{-s}}, \qquad \text{for all } s\in\mathbb{C}. \tag{4.2.71}$$

Hence, the set of poles of the function f (which can be uniquely meromorphically extended to the whole complex plane), is given by

$$\mathscr{P}(f) = \log_2(N+1) + \frac{2\pi}{\log 2}\,\mathrm{i}\mathbb{Z}. \tag{4.2.72}$$

Furthermore, the set of poles of the distance zeta function of the *relative N-plex* $(\partial\Omega_{N,0},\Omega_{N,0})$ is given by

$$\mathscr{P}(\zeta_{\partial\Omega_{N,0},\Omega_{N,0}}) = \{0,1,\dots,N-1\}, \tag{4.2.73}$$

while $\zeta_{\partial\Omega_{N,0},\Omega_{N,0}}(s)\neq 0$ for all $s\in\mathbb{C}\setminus\mathscr{P}(\zeta_{\partial\Omega_{N,0},\Omega_{N,0}})$. Both in (4.2.72) and (4.2.73), all of the poles are simple. Consequently, in light of (4.2.70), we conclude that the set of poles of $\zeta_{\partial\Omega_N,\Omega_N}$, i.e., the complex dimensions of the *inhomogeneous Sierpiński N-gasket* (A_N,Ω_N), is given by

$$\mathscr{P}(\zeta_{A_N,\Omega_N}) = \{0,1,\dots,N-1\}\cup\Big\{\log_2(N+1) + \frac{2\pi}{\log 2}\,\mathrm{i}\mathbb{Z}\Big\}, \tag{4.2.74}$$

with each nonreal complex dimension $\omega_k := \log_2(N+1) + \frac{2\pi}{\log 2}\mathbb{i}k$ (with $k \in \mathbb{Z}\setminus\{0\}$) being simple.[16] In particular, the set of principal complex dimensions of (A_N, Ω_N) is given by[17]

$$\dim_{PC}(A_N,\Omega_N) = \begin{cases} \log_2 3 + \frac{2\pi}{\log 2}\,\mathbb{i}\mathbb{Z}, & \text{for } N=2,\\ 2+\frac{2\pi}{\log 2}\,\mathbb{i}\mathbb{Z}, & \text{for } N=3,\\ \{N-1\}, & \text{for } N\geq 4, \end{cases} \tag{4.2.75}$$

while the (upper) box dimension of (A_N, Ω_N) is given by

$$\overline{\dim}_B(A_N,\Omega_N) = \max\left\{\log_2(N+1), N-1\right\} \tag{4.2.76}$$

and so

$$\overline{\dim}_B(A_N,\Omega_N) = \begin{cases} \log_2 3, & \text{for } N=2,\\ N-1, & \text{for } N\geq 3, \end{cases} \tag{4.2.77}$$

which extends the well-known results for $N = 2$ and 3, corresponding to the usual Sierpiński gasket in $\mathbb{R}^2$ and the tetrahedral gasket in $\mathbb{R}^3$, respectively. Namely, their respective relative box dimensions are equal to $\log_2 3$ and 2.

It can be readily shown that in this case, $\dim_B(A_N,\Omega_N)$ and $\dim_B A_N$ exist and

$$\overline{\dim}_B(A_N,\Omega_N) = \dim_B(A_N,\Omega_N) = \dim_B A_N = \dim_H A_N, \tag{4.2.78}$$

where as before, $\dim_H(\cdot)$ denotes the Hausdorff dimension. More generally, it is easy to see that $\dim_{PC}(A_N,\Omega_N) = \dim_{PC} A_N$, where the equality holds between multisets, that is, counting multiplicities. See also Remark 4.2.27 below.

Moreover, it can also be easily checked (and is essentially known, at least for $N = 2$) that (A_N,Ω_N) is Minkowski nondegenerate if $N \neq 3$:[18]

$$0 < \mathscr{M}_*(A_N,\Omega_N) \leq \mathscr{M}^*(A_N,\Omega_N) < \infty. \tag{4.2.79}$$

In the special case when $N = 3$, due to the factorization formula (4.2.70), ζ_{A_3,Ω_3} has a double pole at $s = 2$ and it can be shown by some of the methods of Chapter 5 (see, especially, Theorem 5.3.21) that in this case, (A_3,Ω_3) is Minkowski degenerate with $\mathscr{M}(A_3,\Omega_3) = +\infty$, but that it is also *h-Minkowski measurable* where the gauge function h is given by $h(t) := \log t^{-1}$ for all $t \in (0,1)$. (For an introduction to gauge functions and gauge Minkowski contents, see the beginning of Subsection 4.5.1 and also Definition 6.1.4 below.) In particular, since $D = \dim_B(A_N,\Omega_N)$ exists and

---

[16] Note that it could happen that $\omega_0 = \log_2(N+1)$ is equal to an integer $m \in \{0,1,\ldots,N-1\}$, which occurs if and only if $N = 2^m - 1$ with $m \in \mathbb{N}\setminus\{1\}$ and $N \geq 3$ or if $N = 1$ (the trivial case of the unit interval discussed in Example 5.5.1 below). In that situation (when $N \geq 3$), $\omega_0 = \sigma_0$ (the similarity dimension of A_N and (A_N,Ω_N), to be discussed further on) is a *double* pole of ζ_{A_N,Ω_N}.

[17] Recall that, by definition, $\dim_{PC}(A_N,\Omega_N) := \mathscr{P}_c(\zeta_{A_N,\Omega_N})$, the set of principal complex dimensions of the RFD (A_N,Ω_N).

[18] The truth of this statement can also be deduced from the methods and results of Chapter 5 below, especially, in Sections 5.3–5.5, including Example 5.5.12 and Subsection 5.5.6.

$\mathscr{M}_*(A_N, \Omega_N) > 0$, the hypothesis of part (c) of Theorem 4.1.7 (and hence, also of part (ii) of Corollary 4.1.10) are satisfied and so, in light of the factorization formula (4.2.70), a moment's reflection shows that

$$\begin{aligned} D &:= \dim_B(A_N, \Omega_N) = \max\{\log_2(N+1), N-1\} \\ &= D(\zeta_{A_N,\Omega_N}) = D_{\mathrm{hol}}(\zeta_{A_N,\Omega_N}), \end{aligned} \tag{4.2.80}$$

as was claimed in Equation (4.2.76) above.

Note that in (4.2.76) and (4.2.80), $\log_2(N+1)$ stands for the *similarity dimension* σ_0 of the self-similar RFD or spray (A_N, Ω_N) (or, equivalently, of the *inhomogeneous* self-similar set A_N), while $N-1$ refers to the (inner) dimension of the boundary $\partial\Omega_{N,0}$ of the generator (the N-plex $\Omega_{N,0}$), i.e., of the RFD $(A_{N,0}, \Omega_{N,0})$.

In the sequel, the function f appearing in Equations (4.2.70)–(4.2.72) will often be called the *scaling zeta function* of the RFD (A, Ω) and denoted by $\zeta_{\mathfrak{S}}$; see, e.g., Subsection 5.5.6 or Section 6.2. Therefore, for example, Equation (4.2.70) can be rewritten as follows (using the abbreviated notation ζ_{A_N,Ω_N} and $\zeta_{A_{N,0},\Omega_{N,0}}$):

$$\zeta_{A_N,\Omega_N}(s) = \zeta_{\mathfrak{S}}(s) \cdot \zeta_{A_{N,0},\Omega_{N,0}}(s). \tag{4.2.81}$$

Also, in Equation (4.2.74), and in agreement with the terminology of [LapPe2–3] and [LapPeWi1–2] (see also [Lap-vFr3, Section 13.1]), $\{0, 1, \ldots, N-1\}$ and $\mathscr{P}(\zeta_{\mathfrak{S}}) = \{\log_2(N+1) + \frac{2\pi}{\log 2}\mathrm{i}\mathbb{Z}\}$ are called, respectively, the set of *integer dimensions* and the set of *scaling complex dimensions* of the self-similar RFD (A_N, Ω_N). Note that some points could be common to those two sets, for instance, when $N = 3$, the point $s = 2$, which is therefore a double pole of ζ_{A_N,Ω_N} or, equivalently, a complex dimension of (A_3, Ω_3) of multiplicity two.

Recall that the *classic Sierpiński N-gasket* S_N (used, for example, in [KiLap1], [Ki1], and the relevant references therein),[19] is a standard (or *homogeneous*) self-similar set. Hence, $S := S_N$ satisfies the fixed point equation $S = \cup_{j=1}^{N+1}\Phi_j(S)$, where $\{\Phi_j\}_{j=1}^{N+1}$ are contractive similitudes of $\mathbb{R}^N$ with corresponding fixed points $\{V_j\}_{j=1}^{N+1}$ and scaling ratios $\{r_j\}_{j=1}^{N+1}$, of common value $1/2$: $r_1 = \cdots = r_{N+1} = 1/2$. In particular, S is the unique nonempty compact subset of $\mathbb{R}^N$ which is the solution of that equation. See, e.g., [Hut] or [Fal1, Chapter 9].

By contrast, the inhomogeneous Sierpiński N-gasket RFD (A_N, Ω_N) is a self-similar spray or RFD, but (for $N \geq 3$) A_N is *not* a (classic or homogeneous) self-similar set. Interestingly, however, A_N is an *inhomogeneous* self similar set (in the sense of [BarDemk], see also [Fra1] and Remark 2.1.87 above), as is explained in more detail when $N = 3$ in the second part of the caption of Figure 4.8 above. In particular, when $N \geq 3$, $A := A_N$ satisfies the following *inhomogeneous* fixed point equation:

$$A = \bigcup_{j=1}^{N+1} \Phi_j(A) \cup B, \tag{4.2.82}$$

[19] The classic Sierpiński N-gasket S_N has been used, for example, in [KiLap1], in a work dealing with the spectral analysis of Laplacians on self-similar fractals.

where B is a suitable (nonempty) compact subset of $\mathbb{R}^N$ (described in the caption of Figure 4.8 in the prototypical special case when $N = 3$); for example, B can be chosen to be the boundary of $\Omega_{N,0}$: we can let $B = \partial\Omega_{N,0} = A_{N,0}$. More specifically, A_N is the unique nonempty compact subset A of $\mathbb{R}^N$ satisfying the identity (4.2.82).

Nevertheless, since in the terminology of Definition 4.2.11, (A_N, Ω_N) is a self-similar spray with generator the N-plex $\Omega_{N,0}$ and ratio list $\{r_1 = \cdots = r_{N+1} = 1/2\}$, the self-similar set S_N (the classic Sierpiński N-gasket), the inhomogeneous self-similar set A_N (the inhomogeneous N-gasket) and the self-similar spray (or RFD) (A_N, Ω_N) have the same similarity dimension, σ_0, which is the unique real solution of the Moran equation $\sum_{j=1}^{N+1} r_j^s = 1$; that is, $(N+1)\cdot 2^{-s} = 1$, with $s \in \mathbb{R}$, or, equivalently,

$$\sigma_0 = \log_2(N+1). \tag{4.2.83}$$

Finally, we point out that Equations (4.2.78) and (4.2.80) imply that

$$\dim_B(A_N, \Omega_N) = \max\left\{\sigma_0, \dim_B(A_{N,0}, \Omega_{N,0})\right\}. \tag{4.2.84}$$

(See also Equation (4.2.50) in Theorem 4.2.17.) Therefore, $\dim_B(A_N, \Omega_N)$ is equal to $\sigma_0 = \dim_B A_N$ when $N \le 3$ and is strictly greater than σ_0 when $N \ge 4$. (See also Remark 4.2.27 just below.) We will obtain a natural generalization and application of these results towards the end of Section 5.5.6; see, especially, part (c) of Remark 5.5.26.

It is noteworthy that when $N = 2$, we not only have that $A_2 = S_2$, the classic Sierpiński gasket, but it is also the case that $A_2 = S_2$ is both a (homogeneous or standard) self-similar set *and* an inhomogeneous self-similar set, with respect to the same iterated functions system (or IFS) $\{\Phi\}_{j=1}^3$, comprised of similarity transformations of $\mathbb{R}^2$. Indeed, in the inhomogeneous fixed point equation (4.2.82), with $A := A_2$ and $N = 2$, we can not only choose $B := \emptyset$ (the empty set), but we can also choose $B := \partial A_{2,0}$, the boundary of the unit triangle.

Remark 4.2.27. It is well known (see, e.g., [Hut], [Fal1, Theorem 9.3]), that if a higher-dimensional self-similar set satisfies the open set condition, as is the case (for every $N \ge 2$) of the standard Sierpiński N-gasket S_N but *not* (for any $N \ge 3$) of the inhomogeneous Sierpiński N-gasket A_N (see the discussion following Equation (4.2.80), along with the caption of Figure 4.8), then its Minkowski and Hausdorff dimensions coincide with its similarity dimension; moreover, $\dim_B S_N$ exists. Hence, this means that

$$\dim_B S_N = \dim_H S_N = \sigma_0 = \log_2(N+1), \tag{4.2.85}$$

where σ_0 is the common similarity dimension of the self-similar set S_N, the inhomogeneous N-gasket A_N, and the self-similar RFD (A_N, Ω_N). Hence, when $N \ge 4$, it follows that $\dim_B A_N > \sigma_0 = \dim_B S_N$. There is no contradiction, however, in light of Equations (4.2.76) and (4.2.77), along with the fact that A_N is not a self-similar set (only a nonhomogeneous self-similar set) for such values of N.[20]

[20] It is possible to construct simpler examples in $\mathbb{R}^2$ which also illustrate this situation; see Examples 4.2.33, 4.2.34 and 4.2.35 below.

What is true, in general, under the above hypotheses (see Theorem 4.2.17 and Definition 4.2.20 above) is that the Minkowski dimension of the self-similar RFD is equal to the maximum of the similarity dimension σ_0 and the (inner) Minkowski dimension D_G of the boundary of the generator G, which is assumed to be sufficiently nice (see Subsection 5.5.6 below);[21] here, $G := \Omega_{N,0}$ and hence, $D_G = \dim_B(A_{N,0},\Omega_{N,0}) = N-1$. We will further discuss this issue in Remark 5.5.26 of Subsection 5.5.6, where we will see that the proper counterpart of this situation is case (iii) of part (c) of Remark 5.5.26, namely, when $D_G := \dim_B(\partial G, G) > \sigma_0$. This latter possibility cannot occur in the case of a (standard) self-similar set; see [Fal1, Theorem 9.3]. More specifically, as does not seem to have been observed before, this impossibility is a somewhat surprising consequence of the aforementioned result of Hutchinson in [Hut], as described in [Fal1, Theorem 9.3] and extending to any dimension $N \geq 1$ a corresponding one-dimensional result due to Moran in [Mora].

In closing this remark, we mention that such a problem does not occur when $N = 1$, which is the situation considered in the theory of the complex dimensions of geometric self-similar strings developed, in particular, in [Lap-vFr3, Chapters 2, 3 and Section 8.4]. Indeed, we then have that $D_G < \sigma_0$ since $D_G = 0$ (when $N = 1$) and $\sigma_0 > 0$ (always).

We next explain in more detail how to calculate the complex dimensions (and hence also the principal complex dimensions) of the relative inhomogeneous Sierpiński N-gasket (A_N,Ω_N) in the prototypical case when $N = 3$.

The relative distance zeta function $\zeta_{\partial\Omega_{N,0},\Omega_{N,0}}$ of the N-plex RFD $(\partial\Omega_{N,0},\Omega_{N,0}) = (A_{N,0},\Omega_{N,0})$ can be explicitly computed as follows, in the case when $N = 3$. It is easy to see that the octahedral RFD $(\partial\Omega_{3,0},\Omega_{3,0})$ can be identified with sixteen copies of disjoint RFDs, each of which is congruent to the pyramidal RFD (T,Ω') in $\mathbb{R}^3$, where Ω' is the open (irregular) pyramid with vertices at $O(0,0,0)$, $A(1/4,0,0)$, $B(1/4,1/4,0)$ and $C(0,0,1/(2\sqrt{2}))$, while the triangle $T = \operatorname{conv}(A,B,C)$ is a face of the pyramid. Since for any $(x,y,z) \in \Omega'$, we have

$$d((x,y,z),T) = \frac{1}{\sqrt{3}}\left(\frac{1}{2\sqrt{2}} - \sqrt{2}x - z\right), \tag{4.2.86}$$

we deduce that (recall that $A_{3,0} := \partial\Omega_{3,0}$)

$$\begin{aligned}\zeta_{A_{3,0},\Omega_{3,0}}(s) &= 16\zeta_{T,\Omega'}(s)\\ &= 16\iiint_{\Omega'} d((x,y,z),T)^{s-3}\mathrm{d}x\,\mathrm{d}y\,\mathrm{d}z\\ &= 16\int_0^{1/4}\mathrm{d}x\int_0^x\mathrm{d}y\int_0^{\frac{1}{2\sqrt{2}}-\sqrt{2}x}\left(\frac{\frac{1}{2\sqrt{2}}-\sqrt{2}x-z}{\sqrt{3}}\right)^{s-3}\mathrm{d}z.\end{aligned} \tag{4.2.87}$$

[21] If we allow the boundary of G to be fractal, then new interesting phenomena may occur, as was illustrated in Subsection 4.2.2 above.

Evaluating the last integral in (4.2.87), we obtain via a direct computation that

$$\begin{aligned}\zeta_{A_{3,0},\Omega_{3,0}}(s) &= 16\frac{(\sqrt{3})^{3-s}}{s-2}\int_0^{1/4}\Big(\frac{1}{2\sqrt{2}}-\sqrt{2}x\Big)^{s-2}x\,\mathrm{d}x \\ &= 8\frac{(\sqrt{3})^{3-s}}{s-2}\int_0^{1/(2\sqrt{2})}u^{s-2}\Big(\frac{1}{2\sqrt{2}}-u\Big)\,\mathrm{d}u \\ &= \frac{8(\sqrt{3})^{3-s}(2\sqrt{2})^{-s}}{s(s-1)(s-2)},\end{aligned} \tag{4.2.88}$$

for any complex number s such that $\operatorname{Re} s > 2$. Therefore, we deduce from (4.2.70) that the distance zeta function of the thetrahedral RFD in $\mathbb{R}^3$ can be meromorphically extended to the whole complex plane and is given for all $s \in \mathbb{C}$ by

$$\zeta_{A_3,\Omega_3}(s) = \frac{8(\sqrt{3})^{3-s}(2\sqrt{2})^{-s}}{s(s-1)(s-2)(1-4\cdot 2^{-s})}. \tag{4.2.89}$$

It is worth noting that $s = 2$ is the only pole of ζ_{A_3,Ω_3} of order 2, since $s = 2$ is a simple pole of both $(s-2)^{-1}$ and $(2^s-4)^{-1}$. More specifically, since the derivative of $1-4\cdot 2^{-s}$ computed at $s = 2$ is nonzero (and, in fact, is equal to $4\log 2$), then $s = 2$ is a simple zero of $1-4\cdot 2^{-s}$; that is, it is a simple pole of $1/(1-4\cdot 2^{-s})$.

Moreover, it immediately follows from Equation (4.2.89) that

$$\zeta_{A_3,\Omega_3}(s) \sim \frac{1}{(s-2)(1-4\cdot 2^{-s})}. \tag{4.2.90}$$

In particular, as we have already seen in Equation (4.2.75) and recalling that $N := 3$ here, we have

$$\dim_{PC}(A_3,\Omega_3) = 2 + \frac{2\pi}{\log 2}\mathbb{i}\mathbb{Z}. \tag{4.2.91}$$

Since $D := \overline{\dim}_B(A_3,\Omega_3) = \dim_B(A_3,\Omega_3) = 2$ is a simple pole of both $1/(s-2)$ and $1/(2^s-4)$, we conclude that $D = 2$ is the only complex dimension of order two of the RFD (A_3,Ω_3). Consequently, the case of the relative Sierpiński 3-gasket (A_3,Ω_3) reveals a new phenomenon: its relative box dimension $D = 2$ is a complex dimension of order (i.e., multiplicity) two, while all the other complex dimensions of the relative Sierpiński 3-gasket (including the double sequence of nonreal complex dimensions on the critical line of convergence $\{\operatorname{Re} s = 2\}$) are simple. Since, as we have already observed earlier, we have that $\dim_{PC}(A_N,\Omega_N) = \dim_{PC} A_N$ for every $N \geq 2$, we deduce from (4.2.91) and the discussion following it that

$$\dim_{PC} A_3 = 2 + \frac{2\pi}{\log 2}\mathbb{i}\mathbb{Z}, \tag{4.2.92}$$

with $s = \dim_B A_3 = 2$ being the only principal complex dimension of A_3 of order two, all the other complex dimensions being simple.

We challenge the interested reader to use similar arguments as in the case when $N=3$ in order to infer that for any $N\geq 3$, the distance zeta function of the relative N-plex $(\partial\Omega_{N,0},\Omega_{N,0})$ is of the form

$$\zeta_{\partial\Omega_{N,0},\Omega_{N,0}}(s)=\frac{g(s)}{s(s-1)\cdots(s-(N-1))}, \tag{4.2.93}$$

where $g(s)$ is a nonvanishing entire function. (Note that, when $N=3$, this is in agreement with Equation (4.2.88) above.) Therefore, we conclude from Equations (4.2.70) and (4.2.71) above that

$$\zeta_{A_N,\Omega_N}(s)=\frac{g(s)}{s(s-1)\cdots(s-(N-1))(1-(N+1)2^{-s})}. \tag{4.2.94}$$

This extends Equation (4.2.89) to any $N\geq 3$ (really, of the base RFD $(A_{3,0},\Omega_{3,0})$ generating the self-similar RFD (A_3,Ω_3)).

In the case when $N\geq 4$, $D=N-1$ is the only principal complex dimension of the inhomogeneous Sierpiński N-gasket RFD. [Indeed, for $N\geq 4$, we have that $\log_2(N+1)<N-1$ (i.e., $N+1<2^{N-1}$), which can be easily proved, for example, by using mathematical induction on N.] Furthermore, we immediately deduce from Equation (4.2.94) that

$$\zeta_{A_N,\Omega_N}(s)\sim\frac{1}{s-(N-1)}. \tag{4.2.95}$$

Moreover, if $N\geq 4$ is of the form $N=2^q-1$ for some integer $q\geq 3$, then $q=\log_2(N+1)$ (note that it is smaller than $D=N-1$) is the only complex dimension of order two (since it is a simple pole of both $(s-q)^{-1}$ and $(1-(N+1)2^{-s})^{-1}$), while all of the other complex dimensions of (A_N,Ω_N) are simple.

On the other hand, if $N\geq 4$ is not of the form $N=2^q-1$ for any integer $q\geq 3$, then all of the complex dimensions of the inhomogeneous Sierpiński N-gasket RFD are simple.

Roughly speaking, in the case when when $N=3$, the fact that $s=2$ has multiplicity two can be explained geometrically as follows: firstly, $s=2$ is a simple pole of the scaling zeta function $\zeta_{\mathfrak{S}}(s)=1/(1-(N+1)2^{-s})=1/(1-4\cdot 2^{-s})$ of the RFD (A_3,Ω_3),[22] while at the same time, $s=2$ is the simple pole arising from the boundary of the first (deleted) octahedron, which is also 2-dimensional; more specifically, $s=2$ is also a simple pole of $\zeta_{A_{3,0},\Omega_{3,0}}$. Therefore, the double pole of ζ_{A_N,Ω_N} arises both from the (inhomogeneous) self-similarity of the RFD (A_N,Ω_N) (or, equivalently, of the set A_N) and from the special geometry of the boundary of the generator (really, of the base RFD $(A_{N,0},\Omega_{N,0})$ generating the self-similar RFD (A_N,Ω_N)) when $N=3$.

Remark 4.2.28. Since as was noted earlier, $\dim_{PC}A_N=\dim_{PC}(A_N,\Omega_N)$, where the equality holds between multisets, exactly the same comment as above holds about the *principal* complex dimensions of the inhomogeneous N-gasket A_N (instead of

[22] Indeed, the similarity dimension of the 3-gasket A_3 is equal to 2.

the complex dimensions of the inhomogeneous N-gasket RFD (A_N,Ω_N)). For example, if $N\geq 4$ is not of the form 2^q-1 for any integer $q\geq 3$, then all of the complex dimensions of A_N are simple, while otherwise (i.e., if $N=2^q-1$, for some $q\geq 3$), then $s=q$ is the only complex dimension of multiplicity 2 and all the other complex dimensions (including $D=\dim_B A_N=\dim_B(A_N,\Omega)=N-1$) are simple. The multiplicity of $s=q$ arises both from the (inhomogeneous) self-similar structure of A_N and of the geometric structure of the boundary of the generator $\Omega_{N,0}$, $A_{N,0}:=\partial\Omega_{N,0}$, exactly as in the case when $N=3$.

In the case of the relative inhomogeneous Sierpiński 2-gasket (A_2,Ω_2), the value of $s=\log_2 3$ (which is the simple pole arising from the self-similarity of (A_2,Ω_2)) is strictly larger than the dimension $s=1$ of the boundary of the deleted triangle (i.e., of the 2-plex $\Omega_{2,0}$). Moreover, the relative 2-Sierpiński gasket is Minkowski nondegenerate and Minkowski nonmeasurable, while the relative inhomogeneous 3-Sierpiński gasket (A_3,Ω_3) is Minkowski degenerate, with its 2-dimensional Minkowski content being equal to $+\infty$. Its gauge function (a notion introduced in Subsection 6.1.4 of Chapter 6.1.1) can be determined by methods involving the fractal tube formulas developed in Chapter 5.

More specifically, in Chapter 5, we will show that it is possible to find a gauge function (namely, $h(t):=\log t^{-1}$ for all $t\in(0,1)$) relative to which the relative inhomogeneous 3-gasket RFD (A_3,Ω_3) is Minkowski *nondegenerate* and moreover, Minkowski *measurable*; see Theorem 5.4.27. (The same is true for the inhomogeneous 3-gasket A_3.) This should be contrasted with the case of the ordinary (classical) Sierpiński N-gasket S_N, which is Minkowski nondegenerate and Minkowski nonmeasurable (in the usual sense, i.e., with respect to the trivial gauge function $h(t)\equiv 1$ correspoding to a standard power law).

On the other hand, when $N\geq 4$, the dimension $D_G=N-1$ of the boundary of the N-plex $\Omega_{N,0}$ is larger than the similarity dimension $\sigma_0=\log_2(N+1)$ arising from "fractality". Hence, $D_G=N-1$. Since $D=N-1$ is the only complex dimension on the critical line (and it is simple), we conclude that for $N\geq 4$, the RFD (A_N,Ω_N) is Minkowski measurable (see Theorem 5.4.20 in Chapter 5). Thus, the case when $N=3$ is indeed very special among all of the inhomogeneous Sierpiński N-gasket RFDs. These issues will be clarified and revisited, as well as placed in a much broader framework, towards the end of Chapters 5 and 6; see, especially, part (c) of Remark 5.5.26 in Subsection 5.5.6, along with Problems 6.2.32, 6.2.35 and 6.2.36.

Example 4.2.29. (*Relative Sierpiński carpet*). Let A be the Sierpiński carpet contained in the unit square Ω. Let (A,Ω) be the corresponding *relative Sierpiński carpet* (or *Sierpiński carpet RFD*), with Ω being the unit square. (See Figure 2.1 on page 49 for a picture of the standard Sierpiński carpet.) Its distance zeta function $\zeta_{A,\Omega}$ coincides with the distance zeta function of the following relative fractal spray (see the end of Definition 4.2.1):

$$\mathrm{Spray}(\Omega_0,\lambda=1/3,b=8),$$

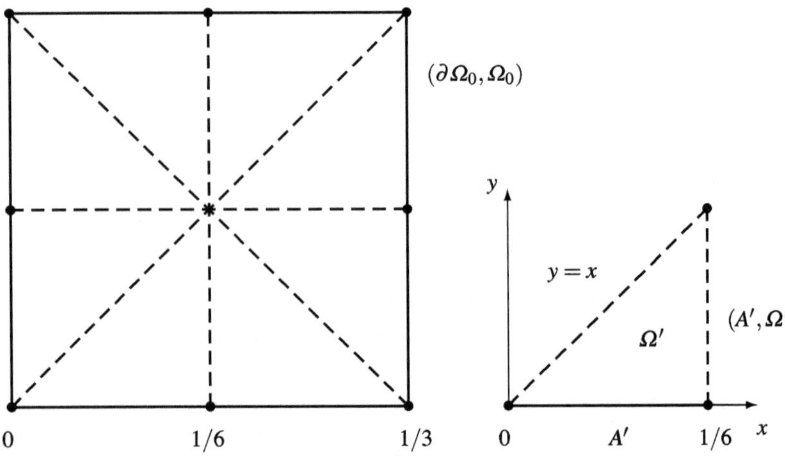

Fig. 4.9 On the left is the base relative fractal drum $(\partial\Omega_0,\Omega_0)$ of the relative Sierpiński carpet (A,Ω) described in Example 4.2.29, where Ω_0 is the associated (open) square with sides $1/3$. The base relative fractal drum $(\partial\Omega_0,\Omega_0)$ can be viewed as the (disjoint) union of eight RFDs, all of which are congruent to the relative fractal drum (A',Ω') depicted on the right. This figure explains Equation (4.2.97); see Lemma 4.2.23.

where Ω_0 is the first deleted open square with sides $1/3$. Similarly as in Example 4.2.24, by using Theorem 4.2.6 and Lemma 4.2.23, we obtain that $\zeta_{A,\Omega}$, the relative distance zeta functions of (A,Ω), has a meromorphic continuation to the entire complex plane given for all $s\in\mathbb{C}$ by

$$\zeta_{A,\Omega}(s)=\frac{8\cdot 6^{-s}}{s(s-1)(1-8\cdot 3^{-s})}. \tag{4.2.96}$$

Indeed, clearly, the base relative fractal drum $(\partial\Omega_0,\Omega_0)$ is the (disjoint) union of eight relative fractal drums, each of which is congruent to a relative fractal drum (A',Ω'), where Ω' is an appropriate isosceles right triangle; see Figure 4.9. We then deduce from Lemma 4.2.23 that

$$\begin{aligned}\zeta_{\partial\Omega_0,\Omega_0}(s)&=8\,\zeta_{A',\Omega'}(s)=8\int_{\Omega'}d((x,y),A')^{s-2}\mathrm{d}x\,\mathrm{d}y\\&=8\int_0^{1/6}\mathrm{d}x\int_0^x y^{s-2}\mathrm{d}y=\frac{8\cdot 6^{-s}}{s(s-1)},\end{aligned} \tag{4.2.97}$$

for all $s\in\mathbb{C}$ with $\mathrm{Re}\,s>1$, and hence, in light of Theorem 4.2.6, that $\zeta_{A,\Omega}(s)$ is given by (4.2.96). Note that, after analytic continuation, we also have

$$\zeta_{\partial\Omega_0,\Omega_0}(s)=\frac{8\cdot 6^{-s}}{s(s-1)},\quad\text{for all } s\in\mathbb{C}. \tag{4.2.98}$$

Since by (4.2.96),

$$\zeta_{A,\Omega}(s) \sim \frac{1}{1-8\cdot 3^{-s}},$$

one deduces from this equivalence that the abscissa of convergence of $\zeta_{A,\Omega}$ is given by $D = \log_3 8 = \dim_B(A,\Omega)$, where the equality follows from Theorem 4.1.7(b) and Remark 4.1.8.

Here, the relative box dimension of A coincides with its usual box dimension, namely, $\log_3 8$. Moreover, the set $\dim_{PC}(A,\Omega) := \mathscr{P}_c(\zeta_{A,\Omega})$ of the relative principal complex dimensions of the Sierpiński carpet A with respect to the unit square Ω is given by

$$\dim_{PC}(A,\Omega) = \log_3 8 + \mathbf{p}\mathbb{i}\mathbb{Z}, \tag{4.2.99}$$

where $\log_3 8 =: D$ is the Minkowski dimension and $\mathbf{p} := 2\pi/\log 3$ is the oscillatory period of the Sierpiński carpet RFD (A,Ω) (as well as of the ordinary Sierpiński carpet). Each principal complex dimension is simple (i.e., is a simple pole of $\zeta_{A,\Omega}$).

Observe that it follows immediately from (4.2.96) that the set $\mathscr{P}(\zeta_{A,\Omega})$ of all relative complex dimensions of the Sierpiński carpet A (with respect to the unit square Ω) is given by

$$\mathscr{P}(\zeta_{A,\Omega}) = \dim_{PC}(A,\Omega) \cup \{0,1\} = (\log_3 8 + \mathbf{p}\mathbb{i}\mathbb{Z}) \cup \{0,1\}, \tag{4.2.100}$$

where $\dim_{PC}(A,\Omega) = \log_3 8 + \mathbf{p}\mathbb{i}\mathbb{Z}$ can be viewed as the set of 'scaling complex dimensions' of the self-similar RFD (A,Ω) and $\{0,1\}$ can be viewed as the set of 'integer dimensions' of (A,Ω) (in the sense of [LapPe2–3] and [LapPeWi1], see also [Lap-vFr3, Section 13.1]). Furthermore, each of these relative complex dimensions is simple (i.e., is a simple pole of $\zeta_{A,\Omega}$). Interestingly, these are exactly the complex dimensions which one would expect to be associated with A, according to the theory developed in [LapPe2–3] and [LapPeWi1–2] (as described in [Lap-vFr3, Section 13.1]) via self-similar tilings (or sprays) and associated tubular zeta functions.

Exactly the same results concerning the principal complex dimensions and the complex dimensions hold for the ordinary Sierpiński carpet A instead of the Sierpiński carpet RFD (A,Ω); in particular, the exact counterparts of (4.2.99) and (4.2.100) hold, with (A,Ω) replaced by A. See Proposition 3.2.1 in Subsection 3.2.1 above.

In light of (4.2.96), the residue of the distance zeta function of the relative Sierpiński carpet (A,Ω) computed at any principal pole $s_k := \log_3 8 + \mathbf{p}\mathbb{i}k$, $k \in \mathbb{Z}$, is given by

$$\operatorname{res}(\zeta_{A,\Omega}, s_k) = \frac{2^{-s_k}}{(\log 3)s_k(s_k-1)}.$$

In particular,

$$|\operatorname{res}(\zeta_{A,\Omega}, s_k)| \sim \frac{2^{-D}}{D\log 3}k^{-2} \quad \text{as} \quad k \to \pm\infty,$$

where $D := \log_3 8$.

Similarly as in the case of the relative Sierpiński gasket (see Proposition 4.2.25), the relative Sierpiński carpet can be viewed as a fractal spray generated by the base RFD appearing in Figure 4.9 on the right.

Proposition 4.2.30 (Relative Sierpiński carpet). *Let (A',Ω') be the RFD defined in Figure 4.9 on the right. Let (A,Ω) be the relative fractal spray generated by the base relative fractal drum (A',Ω'), with scaling ratio $\lambda = 1/3$ and with multiplicities $m_k = 8^k$ for any positive integer k:*

$$(A,\Omega) = \mathrm{Spray}((A',\Omega'),\ \lambda = 1/3,\ m_k = 8^k \text{ for } k \in \mathbb{N}). \tag{4.2.101}$$

(Note that here we assume that the base relative fractal drum (A',Ω') has a multiplicity equal to 8.*) Then, the relative distance zeta function of the relative fractal spray (A,Ω) coincides with the relative distance zeta function of the relative Sierpiński carpet. (See Equation* (4.2.96).*)*

Example 4.2.31. (*Sierpiński N-carpet*). It is easy to generalize the notion of a standard Sierpiński carpet (which is a compact subset of the unit square $[0,1]^2 \subset \mathbb{R}^2$), to the *Sierpiński N-carpet* (or *N-carpet*, for short), defined analogously as a compact subset A of the unit N-dimensional cube $[0,1]^N \subset \mathbb{R}^N$. More specifically, we divide $[0,1]^N$ into the union of 3^N congruent N-dimensional subcubes of length $1/3$ and with disjoint interiors and then remove the middle open subcube. The remaining compact set is denoted by F_1. We then remove the middle open N-dimensional cubes of length $1/3^2$ from the remaining $3^N - 1$ subcubes. The resulting compact subset is denoted by F_2. Proceeding analogously ad infinitum, we obtain a decreasing sequence of compact subsets F_k of $[0,1]^N$, $k \geq 1$. The Sierpiński N-carpet A is then defined by

$$A := \bigcap_{k=1}^{\infty} F_k. \tag{4.2.102}$$

Note that the Sierpiński 1-carpet coincides with the usual ternary Cantor set, while the 2-carpet coincides with the classic Sierpiński carpet; furthermore, the Sierpiński 3-carpet is discussed in [LapRaŽu5, Example 6.10].

It is clear that the *Sierpiński N-carpet RFD* (A,Ω), where A is the standard Sierpiński N-carpet and $\Omega := (0,1)^N$ is the open unit cube of $\mathbb{R}^N$, can be viewed as the following relative fractal spray; see the end of Definition 4.2.1:

$$(A,\Omega) = \mathrm{Spray}\left((\partial\Omega_0,\Omega_0),\lambda = 1/3, b = 3^N - 1\right). \tag{4.2.103}$$

Here, the cube $\Omega_0 = (0,1/3)^N$ is obtained by a suitable translation of the middle open subcube from the first step of the construction of the set A. According to Theorem 4.2.6, we then have that

$$\begin{aligned}\zeta_{A,\Omega}(s) &= f(s)\cdot\zeta_{\partial\Omega_0,\Omega_0}(s)\\ &= \frac{\zeta_{\partial\Omega_0,\Omega_0}(s)}{1-(3^N-1)3^{-s}} \sim \frac{1}{1-(3^N-1)3^{-s}},\end{aligned} \tag{4.2.104}$$

where $f(s) = \zeta_{\mathfrak{S}}(s) = 1/(1-(3^N-1)3^{-s})$, for all $s \in \mathbb{C}$, is the scaling zeta function of the self-similar RFD (A,Ω). Since Ω_0 has a Lipschitz boundary and $\log_{1/\lambda} b = \log_3(3^N-1) \in (N-1,N)$, we deduce from (4.2.104) and from (4.2.12) in Theorem 4.2.6 that the set of principal complex dimensions of the relative Sierpiński N-carpet spray is given by

$$\dim_{PC}(A,\Omega) = \log_3(3^N-1) + \frac{2\pi}{\log 3}\mathrm{i}\mathbb{Z} \tag{4.2.105}$$

and hence,

$$\dim_{PC}(A,\Omega) \subset \{\mathrm{Re}\, s = \log_3(3^N-1)\} \subset \{N-1 < \mathrm{Re}\, s < N\}.$$

In particular, according to Theorem 4.1.7(b), we have that

$$\overline{\dim}_B(A,\Omega) = \log_3(3^N-1). \tag{4.2.106}$$

Furthermore, it can be shown that in the present case of the Sierpiński N-carpet RFD, $\dim_B(A,\Omega)$ and $\dim_B A$ exist and

$$\dim_B(A,\Omega) = \dim_B A = \log_3(3^N-1). \tag{4.2.107}$$

It is easy to see that the set of principal complex dimensions $\dim_{PC} A$ of the Sierpiński N-carpet A in $\mathbb{R}^N$ coincides with the set $\dim_{PC}(A,\Omega)$ appearing in Equation (4.2.105) and that the multiplicities of the complex dimensions are the same (hence, all of the complex dimensions are simple). As simple special cases, for $N=1$ we obtain the set of principal complex dimensions of the ternary Cantor set appearing in Equation (2.1.114) on page 105, or of the usual Sierpiński carpet appearing in Equation (4.2.99), for $N=1$ or $N=2$, respectively.

Since the set of all complex dimensions of the RFD $(\partial\Omega_0,\Omega_0)$ is equal to $\{0,1,\dots,N-1\}$,[23] and hence, $\overline{\dim}_B(\partial\Omega_0,\Omega_0) = \dim_B(\partial\Omega_0,\Omega_0) = N-1$, it follows from Equation (4.2.104) that the set of all complex dimensions of the Sierpiński N-carpet relative fractal spray (A,Ω) is given by

$$\begin{aligned}\mathscr{P}(\zeta_{A,\Omega}) &= \dim_{PC}(A,\Omega) \cup \{0,1,\dots,N-1\} \\ &= \left(\log_3(3^N-1) + \frac{2\pi}{\log 3}\mathrm{i}\mathbb{Z}\right) \cup \{0,1,\dots,N-1\}.\end{aligned} \tag{4.2.108}$$

This concludes for now our study of the relative fractal drum (A,Ω) naturally associated with the N-dimensional Sierpiński carpet.

We will return to this subject in Chapter 5 (Example 5.5.13) when obtaining a corresponding fractal tube formula in the case when $N=3$.

---

[23] Note that the relative zeta function $\zeta_{A,\Omega}$ appearing in Equation (4.2.104) can be meromophically extended in a unique way to the whole complex plane $\mathbb{C}$ since the same can be done with $\zeta_{\partial\Omega_0,\Omega_0}$. See, for example, Equation (4.2.97) dealing with the case when $N=2$.

Remark 4.2.32. It is natural to wonder why the same new phenomena as for the relative inhomogeneous Sierpiński N-gasket (Example 4.2.26) do not occur in the case of the relative Sierpiński N-carpet (A,Ω). In particular, the Minkowski dimension $D := \dim_B(A,\Omega)$ and the similarity dimension σ_0 of the Sierpiński N-carpet RFD (A,Ω) coincide (in fact, $D = \sigma_0 = \log_3(3^N - 1)$, see Equation (4.2.107)). Furthermore, again in light of Equation (4.2.107), D also coincides with the Minkowski dimension of the classic Sierpiński N-carpet A. [Since A is a (homogeneous) self-similar set satisfying the open set condition, we must have that $\dim_B A$ exists and $\dim_B A = \sigma_0$, the common similarity dimension of A and of (A,Ω).] Moreover, $\dim_{PC} A := \mathscr{P}_c(\zeta_A)$ and $\dim_{PC}(A,\Omega) := \mathscr{P}_c(\zeta_{A,\Omega})$ coincide, as multisets. Finally, it is always the case that $\dim_B(A,\Omega) = \max\{\sigma_0, \dim_B(\partial\Omega_0,\Omega_0)\} = \sigma_0$, in agreement with (4.2.106).

All of these statements hold for every $N \geq 1$. The reason, of course, for all these simplifications (compared to the case of the N-gasket RFD in Example 4.2.26, when $N \geq 3$ and, especially, when $N \geq 4$) is that the first component, A, of the self-similar RFD (A,Ω) is precisely the classic Sierpiński N-carpet. Therefore, for every $N \geq 1$, A_N is a self-similar set, in the usual sense, and not just an inhomogeneous self-similar set (as was the case for every $N \geq 3$ of the first component, A_N, of the inhomogeneous self-similar RFD (A_N,Ω_N) in Example 4.2.26).

Example 4.2.33. (*The* $1/2$-*square fractal*). In this planar example, we will further investigate and illustrate the new interesting phenomenon which occurs in the case of the Sierpiński 3-gasket RFD discussed in Example 4.2.26. Namely, we start with the closed unit square $I = [0,1]^2$ in $\mathbb{R}^2$ and subdivide it into 4 smaller squares by taking the centerlines of its sides. We then remove the two diagonal open smaller squares, denoted by G_1 and G_2 in Figure 4.10; so that $G := G_1 \cup G_2$ is our generator in the sense of Definition 4.2.11. Next, we repeat this step with the remaining two closed smaller squares and continue this process, ad infinitum. The $1/2$-*square fractal* is then defined as the set A which remains at the end of the process; see Figure 4.10, where the first 6 iterations are shown. More precisely, the set A is the union of all of the boundaries of the disjoint family of open squares appearing in Figure 4.10 and packed in the unit square I. If we now let $\Omega := (0,1)^2$, we have that (A,Ω) is an example of a self-similar spray (or tiling), in the sense of Definition 4.2.11, with generator $G = G_1 \cup G_2$ and scaling ratios $r_1 = r_2 = 1/2$. Note, however, that A is not a (*homogeneous*) self-similar set in the usual sense (see, e.g., [Fal1, Hut]), defined via iterated function systems (IFSs), but is instead an *inhomogeneous* self-similar set.

More specifically, the set A is the unique nonempty compact subset K of $\mathbb{R}^2$ which is the solution of the inhomogeneous equation

$$K = \bigcup_{j=1}^{2} \Phi_j(K) \cup B, \tag{4.2.109}$$

where Φ_1 and Φ_2 are suitable contractive similitudes of $\mathbb{R}^2$ with fixed points located at the lower left vertex and the upper right vertex of the unit square, respectively, and with a common scaling ratio equal to $1/2$ (i.e., $r_1 = r_2 = 1/2$, where $\{r_j\}_{j=1}^2$ are the scaling ratios of the self-similar RFD (A,Ω)). Furthermore, the nonempty compact set B in Equation (4.2.109) is the union of the left and upper sides of the square G_1 and the right and lower sides of the square G_2; see Figure 4.10. We note that here, the corresponding (classic or homogeneous) self-similar set (i.e., the unique nonempty compact subset of $\mathbb{R}^2$ which is the solution of the homogeneous fixed point equation, $C = \cup_{j=1}^2 \Phi_j(C)$), is the diagonal C of the unit square connecting the lower left and the upper right vertices of the unit square.

Let us now compute the distance zeta function ζ_A of the $1/2$-square fractal. Without loss of generality, we may assume that $\delta > 1/4$; so that we have

$$\zeta_A(s) = \zeta_{A,\Omega}(s) + \zeta_I(s), \tag{4.2.110}$$

where, intuitively, ζ_I denotes the distance zeta function corresponding to the 'outer' δ-neighborhood of A. Clearly, ζ_I is equal to the distance zeta function of the unit square $I := [0,1]^2$; it is straightforward to compute it and show that it has a meromorphic extension to all of $\mathbb{C}$ given by[24]

$$\zeta_I(s) = \frac{4\delta^{s-1}}{s-1} + \frac{2\pi\delta^s}{s}, \tag{4.2.111}$$

for all $s \in \mathbb{C}$.

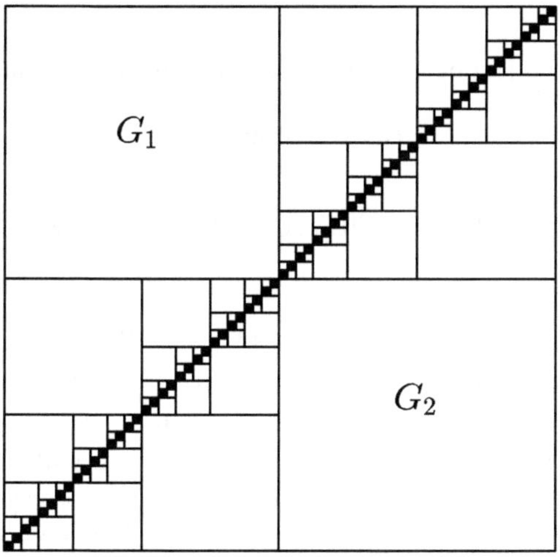

Fig. 4.10 The $1/2$-square fractal A from Example 4.2.33. The first 6 iterations are depicted. Here, $G := G_1 \cup G_2$ is the single generator of the corresponding self-similar spray or RFD (A,Ω), in the sense of Definition 4.2.11. The set A is equal to the complement of the union of the disjoint family of all open squares, with respect to $\Omega = (0,1)^2$. Equivalently, the set A coincides with the closure of the union of the boundaries of all the open squares.

[24] See also the proof of Proposition 3.2.1 where this computation was performed.

Furthermore, by using Theorem 4.2.17, we obtain that

$$\zeta_{A,\Omega}(s) = \frac{\zeta_{\partial G,G}(s)}{1-2\cdot 2^{-s}} = \frac{2^s\zeta_{\partial G,G}(s)}{2^s-2}, \tag{4.2.112}$$

for all $s\in\mathbb{C}$ with $\operatorname{Re} s$ sufficiently large. Next, we compute the distance zeta function of $(\partial G, G)$ by subdividing $G = G_1 \cup G_2$ into 16 congruent triangles (see Figures 4.9 and 4.10) and by using local Cartesian coordinates $(x,y)\in\mathbb{R}^2$ to deduce that

$$\zeta_{\partial G,G}(s) = 16\int_0^{1/4} \mathrm{d}x \int_0^x y^{s-2}\,\mathrm{d}y = \frac{4^{-s}}{s(s-1)},$$

for all $s\in\mathbb{C}$ with $\operatorname{Re} s > 1$. Hence,

$$\zeta_{\partial G,G}(s) = \frac{4^{-s}}{s(s-1)}, \tag{4.2.113}$$

an identity valid initially for all $s\in\mathbb{C}$ such that $\operatorname{Re} s>1$, and then, after meromorphic continuation, for all $s\in\mathbb{C}$. Finally, by combining Equations (4.2.110)–(4.2.113), we conclude that the distance zeta function ζ_A is meromorphic on all of $\mathbb{C}$ and is given by

$$\zeta_A(s) = \frac{2^{-s}}{s(s-1)(2^s-2)} + \frac{4\delta^{s-1}}{s-1} + \frac{2\pi\delta^s}{s}, \tag{4.2.114}$$

for all $s\in\mathbb{C}$.

Consequently (see just below), we have that $\dim_B A$ exists and

$$D(\zeta_A) = \dim_B A = 1,$$
$$\mathscr{P}(\zeta_A) := \mathscr{P}(\zeta_A,\mathbb{C}) = \{0\}\cup(1+\mathbf{p}\mathrm{i}\mathbb{Z}) \tag{4.2.115}$$

and thus

$$\dim_{PC} A := \mathscr{P}_c(\zeta_A) = 1+\mathbf{p}\mathrm{i}\mathbb{Z}, \tag{4.2.116}$$

where the oscillatory period $\mathbf{p}$ of A is given by $\mathbf{p} := \frac{2\pi}{\log 2}$. All of the complex dimensions in $\mathscr{P}(\zeta_A)$ are simple except for $\omega=1$, which is a double pole of ζ_A.

We will revisit this example in Chapter 5 where we will use the distance zeta function of A given by (4.2.114) in order to derive a corresponding fractal tube formula (see Example 5.5.22 in Subsection 5.5.6). For now, we simply mention that it will follow from the results of Chapter 5 (see, especially, Theorem 5.4.30) that $\dim_B A$ exists, $\dim_B A = D(\zeta_A) = 1$ and that A is not Minkowski measurable because of the presence of the double pole of ζ_A at $\omega=1$. On the other hand, we will show that A is h-Minkowski measurable, where the gauge function h is given by $h(t) := \log t^{-1}$ for all $t\in(0,1)$, and by Theorem 5.4.32, the corresponding h-Minkowski content is given by

$$\mathscr{M}^1(A,h) = \zeta_A[1]_{-2} = \frac{1}{4\log 2}, \tag{4.2.117}$$

where $\zeta_A[1]_{-2}$ is the (-2)-nd coefficient in the Laurent series expansion of ζ_A around $s = 1$. Finally, we note that in light of Equation (4.2.115) (and hence, in light of the presence of nonreal complex dimensions), the set A is indeed *fractal*, according to our proposed definition of fractality given in Remark 4.6.24.

Example 4.2.34. (*The* $1/3$-*square fractal*). In the present planar example, we illustrate a situation which is similar to that of the inhomogeneous Sierpiński N-gasket RFD discussed in Example 4.2.26 for $N \geq 4$. Again, we start with the closed unit square $I = [0,1]^2$ in $\mathbb{R}^2$ and subdivide it into 9 smaller congruent squares (similarly as in the case of the Sierpiński carpet). Next, we remove 7 of those smaller squares; that is, we only leave the lower left and the upper right squares (see Figure 4.11). In other words, our generator G (in the sense of Definition 4.2.11) is the (nonconvex) open polygon depicted in Figure 4.11.

As usual, we proceed by iterating this procedure with the two remaining closed squares and then continue this process ad infinitum. The first 4 iterations are depicted in Figure 4.11. The $1/3$-*square fractal* is then defined as the set A which remains at the end of the process. We now let $\Omega := (0,1)^2$, which makes the RFD (A,Ω) a self-similar spray (or tiling), in the sense of Definition 4.2.11, with generator G and scaling ratios $\{r_j\}_{j=1}^2$ such that $r_1 = r_2 = 1/3$. Again, the set A is not a homogeneous self-similar set, but is an inhomogeneous self-similar set.

More specifically, the set A is the unique nonempty compact subset K of $\mathbb{R}^2$ which is the solution of the inhomogeneous fixed point equation

$$K = \bigcup_{j=1}^{2} \Phi_j(K) \cup B, \tag{4.2.118}$$

where Φ_1 and Φ_2 are contractive similitudes of $\mathbb{R}^2$ with fixed points located at the lower left vertex and the upper right vertex of the unit square, respectively, and with a common scaling ratio equal to $1/3$ (i.e., $r_1 = r_2 = 1/3$). Furthermore, the nonempty compact set B in Equation (4.2.118) is equal to the boundary of G without the part belonging to the boundary of the two smaller squares which are left behind in the first iteration; see Figure 4.11. We also observe that here, the corresponding (classic or homogeneous) self-similar set generated by the IFS consisting of Φ_1 and Φ_2, is the ternary Cantor set located along the diagonal of the unit square.

We now proceed by computing the distance zeta function ζ_A of the $1/3$-square fractal. Without loss of generality, we may assume that $\delta > 1/4$; so that we have

$$\zeta_A(s) = \zeta_{A,\Omega}(s) + \zeta_I(s), \tag{4.2.119}$$

where, as before in Example 4.2.33, ζ_I denotes the distance zeta function corresponding to the 'outer' δ-neighborhood of A and coincides with the distance zeta function of the unit square $I := [0,1]^2$. Recall that ζ_I was computed in Example 4.2.33 and is given by Equation (4.2.111) for all $s \in \mathbb{C}$.

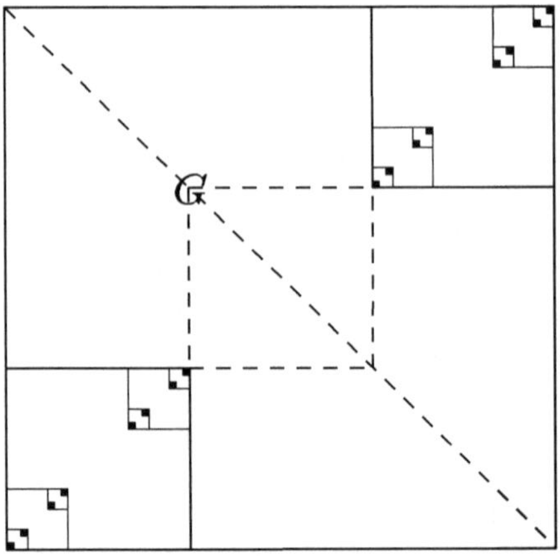

Fig. 4.11 The $1/3$-square fractal A from Example 4.2.34. The first 4 iterations are depicted. Here, G is the single generator of the corresponding self-similar spray or RFD (A,Ω), in the sense of Definition 4.2.11 or Definition 4.2.20. The set A is equal to the complement of the union of the disjoint family of all the open 8-gons, with respect to the open square $\Omega=(0,1)^2$. The largest 8-gon is equal to the union of two open squares indicated by dashed sides of length $2/3$, while each of the next two smaller 8-gons is obtained by scaling the first one by the factor $1/3$. Any of the 2^k 8-gons of the k-th generation is obtained by scaling the first one by the factor $1/3^{k-1}$, for any $k\in\mathbb{N}$. Equivalently, A coincides with the closure of the union of the boundaries of all the 8-gons.

Furthermore, by using Theorem 4.2.17, we obtain that

$$\zeta_{A,\Omega}(s)=\frac{\zeta_{\partial G,G}(s)}{1-2\cdot 3^{-s}}=\frac{3^s\zeta_{\partial G,G}(s)}{3^s-2}, \tag{4.2.120}$$

for all $s\in\mathbb{C}$ with $\operatorname{Re}s$ sufficiently large.

Next, we compute the distance zeta function of $(\partial G,G)$ by subdividing G into 14 congruent triangles denoted by G_i, for $i=1,\dots,14$ (see Figure 4.11). Therefore, by symmetry, we obtain the following functional equation:

$$\zeta_{\partial G,G}(s)=12\zeta_{\partial G,G_1}(s)+2\zeta_{\partial G,G_{13}}, \tag{4.2.121}$$

valid initially for all $s\in\mathbb{C}$ such that $\operatorname{Re}s$ is sufficiently large.

We use local Cartesian coordinates $(x,y)\in\mathbb{R}^2$ to compute $\zeta_{\partial G,G_1}$ and obtain

$$\zeta_{\partial G,G_1}(s)=\int_0^{1/3}\mathrm{d}x\int_0^x y^{s-2}\,\mathrm{d}y=\frac{3^{-s}}{s(s-1)}.$$

Hence,

$$\zeta_{\partial G,G_1}(s)=\frac{3^{-s}}{s(s-1)}, \tag{4.2.122}$$

an identity valid initially for all $s\in\mathbb{C}$ such that $\operatorname{Re}s>1$, and then, after meromorphic continuation, for all $s\in\mathbb{C}$. In order to compute $\zeta_{\partial G,G_{13}}$, we use local polar coordinates (r,θ) and deduce that

$$\zeta_{\partial G,G_{13}}(s) = \int_0^{\pi/2} \mathrm{d}\theta \int_0^{3^{-1}(\sin\theta+\cos\theta)^{-1}} r^{s-1}\,\mathrm{d}r = \frac{3^{-s}}{s}\int_0^{\pi/2}(\cos\theta+\sin\theta)^{-s}\,\mathrm{d}\theta, \tag{4.2.123}$$

valid initially for all $s \in \mathbb{C}$ such that $\operatorname{Re} s > 0$ and after meromorphic continuation, for all $s \in \mathbb{C}$. Note that by using Theorem 2.1.45 with $\varphi(\theta) := (\cos\theta + \sin\theta)^{-1}$ for $\theta \in (0, 2\pi)$, it is easy to check that

$$Z(s) := \int_0^{\pi/2} (\cos\theta + \sin\theta)^{-s}\,\mathrm{d}\theta \tag{4.2.124}$$

is an entire function, since it is of the form of the generalized DTI $f(s) := \int_E \varphi(\theta)^s \mathrm{d}\mu(\theta)$, where $E = [0, \pi/2]$, $\varphi(\theta) := (\cos\theta + \sin\theta)^{-1}$ for all $\theta \in E$ is uniformly bounded by positive constants both from above and below, and $\mathrm{d}\mu(\theta) := \mathrm{d}\theta$.

Finally, by combining Equation (4.2.111) and Equations (4.2.119)–(4.2.124), we obtain that ζ_A is given by

$$\zeta_A(s) = \frac{2}{s(3^s-2)}\left(\frac{6}{s-1} + Z(s)\right) + \frac{4\delta^{s-1}}{s-1} + \frac{2\pi\delta^s}{s}, \tag{4.2.125}$$

an identity valid initially for all $s \in \mathbb{C}$ with $\operatorname{Re} s > 1$ and then, after meromorphic continuation, for all $s \in \mathbb{C}$.

Consequently, we deduce that

$$D(\zeta_A) = 1, \qquad \mathscr{P}(\zeta_A) := \mathscr{P}(\zeta_A, \mathbb{C}) \subseteq \{0\} \cup (\log_3 2 + \mathbf{p}\mathrm{i}\mathbb{Z}) \cup \{1\} \tag{4.2.126}$$

and

$$\dim_{PC} A := \mathscr{P}_c(\zeta_A) = \{1\}, \tag{4.2.127}$$

where the oscillatory period $\mathbf{p}$ of A is given by $\mathbf{p} := \frac{2\pi}{\log 3}$. In Equation (4.2.126), we only have an inclusion since, in principle, some of the complex dimensions with real part $\log_3 2$ may be canceled by the zeros of $6/(s-1) + Z(s)$. However, it can be checked numerically that $\log_3 2 \in \mathscr{P}(\zeta_A)$ and that there also exist nonreal complex dimensions with real part $\log_3 2$ in $\mathscr{P}(\zeta_A)$. All of the complex dimensions in $\mathscr{P}(\zeta_A)$ are simple.

We will revisit this example in Subsection 5.5.6 (see Example 5.5.23) where we will obtain a fractal tube formula for the set A from Equation (4.2.125). For now, we simply mention that, $\dim_B A = 1$ and that, by Theorem 5.4.2, A is Minkowski measurable with Minkowski content given by

$$\mathscr{M}^1(A) = \operatorname{res}(\zeta_A, 1) = 16. \tag{4.2.128}$$

We also note that A is indeed *fractal*, according to our proposed definition of fractality (see Remark 4.6.24). More precisely, in light of Equation (4.2.126), it is *strictly subcritically fractal* and *fractal in dimension* $d := \log_3 2$, in the sense of case (ii) of Remark 5.5.15 below. In closing, we also mention that the set A is rectifiable and that its 'length' (i.e., its 1-dimensional Hausdorff measure) is given by

$$H^1(A) = \frac{\mathscr{M}^1(A)}{\omega_1} = 8, \tag{4.2.129}$$

which can, of course, be easily checked directly. Here, $\omega_1 = 2$ is the volume of the 1-dimensional ball of radius 1.

Example 4.2.35. (*A self-similar fractal nest*). In the final planar example of this subsection, we investigate the case of a 'self-similar fractal nest'.[25] The set A which we now define is an inhomogeneous self-similar set. Similarly as in Example 4.2.34, the set A will be *fractal* in the sense of our proposed definition of fractality given in Remark 4.6.24 and, moreover, will be *strictly subcritically fractal* in the sense of Remark 5.5.15.

Let $a \in (0,1)$ be a real parameter. We define the set A as the union of concentric circles with center at the origin and of radius a^k for $k \in \mathbb{N}_0$ (see Figure 4.12). Furthermore, let G be the open annulus such that ∂G consists of the circles of radius 1 and a, as depicted in Figure 4.12, and let $\Omega := B_1(0)$. We can now consider the RFD (A,Ω) as a self-similar spray with generator G, in the sense of Definition 4.2.11.

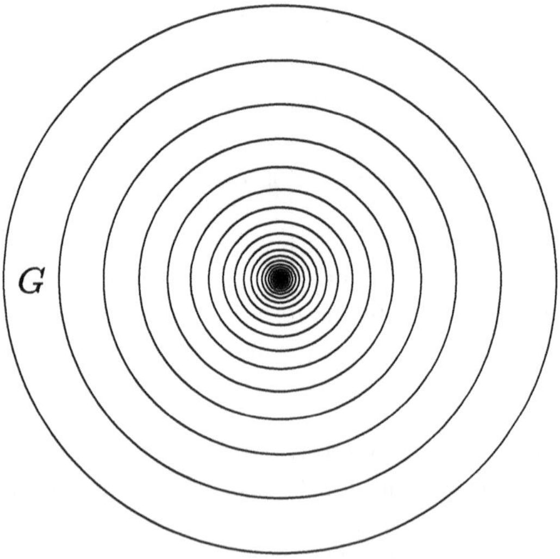

Fig. 4.12 The self-similar fractal nest from Example 4.2.35.

[25] As we shall see, throughout this example, the use of the adjective "self-similar" is somewhat abusive since only one similarity transformation is involved in order to define A.

We note that even though (A,Ω) is a fractal spray, with a single generator G, it is not (strictly speaking) self-similar in the traditional sense because it only has one scaling ratio $r=a$ (associated with a single contractive similitude). However, we will continue using this abuse of language throughout this example. Also, a moment's reflection reveals that this fact does not affect any of the conclusions relevant to the distance zeta function of such an RFD. Namely, we obviously have that

$$(A,\Omega)=(\partial G,G)\sqcup(aA,a\Omega); \tag{4.2.130}$$

so that

$$\zeta_{A,\Omega}(s)=\zeta_{\partial G,G}(s)+\zeta_{aA,a\Omega}(s), \tag{4.2.131}$$

for all $s\in\mathbb{C}$ such that $\operatorname{Re}s$ is sufficiently large. Furthermore, by using the scaling property of the relative distance zeta function (see Theorem 4.1.40), we conclude that

$$\zeta_{A,\Omega}(s)=\frac{\zeta_{\partial G,G}(s)}{1-a^s}, \tag{4.2.132}$$

again for all $s\in\mathbb{C}$ such that $\operatorname{Re}s$ is sufficiently large.

Next, we compute the distance zeta function of the generator by using polar coordinates (r,θ):

$$\begin{aligned}\zeta_{\partial G,G}(s)&=\int_0^{2\pi}\mathrm{d}\theta\int_a^{(1+a)/2}(r-a)^{s-2}r\,\mathrm{d}r\\&\quad+\int_0^{2\pi}\mathrm{d}\theta\int_{(1+a)/2}^1(1-r)^{s-2}r\,\mathrm{d}r\\&=\frac{2^{2-s}\pi(1+a)(1-a)^{s-1}}{s-1},\end{aligned} \tag{4.2.133}$$

an identity valid, after meromorphic continuation, for all $s\in\mathbb{C}$.

Equation (4.2.133) combined with Equation (4.2.132) now yields that $\zeta_{A,\Omega}$ is meromorphic on all of $\mathbb{C}$ and is given for all $s\in\mathbb{C}$ by

$$\zeta_{A,\Omega}(s)=\frac{2^{2-s}\pi(1+a)(1-a)^{s-1}}{(s-1)(1-a^s)}. \tag{4.2.134}$$

Finally, we fix an arbitrary $\delta>(1-a)/2$ and observe that for the distance zeta function of A, we have

$$\zeta_A(s)=\zeta_{A,\Omega}(s)+\zeta_{A,B_{1+\delta}(0)\setminus\Omega}(s), \tag{4.2.135}$$

for all $s\in\mathbb{C}$ with $\operatorname{Re}s$ sufficiently large. Furthermore, we have that

$$\zeta_{A,B_{1+\delta}(0)\setminus\Omega}(s)=\int_0^{2\pi}\mathrm{d}\theta\int_1^{1+\delta}(r-1)^{s-2}r\,\mathrm{d}r=\frac{2\pi\delta^{s-1}}{s-1}+\frac{2\pi\delta^s}{s}, \tag{4.2.136}$$

where the last equality is valid, initially, for all $s\in\mathbb{C}$ such that $\operatorname{Re}s>1$, and then, after meromorphic continuation, for all $s\in\mathbb{C}$.

Combining now the above equation with (4.2.135), we finally obtain that ζ_A is meromorphic on all of $\mathbb{C}$ and is given by

$$\zeta_A(s) = \frac{2^{2-s}\pi(1+a)(1-a)^{s-1}}{(s-1)(1-a^s)} + \frac{2\pi\delta^{s-1}}{s-1} + \frac{2\pi\delta^s}{s}, \tag{4.2.137}$$

for all $s \in \mathbb{C}$.

Consequently (see also Theorem 5.4.30 below), we have that $\dim_B A$ exists and

$$\begin{gathered} D(\zeta_A) = \dim_B A = 1, \\ \mathscr{P}(\zeta_A) := \mathscr{P}(\zeta_A, \mathbb{C}) = \mathbf{p}\mathrm{i}\mathbb{Z} \cup \{1\} \end{gathered} \tag{4.2.138}$$

and

$$\dim_{PC} A := \mathscr{P}_c(\zeta_A) = \{1\}, \tag{4.2.139}$$

where the oscillatory period $\mathbf{p}$ of A is given by $\mathbf{p} := \frac{2\pi}{\log a^{-1}}$ and all of the complex dimensions in $\mathscr{P}(\zeta_A)$ are simple. We will also revisit this example in Subsection 5.5.6 (see Example 5.5.24 below) where its fractal tube formula will be derived directly from Equation (4.2.137) and the results of Chapter 5. Here, we simply mention that $\dim_B A$ exists (which is also easy to check directly), $\dim_B A = 1$ and that, according to Theorem 5.4.2, A is Minkowski measurable, with Minkowski content given by

$$\mathscr{M}^1(A) = \operatorname{res}(\zeta_A, 1) = \frac{4\pi}{1-a}. \tag{4.2.140}$$

We also note that the set A is rectifiable and that its 'length' (really, its 1-dimensional Hausdorff measure) is given by

$$H^1(A) = \frac{\mathscr{M}^1(A)}{\omega_1} = \frac{2\pi}{1-a}, \tag{4.2.141}$$

where, as before, $\omega_1 = 2$ is the volume of the 1-dimensional ball of radius 1. Of course, formula (4.2.141) can also be easily recovered via a direct computation.

In closing, we mention that A is indeed *fractal* according to our proposed definition of fractality (see Remark 4.6.24). More specifically, in light of Equation (4.2.138), A is *strictly subcritically fractal* and *fractal in dimension* $d := 0$, in the sense of Remark 5.5.15 below.

The following example can be viewed as an analog of Example 4.2.35 (the self-similar fractal nest) in the one-dimensional Euclidean space $\mathbb{R}$.

Example 4.2.36. (*The geometric progression fractal string*). Fix $a \in (0,1)$, which will play the role of a parameter. Let $\mathscr{L} = (\ell_k)_{k\geq 0}$ be defined as the geometric sequence with progression a; i.e.,

$$\ell_k := a^k, \quad \text{for all} \quad k \geq 0. \tag{4.2.142}$$

The geometric zeta function of this fractal string is given by

$$\zeta_{\mathscr{L}}(s) = \sum_{k=0}^{\infty} (a^k)^s = \frac{1}{1-a^s}, \tag{4.2.143}$$

valid for all $s \in \mathbb{C}$ with $\operatorname{Re} s > 0$. Upon meromorphic continuation, we see at once that $\zeta_{\mathscr{L}}$ is meromorphic on all of $\mathbb{C}$ and is given by

$$\zeta_{\mathscr{L}}(s) = \frac{1}{1-a^s}, \tag{4.2.144}$$

for all $s \in \mathbb{C}$.

Let $A_{\mathscr{L}}$ be the bounded subset of the real line generated by $\mathscr{L}$; i.e.,

$$A_{\mathscr{L}} := \left\{ a_k := \sum_{j \geq k} \ell_j : k \geq 0 \right\}. \tag{4.2.145}$$

Then, by means of Proposition 5.5.4 of Chapter 5 below (see also Example 2.1.58 and Equation (5.5.15)), we deduce that for any fixed $\delta > 1/2$, its distance zeta function $\zeta_{A_{\mathscr{L}}}$ is meromorphic on all of $\mathbb{C}$ and given by

$$\zeta_{A_{\mathscr{L}}}(s) = \frac{2^{1-s}}{s(1-a^s)} + \frac{2\delta^s}{s}, \tag{4.2.146}$$

for all $s \in \mathbb{C}$. Here, the term $2\delta^s/s$ corresponds to the 'outer' δ-neighborhood of $A_{\mathscr{L}}$, i.e., the left δ-neighborhood of the point 0 and the right δ-neighborhood of the point $a_0 = 1/(1-a)$.

We now see that the set of complex dimensions of $A_{\mathscr{L}}$ (or, equivalently, of $\mathscr{L}$) coincides with the set of principal complex dimensions of $A_{\mathscr{L}}$ (i.e., of $\mathscr{L}$); that is,

$$\mathscr{P}(\zeta_{A_{\mathscr{L}}}) = \dim_{PC} A_{\mathscr{L}} = \mathbf{p}\mathbb{i}\mathbb{Z}, \tag{4.2.147}$$

where $\mathbf{p} := 2\pi/\log a^{-1}$. Furthermore, all of the complex dimensions are simple, except for

$$D := D(\zeta_{A_{\mathscr{L}}}) = \dim_B A_{\mathscr{L}} = 0, \tag{4.2.148}$$

which has multiplicity two. See Remark 4.2.37 below for a justification of this claim.

In Example 5.5.25 of Chapter 5, we will use Equation (4.2.146) in order to obtain an exact fractal tube formula for the set $A_{\mathscr{L}} \subset \mathbb{R}$. For now, we mention that the presence of the double pole of $\zeta_{A_{\mathscr{L}}}(s)$ at $s = 0$ implies that the set $A_{\mathscr{L}}$ is not Minkowski measurable, since $\mathscr{M}^0(A_{\mathscr{L}}) = +\infty$, but that $A_{\mathscr{L}}$ is h-Minkowski measurable with respect to the gauge function h defined by $h(t) := \log t^{-1}$ for all $t \in (0,1)$, and that its h-Minkowski content is given by

$$\mathscr{M}^0(A_{\mathscr{L}}, h) = \frac{2}{\log a^{-1}}. \tag{4.2.149}$$

In closing this example, we point out that the geometric progression string $\mathscr{L}$ (or, equivalently, its canonical geometric realization $A_{\mathscr{L}}$, as well as any of its geometric realizations $\Omega \subset \mathbb{R}$ as bounded open sets of $\mathbb{R}$) is indeed an example of a *fractal* set, according to our proposed definition of fractality (see Remark 4.6.24), due to the presence of nonreal complex dimensions. More specifically, in light of Equation (4.2.147), $A_{\mathscr{L}}$ (or, equivalently, $\mathscr{L}$) is *critically fractal*; i.e., it is *fractal in dimension* $d := D = 0$, in the sense of Remark 5.5.15. Finally, we note that although $A_{\mathscr{L}}$ is h-Minkowski measurable, its intrinsic geometry still possesses geometric oscillations of order $O(t)$ in the fractal tube formula of $A_{\mathscr{L}}$, as will be shown in Example 5.5.25.

Remark 4.2.37. The fact that the complex dimension 0 of $A_{\mathscr{L}}$ has multiplicity two (and not one, as might naively be expected) follows from the following relation between $\zeta_{\mathscr{L}}$ and $\zeta_{A_{\mathscr{L}}}$ (see Equation (5.5.15) or, more generally, Equation (5.5.16) in Subsection 5.5.2 below):

$$\zeta_{A_{\mathscr{L}}}(s) = \frac{2^{1-s}}{s}\zeta_{\mathscr{L}}(s), \tag{4.2.150}$$

valid (in the present case of Example 4.2.36) for all $s \in \mathbb{C}$. Since 0 is a simple pole of $\zeta_{\mathscr{L}}$ (in light of Equation (4.2.144)), it is now apparent that 0 is a double pole of $\zeta_{A_{\mathscr{L}}}$, as claimed. For the same reason, the nonzero poles of $\zeta_{\mathscr{L}}$ and $\zeta_{A_{\mathscr{L}}}$ are the same, and have the same multiplicity.

Remark 4.2.38. When $a := p^{-1}$, where p is a prime number, Example 4.2.36 (the geometric progression string) reduces to the p-th Euler string $\mathscr{L}_p$, studied in [Lap-vFr3, esp., Subsection 4.2.1] (see also [HerLap1–5] and, in the p-adic setting, [LapLu2–3, LapLu-vFr1–2]) and whose geometric zeta function $\zeta_{\mathscr{L}_p}(s)$ coincides with the p-th local Euler factor $(1-p^{-s})^{-1}$, in agreement with (4.2.144) where we have set $a := p^{-1}$.

4.3 Spectral Zeta Functions of Fractal Drums and Their Meromorphic Extensions

We review here some of the known results concerning the spectral asymptotics of (relative) fractal drums, with emphasis on the leading asymptotic behavior of the spectral counting function (or, equivalently, of the eigenvalues), along with a corresponding sharp remainder estimate (obtained in [Lap1] and expressed in terms of the upper box dimension of the boundary).

We then apply these results, along with some results obtained in Section 2.3 of this monograph, in order to show that the spectral zeta functions of these fractal drums have a (nontrivial) meromorphic extension. This fact was already observed in [Lap2–3] by other means, but also by using the error estimates of [Lap1].

Moreover, we show the optimality (or sharpness) of the corresponding upper bound for the abscissa of meromorphic continuation of the spectral zeta function of the fractal drum. This latter result is new and makes use in an essential way of our results obtained later on in this chapter, especially in Sections 4.5 and 4.6.

4.3.1 Spectral Zeta Functions of Fractal Drums in $\mathbb{R}^N$

Let (A,Ω) be a given RFD in $\mathbb{R}^N$. In particular, this means that $|\Omega| < \infty$. We consider the corresponding *Dirichlet eigenvalue problem*, defined on the (possibly disconnected) open set $\Omega_A := \Omega \setminus \overline{A}$.[26] It consists in finding all ordered pairs $(\mu, u) \in \mathbb{C} \times H_0^1(\Omega_A)$ such that $u \neq 0$ and

$$\begin{cases} -\Delta u = \mu u, & \text{in } \Omega_A, \\ u = 0, & \text{on } \partial(\Omega_A), \end{cases} \tag{4.3.1}$$

in the variational sense (see, e.g., [LioMag], [Bre], along with [Lap1] and the relevant references therein). Here, $H_0^1(\Omega_A) := W_0^{1,2}(\Omega_A)$ is the standard *Sobolev space* (see, e.g., [Bre], [GilTru] or [MitŽu]), and $\Delta u = \sum_{k=1}^N \frac{\partial^2 u}{\partial x_k^2}$, where Δ is the Laplace operator. Recall that the Hilbert space $H_0^1(\Omega_A)$ is defined as the completion of $C_0^\infty(\Omega_A)$ (the space of infinitely differentiable complex-valued functions with compact support in Ω_A) under the Sobolev norm

$$\|u\| = \left(\int_{\Omega_A} |u(x)|^2 \, \mathrm{d}x + \int_{\Omega_A} |\nabla u(x)|^2 \, \mathrm{d}x \right)^{1/2} \tag{4.3.2}$$

and the associated inner product.

Equation (4.3.1) is, by definition, interpreted as follows: the scalar μ is an eigenvalue of $-\Delta$ if there exists $u \neq 0$, $u \in H_0^1(\Omega_A)$, such that

$$\int_{\Omega_A} \nabla u(x) \cdot \overline{\nabla \varphi(x)} \, \mathrm{d}x = \mu \int_{\Omega_A} u(x)\, \overline{\varphi(x)} \, \mathrm{d}x,$$

for all $\varphi \in C_0^\infty(\Omega_A)$ (or, equivalently, for all $\varphi \in H_0^1(\Omega_A)$).[27] This is the usual variational formulation of the Dirichlet eigenvalue problem on a bounded open set Ω_A with possibly nonsmooth (or even fractal) boundary. As it turns out, in order for (4.3.1) to be satisfied, μ must be real and even positive.

Throughout Section 4.3, we could assume equivalently that the relative fractal drum (A,Ω) is of the form of a standard fractal drum $(\partial\Omega_0, \Omega_0)$. Indeed, it suffices

[26] For example, if Ω is the unit equilateral triangle and A is the Sierpiński gasket, then Ω_A is the union of a disjoint countable family of open triangles; see Figure 4.5 on page 275.

[27] In the case of Neumann boundary conditions, both $H_0^1(\Omega_A)$ and $C_0^\infty(\Omega_A)$ will be replaced by the Sobolev space $H^1(\Omega_A)$, as will be discussed further on. Also, we must then assume that Ω is a suitable bounded open subset of $\mathbb{R}^N$; see the discussion at the end of this section on pages 343–344.

to apply the quoted results (from [Lap1], for example), to the ordinary fractal drum $(\partial\Omega_0, \Omega_0)$, which is precisely what we will do, implicitly.

The following lemma describes the boundary of $\Omega_A := \Omega \setminus \overline{A}$. As we see, the subset $\overline{A} \setminus \Omega$ of $\overline{A}$ does not have any influence on Ω_A. In particular, if A and Ω are disjoint, then $\Omega_A = \Omega$. For example, $\Omega_{\partial\Omega} = \Omega$. Here and in the sequel of Section 4.3, in order to avoid trivial statements, we assume implicitly that all of the open sets $\Omega \subset \mathbb{R}^N$ are nonempty.

Lemma 4.3.1. *Let (A, Ω) be a relative fractal drum in $\mathbb{R}^N$. If the closure of A does not possess any interior points,*[28] *then $\partial(\Omega_A) = \partial\Omega \cup (\overline{A} \cap \overline{\Omega})$. In particular, if $A \subseteq \overline{\Omega}$, then $\partial(\Omega_A) = \partial\Omega \cup \overline{A}$.*

It is well known that the (eigenvalue) spectrum of the Dirichlet eigenvalue problem (4.3.1) is discrete and consists of an infinite and divergent sequence $(\mu_k)_{k\geq 1}$ of positive numbers (called eigenvalues), without accumulation point (except $+\infty$) and which can be written in nondecreasing order according to multiplicity as follows:

$$0 < \mu_1 \leq \mu_2 \leq \cdots \leq \mu_k \leq \ldots, \qquad \lim_{k\to\infty} \mu_k = +\infty.$$

Furthermore, each of the eigenvalues μ_k is of finite multiplicity. Moreover, if Ω_A is connected, then the first (or 'principal') eigenvalue μ_1 is of multiplicity one (i.e., $\mu_1 < \mu_2$); see [GilTru]. Because the Laplace operator is symmetric, the algebraic and geometric multiplicities of each of its eigenvalues coincide. We say for short that the sequence of eigenvalues $(\mu_k)_{k\geq 1}$ *corresponds* to the relative fractal drum (A, Ω).

Remark 4.3.2. (a) For the present Dirichlet problem (4.3.1), the discreteness of the spectrum, along with the finiteness of the multiplicity of each (necessarily positive) eigenvalue, follows from the fact that for any open subset Ω of $\mathbb{R}^N$ which is bounded (or, more generally, of finite volume), $H_0^1(\Omega)$ is compactly embedded into $H^1(\Omega)$ and hence, into the Lebesgue space $L^2(\Omega_A)$. (See, e.g., [EdmEv].) Recall that the Sobolev space $H^1(\Omega_A) := W^{1,2}(\Omega_A)$ (which is used to formulate the variational Neumann eigenvalue problem) is the space of all functions $u \in L^2(\Omega_A)$ with distributional (or 'weak') gradient $\nabla u \in [L^2(\Omega_A)]^N$. Like $H_0^1(\Omega_A)$, $H^1(\Omega_A)$ is a complex Hilbert space for the Sobolev norm $\|\cdot\|$ defined by (4.3.2) and the associated inner product.

(b) In contrast, for the Neumann problem, which will be briefly discussed towards the end of Subsection 4.3.2, even the discreteness of the spectrum does not always hold (for very rough boundaries) and even when it holds, the counterpart of Weyl's asymptotic formula (Equation (4.3.13) below) need not be verified. (See [Mét2–3].) This is why, following [Lap1], appropriate assumptions will be made on Ω in our discussion of the Neumann eigenvalue problem (or, more generally, of mixed Dirichlet-Neumann boundary conditions) towards the end of Subsection 4.3.2.

[28] It is easy to see that this condition is satisfied if $\overline{\dim}_B A < N$; see page 32.

Definition 4.3.3. The *spectrum of a relative fractal drum* (A,Ω) in $\mathbb{R}^N$, denoted by $\sigma(A,\Omega)$, is defined as the sequence of the square roots of the eigenvalues of the boundary value problem (4.3.1); that is,

$$\sigma(A,\Omega) := (\mu_k^{1/2})_{k\geq 1}. \tag{4.3.3}$$

Physically, the values of $\mu_k^{1/2}$, $k \in \mathbb{N}$, are interpreted as the (normalized)[29] *frequencies of the relative fractal drum.* The eigenvalues are scaled here with the exponent $1/2$, for technical (as well as physical) reasons (and because the Laplacian is a second order linear partial differential operator). See, for example, Lemma 4.3.6.

Definition 4.3.4. The *spectral zeta function* $\zeta^*_{A,\Omega}$ *of a relative fractal drum* (A,Ω) in $\mathbb{R}^N$ is given by

$$\zeta^*_{A,\Omega}(s) := \sum_{k=1}^{\infty} \mu_k^{-s/2}, \tag{4.3.4}$$

for all $s \in \mathbb{C}$ with $\operatorname{Re} s$ sufficiently large.

Example 4.3.5. The spectral zeta function of a fractal string $\mathscr{L} = (\ell_j)_{j\geq 1}$, where $\mathscr{L}$ is viewed as a relative fractal drum $(A_{\mathscr{L}}, \Omega_{\mathscr{L}})$, is given by

$$\zeta^*_{\mathscr{L}}(s) = \sum_{k,j=1}^{\infty} (k \cdot \ell_j^{-1})^{-s} = \zeta(s) \cdot \zeta_{\mathscr{L}}(s),$$

where $\zeta = \zeta_R$ is the Riemann zeta function and $\zeta_{\mathscr{L}}$ is the geometric zeta function of $\mathscr{L}$; see [Lap2–3], [LapMa2] and [Lap-vFr3, Section 1.3]. Hence, by analytic continuation (and since ζ is meromorphic on all of $\mathbb{C}$), we have

$$\zeta^*_{\mathscr{L}}(s) = \zeta(s) \cdot \zeta_{\mathscr{L}}(s), \tag{4.3.5}$$

in every domain $U \subset \mathbb{C}$ to which $\zeta_{\mathscr{L}}$ can be meromorphically continued.

The above definition of the spectrum $\sigma(A,\Omega)$ and of the spectral zeta function $\zeta^*_{A,\Omega}$ of a relative fractal drum is in agreement with the definition of the spectrum of a bounded fractal string $\mathscr{L} = (\ell_j)_{j\geq 1}$ given in [Lap-vFr3, p. 2] or more generally, of a fractal drum (see, e.g., [Lap1–3]). (See also [Lap-vFr3, Equation (1.45), p. 29] and [Lap-vFr3, Appendix B], along with the relevant references therein, including [Gilk] and [See1].) Note that the sequence

$$\mathscr{L}(A,\Omega) := (\mu_k^{-1/2})_{k\geq 1}, \tag{4.3.6}$$

which consists of the reciprocal frequencies in $\sigma(A,\Omega)$, is also a fractal string (possibly unbounded, i.e., $\sum_{k=1}^{\infty} \mu_k^{-1/2} = +\infty$). As we see, the spectral zeta function of a relative fractal drum (A,Ω) is by definition equal to the geometric zeta function of

[29] When $N = 1$, see [Lap-vFr3, footnote 1 on page 2].

the fractal string $\mathscr{L}(A,\Omega)$. It is clear that $D(\zeta^*_{A,\Omega}) \geq 0$. Furthermore, since $\zeta^*_{A,\Omega}$ is a Dirichlet series with positive coefficients (and the spectrum of (A,Ω) is infinite), we also have $D(\zeta^*_{A,\Omega}) = D_{\text{hol}}(\zeta^*_{A,\Omega})$; see Subsection 2.1.3.

We note that the usual definition of the spectrum involves the sequence of eigenvalues $(\mu_k)_{k\geq 1}$ rather than the sequence of their square roots $(\mu_k^{1/2})_{k\geq 1}$, as in Equation (4.3.3). We prefer the definition of the spectrum $\sigma(A,\Omega)$ given in Equation (4.3.3) and hence, the use of the exponent $-s/2$ (rather than of $-s$) in the definition of the spectral zeta function $\zeta^*_{A,\Omega}$ in Equation (4.3.4) since, in this case, Lemma 4.3.6, Proposition 4.3.10 and Theorem 4.3.17 below take a more elegant form. See [Lap2–3] and [Lap-vFr3, p. 29 and Appendix B], and compare, for example, with [Gilk] and [Seel].

The spectrum of a relative fractal drum has an interesting (but elementary) scaling property, which we now state.

Lemma 4.3.6. *Let $\sigma(A,\Omega)$ be the spectrum of a relative fractal drum (A,Ω) in $\mathbb{R}^N$. If λ is any fixed positive real number, then*

$$\sigma(\lambda A,\lambda\Omega) = \lambda^{-1}\sigma(A,\Omega); \tag{4.3.7}$$

that is, $\sigma(\lambda A,\lambda\Omega) = (\lambda^{-1}\mu_k^{1/2})_{k\geq 1}$, where $(\mu_k)_{k\geq 1}$ is the sequence of eigenvalues of problem (4.3.1) on Ω_A. Equivalently, $\mathscr{L}(\lambda A,\lambda\Omega) = \lambda\,\mathscr{L}(A,\Omega)$; see Equation (4.3.6).

Proof. It is easy to see that if μ_k is an eigenvalue corresponding to $-\Delta$, with respect to the domain $\Omega_A = \Omega\setminus\overline{A}$, generated by the relative fractal drum (A,Ω), then $\lambda^{-2}\mu_k$ is an eigenvalue corresponding to the operator $-\Delta$ with respect to the domain $(\lambda\Omega)_{\lambda A}$, generated by $(\lambda A,\lambda\Omega)$. Indeed, if $u_k \in H^1_0(\Omega_A)$ is such that $-\Delta u_k = \mu_k u_k$, $u_k \neq 0$, then for $v_k(y) := u_k(x/\lambda)$, where $y \in (\lambda\Omega)_{\lambda A}$, we have

$$-\Delta v_k(y) = \frac{\mu_k}{\lambda^2} v_k(y).$$

In other words, the sequence of eigenvalues of $-\Delta$ on $(\lambda\Omega)_{\lambda A}$ is equal to $(\mu_k\lambda^{-2})_{k\geq 1}$. (This claim can also be checked directly by using the aforementioned variational formulation of the eigenvalue problem (4.3.1).) Therefore, by Definition 4.3.3,

$$\sigma(\lambda A,\lambda\Omega) = (\lambda^{-1}\mu_k^{1/2})_{k\geq 1} = \lambda^{-1}\sigma(A,\Omega).$$

This completes the proof of the lemma. □

An immediate consequence of Lemma 4.3.6 is the following scaling result for the spectral zeta functions of RFDs.

Proposition 4.3.7 (Scaling property of spectral zeta functions). *Let (A,Ω) be a relative fractal drum in $\mathbb{R}^N$. Then for any $\lambda > 0$, and for all $s \in \mathbb{C}$ such that $\operatorname{Re} s > \overline{\dim}_B(A,\Omega)$, we have*

$$\zeta^*_{\lambda A,\lambda\Omega}(s) = \lambda^s \zeta^*_{A,\Omega}(s).$$

The following result represents a partial extension in the present context of Example 4.3.5 (see also [Lap-vFr3, Theorem 2.1]), in the special case of fractal strings, or of the corresponding result for fractal sprays in [Lap2–3] and [LapPo3]. Its proof is similar to that of Theorem 4.2.5.

Theorem 4.3.8. *Let (A_0,Ω_0) be a base RFD in $\mathbb{R}^N$, and let $\mathscr{L} = (\lambda_j)_{j\geq 1}$ be a nonincreasing sequence of positive numbers tending to zero (and repeated according to multiplicities), i.e., a (not necessarily bounded) fractal string. Assume that (A_j,Ω_j), $j \geq 1$, is a disjoint sequence of RFDs, each of which is obtained by a rigid motion of $\lambda_j(A_0,\Omega_0) = (\lambda_j A_0, \lambda_j \Omega_0)$. Let $(A,\Omega) = \bigcup_{j\geq 1}(A_j,\Omega_j)$ be the corresponding relative fractal spray, generated by (A_0,Ω_0) and $\mathscr{L}$; that is, $(A,\Omega) = (A_0,\Omega_0)\otimes\mathscr{L}$. Then, assuming that $s\in\mathbb{C}$ is such that $\mathrm{Re}\, s > \max\{D(\zeta^*_{A_0,\Omega_0}), \overline{\dim}_B\mathscr{L}\}$, we have*

$$\zeta^*_{A,\Omega}(s) = \zeta^*_{A_0,\Omega_0}(s)\cdot\zeta_{\mathscr{L}}(s), \tag{4.3.8}$$

where $\zeta_{\mathscr{L}}$ is the geometric zeta function of $\mathscr{L}$ (see Equation (2.1.71) of Subsection 2.1.4). In particular, for all $s\in\mathbb{C}$ with $\mathrm{Re}\, s$ sufficiently large, we have

$$\zeta^*_{A,\Omega}(s) = \sum_{k=1}^{\infty}(\mu_k^{(0)})^{-s/2}\sum_{j=1}^{\infty}\lambda_j^s, \tag{4.3.9}$$

where $(\mu_k^{(0)})_{k\geq 1}$ is the sequence of eigenvalues corresponding to the relative fractal drum (A_0,Ω_0). Furthermore, by the principle of analytic continuation, Equation (4.3.8) continues to hold on any domain to which $\zeta_{\mathscr{L}}$ and $\zeta^*_{A_0,\Omega_0}$ can both be meromorphically continued. (A similar comment applies to Equation (4.3.11) below.)

Moreover,

$$D(\zeta^*_{A,\Omega}) = \max\{D(\zeta^*_{A_0,\Omega_0}), \overline{\dim}_B\mathscr{L}\}. \tag{4.3.10}$$

In particular, if $\lambda_j = \lambda^j$ for some fixed $\lambda\in(0,1)$, and each λ^j is of multiplicity b^j, where $b\in\mathbb{N}$, $b\geq 2$, then for $\mathrm{Re}\, s > D(\zeta^*_{A_0,\Omega_0})$

$$\zeta^*_{A,\Omega}(s) = \frac{b\lambda^s}{1-b\lambda^s}\sum_{k=1}^{\infty}(\mu_k^{(0)})^{-s/2} = \frac{b\lambda^s}{1-b\lambda^s}\zeta^*_{A_0,\Omega_0}(s), \tag{4.3.11}$$

and

$$D(\zeta^*_{A,\Omega}) = \max\{D(\zeta^*_{A_0,\Omega_0}), \log_{1/\lambda} b\}.$$

Remark 4.3.9. In the case of fractal sprays (and of fractal strings, in particular), the factorization formula (4.3.8) was first observed in [Lap2–3]. In the special case of fractal strings, it has proved to be very useful; see, especially, [Lap2–3, LapPo1–3, LapMa1–2, HeLap, Lap-vFr1–3, Tep1–2, LalLap1–2, HerLap1–5]. See also, e.g., [Lap-vFr3, Sections 1.4 and 1.5] and [Lap-vFr3, Chapters 6, 9, 10 and 11], both for the case of fractal strings and (possibly generalized or even virtual) fractal sprays.

4.3.2 *Meromorphic Extensions of Spectral Zeta Functions of Fractal Drums*

It is well known that if Ω_0 is any (nonempty) bounded open subset of $\mathbb{R}^N$, and $\sigma(\partial\Omega_0,\Omega_0) = \big((\mu_k^{(0)})^{1/2}\big)_{k\geq 1}$ (that is, $(\mu_k^{(0)})_{k\geq 1}$ is the sequence of eigenvalues of $-\Delta$ with zero (or Dirichlet) boundary data on $\partial\Omega_0$, counting the multiplicities of the eigenvalues), then the following classical asymptotic result holds, known as *Weyl's law* [Wey1–2]:

$$\mu_k^{(0)} \sim \frac{4\pi^2}{(\omega_N|\Omega_0|)^{2/N}} \cdot k^{2/N} \quad \text{as} \quad k \to \infty, \tag{4.3.12}$$

where $\omega_N = \pi^{N/2}/(N/2)!$ is the volume of the unit ball in $\mathbb{R}^N$.[30] We recall that here, consistent with the notation introduced on page 41, the symbol $\sim$ means that the ratio of the left and right sides of (4.3.12) tends to 1 as $k \to \infty$.

The main result of this subsection is stated in Theorem 4.3.17. Its proof is based on the asymptotic result due to the first author, stated in Theorem 4.3.11, combined with Proposition 4.3.10.

The asymptotic result stated in Equation (4.3.12) was obtained by Hermann Weyl in 1912 for piecewise smooth boundaries, in [Wey1–2]. It has since then been extended to a variety of settings (for example, to smooth, compact Riemannian manifolds with or without boundary, various boundary conditions, broader classes of elliptic operators, fractal boundaries, etc.). See, for example, the well-known treatises by Courant and Hilbert [CouHil, Section VI.4] and by Reed and Simon [ReeSim1], along with [Hö3] and the introduction of [Lap1], as well as [Lap2–3] and [Lap-vFr3, Section 12.5 and Appendix B]. It has been extended by G. Métivier in [Mét1–3] during the 1970s to arbitrary bounded subsets of $\mathbb{R}^N$ (in the present case of Dirichlet boundary conditions). Independently and at about the same time, this latter result was also obtained by M. Sh. Birman and M. Z. Solomyak in [BiSo]. Furthermore, in this general setting (for example), sharp error estimates, expressed in terms of the upper Minkowski (or box) dimension of the boundary of Ω_0, were obtained by the first author in the early 1990s in [Lap1]; see Theorem 4.3.11 below, along with the comments following Theorem 4.3.17 and Remark 4.3.23 for further extensions about other boundary conditions and higher-order elliptic operators, with possibly variable coefficients.

In the following result, we consider a class of bounded open subsets Ω_0 of $\mathbb{R}^N$ such that the corresponding sequence of eigenvalues $(\mu_k^{(0)})_{k\geq 1}$ satisfies an asymptotic condition involving the error term as well:

$$\mu_k^{(0)} = \frac{4\pi^2}{(\omega_N|\Omega_0|)^{2/N}} \cdot k^{2/N} + O(k^{\gamma}) \quad \text{as} \quad k \to \infty. \tag{4.3.13}$$

[30] For odd N, we have $(N/2)! = \frac{N}{2}(\frac{N}{2}-1)\cdots\frac{1}{2}$, since $(N/2)! := \Gamma(\frac{N}{2}+1)$, where Γ is the classic gamma function.

Here, we assume that $\gamma \in (-\infty, 2/N)$. It will also be convenient to use the following short-hand notation: $\zeta^*_{\Omega_0} = \zeta^*_{\partial\Omega_0,\Omega_0}$, and more generally, $\zeta^*_{\Omega_0} = \zeta^*_{A_0,\Omega_0}$, provided A_0 and Ω_0 are disjoint. We say for brevity that $\zeta^*_{\Omega_0}$ is the *spectral zeta function of the bounded open subset* Ω_0 *of* $\mathbb{R}^N$.

Proposition 4.3.10. *Assume that* Ω_0 *is an arbitrary bounded open subset of* $\mathbb{R}^N$ *such that the corresponding sequence of eigenvalues of* $-\Delta$*, with zero (or Dirichlet) boundary data on* $\partial\Omega_0$*, counting the multiplicities of the eigenvalues, satisfies the asymptotic condition* (4.3.13)*, where* $\gamma < 2/N$*. Then the spectral zeta function*

$$\zeta^*_{\Omega_0}(s) = \sum_{k=1}^{\infty} (\mu_k^{(0)})^{-s/2} \tag{4.3.14}$$

possesses a (necessarily unique) meromorphic extension (at least) to the open half-plane

$$\{\operatorname{Re} s > N - (2 - \gamma N)\}. \tag{4.3.15}$$

In other words, $D_{\rm mer}(\zeta^*_{\Omega_0}) \le N - (2 - \gamma N)$*. As we see, the meromorphic extension vertical strip, to the left of the vertical line* $\{\operatorname{Re} s = N\}$*, is of width at least* $2 - \gamma N$*.*

The only pole of $\zeta^*_{\Omega_0}$ *in this half-plane is* $s = N$*, and in particular,* $D(\zeta^*_{\Omega_0}) = N$*. Furthermore, it is simple and*

$$\operatorname{res}(\zeta^*_{\Omega_0}, N) = \frac{N\omega_N}{(2\pi)^N} |\Omega_0|. \tag{4.3.16}$$

Proof. Letting $C := 4\pi^2(\omega_N|\Omega_0|)^{-2/N}$, we have that $\mu_k^{(0)} = C \cdot k^{2/N} + d_k$, where $d_k = O(k^\gamma)$ as $k \to \infty$, and hence,

$$\zeta^*_{\Omega_0}(s) = \sum_{k=1}^{\infty} (C \cdot k^{2/N} + d_k)^{-s/2}.$$

To prove the proposition, it suffices to apply Theorem 2.3.12 with $a = 2/N$, $\gamma < a$ and $s_1 = s/2$. Indeed, we obtain that $\zeta^*_{\Omega_0}(s)$ possesses a unique meromorphic extension (at least) to the open half-plane $\{\operatorname{Re} \frac{s}{2} > \frac{N}{2} - (1 - \frac{\gamma}{a})\}$, or, equivalently, to the open half-plane $\{\operatorname{Re} s > N - (2 - \gamma N)\}$, as claimed in (4.3.15). Furthermore, according to the same theorem, the residue of $\zeta^*_{\Omega_0}(2s) = \sum_{k=1}^{\infty} (\mu_k^{(0)})^{-s}$ at $s = a = 2/N$ is equal to $(1/a)C^{-1/a} = (N/2)C^{-N/2}$. Hence, the residue of $\zeta^*_{\Omega_0}(s)$ at $s = N$ can be obtained as follows:

$$\begin{aligned}\operatorname{res}(\zeta^*_{\Omega_0}, N) &= \lim_{s \to N} (s - N)\zeta^*_{\Omega_0}(s) = \lim_{2s \to N} (2s - N)\zeta^*_{\Omega_0}(2s) \\ &= 2 \lim_{s \to N/2} \left(s - \frac{N}{2}\right)\zeta^*_{\Omega_0}(2s) = 2\frac{N}{2}C^{-N/2} = \frac{N\omega_N}{(2\pi)^N}|\Omega_0|,\end{aligned}$$

where in the next-to-last equality, we have used Equation (2.3.18) from Theorem 2.3.12. This completes the proof of Proposition 4.3.10. □

In practice, in light of the remainder estimates of [Lap1] recalled in Theorem 4.3.11 and in Corollary 4.3.14 below, we will apply Proposition 4.3.10 under the assumption that $\tilde{D} := \overline{\dim}_B(\partial\Omega_0, \Omega_0) < N$ and $\gamma \in [(2+\tilde{D}-N)/N, 2/N)$.

In light of Equation (4.3.16), under the hypotheses of Proposition 4.3.10, the residue of the spectral zeta function $\zeta^*_{\Omega_0}$ computed at $s = N$ is proportional to the N-dimensional Lebesgue measure (volume) of Ω_0; see (4.3.16). As we see, this result is of a similar nature as Equation (2.2.4) in Theorem 2.2.3. Moreover, the volume of Ω_0 can be explicitly computed by using the spectral zeta function:

$$|\Omega_0| = \frac{(2\pi)^N}{N\omega_N} \operatorname{res}(\zeta^*_{\Omega_0}, N). \tag{4.3.17}$$

Theorem 4.3.8, combined with Proposition 4.3.10, generalizes [Lap-vFr3, Theorem 1.19] to the N-dimensional case. See also Theorem 4.3.17 below, which relies on Theorem 4.3.11 (or, equivalently, on Corollary 4.3.14) and provides explicit conditions under which Equation (4.3.13) holds, and hence Proposition 4.3.10 can be applied.

It is clear that the claim of Proposition 4.3.10 is true if in (4.3.13) we replace $O(k^\gamma)$ by $O(k^{(\gamma)})$ as $k \to \infty$. For example, we may have $O(k^\gamma \log k)$ as $k \to \infty$ in (4.3.13).

Let $(\mu_k^{(0)})_{k\geq 1}$ be the sequence of eigenvalues of $-\Delta$, where Δ is the Dirichlet Laplacian, associated with a given bounded open subset Ω_0 of $\mathbb{R}^N$. In what follows, we denote by

$$N_v(\mu) := \#\{k \in \mathbb{N} : \mu_k^{(0)} \leq \mu\}, \qquad \text{for} \quad \mu > 0, \tag{4.3.18}$$

the *eigenvalue counting function of the fractal drum*, taking into account the multiplicities. It is also called the *spectral counting function* in the literature; see, e.g., [Lap1–5], [Lap-vFr1–3] and the relevant references therein.

In the proof of Theorem 4.3.17 below, we shall need the following significant result (see [Lap1, Equation (1.8), Theorems 1.1 and 2.3]), which provides a partial resolution of the *modified Weyl–Berry conjecture*. See [Lap1, Corollary 2.1], as well as [Lap1, Theorems 2.1 and 2.3], along with the comments following Theorem 4.3.17 and Remark 4.3.23, for a more general statement involving positive uniformly elliptic linear differential operators (with variable and possibly nonsmooth coefficients) and mixed Dirichlet–Neumann boundary conditions.

Theorem 4.3.11 (Lapidus, [Lap1]). *Let Ω_0 be an arbitrary (nonempty) bounded open subset of $\mathbb{R}^N$. Let $\tilde{D} := \overline{\dim}_B(\partial\Omega_0, \Omega_0)$ denote the upper relative Minkowski (or box) dimension of Ω_0, with respect to $\partial\Omega_0$. Then we have the following remainder estimates:*

(*i*) *If $\tilde{D} \in (N-1, N]$, then for any $d > \tilde{D}$,*

$$N_v(\mu) = (2\pi)^{-N}\omega_N|\Omega_0| \cdot \mu^{N/2} + O(\mu^{d/2}) \quad \text{as} \quad \mu \to +\infty. \tag{4.3.19}$$

(ii) If $\tilde{D} = N-1$, then for any $d > \tilde{D}$,

$$N_\nu(\mu) = (2\pi)^{-N}\omega_N|\Omega_0|\cdot\mu^{N/2} + O(\mu^{d/2}\log\mu) \quad \text{as} \quad \mu \to +\infty. \tag{4.3.20}$$

Moreover, in both cases (i) and (ii), the choice $d = \tilde{D}$ is allowed, provided

$$\mathscr{M}^{*\tilde{D}}(\partial\Omega_0, \Omega_0) < \infty;$$

that is, Ω_0 has finite upper Minkowski content, relative to $\partial\Omega_0$ (i.e., it has finite inner Minkowski content).

Remark 4.3.12. (*a*) For Dirichlet boundary conditions, Theorem 4.3.11 just above (and hence also, Corollary 4.3.14 below) remains valid without change for an arbitrary (and possibly unbounded as well as disconnected) nonempty open set Ω_0 with finite volume: $|\Omega_0| < \infty$.

(*b*) Note that in [Lap1–3], $\tilde{D}$ is referred to as the *inner Minkowski dimension* of Ω_0. It is known that since Ω_0 is a (nonempty) bounded open set, we have $N-1 \leq \tilde{D} \leq N$; see [Lap1, Section 3]. Also, in [Lap1], the case when $\tilde{D} = N-1$ is referred to as the '*nonfractal case*' (or the least fractal case), and the case when $\tilde{D} \in (N-1, N]$ is referred to as the '*fractal case*'. Finally, note that in the *most fractal case* when $\tilde{D} = N$, the error estimate (4.3.19) is still valid, but is uninformative; indeed, even when $d := \tilde{D} = N$, the 'error term' is then of the same order as the 'leading term' in (4.3.19).

(*c*) According to the notation introduced in Remark 2.3.4, this condition can be written more succinctly in the following form:

$$N_\nu(\mu) = (2\pi)^{-N}\omega_N|\Omega_0|\cdot\mu^{N/2} + O(\mu^{(\tilde{D}/2)}) \quad \text{as} \quad \mu \to \infty. \tag{4.3.21}$$

A similar comment applies to the error estimate (4.3.20).

Various aspects of the study of the (possibly modified) Weyl–Berry conjecture are discussed in the introduction of [Lap1], in [Lap3] and, more recently, in a brief survey given in [Lap-vFr3, Section 12.5.1]. See also [Berr1–2], [BroCar], [Lap1–3], [LapPo1–3], [Cae], [vBGilk], [HamLap], [FlVa], [Ger], [GerSc], [MolVai] and the references therein. The result stated in case (*ii*) of Theorem 4.3.11, that is, in the nonfractal case when $\tilde{D} = N-1$, and under the additional assumption that $\mathscr{M}^{*(N-1)}(\partial\Omega)$ is finite, was already obtained in Métivier's work [Mét3, Theorem 6.1 on page 191]; see also [Mét1–2]. Métivier stated his result without the explicit use of box (that is, Minkowski) dimension or Minkowski content. See [Lap-vFr3, Section 12.5] for a more complete list of references. Results concerning the partition function (the trace of the heat semigroup) of the Dirichlet Laplacian have been obtained by Brossard and Carmona [BroCar]. The main estimate in [BroCar] is now a consequence of the results of [Lap1] stated in Theorem 4.3.9, but the converse is not true. Indeed, as is well known, beyond the leading term, the spectral asymptotics for the trace of the heat semigroup do not imply corresponding asymptotics for the eigenvalue counting function (or, equivalently, for the eigenvalues themselves). In fact, when they hold, the pointwise estimates for the eigenvalue counting function are considerably more difficult to prove.

Remark 4.3.13. (*a*) As was mentioned earlier, the first general result concerning the leading term of the asymptotic expansion of the eigenvalues is due to Hermann Weyl in [Wey1–2] towards the beginning of the 20th century, in the case of a sufficiently smooth boundary. Eventually, this result was extended in the 1970s by Guy Métivier in [Mét1–3] (see also [BiSo]) for an arbitrary bounded open set (and for the Dirichlet Laplacian or more general elliptic operators and boundary conditions). The error estimate (4.3.20) is due to Courant in the case of a piecewise smooth boundary (and hence, $\tilde{D} = N-1$), a very special case of (*ii*) in Theorem 4.3.11. An elementary concrete example of that situation can be found in the monograph by Courant and Hilbert [CouHil, p. 431], where in the case when Ω_0 is a rectangle in the plane, with sides a and b, it is shown that the counting function of the associated sequence of eigenvalues of the Dirichlet Laplacian $-\Delta$ satisfies

$$N_V(\mu) = \frac{ab}{4\pi} \cdot \mu + O(\sqrt{\mu}) \quad \text{as} \quad \mu \to +\infty.$$

In fact, an equivalent number-theoretic formulation of this result was already known to Gauss in 1834; see [Gau].

(*b*) In case (*ii*) of Theorem 4.3.11, the remainder estimate (4.3.20) is known to hold without the logarithmic term (i.e., Equation (4.3.19) holds with $d = N-1$ as well as $\mathcal{M}^{*\tilde{D}}(\partial\Omega_0, \Omega_0) < \infty$) if the boundary of Ω_0 is (sufficiently) smooth, or more generally, for sufficiently smooth compact Riemannian manifolds with or without boundary. (By "smooth" here, we mean C^r, that is, r times continuously differentiable, with the positive integer $r \geq 2$ large enough.) See [Hö2–3], [Lap-vFr3, Appendix B] and the introduction of [Lap1] as well as the many references therein, describing, in particular, the results of Hörmander [Hö1], Seeley [See2–3] and Pham The Lai [Ph]. See also [Lap-vFr3, Remark B.1 of Appendix B].

From Theorem 4.3.11 it is possible to derive a result, also due to the first author, regarding the error term for the leading asymptotics of the eigenvalues of the Dirichlet Laplacian $-\Delta$, associated with bounded open sets in $\mathbb{R}^N$. As is well known by the experts in spectral theory, the statement of Corollary 4.3.14 is equivalent to that of Theorem 4.3.11 (by means of a standard Abelian/Tauberian argument, for example); furthermore, Corollary 4.3.14 can be deduced from Theorem 4.3.11 by means of the converse of a Tauberian theorem, called an Abelian theorem in [Sim], for example; see [Lap1, Appendix A] and [Sim] for a closely related situation. In order to keep this part of the exposition essentially self-contained, we provide (at least in a special case) a different proof, based on the elementary Lemma 4.3.15 below.

Corollary 4.3.14 ([Lap1]). *Let Ω_0 be an arbitrary (nonempty) bounded open subset of $\mathbb{R}^N$. As before, we let $\tilde{D} := \overline{\dim}_B(\partial\Omega_0, \Omega_0)$, and let $(\mu_k^{(0)})_{k\geq 1}$ be the sequence of* *eigenvalues of $-\Delta$, where Δ is the Dirichlet Laplacian on Ω_0. Then the following conclusions hold*:

(*i*) *If $\tilde{D} \in (N-1,N]$, then for any $d > \tilde{D}$,*

$$\mu_k^{(0)} = \frac{4\pi^2}{(\omega_N|\Omega_0|)^{2/N}} \cdot k^{2/N} + O(k^{(2+d-N)/N}) \quad \text{as} \quad k \to \infty. \tag{4.3.22}$$

(*ii*) *If $\tilde{D} = N-1$, then for any $d > \tilde{D}$,*

$$\mu_k^{(0)} = \frac{4\pi^2}{(\omega_N|\Omega_0|)^{2/N}} \cdot k^{2/N} + O(k^{(2+d-N)/N} \log k) \quad \text{as} \quad k \to \infty. \tag{4.3.23}$$

Moreover, in each of the cases (*i*) *and* (*ii*), *the choice of $d = \tilde{D}$ is allowed, provided $\mathscr{M}^{*\tilde{D}}(\partial\Omega_0, \Omega_0) < \infty$.*

More succinctly, according to the notation introduced in Remark 2.3.4, we can rewrite (4.3.22) in the following equivalent manner:

$$\mu_k^{(0)} = \frac{4\pi^2}{(\omega_N|\Omega_0|)^{2/N}} \cdot k^{2/N} + O(k^{\left(\frac{2+\tilde{D}-N}{N}\right)}) \quad \text{as} \quad k \to \infty. \tag{4.3.24}$$

A similar comment applies to the remainder term in (4.3.23).

Postponing the proof of Corollary 4.3.14 for a while, we first state and prove an auxilliary technical result.

Lemma 4.3.15. *Let $c > 0$, $m > 0$ and $\alpha \in (-\infty, m)$ be given real numbers. Assume that $(\mu_k)_{k\geq 1}$ is a sequence of positive real numbers satisfying the following condition:*[31]

$$c \cdot \mu_k^m + O(\mu_k^\alpha) = k \quad \text{as} \quad k \to \infty. \tag{4.3.25}$$

Then

$$\mu_k = c^{-1/m} \cdot k^{1/m} + O(k^{\frac{\alpha+1}{m}-1}) \quad \text{as} \quad k \to \infty. \tag{4.3.26}$$

Proof. Step 1: Let us first prove the lemma for $m = 1$. Note that in this case, we have $\alpha < 1$. Without loss of generality, we may assume that $c = 1$; otherwise, we introduce a new sequence $\mu_k' = c\mu_k$. In this case, by the assumption made in the lemma, there exists a positive real number C such that $|k - \mu_k| \leq C\mu_k^\alpha$ for all positive integers k. Since this implies that $k \leq \mu_k + C\mu_k^\alpha$ for all $k \geq 1$, then, clearly,

$$\lim_{k\to\infty} \mu_k = +\infty.$$

Therefore, from

$$\left|\frac{k}{\mu_k} - 1\right| \leq C\mu_k^{\alpha-1}, \tag{4.3.27}$$

[31] Here, we write μ_k^m instead of $(\mu_k)^m$, for example; see also, (4.3.50) below, for instance.

and using $\alpha - 1 < 0$, we conclude that $\lim_{k\to\infty} \frac{k}{\mu_k} = 1$. In particular, there exists a positive constant C_1 such that $\mu_k \leq C_1 k$ for all positive integers k. Hence,

$$|k - \mu_k| \leq C\mu_k^\alpha \leq CC_1^\alpha k^\alpha;$$

that is, $\mu_k = k + O(k^\alpha)$ as $k \to \infty$, which proves the lemma for $m = 1$.

Step 2: We now consider the case when $c \cdot \mu_k^m + O(\mu_k^\alpha) = k$ as $k \to \infty$, with $m > 0$ and $m \neq 1$. Letting $\lambda_k := \mu_k^m$ for every $k \geq 1$, we obtain $c \cdot \lambda_k + O(\lambda_k^{\alpha/m}) = k$ as $k \to \infty$. By Step 1, we then conclude that

$$\lambda_k = \frac{1}{c} \cdot k + O(k^{\alpha/m}) \quad \text{as} \quad k \to \infty.$$

Therefore,

$$\begin{aligned}\mu_k &= \left(\frac{1}{c} \cdot k + O(k^{\alpha/m})\right)^{1/m} = c^{-1/m} k^{1/m} \left(1 + O(k^{\frac{\alpha}{m}-1})\right)^{1/m} \\ &= c^{-1/m} k^{1/m} \left(1 + O(k^{\frac{\alpha}{m}-1})\right) \quad \text{as} \quad k \to \infty,\end{aligned}$$

where in the last equality we have used the fact that $\alpha < m$. This concludes the proof of the lemma. □

Remark 4.3.16. Lemma 4.3.15 permits a slight generalization. If instead of condition (4.3.25), we assume that

$$c \cdot \mu_k^m + O(\mu_k^\alpha) = k + O(k^\beta) \quad \text{as} \quad k \to \infty, \tag{4.3.28}$$

where $\beta < 1$, then (retaining the remaining conditions in the lemma) we have that:

$$\mu_k = c^{-1/m} \cdot k^{1/m} + O(k^{\frac{1}{m} + \max\{\frac{\alpha}{m}, \beta\} - 1}) \quad \text{as} \quad k \to \infty. \tag{4.3.29}$$

This conclusion is obtained by an easy modification of the proof of Lemma 4.3.15.

We are now ready to prove Corollary 4.3.14 (in a special case).

Proof of Corollary 4.3.14. Let us first assume that $\tilde{D} \in (N-1, N]$; that is, let us assume that we are in case (*i*) of the corollary.

For simplicity, we assume that $\tilde{D} \in (N-1, N)$ and that the eigenvalues all have multiplicity one. (The case when $\tilde{D} = N$ is of no interest while the case when $\tilde{D} = N-1$ can be dealt with similarly.) For the general case when the eigenvalues may have multiplicities larger than one, it would be best to work directly with the eigenvalue counting function (and, hence, to use Theorem 4.3.11 instead of Corollary 4.3.14), as is standard and done in [Lap2–3]. See the comment following the proof of Theorem 4.3.17 below.

From the definition of the counting function, we obviously have that $N_\nu(\mu_k^{(0)}) = k$, for all $k \geq 1$. By using Theorem 4.3.11(i), we obtain that

$$(2\pi)^{-N}\omega_N|\Omega_0| \cdot \mu_k^{N/2} + O(\mu_k^{d/2}) = k \quad \text{as} \quad k \to \infty.$$

Now, if we set $m = N/2$ and $\alpha = d/2$, Lemma 4.3.15 immediately implies claim (i) in the corollary. The proof of case (ii) is similar. □

Combining Corollary 4.3.14 with Proposition 4.3.10, we deduce the main result of this section, already obtained by the first author in [Lap2–3]. As in [Lap2–3], it makes an essential use (via Corollary 4.3.14) of the key remainder estimate obtained in [Lap1]. Furthermore, it is stated a little bit more precisely than in [Lap2–3] and makes use of the notation introduced in Subsection 2.1.5. It shows that $D_{\rm mer}(\zeta^*_{\Omega_0})$, the abscissa of meromorphic continuation of $\zeta^*_{\Omega_0}$, does not exceed the upper box dimension of the boundary $\partial\Omega_0$ relative to Ω_0, denoted (as above) $\overline{\dim}_B(\partial\Omega_0, \Omega_0)$ and called the *inner Minkowski dimension of* $\partial\Omega_0$ in [Lap1].

Theorem 4.3.17 (Lapidus, [Lap2–3]). *Let Ω_0 be an arbitrary (nonempty) bounded open subset of $\mathbb{R}^N$ such that $\overline{\dim}_B(\partial\Omega_0, \Omega_0)\} < N$. Then the spectral zeta function $\zeta^*_{\Omega_0}$ of Ω_0 is holomorphic in the open half-plane $\{\mathrm{Re}\, s > N\}$ and $D_{\rm hol}(\zeta^*_{\Omega_0}) = N$. Furthermore, $\zeta^*_{\Omega_0}$ can be (uniquely) meromorphically extended from $\{\mathrm{Re}\, s > N\}$ to (at least) $\{\mathrm{Re}\, s > \overline{\dim}_B(\partial\Omega_0, \Omega_0)\}$. In other words,*[32]

$$D_{\rm mer}(\zeta^*_{\Omega_0}) \leq \overline{\dim}_B(\partial\Omega_0, \Omega_0). \tag{4.3.30}$$

Moreover, $s = N$ is the only pole of $\zeta^*_{\Omega_0}$ in the half-plane $\{\mathrm{Re}\, s > \overline{\dim}_B(\partial\Omega_0, \Omega_0)\}$; it is a simple pole and

$$\mathrm{res}(\zeta^*_{\Omega_0}, N) = \frac{N\omega_N}{(2\pi)^N}|\Omega_0|. \tag{4.3.31}$$

Proof. Let us prove the statement regarding the meromorphicity of $\zeta^*_{\Omega_0}$. [The proof of the statement regarding the holomorphicity of $\zeta^*_{\Omega_0}$ in $\{\mathrm{Re}\, s > N\}$ is left as an easy exercise for the interested reader. (Actually, as will be explained further below, it follows from the known properties of generalized Dirichlet series with positive coefficients recalled in Subsection 2.1.3.)] Let us set $\gamma = \frac{2+d-N}{N}$. Since we then have $N - (2 - \gamma N) = d$, using Proposition 4.3.10 applied to the sequence $(\mu_k^{(0)})_{k\geq 1}$ in Corollary 4.3.14, we conclude that $\zeta^*_{\Omega_0}$ can be meromorphically extended in a unique way to the half-plane $\{\mathrm{Re}\, s > d\}$. This property holds for any $d > \tilde{D} := \overline{\dim}_B(\partial\Omega_0, \Omega_0)$; hence, the function $\zeta^*_{\Omega_0}$ can be meromophically extended to the half-plane $\{\mathrm{Re}\, s > \tilde{D}\}$.

Finally, assume that $\tilde{D} \in (N-1, N)$, for simplicity. Then, since $\tilde{D} < N$, we see that the meromorphic continuation of $\zeta^*_{\Omega_0}$ must have a (simple) pole at $s = N$. Indeed, since $\zeta^*_{\Omega_0}$ is initially given by a (generalized) Dirichlet series with positive

[32] Recall that, by definition, $\{\mathrm{Re}\, s > D_{\rm mer}(\zeta^*_{\Omega_0})\}$ is the largest open right half-plane to which $\zeta^*_{\Omega_0}$ can be meromorphically extended.

coefficients, $\zeta_{\Omega_0}^*$ must have a singularity at $s = N$; but since $\zeta_{\Omega_0}^*$ can be meromorphically continued to a connected open neighborhood of $s = N$ (and in light of either (4.3.19) or (4.3.22)), this singularity must be a simple pole of $\zeta_{\Omega_0}^*$. The value of the residue given in (4.3.31) follows from (4.3.16) in Proposition 4.3.10. This concludes the proof of the theorem. □

Alternatively, Theorem 4.3.17 follows easily from Theorem 4.3.11 (via standard arguments, well known to the experts in spectral theory) by proceeeding as follows, which is the method used in [Lap2–3]. First, observe that, as is well known, the spectral zeta function (essentially) coincides with the Mellin transform of the spectral counting function, at least for $\operatorname{Re} s > \tilde{D}$, where $\tilde{D} := \overline{\dim}_B(\partial\Omega_0, \Omega_0)$. Then, use the remainder estimate for the eigenvalue counting function (see Theorem 4.3.11 above, from [Lap1]), along with a suitable (and standard) Abelian theorem[33] (the converse of a Tauberian theorem, in the terminology of [Sim]) or simply, a direct analysis of the corresponding integral (essentially, the Mellin transform of the spectral counting function N_ν) in order to deduce that $\zeta_{\Omega_0}^*$ admits a meromorphic extension to the open half-plane $\{\operatorname{Re} s > \tilde{D}\}$, as desired. More specifically, one can use Theorem 2.1.47 about the holomorphicity of integrals depending analytically on a parameter, along with Theorem 4.3.11, to deduce that the spectral zeta function $\zeta_{\Omega_0}^*$ admits a meromorphic continuation to $\{\operatorname{Re} s > \tilde{D}\}$, with a single, simple pole located at $s = N$ (thus, the meromorphic continuation is holomorphic for $\operatorname{Re} s > \tilde{D}$ except at $s = N$).

The proof of Theorem 4.3.17 provided above, just after the statement of Theorem 4.3.17, presents the advantage of being elementary (assuming, of course, the results of Theorem 4.3.11, which are not at all elementary). However, at least for now, it is only valid under special assumptions on the multiplicities of the eigenvalues (see Lemma 4.3.15 and Remark 4.3.16 on pages 329 and 330), whereas the aforementioned proof (from [Lap3]) is valid in full generality since it directly relies on Theorem 4.3.11 rather than on Corollary 4.3.14.

We next state an easy but useful consequence of Theorem 4.3.17. At this stage, the reader may wish to review some of the relevant notation introduced in Section 2.1.

Corollary 4.3.18. *Under the same hypotheses as in Theorem 4.3.17, we have (with the notation introduced in Section 2.1)*

$$D(\zeta_{\Omega_0}^*) = D_{\mathrm{hol}}(\zeta_{\Omega_0}^*) = N \tag{4.3.32}$$

and so

$$\Pi(\zeta_{\Omega_0}^*) = \mathscr{H}(\zeta_{\Omega_0}^*) = \{\operatorname{Re} s > N\},^{34} \tag{4.3.33}$$

---

[33] See, e.g., [Sim] or [Lap1, Theorem A in Appendix A] for the case of the Laplace transform instead of the Mellin transform. Of course, a simple change of variable of the form $x = e^t$ then converts the (additive) Laplace transform to the (multiplicative) Mellin transform.

[34] That is, $\Pi(\zeta_{\Omega_0}^*)$, the half-plane of (absolute) convergence of $\zeta_{\Omega_0}^*$, coincides with $\mathscr{H}(\zeta_{\Omega_0}^*)$, the half-plane of holomorphic continuation of $\zeta_{\Omega_0}^*$.

whereas

$$D_{\mathrm{mer}}(\zeta^*_{\Omega_0}) < D_{\mathrm{hol}}(\zeta^*_{\Omega_0}). \tag{4.3.34}$$

Proof. The second equality in (4.3.32), $D_{\mathrm{hol}}(\zeta^*_{\Omega_0}) = N$, holds because (by the second part of Theorem 4.3.17 and since $\overline{\dim}_B(\partial\Omega_0,\Omega_0) < N$) the meromorphic continuation of $\zeta^*_{\Omega_0}$ has a pole at $s = N$, so that $\{\mathrm{Re}\, s > N\}$ is the largest open right half-plane on which $\zeta^*_{\Omega_0}$ is holomorphic: $\mathscr{H}(\zeta^*_{\Omega_0}) = \{\mathrm{Re}\, s > N\}$. Furthermore, the first equality in (4.3.32), $D(\zeta^*_{\Omega_0}) = D_{\mathrm{hol}}(\zeta^*_{\Omega_0})$, holds because $\zeta^*_{\Omega_0}$ is initially given by a (generalized) Dirichlet series with positive coefficients; see Equation (4.3.4) of Definition 4.3.4 above, along with Subsection 2.1.3.1. This proves Equation (4.3.32) and hence also Equation (4.3.33).

Finally, we note that clearly, in light of the second equality in (4.3.32) and of the inequality (4.3.30) in Theorem 4.3.17, the claimed strict inequality (4.3.34) holds since, by hypothesis, we have that $\overline{\dim}_B(\partial\Omega_0,\Omega_0) < N$. This concludes the proof of the corollary. □

For the sake of brevity, let $\tilde{D} := \overline{\dim}_B(\partial\Omega_0,\Omega_0)$, in the sequel. It is noteworthy that the estimates obtained in [Lap1] (and recalled, in particular, in Equations (4.3.19) and (4.3.22)) are best possible (i.e., sharp), in general, in the most important case of a fractal drum for which $N > \tilde{D} > N-1$ and $d = \tilde{D}$, with $\mathscr{M}^{*\tilde{D}}(\partial\Omega_0,\Omega_0) < \infty$;[35] that is, in case (*i*) of Theorem 4.3.11 and of Corollary 4.3.14 (as well as for an open set Ω_0 satisfying $\mathscr{M}^{*\tilde{D}}(\partial\Omega_0,\Omega_0) < \infty$), respectively, the estimates (4.3.19) and (4.3.22) are sharp. See [Lap1, Examples 5.1 and 5.1'] for a one-parameter family of examples $\{\Omega_{0,\alpha}\}_{\alpha>0}$ (based on the α-string, often used in the present book) for which $\tilde{D}$ takes all possible values (as α varies in $(0,+\infty)$) in the allowed open interval $(N-1,N)$ and the error estimates (4.3.19) and (4.3.22) are sharp, with $d := \tilde{D}$; furthermore, each open set $\Omega_{0,\alpha}$ is Minkowski measurable and, in particular, is Minkowski nondegenerate (hence, the condition $\mathscr{M}^{*\tilde{D}}(\partial\Omega_0,\Omega_0) < \infty$ is satisfied).

More specifically, for $\alpha > 0$, let $V_\alpha := \bigcup_{j=1}^{\infty}((j+1)^{-\alpha}, j^{-\alpha})$ denote the α-string. Then, given $N \geq 2$, let $\Omega_{0,\alpha} := V_\alpha \times (0,1)^{N-1}$; so that the bounded open set $\Omega_{0,\alpha}$ is the 'fractal comb' obtained as the disjoint union of the 'teeth' $((j+1)^{-\alpha}, j^{-\alpha}) \times (0,1)^{N-1}$. According to the results of [Lap1, Examples 5.1 and 5.1'] along with [Lap1, Appendix C], for each $\alpha > 0$,

$$\tilde{D} := \dim_B(\partial\Omega_{0,\alpha},\Omega_{0,\alpha}) = (N-1) + (\alpha+1)^{-1} \tag{4.3.35}$$

exists, and the relative fractal drum $(\partial\Omega_{0,\alpha},\Omega_{0,\alpha})$ is Minkowski measurable with Minkowski content

$$\mathscr{M}^{\tilde{D}}(\partial\Omega_{0,\alpha},\Omega_{0,\alpha}) = \frac{2^{1-\tilde{D}}\alpha^{\tilde{D}}}{1-\tilde{D}}.$$

[35] Recall from [Lap1, Corollary 3.2] that (since Ω_0 is a nonempty, bounded and open subset of $\mathbb{R}^N$) $\tilde{D} = \overline{\dim}_B(\partial\Omega_0,\Omega_0)$ always satisfies the following inequality: $N-1 \leq \tilde{D} \leq N$.

Clearly, in light of (4.3.35), $\tilde{D}$ ranges through all of $(N-1,N)$ as α ranges through $(0,+\infty)$. Furthermore, it can be shown by a direct computation (see [Lap1], *loc. cit.*) that the error estimates (4.3.19) and (4.3.22) hold with $d = \tilde{D}$ and cannot be improved. Actually, much more is true in this case, although it is not necessary to know about it for the present argument. Indeed, in light of later results obtained in [LapPo1–2] about the spectral asymptotics of Minkowski measurable fractal strings, one can even show that the error term in (4.3.19) can be replaced by an explicitly computable monotonic (asymptotic) second term, proportional to $\mathscr{M}^{\tilde{D}}(\partial\Omega_0,\Omega_0)\,\mu^{\tilde{D}/2}$ and with the implied constant of proportionality involving the positive number $-\zeta(\tilde{d})$, where $\tilde{d} := 1/(\alpha+1) \in (0,1)$ and ζ denotes the Riemann zeta function; and analogously for (4.3.22). (See [LapPo2–3] and [Lap-vFr3, Subsections 6.5.1 and 8.1.2].)

We note that the open sets $\Omega_{0,\alpha}$ constructed in [Lap1] are not connected. However, much as in [BroCar] and [FlVa], one can open appropriately small gates in each of the 'teeth' of the 'fractal combs' $\Omega_{0,\alpha}$ in order to obtain a one-parameter family $\{\Omega'_{0,\alpha}\}_{\alpha>0}$ of *connected* (and even *simply connected*) open subsets of $\mathbb{R}^N$ (with $N \geq 2$ arbitrary) having the same properties as the family $\{\Omega_{0,\alpha}\}_{\alpha>0}$. More specifically, each domain $\Omega'_{0,\alpha}$ is Minkowski measurable, with

$$\dim_B \Omega'_{0,\alpha} = \dim_B \Omega_{0,\alpha} = (N-1)+(\alpha+1)^{-1} \tag{4.3.36}$$

taking all possible values in $(N-1,N)$, as α varies in the interval $(0,+\infty)$, and for the Dirichlet Laplacian on $\Omega'_{0,\alpha}$, both of the remainder estimates (4.3.19) and (4.3.22) are best possible (with $d := \tilde{D}$ and $\Omega_0 = \Omega_{0,\alpha}$ or $\Omega'_0 = \Omega'_{0,\alpha}$, respectively).

We leave it to the interested reader to verify that the exact same conclusion as above can be reached (for the same two families of examples) in the case of Neumann (instead of the Dirichlet) Laplacian. In this case, we must replace $\tilde{D}$ by $D := \overline{\dim}_B(\partial\Omega_0)$ (as was done in [Lap1] when dealing with Neumann boundary conditions), and, of course, exclude the zero eigenvalue in the original definition (4.3.14) of the corresponding spectral zeta function (which we continue to denote by $\zeta^*_{\Omega_0}$, for simplicity). Observe that for these examples, it is easy to check that $\dim_B(\partial\Omega_0)$ exists and $\tilde{D} = D = \dim_B(\partial\Omega_0)$.

We could naturally be tempted to use the same one-parameter families $\{\Omega_{0,\alpha}\}_{\alpha>0}$ and $\{\Omega'_{0,\alpha}\}_{\alpha>0}$ of open sets and simply connected domains, respectively, along with some of the results of [LapPo2] concerning the modified Weyl–Berry conjecture (in dimension one) to solve the following open problem (Problem 4.3.20), to which we will provide a partial answer in Theorem 4.3.21 below. However, this is not possible, as will be explained in the next remark in the case of this first family.

Remark 4.3.19. To see why the one-parameter family $\{\Omega_{0,\alpha}\}_{\alpha>0}$ cannot be used to resolve part (i) of Problem 4.3.20 below, one can reason as follows (in the case of the Dirichlet Laplacian). First of all, since $\Omega_{0,\alpha} = V_\alpha \times (0,1)^{N-1}$, where $V_\alpha = \cup_{j=1}^{\infty}((j+1)^{-\alpha}, j^{-\alpha})$ is the α-string, we have that the eigenvalues of $\Omega_{0,\alpha}$ are the sums of the eigenvalues of V_α and of those of $(0,1)^{N-1}$; therefore, similarly, the poles of $\zeta^*_{\Omega_0}$ are the sums of the poles of $\zeta^*_{V_\alpha}$ and those of $\zeta^*_{(0,1)^{N-1}}$.

Also, the spectral zeta function of a cube ($(0,1)^{N-1}$, in this case) can be expressed as a linear combination of Epstein zeta functions; it therefore admits a meromorphic extension to all of $\mathbb{C}$, with poles which are all simple and located on the real axis at $\{1,2,\ldots,N-1\}$. Furthermore, according to the classic formula for the spectal zeta function of a fractal string ([Lap2–3], [Lap-vFr3, Theorem 1.10]), we have

$$\zeta^*_{V_\alpha}(s) = \zeta(s)\cdot\zeta_{V_\alpha}(s), \tag{4.3.37}$$

where $\zeta = \zeta_R$ is the Riemann zeta function and ζ_{V_α} is the geometric zeta function of the α-string. Now, by [Lap-vFr3, Theorem 6.21], ζ_{V_α} has a meromorphic extension to all of $\mathbb{C}$ (with simple poles located at $\tilde{d}$ and in (a subset of) $\{-\tilde{d},-2\tilde{d},-3\tilde{d},\ldots\}$), where $\tilde{d} := \dim_B(\partial V_\alpha, V_\alpha) = 1/(\alpha+1)$. Hence, in light of (4.3.37), $\zeta^*_{V_\alpha}$ is meromorphic in all of $\mathbb{C}$ (with one more pole than ζ_{V_α}, namely, the simple pole of ζ_{V_α} at 1). Therefore, $\zeta^*_{0,\alpha}$ can also be meromorphically extended to all of $\mathbb{C}$ (with poles which are all simple, with the exception of $s=1$, which is double, and located on the real axis). We conclude that $D_{\text{mer}}(\zeta^*_{0,\alpha}) = -\infty$ for every $\alpha > 0$, whereas $\tilde{D} := \dim_B(\partial\Omega_{0,\alpha},\Omega_{0,\alpha}) = N-1+(\alpha+1)^{-1}$ sweeps out the interval $(N-1,N)$ as α ranges through $(0,+\infty)$. Therefore, inequality (4.3.30) is strict, in this case, and is in fact, as far as possible from being an equality.

We expect that the following open problem has a positive answer in every dimension $N \geq 1$. (We will show in Theorem 4.3.21 and the ensuing comment, Remark 4.3.23, that this is so both for the Dirichlet and Neumann Laplacians.) In the sequel, we assume implicitly that $\tilde{D} := \dim_B(\partial\Omega_0,\Omega_0) < N$.

Problem 4.3.20. *(i) Determine whether the inequality* (4.3.30) *in Theorem 4.3.17 is sharp; that is, find a bounded open set $\Omega_0 \subset \mathbb{R}^N$ for which*

$$D_{\text{mer}}(\zeta^*_{\Omega_0}) = \overline{\dim}_B(\partial\Omega_0,\Omega_0)$$

for the Dirichlet Laplacian on Ω_0.

(ii) More generally, address the exact counterpart of this problem for higher order elliptic operators (see inequality (4.3.56) *below) and/or for Neumann (or, more generally, for mixed Dirichlet–Neumann) boundary conditions instead of for Dirichlet boundary conditions (see the comment following the statement of this problem); that is, find a bounded open set $\Omega_0 \subset \mathbb{R}^N$ for which $D_{\text{mer}}(\zeta^*_{\Omega_0}) = \overline{\dim}_B(\partial\Omega_0,\Omega_0)$.*

(iii) Either in the setting of (i), or, more generally, in the setting of (ii), find a one-parameter family of bounded open sets solving (i) (or, more generally, (ii)) in the affirmative and for which the dimension $\tilde{D} := \overline{\dim}_B(\partial\Omega_0,\Omega_0)$ takes all the possible values in $(N-1,N)$, as the parameter of the family varies. Furthermore, when $N \geq 2$, find such a family consisting of connected (or even simply connected) open sets.

As before, in the case of Neumann (or, more generally, mixed Dirichlet–Neumann) boundary conditions, we must assume that Ω is a suitable bounded open subset of $\mathbb{R}^N$ (see pages 343–344 at the very end of this section) and replace $\tilde{D} := \overline{\dim}_B(\partial\Omega_0, \Omega_0)$ by $D := \overline{\dim}_B(\partial\Omega_0)$. (We have $\tilde{D} \le D$ and so $N-1 \le D \le N$.) Furthermore, we then let $N(\mu)$ denote the number of (strictly) positive eigenvalues which do not exceed μ (since 0 is always an eigenvalue of the Neumann problem) and similarly exclude the eigenvalue 0 in the original definition of $\zeta^*_{\Omega_0}$, given in Equation (4.3.14), for example (or, more generally, when $m \ge 1$, by (4.3.50) below).

The next theorem is new and provides a partial solution to Problem 4.3.20. It actually answers part (*i*) of the problem in the affirmative. We expect (but do not want to claim) that a suitable modification and/or extension of the construction can be used to solve part (*iii*) as well, at least for the Dirichlet and Neumann Laplacians. It is noteworthy that our construction makes an essential use of aspects of classic fractal string theory and of the theory developed in this book (especially in Sections 4.4–4.6 of the present chapter).

Theorem 4.3.21. *There is an explicitly constructible bounded open subset of* $\mathbb{R}^N$ *solving part* (*i*) *of Problem 4.3.20 in the affirmative* (*and for which* $N-1 < \tilde{D} < N$). *Actually, this open set has a maximally hyperfractal* (*and transcendentally* ∞*-quasiperiodic*) *boundary, in the sense of Section 4.6* (*specifically, of Definition 4.6.23*(*iii*) *and Definition 4.6.7*(*a*)) *below. Equivalently, the associated relative fractal drum* $(\partial\Omega_0, \Omega_0)$ *is maximally hyperfractal* (*and transcendentally* ∞*-quasiperiodic*).

Proof. The proof parallels in part the reasoning outlined in Remark 4.3.19 above. It relies, however, in an essential way on the concepts introduced and the results obtained in Section 4.6 below.

More specifically, assume for now that $N = 1$ and let $\mathscr{L}$ be the (effectively constructible) bounded fractal string obtained in Corollary 4.6.17 of Section 4.6 below. Here, $\mathscr{L}$ is viewed as a relative fractal drum $(\partial V_0, V_0)$, with V_0 a bounded open subset of $\mathbb{R}$. By construction, we have that $(\partial V_0, V_0)$ is transcendentally ∞-quasiperiodic (see Definition 4.6.7(*a*)) and maximally hyperfractal (see Definition 4.6.23(iii)); so that (with $\tilde{d} := \overline{\dim}(\partial V_0, V_0)$) all of the points of the critical line $\{\mathrm{Re}\, s = \tilde{d}\}$ are nonisolated singularities of the geometric zeta function $\zeta_{\mathscr{L}}$ of $\mathscr{L}$. Furthermore, we have

$$\tilde{d} := \overline{\dim}(\partial V_0, V_0) = D(\zeta_{\mathscr{L}}) = D_{\mathrm{mer}}(\zeta_{\mathscr{L}}). \tag{4.3.38}$$

(See part (*a*) of Corollary 4.6.17 below.) Then, in light of the (the counterpart for V_0) of the factorization formula (4.3.37) above, the spectral zeta function $\zeta^*_{V_0}$ also satisfies

$$D_{\mathrm{mer}}(\zeta^*_{V_0}) = \tilde{d} = \overline{\dim}_B(\partial V_0, V_0). \tag{4.3.39}$$

Indeed, the critical line $\{\mathrm{Re}\, s = \tilde{d}\}$ consists entirely of nonisolated singularities of $\zeta^*_{V_0}$. (Also, $\zeta^*_{V_0}$ has a single, simple pole at $s = 1$.) This takes care of the $N = 1$ case.

Next, given a fixed integer $N \ge 2$, let $\Omega_0 := V_0 \times (0,1)^{N-1}$, viewed as a bounded open subset of $\mathbb{R}^N$ (or rather, as the relative fractal drum $(\partial\Omega_0, \Omega_0)$ of $\mathbb{R}^N$). Then, just as in Remark 4.3.19, note that the principal poles / singularities of $\zeta^*_{\Omega_0}$ are the sums of the principal poles / singularities of $\zeta^*_{V_0}$ and the principal pole of $\zeta_{(0,1)^{N-1}}$, which is equal to $N-1$. (Note that the poles of $\zeta_{(0,1)^{N-1}}$ are all simple and located on the real axis, at $\{1,\ldots,N-1\}$.) Therefore, we deduce that $\zeta^*_{\Omega_0}$ has a single (simple) pole at $s=N$ $(=(N-1)+1)$ and that the critical line $\{\mathrm{Re}\, s=\tilde{D}\}$ consists entirely of singularities of $\zeta^*_{\Omega_0}$. Here,

$$\tilde{D}=(N-1)+\tilde{d}=\overline{\dim}_B(\partial\Omega_0,\Omega_0). \tag{4.3.40}$$

It follows that (much as in the $N=1$ case above) we must have

$$\tilde{D}=D_{\mathrm{mer}}(\zeta^*_{\Omega_0}), \tag{4.3.41}$$

as desired. In light of Equations (4.3.40) and (4.3.41), this completes the proof of the theorem. □

We deduce from the above discussion and known properties of the quantities involved that for the bounded open set Ω_0 of Theorem 4.3.21, the abscissa of meromorphic continuation of the spectral zeta function $\zeta^*_{\Omega_0}$ does not only coincide with $\tilde{D}:=\overline{\dim}_B(\partial\Omega_0,\Omega_0)$ (as is stated in Theorem 4.3.21 above), but also coincides with the abscissae of meromorphic, holomorphic and (absolute) convergence of the fractal (i.e., distance and tube) zeta functions of Ω_0, as is stated in the next result. Note that it follows from Theorem 4.3.21 that $D_{\mathrm{hol}}(\zeta^*_{\Omega_0})=D(\zeta^*_{\Omega_0})=N>\tilde{D}$.

Corollary 4.3.22. *For the example discussed in Theorem 4.3.21, we have*

$$\begin{aligned} D_{\mathrm{mer}}(\zeta^*_{\Omega_0}) &= \overline{\dim}_B(\partial\Omega_0,\Omega_0) =: \tilde{D} \\ &= D_{\mathrm{mer}}(f)=D_{\mathrm{hol}}(f)=D(f), \end{aligned} \tag{4.3.42}$$

for all $f\in\{\zeta_{\partial\Omega_0,\Omega_0},\tilde{\zeta}_{\partial\Omega_0,\Omega_0}\}$.

Proof. In light of Theorem 4.3.21, all we have to prove are the last three equalities of (4.3.42) and the equality $D(f)=\tilde{D}$. Now, these inequalities follow by combining the relevant result of Subsections 2.1.2, 2.1.3 and 4.1.1 (see part (*b*) of Theorem 2.1.11, Proposition 2.2.19 and part (*b*) of Theorem 4.1.7). □

Remark 4.3.23. The use of the same geometric example as before in the proof of Theorem 4.3.21, $\Omega_0=V_0\times(0,1)^{N-1}$, and an entirely similar (but slightly simpler) argument, show that the exact counterpart of Theorem 4.3.21 and Corollary 4.3.22 holds for the Neumann (instead of the Dirichlet) Laplacian. Recall that in that case, we must exclude the eigenvalue 0 in the original definition (4.3.14) of the spectral zeta function. We leave the easy verification as an exercise for the interested reader.

Thus far, in connection with the remainder estimates for the leading spectral asymptotics (see Theorems 4.3.11 and 4.3.17 along with Corollaries 4.3.14 and 4.3.18), we have restricted ourselves to discussing the Dirichlet Laplacian $-\Delta$, although the Neumann Laplacian can also be discussed, as well as general positive uniformly elliptic linear differential operators (with variable and possibly nonsmooth coefficients) of order $2m$ (with $m\geq 1$) and of the form

$$\mathscr{A} = \sum_{|\alpha|\leq m, |\beta|\leq m} (-1)^{|\alpha|} D^{\alpha}(a_{\alpha\beta}(x)D^{\beta}), \tag{4.3.43}$$

described in [Lap1, Section 2.2]. We use here the standard multi-index notation: for example, $\alpha := (\alpha_1, \dots, \alpha_N) \in (\mathbb{N} \cup \{0\})^N$, $|\alpha| := \alpha_1 + \cdots + \alpha_N$ and

$$D^{\alpha} := \frac{\partial^{\alpha_1}}{\partial x_1^{\alpha_1}} \cdots \frac{\partial^{\alpha_N}}{\partial x_N^{\alpha_N}}.$$

All of these extensions are obtained in [Lap1]; see Theorem 2.1 and its corollaries in [Lap1]. In the latter case, the assumed assymptotic expansion of the eigenvalues of $\mathscr{A}$ in the corresponding version of Proposition 4.3.10, and implied (or, actually, equivalent to) by [Lap1, Theorem 2.1], should be replaced by

$$\mu_k^{(0)} = (\mu'_{\mathscr{A}}(\Omega_0))^{-2m/N} \cdot k^{2m/N} + O(k^{\gamma}), \quad \text{as} \quad k \to \infty, \tag{4.3.44}$$

where $\mu'_{\mathscr{A}}(\Omega_0)$ is the "Browder–Gårding measure" of Ω_0 defined, for example, in [Hö3] or in [Lap1, Equation (2.18*a*) in Section 2.2] in terms of the (positive definite, unbounded) quadratic form associated with $\mathscr{A}$ and

$$\gamma := \frac{2m + d - N}{N}, \tag{4.3.45}$$

with $d > \tilde{D}$ arbitrary and (as before) $\tilde{D} := \overline{\dim}_B(\partial\Omega_0, \Omega_0)$, the upper Minkowski dimension of the relative fractal drum $(\partial\Omega_0, \Omega_0)$. Furthermore, we may also take $d = \tilde{D}$ provided $\mathscr{M}^{*\tilde{D}}(\partial\Omega_0, \Omega_0) < \infty$. We note that the remainder estimate (4.3.44) actually holds in the above form in the 'fractal case' when $\tilde{D} > N - 1$ (or, equivalently, when $\tilde{D} \in (N-1, N]$, since we always have $\tilde{D} \in [N-1, N]$). Furthermore, in the nonfractal case when $\tilde{D} = N - 1$, we must replace $O(k^{\gamma})$ by $O(k^{\gamma} \log k)$ on the right-hand side of (4.3.44). Here and in the sequel, and as was mentioned earlier, we should replace $\tilde{D}$ by D, where $D := \overline{\dim}_B(\partial\Omega_0)$, the upper Minkowski dimension of the boundary $\partial\Omega_0$, in the case of Neumann (or, more generally, mixed Dirichlet–Neumann) boundary conditions. We should also assume that Ω is a suitable bounded open subset of $\mathbb{R}^N$; see the discussion on pages 343–344 at the very end of this section.

Remark 4.3.24. For Neumann boundary conditions, and for example, for the Neumann Laplacian, one must also use the weak (or variational) formulation of the classic eigenvalue problem $-\Delta u = \mu u$ in Ω_0, with $\partial u/\partial n = 0$ on $\partial\Omega_0$, where $\partial u/\partial n$ stands for the normal derivative of u along $\partial\Omega_0$. However, since $\partial\Omega_0$ is irregular (and hence, $\partial u/\partial n$ is not defined, in general), one must now use the Sobolev space $H^1(\Omega_0) := W^{1,2}(\Omega_0)$ instead of $H_0^1(\Omega_0) := W_0^{1,2}(\Omega_0)$, which was used to formulate the Dirichlet eigenvalue problem; see the discussion following Equation (4.3.1) in Subsection 4.3.1, along with references [LioMag], [Bre] and [Lap1]. (*Neumann boundary conditions* are sometimes referred to as *natural boundary conditions* in the physics and applied mathematics literature, because they are automatically satisfied once the problem has been written in variational form.) An entirely analogous comment applies to general, uniformly elliptic, positive self-adjoint operators of the form (4.3.43); see, e.g., the aforementioned references.

Recall that the *Browder–Gårding measure* $\mu'_{\mathscr{A}}(\mathrm{d}x) := \mu'_{\mathscr{A}}(x)\,\mathrm{d}x$ is the absolutely continuous measure on $\mathbb{R}^N$ (with respect to the Lebesgue measure on $\mathbb{R}^N$), with density $\mu'_{\mathscr{A}}(x)$ given (for a.e. $x \in \Omega_0$) as follows (with $|\cdot| = |\cdot|_N$ denoting the N-dimensional volume or measure, as usual):

$$\mu'_{\mathscr{A}}(x) := (2\pi)^{-N} |\{\xi \in \mathbb{R}^N : a'(x,\xi) < 1\}|, \tag{4.3.46}$$

where $a'(x,\xi)$ denotes the *leading symbol of the quadratic form a* associated with the opeator $\mathscr{A}$ given by (4.3.43):

$$a'(x,\xi) := \sum_{|\alpha|=m,\,|\beta|=m} a_{\alpha\beta}(x)\,\xi^{\alpha+\beta}, \tag{4.3.47}$$

with $\xi^{\kappa} := \xi_1^{\kappa_1}\cdots\xi_N^{\kappa_N}$ for $\kappa = (\kappa_1,\ldots,\kappa_N) \in (\mathbb{N}\cup\{0\})^N$ and $\xi = (\xi_1,\ldots,\xi_N) \in \mathbb{R}^N$ (as well as with $x \in \Omega_0$). So that

$$\mu'_{\mathscr{A}}(\Omega_0) = \int_{\Omega_0} \mu'_{\mathscr{A}}(x)\,\mathrm{d}x, \tag{4.3.48}$$

with $\mu'_{\mathscr{A}}(x)$ given by (4.3.46) and (4.3.47) just above.

Physically, and in light of (4.3.46)–(4.3.48), $\mu'_{\mathscr{A}}(\Omega_0)$ can be interpreted as an integral in the phase space $\mathbb{R}^{2N}$. In fact, it is well known that in the special case when $\mathscr{A}$ is a Schrödinger-type operator, the corresponding Weyl term (namely, the leading term in Equation (4.3.49) below) can be viewed as a volume in phase space (with the eigenvalue parameter μ being thought of as an energy), in agreement with the semiclassical limit of quantum mechanics (see, e.g., [ReeSim1] and [Sim], along with the relevant references therein).

We have just stated, in the remainder estimate (4.3.44), the analog (obtained in [Lap1]) of case (*i*) of Corollary 4.3.14 above. (Observe that when $m = 1$ and in light of (4.3.45), estimate (4.3.44) does reduce to estimate (4.3.22) of Corollary 4.3.14.) Now, in the nonfractal case (or 'least fractal case', still following the terminology of [Lap1]) where $\tilde{D} = N-1$, the exact analog of part (ii) of Corollary 4.3.14 also holds. More specifically, still according to [Lap1, Theorem 2.1 and its corollaries], the precise counterpart of estimate (4.3.44) holds, with $O(k^{\gamma})$ replaced by $O(k^{\gamma}\log k)$, exactly as in estimate (4.3.23) of part (*ii*) of Corollary 4.3.14 (which corresponds to the case when $m = 1$).

Observe that if $N(\mu)$ denotes the eigenvalue counting function of the operator $\mathscr{A}$, the asymptotic remainder estimate (4.3.44) can be written equivalently as follows:

$$N(\mu) = \mu'_{\mathscr{A}}(\Omega_0)\,\mu^{N/2m} + R(\mu), \tag{4.3.49}$$

where the error term $R(\mu)$ is given by $R(\mu) := O(\mu^{d/2m})$ in the fractal case when $\tilde{D} > N-1$ and $R(\mu) := O(\mu^{d/2m}\log\mu)$ in the nonfractal case when $\tilde{D} = N-1$. Here, $d \in (\tilde{D}, N]$ is arbitrary and if $\mathscr{M}^{*\tilde{D}}(\partial\Omega_0,\Omega_0) < \infty$, we may choose $d = \tilde{D}$ as well. [And, similarly, with $D = \overline{\dim}_B(\partial\Omega_0)$ instead of $\tilde{D} = \overline{\dim}_B(\partial\Omega_0, \Omega_0)$ and with $\mathscr{M}^{*\tilde{D}}(\partial\Omega_0)$ instead of $\mathscr{M}^{*\tilde{D}}(\partial\Omega_0, \Omega_0)$, for Neumann or, more generally, for mixed Dirichlet–Neumann (instead of Dirichlet) boundary conditions.] As was observed before, when $\tilde{D} = N - 1$, then $O(k^\gamma)$ must be replaced by $O(k^\gamma \log k)$ on the right-hand side of (4.3.44).

Note that the value (4.3.45) of the exponent γ, appearing in (4.3.44), corresponds to letting $m' := N/2m$ and $\alpha' := d/2m$ (instead of m and α, respectively) in (4.3.25) of Lemma 4.3.15. See Equation (4.3.49) (which we cited from [Lap1, Theorem 2.1]) and recall that $N(\mu_k^{(0)}) = k$ for all $k \geq 1$.

Next, we consider the consequences of the above error estimates ((4.3.44) or, equivalently, (4.3.49)) for the spectral zeta function $\zeta^*_{\Omega_0} := \zeta^*_{\mathscr{A},\Omega_0}$ of the uniformly elliptic operator $\mathscr{A}$ of order $2m$, defined (for $s \in \mathbb{C}$ with $\operatorname{Re} s$ sufficiently large) by

$$\zeta^*_{\Omega_0}(s) := \sum_{k=1}^{\infty} (\mu_k^{(0)})^{-s/2m}. \tag{4.3.50}$$

Observe that since $\mathscr{A}$ is of order $2m$, the (normalized) 'frequencies' of the corresponding drum are given by $\nu_k := (\mu_k^{(0)})^{-1/2m}$, so that $\zeta^*_{\Omega_0}(s) := \sum_{k=1}^{\infty} (\nu_k)^{-s}$, exactly as was done in Definition 4.3.4 when $m = 1$; see Equations (4.3.3) and (4.3.4). Indeed, note that for $m = 1$, Equation (4.3.50) reduces to (4.3.14).

The following result generalizes Theorem 4.3.17 to the present context. We point out that thanks to our definition of $\zeta^*_{\Omega_0}$ in Equation (4.3.50) just above, Theorem 4.3.25 and its consequences (stated, in particular, in Equation (4.3.57) below) take a form which is essentially identical to their counterpart in Theorem 4.3.21 (and in Corollary 4.3.22), for which $m = 1$ and $\mathscr{A}$ is the Laplace operator.

Theorem 4.3.25. *Assume that Ω_0 is a bounded open subset of $\mathbb{R}^N$ such that*

$$\overline{\dim}_B(\partial\Omega_0, \Omega_0) < N.$$

Let $\mathscr{A}$ be a positive uniformly elliptic self-adjoint operator of order $2m$, as described in [Lap1, Section 2.2]. *Then the corresponding spectral zeta function $\zeta^*_{\Omega_0} := \zeta^*_{\mathscr{A},\Omega_0}$, defined by* (4.3.50), *possesses a (necessarily unique) meromorphic extension (at least) to the open half-plane*

$$\{\operatorname{Re} s > \overline{\dim}_B(\partial\Omega_0, \Omega_0)\}.$$

In other words,

$$D_{\text{mer}}(\zeta^*_{\Omega_0}) \leq \overline{\dim}_B(\partial\Omega_0, \Omega_0). \tag{4.3.51}$$

The only pole of $\zeta^*_{\Omega_0}$ in the above half-plane is $s = N$, and hence, in particular, $D(\zeta^*_{\Omega_0}) = N$. Furthermore, it is a simple pole and

$$\operatorname{res}(\zeta^*_{\Omega_0}, N) = N \mu'_{\mathscr{A}}(\Omega_0). \tag{4.3.52}$$

In other words, the residue of the spectral zeta function of the operator $\mathscr{A}$, computed at $s = N$, is proportional to the Browder–Gårding measure of Ω_0.

Proof. Let $\tilde{D} := \overline{\dim}_B(\partial\Omega_0, \Omega_0) < N$. According to [Lap1, Theorem 2.1, case (*i*)], assuming $\tilde{D} > N-1$ the sequence of eigenvalues corresponding to the operator $\mathscr{A}$ satisfies (4.3.49), or, equivalently, condition (4.3.44) with γ defined by (4.3.45) for any $d > \tilde{D}$. Let us fix an arbitrary number $d \in (\tilde{D}, N)$.

First of all, applying Theorem 2.3.12, with $a = 2m/N$, to the sequence of eigenvalues of $\mathscr{A}$ satisfying (4.3.44), we immediately obtain (much as in the proof of Proposition 4.3.10) that $s = N$ is a simple pole. (Note that, since $d < N$, then $\gamma < a$, as required in Theorem 2.3.12.) Furthermore, using (2.3.17) from Theorem 2.3.12, we see that $\zeta^*_{\Omega_0}$ can be meromophically extended at least to the open set of all complex numbers s such that

$$\operatorname{Re}\frac{s}{2m} > \frac{N}{2m} - \left(1 - \frac{\gamma N}{2m}\right) = \frac{d}{2m},$$

that is, to the open half-plane $\{\operatorname{Re} s > d\}$. Since $d > \tilde{D}$ can be chosen arbitrarily close to $\tilde{D}$, we deduce that $D_{\mathrm{mer}}(\zeta^*_{\Omega_0}) \le \tilde{D}$.

Finally, the residue of the spectral zeta function $\zeta^*_{\Omega_0}$ at $s = N$ can then be computed as follows (much in the same way as in the proof of Proposition 4.3.10):

$$\begin{aligned}\operatorname{res}(\zeta^*_{\Omega_0}, N) &= \lim_{s\to N}(s-N)\zeta^*_{\Omega_0}(s) = \lim_{2ms\to N}(2ms-N)\zeta^*_{\Omega_0}(2ms)\\ &= 2m \lim_{s\to N/2m}\left(s - \frac{N}{2m}\right)\zeta^*_{\Omega_0}(2ms) = 2m\frac{N}{2m}\cdot C^{-N/2m}\\ &= N\mu'_{\mathscr{A}}(\Omega_0),\end{aligned} \tag{4.3.53}$$

where in the next-to-last equality, we have used Equation (2.3.18) from Theorem 2.3.12 with $C := \mu'_{\mathscr{A}}(\Omega_0)^{-2m/N}$ and $a := 2m/N$.

In the case when $\tilde{D} = N-1$, we use [Lap1, Theorem 2.1, case (*ii*)], which can be stated equivalently as follows (using, for example, Lemma 4.3.15):

$$\mu_k^{(0)} = (\mu'_{\mathscr{A}}(\Omega_0))^{-2m/N}\cdot k^{2m/N} + O(k^\gamma \log k), \quad \text{as} \quad k\to\infty.$$

Now, we can proceed analogously as in the above case when $\tilde{D} > N-1$. This completes the proof of the theorem. □

As we see, assuming that the hypotheses of Theorem 4.3.25 are satisfied, the Browder–Gårding measure of Ω_0 can be recovered by using the spectral zeta function $\zeta^*_{\Omega_0}$ in the following manner:

$$\mu'_{\mathscr{A}}(\Omega_0) := \frac{1}{N}\operatorname{res}(\zeta^*_{\Omega_0}, N). \tag{4.3.54}$$

Let us now assume that $\tilde{D} < N$ in order for the analog of Weyl's asymptotic estimate to hold (in light of (4.3.44), or, equivalently, (4.3.49)); that is, in order for the error term to be negligible compared to the leading term in (4.3.44) and (4.3.49). It then follows from the above discussion (that is, from estimate (4.3.44) or (4.3.49) when $\tilde{D} > N-1$ or from its counterpart when $\tilde{D} = N-1$) that $\zeta^*_{\Omega_0}$ is holomorphic in the open half-plane $\{\mathrm{Re}\, s > N\}$ and can be (uniquely) meromorphically extended to the (strictly) larger open half-plane $\{\mathrm{Re}\, s > \tilde{D}\}$, with a single (simple) pole at $s = N$ in that half-plane. (This statement is true for any value of $\tilde{D}$ in $[N-1, N)$, whether or not $\mathscr{M}^{*\tilde{D}}(\partial\Omega_0, \Omega_0)$ is finite.) Consequently, we deduce that the abscissa of (absolute) convergence of $\zeta^*_{\Omega_0}$, defined by (4.3.50), satisfies the following identity:

$$D_{\mathrm{hol}}(\zeta^*_{\Omega_0}) = D(\zeta^*_{\Omega_0}) = N, \tag{4.3.55}$$

whereas the abscissa of meromorphic continuation of $\zeta^*_{\Omega_0}$ satisfies the inequality

$$D_{\mathrm{mer}}(\zeta^*_{\Omega_0}) \leq \tilde{D}. \tag{4.3.56}$$

(Observe that when $m = 1$, inequality (4.3.56) formally looks exactly like inequality (4.3.30) of Theorem 4.3.17.) In particular, (since $\tilde{D} < N$, by assumption) we have that

$$D_{\mathrm{mer}}(\zeta^*_{\Omega_0}) < D_{\mathrm{hol}}(\zeta^*_{\Omega_0}).$$

As is noted in [Lap2–3], this latter result (in inequality (4.3.56)) follows from the analog of Theorem 4.3.11 (and Corollary 4.3.14) corresponding to uniformly elliptic differential operators $\mathscr{A}$ of order $2m$, which is obtained in [Lap1, Theorem 2.1 and Corollary 2.2]. See the precise definition of the spectrum and the domain of the operator $\mathscr{A}$ given in [Lap1, Section 2.2]; see also [LioMag] or [Mét1].

In addition, much as in Corollary 4.3.22 (where $m = 1$), we have the following identity (concerning not only the spectral zeta function but also the fractal zeta functions of $(\partial\Omega_0, \Omega_0)$):

$$\begin{aligned} D_{\mathrm{mer}}(\zeta^*_{\Omega_0}) &= \overline{\dim}_B(\partial\Omega_0, \Omega_0) =: \tilde{D} \\ &= D_{\mathrm{mer}}(f) = D_{\mathrm{hol}}(f) = D(f), \end{aligned} \tag{4.3.57}$$

for all $f \in \{\zeta_{\partial\Omega_0,\Omega_0}, \tilde{\zeta}_{\partial\Omega_0,\Omega_0}\}$. And analogously for Neumann or, more generally, mixed Dirichlet–Neumann boundary conditions, except with $D := \overline{\dim}_B(\partial\Omega_0)$ instead of $\tilde{D} := \overline{\dim}_B(\partial\Omega_0, \Omega_0)$ and with $\partial\Omega_0$ instead of the relative fractal drum $(\partial\Omega_0, \Omega_0)$.

Recall that the sharpness of inequality (4.3.56) is addressed in Problem 4.3.20, and that for the Dirichlet Laplacian and in the most important case when $\tilde{D} \in (N-1, N)$, it is established in Theorem 4.3.21 and Corollary 4.3.22 above (which relies in an essential way on the results of Sections 4.5–4.6 below). In light of Remark 4.3.23, the counterpart of the latter statement is also true for the Neumann Laplacian (with $\tilde{D}$ replaced by D, as usual).

In the case of Neumann, or more generally, of mixed Dirichlet–Neumann boundary conditions, it follows from the results of [Lap1] (and [Lap2–3]) that Theorem 4.3.11, Corollary 4.3.14, and hence also Theorem 4.3.17 still hold (along with their more general counterparts for positive uniformly elliptic operators of order $2m$) provided that Ω_0 is assumed to be a bounded open set of $\mathbb{R}^N$ satisfying the *extension property* (explicited in the next paragraph) and $\tilde{D} = \overline{\dim}_B(\partial\Omega_0, \Omega_0)$ (the upper, inner Minkowski dimension of $\partial\Omega_0$) is replaced by $D = \overline{\dim}_B(\partial\Omega_0)$, the upper Minkowski (or box) dimension of $\partial\Omega_0$ in the statement of Theorem 4.3.11, Corollary 4.3.14 and Theorem 4.3.17, as well as Theorem 4.3.21 and Corollary 4.3.22 (which rely on results of Section 4.6 below and Theorem 4.3.25 along with Equation (4.3.56)).[36] See, in particular, [Lap1, Theorem 2.3 and Corollary 2.2].

Recall that the open set $\Omega_0 \subseteq \mathbb{R}^N$ is said to satisfy the *extension property* if every function in the Sobolev space $H^1(\Omega_0) := W^{1,2}(\Omega_0)$ can be extended to a function in $H^1(\mathbb{R}^N) := W^{1,2}(\mathbb{R}^N)$, and the resulting extension operator is a bounded linear operator. For example, a bounded domain Ω_0 in $\mathbb{R}^N$ satisfies the extension property if its boundary $\partial\Omega_0$ is of class C^1; see, e.g., [Bre, Théorème IX.7]. Note that, in this latter case, $\dim_B(\partial\Omega_0, \Omega_0) = N - 1$ and $\mathcal{M}^{*D}(\partial\Omega_0, \Omega_0) < \infty$.

Alternatively, the aforementioned results of [Lap1] imply that (still for Neumann or mixed Dirichlet–Neumann boundary conditions) instead of satisfying the extension property, Ω_0 can be assumed to satisfy the so-called (C')-*condition* [Lap1, Definition 2.2] (which is satisfied, for example, if Ω_0 is locally Lipschitz, or satisfies either a 'segment condition', a 'cone condition', or else is an open set with cusp; see [Mét2–3] or [Lap1, Examples 2.1 and 2.2]), in which case we are necessarily in case (*ii*) of the counterparts of Theorem 4.3.11 and Corollary 4.3.14 (see, especially, Equation (4.3.49) and the text following it), with $D\ (:= \overline{\dim}_B(\partial\Omega_0, \Omega_0)) = N - 1$ and $\mathcal{M}^{*D}(\partial\Omega_0, \Omega_0) < \infty$.

Recall that (as is proved by Jones in [Jon] and discussed in [Lap1, Example 4.2]; see also [Maz]) in two dimensions (i.e., when $N = 2$), a simply connected domain Ω_0 satisfies the *extension property* (or is an *extension domain*) if and only if it is a *quasidisk* (i.e., a Jordan curve which is the quasiconformal image of the unit disk in $\mathbb{R}^2$). The boundary $\partial\Omega_0$ of a quasidisk is called a *quasicircle*, and the property of being a quasicircle can be characterized geometrically by a *chord-arc condition.* Furthermore, a quasicircle can have any Hausdorff dimension between 1 and 2. See [Maz] and [Pom], along with the relevant references therein, for a detailed discussion of quasidiscs, quasicircles and extension domains. The class of quasicircles includes the classic Koch snowflake curve and its natural generalizations, as well as the Julia sets associated with the quadratic maps $z \mapsto z^2 + c$ $(z \in \mathbb{C})$, provided the parameter $c \in \mathbb{C}$ is sufficiently small. Therefore, the Koch snowflake domain (and

---

[36] It is clear from the definitions that $\tilde{D} \leq D$, and it can also be shown (since Ω_0 is open and bounded) that $N - 1 \leq \tilde{D} \leq D \leq N$; see [Lap1, Corollary 3.2]. Furthermore, there are natural examples of planar domains for which $\tilde{D} < D$; see [Lap1, Note added in proof, p. 525] and the relevant reference therein, [Tri2].

its generalizations) and the bounded simply connected domains having for boundary the aforementioned Julia sets, are natural examples of quasidisks and hence, of extension domains.

In higher dimensions, *extension domains* (i.e., domains of $\mathbb{R}^N$ satisfying the (Sobolev) extension property) are more difficult to characterize. However, it has been shown by Hajlasz, Koskela and Tuominen in [HajKosTu1–2] that a bounded domain $\Omega_0 \subset \mathbb{R}^N$ is an extension domain if and only if it satisfies a certain functional analytic condition and the following *measure density condition*; see [HajKosTu1, Theorem 5]. The set $\Omega_0 \subseteq \mathbb{R}^N$ is said to satisfy the *measure density condition* (or to be a *lower Ahlfors regular N-set*) if there exists a positive constant M such that

$$|\Omega_0 \cap B_r(x)| \geq M r^N,$$

for all $x \in \Omega_0$ and all $0 < r \leq 1$, where $B_r(x)$ denotes the open ball of center x and radius r in $\mathbb{R}^N$; see [HajKosTu1].

Finally, we note that for Neumann boundary conditions, the above results concerning spectral asymptotics and spectral zeta functions also extend to higher order uniformly elliptic self-adjoint operators (with variable coefficients), under the analogous hypotheses and with the same changes as those indicated above; see [Lap1].

Remark 4.3.26. It is noteworthy that when the extension property (or else the (C')-condition) is not satisfied, the continuous embedding of $H^1(\Omega_0)$ into $L^2(\Omega_0)$ need not be compact and, hence, the spectrum of the Neumann Laplacian may not be discrete. Actually, even when this spectrum is discrete, there are explicit examples of bounded open sets for which the leading spectral asymptotics of the Neumann Laplacian does not satisfy Weyl's classic law (4.3.12), and hence, let alone the corresponding remainder estimate (4.3.22) (or, equivalently, (4.3.19)). See, e.g., [Mét1–3], [Lap1] and the relevant references therein.

A similar comment can be made about more general uniformly elliptic operators of order $2m$, with Neumann (or, more generally, Dirichlet-Neumann) boundary conditions and with $H^m(\Omega_0)$ instead of $H^1(\Omega_0)$.

4.4 Further Examples of Relative Distance Zeta Functions

The aim of this section is to introduce several classes of RFDs and to study their associated fractal zeta functions. We will focus here on the distance zeta functions, although the corresponding tube zeta functions could be studied as well, either directly or by using the functional equation (4.5.2) below. Of special interest are the unbounded geometric chirps, associated with the standard geometric chirps occurring, for example, in the oscillation theory of differential equations. We also compute the relative distance function of the Cartesian product of fractal strings.

4.4.1 Relative Distance Zeta Functions of Unbounded Geometric Chirps

The following example and result (namely, Example 4.4.1 and Proposition 4.4.3) deal with unbounded geometric chirps; see Figures 4.13, 4.14 and 4.15. Also, refer to Section 3.6 for the case of bounded geometric chirps.

Example 4.4.1. Let A be an (α,β)-geometric chirp, for $\alpha\in(-1,0)$ and $\beta>0$; i.e., A is a union of vertical segments at $x=k^{-1/\beta}$ of length $k^{-\alpha/\beta}$ for $k\in\mathbb{N}$; see (3.6.1). For Ω we take the union of open rectangles R_k for $k\in\mathbb{N}$, where R_k has a base of length $k^{-1/\beta}-(k+1)^{-1/\beta}$ and height $k^{-\alpha/\beta}$; see Figure 4.15. The associated unbounded geometric chirp RFD (A,Ω) approximates the graph of the function

$$x\mapsto x^{\alpha}\sin(\pi x^{-\beta}),\quad \text{for all } x\in(0,1).$$

The relative distance zeta function of (A,Ω) is then given by

$$\zeta_{A,\Omega}(s)=\frac{1}{2^{s-2}(s-1)}\sum_{k=1}^{\infty}k^{-\alpha/\beta}\left(k^{-1/\beta}-(k+1)^{-1/\beta}\right)^{s-1}. \tag{4.4.1}$$

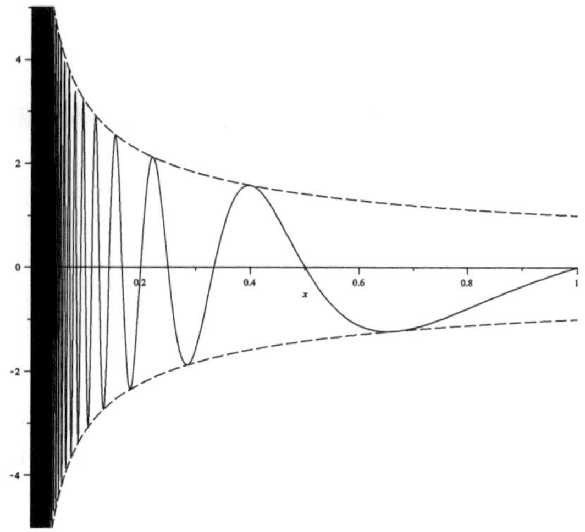

Fig. 4.13 The unbounded $(-1/2,1)$-chirp; the graph of $f(x)=x^{-1/2}\sin(\pi x^{-1})$, $0<x<1$, is fractal near $x=0$. We expect that $\overline{\dim}_B(A,\Omega)=7/4$, as for the related geometric chirp in Proposition 4.4.3(a) and depicted in Figure 4.15 on page 347.

It can be shown that it has a singularity at $s=2-\frac{1+\alpha}{1+\beta}$ and is holomorphic in the open right half-plane $\{\operatorname{Re}s>2-\frac{1+\alpha}{1+\beta}\}$. (See Remark 4.4.2 just below for a justification of this claim.) We conclude from Theorem 4.1.7 that $\overline{\dim}_B(A,\Omega)=2-\frac{1+\alpha}{1+\beta}$, which is Tricot's formula in the case when α is negative and β positive. We note that the original Tricot formula was obtained for $0 < \alpha < \beta$ and can be found in [Tri3, p. 122].

Remark 4.4.2. We provide here a short heuristic proof of the above claim (compare with the proof of Equation (3.6.3) in Subsection 3.6.1 above). Using the Lagrange mean value theorem, we approximate the difference $k^{-1/\beta} - (k+1)^{-1/\beta}$ (where $k \in \mathbb{N}$) by $k^{-\frac{1}{\beta}-1}$. The Dirichlet series on the right-hand side of Equation (4.4.1) then becomes

$$\sum_{k=1}^{\infty} k^{-\alpha/\beta}\left(k^{-\frac{1}{\beta}-1}\right)^{s-1} = \sum_{k=1}^{\infty} k^{-\left(\frac{\alpha}{\beta}+(\frac{1}{\beta}+1)(s-1)\right)}.$$

It converges absolutely if and only if $\frac{\alpha}{\beta} + (\frac{1}{\beta}+1)(\mathrm{Re}\, s - 1) > 1$; that is, when $\mathrm{Re}\, s > 2 - \frac{1+\alpha}{1+\beta}$. This heuristic proof can be easily made precise using Cahen's classical result stated in Theorem 2.1.27. We leave the details as a simple exercise for the

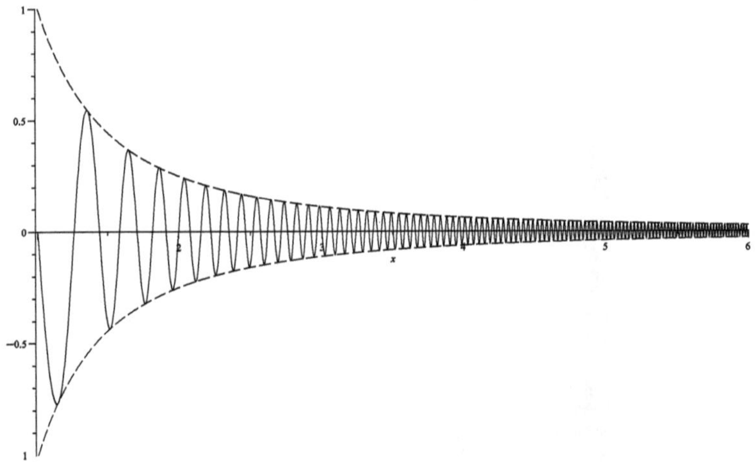

Fig. 4.14 The unbounded $(-2,-3)$-chirp; the graph of $f(x) = x^{-2}\sin(\pi x^3)$, $x > 1$, is fractal near $x = \infty$. We expect that $\overline{\dim}_B(A,\Omega) = 3/2$, as for the related geometric chirp in Proposition 4.4.3(b).

interested reader. (A different proof of the claim as well as additional information can be found in Example 5.5.19 in Subsection 5.5.5 below.)

Proposition 4.4.3. *Let A be an* (α,β)*-geometric chirp defined by (3.6.1), and assume that one of the following conditions holds*:

(a) $-1 < \alpha < 0 < \beta$ and $\Omega = \{(x,y) \in \mathbb{R}^2 : x \in (0,1),\ 0 < x^\alpha < y\}$,

(b) $\beta < \alpha < -1$ and $\Omega = \{(x,y) \in \mathbb{R}^2 : x \in (1,+\infty),\ 0 < x^\alpha < y\}$.

Then

$$\overline{\dim}_B(A,\Omega) = 2 - \frac{1+\alpha}{1+\beta},$$

and moreover, this value coincides with $\dim_{PC}(A,\Omega)$.

Proof. Computing the relative distance zeta function of the (α,β)-geometric chirp from Example 4.4.1 with respect to the 'outer' rectangles and using Lemma 4.1.15, we obtain the result in case (a). We can use the same technique in case (b), due to the fact that $\beta < -1$. □

We note that in Example 4.4.1 and Proposition 4.4.3, we can replace $\overline{\dim}_B(A,\Omega)$ by $d = \dim_B(A,\Omega)$. This can be seen by direct computation: indeed, there exist positive constants c_1 and c_2 such that $c_1\delta^{2-d} \le |A_\delta \cap \Omega| \le c_2\delta^{2-d}$. Therefore, $c_1 \le \mathscr{M}_*^d(A,\Omega) \le \mathscr{M}^{*d}(A,\Omega) \le c_2$.

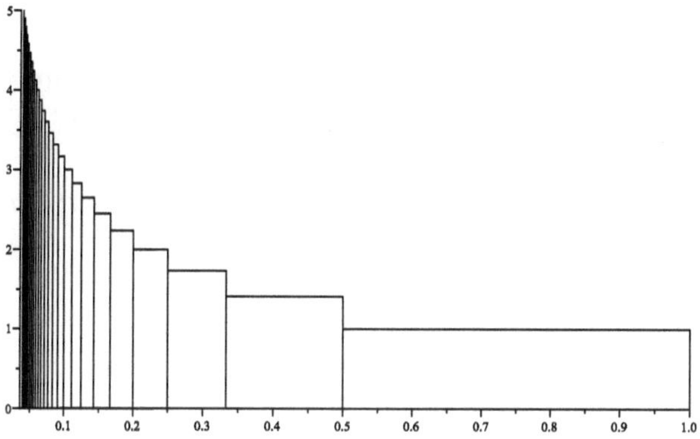

Fig. 4.15 Approximation of the unbounded chirp in Figure 4.13 on page 345, using rectangles. Here, $\alpha = -1/2$ and $\beta = 1$; hence, $\overline{\dim}_B A = 7/4$. In the corresponding RFD (A,Ω), the set A is defined as the union of the vertical segments while the open set Ω is defined as the union of the open rectangles.

In the case when $-1 < \alpha < 0$ and $\beta < 0$, we have $\dim_B(A,\Omega) = 1$, which complements Proposition 4.4.3(a). Analogously, in the case when $\alpha < -1$ and $\alpha \le \beta$, $\beta \ne 0$, we have $\dim_B(A,\Omega) = 1$, which complements Proposition 4.4.3(b).

In the work of the second and third authors with V. Županović [RaŽuŽup], a different approach to the study of the fractal properties of unbounded sets at infinity in $\mathbb{R}^N$ has been undertaken, instead of using the relative box dimensions. If A is an unbounded set which does not possess the origin as its accumulation point, then it is natural to define the *box dimension of A at infinity* as the usual box dimension of $A^{-1} = \{x/|x|^2 : x \in A\}$. Here, A^{-1} is the geometric inversion of A, which under the stated condition is clearly a bounded set. This tool has been applied to the study of the Hopf bifurcation of several polynomial dynamical systems at infinity. In his thesis [Ra1] and in [Ra2], the second author has significantly expanded these ideas. In particular, in [Ra1–2], the notions of Minkowski contents and box dimensions of unbounded open sets with respect to infinity have been introduced and studied, as well as the associated classes of fractal zeta functions, thereby extending to (suitable) unbounded sets $A \subseteq \mathbb{R}^N$ the theory developed in this book and in [LapRaŽu1–8].

4.4.2 Relative Zeta Functions of Cartesian Products of Fractal Strings

In Theorem 3.6.5, we have computed a representative of the zeta function of E relative to E_δ, where E is the boundary of the Cartesian product of two fractal strings $\mathscr{L} = (\ell_j)_{j\geq 1}$ and $\mathscr{M} = (m_k)_{k\geq 1}$. If we consider the zeta function of E relative to the rectangle $\Omega = (0,a_1) \times (0,b_1)$, then we deduce from the proof of Theorem 3.6.5 that $D(s) \equiv 0$ in (3.6.29), which yields the following explicit result.

Theorem 4.4.4. *Assume that the hypotheses of Theorem 3.6.5 are satisfied. Then, for E given as in Theorem 3.6.5 and for $\Omega = (0,a_1) \times (0,b_1)$, we have*

$$\zeta_{E,\Omega}(s) = \frac{2^{2-s}}{s-1} \sum_{j,k=1}^{\infty} \left[|\ell_j - m_k| \min\{\ell_j, m_k\}^{s-1} + \frac{2}{s} \min\{\ell_j, m_k\}^s \right].$$

Furthermore,

$$\overline{\dim}_B(E,\Omega) = \overline{\dim}_B E = 1 + \max\{\overline{\dim}_B \mathscr{L}, \overline{\dim}_B \mathscr{M}\} \tag{4.4.2}$$

is the abscissa of convergence of $\zeta_{E,\Omega}(s)$, and $\zeta_{E,\Omega}(s) \to +\infty$ as $\mathbb{R} \ni s \to \overline{\dim}_B E$ from the right.

Let us consider the product of three fractal strings. Assume that $\mathscr{L} = (\ell_j)_{j\geq 1}$, $\mathscr{M} = (m_k)_{k\geq 1}$, and $\mathscr{N} = (n_r)_{r\geq 1}$, with $a_1 := \sum_j \ell_j$, $b_1 := \sum_k m_k$ and $c_1 := \sum_r n_r$. We identify the strings with three standard families of open intervals, $\mathscr{L} = (I_j)_{j\geq 1}$, $\mathscr{M} = (J_k)_{k\geq 1}$ and $\mathscr{N} = (K_r)_{r\geq 1}$. Furthermore, for any ordered triple $L = (\ell_j, m_j, n_r)$ in $\mathscr{L} \times \mathscr{M} \times \mathscr{N}$, we define its nondecreasing permutation $(M_1(L), M_2(L), M_3(L))$ by $\{\ell_j, m_j, n_r\} = \{M_1(L), M_2(L), M_3(L)\}$ and $M_1(L) \leq M_2(L) \leq M_3(L)$. Note that then, $M_1(L) = \min L$ and $M_3(L) = \max L$. (Further, observe that we are really working here with a Cartesian product of multisets, rather than of ordinary sets. Recall that a multiset is simply a set with multiplicities. For example, $\mathscr{L}$, $\mathscr{M}$ and $\mathscr{N}$ are multisets since, for instance, scales ℓ_j in the sequence $\mathscr{L}$ may have finite multiplicities.) In this context, we obtain the following result.

Theorem 4.4.5. *Let $\mathscr{L} = (I_j)_{j\geq 1}$, $\mathscr{M} = (J_k)_{k\geq 1}$ and $\mathscr{N} = (n_r)_{r\geq 1}$ be three fractal strings, and define the set $E = \partial(\mathscr{L} \times \mathscr{M} \times \mathscr{N})$. Let $\Omega = (0,a_1) \times (0,b_1) \times (0,c_1)$. Then, with the notation introduced just above, we have*

$$\zeta_{E,\Omega}(s) = \frac{2^{3-s}}{s-2}\sum_L \Big[(M_3(L)-M_1(L))(M_2(L)-M_1(L))(M_1(L))^{s-2} \\ +\frac{2}{s-1}(M_3(L)+M_2(L)-2M_1(L))(M_1(L))^{s-1} \\ +\frac{4}{s(s-1)}(M_1(L))^s\Big], \tag{4.4.3}$$

where the summation runs over all ordered triples $L=(\ell_j,m_j,n_r)$ from $\mathscr{L}\times\mathscr{M}\times\mathscr{N}$. Furthermore, the abscissa of convergence $D(\zeta_{E,\Omega})$ of $\zeta_{E,\Omega}$ is given by

$$\overline{\dim}_B(E,\Omega)=\overline{\dim}_B E = 2+\max\{\overline{\dim}_B\mathscr{L},\overline{\dim}_B\mathscr{M},\overline{\dim}_B\mathscr{N}\}, \tag{4.4.4}$$

and $\zeta_{E,\Omega}(s)\to+\infty$ as $\mathbb{R}\ni s\to\overline{\dim}_B E$ from the right.

Proof. The set Ω is a countable disjoint union of open parallelepipeds. Let R be a typical parallelepiped with sides $n\le m\le\ell$. We split R into 16 prisms (8 of them being pairwise isometric and having width $m-n$, and the rest with side $\ell-n$), 32 isometric tetrahedra, and two isometric parallelepipeds with sides $n/2$, $m-n$, $\ell-n$, placed at the center of R. We have to integrate the function $d(x,\partial R)^{s-3}$ over these sets. The integral over each prism of width $m-n$ is equal to

$$\frac{2^{1-s}}{(s-1)(s-2)}(m-n)n^{s-1},$$

(and analogously for the prism of width $\ell-n$). The integral over each tetrahedron is equal to

$$\frac{2^{-s}}{s(s-1)(s-2)}n^s,$$

while the integral over each parallelepiped is equal to

$$\frac{2^{2-s}}{s-2}(m-n)(\ell-n)n^{s-2}.$$

From this, the claim follows easily. We omit the details. The dimension result is an immediate consequence of the finite stability of the upper box dimension and the fact that

$$E=(\overline{A}\times[0,b_1]\times[0,c_1])\cup([0,a_1]\times\overline{B}\times[0,c_1])\cup([0,a_1]\times[0,b_1]\times\overline{C}),$$

where $A=(a_j)_{j\ge1}$ with $a_j:=\sum_{k\ge j}\ell_k$, and analogously for $B=(b_k)_{k\ge1}$ and $C=(c_r)_{r\ge1}$. □

Note that the relative distance zeta function of the set $E=\partial(\mathscr{L}\times\mathscr{M})$ generated by two fractal strings is represented by a double sum (see Theorem 4.4.4), while the relative distance zeta function of $E=\partial(\mathscr{L}\times\mathscr{M}\times\mathscr{N})$ is equal to a triple sum (see Theorem 4.4.5) taken over the indices (j,k,r), since $L=(\ell_j,m_k,n_r)$. Analogously, the relative distance zeta function of the set $E = \partial(\mathscr{L}_1 \times \cdots \times \mathscr{L}_N)$ generated by N fractal strings $\mathscr{L}_i$, $i = 1,\dots,N$, can then be computed, and it is equal to an N-tuple sum. Furthermore, we then have

$$\overline{\dim}_B(E,\Omega) = \overline{\dim}_B E = N - 1 + \max\{\overline{\dim}_B \mathscr{L}_i : i = 1,\dots,N\}.$$

4.5 Meromorphic Extensions of Relative Zeta Functions and Applications

If we consider a class of RFDs with a prescribed value D for the abscissa of convergence of the associated distance relative zeta functions, it is of interest to know the corresponding values of the abscissa of meromorphic continuation $D_{\mathrm{mer}} = D - \alpha$. Clearly, we have $\alpha \geq 0$. Intuitively, the smaller α, the more complex the (fractal) nature of the relative fractal drum. This can be considered even as a definition for comparing the complexity of different RFDs in the class. The most complex, then, is the subclass of relative fractal drums for which the abscissa of meromorphic continuation is equal to D; that is, $\alpha = 0$. And among these, the most complex are the relative fractal drums for which the set of nonisolated singularities is equal to the *whole* critical line $\{\mathrm{Re}\, s = D\}$. Indeed, there cannot be more complexity than that, at least from the present point of view of the higher-dimensional theory of complex fractal dimensions. We call them *maximally hyperfractal drums*; see Definition 4.6.23.

This section is organized as follows. We first study the problem of determining an upper bound for the abscissa of meromorphic extension of the distance (or tube) zeta function for a class of RFDs; see Theorems 4.5.1 and 4.5.2. Furthermore, we construct a class of RFDs for which the abscissa of meromorphic continuation can be explicitly computed. We also construct an explicit class of maximally hyperfractal drums (A,Ω); see Theorem 4.5.8. As a consequence, we then construct (in Section 4.6) a class of maximally hyperfractal strings $\mathscr{L}$, which in turn generate maximally hyperfractal sets $A = A_{\mathscr{L}}$ on the real line; see Corollary 4.6.17.

4.5.1 Meromorphic Extensions of Zeta Functions of Relative Fractal Drums

By analogy with (2.2.8), we introduce the *relative tube zeta function* associated with the relative fractal drum (A,Ω) in $\mathbb{R}^N$. It is defined by

$$\tilde{\zeta}_{A,\Omega}(s) = \int_0^\delta t^{s-N-1} |A_t \cap \Omega| \, dt, \tag{4.5.1}$$

for all $s \in \mathbb{C}$ with $\mathrm{Re}\, s$ sufficiently large, where $\delta > 0$ is fixed. As we see, it involves the *relative tube function* $t \mapsto |A_t \cap \Omega|$. Its abscissa of convergence is given by $D(\tilde{\zeta}_{A,\Omega}) = \overline{\dim}_B(A,\Omega)$. This follows from the following fundamental identity or *functional equation*, which connects the relative tube zeta function $\tilde{\zeta}_{A,\Omega}$ and the relative distance zeta function $\zeta_{A,\Omega}$, defined by (4.1.1):

$$\zeta_{A,A_\delta\cap\Omega}(s) = \delta^{s-N}|A_\delta\cap\Omega| + (N-s)\tilde{\zeta}_{A,\Omega}(s). \tag{4.5.2}$$

This identity is analogous to (2.2.1) and (2.2.23). Its proof is analogous to that of Theorem 2.2.1, using the known identity

$$\int_{A_\delta\cap\Omega} d(x,A)^{-\gamma}\,\mathrm{d}x = \delta^{-\gamma}|A_\delta\cap\Omega| + \gamma\int_0^\delta t^{-\gamma-1}|A_t\cap\Omega|\,\mathrm{d}t, \tag{4.5.3}$$

where $\gamma > 0$; see Lemma 2.1.4, [Žu2, Theorem 2.9(a)], or a more general form provided in [Žu4, Lemma 3.1].

It follows from the identity (4.5.2) that the analog of Proposition 2.2.19 and of Equation (2.2.50) holds in the present more general context. More specifically, provided that $\overline{\dim}_B(A,\Omega) < N$, the relative tube zeta function $\tilde{\zeta}_{A,\Omega}$ and the relative distance zeta function $\zeta_{A,A_\delta\cap\Omega}$ can be (uniquely) meromorphically extended to exactly the same domain $U \subseteq \mathbb{C}$ (chosen to be a connected open neighborhood of a given window $\boldsymbol{W}$, say), when it is possible. Furthermore, the relative fractal drum (A,Ω) has exactly the same visible complex dimensions (and with the same multiplicities), as measured from the point of view of either of these two fractal zeta functions:

$$\mathscr{P}(\tilde{\zeta}_{A,\Omega},\boldsymbol{W}) = \mathscr{P}(\zeta_{A,A_\delta\cap\Omega},\boldsymbol{W}). \tag{4.5.4}$$

In particular, we have

$$\overline{\dim}_B(A,\Omega) = D(\tilde{\zeta}_{A,\Omega}) = D(\zeta_{A,A_\delta\cap\Omega}), \tag{4.5.5}$$

$$\tilde{\zeta}_{A,\Omega} \sim \zeta_{A,A_\delta\cap\Omega}, \tag{4.5.6}$$

and so

$$\dim_{PC}(A,\Omega) = \mathscr{P}_c(\tilde{\zeta}_{A,\Omega}) = \mathscr{P}_c(\zeta_{A,A_\delta\cap\Omega}). \tag{4.5.7}$$

Finally, if $\omega \in U$ is a simple pole of $\tilde{\zeta}_{A,\Omega}$, then it is also a simple pole of $\zeta_{A,A_\delta\cap\Omega}$ and we have

$$\operatorname{res}(\tilde{\zeta}_{A,\Omega},\omega) = \frac{1}{N-\omega}\operatorname{res}(\zeta_{A,A_\delta\cap\Omega},\omega); \tag{4.5.8}$$

that is, the counterpart of Equation (2.2.50) holds in this context. Note that as was the case for ordinary fractal sets A, these residues are independent of the choice of $\delta > 0$.

We next consider a class of RFDs (A,Ω) such that both $D = \dim_B(A,\Omega)$ and $\mathscr{M}^D(A,\Omega)$ exist, but the relative Minkowski content is *degenerate* in the sense that $\mathscr{M}^D(A,\Omega) = +\infty$. In general, we may also have $\mathscr{M}^D(A,\Omega) = 0$, but we do not treat this case here.

We shall treat these two cases by using the following assumption on the asymptotics of the relative tube function $t \mapsto |A_t\cap\Omega|$:

$$|A_t \cap \Omega| = t^{N-D} h(t)(\mathscr{M} + O(t^\alpha)) \quad \text{as} \quad t \to 0^+, \tag{4.5.9}$$

where $\mathscr{M} > 0$, $\alpha > 0$ and $D \leq N$ are given in advance. Here, we assume that the function $h(t)$ has a sufficiently slow growth near the origin, in the sense that for any $c > 0$, $h(t) = O(t^c)$ as $t \to 0^+$. Typical examples of such functions are $h(t) = (\log t^{-1})^m$, $m \geq 1$, or more generally,

$$h(t) = \big(\underbrace{\log \ldots \log}_{n}(t^{-1})\big)^m$$

(the m-th power of the n-th iterated logarithm of t^{-1}, $n \geq 1$), and in these cases we obviously have $\mathscr{M}^D(A,\Omega) = +\infty$. For this and other examples, see [HeLap]. The function $t \mapsto t^D h(t)^{-1}$ is usually called the *gauge function*, but for the sake of simplicity, we shall rather use this name for the function $h(t)$ only.

Assuming that a relative fractal drum (A,Ω) in $\mathbb{R}^N$ is such that $D = \dim_B(A,\Omega)$ exists, and $\mathscr{M}_*^D(A,\Omega) = 0$ or $+\infty$ (or $\mathscr{M}^{*D}(A,\Omega) = 0$ or $+\infty$), it is natural to define as follows a new class of relative lower and upper Minkowski contents of (A,Ω), associated with a suitably chosen gauge function $h(t)$:

$$\begin{aligned} \mathscr{M}_*^D(A,\Omega,h) &= \liminf_{t\to 0^+} \frac{|A_t \cap \Omega|}{t^{N-D}h(t)}, \\ \mathscr{M}^{*D}(A,\Omega,h) &= \limsup_{t\to 0^+} \frac{|A_t \cap \Omega|}{t^{N-D}h(t)}. \end{aligned} \tag{4.5.10}$$

The aim is to find an *explicit* gauge function so that these two contents are in $(0,+\infty)$, and the functions $r \mapsto \mathscr{M}_*^r(A,\Omega,h)$ and $r \mapsto \mathscr{M}^{*r}(A,\Omega,h)$, $r \in \mathbb{R}$, defined exactly as in (4.5.10), except for D replaced with r, have a jump from $+\infty$ to 0 when r crosses the value of D. In this generality, the above *gauge relative Minkowski contents* have been introduced in [Žu4], motivated by [HeLap].

In Equation (4.5.10) above, $\mathscr{M}_*^D(A,\Omega,h)$ (resp., $\mathscr{M}^{*D}(A,\Omega,h)$) is called the *lower* (resp., *upper*) h-Minkowski content of (A,Ω). Furthermore, much as in the usual case when $h \equiv 1$, the RFD (A,Ω) is said to be *h-Minkowski nondegenerate* if

$$0 < \mathscr{M}_*^D(A,\Omega,h) \leq \mathscr{M}^{*D}(A,\Omega,h) < \infty.$$

If for some gauge function h, say, we have that $\mathscr{M}^D(A,\Omega,h) \in (0,+\infty)$ (which means, as usual, that $\mathscr{M}_*^D(A,\Omega,h) = \mathscr{M}^{*D}(A,\Omega,h)$ and that this common value, denoted by $\mathscr{M}^D(A,\Omega,h)$, lies in $(0,+\infty)$), we say (as in [HeLap]) that the fractal drum (A,Ω) is *h-Minkowski measurable*, with *h-Minkowski content* $\mathscr{M}^D(A,\Omega,h)$.

It is easy to see that the counterparts of Theorems 2.3.18 and 2.3.25 also hold in the present context of relative fractal drums. It suffices to replace $|A_t|$ by $|A_t \cap \Omega|$, $\tilde{\zeta}_A(s)$ by $\tilde{\zeta}_{A,\Omega}(s)$, and $\dim_B A$ by $\dim_B(A,\Omega)$. Below, we state and prove two results dealing with RFDs with associated gauge functions; see Theorems 4.5.1 and 4.5.2. In both of these theorems, we have $\mathscr{M}^D(A,\Omega) = +\infty$. As we shall see, certain gauge functions generate higher-order poles of fractal zeta functions. The presence of a nonstandard gauge function may also force the tube (or distance) zeta function to have a partial natural boundary along the critical line $\{\text{Re}\, s = D\}$ (i.e., not to have a meromorphic continuation beyond the open half-plane $\{\text{Re}\, s > D\}$ of holomorphicity). One could then try to view the fractal zeta function as an analytic function on an appropriate Riemann surface. However, we will not investigate this interesting topic here.

In what follows, we denote the Laurent expansion of a meromorphic extension (assumed to exist) of the relative tube zeta function $\tilde{\zeta}_{A,\Omega}$ to a connected open neighborhood of $s = D$ (more specifically, an open punctured disk centered at $s = D$) by

$$\tilde{\zeta}_{A,\Omega}(s) = \sum_{j=-\infty}^{\infty} c_j (s-D)^j, \tag{4.5.11}$$

where, of course, $c_j = 0$ for all $j \ll 0$ (that is, there exists $j_0 \in \mathbb{Z}$ such that $c_j = 0$ for all $j < j_0$).

The following theorem shows that, in order to obtain a meromorphic extension of the zeta function to the left of the abscissa of convergence, it is important to have some information about the second term in the asymptotic expansion of the relative tube function $t \mapsto |A_t \cap \Omega|$ near $t = 0$. We will provide two proofs of this result, because they each highlight different aspects of the issues involved. See Theorem 5.4.29 in Chapter 5 below (as well as Theorem 5.4.30) for a partial converse of Theorem 4.5.1.

Theorem 4.5.1 (Minkowski measurable RFDs). *Let (A,Ω) be a relative fractal drum in $\mathbb{R}^N$, such that (4.5.9) holds for some $D \le N$, $\mathscr{M} > 0$, $\alpha > 0$ and with $h(t) := (\log t^{-1})^m$ for all $t \in (0,1)$, where m is a nonnegative integer. Then the relative fractal drum (A,Ω) is h-Minkowski measurable, $\dim_B(A,\Omega) = D$, and $\mathscr{M}^D(A,\Omega,h) = \mathscr{M}$. Furthermore, the relative tube zeta function $\tilde{\zeta}_{A,\Omega}(s)$ has for abscissa of convergence $D(\tilde{\zeta}_{A,\Omega}) = D$, and it possesses a (necessarily unique) meromorphic extension (at least) to the half-plane $\{\text{Re}\, s > D - \alpha\}$; that is,*

$$D_{\text{mer}}(\tilde{\zeta}_{A,\Omega}) \le D - a. \tag{4.5.12}$$

Moreover, $s = D$ is the unique pole in this half-plane, and it is of order $m+1$. In addition, the coefficients of the Laurent series expansion (4.5.11) corresponding to the principal part of $\tilde{\zeta}_{A,\Omega}(s)$ at $s = D$ are given by

$$\begin{aligned} c_{-m-1} &= m!\mathscr{M}, \\ c_{-m} &= \cdots = c_{-1} = 0 \quad (\text{provided } m \ge 1.) \end{aligned} \tag{4.5.13}$$

If $m = 0$, then D is a simple pole of $\tilde{\zeta}_{A,\Omega}$ and we have that

$$\text{res}(\tilde{\zeta}_{A,\Omega}, D) = \mathscr{M}. \tag{4.5.14}$$

Proof. (First proof of Theorem 4.5.1.) Let us set

$$\begin{aligned}\zeta_1(s) &= \mathscr{M} z_m(s), \quad z_m(s) = \int_0^\delta t^{s-D-1}(\log t^{-1})^m \mathrm{d}t,\\ \zeta_2(s) &= \int_0^\delta t^{s-N-1}(\log t^{-1})^m O(t^{N-D+\alpha})\,\mathrm{d}t.\end{aligned} \tag{4.5.15}$$

Since $\tilde{\zeta}_{A,\Omega}(s) = \zeta_1(s) + \zeta_2(s)$, we can proceed as follows. It is easy to see that for each $\varepsilon > 0$, we have $(\log t^{-1})^m = O(t^{-\varepsilon})$ as $t \to 0^+$; hence,

$$|\zeta_2(s)| \le \int_0^\delta O(t^{\mathrm{Re}\,s-1-D+(\alpha-\varepsilon)})\,\mathrm{d}t.$$

Then, since the integral is well defined for all $s \in \mathbb{C}$ with $\mathrm{Re}\,s > D - (\alpha-\varepsilon)$, in the same way as in the proof of Theorem 2.3.18, we deduce that $D(\zeta_2) \le D - (\alpha - \varepsilon)$. Letting $\varepsilon \to 0^+$, we obtain the following desired inequality: $D(\zeta_2) \le D - \alpha$.

By means of the change of variable $\tau := \log t^{-1}$ (for $0 < t \le \delta$), it is easy to see that

$$z_m(s) = \int_{\log \delta^{-1}}^{+\infty} \mathrm{e}^{-\tau(s-D)} \tau^m \mathrm{d}\tau. \tag{4.5.16}$$

Integration by parts yields the following recursion equation, where we have to assume (at first) that $\mathrm{Re}\,s > D$:

$$z_m(s) = \frac{1}{s-D}\left((\log \delta^{-1})^m \delta^{s-D} + m z_{m-1}(s)\right), \quad m \ge 1, \tag{4.5.17}$$

and $z_0(s) := (s-D)^{-1}\delta^{s-D}$. Since $D(\zeta_2) \le D - \alpha$, it is clear that the coefficients c_j, $j < 0$, of the Laurent series expansion (4.5.11) of $\tilde{\zeta}_{A,\Omega}(s) = \zeta_1(s) + \zeta_2(s)$ in a neighborhood of $s = D$ do not depend on $\delta > 0$. Indeed, changing the value of $\delta > 0$ to $\delta_1 > 0$ in (4.5.1) amounts to adding $\int_\delta^{\delta_1} t^{s-N-1}|A_t \cap \Omega|\,\mathrm{d}t$, which is an entire function of s. Therefore, without loss of generality, we may take $\delta = 1$ in (4.5.17):

$$z_m(s) = \frac{m}{s-D} z_{m-1}(s) = \cdots = \frac{m!}{(s-D)^m} z_0(s) = \frac{m!}{(s-D)^{m+1}}. \tag{4.5.18}$$

In this way, we obtain that

$$\zeta_1(s) = \frac{m!}{(s-D)^{m+1}} \mathscr{M}, \tag{4.5.19}$$

and we can meromorphically extend the definition of ζ_1 from the half-plane $\{\mathrm{Re}\,s > D\}$ to the entire complex plane. The claim then follows from Lemma 2.3.5. $\square$

Proof. (Second proof of Theorem 4.5.1.) Let us define z_0 by

$$z_0(s) = \int_0^\delta t^{s-N-1} \frac{|A_t \cap \Omega|}{(\log t^{-1})^m} \, \mathrm{d}t, \tag{4.5.20}$$

where $\operatorname{Re} s > N - D$. As we see, $z_0(s) = \zeta_1(s)$, where $\zeta_1(s)$ is defined as in the proof of Theorem 2.3.25, except with $|A_t|$ replaced by $|A_t \cap \Omega|$, and $\zeta_0(s)$ has all the properties stated in this theorem for $\zeta_A(s)$. It is easy to see that

$$z_0'(s) = -\int_0^\delta t^{s-N-1} \frac{|A_t \cap \Omega|}{(\log t^{-1})^{m-1}} \, \mathrm{d}t.$$

Therefore, proceeding inductively, we obtain that (still for $\operatorname{Re} s > N - D$)

$$z_0^{(m)}(s) = (-1)^m \int_0^\delta t^{s-N-1} |A_t \cap \Omega| \, \mathrm{d}t = (-1)^m \zeta_{A,\Omega}(s). \tag{4.5.21}$$

We conclude that $\zeta_{A,\Omega}(s)$ and $z_0(s)$ possess the same meromorphic extensions. By using Theorem 2.3.18, we see that $z_0(s) = \zeta_1(s)$ can be meromorphically extended to $\{\operatorname{Re} s > D - \alpha\}$, and therefore the same holds for $\tilde{\zeta}_{A,\Omega}(s)$. The remaining claims follow from the fact that $\tilde{\zeta}_{A,\Omega}(s) = (-1)^m z_0^{(m)}(s)$. $\square$

Next, we consider a class of Minkowski nonmeasurable RFDs with associated gauge functions. The following result is a partial generalization of Theorem 2.3.25, which corresponds to the case when $m = 0$.

Theorem 4.5.2 (Minkowski nonmeasurable RFDs). *Let (A, Ω) be a relative fractal drum in $\mathbb{R}^N$, such that there exist $D \le N$, a nonconstant periodic function $G : \mathbb{R} \to \mathbb{R}$ with minimal period $T > 0$, and a nonnegative integer m, satisfying*

$$|A_t \cap \Omega| = t^{N-D} (\log t^{-1})^m \left(G(\log t^{-1}) + O(t^\alpha) \right) \quad \text{as} \quad t \to 0^+. \tag{4.5.22}$$

Then $\dim_B(A, \Omega)$ exists and $\dim_B(A, \Omega) = D$, G is continuous, and

$$\mathscr{M}_*^D(A, \Omega, h) = \min G, \quad \mathscr{M}^{*D}(A, \Omega, h) = \max G,$$

where $h(t) := (\log t^{-1})^m$ for all $t \in (0, 1)$. Furthermore, the tube zeta function $\tilde{\zeta}_{A,\Omega}$ has for abscissa of convergence $D(\tilde{\zeta}_{A,\Omega}) = D$, and it possesses a (necessarily unique) meromorphic extension (at least) to the half-plane $\{\operatorname{Re} s > D - \alpha\}$; that is,

$$D_{\mathrm{mer}}(\tilde{\zeta}_{A,\Omega}) \le D - \alpha. \tag{4.5.23}$$

Moreover, all of its poles located in this half-plane are of order $m + 1$, and the set of poles $\mathscr{P}(\tilde{\zeta}_{A,\Omega})$ is contained in the vertical line $\{\operatorname{Re} s = D\}$. More precisely,

$$\begin{aligned} \mathscr{P}(\tilde{\zeta}_{A,\Omega}) &= \mathscr{P}_c(\tilde{\zeta}_{A,\Omega}) \\ &= \left\{ s_k = D + \frac{2\pi}{T} k \mathrm{i} \in \mathbb{C} : \hat{G}_0\left(\frac{k}{T}\right) \neq 0, \, k \in \mathbb{Z} \right\}, \end{aligned} \tag{4.5.24}$$

where $s_0 = D \in \mathscr{P}(\tilde{\zeta}_{A,\Omega})$ *and* $\hat{G}_0$ *is the Fourier transform of* G_0 *(as given by (2.3.29)). The poles come in complex conjugate pairs; that is, if* s_k *is a pole, then* s_{-k} *is a pole as well (reality principle, see Remark 2.3.28).*

In addition, if $\tilde{\zeta}_{A,\Omega}(s) = \sum_{j=-\infty}^{\infty} c_j^{(k)}(s-s_k)^j$ *is the Laurent expansion of the tube zeta function in a neighborhood of* $s = s_k$*, for a given* $k \in \mathbb{Z}$*, then*

$$\begin{aligned} c_j^{(k)} &= 0 \quad \textit{for} \quad j < 0 \textit{ and } j \neq -m-1 \\ c_{-m-1}^{(k)} &= \frac{m!}{T}\hat{G}_0\left(\frac{k}{T}\right), \end{aligned} \tag{4.5.25}$$

where G_0 *is the restriction of* G *to the interval* $[0,T]$*, and* $\hat{G}_0$ *is given by (2.3.29), as above. Also,*

$$|c_{-m-1}^{(k)}| \le \frac{m!}{T}\int_0^T G(\tau)\,\mathrm{d}\tau, \quad \lim_{k\to\infty} c_{-m-1}^{(k)} = 0. \tag{4.5.26}$$

In particular, for $k = 0$*, that is, for* $s_0 = D$*, we have*

$$\begin{aligned} c_{-m-1}^{(0)} &= \frac{m!}{T}\int_0^T G(\tau)\,\mathrm{d}\tau \\ m!\,\mathscr{M}_*^D(A,\Omega,h) < c_{-m-1}^{(0)} &< m!\,\mathscr{M}^{*D}(A,\Omega,h). \end{aligned} \tag{4.5.27}$$

If $m = 0$ *(i.e.,* $h(t) = 1$ *for all* $t \in (0,1)$*), then* D *is a simple pole of* $\tilde{\zeta}_{A,\Omega}$ *and we have that*

$$\operatorname{res}(\tilde{\zeta}_{A,\Omega}, D) = \frac{1}{T}\int_0^T G(\tau)\,\mathrm{d}\tau = \widetilde{\mathscr{M}} \tag{4.5.28}$$

and

$$\mathscr{M}_*^D(A,\Omega) < \operatorname{res}(\tilde{\zeta}_{A,\Omega}, D) < \mathscr{M}^{*D}(A,\Omega), \tag{4.5.29}$$

where $\widetilde{\mathscr{M}} = \widetilde{\mathscr{M}}^D(A,\Omega)$ *denotes the average Minkowski content of* (A,Ω)*. (See Remark 4.5.3 below.)*

Proof. For $m \in \mathbb{N}_0$, let us define z_m by

$$z_m(s) = \int_0^\delta t^{s-D-1}(\log t^{-1})^m G(\log t^{-1})\,\mathrm{d}t.$$

The function $z_0(s)$ is the exact counterpart of $\zeta_1(s)$ from the proof of Theorem 2.3.25, with $|A_t|$ replaced by $|A_t \cap \Omega|$ and where, much as in that proof, $\tilde{\zeta}_{A,\Omega} = \zeta_1 + \zeta_2$ and ζ_2 is an entire function. It is easy to see that $z_m(s) = (-1)^m z_0^{(m)}(s)$, therefore, the functions $z_m(s)$ and $z_0(s)$ have the same meromorphic extension, and the same sets of poles. This proves that $\tilde{\zeta}_{A,\Omega}(s)$ can be meromorphically extended from $\{\operatorname{Re} s > D\}$ to the half-plane $\{\operatorname{Re} s > D - \alpha\}$. The set of poles (complex dimensions of (A,Ω)) of the relative zeta function of (A,Ω), contained in this half-plane, is given by

$$\begin{aligned}\mathscr{P}(\tilde{\zeta}_{A,\Omega}) &= \mathscr{P}(z_m) = \mathscr{P}(z_0)\\ &= \left\{ s_k = D + \frac{2\pi}{T} k\mathrm{i} \in \mathbb{C} : \hat{G}_0\left(\frac{k}{T}\right) \neq 0,\ k \in \mathbb{Z} \right\}.\end{aligned}$$

Each of these poles is simple. Furthermore, if

$$z_0(s) = \sum_{j=-1}^{\infty} a_j^{(k)} (s - s_k)^j$$

is the Laurent series of $z_0(s)$ in a neighborhood of $s = s_k$, then

$$z_0^{(m)}(s) = (-1)^m m!\, a_{-1}^{(k)} (s - s_k)^{-m-1} + \sum_{j=0}^{\infty} \frac{(m+j)!}{j!} a_{m+j}^{(k)} (s - s_k)^j.$$

Hence,

$$c_{-m-1}^{(k)} = m!\, a_{-1}^{(k)} = m! \frac{1}{T} \hat{G}_0\left(\frac{k}{T}\right),$$

where, in the last equality, we have used (2.3.33). The remaining claims are easily deduced from the corresponding ones in Theorem 2.3.25. □

Remark 4.5.3. In Equation (4.5.28), $\widetilde{\mathscr{M}} = \widetilde{\mathscr{M}}^D(A,\Omega)$, the *average Minkowski content* of (A,Ω), is defined as the multiplicative Cesàro average of $t^{-(N-D)}|A_t \cap \Omega|$:

$$\widetilde{\mathscr{M}}^D(A,\Omega) := \lim_{\tau \to +\infty} \frac{1}{\log \tau} \int_{1/\tau}^{1} \frac{|A_t \cap \Omega|}{t^{N-D}} \frac{\mathrm{d}t}{t}, \tag{4.5.30}$$

provided the limit exists in $[0,+\infty]$. See Section 2.4, Equation (4.5.9) and compare with [Lap-vFr3, Definition 8.29, Equation (8.55)].

Remark 2.3.19 also applies to Theorems 4.5.1 and 4.5.2. This means that in the statements of these theorems, we may have more general functions, which are $O(t^{\alpha)})$ (instead of $O(t^{\alpha})$) as $t \to 0^+$, like $t^\alpha \log(1/t)$.

Remark 4.5.4. In light of the functional equation (4.5.2) connecting $\zeta_{A,\Omega}$ and $\tilde{\zeta}_{A,\Omega}$, Theorems 4.5.1 and 4.5.2 also hold for relative *distance* zeta functions (instead of relative tube zeta functions), provided $D < N$, and in that case, all of the expressions for the residues and the Laurent coefficients are multiplied by $N - D$.

Example 4.5.5. (*Torus relative fractal drum*). Let Ω be an open torus in $\mathbb{R}^3$ defined by two radii r and R, where $0 < r < R < \infty$, and let $A := \partial\Omega$ be its topological boundary. In order to compute the tube zeta function of the *torus RFD* (A,Ω), we first compute its tube function. Let $\delta \in (0,r)$ be fixed. Using Cavalieri's principle, we obtain that

$$|A_t \cap \Omega|_3 = 2\pi R \left(r^2 - (r-t)^2\right) = 2\pi R(2rt - t^2), \tag{4.5.31}$$

where $t \in (0,\delta)$, from which it follows that

$$\tilde{\zeta}_{A,\Omega}(s) := \int_0^\delta t^{s-4}|A_t \cap \Omega|_3\,dt = 2\pi R\Big(2r\frac{\delta^{s-2}}{s-2} - \frac{\delta^{s-1}}{s-1}\Big) \tag{4.5.32}$$

for all $s \in \mathbb{C}$ such that $\operatorname{Re} s > 2$. The right-hand side defines a meromorphic function on the entire complex plane; so that, by the principle of analytic continuation, $\tilde{\zeta}_{A,\Omega}$ can be (uniquely) meromorphically extended to the whole of $\mathbb{C}$. In particular, we see that the multiset of complex dimensions of the torus RFD (A,Ω) is given by $\mathscr{P}(A,\Omega) = \{1,2\}$. Each of the complex dimensions 1 and 2 is simple. In particular, we have that

$$\dim_{PC}(A,\Omega) = \{2\} \quad \text{and} \quad \operatorname{res}(\tilde{\zeta}_{A,\Omega},2) = 4\pi Rr. \tag{4.5.33}$$

Also, $\overline{\dim}_B(A,\Omega) = D(\tilde{\zeta}_A) = 2$. From Equation (4.5.14) appearing in Theorem 4.5.1 below, we conclude that the 2-dimensional Minkowski content of the torus RFD (A,Ω) is given by

$$\mathscr{M}^2(A,\Omega) = 4\pi Rr. \tag{4.5.34}$$

Since $|A_t|_3 = 2\pi R\big((r+t)^2 - (r-t)^2\big)$, we can also easily compute the 'ordinary' tube zeta function $\tilde{\zeta}_A$ of the torus surface A in $\mathbb{R}^3$:

$$\tilde{\zeta}_A(s) = 8\pi Rr\frac{\delta^{s-2}}{s-2} \tag{4.5.35}$$

for all $s \in \mathbb{C}$. In particular, $\operatorname{res}(\tilde{\zeta}_A,2) = 8\pi Rr$. Using Equations (4.5.2) and (2.2.1), from (4.5.35) we obtain the corresponding distance zeta functions for all $s \in \mathbb{C}$:

$$\zeta_{A,\Omega}(s) = 2\pi R\Big(2r\frac{\delta^{s-2}}{s-2} - \frac{2\delta^{s-1}}{s-1}\Big), \quad \zeta_A(s) = 8\pi Rr\frac{\delta^{s-2}}{s-2}. \tag{4.5.36}$$

Also,

$$\mathscr{P}(\zeta_{A,\Omega}) = \mathscr{P}(\tilde{\zeta}_{A,\Omega}) = \{1,2\}$$

and

$$\mathscr{P}_c(\zeta_{A,\Omega}) = \mathscr{P}_c(\tilde{\zeta}_{A,\Omega}) = \{2\}$$

(with each pole 1 and 2 being simple) and

$$\overline{\dim}_B(A,\Omega) = D(\zeta_{A,\Omega}) = D(\tilde{\zeta}_{A,\Omega}) = 2.$$

Furthermore, we see that $\operatorname{res}(\zeta_{A,\Omega},2) = 4\pi Rr$ and $\operatorname{res}(\zeta_A,2) = 8\pi Rr$, in agreement with Equation (4.5.8), while

$$\mathscr{P}(\zeta_A) = \mathscr{P}(\tilde{\zeta}_A) = \mathscr{P}_c(\zeta_A) = \mathscr{P}_c(\tilde{\zeta}_A) = \{2\}.$$

One can easily extend the example of the 2-torus to any (smooth) closed, compact submanifold of $\mathbb{R}^N$ (and, in particular, of course, to the n-torus, with $n \geq 2$).

This can be done by using Federer's tube formula [Fed1] for sets of positive reach, which extends and unifies Weyl's tube formula [Wey3] for smooth compact submanifolds of $\mathbb{R}^N$ and Steiner's formula (obtained by Steiner and his successors [Stein]) for compact convex subsets of $\mathbb{R}^N$. The global form of Federer's tube formula expresses the volume of t-neighborhoods of a (compact) set of positive reach $A \subset \mathbb{R}^N$ as a polynomial of degree at most N in t, whose coefficients are (essentially) the so-called *Federer's curvatures* and which generalize Weyl's curvatures in [Wey3] (see [BergGos] for an exposition) and Steiner's curvatures in [Stein] (see [Schn2, Chapter 4] for a detailed exposition) in the case of compact submanifolds of $\mathbb{R}^N$ and compact convex sets, respectively.

Recall from [Fed1] that a closed subset A of $\mathbb{R}^N$ is said to be of *positive reach* if there exists $\delta_0 > 0$ such that every point $x \in \mathbb{R}^N$ within a distance less than δ_0 from A has a unique metric projection onto A; see [Fed1] and, e.g., [Schn2]. The *reach* of A, denoted by $\operatorname{reach}(A)$, is then defined as the supremum of all such positive numbers δ_0. Clearly, every closed convex subset of $\mathbb{R}^N$ is of infinite (and hence, positive) reach. Furthermore, if $A \subset \mathbb{R}^2$ is an arc of a circle of radius r, then the reach of A is equal to r.

In the present context, for a compact set of positive reach $A \subset \mathbb{R}^N$, it is easy to deduce from the tube formula in [Fed1] an explicit expression for $\tilde{\zeta}_A$.[37]

Theorem 4.5.6. *Let A be a (nonempty) compact set of positive reach in $\mathbb{R}^N$. Then, for any δ such that $0 < \delta < \operatorname{reach}(A)$, we have that*

$$\tilde{\zeta}_A(s) := \tilde{\zeta}_A(s;\delta) = \sum_{k=0}^{N} c_k \frac{\delta^{s-k}}{s-k}, \tag{4.5.37}$$

where $|A_t| = \sum_{k=0}^{N} c_k t^{N-k}$ for all $t \in (0,\delta)$ and the coefficients c_k are the (normalized) Federer curvatures. (From the functional equation (4.5.2) above, one then deduces at once a corresponding explicit expression for $\zeta_A(s) := \zeta_A(s;\delta)$.)

Hence, $\dim_B A$ exists and

$$D := D(\tilde{\zeta}_A) = D(\zeta_A) = \dim_B A = \max\{k \in \{0,1,\dots,N\} : c_k \neq 0\} \tag{4.5.38}$$

and[38]

$$\mathscr{P} := \mathscr{P}(\tilde{\zeta}_A) = \mathscr{P}(\zeta_A) \subseteq \{0,1,\dots,N\}. \tag{4.5.39}$$

In fact,

$$\mathscr{P} = \big\{k \in \{0,1,\dots,N\} : c_k \neq 0\big\} \subseteq \{k_0,\dots,D\}, \tag{4.5.40}$$

where $k_0 := \min\big\{k \in \{0,1,\dots,D\} : c_k \neq 0\big\}$. Furthermore, each of the complex dimensions of A is simple.

---

[37] Relative versions of Theorem 4.5.6 are also possible, but we will not consider them here.

[38] More precisely, the second equality in Equation (4.5.39) holds only if $D < N$.

Finally, if A is such that its affine hull is all of $\mathbb{R}^N$ (which is the case when the interior of A is nonempty and, in particular, if A is a convex body), then $D = N$, while if A is a (smooth) compact d-dimensional submanifold (with $0 \le d \le N$), then $D = d$.

For the 2-torus $A \subset \mathbb{R}^3$, we have $N = 3$, $D = 2$ (since the Euler characteristic of A is equal to zero), $c_2 \neq 0$,[39] $c_1 = 0$, and hence, $c_0 = 0$, $k_0 = 2$ and $\mathscr{P} = \{2\}$, as was also found in the last displayed equation of Example 4.5.5 via a direct computation.

We note that much more general tube formulas, called (as in [Lap-vFr2–3] and [LapPeWi1–2]) "fractal tube formulas", are obtained in [LapRaŽu5] (as announced in [LapRaŽu4]), as well as in Chapter 5 below, for arbitrary bounded sets (and even more generally, RFDs) in $\mathbb{R}^N$, under mild growth assumptions on the associated fractal zeta functions.

4.5.2 Precise Meromorphic Extensions of Zeta Functions of Countable Unions of Relative Fractal Drums

In Theorem 4.5.8, we construct a class of RFDs in $\mathbb{R}$, with prescribed values of the abscissa of meromorphic continuation of the corresponding zeta functions. This will enable us to construct a class of bounded sets A, with prescribed values of the abscissa of meromorphic continuation of the associated distance or tube zeta functions; see Theorem 4.5.20. The construction makes use of the generalized Cantor sets $C^{(a)}$ introduced in Example 2.2.6.

Definition 4.5.7. Let (A_j, Ω_j), $j \ge 1$, be a given sequence of RFDs in $\mathbb{R}^N$, where $(\Omega_j)_{j\ge1}$ is a disjoint sequence of open subsets of $\mathbb{R}^N$. We define the *union of the relative fractal drums*

$$(A, \Omega) = \bigcup_{j=1}^{\infty} (A_j, \Omega_j),$$

by $A := \cup_{j=1}^{\infty} A_j$ and $\Omega := \cup_{j=1}^{\infty} \Omega_j$, assuming that there exists $\delta > 0$ such that $\Omega \subset A_\delta$ and $|\Omega| < \infty$.

Theorem 4.5.8. *Let $D \in (0,1)$ and $\alpha \in (0, D)$ be prescribed. Let (A, Ω) be a relative fractal drum, defined by $(A, \Omega) = \cup_{j\ge1}(A_j, \Omega_j)$, where $(\Omega_j)_{j\ge1}$ is a family of disjoint open intervals in $\mathbb{R}$, $|\Omega_j| = 2^{-j}$, $A^{(j)} = 2^{-j} C^{(a_j)} + \inf \Omega_j$, and $C^{(a_j)}$ are generalized Cantor sets described in Example 2.2.6, with $a_j \in (0, 1/2)$, $j \ge 1$. Assume that $a_1 = 2^{-1/D}$, and let $(a_j)_{j\ge2}$ be an increasing sequence of positive numbers converging to $2^{-1/(D-\alpha)}$ as $j \to \infty$.*

[39] Note that c_2 is just proportional to the area of the 2-torus, with the proportionality constant being a standard positive constant.

Then, for the relative tube zeta function $\tilde{\zeta}_{A,\Omega}$ *of* (A,Ω)*, we have:*

$$D(\tilde{\zeta}_{A,\Omega}) = D, \quad D_{\mathrm{mer}}(\tilde{\zeta}_{A,\Omega}) = D - \alpha. \tag{4.5.41}$$

(See Definition 2.1.53.) Analogous result holds for the distance zeta function:

$$D(\zeta_{A,\Omega}) = D, \quad D_{\mathrm{mer}}(\zeta_{A,\Omega}) = D - \alpha. \tag{4.5.42}$$

The set of poles of these zeta functions, contained in $\{\mathrm{Re}\, s > D - \alpha\}$*, coincides with the set* $\dim_{PC}(A,\Omega) = \mathscr{P}_c(\zeta_{A,\Omega})$ *of principal complex dimensions of the relative fractal drum* (A,Ω)*:*

$$\dim_{PC}(A,\Omega) = D + \frac{2\pi}{\log(1/a_1)}\mathrm{i}\mathbb{Z}. \tag{4.5.43}$$

In particular, the oscillatory period of the RFD (A,Ω) *is given by* $\mathbf{p} = 2\pi D/\log 2$.[40]

In order to prove Theorem 4.5.8, we shall need the following technical lemma.

Lemma 4.5.9. *Let* $(A_j,\Omega_j)_{j\ge1}$ *be a sequence of RFDs in* $\mathbb{R}^N$ *such that the family of open sets* $(\Omega_j)_{j\ge1}$ *is disjoint. Consider their union* $(A,\Omega) = \cup_{j=1}^{\infty}(A_j,\Omega_j)$*, as introduced in Definition 4.5.7, and assume that* $|\Omega| < \infty$*. If*

$$\partial\Omega_j \subseteq A_j \quad \textit{for all} \quad j \in \mathbb{N}, \tag{4.5.44}$$

then

$$|A_t \cap \Omega| = \sum_{j=1}^{\infty} |(A_j)_t \cap \Omega_j|. \tag{4.5.45}$$

In particular,

$$\tilde{\zeta}_{A,\Omega}(s) = \sum_{j=1}^{\infty} \tilde{\zeta}_{A_j,\Omega_j}(s) \tag{4.5.46}$$

for all $s \in \mathbb{C}$ *such that* $\mathrm{Re}\, s > \overline{\dim}_B(A,\Omega)$.

Proof. For any $j \neq k$ and $a \in A_j$, since $a \notin \Omega_j$, we obviously have $B_t(a) \cap \Omega_k \subset (\partial A_k)_t \cap \Omega_k$. Taking the union over all $a \in A_j$, we obtain

$$(A_j)_t \cap \Omega_k \subseteq (\partial\Omega_k)_t \cap \Omega_k.$$

Using (4.5.44), we see that $(A_j)_t \cap \Omega_k \subseteq (A_k)_t \cap \Omega_k$, and hence,

$$\begin{aligned} A_t \cap \Omega &= \Big(\bigcup_{j=1}^{\infty}(A_j)_t\Big) \cap \Big(\bigcup_{k=1}^{\infty}\Omega_k\Big) \\ &= \bigcup_{j,k=1}^{\infty}((A_j)_t) \cap \Omega_k) = \bigcup_{k=1}^{\infty}(A_k)_t \cap (\Omega_k)). \end{aligned}$$

[40] Compare with Equation (2.2.17) on page 117. It is interesting to note that $\mathbf{p} \to 0^+$ as $D \to 0^+$; see Figure 4.16.

Since the family $(\Omega_k)_{k\geq 1}$ is disjoint, this implies (4.5.45). From this we conclude that for any positive real number s such that $s > \overline{\dim}_B(A,\Omega)$,

$$\begin{aligned}\tilde{\zeta}_{A,\Omega}(s) &= \int_0^\delta t^{s-N-1}|A_t\cap\Omega|\,\mathrm{d}t = \int_0^\delta t^{s-N-1}\Big(\sum_{k=1}^\infty |(A_k)_t\cap\Omega_k|\Big)\,\mathrm{d}t \\ &= \sum_{k=1}^\infty \int_0^\delta t^{s-N-1}|(A_k)_t\cap\Omega_k|\,\mathrm{d}t = \sum_{k=1}^\infty \tilde{\zeta}_{A_k,\Omega_k}(s).\end{aligned}$$

Hence, (4.5.46) holds for s real such that $s > \overline{\dim}_B(A,\Omega)$. But now, using the principle of analytic continuation, we can extend this identity to the open half-plane $\{\mathrm{Re}\, s > \overline{\dim}_B(A,\Omega)\}$, as desired. □

We shall also need the following technical lemma.

Lemma 4.5.10. *Let $D \in (0,1)$ and $\alpha \in (0,D)$ be given. Assume that $(T_j)_{j\geq 1}$ is a decreasing sequence of positive real numbers, converging to a limit which is less than $\pi/(D-\alpha)$. Let $(D_j)_{j\geq 1}$ be a bounded sequence of positive real numbers, and $(G_j)_{j\geq 1}$ a bounded sequence of periodic functions (with G_j being T_j-periodic, for each $j\geq 1$).*

Then, the sequence of functions $(z_j)_{j\geq 1}$, defined by (see (2.3.42) *and* (2.3.43)*)*

$$z_j(s) = \frac{\mathrm{e}^{T_j(s-D_j)}}{\mathrm{e}^{T_j(s-D_j)}-1}\int_{\log\delta^{-1}}^{\log\delta^{-1}+T_j} \mathrm{e}^{-\tau(s-D_j)}G_j(\tau)\,\mathrm{d}\tau \quad \text{for } j\geq 1, \tag{4.5.47}$$

is locally uniformly bounded on $\{0 < \mathrm{Re}\, s < D-\alpha+\varepsilon\}\setminus\overline{\mathscr{S}}$, where

$$\mathscr{S} := \bigcup_{j=1}^\infty \Big(D_j + \frac{2\pi}{T_j}\mathrm{i}\mathbb{Z}\Big) \tag{4.5.48}$$

and ε is a sufficiently small positive real number; that is, for each s_0 in the connected open set $\{0 < \mathrm{Re}\, s < D-\alpha+\varepsilon\}\setminus\overline{\mathscr{S}}$, there exists $M>0$ and a neighborhood $N = N(s_0)$ of s_0 such that $|z_j(s)|\leq M$ for all $j\in\mathbb{N}$ and $s\in N(s_0)$.[41]

Proof. There exists $k_0\in\mathbb{N}$ such that $T_j < \pi/(D-\alpha)$ for all $j\geq k_0$. Therefore, we can assume without loss of generality that $k_0 = 1$; that is, $T_j<\pi/(D-\alpha)$ for all $j\geq 1$. The sequences of real numbers $(T_j)_{j\geq 1}$ and $(D_j)_{j\geq 1}$ are bounded, as well as the sequence of functions $(G_j)_{j\geq 1}$. In light of (4.5.47), it suffices to prove that for any fixed complex number s_0 there exist a neighborhood $N(s_0)$ of s_0, and a positive number c, such that

$$|\mathrm{e}^{T_j(s-D_j)}-1|\geq c, \quad \text{for all } j\in\mathbb{N} \text{ and } s\in N(s_0). \tag{4.5.49}$$

Let us first fix $s_0 \in \{0<\mathrm{Re}\, s < D-\alpha+\varepsilon\}\setminus\mathscr{S}$. Furthermore, let us choose $s_{jk} = s_{jk}(s_0)\in\mathscr{P}_j$ which is closest to s_0. Let $R := d(s_0,S) = d(s_0,s_{jk})$. Then

[41] See also Figure 4.16 and the discussion surrounding Equations (4.5.58)–(4.5.60) below.

$$\begin{aligned}|e^{T_j(s_0-D_j)}-1| &= |e^{T_j(s_0-D_j)}-e^{T_j(s_{jk}-D_j)}| \\ &= |e^{T_j(s_{jk}-D_j)}|\,|e^{T_j(s_0-s_{jk})}-1| \\ &= |e^{T_j(s_0-s_{jk})}-1|.\end{aligned}$$

Let us write $s_0 - s_{jk} = Re^{i\varphi}$, and

$$\begin{aligned} w_j &:= e^{T_j(s_0-s_{jk})} = \exp(T_jRe^{i\varphi}) \\ &= e^{T_jR\cos\varphi}e^{iT_jR\sin\varphi} =: r_je^{i\psi_j}, \end{aligned}$$

where we have set

$$r_j = e^{T_jR\cos\varphi} \qquad \psi_j = T_jR\sin\varphi.$$

We assume without loss of generality that $R < D-\alpha+\varepsilon$, since it suffices to consider $0 < \operatorname{Re} s_0 < D-\alpha+\varepsilon$. We would like to estimate the value of $|w_j - 1|$ from below. Let us fix $\varphi_0 \in (0,\pi/2)$, and consider the following two cases:

Case 1: Assume that

$$\varphi \in (-\pi,\pi] \setminus \{(\frac{\pi}{2}-\varphi_0,\frac{\pi}{2}+\varphi_0)\cup(-\frac{\pi}{2}-\varphi_0,-\frac{\pi}{2}+\varphi_0)\}.$$

We consider the following two subcases:

(a) Assume that $\varphi \in [-\frac{\pi}{2}+\varphi_0,\frac{\pi}{2}-\varphi_0]$. Then

$$r_j = e^{T_jR\cos\varphi} \geq e^{T_jR\cos(\frac{\pi}{2}-\varphi_0)} = e^{T_jR\sin\varphi_0} > 1.$$

Hence,

$$|w_j-1| \geq |w_j|-1 = r_j-1 = e^{T_jR\sin\varphi_0}-1 \geq e^{T_0R\sin\varphi_0}-1 > 0. \tag{4.5.50}$$

(b) Assume that $\varphi \in (-\pi,-\frac{\pi}{2}+\varphi_0]\cup[\frac{\pi}{2}+\varphi_0,\pi]$. Then

$$r_j = e^{T_jR\cos\varphi} \leq e^{T_jR\cos(\frac{\pi}{2}+\varphi_0)} = e^{-T_jR\sin\varphi_0} \leq e^{-T_0R\sin\varphi_0} < 1.$$

Hence,

$$|w_j-1| \geq 1-|w_j| = 1-r_j \geq 1-e^{-T_0R\sin\varphi_0} > 0. \tag{4.5.51}$$

Case 2: Assume that $\varphi \in (\frac{\pi}{2}-\varphi_0,\frac{\pi}{2}+\varphi_0)\cup(-\frac{\pi}{2}-\varphi_0,-\frac{\pi}{2}+\varphi_0)$. Then we have

$$\psi_j = T_jR\sin\varphi \geq T_jR\sin(\frac{\pi}{2}-\varphi_0) = T_jR\cos\varphi_0 \geq T_0R\cos\varphi_0 > 0.$$

Since $T_0(D-\alpha+\varepsilon) < \pi$ for $\varepsilon > 0$ small enough, then for any j large enough we have

$$0 < \psi_j = T_jR\sin\varphi \leq T_jR \leq T_0(D-\alpha+\varepsilon) < \pi;$$

that is,

$$\psi_j \in [T_0 R \cos\varphi_0, T_0(D-\alpha+\varepsilon)] \subset (0,\pi),$$

and therefore,

$$\sin\psi_j \geq \min\{\sin(T_0 R\cos\varphi_0), \sin(T_0(D-\alpha+\varepsilon))\} = \sin(T_0 R\cos\varphi_0) > 0,$$

since $R < D-\alpha+\varepsilon$, for $\varepsilon > 0$ small enough.

If we consider a triangle with vertices at the points 0, 1 and w_j with respect to the (r_j, ψ_j)-polar system (the origin 0 of which is the point s_{jk}), it is clear that the length of the side of the triangle joining 1 with w_j is not smaller than the length of the height of the triangle drawn from 1 to the opposite side connecting 0 and w_j; that is,

$$|w_j - 1| \geq \sin\psi_j > 0.$$

Therefore,

$$|w_j - 1| \geq \sin(T_0 R\cos\varphi_0) > 0. \tag{4.5.52}$$

Making use of (4.5.50), (4.5.51) and (4.5.52), we obtain that

$$|\mathrm{e}^{T_j(s_0-D_j)} - 1| \geq g(s_0),$$

where

$$g(s_0) = \min\{\mathrm{e}^{T_0 d(s_0,S)\sin\varphi_0} - 1, 1-\mathrm{e}^{-T_0 d(s_0,S)\sin\varphi_0}, \sin(T_0 d(s_0,S)\cos\varphi_0)\}.$$

If we take s in a sufficiently small neighborhood $N(s_0)$ of s_0, such that $d(s,S) \geq d_0 > 0$ for some positive constant d_0, then the same type of inequality holds for all $s \in N(s_0)$:

$$|\mathrm{e}^{T_j(s-D_j)} - 1| \geq g(s),$$

where $\varphi_0 \in (0,\pi/2)$ is a fixed angle. The desired constant c is obtained as the infimum of the right-hand side over $s \in N(s_0)$:

$$|\mathrm{e}^{T_j(s-D_j)} - 1| \geq c := \inf_{s\in N(s_0)} g(s) \quad \text{for all } j \in \mathbb{N} \text{ and } s \in N(s_0). \tag{4.5.53}$$

More explicitly, if we let $d_0 := d(N(s_0),S) = \inf_{s\in N(s_0)} d(s,S) > 0$, then we may take

$$c := \min\{\mathrm{e}^{T_0 d_0 \sin\varphi_0} - 1, 1-\mathrm{e}^{-T_0 d_0\sin\varphi_0}, \sin(T_0 d_0\cos\varphi_0)\}. \tag{4.5.54}$$

This concludes the proof of Lemma 4.5.10. □

Proof of Theorem 4.5.8. Note that $|\Omega| = \sum_{j=1}^{\infty} 2^{-j} < \infty$, and $\Omega \subset A_\delta$ for any $\delta > 1/2$. The first equality in (4.5.41) follows from Theorem 4.1.7.

In order to prove the second equality in (4.5.41), we first find a periodic function $G(\tau)$ and $f(t) = O(t^\alpha)$ as $t \to 0^+$, such that

$$|A_t \cap \Omega| = t^{1-D}\left(G\left(\log\frac{1}{t}\right) + f(t)\right).$$

Since $(A_j)_t \subset \overline{\Omega}_j$, where $(A_j)_t$ denotes the t-neighborhood of A_j, we have

$$|(A_j)_t \cap \Omega_j| = 2^{-j} t^{1-D_j}\left(G_j\left(\log\frac{1}{t}\right) - 2t^{D_j}\right). \tag{4.5.55}$$

We note that this identity (called a *fractal tube formula* in [Lap-vFr3, Chapter 8]) is obtained in the same manner as in [Lap-vFr3, Equation (1.11)]; therefore, we will not repeat its proof. Here, $D_j = \dim_B(A_j, \Omega_j) = \log_{1/a_j} 2$, each function G_j is T_j-periodic, where $T_j := \log(1/a_j)$, and $G_j(\tau) \in [\mathscr{M}_*^{D_j}(A_j), \mathscr{M}^{*D_j}(A_j)]$ for every $\tau \in [0, T_j]$ (or equivalently, for all $\tau \in \mathbb{R}$), and the values of the Minkowski contents are given in (2.2.12). Let $D_1 := D$, and note that the sequence $(D_j)_{j\geq 2}$ is monotonically increasing in $(0, D-\alpha)$, and converging to $D-\alpha$.

Using Lemma 4.5.9 and (4.5.55), we obtain

$$\begin{aligned} |A_t \cap \Omega| &= \sum_{j=1}^{\infty} 2^{-j} t^{1-D_j}\left(G_j\left(\log\frac{1}{t}\right) - 2t^{D_j}\right) \\ &= t^{1-D}\left(2^{-1}G_1\left(\log\frac{1}{t}\right) + f(t)\right), \end{aligned} \tag{4.5.56}$$

where

$$f(t) := -t^D + \sum_{j=2}^{\infty} 2^{-j} t^{D-D_j}\left(G_j\left(\log\frac{1}{t}\right) - 2t^{D_j}\right).$$

Since $D - D_j > \alpha$ and $t < 1$, we have

$$|f(t)| \leq t^D + \sum_{j\geq 2} 2^{-j} t^{\alpha}(M+2) = (t^{D-\alpha} + M + 2)t^{\alpha} \leq (M+3)t^{\alpha},$$

where for every $\tau \in [0, T_j]$ (i.e., for every $\tau \in \mathbb{R}$),

$$\begin{aligned} 0 < G_j(\tau) \leq M &:= \sup_{j\geq 2} \mathscr{M}^{*D_j}(A^{(j)}) = \sup_{j\geq 2} 2(1-a_j)\left(\frac{1}{2} - a_j\right)^{D_j - 1} \\ &< 2(1-a_2)\left(\frac{1}{2} - 2^{-1/(D-\alpha)}\right)^{D_2 - 1}, \end{aligned}$$

since both $(a_j)_{j\geq 2}$ and $(D_j)_{j\geq 2}$ are increasing sequences; see (2.2.12). Therefore, $f(t) = O(t^{\alpha})$ as $t \to 0^+$, and we conclude from Theorem 4.5.2 that $D_{\mathrm{mer}}(\tilde{\zeta}_{A,\Omega}) \leq D - \alpha$.

To show the equality, it suffices to prove that $s = D - \alpha$ is a singularity which is not a pole of a meromorphic extension of $\tilde{\zeta}_{A,\Omega}$. More precisely, we show that $D - \alpha$ is a nonisolated singularity of a meromorphic extension of $\tilde{\zeta}_{A,\Omega}$, to which a sequence of distinct poles $(D_j)_{j\geq 2}$ converges from the left. Using the first equality in (4.5.56), we obtain the following identity valid on $\{\mathrm{Re}\, s > D\}$:

$$\begin{aligned}\tilde{\zeta}_{A,\Omega}(s) &= \int_0^\delta t^{s-2}|A_t \cap \Omega|\,\mathrm{d}t \\ &= \sum_{j\geq 1} 2^{-j}\int_0^\delta t^{s-D_j-1}G_j(\log t^{-1})\,\mathrm{d}t - 2\sum_{j\geq 1} 2^{-j}\frac{\delta^s}{s} \\ &= \sum_{j\geq 1} 2^{-s}z_j(s) - 2\frac{\delta^s}{s}.\end{aligned} \tag{4.5.57}$$

The functions $z_j(s)$ have meromorphic extensions to the entire complex plane; see the proof of Theorem 4.5.2. Furthermore, since

$$T_0 := \lim_{j\to\infty} T_j = \lim_{j\to\infty} \log(1/a_j) = \frac{\log 2}{D-\alpha} < \frac{\pi}{D-\alpha},$$

Lemma 4.5.10 shows us that the last series appearing in (4.5.57) converges to a function which is holomorphic on the connected open set

$$\{0 < \mathrm{Re}\, s < D - \alpha + \varepsilon\} \setminus \mathscr{S}_1,$$

for arbitrarily small $\varepsilon > 0$, where $\mathscr{S}_1$ is the set of singularities of $\tilde{\zeta}_{A,\Omega}(s)$ contained in the open right half-plane $\{\mathrm{Re}\, s > 0\}$. More specifically, $\mathscr{S}_1 = \overline{\mathscr{S}}$ (the closure of $\mathscr{S}$ in $\mathbb{C}$) is the closed subset of $\mathbb{C}$ given by

$$\mathscr{S}_1 = \mathscr{S} \cup \mathscr{A}, \tag{4.5.58}$$

where

$$\mathscr{S} := \bigcup_{j=2}^{\infty}\left(D_j + \frac{2\pi}{T_j}\mathrm{i}\mathbb{Z}\right) \tag{4.5.59}$$

and

$$\mathscr{A} := D - \alpha + \mathbf{p}\mathrm{i}\mathbb{Z}. \tag{4.5.60}$$

Here, $\mathscr{S}$ is the set of poles of $\tilde{\zeta}_{A,\Omega}(s)$ in $\{0 < \mathrm{Re}\, s < D - \alpha + \varepsilon\}$,[42] $\mathscr{A}$ is the set of nonisolated singular points (the accumulation points of $\mathscr{S}$) and $\mathbf{p} := 2\pi/T_0$. Therefore, the function $\tilde{\zeta}_{A,\Omega}(s)$, defined by the last expression in (4.5.57), possesses a holomorphic extension to $G = \{0 < \mathrm{Re}\, s < D - \alpha + \varepsilon\} \setminus \mathscr{S}_1$ (note that, as was stated earlier, G is a domain, that is, an open and connected subset of $\mathbb{C}$). In particular, $D - \alpha$ is a singularity of $\tilde{\zeta}_{A,\Omega}(s)$ which is not a pole. This proves that

$$D_{\mathrm{mer}}(\tilde{\zeta}_{A,\Omega}) = D - \alpha.$$

The analogous claims (made in the statement of the theorem) for relative distance zeta functions follow from (4.5.2). □

---

[42] The set $\mathscr{S} := \bigcup_{j\geq 2}(D_j + \frac{2\pi}{T_j}\mathrm{i}\mathbb{Z})$ in Equation (4.5.59) corresponds to the set $\mathscr{S}$ in Equation (4.5.48) of Lemma 4.5.10.

In connection with Equations (4.5.58)–(4.5.60), we point out that in (4.5.58), if we let $\mathscr{P}_j := D_j + \frac{2\pi}{T_j}\mathrm{i}\mathbb{Z}$ for each $j \geq 1$, then we have that $\mathscr{P}_j \to \mathscr{A}$ in the Hausdorff metric, as $j \to \infty$; see Figure 4.16. Note that the sequence $(T_j)_{j\geq 2}$ is decreasing, since a_j is increasing, and hence, the sequence $\mathbf{p}_j := \frac{2\pi}{T_j}$ of *oscillatory quasiperiods* of (A,Ω) is increasing. Also,

$$\mathbf{p}_j \to \mathbf{p} = \frac{2\pi}{T_0} \quad \text{as} \quad j \to \infty,$$

where $T_0 := \frac{\log 2}{D-\alpha}$. It is also interesting to note that, although $\mathrm{Mer}\,(\tilde{\zeta}_{A,\Omega}) = \{\mathrm{Re}\,s > D-\alpha\}$, the tube zeta function $\tilde{\zeta}_{A,\Omega}$ is meromorphic on $\{\mathrm{Re}\,s > 0\} \setminus \mathscr{A}$. Here, the set $\mathrm{Mer}\,\tilde{\zeta}_{A,\Omega}$ is the half-plane of meromorphic continuation introduced in Definition 2.1.53.

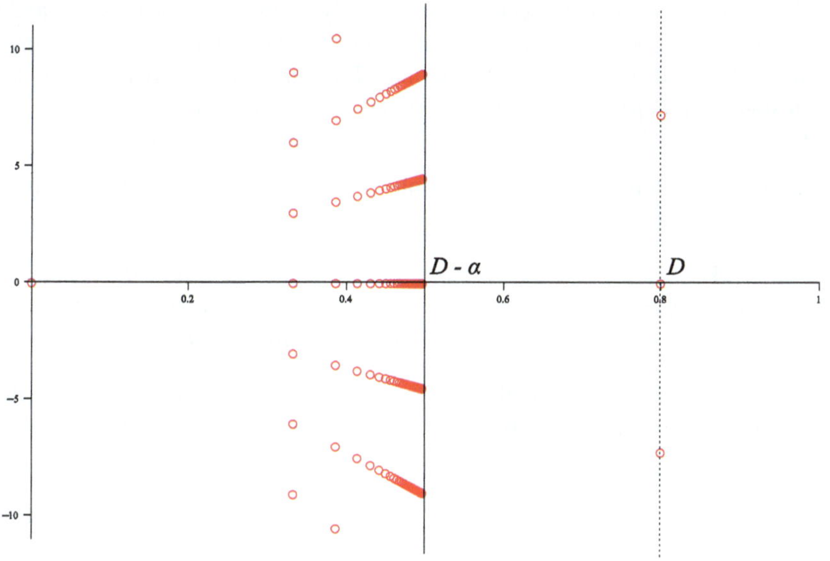

Fig. 4.16 An interesting set of complex dimensions: a sketch of the set $\mathscr{S} := \bigcup_{j\geq 2}(D_j + \frac{2\pi}{T_j}\mathrm{i}\mathbb{Z})$ of the poles of the tube zeta function (or "complex dimensions") of the relative fractal drum (A,Ω) from Theorem 4.5.8, with parameters $D = 4/5$, $\alpha = 3/10$ and the sequence $(a_j)_{j\geq 2}$ defined as $a_j = j/(4(j+1))$ for $j \geq 2$. Here, $D(\tilde{\zeta}_{A,\Omega}) = 4/5$, $D_{\mathrm{mer}}(\tilde{\zeta}_{A,\Omega}) = D - \alpha = 1/2$ and $\mathscr{A} = 2^{-1} + 4\pi(\log 2)\mathrm{i}\mathbb{Z}$ is the set of nonisolated singularities of $\tilde{\zeta}_{A,\Omega}$. (See Equations (4.5.58)–(4.5.60) and the discussion surrounding it.) It is easy to check that the set $\mathscr{S}$ appearing in Equations (4.5.58) and (4.5.59) of the proof of Theorem 4.5.8 is contained in a union of countably many rays emanating from the origin. The dotted vertical line is the critical line $\{\mathrm{Re}\,s = D\}$ of $\tilde{\zeta}_{A,\Omega}$, and to the left of it, the solid vertical line $\{\mathrm{Re}\,s = D-\alpha\}$ is the meromorphy critical line of $\tilde{\zeta}_{A,\Omega}$. It is worth pointing out that in the light of some of the results obtained in Chapters 4 and 5, we will suggest to extend the notion of "complex dimensions" from poles to nonremovable singularities of the associated fractal zeta function (here, $\tilde{\zeta}_{A,\Omega}$).

4.5.3 Precise Meromorphic Extensions of Zeta Functions of Countable Unions of Fractal Strings

In the sequel, an important role is played by the notion of countable union of a sequence of fractal strings, which we now introduce. It extends Definition 3.1.19, in which we have defined the union of two fractal strings.

Definition 4.5.11. Let $\mathscr{L}_j = (\ell_{jk})_{k\geq 1}$, $j \geq 1$, be a sequence of fractal strings in $\mathbb{R}$. The (disjoint) *union of fractal strings*, denoted by

$$\mathscr{L} = \bigsqcup_{j=1}^{\infty} \mathscr{L}_j, \tag{4.5.61}$$

is a new fractal string, defined as the multiset consisting of all $l \in \cup_{j=1}^{\infty} \mathscr{L}_j$, with the multiplicity of l equal to the sum of its multiplicities in all $\mathscr{L}_j$, $j \in \mathbb{N}$. Here, we assume that each $l \in \mathscr{L}$ belongs to at most finitely many fractal strings $\mathscr{L}_j$, and that $\mathscr{L}$ is a sequence of positive numbers converging to zero. Without these assumptions, the union of fractal strings is not well defined. Furthermore, we say that *the union of fractal strings is disjoint*, if for any two indices $j, j' \in \mathbb{N}$, the assumption that $j < j'$ implies that $\mathscr{L}_j \cap \mathscr{L}_{j'} = \emptyset$, where $\mathscr{L}_j$ and $\mathscr{L}_{j'}$ are viewed as ordinary sets.

The following lemma provides a simple construction of well defined countable unions of fractal strings.

Lemma 4.5.12. *Let $\mathscr{L}_j = (\ell_{jk})_{k\geq 1}$, $j \geq 1$, be a sequence of bounded fractal strings. If the sequence of the first elements of the fractal strings converges to zero (that is, if $\ell_{j1} \to 0^+$ as $j \to \infty$), then $\mathscr{L} := \sqcup_{j=1}^{\infty} \mathscr{L}_j$ is a well-defined fractal string.*

Proof. To show that any given element $\ell = \ell_{jk} \in \mathscr{L}$ is of finite multiplicity, it suffices to take $j_0 \in \mathbb{N}$ large enough, $j_0 = j_0(j,k)$, such that $\ell_{jk} > \ell_{j_0 1}$ (this is possible since $\ell_{j_0 1} \to 0^+$ as $j_0 \to \infty$). Then we have

$$\ell_{jk} \in \mathscr{L}_1 \sqcup \cdots \sqcup \mathscr{L}_{j_0 - 1}, \quad \ell_{jk} \notin \bigsqcup_{n=j_0}^{\infty} \mathscr{L}_n, \tag{4.5.62}$$

and hence, the multiplicity of ℓ_{jk} in $\mathscr{L}$ is equal to the sum of the multiplicities of this element in finitely many fractal strings, namely, $\mathscr{L}_1, \ldots, \mathscr{L}_{j_0 - 1}$.

We now show that $\mathscr{L}$ can be ordered as a nonincreasing sequence of positive real numbers $(\ell_m)_{m\geq 1}$, converging to zero. To see this, consider the following sequence of sets

$$\triangle\mathscr{L}_j := \{\ell_{j'k'} : \ell_{j+1,1} \leq \ell_{j'k'} < \ell_{j1}\}. \tag{4.5.63}$$

Here, we assume without loss of generality that the sequence $(\ell_{j1})_{j\geq 1}$ is nonincreasing. Therefore, the sets $\triangle\mathscr{L}_j$ are finite and pairwise disjoint. Furthermore, since ℓ_j converges to zero as $j \to \infty$, we have that

$$\mathscr{L} = \bigsqcup_{j=1}^{\infty} \triangle\mathscr{L}_j. \tag{4.5.64}$$

Here, the union of finite multisets is defined similarly to the union of fractal strings in Definition 4.5.11. Also note that $\min \triangle\mathscr{L}_j = \ell_{j+1,1} > \max \triangle\mathscr{L}_{j+1}$. The desired sequence $\mathscr{L} = (\ell_m)_{m\geq 1}$ is then obtained so that we first order $\triangle\mathscr{L}_1$ as a nonincreasing finite sequence, then continue with $\triangle\mathscr{L}_2$, and so on. □

Lemma 4.5.13. *Let $\mathscr{L}_j$, $j \geq 1$, be a sequence of fractal drums associated with (generalized) Cantor RFDs (A_j, Ω_j) in $\mathbb{R}$, where $(\Omega_j)_{\geq 1}$ is a disjoint family of unit intervals in $\mathbb{R}$, $|\Omega_j| = 2^{-j}$, $A_j = 2^{-j}C^{(a_j)} + \inf \Omega_j$, and $a_j \in (0, 1/2)$ for each $j \geq 1$. Then $\mathscr{L} := \sqcup_{j=1}^{\infty} \mathscr{L}_j$ is a well-defined fractal string.*

Proof. Recall that $\mathscr{L}_j = (\ell_{jk})_{k\geq 1}$ is defined by $\ell_{jk} = |I_{jk}|$, where $(I_{jk})_{k\geq 1}$ is the family of connected components (open intervals) of $\overline{\Omega}_j \setminus A_j = \cup_{k\geq 1} I_{jk}$. We have

$$\ell_{j1} = 2^{-j}(1 - 2a_j) < 2^{-j}, \tag{4.5.65}$$

and hence, $\ell_{j1} \to 0^+$ as $j \to \infty$. The claim now follows from Lemma 4.5.12. □

In the following lemma, we construct a disjoint union of fractal strings, in the sense of Definition 4.5.11. It admits many variations, which we do not discuss here. By $(p_j)_{j\geq 1}$ we denote the usual sequence of prime numbers: $(2, 3, 5, 7, 11, \ldots)$. We construct a disjoint sequence of fractal drums $\mathscr{L}_j = (\ell_{jk})_{k\geq 1}$, $j \in \mathbb{N}$, associated with generalized Cantor sets $C^{(a_j)}$ (see Example 2.2.6), with a suitable choice of the numbers $a_j \in (0, 1/2)$.

Lemma 4.5.14. *Let $\mathscr{L}_j = (\ell_{jk})_{k\geq 1}$, $j \geq 2$, be a sequence of (scaled) Cantor strings, generated by relative fractal drums (A_j, Ω_j), $j \geq 2$, where $(\Omega_j)_{j\geq 1}$ is a disjoint family of intervals in $\mathbb{R}$, $A_j = 2^{-j}C^{(a_j)} + \inf \Omega_j$, and Ω_j is an open interval such that $|\Omega_j| = 2^{-j}$ for each $j \geq 2$. Assume that*

$$a_j = \frac{n_j}{p_j} \quad \textit{for} \quad j \geq 2, \tag{4.5.66}$$

where p_j is the j-th prime number,[43] *and $n_j \in \mathbb{N}$ is such that $n_j < \frac{1}{2}p_j$. Then the fractal string $\mathscr{L} := \sqcup_{j=2}^{\infty} \mathscr{L}_j$ is well defined, and moreover, the union of fractal strings is disjoint. In other words, for every $j, k \geq 1$, each value $\ell_{jk} \in \mathscr{L}$ occurs with the multiplicity 2^{k-1} in $\mathscr{L}$, the same multiplicity as in $\mathscr{L}_j$.*

Proof. From the construction of the Cantor string $\mathscr{L}_j$, we know that

$$\ell_{jk} = 2^{-j} a_j^{k-1}(1 - 2a_j). \tag{4.5.67}$$

---

[43] Here, the sequence of prime numbers is written in increasinig order: $p_1 < p_2 < \cdots < p_j < \ldots$, with $p_j \to \infty$ as $j \to \infty$.

Assume, contrary to the claim, that there exists a pair of indices $j < j'$, such that $\mathscr{L}_j \cap \mathscr{L}_{j'} \neq \emptyset$. In other words, $\ell_{jk} = \ell_{j'k'}$ for some $k, k' \in \mathbb{N}$; that is,

$$2^{-j} a_j^{k-1}(1-2a_j) = 2^{-j'} a_{j'}^{k'-1}(1-2a_{j'}).$$

Using $a_j = n_j/p_j$ and $a_{j'} = n_{j'}/p_{j'}$, we obtain

$$2^{j'-j} n_j^{k-1}(p_j - 2n_j) p_{j'}^{k'} = n_{j'}^{k'-1}(p_{j'} - 2n_{j'}) p_j^k.$$

However, this is impossible since the prime number $p_{j'}$ divides the left-hand side, but not the right-hand side. Indeed, $p_{j'}$ divides neither $n_{j'}$, nor $p_{j'} - 2n_{j'}$, nor p_j. □

Lemma 4.5.15. *Assume that the union $\mathscr{L} = \sqcup_{j=1}^{\infty} \mathscr{L}_j$ of a sequence of fractal strings $(\mathscr{L}_j)_{j\geq 1}$ is well defined, and that it is bounded. Then*

$$\zeta_{\mathscr{L}}(s) = \sum_{j=1}^{\infty} \zeta_{\mathscr{L}_j}(s) \tag{4.5.68}$$

for all $s \in \mathbb{C}$ such that $\operatorname{Re} s > D(\zeta_{\mathscr{L}})$. *Furthermore,* $D(\zeta_{\mathscr{L}}) \geq \sup_{j\geq 1} D(\zeta_{\mathscr{L}_j})$.

Proof. If $\mathscr{L}_j = (\ell_{jk})_{k\geq 1}$, then clearly, $\mathscr{L} = (\ell_{jk})_{j,k\geq 1}$. We have

$$\begin{aligned} \zeta_{\mathscr{L}}(s) &= \sum_{j,k\geq 1} \ell_{jk}^s \quad \text{on} \quad \{\operatorname{Re} s > D(\zeta_{\mathscr{L}})\}, \\ \zeta_{\mathscr{L}_j}(s) &= \sum_{k=1}^{\infty} \ell_{jk}^s \quad \text{on} \quad \{\operatorname{Re} s > D(\zeta_{\mathscr{L}_j})\}, \end{aligned} \tag{4.5.69}$$

for all $j \in \mathbb{N}$. The identity (4.5.68) now follows, as we now explain. Indeed, the two series appearing in (4.5.69) are absolutely convergent, and $D(\zeta_{\mathscr{L}_j}) \leq D(\zeta_{\mathscr{L}})$, for all $j \geq 1$, since $(A_j, \Omega_j) \subseteq (A, \Omega)$ implies that

$$D(\zeta_{\mathscr{L}_j}) = \overline{\dim}_B(A_j, \Omega_j) \leq \overline{\dim}_B(A, \Omega) = D(\zeta_{\mathscr{L}}).$$

□

Definition 4.5.16. Assume that Ω is a bounded interval in $\mathbb{R}$, and $A \subseteq \overline{\Omega}$ is such that A is closed (in $\mathbb{R}$), $|A| = 0$ and $\partial\Omega \subseteq A$. We say that a *fractal string $\mathscr{L}$ is associated with a given relative fractal drum* (A, Ω) in $\mathbb{R}$ if $\mathscr{L} = (\ell_k)_{k\geq 1}$, where $\ell_k := |J_k|$ for each $k \geq 1$, and $(J_k)_{k\geq 1}$ is the disjoint family of all the connected components (i.e., open intervals) of the open set $\Omega \setminus A \subseteq \mathbb{R}$.

Proposition 4.5.17. *Let (A_j, Ω_j) be a sequence of RFDs in $\mathbb{R}$ such that $(\Omega_j)_{j\geq 1}$ is a family of disjoint open intervals, and $\partial\Omega_j \subset A_j \subset \overline{\Omega}_j$, A_j is closed (in $\mathbb{R}$) and $|A_j| = 0$ for each $j \in \mathbb{N}$. Let $(A, \Omega) = \cup_{j=1}^{\infty}(A_j, \Omega_j)$, $|\Omega| < \infty$, and let $\mathscr{L}_j$ be the fractal strings associated with the RFDs (A_j, Ω_j), $j \in \mathbb{N}$. Assume that the sequence $\mathscr{L}_j = (\ell_{jk})_{k\geq 1}$ is nonincreasing for each $j \geq 1$, and is such that the union $\mathscr{L} :=$*

$\sqcup_{j=1}^{\infty}\mathscr{L}_j$ *of fractal strings is well defined (see Definition 4.5.11). If* $\delta > \frac{1}{2}\sup_{j\geq 1}\ell_{j1}$, *then*

$$\zeta_{\mathscr{L}}(s) = s(2\delta)^{s-1}|\Omega| + s(1-s)2^{s-1}\tilde{\zeta}_{A,\Omega}(s), \tag{4.5.70}$$

for every complex number s *in the open half-plane* $\{\operatorname{Re} s > D(\zeta_{A,\Omega})\}$.

Proof. It suffices to prove (4.5.70) on $\{\operatorname{Re} s > D(\zeta_{A,\Omega})\}$. For any such s, we have that

$$\zeta_{A_j,\Omega_j}(s) = \frac{2^{1-s}}{s}\zeta_{\mathscr{L}_j}(s). \tag{4.5.71}$$

This follows from (2.1.84), dropping the second term on the right-hand side, since we deal here with relative zeta functions. Furthermore, using (4.5.2) with $N = 1$, we have

$$\zeta_{A_j,\Omega_j}(s) = \delta^{s-1}|(A_j)_\delta \cap \Omega_j| + (1-s)\tilde{\zeta}_{A_j,\Omega_j}(s). \tag{4.5.72}$$

Note that since $\delta > \frac{1}{2}\ell_{j1}$ for all $j \in \mathbb{N}$, we have $(A_j)_\delta \cap \Omega_j = \Omega_j$. Therefore, we conclude from (4.5.71) and (4.5.72) that for each $j \geq 1$,

$$\zeta_{\mathscr{L}_j}(s) = s(2\delta)^{s-1}|\Omega_j| + s(1-s)2^{s-1}\tilde{\zeta}_{A_j,\Omega_j}(s).$$

The claim now follows by summing up over $j \geq 1$, and using Lemma 4.5.15. □

From previous considerations, it is easy to deduce the following result.

Corollary 4.5.18. *Let* (A_j, Ω_j) *be a sequence of RFDs in* $\mathbb{R}$, *such that* $(\Omega_j)_{j\geq 1}$ *is a disjoint family of open intervals, and* $|\Omega_j| = 2^{-j}$, $A_j = 2^{-j}C^{(a_j)} + \inf \Omega_j$. *Let* $\mathscr{L}_j$, $j \geq 1$, *be a sequence of fractal strings associated with RFDs* (A_j, Ω_j). *Then the fractal string* $\mathscr{L} := \sqcup_{j=1}^{\infty}\mathscr{L}_j$ *is well defined, and*

(*a*) *for every complex number* s *in* $\{\operatorname{Re} s > D(\zeta_{\mathscr{L}})\}$,

$$\zeta_{\mathscr{L}}(s) = \sum_{j=1}^{\infty} 2^{-js}\frac{(1-2a_j)^s}{1-2a_j^s}; \tag{4.5.73}$$

(*b*) *the distance zeta function of the relative fractal drum* $(A, \Omega) = \cup_{j=1}^{\infty}(A_j, \Omega_j)$ *is given by*

$$\zeta_{A,\Omega}(s) = \frac{2^{1-s}}{s}\sum_{j=1}^{\infty} 2^{-js}\frac{(1-2a_j)^s}{1-2a_j^s}, \tag{4.5.74}$$

on $\{\operatorname{Re} s > D(\zeta_{A,\Omega})\}$.

Proof. The fractal string $\mathscr{L}$ is well defined due to Lemma 4.5.12. The claim in part (*a*) follows from

$$\zeta_{\mathscr{L}_j}(s) = \sum_{k=1}^{\infty} 2^{k-1}(2^{-j}a_j^{k-1}(1-2a))^{js} = 2^{-js}\frac{(1-2a_j)^s}{1-2a_j^s},$$

by using Lemma 4.5.15. In order to prove part (*b*), it suffices to use part (*a*), along with (4.5.71). □

We next state the main result of this section, in which we construct a set A with a given specified value of the abscissa of meromorphic continuation of A. In order to do so, it will be convenient to use the following definition.

Definition 4.5.19. Let $\mathscr{L} = (\ell_{jk})_{j,k\geq 1}$ be a fractal string; that is, $\mathscr{L}$ is representable in the form $\mathscr{L} = (m_i)_{i\geq 1}$, where the sequence $(m_i)_{i\geq 1}$ is a nonincreasing reordering of $(\ell_{jk})_{j,k\geq 1}$. Then, the sequence $A = (a_j)_{j\geq 1}$ of positive real numbers is said to be *associated with the fractal string* $\mathscr{L}$ if $a_j := \sum_{i\geq j} m_i$ for each $j \geq 1$.

Theorem 4.5.20. *Let $D \in (0,1)$ and $\alpha \in (0,D)$ be given. Let $\mathscr{L} = (\ell_{jk})_{j,k=1}^{\infty}$ be a bounded fractal string defined as follows. For $j = 1$, we let $\ell_{1k} := a_1^{k-1}(1-2a_1)$, $k \geq 1$, where $a_1 := 2^{-1/D}$. For $j \geq 2$ and $k \geq 1$, we let*

$$\ell_{jk} := 2^{-j} a_j^{k-1}(1-2a_j). \tag{4.5.75}$$

Assume that the sequence $(a_j)_{j\geq 1}$ is increasing and converges to $2^{-1/(D-\alpha)}$ as $j \to \infty$. Then

$$D(\zeta_{\mathscr{L}}) = D, \quad D_{\text{mer}}(\zeta_{\mathscr{L}}) = D - \alpha. \tag{4.5.76}$$

Furthermore, if $A = A_{\mathscr{L}} := \{a_j : j \geq 1\} \subseteq (0,+\infty)$ is the bounded subset of $\mathbb{R}$ associated with the fractal string $\mathscr{L}$, then the same conclusion holds for the distance and tube zeta functions of A:

$$\begin{aligned} D(\zeta_A) = D(\tilde{\zeta}_A) &= D, \\ D_{\text{mer}}(\zeta_A) = D_{\text{mer}}(\tilde{\zeta}_A) &= D - \alpha. \end{aligned} \tag{4.5.77}$$

Moreover, $\dim_{PC} A = D + \frac{2\pi}{T_1}\mathrm{i}\mathbb{Z}$.

Proof. For $j = 1$, the associated fractal string $\mathscr{L}_1 = (\ell_{1k})_{k\geq 1}$ is the Cantor string generated by $A_1 = C^{(a_1)}$. We have $a_1 = 2^{-1/D} < 1/2$, so that the Cantor set $C^{(a_1)}$ is well defined. Furthermore, the box dimension of $C^{(a_1)}$ is given by $\log_{1/a_1} 2 = D$.

For $j \geq 2$, we have $a_j < 2^{-1/(D-\alpha)} < 2^{-1/(1-\alpha)} < 1/2$, so that the (scaled) Cantor sets $A_j = 2^{-j}C^{(a_j)} + \inf \Omega_j$, where $(\Omega_j)_{j\geq 1}$ is a family of disjoint open intervals in $\mathbb{R}$, $|\Omega_j| = 2^{-j}$, are also well defined. Lemma 4.5.13 then implies that the union of fractal strings $\mathscr{L} := \sqcup_{j=1}^{\infty} \mathscr{L}_j$ is well defined, where $\mathscr{L}_j$ are fractal strings associated with (A_j, Ω_j).

We have that

$$T_0 := \lim_j \log(1/a_j) = \log 2^{1/(D-\alpha)} = \frac{\log 2}{D-\alpha} < \frac{\pi}{D-\alpha},$$

so that Lemma 4.5.10 applies. The claim (4.5.76) follows from Theorem 4.5.8. The claims in (4.5.77) follow from Proposition 4.5.17 and (4.1.1), connecting the zeta function of a fractal string $\mathscr{L}$, the distance zeta function of the associated set $A = A_{\mathscr{L}}$, and the tube zeta function of A. □

The set A in Theorem 4.5.20 can be effectively constructed as a set associated with the fractal string $\mathscr{L} = \sqcup_{j=1}^{\infty} \mathscr{L}_j$, where each $\mathscr{L}_j$ is associated with a relative Cantor drum (A_j, Ω_j), described in the proof.

Theorem 4.5.20 shows, in particular, that our main results on the meromorphic extension of distance and tube zeta functions, obtained in Section 2.3, are in general optimal. We plan to study other applications and examples of relative zeta functions in a later work.

4.6 Transcendentally ∞-Quasiperiodic Relative Fractal Drums

One of the new notions explored and used in a key manner in this section is that of 'transcendentally quasiperiodic relative fractal drums', for which the corresponding quasiperiods are algebraically independent; see Section 4.6.1. It enables us, in particular, to construct bounded sets, fractal strings and RFDs that are 'maximally hyperfractal' (in the sense of the new Definition 4.6.23 below); that is, for which the corresponding fractal zeta function has nonisolated singularities at every point of the critical line $\{\mathrm{Re}\, s = D\}$ —and hence, for which the critical line is a (meromorphic) natural boundary (in the sense of part (*ii*) of Definition 1.3.8 in Subsection 1.3.2). The complexity or 'fractality' of the resulting geometric objects is therefore most extreme.

4.6.1 Quasiperiodic Relative Fractal Drums With Infinitely Many Algebraically Independent Quasiperiods

Here, we describe a general construction of quasiperiodic fractal drums possessing infinitely many algebraically incommensurable periods. It is based on properties of generalized Cantor sets, and on Baker's Theorem 3.1.14 from transcendental number theory; see [Ba, Theorem 2.1].

Let $m \geq 2$ be a given integer and $D \in (0,1)$ a given real number. Then for $a > 0$ defined by $a = m^{-1/D}$, we have $am = m^{1-1/D} < 1$, and hence, the generalized Cantor set $A = C^{(m,a)}$ is well defined (see Definition 3.1.1), and $\dim_B A = \log_{1/a} m = D$.

Definition 4.6.1. A finite set of real numbers is said to be *rationally* (*resp., algebraically*) *linearly independent* or simply, *rationally* (*resp., algebraically*) *independent*, if it is linearly independent over the field of rational (resp., algebraic) real numbers.

Definition 4.6.2. A sequence $(T_i)_{i\geq 1}$ of real numbers is said to be *rationally* (*resp., algebraically*) *linearly independent*, if any of its finite subsets is rationally (resp., algebraically) independent.

Definition 4.6.3. Let $m \geq 2$ be a positive integer. Let $\mathbf{p} = (p_i)_{i\geq 1}$ be the sequence of all prime numbers, arranged in increasing order; that is,

$$\mathbf{p} = (2, 3, 5, 7, 11, \ldots).$$

We then define the *exponent sequence* $\mathbf{e} = \mathbf{e}(m) := (\alpha_i)_{i\geq 1}$ *associated with* m, where $\alpha_i \geq 0$ is the multiplicity of p_i in the factorization of m. We also let

$$\mathbf{p}^{\mathbf{e}} := \prod_{\{i\geq 1:\alpha_i>0\}} p_i^{\alpha_i}. \tag{4.6.1}$$

The set of all sequences $\mathbf{e}$ with components in $\mathbb{N}_0 := \mathbb{N} \cup \{0\}$, such that all but at most finitely many components are equal to zero, is denoted by $(\mathbb{N}_0)_c^\infty$.

With this definition, for any integer $m \geq 2$, we obviously have $m = \mathbf{p}^{\mathbf{e}(m)}$. Conversely, any $\mathbf{e} \in (\mathbb{N}_0)_c^\infty$ defines a unique integer $m \geq 2$ such that $m = \mathbf{p}^{\mathbf{e}}$.

Definition 4.6.4. Given an exponent vector $\mathbf{e} = (\alpha_i)_{i\geq 1} \in (\mathbb{N}_0)_c^\infty$, we define the *support of* $\mathbf{e}$ as the set of all indices $i \in \mathbb{N}$ for which $\alpha_i > 0$, and we write

$$S(\mathbf{e}) = \operatorname{supp}(\mathbf{e}) = \{i \geq 1 : \alpha_i > 0\}. \tag{4.6.2}$$

The *support of an integer* $m \geq 2$ is denoted by $\operatorname{supp} m$ and defined by $\operatorname{supp} m := \operatorname{supp} \mathbf{e}(m)$.

The following definition is motivated by Theorem 3.1.15.

Definition 4.6.5. We say that a set $\{\mathbf{e}_i : i \geq 1\}$ of exponent vectors is *rationally linearly independent* if any of its finite subsets is linearly independent over $\mathbb{Q}$. We then say for short that the exponent vectors are rationally independent.

The following two definitions, Definition 4.6.6 and Definition 4.6.7, refine and extend the definition of n-quasiperiodic function and set (Definition 3.1.9 and Definition 3.1.11, respectively).

Definition 4.6.6. We say that a function $G : \mathbb{R} \to \mathbb{R}$ is ∞*-quasiperiodic*, if it is of the form

$$G(\tau) = H(\tau, \tau, \ldots),$$

where $H : \ell^\infty(\mathbb{R}) \to \mathbb{R}$,[44] $H = H(\tau_1, \tau_2, \ldots)$ is a function which is T_j-periodic in its j-th component, for each $j \in \mathbb{N}$, with $T_j > 0$ as minimal periods, and such that the set of periods

$$\{T_j : j \geq 1\} \tag{4.6.3}$$

is *rationally* independent. We say that the *order of quasiperiodicity* of the function G is equal to infinity (or that the function G is ∞*-quasiperiodic*).

In addition, we say that G is

(a) transcendentally quasiperiodic of infinite order (or *transcendentally* ∞*-quasiperiodic*) if the periods in (4.6.3) are *algebraically* independent;

---

[44] Here, $\ell^\infty(\mathbb{R})$ stands for the usual Banach space of bounded sequences $(\tau_j)_{j\geq 1}$ of real numbers, endowed with the norm $\|(\tau_j)_{j\geq 1}\|_\infty := \sup_{j\geq 1} |\tau_j|$.

(*b*) *algebraically quasiperiodic of infinite order* (or *algebraically ∞-quasiperiodic*) of infinite order if the periods in (4.6.3) are *rationally* independent and algebraically dependent.

We say that a sequence $(T_i)_{i\geq 1}$ of real numbers is *algebraically dependent* of infinite order if there exists a finite subset J of $\mathbb{N}$ such that $(T_i)_{i\in J}$ is algebraically dependent (that is, linearly dependent over the field of algebraic numbers). Recall that a finite set of real numbers $\{T_1,\dots,T_k\}$ is said to be *algebraically dependent* if there exist k algebraic real numbers $\lambda_1,\dots,\lambda_k$, not all of them equal to zero, such that $\lambda_1 T_1+\cdots+\lambda_k T_k=0$.

Definition 4.6.7. Let (A,Ω) be a relative fractal drum in $\mathbb{R}^N$ having the following tube formula:

$$|A_t\cap\Omega| = t^{N-D}(G(\log t^{-1})+o(1)) \quad \text{as} \quad t\to 0^+, \tag{4.6.4}$$

where $D\leq N$,[45] and G is a nonnegative function such that

$$0<\liminf_{\tau\to+\infty} G(\tau)\leq \limsup_{\tau\to+\infty} G(\tau)<\infty.$$

(Note that it then follows that $\dim_B(A,\Omega)$ exists and is equal to D. Moreover, $\mathscr{M}_*^D(A,\Omega)=\liminf_{\tau\to+\infty} G(\tau)$ and $\mathscr{M}^{*D}(A,\Omega)=\limsup_{\tau\to+\infty} G(\tau)$.)

We then say that the *relative fractal drum* (A,Ω) *in* $\mathbb{R}^N$ *is quasiperiodic* and of *infinite order of quasiperiodicity* (or, in short, *∞-quasiperiodic*) if the function $G=G(\tau)$ is ∞-quasiperiodic; see Definition 4.6.6.

In addition, (A,Ω) is said to be

(*a*) a *transcendentally ∞-quasiperiodic relative fractal drum* if the corresponding function G is transcendentally ∞-quasiperiodic;

(*b*) an *algebraically ∞-quasiperiodic relative fractal drum* if the corresponding function G is algebraically ∞-quasiperiodic.

Definition 4.6.8. We say that a relative fractal drum (A,Ω) is *n-quasiperiodic*, where $n\geq 2$, if the function G appearing in Definition 4.6.7 is n-quasiperiodic; see Definition 3.1.9. Likewise, one can define *transcendentally n-quasiperiodic relative fractal drums* and *algebraically n-quasiperiodic relative fractal drums.*

In light of Definitions 4.6.7 and 4.6.8, we see that each n-quasiperiodic relative fractal drum, where $n\in(\mathbb{N}\setminus\{1\})\cup\{\infty\}$, is either transcendentally n-quasiperiodic or algebraically n-quasiperiodic. In other words, the family $\mathscr{D}_{\rm qp}(n)$ of n-quasiperiodic RFDs is equal to the disjoint union of the family $\mathscr{D}_{\rm tqp}(n)$ of transcendentally n-quasiperiodic RFDs and the family $\mathscr{D}_{\rm aqp}(n)$ of algebraically n-quasiperiodic RFDs:

$$\mathscr{D}_{\rm qp}(n)=\mathscr{D}_{\rm tqp}(n)\cup\mathscr{D}_{\rm aqp}(n), \quad \text{for} \quad n\in(\mathbb{N}\setminus\{1\})\cup\{\infty\}.$$

[45] Here, D may be negative as well; see Proposition 4.1.35.

Letting

$$\mathscr{D}_{\rm qp} := \bigcup_{n\geq 2} \mathscr{D}_{\rm qp}(n), \quad \mathscr{D}_{\rm tqp} := \bigcup_{n\geq 2} \mathscr{D}_{\rm tqp}(n), \quad \mathscr{D}_{\rm aqp} := \bigcup_{n\geq 2} \mathscr{D}_{\rm aqp}(n)$$

and

$$\overline{\mathscr{D}}_{\rm qp} := \mathscr{D}_{\rm qp} \cup \mathscr{D}_{\rm qp}(\infty), \quad \overline{\mathscr{D}}_{\rm tqp} := \mathscr{D}_{\rm tqp} \cup \mathscr{D}_{\rm tqp}(\infty), \quad \overline{\mathscr{D}}_{\rm aqp} := \mathscr{D}_{\rm aqp} \cup \mathscr{D}_{\rm aqp}(\infty),$$

we have that

$$\overline{\mathscr{D}}_{\rm qp} = \overline{\mathscr{D}}_{\rm tqp} \cup \overline{\mathscr{D}}_{\rm aqp}.$$

Theorem 4.6.9 below shows that the family $\mathscr{D}_{\rm tqp}(\infty)$ is nonempty. Moreover, the family $\mathscr{D}_{\rm aqp}(n)$ of algebraically n-quasiperiodic RFDs is nonempty for any $n \geq 2$, as well as for $n = \infty$, as shown by Radunović in [Ra1].

As we know, the family of bounded fractal strings can be naturally embedded into the family of bounded subsets of $\mathbb{R}$, while the family of bounded subsets of $\mathbb{R}^N$ can be naturally embedded into the family of RFDs. Therefore, we have the following natural embeddings

$$\mathscr{L}_{\rm qp}(n) \subset \mathscr{S}_{\rm qp}(n) \subset \mathscr{D}_{\rm qp}(n). \tag{4.6.5}$$

It is clear that we can define the families $\mathscr{L}_{\rm qp}(\infty)$ and $\mathscr{S}_{\rm qp}(\infty)$, much as we have defined $\mathscr{D}_{\rm qp}(\infty)$ above. In light of the embedding (4.6.5), we then have

$$\mathscr{L}_{\rm qp}(\infty) \subset \mathscr{S}_{\rm qp}(\infty) \subset \mathscr{D}_{\rm qp}(\infty),$$

and analogously

$$\mathscr{L}_{\rm tqp}(\infty) \subset \mathscr{S}_{\rm tqp}(\infty) \subset \mathscr{D}_{\rm tqp}(\infty),$$
$$\mathscr{L}_{\rm aqp}(\infty) \subset \mathscr{S}_{\rm aqp}(\infty) \subset \mathscr{D}_{atp}(\infty).$$

Theorem 4.6.9 below shows that the family $\mathscr{L}_{\rm tqp}(\infty)$ is nonempty. Therefore, the families $\mathscr{S}_{\rm tqp}(\infty)$ and $\mathscr{D}_{\rm tqp}(\infty)$ are nonempyt as well.

The following result can be considered as a fractal set-theoretic interpretation of Baker's theorem [Ba, Theorem 2.1], i.e., of Theorem 2.11, from transcendental number theory. It provides a construction of a transcendentally ∞-quasiperiodic relative fractal drum. In particular, this drum possesses infinitely many algebraically incommensurable quasiperiods T_i. In our construction, we use the two-parameter family of generalized Cantor sets $C^{(m,a)}$ described in Definition 3.1.1 and whose basic properties are described in Proposition 3.1.2.

Theorem 4.6.9. *Let $D \in (0,1)$ be a given real number, and let $(m_i)_{i\geq 1}$ be a sequence of integers, $m_i \geq 2$. For each $i \geq 1$, define $a_i = m_i^{-1/D}$, and let $C^{(m_i,a_i)}$ be the corresponding generalized Cantor set (see Definition 3.1.1). Assume that $(\Omega_i)_{i\geq 1}$ is a family of disjoint open intervals on the real line such that $|\Omega_i| \leq C_1 m_i^{1-1/D} c_i^{1/D}$ for each $i \geq 1$, where the sequence $(c_i)_{i\geq 1}$ of positive real numbers is summable, and $C_1 > 0$. Let*

$$(A,\Omega) := \bigcup_{i\geq 1}(A_i,\Omega_i), \quad \text{where} \quad A_i := |\Omega_i|\, C^{(m_i,a_i)} + \inf\Omega_i, \quad \text{for all} \quad i \geq 1.$$[46]

Assume that the sequence of real numbers

$$(\log m_1, \ldots, \log m_n, \ldots) \text{ is rationally independent.} \tag{4.6.6}$$

Then the sequence of real numbers

$$\left(\frac{1}{D}, T_1, T_2, \ldots\right) \tag{4.6.7}$$

is algebraically independent (that is, lineary independent over the field of algebraic numbers). In other words, the relative fractal drum (A,Ω) is transcendentally quasiperiodic with infinite order of quasiperiodicity. More specifically, its sequence $(T_i)_{i\geq 1}$ of quasiperiods is given by $T_i := \log(1/a_i) = (\log m_i)/D$, for every $i \geq 1$. Furthermore,

$$D(\zeta_{A,\Omega}) = D_{\mathrm{mer}}(\zeta_{A,\Omega}), \tag{4.6.8}$$

and moreover, all of the points on the critical line $\{\mathrm{Re}\, s = D\}$ are nonisolated singularities of $\zeta_{A,\Omega}$; in other words, the relative fractal drum (A,Ω) is also maximally hyperfractal (in the sense of Definition 4.6.23(iii) below and the comment following it).

Finally, the relative fractal drum (A,Ω) is Minkowski nondegenerate, in the sense that

$$0 < \mathscr{M}_*^D(A,\Omega) \leq \mathscr{M}^{*D}(A,\Omega) < \infty.$$

Theorem 4.6.9 admits a partial extension. If instead of condition (4.6.6) we assume that $m_i \to \infty$ as $i \to \infty$, then (4.6.8) still holds, and, moreover, all the points of the critical line are nonisolated singularities of ζ_A. Furthermore, the fractal drum (A,Ω) is Minkowski nondegenerate.

We shall need the following lemma, which states a simple scaling property of the tube functions and Minkowski contents of RFDs. We note that the identity (4.6.10) below yields a partial extension of [Žu4, Proposition 4.4.]. Compare also with the scaling property of the corresponding distance zeta function $\zeta_{A,\Omega}$, obtained in Theorem 4.1.40.

Lemma 4.6.10. (*a*) *Let (A,Ω) be a relative fractal drum in $\mathbb{R}^N$. Then for any fixed $\lambda > 0$, and for all $t > 0$, we have that*

$$(\lambda A)_t \cap \lambda\Omega = \lambda(A_{t/\lambda} \cap \Omega), \quad |(\lambda A)_t \cap \lambda\Omega| = \lambda^N |A_{t/\lambda} \cap \Omega|. \tag{4.6.9}$$

Furthermore, for any real parameter $r \in \mathbb{R}$, we have the following scaling (or homogeneity) properties of relative Minkowski contents:

$$\mathscr{M}^{*r}(\lambda A, \lambda\Omega) = \lambda^r \mathscr{M}^{*r}(A,\Omega), \quad \mathscr{M}_*^r(\lambda A, \lambda\Omega) = \lambda^r \mathscr{M}_*^r(A,\Omega). \tag{4.6.10}$$

---

[46] Note that here, $|\Omega_i|$ plays the role of the scaling factor of the generalized Cantor set $C^{(m_i,a_i)}$.

(b) If A is a generalized Cantor set $C^{(m,a)}$ (see Proposition 3.1.2), then

$$|(\lambda C^{(m,a)})_t \cap (0,\lambda)| = t^{1-D}(G_\lambda(\log t^{-1}) - 2t^D),$$

where

$$G_\lambda(\tau) := \lambda^D G(\tau + \log\lambda)$$

and G is the T-periodic function defined in Equation (3.1.3) *of Proposition 3.1.2.*

Proof. We shall establish parts (a) and (b) separately.

(a) Scaling the set $A_t \cap \Omega$ by the factor λ, we obtain $\lambda(A_t \cap \Omega)$. On the other hand, the same result is then obtained as the intersection of the scaled sets $(\lambda A)_{\lambda t}$ and $\lambda\Omega$; that is,

$$\lambda(A_t \cap \Omega) = (\lambda A)_{\lambda t} \cap \lambda\Omega.$$

The first equality in (4.6.9) now follows by replacing t with t/λ. The second one is an immediate consequence of the first one. We also have

$$\begin{aligned}\mathscr{M}^{*r}(\lambda A, \lambda\Omega) &= \limsup_{t\to 0^+} \frac{|(\lambda A)_t \cap \lambda\Omega|}{t^{N-r}} = \lambda^N \limsup_{t\to 0^+} \frac{|(A)_{t/\lambda} \cap \Omega|}{t^{N-r}} \\ &= \lambda^N \limsup_{\tau\to 0^+} \frac{|(A)_\tau \cap \Omega|}{(\lambda\tau)^{N-r}} = \lambda^r \mathscr{M}^{*r}(A,\Omega).\end{aligned}$$

The second equality in (4.6.10) is proved in the same way, but by now using the lower limit instead of the upper limit.

(b) In the case of the generalized Cantor set, we use (4.6.9) with $N := 1$ along with Proposition 3.1.2:

$$\begin{aligned}|(\lambda C^{(m,a)})_t \cap (0,\lambda) &= \lambda|C^{(m,a)}_{t/\lambda} \cap (0,1)| = \lambda\left(\frac{t}{\lambda}\right)^{1-D}\left(G\left(\log\frac{1}{t/\lambda}\right) - 2(t/\lambda)^D\right) \\ &= t^{1-D}\left(\lambda^D G(\log\lambda + \log t^{-1}) - 2t^D\right).\end{aligned}$$

This completes the proof of the lemma. □

Relative tube zeta functions have a scaling property which is analogous to that obtained in Proposition 2.2.22 for the tube zeta functions of bounded sets. We leave the proof to the interested reader. It suffices to use Lemma 4.6.10(a).

Proposition 4.6.11 (Scaling property of relative tube zeta functions). *Let (A,Ω) be a relative fractal drum and let $\delta > 0$. Let us denote by $\tilde{\zeta}_{A,\Omega}(s;\delta)$ the associated relative fractal zeta function defined by Equation* (4.5.1). *Then, for any $\lambda > 0$, we have $D(\tilde{\zeta}_{\lambda A,\lambda\Omega}) = D(\tilde{\zeta}_{A,\Omega};\delta) = \overline{\dim}_B(A,\Omega)$ and*

$$\tilde{\zeta}_{\lambda A,\lambda\Omega}(s;\lambda\delta) = \lambda^s \tilde{\zeta}_{A,\Omega}(s;\delta), \tag{4.6.11}$$

for all $s \in \mathbb{C}$ *such that* $\operatorname{Re} s > \overline{\dim}_B(A,\Omega)$. *Furthermore, if* $\omega \in \mathbb{C}$ *is a simple pole of* $\tilde{\zeta}_{A,\Omega}(s;\delta)$, *meromophically extended to a connected open neighborhood of the critical line* $\{\operatorname{Re} s = \overline{\dim}_B(A,\Omega)\}$ *(as usual, we keep the same notation for the extended function), then*

$$\operatorname{res}(\tilde{\zeta}_{\lambda A,\lambda\Omega},\omega) = \lambda^{\omega} \operatorname{res} \tilde{\zeta}_{A,\Omega}(\,\cdot\,;\delta),\omega). \tag{4.6.12}$$

One proof of Proposition 4.6.11 would rely on the functional equation (4.5.2) combined with Theorem 4.1.40, the scaling property of the distance zeta function.

In the proof of Theorem 4.6.9, we shall use the following simple fact. If a function $G(\tau) = H(\tau,\tau,\dots)$ is transcendentally quasiperiodic with respect to a sequence of quasiperiods $(T_i)_{i\geq 1}$, it is clear that for any fixed sequence of real numbers $\mathbf{d} := (d_i)_{i\geq 1}$, the corresponding function

$$G_{\mathbf{d}}(\tau) = H(d_1+\tau, d_2+\tau,\dots)$$

is quasiperiodic with respect to the same sequence of quasiperiods.

Proof of Theorem 4.6.9. The proof is divided into three steps.

Step 1: First of all, note that the generalized Cantor sets $C^{(m_i,a_i)}$ are well defined, since $m_i a_i = m_i^{1-1/D} < 1$ for each $i \geq 1$; see Definition 3.1.1. Furthermore,

$$|\Omega| = \sum_{i=1}^{\infty} |\Omega_i| \leq C_1 \sum_{i=1}^{\infty} m_i^{1-1/D} c_i^{1/D} \leq C_1 \sum_{i=1}^{\infty} c_i^{1/D} \leq C_1 \sum_{i=1}^{\infty} c_i < \infty,$$

where we have assumed without loss of generality that $c_i \leq 1$ for all $i \geq 1$. Using Lemma 4.6.10, we have

$$\begin{aligned}|A_t \cap \Omega| &= \sum_{i=1}^{\infty} |(A_i)_t \cap \Omega_i| = t^{1-D} \sum_{i=1}^{\infty} |\Omega_i|^D \left(G_i\Big(\log|\Omega_i| + \log\frac{1}{t}\Big) - 2t^D \right) \\ &= t^{1-D}\Big(G\Big(\log\frac{1}{t}\Big) - 2|\Omega| t^D \Big),\end{aligned}$$

where

$$G(\tau) := \sum_{i=1}^{\infty} |\Omega_i|^D G_i(\log|\Omega_i| + \tau)$$

and the functions $G_i = G_i(\tau)$ are T_i-periodic, with $T_i := \log(1/a_i)$, for all $i \geq 1$. This shows that $G(\tau) = H(\tau,\tau,\dots)$, where

$$H((\tau_i)_{i\geq 1}) := \sum_{i=1}^{\infty} |\Omega_i|^D G_i(\log|\Omega_i| + \tau_i).$$

Note that the last series is well defined, and that so is the series defining $G(\tau)$. Indeed, letting $\mathcal{M}_i = \mathcal{M}^{*D}(C^{(m_i,a_i)})$ and using Proposition 3.1.2, we see that

$$0 < G_i(\tau) \leq \mathcal{M}_i = \left(\frac{2(m_i-1)}{1-m_i a_i}\right)^{1-D} \frac{m_i}{m_i-1}(1-a_i) \leq C m_i^{1-D}, \tag{4.6.13}$$

where C is a positive constant independent of i, since $m_i \to \infty$ and $m_i a_i \to 0$ as $i \to \infty$. Therefore,

$$\sum_{i=1}^{\infty} |\Omega_i|^D G_i(\tau_i) \le \sum_{i=1}^{\infty} (C_1^D m_i^{D-1} c_i)\,(C m_i^{1-D}) = C C_1^D \sum_{i=1}^{\infty} c_i < \infty.$$

In particular,

$$\mathscr{M}^{*D}(A,\Omega) \le C C_1^D \sum_{i=1}^{\infty} c_i < \infty.$$

On the other hand, since $(A_1,\Omega_1) \supset (A,\Omega)$, we can use Lemma 4.6.10(a) (with $r := D$) and Proposition 3.1.2 to obtain that

$$\mathscr{M}_*^D(A,\Omega) \ge \mathscr{M}_*^D(A_1,\Omega_1) = |\Omega_1|^D \mathscr{M}_*^D(C^{(m_1,a_1)}) = |\Omega_1|^D \frac{1}{D}\left(\frac{2D}{1-D}\right)^{1-D} > 0.$$

Step 2: Let n be any fixed positive integer. Since the set of real numbers

$$\{\log m_1, \dots, \log m_n\}$$

is rationally independent, we conclude from Baker's theorem (see Theorem 3.1.14 above or [Ba, Theorem 2.1]) that the set of real numbers $\{1, \log m_1, \dots, \log m_n\}$ is algebraically independent. Dividing all of these numbers by D, and using $D = (\log m_i)/T_i$, where $T_i = \log(1/a_i)$ for all i (see Proposition 3.1.2), we deduce that

$$\left\{\frac{1}{D}, \frac{\log m_1}{D}, \dots, \frac{\log m_n}{D}\right\} = \left\{\frac{1}{D}, T_1, \dots, T_n\right\}$$

is algebraically independent as well. Since n is arbitrary, this proves that the relative fractal drum (A,Ω) is transcendentally ∞-quasiperiodic, in the sense of Definition 4.6.7.

Step 3: To prove the last claim, note that the critical line $\{\operatorname{Re} s = D\}$ contains the union of the set of poles $\mathscr{P}_i := \mathscr{P}(\tilde{\zeta}_{A_i,\Omega_i}, \mathbb{C}) = D + \mathbf{p}_i \mathrm{i}\mathbb{Z}$ of the tube zeta functions $\tilde{\zeta}_{A_i,\Omega_i}$, $i \ge 1$. Since the integers m_i are all distinct, we have that $m_i \to \infty$ as $i \to \infty$, and therefore, $\mathbf{p}_i = 2\pi/T_i = 2\pi D/\log m_i \to 0$. This proves that the union $\cup_{i\ge1}\mathscr{P}_i$, as a set of nonisolated singularities of $\tilde{\zeta}_{A,\Omega} = \sum_{i\ge1}\tilde{\zeta}_{A_i,\Omega_i}$ (see Lemma 4.5.9), is dense in the critical line $\{\operatorname{Re} s = D\}$. Since we have a dense set of nonisolated singularities of $\zeta_{A,\Omega}$ along the critical line, then in fact, each point on the line is a nonisolated singularity. Indeed, reasoning by contradiction, if any point (say, s_0) on the critical line is a removable singularity, then there is a punctured connected open neighborhood of s_0 in which the fractal zeta function $\zeta_{A,\Omega}$ is holomorphic, and hence, the same is true along the corresponding punctured open interval (along the critical line) containing the singularity s_0, which is impossible. (For more details, see the proof of Lemma 4.6.12 just below.) It follows, in particular, that (4.6.8) holds, as desired.

We note that the above argument can be summarized as follows: The set of nonisolated singularities along $L := \{\operatorname{Re} s = D\}$ is closed in the critical line L. Since the latter set is already known to be dense in L, it follows that it must be all of L. This argument is the content of Lemma 4.6.12 just below. □

At the end of the proof of Step 3 of Theorem 4.6.9, we have used the following lemma.[47]

Lemma 4.6.12. *In Step 3 of the proof of Theorem 4.6.9 above, the set of nonisolated singularities of $\zeta_{A,\Omega}$ along the critical line $L := \{\operatorname{Re} s = D\}$ is both closed and dense in, and therefore coincides with, L.*

Proof. We already know from the first part of Step 3 of the proof of Theorem 4.6.9 that the set of nonisolated singularities of $\zeta_{A,\Omega}$ along L is dense in L. Therefore, all we need to show is that it is also closed in L. Equivalently, we must show that the set of removable singularities of $\zeta_{A,\Omega}$ along L is open in L.

For this purpose, assume that there exists $s_0 \in L$ which is a removable singularity of $\zeta_{A,\Omega}$. By definition, this means that there exists an open disk $U := B_\rho(s_0)$ in $\mathbb{C}$ centered at s_0 and such that $\zeta_{A,\Omega}$ is holomorphic in the punctured disk $U \setminus \{s_0\}$. (Upon resolution of the singularity at s_0, we could take all of U instead and hence, all of I just below.) Therefore, if $I := U \cap L$ is the corresponding open interval along the critical line $L := \{\operatorname{Re} s = D\}$, then $\zeta_{A,\Omega}$ cannot have any nonremovable singularity in the punctured interval $I \setminus \{s_0\}$, and therefore consists entirely of removable singularities.

This establishes the fact that the set of removable singularities along L is open in L, and thereby concludes the proof of the lemma. In particular, we have shown that every point of the line L is a nonisolated nonremovable singularity of $\zeta_{A,\Omega}$; i.e, L is a natural barrier for $\zeta_{A,\Omega}$. More specifically, L is a (meromorphic) natural boundary for $\zeta_{A,\Omega}$, in the sense of part (*ii*) of Definition 1.3.8 in Subsection 1.3.2. □

It is noteworthy that the sequence $\mathscr{M}^{*D}(C^{(m_i,a_i)},(0,1))$ appearing in Theorem 4.6.9 is divergent. More precisely, it is easy to deduce from the equality in (4.6.13) that

$$\mathscr{M}^{*D}(C^{(m_i,a_i)},(0,1)) \sim (2m_i)^{1-D} \quad \text{as} \quad i \to \infty.$$

The conditions of Theorem 4.6.9 are satisfied if, for example, $m_i := p_i$ for every $i \geq 1$ (that is, $(m_i)_{i\geq 1}$ is the sequence of prime numbers $(p_i)_{i\geq 1}$, written in increasing order), and if $C_1 := 1$ and $c_i := 2^{-i}$ for all $i \geq 1$. More general choices of the sequence $(m_i)_{\geq 1}$ can be found in Theorem 4.6.13 below; see also Remark 4.6.15.

Theorem 4.6.9 shows that a result about the meromorphic extensions of distance relative zeta functions, obtained in Theorem 4.5.2 for a class of Minkowski nonmeasurable RFDs satisfying a periodicity condition, cannot be extended to transcendentally quasiperiodic RFDs with infinitely many quasiperiods. For quasiperiodic sets and RFDs with finitely many quasiperiods, such extensions are also possible. See, for example, Theorem 2.3.43 and its obvious extension to the context of RFDs.

[47] It will be apparent to the reader that, in the statement of Lemma 4.6.12, the closedness statement is of a general nature.

4.6.2 Hyperfractals and Transcendentally ∞-Quasiperiodic Fractal Strings and Sets

The following result provides some sufficient conditions on the sequence $(m_i)_{i\geq 1}$, for the rational independence to hold in condition (4.6.6). It complements Theorem 3.1.15.

Theorem 4.6.13. *Let $m_i \geq 2$ be given integers, $i \geq 1$, and let $S_i := \operatorname{supp}(m_i)$ be their corresponding supports (see Definition 4.6.4). Assume that*

$$i \mapsto \max S_i \quad \text{is increasing.} \tag{4.6.14}$$

Let $D \in (0,1)$, and define the relative fractal drum $(A,\Omega) = \cup_{i=1}^{\infty}(A_i,\Omega_i)$, where $A_i := 2^{-i}C^{(m_i,a_i)} + \inf\Omega_i$, $a_i := m_i^{-1/D}$, and the family of open intervals $(\Omega_i)_{i\geq 1}$ is disjoint, with $|\Omega_i| := 2^{-i}$ for all $i \geq 1$. Then the relative fractal drum (A,Ω) is transcendentally quasiperiodic and with infinite order of quasiperiodicity. Furthermore,

$$D(\zeta_{A,\Omega}) = D_{\mathrm{mer}}(\zeta_{A,\Omega}) = D_{\mathrm{hol}}(\zeta_{A,\Omega}), \tag{4.6.15}$$

and moreover, all of the points on the critical line $\{\operatorname{Re} s = D\}$ are nonisolated singularities of $\zeta_{A,\Omega}$.

In order to prove this result, we shall use the following auxiliary lemma.

Lemma 4.6.14. *Let $(m_i)_{i\geq 1}$ be a sequence of integers, $m_i \geq 2$, such that the sequence of the associated exponent vectors $(\mathbf{e}(m_i))_{i\geq 1}$ is rationally linearly independent. Then the sequence $(\log m_i)_{i\geq 1}$ is rationally linearly independent as well.*

Proof. Let n be a fixed positive integer. Since the vectors $\mathbf{e}(m_1),\dots,\mathbf{e}(m_n)$ are rationally linearly independent, then using Steps 1 and 2 of the proof of Theorem 3.1.15, we conclude that the numbers $\log m_1,\dots,\log m_n$ are rationally linearly independent as well. □

Proof of Theorem 4.6.13. Condition (4.6.14) ensures that any pair of numbers m_i and m_j, $i \neq j$, has different corresponding sets of prime factors. From this we can easily conclude that the exponent vectors $\mathbf{e}(m_i)$, $i \geq 1$, are rationally linearly independent. Indeed, assume that

$$k_1\mathbf{e}(m_1) + \cdots + k_n\mathbf{e}(m_n) = 0, \tag{4.6.16}$$

for some $n \in \mathbb{N}$, where the coefficients k_i are integers. Let $s_i := \max S_i$. By looking at the s_n-th component of (4.6.16), we immediately obtain that $k_n = 0$. We then apply the same reasoning to the s_{n-1}-th component in order to obtain $k_{n-1} = 0$, and so on.

Using Lemma 4.6.14, we conclude that the sequence of integers $(\log m_i)_{i\geq 1}$ is rationally linearly independent. The claim then follows from Theorem 4.6.9. □

Remark 4.6.15. It is easy to see that condition (4.6.14) in Theorem 4.6.13 can be relaxed. More specifically, it suffices to assume that $i \mapsto \max S_i$ be injective. Indeed, if the map is injective, then each member of the set $\{\max S_i : i \in \mathbb{N}\}$ has multiplicity 1, and after a suitable permutation, we can obtain (4.6.14).[48]

In Theorems 4.6.9 and 4.6.13, we have constructed a transcendentally quasiperiodic relative fractal drum (A,Ω) with infinite order of quasiperiodicity. In particular, (A,Ω) has infinitely many algebraically incommensurable quasiperiods $T_i = \frac{1}{D}\log m_i$, $i \geq 1$.

The following corollary (Corollary 4.6.17) shows that there exist bounded fractal strings $\mathscr{L} = (\ell_j)_{j\geq 1}$ with infinitely many algebraically incommensurable quasiperiods (i.e., with infinitely many incommensurable quasifrequencies). We see from the proof of this result that $\mathscr{L}$ can be effectively constructed.

Definition 4.6.16. As we know, any bounded fractal string $\mathscr{L} = (\ell_j)_{j\geq 1}$ can be naturally identified with a relative fractal drum $(A_{\mathscr{L}},\Omega_{\mathscr{L}})$ in $\mathbb{R}$, where

$$A_{\mathscr{L}} := \Big\{a_k := \sum_{j\geq k} \ell_j : k \geq 1\Big\}, \quad \Omega_{\mathscr{L}} := \bigcup_{k=1}^{\infty} (a_{k+1}, a_k),$$

with $|\Omega_{\mathscr{L}}| = \sum_{j=1}^{\infty} \ell_j < \infty$. Let $n \in \mathbb{N} \cup \{\infty\}$ be fixed. We say that *a bounded fractal string* $\mathscr{L} = (\ell_j)_{j\geq 1}$ *is n-quasiperiodic* if the corresponding relative fractal drum $(A_{\mathscr{L}},\Omega_{\mathscr{L}})$ is n-quasiperiodic. The *order of quasiperiodicity of a bounded fractal string* $\mathscr{L}$ is defined as the order of quasiperiodicity of the corresponding relative fractal drum $(A_{\mathscr{L}},\Omega_{\mathscr{L}})$; see Definitions 4.6.7 and 4.6.8.

In addition, we say that *a bounded fractal string* $\mathscr{L} = (\ell_j)_{j\geq 1}$ *is transcendentally* (resp., *algebraically*) *∞-quasiperiodic* if the corresponding relative fractal drum $(A_{\mathscr{L}},\Omega_{\mathscr{L}})$ is transcendentally (resp., algebraically) ∞-quasiperiodic.

The family $\mathscr{L}_{\rm qp}$ of all ∞-quasiperiodic fractal strings is the disjoint union of the family $\mathscr{L}_{\rm aqp}$ of algebraically ∞-quasiperiodic fractal strings and the family $\mathscr{L}_{\rm tqp}$ of transcendentally ∞-quasiperiodic fractal strings:

$$\mathscr{L}_{\rm qp}(\infty) = \mathscr{L}_{\rm aqp}(\infty) \cup \mathscr{L}_{\rm tqp}(\infty).$$

If we let

$$\overline{\mathscr{L}}_{\rm qp} := \mathscr{L}_{\rm qp} \cup \mathscr{L}_{\rm qp}(\infty), \quad \overline{\mathscr{L}}_{\rm tqp} := \mathscr{L}_{\rm tqp} \cup \mathscr{L}_{\rm tqp}(\infty), \quad \overline{\mathscr{L}}_{\rm aqp} := \mathscr{L}_{\rm aqp} \cup \mathscr{L}_{\rm aqp}(\infty),$$

where $\mathscr{L}_{\rm qp}$, $\mathscr{L}_{\rm tqp}$ and $\mathscr{L}_{\rm aqp}$ are defined by (3.1.30) on page 202, then

$$\overline{\mathscr{L}}_{\rm qp} = \overline{\mathscr{L}}_{\rm tqp} \cup \overline{\mathscr{L}}_{\rm aqp}.$$

We expect that the family $\overline{\mathscr{L}}_{\rm aqp}$ is nonempty.

---

[48] We wish to thank Tomislav Šikić for this remark.

Corollary 4.6.17. (*a*) *There exists an effectively constructible bounded fractal string* $\mathscr{L} = (\ell_j)_{j\geq 1}$ *in* $\mathbb{R}$ *which is transcendentally* ∞*-quasiperiodic* (*see Definition* 4.6.7), *such that*

$$D(\zeta_{\mathscr{L}}) = D_{\rm hol}(\zeta_{\mathscr{L}}) = D_{\rm mer}(\zeta_{\mathscr{L}}) \tag{4.6.17}$$

and all of the points on the critical line $\{\operatorname{Re} s = D\}$ *are nonisolated singularities of the geometric zeta function* $\zeta_{\mathscr{L}}$; *in other words, the fractal string* $\mathscr{L}$ *is also maximally hyperfractal* (*in the sense of Definition* 4.6.23(*iii*) *below and the comment following it*).

(*b*) *In particular, there exists an effectively constructible bounded subset* A_0 *of* $\mathbb{R}$, *which is transcendentally* ∞*-quasiperiodic, such that*

$$D(\zeta_{A_0}) = D_{\rm hol}(\zeta_{\mathscr{L}}) = D_{\rm mer}(\zeta_{A_0}) \tag{4.6.18}$$

and all of the points on the critical line $\{\operatorname{Re} s = D\}$ *are nonisolated singularities of the distance zeta function* ζ_{A_0} (*as well as of the tube zeta function* $\tilde{\zeta}_{A_0}$); *in other words, the bounded set* A_0 *is also maximally hyperfractal* (*in the sense of Definition* 4.6.23(*iii*) *below and the comment following it*).

Proof. (*a*) It suffices to note that each relative subdrum (A_i, Ω_i) of (A, Ω), defined in Theorem 4.6.13, can be viewed as a fractal string $\mathscr{L}_i$ (i.e., Cantor's string) associated with a generalized Cantor set $A_i = C^{(m_i, a_i)}$. Therefore, the relative fractal drum $(A, \Omega) = \cup_{i\geq 1}(A_i, \Omega_i)$ can be viewed as a bounded fractal string $\mathscr{L} = \sqcup_{i\geq 1}\mathscr{L}_i$.

(*b*) To prove this, it suffices to associate a new RFD (A_0, Ω_0) to the fractal string $\mathscr{L}$ from (*a*). Its construction can be found in Definition 4.6.16. □

Remark 4.6.18. Note that the set A_0 in Corollary 4.6.17(b) does not coincide with the set A from the relative fractal drum (A, Ω), associated with the fractal string $\mathscr{L}$. Indeed, A is a union of a countable family of Cantor sets (therefore, an uncountable set), whereas A_0 is a decreasing sequence of positive real numbers converging to zero. Here, A_0 is generated by the union of a sequence of generalized Cantor strings $\mathscr{L}_i$, $i \geq 1$, and each $\mathscr{L}_i$ is generated by a generalized relative Cantor drum.

Remark 4.6.19. For $N \geq 2$, one can readily extend Corollary 4.6.17 to obtain an explicitly constructible maximally hyperfractal and transcendentally ∞-quasiperiodic fractal spray in $\mathbb{R}^N$, and correspondingly, a bounded subset A of $\mathbb{R}^N$ having those same exact properties. Indeed, it suffices to proceed exactly as in the passage from Example 5.1 to Example 5.1' in [Lap1]. Namely, for example, if $A_0 \subset \mathbb{R}$ is the bounded set obtained in part (*b*) of Corollary 4.6.17, simply let $A := A_0 \times [0,1]^{N-1}$, now viewed as a bounded subset of $\mathbb{R}^N$. (See also Subsection 4.6.4.)

Remark 4.6.20. There is a classic example of a function which is holomorphic on the open unit disk in $\mathbb{C}$ and is such that each of its points on the boundary is a nonisolated singularity. See Problem 6.2.18 on page 558.

Example 4.6.21. Concerning Lemma 4.6.14, there are many other ways to ensure the rational independence of $\log m_i$, $i \geq 1$. For example, if $m_1, \ldots, m_n$ are the positive integers defined by

$$\begin{aligned} m_1 &= p_{j_1}^k p_{j_2} \cdots p_{j_n} \\ m_2 &= p_{j_1} p_{j_2}^k \cdots p_{j_n} \\ &\vdots \\ m_n &= p_{j_1} p_{j_2} \cdots p_{j_n}^k, \end{aligned}$$

where $k \geq 2$ is a fixed integer, then these integers have identical supports, and their exponent vectors are given by

$$\begin{aligned} \mathbf{e}(m_1) &= (k, 1, 1, \ldots, 1), \\ \mathbf{e}(m_2) &= (1, k, 1, \ldots, 1), \\ &\vdots \\ \mathbf{e}(m_n) &= (1, 1, \ldots, 1, k), \end{aligned}$$

where we have truncated the exponent vectors outside of their supports. It is easy to see that these vectors are rationally linearly independent. Indeed, we have $a := \frac{1}{k+n-1}(\mathbf{e}(m_1) + \cdots + \mathbf{e}(m_n)) = (1, 1, \ldots, 1)$, and therefore, the vectors

$$\frac{1}{k-1}(\mathbf{e}(m_1) - a), \ldots, \frac{1}{k-1}(\mathbf{e}(m_n) - a)$$

form the standard basis of $\mathbb{Q}^n$.

Example 4.6.22. Let $(P_j)_{j \geq 1}$ be a partition of the set of all prime numbers, such that each set P_j is finite. Applying the construction from Example 4.6.21 on each P_j, with $k = k_j \geq 2$, we obtain an infinite sequence of integers $(m_i)_{i \geq 1}$ such that the associated sequence $(\mathbf{e}(m_i))_{i \geq 1}$ of their exponent vectors is rationally linearly independent.

Alternatively, we can also use the constructions from Remark 4.6.21 and from (4.6.14) intermittently, applied on the elements of the sequence of sets $(P_j)_{j \geq 1}$.

In Subsection 4.6.4, we will extend the construction carried out in the present subsection to obtain maximally hyperfractal sets in $\mathbb{R}^N$ of arbitrarily prescribed dimension $D \in (N-1, N)$, for any $N \geq 1$.

4.6.3 Fractality, Hyperfractality and Complex Dimensions

The following definition is closely related to the the notion of fractality (given in [Lap-vFr3], Sections 12.1.1 and 12.1.2, including Figures 12.1–12.3), as will be explained in Remark 4.6.24 below. At this point, the reader may wish to review the definition of a (*meromorphic*) *partial natural boundary* and that of a (*meromorphic*) *natural boundary* (and correspondingly, of a *partial domain of meromorphy* and of a *domain of meromorphy*) given, respectively, in part (*i*) and in part (*ii*) of Definition 1.3.8 of Subsection 1.3.2 on page 39 (and as strengthened in Remark 1.3.9).

Definition 4.6.23. (*Hyperfractality*). Let A be a bounded subset of $\mathbb{R}^N$ and let $D := \overline{\dim}_B A$. Then:

(*i*) The set A is a *hyperfractal* (or is *hyperfractal*) if there is a screen $\boldsymbol{S}$ (see page 95 above or Definition 5.1.1 on page 411 below) which is a (meromorphic) partial natural boundary for the associated tube (or equivalently, if $D < N$, distance) zeta function of A. This means, in particular, that the fractal zeta function cannot be meromorphically continued to any connected open neighborhood of S (or, equivalently, of the associated window $\boldsymbol{W}$); see Definition 1.3.8(*i*) for the precise definition of a (meromorphic) partial natural boundary. (See also both parts of Remark 1.3.9.) Equivalently, the interior $\mathring{\boldsymbol{W}}$ of the window is a partial domain of meromorphy for the fractal zeta function of A.

(*ii*) The set A is a *strong hyperfractal* (or is *strongly hyperfractal*) if the critical line $\{\mathrm{Re}\, s = D\}$ is a (meromorphic) partial natural boundary of the associated fractal zeta function; that is, if we can choose $S = \{\mathrm{Re}\, s = D\}$ in (*i*).[49] Equivalently, the open right half-plane $\{\mathrm{Re}\, s > D\}$ is a partial domain of meromorphy for $\tilde{\zeta}_A$ (or equivalently, if $D < N$, for ζ_A), also in the sense of Definition 1.3.8(*i*).

(*iii*) Finally, the set A is *maximally hyperfractal* if it is strongly hyperfractal and every point of the critical line $\{\mathrm{Re}\, s = D\}$ is a nonisolated singularity of the fractal zeta function of A. In that case, the critical line $\{\mathrm{Re}\, s = D\}$ is a meromorphic natural boundary of the fractal zeta function; see Definition 1.3.8(*ii*) for the precise definition of a partial natural boundary. In short, the fractal zeta function of A cannot be extended meromorphically (and, a fortiori, holomorphically) to any punctured (and connected) open neighborhood of s, given any point s of the critical line. Equivalently, the open right half-plane $\{\mathrm{Re}\, s > D\}$ is a domain of meromorphy for $\tilde{\zeta}_A$ (or equivalently, if $D < N$, for ζ_A), also in the sense of Definition 1.3.8(*ii*)

An analogous definition can be provided (in the obvious manner) where instead of A, we have a fractal string $\mathscr{L} = (\ell_j)_{j \geq 1}$ in $\mathbb{R}$ or, more generally, a relative fractal drum (A, Ω) in $\mathbb{R}^N$.

Remark 4.6.24. (*Complex dimensions and the definition of fractality*). In [Lap-vFr1–3], a geometric object is said to be "fractal" if the associated zeta function has at least one nonreal complex pole (with positive real part); i.e., the object has at least one nonreal complex dimension.[50] (See [Lap-vFr3, Sections 12.1

[49] Recall from Theorem 2.1.11(*a*) that since $D = \overline{\dim}_B A$, the fractal zeta function ζ_A is holomorphic (and hence, meromorphic) in the window $\boldsymbol{W} = \{\mathrm{Re}\, s > D\}$, in that case.

[50] Then, clearly, it has at least two nonreal complex conjugate complex dimensions.

and 12.2] for a detailed discussion.) In [Lap-vFr2, Lap-vFr3], in order, in particular, to take into account some possible situations pertaining to random fractals (see [HamLap], partly described in [Lap-vFr3, Section 13.4]), the definition of fractality (within the context of the theory of complex dimensions) was extended so as to allow for the case described in part (i) of Definition 4.6.23 just above, namely, the existence of a partial natural boundary along a screen. See [Lap-vFr3, Subsection 13.4.3].

We note that in [Lap-vFr3] (and the other aforementioned references), the term "hyperfractal" was not used to refer to case (i) (or to any other situation). More important, except for fractal strings and in very special higher-dimensional situations (such as suitable fractal sprays), one did not have to our disposal (as we now do, thanks to the general theory developed in this book and in [LapRaŽu1–8]) a general definition of "fractal zeta function" associated with an arbitrary bounded subset of $\mathbb{R}^N$, for every $N \geq 1$. Therefore, we can now define the "fractality" of any bounded subset of $\mathbb{R}^N$ (including Julia sets and the Mandelbrot set) and, more generally, of any relative fractal drum, by the presence of a nonreal complex dimension or else by the "hyperfractality" (in the sense of part (i) of Definition 4.6.23) of the geometric object under consideration. Here, "complex dimension" is understood as a (visible) pole of the associated fractal zeta function (the distance or tube zeta function of a bounded subset or a relative fractal drum of $\mathbb{R}^N$, or else, as was the case in most of [Lap-vFr3], the geometric zeta function of a fractal string).

Much as in [Lap-vFr1–3] and [Lap3–8], this terminology (concerning fractality, hyperfractality, and complex dimensions), can be extended to 'virtual geometries', as well as to (absolute or) relative fractal drums, noncommutative geometries, dynamical systems, and arithmetic geometries, via suitably associated 'fractal zeta functions', be they absolute or relative distance or tube zeta functions, spectral zeta functions, dynamical zeta functions, or arithmetic zeta functions (or their logarithmic derivatives thereof).

As we have seen in Theorem 4.6.9 and Corollary 4.6.17, there exist bounded sets A_0, fractal strings $\mathscr{L}$ and RFDs (A,Ω), that are maximally hyperfractal. In other words, all the points on the critical line $\{\mathrm{Re}\, s = D\}$ are nonisolated singularities of the corresponding zeta functions. (See Problem 6.2.20.) Furthermore, the construction provided in Subsection 4.6.4 below will show that for any integer $n \geq 1$ there exists a maximally hyperfractal bounded subset of $\mathbb{R}^N$, of arbitrary prescribed dimension $D \in (N-1,N)$; see Corollary 4.6.28. In addition, we recall that in Example 3.3.7, we have constructed a fractal string $\mathscr{L}_\infty$ whose associated fractal zeta function has a countable set of essential singularities on the critical line; see Equation (3.3.32). Such a fractal string is therefore strongly hyperfractal, in the sense of part (ii) of Definition 4.6.23 (and as strengthened in part (b) of Remark 1.3.9). It is worth pointing out that this construction was generalized to a whole class of strongly hyperfractal RFDs which are not maximally hyperfractal; see Example 4.2.10 of Subsection 4.2.2 above.

Corollary 4.6.17 provides a partial answer to a part of [Lap-vFr3, Problem 13.146, p. 473] (building on open problems proposed toward the end of [HamLap]). Note that in Corollary 4.6.17(b) we have constructed a (deterministic) hyperfractal A_0 on the real line, which is just a bounded countable set on the real line (more precisely, a bounded decreasing sequence converging to zero; see Remark 4.6.18). In this sense, A_0 may be viewed as being fairly simple. Recall, however, that it has been (effectively) constructed by means of a countable family of generalized Cantor sets, and in this sense, this sequence (as well as the corresponding hyperfractal string) is extremely complex. Also, we stress that in this construction, we do not use any random fractal sets. Random fractal strings, along with the associated random zeta functions and complex dimensions, are the object of the work of Ben Hambly and the first author in [HamLap], which is surveyed in [Lap-vFr3, Section 13.4] where the aforementioned open problem can be found.

In short, the latter problem asks whether almost surely, and within a suitably defined class of random fractals, the associated (pointwise) random fractal zeta functions have a (meromorphic) partial natural boundary. Our present work now enables us to give a proper meaning to the notion of 'fractal zeta functions' in higher dimensions, and hence to adapt it to random fractals (by naturally extending the notions introduced for random fractal strings in [HamLap]). Moreover, in the deterministic setting, the examples constructed here indicate that in some cases, one can obtain a much stronger conclusion; namely, the partial natural boundary can consist solely of nonisolated singularities. In turn, in the random setting, one may complete the above open problem (from [HamLap] and [Lap-vFr3]) by asking whether, almost surely, the random fractals within a suitable class are *maximally hyperfractal*, and hence, admit the critical line as a (meromorphic) natural boundary (for the associated fractal zeta function).

Given $d \in \mathbb{R}$ such that $d \le D$, Definition 4.6.23 (or its obvious conterpart for a fractal string $\mathscr{L}$ or a relative fractal drum (A,Ω)) can be extended as follows. In the analog of case (i), A is said to be *hyperfractal* (respectively, *strictly hyperfractal*) *in dimension* d if the screen $\boldsymbol{S}$ can be chosen so that $\sup S = d$ (respectively, $\max S$ exists and $\max S = d$).[51] In the analog of case (ii), A is said to be *strongly hyperfractal in dimension* d if the vertical line $\{\mathrm{Re}\, s = d\}$ is a (meromorphic) partial natural boundary of the associated zeta function (that is, if we can choose $S = \{\mathrm{Re}\, s = d\}$ in the counterpart of (i)). Finally, A is said to be *maximally hyperfractal in dimension* d if it is strongly hyperfractal in dimension d and every point of the vertical line $\{\mathrm{Re}\, s = d\}$[52] is a nonisolated singularity of the zeta function. Therefore, $\{\mathrm{Re}\, s = D\}$ is a (meromorphic) natural boundary (for the associated fractal zeta function).

It would be interesting to consider the following open problem, which complements in a different direction the problem about random fractals stated above in the discussion following Remark 4.6.24. Namely, one may ask whether given a (deterministic or random) relative fractal drum which is hyperfractal or even, maximally

[51] As in [Lap-vFr2], we adopt the following notation: $\sup S := \sup_{t\in\mathbb{R}} S(t)$, and similarly for $\max S$ (when it exists). See the definition of a screen on page 95.

[52] Except possibly for some points in a small neighborhood of d in that line.

hyperfractal (with respect to the standard power law gauge function, $h \equiv 1$), one can sometime find another (non power law) gauge function h (in the sense of Definition 6.1.4 of Section 6.1 below) for which the associated zeta function no longer has a partial natural boundary. We note that in order to address this problem, one should be ready to work with analytic functions on suitable Riemann surfaces rather than just on $\mathbb{C}$ or on the Riemann sphere $\widetilde{\mathbb{C}} := \mathbb{C} \cup \{\infty\}$. See [EsLapRRo] and [ElLapMacRo] where a related open problem is raised in connection with the 'multifractal zeta functions' of [LapRo1, LapLéRo, LéMen, ElLapMacRo]. We plan to develop this point of view in a later work, especially in connection with the results of Chapter 5 below, on fractal tube formulas and Minkowski measurability criteria.

We will pursue the discussion of fractality in Subsection 5.5.4, in connection with the devil's staircase (the graph of the Cantor function) and a suitable version thereof studied in Example 5.5.14. See, especially, Remark 5.5.15 and the comments surrounding it. We will also revisit this issue (the notion of fractality and the related notions of critical fractality, subcritical fractality, and more general fractality in dimension $d \in \mathbb{R}$, all introduced in Subsection 5.5.4) in various places, including in Subsection 5.5.6, when discussing Example 5.5.22 (the 1/2-square fractal), Example 5.5.23 (the 1/3-square fractal), and Example 5.5.25 (the geometric progression fractal string).

4.6.4 Maximal Hyperfractals in Euclidean Spaces

The aim of this subsection is to show that, given a maximal hyperfractal set A in $\mathbb{R}^N$, the sets of the form $A \times [0,1]^m$ will also be maximally hyperfractal for any positive integer m. The main result is stated in Theorem 4.6.27 below. It will enable us, in particular, to obtain an N-dimensional analog of part (b) of Corollary 4.6.17 above; see Corollary 4.6.28 below.

Lemma 4.6.25. *Assume that $f = f(s)$ is a Dirichlet-type integral (DTI) such that $D_{\rm hol}(f) \in \mathbb{R}$ and the corresponding critical line of holomorphic continuation $\{\operatorname{Re} s = D_{\rm hol}(f)\}$ consists entirely of nonisolated singularities. Assume that a ($\mathbb{C}$-valued) function $g = g(s)$ is holomorphic on the open right half-plane $\{\operatorname{Re} s > \alpha\}$, where $\alpha \in \mathbb{R} \cup \{-\infty\}$ and $\alpha < D_{\rm hol}(f)$. Then $D_{\rm hol}(f+g) = D_{\rm hol}(f)$ and hence, the critical line $\{\operatorname{Re} s = D_{\rm hol}(f+g)\}$ of holomorphic continuation corresponding to the function $f+g$ also consists entirely of nonisolated singularities.*

Proof. Since f is holomorphic on $\{\operatorname{Re} s > D_{\rm hol}(f)\}$, and by definition, the holomorphicity lower bound $D_{\rm hol}(f)$ is optimal (i.e., it is the infimum of all $\beta \in \mathbb{R}$ such that f is holomorphic on $\{\operatorname{Re} s > \beta\}$), it then follows that $D_{\rm hol}(f) = D_{\rm hol}(f+g)$; indeed, by hypothesis, g is holomorphic on the open right half-plane $\{\operatorname{Re} s > \alpha\}$ containing $\{\operatorname{Re} s > D_{\rm hol}(f)\}$.

In order to prove the second claim, we argue by contradiction and assume that some complex number s_0 with $\operatorname{Re} s_0 = D_{\rm hol}(f+g)$ is a removable singularity of

$f+g$. Then, since g is holomorphic at s_0 (because $\alpha < D_{\rm hol}(g)$), it would follow that s_0 is a removable singularity of the function $f = (f+g) - g$ as well. However, this would contradict the assumption according to which the holomorphy critical line $\{\mathrm{Re}\, s = D_{\rm hol}(f)\}$ consists of nonisolated singularities. □

Remark 4.6.26. Actually, a slightly more general result holds. Indeed, it suffices to assume that the function $g = g(s)$ appearing in Lemma 4.6.25 is holomorphic on an open subset of $\mathbb{C}$ containing the *closed* right half-plane $\{\mathrm{Re}\, s \geq D_{\rm hol}(f)\}$.

Theorem 4.6.27. *Assume that A is a maximally hyperfractal subset of $\mathbb{R}^N$ and let d be a positive integer. Then the set $A \times [0,1]^d$ is also maximally hyperfractal.*

Proof. By part (a) of Theorem 2.2.32, we can write

$$\zeta_{A\times[0,1]^d}(s) = \zeta_A(s-d) + g(s) \tag{4.6.19}$$

for all $s \in \mathbb{C}$ with $\mathrm{Re}\, s > \overline{\dim}_B A + d$, where

$$g(s) := \sum_{k=1}^{d} \binom{d}{k} \zeta_A(s-d+k)$$

is holomorphic on $\{\mathrm{Re}\, s > \overline{\dim}_B A + d - 1\}$. By hypothesis, the critical line of holomorphic continuation of the function $f(s) := \zeta_A(s-d)$ is the vertical line $\{\mathrm{Re}\, s = \overline{\dim}_B A + d\}$ and consists entirely of nonisolated singularities. On the other hand, the function $g(s)$ is holomorphic on $\{\mathrm{Re}\, s > \alpha := \overline{\dim}_B A + d - 1\}$, since this is the case of the functions $\zeta_A(s-d+k)$ for $k = 1,2,\ldots,d$. (Here, we have also used the easily verified fact that $\overline{\dim}_B A$ does not depend on N, the embedding dimension; see also Proposition 4.7.6 below for a more general context.) Since $\alpha < \overline{\dim}_B A + d$, the claim now follows from Lemma 4.6.25. □

The identity (4.6.19) implies that

$$\zeta_{A\times[0,1]^d}(s) \sim \zeta_A(s-d), \tag{4.6.20}$$

which we call the *shift property* of the distance zeta function with respect to the Cartesian product of A with the d-dimensional cube $[0,1]^d$. Furthermore, the set $A \times [0,1]^d$ is called the *fractal grill* generated by A.

Corollary 4.6.28. *Let N be any positive integer. Then, for any $D \in (N-1,N)$, there is an explicitly constructible maximally hyperfractal subset A of $\mathbb{R}^N$ such that* $\dim_B A = D$.

Proof. Let $A_{\mathscr{L}}$ be a maximally hyperfractal set in $\mathbb{R}$ of the sort constructed in part (b) of Corollary 4.6.17 above. It then suffices to let $A := A_{\mathscr{L}} \times [0,1]^{N-1}$ and to apply Theorem 4.6.27 to the set $A_{\mathscr{L}} \subset \mathbb{R}$ instead of A and with $d = N-1$. □

Actually, by considering $A_{\mathscr{L}} \times [0,1]^d$, the Cartesian product of $A_{\mathscr{L}}$ by $[0,1]^d$, with $1 \leq d \leq N-1$, the same proof as the one just above shows that in the statement of Corollary 4.6.28, we may assume that $\overline{\dim}_B A \in (d,N)$, for any $d = 1,\ldots,N-1$.

4.7 Complex Dimensions and Embeddings Into Higher-Dimensional Spaces

In this section, we obtain useful results concerning relative fractal drums and bounded subsets of $\mathbb{R}^N$ embedded into higher-dimensional spaces. In particular, we show that the complex dimensions (and their multiplicities) of a bounded set (or, more generally, of a relative fractal drum) are independent of the dimension of the ambient space. (See Theorem 4.7.3 and Theorem 4.7.10, respectively.) In addition, we apply some of these results in order to calculate the complex dimensions of the Cantor dust.

4.7.1 Embeddings Into Higher Dimensions in the Case of Bounded Sets

We begin this subsection by stating a result which (along with the subsequent result, Theorem 4.7.2) will be key to the developments in this section. We first work with bounded sets, in Subsection 4.7.1, and then with general RFDs, in Subsection 4.7.2.

Proposition 4.7.1. *Let $A \subseteq \mathbb{R}^N$ be a bounded set and let $\overline{D} := \overline{\dim}_B A$. Then, for the tube zeta functions of A and $A \times \{0\} \subseteq \mathbb{R}^{N+1}$, the following equality holds:*

$$\tilde{\zeta}_{A\times\{0\}}(s;\delta) = 2\int_0^{\pi/2} \frac{\tilde{\zeta}_A(s;\delta\sin\tau)}{\sin^{s-N-1}\tau}\,\mathrm{d}\tau, \tag{4.7.1}$$

for all $s \in \mathbb{C}$ such that $\mathrm{Re}\, s > \overline{D}$.

Proof. First of all, it is well known and easy to check directly from the definitions (see Equations (1.3.1) and (1.3.4)) that $\overline{\dim}_B(A\times\{0\}) = \overline{\dim}_B A$, from which we conclude that the tube zeta functions of A and $A\times\{0\}$ are both holomorphic in the right half-plane $\{\mathrm{Re}\, s > \overline{D}\}$. Furthermore, we use the fact (see [Res, Proposition 6]) that for every $t > 0$, we have

$$|(A\times\{0\})_t|_{N+1} = 2\int_0^t |A_{\sqrt{t^2-u^2}}|_N\,\mathrm{d}u, \tag{4.7.2}$$

where as before, $|\cdot|_N$ denotes the N-dimensional Lebesgue measure. After having made the change of variable $u := t\cos v$, this yields

$$|(A\times\{0\})_t|_{N+1} = 2t\int_0^{\pi/2} |A_{t\sin v}|_N \sin v\,\mathrm{d}v. \tag{4.7.3}$$

Finally, for the tube zeta function of $A\times\{0\}$, we can write successively:

$$\begin{aligned}\tilde{\zeta}_{A\times\{0\}}(s;\delta)&=\int_0^\delta t^{s-N-2}|(A\times\{0\})_t|_{N+1}\,\mathrm{d}t\\&=2\int_0^\delta t^{s-N-1}\,\mathrm{d}t\int_0^{\pi/2}|A_{t\sin v}|_N\sin v\,\mathrm{d}v\\&=2\int_0^{\pi/2}\sin v\,\mathrm{d}v\int_0^\delta t^{s-N-1}|A_{t\sin v}|_N\,\mathrm{d}t\\&=2\int_0^{\pi/2}\sin^{N+1-s}v\,\mathrm{d}v\int_0^{\delta\sin v}\tau^{s-N-1}|A_\tau|_N\,\mathrm{d}\tau\\&=2\int_0^{\pi/2}\frac{\tilde{\zeta}_A(s;\delta\sin v)}{\sin^{s-N-1}v}\,\mathrm{d}v,\end{aligned}$$

where we have used the Fubini–Tonelli theorem in order to justify the interchange of integrals (in the third equality), as well as made another change of variable (in the fourth equality), namely, $\tau:=t\sin v$. This completes the proof of the proposition. □

Theorem 4.7.2. *Let $A\subseteq\mathbb{R}^N$ be a bounded set and let $\overline{D}:=\overline{\dim}_BA$. Then, we have the following equality between $\tilde{\zeta}_A$, the tube zeta function of A, and $\tilde{\zeta}_{A_M}$, the tube zeta function of $A_M:=A\times\{0\}\cdots\times\{0\}\subseteq\mathbb{R}^{N+M}$, with $M\in\mathbb{N}$ arbitrary:*

$$\tilde{\zeta}_{A_M}(s;\delta)=\frac{(\sqrt{\pi})^M\,\Gamma\left(\frac{N-s}{2}+1\right)}{\Gamma\left(\frac{N+M-s}{2}+1\right)}\tilde{\zeta}_A(s;\delta)+E(s;\delta),\tag{4.7.4}$$

initially valid for all $s\in\mathbb{C}$ such that $\operatorname{Re}s>\overline{D}$. *Here, the* error function $E(s):=E(s;\delta)$ *(initially defined in the case when $M=1$ by the integral on the right-hand side of Equation* (4.7.7) *below) admits a meromorphic extension to all of $\mathbb{C}$. The possible poles (in $\mathbb{C}$) of $E(s;\delta)$ are located at $s_k:=N+2+2k$ for every $k\in\mathbb{N}_0$, and all of them are simple. (It follows that $\tilde{\zeta}_A$ is well defined at each s_k.) Moreover, we have that for each $k\in\mathbb{N}_0$,*[53]

$$\operatorname{res}(E(\cdot;\delta),s_k)=\frac{(-1)^{k+1}(\sqrt{\pi})^M}{k!\,\Gamma\left(\frac{M}{2}-k\right)}\tilde{\zeta}_A(s_k;\delta).\tag{4.7.5}$$

More specifically, if M is even, then all of the poles s_k of $E(s;\delta)$ for $k\geq M/2$ are canceled; i.e., the corresponding residues in (4.7.5) *are equal to zero. On the other hand, if M is odd, there are no such cancellations and all of the residues in* (4.7.5) *are nonzero; so that all the s_k's are simple poles of $E(s;\delta)$ in that case.*

[53] We refer to Theorem 4.7.3 for more precise information about the domain of validity of the approximate functional equation (4.7.4), and to Corollary 4.7.4 for information about the relationship between the (visible) poles of $\tilde{\zeta}_A$ and $\tilde{\zeta}_{A_M}$.

Proof. We will prove the theorem in the case when $M = 1$. The general case when $M \in \mathbb{N}$ then follows immediately by induction. From Proposition 4.7.1 we have that for $\mathrm{Re}\, s > \overline{\dim}_B A$, formula (4.7.1) holds. In turn, this latter identity can be written as

$$\begin{aligned}\tilde{\zeta}_{A\times\{0\}}(s;\delta) &= 2\tilde{\zeta}_A(s;\delta)\int_0^{\pi/2}\frac{\mathrm{d}\tau}{\sin^{s-N-1}\tau}\\ &\quad -2\int_0^{\pi/2}\frac{\mathrm{d}\nu}{\sin^{s-N-1}\nu}\int_{\delta\sin\nu}^{\delta}\tau^{s-N-1}|A_\tau|_N\,\mathrm{d}\tau\\ &= \tilde{\zeta}_A(s;\delta)\cdot \mathrm{B}\left(\frac{N-s}{2}+1,\frac{1}{2}\right)+E(s;\delta),\end{aligned} \tag{4.7.6}$$

where $\mathrm{B}(u,v)$ denotes the *Euler beta function* and

$$E(s;\delta) := -2\int_0^{\pi/2}\frac{\mathrm{d}\nu}{\sin^{s-N-1}\nu}\int_{\delta\sin\nu}^{\delta}\tau^{s-N-1}|A_\tau|_N\,\mathrm{d}\tau. \tag{4.7.7}$$

By using the functional equation which links the beta function with the gamma function (namely, $\mathrm{B}(x,y) = \Gamma(x)\Gamma(y)/\Gamma(x+y)$ for all $x,y>0$ and hence, upon meromorphic continuation, for all $x,y\in\mathbb{C}$), we obtain that (4.7.4) holds (with $M=1$) for all $s\in\mathbb{C}$ such that $\mathrm{Re}\, s > \overline{\dim}_B A$.

By looking at the expression for $E(s;\delta)$ in (4.7.7), we see that the integrand is holomorphic for every $\nu\in(0,\pi/2)$ since the integral $\int_{\delta\sin\nu}^{\delta}\tau^{s-N-1}|A_\tau|_N\,\mathrm{d}\tau$ is equal to $\tilde{\zeta}_A(s;\delta)-\tilde{\zeta}_A(s;\delta\sin\nu)$, which is an entire function. Furthermore, if we assume that $\mathrm{Re}\, s < N+1$, then since $\tau\mapsto\tau^{\mathrm{Re}\, s-N-1}$ is decreasing, we have the following estimate:

$$\begin{aligned}|E(s;\delta)| &\le 2\int_0^{\pi/2}\sin^{N+1-\mathrm{Re}\, s}\nu\,\mathrm{d}\nu\int_{\delta\sin\nu}^{\delta}\tau^{\mathrm{Re}\, s-N-1}|A_\tau|_N\,\mathrm{d}\tau\\ &\le 2|A_\delta|_N\int_0^{\pi/2}\sin^{N+1-\mathrm{Re}\, s}\nu\,\mathrm{d}\nu\int_{\delta\sin\nu}^{\delta}\tau^{\mathrm{Re}\, s-N-1}\,\mathrm{d}\tau\\ &\le 2\delta^{\mathrm{Re}\, s-N-1}|A_\delta|_N\int_0^{\pi/2}\sin^{N+1-\mathrm{Re}\, s}\nu\,\sin^{\mathrm{Re}\, s-N-1}\nu\int_{\delta\sin\nu}^{\delta}\mathrm{d}\tau\\ &= 2\delta^{\mathrm{Re}\, s-N}|A_\delta|_N\int_0^{\pi/2}(1-\sin\nu)\,\mathrm{d}\nu\\ &= 2\delta^{\mathrm{Re}\, s-N}|A_\delta|_N\left(\frac{\pi}{2}-1\right).\end{aligned} \tag{4.7.8}$$

Hence,

$$|E(s;\delta)| \le 2\delta^{\mathrm{Re}\, s-N}|A_\delta|\left(\frac{\pi}{2}-1\right). \tag{4.7.9}$$

We conclude from this inequality that for $s_0\in\{\mathrm{Re}\, s<N+1\}$, the condition $(3')$ of Remark 2.1.48 is satisfied, which implies, in light of Theorem 2.1.47, that $E(s;\delta)$ is holomorphic on the open half-plane $\{\mathrm{Re}\, s<N+1\}$.

On the other hand, we know that both of the tube zeta functions $\tilde{\zeta}_A$ and $\tilde{\zeta}_{A_M}$ are holomorphic on $\{\mathrm{Re}\, s>\overline{\dim}_B A\}\supseteq\{\mathrm{Re}\, s>N\}$. The fact that $E(s;\delta)$ is meromorphic on $\mathbb{C}$, as well as the statement about its poles, now follows from Equation (4.7.4) (with $M = 1$) and the fact that the gamma function is nowhere vanishing in $\mathbb{C}$. (In fact, $1/\Gamma(s)$ is an entire function with zeros at the nonpositive integers.) More specifically, the locations of the poles of $E(s;\delta)$ must coincide with the locations of the poles $s_k = N+2+2k$, for $k \in \mathbb{N}_0$, of $\Gamma((N-s)/2+1)$ since the left-hand side of (4.7.4) is holomorphic on $\{\mathrm{Re}\, s > \overline{\dim}_B A\}$ and because $\tilde{\zeta}_A(s_k) > 0$ (since it is defined as the integral of a positive function). Note that since $N \geq \overline{D}$, we have $s_k > \overline{D}$, and hence, $\tilde{\zeta}_A$ is well defined at s_k, for each $k \in \mathbb{N}_0$.

Finally, by multiplying (4.7.4) by $(s - s_k)$, taking the limit as $s \to s_k$ and then using the fact that the residue of the gamma function at $-k$ is equal to $(-1)^k/k!$, we deduce that (4.7.5) holds, as desired.

Furthermore, if M is odd, there are no cancellations between the poles of the numerator and of the denominator in (4.7.4) since an integer cannot be both even and odd; i.e., the residues are nonzero for each $k \in \mathbb{N}_0$. On the other hand, if M is even, then it is clear that all of the residues at s_k for $k \geq M/2$ are equal to zero; i.e., the corresponding poles at s_k cancel out with the poles of the denominator in (4.7.4). □

Theorem 4.7.2 has as an important consequence, namely, the fact that the notion of complex dimensions does not depend on the dimension of the ambient space.

Theorem 4.7.3. *Let $A \subseteq \mathbb{R}^N$ be a bounded set and A_M be its embedding into $\mathbb{R}^{N+M}$, with $M \in \mathbb{N}$ arbitrary. Then, the tube zeta function $\tilde{\zeta}_A$ of A has a meromorphic extension to a given connected open neighborhood U of the critical line $\{\mathrm{Re}\, s = \overline{\dim}_B A\}$ if and only if the analogous statement is true for the tube zeta function $\tilde{\zeta}_{A_M}$ of A_M. Furthermore, in that case, the approximate functional equation* (4.7.4) *remains valid for all $s \in U$. In addition, the multisets*[54] *of the poles of $\tilde{\zeta}_A$ and $\tilde{\zeta}_{A_M}$ located in U coincide; i.e., $\mathscr{P}(\tilde{\zeta}_A, U) = \mathscr{P}(\tilde{\zeta}_{A_M}, U)$.*[55] *Consequently, neither the values nor the multiplicities of the complex dimensions of A depend on the dimension of the ambient space.*

Proof. This is a direct consequence of Theorem 4.7.2 and the principle of analytic continuation. More precisely, identity (4.7.4) is valid for all $s \in \mathbb{C}$ such that $\mathrm{Re}\, s > \overline{\dim}_B A$ and the function $E(s;\delta)$ is meromorphic on all of $\mathbb{C}$. Furthermore, according to Theorem 4.7.2, the poles of $E(s;\delta)$ belong to $\{\mathrm{Re}\, s \geq N+2\}$, which implies that the function $s \mapsto E(s;\delta)$ is holomorphic on $\{\mathrm{Re}\, s < N+2\}$. Identity (4.7.4) then remains valid if any of the two zeta functions involved (namely, $\tilde{\zeta}_A$ or $\tilde{\zeta}_{A_M}$) has a meromorphic continuation to some connected open neighborhood of the critical line $\{\mathrm{Re}\, s = \overline{\dim}_B A\}$. This completes the proof of the theorem. □

Corollary 4.7.4. *Let $A \subseteq \mathbb{R}^N$ be a bounded set (with $\overline{D} := \overline{\dim}_B A$) such that its tube zeta function $\tilde{\zeta}_A$ has a meromorphic continuation to a connected open neighborhood U of the critical line $\{\mathrm{Re}\, s = \overline{\dim}_B A\}$. Furthermore, suppose that $s = \overline{D}$ is a simple*

[54] In these multisets, each pole is counted according to its multiplicity.

[55] Recall that the bounded sets A and A_M have the same upper Minkowski dimension, $\overline{\dim}_B A = \overline{\dim}_B A_M$, and hence, the same critical line $\{\mathrm{Re}\, s = \overline{\dim}_B A\}$.

pole of $\tilde{\zeta}_A$. *Let* $A_M \subseteq \mathbb{R}^{N+M}$ *be the canonical embedding of* A *into* $\mathbb{R}^{N+M}$, *with* $M \in \mathbb{N}$ *arbitrary, as in Theorem 4.7.2. Then*

$$\operatorname{res}(\tilde{\zeta}_{A_M}, \overline{D}) = \frac{(\sqrt{\pi})^M \, \Gamma\left(\frac{N-\overline{D}}{2}+1\right)}{\Gamma\left(\frac{N+M-\overline{D}}{2}+1\right)} \operatorname{res}(\tilde{\zeta}_A, \overline{D}). \tag{4.7.10}$$

We point out here that the above corollary is compatible with the dimensional invariance of the normalized Minkowski content, obtained in [Kne] (see also [Res]). More specifically, if in the above corollary, we assume, in addition, that $\overline{D}$ is the only pole of the tube zeta function of A on the critical line $\{\operatorname{Re} s = \overline{D}\}$ (i.e., $\overline{D}$ is the only complex dimension of A with real part $\overline{D}$), then, according to Theorem 5.4.2 of Chapter 5 below (the "sufficient condition for Minkowski measurability"), A and $A \times \{0\}$ are Minkowski measurable with Minkowski dimension $D := \overline{D}$ and have respective Minkowski contents satisfying the following identity:

$$\frac{\mathscr{M}^D(A)}{\pi^{\frac{D-N}{2}}\,\Gamma\left(\frac{N-D}{2}+1\right)} = \frac{\mathscr{M}^D(A \times \{0\})}{\pi^{\frac{D-N-1}{2}}\,\Gamma\left(\frac{N+1-D}{2}+1\right)}. \tag{4.7.11}$$

4.7.2 Embeddings Into Higher Dimensions in the Case of Relative Fractal Drums

The observations made in the previous subsection in the context of bounded subsets of $\mathbb{R}^N$ can also be extended to the more general context of relative fractal drums (RFDs) in $\mathbb{R}^N$. More specifically, let (A, Ω) be a relative fractal drum in $\mathbb{R}^N$ and let

$$(A \times \{0\}, \Omega \times (-1,1))$$

be its natural embedding into $\mathbb{R}^{N+1}$. We want to connect the relative tube zeta functions of these two RFDs; the following lemma will be needed for this purpose.

Lemma 4.7.5. *Let* (A, Ω) *be a relative fractal drum in* $\mathbb{R}^N$ *and fix* $\delta \in (0,1)$. *Then we have*

$$\left|(A \times \{0\})_\delta \cap (\Omega \times (-1,1))\right|_{N+1} = 2\int_0^\delta |A_{\sqrt{\delta^2-u^2}} \cap \Omega|_N \, du. \tag{4.7.12}$$

Proof. We proceed much as in the proof of [Res, Proposition 6]. Namely, if we let $(x,y) \in \mathbb{R}^N \times \mathbb{R} \equiv \mathbb{R}^{N+1}$ and define

$$V := \{(x,y) : d_{N+1}((x,y), A \times \{0\}) \le \delta\} \cap \{(x,y) : x \in \Omega, |y| \le 1\}, \tag{4.7.13}$$

where $(x,y) \in \mathbb{R}^N \times \mathbb{R} \simeq \mathbb{R}^{N+1}$ and for any $k \in \mathbb{N}$, d_k denotes the Euclidean distance in $\mathbb{R}^k$. It is clear that the following equality holds:

$$d_{N+1}((x,y), A \times \{0\}) = \sqrt{d_N(x,A)^2 + y^2}.$$

This implies that for a fixed $y \in [-\delta, \delta] \subset \mathbb{R}$, we have

$$\begin{aligned} V_y :&= \{x \in \mathbb{R}^N : d_{N+1}((x,y), A \times \{0\}) \le \delta\} \\ &= \left\{x \in \mathbb{R}^N : d_N(x,A) \le \sqrt{\delta^2 - y^2}\right\}. \end{aligned} \tag{4.7.14}$$

(Note that if $|y| > \delta$, then V_y is empty.) Finally, Fubini's theorem implies that

$$\begin{aligned} \big|(A \times \{0\})_\delta \cap (\Omega \times (-1,1))\big|_{N+1} &= \int_V \mathrm{d}x\,\mathrm{d}y \\ &= \int_{-\delta}^{\delta} \mathrm{d}y \int_{V_y \cap \{x \in \mathbb{R}^N : x \in \Omega\}} \mathrm{d}x \\ &= 2\int_0^{\delta} |A_{\sqrt{\delta^2 - y^2}} \cap \Omega|_N \,\mathrm{d}y, \end{aligned}$$

which completes the proof of the lemma. □

The above lemma will eventually yield (in Theorem 4.7.10 below) an RFD analog of Proposition 4.7.1 from Subsection 4.7.1 above. First, however, we will show that the upper and lower relative box dimensions of an RFD are independent of the ambient space dimension.

Proposition 4.7.6. *Let (A,Ω) be an RFD in $\mathbb{R}^N$ and let*

$$(A,\Omega)_M := (A_M, \Omega \times (-1,1)^M) \tag{4.7.15}$$

be its embedding into $\mathbb{R}^{N+M}$, for some $M \in \mathbb{N}$. Then we have that

$$\overline{\dim}_B(A,\Omega) = \overline{\dim}_B(A,\Omega)_M \tag{4.7.16}$$

and

$$\underline{\dim}_B(A,\Omega) = \underline{\dim}_B(A,\Omega)_M. \tag{4.7.17}$$

Proof. We only prove the proposition in the case when $M = 1$, from which the general result then easily follows by induction. It is clear that for $0 < \delta < 1$, we have

$$\begin{aligned} (A \times \{0\})_\delta \cap (\Omega \times (-1,1)) &\subseteq (A \times \{0\})_\delta \cap (\Omega \times (-\delta,\delta)) \\ &\subseteq (A_\delta \cap \Omega) \times (-\delta,\delta); \end{aligned}$$

so that

$$|(A \times \{0\})_\delta \cap (\Omega \times (-1,1))|_{N+1} \le 2\delta |A_\delta \cap \Omega|_N. \tag{4.7.18}$$

This observation, in turn, implies that for every $r \in \mathbb{R}$, we have

$$\frac{|(A\times\{0\})_\delta \cap (\Omega\times(-1,1))|_{N+1}}{\delta^{N+1-r}} \leq \frac{2|A_\delta \cap \Omega|_N}{\delta^{N-r}}. \tag{4.7.19}$$

Furthermore, by successively taking the upper and lower limits as $\delta \to 0^+$ in Equation (4.7.19) just above, we obtain the following inequalities involving the r-dimensional upper and lower relative Minkowski contents, respectively:

$$\mathcal{M}^{*r}(A,\Omega)_1 \leq 2\mathcal{M}^{*r}(A,\Omega) \quad \text{and} \quad \mathcal{M}_*^r(A,\Omega)_1 \leq 2\mathcal{M}_*^r(A,\Omega). \tag{4.7.20}$$

In light of the definition of the relative upper and lower box (or Minkowski) dimensions (see Equation (4.1.4) and Equation (4.1.6), along with the text surrounding them), we deduce that

$$\overline{\dim}_B(A,\Omega)_1 \leq \overline{\dim}_B(A,\Omega) \quad \text{and} \quad \underline{\dim}_B(A,\Omega)_1 \leq \underline{\dim}_B(A,\Omega). \tag{4.7.21}$$

On the other hand, for geometric reasons, we have that

$$(A_{\delta/2} \cap \Omega) \times \left(-\frac{\delta\sqrt{3}}{2}, \frac{\delta\sqrt{3}}{2}\right) \subseteq (A\times\{0\})_\delta \cap (\Omega\times(-1,1));$$

so that

$$\delta\sqrt{3}|A_{\delta/2} \cap \Omega|_N \leq |(A\times\{0\})_\delta \cap (\Omega\times(-1,1))|_{N+1}. \tag{4.7.22}$$

Much as before, this inequality implies that for every $r \in \mathbb{R}$, we have

$$\frac{\sqrt{3}|A_{\delta/2} \cap \Omega|_N}{2^{N-r}(\delta/2)^{N-r}} \leq \frac{|(A\times\{0\})_\delta \cap (\Omega\times(-1,1))|_{N+1}}{\delta^{N+1-r}} \tag{4.7.23}$$

and by successively taking the upper and lower limits as $\delta \to 0^+$, we obtain that

$$\frac{\sqrt{3}\mathcal{M}^{*r}(A,\Omega)}{2^{N-r}} \leq \mathcal{M}^{*r}(A,\Omega)_1 \quad \text{and} \quad \frac{\sqrt{3}\mathcal{M}_*^r(A,\Omega)}{2^{N-r}} \leq \mathcal{M}_*^r(A,\Omega)_1. \tag{4.7.24}$$

Finally, this completes the proof because (again in light of Equation (4.1.4) and Equation (4.1.6), along with the text surrounding them), (4.7.24) implies the reverse inequalities for the upper and lower relative box dimensions in (4.7.21). □

Remark 4.7.7. Observe that it follows from Proposition 4.7.6 (combined with part (*b*) of Theorem 4.1.7) that the RFDs (A,Ω) and $(A,\Omega)_M$ have the same upper Minkowski dimension, $\overline{\dim}_B(A,\Omega) = \overline{\dim}_B(A,\Omega)_M$, and hence, the same critical line $\{\mathrm{Re}\, s = \overline{\dim}_B(A,\Omega)\}$. This fact will be used implicitly in the statement of Proposition 4.7.8 as well as in the statements of Theorems 4.7.9 and 4.7.10 just below.

We can now state the desired results for embedded RFDs and their relative fractal zeta functions. In light of Lemma 4.7.5 and Proposition 4.7.6, the proofs follow the same steps as in the corresponding results established in Subsection 4.7.1 about bounded subsets of $\mathbb{R}^N$ (namely, Proposition 4.7.1 and Theorem 4.7.2, respectively), and for this reason, we will omit them.

Proposition 4.7.8. *Fix $\delta \in (0,1)$ and let (A,Ω) be an RFD in $\mathbb{R}^N$, with $\overline{D} := \overline{\dim}_B(A,\Omega)$. Then, for the relative tube zeta functions of (A,Ω) and $(A,\Omega)_1 := (A \times \{0\}, \Omega \times (-1,1))$, the following equality holds:*

$$\tilde{\zeta}_{A\times\{0\},\Omega\times(-1,1)}(s;\delta) = 2\int_0^{\pi/2} \frac{\tilde{\zeta}_{A,\Omega}(s;\delta\sin\tau)}{\sin^{s-N-1}\tau}\,d\tau, \tag{4.7.25}$$

for all $s \in \mathbb{C}$ such that $\operatorname{Re} s > \overline{D}$.

Theorem 4.7.9. *Fix $\delta \in (0,1)$ and let (A,Ω) be an RFD in $\mathbb{R}^N$, with $\overline{D} := \overline{\dim}_B(A,\Omega)$. Then, we have the following equality between $\tilde{\zeta}_{A,\Omega}$, the tube zeta function of (A,Ω), and $\tilde{\zeta}_{A_M,\Omega\times(-1,1)^M}$, the tube zeta function of the relative fractal drum $(A,\Omega)_M := (A_M, \Omega\times(-1,1)^M)$ in $\mathbb{R}^{N+M}$, where $M \in \mathbb{N}$ is arbitrary:*

$$\tilde{\zeta}_{A_M,\Omega\times(-1,1)^M}(s;\delta) = \frac{\left(\sqrt{\pi}\right)^M \Gamma\left(\frac{N-s}{2}+1\right)}{\Gamma\left(\frac{N+M-s}{2}+1\right)}\tilde{\zeta}_{A,\Omega}(s;\delta) + E(s;\delta), \tag{4.7.26}$$

initially valid for all $s \in \mathbb{C}$ such that $\operatorname{Re} s > \overline{D}$.[56] *Here, the* error function $E(s) := E(s;\delta)$ *is meromorphic on all of $\mathbb{C}$. Furthermore, the possible poles (in $\mathbb{C}$) of $E(s;\delta)$ are located at $s_k := N+2+2k$ for every $k \in \mathbb{N}_0$, and all of them are simple. (It follows that $\tilde{\zeta}_A$ is well defined at each s_k.) Moreover, we have that for each $k \in \mathbb{N}_0$,*

$$\operatorname{res}(E(\cdot\,;\delta), s_k) = \frac{(-1)^{k+1}\left(\sqrt{\pi}\right)^M}{k!\,\Gamma\left(\frac{M}{2}-k\right)}\tilde{\zeta}_{A,\Omega}(s_k;\delta). \tag{4.7.27}$$

More specifically, if M is even, then all of the poles s_k of $E(s;\delta)$ for $k \geq M/2$ are canceled; i.e., the corresponding residues in (4.7.27) *are equal to zero. On the other hand, if M is odd, there are no such cancellations and all of the residues in* (4.7.27) *are nonzero; so that all of the s_k's are simple poles of $E(s;\delta)$ in that case.*

We deduce at once from Theorem 4.7.9 the following key result about the invariance of the complex dimensions of a relative fractal drum with respect the dimension of the ambient space. This result extends Theorem 4.7.3 to general RFDs.

Theorem 4.7.10. *Let (A,Ω) be an RFD in $\mathbb{R}^N$ and let the RFD $(A,\Omega)_M := (A_M, \Omega\times(-1,1)^M)$ be its embedding into $\mathbb{R}^{N+M}$, for some arbitrary $M \in \mathbb{N}$. Then, the tube zeta function $\tilde{\zeta}_{A,\Omega}$ of (A,Ω) has a meromorphic extension to a given connected open neighborhood U of the critical line* $\{\operatorname{Re} s = \overline{\dim}_B(A,\Omega)\}$ *if and only if*

---

[56] See Theorem 4.7.10 for more precise information about the domain of validity of the approximate functional equation (4.7.26).

the analogous statement is true for the tube zeta function $\tilde{\zeta}_{(A,\Omega)_M} := \tilde{\zeta}_{A_M,\Omega\times(-1,1)^M}$ *of* $(A,\Omega)_M$. *(See Remark 4.7.7 just above.) Furthermore, in that case, the approximate functional equation* (4.7.26) *remains valid for all* $s \in U$. *In addition, the multisets of the poles of* $\tilde{\zeta}_{A,\Omega}$ *and* $\tilde{\zeta}_{(A,\Omega)_M}$ *belonging to* U *coincide; i.e.,*

$$\mathscr{P}(\tilde{\zeta}_{A,\Omega}, U) = \mathscr{P}(\tilde{\zeta}_{(A,\Omega)_M}, U). \tag{4.7.28}$$

Consequently, neither the values nor the multiplicities of the complex dimensions of the RFD (A,Ω) *depend on the dimension of the ambient space.*

Remark 4.7.11. In the above discussion about embedding RFDs into higher-dimensional spaces, we can also make similar observations if we embed (A,Ω) as a 'one-sided' RFD, for example of the form $(A\times\{0\}, \Omega\times(0,1))$, a fact which can be more useful when decomposing a relative fractal drum into a union of relative fractal subdrums in order to compute its distance (or tube) zeta function.[57] This observation follows immediately from the above results for 'two-sided' embeddings of RFDs since, by symmetry, we have

$$\tilde{\zeta}_{A\times\{0\},\Omega\times(-1,1)}(s) = 2\,\tilde{\zeta}_{A\times\{0\},\Omega\times(0,1)}(s). \tag{4.7.29}$$

We note that when using the above formulas, one only has to be careful to take into account the factor 2. Furthermore, we can also embed (A,Ω) as

$$(A\times\{0\}, \Omega\times(-\alpha,\alpha)) \quad \text{or} \quad (A\times\{0\}, \Omega\times(0,\alpha)), \tag{4.7.30}$$

for some $\alpha > 0$, but in that case, the corresponding formulas will only be valid for all $\delta \in (0,\alpha)$.

We could now use the functional equation (2.2.23) connecting the tube and distance zeta functions, in order to translate the above results in terms of $\zeta_{A,\Omega} := \zeta_{A,\Omega}(\,\cdot\,;\delta)$, the (relative) distance zeta function of the RFD (A,Ω). However, we will instead use another approach because it gives some additional information about the resulting error function. More specifically, consider the *Mellin zeta function of a relative fractal drum*, to be introduced and studied in Section 5.4 below (see Definition 5.4.6). Here, we state some of its properties (see Theorems 5.4.7, 5.4.9 and 5.4.10) which will be needed in the following discussion.

The Mellin zeta function of an RFD (A,Ω) with $\overline{\dim}_B(A,\Omega) < N$ is initially defined by

$$\zeta^{\mathfrak{M}}_{A,\Omega}(s) = \int_0^{+\infty} t^{s-N-1}|A_t\cap\Omega|\,\mathrm{d}t, \tag{4.7.31}$$

for all $s \in \mathbb{C}$ located in a suitable vertical strip. In fact, in light of Theorem 5.4.7, the above Lebesgue integral is absolutely convergent (and hence, convergent) for all $s \in \mathbb{C}$ such that $\mathrm{Re}\, s \in (\overline{\dim}_B(A,\Omega), N)$. Moreover, the relative distance and Mellin zeta functions of (A,Ω) are connected by the functional equation

[57] See Subsection 4.2.3 for examples of such decompositions in the case of the relative Sierpiński gasket and carpet, as well as of their higher-dimensional analogs.

$$\zeta_{A,\Omega}(s) = (N-s)\zeta_{A,\Omega}^{\mathfrak{M}}(s), \tag{4.7.32}$$

on every open connected set $U \subseteq \mathbb{C}$ to which any of the two zeta functions has a meromorphic continuation. Observe that in (4.7.32), the parameter δ is absent. Indeed, this means implicitly that the functional equation (4.7.32) is valid only for the parameters $\delta > 0$ for which $\Omega \subseteq A_\delta$ is satisfied; that is, when the equality $\zeta_{A,\Omega}(s;\delta) = \int_\Omega d(x,A)^{s-N}\,\mathrm{d}x$ is satisfied.

We will now embed the relative fractal drum (A,Ω) of $\mathbb{R}^N$ into $\mathbb{R}^{N+1}$ as

$$(A\times\{0\},\Omega\times\mathbb{R}).$$

Strictly speaking, this is not a relative fractal drum in $\mathbb{R}^{N+1}$ since there does not exist $\delta > 0$ such that $\Omega\times\mathbb{R} \subseteq (A\times\{0\})_\delta$. On the other hand, observe that Lemma 4.7.5 is now valid for every $\delta > 0$; that is,

$$\big|(A\times\{0\})_\delta \cap (\Omega\times\mathbb{R})\big|_{N+1} = 2\int_0^\delta |A_{\sqrt{\delta^2-u^2}}\cap\Omega|_N\,\mathrm{d}u. \tag{4.7.33}$$

Proposition 4.7.12. *Let (A,Ω) be an RFD in $\mathbb{R}^N$ such that $\overline{\dim}_B(A,\Omega) < N$. Then the function $F = F(s)$, defined by the integral*

$$F(s) := \int_0^{+\infty} t^{s-N-2}\big|(A\times\{0\})_t \cap (\Omega\times\mathbb{R})\big|_{N+1}\,\mathrm{d}t, \tag{4.7.34}$$

is holomorphic inside the vertical strip $\{\overline{\dim}_B(A,\Omega) < \mathrm{Re}\,s < N\}$.

Proof. We split the integral into two integrals: $F(s) = \int_0^1 + \int_1^{+\infty}$. According to Proposition 4.7.6, the first integral,

$$\begin{aligned}&\int_0^1 t^{s-N-2}\big|(A\times\{0\})_t \cap (\Omega\times\mathbb{R})\big|_{N+1}\,\mathrm{d}t\\ &=\int_0^1 t^{s-N-2}\big|(A\times\{0\})_t \cap (\Omega\times(-1,1))\big|_{N+1}\,\mathrm{d}t,\end{aligned}$$

defines a holomorphic function on the right half-plane $\{\mathrm{Re}\,s > \overline{\dim}_B(A,\Omega)\}$.

In order to deal with the second integral, we observe that

$$\big|(A\times\{0\})_t \cap (\Omega\times\mathbb{R})\big|_{N+1} \le 2t|\Omega|_N,$$

and consequently, deduce that

$$\left|\int_1^{+\infty} t^{s-N-2}\big|(A\times\{0\})_t \cap (\Omega\times\mathbb{R})\big|_{N+1}\,\mathrm{d}t\right| \le 2|\Omega|_N\int_1^{+\infty} t^{\mathrm{Re}\,s-N-1}\,\mathrm{d}t = \frac{2|\Omega|_N}{N-\mathrm{Re}\,s},$$

for all $s\in\mathbb{C}$ such that $\mathrm{Re}\,s < N$. In light of Theorem 2.1.47 and Remark 2.1.48, the latter inequality implies that the integral over $(1,+\infty)$ defines a holomorphic function on the left half-plane $\{\mathrm{Re}\, s < N\}$. Therefore, it follows that $F(s)$ is holomorphic in the vertical strip $\{\overline{\dim}_B(A,\Omega) < \mathrm{Re}\, s < N\}$ and the proof of the proposition is complete. □

In light of the above proposition, we continue to use the convenient notation $\zeta^{\mathfrak{M}}_{A\times\{0\},\Omega\times\mathbb{R}}$ for the integral appearing on the right-hand side of (4.7.34) although, as was noted earlier, $(A\times\{0\},\Omega\times\mathbb{R})$ is not technically a relative fractal drum in $\mathbb{R}^{N+1}$; see Remark 4.7.11 above. The following result is the counterpart of Theorem 4.7.2 in the present, more general context.

Theorem 4.7.13. *Let (A,Ω) be a relative fractal drum in $\mathbb{R}^N$ such that $\overline{D} := \overline{\dim}_B(A,\Omega) < N$. Then, for every $a > 0$, the following approximate functional equation holds:*

$$\zeta_{A\times\{0\},\Omega\times(-a,a)}(s) = \frac{\sqrt{\pi}\,\Gamma\left(\frac{N-s}{2}\right)}{\Gamma\left(\frac{N+1-s}{2}\right)}\zeta_{A,\Omega}(s) + E(s;a), \tag{4.7.35}$$

initially valid for all $s\in\mathbb{C}$ such that $\mathrm{Re}\, s > \overline{D}$. Here, the error function $E(s) := E(s;a)$ *is initially given (for all $s\in\mathbb{C}$ such that* $\mathrm{Re}\, s < N$*) by*

$$E(s;a) := (s-N-1)\int_a^{+\infty} t^{s-N-2}|(A\times\{0\})_t\cap\Omega\times(\mathbb{R}\setminus(-a,a))|_{N+1}\,\mathrm{d}t, \tag{4.7.36}$$

and admits a meromorphic extension to all of $\mathbb{C}$, with a set of simple poles equal to $\{N+2k : k\in\mathbb{N}_0\}$.

Moreover, Equation (4.7.35) *remains valid on any connected open neighborhood of the critical line $\{\mathrm{Re}\, s = \overline{D}\}$ to which $\zeta_{A,\Omega}$ (or, equivalently, $\zeta_{A\times\{0\},\Omega\times(-a,a)}$) can be meromorphically continued.*

Proof. In a completely analogous way as in the proof of Theorem 4.7.2, we obtain that

$$\tilde{\zeta}_{A\times\{0\},\Omega\times\mathbb{R}}(s;\delta) = \frac{\sqrt{\pi}\,\Gamma\left(\frac{N-s}{2}+1\right)}{\Gamma\left(\frac{N+1-s}{2}+1\right)}\tilde{\zeta}_{A,\Omega}(s;\delta) + \tilde{E}(s;\delta), \tag{4.7.37}$$

now valid for all $\delta > 0$ (see Equation (4.7.33) above and the discussion preceding it). Furthermore, the error function $\tilde{E}(s) := \tilde{E}(s;\delta)$ is holomorphic on $\{\mathrm{Re}\, s < N+1\}$ and

$$|\tilde{E}(s,\delta)| \le 2\delta^{\mathrm{Re}\, s-N}|A_\delta\cap\Omega|_N\left(\frac{\pi}{2}-1\right) \tag{4.7.38}$$

for all $s\in\mathbb{C}$ such that $\mathrm{Re}\, s < N+1$. See the proof of Theorem 4.7.2 and Equation (4.7.8) in order to derive the above estimate. The estimate (4.7.38) now implies that the sequence of holomorphic functions $\tilde{E}(\,\cdot\,;n)$ tends to 0 as $n\to\infty$, uniformly on every compact subset of $\{\mathrm{Re}\, s < N\}$, since $|A_n\cap\Omega| = |\Omega|$ for all n sufficiently large. Furthermore, we also have that $\tilde{\zeta}_{A,\Omega}(\,\cdot\,;n)\to\zeta^{\mathfrak{M}}_{A,\Omega}$ and

$$\tilde{\zeta}_{A\times\{0\},\Omega\times\mathbb{R}}(s;n)\to\zeta^{\mathfrak{M}}_{A\times\{0\},\Omega\times\mathbb{R}} \quad \text{as} \quad n\to\infty, \tag{4.7.39}$$

uniformly on every compact subset of $\{\overline{D} < \mathrm{Re}\, s < N\}$. This implies that by taking the limit in (4.7.37) as $\delta \to +\infty$, we obtain the following functional equality between holomorphic functions:

$$\zeta^{\mathfrak{M}}_{A\times\{0\},\Omega\times\mathbb{R}}(s) = \frac{\sqrt{\pi}\,\Gamma\left(\frac{N-s}{2}+1\right)}{\Gamma\left(\frac{N+1-s}{2}+1\right)}\zeta^{\mathfrak{M}}_{A,\Omega}(s), \tag{4.7.40}$$

valid in the vertical strip $\{\overline{D} < \mathrm{Re}\, s < N\}$. We can obtain this equality even more directly by applying Lebesgue's dominated convergence theorem to a counterpart of (4.7.25).

Moreover, according to (4.7.32) and (4.7.40), we have the functional equation

$$\zeta^{\mathfrak{M}}_{A\times\{0\},\Omega\times\mathbb{R}}(s) = \frac{2\sqrt{\pi}\,\Gamma\left(\frac{N-s}{2}\right)}{\Gamma\left(\frac{N+1-s}{2}+1\right)}\zeta_{A,\Omega}(s), \tag{4.7.41}$$

from which we deduce that the right-hand side admits a meromorphic extension to the right half-plane $\{\mathrm{Re}\, s > \overline{D}\}$, with simple poles located at the simple poles of $\Gamma((N-s)/2)$; that is, at $s_k := N+2k$ for all $k \in \mathbb{N}_0$. (Observe that in the above ratio of gamma functions, there are no cancellations between the poles of the numerator and of the denominator; indeed, an integer cannot be both even and odd.) From this we conclude that by the principle of analytic continuation, the same property also holds for the left-hand side of (4.7.41) and, furthermore, the left-hand side has a meromorphic extension to any domain $U \subseteq \mathbb{C}$ to which the right-hand side can be meromorphically extended.

In order to complete the proof of the theorem, we now observe that for any $a > 0$, since

$$\begin{aligned}\big|(A\times\{0\})_t \cap (\Omega\times\mathbb{R})\big| &= \big|(A\times\{0\})_t \cap (\Omega\times(-a,a))\big| \\ &\quad + \big|(A\times\{0\})_t \cap (\Omega\times(\mathbb{R}\setminus(-a,a)))\big|,\end{aligned}$$

the left-hand side of (4.7.41) can be split into two parts, as follows:

$$\begin{aligned}\zeta^{\mathfrak{M}}_{A\times\{0\},\Omega\times\mathbb{R}}(s) &= \zeta^{\mathfrak{M}}_{A\times\{0\},\Omega\times(-a,a)}(s) \\ &\quad + \int_a^{+\infty} t^{s-N-2}\big|(A\times\{0\})_t \cap (\Omega\times(\mathbb{R}\setminus(-a,a)))\big|\,dt \\ &= \frac{\zeta_{A\times\{0\},\Omega\times(-a,a)}(s)}{N+1-s} - \frac{E(s;a)}{N+1-s}.\end{aligned}$$

We then combine this observation with (4.7.41) to obtain (4.7.35). From the theory developed in this chapter (see Theorem 4.1.7), we know that $\zeta_{A\times\{0\},\Omega\times(-a,a)}(s)$ is holomorphic on the open right half-plane $\{\mathrm{Re}\, s > \overline{D}\}$. Furthermore, much as in the proof of Proposition 4.7.12, we can show that $E(s) := E(s;a)$ defines a holomorphic function on the open left half-plane $\{\mathrm{Re}\, s < N\}$. This fact, along with the functional equation (4.7.35), now ensures that $E(s;a)$ admits a meromorphic continuation to all of $\mathbb{C}$, with a set of simple poles equal to $\{N+2k : k \in \mathbb{N}_0\}$. (Note that $\zeta_{A,\Omega}(s) > 0$

for all $s \in [N, +\infty)$, which implies that there are no zero-pole cancellations on the right-hand side of (4.7.35).) This completes the proof of Theorem 4.7.13. $\square$

We note that in Example 4.7.15 below, we actually want to embed (A, Ω) into $\mathbb{R}^{N+1}$, as $(A \times \{0\}, \Omega \times (0,a))$ for some $a > 0$. By looking at the proof of the above theorem and using a suitable symmetry argument, we can obtain the following result, which deals with this type of embedding.

Theorem 4.7.14. *Let (A, Ω) be a relative fractal drum in $\mathbb{R}^N$ such that $\overline{D} := \overline{\dim}_B(A, \Omega) < N$. Then, the following approximate functional equation holds:*

$$\zeta_{A\times\{0\},\Omega\times(0,a)}(s) = \frac{\sqrt{\pi}\,\Gamma\left(\frac{N-s}{2}\right)}{2\Gamma\left(\frac{N+1-s}{2}\right)}\zeta_{A,\Omega}(s) + E(s;a), \tag{4.7.42}$$

initially valid for all $s \in \mathbb{C}$ such that $\operatorname{Re} s > \overline{D}$. Here, the error function $E(s) := E(s;a)$ *is initially given (for all $s \in \mathbb{C}$ such that* $\operatorname{Re} s < N$*) by*

$$E(s;a) := (s-N-1)\int_a^{+\infty} t^{s-N-2}|(A\times\{0\})_t \cap \Omega \times (\mathbb{R}\setminus(0,a))|_{N+1}\,\mathrm{d}t, \tag{4.7.43}$$

and admits a meromorphic continuation to all of $\mathbb{C}$, with a set of simple poles equal to $\{N+2k : k \in \mathbb{N}_0\}$.

Moreover, Equation (4.7.42) *remains valid on any connected open neighborhood of the critical line $\{\operatorname{Re} s = \overline{D}\}$ to which $\zeta_{A,\Omega}$ (or, equivalently, $\zeta_{A\times\{0\},\Omega\times(0,a)}$) can be meromorphically continued.*

Example 4.7.15. (*Complex dimensions of the Cantor dust RFD*). In this example, we will consider the relative fractal drum consisting of the Cantor dust contained in $[0,1]^2$ and compute its distance zeta function. More precisely, let $A := C^{(1/3)} \times C^{(1/3)}$ be the Cantor dust (i.e., the Cartesian product of the ternary Cantor set $C := C^{1/3}$ by itself; see Figure 1.2 of Subsection 1.1) and let $\Omega := (0,1)^2$. We will not obtain an explicit formula in a closed form but we will instead use Theorem 4.7.14 in order to deduce that the distance zeta function of the Cantor dust has a meromorphic continuation to all of $\mathbb{C}$.

More interestingly, we will also show that the set of complex dimensions of the Cantor dust is a subset of the union of a periodic set contained in the critical line $\{\operatorname{Re} s = \log_3 4\}$ and the set of complex dimensions of the Cantor set (which is a periodic set contained in the critical line $\{\operatorname{Re} s = \log_3 2\}$). *This fact is significant because it shows that in this case, the distance (or tube) zeta function also detects the 'lower-dimensional' fractal nature of the Cantor dust.*

Note that, as is well known, the Minkowski dimension of the RFD (or Cantor string) $(C,(0,1))$ is given by $\dim_B(C,(0,1)) = \log_3 2$ (see [Lap-vFr1, Subsection 1.2.2] or Equation (2.2.17) in Example 2.2.6 above). Furthermore, it will follow from the discussion below that, as might be expected since $(A,\Omega) = (C,(0,1)) \times (C,(0,1))$,

$$\dim_B(A,\Omega) = 2\dim_B(C,(0,1)) = \log_3 4. \tag{4.7.44}$$

Consequently, it follows that the critical line of the RFD in $\mathbb{R}$ ('Cantor string') $(C,(0,1))$ is the vertical line $\{\mathrm{Re}\, s = \log_3 2\}$, while the critical line of the RFD in $\mathbb{R}^2$ ('Cantor dust') (A,Ω) is the vertical line $\{\mathrm{Re}\, s = \log_3 4\}$, as was stated in the previous paragraph.

The construction of the RFD (A,Ω) can be carried out by beginning with the unit square and removing the open middle-third 'cross', and then iterating this procedure ad infinitum. (See Figure 1.2 on page 10.) This procedure implies that we can subdivide the Cantor dust into a countable union of RFDs which are scaled down versions of two base (or generating) RFDs, denoted by (A_1,Ω_1) and (A_2,Ω_2). The first one of these base RFDs, (A_1,Ω_1), is defined by $\Omega_1 := (0,1/3)^2$ and by A_1 being the union of the four vertices of the closure of Ω_1 (namely, of the square $[0,1/3]^2$). Furthermore, the second base RFD, (A_2,Ω_2), is defined by $\Omega_2 := (0,1/3)\times(0,1/6)$ and by A_2 being the ternary Cantor set contained in $[0,1/3]\times\{0\}$.

At the n-th step of the iteration, we have exactly 4^{n-1} RFDs of the type $(a_nA_1,a_n\Omega_1)$ and $8\cdot 4^{n-1}$ RFDs of the type $(a_nA_2,a_n\Omega_2)$, where $a_n := 3^{-n}$ for each $n\in\mathbb{N}$. This observation, together with the scaling property of the relative distance zeta function (see Theorem 4.1.40), yields successively (for all $s\in\mathbb{C}$ with $\mathrm{Re}\, s$ sufficiently large):

$$\begin{aligned}\zeta_{A,\Omega}(s) &= \sum_{n=1}^{\infty} 4^{n-1}\zeta_{a_nA_1,a_n\Omega_1}(s) + 8\sum_{n=1}^{\infty} 4^{n-1}\zeta_{a_nA_2,a_n\Omega_2}(s)\\ &= \left(\zeta_{A_1,\Omega_1}(s) + 8\zeta_{A_2,\Omega_2}(s)\right)\sum_{n=1}^{\infty} 4^{n-1}\cdot 3^{-ns}\\ &= \frac{1}{3^s-4}\left(\zeta_{A_1,\Omega_1}(s) + 8\zeta_{A_2,\Omega_2}(s)\right).\end{aligned} \tag{4.7.45}$$

Moreover, for the relative distance zeta function of (A_1,Ω_1), we have

$$\begin{aligned}\zeta_{A_1,\Omega_1}(s) &= 8\int_0^{1/6} \mathrm{d}x \int_0^x \left(\sqrt{x^2+y^2}\right)^{s-2} \mathrm{d}y\\ &= 8\int_0^{\pi/4} \mathrm{d}\theta \int_0^{1/6\cos\theta} r^{s-1}\,\mathrm{d}r\\ &= \frac{8}{6^s s}\int_0^{\pi/4} \cos^{-s}\theta\,\mathrm{d}\varphi = \frac{8I(s)}{6^s s},\end{aligned} \tag{4.7.46}$$

where $I(s) := \int_0^{\pi/4}\cos^{-s}\theta\,\mathrm{d}\theta$ is easily seen to be an entire function (by means of Theorem 2.1.45 with $\varphi(\theta) := \cos^{-1}\theta$ for $\theta\in(0,\pi/4)$).[58] Consequently, $\zeta_{A,\Omega}$ admits a meromorphic continuation to all of $\mathbb{C}$ and we have

$$\zeta_{A,\Omega}(s) = \frac{8}{3^s-4}\left(\frac{I(s)}{6^s s} + \zeta_{A_2,\Omega_2}(s)\right), \tag{4.7.47}$$

[58] In fact, $I(s) = 2^{-1}\mathrm{B}_{1/2}(1/2,(1-s)/2)$, where $\mathrm{B}_x(a,b) := \int_0^x t^{a-1}(1-t)^{b-1}\,\mathrm{d}t$ is the *incomplete beta function.*

for all $s \in \mathbb{C}$. Furthermore, let $\zeta_{C,(0,1)}$ be the relative distance zeta function of the Cantor middle-third set constructed inside $[0,1]$; see Example 5.5.3 in Chapter 5 below. From Theorem 4.7.14 and the scaling property of the relative distance zeta function (Theorem 4.1.40), we now deduce that

$$\begin{aligned}\zeta_{A_2,\Omega_2}(s) &= \frac{\sqrt{\pi}\,\Gamma\left(\frac{1-s}{2}\right)}{2\Gamma\left(\frac{2-s}{2}\right)}\zeta_{3^{-1}C,3^{-1}(0,1)}(s) + E(s;6^{-1}) \\ &= \frac{\Gamma\left(\frac{1-s}{2}\right)}{\Gamma\left(\frac{2-s}{2}\right)}\frac{\sqrt{\pi}}{6^s s(3^s-2)} + E(s;6^{-1}),\end{aligned} \tag{4.7.48}$$

where $E(s;6^{-1})$ is meromorphic on all of $\mathbb{C}$ with a set of simple poles equal to $\{2k+1 : k \in \mathbb{N}_0\}$; so that for all $s \in \mathbb{C}$, we have

$$\zeta_{A,\Omega}(s) = \frac{8}{s(3^s-4)}\left(\frac{I(s)}{6^s} + \frac{\Gamma\left(\frac{1-s}{2}\right)}{\Gamma\left(\frac{2-s}{2}\right)}\frac{\sqrt{\pi}}{6^s s(3^s-2)} + E(s;6^{-1})\right). \tag{4.7.49}$$

Formula (4.7.49) implies that $\mathscr{P}(\zeta_{A,\Omega})$, the set of all complex dimensions (in $\mathbb{C}$) of the 'relative' Cantor dust, is a subset of

$$\left(\log_3 4 + \frac{2\pi}{\log 3}\mathrm{i}\mathbb{Z}\right) \cup \left(\log_3 2 + \frac{2\pi}{\log 3}\mathrm{i}\mathbb{Z}\right) \cup \{0\} \tag{4.7.50}$$

and consists of simple poles of $\zeta_{A,\Omega}$. Of course, we know that $\log_3 4 \in \mathscr{P}(\zeta_{A,\Omega})$, but we can only conjecture that the other poles on the critical line $\{\mathrm{Re}\, s = \log_3 4\}$ are in $\mathscr{P}(\zeta_{A,\Omega})$ since it may happen that there are zero-pole cancellations in (4.7.49). On the other hand, since it is known that the Cantor dust is not Minkowski measurable (see [FaZe]), we can deduce from the sufficient condition for Minkowski measurability obtained in Theorem 5.4.2 of Chapter 5 below that there must exist at least two other (necessarily nonreal) poles $s_{\pm k_0} = \log_3 4 \pm \frac{2k_0\pi\mathrm{i}}{\log 3}$ of $\zeta_{A,\Omega}$, for some $k_0 \in \mathbb{N}$.[59] From (4.7.49) we cannot even claim that $0 \in \mathscr{P}(\zeta_{A,\Omega})$ for sure, but we can see that all of the principal complex dimensions of the Cantor set are elements of $\mathscr{P}(\zeta_{A,\Omega})$; i.e., $\log_3 2 + \frac{2\pi}{\log 3}\mathrm{i}\mathbb{Z} \subseteq \mathscr{P}(\zeta_{A,\Omega})$. We conjecture that we also have $\log_3 4 + \frac{2\pi}{\log 3}\mathrm{i}\mathbb{Z} \subseteq \mathscr{P}(\zeta_{A,\Omega})$; that is, we conjecture that $\mathscr{P}_c(\zeta_{A,\Omega}) = \log_3 4 + \frac{2\pi}{\log 3}\mathrm{i}\mathbb{Z}$.

The above example can be easily generalized to Cartesian products of any finite number of generalized Cantor sets, in which case we conjecture that the set of complex dimensions of the product is contained in the union of sets of complex dimensions of each of the factors, modulo any zero-pole cancellations which may occur. In light of this and other similar examples, it would be interesting to obtain some results about zero-free regions for fractal zeta functions. We leave this problem as a possible subject for future investigations.

---

[59] Indeed, according to Theorem 5.4.2, $D := \log_3 4$ cannot be the only complex dimension of (A,Ω) on the critical line $\{\mathrm{Re}\, s = D\}$ since otherwise, the Cantor dust would be Minkowski measurable, which is a contradiction.

Chapter 5
Fractal Tube Formulas and Complex Dimensions

> *There exist a limited number of very simple fundamental relationships that together constitute the schema by means of which the remaining theorems can be developed logically and without difficulty.*
>
> Jakob Steiner (1796–1863)

Abstract In this chapter, we reconstruct information about the geometry of relative fractal drums (and, consequently, compact sets) in $\mathbb{R}^N$ from their associated fractal zeta functions. Roughly speaking, given a relative fractal drum (A,Ω) in $\mathbb{R}^N$ (with $N \geq 1$ arbitrary), we derive an asymptotic formula for its relative tube function $t \mapsto |A_t \cap \Omega|$ as $t \to 0^+$, expressed as a sum taken over its complex dimensions of the residues of its (suitably modified and meromorphically extended) fractal zeta function. The resulting asymptotic formulas are called *fractal tube formulas* and are valid either pointwise or distributionally, as well as with or without an error term, depending on the growth properties of the associated fractal zeta functions. We note that these fractal tube formulas are expressed either in terms of the tube zeta function $\tilde{\zeta}_{A,\Omega}$ or, more interestingly, in terms of the distance zeta function $\zeta_{A,\Omega}$. The results of this chapter generalize to higher dimensions and arbitrary relative fractal drums the corresponding ones obtained previously for fractal strings by the first author and M. van Frankenhuijsen. We illustrate these results by obtaining fractal tube formulas for a number of well-known fractal sets, including the Sierpiński gasket and 3-carpet along with higher-dimensional analogs, a version of the graph of the Cantor function (i.e., of the devil's staircase), fractal strings, fractal sprays, self-similar sprays and tilings, as well as certain non self-similar fractals, such as fractal nests and unbounded geometric chirps. We also apply these results in an essential way in order to obtain and establish a Minkowski measurability criterion for a large class of relative fractal drums (and, in particular, of bounded sets) in $\mathbb{R}^N$, with $N \geq 1$ arbitrary. More specifically, under appropriate hypotheses, a relative fractal drum (and, in particular, a bounded set) in $\mathbb{R}^N$ of (upper) Minkowski dimension D is shown to be Minkowski measurable if and only if its only complex dimension with real part equal to D is D itself, and D is simple. We also discuss the notion of fractality defined in our context as the presence of at least one nonreal complex dimension. We show, in particular, that as is expected and intuitive, (a variant of) the Cantor graph (or devil's staircase) is "fractal" in our sense, whereas as is well known, it is not "fractal" in Mandelbrots's sense.

M.L. Lapidus et al., *Fractal Zeta Functions and Fractal Drums*,
Springer Monographs in Mathematics, DOI 10.1007/978-3-319-44706-3_5

Key words: Mellin transform, fractal set, fractal string, relative fractal drum (RFD), complex dimensions of an RFD, box dimension, fractal zeta functions, distance zeta function, tube zeta function, Minkowski content, Minkowski measurable set, Minkowski measurability criterion, fractal tube formulas, residue, meromorphic extension, gauge-Minkowski measurability, singularities of fractal zeta functions.

In this chapter, we reconstruct information about the geometry of relative fractal drums (and, consequently, compact sets) in $\mathbb{R}^N$ from their associated fractal zeta functions. Roughly speaking, given a relative fractal drum (A,Ω) in $\mathbb{R}^N$ (with $N \geq 1$ arbitrary), we derive an asymptotic formula for its relative tube function $t \mapsto |A_t \cap \Omega|$ as $t \to 0^+$, expressed as a sum taken over its complex dimensions of the residues of its (suitably modified and meromorphically extended) fractal zeta function. The resulting asymptotic formulas are called *fractal tube formulas* and are valid either pointwise or distributionally, as well as with or without an error term, depending on the growth properties of the associated fractal zeta functions. We note that these fractal tube formulas are expressed either in terms of the tube zeta function (see Sections 5.1 and 5.2) or, more interestingly, in terms of the distance zeta function (see Section 5.3).

The results of this chapter generalize to higher dimensions the corresponding ones obtained previously for fractal strings by the first author and M. van Frankenhuijsen (see [Lap-vFr3, Chapters 5 and 8]). We illustrate these results by obtaining fractal tube formulas for a number of well-known fractal sets, including the Sierpiński gasket and 3-carpet along with higher-dimensional analogs, a version of the graph of the Cantor function (i.e., of the devil's staircase), fractal strings, fractal sprays, self-similar sprays and tilings, as well as certain non self-similar fractals, such as fractal nests and unbounded geometric chirps. We also apply our fractal tube formulas in an essential way in order to obtain and establish a Minkowski measurability criterion for a large class of relative fractal drums (and, in particular, of bounded sets) in $\mathbb{R}^N$, with $N \geq 1$ arbitrary; see Section 5.4. More specifically, under appropriate hypotheses, a relative fractal drum (and, in particular, a bounded set) in $\mathbb{R}^N$ of Minkowski dimension D is shown to be Minkowski measurable if and only if its only complex dimension with real part equal to D is D itself, and D is simple. Again, this criterion generalizes to higher dimensions the corresponding one already obtained for fractal strings (see [Lap-vFr3, Theorem 8.15]).

In closing this introduction to Chapter 5, it may be helpful to the readers to point out the relationship between aspects of our present work on fractal tube formulas and the classic Steiner tube formula [Stein], as generalized in various ways by many authors (including Minkowski [Mink], Weyl [Wey3] and later, Federer [Fed1–2]) and as stated in the case of compact convex sets in [Schn2, Theorem 4.2.1].[1]

Let A be a compact convex subset of $\mathbb{R}^N$ (with $N \geq 1$) and let B^k denote the k-dimensional unit ball of $\mathbb{R}^k$ (for any integer $k \geq 1$) with k-dimensional volume (or Lebesgue measure) denoted by $|B^k|_k$. We also let $|B^0|_0 := 1$. Note that for $t \geq 0$, the

[1] Our exposition of this material closely follows part of [Lap-vFr3, Subsection 13.1.3]; see also [LapPe2–3] and [LapPeWi1].

t-neighborhood (or t-parallel body) of A can be written as $A_t = A + tB^N$. Then its volume $V_A(t) := |A_t|_N$ can be expressed as a polynomial of degree $\leq N$ (exactly N if $|A|_N > 0$; e.g., if A has nonempty interior) in the variable t:

$$V_A(t) = \sum_{k=0}^{N} \mu_k(A)|B^{N-k}|_{N-k}t^{N-k}, \tag{5.0.1}$$

where for $k = 0, 1, \ldots, N$, $\mu_k(A)$ denotes the k-th *intrinsic volume* of A.

Up to some suitable normalizing and multiplicative constant (depending on k, for each $k \in \{0, 1, \ldots, N\}$, the k-th intrinsic volume $\mu_k(A)$) coincides with the k-th *total curvature* of A or the so-called $(N-k)$-th *Quermassintegral* of A. Moreover, still for $k \in \{0, 1, \ldots, N\}$, $\mu_k(A)$ can be interpreted either combinatorially and algebraically in terms of appropriate valuations (see [KlRot]) or (in a closely related context) within the framework of integral geometry, as the average measure of orthogonal projections to $(N-k)$-dimensional subspaces of Euclidean space $\mathbb{R}^N$. See, e.g., [Schn2] and [KlRot, Chapter 7]. (This latter geometric interpretation was already implicit in Steiner's original work [Stein] and that of his immediate successors, where $N = 2$ or $N = 3$.)

To make a long and beautiful story short, let us simply mention here that (up to a suitable normalizing multiplicative constant) $\mu_0(A)$ corresponds to the *Euler characteristic*,[2] $\mu_1(A)$ to the so-called *mean width*, $\mu_{N-1}(A)$ is the *surface area* and $\mu_N(A)$ the (N-dimensional) *volume* of A (i.e., $\mu_N(A) = |A|_N = |A|$, in our notation).

Finally, let us point out that the intrinsic volumes μ_k have the following algebraic and geometric properties:

(*i*) Each μ_k is homogeneous of degree k; i.e., for all $\lambda > 0$,

$$\mu_k(\lambda A) = \lambda^k \mu_k(A), \tag{5.0.2}$$

and

(*ii*) each μ_k is rigid motion invariant; more specifically, for any affine isometry (i.e., displacement) R of $\mathbb{R}^N$, we have that

$$\mu_k\big(R(A)\big) = \mu_k(A). \tag{5.0.3}$$

Remarkably, for any (visible) complex dimension ω of a bounded subset A of $\mathbb{R}^N$ (or, more generally, of an RFD (A, Ω) of $\mathbb{R}^N$), the corresponding coefficient of our fractal tube formula (assuming that we are in the case of simple poles), that is, essentially, the residue of the fractal zeta function at $s = \omega$ (see Equation (1.1.2) in the introduction or Equation (5.1.46) in Theorem 5.1.16 below), satisfies entirely analogous homogeneous and geometric invariance properties (with k replaced by

---

[2] In the present case of compact convex sets, μ_0 is always equal to 1. However, in the more general setting of sets of positive reach or of finite unions of such sets, it is $\mathbb{Z}$-valued; see, e.g., [Schn2, Section 3.4] and [Fed1, Zä2].

ω in the counterparts of Equations (5.0.2) and (5.0.3)).[3] Furthermore, of course, the resulting tube formula is no longer a polynomial of degree at most N in the variable t but involves a typically infinite sum over all of the underlying visible complex dimensions of A (or of the RFD (A,Ω)). Moreover, as we shall see in many examples, the coefficients of the fractal tube formula that correspond to the set of (visible) complex dimensions can frequently be naturally decomposed as a set of *integer dimensions* (say, $\omega = k \in \{0,1,\ldots,N\}$) and of *scaling complex dimensions* (say, $\omega \in \mathfrak{D}_{\mathfrak{S}}$). See, especially, the discussion of the Sierpiński gasket and of the 3-dimensional carpet in Subsection 5.5.3, along with that of self-similar sprays in Subsection 5.5.6. Such a situation already arose in the very special but important case of fractal sprays studied in [LapPe2–3] and [LapPeWi1–2]. Of course, if $\mathscr{D}_{\mathfrak{S}}$ happens to be empty (which is certainly the case if A is a compact convex set), then $V_A(t)$ reduces to a polynomial expression of degree $\leq N$ in t and the corresponding tube formula is Steiner-like, much as in Equation (5.0.1) above.

We leave to a later work a further and much more detailed exploration of the possible geometric, algebraic and combinatorial interpretations of our fractal tube formulas (as well as of local versions thereof), in the spirit of the above discussion and particularly, the work of Stein [Stein], Minkowski [Mink] (see also [Schn2]), Weyl [Wey3] (see also [BergGos] and [Gra]), Federer ([Fed2] and especially, his work in [Fed1] on *local tube formulas* and *curvature measures*), Klain and Rota [KlRot] and many other authors; see, e.g., the books [BergGos], [Bla], [Schn2], [Gra], [Lap-vFr1–3], along with the articles [Fu1–2], [HugLasWeil], [KeKom], [Kom], [Kow], [LapPe1–3], [LapPeWi1], [LlWi], [Mil], [Ol1–2], [RatWi1–2], [Schn1], [Sta], [Wi], [WiZä], [Zä1–5] and the many relevant references therein.

5.1 Pointwise Tube Formulas

In this section, given a relative fractal drum (A,Ω) in $\mathbb{R}^N$, with $N \geq 1$ arbitrary, we obtain and derive a corresponding pointwise fractal tube formula, expressed in terms of its complex dimensions. The proof of our pointwise (and later, in Section 5.2, distributional) fractal tube formulas follows many of the same steps as in [Lap-vFr3, Chapters 5 and 8] in the case of the geometric zeta functions of fractal strings, but now in the new and significantly more general context of relative fractal drums (and, in particular, of bounded sets) in $\mathbb{R}^N$, with $N \geq 1$ arbitrary. There are, however, a number of technical differences, related, in particular, to the use of the Mellin transform inversion formula, as well as in later sections (Sections 5.3 and 5.4), of various intermediate zeta functions, called the shell and Mellin zeta functions, in addition to the tube and distance zeta functions.

[3] In light of the definitions, the analog of Equation (5.0.3) obviously holds in our context. For the counterpart of Equation (5.0.2), see Equation (2.1.106) in Proposition 2.1.77, which follows from the scaling property of ζ_A stated in Equation (2.1.105).

For a number of results concerning tube formulas and their generalizations in a variety of settings (including convex bodies, smooth compact submanifolds of Euclidean spaces, compact Riemannian and Lipschitz manifolds, sets of positive reach, semi-algebraic sets, fractal strings and sprays), as well as related topics, we mention, in particular [BergGos, Bla, CheeMüSchr1–2, DemDenKoÜ, DemKoÖÜ, DenKoÖÜ, Fed1, Fu1–2, Gra, HugLasWeil, KeKom, KlRot, Kow, LapLu3, LapLu-vFr1–2, LapPe1–3, LapPeWi1–2, LapRaŽu4–5, Mil, Mink, MitŽu, Ol1–2, RatWi1–2, Schn1–2, Sta, Stein, Wey3, Wi, WiZä, Zä1–5], along with the many relevant references therein. See also the second part of the introduction to this chapter for a brief overview.

5.1.1 Definitions and Preliminaries

We begin by stating several definitions which are already introduced in [Lap-vFr3] in the setting of generalized fractal strings, and adapt them to the setting of relative fractal drums in $\mathbb{R}^N$, for any $N \geq 1$.

Definition 5.1.1. The *screen* $\boldsymbol{S}$ is the graph of a bounded, real-valued Lipschitz continuous function $S(\tau)$, with the horizontal and vertical axes interchanged:

$$\boldsymbol{S} := \{S(\tau) + \mathrm{i}\tau \, : \, \tau \in \mathbb{R}\}. \tag{5.1.1}$$

The Lipschitz constant of S is denoted by $\|S\|_{\mathrm{Lip}}$; so that

$$|S(x) - S(y)| \leq \|S\|_{\mathrm{Lip}} |x - y|, \quad \text{for all } x, y \in \mathbb{R}.$$

Furthermore, the following quantities are associated to the screen:

$$\inf S := \inf_{\tau \in \mathbb{R}} S(\tau) \quad \text{and} \quad \sup S := \sup_{\tau \in \mathbb{R}} S(\tau).$$

As before, given an RFD (A, Ω) in $\mathbb{R}^N$, we denote its upper relative box dimension by $\overline{D} := \overline{\dim}_B(A, \Omega)$; recall that $\overline{D} \leq N$. We always assume, additionally, that $\overline{D} > -\infty$ and the screen $\boldsymbol{S}$ lies to the left of the *critical line* $\{\mathrm{Re}\, s = \overline{D}\}$, i.e., that $\sup S \leq \overline{D}$. Also, in the sequel, we assume that $\inf S > -\infty$ (see, however, Remark 5.1.2 below); hence, we have that

$$-\infty < \inf S \leq \sup S \leq \overline{D}. \tag{5.1.2}$$

Moreover, the *window* $\boldsymbol{W}$ is defined as the part of the complex plane to the right of $\boldsymbol{S}$; that is,

$$\boldsymbol{W} := \{s \in \mathbb{C} : \mathrm{Re}\, s \geq S(\mathrm{Im}\, s)\}. \tag{5.1.3}$$

(Note that $\boldsymbol{W}$ is a closed subset of $\mathbb{C}$ and that $\boldsymbol{S} = \partial \boldsymbol{W}$, the boundary of $\boldsymbol{W}$.) We say that the relative fractal drum (A, Ω) is *admissible* if its relative tube (or distance) zeta function can be meromorphically extended (necessarily uniquely) to an open connected neighborhood of some window $\boldsymbol{W}$, defined as above.

Remark 5.1.2. Occasionally, in the strongly languid case, in the sense of Definition 5.1.4 or Definition 5.3.9 below (and hence, in particular, when the fractal zeta function involved is meromorphic on all of $\mathbb{C}$), it is convenient to implicitly move the screen $\boldsymbol{S}$ to $-$infinity (i.e., to let $S \equiv -\infty$) and thus to choose $\boldsymbol{W} := \mathbb{C}$.

The next definition adapts [Lap-vFr3, Definition 5.2] to the case of relative fractal drums in $\mathbb{R}^N$ (and, in particular, to the case of bounded subsets of $\mathbb{R}^N$).

Definition 5.1.3. (*Languidity*, adapted from [Lap-vFr3]). An admissible relative fractal drum (A,Ω) in $\mathbb{R}^N$ is said to be *languid* if for some fixed $\delta > 0$, its tube zeta function $\tilde{\zeta}_{A,\Omega}(\,\cdot\,;\delta)$ satisfies the following growth conditions:

There exists a real constant κ and a two-sided sequence $(T_n)_{n\in\mathbb{Z}}$ of real numbers such that $T_{-n} < 0 < T_n$ for all $n \geq 1$ and

$$\lim_{n\to\infty} T_n = +\infty, \quad \lim_{n\to\infty} T_{-n} = -\infty \tag{5.1.4}$$

satisfying the following two hypotheses, **L1** and **L2**:[4]

L1 For a fixed real constant $c > \overline{\dim}_B(A,\Omega)$, there exists a positive constant $C > 0$ such that for all $n \in \mathbb{Z}$ and all $\sigma \in (S(T_n),c)$,[5]

$$|\tilde{\zeta}_{A,\Omega}(\sigma + \mathrm{i}T_n;\delta)| \leq C(|T_n|+1)^{\kappa}. \tag{5.1.5}$$

L2 For all $\tau \in \mathbb{R}$, $|\tau| \geq 1$,

$$|\tilde{\zeta}_{A,\Omega}(S(\tau) + \mathrm{i}\tau;\delta)| \leq C|\tau|^{\kappa}, \tag{5.1.6}$$

where C is a positive constant which (without loss of generality) can be chosen to be the same one as in condition **L1**.

Note that hypothesis **L1** is a polynomial growth condition along horizontal segments (necessarily not passing through any singularities of $\tilde{\zeta}_{A,\Omega}(\,\cdot\,;\delta)$), while hypothesis **L2** is a polynomial growth condition along the vertical direction of the screen. These hypotheses will be needed in order to establish the pointwise and distributional tube formulas with error term.

It is noteworthy that there exist RFDs not satisfying condition **L2** (and hence, which are not languid) for some choices of the screen $\boldsymbol{S}$. For a specific example, see [Lap-vFr3, Example 5.32] which is a nonlattice self-similar fractal string, viewed naturally as an RFD (A,Ω), and is such that there is no screen $\boldsymbol{S}$ passing between the critical line $\{\mathrm{Re}\,s = D\}$ (where $D := \dim_B(A,\Omega)$) and the complex dimensions to the left of this line along which the fractal string is languid.

---

[4] Here, unlike in the definition given in [Lap-vFr3], we do not need to assume that $\lim_{n\to+\infty} T_n/|T_{-n}| = 1$.

[5] This is a slight modification of the original definition of languidity from [Lap-vFr3], where c was replaced by $+\infty$; compare with [Lap-vFr3, Definition 5.2, pp. 146–147]. Furthermore, it is clear that if condition **L1** is satisfied for some $c > \overline{\dim}_B(A,\Omega)$, then it is also satisfied for any c_1 such that $\overline{\dim}_B(A,\Omega) < c_1 < c$.

In order to obtain the pointwise and distributional tube formulas without error terms (that is, *exact* tube formulas), we will need a stronger notion of languidity. Accordingly, we introduce the following definition, which adapts to our current more general situation the definition of strong laguidity given in [Lap-vFr3, Definition 5.3].

Definition 5.1.4. (*Strong languidity*, adapted from [Lap-vFr3]). We say that an admissible relative fractal drum (A,Ω) in $\mathbb{R}^N$ is *strongly languid* if for some $\delta > 0$, its tube zeta function satisfies condition **L1** with $S(T_n) \equiv -\infty$ (that is, with $S(T_n)$ replaced by $-\infty$) in (5.1.5); i.e., for every $\sigma < c$ and, additionally, there exists a sequence of screens $\boldsymbol{S}_m\colon \tau \mapsto S_m(\tau) + \mathrm{i}\tau$ for $m \geq 1$, $\tau \in \mathbb{R}$ with $\sup S_m \to -\infty$ as $m \to \infty$ and with a uniform Lipschitz bound, $\sup_{m\geq 1} \|S_m\|_{\mathrm{Lip}} < \infty$, such that the following condition holds:

L2' There exist constants $B, C > 0$ such that for all $\tau \in \mathbb{R}$ and $m \geq 1$,

$$|\tilde{\zeta}_{A,\Omega}(S_m(\tau) + \mathrm{i}\tau; \delta)| \leq C B^{|S_m(\tau)|} (|\tau| + 1)^{\kappa}. \tag{5.1.7}$$

It is clear that hypothesis **L2'** implies hypothesis **L2**; so that a strongly languid relative fractal drum is languid. We also note that if a relative fractal drum is languid for some κ, then it is also languid for any $\kappa_1 > \kappa$. (Observe that for $\tilde{\zeta}_{A,\Omega}$ or, equivalently, the RFD (A,Ω), to be strongly languid, $\tilde{\zeta}_{A,\Omega}$ must admit a meromorphic continuation to all of $\mathbb{C}$; see also Remark 5.1.2 above.)

We will also use the notion of *languid* (or else, *strongly languid*) *relative tube zeta function*, in the obvious sense.

As we shall see throughout this chapter, most of the geometrically interesting examples of RFDs (and, in particular, of bounded sets) in $\mathbb{R}^N$ considered in this monograph are either languid (relative to a suitable screen), in the sense of Definition 5.1.3 above (or of its counterpart for the distance zeta function, in Definition 5.3.9 below) or else, strongly languid, in the sense of Definition 5.1.4 just above (or, again, in the sense of Definition 5.3.9).

Although, as was already explained, the dependence of the tube zeta function $\tilde{\zeta}_{A,\Omega} = \tilde{\zeta}_{A,\Omega}(\,\cdot\,;\delta)$ on $\delta > 0$ is inessential, it is not clear whether the counterpart of this statement is also true for the languidity conditions. More precisely, we show that changing the parameter $\delta > 0$ will preserve languidity, but possibly with a different languidity exponent κ_δ. This is the content of the next proposition.

Proposition 5.1.5. *Let (A,Ω) be a relative fractal drum in $\mathbb{R}^N$. If the relative tube zeta function $\tilde{\zeta}_{A,\Omega}(\,\cdot\,;\delta)$ satisfies the languidity conditions* **L1** *and* **L2** *for some $\delta > 0$ and with languidity exponent $\kappa \in \mathbb{R}$, then so does $\tilde{\zeta}_{A,\Omega}(\,\cdot\,;\delta_1)$ for any $\delta_1 > 0$ and with $\kappa_{\delta_1} := \max\{\kappa, 0\}$.*

Furthermore, an entirely analogous statement is also true in the case when $\tilde{\zeta}_{A,\Omega}(\,\cdot\,;\delta)$ is strongly languid, under the additional assumption that $\delta \geq 1$ and $\delta_1 \geq 1$.

Proof. Without loss of generality, we may assume that $\delta < \delta_1$. Then, the conclusion follows from the fact that $\tilde{\zeta}_{A,\Omega}(\,\cdot\,;\delta_1) = \tilde{\zeta}_{A,\Omega}(\,\cdot\,;\delta) + f(s)$, where f is entire and

$$|f(s)| \le \int_{\delta}^{\delta_1} t^{\operatorname{Re}s-N-1}|A_t\cap\Omega|\,dt \le \begin{cases} |\Omega|\frac{\delta_1^{\operatorname{Re}s-N}-\delta^{\operatorname{Re}s-N}}{\operatorname{Re}s-N}, & \operatorname{Re}s\neq N,\\ |\Omega|(\log\delta_1-\log\delta), & \operatorname{Re}s=N. \end{cases} \tag{5.1.8}$$

Since, clearly, the upper bound on $|f(s)|$ does not depend on $\operatorname{Im}s$, we conclude that f satisfies the languidity conditions **L1** and **L2** with the languidity exponent $\kappa_f := 0$ and for any given window $\boldsymbol{W}$. This observation implies that then, $\tilde{\zeta}_{A,\Omega}(\,\cdot\,;\delta_1)$ is languid for the languidity exponent $\kappa_{\delta_1} := \max\{\kappa, 0\}$ and with the same window as for $\tilde{\zeta}_{A,\Omega}(\,\cdot\,;\delta)$.

The additional assumption for the case of strong languidity is needed since **L1** must then be satisfied for all $\sigma\in(-\infty,c)$, in the notation of Definition 5.1.3, and for this to be achieved we need that $\delta_1>\delta\ge 1$ in (5.1.8); indeed, otherwise, we do not have an upper bound on $|f(s)|$ when $\operatorname{Re}s\to-\infty$. □

Given a fixed open window $\boldsymbol{W}$ and a double sequence $(T_n)_{n\in\mathbb{Z}}$, as in Definition 5.1.3, and a real constant $c>0$, we consider the set

$$\mathscr{F}_{\mathrm{lan}} = \mathscr{F}_{\mathrm{lan}}(\boldsymbol{W},(T_n)_{n\in\mathbb{Z}},c) \tag{5.1.9}$$

of all meromorphic functions $f: U\to\mathbb{C}$ which we call *languid functions*, defined on a domain $U=U(f)\subseteq\mathbb{C}$ containing $\boldsymbol{W}$, and satisfying all of the conditions of Definition 5.1.3, in which the constants appearing there also depend on f (for example, $\kappa=\kappa(f)$, with the condition $c>N$ appearing in **L1** replaced by $c>\sup S$). It is clear that some of the functions $f\in\mathscr{F}_{\mathrm{lan}}(\boldsymbol{W},(T_n)_{n\in\mathbb{Z}})$ are of the form $f=\zeta_{A,\Omega}$, for some RFD (A,Ω). In the following discussion, we will no longer indicate explicitly the dependence of the languidity exponent κ on $\delta>0$ but when necessary, we will instead denote κ by $\kappa(f)$ in order to highlight the dependence on the underlying meromorphic function f.

It is easy to check that the set $\mathscr{F}_{\mathrm{lan}}=\mathscr{F}_{\mathrm{lan}}(\boldsymbol{W},(T_n)_{n\in\mathbb{Z}})$ of languid functions is a vector space and that it is even an algebra with respect to the pointwise multiplication of functions. Furthermore, for any $f,\ g\in\mathscr{F}_{\mathrm{lan}}$ and $\lambda,\mu\in\mathbb{C}$, we have that

$$\kappa(\lambda f+\mu g)\le\max\{\kappa(f),\kappa(g)\} \quad\text{and}\quad \kappa(f\cdot g)\le\kappa(f)+\kappa(g).$$

Any function $f\in\mathscr{F}_{\mathrm{lan}}(\boldsymbol{W},(T_n)_{n\in\mathbb{Z}},c)$ is said to be languid in the set $\boldsymbol{W}\cap\{\operatorname{Re}s<c\}$ (with respect to the double sequence $(T_n)_{n\in\mathbb{Z}}$). All of the examples of languid fractal zeta functions provided in this book belong to the above family $\mathscr{F}_{\mathrm{lan}}(\boldsymbol{W},(T_n)_{n\in\mathbb{Z}},c)$, for a suitable choice of window $\boldsymbol{W}$, sequence $(T_n)_{n\in\mathbb{Z}}$ and of the constant $c>\sup S$. The preceding comment justifies to call the vector space $\mathscr{F}_{\mathrm{lan}}(\boldsymbol{W},(T_n)_{n\in\mathbb{Z}},c)$ the *algebra of languid functions*.

In several of the applications (for example, when dealing with the languidity of Cantor strings of higher order, i.e., obtained as consecutive tensor products of the Cantor string by itself; see Example 4.2.10 along with Subsection 5.4.4), we shall need the following property. (Observe that bounded fractal strings $\mathscr{L}$ can be viewed as relative fractal drums $(\partial\Omega,\Omega)$, where $\Omega\subset\mathbb{R}$ is an open set which is a geometric realization of $\mathscr{L}$.)

Let $\mathscr{L}_1$ and $\mathscr{L}_2$ be two bounded fractal strings such that

$$\zeta_{\mathscr{L}_j} \in \mathscr{F}_{\mathrm{lan}}(\boldsymbol{W}, (T_n)_{n\in\mathbb{Z}}, c), \text{ for } j = 1,2 \text{ and some } c > 1.$$

Then $\zeta_{\mathscr{L}_1\otimes\mathscr{L}_2} \in \mathscr{F}_{\mathrm{lan}}(\boldsymbol{W}, (T_n)_{n\in\mathbb{Z}}, c)$, where $\mathscr{L}_1 \otimes \mathscr{L}_2$ denotes the tensor product of $\mathscr{L}_1$ and $\mathscr{L}_2$. Furthermore, $\kappa_{12} \le \kappa_1 + \kappa_2$, where $\kappa_{12} := \kappa(\mathscr{L}_1 \otimes \mathscr{L}_2)$ and $\kappa_j := \kappa(\mathscr{L}_j)$, for $j = 1,2$.

The above statement follows from the fact that $\zeta_{\mathscr{L}_1\otimes\mathscr{L}_2}(s) = \zeta_{\mathscr{L}_1}(s) \cdot \zeta_{\mathscr{L}_2}(s)$, for all $s \in \mathbb{C}$ with $\operatorname{Re} s$ sufficiently large. Then, upon meromorphic continuation of $\zeta_{\mathscr{L}_1}$ and $\zeta_{\mathscr{L}_2}$ to a domain $U \subseteq \mathbb{C}$ containing the common window $\boldsymbol{W}$, we take $c > N := 1$ as in condition **L1** of Definition 5.1.3 above.

We now return to the main course of our discussion, with the goal of establishing a pointwise tube formula expressed in terms of the tube zeta function $\tilde{\zeta}_{A,\Omega}$.

In order to obtain the relative tube formula expressed in terms of the complex dimensions of the relative fractal drum (A,Ω), we will need to work (for each $k \in \mathbb{N}$) with the k-th primitive (or k-th anti-derivative) function, $V^{[k]} = V^{[k]}(t)$, of the relative tube function $V = V(t)$ vanishing along with its first $(k-1)$ derivatives at $t = 0$. Therefore, we let

$$V(t) = V_{A,\Omega}(t) = V^{[0]}(t) := |A_t \cap \Omega| \tag{5.1.10}$$

and

$$V^{[k]}(t) = V^{[k]}_{A,\Omega}(t) := \int_0^t V^{[k-1]}(\tau)\,\mathrm{d}\tau, \qquad \text{for each } k \in \mathbb{N}. \tag{5.1.11}$$

(Recall that $\mathbb{N} := \{1,2,3,\dots\}$ and $\mathbb{N}_0 := \mathbb{N} \cup \{0\}$.) In the special case of a bounded subset $A \subset \mathbb{R}^N$ (corresponding to the choice of an RFD of the form (A, A_δ), for some $\delta > 0$), we use the analogous notation $V^{[k]}(t) = V^{[k]}_A(t)$ for the k-th primitive function of the tube function $V(t) = V_A(t) := |A_t|$, where $k \in \mathbb{N}_0$.

Furthermore, we recall that for any $s \in \mathbb{C}$, the *Pochhammer* symbol is defined by

$$(s)_0 := 1, \quad (s)_k := s(s+1)\cdots(s+k-1) \tag{5.1.12}$$

for any nonnegative integer k and, more generally, for the purpose of Section 5.2, for every $k \in \mathbb{Z}$ by

$$(s)_k := \frac{\Gamma(s+k)}{\Gamma(s)}, \tag{5.1.13}$$

where Γ denotes the classic gamma function.

It is natural to wonder why we do not simply work with the tube function $V = V^{[0]} = V^{[0]}_{A,\Omega}$ instead of all of its primitives $V^{[k]} = V^{[k]}_{A,\Omega}$ (for any integer $k \ge 0$). For an answer to this question, we refer the curious reader to Remark 5.1.19 at the end of Subsection 5.1.3 below, as well as to the comment preceding it. We also mention that in the distributional setting, we will allow k to be any integer in $\mathbb{Z}$ (rather than in $\mathbb{N} \cup \{0\}$) in the definition of the corresponding k-th 'tube distribution' $\mathscr{V}^{[k]} = \mathscr{V}^{[k]}_{A,\Omega}$; see Definition 5.2.1 at the beginning of Section 5.2. In that setting, the case when $k=-1$ yields the most fundamental fractal tube formula (for the distribution $\mathscr{V}^{[-1]}=\mathscr{V}^{[-1]}_{A,\Omega}$, which can also be viewed as a measure).

Before stating the main relationship connecting $V^{[k]}=V^{[k]}_{A,\Omega}$ and the tube zeta function $\tilde{\zeta}_{A,\Omega}$ of the RFD (A,Ω), valid for any integer $k\geq 0$, we begin by considering the key special case when $k=0$ (so that $V^{[0]}=V=V_{A,\Omega}$). It turns out that $V(t)=|A_t\cap\Omega|$ is essentially equal to the inverse Mellin transform of $\tilde{\zeta}_{A,\Omega}$, as will be seen in Theorem 5.1.7 below. Before stating and proving the latter result, we need to briefly provide some basic information about the Mellin transform and its inverse transform.

First, as an initial motivation for the approach used in this chapter, we note that the tube zeta function coincides with the Mellin transform of a modification of the tube function $t\mapsto|A_t\cap\Omega|$, where as before, $|A_t\cap\Omega|=|A_t\cap\Omega|_N$ denotes the N-dimensional volume of $A_t\cap\Omega\subseteq\mathbb{R}^N$. More specifically, one has that for all $s\in\mathbb{C}$ such that $\operatorname{Re}s>\overline{\dim}_B(A,\Omega)$,

$$\tilde{\zeta}_{A,\Omega}(s;\delta)=\int_0^{+\infty}t^{s-1}\left(\chi_{(0,\delta)}(t)t^{-N}|A_t\cap\Omega|\right)\mathrm{d}t, \tag{5.1.14}$$

where $\chi_{(0,\delta)}$ denotes the characteristic function of the set $(0,\delta)$. Recall that the Mellin transform of a function $f\colon\mathbb{R}\to\mathbb{R}$ is defined by

$$\{\mathfrak{M}f\}(s):=\int_0^{+\infty}t^{s-1}f(t)\,\mathrm{d}t, \tag{5.1.15}$$

where s is a complex number with large enough real part. Furthermore, the Mellin inversion theorem, which we recall here for the sake of completeness, together with Equation (5.1.14), yields an integral expression for the tube function of a given relative fractal drum.

Theorem 5.1.6 (Mellin's inversion theorem, cited from [Tit2, Theorem 28]**).** *Let $f\colon(0,+\infty)\to\mathbb{R}$ be such that for a given $y>0$, $f(t)$ is of bounded variation in a connected open neighborhood of the point $t=y$. Furthermore, assume that the function $t\mapsto t^{c-1}f(t)$ belongs to $L^1(0,+\infty)$, where c is a real number, and define*

$$\{\mathfrak{M}f\}(s):=\int_0^{+\infty}t^{s-1}f(t)\,\mathrm{d}t \tag{5.1.16}$$

for all $s\in\mathbb{C}$ such that $\operatorname{Re}s=c$. Then, for the above value of y, the following inversion formula holds:

$$\frac{1}{2}\big(f(y+0)+f(y-0)\big)=\frac{1}{2\pi\mathrm{i}}\int_{c-\mathrm{i}\infty}^{c+\mathrm{i}\infty}y^{-s}\{\mathfrak{M}f\}(s)\,\mathrm{d}s, \tag{5.1.17}$$

where $f(y+0)$ and $f(y-0)$ denote, respectively, the right and left limits of f at y. Here, on the right-hand side of (5.1.17)*, the contour integral is taken over the vertical line $\{\operatorname{Re}s=c\}$.*

We can now state the announced integral formula connecting the relative tube function of the RFD (A,Ω) and the tube zeta function $\tilde{\zeta}_{A,\Omega}(\,\cdot\,;\delta)$.

Theorem 5.1.7. *Let (A,Ω) be a relative fractal drum in $\mathbb{R}^N$ and fix $\delta > 0$. Then, for any fixed $c > \overline{\dim}_B(A,\Omega)$ and for every $t \in (0,\delta)$, we have*

$$|A_t \cap \Omega| = \frac{1}{2\pi\mathrm{i}} \int_{c-\mathrm{i}\infty}^{c+\mathrm{i}\infty} t^{N-s} \tilde{\zeta}_{A,\Omega}(s;\delta)\,\mathrm{d}s. \tag{5.1.18}$$

Proof. Let $f(t) := \chi_{(0,\delta)}(t) t^{-N} |A_t \cap \Omega|$ and observe that $t \mapsto |A_t \cap \Omega|$ is nondecreasing, and hence, is locally of bounded variation on $(0,+\infty)$. Since the product of two functions of locally bounded variation is also a function of locally bounded variation, we conclude that f is also locally of bounded variation on $(0,+\infty)$. Furthermore, we deduce from Theorem 4.1.7 and from the functional equality (4.5.2) that the integral defining the tube zeta function $\tilde{\zeta}_{A,\Omega}$ in Equation (5.1.14) is absolutely convergent (and hence, convergent) for all $s \in \mathbb{C}$ such that $\operatorname{Re} s > \overline{\dim}_B(A,\Omega)$ or, in other words, $t \mapsto t^{\operatorname{Re} s-1} f(t)$ belongs to $L^1(0,+\infty)$ for such s. Consequently, the Mellin transform $\{\mathfrak{M}f\}(s)$ of f is well defined by Equation (5.1.16) and coincides with $\tilde{\zeta}_{A,\Omega}(s;\delta)$ for $c = \operatorname{Re} s > \overline{\dim}_B(A,\Omega)$; that is, Equation (5.1.14) holds for all $s \in \mathbb{C}$ such that $\operatorname{Re} s > \overline{\dim}_B(A,\Omega)$, as was claimed above. Therefore, by Theorem 5.1.6, we can recover the relative tube function from the relative tube zeta function and for positive $y \neq \delta$, we have

$$\chi_{(0,\delta)}(y) y^{-N} |A_y \cap \Omega| = \frac{1}{2\pi\mathrm{i}} \int_{c-\mathrm{i}\infty}^{c+\mathrm{i}\infty} y^{-s} \tilde{\zeta}_{A,\Omega}(s;\delta)\,\mathrm{d}s, \tag{5.1.19}$$

where $c > \overline{\dim}_B(A,\Omega)$ is arbitrary; that is, (5.1.18) is valid for all $t \in (0,\delta)$, as desired. □

One of the main goals in this chapter will be to express formula (5.1.18) in a more useful and applicable way. More specifically, we will express the right-hand side of (5.1.18) in terms of the relative distance zeta function and as a sum (interpreted in a suitable way) of residues over the complex dimensions of the given relative fractal drum. The resulting identity will be called a "fractal tube formula" (as in [Lap-vFr3]) or simply, a tube formula.

A priori, one would naively expect that Equation (5.1.16) and hence also, Equation (5.1.17), only holds for $c \geq N$. (Indeed, since $f(t) = 0$ for all $t \geq \delta$ and $|A_t \cap \Omega| \leq |\Omega|$, we easily see that $t \mapsto t^{c-1} f(t)$ belongs to $L^1(0,+\infty)$ for $c \geq N$.) The stronger conclusion obtained in Theorem 5.1.7 requires the aforementioned results obtained in Chapter 4 (and whose detailed proofs where given in Chapter 2 in the important special case of bounded subsets of $\mathbb{R}^N$).

The following result is really a corollary of Theorem 5.1.7 but given its importance for the rest of this section, we state it as a separate proposition.

Proposition 5.1.8. *Let (A,Ω) be a relative fractal drum in $\mathbb{R}^N$ and let $\delta > 0$ be fixed. Then for every $t \in (0,\delta)$ and $k \in \mathbb{N}_0$, we have*

$$V_{A,\Omega}^{[k]}(t) = \frac{1}{2\pi \mathrm{i}} \int_{c-\mathrm{i}\infty}^{c+\mathrm{i}\infty} \frac{t^{N-s+k}}{(N-s+1)_k} \tilde{\zeta}_{A,\Omega}(s;\delta)\,\mathrm{d}s, \tag{5.1.20}$$

where $c \in (\overline{\dim}_B(A,\Omega), N+1)$ *is arbitrary.*

Proof. By Theorem 5.1.7, we have the following equalities, valid (pointwise) for all $t \in (0,\delta)$:

$$\begin{aligned} V_{A,\Omega}^{[1]}(t) = \int_0^t V_{A,\Omega}(\tau)\,\mathrm{d}\tau &= \frac{1}{2\pi \mathrm{i}} \int_0^t \int_{c-\mathrm{i}\infty}^{c+\mathrm{i}\infty} \tau^{N-s} \tilde{\zeta}_{A,\Omega}(s;\delta)\,\mathrm{d}s\,\mathrm{d}\tau \\ &= \frac{1}{2\pi \mathrm{i}} \int_{c-\mathrm{i}\infty}^{c+\mathrm{i}\infty} \int_0^t \tau^{N-s} \tilde{\zeta}_{A,\Omega}(s;\delta)\,\mathrm{d}\tau\,\mathrm{d}s \\ &= \frac{1}{2\pi \mathrm{i}} \int_{c-\mathrm{i}\infty}^{c+\mathrm{i}\infty} \frac{t^{N-s+1}}{N-s+1} \tilde{\zeta}_{A,\Omega}(s;\delta)\,\mathrm{d}s, \end{aligned}$$

since $N-c+1>0$. The interchange of the order of integration is justified by combining Lebesgue's dominated convergence theorem and the Fubini–Tonelli theorem. Iterating this calculation $k-1$ more times, we prove the statement of the proposition. □

We adapt the following definition of the truncated screen and window from Section 5.3 of [Lap-vFr3], where it was stated for languid generalized fractal strings, so that it can be used in the same form in the case of relative fractal drums in $\mathbb{R}^N$.

Definition 5.1.9. (*The truncated screen and window*). Given an integer $n \geq 1$ and a languid relative fractal drum in $\mathbb{R}^N$, the *truncated screen* $\boldsymbol{S}_{|n}$ is the part of the screen $\boldsymbol{S}$ restricted to the interval $[T_{-n}, T_n]$, and the *truncated window* $\boldsymbol{W}_{|n}$ is the window $\boldsymbol{W}$ intersected with the horizontal strip between T_{-n} and T_n; i.e.,

$$\boldsymbol{W}_{|n} := \boldsymbol{W} \cap \{s \in \mathbb{C} : T_{-n} \leq \operatorname{Im} s \leq T_n\}. \tag{5.1.21}$$

We then call $\mathscr{P}(\tilde{\zeta}_{A,\Omega}, \boldsymbol{W}_{|n})$ the set of *truncated visible complex dimensions*; i.e., it is the set of visible complex dimensions of (A,Ω) relative to the window $\boldsymbol{W}$ and with imaginary parts between T_{-n} and T_n. Note that since by assumption, there are no poles of $\tilde{\zeta}_{A,\Omega}$ along the screen $\boldsymbol{S}$, we could replace $\boldsymbol{W}_{|n}$ by its interior $\mathring{\boldsymbol{W}}_{|n}$, in the aforementioned notation:

$$\mathscr{P}\left(\tilde{\zeta}_{A,\Omega}, \boldsymbol{W}_{|n}\right) = \mathscr{P}\left(\tilde{\zeta}_{A,\Omega}, \mathring{\boldsymbol{W}}_{|n}\right). \tag{5.1.22}$$

5.1.2 Pointwise Tube Formula with Error Term

We stress that from now on, the phrase "let (A,Ω) be a languid (or strongly languid) relative fractal drum", will implicitly mean that (A,Ω) is admissible for some window $\boldsymbol{W}$ and for some $\delta > 0$, the relative tube zeta function $\tilde{\zeta}_{A,\Omega}(s;\delta)$ of (A,Ω) satisfies the languidity conditions of Definition 5.1.3 (or Definition 5.1.4, respectively).

Let us now derive a 'truncated pointwise tube formula' (Lemma 5.1.10), from which the general pointwise tube formula (Theorem 5.1.11 below) will follow. Note that Lemma 5.1.10 is the counterpart, valid for any $N \geq 1$, of [Lap-vFr3, Lemma 5.9]. Furthermore, recall from the end of Subsection 5.1.1 that for each integer $n \geq 1$, the truncated screen $\boldsymbol{S}_{|n}$ and the associated truncated window $\boldsymbol{W}_{|n}$ were defined in Definition 5.1.9.

Lemma 5.1.10 (Truncated pointwise tube formula). *Let $k \geq 0$ be an integer and (A,Ω) a languid relative fractal drum in $\mathbb{R}^N$ for a fixed $\delta > 0$ and for some fixed languidity exponent $\kappa \in \mathbb{R}$. Furthermore, fix a constant $c \in (\overline{\dim}_B(A,\Omega), N+1)$. Then, for all $t \in (0,\delta)$ and all $n \geq 1$, we have*

$$\begin{aligned} I_n &:= \frac{1}{2\pi\mathrm{i}} \int_{c+\mathrm{i}T_{-n}}^{c+\mathrm{i}T_n} \frac{t^{N-s+k}}{(N-s+1)_k} \tilde{\zeta}_{A,\Omega}(s;\delta)\,\mathrm{d}s \\ &= \sum_{\omega \in \mathscr{P}(\tilde{\zeta}_{A,\Omega}, \boldsymbol{W}_{|n})} \operatorname{res}\left(\frac{t^{N-s+k}}{(N-s+1)_k} \tilde{\zeta}_{A,\Omega}(s;\delta), \omega\right) \\ &\quad + \frac{1}{2\pi\mathrm{i}} \int_{\boldsymbol{S}_{|n}} \frac{t^{N-s+k}}{(N-s+1)_k} \tilde{\zeta}_{A,\Omega}(s;\delta)\,\mathrm{d}s + E_n(t). \end{aligned} \tag{5.1.23}$$

Moreover, assuming that hypothesis **L1** *is fulfilled, we have the following pointwise remainder estimate, valid for all $t \in (0,\delta)$:*

$$|E_n(t)| \leq t^{N+k} K_\kappa \max\left\{T_n^{\kappa-k}, |T_{-n}|^{\kappa-k}\right\}(c - \inf S) \max\left\{t^{-c}, t^{-\inf S}\right\}, \tag{5.1.24}$$

where K_κ is a positive constant depending only on the languidity exponent κ.[6]

Finally, for each point $s = S(\tau) + \mathrm{i}\tau$, where $\tau \in \mathbb{R}$ is such that $|\tau| > 1$, and for all $t \in (0,\delta)$, the integrand over the truncated screen appearing in (5.1.23) *is bounded in absolute value by*

$$C t^{N+k} \max\left\{t^{-\sup S}, t^{-\inf S}\right\} |\tau|^{\kappa-k}, \tag{5.1.25}$$

when hypothesis **L2** *holds, and by*

$$C_\kappa t^{N+k} \max\left\{B^{|\inf S|}, B^{|\sup S|}\right\} \max\left\{t^{-\sup S}, t^{-\inf S}\right\} |\tau|^{\kappa-k}, \tag{5.1.26}$$

when hypothesis **L2'** *holds, with the constant C_κ depending only on κ.*[7]

Proof. Let $\overline{D} := \overline{\dim}_B(A,\Omega)$; for the sake of brevity, we will write $\tilde{\zeta}_{A,\Omega}(s)$ instead of $\tilde{\zeta}_{A,\Omega}(s;\delta)$ throughout the proof. Now, we replace the integral over the segment $[c + \mathrm{i}T_{-n}, c + \mathrm{i}T_n]$ with the integral over the contour Γ consisting of this segment, the truncated screen $\boldsymbol{S}_{|n}$ and the two horizontal segments joining $S(T_{\pm n}) + \mathrm{i}T_{\pm n}$ and $c + \mathrm{i}T_{\pm n}$ (see Figure 5.1). In other words, we have

---

[6] More precisely, K_κ depends only on κ and the constant C occurring in hypothesis **L1**.

[7] Here, the constant C_κ actually depends only on κ and on the constant C appearing in hypothesis **L1**.

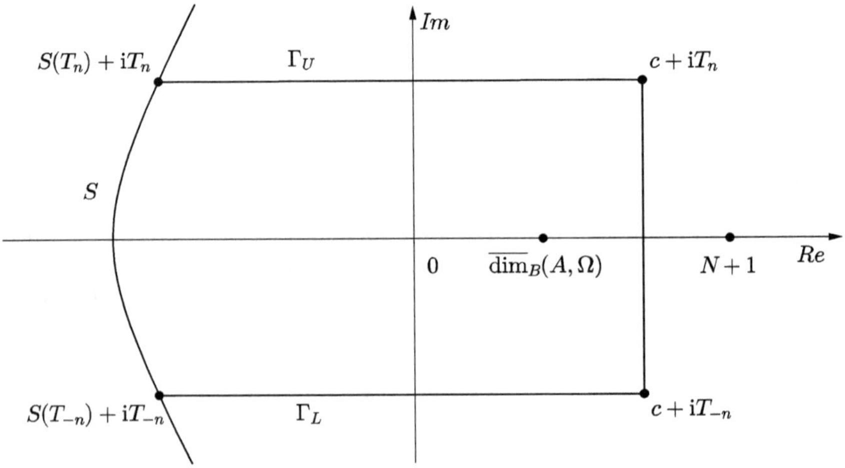

Fig. 5.1 The truncated window $\boldsymbol{W}_{|n}$ and the contour Γ which we use to estimate the integral I_n in Lemma 5.1.10.

$$
\begin{aligned}
I_n &= \frac{1}{2\pi\mathrm{i}} \int_{c+\mathrm{i}T_{-n}}^{c+\mathrm{i}T_n} \frac{t^{N-s+k}}{(N-s+1)_k} \tilde{\zeta}_{A,\Omega}(s)\,\mathrm{d}s \\
&= \frac{1}{2\pi\mathrm{i}} \oint_{\Gamma} \frac{t^{N-s+k}}{(N-s+1)_k} \tilde{\zeta}_{A,\Omega}(s)\,\mathrm{d}s \\
&\quad + \frac{1}{2\pi\mathrm{i}} \int_{\boldsymbol{S}_{|n}} \frac{t^{N-s+k}}{(N-s+1)_k} \tilde{\zeta}_{A,\Omega}(s)\,\mathrm{d}s + E_n(t),
\end{aligned}
$$

where

$$
E_n(t) := \frac{1}{2\pi\mathrm{i}} \int_{\Gamma_L \cup \Gamma_U} \frac{t^{N-s+k}}{(N-s+1)_k} \tilde{\zeta}_{A,\Omega}(s)\,\mathrm{d}s.
$$

Furthermore, the integrand appearing above is meromorphic on the bounded domain having Γ as its boundary and its poles are exactly the poles of the relative tube zeta function since $c \in (\overline{\dim}_B(A,\Omega), N+1)$ ensures that there are no zeros of $(N-s+1)_k$ inside of Γ. Consequently, we deduce from the residue theorem that

$$
\begin{aligned}
I_n &= \sum_{\omega \in \mathscr{P}(\tilde{\zeta}_{A,\Omega}, \boldsymbol{W}_{|n})} \operatorname{res}\left(\frac{t^{N-s+k}}{(N-s+1)_k} \tilde{\zeta}_{A,\Omega}(s), \omega \right) \\
&\quad + \frac{1}{2\pi\mathrm{i}} \int_{\boldsymbol{S}_{|n}} \frac{t^{N-s+k}}{(N-s+1)_k} \tilde{\zeta}_{A,\Omega}(s)\,\mathrm{d}s + E_n(t).
\end{aligned}
$$

To obtain the upper bound on $|E_n(t)|$, we first observe that for $s = \sigma + \mathrm{i}T_n$ we have $|(N-s+1)_k| \geq T_n^k$ and we estimate the integrals over the upper segment Γ_U and the lower segment Γ_L under hypothesis **L1**:

$$\left|\int_{\Gamma_U} \frac{t^{N-s+k}}{(N-s+1)_k} \tilde{\zeta}_{A,\Omega}(s)\,\mathrm{d}s\right| = \left|\int_{\boldsymbol{S}(T_n)}^{c} \frac{t^{N+k-\sigma-\mathrm{i}T_n}}{(N+1-(\sigma+\mathrm{i}T_n))_k} \tilde{\zeta}_{A,\Omega}(\sigma+\mathrm{i}T_n)\,\mathrm{d}\sigma\right|$$
$$\leq t^{N+k}C(T_n+1)^{\kappa}T_n^{-k}\int_{\boldsymbol{S}(T_n)}^{c} t^{-\sigma}\,\mathrm{d}\sigma$$
$$\leq t^{N+k}K_\kappa T_n^{\kappa-k}\big(c-S(T_n)\big)\max\left\{t^{-c},t^{-S(T_n)}\right\},$$

where K_κ is a positive constant such that $C(|T_n|+1)^\kappa \leq K_\kappa |T_n|^\kappa$ for all $n \in \mathbb{Z}$. Furthermore, since $\inf S \leq S(\tau)$ for all $\tau \in \mathbb{R}$, we have

$$\left|\int_{\Gamma_U} \frac{t^{N-s+k}\tilde{\zeta}_{A,\Omega}(s)\,\mathrm{d}s}{(N-s+1)_k}\right| \leq t^{N+k}K_\kappa T_n^{\kappa-k}(c-\inf S)\max\left\{t^{-c},t^{-\inf S}\right\}. \tag{5.1.27}$$

A similar calculation for the integral over the lower line segment yields

$$\left|\int_{\Gamma_L} \frac{t^{N-s+k}\tilde{\zeta}_{A,\Omega}(s)\,\mathrm{d}s}{(N-s+1)_k}\right| \leq t^{N+k}K_\kappa |T_{-n}|^{\kappa-k}(c-\inf S)\max\left\{t^{-c},t^{-\inf S}\right\}. \tag{5.1.28}$$

Therefore, putting (5.1.27) and (5.1.28) together, we obtain the upper bound (5.1.24).[8]

In order to estimate the integrand over the truncated screen $\boldsymbol{S}_{|n}$, we observe that for $s = S(\tau)+\mathrm{i}\tau$ with $|\tau| > 1$, we have

$$\begin{aligned}\left|\frac{t^{N-s+k}}{(N-s+1)_k}\tilde{\zeta}_{A,\Omega}(s)\right| &\leq Ct^{N-S(\tau)+k}|\tau|^{\kappa-k}\\ &\leq Ct^{N+k}\max\left\{t^{-\sup S},t^{-\inf S}\right\}|\tau|^{\kappa-k},\end{aligned} \tag{5.1.29}$$

under hypothesis **L2** and similarly, under hypothesis **L2'**. (Then, C_κ is a constant chosen so that $C(|\tau|+1)^\kappa \leq C_\kappa|\tau|^\kappa$ holds for all τ such that $|\tau| > 1$.) This completes the proof of the lemma. □

We can now state and prove the announced result.

Theorem 5.1.11 (Pointwise fractal tube formula with error term, via $\tilde{\zeta}_{A,\Omega}$). *Let (A,Ω) be a relative fractal drum in $\mathbb{R}^N$ which is languid for some fixed $\delta > 0$ and some fixed languidity exponent $\kappa \in \mathbb{R}$. Furthermore, let $k > \kappa+1$ be a nonnegative integer. Then, the following pointwise fractal tube formula with error term, expressed in terms of the tube zeta function $\tilde{\zeta}_{A,\Omega} := \tilde{\zeta}_{A,\Omega}(\,\cdot\,;\delta)$, is valid for every $t \in (0,\delta)$:*

$$V_{A,\Omega}^{[k]}(t) = \sum_{\omega\in\mathscr{P}(\tilde{\zeta}_{A,\Omega},\boldsymbol{W})} \operatorname{res}\left(\frac{t^{N-s+k}}{(N-s+1)_k}\tilde{\zeta}_{A,\Omega}(s),\omega\right) + \tilde{R}_{A,\Omega}^{[k]}(t). \tag{5.1.30}$$

[8] The constant K_κ in (5.1.24) is actually equal to the present constant K_κ divided by π.

Here, for every $t \in (0,\delta)$, the (pointwise) error term $\tilde{R}^{[k]}_{A,\Omega}$ is given by the absolutely convergent (and hence, convergent) integral

$$\tilde{R}^{[k]}_{A,\Omega}(t) = \frac{1}{2\pi\mathrm{i}} \int_{S} \frac{t^{N-s+k}}{(N-s+1)_k} \tilde{\zeta}_{A,\Omega}(s)\,ds. \qquad (5.1.31)$$

Furthermore, we have the following pointwise error estimate, valid for all $t \in (0,\delta)$:

$$\left|\tilde{R}^{[k]}_{A,\Omega}(t)\right| \leq t^{N+k} \max\{t^{-\sup S}, t^{-\inf S}\} \left(\frac{C(1+\|S\|_{\mathrm{Lip}})}{2\pi(k-\kappa-1)} + C' \right), \qquad (5.1.32)$$

where C is the positive constant appearing in **L1** *and* **L2** *and C' is some suitable positive constant. These constants depend only on the relative fractal drum (A,Ω) and the screen, but not on the value of the nonnegative integer k.*

In particular, we have the following pointwise error estimate:

$$\tilde{R}^{[k]}_{A,\Omega}(t) = O(t^{N-\sup S+k}) \quad \text{as} \quad t \to 0^+. \qquad (5.1.33)$$

Moreover, if $S(\tau) < \sup S$ for all $\tau \in \mathbb{R}$ (i.e., if the screen $\boldsymbol{S}$ lies strictly to the left of the vertical line $\{\mathrm{Re}\,s = \sup S\}$), then we have the following stronger pointwise estimate:

$$\tilde{R}^{[k]}_{A,\Omega}(t) = o(t^{N-\sup S+k}) \quad \text{as} \quad t \to 0^+. \qquad (5.1.34)$$

Before establishing Theorem 5.1.11, we must make the following two comments (in parts (a) and (b) of Remark 5.1.12), which will help to understand the statement of the theorem. *We stress that comments similar to those in Remark 5.1.12 also apply to all other theorems stated below* (in this chapter), *in which a* (typically infinite) *sum over the (visible) complex dimensions appears, either in reference to a pointwise or distributional fractal tube formula.* (See, in particular, Theorems 5.1.13, 5.1.14, 5.1.16, 5.2.2, 5.2.4, 5.2.6, 5.3.11, 5.3.13, 5.3.16, 5.3.17, 5.3.19, 5.3.20, 5.3.21, 5.4.14, along with Corollaries 5.2.12 and 5.3.14.) More specifically, part (b) of Remark 5.1.12 remains valid without change, and likewise for the counterpart of part (a) of Remark 5.1.12 in reference to a (potentially infinite) sum over the (visible) complex dimensions occurring in a pointwise fractal tube formula (such as in Theorem 5.1.13, 5.1.14 and 5.1.16 of Subsection 5.1.2 and in Theorems 5.3.11, 5.3.13, 5.3.16, 5.3.17, along with Corollary 5.3.14 of Subsection 5.3.2 below). Moreover, in the counterpart of part (a) of Remark 5.1.12, when referring to a distributional (rather than a pointwise) fractal tube formula (such as in Theorems 5.2.2, 5.2.4, 5.2.6, and Corollary 5.2.12 of Section 5.2 below or in Theorems 5.3.19, 5.3.20, 5.3.21 of Subsection 5.3.3), the (potentially infinite) sum has to be interpreted as a distributional (rather than pointwise) limit of the partial sums.

Remark 5.1.12. (*a*) The (potentially infinite) sum appearing in (5.1.30) in the above theorem (Theorem 5.1.11) is to be understood as the limit

$$\lim_{n\to\infty} \sum_{\omega\in\mathscr{P}(\tilde{\zeta}_{A,\Omega},\boldsymbol{W}_{|n})} \operatorname{res}\left(\frac{t^{N-s+k}}{(N-s+1)_k}\tilde{\zeta}_{A,\Omega}(s),\omega\right), \tag{5.1.35}$$

where $\boldsymbol{W}_{|n}$ is the truncated window given by Definition 5.1.9 or, in other words, as the pointwise limit of the partial sums over the complex dimensions contained in $\boldsymbol{W}_{|n}$. Furthermore, the existence of this limit follows from the proof of the theorem; that is, the series in (5.1.30) converges pointwise and conditionally. On the other hand, Theorem 5.1.11 does not give any information about the possible absolute convergence of the series in (5.1.30). A similar situation occurs in [Lap-vFr3, Chapters 5 and 8] and, in fact, also in Riemann's original explicit formula for the counting function of the prime numbers (see, e.g., [Edw] or [In]).

(*b*) Moreover, the sum over the set $\mathscr{P}(\tilde{\zeta}_{A,\Omega},\boldsymbol{W})$ in Equation (5.1.30) of Theorem 5.1.11 does not depend on δ since changing the parameter δ has no effect on the residues appearing in (5.1.30). This follows easily from the fact that the principal parts of the meromorphic extension of the relative tube zeta function around any of its poles do not depend on δ (see Subsection 4.5.1). In other words, when applying Theorem 5.1.11, one has to determine that (A,Ω) is languid for some $\delta>0$, but when calculating the sum, one can take any $\delta>0$; that is, in practice, the most convenient one in the particular example one is interested in.

Proof of Theorem 5.1.11. Without loss of generality, let $c\in(\overline{\dim}_B(A,\Omega),N+1)$ be the constant from the languidity condition **L1** of Definition 5.1.3. We will prove the theorem by using Lemma 5.1.10 in order to obtain (5.1.23) and then, by letting $n\to\infty$. We note that $E_n(t)$ tends to zero for $k>\kappa$ at the rate of some negative power of $\min\{T_n,|T_{-n}|\}$. Furthermore, for $k>\kappa+1$, the error term $\tilde{R}^{[k]}(t)_{A,\Omega}$ is absolutely convergent (and hence, pointwise convergent). Indeed, note that, since $\tau\mapsto S(\tau)$ is Lipschitz continuous, it is differentiable almost everywhere and, consequently, the derivative of $\tau\mapsto S(\tau)+\mathrm{i}\tau$ is bounded by $(1+\|S\|_{\mathrm{Lip}})$ for almost all $\tau\in\mathbb{R}$. Moreover, since

$$\int_1^{+\infty}\tau^{\kappa-k}\,\mathrm{d}\tau=\frac{1}{k-\kappa-1}$$

for $k>\kappa+1$, the upper bound (5.1.32) on the error term $\tilde{R}^{[k]}_{A,\Omega}(t)$ now follows from (5.1.25). The positive constant C' in (5.1.32) is the constant which corresponds to the integral over the part of the screen for which $|\tau|<1$; i.e.,

$$C':=\frac{1}{2\pi}\int_{\boldsymbol{S}\cap\{|\operatorname{Im}S|<1\}}\frac{|\tilde{\zeta}_{A,\Omega}(s)|}{|(N-s+1)_k|}\,|\mathrm{d}s|.$$

In the case when the screen stays strictly to the left of the line $\{\operatorname{Re}s=\sup S\}$, we can obtain the better estimate (5.1.34) by using a well-known method; see, e.g., [In, pp. 33–34]. Namely, for any given $\varepsilon>0$, we have to show that (5.1.31) is bounded by $\varepsilon t^{N-\sup S+k}$. For a given $T>0$, we can split the integral appearing on the right-hand side of (5.1.31) into the following two parts. The first one is the integral over the part of the screen for which $|\operatorname{Im} S|>T$, and the second one is the integral over the part of the screen for which $|\operatorname{Im} S|\leq T$. Since the first integral is absolutely convergent, we can choose T sufficiently large so that it is bounded by $\frac{1}{2}\varepsilon t^{N-\sup S+k}$. For the second integral, we observe that the maximum of $S(\tau)$ for all $\tau\in[-T,T]$ is strictly less than $\sup S$; i.e., we can choose $\alpha>0$ such that $S(\tau)<\sup S-\alpha$ for $\tau\in[-T,T]$. This implies that the integral over the part of the screen for which $|\operatorname{Im} S|\leq T$ is of order $O(t^{N-\sup S+k+\alpha})$ as $t\to 0^+$.[9] Hence, for all sufficiently small $t>0$ it is bounded by $\frac{1}{2}\varepsilon t^{N-\sup S+k}$. This proves that $\tilde{R}^{[k]}_{A,\Omega}(t)=o(t^{N-\sup S+k})$ as $t\to 0^+$, as desired, and therefore completes the proof of the theorem. □

5.1.3 Exact Pointwise Tube Formula

In the case of a strongly languid relative fractal drum, we are able to obtain a pointwise formula without an error term. Such an explicit formula is said to be *exact*.

Theorem 5.1.13 (Exact pointwise fractal tube formula via $\tilde{\zeta}_{A,\Omega}$). *Let (A,Ω) be a relative fractal drum in $\mathbb{R}^N$ which is strongly languid for some fixed $\delta>0$ and some fixed languidity exponent $\kappa\in\mathbb{R}$. Furthermore, let $k>\kappa$ be a nonnegative integer. Then, the following exact pointwise fractal tube formula, expressed in terms of the tube zeta function $\tilde{\zeta}_{A,\Omega}:=\tilde{\zeta}_{A,\Omega}(\,\cdot\,;\delta)$, holds for all $t\in(0,\min\{1,\delta,B^{-1}\})$:*

$$V^{[k]}_{A,\Omega}(t)=\sum_{\omega\in\mathscr{P}(\tilde{\zeta}_{A,\Omega},\mathbb{C})}\operatorname{res}\left(\frac{t^{N-s+k}}{(N-s+1)_k}\tilde{\zeta}_{A,\Omega}(s),\omega\right). \tag{5.1.36}$$

Here, B is the positive constant appearing in hypothesis **L2'**.

Proof. For a fixed integer $n\geq 1$, we apply Lemma 5.1.10 with the screen $\boldsymbol{S}_m$ given by hypothesis **L2'**. We first let $m\to\infty$ while keeping n fixed. Since the screens $\boldsymbol{S}_m$ have a uniform Lipschitz bound, if we take $t<\min\{1,B^{-1}\}$, then the sequence of integrals over the truncated screens $\boldsymbol{S}_{m|n}$ converges to 0 as $m\to\infty$.[10] Indeed, let us take m_0 large enough so that $\sup S_m<0$ for every $m\geq m_0$. This is possible since $\sup S_m\to-\infty$ as $m\to\infty$; see hypothesis **L2'** in Definition 5.1.4.

Furthermore, for every $m\geq 1$ and $n\geq 1$, the integral over the truncated screen $\boldsymbol{S}_{m|n}$ is given by

$$I_{n,m}:=\frac{1}{2\pi\mathrm{i}}\int_{\boldsymbol{S}_{m|n}}\frac{t^{N-s+k}}{(N-s+1)_k}\tilde{\zeta}_{A,\Omega}(s)\,\mathrm{d}s \tag{5.1.37}$$

[9] Note that since the screen $\boldsymbol{S}$ avoids the poles of the relative tube zeta function, we have that $\tilde{\zeta}_{A,\Omega}(s)$ is bounded for all $s\in\mathbb{C}$ in the part of the screen $\boldsymbol{S}$ for which $|\operatorname{Im} S|\leq T$.

[10] Here and throughout this proof, $\boldsymbol{S}_{m|n}$ denotes the n-th truncated screen associated with the screen $\boldsymbol{S}_m$, in the sense of Lemma 5.1.10 and Figure 5.1 above.

and, similarly as in the proof of Lemma 5.1.10, we have that the integrand is bounded in absolute value by

$$C_\kappa \max\left\{B^{|\inf S_{m|n}|}, B^{|\sup S_{m|n}|}\right\} t^{N+|\sup S_{m|n}|+k}, \tag{5.1.38}$$

where C_κ is a suitable constant depending only on κ. Here, we use the notation

$$\inf S_{m|n} := \inf_{\tau\in[T_{-n},T_n]} S_m(\tau) \text{ and } \sup S_{m|n} := \sup_{\tau\in[T_{-n},T_n]} S_m(\tau). \tag{5.1.39}$$

We now let $L := \sup_{m\geq 1} \|S_m\|$ be the uniform Lipschitz bound for the sequence of screens $\boldsymbol{S}_m$. Then, the derivative of $\tau \mapsto S_m(\tau) + \mathbb{i}\tau$ is bounded for almost every $\tau \in [T_{-n}, T_n]$ by $(1+L)$.

We must next consider the following two cases: firstly, if $B < 1$, we then have that

$$|I_{n,m}| \leq \frac{C_\kappa(1+L)B^{|\sup S_{m|n}|}}{2\pi}(T_n - T_{-n})t^{N+|\sup S_{m|n}|+k},$$

and, since $t < 1$, we have that $I_{n,m} \to 0$ as $m \to \infty$. Secondly, if $B \geq 1$, we deduce from the Lipschitz condition on S_m that we have

$$\sup S_{m|n} - \inf S_{m|n} \leq L(T_n - T_{-n});$$

i.e.,

$$|\inf S_{m|n}| \leq |\sup S_{m|n}| + L(T_n - T_{-n}),$$

from which we deduce the estimate

$$|I_{n,m}| \leq \frac{C_\kappa(1+L)B^{L(T_n-T_{-n})}}{2\pi}(T_n - T_{-n})(Bt)^{|\sup S_{m|n}|}t^{N+k}.$$

Therefore, $I_{n,m} \to 0$ as $m \to \infty$ since $Bt < 1$.

We now let $E_{n,m}(t)$ be the error function appearing in (5.1.23) for the truncated screen $\boldsymbol{S}_{m|n}$ and we will complete the proof by showing that its iterated limit converges to zero pointwise. For $c \in (\overline{\dim}_B(A,\Omega), N+1)$ and since $0 < t < 1$, we have, much as in the proof of Lemma 5.1.10, that

$$\begin{aligned}\left|\int_{\Gamma_{U_m}} \frac{t^{N-s+k}\tilde{\zeta}_{A,\Omega}(s)\,\mathrm{d}s}{(N-s+1)_k}\right| &\leq t^{N+k}C(T_n+1)^\kappa T_n^{-k}\int_{-\infty}^{c} t^{-\sigma}\,\mathrm{d}\sigma \\ &\leq t^{N+k}K_\kappa T_n^{\kappa-k}\frac{t^{-c}}{\log(1/t)}.\end{aligned} \tag{5.1.40}$$

Here, Γ_{U_m} is the segment connecting $S_m(T_n) + \mathbb{i}T_n$ and $c + \mathbb{i}T_n$. A similar reasoning for the corresponding integral over the lower segment gives us the following upper bound on $|E_{n,m}(t)|$, independent of m:

$$|E_{n,m}(t)| \leq \frac{t^{N-c+k}}{\pi\log(1/t)}K_\kappa \max\{T_n^{\kappa-k}, |T_{-n}|^{\kappa-k}\}.$$

Finally, this inequality, which is valid for all $m \geq 1$ and all $n \geq 1$, implies that for a fixed $k > \kappa$, the iterated limit of $E_{n,m}(t)$ tends to 0 when $m \to \infty$ and then $n \to \infty$; i.e., we have

$$\lim_{n\to\infty}\big(\lim_{m\to\infty} E_{n,m}(t)\big) = 0.$$

This concludes the proof of the theorem. □

Theorems 5.1.11 and 5.1.13 are of most interest in the case when $k = 0$, i.e., when we obtain a pointwise formula for the volume of the relative t-neighborhood $|A_t \cap \Omega|$ in terms of the complex dimensions of (A, Ω). We state this case as a separate theorem.

Theorem 5.1.14 (Pointwise fractal tube formula via $\tilde{\zeta}_{A,\Omega}$; level $k = 0$). *Under the same hypotheses as in Theorem 5.1.11, with languidity exponent $\kappa < -1$ (resp., under the same hypotheses as in Theorem 5.1.13, with languidity exponent $\kappa < 0$) and with $k := 0$, we have the following pointwise formula for the tube function of the relative fractal drum (A, Ω) in $\mathbb{R}^N$:*

$$|A_t \cap \Omega| = \sum_{\omega \in \mathscr{P}(\tilde{\zeta}_{A,\Omega}, \boldsymbol{W})} \operatorname{res}\left(t^{N-s}\tilde{\zeta}_{A,\Omega}(s), \omega\right) + \tilde{R}^{[0]}_{A,\Omega}(t), \tag{5.1.41}$$

valid pointwise for all $t \in (0, \delta)$ and where $\tilde{R}^{[0]}_{A,\Omega}(t)$ is the error term given by formula (5.1.31) *with $k := 0$. Furthermore, we have the following pointwise error estimate*:

$$\tilde{R}^{[0]}_{A,\Omega}(t) = O(t^{N-\sup S}) \quad \text{as} \quad t \to 0^+. \tag{5.1.42}$$

Moreover, if $S(\tau) < \sup S$ for every $\tau \in \mathbb{R}$ (i.e., if the screen $\boldsymbol{S}$ lies strictly to the left of the vertical line $\{\operatorname{Re} s = \sup S\}$), we then have

$$\tilde{R}^{[0]}_{A,\Omega}(t) = o(t^{N-\sup S}) \quad \text{as} \quad t \to 0^+. \tag{5.1.43}$$

Finally, in the special case of Theorem 5.1.13 where (A, Ω) is assumed to be strongly languid, then $\tilde{R}^{[0]}_{A,\Omega}(t) \equiv 0$ and $\boldsymbol{W} := \mathbb{C}$ in (5.1.41)*; so that the pointwise fractal tube formula* (5.1.41) *becomes exact.*

Remark 5.1.15. In the applications, we often have to consider the case when all of the visible complex dimensions are simple. More specifically, if we assume that all of the poles of $\tilde{\zeta}_{A,\Omega}$ visible through the window $\boldsymbol{W}$ (i.e., lying in $\boldsymbol{W}$) are simple, then in the statement of Theorem 5.1.14, the sum over the visible complex dimensions appearing in Equation (5.1.41) can be replaced by the following expression:

$$\sum_{\omega \in \mathscr{P}(\tilde{\zeta}_{A,\Omega}, \boldsymbol{W})} \tilde{c}_\omega t^{N-\omega}, \tag{5.1.44}$$

where for each $\omega \in \mathscr{P}(\tilde{\zeta}_{A,\Omega}, \boldsymbol{W})$, we have

$$\tilde{c}_\omega := \operatorname{res}(\tilde{\zeta}_{A,\Omega}, \omega). \tag{5.1.45}$$

In light of Remark 5.1.15, we obtain the following result, which (even though it is an immediate corollary of Theorem 5.1.14) we state as a separate theorem because of its importance in the applications. (See, especially, Sections 5.4 and 5.5 below.)

Theorem 5.1.16 (Pointwise fractal tube formula via $\tilde{\zeta}_{A,\Omega}$; level $k=0$ and simple poles). *Assume that the hypotheses of Theorem 5.1.14 hold. Suppose, in addition, that all of the visible complex dimensions of the RFD (A,Ω) are simple (i.e., all of the poles of $\tilde{\zeta}_{A,\Omega}$ belonging to the window $\boldsymbol{W}$ are simple). Then, the pointwise fractal tube formula, expressed in terms of the tube zeta function $\tilde{\zeta}_{A,\Omega}$, takes the following simpler form:*

$$|A_t\cap\Omega| = \sum_{\omega\in\mathscr{P}(\tilde{\zeta}_{A,\Omega},\boldsymbol{W})} \operatorname{res}\left(\tilde{\zeta}_{A,\Omega}(s),\omega\right)t^{N-\omega} + \tilde{R}^{[0]}_{A,\Omega}(t), \tag{5.1.46}$$

where the error term $\tilde{R}^{[0]}_{A,\Omega}$ is the same as in Theorem 5.1.14 and hence, satisfies the same estimates [(5.1.42) *or* (5.1.43), *depending on the hypotheses*] *as in Theorem 5.1.14.*

In particular, in the strongly languid case, we have $\tilde{R}^{[0]}(t)\equiv 0$ and $\boldsymbol{W}:=\mathbb{C}$; consequently, (5.1.46) *becomes the following* exact *fractal tube formula, valid pointwise for all $t\in(0,\min\{1,\delta,B^{-1}\})$:*

$$|A_t\cap\Omega| = \sum_{\omega\in\mathscr{P}(\tilde{\zeta}_{A,\Omega},\mathbb{C})} \operatorname{res}\left(\tilde{\zeta}_{A,\Omega}(s),\omega\right)t^{N-\omega}, \tag{5.1.47}$$

where B is the constant appearing in hypothesis **L2'** *of Definition 5.1.4.*

Remark 5.1.17. Naturally, in light of Theorem 5.1.11 and Theorem 5.1.13, the counterpart of Remark 5.1.15 and Theorem 5.1.16 holds for any level k (satisfying the assumptions of the relevant result). For example, provided that all of the complex dimensions visible through $\boldsymbol{W}$ are simple, the exact pointwise fractal tube formula (5.1.36) of Theorem 5.1.13 becomes (for all $t\in(0,\min\{1,\delta,B^{-1}\})$)

$$V^{[k]}_{A,,\Omega}(t) = \sum_{\omega\in\mathscr{P}(\tilde{\zeta}_{A,\Omega},\mathbb{C})} \operatorname{res}\left(\tilde{\zeta}_{A,\Omega}(s),\omega\right)\frac{t^{N-\omega+k}}{(N-\omega+1)_k}, \tag{5.1.48}$$

and similarly for the pointwise fractal tube formula with error term given in (5.1.30) of Theorem 5.1.11.

Note that in light of (5.1.12) and for each $k\in\mathbb{N}_0$, we have (with the obvious convention if $k=0$)

$$(N-s+1)_k = (N-s+1)(N-s+2)\cdots(N-s+k) \tag{5.1.49}$$

and hence, the zeros of $s\mapsto(N-s+1)_k$ are simple and occur precisely at

$$s=N+1,N+2,\dots,N+k. \tag{5.1.50}$$

(Clearly, since $(N-s+1)_0 = 1$, (5.1.49) does not have any zeros if $k=0$.) Consequently, since $\overline{\dim}_B(A,\Omega) \le N$ and k is nonnegative (i.e., $k \in \mathbb{N}_0$) in the present case of pointwise tube formulas, the complex number $(N-\omega+1)_k$ is never equal to zero for $\omega \in \mathscr{P}(\tilde{\zeta}_{A,\Omega}) := \mathscr{P}(\tilde{\zeta}_{A,\Omega}, \mathbb{C})$ (or else for $\omega \in \mathscr{P}(\tilde{\zeta}_{A,\Omega}, \boldsymbol{W})$, in the case of a pointwise tube formula with error term). Moreover, if we work with a distributional tube formula (as will be case in Section 5.2 and part of Section 5.3, for example), the level k is allowed to be negative ($i.e., k \in \mathbb{Z}$). However, in the case of a negative integer k, the function $s \mapsto (N-s+1)_k$ does not have any zeros, but only simple poles located precisely at

$$s = N+1+k, N+2+k, \ldots, N; \tag{5.1.51}$$

so that its reciprocal has simple zeros precisely at those same points. Indeed, note that if $k<0$, then

$$(N-s+1)_k = \frac{\Gamma(N-s+1+k)}{\Gamma(N-s+1)} \tag{5.1.52}$$

and observe that (on the right-hand side of (5.1.52)) the numerator has simple poles located precisely at $s = N+1+k, N+2+k, N+3+k, \ldots$, while the denominator has simple zeros located precisely at $s = N+1, N+2, N+3, \ldots$, which cancel out all of the poles of the numerator except for the ones stated in (5.1.51).

In light of this discussion, for the distributional tube formulas obtained in Section 5.2 and in part of Section 5.3 below, the same comment (concerning the complex dimensions of (A,Ω)) applies as for the pointwise tube formulas; i.e., the complex dimensions of (A,Ω) are never zeros of $s \mapsto (N-s+1)_k$. But we also note that in the distributional case, it may happen that ω is a zero of $s \mapsto (N-s+1)_k^{-1}$, which then cancels out the term corresponding to $t^{N-\omega}$ in Equation (5.1.48).

Remark 5.1.18. The obvious counterpart of Remark 5.1.15, Theorem 5.1.16 and Remark 5.1.17 holds for all of the fractal tube formulas considered in this chapter, whether they are pointwise or distributional formulas, with or without error term, as well as expressed in terms of either $\zeta_{A,\Omega}$ or $\tilde{\zeta}_{A,\Omega}$ or (with the notation of Subsection 5.3.1 or 5.4.2, respectively) $\breve{\zeta}_{A,\Omega}$ or $\zeta^{\mathfrak{M}}_{A,\Omega}$. In the case of $\zeta_{A,\Omega}$, $\breve{\zeta}_{A,\Omega}$ and $\zeta^{\mathfrak{M}}_{A,\Omega}$, one must assume, in addition, that $\overline{D} := \dim_B(A,\Omega) < N$. (See also the second part of Remark 5.1.17 just above, along with Remark 5.3.18 below.)

The following comment will help explain, in part, why we work with (pointwise) fractal tube formulas at level k (i.e., for the k-th primitive $V^{[k]}_{A,\Omega}$ of $V_{A,\Omega} = V^{[0]}_{A,\Omega}$), rather than simply at level 0 (i.e., for the tube function $V := V_{A,\Omega}$ itself). Observe, in addition, that the larger k is, the weaker the assumptions on the growth of the corresponding fractal zeta function (here, in Subsections 5.1.2 and 5.1.3, the tube zeta function $\tilde{\zeta}_{A,\Omega}$) in the statement of the corresponding pointwise fractal tube formula; see, e.g., the hypotheses of Theorem 5.1.11 and of Theorem 5.1.13, in Subsections 5.1.2 and Subsection 5.1.3, respectively. (A similar comment could be made about all of the fractal tube formulas established in this chapter).

Remark 5.1.19. Recall from [Schw, Section VII, I, esp., p. 226], that given a periodic distribution T on $\mathbb{R}$, in order to establish (under suitable, but rather weak, hypotheses) the convergence of its Fourier series, in the sense of distributions, one first integrates the given distribution sufficiently many times (say, k times, with k large enough); so that at the k-th level (i.e., for the k-th antiderivative of T, now viewed as a "nice" pointwise function), one can apply the usual theorems concerning the *pointwise* (and uniform) convergence of Fourier series. One then views the k-th primitive as a distribution and differentiates it k times, along with its convergent Fourier series, in the distributional sense. One therefore concludes that the periodic distribution T is equal (in the sense of distributions) to its (distributionally convergent) Fourier series. An entirely analogous procedure will be used here, in a different but related context, in order to establish the distributional fractal tube formulas from Subsection 5.2 just below, by essentially deducing them from the corresponding pointwise fractal tube formulas obtained in the present subsections (i.e., Subsection 5.1.2 and Subsection 5.1.3). Hence, the importance of having established the pointwise fractal tube formulas in Subsections 5.1.2 and 5.1.3, not only at level $k=1$ (when possible) but (under significantly weaker hypotheses) also at any sufficiently large level $k \geq 1$. The exact same procedure was used in [Lap-vFr1–3] in order to deduce the main distributional explicit formulas from the corresponding pointwise explicit formulas; see, especially, the first proof of Theorem 5.18 along with Remark 5.20 on page 160 of [Lap-vFr3].[11]

5.2 Distributional Tube Formulas

In this section, our goal is to obtain the distributional analogs of Theorems 5.1.11 and 5.1.13 in order to derive a distributional fractal tube formula for $V^{[k]}_{A,\Omega}(t)$, valid for any integer $k \in \mathbb{Z}$ and still expressed in terms of the (visible) poles of the tube zeta function $\tilde{\zeta}_{A,\Omega}$. This will provide us with information (in the sense of Schwartz distributions or generalized functions) about the tube function of a relative fractal drum (A,Ω), no matter for which exponent $\kappa \in \mathbb{R}$ the relative fractal drum (A,Ω) is languid. (See Definition 5.1.3.) More precisely, let $\delta > 0$ and $\mathscr{D}(0,\delta) := C_c^\infty(0,\delta)$ be the space of infinitely differentiable (complex-valued) test functions with compact support contained in $(0,\delta)$. In fact, we will start with a larger space of test functions for which the formulas obtained here will be valid. Namely, let $\mathscr{K}(0,\delta)$ be the set of test functions φ in the class $C^\infty(0,\delta)$, such that for all $m \in \mathbb{Z}$ and $q \in \mathbb{N}$, we have $t^m \varphi^{(q)}(t) \to 0$, as $t \to 0^+$ and also that $(t-\delta)^m \varphi^{(q)}(t) \to 0$ as $t \to \delta^-$, where $\varphi^{(q)}$ denotes the q-th derivative of φ.

Note that $\mathscr{D}(0,\delta) \subseteq \mathscr{K}(0,\delta)$, with a continuous embedding. Hence, we have the following (reverse) inclusion, also with a continuous embedding, between the corresponding spaces of distributions (i.e., the dual spaces):

[11] There will be some significant technical differences in the execution of the method in the present book, but they will not necessarily be pointed out.

$$\mathscr{K}'(0,\delta) \subseteq \mathscr{D}'(0,\delta). \tag{5.2.1}$$

General information about the theory of distributions (or generalized functions) can be found in [Schw, Bre, Foll, Hö2, JohLap, JohLapNi, ReeSim1].

Definition 5.2.1. Let (A,Ω) be a relative fractal drum in $\mathbb{R}^N$ and let $k \in \mathbb{Z}$ be an arbitrary integer. We define the distribution $\mathscr{V}^{[k]} = \mathscr{V}^{[k]}_{A,\Omega}$ on $\mathscr{K}(0,\delta)$ to be the $|k|$-th distributional derivative of $V(t) = |A_t \cap \Omega|$ in case $k < 0$ and the k-th primitive (or k-th anti-derivative) function (considered as a regular distribution in $\mathscr{K}'(0,\delta)$) of $V(t)$ if $k > 0$. For $k = 0$, this is the (regular) distribution generated by the locally integrable function $V(t)$. (Observe that $V = V(t)$ is locally integrable on $(0,+\infty)$ because it is continuous.) More specifically, for any test function $\varphi \in \mathscr{K}(0,\delta)$, we have

$$\langle \mathscr{V}^{[k]}, \varphi\rangle := \int_0^{+\infty} V^{[k]}(t)\varphi(t)\,\mathrm{d}t, \qquad \text{for } k \geq 0, \tag{5.2.2}$$

and

$$\langle \mathscr{V}^{[k]}, \varphi\rangle := (-1)^{|k|}\int_0^{+\infty} V(t)\varphi^{(|k|)}(t)\,\mathrm{d}t, \qquad \text{for } k < 0. \tag{5.2.3}$$

Here, and from now on, for convenience, we always extend the test function $\varphi \in \mathscr{K}(0,\delta)$ to the interval $[\delta,+\infty)$ by letting $\varphi_{|[\delta,+\infty)} \equiv 0$.

Note that it follows from Definition 5.2.1 that for all $k_1, k_2 \in \mathbb{Z}$ such that $k_1 < k_2$, $\mathscr{V}^{[k_1]}$ is the $(k_2 - k_1)$-th distributional derivative of $\mathscr{V}^{[k_2]}$.

Also recall that the extended definition to an arbitrary $k \in \mathbb{Z}$ of the Pochhammer symbol $(s)_k$ initially defined by (5.1.12) for the case when $k \in \mathbb{N}_0$ is given by:

$$(s)_k := \frac{\Gamma(s+k)}{\Gamma(s)}. \tag{5.2.4}$$

Suppose now that $\varphi \in \mathscr{K}(0,\delta)$ is a test function. The decay conditions on φ imply that $t^s\varphi(t)$ is Lebesgue integrable on $(0,\delta)$ for every $s \in \mathbb{C}$ and that $\{\mathfrak{M}\varphi\}(s)$, the Mellin transform of φ, is an entire function of $s \in \mathbb{C}$, as can be directly verified by using Theorem 2.1.47 about the holomorphicity of an integral depending analytically on a parameter (see also [Tit2, Theorem 31]).

Furthermore, let $g(s)$ be a meromorphic function. Then, the residue $\operatorname{res}(g(s),\omega)$ vanishes unless ω is a pole of g. Moreover, for all $k \in \mathbb{Z}$, $N \in \mathbb{N}$ and by choosing a suitable contour Γ around ω, we have

$$\begin{aligned}\int_0^{+\infty}\varphi(t)\operatorname{res}(t^{N-s+k}g(s),\omega)\,\mathrm{d}t &= \int_0^{+\infty}\varphi(t)\frac{1}{2\pi\mathbb{i}}\oint_\Gamma t^{N-s+k}g(s)\,\mathrm{d}s\,\mathrm{d}t\\ &= \frac{1}{2\pi\mathbb{i}}\oint_\Gamma g(s)\int_0^{+\infty}t^{N-s+k}\varphi(t)\,\mathrm{d}t\,\mathrm{d}s\\ &= \operatorname{res}\big(\{\mathfrak{M}\varphi\}(N-s+k+1)\,g(s),\omega\big).\end{aligned}$$

The change of the order of integration is justified by the Fubini–Tonelli theorem since the last integral above is absolutely convergent. In short, for every $\varphi \in \mathscr{K}(0,\delta)$, we have

$$\left\langle \operatorname{res}(t^{N-s+k} g(s),\omega),\varphi\right\rangle = \operatorname{res}\left(\{\mathfrak{M}\varphi\}(N-s+1+k)\,g(s),\omega\right), \tag{5.2.5}$$

where $g(s)$ is a meromorphic function on a connected open neighborhood of $\omega \in \mathbb{C}$ and where $k \in \mathbb{Z}$ and $N \in \mathbb{N}$.

Note that for $k=-1$, the distribution $\mathscr{V}^{[-1]} = \mathscr{V}^{[-1]}_{A,\Omega}$ can be viewed as a positive measure on $(0,+\infty)$; indeed, it is the distributional derivative of the nondecreasing and locally integrable function $t \mapsto V(t) = V_{A,\Omega}(t)$ on $(0,+\infty)$. By analogy with the special case of fractal strings (discussed in [Lap-vFr3, Subsection 6.3.1]), as well as with the mathematical and theoretical physics literature on spectral theory, semiclassical approximation and quantum mechanics, we call it the *geometric density of (volume) states* of the RFD (A,Ω). (Compare with the relevant references in [Lap-vFr3], *ibid.*) From a fundamental point of view, this measure $\mathscr{V}^{[-1]} = \mathscr{V}^{[-1]}_{A,\Omega}$ is the most important 'distributional tube function' and the corresponding fractal tube formulas are the most useful distributional fractal tube formulas. (See also the comment concluding Subsection 5.2.2 on page 437 below.)

5.2.1 Distributional Tube Formula with Error Term

We are now able to state the distributional analog of Theorem 5.1.11; that is, the distributional tube formula with an error term.

Theorem 5.2.2 (Distributional fractal tube formula with error term, via $\tilde{\zeta}_{A,\Omega}$). *Let (A,Ω) be a relative fractal drum in $\mathbb{R}^N$ which is languid for some languidity exponent $\kappa \in \mathbb{R}$ and some $\delta > 0$. Then, for every $k \in \mathbb{Z}$, the distribution $\mathscr{V}^{[k]}_{A,\Omega}$ in $\mathscr{K}'(0,\delta)$ (and hence, also in $\mathscr{D}'(0,\delta)$) is given by the following distributional fractal tube formula, with error term and expressed in terms of the tube zeta function $\tilde{\zeta}_{A,\Omega} := \tilde{\zeta}_{A,\Omega}(\,\cdot\,;\delta)$:*

$$\mathscr{V}^{[k]}_{A,\Omega}(t) = \sum_{\omega \in \mathscr{P}(\tilde{\zeta}_{A,\Omega},\boldsymbol{W})} \operatorname{res}\left(\frac{t^{N-s+k}}{(N-s+1)_k}\tilde{\zeta}_{A,\Omega}(s),\omega\right) + \tilde{\mathscr{R}}^{[k]}_{A,\Omega}(t). \tag{5.2.6}$$

That is, the action of $\mathscr{V}^{[k]}_{A,\Omega}$ on an arbitrary test function $\varphi \in \mathscr{K}(0,\delta)$ is given by

$$\begin{aligned}\langle \mathscr{V}^{[k]}_{A,\Omega},\varphi\rangle = &\sum_{\omega \in \mathscr{P}(\tilde{\zeta}_{A,\Omega},\boldsymbol{W})} \operatorname{res}\left(\frac{\{\mathfrak{M}\varphi\}(N-s+1+k)\,\tilde{\zeta}_{A,\Omega}(s)}{(N-s+1)_k},\omega\right)\\ &+\langle \tilde{\mathscr{R}}^{[k]}_{A,\Omega},\varphi\rangle.\end{aligned} \tag{5.2.7}$$

Here, the (distributional) error term $\tilde{\mathscr{R}}^{[k]}_{A,\Omega}$ is given by the distribution in $\mathscr{K}'(0,\delta)$ defined for all test functions $\varphi \in \mathscr{K}(0,\delta)$ by

$$\langle \tilde{\mathscr{R}}^{[k]}_{A,\Omega}, \varphi \rangle = \frac{1}{2\pi\mathrm{i}} \int_{S} \frac{\{\mathfrak{M}\varphi\}(N-s+1+k)\, \tilde{\zeta}_{A,\Omega}(s)}{(N-s+1)_k}\, \mathrm{d}s. \tag{5.2.8}$$

(The corresponding distributional error estimate for $\tilde{\mathscr{R}}^{[k]}_{A,\Omega}$ will be given in Theorem 5.2.11 of Subsection 5.2.3 below.)

Proof. We begin the proof by fixing $k \in \mathbb{N}_0$ such that $k > \kappa + 1$ and a constant $c \in (\overline{\dim}_B(A,\Omega), N+1)$. (Note that by fixing $c \in (\overline{\dim}_B(A,\Omega), N+1)$, we have ensured that none of the poles of $(N-s+1)_k^{-1}$ are located in the window $\boldsymbol{W}$. Indeed, since

$$(N-s+1)_k^{-1} = \frac{\Gamma(N-s+1)}{\Gamma(N-s+1+k)}, \tag{5.2.9}$$

we can see that the set of its poles is a subset of $\{N+n : n \in \mathbb{N}\}$; see also the discussion provided in Remark 5.1.17 above.) Then, for every test function $\varphi \in \mathscr{K}(0,\delta)$, we have successively:

$$\begin{aligned} \langle V^{[k]}_{A,\Omega}, \varphi \rangle &= \int_0^{+\infty} V^{[k]}_{A,\Omega}(t)\varphi(t)\,\mathrm{d}t \\ &= \frac{1}{2\pi\mathrm{i}} \int_{c-\mathrm{i}\infty}^{c+\mathrm{i}\infty} \int_0^{+\infty} \varphi(t) \frac{t^{N-s+k}}{(N-s+1)_k} \tilde{\zeta}_{A,\Omega}(s)\,\mathrm{d}t\,\mathrm{d}s \\ &= \frac{1}{2\pi\mathrm{i}} \int_{c-\mathrm{i}\infty}^{c+\mathrm{i}\infty} \frac{\{\mathfrak{M}\varphi\}(N-s+1+k)\, \tilde{\zeta}_{A,\Omega}(s)}{(N-s+1)_k}\,\mathrm{d}s. \end{aligned} \tag{5.2.10}$$

Here, the interchange of the order of integration in the second equality is justified by the Fubini–Tonelli theorem since the first integral above is absolutely convergent. (It is easy to see that $|V^{[k]}(t)| \leq |A_t| t^k$, for all $t \in (0,+\infty)$ and $k \geq 0$.) One can now approximate the last integral above in the same way as in Lemma 5.1.10; that is, we approximate it by the following expression:

$$\begin{aligned} &\sum_{\omega \in \mathscr{P}(\tilde{\zeta}_{A,\Omega}, \boldsymbol{W}_{|n})} \operatorname{res}\left(\frac{\{\mathfrak{M}\varphi\}(N-s+1+k)\, \tilde{\zeta}_{A,\Omega}(s)}{(N-s+1)_k} \right) \\ &+ \frac{1}{2\pi\mathrm{i}} \int_{\boldsymbol{S}_{|n}} \frac{\{\mathfrak{M}\varphi\}(N-s+1+k)\, \tilde{\zeta}_{A,\Omega}(s)}{(N-s+1)_k}\,\mathrm{d}s \\ &+ \frac{1}{2\pi\mathrm{i}} \int_{\Gamma_L \cup \Gamma_U} \frac{\{\mathfrak{M}\varphi\}(N-s+1+k)\, \tilde{\zeta}_{A,\Omega}(s)}{(N-s+1)_k}\,\mathrm{d}s. \end{aligned} \tag{5.2.11}$$

Furthermore, in light of (5.2.5), the latter expression is equal to

$$\begin{aligned}&\sum_{\omega\in\mathscr{P}(\tilde{\zeta}_{A,\Omega},\boldsymbol{W}_{|n})}\left\langle \operatorname{res}\left(\frac{t^{N-s+k}}{(N-s+1)_k}\tilde{\zeta}_{A,\Omega}(s),\omega\right),\varphi\right\rangle\\&+\frac{1}{2\pi\mathbb{i}}\int_{\boldsymbol{S}_{|n}}\frac{\{\mathfrak{M}\varphi\}(N-s+1+k)\,\tilde{\zeta}_{A,\Omega}(s)}{(N-s+1)_k}\,\mathrm{d}s\\&+\int_0^{+\infty}E_n(t)\varphi(t)\,\mathrm{d}t,\end{aligned}\tag{5.2.12}$$

where the error term $E_n(t)$ is given as in Lemma 5.1.10 and its proof.

Next, by letting $n\to\infty$, we deduce by the same argument as in Theorem 5.1.11 that the integral of the error function $E_n(t)\varphi(t)$ tends to zero and, similarly, we show that the integral over the truncated screen $\boldsymbol{S}_{|n}$ converges absolutely. Thus, we deduce that

$$\begin{aligned}\langle V_{A,\Omega}^{[k]},\varphi\rangle=&\sum_{\omega\in\mathscr{P}(\tilde{\zeta}_{A,\Omega},\boldsymbol{W})}\operatorname{res}\left(\frac{\{\mathfrak{M}\varphi\}(N-s+1+k)\,\tilde{\zeta}_{A,\Omega}(s)}{(N-s+1)_k},\omega\right)\\&+\langle\tilde{\mathscr{R}}_{A,\Omega}^{[k]},\varphi\rangle,\end{aligned}\tag{5.2.13}$$

where $\tilde{\mathscr{R}}_{A,\Omega}^{[k]}$ is given by its action on test functions as shown in Equation (5.2.8).

Moreover, observe that the expression on the right-hand side of (5.2.13) defines a distribution in $\mathscr{K}'(0,\delta)$ (since $V_{A,\Omega}^{[k]}$ is locally integrable). This concludes the proof of the theorem in the case when $k>\max\{-1,\kappa+1\}$. (Note that it follows from this first part of the proof that (5.2.13) continues to hold if k is replaced by any $k'\in\mathbb{N}_0$ such that $k'\geq k$.)

In the case when $k\leq\kappa+1$ and $k\in\mathbb{Z}$, we choose an integer q such that $k+q>\max\{\kappa+1,-1\}$ and note that by the definition of the distributional derivative (or alternatively, in light of Equations (5.2.2) and (5.2.3) defining $\mathscr{V}_{A,\Omega}^{[k]}$), we have that

$$\langle\mathscr{V}_{A,\Omega}^{[k]},\varphi\rangle=(-1)^q\langle\mathscr{V}_{A,\Omega}^{[k+q]},\varphi^{(q)}\rangle;\tag{5.2.14}$$

see the comment following Definition 5.2.1 above. Finally, in order to complete the proof, we use identity (5.2.14) together with (5.2.13) applied at level $k+q$,[12] along with the following well-known (and easy to verify) fact about the Mellin transform (see Equation (5.1.16) defining $\{\mathfrak{M}f\}(s)$):

$$\{\mathfrak{M}\varphi\}(s)=\frac{(-1)^q}{(s)_q}\{\mathfrak{M}\varphi^{(q)}\}(s+q),\tag{5.2.15}$$

for all $s\in\mathbb{C}$ and $q\in\mathbb{Z}$. We therefore deduce that (5.2.7) holds, with $\langle\tilde{\mathscr{R}}_{A,\Omega}^{[k]},\varphi\rangle$ given by (5.2.8), as required. This concludes the proof of the theorem. □

[12] See the parenthetical comment appearing shortly after Equation (5.2.13).

Remark 5.2.3. Note that in the above proof of Theorem 5.2.2, we have established the fact that the sum over the (visible) complex dimensions appearing in (5.2.6) defines a distribution in $\mathscr{K}'(0,\delta)$ and hence, according to the inclusion (5.2.1), also in $\mathscr{D}'(0,\delta)$. In turn, this fact implies that both terms on the right-hand side of (5.2.6) are, on their own, distributions in $\mathscr{K}'(0,\delta)$. Namely, this is a consequence of the following well-known fact about the convergence of distributions, which itself follows from a suitable generalization of the Hahn–Banach theorem to locally convex topological spaces (see, for example, [Hö2, Theorem 2.1.8, p. 39]):

Let $(\mathscr{T}_n)_{n\geq 1}$ be a sequence of distributions in $\mathscr{D}'(0,\delta)$ such that

$$\langle \mathscr{T}, \varphi \rangle := \lim_{n\to\infty} \langle \mathscr{T}_n, \varphi \rangle$$

exists for every test function $\varphi \in \mathscr{D}(0,\delta)$. Then, $\mathscr{T}$ is a distribution in $\mathscr{D}'(0,\delta)$ and therefore, $\mathscr{T}_n \to \mathscr{T}$ in $\mathscr{D}'(0,\delta)$.

This result, applied to the appropriate sequence of partial sums over the set $\mathscr{P}(\tilde{\zeta}_{A,\Omega}, \boldsymbol{W})$, implies that the sum over the visible complex dimensions in (5.2.6) is indeed a distribution in $\mathscr{D}'(0,\delta)$ and hence, that each term taken separately on the right-hand side of (5.2.6) defines a distribution in $\mathscr{D}'(0,\delta)$.

An entirely analogous comment applies to Theorem 5.2.4 below, with the space of test functions coinciding with $\mathscr{D}(0,\delta_0)$ and thus with the associated space of distributions being equal to $\mathscr{D}'(0,\delta_0)$.

5.2.2 Exact Distributional Tube Formula

In the next theorem, we obtain a distributional analog of the pointwise tube formula without error term stated in Theorem 5.1.13 of Subsection 5.1.3. The resulting formula will be an asymptotic distributional formula; i.e., it will be valid for test functions in $\mathscr{K}(0,\delta)$ that are supported on the left of B^{-1}, where $B > 0$ is the constant appearing in hypothesis **L2'**.

Theorem 5.2.4 (Exact distributional tube formula via $\tilde{\zeta}_{A,\Omega}$). *Let (A,Ω) be a relative fractal drum in $\mathbb{R}^N$ which is strongly languid for some languidity exponent $\kappa \in \mathbb{R}$ and some $\delta > 0$. Furthermore, let $\delta_0 := \min\{1,\delta,B^{-1}\}$. Then, for every $k \in \mathbb{Z}$, the distribution $\mathscr{V}_{A,\Omega}^{[k]}$ in $\mathscr{D}'(0,\delta_0)$ is given by the following exact distributional tube formula in $\mathscr{D}'(0,\delta_0)$, expressed in terms of the tube zeta function $\tilde{\zeta}_{A,\Omega} := \tilde{\zeta}_{A,\Omega}(\,\cdot\,;\delta)$:*

$$\mathscr{V}_{A,\Omega}^{[k]}(t) = \sum_{\omega \in \mathscr{P}(\tilde{\zeta}_{A,\Omega},\mathbb{C})} \operatorname{res}\left(\frac{t^{N-s+k}}{(N-s+1)_k}\tilde{\zeta}_{A,\Omega}(s), \omega\right). \tag{5.2.16}$$

That is, the action of $\mathscr{V}_{A,\Omega}^{[k]}$ *on an arbitrary test function* $\varphi \in \mathscr{D}(0,\delta_0)$ *is given by*

$$\langle \mathscr{V}_{A,\Omega}^{[k]}, \varphi \rangle = \sum_{\omega \in \mathscr{P}(\tilde{\zeta}_{A,\Omega},\mathbb{C})} \operatorname{res}\left(\frac{\{\mathfrak{M}\varphi\}(N-s+1+k)\,\tilde{\zeta}_{A,\Omega}(s)}{(N-s+1)_k}, \omega \right). \tag{5.2.17}$$

Proof. We will prove the theorem by applying Theorem 5.2.2 to the sequence of screens $\boldsymbol{S}_m$ and showing that the corresponding error term tends to zero as $m \to \infty$. By choosing $q \in \mathbb{N}$ such that $k+q > \kappa+1$ and $m \in \mathbb{N}$ such that $\sup S_m < 0$, we deduce from (5.2.6) the following distributional identity, viewed as an equality in $\mathscr{D}'(0,\delta_0)$:

$$\mathscr{V}_{A,\Omega}^{[k+q]}(t) = \sum_{\omega \in \mathscr{P}(\tilde{\zeta}_{A,\Omega},\boldsymbol{W}_m)} \operatorname{res}\left(\frac{t^{N-s+k+q}}{(N-s+1)_{k+q}} \tilde{\zeta}_{A,\Omega}(s), \omega \right) + \tilde{\mathscr{R}}_m^{[k+q]}(t). \tag{5.2.18}$$

We now fix a test function $\varphi \in \mathscr{D}(0,\delta_0)$; since by definition, φ has compact support, there exists $\nu \in (0,1)$ such that the support of φ is contained in $(0, \nu B^{-1}]$. Using this fact, we can estimate the Mellin transform of φ in the following way, for all $s \in \mathbb{C}$ such that $\operatorname{Re} s < 0$:

$$\left|\{\mathfrak{M}\varphi\}(N-s+1+k+q)\right| \le (\nu B^{-1})^{-\operatorname{Re} s} \int_0^{+\infty} t^{N+k+q} |\varphi(t)|\, \mathrm{d}t. \tag{5.2.19}$$

Using this estimate, hypothesis **L2'**, along with the fact that

$$|(N - S_m(\tau) - \mathrm{i}\tau + 1)_{k+q}| \ge \left(\sqrt{1+\tau^2}\right)^{k+q},$$

we estimate the distributional error $\tilde{\mathscr{R}}_m^{[k+q]}$ as follows (we let $|\mathrm{d}s| := |s'(\tau)|\,\mathrm{d}\tau$):

$$\begin{aligned} |\langle \tilde{\mathscr{R}}_m^{[k+q]}, \varphi \rangle| &\le \int_{\boldsymbol{S}_m} |\{\mathfrak{M}\varphi\}(N-s+1+k+q)| \frac{|\tilde{\zeta}_A(s)|}{|(N-s+1)_{k+q}|} |\mathrm{d}s| \\ &\le \tilde{K}(1 + \|S_m\|_{\mathrm{Lip}}) \int_{-\infty}^{+\infty} (B\nu B^{-1})^{|S_m(\tau)|} \frac{(1+|\tau|)^\kappa}{\left(\sqrt{1+\tau^2}\right)^{k+q}}\, \mathrm{d}\tau \\ &\le K \nu^{|\sup S_m|} \int_{-\infty}^{+\infty} \frac{(1+|\tau|)^\kappa}{\left(\sqrt{1+\tau^2}\right)^{k+q}}\, \mathrm{d}\tau, \end{aligned} \tag{5.2.20}$$

with K being a suitable positive constant. The last inequality follows since, according to hypothesis **L2'**, the sequence of screens $(\boldsymbol{S}_m)_{m \ge 1}$ has a uniform Lipschitz bound; see the definition of strong languidity given in Definition 5.1.4. Furthermore, the last integral in the above calculation is convergent since $k+q > \kappa+1$.

Next, by letting $m \to \infty$, we deduce that $\langle \tilde{\mathscr{R}}_m^{[k+q]}, \varphi \rangle \to 0$ since $|\sup S_m| \to \infty$, and thus we conclude that $\tilde{\mathscr{R}}_m^{[k+q]} \to 0$ as $m \to \infty$, in $\mathscr{D}'(0,\delta_0)$. Finally, in light of (5.2.18), we obtain the statement of the theorem for the distribution $\mathscr{V}_{A,\Omega}^{[k+q]}$ in $\mathscr{D}'(0,\delta_0)$; in order to obtain the statement for $\mathscr{V}_{A,\Omega}^{[k]}$ itself, we use the exact same argument as in the proof of Theorem 5.2.2. In particular, in order to obtain (5.2.16) for the distribution $\mathscr{V}_{A,\Omega}^{[k]}$ in $\mathscr{D}'(0,\delta_0)$, we simply differentiate q times the resulting identity for $\mathscr{V}_{A,\Omega}^{[k+q]}$, in the distributional sense. □

Remark 5.2.5. One can see from the proof of Theorem 5.2.4 that the distributional formula (5.2.17) is actually valid for a larger class of test functions. More specifically, it is valid for all $\varphi \in \mathscr{K}(0,\delta_0)$ which have support in $(0,\nu B^{-1}]$, for some $\nu \in (0,1)$, possibly depending on φ.

We next state as a separate theorem the most interesting special case (beside the case when $k=-1$, see the comments just before Subsection 5.2.1 above and at the end of the present subsection below) of the distributional fractal tube formula (with and without an error term); namely, the case when $k=0$ and hence, $\mathscr{V}_{A,\Omega}^{[0]}(t) = |A_t \cap \Omega|$ for all $t>0$ (and as a regular distribution in $\mathscr{D}'(0,\delta_0)$).

Theorem 5.2.6 (Distributional fractal tube formula via $\tilde{\zeta}_{A,\Omega}$; level $k=0$). *Let (A,Ω) be a relative fractal drum in $\mathbb{R}^N$. Under the same hypotheses as in Theorem 5.2.2 (that is, if the relative fractal drum (A,Ω) is languid for some languidity exponent $\kappa \in \mathbb{R}$ and some $\delta > 0$), and with $k := 0$, we have the following distributional equality for the tube function $t \mapsto |A_t \cap \Omega|$ of (A,Ω), expressed in terms of $\tilde{\zeta}_{A,\Omega} := \tilde{\zeta}_{A,\Omega}(\,\cdot\,;\delta)$ and valid in $\mathscr{K}'(0,\delta)$ (and hence, also in $\mathscr{D}'(0,\delta)$):*

$$|A_t \cap \Omega| = \sum_{\omega \in \mathscr{P}(\tilde{\zeta}_{A,\Omega},\boldsymbol{W})} \operatorname{res}\left(t^{N-s}\tilde{\zeta}_{A,\Omega}(s),\omega\right) + \tilde{\mathscr{R}}_{A,\Omega}^{[0]}(t), \tag{5.2.21}$$

where $\tilde{\mathscr{R}}_{A,\Omega}^{[0]}$ is the distribution in $\mathscr{K}'(0,\delta)$ given for all $\varphi \in \mathscr{K}(0,\delta)$ (and, in particular, for all $\varphi \in \mathscr{D}(0,\delta)$) by formula (5.2.8) with $k=0$. (See Remark 5.2.8 and Corollary 5.2.12 below.)

Moreover, under the same hypotheses as in Theorem 5.2.4 (that is, if (A,Ω) is strongly languid for some $\kappa \in \mathbb{R}$ and some $\delta > 0$), then (5.2.21) holds as a distributional equality in $\mathscr{D}'(0,\delta_0)$, where $\delta_0 := \min\{1,\delta,B^{-1}\}$ and with $\tilde{\mathscr{R}}_{A,\Omega}^{[0]} \equiv 0$ and $\boldsymbol{W} := \mathbb{C}$; so that (5.2.21) is an exact *fractal tube formula in this case.*

Remark 5.2.7. Note that when the expression on the left-hand side of the distributional equality (5.2.21) (namely, $t \mapsto |A_t \cap \Omega|$) defines a locally integrable function of the variable t (which may not be the case since this expression is a distribution acting on test functions $\varphi \in \mathscr{K}(0,\delta)$ defined by (5.2.2), or equivalently, by (5.2.7), with $k := 0$), we also have an equality, valid pointwise almost everywhere, between the tube function $t \mapsto |A_t \cap \Omega|$ and the expression on the right-hand side of (5.2.21).[13] (It is easy to check that the tube function $t \mapsto |A_t \cap \Omega|$ is locally integrable, and hence defines a regular distribution, if, for example, either A or Ω is bounded or, more generally, has finite volume in $\mathbb{R}^N$.)

[13] Here, we use the well-known fact according to which a locally integrable function uniquely defines a (regular) distribution, see, e.g., [Schw, JohLap, Hö2, Bre].

Remark 5.2.8. A suitable distributional error estimate for $\tilde{\mathscr{R}}^{[0]}_{A,\Omega}$ will be provided in Corollary 5.2.12 below.

Let $\eta := \mathscr{V}^{[-1]} = \frac{d\mathscr{V}^{[0]}}{dt}$ (in the distributional sense) be the positive measure which we called the *geometric density of (volume) states* of the RFD (A,Ω) on page 431, just before the beginning of Subsection 5.2.1. Then, by applying Theorem 5.2.2 (from Subsection 5.2.1) or Theorem 5.2.4 (from Subsection 5.2.2) at level $k = -1$ (rather than at level $k = 0$), we would obtain the most important form of the distributional fractal tube formula with or withouth error term, respectively. (An entirely ananlogous comment could be made about all of the distributional fractal tube formulas obtained in this chapter.) We leave it as a simple exercise for the reader to write down explicitly the corresponding fractal distributional tube formulas and to express them in terms of the Mellin transform of the test functions φ to which they can be applied. (Compare with the corresponding results of [Lap-vFr3, Subsection 6.3.1] obtained for the geometric and spectral densities of states of fractal strings.)

5.2.3 Estimate for the Distributional Error Term

We would now like to give a distributional estimate for the error term appearing in Theorem 5.2.2, interpreted in the same sense as was done in [Lap-vFr3, Subsection 5.2.4], in the case of the distributional explicit formula for generalized fractal strings. Therefore, let us next introduce the notion of a distributional order of growth (see [EstKa, JaffMey, PiStVi] and also, independently, [Lap-vFr1–2] and [Lap-vFr3, Definition 5.29]).

For a test function $\varphi \in \mathscr{D}(0,+\infty)$ and $a > 0$, we let

$$\varphi_a(t) := \frac{1}{a}\varphi\left(\frac{t}{a}\right). \tag{5.2.22}$$

Note that $\int_0^{+\infty} \varphi_a(t)\,dt = \int_0^{+\infty} \varphi(t)\,dt$, for every $a > 0$. Furthermore, when $a \to 0^+$, the support of φ_a becomes 'narrower' and gets 'closer' to zero, while the amplitude of φ_a tends to infinity in absolute value. On the other hand, when $a \to +\infty$, then the support of φ_a becomes 'wider' and 'escapes' to infinity, while the amplitude of φ_a tends to zero in absolute value.

Definition 5.2.9. Let $\mathscr{R}$ be a distribution in $\mathscr{D}'(0,\delta)$ and let $\alpha \in \mathbb{R}$. We say that $\mathscr{R}$ is of *asymptotic order* at most t^α (resp., less than t^α) as $t \to 0^+$ if applied to an arbitrary test function φ_a in $\mathscr{D}(0,\delta)$, we have that[14]

$$\langle \mathscr{R}, \varphi_a \rangle = O(a^\alpha) \quad (\text{resp., } \langle \mathscr{R}, \varphi_a \rangle = o(a^\alpha)), \quad \text{as} \quad a \to 0^+. \tag{5.2.23}$$

We then write that $\mathscr{R}(t) = O(t^\alpha)$ (resp., $\mathscr{R}(t) = o(t^\alpha)$), as $a \to 0^+$.

---

[14] In formula (5.2.23), the implicit constant may depend on the test function φ.

Remark 5.2.10. We point out that if f is a continuous function such that pointwise, $f(t) = O(t^\alpha)$ or $f(t) = o(t^\alpha)$ as $t \to 0^+$, for some $\alpha \in \mathbb{R}$, then f also satisfies the same asymptotic estimate, in the distributional sense of Definition 5.2.9. Namely, by taking $\varphi \in \mathscr{D}(0,\delta)$, we have

$$\langle f, \varphi_a \rangle = \int_0^{+\infty} f(t)\varphi_a(t)\,\mathrm{d}t = \int_0^{+\infty} f(a\tau)\varphi(\tau)\,\mathrm{d}\tau. \tag{5.2.24}$$

If $f = O(t^\alpha)$, then, since φ has compact support, we can take a sufficiently small so that in the above integral, we have $|f(a\tau)| \le Ca^\alpha\tau^\alpha$, for some positive constant C. In other words, for all positive a sufficiently small, we have

$$|\langle f, \varphi_a \rangle| \le Ca^\alpha \int_0^{+\infty} \tau^\alpha \varphi(\tau)\,\mathrm{d}\tau = K_\varphi a^\alpha, \tag{5.2.25}$$

where the constant K_φ depends only on the test function φ. One reasons analogously in the case when $f(t) = o(t^\alpha)$ as $t \to 0^+$. The same comment can also be made about the asymptotics as $t \to +\infty$. On the other hand, we note that clearly, a distributional asymptotic estimate in the case of regular distributions, does not in general imply the usual pointwise one; see, for example, [PiStVi].

Finally, also observe that for a test function $\varphi \in \mathscr{D}(0,\delta)$ and $a > 0$, the Mellin transform of φ_a satisfies the following scaling identity (see Equation (5.1.16) defining $\{\mathfrak{M}f\}(s)$):

$$\{\mathfrak{M}\varphi_a\}(s) = a^{s-1}\{\mathfrak{M}\varphi\}(s), \tag{5.2.26}$$

for all $s \in \mathbb{C}$.

We can now state the following result about the order of growth of the distributional error term appearing in Theorem 5.2.2. It is the analog in our present context of [Lap-vFr3, Theorem 5.30].

Theorem 5.2.11 (Estimate for the distributional error term). *Assume that the hypotheses of Theorem 5.2.2 are satisfied, for a fixed $k \in \mathbb{Z}$. Then, the distribution $\tilde{\mathscr{R}}^{[k]}_{A,\Omega}(t)$ given by* (5.2.8) *is of asymptotic order at most $t^{N-\sup S+k}$ as $t \to 0^+$; i.e,*

$$\tilde{\mathscr{R}}^{[k]}_{A,\Omega}(t) = O(t^{N-\sup S+k}) \quad \text{as} \quad t \to 0^+, \tag{5.2.27}$$

in the sense of Definition 5.2.9.

Moreover, if $S(\tau) < \sup S$ for all $\tau \in \mathbb{R}$ (that is, if the screen $\boldsymbol{S}$ lies strictly to the left of the vertical line $\{\mathrm{Re}\, s = \sup S\}$), then $\tilde{\mathscr{R}}^{[k]}_{A,\Omega}(t)$ is of asymptotic order less than $t^{N-\sup S+k}$; i.e.,

$$\tilde{\mathscr{R}}^{[k]}_{A,\Omega}(t) = o(t^{N-\sup S+k}) \quad \text{as} \quad t \to 0^+, \tag{5.2.28}$$

also in the sense of Definition 5.2.9.

Proof. Let φ be a test function. Then, the integral defining $\langle \tilde{\mathscr{R}}^{[k]}_{A,\Omega}, \varphi \rangle$ in Equation (5.2.8) converges absolutely. Furthermore, for any $a \in (0,1)$, and by using (5.2.26), we obtain the following estimate:

$$\begin{aligned} |\langle \tilde{\mathscr{R}}^{[k]}_{A,\Omega}, \varphi_a \rangle| &\leq \frac{1}{2\pi} \int_S \frac{|\{\mathfrak{M}\varphi_a\}(N-s+1+k)|}{|(N-s+1)_k|} |\tilde{\zeta}_{A,\Omega}(s)| \, |ds| \\ &= \frac{1}{2\pi} \int_S a^{N-\mathrm{Re}\, s+k} \frac{|\{\mathfrak{M}\varphi\}(N-s+1+k)|}{|(N-s+1)_k|} |\tilde{\zeta}_{A,\Omega}(s)| \, |ds| \\ &\leq \mathrm{const} \cdot a^{N-\sup S+k}; \end{aligned}$$

this proves the first part of the theorem.

In order to establish the second part of the theorem, we use an argument similar to the one used in the proof of estimate (5.1.34) of Theorem 5.1.11. □

In the next corollary of either Theorem 5.2.2 or of Theorem 5.2.6, we consider the important situation when $k=0$, and as a special case, when all of the (visible) complex dimensions are simple (in the spirit of Remark 5.1.15 and Theorem 5.1.16). (We leave it as an exercise for the interested reader to state the corresponding corollary in the key case when $k=-1$.)

Corollary 5.2.12 (Estimate for the distributional error term; level $k=0$). *Under the hypotheses of Theorem 5.2.2, with $k=0$ (or equivalently, under the hypotheses of Theorem 5.2.6 in the languid case), the distributional error term in* (5.2.6) *(or equivalently, in* (5.2.21)*) given by* (5.2.8) *with $k=0$ satisfies the following error estimate:*

$$\tilde{\mathscr{R}}^{[0]}_{A,\Omega}(t) = O(t^{N-\sup S}) \quad \text{as} \quad t \to 0^+ \tag{5.2.29}$$

(resp., $\tilde{\mathscr{R}}^{[0]}_{A,\Omega}(t) = o(t^{N-\sup S})$ as $t \to 0^+$, in the special case when the screen $\boldsymbol{S}$ lies strictly to the left of the line $\{\mathrm{Re}\, s = \sup S\}$).

Consequently, we have that

$$|A_t \cap \Omega| = \sum_{\omega \in \mathscr{P}(\tilde{\zeta}_{A,\Omega}, \boldsymbol{W})} \mathrm{res}\left(t^{N-s} \tilde{\zeta}_{A,\Omega}(s), \omega\right) + O(t^{N-\sup S}) \tag{5.2.30}$$

(resp., Equation (5.2.30) *holds with $o(t^{N-\sup S})$ instead of $O(t^{N-\sup S})$) as $t \to 0^+$, in the distributional sense. Moreover, on the right-hand side of* (5.2.30)*, the sum over the visible complex dimensions of (A,Ω) becomes*

$$\sum_{\omega \in \mathscr{P}(\tilde{\zeta}_{A,\Omega}, \boldsymbol{W})} t^{N-\omega} \mathrm{res}\left(\tilde{\zeta}_{A,\Omega}, \omega\right) \tag{5.2.31}$$

in the important special case when all the visible complex dimensions (i.e., all of the poles of $\tilde{\zeta}_{A,\Omega} = \tilde{\zeta}_{A,\Omega}(\,\cdot\,;\delta)$ lying in $\boldsymbol{W}$) are simple.

5.3 Tube Formulas in Terms of the Relative Distance Zeta Function

In this section, we translate the results from the previous sections in terms of the relative distance zeta functions $\zeta_{A,\Omega} := \zeta_{A,\Omega}(\,\cdot\,;\delta)$ (instead of the tube zeta function $\tilde{\zeta}_{A,\Omega}$, as in Sections 5.1 and 5.2). This is extremely useful in the applications since the relative distance zeta function $\zeta_{A,\Omega}$ of an RFD (A,Ω), can be calculated without knowing its relative tube function $t \mapsto |A_t \cap \Omega|$. Of course, the results will follow, in particular, from the functional equation (4.5.2) which connects these two fractal zeta functions, $\zeta_{A,\Omega}$ and $\tilde{\zeta}_{A,\Omega}$. More precisely, in order to derive the analogous results in terms of the distance zeta function, we will introduce a new fractal zeta function, called the *relative shell zeta function*, which satisfies a more direct functional equation, compared to (4.5.2). For $A \subseteq \mathbb{R}^N$ and $t,\delta > 0$ with $t \leq \delta$, let

$$A_{t,\delta} := A_\delta \setminus \overline{A_t}. \tag{5.3.1}$$

The subset $A_{t,\delta}$ so defined can be thought of as the (t,δ)-*shell* of A.

Stachó has proved in [Sta] that for any bounded set $A \subset \mathbb{R}^N$ and every $t > 0$, we have that $|\partial A_t| = 0$, where ∂A_t denotes the boundary of A_t in $\mathbb{R}^N$ and (as usual) $|\partial A_t|$ denotes its N-dimensional volume. Since any unbounded set in $\mathbb{R}^N$ may be partitioned into a countable union of bounded subsets, the exact same statement is also true for any unbounded subset of $\mathbb{R}^N$. Consequently, for any relative fractal drum (A,Ω) in $\mathbb{R}^N$, we have (for $0 < t \leq \delta$)

$$|A_{t,\delta} \cap \Omega| = |A_\delta \cap \Omega| - |\overline{A_t} \cap \Omega| = |A_\delta \cap \Omega| - |A_t \cap \Omega|. \tag{5.3.2}$$

5.3.1 The Relative Shell Zeta Function

Let $\tilde{\zeta}_{A,\Omega}(\,\cdot\,;\delta)$ be the tube zeta function of the relative fractal drum (A,Ω) in $\mathbb{R}^N$ and assume that $\operatorname{Re} s > N$, then we have

$$\begin{aligned}\tilde{\zeta}_{A,\Omega}(s;\delta) &= \int_0^\delta t^{s-N-1}|A_t \cap \Omega|\,dt\\ &= \int_0^\delta t^{s-N-1}(|A_\delta \cap \Omega| - |A_{t,\delta} \cap \Omega|)\,dt\\ &= \frac{\delta^{s-N}|A_\delta \cap \Omega|}{s-N} - \int_0^\delta t^{s-N-1}|A_{t,\delta} \cap \Omega|\,dt.\end{aligned} \tag{5.3.3}$$

Definition 5.3.1. Let (A,Ω) be an RFD in $\mathbb{R}^N$ and fix $\delta > 0$. We define the *shell zeta function* $\breve{\zeta}_{A,\Omega} := \breve{\zeta}_{A,\Omega}(\,\cdot\,;\delta)$ *of* A *relative to* Ω (or the *relative shell zeta function*) by

$$\breve{\zeta}_{A,\Omega}(s;\delta) := -\int_0^\delta t^{s-N-1}|A_{t,\delta} \cap \Omega|\,dt, \tag{5.3.4}$$

for all $s \in \mathbb{C}$ with $\operatorname{Re} s$ sufficiently large. Here, the integral is taken in the Lebesgue sense.

In light of (5.3.3), we can now easily obtain the following theorem.

Theorem 5.3.2. *Let (A,Ω) be an RFD in $\mathbb{R}^N$ and fix $\delta > 0$. Then, the shell zeta function $\breve{\zeta}_{A,\Omega}(\,\cdot\,;\delta)$ of (A,Ω) is holomorphic on the open right half-plane $\{\operatorname{Re} s > N\}$ and*

$$\frac{\mathrm{d}}{\mathrm{d}s}\breve{\zeta}_{A,\Omega}(s;\delta) = -\int_0^\delta t^{s-N-1}|A_{t,\delta}\cap\Omega|\log t\,\mathrm{d}t, \tag{5.3.5}$$

for all $s \in \mathbb{C}$ such that $\operatorname{Re} s > N$.

Furthermore, for all $s \in \mathbb{C}$ such that $\operatorname{Re} s > N$, $\breve{\zeta}_{A,\Omega}(\,\cdot\,;\delta)$ satisfies the following functional equations, connecting it to the tube and distance zeta functions of (A,Ω), respectively:

$$\breve{\zeta}_{A,\Omega}(s;\delta) = \frac{\delta^{s-N}|A_\delta\cap\Omega|}{s-N} + \tilde{\zeta}_{A,\Omega}(s;\delta) \tag{5.3.6}$$

and

$$\zeta_{A,\Omega}(s;\delta) = (N-s)\breve{\zeta}_{A,\Omega}(s;\delta). \tag{5.3.7}$$

Proof. To prove the holomorphicity of $\breve{\zeta}_{A,\Omega}(\,\cdot\,;\delta)$, one observes that for every real number $\sigma > N$, we have

$$|\breve{\zeta}_{A,\Omega}(\sigma;\delta)| \le |A_\delta\cap\Omega|\int_0^\delta t^{\sigma-N-1}\,\mathrm{d}t < \infty,$$

and one then uses Theorem 2.1.45 which also yields the formula (5.3.5) for the derivative. Formula (5.3.6) is a rewriting of (5.3.3) and by combining it with the functional equation (2.2.23), which connects the relative distance and tube zeta functions, we obtain (5.3.7). □

In light of Theorem 4.1.7, the principle of analytic continuation combined with Equation (5.3.6) (or (5.3.7)) now immediately yields the following properties of the relative shell zeta function.

Theorem 5.3.3. *Let (A,Ω) be a relative fractal drum in $\mathbb{R}^N$ such that $\overline{\dim}_B(A,\Omega) < N$ and fix $\delta > 0$. Then the following properties hold:*

(*a*) *The relative shell zeta function $\breve{\zeta}_{A,\Omega}(s;\delta)$ is meromorphic in the half-plane $\{\operatorname{Re} s > \overline{\dim}_B(A,\Omega)\}$, with a single simple pole at $s = N$. Furthermore,*

$$\operatorname{res}(\breve{\zeta}_{A,\Omega}(\,\cdot\,;\delta),N) = -|A_\delta\cap\Omega|. \tag{5.3.8}$$

(*b*) *If the relative box (or Minkowski) dimension $D := \dim_B(A,\Omega)$ exists, $D < N$, and $\mathscr{M}_*^D(A,\Omega) > 0$, then $\breve{\zeta}_{A,\Omega}(s) \to +\infty$ as $s \in \mathbb{R}$ converges to D from the right.*

Proof. We conclude from the principle of analytic continuation that the functional equalities (5.3.6) and (5.3.7) continue to hold on any open connected set $U \supseteq \{\mathrm{Re}\, s > N\}$ to which any of the three relative zeta functions, $\zeta_{A,\Omega}$, $\tilde{\zeta}_{A,\Omega}$ or $\zeta_{A,\Omega}$, has a holomorphic continuation. In light of this, part (a) follows from the counterpart of Theorem 4.1.7 for the relative tube zeta function and from (5.3.6), while part (b) follows from Theorem 4.1.7 and (5.3.7). □

The following corollary is an immediate consequence of the above theorem, or more precisely, of the functional equation (5.3.6) and the fact that for a given RFD (A,Ω) in $\mathbb{R}^N$ and any fixed $\delta_1, \delta_2 > 0$, the difference $\tilde{\zeta}_{A,\Omega}(s;\delta_1) - \tilde{\zeta}_{A,\Omega}(s;\delta_2)$ is an entire function. (See Proposition 2.2.13, which has an obvious counterpart of RFDs.)

Corollary 5.3.4. *Let (A,Ω) be an RFD in $\mathbb{R}^N$ such that $\overline{\dim}_B(A,\Omega) < N$ and fix $\delta_1, \delta_2 > 0$ such that $\delta_1 < \delta_2$. Then, the difference $\breve{\zeta}_{A,\Omega}(s;\delta_1) - \breve{\zeta}_{A,\Omega}(s;\delta_2)$ is meromorphic on all of $\mathbb{C}$, with a single simple pole at $s = N$ of residue $|A_{\delta_1,\delta_2} \cap \Omega|$.*

The next corollary follows at once from the first part of the proof of Theorem 5.3.3. (Similarly as before, in the sequel, for simplicity, we will often use the short-hand notation $\breve{\zeta}_{A,\Omega}$, $\tilde{\zeta}_{A,\Omega}$ and $\zeta_{A,\Omega}$, respectively, for the shell, tube and distance zeta function of the RFD (A,Ω).)

Corollary 5.3.5. *Let (A,Ω) be an RFD in $\mathbb{R}^N$. Then, the functional equations* (5.3.6) *and* (5.3.7) *continue to hold on any connected open neighborhood $U \subseteq \mathbb{C}$ of the critical line $\{\mathrm{Re}\, s = \overline{\dim}_B(A,\Omega)\}$ to which any of the three relative zeta functions $\breve{\zeta}_{A,\Omega}$, $\tilde{\zeta}_{A,\Omega}$ or $\zeta_{A,\Omega}$ can be meromorphically continued. More specifically, if either $\breve{\zeta}_{A,\Omega}$, $\tilde{\zeta}_{A,\Omega}$ or $\zeta_{A,\Omega}$ has a (necessarily unique) meromorphic continuation on the domain $U \subseteq \mathbb{C}$, then so do the other two fractal zeta functions and the functional equations* (5.3.6) *and* (5.3.7) *continue to hold for all $s \in U$ between the resulting meromorphic extensions of $\breve{\zeta}_{A,\Omega}$, $\tilde{\zeta}_{A,\Omega}$ and $\zeta_{A,\Omega}$.*

Moreover, in light of the obvious counterpart for RFDs of Theorem 2.2.14 and the functional equation (5.3.6), we have the following result.

Theorem 5.3.6. *Assume that (A,Ω) is a Minkowski nondegenerate RFD in $\mathbb{R}^N$, that is, $0 < \mathscr{M}_*^D(A,\Omega) \le \mathscr{M}^{*D}(A,\Omega) < \infty$ (in particular, $\dim_B(A,\Omega) = D$), and $D < N$. If $\tilde{\zeta}_{A,\Omega}(s)$ can be extended meromorphically to a connected neighborhood of $s = D$, then D is necessarily a simple pole of $\breve{\zeta}_{A,\Omega}(s)$, the residue $\mathrm{res}(\breve{\zeta}_{A,\Omega}, D)$ is independent of δ and*

$$\mathscr{M}_*^D(A,\Omega) \le \mathrm{res}(\breve{\zeta}_{A,\Omega}, D) \le \mathscr{M}^{*D}(A,\Omega). \tag{5.3.9}$$

Furthermore, if (A,Ω) is Minkowski measurable, then

$$\mathrm{res}(\breve{\zeta}_{A,\Omega}, D) = \mathscr{M}^D(A,\Omega). \tag{5.3.10}$$

The most useful fact about the relative shell zeta function is that the residues of its meromorphic extension at any of its (simple) poles belonging to the open left half-plane $\{\mathrm{Re}\, s < N\}$ have a simple connection to the residues of the relative tube or distance zeta functions. (See also Corollary 5.3.5 just above.)

Lemma 5.3.7. *Assume that (A,Ω) is an RFD in $\mathbb{R}^N$ such that its tube or distance or shell zeta function is meromorphic on some connected open neighborhood $U \subseteq \mathbb{C}$ of the critical line $\{\mathrm{Re}\, s = \overline{\dim}_B(A,\Omega)\}$. Then, the multisets of poles located in $U\setminus\{N\}$ of each of the three zeta functions, $\breve{\zeta}_{A,\Omega}$, $\tilde{\zeta}_{A,\Omega}$ and $\zeta_{A,\Omega}$, coincide:*

$$\mathscr{P}(\breve{\zeta}_{A,\Omega}, U\setminus\{N\}) = \mathscr{P}(\tilde{\zeta}_{A,\Omega}, U\setminus\{N\}) = \mathscr{P}(\zeta_{A,\Omega}, U\setminus\{N\}). \tag{5.3.11}$$

Moreover, if $\omega \in U\setminus\{N\}$ is a simple pole of one of the three fractal zeta functions $\breve{\zeta}_{A,\Omega}$, $\tilde{\zeta}_{A,\Omega}$ or $\zeta_{A,\Omega}$, then it is also a simple pole of the other two fractal zeta functions and we have

$$\mathrm{res}(\breve{\zeta}_{A,\Omega},\omega) = \mathrm{res}(\tilde{\zeta}_{A,\Omega},\omega) = \frac{\mathrm{res}(\zeta_{A,\Omega},\omega)}{N-\omega}. \tag{5.3.12}$$

Although the shell zeta function, $\breve{\zeta}_{A,\Omega}$, may seem rather artificial in the present context of relative fractal drums, it will prove to be quite useful as a "translation tool" for deriving the tube formulas (originally obtained via the tube zeta function, $\tilde{\zeta}_{A,\Omega}$, in Sections 5.1 and 5.2) in terms of the much more practical geometric distance zeta function, $\zeta_{A,\Omega}$. We note that the shell zeta function originally arose naturally in [Ra1–2], where it was used, in particular, to generalize the present theory of complex dimensions developed in this book and in [LapRaŽu1–8] to the special case of *unbounded sets at infinity* having infinite Lebesgue measure.

5.3.2 *Pointwise Tube Formulas in Terms of the Distance Zeta Function*

Similarly as in the case of the relative tube zeta function of (A,Ω), we observe that $\breve{\zeta}_{A,\Omega}(s) = \{\mathfrak{M}f\}(s)$, where $f(s) := -t^{-N}\chi_{(0,\delta)}(t)|A_{t,\delta}\cap\Omega|$. Note that f is continuous and of bounded variation on $(0,+\infty)$; so that we can apply the Mellin inversion theorem (Theorem 5.1.6), much as in the proof of Theorem 5.1.7, and conclude that

$$|A_{t,\delta}\cap\Omega| = -\frac{1}{2\pi \mathrm{i}}\int_{c-\mathrm{i}\infty}^{c+\mathrm{i}\infty} t^{N-s}\breve{\zeta}_{A,\Omega}(s;\delta)\,\mathrm{d}s, \tag{5.3.13}$$

where $c > N$ is arbitrary and $t \in (0,\delta)$. In light of (5.3.2), the following theorem is an immediate consequence of the identity (5.3.13).

Theorem 5.3.8. *Let (A,Ω) be a relative fractal drum in $\mathbb{R}^N$ and fix $\delta > 0$. Then, for every $t \in (0,\delta)$ and any real number $c > N$, we have*

$$|A_t\cap\Omega| = |A_\delta\cap\Omega| + \frac{1}{2\pi \mathrm{i}}\int_{c-\mathrm{i}\infty}^{c+\mathrm{i}\infty} t^{N-s}\breve{\zeta}_{A,\Omega}(s;\delta)\,\mathrm{d}s. \tag{5.3.14}$$

It is now clear that if the shell zeta function of (A,Ω) satisfies the languidity conditions of Definition 5.1.3, with the constant $c>N$ in the condition **L1**, or the strong languidity conditions of Definition 5.1.4, we can rewrite the results of Sections 5.1 and 5.2 verbatim in terms of the shell zeta function. Note that for this to work, it was crucial that in the truncated pointwise formula of Lemma 5.1.10, we had the freedom to choose any $c\in(\overline{\dim}_B(A,\Omega),N+1)$. Furthermore, observe that the additional pole of the shell zeta function at $s=N$ will cancel out the term $|A_\delta\cap\Omega|$ in (5.3.14) above. More specifically, in the analog of the pointwise formula stated in Theorem 5.1.11 for the relative shell zeta function, we obtain the following pointwise fractal tube formula with error term, expressed in terms of the shell zeta function $\breve{\zeta}_{A,\Omega}:=\breve{\zeta}_{A,\Omega}(\cdot\,;\delta)$:

$$\begin{aligned}V^{[k]}_{A,\Omega}(t)=&\sum_{\omega\in\mathscr{P}(\breve{\zeta}_{A,\Omega},\boldsymbol{W})}\operatorname{res}\left(\frac{t^{N-s+k}}{(N-s+1)_k}\breve{\zeta}_{A,\Omega}(s;\delta),\omega\right)\\&+|A_\delta\cap\Omega|\frac{t^k}{(1)_k}+\breve{R}^{[k]}_{A,\Omega}(t),\end{aligned}\tag{5.3.15}$$

valid pointwise for all $t\in(0,\delta)$. Here, just as in the statement of Theorem 5.1.11, the shell zeta function $\breve{\zeta}_{A,\Omega)}$ of the RFD (A,Ω) is assumed to be languid for some fixed $\delta>0$ and some fixed constant $\kappa\in\mathbb{R}$, as well as with the constant c satisfying $c>N$. Furthermore, the nonnegative integer k is assumed to be such that $k>\kappa+1$ and for every $t\in(0,\delta)$, the error term $\breve{R}^{[k]}_{A,\Omega}$ is given (much as in (5.1.31)) by the absolutely convergent (and hence, convergent) integral

$$\breve{R}^{[k]}_{A,\Omega}(t)=\frac{1}{2\pi\mathrm{i}}\int_S\frac{t^{N-s+k}}{(N-s+1)_k}\breve{\zeta}_{A,\Omega}(s;\delta)\,\mathrm{d}s.\tag{5.3.16}$$

Moreover, it satisfies the exact analog of the pointwise error estimate (5.1.32), valid pointwise for all $t\in(0,\delta)$. Hence, it satisfies (for $\breve{\zeta}_{A,\Omega}$ instead of for $\tilde{\zeta}_{A,\Omega}$) the error estimate (5.1.33) and, in the special case when the screen $\boldsymbol{S}$ lies strictly to the left of the vertical line $\{\operatorname{Re}s=\sup S\}$, it satisfies the exact analog (for $\breve{\zeta}_{A,\Omega}$) of the stronger error estimate (5.1.34).

In addition, by singling out the residue at $s=N$ from the above sum and using Lemma 5.3.7 and Theorem 5.3.3(a), along with the functional equation (5.3.7), we can rewrite the above equation in (5.3.15) as

$$V^{[k]}_{A,\Omega}(t)=\sum_{\omega\in\mathscr{P}(\zeta_{A,\Omega},\boldsymbol{W})}\operatorname{res}\left(\frac{t^{N-s+k}}{(N-s)_{k+1}}\zeta_{A,\Omega}(s;\delta),\omega\right)+R^{[k]}_{A,\Omega}(t),\tag{5.3.17}$$

where the pointwise error term $R^{[k]}_{A,\Omega}$ is now given by the absolutely convergent (and hence, convergent) integral

$$R^{[k]}_{A,\Omega}(t)=\frac{1}{2\pi\mathrm{i}}\int_S\frac{t^{N-s+k}}{(N-s)_{k+1}}\zeta_A(s,\Omega;\delta)\,\mathrm{d}s.\tag{5.3.18}$$

Let us now define the analogs of the languidity conditions of a relative fractal drum in terms of its relative distance zeta function.

Definition 5.3.9. (*d-languidity and strong d-languidity*). We say that a relative fractal drum (A,Ω) in $\mathbb{R}^N$ is *d-languid* (resp., *strongly d-languid*) if it is languid in the sense of Definition 5.1.3 (resp., Definition 5.1.4), but with the relative tube zeta function $\tilde{\zeta}_{A,\Omega} = \tilde{\zeta}_{A,\Omega}(\,\cdot\,;\delta)$ replaced by the relative distance zeta function $\zeta_{A,\Omega} = \zeta_{A,\Omega}(\,\cdot\,;\delta)$ and with the constant c appearing in **L1** satisfying $c > N$.

The following lemma is an immediate consequence of the functional equation (5.3.7).

Lemma 5.3.10. *Let (A,Ω) be a relative fractal drum in $\mathbb{R}^N$ such that $\overline{\dim}_B(A,\Omega) < N$ and which is d-languid for some value $\delta > 0$ and with some d-languidity exponent $\kappa_d \in \mathbb{R}$. Then the shell zeta function $\breve{\zeta}_{A,\Omega}$ of (A,Ω) satisfies the languidity conditions of Definition 5.1.3 for the same value of δ and with the languidity exponent $\kappa := \kappa_d - 1$.*

Furthermore, if (A,Ω) is strongly d-languid with the corresponding constant $B > 0$ and for some d-languidity exponent $\kappa_d \in \mathbb{R}$ and some $\delta > 0$, then the shell zeta function $\breve{\zeta}_{A,\Omega}$ of (A,Ω) satisfies the strong languidity conditions of Definition 5.1.4 with the languidity exponent $\kappa := \kappa_d - 1$ and with the same constant B as well as the same value of δ.

We are now able to state and prove the main theorem of this section, which is the analog for $\zeta_{A,\Omega}$ of Theorem 5.1.11 stated in Section 5.1 in terms of $\tilde{\zeta}_{A,\Omega}$.

Theorem 5.3.11 (Pointwise fractal tube formula with error term, via $\zeta_{A,\Omega}$). *Let (A,Ω) be a relative fractal drum in $\mathbb{R}^N$ which is d-languid for some $\delta > 0$ and with d-languidity exponent $\kappa_d \in \mathbb{R}$. Furthermore, assume that $\overline{\dim}_B(A,\Omega) < N$ and let $k > \kappa_d$ be a nonnegative integer. Then, the following pointwise fractal tube formula, expressed in terms of the distance zeta function $\zeta_{A,\Omega} := \zeta_{A,\Omega}(\,\cdot\,;\delta)$, is valid for every $t \in (0,\delta)$:*

$$V_{A,\Omega}^{[k]}(t) = \sum_{\omega \in \mathscr{P}(\zeta_{A,\Omega}),\boldsymbol{W})} \operatorname{res}\left(\frac{t^{N-s+k}}{(N-s)_{k+1}}\zeta_{A,\Omega}(s),\omega\right) + R_{A,\Omega}^{[k]}(t). \tag{5.3.19}$$

Here, for every $t \in (0,\delta)$, the error term $R_{A,\Omega}^{[k]}$ is given by the absolutely convergent (and hence, convergent) integral

$$R_{A,\Omega}^{[k]}(t) = \frac{1}{2\pi\mathrm{i}}\int_S \frac{t^{N-s+k}}{(N-s)_{k+1}}\zeta_{A,\Omega}(s)\,\mathrm{d}s. \tag{5.3.20}$$

Furthermore, for every $t \in (0,\delta)$, we have

$$|R_{A,\Omega}^{[k]}(t)| \le t^{N+k}\max\{t^{-\sup S}, t^{-\inf S}\}\left(\frac{C(1+\|S\|_{\mathrm{Lip}})}{2\pi(k-\kappa_d)} + C'\right), \tag{5.3.21}$$

where C is the constant appearing in **L1** *and* **L2** *and C' is some suitable positive constant. These constants depend only on the relative fractal drum (A,Ω) and the screen, but not on k.*

In particular, we have the following pointwise error estimate:

$$R^{[k]}_{A,\Omega}(t) = O(t^{N-\sup S+k}) \quad \text{as} \quad t \to 0^+. \tag{5.3.22}$$

Moreover, if $S(\tau) < \sup S$ (i.e., if the screen $\boldsymbol{S}$ lies strictly left of the vertical line $\{\mathrm{Re}\, s = \sup S\}$), then we have the following stronger pointwise error estimate:

$$R^{[k]}_{A,\Omega}(t) = o(t^{N-\sup S+k}) \quad \text{as} \quad t \to 0^+. \tag{5.3.23}$$

Proof. In light of Lemma 5.3.10, we have that $\zeta_{A,\Omega}$, the shell zeta function of (A,Ω), also satisfies the appropriate languidity conditions with languidity exponent $\kappa := \kappa_d - 1$ and for the same value of δ. The theorem now follows much as in the case of the relative tube zeta function $\tilde{\zeta}_{A,\Omega}$; see the proof of Theorem 5.1.11 and the discussion following Theorem 5.3.8. □

Remark 5.3.12. In Theorem 5.3.11, the additional assumption according to which $\overline{\dim}_B(A,\Omega) < N$ is made in order to avoid the situation where $s = N$ is a pole of $\tilde{\zeta}_{A,\Omega}$. We will also assume that this additional hypothesis is satisfied in the statements of all the other theorems involving the relative distance zeta function in the present section.

Theorem 5.3.13 (Exact pointwise fractal tube formula via $\zeta_{A,\Omega}$). *Let (A,Ω) be a relative fractal drum in $\mathbb{R}^N$ which is strongly d-languid for some $\delta > 0$ and with d-languidity exponent $\kappa_d \in \mathbb{R}$. Furthermore, let $k > \kappa_d - 1$ be a nonnegative integer and assume that $\overline{\dim}_B(A,\Omega) < N$. Then, the following exact pointwise fractal tube formula, expressed in terms of the distance zeta function $\zeta_{A,\Omega} := \zeta_{A,\Omega}(\,\cdot\,;\delta)$, holds for every $t \in (0, \min\{1,\delta,B^{-1}\})$:*

$$V^{[k]}_{A,\Omega}(t) = \sum_{\omega \in \mathscr{P}(\zeta_{A,\Omega},\mathbb{C})} \operatorname{res}\left(\frac{t^{N-s+k}}{(N-s)_{k+1}}\zeta_{A,\Omega}(s), \omega\right). \tag{5.3.24}$$

Here, B is the constant appearing in **L2'** *and κ_d is the exponent occurring in the statement of hypotheses* **L1** *and* **L2'**.

Proof. In light of Lemma 5.3.10 and the functional equation (5.3.7), the proof of the theorem parallels that of Theorem 5.3.11, except for the tube zeta function $\tilde{\zeta}_{A,\Omega}(\,\cdot\,;\delta)$ now replaced by the shell zeta function $\zeta_{A,\Omega}(\,\cdot\,;\delta)$. □

In some cases, we will have a relative fractal drum (A,Ω) that is 'almost' strongly d-languid, but not exactly. More precisely, (A,Ω) will satisfy all of the conditions of strong d-languidity except for the condition that **L1** is satisfied for *all* $\sigma < c$. For example, let A be the middle-third Cantor set constructed in $[0,1]$ and let $\Omega = (0,1)$. Then, the relative distance zeta function $\zeta_{A,\Omega}$ is meromorphic on all of $\mathbb{C}$ and given for all $s \in \mathbb{C}$ by (see (2.1.113) from Example 2.1.82):

$$\zeta_{A,\Omega}(s) = \frac{2^{1-s}}{s(3^s-2)}. \tag{5.3.25}$$

As one can easily check, it almost satisfies the strong languidity conditions with $\kappa_d := -1$, where the sequence of screens $\boldsymbol{S}_m$ can be taken as the sequence of vertical lines $\{\mathrm{Re}\, s = -m\}$ for $m \in \mathbb{N}$. The problem here is due to the factor 2^{-s} which tends exponentially fast to $+\infty$ as $\mathrm{Re}\, s \to -\infty$, so that condition **L1** cannot be fulfilled for all $\sigma < c$. In order to obtain a pointwise formula in this and similar cases, we can multiply $\zeta_{A,\Omega}(s)$ by 2^s and then, the resulting function will be strongly d-languid. On the other hand, by the scaling property of the relative distance zeta function (see Theorem 4.1.40), we have that $2^s\zeta_{A,\Omega}(s) = \zeta_{2A,2\Omega}(s)$. Hence, we can state the following corollary dealing with this situation and which will be used repeatedly (most often implicitly) in Sections 5.4 and 5.5.

Corollary 5.3.14 (Exact pointwise fractal tube formula via $\zeta_{A,\Omega}$; scaled version). *Let (A,Ω) be a relative fractal drum in $\mathbb{R}^N$ such that $\overline{\dim}_B(A,\Omega) < N$. Furthermore, assume that there exists a scaling factor $\lambda > 0$ such that $(\lambda A, \lambda\Omega)$ is a strongly d-languid RFD in $\mathbb{R}^N$, for some $\delta > 0$ and with d-languidity exponent $\kappa_d \in \mathbb{R}$. Moreover, let $k > \kappa_d - 1$ be a nonnegative integer. Then, the following exact pointwise fractal tube formula, expressed in terms of the distance zeta function $\zeta_{A,\Omega}$, holds for every $t \in (0, \lambda^{-1}\min\{1,\delta,B^{-1}\})$:*

$$V_{A,\Omega}^{[k]}(t) = \sum_{\omega\in\mathscr{P}(\zeta_{A,\Omega},\mathbb{C})} \operatorname{res}\left(\frac{t^{N-s+k}}{(N-s)_{k+1}}\zeta_{A,\Omega}(s),\omega\right). \tag{5.3.26}$$

Here, B is the constant appearing in **L2'** *(for the function $s \mapsto \zeta_{\lambda A,\lambda\Omega}(s;\delta) = \lambda^s\zeta_{A,\Omega}(s;\delta\lambda^{-1})$) and κ_d is the exponent occurring in the statement of hypotheses* **L1** *and* **L2'**.

Proof. Let us denote by $V_\lambda^{[k]}(\tau)$ the k-th primitive of the function

$$\tau \mapsto |(\lambda A)_\tau \cap \lambda\Omega|.$$

Since we know from Lemma 4.6.10 that $V_\lambda^{[0]}(\tau) = \lambda^N V^{[0]}(\tau/\lambda)$, we deduce that

$$V_\lambda^{[1]}(\tau) = \int_0^\tau V_\lambda^{[0]}(t)\,\mathrm{d}t = \lambda^N\int_0^\tau V_{A,\Omega}^{[0]}(t/\lambda)\,\mathrm{d}t = \lambda^{N+1}\int_0^{\tau/\lambda} V_{A,\Omega}^{[0]}(\xi)\,\mathrm{d}\xi, \tag{5.3.27}$$

or, in other words, $V_\lambda^{[1]}(\tau) = \lambda^{N+1}V_{A,\Omega}^{[1]}(\tau/\lambda)$ and hence, by induction,

$$V_\lambda^{[k]}(\tau) = \lambda^{N+k}V_{A,\Omega}^{[k]}(\tau/\lambda), \tag{5.3.28}$$

for all nonnegative integers k. We now apply Theorem 5.3.13 to the relative fractal drum $(\lambda A, \lambda\Omega)$ and obtain the following fractal tube formula, valid pointwise for all $\tau \in (0, \min\{1,\delta,B^{-1}\})$:

$$V_\lambda^{[k]}(\tau) = \sum_{\omega \in \mathscr{P}(\zeta_{\lambda A,\lambda\Omega},\mathbb{C})} \operatorname{res}\left(\frac{\tau^{N-s+k}}{(N-s)_{k+1}}\zeta_{\lambda A,\lambda\Omega}(s;\delta),\omega\right). \tag{5.3.29}$$

Now combining (5.3.28) with (5.3.29) and the scaling property of the relative distance zeta function (Theorem 4.1.40), we deduce that

$$\lambda^{N+k}V_{A,\Omega}^{[k]}(\tau/\lambda) = \sum_{\omega \in \mathscr{P}(\zeta_{A,\Omega},\mathbb{C})} \operatorname{res}\left(\frac{\tau^{N-s+k}\lambda^s}{(N-s)_{k+1}}\zeta_{A,\Omega}(s;\delta\lambda^{-1}),\omega\right). \tag{5.3.30}$$

Finally, we complete the proof of the corollary by multiplying the above identity by λ^{-N-k} and introducing a new variable $t := \tau/\lambda$. □

Remark 5.3.15. We point out that an analogous corollary can be stated in terms of the relative tube zeta function and the exact pointwise tube formula of Theorem 5.1.13.

The most interesting situation is, of course, the case when we can apply Theorems 5.3.11 and 5.3.13 at the level $k=0$. We now state the corresponding corollaries of these two theorems as a separate (and single) theorem.

Theorem 5.3.16 (Pointwise fractal tube formula via $\zeta_{A,\Omega}$; level $k=0$).

(*i*) *Under the same hypotheses as in Theorem 5.3.11, with $k := 0$, and using the same notation as in that theorem, with $\kappa_d < 0$, the following pointwise fractal tube formula with error term, expressed in terms of the distance zeta function $\zeta_{A,\Omega} := \zeta_{A,\Omega}(\cdot;\delta)$, holds for all $t \in (0,\delta)$:*

$$|A_t \cap \Omega| = \sum_{\omega \in \mathscr{P}(\zeta_{A,\Omega},\boldsymbol{W})} \operatorname{res}\left(\frac{t^{N-s}}{N-s}\zeta_{A,\Omega}(s),\omega\right) + R_{A,\Omega}^{[0]}(t), \tag{5.3.31}$$

where $R_{A,\Omega}^{[0]}(t)$ is the error term given by formula (5.3.20) *with $k := 0$. Furthermore, we have the following pointwise error estimate:*

$$R_{A,\Omega}^{[0]}(t) = O(t^{N-\sup S}) \quad \text{as} \quad t \to 0^+. \tag{5.3.32}$$

Moreover, if $S(\tau) < \sup S$ for every $\tau \in \mathbb{R}$ (i.e., if the screen $\boldsymbol{S}$ lies strictly to the left of the vertical line $\{\operatorname{Re} s = \sup S\}$), we then have the following stronger pointwise error estimate:

$$R_{A,\Omega}^{[0]}(t) = o(t^{N-\sup S}) \quad \text{as} \quad t \to 0^+. \tag{5.3.33}$$

(*ii*) *Finally, under the same hypotheses as in Theorem 5.3.13 or Corollary 5.3.14, with $k := 0$ and $\kappa_d < 1$, and if, in addition $(\lambda A, \lambda\Omega)$ is strongly d-languid for some $\lambda > 0$, then the fractal tube formula* (5.3.31) *holds pointwise for all $t \in (0, \lambda^{-1}\min\{1,\delta,B^{-1}\})$, with $R_{A,\Omega}^{[0]}(t) \equiv 0$ and $\boldsymbol{W} := \mathbb{C}$; so that* (5.3.31) *becomes an exact fractal tube formula in this case.*

The exact analog of Remark 5.1.15, Theorem 5.1.16 and Remark 5.1.17 holds in the present situation, except for the relative tube zeta function $\tilde{\zeta}_{A,\Omega}$ replaced by the relative distance zeta function $\zeta_{A,\Omega}$ of the RFD (A,Ω). In order to avoid too many repetitions, we only state the counterpart of Theorem 5.1.16 in the present context. It is, of course, the corollary of Theorem 5.3.16 corresponding to the level $k=0$.

Theorem 5.3.17 (Pointwise fractal tube formula via $\zeta_{A,\Omega}$; level $k=0$ and the case of simple poles). *Assume that the hypotheses of Theorem 5.3.16 hold. Suppose, in addition, that all of the visible complex dimensions of the RFD (A,Ω) are simple (i.e., all of the poles of $\zeta_{A,\Omega}$ or, equivalently, since $\overline{D} := \overline{\dim}_B(A,\Omega) < N$ here, of $\tilde{\zeta}_{A,\Omega}$, belonging to the window $\boldsymbol{W}$ are simple). Then, the pointwise fractal tube formula* (5.3.31)*, expressed in terms of $\zeta_{A,\Omega}$, takes the following simpler form, valid for all $t \in (0,\delta)$:*

$$|A_t \cap \Omega| = \sum_{\omega \in \mathscr{P}(\zeta_{A,\Omega},\boldsymbol{W})} \frac{t^{N-\omega}}{N-\omega} \operatorname{res}\left(\zeta_{A,\Omega}(s),\omega\right) + R^{[0]}_{A,\Omega}(t), \tag{5.3.34}$$

where the (pointwise) error term $R^{[0]}_{A,\Omega}$ is the same as in Theorem 5.3.11 at level $k=0$ and hence, satisfies the same (pointwise) error estimates [(5.3.32) *or* (5.3.33), *depending on the hypotheses*] *as in Theorem 5.3.16. In particular, in the strongly languid case (i.e., if $(\lambda A,\lambda\Omega)$ is strongly languid for some $\lambda > 0$), we have $R^{[0]}_{A,\Omega} \equiv 0$ and $\boldsymbol{W} := \mathbb{C}$, so that* (5.3.34) *then becomes an* exact *pointwise fractal tube formula, valid for all $t \in (0, \lambda^{-1}\min\{1,\delta,B^{-1}\})$.*

Remark 5.3.18. Note that the hypothesis according to which $\overline{D} := \overline{\dim}_B(A,\Omega) < N$ (see Theorems 5.3.11 and 5.3.13, Corollary 5.3.14, along with Theorems 5.3.16 and 5.3.17) imply that $s=N$ is not a pole of $\zeta_{A,\Omega}$, a fact which is explicitly needed in deducing (5.3.34) from (5.3.31). Indeed, since (by part (a) of Theorem 4.1.7) $\zeta_{A,\Omega}$ is holomorphic in the open right half-plane $\{\operatorname{Re} s > \overline{D}\}$, we then have the following inclusions:

$$\mathscr{P}(\zeta_{A,\Omega},\boldsymbol{W}) \subseteq \{\operatorname{Re} s \le \overline{D}\} \subseteq \{\operatorname{Re} s \le N\}. \tag{5.3.35}$$

5.3.3 Distributional Tube Formulas in Terms of the Distance Zeta Function

Let us now state the distributional analogs of the above results in terms of the relative distance zeta function. The proofs are completely analogous to the ones from Section 5.2 for the case of the relative tube zeta function. Again, we use the relative shell zeta function and the same scaling technique as in the proof of Corollary 5.3.14 (and Theorem 5.3.13) above to obtain the desired results under the hypotheses of d-languidity (Theorem 5.3.19) or of strong d-languidity (Theorem 5.3.20), respectively.

Theorem 5.3.19 (Distributional fractal tube formula with error term, via $\zeta_{A,\Omega}$). *Let (A,Ω) be a d-languid relative fractal drum in $\mathbb{R}^N$ for some $\delta>0$ and d-languidity exponent $\kappa_d\in\mathbb{R}$. Furthermore, assume that $\overline{\dim}_B(A,\Omega)<N$. Then, for every $k\in\mathbb{Z}$, the distribution $\mathscr{V}^{[k]}_{A,\Omega}$ in $\mathscr{K}'(0,\delta)$ (and hence, also in $\mathscr{D}'(0,\delta)$) is given by the following distributional fractal tube formula, with error term and expressed in terms of the distance zeta function $\zeta_{A,\Omega}:=\zeta_{A,\Omega}(\,\cdot\,;\delta)$:*

$$\mathscr{V}^{[k]}_{A,\Omega}(t)=\sum_{\omega\in\mathscr{P}(\zeta_{A,\Omega},\boldsymbol{W})}\operatorname{res}\left(\frac{t^{N-s+k}}{(N-s)_{k+1}}\zeta_{A,\Omega}(s),\omega\right)+\mathscr{R}^{[k]}_{A,\Omega}(t). \tag{5.3.36}$$

That is, the action of $\mathscr{V}^{[k]}_{A,\Omega}(t)$ on an arbitrary test function $\varphi\in\mathscr{K}(0,\delta)$ is given by

$$\begin{aligned}\langle\mathscr{V}^{[k]}_{A,\Omega},\varphi\rangle=&\sum_{\omega\in\mathscr{P}(\zeta_{A,\Omega},\boldsymbol{W})}\operatorname{res}\left(\frac{\{\mathfrak{M}\varphi\}(N-s+1+k)\,\zeta_{A,\Omega}(s)}{(N-s)_{k+1}},\omega\right)\\&+\langle\mathscr{R}^{[k]}_{A,\Omega},\varphi\rangle.\end{aligned} \tag{5.3.37}$$

Here, the distribution $\mathscr{R}^{[k]}_{A,\Omega}$ in $\mathscr{K}'(0,\delta)$ is the distributional error term given for all $\varphi\in\mathscr{K}(0,\delta)$ by

$$\langle\mathscr{R}^{[k]}_{A,\Omega},\varphi\rangle=\frac{1}{2\pi\mathrm{i}}\int_{\boldsymbol{S}}\frac{\{\mathfrak{M}\varphi\}(N-s+1+k)\,\zeta_{A,\Omega}(s)}{(N-s)_{k+1}}\,\mathrm{d}s. \tag{5.3.38}$$

Furthermore, the distribution $\mathscr{R}^{[k]}_{A,\Omega}(t)$ is of asymptotic order at most $t^{N-\sup S+k}$ as $t\to0^+$; i.e,

$$\mathscr{R}^{[k]}_{A,\Omega}(t)=O(t^{N-\sup S+k})\quad\text{as}\quad t\to0^+, \tag{5.3.39}$$

in the sense of Definition 5.2.9.

If, in addition, $S(\tau)<\sup S$ for all $\tau\in\mathbb{R}$ (that is, if the screen $\boldsymbol{S}$ lies strictly to the left of the vertical line $\{\operatorname{Re}s=\sup S\}$), then $\mathscr{R}^{[k]}_{A,\Omega}(t)$ is of asymptotic order less than $t^{N-\sup S+k}$; i.e., still in the sense of Definition 5.2.9, we have that

$$\mathscr{R}^{[k]}_{A,\Omega}(t)=o(t^{N-\sup S+k})\quad\text{as}\quad t\to0^+. \tag{5.3.40}$$

In the case of a (possibly scaled) strongly d-languid relative fractal drum, as before, we obtain a distributional formula without an error term, as stated in the next theorem.

Theorem 5.3.20 (Exact distributional fractal tube formula via $\zeta_{A,\Omega}$). *Let (A,Ω) be a relative fractal drum in $\mathbb{R}^N$ and assume also that $\overline{\dim}_B(A,\Omega)<N$. Furthermore, assume that there exists $\lambda>0$ such that $(\lambda A,\lambda\Omega)$ is strongly d-languid for some $\delta>0$, $\kappa_d\in\mathbb{R}$, and let $\delta_0:=\lambda^{-1}\min\{1,\delta,B^{-1}\}$.[15] Then, for every $k\in\mathbb{Z}$, the distribution $\mathscr{V}^{[k]}_{A,\Omega}$ in $\mathscr{D}'(0,\delta_0)$ is given in terms of $\zeta_{A,\Omega}:=\zeta_{A,\Omega}(\,\cdot\,;\delta)$ by*

[15] Here, B is the constant appearing in condition **L2'** for the function $\zeta_{\lambda A,\lambda\Omega}(s;\delta)$.

$$\mathscr{V}_{A,\Omega}^{[k]}(t) = \sum_{\omega \in \mathscr{P}(\zeta_{A,\Omega},\mathbb{C})} \operatorname{res}\left(\frac{t^{N-s+k}}{(N-s)_{k+1}}\zeta_{A,\Omega}(s), \omega\right). \tag{5.3.41}$$

That is, the action of $\mathscr{V}_{A,\Omega}^{[k]}$ *on an arbitrary test function* $\varphi \in \mathscr{D}(0,\delta_0)$ *is given by*

$$\langle \mathscr{V}_{A,\Omega}^{[k]}(t), \varphi \rangle = \sum_{\omega \in \mathscr{P}(\zeta_{A,\Omega},\mathbb{C})} \operatorname{res}\left(\frac{\{\mathfrak{M}\varphi\}(N-s+1+k)\,\zeta_{A,\Omega}(s)}{(N-s)_{k+1}}, \omega\right). \tag{5.3.42}$$

We conclude this section by stating as a separate (and single) theorem the most interesting special case of Theorems 5.3.19 and 5.3.20, when $k=0$.

Theorem 5.3.21 (Distributional fractal tube formula via $\zeta_{A,\Omega}$; level $k=0$). *Under the same hypotheses as in Theorem 5.3.19, with* $k := 0$, *we have the following distributional equality in* $\mathscr{K}'(0,\delta)$ *for the relative tube function* $t \mapsto |A_t \cap \Omega|$ *of the relative fractal drum* (A,Ω) *in* $\mathbb{R}^N$:

$$|A_t \cap \Omega| = \sum_{\omega \in \mathscr{P}(\zeta_{A,\Omega},\boldsymbol{W})} \operatorname{res}\left(\frac{t^{N-s}}{N-s}\zeta_{A,\Omega}(s), \omega\right) + \mathscr{R}_{A,\Omega}^{[0]}(t), \tag{5.3.43}$$

where $\mathscr{R}_{A,\Omega}^{[0]}(t)$ *is given by* (5.3.38) *for* $k=0$ *and* $\mathscr{R}_{A,\Omega}^{[0]}(t) = O(t^{N-\sup S})$ *as* $t \to 0^+$ *or, if* $S(\tau) < \sup S$ *for all* $\tau \in \mathbb{R}$, *then* $\mathscr{R}_{A,\Omega}^{[0]}(t) = o(t^{N-\sup S})$ *as* $t \to 0^+$.

Moreover, under the same hypotheses as in Theorem 5.3.20, with $k := 0$, *and if* $(\lambda A, \lambda\Omega)$ *is strongly d-languid for some* $\lambda > 0$, *then the analog of* (5.3.43) *holds in* $\mathscr{D}'(0,\delta_0)$, *where* $\delta_0 := \lambda^{-1}\min\{1,\delta,B^{-1}\}$ *and with* $\mathscr{R}_{A,\Omega}^{[0]}(t) \equiv 0$ *and* $\boldsymbol{W} := \mathbb{C}$; *so that we obtain an* exact *distributional fractal tube formula in this case.*

Finally, if, in addition, each visible complex dimension of (A,Ω) *is simple (i.e., if each pole of* $\zeta_{A,\Omega}$ *or, equivalently, of* $\tilde{\zeta}_{A,\Omega}$, *located in* $\boldsymbol{W}$ *is simple), then the sum over the complex dimensions in* (5.3.43) (*or in its analog with* $\boldsymbol{W} := \mathbb{C}$, *for the exact tube formula*) *becomes*

$$\sum_{\omega \in \mathscr{P}(\zeta_{A,\Omega},\boldsymbol{W})} \frac{t^{N-\omega}}{N-\omega} \operatorname{res}\left(\zeta_{A,\Omega}(s), \omega\right). \tag{5.3.44}$$

5.4 A Criterion for Minkowski Measurability

In this section, we obtain, in particular, a necessary and sufficient condition for the Minkowski measurability of a large class of RFDs (A,Ω) in $\mathbb{R}^N$, expressed in terms of the principal poles of their fractal zeta functions. More specifically, *under suitable hypotheses, an* RFD *with Minkowski dimension D is Minkowski measurable if and only if its only complex dimension with real part D is equal to D itself, and D is simple.* (See Theorems 5.4.20 and 5.4.25, along with Remark 5.4.21.) We also obtain a sufficient condition (with weaker hypotheses imposed on the RFD in comparison to the Minkowski measurability criterion of Theorem 5.4.20) for the Minkowski measurability of a relative fractal drum; see Theorem 5.4.2. Furthermore, we establish an upper bound for the upper Minkowski content of an RFD in terms of the residue at $s = \overline{D}$ of its fractal zeta function, where $\overline{D}$ denotes the upper Minkowski dimension of the RFD; see Theorem 5.4.4. Naturally, all of theses results apply, in particular, to bounded subsets A of $\mathbb{R}^N$, with $N \geq 1$ arbitrary, by simply considering the associated RFD (A, A_δ), for any $\delta > 0$ or, even more conveniently, the RFD (A, Ω), where Ω is any open neighborhood of the set A.

5.4.1 A Sufficient Condition for Minkowski Measurability

In this subsection, we show that a sufficient condition for the Minkowski measurability of a relative fractal drum (A, Ω) can be given in terms of its relative tube (or distance) zeta function. This will be a consequence of a well-known Tauberian theorem due to Wiener and Pitt (see [PitWie]) and which generalizes the famous Ikehara Tauberian theorem. The proof of the Wiener–Pitt Tauberian theorem can also be found in [Kor, Chapter III, Lemma 9.1 and Proposition 4.3] or in [Pit, Section 6.1] and in [Dia], where a different proof using a technique due to Bochner is given. We next state this theorem, for the sake of completeness. (Clearly, in either the statement of Theorem 5.4.1 or in the proofs of Theorems 5.4.2 and 5.4.4, the symbol h does not refer to a gauge function, unlike in Subsection 5.4.4, for example.)

Theorem 5.4.1 (The Wiener–Pitt Tauberian theorem, cited from [Kor]**).** *Let $\sigma\colon \mathbb{R} \to \mathbb{R}$ be such that $\sigma(t)$ vanishes for all $t < 0$, is nonnegative for all $t \geq 0$, and such that its Laplace transform*

$$F(s) := \{\mathfrak{L}\sigma\}(s) := \int_0^{+\infty} \mathrm{e}^{-st}\sigma(t)\,\mathrm{d}t \tag{5.4.1}$$

exists for all $s \in \mathbb{C}$ such that $\operatorname{Re} s > 0$. Furthermore, suppose that for some constants $A > 0$ and $\lambda > 0$, the function

$$H(s) := F(s) - \frac{A}{s}, \quad s := x + \mathbb{i}y \quad (x > 0,\, y \in \mathbb{R}), \tag{5.4.2}$$

converges in $L^1(-\lambda, \lambda)$ to a boundary function $H(\mathbb{i}y)$ as $x \to 0^+$. Then, for every fixed real number $h \geq 2\pi/\lambda$, we have that

$$\sigma_h(u) := \frac{1}{h}\int_u^{u+h} \sigma(t)\,\mathrm{d}t \leq CA + o(1) \quad \text{as} \quad u \to +\infty, \tag{5.4.3}$$

for some positive constant $C < 3$.

Moreover, if the above constant λ can be taken to be arbitrarily large, then for every fixed $h > 0$,

$$\sigma_h(u) \to A \quad \text{as} \quad u \to +\infty. \tag{5.4.4}$$

Let us now state the aforementioned result and then prove it by using the above Tauberian theorem.

Theorem 5.4.2 (Sufficient condition for Minkowski measurability). *Let (A,Ω) be a relative fractal drum in $\mathbb{R}^N$ and let $\overline{D} := \overline{\dim}_B(A,\Omega)$. Furthermore, suppose that the relative tube zeta function $\tilde{\zeta}_{A,\Omega}$ of (A,Ω) can be meromorphically extended to a connected open neighborhood $U \subseteq \mathbb{C}$ of the critical line $\{\mathrm{Re}\, s = \overline{D}\}$, with a single pole $\overline{D}$, which is assumed to be simple. Then $D := \dim_B(A,\Omega)$ exists, $D = \overline{D}$ and (A,Ω) is Minkowski measurable with Minkowski content given by*

$$\mathscr{M}^D(A,\Omega) = \operatorname{res}(\tilde{\zeta}_{A,\Omega}, D). \tag{5.4.5}$$

Moreover, if we assume, in addition, that $\overline{D} < N$, then the theorem is also valid if we replace the relative tube zeta function $\tilde{\zeta}_{A,\Omega}$ by the relative distance zeta function $\zeta_{A,\Omega}$ of (A,Ω), and in that case, we have

$$\mathscr{M}^D(A,\Omega) = \frac{\operatorname{res}(\zeta_{A,\Omega}, D)}{N-D}. \tag{5.4.6}$$

Proof. Without loss of generality, for the tube zeta function $\tilde{\zeta}_{A,\Omega}(\,\cdot\,;\delta)$ we may choose $\delta = 1$ and change the variable of integration by letting $u := 1/t$:

$$\begin{aligned}\tilde{\zeta}_{A,\Omega}(s+\overline{D}) &= \int_0^1 t^{s+\overline{D}-1-N}|A_t \cap \Omega|\,\mathrm{d}t \\ &= \int_1^{+\infty} u^{-s-1-\overline{D}+N}|A_{1/u} \cap \Omega|\,\mathrm{d}u \\ &= \int_0^{+\infty} \mathrm{e}^{-sv}\mathrm{e}^{v(N-\overline{D})}|A_{\mathrm{e}^{-v}} \cap \Omega|\,\mathrm{d}v = \{\mathfrak{L}\sigma\}(s),\end{aligned} \tag{5.4.7}$$

where we have made another change of variable in the second to last equality, namely, $v := \log u$, and we have let $\sigma(v) := \mathrm{e}^{v(N-\overline{D})}|A_{\mathrm{e}^{-v}} \cap \Omega|$. Clearly, the definition of the relative tube zeta function of (A,Ω) implies that the residue of $\tilde{\zeta}_{A,\Omega}(s)$ at $s = \overline{D}$ must be real and positive. (Note that, a priori, it should be nonnegative, but since by hypothesis, $\overline{D}$ is a pole of the meromorphic continuation of $\tilde{\zeta}_{A,\Omega}$ to U, the residue at $\overline{D}$ must be different from zero.) Furthermore, since $s = \overline{D}$ is the only pole of $\tilde{\zeta}_{A,\Omega}$ in U, we conclude that

$$H(s) := \tilde{\zeta}_{A,\Omega}(s+\overline{D}) - \frac{\operatorname{res}(\tilde{\zeta}_{A,\Omega}, \overline{D})}{s} \tag{5.4.8}$$

is holomorphic in the connected open neighborhood $U_{\overline{D}} := \{s \in \mathbb{C} : s + \overline{D} \in U\}$ of the vertical line $\{\mathrm{Re}\, s = 0\}$. In other words, we can apply Theorem 5.4.1 (for arbitrarily large $\lambda > 0$, in the notation of that theorem) and conclude that

$$\sigma_h(u) = \frac{1}{h}\int_u^{u+h} \sigma(v)\,\mathrm{d}v \to \operatorname{res}(\tilde{\zeta}_{A,\Omega}, \overline{D}) \quad \text{as} \quad u \to +\infty, \tag{5.4.9}$$

for every $h > 0$.[16] In particular, since $v \mapsto |A_{\mathrm{e}^{-v}} \cap \Omega|$ is nonincreasing, we next consider the following two cases:

Case (a): We assume that $\overline{D} < N$. Hence, we have

$$\begin{aligned}\frac{1}{h}\int_u^{u+h} \mathrm{e}^{v(N-\overline{D})}|A_{\mathrm{e}^{-v}} \cap \Omega|\,\mathrm{d}v &\le \frac{|A_{\mathrm{e}^{-u}} \cap \Omega|}{h}\int_u^{u+h} \mathrm{e}^{v(N-\overline{D})}\,\mathrm{d}v\\ &= \frac{|A_{\mathrm{e}^{-u}} \cap \Omega|}{\mathrm{e}^{-u(N-\overline{D})}}\,\frac{\mathrm{e}^{h(N-\overline{D})}-1}{(N-\overline{D})h}.\end{aligned}$$

By taking the lower limit of both sides as $u \to +\infty$, we obtain that

$$\operatorname{res}(\tilde{\zeta}_{A,\Omega},\overline{D}) \le \mathscr{M}_*^{\overline{D}}(A,\Omega)\frac{\mathrm{e}^{h(N-\overline{D})}-1}{(N-\overline{D})h}. \tag{5.4.10}$$

Since this is true for every $h > 0$, we can deduce by letting $h \to 0^+$ that

$$\operatorname{res}(\tilde{\zeta}_{A,\Omega},\overline{D}) \le \mathscr{M}_*^{\overline{D}}(A,\Omega). \tag{5.4.11}$$

On the other hand, we have

$$\begin{aligned}\frac{1}{h}\int_u^{u+h} \mathrm{e}^{v(N-\overline{D})}|A_{\mathrm{e}^{-v}} \cap \Omega|\,\mathrm{d}v &\ge \frac{|A_{\mathrm{e}^{-(u+h)}} \cap \Omega|}{h}\int_u^{u+h} \mathrm{e}^{v(N-\overline{D})}\,\mathrm{d}v\\ &= \frac{|A_{\mathrm{e}^{-(u+h)}} \cap \Omega|}{\mathrm{e}^{-(u+h)(N-\overline{D})}}\,\frac{1-\mathrm{e}^{-h(N-\overline{D})}}{(N-\overline{D})h}\end{aligned} \tag{5.4.12}$$

and, similarly as before, by taking the upper limit of both sides as $u \to +\infty$, we obtain that

$$\operatorname{res}(\tilde{\zeta}_{A,\Omega},\overline{D}) \ge \mathscr{M}^{*\overline{D}}(A,\Omega)\frac{1-\mathrm{e}^{-h(N-\overline{D})}}{(N-\overline{D})h}. \tag{5.4.13}$$

Since this is true for every $h > 0$, we let $h \to 0^+$ and conclude that

$$\operatorname{res}(\tilde{\zeta}_{A,\Omega},\overline{D}) \ge \mathscr{M}^{*\overline{D}}(A,\Omega). \tag{5.4.14}$$

This latter inequality, combined with (5.4.11), implies that (A,Ω) is $\overline{D}$-Minkowski measurable which, a fortiori, implies that $D = \dim_B(A,\Omega) = \overline{D}$. Furthermore, we also conclude that $\operatorname{res}(\tilde{\zeta}_{A,\Omega},D) = \mathscr{M}^D(A,\Omega)$, the Minkowski content of (A,Ω).

Case (b): We will now assume that $\overline{D} = N$. Therefore, in this case we have

$$|A_{\mathrm{e}^{-(u+h)}} \cap \Omega| = \frac{|A_{\mathrm{e}^{-(u+h)}} \cap \Omega|}{\mathrm{e}^{-(u+h)(N-N)}} \le \frac{1}{h}\int_u^{u+h} |A_{\mathrm{e}^{-v}} \cap \Omega|\,\mathrm{d}v \le \frac{|A_{\mathrm{e}^{-u}} \cap \Omega|}{\mathrm{e}^{-u(N-N)}} = |A_{\mathrm{e}^{-u}} \cap \Omega|.$$

[16] The convergence of $H(s)$ in $L^1(-\lambda,\lambda)$ as $\operatorname{Re} s \to 0^+$, which is required by Theorem 5.4.1, follows easily since the contour integral of H over the rectangle with vertices at $-\lambda\mathrm{i}$, $\operatorname{Re} s - \lambda\mathrm{i}$, $\operatorname{Re} s + \lambda\mathrm{i}$, and $\lambda\mathrm{i}$ is equal to zero.

Then, by taking, respectively, the lower and upper limits as $u \to +\infty$, we obtain that

$$\mathscr{M}^{*N}(A,\Omega) \leq \operatorname{res}(\tilde{\zeta}_{A,\Omega},N) \leq \mathscr{M}_*^N(A,\Omega). \tag{5.4.15}$$

Finally, if $D < N$, then the part of the theorem dealing with the distance (instead of the tube) zeta function of (A,Ω) follows at once from case (a) of the proof for $\tilde{\zeta}_{A,\Omega}$. This is so in light of the functional equation (4.5.2) connecting $\zeta_{A,\Omega}$ and $\tilde{\zeta}_{A,\Omega}$, or more precisely, of the relation between the residues at the simple pole $s = D$ of the two zeta functions which follows from it (namely, we have that $\operatorname{res}(\zeta_{A,\Omega},D) = (N-D)\operatorname{res}(\tilde{\zeta}_{A,\Omega},D)$). This concludes the proof of the theorem. □

Remark 5.4.3. In light of Theorem 5.4.1, the assumptions of Theorem 5.4.2 can be weakened. More precisely, it suffices to assume that for every fixed $\lambda > 0$, the function

$$\tilde{\zeta}_{A,\Omega}(s) - \frac{\operatorname{res}(\tilde{\zeta}_{A,\Omega},\overline{D})}{s-\overline{D}} \tag{5.4.16}$$

(restricted to the vertical line segment $(-\mathrm{i}\lambda,\mathrm{i}\lambda)$ and viewed as a function of $\tau := \operatorname{Im} s \in (-\lambda,\lambda)$), converges in $L^1(-\lambda,\lambda)$ to a boundary function $H(\mathrm{i}\operatorname{Im} s)$ as $\operatorname{Re} s \to \overline{D}^+$. Consequently, $H(\mathrm{i}\tau)$ must then satisfy

$$\int_{-\lambda}^{\lambda} |H(\mathrm{i}\tau)|\,\mathrm{d}\tau < \infty, \tag{5.4.17}$$

for every $\lambda > 0$.

In the case when, besides $\overline{D}$, there are other singularities on the critical line $\{\operatorname{Re} s = \overline{D}\}$ of the relative fractal drum (A,Ω), we can use Theorem 5.4.1 to derive an upper bound for the upper $\overline{D}$-dimensional Minkowski content of (A,Ω), expressed in terms of the residue of its relative tube (or distance) zeta function at $s = \overline{D}$, as we now explain in the next result.

Theorem 5.4.4 (Upper bound for the upper Minkowski content). *Let (A,Ω) be a relative fractal drum in $\mathbb{R}^N$ and let $\overline{D} := \overline{\dim}_B(A,\Omega)$. Furthermore, assume that the relative tube zeta function $\tilde{\zeta}_{A,\Omega}$ of (A,Ω) can be meromorphically extended to a connected open neighborhood U of the critical line $\{\operatorname{Re} s = \overline{D}\}$ and that $\overline{D}$ is a simple pole of its meromorphic continuation to U. Also assume that the critical line $\{\operatorname{Re} s = \overline{D}\}$ contains another pole of $\tilde{\zeta}_{A,\Omega}$, different from $\overline{D}$. Furthermore, let*

$$\lambda_{A,\Omega} := \inf\{|\overline{D}-\omega| : \omega \in \dim_{PC}(A,\Omega) \setminus \{\overline{D}\}\}. \tag{5.4.18}$$

Then, if $\overline{D} < N$, we have the following upper bound for the upper $\overline{D}$-dimensional Minkowski content of (A,Ω), expressed in terms of the residue at $s := \overline{D}$ of the relative tube zeta function of (A,Ω):

$$\mathscr{M}^{*\overline{D}}(A,\Omega) \leq \frac{C\lambda_{A,\Omega}(N-\overline{D})}{2\pi\left(1-\mathrm{e}^{-2\pi(N-\overline{D})/\lambda_{A,\Omega}}\right)} \operatorname{res}(\tilde{\zeta}_{A,\Omega},\overline{D}); \tag{5.4.19}$$

moreover, in the case when $\overline{D} = N$, *we have*

$$\mathscr{M}^{*N}(A,\Omega) \leq C\operatorname{res}(\tilde{\zeta}_{A,\Omega},N), \tag{5.4.20}$$

where (both in (5.4.19), (5.4.20) *just above and in* (5.4.21) *just below) C is a positive constant such that* $C < 3$.

Finally, if $\overline{D} < N$, *we have the following upper bound for the upper $\overline{D}$-dimensional Minkowski content of* (A,Ω), *expressed in terms of the residue at* $s := \overline{D}$ *of the relative distance zeta function of* (A,Ω):

$$\mathscr{M}^{*\overline{D}}(A,\Omega) \leq \frac{C\lambda_{A,\Omega}}{2\pi\left(1-\mathrm{e}^{-2\pi(N-\overline{D})/\lambda_{A,\Omega}}\right)}\operatorname{res}(\zeta_{A,\Omega},\overline{D}). \tag{5.4.21}$$

Proof. We use the same reasoning as in the proof of Theorem 5.4.2, with the only difference residing in the fact that we can now only use the weaker statement (5.4.3) of Theorem 5.4.1 since by hypothesis, there is another pole on the critical line $\{\operatorname{Re} s = \overline{D}\}$, besides $\overline{D}$ itself. More specifically, if $\overline{D} < N$ and $\lambda < \lambda_{A,\Omega}$, then by using (5.4.12) and (5.4.3), we show that for every $h \geq 2\pi/\lambda$, we have

$$C\operatorname{res}(\tilde{\zeta}_{A,\Omega},\overline{D}) \geq \mathscr{M}^{*\overline{D}}(A,\Omega)\frac{1-\mathrm{e}^{-h(N-\overline{D})}}{(N-\overline{D})h}. \tag{5.4.22}$$

Since the right-hand side of (5.4.22) just above is a decreasing function of h, we obtain the best estimate for $h = 2\pi/\lambda$. Furthermore, since this is true for every $\lambda < \lambda_{A,\Omega}$, we obtain (5.4.19) by letting $\lambda \to \lambda_{A,\Omega}^{-}$. Moreover, (5.4.22) is also valid if $\overline{D} = N$, but without the factor that depends on h, by a similar argument as in case (*b*) of the proof of Theorem 5.4.2. Finally, if $\overline{D} < N$, the statement about the relative distance zeta function $\zeta_{A,\Omega}$ follows by the same argument as in case (*a*) of the proof of Theorem 5.4.2, by also using the functional equation (4.5.2) connecting $\zeta_{A,\Omega}$ and $\tilde{\zeta}_{A,\Omega}$. □

Remark 5.4.5. Much as in the case of Theorem 5.4.2 (see Remark 5.4.3), the hypotheses of Theorem 5.4.4 can be weakened. However, we have stated Theorem 5.4.4 in the above form because this is the most common situation which is encountered in our examples of RFDs. For instance, in order to obtain an upper bound for the upper $\overline{D}$-dimensional Minkowski content of (A,Ω), it suffices to assume that the relative tube zeta function $\tilde{\zeta}_{A,\Omega}$ can be holomorphically continued to a punctured disk $B_r(\overline{D}) \setminus \{\overline{D}\}$, centered at $\overline{D}$ and with radius $r > 0$. In that case, (5.4.19) is valid with $\lambda_{A,\Omega}$ replaced by the radius r. Of course, the larger the radius of the disk, the better the upper bound. All one actually needs is the L^1-convergence as $\operatorname{Re} s \to \overline{D}^{+}$ of the relative tube or distance zeta function of (A,Ω) to a boundary function defined on a symmetric vertical interval $(\overline{D} - r\mathbb{i}, \overline{D} + r\mathbb{i})$, similarly as in Remark 5.4.3.

5.4.2 The Relative Mellin Zeta Function

In order to obtain a criterion for Minkowski measurability, we will need to extend to a larger space of test functions the distributional tube formulas derived in the previous sections. It will suffice to extend them to the space $\mathscr{K}(0,+\infty)$.[17] We now observe that in Definition 4.1.2, we have assumed that an RFD (A,Ω) has the property that there exists $\delta > 0$ such that $\Omega \subseteq A_\delta$. If this condition is not fulfilled, we can always replace Ω by $\tilde{\Omega} := A_\delta \cap \Omega$ and work instead with the new RFD $(A,\tilde{\Omega})$ because the fractal properties of this new RFD will be identical to those of (A,Ω). This last statement follows directly from the definition of the relative Minkowski dimension (see page 249). Furthermore, in light of the above discussion, for an RFD (A,Ω) we have that $A_\delta \cap \Omega = \Omega$ for all δ sufficiently large; consequently, for such values of δ, we have that $|A_\delta \cap \Omega| = |\Omega|$, which implies that we can actually redefine the tube zeta function in a way which will be more suitable. More precisely, assume that $\overline{D} := \overline{\dim}_B(A,\Omega) < N$ and recall the functional equation (2.2.23), written in the following integral form:

$$\int_{A_\delta \cap \Omega} d(x,A)^{s-N}\,\mathrm{d}x = \delta^{s-N}|A_\delta \cap \Omega| + (N-s)\int_0^\delta t^{s-N-1}|A_t \cap \Omega|\,\mathrm{d}t, \tag{5.4.23}$$

initially valid for all $s \in \mathbb{C}$ such that $\operatorname{Re} s > \overline{D}$. Furthermore, by taking the complex number s in the vertical strip $\{\operatorname{Re} s > \overline{D}\} \cap \{\operatorname{Re} s < N\}$ and letting $\delta \to +\infty$, we obtain the following equality:

$$\int_\Omega d(x,A)^{s-N}\,\mathrm{d}x = (N-s)\int_0^{+\infty} t^{s-N-1}|A_t \cap \Omega|\,\mathrm{d}t. \tag{5.4.24}$$

As we can see, on the right-hand side of (5.4.24), we have the Mellin transform of the function $t^{-N}|A_t \cap \Omega|$ and this integral is absolutely convergent inside the vertical strip $\{\operatorname{Re} s > \overline{D}\} \cap \{\operatorname{Re} s < N\}$. Indeed, we have that

$$\int_0^{+\infty} t^{s-N-1}|A_t \cap \Omega|\,\mathrm{d}t = \int_0^1 t^{s-N-1}|A_t \cap \Omega|\,\mathrm{d}t + \int_1^{+\infty} t^{s-N-1}|A_t \cap \Omega|\,\mathrm{d}t, \tag{5.4.25}$$

and the integral over $(0,1)$ is equal to $\tilde{\zeta}_{A,\Omega}(s;1)$ and hence, is absolutely convergent on $\{\operatorname{Re} s > \overline{D}\}$, while for the integral over $(1,+\infty)$, we have

$$\begin{aligned}\left|\int_1^{+\infty} t^{s-N-1}|A_t \cap \Omega|\,\mathrm{d}t\right| &\le \int_1^{+\infty} t^{\operatorname{Re} s-N-1}|A_t \cap \Omega|\,\mathrm{d}t \\ &\le |\Omega|\int_1^{+\infty} t^{\operatorname{Re} s-N-1}\,\mathrm{d}t = \frac{|\Omega|}{N-\operatorname{Re} s}.\end{aligned} \tag{5.4.26}$$

[17] Here, $\mathscr{K}(0,+\infty)$ is defined exactly as $\mathscr{K}(0,\delta)$ just before Definition 5.2.1, except for δ replaced by $+\infty$, and in this case, we require that for every $\varphi \in \mathscr{K}(0,+\infty)$, $t^m\varphi^{(q)}(t) \to 0$ as $t \to +\infty$, where $\varphi^{(q)}$ denotes the q-th derivative of φ.

We then conclude from Theorem 2.1.45 (about the holomorphicity of an integral depending analytically on a complex parameter) that the integral on the right-hand side of (5.4.24) defines a holomorphic function on the vertical strip $\{\overline{D} < \mathrm{Re}\, s < N\}$ and upon analytic continuation, that the entire right-hand side of (5.4.24) coincides (within that strip) with the relative distance zeta function $\zeta_{A,\Omega}(s)$; i.e., the identity (5.4.24) holds as an equality between holomorphic functions defined on the open vertical strip $\{\overline{D} < \mathrm{Re}\, s < N\}$.

Moreover, upon further meromorphic continuation (and since, by Theorem 4.1.7, $\zeta_{A,\Omega}$ is holomorphic in the open right half-plane $\{\mathrm{Re}\, s > \overline{D}\}$), we also deduce that if $\zeta_{A,\Omega}$ can be meromorphically continued to a given connected open neighborhood U of the critical line $\{\mathrm{Re}\, s = \overline{D}\}$ (or, equivalently, of the closed half-plane $\{\mathrm{Re}\, s \geq \overline{D}\}$), then with the terminology and notation of Definition 5.4.6 just below, so can the Mellin zeta function $\zeta_{A,\Omega}^{\mathfrak{M}}$. Hence, we deduce that the following functional equation (between meromorphic functions) holds:

$$\zeta_{A,\Omega}(s) = (N-s)\zeta_{A,\Omega}^{\mathfrak{M}}(s), \tag{5.4.27}$$

for all $s \in U$.

Definition 5.4.6. Let (A,Ω) be an RFD in $\mathbb{R}^N$ such that $\overline{\dim}_B(A,\Omega) < N$. We define the *Mellin zeta function* $\zeta_{A,\Omega}^{\mathfrak{M}}$ of (A,Ω) by

$$\zeta_{A,\Omega}^{\mathfrak{M}}(s) := \int_0^{+\infty} t^{s-N-1}|A_t \cap \Omega|\, \mathrm{d}t, \tag{5.4.28}$$

for all $s \in \mathbb{C}$ with $\mathrm{Re}\, s \in (\overline{\dim}_B(A,\Omega), N)$. Here, the integral is taken in the Lebesgue sense.

In the discussion preceding Definition 5.4.6, we have proved a part of the following theorem.

Theorem 5.4.7. *Let (A,Ω) be an RFD in $\mathbb{R}^N$ such that $\overline{\dim}_B(A,\Omega) < N$. Then, the Mellin zeta function $\zeta_{A,\Omega}^{\mathfrak{M}}$, as given by Equation* (5.4.28)*, is holomorphic on the open vertical strip $\{\overline{\dim}_B(A,\Omega) < \mathrm{Re}\, s < N\}$ and*

$$\frac{\mathrm{d}}{\mathrm{d}s}\zeta_{A,\Omega}^{\mathfrak{M}}(s) = \int_0^{+\infty} t^{s-N-1}|A_t \cap \Omega| \log t\, \mathrm{d}t, \tag{5.4.29}$$

for all s in $\{\overline{\dim}_B(A,\Omega) < \mathrm{Re}\, s < N\}$. Furthermore, $\{\overline{\dim}_B(A,\Omega) < \mathrm{Re}\, s < N\}$ is the largest vertical strip (of the form $\{\alpha < \mathrm{Re}\, s < \beta\}$, with $-\infty \leq \alpha < \beta \leq +\infty$) on which the integral on the right-hand side of (5.4.28) *is absolutely convergent (i.e., is a convergent Lebesgue integral).*

Moreover, for all $s \in \mathbb{C}$ such that $\overline{\dim}_B(A,\Omega) < \mathrm{Re}\, s < N$ and for any fixed $\delta > 0$ such that $\Omega \subseteq A_\delta$, $\zeta_{A,\Omega}^{\mathfrak{M}}$ satisfies the following functional equations connecting it to $\tilde{\zeta}_{A,\Omega}$ and $\zeta_{A,\Omega}$, respectively:

$$\zeta_{A,\Omega}^{\mathfrak{M}}(s) = \tilde{\zeta}_{A,\Omega}(s;\delta) + \frac{\delta^{s-N}|\Omega|}{N-s} \tag{5.4.30}$$

and

$$\zeta^{\mathfrak{M}}_{A,\Omega}(s) = \frac{\zeta_{A,\Omega}(s;\delta)}{N-s}. \tag{5.4.31}$$

Remark 5.4.8. We point out that functional equations similar to (5.4.30) and (5.4.31) are also satisfied for every $\delta > 0$ such that $\Omega \not\subseteq A_\delta$, but one then also has to add to the right-hand side of each of the corresponding functional equations a suitable function f which is meromorphic on $\mathbb{C}$ and has a single, simple pole at $s = N$.

Proof of Theorem 5.4.7. We have already proved the first part of the theorem. The optimality of the vertical strip follows directly from (5.4.25) (or, more precisely, from (5.4.27)). Namely, the lower bound $\overline{\dim}_B(A,\Omega)$ is a consequence of the presence of the first integral on the right-hand side of (5.4.25) since the latter integral is equal to $\tilde{\zeta}_{A,\Omega}(s;1)$. Furthermore, the upper bound N is a consequence of the presence of the second integral on the right-hand side of (5.4.25), since that integral is divergent for any real number s such that $s > N$. To see this, let $\delta \geq 1$ be such that $\Omega \subseteq A_\delta$ and make the following observation:

$$\begin{aligned}\int_1^{+\infty} t^{s-N-1}|A_t \cap \Omega|\,dt &\geq \int_\delta^{+\infty} t^{s-N-1}|A_t \cap \Omega|\,dt \\ &= |\Omega| \int_\delta^{+\infty} t^{s-N-1}\,dt = +\infty.\end{aligned} \tag{5.4.32}$$

The functional equation (5.4.31) is already proven, while (5.4.30) can be proven directly by splitting the integral defining $\zeta^{\mathfrak{M}}_{A,\Omega}$ over the intervals $(0,\delta)$ and $(\delta,+\infty)$ or by using the functional equation (4.5.2) connecting the tube and distance zeta functions. □

As a consequence of the functional equations (5.4.31), (5.4.30) and the principle of analytic continuation, we obtain the following two theorems, which follow from the obvious extensions to RFDs of the corresponding ones for the relative distance and tube zeta functions (see Theorems 2.1.11, 2.2.3, 2.2.11 and 2.2.14).

Theorem 5.4.9. *Let (A,Ω) be a relative fractal drum in $\mathbb{R}^N$ such that $\overline{\dim}_B(A,\Omega) < N$. Then the following properties hold:*

(a) The Mellin zeta function $\zeta^{\mathfrak{M}}_{A,\Omega}$ is meromorphic in the half-plane $\{\mathrm{Re}\, s > \overline{\dim}_B(A,\Omega)\}$ with a single, simple pole at $s = N$. Furthermore,

$$\operatorname{res}(\zeta^{\mathfrak{M}}_{A,\Omega}, N) = -|\Omega|. \tag{5.4.33}$$

(b) If the relative box (or Minkowski) dimension $D := \dim_B(A,\Omega)$ exists, and $\mathscr{M}^D_*(A,\Omega) > 0$, then $\zeta^{\mathfrak{M}}_{A,\Omega}(s) \to +\infty$ as $s \in \mathbb{R}$ converges to D from the right.

Proof. By means of the principle of analytic continuation, and since $\overline{\dim}_B(A,\Omega) < N$, we conclude that the functional equalities (5.4.30) and (5.4.31) continue to hold on any connected open neighborhood $U \subseteq \mathbb{C}$ of the vertical strip $\{\overline{\dim}_B(A,\Omega) <$

$\mathrm{Re}\, s < N\}$ to which any (*and hence, all*) of the three relative zeta functions has a holomorphic continuation. (See also the text surrounding Equation (5.4.27).) As a result, part (*a*) follows from the counterpart of Theorem 4.1.7 for the relative tube zeta function (see also Theorem 2.2.11) and (5.4.30), while part (*b*) follows from Theorem 4.1.7 and (5.4.31). □

Furthermore, in light of Theorem 2.2.14 and (5.4.30), one obtains the following result.

Theorem 5.4.10. *Assume that (A,Ω) is a Minkowski nondegenerate RFD in $\mathbb{R}^N$, that is, $0 < \mathcal{M}_*^D(A,\Omega) \le \mathcal{M}^{*D}(A,\Omega) < \infty$ (in particular, $D := \dim_B(A,\Omega)$ exists), and $D < N$. If $\zeta_{A,\Omega}^{\mathfrak{M}}$ can be extended meromorphically to a connected open neighborhood of $s = D$, then D is necessarily a simple pole of $\zeta_{A,\Omega}^{\mathfrak{M}}$, the residue $\mathrm{res}(\zeta_{A,\Omega}^{\mathfrak{M}}, D)$ is independependent of δ and*

$$\mathcal{M}_*^D(A,\Omega) \le \mathrm{res}(\zeta_{A,\Omega}^{\mathfrak{M}}, D) \le \mathcal{M}^{*D}(A,\Omega). \tag{5.4.34}$$

Furthermore, if (A,Ω) is Minkowski measurable, then

$$\mathrm{res}(\zeta_{A,\Omega}^{\mathfrak{M}}, D) = \mathcal{M}^D(A,\Omega). \tag{5.4.35}$$

Lemma 5.4.11. *Assume that (A,Ω) is an RFD in $\mathbb{R}^N$ with $\overline{\dim}_B(A,\Omega) < N$ and such that its tube or distance or Mellin zeta function is meromorphic on some connected open neighborhood U of the vertical strip $\{\overline{\dim}_B(A,\Omega) < \mathrm{Re}\, s < N\}$.*[18] *Then, the multisets of poles located in $U \setminus \{N\}$ of each of these three zeta functions, $\tilde{\zeta}_{A,\Omega}$, $\zeta_{A,\Omega}$ and $\zeta_{A,\Omega}^{\mathfrak{M}}$, coincide:*

$$\mathscr{P}(\tilde{\zeta}_{A,\Omega}, U \setminus \{N\}) = \mathscr{P}(\zeta_{A,\Omega}, U \setminus \{N\}) = \mathscr{P}(\zeta_{A,\Omega}^{\mathfrak{M}}, U \setminus \{N\}). \tag{5.4.36}$$

Moreover, if $\omega \in U \setminus \{N\}$ is a simple pole of any (and hence, of all) of these three zeta functions, then[19]

$$\mathrm{res}\left(\zeta_{A,\Omega}^{\mathfrak{M}}, \omega\right) = \mathrm{res}\left(\tilde{\zeta}_{A,\Omega}, \omega\right) = \frac{\mathrm{res}\left(\zeta_{A,\Omega}, \omega\right)}{N - \omega}. \tag{5.4.37}$$

We can now use the Mellin inversion theorem (Theorem 5.1.6) in order to derive the following inversion formula for the Mellin zeta function.

Theorem 5.4.12. *Let (A,Ω) be an RFD in $\mathbb{R}^N$ such that $\overline{\dim}_B(A,\Omega) < N$. Then, for any $c \in (\overline{\dim}_B(A,\Omega), N)$ and $t > 0$, the following formula is valid pointwise:*

---

[18] Recall from Theorem 4.1.7 and its counterpart for the relative tube zeta function that $\zeta_{A,\Omega}$ and $\tilde{\zeta}_{A,\Omega}$ are holomorphic on the open right-half plane $\{\mathrm{Re}\, s > \overline{\dim}_B(A,\Omega)\}$.

[19] Clearly, in the case when $\omega \in U \setminus \{N\}$ is a multiple pole, an analogous relation holds between the principal parts at ω of $\tilde{\zeta}_{A,\Omega}(s)$, $\zeta_{A,\Omega}^{\mathfrak{M}}(s)$ and the meromorphic function $\zeta_{A,\Omega}(s)/(N-s)$; furthermore, in that case, ω has the same multiplicity for either of these zeta functions.

$$|A_t \cap \Omega| = \frac{1}{2\pi\mathrm{i}} \int_{c-\mathrm{i}\infty}^{c+\mathrm{i}\infty} t^{N-s} \zeta_{A,\Omega}^{\mathfrak{M}}(s)\,\mathrm{d}s. \tag{5.4.38}$$

Proof. The conclusion follows directly from Theorem 5.1.6, along with the fact that the function $t \mapsto t^{-N}|A_t \cap \Omega|$ is continuous and of locally bounded variation on $(0,+\infty)$ and $t \mapsto t^{c-N-1}|A_t \cap \Omega|$ is in $L^1(0,+\infty)$ for every $c \in (\overline{\dim}_B(A,\Omega),N)$. See also the proof of Theorem 5.1.7 since the reasoning here is along the same lines. □

Note that in the above theorem, it is crucial to choose $c \in (\overline{\dim}_B(A,\Omega),N)$ for the hypothesis of the Mellin inversion theorem to be satisfied. In other words, the inversion theorem (Theorem 5.1.6) is no longer applicable in the case when $\overline{\dim}_B(A,\Omega) = N$ since then, we cannot define the Mellin zeta function. Note that this is in contrast with the situation from Sections 5.1 and 5.2 where we have worked with the relative tube zeta function. One can now impose languidity conditions on the Mellin zeta function $\zeta_{A,\Omega}^{\mathfrak{M}}$ and rewrite Sections 5.1 and 5.2 in terms of $\zeta_{A,\Omega}^{\mathfrak{M}}$ since the fact that we have to choose $c \in (\overline{\dim}_B(A,\Omega),N)$ is not a hindrance. Indeed, originally we had the freedom to choose any $c \in (\overline{\dim}_B(A,\Omega),N+1)$ in Proposition 5.1.8. Furthermore, this will ensure that although $s = N$ is always a pole of the Mellin zeta function, it will never be a part of the sum over the residues of $\zeta_{A,\Omega}^{\mathfrak{M}}$ since it is always located strictly to the right of the vertical line $\{\operatorname{Re} s = c\}$ over which we integrate in (5.4.38).

Moreover, one can also derive the corresponding results about the distance zeta function $\zeta_{A,\Omega}$ directly from the Mellin zeta function $\zeta_{A,\Omega}^{\mathfrak{M}}$ and without the use of the shell zeta function $\breve{\zeta}_{A,\Omega}$. However, one has to be careful and always choose δ sufficiently large so that $\Omega \subseteq A_\delta$ in order for the functional equation (5.4.31) to be satisfied. One other issue that is not clear in this context is whether the restriction of having to choose δ large enough for the inclusion $\Omega \subseteq A_\delta$ to hold could increase the 'languidity exponent' κ_d of $\zeta_{A,\Omega}(s)$. This is not the case in all of the examples we will consider, but a general result along these lines has yet to be obtained.

Proposition 5.4.13. *Let (A,Ω) be a relative fractal drum in $\mathbb{R}^N$. If the relative distance zeta function $\zeta_{A,\Omega}(\,\cdot\,;\delta)$ satisfies the languidity conditions* **L1** *and* **L2** *of Definition 5.1.3 for some $\delta > 0$ and with d-languidity exponent $\kappa_d \in \mathbb{R}$, then so does $\zeta_{A,\Omega}(\,\cdot\,;\delta_1)$ for any $\delta_1 > 0$ and with $(\kappa_d)_{\delta_1} := \max\{\kappa_d, 0\}$.*

Furthermore, the analogous statement is also true in the case when $\zeta_{A,\Omega}(\,\cdot\,;\delta)$ is strongly d-languid, under the additional assumption that $\delta \geq 1$ and $\delta_1 \geq 1$.

Proof. Without loss of generality, we may assume that $\delta < \delta_1$. Then, the conclusion follows from the fact that for a given a window $\boldsymbol{W}$ we have $\zeta_{A,\Omega}(s;\delta_1) = \zeta_{A,\Omega}(s;\delta) + g(s)$ for all $s \in \boldsymbol{W}$, where g is defined for all $s \in \mathbb{C}$ and is an entire function. (This fact follows directly from Lemma 2.1.15.) Furthermore, for all $s \in \mathbb{C}$, we have the following upper bound on $|g(s)|$:

$$|g(s)| \leq \int_{(A_{\delta_1} \setminus \overline{A_\delta}) \cap \Omega} d(x,A)^{\operatorname{Re} s - N}\,\mathrm{d}x \leq |\Omega| \max\{\delta^{\operatorname{Re} s - N}, \delta_1^{\operatorname{Re} s - N}\}. \tag{5.4.39}$$

As we can see, the upper bound on $|g(s)|$ does not depend on $\operatorname{Im} s$ and therefore, we conclude that g satisfies the languidity conditions **L1** and **L2** of Definition 5.1.3 with the languidity exponent $\kappa_g = 0$ and for any given window $\boldsymbol{W}$. This observation implies that then, $\zeta_{A,\Omega}(\,\cdot\,;\delta_1)$ is languid with d-languidity exponent $(\kappa_d)_{\delta_1} := \max\{\kappa_d, 0\}$ and for the same window as for $\zeta_{A,\Omega}(\,\cdot\,;\delta)$.

The additional assumption for strong d-languidity is needed since **L1** must then be satisfied for all $\sigma \in (-\infty, c)$, in the notation of Definition 5.1.3. Furthermore, for this condition to be achieved, we need that $\delta_1 > \delta \geq 1$ in (5.4.39) since otherwise, we cannot obtain an upper bound on $|g(s)|$ when $\operatorname{Re} s \to -\infty$. □

In order to avoid unnecessary repetitions, we will not restate all of the theorems of Sections 5.1 and 5.2 in terms of the Mellin zeta function, but we will do so only for the distributional fractal tube formula with error term because the corresponding result will be needed for establishing the Minkowski measurability criterion of Subsection 5.4.3 below; see Theorem 5.4.20 along with Theorem 5.4.15.

Recall that the original motivation for introducing the Mellin zeta function was to obtain a distributional fractal tube formula valid on a larger space of test functions, more precisely, on the space $\mathscr{K}(0,+\infty)$; that is, the space of test functions φ in the class $C^\infty(0,+\infty)$, such that for all $m \in \mathbb{Z}$ and $q \in \mathbb{N}$, we have $t^m \varphi^{(q)}(t) \to 0$, both as $t \to 0^+$ and as $t \to +\infty$. Also note that $\mathscr{D}(0,+\infty) \subseteq \mathscr{K}(0,+\infty)$, and hence, we have the following (reverse) relation between the corresponding spaces of distributions (or dual spaces):

$$\mathscr{K}'(0,+\infty) \subseteq \mathscr{D}'(0,+\infty). \tag{5.4.40}$$

Theorem 5.4.14 (Distributional fractal tube formula with error term, via $\zeta^{\mathfrak{M}}_{A,\Omega}$; level $k=0$). *Let (A,Ω) be a relative fractal drum in $\mathbb{R}^N$ such that $\overline{\dim}_B(A,\Omega) < N$. Furthermore, assume that $\zeta^{\mathfrak{M}}_{A,\Omega}$ satisfies the languidity conditions* **L1** *and* **L2** *of Definition 5.1.3, for some $\kappa \in \mathbb{R}$. Then, the distribution $\mathscr{V}^{[0]}_{A,\Omega}$ in $\mathscr{K}'(0,+\infty)$ is given by the following distributional identity in $\mathscr{K}'(0,+\infty)$:*

$$\mathscr{V}^{[0]}_{A,\Omega}(t) = \sum_{\omega \in \mathscr{P}(\zeta^{\mathfrak{M}}_{A,\Omega}, \boldsymbol{W})} \operatorname{res}\left(t^{N-s}\zeta^{\mathfrak{M}}_{A,\Omega}(s), \omega\right) + \mathscr{R}^{\mathfrak{M}[0]}_{A,\Omega}(t). \tag{5.4.41}$$

That is, the action of $\mathscr{V}^{[0]}_{A,\Omega}$ on an arbitrary test function $\varphi \in \mathscr{K}(0,+\infty)$ is given by

$$\begin{aligned}\langle \mathscr{V}^{[0]}_{A,\Omega}, \varphi \rangle = &\sum_{\omega \in \mathscr{P}(\zeta^{\mathfrak{M}}_{A,\Omega}, \boldsymbol{W})} \operatorname{res}\left(\{\mathfrak{M}\varphi\}(N-s+1)\, \zeta^{\mathfrak{M}}_{A,\Omega}(s), \omega\right) \\ &+ \langle \mathscr{R}^{\mathfrak{M}[0]}_{A,\Omega}, \varphi \rangle.\end{aligned} \tag{5.4.42}$$

Here, the distributional error term $\mathscr{R}^{\mathfrak{M}[0]}_{A,\Omega}$ is the distribution in $\mathscr{K}'(0,+\infty)$ given for all $\varphi \in \mathscr{K}(0,+\infty)$ by

$$\langle \mathscr{R}^{\mathfrak{M}[0]}_{A,\Omega}, \varphi \rangle = \frac{1}{2\pi \mathrm{i}} \int_S \{\mathfrak{M}\varphi\}(N-s+1)\, \zeta^{\mathfrak{M}}_{A,\Omega}(s)\, ds. \tag{5.4.43}$$

Furthermore, the distribution $\mathscr{R}^{\mathfrak{M}[0]}_{A,\Omega}(t)$ *is of asymptotic order at most* $t^{N-\sup S}$ *as* $t \to 0^+$*; i.e.,*

$$\mathscr{R}^{\mathfrak{M}[0]}_{A,\Omega}(t) = O(t^{N-\sup S}) \quad \text{as} \quad t \to 0^+, \tag{5.4.44}$$

in the sense of Definition 5.2.9.

Moreover, if $S(\tau) < \sup S$ *for all* $\tau \in \mathbb{R}$ *(that is, if the screen* $\boldsymbol{S}$ *lies strictly to the left of the vertical line* $\{\operatorname{Re} s = \sup S\}$*), then* $\mathscr{R}^{\mathfrak{M}[0]}_{A,\Omega}(t)$ *is of asymptotic order less than* $t^{N-\sup S}$*; i.e.,*

$$\mathscr{R}^{\mathfrak{M}[0]}_{A,\Omega}(t) = o(t^{N-\sup S}) \quad \text{as} \quad t \to 0^+, \tag{5.4.45}$$

still in the sense of Definition 5.2.9.

5.4.3 Characterization of Minkowski Measurability

Having expanded (in Theorem 5.4.14 of Subsection 5.4.2) the space of test functions for which the distributional tube formula is valid, we can now obtain a necessary condition for the Minkowski measurability of a languid relative fractal drum. We stress that in the statement of the following theorem, the phrase according to which the Mellin zeta function is languid means that $\zeta^{\mathfrak{M}}_{A,\Omega}$ satisfies the languidity conditions of Definition 5.1.3 for some languidity exponent $\kappa \in \mathbb{R}$, with the caveat that in condition **L1** we now assume that $c \in (\overline{\dim}_B(A,\Omega), N)$.

Theorem 5.4.15 (Necessary condition for Minkowski measurability). *Let* (A,Ω) *be a relative fractal drum in* $\mathbb{R}^N$ *such that* $D := \dim_B(A,\Omega)$ *exists,* $D < N$ *and* (A,Ω) *is Minkowski measurable. Furthermore, assume that its Mellin zeta function* $\zeta^{\mathfrak{M}}_{A,\Omega}$ *is languid for some screen* $\boldsymbol{S}$ *passing strictly to the left of the critical line* $\{\operatorname{Re} s = D\}$ *and strictly to the right of all the complex dimensions of* (A,Ω) *with real part strictly less than* D*.*

Then, D *is the only pole of* $\zeta^{\mathfrak{M}}_{A,\Omega}$ *located on the critical line* $\{\operatorname{Re} s = D\}$ *and it is simple.*

Proof. Since (A,Ω) is languid, the hypotheses of Theorem 5.4.10 are satisfied and, therefore, $s = D$ is a simple pole of $\zeta^{\mathfrak{M}}_{A,\Omega}$. Furthermore, also by Theorem 5.4.10, we have that $\mathscr{M} := \mathscr{M}^D(A,\Omega) = \operatorname{res}(\zeta^{\mathfrak{M}}_{A,\Omega}, D)$. It remains to show that D is the only pole located on the critical line. First, we deduce at once from the definition of the Mellin zeta function given in Equation (5.4.28) that $|\zeta^{\mathfrak{M}}_{A,\Omega}(s)| \le \zeta^{\mathfrak{M}}_{A,\Omega}(\operatorname{Re} s)$, for all $s \in \{D < \operatorname{Re} s < N\}$. We conclude from this inequality that if ξ is another pole of $\zeta^{\mathfrak{M}}_{A,\Omega}$ with $\operatorname{Re} \xi = D$, then it must also be simple.

Now, let us denote by $\xi_n := D + \mathrm{i}\gamma_n$, with $\gamma_n \in \mathbb{R}$ and $n \in J$, the potentially infinite sequence of poles of $\zeta^{\mathfrak{M}}_{A,\Omega}$ with real part D (i.e., of principal poles of $\zeta^{\mathfrak{M}}_{A,\Omega}$).[20] Here,

[20] As is well known, a meromorphic function on a domain of $\mathbb{C}$ can only have an at most countable number of poles. This follows from the fact that any compact subset $K \subseteq \mathbb{C}$ may contain only finitely many poles (since otherwise, there would be a limit point of poles in K) and any domain of $\mathbb{C}$ is contained in a countable union of compact sets.

$J \subseteq \mathbb{N}_0$ is a finite or infinite subset of $\mathbb{N}_0$, $0 \in J$, and we use the convention according to which $\gamma_0 := 0$ and hence, $\xi_0 := D$. Since D is simple (i.e., a simple pole of $\zeta_{A,\Omega}^{\mathfrak{M}}$), we must have $\gamma_n \neq 0$ for all $n \in J \setminus \{0\}$.

Observe that in light of the argument given in the first part of the proof, each principal pole ξ_n is then also *simple*, for every $n \in J \setminus \{0\}$; so that we can let $a_n := \operatorname{res}(\zeta_{A,\Omega}^{\mathfrak{M}}, \xi_n)$, for every $n \in J$. Furthermore, as was established at the beginning of the proof, we have $a_0 = \operatorname{res}(\zeta_{A,\Omega}^{\mathfrak{M}}, D) = \mathscr{M}$, the Minkowski content of (A,Ω). (Recall that the RFD (A,Ω) is assumed to be Minkowski measurable.)

Next, we will show that $J \setminus \{0\}$ is empty and therefore, that D is the only principal pole of $\zeta_{A,\Omega}^{\mathfrak{M}}$ (and is simple), as desired. For this purpose, we reason by contradiction and assume that $J \setminus \{0\}$ is nonempty. Then, in light of Theorem 5.4.14 (the distributional fractal tube formula at level $k = 0$ via $\zeta_{A,\Omega}^{\mathfrak{M}}$) applied with the stronger error estimate given by (5.4.45) and for the same choice of screen $\boldsymbol{S}$ as assumed to exist in the statement of that theorem (and which also exists, by the hypotheses of the present theorem), we have that

$$\begin{aligned} |A_t \cap \Omega| &= \sum_{n \in J} a_n t^{N-\xi_n} + o(t^{N-D}) \\ &= \mathscr{M} t^{N-D} + t^{N-D} \sum_{n \in J \setminus \{0\}} a_n t^{-\mathrm{i}\gamma_n} + o(t^{N-D}) \quad \text{as} \quad t \to 0^+, \end{aligned} \tag{5.4.46}$$

in the distributional sense since, by assumption, the screen $\boldsymbol{S}$ lies strictly to the left of the critical line $\{\operatorname{Re} s = D\}$.

On the other hand, since (A,Ω) is Minkowski measurable, we know that its relative tube function satisfies

$$|A_t \cap \Omega| = \mathscr{M} t^{N-D} + o(t^{N-D}) \quad \text{as} \quad t \to 0^+, \tag{5.4.47}$$

in the usual pointwise sense and hence also, in the distributional sense. Combining (5.4.46) with (5.4.47) yields that

$$\sum_{n \in J \setminus \{0\}} a_n t^{-\mathrm{i}\gamma_n} = o(1) \quad \text{as} \quad t \to 0^+, \tag{5.4.48}$$

in the distributional sense. After a (distributional) change of variable $\tau := \log t$ (note that $\tau \in C^\infty(0,+\infty)$), the uniqueness theorem for almost periodic distributions (see [Schw, Section VI.9.6, p. 208]) can be applied and now implies that (5.4.48) can only be true if $a_n = 0$ for all $n \in J \setminus \{0\}$; that is, only if $J \setminus \{0\}$ is empty (which contradicts our assumption) or, equivalently, only if there are no other poles on the critical line, except for $s = D$, as we needed to show. □

Remark 5.4.16. The above theorem can also be stated in terms of the relative tube and distance zeta functions of (A,Ω). This claim follows from the fact that the functional equations (5.4.30) and (5.4.31) which connect the relative tube zeta function, the relative distance zeta function and the Mellin zeta function of (A,Ω), along with Propositions 5.1.5 and 5.4.13, imply that if the languidity conditions **L1** and **L2** of Definition 5.1.3 are satisfied by the tube or distance zeta function, then they are also satisfied by the Mellin zeta function, with a possibly different languidity exponent. We can still, however, apply Theorem 5.4.14.

Remark 5.4.17. We point out that Theorem 5.4.15 (more precisely, its counterpart for the relative tube zeta function) is at the same time more and less general than Theorem 2.3.18 (more precisely, than its counterpart for RFDs). Indeed, in the counterpart of Theorem 2.3.18, we would assume that for some $\alpha > 0$, the tube function of (A,Ω) satisfies

$$|A_t \cap \Omega| = t^{N-D}(\mathscr{M} + O(t^{\alpha})) \quad \text{as} \quad t \to 0^+, \tag{5.4.49}$$

pointwise, which is a stronger assumption than the mere hypothesis of Minkowski measurability of (A,Ω). Then, the conclusion that D is the only complex dimension of (A,Ω) with real part D and that it is simple follows without the additional assumptions of Theorem 5.4.15.

On the other hand, in (the counterpart of) Theorem 5.4.15, we make a weaker hypothesis about the tube function $t \mapsto |A_t \cap \Omega|$ of (A,Ω); that is, we only assume that (A,Ω) is Minkowski measurable (with Minkowski dimension $\dim_B(A,\Omega) < N$) but we must make a stronger assumption on the fractal zeta function in order to draw the same conclusion about the complex dimensions of (A,Ω). This tradeoff is desirable since, in general, we want to draw as much information as possible about the geometry of RFDs directly from their fractal zeta functions, more precisely, from their distance zeta functions. In light of this, it would be of great interest to find out whether the languidity hypothesis or the conditions on the screen in Theorem 5.4.15 can be weakened.

Remark 5.4.18. It clearly follows from the proof of Theorem 5.4.15 that it would suffice to assume in the statement of that theorem that the RFD (A,Ω) is Minkowski measurable *in the distributional sense* (which specifically means in the present context that Equation (5.4.47) holds as a distributional identity in $\mathscr{K}'(0,+\infty)$, with $\mathscr{M} \in (0,+\infty)$).

The previous remark motivates us to introduce the following definition.

Definition 5.4.19. (*Weak vs. strong Minkowski measurability*).

(*i*) A relative fractal drum (A,Ω) such that $D := \dim_B(A,\Omega)$ exists is said to be *Minkowski measurable, in the distributional sense* (or *weakly Minkowski measurable*, in short) if there exist a constant $\mathscr{M} \in (0,+\infty)$ such that, in the sense of distributions,

$$\lim_{t\to 0^+} t^{-(N-D)}|A_t \cap \Omega| = \mathscr{M}, \text{ in } \mathscr{K}'(0,+\infty); \tag{5.4.50}$$

i.e., for every $\varphi \in \mathscr{K}(0,+\infty)$,[21]

[21] For the definition of the spaces $\mathscr{K}'(0,+\infty)$ and $\mathscr{K}'(0,+\infty)$, see the discussion preceding Equation (5.4.40).

$$\lim_{a\to 0^+}\int_0^{+\infty} t^{-(N-D)}|A_t\cap\Omega|\varphi_a(t)\,dt = \mathscr{M}\lim_{a\to 0^+}\int_0^{+\infty}\varphi_a(t)\,dt = \mathscr{M}\int_0^{+\infty}\varphi(t)\,dt. \tag{5.4.51}$$

Here, as before, φ_a is the scaled version of φ defined by (5.2.22).[22] Then, $\mathscr{M}$ is called the *weak Minkowski content* of the RFD (A,Ω).

(ii) Much as in part (i) of this definition, we can say that a relative fractal drum (A,Ω) is *strongly Minkowski measurable* (with *strong Minkowski content* $\mathscr{M}$) if it is Minkowski measurable in the usual (pointwise) sense of Subsection 4.1.1; i.e., if there exists a constant $\mathscr{M}\in(0,+\infty)$ such that

$$\lim_{t\to 0^+} t^{-(N-D)}|A_t\cap\Omega| = \mathscr{M}, \quad \text{in } \mathbb{R}. \tag{5.4.52}$$

Clearly, if (A,Ω) is strongly Minkowski measurable, it is also weakly Minkowski measurable and then, the strong and weak Minkowski contents of (A,Ω) coincide.[23] We note that we could similarly distinguish between weak and strong (or ordinary) Minkowski nondegeneracy, for example, although this definition will not be needed in the sequel.

We stress that the notion of Minkowski measurability being characterized in all of the criteria stated below in the remainder of this subsection (namely, Theorem 5.4.20, Theorem 5.4.25 and Corollary 5.4.26) *is always the notion of strong (or ordinary) Minkowski measurability*, in the sense of part (ii) of Definition 5.4.19 above (as opposed to that of weak Minkowski measurability, introduced in part (i) of Definition 5.4.19 and which, according to Remark 5.4.18 could be used in the statement of Theorem 5.4.15, the necessary condition for Minkowski measurability).

Finally, we can now state the announced Minkowski measurability criterion, the proof of which follows directly from Theorems 5.4.2 and 5.4.15.

Theorem 5.4.20 (Minkowski measurability criterion in terms of $\zeta_{A,\Omega}$). *Let (A,Ω) be a relative fractal drum in $\mathbb{R}^N$ such that $D := \dim_B(A,\Omega)$ exists and $D<N$. Furthermore, assume that (A,Ω) is d-languid for a screen* **S** *passing* strictly *between the critical line $\{\mathrm{Re}\,s = D\}$ and all the complex dimensions of (A,Ω) with real part strictly less than D. Then the following statements are equivalent:*

(a) *The RFD (A,Ω) is (strongly) Minkowski measurable.*

(b) *D is the only pole of the relative distance zeta function $\zeta_{A,\Omega}$ located on the critical line $\{\mathrm{Re}\,s = D\}$ and it is simple.*

---

[22] The second equality in (5.4.51) follows directly from the definition of φ_a since an elementary change of variable shows that the first integral on the right-hand side of (5.4.51) does not depend on a.

[23] More specifically, the *weak Minkowski content* is the regular distribution defined by (5.4.51) and associated with the constant function $\mathscr{M}$, the *strong* (or *ordinary*) *Minkowski content* of (A,Ω).

Remark 5.4.21. The above criterion is also valid if in part (*b*) of Theorem 5.4.20, we replace $\zeta_{A,\Omega}$ with the relative tube zeta function $\tilde{\zeta}_{A,\Omega}$, the Mellin zeta function $\zeta^{\mathfrak{M}}_{A,\Omega}$ or the relative shell zeta function $\breve{\zeta}_{A,\Omega}$. In this case, it suffices to assume that the chosen fractal zeta function satisfies the usual languidity conditions of Definition 5.1.3 (along with the condition from Theorem 5.4.20 about the existence of a suitable screen). In fact, if we state the theorem in terms of the relative tube zeta function $\tilde{\zeta}_{A,\Omega}$, we may omit the condition that $\dim_B(A,\Omega) < N$, as we shall see in Theorem 5.4.25 below.

Remark 5.4.22. Theorem 5.4.20 extends to RFDs in $\mathbb{R}^N$, with $N \geq 1$ arbitrary, the Minkowski measurability criterion for fractal strings obtained in [Lap-vFr3, Theorem 8.15 of Section 8.3]. More specifically, the latter criterion corresponds to the $N = 1$ case of Theorem 5.4.20. We note that in [Lap-vFr3], the criterion was formulated in terms of the principal complex dimensions of the underlying fractal string (interpreted as the poles of the associated geometric zeta function with real part equal to D, the Minkowski dimension of the string). However, in light of the results of Subsection 2.1.4 (see, especially, Proposition 2.1.59 and Corollary 2.1.61), they can now be restated equivalently in terms of the principal poles of the distance zeta function of the corresponding relative fractal drum $(\partial\Omega,\Omega)$, where Ω is any geometric realization of the fractal string. We mention that in the statement of [Lap-vFr3, Theorem 8.15 of Section 8.3], the fact that the fractal string was weakly (rather than strongly) Minkowski measurable should have been stressed more explicitly. In this regard, we note that Theorem 5.4.20 shows that in [Lap-vFr3] (*ibid*), we can now replace weak by strong (or ordinary) Minkowski measurability, which is the best possible result.

On the other hand, in [Lap-vFr3, Theorems 8.23 and 8.36 of Section 8.4], the characterization of Minkowski measurability obtained for self-similar strings was stated in terms of the strong (i.e., ordinary) Minkowski measurability of the fractal strings.[24] We do not consider the N-dimensional counterpart of such a situation here, although this would certainly be of interest. (See, however, Subsection 5.5.6 below, where we obtain an appropriate analog for self-similar sprays of this characterization and discuss its potential N-dimensional counterpart for self-similar sets; see also Problems 6.2.35 and 6.2.36, along with the text surrounding them.)

Beside the obvious difficulty of computing the distance (or another fractal) zeta function of a (suitable) 'self-similar RFD' in $\mathbb{R}^N$ (and, in particular, of a compact self-similar set in $\mathbb{R}^N$ satisfying the open set condition, say), an important remaining issue is to remove (as was done in [Lap-vFr3, Section 8.4] when $N = 1$) the condition concerning the existence of an appropriate screen $\boldsymbol{S}$ (as is assumed in Theorem 5.4.20, as well as in Theorem 5.4.25 and Corollary 5.4.26 below). Indeed, in the lattice case, this condition is clearly always satisfied as long as the 'base RFD'

[24] Namely, a self-similar string (of Minkowski dimension $D \in (0,1)$) is (strongly) Minkowski measurable if and only if it is a nonlattice string (i.e., the multiplicative group generated by its distinct scaling ratios is not of rank 1) or, equivalently, if its Minkowski dimension D is the only complex dimension of real part D. (It is known from [Lap-vFr3, Theorems 2.16 and 3.6]) that for a self-similar string, D is always simple.)

is 'nice enough' (see Theorem 4.2.17 where the distance zeta function for 'self-similar RFDs' was computed). On the other hand, it is shown in [Lap-vFr3] that in the nonlattice case, there are examples of nonlattice self-similar strings (and hence, of 'self-similar RFDs' and sets in $\mathbb{R}^N$) for which it is not fulfilled. (See [Lap-vFr3, Example 5.32] showing that for a given self-similar string, it is not always possible to choose a screen $\boldsymbol{S}$ passing strictly between the critical line $\{\mathrm{Re}\, s = D\}$ and the complex dimensions to the left of this line and along which the RFD is languid.)

However, we also stress that this issue regarding the nonlattice case is, a priori, occurring only in one direction of the desired (Minkowski measurability) characterization theorem; that is, in the direction which aims at proving that, under suitable hypotheses, a Minkowski measurable 'self-similar RFD' (with a 'nice enough' base RFD or generator) is always nonlattice.[25] Indeed, for the other direction, we do not need the restrictive assumption about the existence of a suitable screen along which the RFD is languid since the desired Minkowski measurability conclusion should follow from the sufficient condition provided in Theorem 5.4.2. More specifically, in light of Theorem 4.2.17, and under appropriate hypotheses,[26] we expect to draw the conclusion that a nonlattice self-similar RFD is always Minkowski measurable since the only pole on the critical line is its Minkowski dimension (which is equal to the maximum of the inner Minkowski dimension of the boundary of the generator and the unique real solution of their associated complexified Moran equation, see Equation (5.5.186) below), and it is simple. Of course, the above potential "theorem" should be more precisely stated, with the expressions of 'self-similar RFD', 'nice enough' and 'base RFD' (or 'generator') being unambiguously specified. We leave this task for a future work. (See also the end of Subsection 5.5.6 below where these issues are addressed and essentially resolved in the important special case of self-similar sprays, under mild assumptions.)

In the next result (Corollary 5.4.23), which follows from a combination of Theorem 5.4.2 and Theorem 5.4.20, we recover the aforementioned characterization of the Minkowski measurability of self-similar fractal strings (with possibly multiple gaps,[27] in the sense of [Lap-vFr3, Chapters 2 and 3]), obtained in [Lap-vFr3, Sec-

25 Actually, this is only a problem at first glance. In fact, a moment's reflection shows that it suffices to reason by contradiction in order to resolve this problem. Indeed, in the lattice case, we can always find a suitable screen satisfying the required hypotheses of Theorem 5.4.20 and, consequently, conclude that a lattice RFD is not Minkowski measurable and reach a contradiction after an application of Theorem 5.4.20. We will proceed exactly in this manner in the proof of Corollary 5.4.23 (where $N = 1$), as well as in Subsection 5.5.6, Remark 5.5.26(c), when dealing in a similar manner with (higher-dimensional) self-similar sprays.

26 In particular, we assume that the generators of the associated self-similar tilings (or sprays) are pluriphase (in the sense of [LapPe3] and [LapPeWi1], along with [Lap-vFr3, Section13.1]) and (as will be done in Subsection 5.5.6 for self-similar sprays in case (i) of part (c) and part (a) of Remark 5.5.26) that the Minkowski dimension of their boundary is strictly smaller than their similarity dimension (or, equivalently, than the similarity dimension of the associated self-similar tiling or spray).

27 We note that the general case of multiple gaps precisely corresponds to the general case of multiple generators for self-similar sprays; see, especially, part (b) of Remark 5.5.26 in Subsection 5.5.6.

tion 8.4, esp., Theorems 8.23, 8.25 and 8.36, along with Corollary 8.27]. In the proof of the next corollary, we will use the known fact (established in [Lap-vFr3, Theorems 2.16 and 3.6]) that a self-similar string is nonlattice if and only if its only scaling complex dimension (i.e., the only pole of its geometric or scaling zeta function, introduced in Remark 5.5.20 of Subsection 5.5.6) with real part σ_0 (the similarity dimension of the string) is σ_0 itself. Note that this last statement also uses the fact (established in a part of [Lap-vFr3, Corollary 8.27]) according to which a lattice self-similar string (with multiple gaps) has infinitely many principal scaling complex dimensions (i.e., potential poles of the geometric or scaling zeta function with real part σ_0 and with nonzero residues) of the form $\sigma_0 + \mathbb{i}k\mathbf{p}$, where $\mathbf{p} := 2\pi/\log r^{-1}$, and $r \in (0,1)$ is the single generator of the multiplicative group (of rank one) generated by the underlying distinct scaling ratios; therefore, it has at least one nonreal principal complex dimension. (This latter fact is easy to check in the case of a single gap; then, the set of principal scaling complex dimensions is all of $\sigma_0 + \mathbf{p}\mathbb{i}\mathbb{Z}$.) Finally, we recall from [Lap-vFr3, Chapters 2 and 3] that σ_0 (and hence, each of the other principal scaling complex dimensions) is always simple, either in the lattice case or in the nonlattice case. This latter fact is also easy to check directly from the definitions.

In the following corollary of Theorem 5.4.20 (combined with the aforementioned results in [Lap-vFr3, Chapters 2 and 3]), Ω denotes an arbitrary geometric realization of a (nontrivial) bounded self-similar fractal string $\mathcal{L} := (\ell_j)_{j=1}^{\infty}$, as a bounded open subset of $\mathbb{R}$; see Subsection 2.1.4. Furthermore, $\partial\Omega$ denotes its boundary (in $\mathbb{R}$) and $(\partial\Omega,\Omega)$ is the associated relative fractal drum (or RFD) in $\mathbb{R}$. We note that in [Lap-vFr3], the term RFD (or 'relative fractal drum') was not used but that an equivalent notion was used instead in the present situation of fractal strings.

Corollary 5.4.23 (Characterization of the Minkowski measurability of self-similar strings, [Lap-vFr3, Section 8.4]**).** *Let $(\partial\Omega,\Omega)$ (or $\mathcal{L}$) be a (nontrivial, bounded) self-similar fractal string, with (upper) Minkowski dimension $D < 1$; so that $D = \sigma_0$, its similarity dimension. Then the following statements are equivalent*:

(i) The RFD $(\partial\Omega,\Omega)$ is Minkowski measurable.

(ii) The self-similar string $\mathcal{L}$ (or, equivalently, the self-similar RFD $(\partial\Omega,\Omega)$) is nonlattice.

(iii) The only principal scaling complex dimension of $(\partial\Omega,\Omega)$ is $D = \sigma_0$.

Proof. We already know from the discussion preceding the statement of the corollary that (ii) and (iii) are equivalent, based on the results of [Lap-vFr3, Chapters 2, 3 and 8]; see, especially, [Lap-vFr3, Theorems 8.23 and 8.36].

Next, we show that (i) and (iii) are equivalent. Note that since $D = \sigma_0$ and $\sigma_0 > 0$, we have that $D \in (0,1)$.

First, observe that (since σ_0 is always simple) the fact that (iii) implies (i) follows from Theorem 5.4.2, the sufficient condition for the Minkowski measurability of an RFD. Observe that in order to verify that the hypotheses of Theorem 5.4.2 are satisfied by the RFD $(\partial\Omega,\Omega)$ and its distance zeta function $\zeta_{\partial\Omega,\Omega}$, we use the fact that the scaling (and thus, geometric) zeta function $\zeta_{\mathfrak{S}}$ of a self-similar string is strongly languid with exponent $\kappa := 0$ (and hence, also for any $\kappa \geq 0$), as is shown in [Lap-vFr3, Section 6.4, just above Remark 6.12], combined with the key relation

$$\zeta_{\partial\Omega,\Omega}(s) = \frac{2^{1-s}\zeta_{\mathscr{L}}(s)}{s}. \tag{5.4.53}$$

See Equation (5.5.16) in Subsection 5.5.2 below, where it is proved for any fractal string $\mathscr{L}$; here, $\zeta_{\mathscr{L}} = \zeta_{\mathfrak{S}}$. Therefore, the distance zeta function $\zeta_{\partial\Omega,\Omega}$ is strongly d-languid with exponent $\kappa_d := -1$. Consequently (and assuming that (*iii*) holds), the hypotheses of Theorem 5.4.2 are satisfied and so, it follows that $(\partial\Omega,\Omega)$ is Minkowski measurable; i.e., (*i*) holds, as desired.

Now, all that remains to show is that (*i*) implies (*iii*). More explicitly, we need to show that the fact that $(\partial\Omega,\Omega)$ is Minkowski measurable, implies that $\mathscr{L}$ (or, equivalently, $(\partial\Omega,\Omega)$) is nonlattice. For this purpose, we reason by contradiction. Namely, we assume that (*i*) holds (i.e., $(\partial\Omega,\Omega)$ is Minkowski measurable) but that $\mathscr{L}$ is a *lattice* (self-similar) string. Since $\mathscr{L}$ is lattice,[28] its scaling complex dimensions are located (and periodically distributed) on finitely many vertical lines (possibly on a single such line), the right most of which is the vertical line $\{\mathrm{Re}\, s = \sigma_0\}$, the critical line (since $\sigma_0 = D$). Therefore, we can obviously choose, as is required by the hypotheses of Theorem 5.4.20 (the Minkowski measurability criterion), a screen $\boldsymbol{S}$ passing strictly between the vertical line $\{\mathrm{Re}\, s = \sigma_0 = D\}$ and all the complex dimensions (i.e., the poles of $\zeta_{\partial\Omega,\Omega}$) with real part strictly less than $D = \sigma_0$. In light of Equation (5.4.53), it suffices to let S be any vertical line $\{\mathrm{Re}\, s = \Theta\}$, where $\Theta \in (\max\{0,\sigma_1\},\sigma_0)$ and σ_1 is the abscissa of the second to last (right most) vertical line on which the scaling complex dimensions of $\mathscr{L}$ (or of $(\partial\Omega,\Omega)$) are located. (If σ_1 does not exist, then we can choose any $\Theta \in (0,\sigma_0)$.) The fact that the strong d-languidity assumption is satisfied by $(\partial\Omega,\Omega)$ is explained in the previous step of the proof. The corresponding argument is valid for any self-similar string.

We may therefore apply Theorem 5.4.20 and deduce from the fact that the RFD $(\partial\Omega,\Omega)$ is Minkowski measurable that $D = \sigma_0$ must be its only complex dimension of real part $D = \sigma_0$.[29] This contradicts the fact that $\mathscr{L}$ is a lattice string, and hence has infinitely many (and thus at least two complex conjugate) nonreal principal scaling complex dimensions. We deduce from this contradiction that $\mathscr{L}$ must be a nonlattice string (i.e., (*ii*) holds) and hence (since (*ii*) and (*iii*) are equivalent), that (*iii*) holds, as desired.

This concludes the proof of the corollary. □

Remark 5.4.24. In Subsection 5.5.6, by using an analogous method, we will extend Corollary 5.4.23 to higher dimensions, that is, to a large class of self-similar sprays in $\mathbb{R}^N$, with $N \geq 1$ arbitrary. In the general case (and under some mild assumptions), Minkowski measurability will have to be replaced by 'possibly subcritical

[28] In the case of a self-similar string with multiple gaps, this means that the multiplicative group generated by both the distinct scaling ratios and gaps is of rank 1; see [Lap-vFr3, Chapters 2 and 3].
[29] Note that, in light of (5.4.53) and since $\sigma_0 > 0$, it follows that the principal complex dimensions of $(\partial\Omega,\Omega)$ and the principal scaling complex dimensions coincide: $\mathscr{P}_c(\zeta_{\partial\Omega,\Omega}) = \mathscr{P}_c(\zeta_{\mathscr{L}})$.

Minkowski measurability', in a sense to be explained there. (See especially, part (c) of Remark 5.5.26.)

Next, we give the counterpart of Theorem 5.4.20, but now expressed in terms of the tube zeta function $\tilde{\zeta}_{A,\Omega}$ (instead of the distance zeta function $\zeta_{A,\Omega}$) and with the restriction $\dim_B(A,\Omega) < N$ removed in this case.

Theorem 5.4.25 (Minkowski measurability criterion in terms of $\tilde{\zeta}_{A,\Omega}$). *Let (A,Ω) be a relative fractal drum in $\mathbb{R}^N$ such that $D := \dim_B(A,\Omega)$ exists. Furthermore, assume that (A,Ω) is languid for a screen $\boldsymbol{S}$ passing strictly to the left of the critical line $\{\mathrm{Re}\, s = D\}$ and strictly to the right of all the complex dimensions of (A,Ω) with real part strictly less than D. Then the following statements are equivalent:*

(a) The RFD (A,Ω) is (strongly) Minkowski measurable.

(b) D is the only pole of the relative tube zeta function $\tilde{\zeta}_{A,\Omega}$ located on the critical line $\{\mathrm{Re}\, s = D\}$, and it is simple.

Proof. First of all, if $D = \dim_B(A,\Omega) < N$, then, again, the conclusion of the theorem follows from Theorems 5.4.2 and 5.4.15 together with Remark 5.4.16.

In the case when $D = N$, we will embed (A,Ω) into $\mathbb{R}^{N+1}$, as was done in Subsection 4.7.2, and then use Theorem 4.7.9. The fact that (b) implies (a) is a consequence of Theorem 5.4.2 since there are no restrictions of the type $\dim_B(A,\Omega) < N$ in the hypotheses of that theorem. Actually, it follows directly from the definition of the relative Minkowski content that $\dim_B(A,\Omega) = N$ implies that $\mathscr{M}^N(A,\Omega)$ exists and $\mathscr{M}^N(A,\Omega) = |\overline{A} \cap \Omega|$.

In order to prove that (a) implies (b), we embed (A,Ω) into $\mathbb{R}^{N+1}$ as $(A,\Omega)_1 := (A \times \{0\}, \Omega \times (-\delta,\delta))$, for some suitable $\delta > 0$, and conclude from Theorem 4.7.9 that the relative tube zeta functions of the RFDs (A,Ω) and $(A,\Omega)_1$ are connected by the approximate functional equation (4.7.26) from Theorem 4.7.9 (for $M = 1$); that is,

$$\tilde{\zeta}_{A\times\{0\},\Omega\times(-\delta,\delta)}(s;\delta) = \frac{\sqrt{\pi}\,\Gamma\left(\frac{N-s}{2}+1\right)}{\Gamma\left(\frac{N+1-s}{2}+1\right)}\tilde{\zeta}_{A,\Omega}(s;\delta) + E(s;\delta). \tag{5.4.54}$$

Here, $\delta > 0$ is chosen such that $\tilde{\zeta}_{A,\Omega}(\,\cdot\,;\delta)$ satisfies the languidity hypothesis of the theorem. We will now show that $\tilde{\zeta}_{A\times\{0\},\Omega\times(-\delta,\delta)}(\,\cdot\,;\delta)$ satisfies the needed languidity conditions of Definition 5.1.3 and use Theorem 5.4.15 to conclude the proof. The error function $E(\,\cdot\,;\delta)$ is holomorphic on the open left half-plane $\{\mathrm{Re}\, s < N+1\}$ and bounded by $2\delta^{\mathrm{Re}\, s-N}|A_\delta \cap \Omega|_N \left(\frac{\pi}{2}-1\right)$ (see the proof of Theorem 4.7.2 and Equation (4.7.8)). In other words, $E(\,\cdot\,;\delta)$ is languid (with a languidity exponent equal to 0). Furthermore, for any $a,b \in \mathbb{C}$ such that $\mathrm{Re}(b-a) > 0$, we have the following pointwise asymptotic expansion:

$$\frac{\Gamma(z+a)}{\Gamma(z+b)} \sim z^{a-b}\sum_{n=0}^{\infty}\frac{(-1)^n B_n^{(a-b+1)}(a)}{n!}\frac{\Gamma(b-a+n)}{\Gamma(b-a)}\frac{1}{z^n} \quad \text{as} \quad |z| \to +\infty, \tag{5.4.55}$$

in the sector $|\arg z| < \pi$.[30] Substituting $z := \frac{N-s}{2}+1$, $a := 0$ and $b := 1/2$ into Equation (5.4.55), we obtain that

$$\frac{\Gamma\left(\frac{N-s}{2}+1\right)}{\Gamma\left(\frac{N+1-s}{2}+1\right)} \sim (N-s+2)^{-\frac{1}{2}} \sum_{n=0}^{\infty} \frac{(2n)!\sqrt{2}(-1)^n B_n^{(1/2)}(0)}{2^n (n!)^2 (N-s+2)^n} \quad \text{as} \quad |s| \to +\infty, \tag{5.4.56}$$

for all $s \in \mathbb{C} \setminus [N+2,+\infty)$.[31] In particular, we have that

$$\frac{\Gamma\left(\frac{N-s}{2}+1\right)}{\Gamma\left(\frac{N+1-s}{2}+1\right)} = O(|s|^{-1/2}) \quad \text{as} \quad |s| \to +\infty, \tag{5.4.57}$$

for all $s \in \mathbb{C} \setminus [N+2,+\infty)$, from which we conclude that the product of this ratio of gamma functions with the relative tube zeta function $\tilde{\zeta}_{A,\Omega}(\,\cdot\,;\delta)$ is languid with a languidity exponent no greater than $\kappa - 1/2$, where κ is the languidity exponent of $\tilde{\zeta}_{A,\Omega}(\,\cdot\,;\delta)$. This fact, along with Equation (5.4.54) and the languidity of $E(\,\cdot\,;\delta)$, implies that $\tilde{\zeta}_{A\times\{0\},\Omega\times(-\delta,\delta)}(\,\cdot\,;\delta)$ is languid with the same choice of a double sequence $(T_n)_{n\in\mathbb{Z}\setminus\{0\}}$ and the screen $\boldsymbol{S}$ as for $\tilde{\zeta}_{A,\Omega}(\,\cdot\,;\delta)$ and with a languidity exponent no greater than $\max\{\kappa - 1/2, 0\}$.

On the other hand, if (A,Ω) is Minkowski measurable, then this is also true for the embedded RFD $(A\times\{0\},\Omega\times(-\delta,\delta))$. In light of Lemma 4.7.5, this fact follows in a completely analogous way as in the case of bounded subsets of $\mathbb{R}^N$ which was proven in [Kne] (see also [Res]) and extended to RFDs in Subsection 4.7.2 above. We now conclude the proof by invoking Theorem 5.4.15, or rather, its counterpart expressed in terms of the relative tube zeta function (see Remark 5.4.16). □

In the next corollary of Theorem 5.4.20 and 5.4.25, and in light of Lemma 5.4.11 and Remark 5.4.16, we can indifferently interpret the (principal) complex dimensions of the RFD (A,Ω) as being the (principal) poles of either the distance, tube, shell or Mellin zeta function of (A,Ω). This is the reason why we assume that the hypotheses of both Theorems 5.4.20 and 5.4.25 are satisfied; i.e., we assume that $\dim_B(A,\Omega) < N$ in order to avoid the situation when N is a pole of the tube zeta function but is not a pole of the distance zeta function, which may happen.

Corollary 5.4.26 (Characterization of Minkowski measurability in terms of the complex dimensions). *Let (A,Ω) be a relative fractal drum in $\mathbb{R}^N$, with $N \geq 1$ arbitrary, such that $D := \dim_B(A,\Omega)$ exists and $D < N$. Assume also that any of its fractal zeta functions (specifically, $\zeta_{A,\Omega}$ or $\tilde{\zeta}_{A,\Omega}$, respectively) satisfies the hypotheses of Theorem 5.4.20 (or of Theorem 5.4.25, respectively) concerning the languidity and the screen. Then, the following statements are equivalent:*

---

[30] Here, $B_n^{(\sigma)}(x)$ is the n-th generalized Bernoulli polynomial (see, e.g., [SriTod] for the exact definition and an explicit expression). See also [Tem, Subsection 3.6.2] for this result on the asymptotics of ratios of gamma functions.

[31] We have used here the classic identity $\Gamma(1/2) = \sqrt{\pi}$ and, more generally, $\Gamma(1/2+n) = \frac{(2n)!}{4^n n!}\sqrt{\pi}$, for every $n \in \mathbb{N}_0$.

(a) The RFD (A,Ω) is (strongly) Minkowski measurable.

(b) D is the only complex dimension of the RFD (A,Ω) with real part equal to D (i.e., located on the critical line $\{\operatorname{Re} s = D\}$), and it is simple.

5.4.4 h-Minkowski Measurability and Optimal Tube Function Asymptotic Expansion

In this subsection, we first obtain a very general result (Theorem 5.4.27) about generating h-Minkowski measurable RFDs, where $h(t) := (\log t^{-1})^{m-1}$ for all $t \in (0,1)$ and m is a positive integer, by using only some information about the principal poles and their multiplicities. Its proof rests on the use of the pointwise tube formula (Theorem 5.1.13). Theorem 5.4.27 is in fact a partial converse of Theorem 4.5.1. Especially important is the asymptotic expansion of the tube function stated in Equation (5.4.60), from which it is possible to deduce the optimal tube function asymptotic expansion for a class of h-Minkowski measurable RFDs, as stated in Theorem 5.4.29.

We invite the reader to review the definition of a gauge function h provided in Subsection 4.5.1 above (in the text surrounding Equation (4.5.10)) as well as, in particular, of the corresponding notion of Minkowski h-measurability. (See also Section 6.1 below, with the obvious changes in notation, such as the bounded set A being replaced by the RFD (A,Ω).) The notion of Minkowski h-measurability is motivated, geometrically and physically, by the study of non power law scaling behavior which arises in many natural examples. See [HeLap] and the relevant references therein.

Theorem 5.4.27 (Generating h-Minkowski measurable RFDs). *Let (A,Ω) be a relative fractal drum in $\mathbb{R}^N$ which is languid with languidity exponent $\kappa < -1$ or such that $(\lambda A, \lambda\Omega)$ is strongly languid for some $\lambda > 0$ with languidity exponent $\kappa < 0$, for a screen $\boldsymbol{S}$ passing strictly between the critical line $\{\operatorname{Re} s = \overline{\dim}_B(A,\Omega)\}$ and all the complex dimensions of (A,Ω) with real part strictly less than $\overline{D} := \overline{\dim}_B(A,\Omega)$. Furthermore, suppose that $\overline{D}$ is the only pole of its relative tube zeta function $\tilde{\zeta}_{A,\Omega}$ with real part equal to $\overline{D}$ of order $m \geq 1$ and, additionally, that there exists (at most) finitely many nonreal poles of $\tilde{\zeta}_{A,\Omega}$ with real part $\overline{D}$. Moreover, assume that the multiplicity of each of those nonreal poles is of order strictly less than m. Then, $\dim_B(A,\Omega)$ exists and is equal to $D := \overline{D}$. Moreover, $\mathcal{M}^D(A,\Omega)$ exists and is equal to $+\infty$; hence, (A,Ω), is Minkowski degenerate.*

Finally, an appropriate gauge function for (A,Ω) is $h(t) := (\log t^{-1})^{m-1}$, for all $t \in (0,1)$, and we have that, relative to *h, the RFD (A,Ω) is not only Minkowski nondegenerate but is also Minkowski measurable with Minkowski content given by*

$$\mathcal{M} := \mathcal{M}^D(A,\Omega,h) = \frac{\tilde{\zeta}_{A,\Omega}[D]_{-m}}{(m-1)!}, \tag{5.4.58}$$

where $\tilde{\zeta}_{A,\Omega}[D]_{-m}$ *denotes the coefficient corresponding to* $(s-D)^{-m}$ *in the Laurent expansion of* $\tilde{\zeta}_{A,\Omega}$ *around* $s=D$. *Moreover, if there is at least one nonreal complex dimension on the critical line* $\{\mathrm{Re}=D\}$, *then the tube function* $t\mapsto|A_t\cap\Omega|$ *has the following pointwise asymptotic estimate:*

$$|A_t\cap\Omega| = t^{N-D}h(t)\big(\mathscr{M}+O((\log t^{-1})^{-1})\big) \quad \text{as} \quad t\to 0^+, \tag{5.4.59}$$

while if D *is the unique pole of* $\tilde{\zeta}_{A,\Omega}$ *on the critical line (i.e., the unique principal complex dimension of* (A,Ω)*), we have the following sharper asymptotic estimate:*

$$|A_t\cap\Omega| = t^{N-D}h(t)\big(\mathscr{M}+O(t^{D-\sup S})\big) \quad \text{as} \quad t\to 0^+. \tag{5.4.60}$$

Proof. Let $\omega_0:=\overline{D}:=\overline{\dim}_B(A,\Omega)$,[32] and let $\omega_j:=\overline{D}+\mathrm{i}\gamma_j$, where $\gamma_j\in\mathbb{R}\setminus\{0\}$ for $j\in J$ and J is a finite and symmetric subset of $\mathbb{Z}\setminus\{0\}$. That is, $\{\omega_j\}_{j\in J}$ is the (finite) set of all the other poles of $\tilde{\zeta}_{A,\Omega}$ located on the critical line $\{\mathrm{Re}\,s=\overline{D}\}$, i.e., with real part $\overline{D}$. We also let $\gamma_0:=0$ and $m_0:=m$, in order to be consistent with the notation introduced just below. Furthermore, for each $j\in J$, let m_j be the multiplicity of ω_j and then, by hypothesis of the theorem, we have that $m_j<m$ for every $j\in J$. By Theorem 5.1.13 and since the screen $\boldsymbol{S}$ is strictly to the right of all the other complex dimensions of (A,Ω) with real part strictly less than $\overline{D}$ and strictly to the left of the critical line $\{\mathrm{Re}\,s=\overline{D}\}$, we obtain a pointwise tube formula for (A,Ω) with an error term which is of strictly higher asymptotic order as $t\to 0^+$ than the term corresponding to the residue at $s=\overline{D}$; that is, we have the following pointwise tube formula with error term:

$$|A_t\cap\Omega| = \sum_{j\in J\cup\{0\}} \mathrm{res}(t^{N-s}\tilde{\zeta}_{A,\Omega}(s),\omega_j)+O(t^{N-\sup S}) \quad \text{as}\ t\to 0^+. \tag{5.4.61}$$

We next consider the Taylor expansion of t^{N-s} around $s=\omega_j$ (for each $j\in J\cup\{0\}$):

$$t^{N-s}=t^{N-\omega_j}\,\mathrm{e}^{(s-\omega_j)\log t^{-1}}=t^{N-\omega_j}\sum_{n=0}^{\infty}\frac{(\log t^{-1})^n}{n!}(s-\omega_j)^n; \tag{5.4.62}$$

we then multiply it by the Laurent expansion of $\tilde{\zeta}_{A,\Omega}(s)$ around $s=\omega_j$ and extract the residue of this product in order to deduce that

$$\mathrm{res}(t^{N-s}\tilde{\zeta}_{A,\Omega}(s),\omega_j)=t^{N-\omega_j}\sum_{n=0}^{m_j-1}\frac{(\log t^{-1})^n}{n!}\tilde{\zeta}_{A,\Omega}[\omega_j]_{-n-1}. \tag{5.4.63}$$

In light of this identity and of (5.4.61), we conclude that $\dim_B(A,\Omega)$ exists and is equal to $\overline{D}$. Furthermore, since $m_j<m_0=m$, we conclude that the highest power of $\log t^{-1}$ appearing in the fractal tube formula (5.4.61) is $m-1$, and that it appears only in the sum (5.4.63) when $j=0$. Therefore, if we choose $h(t):=(\log t^{-1})^{m-1}$

[32] It will follow from the proof that $\dim_B(A,\Omega)$ exists and that $\overline{D}=\dim_B(A,\Omega)$ and hence, is equal to D.

for all $t \in (0,1)$ as our gauge function, the statements about the Minkowski content and the gauge Minkowski content (in the usual sense as well as with respect to h) now also follow from the fractal tube formula (5.4.61).

We easily deduce Equations (5.4.59) and (5.4.60) from (5.4.61) by rewriting (5.4.63) as follows:

$$\operatorname{res}(t^{N-s}\tilde{\zeta}_{A,\Omega}(s),\omega_j) = t^{N-D}h(t)\sum_{n=0}^{m_j-1} t^{D-\omega_j}\frac{(\log t^{-1})^{n-m+1}}{n!}\tilde{\zeta}_{A,\Omega}[\omega_j]_{-n-1}. \tag{5.4.64}$$

Indeed, for $j=0$, we have that $m_0 = m$ and $\omega_0 = D$; so that the term on the right-hand side of (5.4.64) corresponding to $n = m-1$ is equal to $\frac{\tilde{\zeta}_{A,\Omega}[\omega_j]_{-m}}{(m-1)!}$ (i.e., to $\mathscr{M}$, the h-Minkowski content of (A,Ω); see Equation (5.4.58) above), while for any $n \in \{0,\dots,m-2\}$ (if this set is nonempty, i.e., if $m \geq 2$), we are left with a function which is $O((\log t^{-1})^{-1})$ as $t \to 0^+$ (if $m=1$, the corresponding function is absent; i.e., it is equal to zero). Equations (5.4.59) and (5.4.60) then follow because for $j \neq 0$ ($j \in J$), we have that $|t^{D-\omega_j}| = 1$ (since $D - \omega_j$ is a purely imaginary complex number) and $(\log t^{-1})^{n-m+1} = O((\log t^{-1})^{-1})$ as $t \to 0^+$.

This concludes the proof of the theorem. □

Remark 5.4.28. In light of the proof of Theorem 5.4.27, the error term $O((\log t^{-1})^{-1})$ in Equation (5.4.59) can be slightly improved to $O((\log t^{-1})^{n-m+1})$, where n is the largest positive integer such that $n < m-1$ and for which there exists $j \in J$ such that $\tilde{\zeta}_{A,\Omega}[\omega_j]_{-n-1} \neq 0$.

The further to the left we can meromophically extend the tube zeta function $\tilde{\zeta}_{A,\Omega}$ of a given RFD (A,Ω) (meaning, the smaller the value of $\sup S$, for a given screen $\boldsymbol{S}$ relative to which $\tilde{\zeta}_{A,\Omega}$ is admissible), the sharper the estimate (5.4.60) in Theorem 5.4.27. Therefore, if we denote by $\mathscr{S}(A,\Omega)$ the family of all possible screens $\boldsymbol{S}$ such that the tube zeta function $\tilde{\zeta}_{A,\Omega}$ admits a meromorphic extension to a connected open neighborhood of the corresponding window $\boldsymbol{W} = \boldsymbol{W}(\boldsymbol{S})$, it is natural to define the following (extended) real number:

$$\operatorname{or}(A,\Omega) := \inf_{S \in \mathscr{S}(A,\Omega)} \sup S \in [-\infty, \overline{\dim}_B(A,\Omega)], \tag{5.4.65}$$

which we call the *order* of the RFD (A,Ω). It is clear that

$$\operatorname{or}(A,\Omega) \leq D_{\text{mer}}(\tilde{\zeta}_{A,\Omega}), \tag{5.4.66}$$

where $D_{\text{mer}}(\tilde{\zeta}_{A,\Omega})$ is the abscissa of meromorphic continuation of the tube zeta function $\tilde{\zeta}_{A,\Omega}$. In light of Equation (5.4.60) in Theorem 5.4.27, we can then deduce the following significant conclusion.

Theorem 5.4.29 (Optimal tube function asymptotic expansion). *Let (A,Ω) be a relative fractal drum such that the conditions of Theorem 5.4.27 are satisfied, with $D := D(\tilde{\zeta}_{A,\Omega})$ being the unique pole of $\tilde{\zeta}_{A,\Omega}$ in the open right half-plane*

$\{\operatorname{Re} s > D_{\text{mer}}(\tilde{\zeta}_{A,\Omega})\}$, *of order* $m \geq 1$. *Then,* (A,Ω) *is h-Minkowski measurable with h-Minkowski content* $\mathscr{M}(A,\Omega,h) = \mathscr{M}$. *Furthermore, for any positive real number* ε, *the tube function* $t \mapsto |A_t \cap \Omega|$ *has the following pointwise asymptotic expansion, with error term*:

$$|A_t \cap \Omega| = t^{N-D} h(t) \big(\mathscr{M} + O(t^{D - D_{\text{mer}}(\tilde{\zeta}_{A,\Omega}) - \varepsilon})\big) \quad \text{as} \quad t \to 0^+, \tag{5.4.67}$$

where $h(t) := (\log t^{-1})^{m-1}$ *for all* $t \in (0,1)$.

Moreover, the asymptotic formula in Equation (5.4.67) *is optimal; that is, the exponent* $D - D_{\text{mer}}(\tilde{\zeta}_{A,\Omega})$ *appearing in the error term cannot be replaced by a larger number. In addition, the order of the RFD* (A,Ω) *is equal to the abscisssa of meromorphic continuation of the corresponding tube zeta function* $\tilde{\zeta}_{A,\Omega}$*; i.e.,*

$$\operatorname{or}(A,\Omega) = D_{\text{mer}}(\tilde{\zeta}_{A,\Omega}). \tag{5.4.68}$$

Proof. First of all, by using Theorem 5.4.27 and since for any $\varepsilon > 0$ there exists $S \in \mathscr{S}$ such that $\sup S > \operatorname{or}(A,\Omega) + \varepsilon$, we have that

$$|A_t \cap \Omega| = t^{N-D} h(t) \big(\mathscr{M} + O(t^{D - \operatorname{or}(A,\Omega) - \varepsilon})\big) \quad \text{as} \quad t \to 0^+. \tag{5.4.69}$$

Equation (5.4.67) then follows from (5.4.66).

In order to establish the *optimality* of the asymptotic formula in Equation (5.4.67), we reason by contradiction. Assume that we have

$$|A_t \cap \Omega| = t^{N-D} h(t) \big(\mathscr{M} + O(t^{\alpha})\big) \quad \text{as} \quad t \to 0^+, \tag{5.4.70}$$

for some real number

$$\alpha > D - D_{\text{mer}}(\tilde{\zeta}_{A,\Omega}). \tag{5.4.71}$$

(Here, in the statement of (5.4.71), we can omit ε since without loss of generality, we can always choose a smaller real number α satisfying the same strict inequality.) In light of Theorem 4.5.1 (applied with $m-1$ instead of m), we conclude that $\tilde{\zeta}_{A,\Omega}$ can be meromophically extended (at least) to the open right half-plane $\{\operatorname{Re} s > D - \alpha\}$. It then follows from the definition of the abscissa of meromorphic continuation that $D_{\text{mer}}(\tilde{\zeta}_{A,\Omega}) \leq D - \alpha$. On the other hand, we deduce from (5.4.71) that $D - \alpha < D_{\text{mer}}(\tilde{\zeta}_{A,\Omega})$, which contradicts the previous inequality.

Finally, in order to prove Equation (5.4.68), it suffices to note that, due to the optimality proved just above, we must have that $D - \operatorname{or}(A,\Omega) \leq D - D_{\text{mer}}(\tilde{\zeta}_{A,\Omega})$ (see Equations (5.4.69) and (5.4.67)); i.e., $\operatorname{or}(A,\Omega) \geq D_{\text{mer}}(\tilde{\zeta}_{A,\Omega})$, which together with (5.4.66), implies the desired equality (5.4.68). This completes the proof of the theorem. ☐

In other words, assuming that the conditions of Theorem 5.4.29 are satisfied, we have shown that the larger the difference

$$\alpha(A,\Omega) := D(\tilde{\zeta}_{A,\Omega}) - D_{\text{mer}}(\tilde{\zeta}_{A,\Omega}) \tag{5.4.72}$$

between the abscissa of (absolute) convergence and the abscissa of meromorphic continuation of the tube zeta function $\tilde{\zeta}_{A,\Omega}$ of a given RFD (A,Ω), the better the asymptotic estimate (5.4.67) of the tube function $t \mapsto |A_t \cap \Omega|$ when $t \to 0^+$.

Our next result, Theorem 5.4.30, can be viewed as the converse of Theorem 4.5.1, about the existence of meromorphic extensions of the tube (or distance) zeta functions of suitable Minkowski measurable RFDs. In addition, *it shows that, in some precise sense, these results are optimal.*

Theorem 5.4.30. *Let (A,Ω) be an RFD such that the conditions of Theorem 5.4.27 are satisfied. Furthermore, assume that there exists a positive real number α such that the relative tube zeta function $\tilde{\zeta}_{A,\Omega}$ can be meromorphically extended to the open right half-plane $\{\mathrm{Re}\, s > D-\alpha\}$, with $D := D(\tilde{\zeta}_{A,\Omega})$ being the unique pole of $\tilde{\zeta}_{A,\Omega}$ in this right half-plane, of order $m \geq 1$. Then, the tube function $t \mapsto |A_t \cap \Omega|$ has the following pointwise asymptotic expansion, with error term of order α:*

$$|A_t \cap \Omega| = t^{N-D} h(t)\big(\mathscr{M} + O(t^{\alpha})\big) \quad \text{as} \quad t \to 0^+, \tag{5.4.73}$$

where the gauge function h is given by $h(t) := (\log t^{-1})^{m-1}$ for all $t \in (0,1)$ and $\mathscr{M} = \mathscr{M}(A,\Omega,h)$ is the h-Minkowski content of (A,Ω).

Moreover, if we let

$$r(A,\Omega) := \sup\{\mathrm{Re}\, s : s \in \mathscr{P}(\tilde{\zeta}_{A,\Omega}) \setminus \{D\}\}, \tag{5.4.74}$$

then the tube function $t \mapsto |A_t \cap \Omega|$ has the following poinntwise asymptotic expansion, with error term, for any positive real number ε:

$$|A_t \cap \Omega| = t^{N-D} h(t)\big(\mathscr{M} + O(t^{D-r(A,\Omega)-\varepsilon})\big) \quad \text{as} \quad t \to 0^+, \tag{5.4.75}$$

Proof. Equation (5.4.73) follows from Equation (5.4.59) of Theorem 5.4.27, by choosing the screen $\boldsymbol{S}$ to be the vertical line $\{\mathrm{Re}\, s = D-\alpha\}$; that is, $S(x) := D-\alpha$ for all $x \in \mathbb{R}$. Indeed, in this case, we have $D - \sup S = D-(D-\alpha) = \alpha$.

Equation (5.4.75) follows easily from (5.4.73) by letting $\alpha := D - r(A,\Omega) - \varepsilon$, for $\varepsilon > 0$ small enough. Indeed, for such an ε, $s = D$ is the only pole in the open right half-plane $\{\mathrm{Re}\, s > r(A,\Omega)+\varepsilon = D-\alpha\}$. □

In light of the functional equation (4.5.2) connecting the relative distance zeta function with the relative tube zeta function, it is clear that when $\overline{\dim}_B A < N$, the value of the order $\alpha(A,\Omega)$ can be analogously defined by using the relative distance zeta function $\zeta_{A,\Omega}$ instead of the relative tube zeta function $\tilde{\zeta}_{A,\Omega}$ in Equation (5.4.72). Furthermore, $D(\zeta_{A,\Omega}) = D(\tilde{\zeta}_{A,\Omega}) = \overline{\dim}_B(A,\Omega)$, $D_{\mathrm{mer}}(\zeta_{A,\Omega}) = D_{\mathrm{mer}}(\tilde{\zeta}_{A,\Omega})$, and the analog of Theorem 5.4.29 can be easily stated and proved in the case of the relative distance zeta function of a given RFD (A,Ω), instead of the relative tube zeta function.

Remark 5.4.31. It may be that the conclusion of Theorem 5.4.27 (and of Theorem 5.4.32 below, respectively) is also true in the case when there exists an infinite sequence of nonreal complex dimensions of (A,Ω) with real part $\overline{D}$ such that each of them has multiplicity strictly less than that of $\overline{D}$.[33] The fractal tube formula (5.4.61) ((5.4.78), respectively) also holds pointwise in this case, but in order to obtain the conclusion about $\dim_B(A,\Omega)$ and the h-Minkowski measurability of (A,Ω), we have to be able to justify the interchange of the limit as $t\to 0^+$ and the infinite sum which appears in this case in Equation (5.4.61) (respectively, (5.4.78)). A priori, we do not have such a justification to our disposal without making additional assumptions on the nature of the convergence of the sum in (5.4.61) (respectively, (5.4.78)).

It would be interesting to try to extend the above result and obtain a type of gauge Minkowski measurability criterion, in the likes of Theorem 5.4.20. (Some of the results obtained in [HeLap] may be useful for this purpose.)[34] See Theorem 4.5.1 for a partial converse of the above theorem in the case when the relative tube function satisfies the following pointwise asymptotic expansion, with error term:

$$|A_t\cap\Omega| = t^{N-D}(\log t^{-1})^{m-1}(\mathscr{M}+O(t^{\alpha})) \quad \text{as} \quad t\to 0^+, \tag{5.4.76}$$

where $m\in\mathbb{N}$ and $\alpha>0$.

As always, we can reformulate the above theorems in terms of the distance (instead of the tube) zeta function. As an example, we state the counterpart for $\zeta_{A,\Omega}$ of Theorem 5.4.27.

Theorem 5.4.32. *Let (A,Ω) be a relative fractal drum in $\mathbb{R}^N$ such that $\overline{\dim}_B(A,\Omega)<N$. Also assume that (A,Ω) is d-languid with $\kappa_d<0$ or is such that $(\lambda A,\lambda\Omega)$ is strongly d-languid for some $\lambda>0$ with $\kappa_d<1$, for a screen $\boldsymbol{S}$ passing strictly between the critical line $\{\mathrm{Re}\, s=\overline{\dim}_B(A,\Omega)\}$ and all the complex dimensions of (A,Ω) with real part strictly less than $\overline{D}:=\overline{\dim}_B(A,\Omega)$. Furthermore, suppose that $\overline{D}$ is the only pole of the relative distance zeta function $\zeta_{A,\Omega}$ with real part equal to $\overline{D}$ of order $m\geq 1$ and, additionally, that there exists (at most) finitely many nonreal poles of $\zeta_{A,\Omega}$ with real part $\overline{D}$. Moreover, assume that the multiplicity of each of those nonreal poles is of order strictly less than m. Then, $\dim_B(A,\Omega)$ exists and is equal to $D:=\overline{D}$. Also, $\mathscr{M}^D(A,\Omega)$ exists and is equal to $+\infty$; hence, (A,Ω), is Minkowski degenerate.*

In addition, an appropriate gauge function for (A,Ω) is $h(t):=(\log t^{-1})^{m-1}$ for all $t\in(0,1)$ and we have that, relative to *h, and in the terminology of Definition 6.1.4 of Section 6.1 below, the RFD (A,Ω) is not only h-Minkowski nondegenerate but is also h-Minkowski measurable, with h-Minkowski content given by*

---

[33] See Example 5.5.22 where we are in such a situation and the conclusion of Theorem 5.4.32 holds.

[34] Recall, in particular, that in [HeLap], a gauge Minkowski measurability criterion was obtained for fractal strings, extending to the case of non power laws the one obtained (when $h\equiv 1$) in [LapPo1–2]. This criterion does not involve the notion of complex dimensions and is stated only in terms of the underlying gauge function h and the asymptotic behavior of the lengths of the string.

$$\mathscr{M}^D(A,\Omega,h) = \frac{\zeta_{A,\Omega}[D]_{-m}}{(N-D)(m-1)!}, \tag{5.4.77}$$

where $\zeta_{A,\Omega}[D]_{-m}$ denotes the coefficient corresponding to $(s-D)^{-m}$ in the Laurent expansion of $\zeta_{A,\Omega}$ around $s = D$.

Finally, the exact same conclusions as in Theorem 5.4.27 hold concerning the asymptotic expansion of $|A_t \cap \Omega|$ in either (5.4.59) *or* (5.4.60)*, but with $\zeta_{A,\Omega}$ in place of $\tilde{\zeta}_{A,\Omega}$ in the respective hypotheses.*

Proof. We will prove the theorem in the special case when $\overline{D}$ is the only pole with real part equal to $\overline{D}$. The general case then follows analogously as in the proof of Theorem 5.4.27. Let $\overline{D} := \overline{\dim}_B(A,\Omega)$. By Theorem 5.3.16, we have the following pointwise asymptotic tube formula, with error term:

$$|A_t \cap \Omega| = \operatorname{res}\left(\frac{t^{N-s}}{N-s}\zeta_{A,\Omega}(s), \overline{D}\right) + O(t^{N-\sup S}) \quad \text{as } t \to 0^+. \tag{5.4.78}$$

Furthermore, we expand $(N-s)^{-1}$ into a Taylor series around $s = \overline{D}$:

$$\frac{1}{N-s} = \sum_{n=0}^{\infty} \frac{(-1)^n (s-\overline{D})^n}{n!(N-\overline{D})^{n+1}}; \tag{5.4.79}$$

we then multiply the resulting Taylor series by (5.4.62) in order to obtain the following Taylor expansion of $t^{N-s}/(N-s)$ around $s = \overline{D}$:

$$\frac{t^{N-s}}{N-s} = \sum_{n=0}^{\infty} (s-\overline{D})^n \sum_{k=0}^{n} \frac{(-1)^{n-k}(\log t^{-1})^k}{k!(n-k)!(N-\overline{D})^{n-k+1}}. \tag{5.4.80}$$

We next multiply the above Taylor series (in (5.4.80)) by the Laurent expansion of $\zeta_{A,\Omega}(s)$ around $s = \overline{D}$ and extract the residue of the resulting product to deduce that

$$\operatorname{res}\left(\frac{t^{N-s}}{N-s}\zeta_{A,\Omega}(s), \overline{D}\right) = t^{N-\overline{D}} \sum_{n=0}^{m-1} \sum_{k=0}^{n} \frac{(-1)^{n-k}(\log t^{-1})^k \zeta_{A,\Omega}[\overline{D}]_{-n-1}}{k!(n-k)!(N-\overline{D})^{n-k+1}}.$$

We then complete the proof of the theorem by reasoning analogously as in the proof of Theorem 5.4.27. □

5.5 Examples and Applications

In this section, we illustrate the theory of fractal tube formulas developed in Sections 5.1–5.3 (along with the associated Minkowski measurability criterion obtained in Section 5.4) by means of several examples of bounded (fractal) sets and relative fractal drums. These examples include the line segment and the $(N-1)$-dimensional sphere (Subsection 5.5.1), the recovery of the known tube formulas (from [Lap-vFr3]) for fractal strings (Subsection 5.5.2), the Sierpiński gasket and the 3-dimensional Sierpiński carpet, along with the inhomogeneous higher-dimensional N-gasket RFDs, with $N \geq 3$ (Subsection 5.5.3), a suitable version of the Cantor graph (the 'devil's staircase') and an associated discussion of 'fractality' expressed in terms of the presence of nonreal complex dimensions (Subsection 5.5.4), two families of examples which are not self-similar, namely, fractal nests and unbounded geometric chirps (Subsection 5.5.5), as well as, finally, the recovery and significant extensions of the known fractal tube formulas (from [LapPe2–3, LapPeWi1–2]) for self-similar sprays (Subsection 5.5.6).

5.5.1 The Line Segment and the Sphere

We begin by considering the trivial example of the unit interval in $\mathbb{R}$, which illustrates the case when we cannot use the distance zeta function in order to recover the tube formula, since $D = N = 1$.

Example 5.5.1. Let $I = [0,1]$ be the unit interval in $\mathbb{R}$. Then the meromorphic continuations to $\mathbb{C}$ of its distance and tube zeta functions are respectively given by

$$\zeta_I(s) = \frac{2\delta^s}{s} \quad \text{and} \quad \tilde{\zeta}_I(s) = \frac{2\delta^s}{s} + \frac{\delta^{s-1}}{s-1}, \quad \text{for all } s \in \mathbb{C}. \tag{5.5.1}$$

As we can see, the distance zeta function fails to provide information about the Minkowski content in this case, because the pole at $s = 1$ is canceled by means of the functional equation (2.2.23). On the other hand, it is clear that $\tilde{\zeta}_I$ is strongly languid if we choose $\delta > 1$ for $\kappa := -1$ and a sequence of screens consisting of the vertical lines $\{\mathrm{Re}\, s = -m\}$, where $m \in \mathbb{N}$. We then recover from Theorem 5.1.14 the following exact pointwise tube formula:

$$|I_t| = t^{N-0}\operatorname{res}(\tilde{\zeta}_I, 0) + t^{N-1}\operatorname{res}(\tilde{\zeta}_I, 1) = 2t + 1, \tag{5.5.2}$$

initially valid for all $t \in (0, \delta)$. Actually, since $\delta > 1$ may be taken arbitrary large, the exact tube formula (5.5.2) is valid for all $t > 0$. Note that, of course, it is immediate to check directly that the tube formula (5.5.2) holds for all $t > 0$.

Next, let us look at the example of the $(N-1)$-dimensional sphere in $\mathbb{R}^N$, for which the tube zeta function has been explicitly calculated in Example 2.2.21.

Example 5.5.2. Let $B_R(0)$ be the ball of $\mathbb{R}^N$ centered at the origin and with radius $R > 1$; furthermore, let $A := \partial B_R(0)$ be its boundary, i.e., the $(N-1)$-dimensional sphere of radius R. Then, for a fixed $\delta \in (0, R)$, the tube zeta function of A is meromorphic on $\mathbb{C}$ and given by

$$\tilde{\zeta}_A(s) = \omega_N \sum_{k=0}^{N} (1 - (-1)^k) R^{N-k} \binom{N}{k} \frac{\delta^{s-N+k}}{s-(N-k)}, \quad \text{for all } s \in \mathbb{C}. \tag{5.5.3}$$

Moreover, we have shown in Example 2.2.21 that $\dim_B A = D(\tilde{\zeta}_A) = N-1$ and that, in addition, the set of complex dimensions of A is given by

$$\begin{aligned}\mathscr{P}(\tilde{\zeta}_A) := \mathscr{P}(\tilde{\zeta}_A,\mathbb{C}) &= \left\{N-(2j+1) : j = 0,1,2,\ldots,\left\lfloor\frac{N-1}{2}\right\rfloor\right\} \\ &= \left\{N-1, N-3, \ldots, N-\left(2\left\lfloor\frac{N-1}{2}\right\rfloor+1\right)\right\}.\end{aligned} \tag{5.5.4}$$

Also, $\mathscr{P}(\tilde{\zeta}_A) = \mathscr{P}(\zeta_A)$.

The residue of the tube zeta function $\tilde{\zeta}_A$ at any of its poles $m \in \mathscr{P}(\tilde{\zeta}_A)$ is given by

$$\operatorname{res}(\tilde{\zeta}_A, m) = 2\omega_N \binom{N}{m} R^m. \tag{5.5.5}$$

Observe that, by choosing $\delta = 1$, we have that $\tilde{\zeta}_A$ is strongly languid with $\kappa = -1$. More specifically, we may take the sequence of screens $\boldsymbol{S}_m$ as the sequence of vertical lines $\{\operatorname{Re} s = -m\}$, with $m \in \mathbb{N}$. Then, in light of Theorem 5.1.14 (the strongly languid case), we recover the following well-known tube formula of A; i.e., for all $t \in (0,1)$, we have successively:

$$\begin{aligned}|A_t| &= \sum_{\omega \in \mathscr{P}(\tilde{\zeta}_A)} t^{N-\omega} \operatorname{res}(\tilde{\zeta}_A, \omega) \\ &= 2\omega_N \sum_{j=0}^{\lfloor\frac{N-1}{2}\rfloor} \binom{N}{2j+1} t^{2j+1} R^{N-(2j+1)} \\ &= \omega_N \sum_{k=0}^{N} \binom{N}{k} (1-(-1)^k) t^k R^{N-k} \\ &= \omega_N \left((R+t)^N - (R-t)^N\right).\end{aligned} \tag{5.5.6}$$

We refer to Theorem 4.5.6 on page 359 and the discussion surrounding it for a large class of additional examples of exact (pointwise) tube formulas for bounded sets and relative fractal drums in $\mathbb{R}^N$ associated with compact sets of positive reach, including compact convex sets in $\mathbb{R}^N$ and compact smooth submanifolds of $\mathbb{R}^N$. Recall that Theorem 4.5.6 relied in a key manner on the tube formula obtained by H. Federer in [Fed1].

5.5.2 *Tube Formulas for Fractal Strings*

In the present subsection, we apply our general theory of fractal tube formulas for relative fractal drums (and, in particular, for bounded sets) in $\mathbb{R}^N$ to the one-dimensional case (i.e., $N = 1$) in order to recover the known (pointwise and distributional) fractal tube formulas for fractal strings obtained in [Lap-vFr3]. [Completely analogously, we could obtain fractal tube formulas for bounded closed (or, equivalently, compact) subsets of the real line.] We begin by discussing the prototypical example of the Cantor string (viewed as an RFD), in Example 5.5.3, and further illustrate our results by means of two well-known examples, namely, the Fibonacci string (in Example 5.5.9) and the a-string (in Example 5.5.10). Along the way, we discuss the case of general fractal strings as well as the associated fractal tube formulas.

Example 5.5.3. (*The standard ternary Cantor set and string*). Let C be the standard ternary Cantor set in $[0,1]$ and fix $\delta \geq 1/6$. Then, it is easy to deduce from the discussion in Example 2.1.82 that the 'absolute' distance zeta function of C is meromorphic in all of $\mathbb{C}$ and given by

$$\zeta_{C,C_\delta}(s) = \frac{2^{1-s}}{s(3^s-2)} + \frac{2\delta^s}{s}, \quad \text{for all } s \in \mathbb{C}, \tag{5.5.7}$$

where the term $2\delta^s/s$ corresponds to the integral over the 'outer' neighborhood of the two endpoints 0 and 1. Consequently, the relative distance zeta function of $(C,(0,1))$ is also meromorphic on all of $\mathbb{C}$ and given by

$$\zeta_{C,(0,1)}(s) = \frac{2^{1-s}}{s(3^s-2)}, \quad \text{for all } s \in \mathbb{C}. \tag{5.5.8}$$

Hence, in light of (5.5.7) and (5.5.8), the sets of complex dimensions of the Cantor set C and of the Cantor string $(C,(0,1))$, viewed as an RFD, coincide:

$$\mathscr{P}(\zeta_C) = \mathscr{P}(\zeta_{C,(0,1)}) = \{0\} \cup \left(\log_3 2 + \frac{2\pi}{\log 3}\mathrm{i}\mathbb{Z}\right). \tag{5.5.9}$$

In (5.5.9), each of the complex dimensions is simple. Furthermore, the Minkowski dimension $D := \dim_B(C,(0,1))$ of the Cantor string exists and $D = \log_3 2$, the Minkowski dimension of the Cantor set, which also exists. Furthermore, $\mathbf{p} := \frac{2\pi}{\log 3}$ is the oscillatory period of the Cantor set (or string), viewed as a *lattice* self-similar set (or string); see [Lap-vFr3, Chapter 2, esp., Subsection 2.3.1 and Section 2.4].

It is clear that $(\lambda C, \lambda(0,1))$ is strongly d-languid for $\kappa_d := -1$, any $\lambda \geq 2$ and a sequence of screens consisting of the vertical lines $\{\mathrm{Re}\, s = -m\}$ for $m \in \mathbb{N}$, along with the constant $B_\lambda := 2/\lambda$ in the strong languidity condition **L2'**.[35] Theorem 5.3.16 (or, really, Theorem 5.3.17 since all of the complex dimensions of the RFD are simple) then enables us to recover the following exact pointwise fractal formula for the inner t-neighborhood of C, whch is valid for all $t \in (0, \min\{1/\lambda, 1/2\}) = (0, 1/2)$:

[35] Without loss of generality, we can fix $\delta \geq 1$, here.

$$\begin{aligned}
|C_t \cap (0,1)| &= \sum_{\omega \in \mathscr{P}(\zeta_{C,(0,1)})} \operatorname{res}\left(\frac{t^{1-s}}{1-s}\zeta_{C,(0,1)}(s), \omega\right) \\
&= \sum_{\omega \in \mathscr{P}(\zeta_{C,(0,1)})} \frac{t^{1-\omega}}{1-\omega} \operatorname{res}\left(\zeta_{C,(0,1)}, \omega\right) \\
&= \frac{1}{2\log 3} \sum_{k=-\infty}^{+\infty} \frac{(2t)^{1-\omega_k}}{(1-\omega_k)\omega_k} - 2t \\
&= \frac{(2t)^{1-D}}{2\log 3} \sum_{k=-\infty}^{+\infty} \frac{(2t)^{-\mathbb{i}k\mathbf{p}}}{(1-\omega_k)\omega_k} - 2t \\
&= t^{1-D} G\left(\log_3 (2t)^{-1}\right) - 2t,
\end{aligned} \tag{5.5.10}$$

where $\omega_k := D + \mathbb{i}k\mathbf{p}$ for each $k \in \mathbb{Z}$, $D := \dim_B(C,(0,1)) = \log_3 2$ (as above), and $\mathbf{p} := \frac{2\pi}{\log 3}$ denote, respectively, the relative Minkowski dimension and the 'oscillatory period' of the Cantor string RFD $(C,(0,1))$ in $\mathbb{R}$ (or, equivalently, of the Cantor string $\mathscr{L}_{CS}$). Furthermore, G is the positive, nonconstant 1-periodic function, which is bounded away from zero and infinity and given by the following Fourier series expansion:

$$G(x) := \frac{2^{-D}}{\log 3} \sum_{k \in \mathbb{Z}} \frac{e^{2\pi \mathbb{i} kx}}{\omega_k (1-\omega_k)}. \tag{5.5.11}$$

In (5.5.10), the second equality follows from the fact that all of the complex dimensions of $(C,(0,1))$ are simple (see also Theorem 5.3.17 above), while the third equality is obtained by computing the residues of $\zeta_{C,(0,1)}$ at each $s := \omega_k$ (for $k \in \mathbb{Z}$) and at $s = 0$; in particular, we have that

$$\operatorname{res}\left(\zeta_{C,(0,1)}, \omega_k\right) = \frac{2^{-\omega_k}}{\omega_k \log 3}, \quad \text{for all } k \in \mathbb{Z}. \tag{5.5.12}$$

Of course, the above exact pointwise fractal tube formula (5.5.10) coincides with the one obtained by a direct computation for the Cantor string (see [Lap-vFr3, Subsection 1.1.2]) or from the general theory of fractal tube formulas for fractal strings (see [Lap-vFr3, Chapter 8, esp., Sections 8.1 and 8.2]) and, in particular, for self-similar strings (see, especially, [Lap-vFr3, Subsection 8.4.1, Example 8.2.2]).[36] Note that the 'absolute' tube function $|C_t|$ has the same expression as in (5.5.10) above but now without the term $-2t$, which is in accordance with (5.5.7).

Finally, observe that, in agreement with the lattice case of the general theory of self-similar strings developed in [Lap-vFr3, Chapters 2–3, and Section 8.4], we can rewrite the pointwise fractal tube formula (5.5.10) as follows (with $D := \dim_B C = \log_2 3$):

$$t^{-(1-D)} V_{C,(0,1)}(t) = t^{-(1-D)} |C_t \cap (0,1)| = G\left(\log_3 (2t)^{-1}\right) + o(1), \tag{5.5.13}$$

---

[36] *Caution*: in [Lap-vFr1, Subsection 8.4], the Cantor string is defined slightly differently, and hence, C is replaced by $3^{-1}C$.

where G is given by (5.5.11). Therefore, since G is periodic and nonconstant, it is clear that $t^{-(1-D)}V_{C,(0,1)}(t)$ cannot have a limit as $t \to 0^+$. It follows that the Cantor string RFD $(C,(0,1))$ (or, equivalently, the Cantor string $\mathscr{L}_{CS}$) is *not* Minkowski measurable but (since G is also bounded away from zero and infinity) is Minkowski nondegenerate. (This was first proved in [LapPo1–2] via a direct computation, leading to the precise values of $\mathscr{M}_*$ and $\mathscr{M}^*$, and reproved in [Lap-vFr3, Subsection 8.4.2] by using either the pointwise fractal tube formulas or a self-similar fractal string analog of the Minkowski measurability criterion; i.e., of the $N = 1$ case of Theorem 5.4.20; see Remark 5.4.22 and, especially, Corollary 5.4.23.) Note that, of course, as was alluded to just above, we can also deduce the Minkowski nonmeasurability of $(C,(0,1))$ from the $N = 1$ case of Theorem 5.4.20. Indeed, $D := \dim_B(C,(0,1)) = \log_3 2 < 1$, although it is simple, is *not* the only complex dimension of the RFD $(C,(0,1))$ (or, equivalently, of the Cantor string) with real part equal to D, since G is nonconstant. In addition, the remaining hypotheses of Theorem 5.4.20 are clearly satisfied.

The above example demonstrates how the theory developed in this chapter generalizes (to arbitrary dimensions $N \geq 1$) the corresponding one for fractal strings developed in [Lap-vFr3, Chapter 8].[37] More generally, the following result provides a general connection between the geometric zeta function of a nontrivial fractal string $\mathscr{L} = (\ell_j)_{j\geq 1}$ and the (relative) distance zeta function of the bounded subset of $\mathbb{R}$ given by

$$A_{\mathscr{L}} := \left\{ a_k := \sum_{j\geq k} \ell_j : k \geq 1 \right\} \tag{5.5.14}$$

or, more specifically, of the RFD $(A_{\mathscr{L}},(0,\ell))$. See also Remark 5.5.5 and Equation (5.5.16) below.

Proposition 5.5.4. *Let $\mathscr{L} = (\ell_j)_{j\geq 1}$ be a nontrivial bounded fractal string and let $\ell := \zeta_{\mathscr{L}}(1) = \sum_{j=1}^{\infty} \ell_j$ denote its total length. Then, for every $\delta \geq \ell_1/2$, we have the following functional equation for the distance zeta function of the relative fractal drum $(A_{\mathscr{L}},(0,\ell))$:*

$$\zeta_{A_{\mathscr{L}},(0,\ell)}(s;\delta) = \frac{2^{1-s}}{s}\zeta_{\mathscr{L}}(s), \tag{5.5.15}$$

valid on any connected open neighborhood $U \subseteq \mathbb{C}$ of the critical line $\{\operatorname{Re} s = \overline{\dim}_B(A_{\mathscr{L}},(0,\ell))\}$ to which any (and hence, each) of the two fractal zeta functions $\zeta_{A_{\mathscr{L}},(0,\ell)}$ and $\zeta_{\mathscr{L}}$ possesses a meromorphic continuation.[38]

[37] One should somewhat qualify this statement, however, because the higher-dimensional counterpart of the theory of fractal tube formulas for self-similar strings developed in [Lap-vFr3, Section 8.4] is not developed in this book in the general case of self-similar RFDs (and, for example, of self-similar sets satisfying the open set condition), except in the special case of self-similar sprays discussed in Subsection 5.5.6 below.

[38] If we do not require that $\delta \geq \ell_1/2$, then we have that $\zeta_{A_{\mathscr{L}}}(s;\delta) = 2^{1-s}s^{-1}\zeta_{\mathscr{L}}(s) + v(s)$, where v is holomorphic on $\{\operatorname{Re} s > 0\}$. On the other hand, in order to apply the theory, we may restrict ourselves to the case when $\delta \geq \ell_1/2$.

Furthermore, if $\zeta_{\mathscr{L}}$ is languid for some languidity exponent $\kappa_{\mathscr{L}} \in \mathbb{R}$, then $\zeta_{A_{\mathscr{L}},(0,\ell)}(\cdot\,;\delta)$ is d-languid for the d-languidity exponent $\kappa_d := \kappa_{\mathscr{L}} - 1$, with any $\delta \geq \ell_1/2$.

Moreover, if $\zeta_{\mathscr{L}}$ is strongly languid, then so is $\zeta_{\lambda A_{\mathscr{L}},(0,\lambda\ell)}(\cdot\,;\delta\lambda)$ for any $\lambda \geq 2$ and any $\delta \geq \ell_1/2$.

Proof. The functional equation (5.5.15) is already derived in Example 2.1.58. More precisely, it follows from Equation (2.1.84) and the principle of analytic continuation. Furthermore, the statements about the languidity follow directly from the definition. □

Remark 5.5.5. There is nothing special about the bounded set $A_{\mathscr{L}} \subset \mathbb{R}$ associated with $\mathscr{L}$, as was already pointed out in Corollary 2.1.61 and the comments surrounding it (see also Remark 5.5.6 below). In fact, in the statement of Proposition 5.5.4, we could replace $A_{\mathscr{L}}$ with $\partial\Omega$, where the bounded open set $\Omega \subset \mathbb{R}$ is an arbitrary geometric realization of the fractal string $\mathscr{L}$. Similarly, in recovering the fractal tube formulas for fractal strings obtained in [Lap-vFr3, Chapter 8], one can use $\zeta_{\partial\Omega,\Omega} := \zeta_{\partial\Omega,\Omega}(\cdot\,;\delta)$ instead of $\zeta_{A_{\mathscr{L}},(0,\ell)} := \zeta_{A_{\mathscr{L}},(0,\ell)}(\cdot\,;\delta)$. This is precisely what we will do in the subsequent discussion.

Let $\partial\Omega$ be the boundary of Ω, where the bounded open set $\Omega \subset \mathbb{R}$ is any geometric realization of the bounded (nontrivial) fractal string $\mathscr{L}$ such that $\overline{\dim}_B(\partial\Omega,\Omega) < 1$. Then, under suitable hypotheses (namely, we assume that either $\zeta_{\partial\Omega,\Omega}$ or $\zeta_{\mathscr{L}}$ has a meromorphic continuation to a connected open neighborhood U of the critical line $\{\mathrm{Re}\, s = \overline{\dim}_B(\partial\Omega,\Omega)\}$), we have (much as in (5.5.15) above) the following key *functional equation connecting the distance zeta function $\zeta_{\partial\Omega,\Omega}$ of the RFD $(\partial\Omega,\Omega)$ and the geometric zeta function $\zeta_{\mathscr{L}}$ of the fractal string $\mathscr{L} := (\ell_j)_{j=1}^{\infty}$* (see also part (*ii*) of Corollary 2.1.61 on page 92 for more detailed information):[39]

$$\zeta_{\partial\Omega,\Omega}(s) = \frac{2^{1-s}\zeta_{\mathscr{L}}(s)}{s}, \tag{5.5.16}$$

valid for all $s \in U$. Of course, it then follows that each of the two fractal zeta functions $\zeta_{\partial\Omega,\Omega}$ and $\zeta_{\mathscr{L}}$ has a unique meromorphic continuation to all of U.

Consequently, by choosing $U := \mathring{\boldsymbol{W}}$ to be the interior of a suitable window $\boldsymbol{W}$ (with an associated screen $\boldsymbol{S}$), we deduce from the results of Section 5.3 that the tube function

$$V_{\mathscr{L}}(t) := |\{x \in \Omega : d(x,\partial\Omega) < t\}|_1 = V_{\partial\Omega,\Omega}(t) \tag{5.5.17}$$

[39] We note that the functional equation (5.5.16) is valid, without any hypothesis on the bounded fractal string $\mathscr{L}$ (or on its distance and geometric zeta functions), for all $s \in \mathbb{C}$ with $\mathrm{Re}\, s$ sufficiently large (namely, for $\mathrm{Re}\, s > \overline{D}$, where $\overline{D} := \overline{\dim}_B(\partial\Omega,\Omega) = D(\zeta_{\partial\Omega,\Omega}) = D(\zeta_{\mathscr{L}})$). Its proof is provided in the proof of part (*ii*) of Corollary 2.1.61 on page 92.

can be expressed via the following fractal tube formula (with or without error term and pointwise or distributionally, depending on the assumptions), for every $\delta \geq \ell_1/2$:[40]

$$\begin{aligned} V_{\mathscr{L}}(t) &= V_{\partial\Omega,\Omega}(t) \\ &= \sum_{\omega \in \mathscr{P}(\zeta_{\partial\Omega,\Omega}, \boldsymbol{W})} \operatorname{res}\left(\frac{t^{1-s}}{1-s}\zeta_{\partial\Omega,\Omega}(s), \omega\right) + R^{[0]}_{\partial\Omega,\Omega}(t) \\ &= \sum_{\omega \in \mathscr{P}(\zeta_{\partial\Omega,\Omega}, \boldsymbol{W})} \operatorname{res}\left(\frac{(2t)^{1-s}}{s(1-s)}\zeta_{\mathscr{L}}(s), \omega\right) + R^{[0]}_{\partial\Omega,\Omega}(t), \end{aligned} \tag{5.5.18}$$

where, in the languid case, we have the error estimate $R^{[0]}_{\partial\Omega,\Omega}(t) = O(t^{1-\sup S})$ as $t \to 0^+$ or $R^{[0]}_{\partial\Omega,\Omega}(t) = o(t^{1-\sup S})$ as $t \to 0^+$ (also depending on the hypotheses; more specifically, in order to obtain the better error estimate, we also have to assume that the screen $\boldsymbol{S}$ is strictly to the left of the vertical line $\{\operatorname{Re} s = \sup S\}$), or else, $R^{[0]}_{\partial\Omega,\Omega}(t) \equiv 0$ and $\boldsymbol{W} := \mathbb{C}$ in the strongly languid case. Here, $\mathscr{P}(\zeta_{\partial\Omega,\Omega}, \boldsymbol{W})$ denotes the set of visible complex dimensions of $(\partial\Omega, \Omega)$, visible through a given window $\boldsymbol{W}$ (with an associated screen $\boldsymbol{S}$), and in light of the counterpart for the RFD $(\partial\Omega, \Omega)$ of Equation (5.5.15) along with Remark 5.5.5, we have that

$$\mathscr{P}(\zeta_{\partial\Omega,\Omega}, \boldsymbol{W} \setminus \{0\}) = \mathscr{P}(\zeta_{\mathscr{L}}, \boldsymbol{W} \setminus \{0\}), \tag{5.5.19}$$

where the equality holds between multisets. Furthermore, if $0 \in \boldsymbol{W}$ and if $\zeta_{\mathscr{L}}(0)$ is defined and not equal to zero (i.e., if $\zeta_{\mathscr{L}}(0) \neq 0$), then, $0 \in \mathscr{P}(\zeta_{\partial\Omega,\Omega}, \boldsymbol{W})$ and it has multiplicity one. On the other hand, if $0 \in \mathscr{P}(\zeta_{\mathscr{L}}, \boldsymbol{W})$ and is a pole of multiplicity m for some $m \in \mathbb{N}$, then, $0 \in \mathscr{P}(\zeta_{\partial\Omega,\Omega}, \boldsymbol{W})$ and it has multiplicity $m+1$. In other words, we have the following equality between multisets:

$$\mathscr{P}(\zeta_{\partial\Omega,\Omega}, \boldsymbol{W}) = \mathscr{P}(\zeta_{\mathscr{L}}, \boldsymbol{W}) \cup \{0\}_{0 \in \boldsymbol{W},\ \zeta_{\mathscr{L}}(0) \neq 0}, \tag{5.5.20}$$

where $\{0\}_{0 \in \boldsymbol{W},\ \zeta_{\mathscr{L}}(0) \neq 0}$ is equal to $\{0\}$ if $0 \in \boldsymbol{W}$ and $\zeta_{\mathscr{L}}(0) \neq 0$, and to the empty set otherwise.

If, in addition, each of the visible complex dimensions of $(\partial\Omega, \Omega)$ (i.e., each pole of $\zeta_{\partial\Omega,\Omega}$ in $\boldsymbol{W}$) is simple, then (in light of (5.5.18)) the fractal tube formula (5.5.18) takes the following simpler form:

$$\begin{aligned} V_{\mathscr{L}}(t) &= V_{\partial\Omega,\Omega}(t) \\ &= \sum_{\omega \in \mathscr{P}(\zeta_{\mathscr{L}}, \boldsymbol{W})} \frac{(2t)^{1-\omega}}{\omega(1-\omega)} \operatorname{res}(\zeta_{\mathscr{L}}(s), \omega) + \{2t\zeta_{\mathscr{L}}(0)\}_{0 \in \boldsymbol{W}} \\ &\quad + R^{[0]}_{\partial\Omega,\Omega}(t), \end{aligned} \tag{5.5.21}$$

[40] Namely, we are assuming here either the hypotheses of Theorem 5.3.16 (i.e., of Theorem 5.3.11 at level $k = 0$), for the pointwise tube formula, or else, the hypotheses of Theorem 5.3.21 (i.e., of Theorem 5.3.19 at level $k = 0$), for the distributional tube formula.

where the (pointwise or distributional) error term $R^{[0]}_{\partial\Omega,\Omega}(t)$ is estimated as above (in the languid case) or else, $R^{[0]}_{\partial\Omega,\Omega}(t) \equiv 0$ and $\boldsymbol{W} := \mathbb{C}$ (in the strongly languid case). Here, the term $\{2t\zeta_{\mathscr{L}}(0)\}_{0\in\boldsymbol{W}}$ is equal to zero if $0 \notin \boldsymbol{W}$ and to $2t\zeta_{\mathscr{L}}(0)$ if $0 \in \boldsymbol{W}$. If, however, 0 is a simple, visible pole of $\zeta_{\mathscr{L}}$, then we should replace $\{2t\zeta_{\mathscr{L}}(0)\}_{0\in\boldsymbol{W}}$ on the right-hand side of (5.5.21) by the term

$$2t(1-\log(2t))\operatorname{res}(\zeta_{\mathscr{L}},0) + 2t\zeta_{\mathscr{L}}[0]_0, \tag{5.5.22}$$

where $\zeta_{\mathscr{L}}[0]_0$ stands for the constant term in the Laurent series expansion of $\zeta_{\mathscr{L}}$ around $s = 0$. This is in agreement with [Lap-vFr3, Corollary 8.3] (resp., [Lap-vFr3, Corollary 8.10]) in the case of a distributional (resp., pointwise) fractal tube formula.

Note that in light of (5.5.19), formula (5.5.18) can be rewritten as follows, in terms of the set $\mathscr{P}(\zeta_{\mathscr{L}},\boldsymbol{W})$ of all visible poles of $\zeta_{\mathscr{L}}$ (see also Remark 5.5.7 below):

$$\begin{aligned} V_{\mathscr{L}}(t) &= V_{\partial\Omega,\Omega}(t) \\ &= \sum_{\omega\in\mathscr{P}(\zeta_{\mathscr{L}},\boldsymbol{W})} \operatorname{res}\left(\frac{(2t)^{1-s}}{s(1-s)}\zeta_{\mathscr{L}}(s),\omega\right) \\ &\quad + \{2t\zeta_{\mathscr{L}}(0)\}_{0\in\boldsymbol{W}\setminus\mathscr{P}(\zeta_{\mathscr{L}},\boldsymbol{W})} + R^{[0]}_{\partial\Omega,\Omega}(t), \end{aligned} \tag{5.5.23}$$

which is in agreement with [Lap-vFr3, Theorem 8.1] (resp., [Lap-vFr3, Theorem 8.7]) in the case of a distributional (resp., pointwise) fractal tube formula.

Naturally, $\mathscr{P}(\zeta_{\mathscr{L}},\boldsymbol{W})$ is viewed as a multiset; that is, on the right-hand side of (5.5.19) or (5.5.20), each visible 'scaling complex dimension' $\omega \in \mathscr{P}(\zeta_{\mathscr{L}},\boldsymbol{W})$ (i.e., each visible pole of the geometric zeta function $\zeta_{\mathscr{L}}$) occurs according to its multiplicity. An entirely analogous comment can be made about the multiset $\mathscr{P}(\zeta_{\partial\Omega,\Omega},\boldsymbol{W})$ and the associated visible complex dimensions $\omega \in \mathscr{P}(\zeta_{\partial\Omega,\Omega},\boldsymbol{W})$.

Remark 5.5.6. As was first observed in [LapPo1–2] and as can be easily checked via a direct computation, $V_{\partial\Omega,\Omega}$ depends only on the fractal string $\mathscr{L} = (\ell_j)_{j=1}^{\infty}$ and *not* on the chosen geometric representation of $\mathscr{L}$ via a bounded open set $\Omega \subset \mathbb{R}$. (See also [Lap-vFr3, Equation (8.1), p. 238].) Hence, we may use the notation $V_{\partial\Omega,\Omega} = V_{\mathscr{L}}$. More specifically, a moment's reflection reveals that for every $t > 0$, we have that

$$V_{\partial\Omega,\Omega}(t) = \sum_{\ell_j\geq 2t} 2t + \sum_{\ell_j<2t} \ell_j, \tag{5.5.24}$$

which clearly depends only on the fractal string $\mathscr{L} = (\ell_j)_{j=1}^{\infty}$.

Remark 5.5.7. In [Lap-vFr3], the elements of $\mathscr{P}(\zeta_{\mathscr{L}},\boldsymbol{W})$ are called the (visible) complex dimensions of $\mathscr{L}$. In the present book, the relationship with the actual (visible) complex dimension of the RFD $(\partial\Omega,\Omega)$ (i.e., the visible poles of $\zeta_{\partial\Omega,\Omega}$) is given by Equations (5.5.19) and (5.5.20), along with the text surrounding them. Much as in [LaPe2–3, LapPeWi1–2] and [Lap-vFr3, Section 13.1], we propose to refer to the elements of $\mathscr{P}(\zeta_{\mathscr{L}},\boldsymbol{W})$ (i.e., to the visible poles of the geometric zeta function $\zeta_{\mathscr{L}}$) as the visible *scaling complex dimensions* of the fractal string $\mathscr{L}$.

Similarly, $\zeta_{\mathscr{L}}$ will also be occasionally referred to as the *scaling zeta function* of $\mathscr{L}$ (or rather, of the associated RFD $(\partial\Omega,\Omega)$) and denoted by $\zeta_{\mathfrak{S}}$; see, especially, Subsection 5.5.6 and the text surrounding Problems 6.2.35, 6.2.36 and 6.2.38.

Remark 5.5.8. We leave it as an easy exercise for the interested reader to use the counterpart for the RFD $(\partial\Omega,\Omega)$ of the functional equation (5.5.15) in Proposition 5.5.4 in order to express the languidity, as well as the strong languidity conditions, in terms of the geometric zeta function $\zeta_{\mathscr{L}}$ instead of the distance zeta function $\zeta_{\partial\Omega,\Omega}$. Furthermore, the reader can easily check that the results of Example 5.5.3 concerning the Cantor string

$$\mathscr{L} := \left(\frac{1}{3},\frac{1}{9},\frac{1}{9},\frac{1}{27},\frac{1}{27},\frac{1}{27},\frac{1}{27},\dots\right) \tag{5.5.25}$$

(see, especially, Equation (5.5.10)) are compatible with both (5.5.21) and (5.5.23). Indeed, in light of (5.5.8) and (5.5.15), we have (for all $s \in \mathbb{C}$)

$$\zeta_{CS}(s) = \frac{1}{3^s - 2}, \tag{5.5.26}$$

from which it follows that $\zeta_{CS}(0) = -1$ and (with $\boldsymbol{W} := \mathbb{C}$) the term $\{2t\zeta_{\mathscr{L}}(0)\}_{0\in\boldsymbol{W}}$ in both (5.5.21) and (5.5.23) becomes $-2t$, in agreement with (5.5.10).

Let us now apply Proposition 5.5.4 (along with Remark 5.5.5) and the above discussion in order to recover the formula of the tubular volume of the boundary of a well-known fractal string studied in [Lap-vFr3, Subsection 2.3.2].

Example 5.5.9. (*The Fibonacci string*). Let Fib be the Fibonacci string (with total length 4) where the sequence of *distinct* lengths is given by $\ell_j := 2^{-j}$, for $j \in \mathbb{N}_0$, and each length ℓ_j has multiplicity F_{j+1}. Here, for each $n \in \mathbb{N}_0$, F_n denotes the n-th Fibonacci number defined by the following recursion formula:

$$F_{n+1} = F_n + F_{n-1} \ \text{ for all } \ n \geq 1, \quad \text{and} \quad F_0 := 0,\ F_1 := 1. \tag{5.5.27}$$

Then, for the geometric zeta function of the Fibonacci string, we have (see [Lap-vFr3, Equation (2.20)])

$$\zeta_{\mathrm{Fib}}(s) = \frac{1}{1 - 2^{-s} - 4^{-s}} \quad \text{for all } s \in \mathbb{C}, \tag{5.5.28}$$

and we deduce from Proposition 5.5.4 that

$$\zeta_{A_{\mathrm{Fib}},(0,4)}(s;1) = \frac{2^{1-s}}{s(1-2^{-s}-4^{-s})} = \frac{2^{s+1}}{s(4^s-2^s-1)}, \tag{5.5.29}$$

also for all $s \in \mathbb{C}$. Therefore, one can easily check that the set of complex dimensions of the RFD $(A_{\mathrm{Fib}},(0,4))$ consists solely of *simple* poles of $\zeta_{A_{\mathrm{Fib}},(0,4)} := \zeta_{A_{\mathrm{Fib}},(0,4)}(\,\cdot\,;1)$ and is given by

$$\mathscr{P}(\zeta_{A_{\mathrm{Fib}}}) := \mathscr{P}(\zeta_{A_{\mathrm{Fib}},(0,4)}) = \left(-D + \frac{\mathbf{p}\mathbb{i}}{2} + \mathbf{p}\mathbb{i}\mathbb{Z}\right) \cup \{0\} \cup (D + \mathbf{p}\mathbb{i}\mathbb{Z}), \qquad (5.5.30)$$

where $D := \dim_B(A_{\mathrm{Fib}}, (0,4)) = \log_2 \phi$, with $\phi = (1+\sqrt{5})/2$ being the *golden mean*, and with oscillatory period $\mathbf{p} := 2\pi/\log 2$. (Here, Fib is viewed as a lattice self-similar string; see [Lap-vFr3, Chapter 2, esp., Subsection 2.3.2].) Similarly as in Example 5.5.3, one checks that we can apply Theorem 5.3.16 with any $\lambda \geq 1/2$ and a corresponding $B_\lambda := 1/(2\lambda)$ in order to recover the following exact pointwise fractal tube formula, valid for all $t \in (0,1)$:

$$\begin{aligned}
V_{\mathrm{Fib}}(t) &:= V_{A_{\mathrm{Fib}},(0,4)}(t) = |(A_{\mathrm{Fib}})_t \cap (0,4)| \\
&= \sum_{\omega \in \mathscr{P}(\zeta_{A_{\mathrm{Fib}}})} \operatorname{res}\left(\frac{t^{1-s}}{1-s}\zeta_{A_{\mathrm{Fib}},(0,4)}(s), \omega\right) \\
&= \sum_{\omega \in \mathscr{P}(\zeta_{A_{\mathrm{Fib}}})} \frac{t^{1-\omega}}{1-\omega} \operatorname{res}\left(\zeta_{A_{\mathrm{Fib}},(0,4)}, \omega\right) \qquad (5.5.31)\\
&= \frac{(2t)^{1-D}\phi}{\sqrt{5}\log 2} \sum_{k=-\infty}^{+\infty} \frac{(2t)^{-\mathbb{i}k\mathbf{p}}}{(1-D-\mathbb{i}k\mathbf{p})(D+\mathbb{i}k\mathbf{p})} - 2t \\
&\quad + \frac{(2t)^{1+D}(\phi-1)}{\sqrt{5}\log 2} \sum_{k=-\infty}^{+\infty} \frac{(2t)^{-\mathbb{i}\mathbf{p}/2-\mathbb{i}k\mathbf{p}}}{(1+D-\mathbb{i}\mathbf{p}/2-\mathbb{i}k\mathbf{p})(-D+\mathbb{i}\mathbf{p}/2+\mathbb{i}k\mathbf{p})}.
\end{aligned}$$

Of course, the above formula coincides with the formula derived in [Lap-vFr3, Subsection 2.3.2]. It is also consistent with the discussion (of fractal tube formulas for fractal strings) following Proposition 5.5.4 above, in the strongly languid case (hence, with $\boldsymbol{W} := \mathbb{C}$ and $R^{[0]}_{A_{\mathrm{Fib}},(0,4)}(t) \equiv 0$) and in the pointwise case. (See, in particular, Equation (5.5.21), where we have set $R^{[0]}_{A_{\mathrm{Fib}},(0,4)}(t) \equiv 0$.)

Much as we did for the Cantor string in Equations (5.5.10) and (5.5.11), it is now immediate to rewrite (5.5.31) in the following form, which is consistent with the general theory of (exact pointwise) fractal tube formulas for lattice self-similar strings developed in [Lap-vFr3, Subsection 8.4.2, esp., Theorem 8.4.2]:

$$V_{\mathrm{Fib}}(t) = (2t)^{1-D} G_1\left(\log_2 (2t)^{-1}\right) - 2t + (2t)^{1+D} G_2\left(\log_2 (2t)^{-1}\right), \qquad (5.5.32)$$

where G_1 and G_2 are explicitly known nonconstant 1-periodic functions on $\mathbb{R}$; furthermore, G_1 and $|G_2|$ are bounded away from zero and infinity.

Example 5.5.10. (*The a-string*). For a given $a > 0$, the a-string $\mathscr{L}_a$ can be realized as the bounded open set $\Omega_a \subset \mathbb{R}$ obtained by removing the points j^{-a} for $j \in \mathbb{N}$ from the interval $(0,1)$; that is,

$$\Omega_a = \bigcup_{j=1}^{\infty} \left((j+1)^{-a}, j^{-a}\right), \qquad (5.5.33)$$

so that the sequence of lengths of $\mathscr{L}_a$ is defined by

$$\ell_j = j^{-a} - (j+1)^{-a}, \text{ for } j = 1,2,\ldots, \tag{5.5.34}$$

and $\partial\Omega_a = \{j^{-a} : j \geq 1\} \cup \{0\} = A_{\mathscr{L}_a} \cup \{0\}$. Hence, its geometric zeta function is given (for all $s \in \mathbb{C}$ such that $\operatorname{Re} s > \dim_B \mathscr{L}_a$) by

$$\zeta_{\mathscr{L}_a}(s) = \sum_{j=1}^{\infty} \ell_j^s = \sum_{j=1}^{\infty} \left(j^{-a} - (j+1)^{-a}\right)^s.$$

It then follows from Proposition 5.5.4 that for $\delta > (1 - 2^{-a})/2$, its distance zeta function is given by (see Remark 5.5.11 at the end of this subsection)

$$\zeta_{A_{\mathscr{L}_a},(0,1)}(s;\delta) = \frac{\zeta_{\mathscr{L}_a}(s)}{2^{s-1}s} = \frac{1}{2^{s-1}s} \sum_{j=1}^{\infty} \left(j^{-a} - (j+1)^{-a}\right)^s, \tag{5.5.35}$$

where the second equality holds for all $s \in \mathbb{C}$ such that $\operatorname{Re} s > \dim_B \mathscr{L}_a$ while the first equality holds for all $s \in \mathbb{C}$ (since, as will be recalled just below, $\zeta_{\mathscr{L}_a}$ and hence also $\zeta_{A_{\mathscr{L}_a},(0,1)}$, admits a meromorphic extension to all of $\mathbb{C}$).

Furthermore, the properties of the geometric zeta function $\zeta_{\mathscr{L}_a}$ of the a-string are well-known (see [Lap-vFr3, Theorem 6.21]). Namely, $\zeta_{\mathscr{L}_a}$ has a meromorphic continuation to the whole of $\mathbb{C}$ and its poles in $\mathbb{C}$ are located at

$$D := \dim_B \mathscr{L}_a = \dim_B A_{\mathscr{L}_a} = \frac{1}{a+1} \tag{5.5.36}$$

and at (a subset of) $\{-\frac{m}{a+1} : m \in \mathbb{N}\}$. Moreover, all of its poles are simple and $\operatorname{res}(\zeta_{\mathscr{L}_a}, D) = Da^D$.[41] In addition, for any screen $\boldsymbol{S}$ not passing through a pole, the function $\zeta_{\mathscr{L}_a}$ satisfies **L1** and **L2** with $\kappa := \frac{1}{2} - (a+1)\inf S$, if $\inf S \leq 0$ and $\kappa := \frac{1}{2}$ if $\inf S \geq 0$. From these facts and Equation (5.5.35), we conclude that the set $A_{\mathscr{L}_a}$ is d-languid with $\kappa_d := -\frac{1}{2} - (a+1)\inf S$ if $\inf S \leq 0$ and with $\kappa_d := -\frac{1}{2}$ if $\inf S \geq 0$. For each $M \in \mathbb{N}_0$, where (as before) $\mathbb{N}_0 := \mathbb{N} \cup \{0\}$, we can now choose the screen $\boldsymbol{S}_M$ to be some vertical line between $-\frac{M+1}{1+a}$ and $-\frac{M+2}{1+a}$ and let $\boldsymbol{W}_M$ be the corresponding window. Applying Theorem 5.3.21, we now obtain the following asymptotic distributional formula for the tube function $t \mapsto |(A_{\mathscr{L}_a})_t \cap (0,1)|$ when $t \to 0^+$:

$$\begin{aligned} |(A_{\mathscr{L}_a})_t \cap (0,1)| = &\sum_{\omega \in \mathscr{P}(\zeta_{A_{\mathscr{L}_a}}, \boldsymbol{W}_M)} \operatorname{res}\left(\frac{t^{1-s}}{1-s}\zeta_{A_{\mathscr{L}_a,(0,1)}}(s;\delta), \omega\right) \\ &+ O(t^{1-\sup S_M}). \end{aligned} \tag{5.5.37}$$

---

[41] In [Lap-vFr3, Theorem 6.21], it is stated that $\operatorname{res}(\zeta_{\mathscr{L}_a}, D) = a^D$, which is a misprint. More specifically, in the proof of that theorem, the source of the misprint is the fact that the residue of $\zeta((a+1)s)$ at $s = 1/(a+1)$ is equal to $1/(a+1)$ and not to 1. Here, ζ is the Riemann zeta function.

More specifically, since we know that all the poles are simple and $\zeta_{\mathscr{L}_a}(0) = -1/2$ (see [Lap-vFr3, p. 205]), we have that

$$\begin{aligned} &\operatorname{res}(\zeta_{A_{\mathscr{L}_a}}, D) = 2^{1-D} D^{-1} \operatorname{res}(\zeta_{\mathscr{L}_a}, D) = 2^{1-D} a^D, \\ &\operatorname{res}(\zeta_{A_{\mathscr{L}_a}}, 0) = 2\zeta_{\mathscr{L}_a}(0) = -1. \end{aligned} \tag{5.5.38}$$

Consequently, and in agreement with the discussion following Proposition 5.5.4 in the special case of simple complex dimensions (see, especially, Equation (5.5.21) above), we have that

$$\begin{aligned} |(A_{\mathscr{L}_a})_t \cap (0,1)| = \frac{2^{1-D} a^D}{1-D} t^{1-D} - t - \sum_{m=1}^{M} \frac{\operatorname{res}(\zeta_{\mathscr{L}_a}, -mD)\,(2t)^{1+mD}}{(1+mD)mD} \\ + O(t^{1+(M+1)D}), \quad \text{as} \quad t \to 0^+, \end{aligned} \tag{5.5.39}$$

where the sum is interpreted as being equal to 0 if $M = 0$. In particular, $\dim_B A_{\mathscr{L}_a} = D$ (as was stated above), and, according to Theorem 5.4.20 (the Minkowski measurability criterion), the a-string is Minkowski measurable with Minkowski content given by

$$\mathscr{M}^D(A_{\mathscr{L}_a}) = \frac{2^{1-D} a^D}{1-D}, \tag{5.5.40}$$

as was first established in [Lap1, Example 5.1] and later reproved in [LapPo1–2] via a general Minkowski measurability criterion for fractal strings (expressed in terms of the asymptotic behavior of $(\ell_j)_{j=1}^{\infty}$, here, $\ell_j \sim a j^{-1/D}$ as $j \to \infty$) and then, in [Lap-vFr1–3] (via the theory of complex dimensions of fractal strings, specifically, via the special case of Theorem 5.4.20 when $N = 1$). We point out that (5.5.39) coincides with the 'inner' tube formula of the a-string (see [Lap-vFr3, Subsection 8.1.2]).[42] Furthermore, by choosing a screen to the right of $-D/2$, we conclude that (5.5.39) is actually valid pointwise since then, $\kappa_d < 0$ (see Theorem 5.3.16).

Remark 5.5.11. Throughout the discussion provided in Example 5.5.10, and without affecting any of the results, we could have replaced the RFD $(A_{\mathscr{L}_a}, (0,1))$ by the equivalent RFD $(\partial\Omega_a, \Omega_a)$, where Ω_a is defined by (5.5.33) and, more generally, by the RFD $(\partial\Omega, \Omega)$, where Ω is an arbitrary geometric realization of the fractal string $\mathscr{L}_a$. Indeed, as we know from Subsection 5.5.2 (see, especially, Remark 5.5.5), all of the results obtained here are independent of the choice of the geometric realization of the fractal string $\mathscr{L}_a$.

[42] More precisely, the two expressions coincide after we have taken into account the misprint mentioned in footnote 41 on page 490 and added the term $2\zeta_{\mathscr{L}}(0)$ which seems to have been forgotten in [Lap-vFr3].

5.5.3 The Sierpiński Gasket and 3-Carpet

In this subsection, we provide an exact, pointwise fractal tube formula for the Sierpiński gasket (Example 5.5.12) and for a three-dimensional analog of the Sierpiński carpet (Example 5.5.13). Naturally, although the required computation is somewhat more complicated, one could similarly derive from our general results in Section 5.3 exact, pointwise fractal tube formulas for the N-dimensional analogs of the Sierpiński gasket and carpet, with $N \geq 2$ arbitrary. We leave it to the interested reader to carry out the corresponding detailed computations and to imagine other (two- or higher-dimensional) examples of self-similar fractal sets or self-similar RFDs which can be dealt with explicitly within the present general theory of (higher-dimensional) fractal tube formulas.[43] The example of the Sierpiński 3-carpet discussed in detail in Example 5.5.13 below should give a good idea as to how to proceed in other, related situations, including especially for the higher-dimensional inhomogeneous N-gasket RFDs (with $N \geq 4$) discussed in Example 4.2.26 and for other self-similar RFDs which can also be dealt with within the general theory of fractal tube formulas and their applications developed in this chapter.

Example 5.5.12. (The Sierpiński gasket). Let A be the Sierpiński gasket in $\mathbb{R}^2$, constructed in the usual way inside the unit triangle. Furthermore, we assume without loss of generality that $\delta > 1/4\sqrt{3}$, so that A_δ be simply connected. Then, the distance zeta function ζ_A of the Sierpiński gasket is meromorphic on the whole complex plane and is given by

$$\zeta_A(s;\delta) = \frac{6(\sqrt{3})^{1-s}2^{-s}}{s(s-1)(2^s-3)} + 2\pi\frac{\delta^s}{s} + 3\frac{\delta^{s-1}}{s-1}, \tag{5.5.41}$$

for all $s \in \mathbb{C}$ (see Proposition 3.2.3 in Subsection 3.2.1). In particular, the set of complex dimensions of the Sierpiński gasket is given by

$$\mathscr{P}(\zeta_A) := \mathscr{P}(\zeta_A, \mathbb{C}) = \{0,1\} \cup \left(\log_2 3 + \frac{2\pi}{\log 2}\mathbb{i}\mathbb{Z}\right), \tag{5.5.42}$$

with each complex dimension being simple.

By letting $\omega_k := \log_2 3 + \mathbb{i}k\mathbf{p}$ (for each $k \in \mathbb{Z}$) and $\mathbf{p} := 2\pi/\log 2$, we have that

$$\operatorname{res}(\zeta_A, \omega_k) = \frac{6(\sqrt{3})^{1-\omega_k}}{4^{\omega_k}(\log 2)\omega_k(\omega_k - 1)} \quad \text{for all } k \in \mathbb{Z}, \tag{5.5.43}$$

$$\operatorname{res}(\zeta_A, 0) = 3\sqrt{3} + 2\pi, \quad \text{and} \quad \operatorname{res}(\zeta_A, 1) = 0. \tag{5.5.44}$$

---

[43] The authors have recently obtained an explicit fractal tube formula for the Koch drum (or the Koch snowflake RFD), by using the general theory developed in this chapter. This important example should be discussed in a later work and its conclusions compared with those of [LapPe1] (as discussed in [Lap-vFr3, Subsection 12.2.1]).

Similarly as in Examples 5.5.3 and 5.5.9, one can check that $\zeta_{\lambda A}(\cdot\,;\delta\lambda)$ is strongly languid with $\kappa_d := -1$ for every $\delta \geq 1/2\sqrt{3}$ and any $\lambda \geq 2\sqrt{3}$; so that we can apply Theorem 5.3.16 (or, more specifically, its corollary given in Theorem 5.3.17 at level $k = 0$ and in the case of simple poles) in order to obtain the following exact pointwise fractal tube formula:

$$\begin{aligned}|A_t| &= \sum_{\omega\in\mathscr{P}(\zeta_A)} \operatorname{res}\left(\frac{t^{2-s}}{2-s}\zeta_A(s;\delta),\omega\right)\\ &= t^{2-\log_2 3}\frac{6\sqrt{3}}{\log 2}\sum_{k=-\infty}^{+\infty}\frac{(4\sqrt{3})^{-\omega_k}t^{-\mathrm{i}k\mathbf{p}}}{(2-\omega_k)(\omega_k-1)\omega_k}+\left(\frac{3\sqrt{3}}{2}+\pi\right)t^2\\ &= t^{2-D}G(\log_2 t^{-1})+\left(\frac{3\sqrt{3}}{2}+\pi\right)t^2,\end{aligned}$$

valid for all $t \in (0, 1/2\sqrt{3})$. (Here, G is a positive, nonconstant 1-periodic function, which is bounded away from zero and infinity and is given explicitly by the convergent Fourier series $G(x) := \frac{6\sqrt{3}}{\log 2}\sum_{k=-\infty}^{+\infty}\frac{(4\sqrt{3})^{-\omega_k}\exp(2\pi\mathrm{i}kx)}{(2-\omega_k)(\omega_k-1)\omega_k}$, for all $x \in \mathbb{R}$.) Note that this fractal tube formula coincides with the one obtained in [LapPe3] and [LapPeWi1] and, more recently, via a different (but related) technique in [DenKoÖÜ].

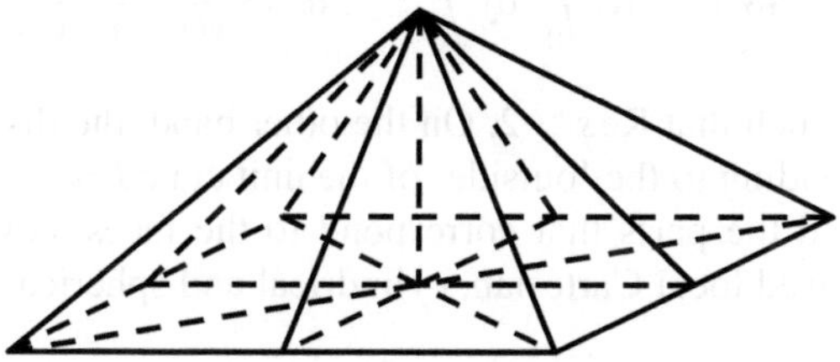

Fig. 5.2 The pairwise congruent pyramids into which we subdivide the cube A_1 from Example 5.5.13. Eight of them, corresponding to one face of A_1, are shown here.

Example 5.5.13. (*The* 3-*carpet*). Let A be the three-dimensional analog of the Sierpiński carpet. More specifically, we construct A by dividing the closed unit cube of $\mathbb{R}^3$ into 27 pairwise congruent cubes and remove the open middle cube. Then, we iterate this step with each of the 26 remaining smaller closed cubes; and so on, ad infinitum. By choosing $\delta > 1/6$, we have that A_δ is simply connected. Let us now calculate the distance zeta function ζ_A of the three-dimensional carpet A. Note that

$$\zeta_A(s;\delta) = \zeta_{A,I}(s) + \zeta_{A,A_\delta\setminus I}(s),$$

where I denotes the closed unit cube in $\mathbb{R}^3$. Let us denote by B_1 the open unit cube of side $1/3$ removed in the first step of the construction; so that we have the following equalities:

$$\zeta_{A,I}(s) = \zeta_{A,B_1}(s) + \zeta_{A,I\setminus B_1}(s) = \zeta_{\partial B_1,B_1}(s) + 26\,\zeta_{3^{-1}A,3^{-1}I}(s), \tag{5.5.45}$$

for all $s \in \mathbb{C}$ with $\operatorname{Re} s$ sufficiently large. The first equality is obvious, while the second equality in (5.5.45) follows from the self-similarity of A. More precisely, this equality follows since the relative fractal drum $(A, I \setminus B_1)$ consists of 26 copies of (A, I) scaled down by 3^{-1}. Hence, by the scaling property of the relative distance zeta function (see Theorem 4.1.40), we have that

$$\zeta_{A,I}(s) = \zeta_{\partial B_1,B_1}(s) + 26 \cdot 3^{-s} \zeta_{A,I}(s),$$

which yields

$$\zeta_{A,I}(s) = \frac{\zeta_{\partial B_1,B_1}(s)}{1 - 26 \cdot 3^{-s}}, \tag{5.5.46}$$

for all $s \in \mathbb{C}$ with $\operatorname{Re} s$ sufficiently large. The distance zeta function $\zeta_{\partial B_1,B_1}$ can be easily calculated by dividing the cube B_1 into 48 pairwise congruent pyramids (see Figure 5.2) and then integrating in local Cartesian coordinates $(x,y) \in \mathbb{R}^2$ over each resulting pyramid:

$$\zeta_{\partial B_1,B_1}(s) = 48 \int_0^{1/6} \mathrm{d}x \int_0^x \mathrm{d}y \int_0^y z^{s-3}\,\mathrm{d}z = \frac{48 \cdot 6^{-s}}{s(s-1)(s-2)}, \tag{5.5.47}$$

valid for all $s \in \mathbb{C}$ such that $\operatorname{Re} s > 2$. On the other hand, the distance zeta function $\zeta_{A,A_\delta \setminus I}(s)$ corresponding to the 'outside' of the unit cube I is easy to calculate once we have subdivided the parts that correspond to the faces, edges and vertices of the unit cube and used local Cartesian, cylindrical and spherical coordinates in $\mathbb{R}^3$, respectively:

$$\begin{aligned}\zeta_{A,A_\delta\setminus I}(s) &= 6\int_0^1 \mathrm{d}x \int_0^1 \mathrm{d}y \int_0^\delta z^{s-3}\,\mathrm{d}z + 12\int_0^{\pi/2} \mathrm{d}\varphi \int_0^\delta r^{s-2}\,\mathrm{d}r \int_0^1 \mathrm{d}z \\ &\quad + 8\int_0^{\pi/2} \sin\theta\,\mathrm{d}\theta \int_0^{\pi/2} \mathrm{d}\varphi \int_0^\delta r^{s-1}\,\mathrm{d}r \\ &= \frac{6\delta^{s-2}}{s-2} + \frac{6\pi\delta^{s-1}}{s-1} + \frac{4\pi\delta^s}{s},\end{aligned} \tag{5.5.48}$$

again valid for all $s \in \mathbb{C}$ such that $\operatorname{Re} s > 2$. From the above calculation and from (5.5.46) together with (5.5.47), we deduce that ζ_A can be meromorphically continued to all of $\mathbb{C}$ and is then given by

$$\zeta_A(s) := \zeta_A(s,\delta) = \frac{48 \cdot 2^{-s}}{s(s-1)(s-2)(3^s - 26)} + \frac{4\pi\delta^s}{s} + \frac{6\pi\delta^{s-1}}{s-1} + \frac{6\delta^{s-2}}{s-2}, \tag{5.5.49}$$

for every $s \in \mathbb{C}$.

It follows that the set of complex dimensions of the 3-carpet A is given by

$$\mathscr{P}(\zeta_A) := \mathscr{P}(\zeta_A, \mathbb{C}) = \{0,1,2\} \cup \left(\log_3 26 + \mathbf{p}\mathbb{i}\mathbb{Z}\right), \tag{5.5.50}$$

where $D := \log_3 26$ $(= D(\zeta_A))$ is the Minkowski (or box) dimension of the 3-carpet A and $\mathbf{p} := 2\pi/\log 3$ is the oscillatory period of A (viewed as a lattice self-similar set). In (5.5.50), each of the complex dimensions is simple. Furthermore, a routine computation shows that

$$\operatorname{res}(\zeta_A, 0) = 4\pi - \frac{24}{25}, \quad \operatorname{res}(\zeta_A, 1) = 6\pi + \frac{24}{23}, \quad \operatorname{res}(\zeta_A, 2) = \frac{96}{17} \tag{5.5.51}$$

and, by letting $\omega_k := \log_3 26 + \mathbb{i}k\mathbf{p}$ (for all $k \in \mathbb{Z}$),

$$\operatorname{res}(\zeta_A, \omega_k) = \frac{24}{13 \cdot 2^{\omega_k} \omega_k (\omega_k - 1)(\omega_k - 2) \log 3}. \tag{5.5.52}$$

One also easily checks that the hypotheses of Theorem 5.3.16 (or, really, of Theorem 5.3.17 since all of the complex dimensions in (5.5.50) are simple) are satisfied for every $\delta \geq 1/2$ and any scaling factor $\lambda \geq 2$, and thus we obtain the following exact pointwise tube formula, valid for all $t \in (0, 1/2)$:

$$\begin{aligned} |A_t| = {} & \frac{24 t^{3-\log_3 26}}{13 \log 3} \sum_{k=-\infty}^{+\infty} \frac{2^{-\omega_k} t^{-\mathbb{i}k\mathbf{p}}}{(3-\omega_k)(\omega_k - 1)(\omega_k - 2)\omega_k} \\ & + \left(6 - \frac{6}{17}\right) t + \left(3\pi + \frac{12}{23}\right) t^2 + \left(\frac{4\pi}{3} - \frac{8}{25}\right) t^3. \end{aligned} \tag{5.5.53}$$

In particular, we conclude that $D := \dim_B A = \log_3 26$ (as was noted before) and, by Theorem 5.4.20, that the three-dimensional Sierpiński carpet is not Minkowski measurable, which is expected (see [Lap3]). We also point out that the part $6t + 3\pi t^2 + 4\pi t^3/3$ from the above Equation (5.5.53) is exactly equal to $|I_t| - |I|$, where I is the closed unit cube of $\mathbb{R}^3$.

Finally, we note that clearly, the first term on the right-hand side of (5.5.53) can be rewritten in the following form (still with $D := \dim_B A = \log_3 26$):

$$t^{3-D} G(\log_3 t^{-1}), \tag{5.5.54}$$

where G is a positive, nonconstant 1-periodic function which is bounded away from zero and infinity and is given explicitly by the convergent Fourier series $G(x) := \frac{24}{13 \log 3} \sum_{k=-\infty}^{+\infty} \frac{2^{-\omega_k} \exp(2\pi \mathbb{i} k x)}{(3-\omega_k)(\omega_k - 1)(\omega_k - 2)\omega_k}$, for all $x \in \mathbb{R}$. Therefore, also as expected (see [Lap3]), the 3-carpet is Minkowski nondegenerate: $0 < \mathscr{M}_*(A) < \mathscr{M}^*(A) < \infty$.

Of course, exactly the same comment as above about the Minkowski nonmeasurability and the Minkowski nondegeneracy could have been made about the Sierpiński gasket discussed in Example 5.5.12.

5.5.4 A Relative Fractal Drum Generated by the Cantor Function

The example dicussed in this subsection, namely, a version of the Cantor graph (or "devil's staircase", in the terminology of [Man1]) plays an important role in showing

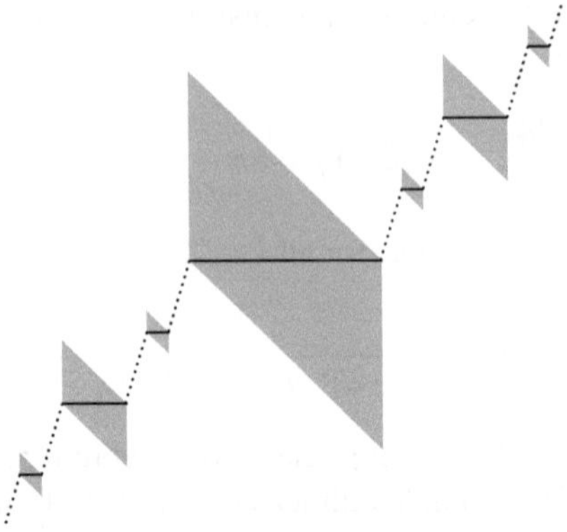

Fig. 5.3 The third step in the construction of the Cantor graph relative fractal drum (A, Ω) from Example 5.5.14. One can see, in particular, the sets B_k, $\triangle_k$ and $\tilde{\triangle}_k$ for $k = 1,2,3$.

why the notion of complex dimensions gives a lot more information than the mere (Minkowski or Hausdorff) fractal dimension, as will be explained below in relation to the elusive notion of "fractality".

We invite the interested reader to review the discussion of the classic Cantor graph provided in the introduction to this book in Remark 1.2.1 of Section 1.2, along with Figures 1.5, 1.6 and 1.7 on pages 25–28.

Example 5.5.14. (*The Cantor graph RFD*). In this example, we compute the distance zeta function of the RFD (A, Ω) in $\mathbb{R}^2$, where A is the graph of the Cantor function (i.e., the Cantor graph) and Ω is the union of triangles $\triangle_k$ that lie above and the triangles $\tilde{\triangle}_k$ that lie below each of the horizontal parts of the graph denoted by B_k. (At each step of the construction there are 2^{k-1} pairwise congruent triangles $\triangle_k$ and $\tilde{\triangle}_k$.) Each of these triangles is isosceles, has for one of its sides a horizontal part of the Cantor graph, and has a right angle at the left end of B_k, in the case of $\triangle_k$, or at the right end of B_k, in the case of $\tilde{\triangle}_k$. See Figure 5.3.

For obvious geometric reasons and by using the scaling property (see Theorem 4.1.40) of the relative distance zeta function of the resulting RFD (A, Ω), called the *Cantor graph RFD*, we then have the following identity:

$$\begin{aligned}
\zeta_{A,\Omega}(s) &= \sum_{k=1}^{\infty} 2^k \zeta_{B_k,\triangle_k}(s) \\
&= \sum_{k=1}^{\infty} 2^k \zeta_{3^{-k}B_1,3^{-k}\triangle_1}(s) \qquad (5.5.55) \\
&= \zeta_{B_1,\triangle_1}(s) \sum_{k=1}^{\infty} \frac{2^k}{3^{ks}} = \frac{2\zeta_{B_1,\triangle_1}(s)}{3^s-2},
\end{aligned}$$

valid for all $s \in \mathbb{C}$ with $\operatorname{Re} s$ sufficiently large. Here, $(B_1, \triangle_1)$ is the relative fractal drum described above with two perpendicular sides of length equal to 1. It is straightforward to compute its relative distance zeta function:

$$\zeta_{B_1,\triangle_1}(s) = \int_0^1 \mathrm{d}x \int_0^x y^{s-2}\,\mathrm{d}y = \frac{1}{s(s-1)}, \qquad (5.5.56)$$

valid, initially, for all $s \in \mathbb{C}$ such that $\operatorname{Re} s > 1$ and then, upon meromorphic continuation, for all $s \in \mathbb{C}$. This fact, combined with the last equality of Equation (5.5.55), gives us the distance zeta function of (A, Ω), which is clearly meromorphic on all of $\mathbb{C}$:

$$\zeta_{A,\Omega}(s) = \frac{2}{s(3^s-2)(s-1)}, \quad \text{for all } s \in \mathbb{C}. \qquad (5.5.57)$$

We therefore deduce that the set of complex dimensions of the RFD (A, Ω) is given by

$$\mathscr{P}(\zeta_{A,\Omega}) := \mathscr{P}(\zeta_{A,\Omega}, \mathbb{C}) = \{0,1\} \cup \left(\log_3 2 + \frac{2\pi}{\log 3}\mathrm{i}\mathbb{Z}\right), \qquad (5.5.58)$$

with each complex dimension being simple.

We conclude from Theorem 5.4.2 that $\dim_B(A, \Omega) = 1$ and that the RFD (A, Ω) is Minkowski measurable. Moreover, one also deduces from Theorem 5.4.2 that the (one-dimensional) Minkowski content of (A, Ω) is given by

$$\mathscr{M}^1(A, \Omega) = \frac{\operatorname{res}(\zeta_{A,\Omega}, 1)}{2-1} = 2, \qquad (5.5.59)$$

which coincides with the length of the Cantor graph (i.e., the graph of the Cantor function, also called the devil's staircase in [Man1]).

In the sequel, we associate the RFD $(A, A_{1/3})$ in $\mathbb{R}^2$ to the classic Cantor graph.[44] We do not know if (5.5.58) coincides with the set of complex dimensions of the 'full' graph of the Cantor function (i.e., the original devil's staircase), or equivalently, the RFD $(A, A_{1/3})$, but we expect that this is indeed the case since (A, Ω) is a 'relative fractal subdrum' of $(A, A_{1/3})$. Moreover, it clearly follows from the construction of (A, Ω) that for the distance zeta function of the RFD $(A, A_{1/3})$ associated with the graph of the Cantor function, we have

[44] Recall that the classic Cantor graph (or 'full Cantor graph') was discussed at some length in the introduction (Chapter 1), on pages 25–28; see, especially, Remark 1.2.1 along with Figures 1.5, 1.6 and 1.7.

$$\zeta_{A,A_{1/3}}(s) = \zeta_{A,\Omega}(s) + \zeta_{A,A_{1/3}\setminus\Omega}(s). \tag{5.5.60}$$

In order to prove that $\mathscr{P}(\zeta_{A,\Omega})$, given by (5.5.58), is a subset of the complex dimensions of the 'full' Cantor graph, it would therefore remain to show that $\zeta_{A,A_{1/3}\setminus\Omega}(s)$ has a meromorphic continuation to some connected open neigborhood U of the critical line $\{\mathrm{Re}\, s = 1\}$ such that U contains the set of complex dimensions of (A,Ω), as given by (5.5.58), and that there are no pole-pole cancellations in the right-hand side of (5.5.60).

One easily checks that $\lambda^s\zeta_{A,\Omega}(s;1/3)$ is strongly d-languid for any $\lambda \geq 1$, with $\kappa_d := -2$, and thus we can apply Theorem 5.3.16 in order to obtain the following exact pointwise fractal tube formula for the RFD (A,Ω), valid for all $t \in (0,1)$:

$$\begin{aligned} V_{A,\Omega}(t) := |A_t \cap \Omega| &= \sum_{\omega\in\mathscr{P}(\zeta_{A,\Omega})} \operatorname{res}\left(\frac{t^{2-s}}{2-s}\zeta_{A,\Omega}(s),\omega\right) \\ &= \sum_{\omega\in\mathscr{P}(\zeta_{A,\Omega})} \frac{t^{2-\omega}}{2-\omega}\operatorname{res}\left(\zeta_{A,\Omega}(s),\omega\right) \\ &= 2t + \frac{t^{2-\log_3 2}}{\log 3}\sum_{k=-\infty}^{+\infty}\frac{t^{-\mathbb{i}k\mathbf{p}}}{(2-\omega_k)(\omega_k-1)\omega_k} + t^2 \\ &= 2t^{2-D_{CF}} + t^{2-D_{CS}}G_{CF}\left(\log_3 t^{-1}\right) + t^2, \end{aligned} \tag{5.5.61}$$

where $\omega_k := \log_3 2 + \mathbb{i}k\mathbf{p}$ (for each $k \in \mathbb{Z}$), $D_{CF} = \dim_B(A,\Omega) = 1$, $D_{CS} = \log_3 2$ and $\mathbf{p} := 2\pi/\log 3$.

In the last line of (5.5.61), G_{CF} is a nonconstant 1-periodic function on $\mathbb{R}$, which is bounded away from zero and infinity. It is given by the following convergent (and even, absolutely convergent) Fourier series:

$$G_{CF}(x) := \frac{1}{\log 3}\sum_{k=-\infty}^{+\infty}\frac{\mathrm{e}^{2\pi\mathbb{i}kx}}{(2-\omega_k)(\omega_k-1)\omega_k}, \quad \text{for all } x \in \mathbb{R}. \tag{5.5.62}$$

Note that in order to obtain the third equality in (5.5.61), and hence also the above expression for G_{CF} given in (5.5.62), we have used the fact that (in light of (5.5.57) and (5.5.58))

$$\operatorname{res}\left(\zeta_{A,\Omega}(s),\omega_k\right) = \frac{1}{\log 3(\omega_k-1)\omega_k}, \tag{5.5.63}$$

for all $k \in \mathbb{Z}$.

It follows from (5.5.61) and (5.5.62) that even though this version of the Cantor graph, described by the RFD (A,Ω), is Minkowski measurable and hence does not have any oscillations of leading order, it has *oscillations of lower order*, corresponding to the complex dimensions of the Cantor set (or string) of the form $D_{CS} + \mathbb{i}k\mathbf{p}$, with $k \in \mathbb{Z}$ (see Example 5.5.3, especially, Equation (5.5.9)); that is, it has *subcritical oscillations*, of order $2 - D_{CS} \approx 1.3691$, where $D_{CS} := \log_3 2$ is the Minkowski dimension of the Cantor set (or string). In fact, in light of the pointwise fractal tube formula (5.5.61) and since the RFD (A,Ω) has Minkowski content

$\mathscr{M}_{CF} := \mathscr{M}(A,\Omega) = 2$ (see Equation (5.5.59) above), as well as Minkowski dimension $D_{CF} := \dim_B(A,\Omega) = 1$, we have that

$$\begin{aligned} 0 &< \liminf_{t\to 0^+} t^{-(2-D_{CS})} \left| \mathscr{M}_{CF} t^{2-D_{CF}} - V_{A,\Omega}(t) \right| \\ &< \limsup_{t\to 0^+} t^{-(2-D_{CS})} \left| \mathscr{M}_{CF} t^{2-D_{CF}} - V_{A,\Omega}(t) \right| < \infty. \end{aligned} \tag{5.5.64}$$

Hence, we see that even though the *leading term* (as $t \to 0^+$) in the fractal tube formula (5.5.61) is of order $2 - D_{CF} = 1$ (i.e., of order $t^{2-D_{CF}} = t$), determined by the Minkowski dimension $D_{CF} = 1$ of (A,Ω), as should be case, and is *monotonic* (and therefore, *nonoscillatory*), the *asymptotic second term*, $h(t) := t^{2-D_{CS}} G(\log_3 t^{-1})$, is of order $2 - D_{CS}$, determined by the Minkowski dimension $D_{CS} = \log_3 2$ of the *Cantor set* (or string), and is *oscillatory* (in fact, multiplicatively periodic, or "log-periodic", to use the physicists' terminology).

Remark 5.5.15. (*Critical vs subcritical fractals*). Recall from Remark 4.6.24 that a geometric object is said to be "fractal" if it has at least one nonreal complex dimension (or if its fractal zeta function has a (meromorphic) partial natural boundary along a suitable screen, in which case it is said to be "hyperfractal"). (See [Lap-vFr3, Sections 12.1 and 12.2], along with [Lap-vFr3, Subsection 13.4.3], as adapted and extended to our general higher-dimensional theory of complex dimensions in Subsection 4.6.3 above, especially, in Definition 4.6.23, Remark 4.6.24 and the comments following it.) Accordingly, the present version of the Cantor graph (i.e., the RFD (A,Ω) from Example 5.5.14 just above) is "fractal" in this sense.

In addition, following [Lap-vFr1–3] (see, especially, [Lap-vFr3, Section 3.7]), given $d \in \mathbb{R}$ (with $d \le N$), we say that a geometric object is *fractal in dimension* d if it has at least one *nonreal* (visible) complex dimension of real part d.[45] (Automatically, it will then have at least one pair of nonreal complex conjugate complex dimensions of real part d.) If the object in question is an RFD (A,Ω) (and, in particular, a bounded set A) in $\mathbb{R}^N$, with upper (relative) Minkowski dimension $\overline{\dim}_B(A,\Omega)$ (or, in particular $\overline{\dim}_B A$) denoted by $\overline{D}$, then we can distinguish between the following two different and interesting cases:[46]

(*i*) (*Critical case*). The RFD (A,Ω) is fractal in dimension $d := \overline{D}$, in which case (A,Ω) is said to be *critically fractal*. Indeed, under suitable hypotheses, it then follows from the fractal tube formulas of Sections 5.1–5.3 that it has at least one nonreal complex dimension on the critical line $\{\mathrm{Re}\, s = \overline{D}\}$, thereby giving rise to *geometric oscillations of leading order*.

(*ii*) (*Subcritical case*). The RFD (A,Ω) is not fractal in dimension $\overline{D}$ (i.e., it does not have any nonreal principal complex dimension), but it is fractal in some dimen-

---

45 We allow here the number d to be nonpositive, since it enables us to deal with a broader class of potential fractals.

46 We assume here implicitly that the fractal zeta function of (A,Ω) under consideration has a meromorphic extension to a connected open neighborhood of the critical line $\{\mathrm{Re}\, s = \overline{\dim}_B(A,\Omega)\}$, say, to the interior of a window $\boldsymbol{W}$ with associated screen $\boldsymbol{S}$ such that $\sup S < \overline{D} := \overline{\dim}_B(A,\Omega)$. We also assume that $\overline{D} \in \mathbb{R}$; i.e. (since $\overline{D} \le N$), $\overline{D} \neq -\infty$.

sion $d < \overline{D}$. The RFD (A,Ω) is then said to be *subcritically fractal*. (Sometimes, we will also say that (A,Ω) is *"strictly subcritically fractal"* in order to emphasize the fact that $d < D$, and we will say that (A,Ω) is *"possibly subcritically fractal"* in order to indicate that $d \leq D$ instead of $d < D$.)

[Other cases are possible, such as (A,Ω) being hyperfractal, even in case (i) or (ii), or else (A,Ω) being *nonfractal*; that is, neither having a nonreal (visible) complex dimension nor being hyperfractal. However, we are not concerned with these situations in the present context.]

Given an RFD (A,Ω), we define $\alpha \in \mathbb{R} \cup \{-\infty\}$, the *subcriticality index* of (A,Ω), via the following formula:

$$\alpha = \alpha_{A,\Omega} := \sup\{d \in \mathbb{R} : (A,\Omega) \text{ is fractal in dimension } d\}. \tag{5.5.65}$$

By convention, we let $\alpha_{A,\Omega} = -\infty$ if (A,Ω) is not fractal in dimension d, for any $d \in \mathbb{R}$. Clearly, we always have $\alpha_{A,\Omega} \leq \overline{D} \leq N$.

We note that even if (A,Ω) is subcritically fractal, it could happen that $\alpha_{A,\Omega} = \overline{D} := \overline{\dim}_B(A,\Omega)$. This is the case, for instance, if $(A,\Omega) := (\partial\Omega,\Omega)$ is a generic, nonlattice self-similar string, in the sense of [Lap-vFr3, Subsection 3.2.1].[47] Then, as was conjectured in [Lap-vFr3, Subsection 3.7.1] (as well as, more specifically, in reference [Lap–vF6] of [Lap-vFr3]) and later proved in [MorSepVi1], the *set of dimensions of fractality* of (A,Ω) (i.e., the set of real numbers d such that (A,Ω) is fractal in dimension d) is dense in some compact interval of the form $[D_*,\overline{D}]$, with $D_* \in \mathbb{R}$ and $D_* < \overline{D}$. As a result, in light of (5.5.65), it follows that $\alpha_{A,\Omega} = \overline{D}$. However, (A,Ω) is not critically fractal (because according to [Lap-vFr3, Theorem 2.16], a (generic) nonlattice string does not have any nonreal complex dimensions of real part $\overline{D}$), even though it is subcritically fractal in dimension $d < \overline{D}$ for a dense (and countable) set of real numbers d in $[D_*,\overline{D}]$.

We now return to the RFD considered in Example 5.5.14 (that is, the version of the Cantor graph denoted by (A,Ω)), and we refer to Remark 5.5.15 just above for the appropriate terminology and definitions. As we have seen, (A,Ω) is fractal. More specifically, *it is not critically fractal* (because its only complex dimension of real part D_{CF} $(= \overline{D} = \dim_B(A,\Omega)) = 1$ is 1 itself, the Minkowski dimension of the Cantor graph, and it is simple) *but it is strictly subcritically fractal*. In fact, it is subcritically fractal in a single dimension, namely, in dimension $d = D_{CS} = \log_3 2$, the Minkowski dimension of the Cantor set. Consequently, in light of (5.5.65), the subcriticality index of (A,Ω) is given by

$$\alpha_{A,\Omega} = D_{CS} = \log_3 2, \tag{5.5.66}$$

and it is attained.

---

[47] Recall from [Lap-vFr3, Chapters 2–3] that a self-similar string with distinct scaling ratios $\rho_1,\ldots,\rho_n$ in $(0,1)$ is said to be *lattice* (resp., *nonlattice*) if the rank of the group generated by $\rho_1,\ldots,\rho_n$ (viewed as a multiplicative subgroup of $(0,+\infty)$) is equal to 1 (resp., > 1), and *generic nonlattice* if the rank is equal to n, the maximal possible rank, and $n > 1$.

We expect the same result to hold for the devil's staircase itself (i.e., the 'full' graph of the Cantor function), represented by the RFD $(A, A_{1/3})$ and of which (A, Ω) is a 'relative fractal subdrum', as was explained above. Clearly, in light of (5.5.60) and (5.5.58), we have the following inclusions (between multisets):

$$\begin{aligned}\mathscr{P}(\zeta_{A,A_{1/3}}) &\subseteq \mathscr{P}(\zeta_{A,\Omega}) \cup \mathscr{P}(\zeta_{A,A_{1/3}\setminus\Omega}) \\ &\subseteq \{0,1\} \cup \left\{ D_{CS} + \frac{2\pi}{\log 3}\mathrm{i}\mathbb{Z} \right\}.\end{aligned} \tag{5.5.67}$$

Also, we know for a fact that $\dim_B(A, A_{1/3})$ exists and

$$D(\zeta_{A,A_{1/3}}) = \dim_B(A, A_{1/3}) = 1; \tag{5.5.68}$$

so that

$$\dim_{PC}(A, A_{1/3}) := \mathscr{P}_c(\zeta_{A,A_{1/3}}) = \{1\}. \tag{5.5.69}$$

(Thus, we also have that $\{1\} \subseteq \mathscr{P}(\zeta_{A,A_{1/3}})$ in (5.5.67).) Note that (5.5.68) (and hence, (5.5.69)) follows from the rectifiability of the devil's staircase, combined with a well-known result in [Fed2] and with part (*b*) of Theorem 4.1.7.

As was mentioned earlier in the discussion of Example 5.5.14 (and was predicted in [Lap-vFr3, Subsections 12.1.2 and 12.3.2], based on an 'approximate tube formula'), we expect that $\mathscr{P}(\zeta_{A,A_{1/3}}) = \mathscr{P}(\zeta_{A,\Omega})$, as given by (5.5.58), and hence, that we actually have equalities instead of inclusions in (5.5.67), even equalities between multisets. If so, then the 'full' Cantor graph $(A, A_{1/3})$ is fractal, not critically fractal, but (strictly) subcritically fractal in the single dimension $d := D_{CS} = \log_3 2$.

Clearly, both (A, Ω) and $(A, A_{1/3})$ should be fractal for any proper definition of fractality. This would completely resolve the following apparent paradox: the RFD $(A, A_{1/3})$ is not "fractal" according to Mandelbrot's original definition of fractality given in [Man1],[48] even though everyone feels and expects it to be "fractal" simply after having glanced at the 'full' Cantor graph $(A, A_{1/3})$ (the 'devil's staircase' in the sense of [Man1]). The same is true for the 'partial' Cantor graph (A, Ω), for which we can now rigorously prove that it is "fractal" (in the sense of the present theory of complex dimensions) even though it is only (strictly) subcritically fractal, which may explain, in hindsight, why some practitioners refer to it as a "borderline fractal" (see, e.g., [PeitJüSa]).

We conclude this discussion by quoting (as in [Lap-vFr3, p. 335]) Mandelbrot [Man1, p. 82] writing about the devil's staircase (the 'full' Cantor graph, depicted in [Man1, Plate 83, p. 83]):

---

48 Indeed, Mandelbrot's definition, given in [Man1, p. 15], can be stated as follows: A geometric object is "fractal" if its Hausdorff dimension is strictly greater than (i.e., is not equal to) its topological dimension. However, note that the Hausdorff, Minkowski and topological dimensions coincide and are equal to 1 in the case of (either the 'full' or the 'partial') Cantor graph. If, in addition, we replaced "Hausdorf dimension" by (relative, upper) "Minkowski dimension" in the above definition and we interpreted the topological dimension in the obvious way, we would also reach the analogous conclusion for both $(A, A_{1/3})$ and (A, Ω), which therefore would still not be fractal according to this modified Mandelbrot definition.

> *One would love to call the present curve a fractal, but to achieve this goal we would have to define fractals less stringently, on the basis of notions other than D* [the Hausdorff dimension] *alone.*

Thanks to the higher-dimensional theory of complex dimensions of fractals and the associated fractal tube formulas developed in this book and in [LapRaŽu1–8], building on the corresponding theory for fractal strings developed in [Lap-vFr1–3], we are now tentatively close to having resolved this apparent paradox, which has long puzzled the first author and was one of the key motivations for the development of the mathematical theory of complex dimensions. Furthermore, if we use the 'partial' Cantor graph (A,Ω) as a suitable substitute for the 'full' Cantor graph, viewed as the RFD $(A,A_{1/3})$, the corresponding paradox is indeed completely resolved here. We invite the interested reader to extend the conclusions of the present example (i.e., Example 5.5.14) from (A,Ω) to $(A,A_{1/3})$, and thereby, to fully prove the conjectures and statements made in [Lap-vFr3, Subsection 12.1.2], as well as here, about the devil's staircase itself.

5.5.5 Fractal Nests and Unbounded Geometric Chirps

In this subsection, we apply our general fractal tube formulas to several families of fractal nests (Example 5.5.16) and of (unbounded) geometric chirps (Example 5.5.19). Both of these families are examples of fractal sets which are *not* self-similar or, more generally, 'self-alike' in any sense. We also draw some useful conclusions about the interesting new situation when the fractal zeta function of the RFD (A,Ω) has a pole of order m (with $m\in\mathbb{N}$, $m\geq 2$) located at $D:=\dim_B(A,\Omega)$ and all the other poles with real part D are of order strictly less than m. More specifically, under suitable additional hypotheses, we show that the RFD (A,Ω) is then Minkowski degenerate, with Minkowski content $\mathscr{M}(A,\Omega)=+\infty$, but is nevertheless *gauge Minkowski measurable* with respect to the *gauge function* h given by $h(t):=(\log t^{-1})^{m-1}$, for all $t\in(0,1)$. (See Theorems 5.4.27 and 5.4.32 for this result; also, for an introduction to gauge functions, see the beginning of Subsection 4.5.1, along with Definition 6.1.4.)

Example 5.5.16. (Fractal nests). We let $\mathscr{L}=(\ell_j)_{j\geq 1}$ be a bounded fractal string and, as before, let $A_{\mathscr{L}}=\{a_k:k\in\mathbb{N}\}\subset\mathbb{R}$, with $a_k:=\sum_{j\geq k}\ell_j$ for each $k\geq 1$. Furthermore, consider now $A_{\mathscr{L}}$ as a subset of the x_1-axis in $\mathbb{R}^2$ and let A be the planar set obtained by rotating $A_{\mathscr{L}}$ around the origin; i.e., A is a union of concentric circles of radii a_k and center at the origin (see Figure 3.2). For $\delta>\ell_1/2$, the distance zeta function of A is given (for $\operatorname{Re}s$ sufficiently large) by

$$\zeta_A(s)=\frac{2^{2-s}\pi}{s-1}\sum_{j=1}^{\infty}\ell_j^{s-1}(a_j+a_{j+1})+\frac{2\pi\delta^s}{s}+\frac{2\pi a_1\delta^{s-1}}{s-1}; \tag{5.5.70}$$

see Equation (3.5.1) from Example 3.5.1. The last two terms in the above formula correspond to the annulus $a_1 < r < a_1 + \delta$ and we will neglect them; that is, without affecting the final outcome, we will only consider the relative distance zeta function $\zeta_{A,\Omega}$, with $\Omega := B_{a_1}(0)$.[49] Furthermore, since $a_{j+1} = a_j - \ell_j$ for each $j \geq 1$, we have

$$\begin{aligned}\zeta_{A,\Omega}(s) &= \frac{2^{2-s}\pi}{s-1}\sum_{k=1}^{\infty}\ell_j^{s-1}(2a_j - \ell_j)\\ &= \frac{2^{3-s}\pi}{s-1}\sum_{j=1}^{\infty}a_j\ell_j^{s-1} - \frac{2^{2-s}\pi}{s-1}\sum_{j=1}^{\infty}\ell_j^s \\ &= \frac{2^{3-s}\pi}{s-1}\zeta_1(s) - \frac{2^{2-s}\pi}{s-1}\zeta_{\mathscr{L}}(s),\end{aligned} \tag{5.5.71}$$

where we have denoted by $\zeta_1 = \zeta_1(s)$ the first of the two sums appearing after the second equality and where $\zeta_{\mathscr{L}} = \zeta_{\mathscr{L}}(s)$ is the geometric zeta function of the fractal string $\mathscr{L}$.

Let us next consider an interesting special case of the fractal nest above; that is, the relative fractal drum (A_a, Ω) corresponding to the a-string $\mathscr{L} := \mathscr{L}_a$, with $a > 0$; so that $\ell_j := j^{-a} - (j+1)^{-a}$ for all $j \geq 1$ and hence, $a_j = j^{-a}$ for every $j \geq 1$. In this case, we have that (for $\operatorname{Re} s$ large enough)

$$\zeta_{A_a,\Omega}(s) = \frac{2^{3-s}\pi}{s-1}\sum_{j=1}^{\infty}j^{-a}\ell_j^{s-1} - \frac{2^{2-s}\pi}{s-1}\zeta_{\mathscr{L}}(s). \tag{5.5.72}$$

Since the geometric zeta function $\zeta_{\mathscr{L}} = \zeta_{\mathscr{L}_a}$ has already been analyzed in Example 5.5.10 (based on the results of [Lap-vFr3, Subsection 6.5.1]), we will now do the same for the zeta function ζ_1 by means of a technique analogous to the one used in the proof of [Lap-vFr3, Theorem 6.21]. Here, $\zeta_1(s)$ is initially defined by the following Dirichlet series (still with $\ell_j := j^{-a} - (j+1)^{-a}$, for all $j \geq 1$):

$$\zeta_1(s) = \sum_{j=1}^{\infty}j^{-a}\ell_j^{s-1}, \tag{5.5.73}$$

for all $s \in \mathbb{C}$ with $\operatorname{Re} s$ sufficiently large. Hence, we have $\zeta_1(s) = \zeta_{\mathscr{L},-a}(s-1)$, in the notation of the next theorem, and $\zeta_1(s) = \zeta_{\mathscr{L},-a,1}(s)$, in the notation of Corollary 5.5.18 following it.

Theorem 5.5.17. *Let $a > 0$, $b \in \mathbb{R}$, and let $\mathscr{L} = \mathscr{L}_a$ be the a-string with lengths ℓ_j given by (5.5.34); i.e., $\ell_j = j^{-a} - (j+1)^{-a}$ for all $j \geq 1$. Then, the Dirichlet series $\zeta_{\mathscr{L},b}(s) := \sum_{j=1}^{\infty} j^b \ell_j^s$ (defined initially for all $s \in \mathbb{C}$ with* Re *s sufficiently large) has a meromorphic continuation to all of $\mathbb{C}$. The poles of $\zeta_{\mathscr{L},b}$ are located at*

$$C := D(\zeta_{\mathscr{L},b}) = \frac{b+1}{a+1} \tag{5.5.74}$$

[49] Here, for $r > 0$, $B_r(x)$ denotes the open ball of radius r and with center at x.

and in (a subset of) $\left\{\frac{b-m}{a+1} : m \in \mathbb{N}_0\right\} \setminus \{0\}$, *and they are all simple.*[50] *In particular, we have the following inclusions:*[51]

$$\begin{aligned}\left\{\frac{b+1}{a+1}\right\} &\subseteq \mathscr{P}(\zeta_{\mathscr{L},b}) := \mathscr{P}(\zeta_{\mathscr{L},b}, \mathbb{C}) \\ &\subseteq \left\{\frac{b+1}{a+1}\right\} \cup \left(\left\{\frac{b-m}{a+1} : m \in \mathbb{N}_0\right\} \setminus \{0\}\right).\end{aligned} \tag{5.5.75}$$

Furthermore, the residue of $\zeta_{\mathscr{L},b}$ *at* $C = \frac{b+1}{a+1}$ *is equal to* $\frac{a^C}{a+1}$; *so that* $\frac{b+1}{a+1}$ *is always a (necessarily simple) pole of* $\zeta_{\mathscr{L},b}$.

Moreover, for any screen $\mathbf{S}_\sigma$ *chosen to be a vertical line* $\{\mathrm{Re}\, s = \sigma\}$, *with* $\sigma \in \mathbb{R} \setminus \mathscr{P}(\zeta_{\mathscr{L},b})$, *the zeta function* $\zeta_{\mathscr{L},b}$ *satisfies the languidity conditions* **L1** *and* **L2**, *with* $\kappa := \frac{1}{2} + b - (a+1)\sigma$ *if* $\sigma \le \frac{b}{a+1}$ *and* $\kappa := \frac{1}{2}(1 + b - (a+1)\sigma)$ *if* $\sigma \in \left[\frac{b}{a+1}, \frac{b+1}{a+1}\right]$.

Finally, we have that $\zeta_{\mathscr{L},b}(0) = \zeta(-b)$ *for all* $b \in \mathbb{R} \setminus \{-1\}$, *where* ζ *is the Riemann zeta function.*

Proof. We begin by computing the first term of an asymptotic expansion of ℓ_j:

$$\ell_j = j^{-a} - (j+1)^{-a} = a \int_j^{j+1} x^{-a-1}\, dx = a j^{-a-1} + H(j), \tag{5.5.76}$$

where $j \ge 1$ and $H(j) := a \int_j^{j+1} (x^{-a-1} - j^{-a-1})\, dx$. We next introduce a new variable $t := x/j - 1$ and let

$$h_j := a^{-1} j^{a+1} H(j) = j \int_0^{1/j} \left((1+t)^{-a-1} - 1\right) dt. \tag{5.5.77}$$

Note that $h_j = O(1/j)$ as $j \to \infty$. By now choosing an integer $M \ge 0$, we have

$$\begin{aligned} j^b \ell_j^s &= j^b \left(a j^{-a-1}(1+h_j)\right)^s \\ &= a^s j^{b-s(a+1)} \left(\sum_{n=0}^{M} \binom{s}{n} h_j^n + O\left(\frac{(|s|+1)^{M+1}}{j^{M+1}}\right)\right) \quad \text{as } j \to \infty, \end{aligned} \tag{5.5.78}$$

where we have let

$$\binom{s}{n} := \frac{(s-n+1)_n}{n!}, \quad \text{for all } s \in \mathbb{C} \text{ and } n \in \mathbb{N}_0. \tag{5.5.79}$$

---

[50] Here, as usual, we let $\mathbb{N}_0 := \mathbb{N} \cup \{0\}$.

[51] For 'generic' values of a and b, the second inclusion in (5.5.75) should be an equality while for 'most' values of those parameters, $\mathscr{P}(\zeta_{\mathscr{L},b})$ should at least contain an infinite subset of $\{\frac{b-m}{a+1} : m \in \mathbb{N}_0\}$. However, this informal comment will not be needed in the sequel and the corresponding conjecture has not been proved or even precisely formulated.

(Clearly, $\binom{s}{n}$ is a natural generalization of the usual binomial coefficient to an arbitrary value of the parameter $s \in \mathbb{C}$.) We thus obtain the following identity:

$$\zeta_{\mathscr{L},b}(s) = \sum_{n=0}^{M} a^s \binom{s}{n} \sum_{j=0}^{\infty} h_j^n j^{b-s(a+1)} + f(s), \tag{5.5.80}$$

where $f(s)$ is defined and holomorphic on the open half-plane $\{\operatorname{Re} s > \frac{b-M}{a+1}\}$. Furthermore, the first term (i.e., the term corresponding to $n = 0$ in the above sum) is equal to $a^s \zeta((a+1)s - b)$, where ζ is the Riemann zeta function, and thus has a single, simple pole at $s = C := D(\zeta_{\mathscr{L},b}) = \frac{b+1}{a+1}$.[52] In order to compute the residue of $a^s \zeta((a+1)s-b)$ at $s = \frac{b+1}{a+1}$, we use the fact that the principal part of the Riemann zeta function at $s = 1$ is equal to $1/(s-1)$ and consequently,

$$\lim_{s \to C} (s-C) a^s \zeta((a+1)s-b) = \lim_{s \to C} a^s \frac{s-C}{(a+1)s-b-1} = \frac{a^{\frac{b+1}{a+1}}}{a+1}. \tag{5.5.81}$$

A well-known result (due to Lindelöf) about the growth of the Riemann zeta function along vertical lines (see, e.g., [Edw, Section 9.2]) implies that the first term in (5.5.80) grows along the vertical lines $\{\operatorname{Re} s = \sigma\}$, for some $\sigma \in \mathbb{R}$, as $(|t|+1)^{\frac{1}{2}+b-\sigma(a+1)}$ if $\sigma < \frac{b}{a+1}$, as $(|t|+1)^{\frac{1}{2}(b+1-(a+1)\sigma)}$ if $\sigma \in \left[\frac{b}{a+1}, \frac{b+1}{a+1}\right]$, and is bounded from above by a constant (possibly depending on σ) if $\sigma > \frac{b+1}{a+1}$.

It now remains to analyze the functions

$$\sum_{j=1}^{\infty} h_j^n j^{b-(a+1)s}, \tag{5.5.82}$$

for each $n \geq 1$.

Let us fix $M \in \mathbb{N}_0$, for now. Then, the asymptotic expansion $(1+t)^{-a-1} = \sum_{m=0}^{M} \binom{-a-1}{m} t^m + O(t^{M+1})$ as $t \to 0^+$, together with (5.5.77), yields

$$\begin{aligned} h_j &= j \int_0^{1/j} \sum_{m=1}^{M} \binom{-a-1}{m} t^m \, \mathrm{d}t + O(j^{-M-1}) \\ &= -\frac{1}{a} \sum_{m=1}^{M} \binom{-a}{m+1} j^{-m} + O(j^{-M-1}) \quad \text{as } j \to \infty. \end{aligned} \tag{5.5.83}$$

We proceed by taking the n-th power of the above expansion to obtain an asymptotic expansion for h_j^n and substitute this into (5.5.82). It enables us to express each of the functions in (5.5.82) as a sum of constant multiples of $\zeta(m+(a+1)s-b)$, for $n \leq m \leq M$, and of a remainder term of order $O(j^{-M-1})$. Since $\zeta(m+(a+1)s-b)$ has a simple pole at $s = \frac{b+1-m}{a+1}$ and in view of (5.5.80), we conclude that $\zeta_{\mathscr{L},b}(s)$

[52] See, e.g., [Tit3] or [Edw] for the relevant properties of the Riemann zeta function. Recall, in particular, that ζ has a meromorphic continuation to all of $\mathbb{C}$ with a single, simple pole at $s = 1$ (with residue 1) and that it is initially defined by the Dirichlet series $\zeta(s) = \sum_{j=1}^{\infty} j^{-s}$ for all $s \in \mathbb{C}$ with $\operatorname{Re} s > 1$.

has a meromorphic continuation to the open right half-plane $\{\operatorname{Re} s > \frac{b+1-M}{1+a}\}$, with simple poles at $s = \frac{b+1-m}{1+a}$ for $m = 0,1,2,\ldots,M$. To be more specific, some of these potential poles of $\zeta_{\mathscr{L},b}$ may not actually be poles (due to cancellations), depending on the choice of the parameters a and b. (See, however, the unproven assertion in footnote 51 in this subsection.) Furthermore, 0 is never a pole of $\zeta_{\mathscr{L},b}$, since we can see from (5.5.80) that it is canceled by the factor $\binom{s}{m}$ for $m \geq 1$. Moreover, since M is arbitrary, we conclude that $\zeta_{\mathscr{L},b}$ has a meromorphic continuation to all of $\mathbb{C}$. Next, note that for each integer $m \geq 1$, the growth of $\zeta(m+(a+1)s-b)$ is dominated by the growth of the first term $a^s\zeta((a+1)s-b)$ and therefore, we have proved the statement about the languidity of $\zeta_{\mathscr{L},b}$.

Finally, the last statement of the theorem follows from an application of the principle of analytic continuation since we deduce directly from the definition of $\zeta_{\mathscr{L},b}$ that $\zeta_{\mathscr{L},b}(0) = \zeta(-b)$ for all $b \in \{\operatorname{Re} s < -1\}$. □

In order to complete the present discussion of the example of the fractal nests, as well as in preparation for the example of the unbounded geometric chirps (Example 5.5.19 below), we will need the following simple consequence of the above theorem.

Corollary 5.5.18. *Let $a > 0$, $b \in \mathbb{R}$, $\tau \in \mathbb{R}$ and let $\mathscr{L} := \mathscr{L}_a$ be the a-string with lengths ℓ_j given by (5.5.34). Then, the Dirichlet series $\zeta_{\mathscr{L},b,\tau}(s) := \sum_{j=1}^{\infty} j^b \ell_j^{s-\tau}$ (initially defined for all $s \in \mathbb{C}$ with $\operatorname{Re} s$ sufficiently large) has a meromorphic continuation to all of $\mathbb{C}$. The poles of $\zeta_{\mathscr{L},b,\tau}$ are located at*

$$D(\zeta_{\mathscr{L},b,\tau}) = \frac{b+1}{a+1} + \tau \tag{5.5.84}$$

and in (a subset of) $\{\frac{b-m}{a+1} + \tau : m \in \mathbb{N}_0\} \setminus \{\tau\}$, and they are all simple. In particular, we have the following inclusions:[53]

$$\begin{aligned}\left\{\frac{b+1}{a+1} + \tau\right\} &\subseteq \mathscr{P}(\zeta_{\mathscr{L},b,\tau}) := \mathscr{P}(\zeta_{\mathscr{L},b,\tau}, \mathbb{C}) \\ &\subseteq \left\{\frac{b+1}{a+1} + \tau\right\} \cup \left(\left\{\frac{b-m}{a+1} + \tau : m \in \mathbb{N}_0\right\} \setminus \{\tau\}\right).\end{aligned} \tag{5.5.85}$$

Furthermore, the residue of $\zeta_{\mathscr{L},b,\tau}$ at $\frac{b+1}{a+1} + \tau$ is equal to $\frac{a^{(b+1)/(a+1)}}{a+1}$; so that $D(\zeta_{\mathscr{L},b,\tau}) = \frac{b+1}{a+1} + \tau$ is always a (necessarily simple) pole of $\zeta_{\mathscr{L},b,\tau}$.

Moreover, for any screen $\mathbf{S}_\sigma$ chosen to be a vertical line $\{\operatorname{Re} s = \sigma\}$, with $\sigma \in \mathbb{R} \setminus \mathscr{P}(\zeta_{\mathscr{L},b,\tau})$, the zeta function $\zeta_{\mathscr{L},b,\tau}$ satisfies the languidity conditions **L1** *and* **L2**, *with $\kappa := \frac{1}{2} + b - (a+1)\sigma$ if $\sigma \leq \frac{b}{a+1} + \tau$ and $\kappa := \frac{1}{2}(1 + b - (a+1)\sigma)$ if $\sigma \in \left[\frac{b}{a+1} + \tau, \frac{b+1}{a+1} + \tau\right]$.*

Finally, we have that $\zeta_{\mathscr{L},b,\tau}(\tau) = \zeta(-b)$ for all $b \in \mathbb{R} \setminus \{-1\}$.

---

[53] A comment entirely analogous to the one made in footnote 51 on page 504 holds relative to 'generic' (or else 'most') values of the parameters a, b and τ. (Recall that $\mathscr{L} = \mathscr{L}_a$, so that $\zeta_{\mathscr{L},b,\tau}$ depends on a, b and τ.)

Proof. Since $\zeta_{\mathscr{L},b,\tau}(s) = \zeta_{\mathscr{L},b}(s-\tau)$, this an immediate consequence of Theorem 5.5.17. $\square$

Let us now return to Example 5.5.16, where the distance zeta function of (A_a,Ω) is given by (5.5.72); see also (5.5.73) and the brief discussion following it. We therefore deduce from Corollary 5.5.18 and the discussion of $\zeta_{\mathscr{L}} = \zeta_{\mathscr{L}_a}$ in Example 5.5.16, combined with an application of the principle of analytic continuation, that $\zeta_{A_a,\Omega}$ is meromorphic on all of $\mathbb{C}$ and is given for all $s \in \mathbb{C}$ by

$$\zeta_{A_a,\Omega}(s) = \frac{2^{3-s}\pi}{s-1}\zeta_{\mathscr{L},-a,1}(s) - \frac{2^{2-s}\pi}{s-1}\zeta_{\mathscr{L}}(s). \tag{5.5.86}$$

Moreover, the set of complex dimensions of (A_a,Ω) satisfies the inclusion

$$\begin{aligned}\mathscr{P}(\zeta_{A_a,\Omega}) &:= \mathscr{P}(\zeta_{A_a,\Omega},\mathbb{C}) \\ &\subseteq \left\{1, \frac{2}{a+1}, \frac{1}{a+1}\right\} \cup \left\{-\frac{m}{a+1} : m \in \mathbb{N}\right\}.\end{aligned} \tag{5.5.87}$$

We do not have an equality here, due to the possibility of zero-pole cancellations. Furthermore, if $a \neq 1$, all of the above (potential) complex dimensions are simple. Moreover, we are certain that $\frac{2}{a+1}$ is always a complex dimension of (A_a,Ω) since it is never canceled, as a pole. In fact, by letting $D := \frac{2}{a+1}$, we have for all positive $a \neq 1$ that

$$\operatorname{res}\left(\zeta_{A_a,\Omega}, D\right) = \frac{2^{2-D}D\pi}{D-1}a^{D-1}. \tag{5.5.88}$$

We now conclude from Theorem 5.4.2 (and part (b) of Theorem 4.1.7) that if $a \in (0,1)$, $\dim_B(A_a,\Omega) = D(\zeta_{A_a,\Omega}) = D$ and (A_a,Ω) is Minkowski measurable with Minkowski content given by

$$\mathscr{M}^D(A_a,\Omega) = \frac{2^{2-D}D\pi}{(2-D)(D-1)}a^{D-1}. \tag{5.5.89}$$

Furthermore, it also follows from Theorem 5.4.2 that if $a > 1$, we have that $\dim_B(A_a,\Omega) = 1$ and the corresponding residue is given by

$$\operatorname{res}(\zeta_{A_a,\Omega}, 1) = 4\pi\zeta_{\mathscr{L},-a,1}(1) - 2\pi\zeta_{\mathscr{L}}(1) = 4\pi\zeta(a) - 2\pi. \tag{5.5.90}$$

Therefore, still for $a > 1$, the RFD (A_a,Ω) is Minkowski measurable with Minkowski content given by

$$\mathscr{M}^1(A_a,\Omega) = 4\pi\zeta(a) - 2\pi; \tag{5.5.91}$$

note that $\mathscr{M}^1(A_a,\Omega)$ is positive since $\zeta(a) > 1$ for $a > 1$; so that

$$2\pi < \mathscr{M}^1(A_a,\Omega) < \infty.$$

In the critical case when $a=1$, we have that $s=1$ is a pole of second order (i.e., of multiplicity two) of $\zeta_{A_1,\Omega}$ and since it is a simple pole of $\zeta_{\mathscr{L},-1,1}$, we deduce from (5.5.86) that

$$\operatorname{res}(\zeta_{A_1,\Omega},1)=4\pi\zeta_{\mathscr{L},-1,1}[1]_0-2\pi, \tag{5.5.92}$$

where for each $m\in\mathbb{Z}$, $\zeta_{\mathscr{L},-1,1}[\omega]_m$ stands for the m-th coefficient in the Laurent series expansion of $\zeta_{\mathscr{L},-1,1}$ around $s=\omega$. We conclude that in this case (i.e., when $a=1$), by Theorem 2.2.3 (and part (b) of Theorem 4.1.7), the RFD (A_1,Ω) must be Minkowski degenerate with $\dim_B(A_1,\Omega)=D(\zeta_{A_1,\Omega})=1$.[54] We can also compute the coefficient corresponding to $(s-1)^{-2}$ in the Laurent expansion of $\zeta_{A_1,\Omega}$ around $s=1$, by using Corollary 5.5.18:

$$\zeta_{A_1,\Omega}[1]_{-2}=4\pi\operatorname{res}(\zeta_{\mathscr{L},-1,1},1)=2\pi. \tag{5.5.93}$$

Assume now that $a\neq 1$. For $M\in\mathbb{N}\cup\{0\}$, as before, we choose the screen $\boldsymbol{S}_M$ to be some vertical line between $-\frac{M+1}{1+a}$ and $-\frac{M+2}{1+a}$, and let $\boldsymbol{W}_M$ be the corresponding window. After having applied Theorem 5.3.21, we then obtain the following asymptotic distributional formula for the tube function $V(t):=|(A_a)_t\cap\Omega|$, as $t\to 0^+$:

$$\begin{aligned}V(t)=&\frac{2^{2-D}D\pi}{(2-D)(D-1)}a^{D-1}t^{2-D}+(4\pi\zeta(a)-2\pi)t\\&+\frac{\operatorname{res}\left(\zeta_{A_a,\Omega},\frac{1}{a+1}\right)t^{2-\frac{1}{a+1}}}{2-\frac{1}{a+1}}\\&+\sum_{m=1}^{M}\frac{\operatorname{res}\left(\zeta_{A_a,\Omega},-\frac{m}{a+1}\right)t^{2+\frac{m}{a+1}}}{2+\frac{m}{a+1}}+O\big(t^{2+\frac{M+1}{a+1}}\big)\quad\text{as}\quad t\to 0^+,\end{aligned} \tag{5.5.94}$$

where the sum is interpreted as being equal to 0 if $M=1$. By choosing as a screen a vertical line $\{\operatorname{Re} s=\sigma\}$, with $\sigma>-\frac{1}{2(a+1)}$, we obtain a pointwise fractal tube formula with a pointwise error term of order $O(t^{2-\sigma})$; indeed, in light of Corollary 5.5.18, we have that $\kappa_d<0$ and hence, we can apply part (i) of Theorem 5.3.16. This pointwise formula is still given by (5.5.94) but now interpreted pointwise and valid for all $t>0$. It is actually initially valid for all $t\in(0,\delta)$ but since $\delta>\ell_1/2$ may be taken arbitrary large, we conclude that it is valid for all $t>0$. Of course, we actually do not know much about the above error term when t is not close to zero, but that does not matter because we are not interested in the values of $V(t)=|(A_a)_t\cap\Omega|$ for large t. (Note also that clearly, $|(A_a)_t\cap\Omega|=|\Omega|=|B_1(0)|=\pi$, for all t sufficiently large.)

54 Actually, it can also be shown directly that $\mathscr{M}^1(A_1,\Omega)$ exists in this case and is equal to $+\infty$; see Example 3.5.1.

Let us next consider the critical case when $a=1$. Choose a screen given by the vertical line $\{\mathrm{Re}\, s=\sigma\}$, with $\sigma\in(-3/4,-1/2)$; we then obtain the following pointwise fractal tube formula with error term:

$$\begin{aligned} V(t) &= \operatorname{res}\left(\frac{t^{2-s}}{2-s}\zeta_{A_1,\Omega}(s),1\right)+\frac{2}{3}\operatorname{res}\left(\zeta_{A_1,\Omega},\frac{1}{2}\right)t^{\frac{3}{2}} \\ &+\frac{2}{5}\operatorname{res}\left(\zeta_{A_1,\Omega},-\frac{1}{2}\right)t^{\frac{5}{2}}+O(t^{2-\sigma}) \quad \text{as} \quad t\to 0^+. \end{aligned} \tag{5.5.95}$$

We expand the function $t^{2-s}/(2-s)$ into a Taylor series around $s=1$, as follows:

$$\frac{t^{2-s}}{2-s}=t\sum_{n=0}^{\infty}(s-1)^n\sum_{k=0}^{n}\frac{(-1)^{n-k}(\log t^{-1})^k}{k!(n-k)!}. \tag{5.5.96}$$

We then deduce from (5.5.92) and (5.5.93) that

$$\operatorname{res}\left(\frac{t^{2-s}}{2-s}\zeta_{A_1,\Omega}(s),1\right)=2\pi t\log t^{-1}+4\pi t(\zeta_{\mathscr{L},-1,1}[1]_0-1); \tag{5.5.97}$$

so that (still pointwise)

$$V(t)=2\pi t\log t^{-1}+4\pi t(\zeta_{\mathscr{L},-1,1}[1]_0-1)+o(t) \quad \text{as} \quad t\to 0^+. \tag{5.5.98}$$

The above tube formula is in agreement with the fact that (A_1,Ω) is Minkowski degenerate but it is also clear that one can choose the function $h(t):=\log t^{-1}$, for all $t\in(0,1)$, as an appropriate gauge function (see the beginning of Subsection 4.5.1 for an introduction to gauge functions, along with Definition 6.1.4 below). More precisely, one then has that $\mathscr{M}^1(A_1,\Omega,h)$, the *gauge relative Minkowski content* of (A_1,Ω), is well defined and

$$\mathscr{M}^1(A_1,\Omega,h)=\lim_{t\to 0^+}\frac{|(A_1)_t\cap\Omega|}{t\,h(t)}=2\pi. \tag{5.5.99}$$

In particular, the RFD (A,Ω) is h-Minkowski measurable, in the sense of Definition 6.1.6 (extended to RFDs in the obvious way). The choice of the above gauge function h is connected with the fact that when $a=1$, $\zeta_{A,\Omega}$ possesses a (unique) pole of order two on the critical line, located at $s=1$. (See Subsection 5.4.4 above.) Alternatively, one may also view this situation as a kind of "merging" of two simple complex dimensions of the RFD (A_a,Ω), namely, 1 and $2/(a+1)$, into a single complex dimension of order two (located at $s=1$) as $a\to 1$.

Example 5.5.19. (*Unbounded geometric chirps*). In this example, we consider and study a type of unbounded geometric chirp. A standard *geometric* (α,β)-*chirp*, with positive parameters α and β, is a simple geometric approximation of the graph of the function $f(x):=x^{\alpha}\sin(\pi x^{-\beta})$, for all $x\in(0,1)$. (See Example 4.4.1 and Proposition 4.4.3, along with Figures 4.13, 4.14 and 4.15 on pages 345–347.)

If the parameters α and β satisfy the inequalities $-1<\alpha<0<\beta$, as we shall assume from now on, we obtain an example of an unbounded chirp function f which we approximate by the unbounded geometric (α,β)-chirp. More specifically, let $A_{\alpha,\beta}$ be the union of vertical segments with abscissae $x:=j^{-1/\beta}$ and of lengths $j^{-\alpha/\beta}$, for every $j\in\mathbb{N}$. Furthermore, define Ω as a union of the rectangles R_j for $j\in\mathbb{N}$, where R_j has a base of length $j^{-1/\beta}-(j+1)^{-1/\beta}$ and height $j^{-\alpha/\beta}$; see Figure 4.15. The relative distance zeta function of (A,Ω) is computed in Example 4.4.1 and is given by

$$\begin{aligned}\zeta_{A_{\alpha,\beta},\Omega}(s)&=\frac{2^{2-s}}{(s-1)}\sum_{j=1}^{\infty}j^{-\alpha/\beta}\left(j^{-1/\beta}-(j+1)^{-1/\beta}\right)^{s-1}\\&=\frac{2^{2-s}}{(s-1)}\zeta_{\mathscr{L},-\alpha/\beta,1}(s),\end{aligned}\tag{5.5.100}$$

where $\mathscr{L}$ is the β^{-1}-string. In light of Corollary 5.5.18, we conclude that $\zeta_{A_{\alpha,\beta},\Omega}$ has a meromorphic continuation to all of $\mathbb{C}$ and

$$\mathscr{P}(\zeta_{A_{\alpha,\beta},\Omega})\subseteq\left\{1,2-\frac{1+\alpha}{1+\beta}\right\}\cup\{D_m:m\in\mathbb{N}\},\tag{5.5.101}$$

where $D_m:=2-\frac{1+\alpha+m\beta}{1+\beta}$. Let $D:=2-\frac{1+\alpha}{1+\beta}$. Also, by the same corollary and from (5.5.100), we have that both 1 and D are simple poles of $\zeta_{A_{\alpha,\beta}}$. Furthermore, we have that $D>1$ and, consequently, $\dim_B(A_{\alpha,\beta},\Omega)=D$ and the RFD $(A_{\alpha,\beta},\Omega)$ is Minkowski measurable with Minkowski content given by

$$\begin{aligned}\mathscr{M}^D(A_{\alpha,\beta},\Omega)&=\frac{\operatorname{res}(\zeta_{A_{\alpha,\beta},\Omega},D)}{2-D}=\frac{2^{2-D}}{(2-D)(D-1)}\frac{\beta^{\frac{1+\alpha}{1+\beta}}}{1+\beta}\\&=\frac{(2\beta)^{2-D}}{(2-D)(D-1)(1+\beta)}.\end{aligned}\tag{5.5.102}$$

Moreover, the residue at $s=1$ is given by

$$\operatorname{res}(\zeta_{A_{\alpha,\beta},\Omega},1)=2\zeta_{\mathscr{L},-\alpha/\beta,1}(1)=2\zeta\left(\frac{\alpha}{\beta}\right).\tag{5.5.103}$$

It follows that $s=1$ is indeed a simple pole of $\zeta_{A_{\alpha,\beta},\Omega}$.

Similarly as in previous examples, for $M\in\mathbb{N}\cup\{0\}$, we choose the screen $\boldsymbol{S}_M$ to be a vertical line $\{\operatorname{Re}s=\sigma\}$, for some real number σ lying strictly between $2-\frac{1+\alpha+(M+1)\beta}{1+\beta}$ and $2-\frac{1+\alpha+(M+2)\beta}{1+\beta}$ (so that $\zeta_{A_{\alpha,\beta},\Omega}$ is d-languid in the associated window), and let $\boldsymbol{W}_M$ be the corresponding window. From Theorem 5.3.21, we then obtain the following asymptotic distributional formula for the tube function $V(t):=|(A_{\alpha,\beta})_t\cap\Omega|$:

$$
\begin{aligned}
V(t) = {} & \frac{(2\beta t)^{2-D}}{(2-D)(D-1)(1+\beta)} + \frac{t^{2-D_1}\operatorname{res}(\zeta_{A_{\alpha,\beta},\Omega}, D_1)}{2-D_1} \\
& + 2t\,\zeta\Big(\frac{\alpha}{\beta}\Big) + \sum_{m=2}^{M} \frac{t^{2-D_m}\operatorname{res}(\zeta_{A_{\alpha,\beta},\Omega}, D_m)}{2-D_m} \\
& + O(t^{2-D_{M+1}}) \quad \text{as} \quad t \to 0^+.
\end{aligned}
\tag{5.5.104}
$$

Note that the second noninteger complex dimension, namely, $D_1 = 1 - \frac{\alpha}{1+\beta}$, is also greater than 1. Finally, by choosing as a screen a vertical line to the right of $-\frac{2\alpha+\beta}{1+\beta}$, we actually obtain a pointwise formula still given by (5.5.104) above; indeed, we then have $\kappa_d < 0$, so that we can apply part (*i*) of Theorem 5.3.16.

5.5.6 Tube Formulas and Minkowski Measurability Criteria for Self-Similar Sprays

We conclude this section by explaining how the results of this chapter may also be applied to recover and significantly extend, as well as place within a general conceptual framework, the tube formulas for self-similar sprays generated by an arbitrary open set $G \subset \mathbb{R}^N$ of finite N-dimensional Lebesgue measure. (See, especially, [LapPe2–3] extended to a significantly more general setting in [LapPeWi1], along with the exposition of those results given in [Lap-vFr3, Section 13.1]; see also [DenKoÖÜ] for another, but related, proof of some of those results.)

Recall from Definition 4.2.11 that a self-similar spray (with a single generator G, assumed to be bounded and open) is defined as a collection $(G_k)_{k\in\mathbb{N}}$ of pairwise disjoint (bounded) open sets $G_k \subset \mathbb{R}^N$, with $G_0 := G$ and such that for each $k \in \mathbb{N}$, G_k is a scaled copy of G by some factor $\lambda_k > 0$. (We let $\lambda_0 := 1$.) The associated scaling sequence $(\lambda_k)_{k\in\mathbb{N}}$ is obtained from a ratio list $\{r_1, r_2, \dots, r_J\}$, with $0 < r_j < 1$ for each $j = 1, \dots, J$ and such that $\sum_{j=1}^{J} r_j^N < 1$, by considering all possible words built out of the scaling ratios r_j; see Equation (4.2.31). Here, as in Definition 4.2.11, $J \geq 2$ and the scaling ratios $r_1, \dots, r_J$ are repeated according to their multiplicities.

Let us next assume that (A, Ω) is the self-similar spray considered as a relative fractal drum and defined as $A := \partial(\sqcup_{k=0}^{\infty} G_k)$ and $\Omega := \sqcup_{k=0}^{\infty} G_k$, with $\overline{\dim}_B(\partial G, G) < N$. Then, by Theorem 4.2.17, we have the following key formula, called a *factorization formula*, for its associated distance zeta function $\zeta_{A,\Omega}$, expressed as follows in terms of the distance zeta function of the boundary of the generator (relative to the generator), $\zeta_{\partial G,G}$, and the scaling ratios $\{r_j\}_{j=1}^{J}$:

$$
\zeta_{A,\Omega}(s) = \frac{\zeta_{\partial G}(s, G)}{1 - \sum_{j=1}^{J} r_j^s}. \tag{5.5.105}
$$

(See also part (a) of Remark 5.5.26 below.)[55] It now suffices to assume that the relative distance zeta function $\zeta_{\partial G,G}$ of the generating relative fractal drum $(\partial G, G)$ satisfies suitable languidity conditions in order to apply (at level $k=0$) the fractal tube formulas of Sections 5.1–5.3 and to obtain a pointwise or distributional formula, with or without error term, for the 'inner' volume of $\sqcup_{k=0}^{\infty} G_k$:[56]

$$\begin{aligned} V_{A,\Omega}(t) &:= |A_t \cap \Omega| \\ &= \sum_{\omega \in (\mathfrak{D} \cap \boldsymbol{W}) \cup \mathscr{P}(\zeta_{\partial G,G}, \boldsymbol{W})} \operatorname{res}\left(\frac{t^{N-s} \zeta_{\partial G,G}(s)}{(N-s)\left(1 - \sum_{j=1}^{J} r_j^s\right)}, \omega \right) \\ &\quad + R_{A,\Omega}(t), \end{aligned} \tag{5.5.106}$$

where $\mathfrak{D}$ denotes the set of solutions in $\mathbb{C}$ of $\sum_{j=1}^{J} r_j^s = 1$, the complexified Moran equation, and $R_{A,\Omega} := R_{A,\Omega}^{[0]}$ is a pointwise or distributional error term (or else $R_{A,\Omega}(t) \equiv 0$ and $\boldsymbol{W} := \mathbb{C}$, in the case of an exact tube formula, provided $\zeta_{\partial G,G}$ is strongly d-languid), depending on the d-languidity growth conditions satisfied by $\zeta_{\partial G,G}$.

In the d-languid (but not necessarily strongly d-languid) case, $R_{A,\Omega} = R_{A,\Omega}^{[0]}$ satisfies the following (pointwise or distributional) error estimate (at level $k=0$):

$$R_{A,\Omega}(t) = O(t^{N - \sup S}) \quad \text{as} \quad t \to 0^+, \tag{5.5.107}$$

where S is the screen associated to the window $\boldsymbol{W}$.

Remark 5.5.20. Observe that in the notation also used in part (b) of Remark 5.5.26 below (and towards the end of Chapter 6), we can rewrite Equation (5.5.105) as follows:

$$\zeta_{A,\Omega}(s) = \zeta_{\mathfrak{S}}(s) \cdot \zeta_{\partial G,G}(s), \tag{5.5.108}$$

where the geometric zeta function $\zeta_{\mathfrak{S}}$ of the associated self-similar string (with scaling ratios $\{r_j\}_{j=1}^{J}$ and a single gap length, equal to one, in the terminology of [Lap-vFr3, Chapters 2 and 3]) is meromorphic in all of $\mathbb{C}$ and given for all $s \in \mathbb{C}$ by

$$\zeta_{\mathfrak{S}}(s) = \frac{1}{1 - \sum_{j=1}^{J} r_j^s}. \tag{5.5.109}$$

In general, given a connected open set $U \subseteq \mathbb{C}$ containing the vertical line $\{\operatorname{Re} s = \dim_B(A,\Omega)\}$, $\zeta_{A,\Omega}$ is meromorphic in U if and only if $\zeta_{\partial G,G}$ is; furthermore, in that case, the *factorization formula* (5.5.108) then holds for all $s \in U$. We note that in the

[55] For the case of multiple generators, see Equation (5.5.172) in part (b) of Remark 5.5.26.

[56] Here and throughout the rest of this subsection, we use the notation $V_{A,\Omega}$, consistent with the statement of a pointwise tube formula. In the case of the distributional tube formulas, we should use instead the notation $\mathscr{V}_{A,\Omega}$. (And analogously for the error term $R_{A,\Omega}(t)$ in (5.5.106), which should then be denoted by $\mathscr{R}_{A,\Omega}(t)$, in the distributional case.) For notational simplicity, however, we will not do so in this discussion.

sequel (see Remark 5.5.26 below) and following [LapPe2–3] and [LapPeWi1–2], we will often refer to $\zeta_{\mathfrak{S}}$ as the *scaling zeta function* of the self-similar spray (A,Ω) and to its poles in $\mathbb{C}$ (composing the multiset $\mathfrak{D}$) as the *scaling complex dimensions* of (A,Ω). In the present case, they are the solutions (counting multiplicities) of the *complexified Moran equation* $\sum_{j=1}^{J} r_j^s = 1$. We will also sometimes write $\mathfrak{D}_{\mathfrak{S}}$ instead of $\mathfrak{D}$, so that $\mathfrak{D}_{\mathfrak{S}} := \mathfrak{D}$; hence, similarly, $\mathfrak{D}_{\mathfrak{S}} \cap \boldsymbol{W} = \mathfrak{D} \cap \boldsymbol{W}$, the set of *visible scaling complex dimensions* of (A,Ω), denotes the set of poles of $\zeta_{\mathfrak{S}}$ visible through the window $\boldsymbol{W}$. See Equations (5.5.106) above and (5.5.175) below.

Typically, we will work with generators such that $\zeta_{\partial G,G}$ is strongly d-languid and consequently, since $\zeta_{\mathfrak{S}}$ (as given by (5.5.109)) is strongly d-languid (after a possible scaling by an appropriate scaling factor $\lambda_{\mathfrak{S}} > 0$; see Corollary 5.3.14 and the discussion preceding it), $\zeta_{A,\Omega}$ will be strongly d-languid (also after a possible scaling by the same scaling factor $\lambda_{\mathfrak{S}}$) and given by the factorization formula (5.5.108) (or (5.5.105)), for all $s \in \mathbb{C}$. As a result, unless we need to work with a 'truncated tube formula' (corresponding to a fractal tube formula with error term associated with a suitable screen $\boldsymbol{S}$), we will be able to obtain an *exact* fractal tube formula, as we will now see.

Assume next that the generator G is *monophase* (in the sense of [LapPe2–3] and [LapPeWi1–2]); that is, the volume of its 'inner' t-neighborhood is given by a polynomial $\sum_{i=0}^{N-1} \kappa_i t^{N-i}$ for all $t \in \mathbb{R}$ such that $0 < t < g$. Here, g is the *inradius* of G, i.e., the supremum of the radii of all the balls which are contained in G. Since then,

$$V_{\partial G,G}(t) := |(\partial G)_t \cap G| = \sum_{i=0}^{N-1} \kappa_i t^{N-i}, \tag{5.5.110}$$

for $0 < t < g$, we can explicitly calculate the relative tube zeta function of G, as follows (initially, for all $s \in \mathbb{C}$ with $\operatorname{Re} s$ sufficiently large):

$$\tilde{\zeta}_{\partial G,G}(s;g) = \int_0^g t^{s-N-1} \sum_{i=0}^{N-1} \kappa_i t^{N-i}\, \mathrm{d}t = \sum_{i=0}^{N-1} \frac{\kappa_i g^{s-i}}{s-i}. \tag{5.5.111}$$

It is obviously meromorphic on all of $\mathbb{C}$ and still given by (5.5.111) for all $s \in \mathbb{C}$.

Using the functional equation which connects the relative tube and distance zeta functions (see Equation (4.5.2)), we now obtain the following explicit expression for the relative distance zeta function of the generator G:

$$\begin{aligned} \zeta_{\partial G,G}(s) := \zeta_{\partial G,G}(s;g) &= g^{s-N}|(\partial G)_g \cap G| + (N-s)\tilde{\zeta}_{\partial G,G}(s;g) \\ &= g^{s-N}|G| + (N-s)\sum_{i=0}^{N-1} \frac{\kappa_i g^{s-i}}{s-i} \\ &= (N-s)\sum_{i=0}^{N} \frac{\kappa_i g^{s-i}}{s-i}, \end{aligned} \tag{5.5.112}$$

where we have let $\kappa_N := -|G|$.

Consequently, by substituting (5.5.112) into (5.5.106), we recover (and significantly extend as well as place within the broader framework of the theory of fractal tube formulas via fractal zeta functions) a well-known result obtained in [LapPe3] and more generally in [LapPeWi1], as well as more recently, via a different (but related) technique in [DenKoÖÜ]:

$$V_{A,\Omega}(t) := |A_t \cap \Omega| = \sum_{\omega \in \mathfrak{D} \cup \{0,1,\ldots,N-1\}} \operatorname{res}\left(t^{N-s} \frac{\sum_{i=0}^{N} \kappa_i \frac{g^{s-i}}{s-i}}{\left(1 - \sum_{j=1}^{J} r_j^s\right)}, \omega \right). \qquad (5.5.113)$$

This is an exact pointwise fractal tube formula. Indeed, after an appropriate scaling by a factor $\lambda_G > 0$, $\zeta_{\partial G,G}$ is shown to be strongly d-languid with $(\kappa_d)_G := 0$ for a suitable infinite sequence of vertical lines $\{\operatorname{Re} s = \alpha_m\}$, $m \geq 1$ with $\alpha_m \in \mathbb{R}$ and $\alpha_m \to -\infty$ as $m \to \infty$. Also, it is easy to check (after an appropriate scaling by a factor $\lambda_{\mathfrak{S}} > 0$) that $\zeta_{\mathfrak{S}}(s) = (1 - \sum_{j=1}^{J} r_j^s)^{-1}$ is strongly d-languid, with $(\kappa_d)_{\mathfrak{S}} := 0$ (see [Lap-vFr3, Equation (6.36), p. 195]). Hence (after a suitable scaling by $\lambda := \lambda_{A,\Omega}$, depending on both λ_G and $\lambda_{\mathfrak{S}}$), we deduce from the factorization formula (5.5.105) (or, equivalently, (5.5.108)) that $\zeta_{A,\Omega}$ is strongly d-languid, with exponent $(\kappa_d)_{A,\Omega} := 0$ for this same sequence of vertical lines $\{\operatorname{Re} s = \alpha_m\}$, $m \geq 1$. We can therefore conclude from Theorem 5.3.16 that the tube formula (5.5.113) is valid pointwise and without an error term in this case, for all positive t sufficiently small.[57]

If needed, one can also obtain a corresponding 'truncated' pointwise fractal tube formula (with error term), relative to a suitable screen, in the spirit of [Lap-vFr3, Corollary 8.27 and Subsection 8.4.4].

A completely analogous reasoning can be used for the case of *pluriphase* generators G for which the 'inner' tubular volume is given as a piecewise polynomial. In a future work, we plan to investigate for which classes of generators the tube formula (5.5.106) can be applied pointwise or distributionally. It is clearly a very large class, corresponding to essentially all of the self-similar sprays (and hence, also all of the self-similar tilings, in the sense of [Pe, LapPe2–3, LapPeWi1–2, PeWi]) of interest, including (in light of the results of [KoRati], proving a conjecture of [LapPe2–3]) those with generators that are convex polyhedra (or polytopes), under mild assumptions.

Remark 5.5.21. We point out that in [LapPeWi1], which (prior to the present work and that in [LapRaŽu5]) was the paper containing the most elaborate results concerning the fractal tube formulas for self-similar sprays (and other fractal sprays), a considerable amount of effort was required to obtain analogous (but less general) fractal tube formulas, with or without error term. Furthermore, the 'tubular zeta functions' used in [LapPe2–3] and, in the more general context of [LapPeWi1], were introduced in an ad hoc manner. Here, by contrast, both the fractal tube formulas and the fractal zeta functions (in the present situation, the distance zeta functions)

---

[57] For an exact interval within which the fractal tube formula is valid, one has to explicitly calculate the scaling factors λ_G and $\lambda_{\mathfrak{S}}$, which we leave as an exercise for the interested reader.

occurring in the corresponding tube formulas follow naturally from the general theory developed in this book, and in particular, in the present chapter.

We close this comment by mentioning that the interested reader can find in [LapPe2–3], [LapPeWi1–2], as well as in the exposition given in [Lap-vFr3, Subsection 13.1.4], a number of concrete examples illustrating the fractal tube formulas for self-similar sprays (or tilings). These examples include the Cantor tiling, the Koch tiling, the Sierpiński gasket and carpet tilings, along with the pentagasket tiling (see, e.g., [Lap-vFr3, Example 13.33]), which is an interesting and natural example of a self-similar spray with multiple generators; see part (b) of Remark 5.5.26 below for the relevance of this latter example. In all of these examples, the underlying generators of the fractal sprays (or tilings) are convex polygons and therefore, satisfy the required assumptions.

The next four examples respectively complement Examples 4.2.33, 4.2.34, 4.2.35 and 4.2.36 from Subsection 4.2.3. We obtain here their corresponding fractal tube formulas and illustrate the interesting situations which may arise, in particular, for self-similar sprays (or tilings), or more generally, for self-similar RFDs. These examples enable us, in particular, to further illustrate our proposed definition of (critical and subcritical) fractality (see Remark 4.6.24). Accordingly, the sets and RFDs considered in these examples are indeed fractal, in that sense, and their fractality reflects their intrinsic geometric oscillations, as is made evident by the corresponding fractal tube formulas.

Example 5.5.22. (*Fractal tube formula for the* $1/2$*-square fractal*). Let us consider the $1/2$-square fractal A from Example 4.2.33 and depicted in Figure 4.10. Its distance zeta function was obtained in Example 4.2.33 (see Equation (4.2.114)), where it was shown to be meromorphic on all of $\mathbb{C}$ and given by

$$\zeta_A(s) = \frac{2^{-s}}{s(s-1)(2^s-2)} + \frac{4}{s-1} + \frac{2\pi}{s}, \tag{5.5.114}$$

for every $s \in \mathbb{C}$. In (5.5.114), without loss of generality, we have chosen $\delta := 1$. Furthermore, as was discussed in Example 4.2.33, it follows at once from (5.5.114) that

$$\begin{aligned} D(\zeta_A) &= 1, \\ \mathscr{P}(\zeta_A) := \mathscr{P}(\zeta_A, \mathbb{C}) &= \{0\} \cup (1 + \mathbf{p}\mathrm{i}\mathbb{Z}) \end{aligned} \tag{5.5.115}$$

and

$$\dim_{PC} A := \mathscr{P}_c(\zeta_A) = \{1\}, \tag{5.5.116}$$

where the oscillatory period $\mathbf{p}$ of A is given by $\mathbf{p} := \frac{2\pi}{\log 2}$ and all of the complex dimensions in $\mathscr{P}(\zeta_A)$ are simple, except for $\omega_0 := 1$ which is a double pole of ζ_A.

One easily sees that λA is strongly d-languid for $\kappa_d := -1$, for any $\lambda \geq 2$ and for a sequence of screens consisting of the vertical lines $\{\mathrm{Re}\, s = -m\}$, $m \in \mathbb{N}$, along with the choice of the constant $B_\lambda := 2/\lambda$ in the strong languidity condition **L2'**. Therefore, we can use Theorem 5.3.16 in order to obtain the following exact pointwise fractal tube formula, valid for all $t \in (0, \min\{1/\lambda, 1/2\}) = (0, 1/2)$:

$$
\begin{aligned}
V_A(t) := |A_t| &= \sum_{\omega \in \mathscr{P}(\zeta_A)} \operatorname{res}\left(\frac{t^{2-s}}{2-s}\zeta_A(s), \omega\right) \\
&= \operatorname{res}\left(\frac{t^{2-s}}{2-s}\zeta_A(s), 1\right) + \sum_{\omega \in \mathscr{P}(\zeta_A)\setminus\{1\}} \frac{t^{2-\omega}}{2-\omega} \operatorname{res}\left(\zeta_A, \omega\right).
\end{aligned}
\tag{5.5.117}
$$

We now let $\omega_k := 1 + \mathbb{i}\mathbf{p}k$ for each $k \in \mathbb{Z}$ and note that, in light of (5.5.114),

$$
\operatorname{res}(\zeta_A, 0) = 1 + 2\pi \quad \text{and} \quad \operatorname{res}(\zeta_A, \omega_k) = \frac{4^{-\mathbb{i}\mathbf{p}k}}{4\omega_k(\omega_k - 1)}, \tag{5.5.118}
$$

for every $k \in \mathbb{Z} \setminus \{0\}$.

In order to compute the residue at $\omega_0 = 1$ in (5.5.117), we reason analogously as in the proof of Theorem 5.4.32 (see Equation (5.4.80) and the comment following it) to conclude that

$$
\begin{aligned}
\operatorname{res}\left(\frac{t^{2-s}}{2-s}\zeta_A(s), 1\right) &= t \sum_{n=0}^{1} \sum_{k=0}^{n} \frac{(-1)^{n-k}(\log t^{-1})^k \zeta_A[1]_{-n-1}}{k!(n-k)!} \\
&= t \left(\zeta_A[1]_{-1} - \zeta_A[1]_{-2} + \zeta_A[1]_{-2} \log t^{-1}\right).
\end{aligned}
\tag{5.5.119}
$$

The Laurent series expansion of ζ_A around $s = 1$ is given by

$$
\zeta_A(s) = \frac{1}{4\log 2\,(s-1)^2} + \frac{29\log 2 - 2}{8\log 2\,(s-1)} + O(1), \tag{5.5.120}
$$

so that

$$
\zeta_A[1]_{-2} = \frac{1}{4\log 2} \quad \text{and} \quad \zeta_A[1]_{-1} = \frac{29\log 2 - 2}{8\log 2}, \tag{5.5.121}
$$

which, combined with (5.5.119), yields

$$
\operatorname{res}\left(\frac{t^{2-s}}{2-s}\zeta_A(s), 1\right) = \frac{1}{4\log 2} t \log t^{-1} + \frac{29\log 2 - 4}{8\log 2} t. \tag{5.5.122}
$$

Finally, we obtain the following exact fractal tube formula for the $1/2$-square fractal A, valid for all $t \in (0, 1/2)$:

$$
\begin{aligned}
V_A(t) := |A_t| &= \frac{1}{4\log 2} t \log t^{-1} + \frac{29\log 2 - 4}{8\log 2} t \\
&\quad + t \sum_{k \in \mathbb{Z}\setminus\{0\}} \frac{(4t)^{-\mathbb{i}\mathbf{p}k}}{4\omega_k(\omega_k - 1)(2 - \omega_k)} + \frac{1+2\pi}{2} t^2 \\
&= \frac{1}{4\log 2} t \log t^{-1} + t\, G\left(\log_2 (4t)^{-1}\right) + \frac{1+2\pi}{2} t^2,
\end{aligned}
\tag{5.5.123}
$$

where G is a nonconstant 1-periodic function on $\mathbb{R}$, which is bounded away from zero and infinity. It is given by the following absolutely convergent (and hence, convergent) Fourier series:

$$G(x) := \frac{29\log 2 - 4}{8\log 2} + \frac{1}{4}\sum_{k\in\mathbb{Z}\setminus\{0\}} \frac{\mathrm{e}^{2\pi \mathrm{i} kx}}{(2-\omega_k)(\omega_k - 1)\omega_k}, \quad \text{for all } x \in \mathbb{R}. \tag{5.5.124}$$

To conclude our discussion of this example, we note that it is now clear from the fractal tube formula (5.5.123) for the $1/2$-square fractal that $\dim_B A = 1$ and that A is Minkowski degenerate with (ordinary) Minkowski content $\mathscr{M}^1(A) = +\infty$. On the other hand, we deduce from a direct computation that A is h-Minkowski measurable with $h(t) := \log t^{-1}$ (for all $t \in (0,1)$) and with h-Minkowski content given by

$$\mathscr{M}^1(A,h) = \frac{1}{4\log 2}. \tag{5.5.125}$$

Finally, although $D := \dim_B A = 1$ (which is also the topological dimension of A) and hence, A would not be considered fractal in the classical sense, we also see from (5.5.123) that the nonreal complex dimensions of A with real part equal to D give rise to (intrinsic) geometric oscillations of order t^{2-D} (or simply, $2-D$) in its fractal tube formula. More specifically, according to our proposed definition of fractality given in Remark 4.6.24 and further refined in Remark 5.5.15 (case (i)) above, A is *critically fractal* in dimension $d := D = \dim_B A = 1$. (See also the concluding comments concerning the example of the Cantor graph in Subsection 5.5.4, especially on pages 499–502.)

Example 5.5.23. (*Fractal tube formula for the* $1/3$-*square fractal*). Let us now consider the $1/3$-square fractal A from Example 4.2.34 and depicted in Figure 4.11. Its distance zeta function was obtained in Example 4.2.34 (see Equation (4.2.125)), where it was shown to be meromorphic on all of $\mathbb{C}$ and given by

$$\zeta_A(s) = \frac{2}{s(3^s-2)}\left(\frac{6}{s-1} + Z(s)\right) + \frac{4}{s-1} + \frac{2\pi}{s}, \tag{5.5.126}$$

for all $s \in \mathbb{C}$. Here, the entire function Z is given by

$$Z(s) := \int_0^{\pi/2} (\cos\varphi + \sin\varphi)^{-s}\,\mathrm{d}\varphi \tag{5.5.127}$$

and, without loss of generality, we have chosen $\delta := 1$. Furthermore, as was discussed in Example 4.2.34, it follows at once from (5.5.126) that

$$D(\zeta_A) = 1 \tag{5.5.128}$$

and

$$\mathscr{P}(\zeta_A) := \mathscr{P}(\zeta_A, \mathbb{C}) \subseteq \{0\} \cup (\log_3 2 + \mathbf{p}\mathrm{i}\mathbb{Z}) \cup \{1\}, \tag{5.5.129}$$

where the oscillatory period $\mathbf{p}$ of A is given by $\mathbf{p} := \frac{2\pi}{\log 3}$ and all of the complex dimensions in $\mathscr{P}(\zeta_A)$ are simple. In Equation (5.5.129), we only have an inclusion since, at least in principle, some of the complex dimensions with real part $\log_3 2$ may be canceled by the zeros of $6/(s-1)+Z(s)$. However, it can be checked numerically that there exist nonreal complex dimensions with real part $\log_3 2$ in $\mathscr{P}(\zeta_A)$. Moreover, observe that we have

$$|Z(s)| \le \begin{cases} 2^{-\operatorname{Re} s/2-1}\pi & \text{if} \quad \operatorname{Re} s < 0, \\ \pi/2 & \text{if} \quad \operatorname{Re} s \ge 0, \end{cases} \tag{5.5.130}$$

from which we conclude that that λA is strongly d-languid for $\kappa_d := -1$, any $\lambda \ge \sqrt{2}$ and a sequence of screens consisting of the vertical lines $\{\operatorname{Re} s = -m\}$, $m \in \mathbb{N}$, along with the constant $B_\lambda := \sqrt{2}/\lambda$ in the strong languidity condition **L2'**. Therefore, we can use Theorem 5.3.16 to obtain the following exact pointwise fractal tube formula, valid for all $t \in (0, \min\{1/\lambda, 1/\sqrt{2}\}) = (0, 1/\sqrt{2})$:

$$\begin{aligned} V_A(t) := |A_t| &= \sum_{\omega \in \mathscr{P}(\zeta_A)} \operatorname{res}\left(\frac{t^{2-s}}{2-s}\zeta_A(s), \omega\right) \\ &= \sum_{\omega \in \mathscr{P}(\zeta_A)} \frac{t^{2-s}}{2-s} \operatorname{res}\left(\zeta_A(s), \omega\right) \\ &= 16t + \frac{t^{2-\log_3 2}}{\log 3} \sum_{k=-\infty}^{+\infty} \frac{(3t)^{-\mathrm{i}\mathbf{p}k}}{\omega_k(2-\omega_k)}\left(\frac{6}{\omega_k - 1} + Z(\omega_k)\right) \\ &\quad + \frac{12+\pi}{2}t^2 \\ &= 16t + t^{2-\log_3 2} G\left(\log_3 (3t)^{-1}\right) + \frac{12+\pi}{2}t^2. \end{aligned} \tag{5.5.131}$$

Here, we have used the fact that

$$\operatorname{res}\left(\zeta_A, 1\right) = 16, \quad \operatorname{res}\left(\zeta_A, 0\right) = 12 + \pi \tag{5.5.132}$$

and

$$\operatorname{res}\left(\zeta_A, \omega_k\right) = \frac{3^{-\mathrm{i}\mathbf{p}k}}{(\log 3)\omega_k}\left(\frac{6}{\omega_k - 1} + Z(\omega_k)\right), \tag{5.5.133}$$

where we have let $\omega_k := \log_3 2 + \mathrm{i}\mathbf{p}k$ for each $k \in \mathbb{Z}$. It can be checked numerically that $\operatorname{res}(\zeta_A, \omega_k) \neq 0$ (at least) for $k = -1, 0, 1$ and we conjecture that this is also true for all $k \in \mathbb{Z}$.[58] However, the fact that $\operatorname{res}(\zeta_A, \omega_k) \neq 0$ for $k = -1, 0, 1$ suffices to deduce that the function G in the last line of (5.5.131) is a *nonconstant* 1-periodic function on $\mathbb{R}$, which is bounded away from zero and infinity and is given by the following absolutely convergent (and hence, convergent) Fourier series:

---

[58] We caution the reader that we do not have a rigorous proof of this last statement.

$$G(x) := \frac{1}{\log 3} \sum_{k=-\infty}^{+\infty} \frac{e^{2\pi \mathrm{i} kx}}{(2-\omega_k)\omega_k} \left(\frac{6}{\omega_k - 1} + Z(\omega_k) \right), \quad \text{for all } x \in \mathbb{R}. \tag{5.5.134}$$

In conclusion, we observe that it is clear from the fractal tube formula (5.5.131) that $\dim_B A = 1$ and A is Minkowski measurable, with Minkowski content given by

$$\mathscr{M}^1(A) = 16. \tag{5.5.135}$$

Moreover, since the set A is rectifiable, we have that $H^1(A) = \mathscr{M}^1(A)/2 = 8$, which can, of course, also be computed directly.[59] (Here, as before, $H^1(A)$ denotes the 1-dimensional Hausdorff measure of A.) On the other hand, although $D := \dim_B A = 1$ (which also coincides with the topological dimension of A) and thus A would not be considered fractal in the classical sense, we also see from (5.5.131) that the nonreal complex dimensions of A with real part equal to $\log_3 2$ give rise to (intrinsic) geometric oscillations of order $t^{2-\log_3 2}$ (or simply, $2 - \log_3 2$) in its fractal tube formula. Therefore, according to our proposed definition of fractality given in Remark 4.6.24 and further refined in Remark 5.5.15 (case (i)) above, the $1/3$-square fractal A is fractal; more precisely, it is *strictly subcritically fractal* in dimension $d := \log_3 2$. (See also the discussion at the end of Subsection 5.5.4, especially on pages 499–502.)

Example 5.5.24. (*Fractal tube formula for the self-similar fractal nest*). Let us now consider the self-similar fractal nest A from Example 4.2.35 and depicted in Figure 4.12. Its distance zeta function was obtained in Example 4.2.35 (see Equation (4.2.137)), where it was shown to be meromorphic on all of $\mathbb{C}$ and given by

$$\zeta_A(s) = \frac{2^{2-s}\pi(1+a)(1-a)^{s-1}}{(s-1)(1-a^s)} + \frac{2\pi}{s-1} + \frac{2\pi}{s}, \tag{5.5.136}$$

for all $s \in \mathbb{C}$; here, without loss of generality, we have chosen $\delta := 1$. Recall that $a \in (0,1)$ is a real parameter. Also recall that

$$D(\zeta_A) = 1 \tag{5.5.137}$$

and

$$\mathscr{P}(\zeta_A) := \mathscr{P}(\zeta_A, \mathbb{C}) = \mathbf{p}\mathrm{i}\mathbb{Z} \cup \{1\}, \tag{5.5.138}$$

where the oscillatory period $\mathbf{p}$ of A is given by $\mathbf{p} := \frac{2\pi}{\log a^{-1}}$ and all of the complex dimensions in $\mathscr{P}(\zeta_A)$ are simple.

It is now easy to check that λA is strongly d-languid with $\kappa_d := -1$, for any $\lambda \geq 2$ if $a \in (0, 1/2]$ or for any $\lambda \geq 2(1-a)/a$ if $a \in (1/2, 1)$ and (in both cases) for a sequence of screens consisting of vertical lines $\{\mathrm{Re}\, s = -m\}$, $m \in \mathbb{N}$, in the strong languidity condition **L2'**. Furthermore, once again, we can use Theorem 5.3.16 in

[59] In order to obtain the length $H^1(A)$ of A, we divide its Minkowski content by the volume of the unit ball of $\mathbb{R}^{N-1} = \mathbb{R}$ (since here $N = 2$), which is equal to 2.

order to obtain the following exact pointwise fractal tube formula, which is valid for all $t \in (0, \min\{1/2, a/2(1-a)\})$:

$$\begin{aligned} V_A(t) := |A_t| &= \sum_{\omega \in \mathscr{P}(\zeta_A)} \operatorname{res}\left(\frac{t^{2-s}}{2-s}\zeta_A(s), \omega\right) \\ &= \sum_{\omega \in \mathscr{P}(\zeta_A)} \frac{t^{2-s}}{2-s} \operatorname{res}(\zeta_A(s), \omega) \\ &= \frac{4\pi}{1-a}t + \left(\pi + \frac{4\pi(1+a)}{(\log a^{-1})(1-a)} \sum_{k=-\infty}^{+\infty} \frac{\left(\frac{2t}{1-a}\right)^{-\mathbb{i}\mathbf{p}k}}{(\omega_k - 1)(2-\omega_k)}\right) t^2 \\ &= \frac{4\pi}{(1-a)}t + t^2 G\left(\log_{a^{-1}}\left(\frac{2t}{1-a}\right)\right). \end{aligned} \tag{5.5.139}$$

Here, we have used the fact that

$$\operatorname{res}(\zeta_A, 1) = \frac{4\pi}{1-a}, \quad \operatorname{res}(\zeta_A, 0) = 2\pi + \frac{4\pi(1+a)}{(\log a)(1-a)} \tag{5.5.140}$$

and

$$\operatorname{res}(\zeta_A, \omega_k) = \frac{4\pi(1+a)}{(\log a^{-1})(\omega_k - 1)} \left(\frac{2}{1-a}\right)^{-\mathbb{i}\mathbf{p}k}, \tag{5.5.141}$$

where we have let $\omega_k := \mathbb{i}\mathbf{p}k$ for each $k \in \mathbb{Z}$. Furthermore, the function G appearing in the last line of (5.5.139) is a nonconstant 1-periodic function on $\mathbb{R}$, which is bounded away from zero and infinity and is given by the following absolutely convergent (and hence, convergent) Fourier series:

$$G(x) := \pi + \frac{4\pi(1+a)}{(\log a^{-1})(1-a)} \sum_{k=-\infty}^{+\infty} \frac{\mathrm{e}^{2\pi\mathbb{i}kx}}{(2-\omega_k)(\omega_k - 1)}, \quad \text{for all } x \in \mathbb{R}. \tag{5.5.142}$$

It clearly follows from the fractal tube formula (5.5.139) that $\dim_B A = 1$ and A is Minkowski measurable with Minkowski content given by

$$\mathscr{M}^1(A) = \frac{4\pi}{1-a}. \tag{5.5.143}$$

Furthermore, since the set A is rectifiable, we have that $H^1(A) = \mathscr{M}^1(A)/2 = 2\pi/(1-a)$, which, of course, can also be easily checked via a direct computation.

Finally, we conclude this example by observing that although $D := \dim_B A = 1$ (which is also the topological dimension of A) and thus A would not be considered fractal in the classical sense, we also see from (5.5.139) that the nonreal complex dimensions of A with real part equal to 0 give rise to (intrinsic) geometric oscillations of order t^2 (or simply, 2) in its fractal tube formula. Consequently, according to our proposed definition of fractality given in Remark 4.6.24 and further refined in Remark 5.5.15 (case (i)) above, the self-similar fractal nest A is indeed fractal;

more precisely, it is *strictly subcritically fractal* in dimension $d := 0$. (See also the discussion closing Subsection 5.5.4 above.)

Example 5.5.25. (*Fractal tube formula for the geometric progression string*). We now consider the geometric progression fractal string $\mathscr{L}$ from Example 4.2.36; more specifically, we consider its geometric realization $A_{\mathscr{L}}$ in $\mathbb{R}$. Its distance zeta function was obtained in Example 4.2.36 (see Equation (4.2.146)), where it was shown to be meromorphic on all of $\mathbb{C}$ and given by

$$\zeta_{A_{\mathscr{L}}}(s) = \frac{2^{1-s}}{s(1-a^s)} + \frac{2}{s}, \tag{5.5.144}$$

for all $s \in \mathbb{C}$.[60] Recall that $a \in (0,1)$ is a real parameter. Also recall that (as is obvious in light of Equation (5.5.144))

$$D(\zeta_{A_{\mathscr{L}}}) = 0 \tag{5.5.145}$$

and

$$\mathscr{P}(\zeta_{A_{\mathscr{L}}}) = \mathbf{p}\mathbb{i}\mathbb{Z}, \tag{5.5.146}$$

where the oscillatory period $\mathbf{p}$ of $A_{\mathscr{L}}$ is given by $\mathbf{p} := \frac{2\pi}{\log a^{-1}}$ and all of the complex dimensions in $\mathscr{P}(\zeta_A)$ are simple, except for $\omega_0 := 0$, which has multiplicity two. It is now easy to check that $\lambda A_{\mathscr{L}}$ is strongly d-languid for $\kappa_d := -1$, any $\lambda \geq 2a$ and a sequence of screens consisting of the vertical lines $\{\mathrm{Re}\, s = -m\}$, $m \in \mathbb{N}$, in the strong languidity condition **L2'**. Therefore, once more, we can use Theorem 5.3.16 in order to obtain the following exact pointwise fractal tube formula, valid for all $t \in (0, 1/(2a))$:

$$\begin{aligned} V_{A_{\mathscr{L}}}(t) := |(A_{\mathscr{L}})_t| &= \sum_{\omega \in \mathscr{P}(\zeta_A)} \operatorname{res}\left(\frac{t^{1-s}}{1-s}\zeta_{A_{\mathscr{L}}}(s), \omega\right) \\ &= \operatorname{res}\left(\frac{t^{1-s}}{1-s}\zeta_{A_{\mathscr{L}}}(s), 0\right) \\ &\quad + \sum_{\omega \in \mathscr{P}(\zeta_{A_{\mathscr{L}}}) \setminus \{0\}} \frac{t^{1-\omega}}{1-\omega} \operatorname{res}\left(\zeta_{A_{\mathscr{L}}}, \omega\right). \end{aligned} \tag{5.5.147}$$

We now let $\omega_k := \mathbb{i}\mathbf{p}k$ for each $k \in \mathbb{Z}$ and note that:

$$\operatorname{res}\left(\zeta_{A_{\mathscr{L}}}, \omega_k\right) = \frac{2^{1-\omega_k}}{(\log a^{-1})\omega_k}, \tag{5.5.148}$$

for every $k \in \mathbb{Z} \setminus \{0\}$.

In order to calculate the residue of $\zeta_{A_{\mathscr{L}}}$ at $\omega_0 = 0$ in (5.5.147), we reason analogously as in the proof of Theorem 5.4.32 (see Equation (5.4.80) and the comment following it) in order to deduce that

[60] Here, without loss of generality, we have fixed $\delta := 1$.

$$\begin{aligned}\operatorname{res}\left(\frac{t^{1-s}}{1-s}\zeta_{A_{\mathscr{L}}}(s),0\right) &= t\sum_{n=0}^{1}\sum_{k=0}^{n}\frac{(-1)^{n-k}(\log t^{-1})^k\zeta_{A_{\mathscr{L}}}[0]_{-n-1}}{k!(n-k)!} \\ &= t\left(\zeta_{A_{\mathscr{L}}}[0]_{-1}-\zeta_{A_{\mathscr{L}}}[0]_{-2}+\zeta_{A_{\mathscr{L}}}[0]_{-2}\log t^{-1}\right).\end{aligned} \tag{5.5.149}$$

The Laurent series expansion of ζ_A around $s=0$ is given by

$$\zeta_{A_{\mathscr{L}}}(s)=\frac{2}{(\log a^{-1})s^2}+\frac{3-\log_{a^{-1}}4}{s}+O(1). \tag{5.5.150}$$

Hence,

$$\zeta_{A_{\mathscr{L}}}[0]_{-2}=\frac{2}{\log a^{-1}} \quad \text{and} \quad \zeta_{A_{\mathscr{L}}}[0]_{-1}=3-\log_{a^{-1}}4, \tag{5.5.151}$$

which combined with Equation (5.5.149) yields

$$\operatorname{res}\left(\frac{t^{1-s}}{1-s}\zeta_{A_{\mathscr{L}}}(s),0\right)=\frac{2}{\log a^{-1}}t\log t^{-1}+\left(3-\frac{\log 4-2}{\log a^{-1}}\right)t. \tag{5.5.152}$$

Finally, we obtain the following exact pointwise fractal tube formula for $A_{\mathscr{L}}$, valid for all $t\in(0,1/(2a))$:

$$\begin{aligned}V_{A_{\mathscr{L}}}(t):=|(A_{\mathscr{L}})_t| &= \frac{2}{\log a^{-1}}t\log t^{-1}+\left(3-\frac{\log 4-2}{\log a^{-1}}\right)t \\ &\quad+\frac{2t}{\log a^{-1}}\sum_{k\in\mathbb{Z}\setminus\{0\}}\frac{(2t)^{-\mathrm{i}\mathbf{p}k}}{\omega_k(1-\omega_k)} \\ &= \frac{2}{\log a^{-1}}t\log t^{-1}+t\,G\left(\log_{a^{-1}}(2t)^{-1}\right),\end{aligned} \tag{5.5.153}$$

where G is a nonconstant 1-periodic function on $\mathbb{R}$, which is bounded away from zero and infinity and given by the following absolutely convergent (and hence, convergent) Fourier series:

$$G(x):=\left(3-\frac{\log 4-2}{\log a^{-1}}\right)+\frac{2}{\log a^{-1}}\sum_{k\in\mathbb{Z}\setminus\{0\}}\frac{\mathrm{e}^{2\pi\mathrm{i}kx}}{\omega_k(1-\omega_k)}, \quad \text{for all } x\in\mathbb{R}. \tag{5.5.154}$$

It is now clear from the fractal tube formula (5.5.153) for $A_{\mathscr{L}}$ that, as was already stated, $\dim_B A_{\mathscr{L}}=0$ (which is also the topological dimension of A) and $A_{\mathscr{L}}$ is Minkowski degenerate with (ordinary) Minkowski content $\mathscr{M}^0(A_{\mathscr{L}})=+\infty$. On the other hand, $A_{\mathscr{L}}$ is h-Minkowski measurable, with $h(t):=\log t^{-1}$ for all $t\in(0,1)$, and its h-Minkowski content is given by

$$\mathscr{M}^0(A_{\mathscr{L}},h)=\frac{2}{\log a^{-1}}. \tag{5.5.155}$$

We conclude this example by pointing out that, although $D := \dim_B A_{\mathscr{L}} = 0$ (which is also the topological dimension of A) and hence, much as in Examples 5.5.22, 5.5.23 and 5.5.24, $A_{\mathscr{L}}$ would not be considered fractal in the classical sense, we also see from (5.5.153) that the nonreal complex dimensions of $A_{\mathscr{L}}$ with real part equal to D give rise to (intrinsic) geometric oscillations of order t (or simply, 1) in its fractal tube formula (even though $A_{\mathscr{L}}$ is h-Minkowski measurable). More specifically, according to our proposed definition of fractality given in Remark 4.6.24 and further refined in Remark 5.5.15 (case (i)) above, $A_{\mathscr{L}}$ (or, equivalently, $\mathscr{L}$) is *critically fractal* in dimension $d := D = \dim_B A_{\mathscr{L}} = 0$. (See the concluding comments of Subsection 5.5.4, on pages 499–502 for a closely related situation.)

The next remark (Remark 5.5.26) is extensive and is composed of three parts, labeled (a), (b) and (c). It addresses various aspects and generalizations of the above discussion. Part (a) discusses the potential generalizations of the above fractal tube formulas to *self-similar sets*, rather than to self-similar sprays (or tilings). Moreover, part (b) provides explicitly the easy extension of the above fractal tube formulas to self-similar sprays with *multiple generators*. Finally, part (c) discusses the characterization of the Minkowski measurability (or, more generally, the possibly subcritical Minkowski measurability) of a large class of self-similar sprays and self-similar sets. (We could analogously discuss the general applications of fractal tube formulas and Minkowski measurability criteria to not necessarily self-similar fractal sprays. However, for the sake of clarity and brevity, we will not do so here and simply mention that, under appropriate assumptions, the corresponding discussion would parallel the one provided in the relevant portions of parts (a) and (b) of Remark 5.5.26 just below.)

Remark 5.5.26. (a) (*Towards fractal tube formulas for self-similar sets*). We also expect that under suitable hypotheses (see below), and by using (or appropriately extending) the corresponding results of the present subsection about self-similar sprays, we can obtain (pointwise or distributional) fractal tube formulas, with or without error term, for a large class of self-similar sets (satisfying the open set condition). A key step for deriving such results should consist in obtaining an approximate functional equation connecting the distance zeta functions of the given self-similar set F and of the associated self-similar tiling (or spray) (A,Ω), viewed as a self-similar RFD.

More specifically, we expect that, under suitable assumptions, we have

$$\zeta_F(s) = \zeta_{A,\Omega}(s) + \zeta_{O,\mathrm{out}}(s) + f(s), \tag{5.5.156}$$

where f is a function which is holomorphic on some open right half-plane $\{\mathrm{Re}\, s > \eta\}$, with $-\infty \le \eta < \min\{D_G, \sigma_0\}$. Here, D_G and σ_0 stand, respectively, for the (upper) Minkowski dimension of the generator G of the self-similar spray (or tiling), and for the *similarity dimension* of F (or, equivalently, of (A,Ω)); that is, $\sigma_0 \in (0,N)$ is the unique real solution of the Moran equation $\sum_{j=1}^{J} r_j^s = 1$.

Furthermore, $\zeta_{O,\mathrm{out}} := \zeta_{K,K^c}$ is the distance zeta function of a relatively "simple" RFD (denoted in an interesting special case by (K,K^c) below, with $K := \overline{O}$ and

$K^c := \mathbb{R}^N \setminus O$) taking into account the contributions of the *outer* neighborhoods of $K = \overline{O}$, where O is an admissible open set, in a sense to be explained below. For the purpose of the present discussion, we assume that $\overline{O}$, the closure of O in $\mathbb{R}^N$, is the convex hull of F, but significantly more general situations will be considered below.

Moreover, (A,Ω) is the canonical self-similar tiling associated with the self-similar set F, in the sense of [Pe] to be briefly described below (and as used in [LapPe2–3], [LapPeWi1–2], as well as described in [Lap-vFr3, Section 13.1]) and viewed as a self-similar RFD. It is a special case of a self-similar spray (in the sense of [Lap3, LapPo3]) having a natural geometric meaning. In particular, the bounded open set Ω is the complement of F in the (necessarily closed) convex hull of F (or, more generally, in the closure $\overline{O}$ of the admissible open set O, see below) of countably many scaled copies (the 'tiles') of the generator G of the fractal spray (or tiling).[61] In most situations of interest, the tiles are simple polyhedra (equivalently, the generator is a simple polyhedron).

We assume that the self-similar tiling is nontrivial, which is known to imply that the self-similar set F satisfies the open set condition (in the sense of [Hut]; see also, e.g., [Fal1]) and that $\dim_B(A,\Omega) < N$ (which will be implied by our other hypotheses). We also assume that the generator G is sufficiently 'nice', for instance, monophase or, more generally, pluriphase (in the sense of [LapPe2–3], [LapPeWi1–2]), as discussed earlier in the present subsection. For example, under mild nondegeneracy hypotheses, G can be a convex polytope of $\mathbb{R}^N$; see the main result in [KoRati], which resolved and specified a conjecture in [LapPe2] and [LapPe3]. (More information about self-similar tilings, their generators, the open set condition, and the nontriviality condition, is provided towards the end of the present discussion, i.e., of part (a) of this remark.)

Note that it is well known in the literature on fractal geometry (see, e.g., [Hut] and Theorem 9.3, page 140 of the third edition of [Fal1]) that since the self-similar set F satisfies the open set condition and letting $D_F := \dim_B F$, we have that the box (i.e., Minkowski) dimension D_F of F exists and $\sigma_0 = \dim_H F = D_F$ $(= D(\zeta_F)$, by Theorem 2.1.11 and Corollary 2.1.20 above). We caution the reader, however, that the same statement does not hold, in general, for a self-similar RFD (which is not, necessarily, a self-similar set), as the example of the inhomogeneous Sierpiński N-gasket RFD (and similar examples; see Examples 4.2.33, 4.2.34 and 4.2.35 from Subsection 4.2.3, revisited from the point of view of the fractal tube formulas in Examples 5.5.22, 5.5.23 and 5.5.24, respectively) shows. More specifically, for the inhomogeneous Sierpiński N-gasket RFD (A_N,Ω_N), when $N \geq 4$, the open set condition is satisfied by the corresponding self-similar set F but we have

$$D_{A_N,\Omega_N} = D_G = N-1 > \sigma_0 = D_F = \log_2(N+1), \tag{5.5.157}$$

where G is the single generator of the inhomogeneous N-gasket RFD. Here, the corresponding self-similar set $F := S_3$ is the well-known Sierpiński tetrahedron (or

[61] For notational simplicity, we assume here that there is a single generator. Entirely analogous results are expected in the case of finitely many generators, as is explained at the end of part (c) of this remark, based on part (b) below.

pyramid) S_3, in the case when $N=3$ and a higher-dimensional analogue $F:=S_N$ when $N\geq 4$.[62] (See Example 4.2.26, along with the discussion surrounding Equation (5.5.185).) We caution the reader, however, that we can have simultaneously $D_F=\sigma_0$ and $D_{A_N,\Omega_N}=D_{A_N}=\max\{D_G,\sigma_0\}>\sigma_0$. Indeed, as was discussed in Example 4.2.26 and will be further discussed below, A_N is an *inhomogeneous* self-similar set, *not* an ordinary (or homogeneous) self-similar set. This may be a somewhat confusing point worth being highlighted. The first author (Michel Lapidus) wishes to thank Michael Barnsley and Martina Zähle for their helpful queries about this issue.

What should be true, in general, is that for a self-similar set F satisfying the open set condition (and under some other mild conditions on the generator G), we have that $D_G\leq\sigma_0$ and therefore (in light of Corollary 2.1.20, in particular),

$$\begin{aligned} D_F &:= \dim_B F = D(\zeta_F) = D_{A,\Omega} = D(\zeta_{A,\Omega}) \\ &= \dim_B(A,\Omega) = \max\{D_G,\sigma_0\} = \sigma_0. \end{aligned} \tag{5.5.158}$$

In the present discussion, we do not need to know that (or when) Equation (5.5.158) holds, but we need to be aware of the following fact, which is a consequence of the factorization formula (5.5.105) (or, equivalently, (5.5.108)) combined with parts (b) and (c) of Theorem 4.1.7:[63]

$$D_{A,\Omega} := \dim_B(A,\Omega) = D(\zeta_{A,\Omega}) = D_{\rm hol}(\zeta_{A,\Omega}) = \max\{D_G,\sigma_0\}, \tag{5.5.159}$$

with (again by these same results) $D_G=D(\zeta_{\partial G,G})$. (Clearly, we have that $\sigma_0<N$ since, by hypothesis, $\sum_{j=1}^J r_j^N<1$; also, we have that $\sigma_0>0$ since $J\geq 2$.)

The hypotheses under which we expect these results to hold, include, for example, that the generator G is pluriphase (for instance, a polytope, under mild assumption, by [KoRati]).

We mention that other, more general, choices of self-similar tilings associated with a self-similar set F and a corresponding open set O satisfying the open set condition (OSC, in short) can be made, as in [PeWi] (and as used in [LapPeWi1–2] or described in [Lap-vFr3, Section 13.1]); they give rise to different, although related, self-similar RFDs but the corresponding results are expected to be analogous. In fact, this extra flexibility is likely to be essential for making the present ideas more specific and to correctly implement them.

These more general self-similar tilings are associated with 'feasible' (bounded, nonempty) open sets O; i.e., the OSC is satisfied in the following sense: O is a nonempty open subset such that

$$\Phi_j(O)\subseteq O \quad \text{for} \quad j=1,\ldots,J \tag{5.5.160}$$

[62] More precisely, the self-similar set $F:=S_N$ is the unique fixed point of the IFS consisting of $(N+1)$ contractive similitudes of $\mathbb{R}^N$ of scaling factor equal to 1/2, as described in Example 4.2.26.

[63] We assume here implicitly that when $D_G\neq\sigma_0$, the generator G (and hence also, the RFD (A,Ω)) has positive lower Minkowski content, which is the case in most situations of interest.

and

$$\Phi_j(O)\cap\Phi_{j'}(O)=\emptyset \quad \text{for} \quad j\neq j', \quad \text{with} \quad j,j'\in\{1,\ldots,J\}, \tag{5.5.161}$$

where $\{\Phi_j\}_{j=1}^J$ are J contractive similitudes of $\mathbb{R}^N$ with scaling ratios $\{r_j\}_{j=1}^J$ defining the self-similar set in the usual way:

$$F=\bigcup_{j=1}^J \Phi_j(F)=:\Phi(F). \tag{5.5.162}$$

Furthermore, O is 'acceptable', in the sense that the associated self-similar tiling is nontrivial:

$$O\not\subseteq\Phi(\overline{O}), \tag{5.5.163}$$

where $\overline{O}$ denotes the closure of O in $\mathbb{R}^N$. The associated self-similar tiling $\mathfrak{t}:=\mathfrak{t}(O)$ is then given by

$$\mathfrak{t}=\mathfrak{t}(O):=\{\Phi_w(G)\,:\,w\in\mathfrak{W}\}, \tag{5.5.164}$$

where $\mathfrak{W}:=\cup_{k=0}^{\infty}\{1,\ldots,J\}^k$ is the set of all finite words based on the alphabet $\{1,\ldots,J\}$ and for each $w=(w_1,\ldots,w_k)\in\mathfrak{W}$, $\Phi_w:=\Phi_{w_1}\circ\cdots\circ\Phi_{w_k}$ is the appropriate composition of the maps $\Phi_{w_1},\ldots,\Phi_{w_k}$. Here, the *generator* G of the self-similar tiling $\mathfrak{t}$ (or, equivalently, of the associated self-similar RFD (A,Ω)) is the (bounded) connected open set of $\mathbb{R}^N$ defined by $G:=O\setminus\Phi(\overline{O})$, the elements of O which are not in $\Phi(\overline{O})$. Note that according to the nontriviality condition, G is nonempty. Also, for notational simplicity, we assume in the present discussion that there is a single generator. (In general, the open set $O\setminus\Phi(\overline{O})$ need not be connected. Then, the generators $\{G^{(q)}\}_{q=1}^Q$ of $\mathfrak{t}$, where Q is assumed to be a finite number, for simplicity, are the bounded open sets which are defined as the connected components of the open set $O\setminus\Phi(\overline{O})$.)

It is shown in [PeWi] (extending [Pe] to this more general situation) that the *tiles* $\Phi_w(G)$ (for $w\in\mathfrak{W}$) in $\mathfrak{t}=\mathfrak{t}(O)$ are pairwise disjoint (bounded) open sets, and that the closure of their union coincides with $\overline{O}$, the closure of O:[64]

$$\overline{O}=\overline{\bigsqcup_{w\in\mathfrak{W}}\Phi_w(G)}, \tag{5.5.165}$$

thereby, justifying the use of the term "self-similar tiling" in this context.

Let us close this discussion by giving one class of examples for which a relation of the type (5.5.156) is satisfied, in a strong sense (since the counterpart of f is then identically equal to zero) and for the tube (instead of the distance) zeta

[64] We follow here the discussion of self-similar tilings provided in [LapPeWi2]. If, as in part (*b*) of this remark on page 528, we have multiple generators, $\{G^{(q)}\}_{q=1}^Q$, rather than a single generator G, then (5.5.165) should be replaced by the following equality: $\overline{O}=\overline{\bigsqcup_{q=1}^Q\bigsqcup_{w\in\mathfrak{W}}\Phi_w(G^{(q)})}$, where $\overline{B}$ denotes the closure in $\mathbb{R}^N$ of $B\subseteq\mathbb{R}^N$. We note that the $N=1$ case precisely corresponds to the self-similar strings with multiple gaps discussed in [Lap-vFr3, Chapter 2, esp., Section 12.1].

functions. (As a consequence of the functional equations (2.2.23) and (4.5.2), it will then follow that (5.5.156) actually holds for the distance zeta functions, with f being an explicitly computable entire function.) This is directly connected to a result in [PeWi] and to the hypotheses made in [LapPeWi2, Section 5] in order to transfer results (such as Minkowski measurability results) from self-similar tilings to certain self-similar sets. The main geometric condition about the self-similar tiling $\mathfrak{t} = \mathfrak{t}(O)$ required here is the following *compatibility condition*. Namely, the bounded open set O (in addition to being 'admissible', i.e., feasible and acceptable) is such that $\partial O \subset F$, where ∂O denotes the boundary of O in $\mathbb{R}^N$. This condition is satisfied, for example, by the Sierpiński gasket and the Sierpiński carpet tilings, but not by the Koch tiling (see, e.g., [LapPeWi2, Figure 2.1, p. 189]). Then, if we consider the compact subset of $\mathbb{R}^N$ defined by $K := \overline{O}$, it follows from the compatibility condition that for every $t > 0$, we have that

$$|F_t| = V(\mathfrak{t}, t) + |K_t \setminus K|, \tag{5.5.166}$$

where $V(\mathfrak{t}, t)$ is the volume of the inner t-neighborhood of $\mathfrak{t} = \mathfrak{t}(O)$ (the disjoint union of the inner t-neighborhoods of the tiles of $\mathfrak{t}$), just as in the general theory of fractal sprays [Lap2–3, LapPo3, Lap-vFr1–3]), and

$$|K_t \setminus K| = |K_t| - |K| = |K_t \cap K^c|, \tag{5.5.167}$$

with $K^c := \mathbb{R}^N \setminus K$ (the complement of K in $\mathbb{R}^N$). Here, $K_t \setminus K$ is the 'outer' (or 'excised') t-neighborhood of K. Note that therefore, in light of (5.5.167),

$$|K_t \setminus K| = |K_t \cap K^c| = V_{K,K^c} \tag{5.5.168}$$

is the tube function of the RFD (K, K^c).

It follows at once from (5.5.166) and (5.5.168) that

$$|F_t| = V(\mathfrak{t}, t) + V_{K,K^c}(t) = V_{A,\Omega}(t) + V_{K,K^c}(t), \tag{5.5.169}$$

where we have identified the self-similar tiling with the RFD (A, Ω) given by the self-similar spray defined by $\mathfrak{t}$ and hence, generated by G. (For notational simplicity, we write $|\cdot|$ instead of $|\cdot|_N$ in Equations (5.5.166)–(5.5.169).) Observe that in light of (5.5.169), $|F_t|$ is the sum of an *inner contribution* (corresponding to the RFD (A, Ω), associated with the tiles of $\mathfrak{t} = \mathfrak{t}(O)$) and an *outer contribution* (corresponding to the RFD (K, K^c), where $K := \overline{O}$).

In light of the definition of the tube zeta function of an RFD (A, Ω) and since, by construction, $V_{A,\Omega} = |A_t \cap \Omega|$ and $V_{K,K^c} = |K_t \cap K^c|$, we deduce the following key identity between the corresponding tube zeta functions:

$$\tilde{\zeta}_F(s) = \tilde{\zeta}_{A,\Omega}(s) + \tilde{\zeta}_{K,K^c}(s), \tag{5.5.170}$$

for all $s \in \mathbb{C}$ with $\operatorname{Re} s$ sufficiently large, and then, as usual, after meromorphic continuation, for all $s \in U$, where $U \subseteq \mathbb{C}$ is a domain to which both $\tilde{\zeta}_{A,\Omega}$ and $\tilde{\zeta}_{K,K^c}$ can be meromorphically continued. Since, according to our results about self-similar tilings (sprays) obtained in the first part of this subsection, $\zeta_{A,\Omega}$ and hence also, $\tilde{\zeta}_{A,\Omega}$, can be meromorphically continued to all of $\mathbb{C}$ (and is explicitly known in terms of $\zeta_{\partial G,G}$ and the scaling ratios $\{r_j\}_{j=1}^J$, see Equation (5.5.105) along with the functional Equation (4.5.2)), the identity (5.5.170) is valid for all $s \in U$, where U is any connected open subset of $\mathbb{C}$ to which $\tilde{\zeta}_{K,K^c}$ can be meromorphically continued.

We leave it as an exercise for the interested reader to deduce from (5.5.170) and a repeated application of the functional equation (4.5.2) (connecting the relative distance and tube zeta functions) a corresponding identity for the distance zeta functions ζ_F, $\zeta_{A,\Omega}$ and $\zeta_{O,\mathrm{out}} := \zeta_{K,K^c}$. In fact, as desired, one obtains an expression which is exactly of the form (5.5.156), where the 'error function' $f = f(s)$ is an entire function (which is explicitly computable and depends on the unimportant choice of $\delta > 0$ occurring in the functional equation (4.5.2)).

In the more general situation when the compatibility condition $\partial O \subset F$ is not necessarily satisfied, we expect that, under suitable hypotheses, we have a relation of the form

$$\tilde{\zeta}_F(s) = \tilde{\zeta}_{A,\Omega} + \tilde{\zeta}_{K,K^c} + g(s), \tag{5.5.171}$$

where g is a holomorphic function of the above type; namely, g is holomorphic on some open right half-plane $\{\operatorname{Re} s > \alpha\}$, with $\alpha \in \mathbb{R} \cup \{-\infty\}$ small enough (say, $\alpha < \min\{D_{A,\Omega}, D_{K,K^c}\}$). Then, exactly as above, by repeatedly using the functional equation (4.5.2), one would obtain an expression of the desired form (5.5.156), for the distance zeta function ζ_F itself.

This last discussion clearly suggests that, in general, the situation is somewhat complicated but that it might be best apprehended by first working with (absolute and relative) tube (instead of distance) zeta functions, then deducing the corresponding identity (5.5.156) between the distance zeta functions (via a repeated use of the functional equation (4.5.2)) and finally applying one of the appropriate tube formulas (expressed in terms of the distance zeta functions) from Section 5.3 in order to obtain the tube formula for $V_{O,\mathrm{out}} := V_{K,K^c}$.

Alternatively, one may wish to directly apply the tube formulas obtained in Sections 5.1 and 5.2 (rather than in Section 5.3). Finally, we note that in light of the functional equation connecting $\zeta_{A,\Omega}$ and $\tilde{\zeta}_{A,\Omega}$ (see Equation (4.5.2)) and in light of Equation (5.5.105) (or, equivalently, (5.5.108)) giving the explicit expression of $\zeta_{A,\Omega}$ (namely, $\zeta_{A,\Omega}(s) = \zeta_{\partial G,G}(s)/(1 - \sum_{j=1}^J r_j^s)$), we have in our possession an explicit expression for $\tilde{\zeta}_{A,\Omega}(s)$. We expect that in many cases, $\tilde{\zeta}_{K,K^c}(s)$ can also be computed or, at least, appropriately approximated (modulo a suitable holomorphic function, of the above type).

(*b*) (*Fractal tube formulas for self-similar sprays with multiple generators*). Moreover, we mention that all the results of the present subsection easily extend to self-similar sprays (and tilings) with finitely many (rather than a single) generators. For example, if $G^{(1)}, \ldots, G^{(Q)}$ denote these Q generators, with $Q \in \mathbb{N}$, Equation (5.5.105) simply becomes

$$\begin{aligned}\zeta_{A,\Omega}(s) &= \sum_{q=1}^{Q} \zeta_{A^{(q)},\Omega^{(q)}}(s) = \frac{\sum_{q=1}^{Q} \zeta_{\partial G^{(q)},G^{(q)}}(s)}{1-\sum_{j=1}^{J} r_j^s} \\ &= \zeta_{\mathfrak{S}}(s) \sum_{q=1}^{Q} \zeta_{\partial G^{(q)},G^{(q)}}(s),\end{aligned} \tag{5.5.172}$$

where (A,Ω) denotes the self-similar spray (viewed as an RFD) with Q generators $G^{(1)},\dots,G^{(Q)}$ and scaling ratios $\{r_j\}_{j=1}^{J}$, and for each $q=1,\dots,Q$, $(A^{(q)},\Omega^{(q)})$ stands for the self-similar spray with the single generator $G^{(q)}$ and the same scaling ratios (or ratio list) $\{r_j\}_{j=1}^{J}$. Furthermore, in the last equality of (5.5.172), $\zeta_{\mathfrak{S}}$ is the scaling zeta function of the self-similar spray (A,Ω) given by (5.5.109). Moreover, with σ_0 $(=D(\zeta_{\mathfrak{S}})=D_{\text{hol}}(\zeta_{\mathfrak{S}})$, by Theorem 4.1.7) still denoting the similarity dimension, (5.5.159) becomes (due to possible cancellations between the zeros of the numerator and the denominator, in the last two terms of (5.5.172))[65]

$$\begin{aligned}D_{A,\Omega} &:= \dim_B(A,\Omega) = D(\zeta_{A,\Omega}) = D_{\text{hol}}(\zeta_{A,\Omega}) \\ &\leq \max\left\{ \max_{q=1,\dots,Q} \{D_{G^{(q)}}\}, \sigma_0 \right\}.\end{aligned} \tag{5.5.173}$$

Similarly, exactly as in [LapPe2–3] and [LapPeWi1–2], the fractal tube formula (of a given type) associated with the RFD (or self-similar spray) (A,Ω) is simply obtained by adding the corresponding fractal tube formulas (of the same type) associated with the RFDs (or self-similar sprays) $(A^{(q)},\Omega^{(q)})$, for each $q=1,\dots,Q$, and likewise for the corresponding error terms and error estimates (when applicable). [Naturally, this observation can also be applied to the case of self-similar sets discussed in the previous comment (see part (a) of this remark), when the corresponding self-similar tilings have multiple generators (rather than a single generator).]

More specifically, the counterpart of (5.5.113) would become

$$V_{A,\Omega}(t) = \sum_{q=1}^{Q} V_{A^{(q)},\Omega^{(q)}}(t), \tag{5.5.174}$$

where, for each $q=1,\dots,Q$, $V_{A^{(q)},\Omega^{(q)}}(t)$ is given by the fractal tube formula (5.5.106), but with (A,Ω) replaced by $(A^{(q)},\Omega^{(q)})$ and likewise, with $(\partial G,G)$ replaced by $(\partial G^{(q)},G^{(q)})$. Therefore, we have the following (pointwise or distributional) formula (with or without error term):

$$\begin{aligned}V_{A,\Omega}(t) &:= |A_t \cap \Omega| \\ &= \sum_{q=1}^{Q} \sum_{\omega \in (\mathfrak{D}\cap \boldsymbol{W}) \cup \mathscr{P}(\zeta_{\partial G^{(q)},G^{(q)}},\boldsymbol{W})} \operatorname{res}\left(\frac{t^{N-s}\zeta_{\partial G^{(q)},G^{(q)}}(s)}{(N-s)\left(1-\sum_{j=1}^{J} r_j^s\right)}, \omega \right) \\ &\quad + R_{A,\Omega}(t),\end{aligned} \tag{5.5.175}$$

[65] For the third equality in (5.5.173) to hold, we must implicitly assume a mild condition on (A,Ω); namely, that its lower Minkowski content $\mathscr{M}_*(A,\Omega)$ is positive.

where, as before, $\mathfrak{D} = \mathfrak{D}_{\mathfrak{S}}$ is the set of complex solutions of $\sum_{j=1}^{J} r_j^s = 1$ and $R_{A,\Omega}(t) := \sum_{q=1}^{Q} R_{A^{(q)},\Omega^{(q)}}(t)$ is the sum of the error terms associated with each self-similar RFD $(A^{(q)},\Omega^{(q)})$, for $q \in \{1,\ldots,Q\}$.

In the case when each of the generators $G^{(q)}$ $(q = 1,\ldots,Q)$ is monophase, then each $V_{A^{(q)},\Omega^{(q)}}$ is given by the exact tube formula (5.5.113), except for (A,Ω) replaced by $(A^{(q)},\Omega^{(q)})$, g replaced by $g^{(q)}$, the inner radius of $G^{(q)}$, and κ_i replaced by $\kappa_i^{(q)}$ for each $i = 0,\ldots,N$:

$$\begin{aligned} V_{A,\Omega}(t) &:= |A_t \cap \Omega| \\ &= \sum_{q=1}^{Q} \sum_{\omega \in \mathfrak{D} \cup \{0,1,\ldots,N-1\}} \operatorname{res}\left(t^{N-s} \frac{\sum_{i=0}^{N} \kappa_i^{(q)} \frac{(g^{(q)})^{s-i}}{s-i}}{\left(1 - \sum_{j=1}^{J} r_j^s\right)}, \omega \right). \end{aligned} \tag{5.5.176}$$

Furthermore, in the important special case when $\mathfrak{D} = \mathfrak{D}_{\mathfrak{S}}$ (the set of 'scaling complex dimensions' of the self-similar spray, in the terminology of [LapPe2–3], [LapPeWi1–2] or of [Lap-vFr3, Section 13.1]) consists only of simple zeros of the Dirichlet polynomial $1 - \sum_{j=1}^{J} r_j^s$ (or, equivalently, of simple poles of the 'scaling zeta function' $\zeta_{\mathfrak{S}}(s) := (1 - \sum_{j=1}^{J} r_j^s)^{-1}$), and the unique real zero σ_0 of $1 - \sum_{j=1}^{J} r_j^s$ is not an integer,[66] the fractal tube formula (5.5.176) takes the following simpler form:

$$V_{A,\Omega}(t) := |A_t \cap \Omega| = \sum_{\omega \in \mathfrak{D} \cup \{0,1,\ldots,N-1\}} d_\omega t^{N-\omega}, \tag{5.5.177}$$

where

$$d_\omega := \operatorname{res}\left(\zeta_{\mathfrak{S}}, \omega\right) \sum_{q=1}^{Q} \sum_{i=0}^{N} \frac{\kappa_i^{(q)} (g^{(q)})^{\omega - i}}{\omega - i}, \quad \text{if} \quad \omega \in \mathfrak{D}, \tag{5.5.178}$$

or

$$d_\omega := \zeta_{\mathfrak{S}}(\omega) \sum_{q=1}^{Q} \kappa_\omega^{(q)}, \quad \text{if} \quad \omega \in \{0,1,\ldots,N-1\}, \tag{5.5.179}$$

and (as above, and in the terminology of [LapPe2–3], [LapPeWi1–2]) $\zeta_{\mathfrak{S}}$ denotes the 'scaling zeta function' of the spray; namely,

$$\zeta_{\mathfrak{S}}(s) := \frac{1}{1 - \sum_{j=1}^{J} r_j^s}, \quad \text{for all} \quad s \in \mathbb{C}. \tag{5.5.180}$$

Consequently, we obtain the following more explicit form of the fractal tube formula, which shows the respective contributions of the 'scaling complex dimensions' (i.e., the poles of $\zeta_{\mathfrak{S}}$, or, equivalently, the elements of $\mathfrak{D} =: \mathfrak{D}_{\mathfrak{S}}$) and the 'integer complex dimensions' (i.e., the elements of $\{0,\ldots,N-1\}$ that are poles of $\zeta_{\partial G,G}$):

[66] See the example of the Sierpiński 3-gasket discussed in Example 4.2.26 where this situation occurs.

$$
\begin{aligned}
V_{A,\Omega}(t) &:= |A_t \cap \Omega| \\
&= \sum_{\omega \in \mathfrak{D}_{\mathfrak{S}}} \operatorname{res}\left(\zeta_{\mathfrak{S}}, \omega\right) \left(\sum_{q=1}^{Q} \sum_{i=0}^{N} \frac{\kappa_i^{(q)} (g^{(q)})^{\omega - i}}{\omega - i} \right) t^{N-\omega} \\
&\quad + \sum_{i=0}^{N-1} \zeta_{\mathfrak{S}}(i) \left(\sum_{q=1}^{Q} \kappa_i^{(q)} \right) t^{N-i}.
\end{aligned}
\tag{5.5.181}
$$

(Compare with [LapPe3, LapPeWi1] and [Lap-vFr3, Corollary 13.16].)

In closing this comment, we note that the fractal tube formula (5.5.176) (or, equivalently, (5.5.181)) can be immediately extended from the case of monophase to that of pluriphase generators, as a consequence of the general fractal tube formula (5.5.175). Therefore, we can recover and considerably extend all of the results obtained in [LapPe2–3] and [LapPeWi1–2] (see also [DemKoÖÜ] and, especially, [DenKoÖÜ], based on those references), as well as the corresponding results discussed in the exposition provided in [Lap-vFr3, Section 13.1].

Also, we note that, naturally, the discussion of part (a) of this remark can be combined with the present one to extend from single to multiple (sufficiently nice) generators the conjectures regarding the fractal zeta functions of self-similar sets (satisfying the open set condition) and corresponding fractal tube formulas. In particular, we expect that (with the notation of part (a)) Equation (5.5.156) remains true, with $\zeta_{A,\Omega}$ given by the following identity:

$$\zeta_F(s) = \zeta_{A,\Omega}(s) + \zeta_{O,\mathrm{out}}(s) + f(s), \tag{5.5.182}$$

where the complex-valued function f is holomorphic in a suitable open right half-plane $\{\operatorname{Re} s > \alpha\}$, with $\alpha \in \mathbb{R} \cup \{-\infty\}$ sufficiently small, and where (A,Ω) is the canonical self-similar tiling associated with the given self-similar set F. We then conjecture that the corresponding fractal tube formula is given by

$$V_F(t) = V_{A,\Omega}(t) + V_{O,\mathrm{out}}(t) + R_F(t), \tag{5.5.183}$$

where $V_{A,\Omega}$ is still given by Equation (5.5.175), the fractal tube formula for the canonical self-similar tiling (A,Ω) (or an appropriate more general self-similar tiling) associated with the self-similar set F, and the error term R_F can (roughly) be split into three parts,

$$R_F = R_{A,\Omega} + R_{O,\mathrm{out}} + R_f, \tag{5.5.184}$$

the error term $R_{A,\Omega}$ appearing in (5.5.175), $R_{O,\mathrm{out}}$, the error term appearing in the fractal tube formula for the RFD $(O,\mathrm{out}) := (K,K^c)$ described towards the end of part (a) (or a suitable analog thereof), expressed in terms of the distance zeta function $\zeta_{O,\mathrm{out}} := \zeta_{K,K^c}$ (by using the results of Section 5.3), as well as the error term R_f corresponding to the contribution of the holomorphic function f appearing in (5.5.182) (the counterpart of (5.5.156) in the present context).

(*c*) (*Minkowski measurability criteria for self-similar sprays and sets*). Let (A,Ω) be a self similar spray with scaling ratios $\{r_j\}_{j=1}^J$ and (for notational simplicity) a single generator G assumed to be pluriphase (for example, a suitable polytope, by [KoRati]).[67] Then, under mild assumptions, $D = \dim_B(A,\Omega)$ exists, $D < N$ and (A,Ω) is Minkowski nondegenerate. In particular $\mathscr{M}_*(A,\Omega) > 0$. Now, let σ_0 denote the similarity dimension of (A,Ω); namely, σ_0 is the unique real solution of the Moran equation $\sum_{j=1}^J r_j^s = 1$. Then, in light of the factorization formula (5.5.105), (namely, since $\zeta_{\partial G,G}$ is then meromorphic in all of $\mathbb{C}$, $\zeta_{A,\Omega}(s) = \zeta_{\mathfrak{S}}(s)\cdot\zeta_{\partial G,G}(s)$, for all $s\in\mathbb{C}$, where $\zeta_{\mathfrak{S}}(s) := (1-\sum_{j=1}^J r_j^s)^{-1}$) and by Theorem 4.1.7(*c*) (or Corollary 4.1.10(*ii*)), we have that

$$\begin{aligned} D = \dim_B(A,\Omega) &= D(\zeta_{A,\Omega}) = D_{\rm hol}(\zeta_{A,\Omega}) \\ &= \max\left\{\sigma_0, \overline{\dim}_B(\partial G,G)\right\}. \end{aligned} \tag{5.5.185}$$

Actually, in the second equality above, initially, we have $\overline{\dim}_B(A,\Omega) = D(\zeta_{A,\Omega})$, which is a general result for any RFD (A,Ω), by Theorem 4.1.7. The stronger result in our case follows from the sufficiency condition given by Theorem 5.4.2 in the nonlattice case, and by the pointwise fractal tube formula of case (*ii*) of Theorem 5.3.16 in the lattice case (since the generator is assumed to be pluriphase, it is not difficult to check that the strong d-languidity is satisfied with d-languidity exponent $\kappa_d := 0$; see also the discussion following Equation (5.5.113) in the monophase case).[68] In particular, we obtain the following natural generalization of the identity (4.2.80), initially established in the case of the inhomogeneous N-gasket RFD in Example 4.2.26 of Subsection 4.2.3 (in the notation of Example 4.2.26, we have $G := \Omega_{N,0}$):

$$D = \dim_B(A,\Omega) = D(\zeta_{A,\Omega}) = \max\left\{\sigma_0, \overline{\dim}_B(\partial G,G)\right\}. \tag{5.5.186}$$

Note that, under the present assumptions, we have that $\overline{\dim}_B(\partial G,G)$ belongs to $\{0,1,\ldots,N-1\}$ and hence, is an integer. Also, if G is a polytope, for example, $D_G := \dim_B(\partial G,G)$ exists and therefore can be substituted for $\overline{D}_G := \overline{\dim}_B(\partial G,G)$ in (5.5.185) and (5.5.186). For notational simplicity, we will assume in the sequel that $D_G = \dim_B(\partial G,G)$ exists.

Next, *we distinguish the following three cases*: (*i*) $D_G < \sigma_0$; (*ii*) $D_G = \sigma_0$; *and* (*iii*) $D_G > \sigma_0$.

***Case* (i):** $D_G < \sigma_0$. Then, by (5.5.186), $D = \sigma_0$ and all of the poles of $\zeta_{\partial G,G}$ (which are all simple and form a subset of $\{0,1,\ldots,N-1\}$) are located strictly to the left of D. Therefore, in light of the factorization formula (5.5.105) recalled

[67] The assumptions on G could be significantly weakened but we will refrain from doing so here, in order not to complicate the statements unnecessarily.

[68] Actually, the following issue arises in the critical case when $D = \sigma_0 = \overline{\dim}_B(\partial G,G)$ since then, by the factorization formula (5.5.105), D is a *double* pole of $\zeta_{A,\Omega}$. Therefore, in the lattice case, we can apply Theorem 5.4.32, but in the nonlattice case, we do not have a rigorous justification of the second equality in (5.5.185), although we conjecture that it is also true.

above, it follows that the principal complex dimensions of (A,Ω) coincide with the complex solutions of the complexified Moran equation $\sum_{j=1}^{J} r_j^s = 1$ (i.e., with the scaling complex dimensions of (A,Ω)). Recall from [Lap-vFr3, Theorem 3.6] that σ_0 is always a simple pole of $\zeta_{\mathfrak{S}}$ and that, in the nonlattice case, it is the only pole of $\zeta_{\mathfrak{S}}$ located on the vertical line $\{\mathrm{Re}\, s = \sigma_0\}$, while in the lattice case, the poles of $\zeta_{\mathfrak{S}}$ form an infinite subset of $\sigma_0 + \mathbf{p}\mathrm{i}\mathbb{Z}$, where $\mathbf{p} := 2\pi/\log(r^{-1})$ is the *oscillatory period*, with $r \in (0,1)$ being the single generator of the multiplicative group (of rank 1) generated by the scaling ratios $r_1,\dots,r_J$. (See also the comments preceding the statements of Corollary 5.4.23 above.)

We deduce, in particular, that since D is simple, *the RFD (A,Ω) has a nonreal complex dimension with real part D $(=\sigma_0)$ if and only if we are in the lattice case, and hence*, in light of Theorems 5.4.2 and 5.4.20, *if and only if (A,Ω) is Minkowski measurable*. More specifically, we reason exactly as in the proof of Corollary 5.4.23 (which corresponds to the case when $N = 1$). Namely, if (A,Ω) is lattice, then it satisfies the hypotheses of Theorem 5.4.15 concerning the languidity and the screen and hence, since in the lattice case, in addition to D there are other poles with real part D, we conclude that (A,Ω) cannot be Minkowski measurable. On the other hand, if (A,Ω) is nonlattice, then the only pole with real part D is D itself and it is simple. Consequently, (A,Ω) satisfies the hypotheses of Theorem 5.4.2 and hence, (A,Ω) is Minkowski measurable.[69]

Therefore, in case (i), (A,Ω) is Minkowski measurable if and only if D is its only principal complex dimension and also, if and only if the self-similar spray (A,Ω) is nonlattice.

This proves (for the case of self-similar sprays) the geometric part of a conjecture of the first author in [Lap3, Conjecture 3, pp. 163–164], in case (i) (and, in particular, in case D is *not* an integer). Note that for a self-similar string (i.e., when $N = 1$), we are always in case (i) and therefore, we have reproved the characterization of the Minkowski measurability for self-similar strings obtained in [Lap-vFr3, Section 8.4, esp., Theorems 8.23 and 8.36] and recovered (via a different method) in Corollary 5.4.23 above.

We note that [Lap3, Conjecture 3] was stated both for the geometry and spectra of self-similar fractal drums, satisfying the open set condition. More specifically, it was stated in terms of *the leading geometric and spectral oscillations* of such drums but not explicitly in terms of complex dimensions. Therefore, from this point of view, our results go beyond the scope of the geometric part of that conjecture.

---

[69] See [Lap-vFr3], Chapters 2 and 3, especially, Theorem 2.16 and Theorem 3.6, for a detailed analysis of the structure of the scaling complex dimensions, in the lattice and nonlattice cases. In particular, in the lattice case, the scaling complex dimensions are periodically distributed along finitely many vertical lines (the right-most of which is $\{\mathrm{Re}\, s = \sigma_0\}$), while in the nonlattice case, they are '*quasiperiodically distributed*' and can be approximated by the scaling complex dimensions of an infinite sequence of self-similar strings, with oscillatory periods $\mathbf{p}_n$ increasing to infinity exponentially fast as $n \to \infty$. Observe that in [Lap-vFr3, Chapter 3], no assumption is made about the underlying scaling ratios (and gaps), so that the corresponding results (and hence also, [Lap-vFr3, Theorems 2.16 and 3.6]) can be applied to self-similar sprays in $\mathbb{R}^N$, for an arbitrary $N \geq 1$.

***Case* (ii):** $D_G = \sigma_0$ and hence, D is an integer. Since D_G and σ_0 are simple poles of $\zeta_{\partial G,G}$ and $\zeta_{\mathfrak{S}}$, respectively, it follows from the factorization formula (5.5.108), $\zeta_{A,\Omega} = \zeta_{\mathfrak{S}} \cdot \zeta_{\partial G,G}$, that D is a double (and hence, a multiple) pole of $\zeta_{A,\Omega}$. Therefore, according to Theorem 5.4.10 (and by using Lemma 5.4.11), the self-similar spray (i.e., the RFD) (A,Ω) *is not Minkowski measurable*, independently of whether or not the self-similar spray is lattice or nonlattice. Under the additional assumptions[70] of Theorem 5.4.32 (the case when $m = 2$ in the notation of that theorem) on the RFD (A,Ω), we conclude in this case that it is h-Minkowski measurable, where the gauge function h is given by $h(t) := \log t^{-1}$ for every $t \in (0,1)$.

***Case* (iii):** $D_G > \sigma_0$. Then, in light of (5.5.186), we have $D = D_G$ (hence, D is an integer). Furthermore, according to [Lap-vFr3, Equation (3.9)] (see also Subsection 2.1.4 above), all of the poles of $\zeta_{\mathfrak{S}}$ have real part $\leq \sigma_0$ and thus, have real part $< D$. In light of the factorization formula (5.5.105), it then follows that the only principal complex dimension of (A,Ω) is $D = D_G$ itself, and it is simple (since D_G is a simple pole of $\zeta_{\partial G,G}$); i.e., $\dim_{PC}(A,\Omega) = \{D_G\}$. (Note that in most cases of interest, we have $D_G = N-1$.) Consequently, by Theorem 5.4.20 (the Minkowski measurability criterion), the RFD (A,Ω) is Minkowski measurable.

Therefore, in case (iii), (A,Ω) *is always Minkowski measurable*, whether or not the self-similar spray (A,Ω) is lattice or nonlattice.

In summary, if D is not an integer, then we must be in case (i). In that situation, (A,Ω) is Minkowski measurable if and only if it is nonlattice (the same conclusion holds in case (i) even if D is an integer, which can happen). In particular, the geometric part of Conjecture 3 of [Lap3, p. 163–164] is true in this case (and, more generally, in case (i), which is the one most often encountered in practice). (See also the comments below about the later results of [KomPeWi].)

If we are not in case (i), then D is an integer and either $\sigma_0 = D_G$ $(= D)$ and hence, (A,Ω) is not Minkowski measurable or else $\sigma_0 < D_G$ $(= D)$, and hence (A,Ω) is Minkowski measurable. Then, clearly, the conclusion of the geometric part of [Lap3, Conjecture 3] fails when we are not in case (i). Note, however, that in case (iii), this fact does not contradict [Lap3, Conjecture 3] because case (iii) cannot occur for self-similar sets satisfying the open set condition, which was the only situation considered in that conjecture. Indeed, in that situation, we have (for any $\delta > 0$)

$$\dim_B(A,\Omega) \leq \dim_B(A,A_\delta) = \dim_B A = \sigma_0,$$

where the last equality is well known and was discussed earlier. (See, e.g., [Fal1, Theorem 9.3].)

Recall from the discussion of the inhomogeneous relative N-gasket (A_N,Ω_N) in Example 4.2.26 of Subsection 4.2.3 that case (i) occurs when $N = 2$ (the usual Sierpiński gasket), case (ii) when $N = 3$, and case (iii) when $N \geq 4$. Therefore, each

[70] These assumptions are always satisfied in the lattice case, whereas in the nonlattice case it may not be possible to choose a suitable screen. Note also that for now, we do not have an analog of Theorem 5.4.2 for h-Minkowski measurability sufficiency; we leave the problem of proving such a sufficiency theorem for a future work. In spite of these two cautionary comments, we believe that the stated h-Minkowski measurability result is true quite generally, in case (ii).

of the cases (i)–(iii) is naturally realized for general self-similar sprays (or RFDs), even though for self similar sets F (satisfying the open set condition), only case (i) or case (ii) can occur because we always have that $D_F = \sigma_0$ for such sets.

We point out that the geometric part of [Lap3, Conjecture 3] has been proved for self-similar sets (rather than for general self-similar sprays or RFDs) satisfying the open set condition, first when $N = 1$ in [Lap-vFr1–3] (see [Lap-vFr3, Section 8.4] and the earlier books [Lap-vFr1–2]), by using the fractal tube formulas for fractal strings, and then, in parallel with the present work, when $N \geq 1$ in [KomPeWi], by using the renewal theorem, in particular. The aforementioned works extend a variety of results previously obtained in [Lap3] (when $N = 1$ and by using the renewal theorem), in [Fal2] (also when $N = 1$, and by using this same theorem as well) and then, in [Gat] (when $N \geq 2$, and also by using the renewal theorem) and in [Lap-vFr1–3] (as mentioned above), as well as later, under some relatively restrictive hypotheses, for self-similar sprays in [LapPeWi2] (when $N \geq 1$ and by using the fractal tube formulas for self-similar sprays of [LapPeWi1], along with techniques from [Lap-vFr3, Section 8.4]).

What was missing in the results of [Lap3, Fal2, Gat] (but not of [Lap-vFr1–3] and of [LapPeWi2]) was to show that lattice self-similar sets are *not* Minkowski measurable (as was the content of part of the aforementioned conjecture of [Lap3]), which is now known to be true when D is not an integer. We note that case (ii) was also considered in [KomPeWi], with the same conclusion as above. In our setting, however, by using the results and methods of Subsection 5.4.4, we could also (under appropriate hypotheses) obtain definite conclusions about the h-Minkowski measurability of (A, Ω) for some suitable gauge function h (namely, $h(t) := \log t^{-1}$, for all $t \in (0,1)$). In the inhomogeneous (rather than homogeneous or strictly) self-similar case, explicit examples of such situations and conclusions are provided in Examples 5.5.22, 5.5.25, as well as in Example 4.2.26 when $N = 3$ (the relative Sierpiński 3-gasket); see Theorem 5.4.27 (along with Theorem 4.5.1).

Finally, we close this discussion with four comments:

First, we expect that the above results about self-similar sprays cases (i), (ii) and (iii) can also be proved by analogous methods for self-similar sets in $\mathbb{R}^N$ (satisfying the open set condition), via a suitable functional equation connecting the fractal zeta functions of the self-similar set and the associated self-similar tiling (or spray), as was suggested earlier in this subsection, or else by some other method based on symmetry considerations and the key scaling properties of fractal zeta functions established in this book.

Second, whether for self-similar sprays or for self-similar sets (and, more generally, for self-similar RFDs), the geometric part of the original conjecture [Lap3, Conjecture 3] can be naturally extended as follows (both in case (i) for which it is now a theorem, and in case (iii)):

Assume that $\sigma_0 \neq D_G$. *Then, the self-similar spray (or set) is (possibly subcritically) Minkowski measurable in dimension* $d := \sigma_0$ *if and only if the only complex dimension with real part* d *is* d *itself.*[71] Note that here, d is simple when $d \neq D_G$.

In the case of self-similar sprays, this conjecture should follow from the results of this chapter, much as in case (i) above, by an appropriate adaptation (to the strictly subcritical case when $D_G > d$) and extension of the statements and proofs of Theorems 5.4.2 and 5.4.20. (Of course, case (i) corresponds to the case when $D_G < d$ and is already proved above.)

We note that there may be one exception to the above conjecture; namely, it could happen that $D_G > d$ and $\zeta_{\partial G,G}(d) = 0$,[72] thereby giving rise to cancellations.[73] Of course, this could only happen for a general self-similar RFD (or spray), but not for a self-similar set.

Recall once again that for self-similar sets (satisfying the open set condition), we have that $D_F = d$ and hence, we can replace "possibly subcritically" by "critically" in the above conjecture, in agreement with the original conjecture made in [Lap3] (and now proved in that case in [KomPeWi], as discussed above). We point out that case (ii) (when $D_G = d$) will be discussed in more detail towards the end of Chapter 6 (see, especially, Problem 6.2.36 and the much more precise conjecture to be stated there), is not yet proved for self-similar sets, but is expected to be true and provable by means of some of the results of the present chapter.

Our third comment is that many of the results and conjectures of this subsection extend naturally to the setting of (*not necessarily self-similar*) fractal sprays, under appropriate assumptions on the generators of the sprays and on the (generalized) fractal strings by which they are scaled. Then, Minkowski measurability (respectively, Minkowski measurability in dimension d) should be equivalent to the fact that the only complex dimension of real part D (respectively, d) is D (respectively, d) itself and D (respectively, d) is simple. This follows exactly along the same lines as above by using, in particular, the factorization formula, $\zeta_{A,\Omega} = \zeta_{\mathfrak{S}} \cdot \zeta_{\partial G,G}$, where (A,Ω) is a fractal spray with a single generator G and $\zeta_{\mathfrak{S}}$ is the geometric zeta function of the fractal string by which G is scaled; in other words, $\zeta_{\mathfrak{S}}$ is the scaling zeta function of the fractal spray. Using our previously introduced terminology,

---

[71] Roughly speaking, "Minkowski measurability in dimension d" means that if $W(t)$ is the part of the fractal tube formula corresponding to the complex dimensions with real part $\leq d$, then the limit of $t^{-(N-d)}W(t)$ (as $t \to 0^+$) exists in $\mathbb{R}$. Note that, according to this definition, the 'd-Minkowski content' may be negative, which cannot happen for the classical Minkowski content; alternatively, one may prefer to replace $W(t)$ by $|W(t)|$ in the above definition, so that the corresponding content be always nonnegative. (One can similarly define the notion of Minkowski nondegeneracy in dimension d.) When $D_G < d$, we have $d = D$ and hence, the notion of Minkowski measurability in dimension d coincides with the usual notion of Minkowski measurability.

[72] Note that this cannot occur when $D_G < d$ because then, we must have that $\zeta_{\partial G,G}(d) > 0$ by definition of $\zeta_{\partial G,G}(s)$ for all $s \in \mathbb{C}$ such that $\operatorname{Re} s > D_G$; that is, $\zeta_{\partial G,G}(d)$ is then given by the (convergent) Lebesgue integral of a positive function (see Equation (4.1.1)).

[73] Theoretically, it could happen that $\zeta_{\partial G,G}$ also cancels some of the other or even all of the complex dimensions with real part equal to σ_0, but we do not have examples of such self-similar RFDs. We do not even have an example for which the cancellation of σ_0 occurs.

this would mean that the corresponding RFD is not fractal in dimension D (respectively, d) and hence, is not critically fractal (respectively, is not subcritically fractal in dimension d). Similarly, under appropriate hypotheses, critical Minkowski nonmeasurability (respectively, subcritical Minkowski nonmeasurability in dimension $d < D$) would be equivalent to critical fractality (respectively, to strict subcritical fractality in dimension $d < D$).

We close this section by a fourth and final comment. Namely, we mention that one could provide many further examples illustrating our fractal tube formulas, as applied to self-similar sprays or, more generally, fractal sprays. These examples would include the Koch curve tiling, the Sierpiński gasket tiling, the pentagasket tiling and the Menger sponge tiling depicted, respectively, in Figures 6.1–6.5 of [LapPeWi1]. We refer to [LapPe2–3] for the corresponding tube formulas. We point out that the pentagasket tiling is of special interest because it is a natural example of a self-similar spray with multiple generators. Recall from Remark 5.5.26(b) that in the case of fractal sprays with multiple (say, Q) generators, it suffices to apply the results of the present subsection to each of the corresponding Q fractal sprays with a single generator, and then to add-up the resulting Q fractal tube formulas.

Other interesting examples include the Cantor carpet, the U-shaped modification of the Sierpiński carpet tiling (which has a generator that is itself "fractal", in our sense), the binary trees, and the Apollonian packings depicted, respectively, in Figures 6.6, 6.9, 6.11 and 6.12 of [LapPeWi1] and whose associated fractal tube formulas are provided and discussed in Subsections 6.1–6.4 of [LapPeWi1].

Moreover, recall from footnote 43 on page 492 that we can now use the results and methods of this chapter (including of this subsection) to also obtain a fractal tube formula for the Koch curve and for the Koch snowflake RFDs, which are important geometric examples that are definitely not fractal sprays.[74] This result remains to be fully explicited and compared with the tube formula obtained for these same examples in [LapPe1] via a direct computation (and without the use of fractal zeta functions and their associated complex dimensions); see [Lap-vFr3, Subsection 12.21] for an exposition of the main result of [LapPe1].

Finally, we recall that our methods apply naturally to fractal sprays which are not necessarily self-similar (such as the last three examples mentioned in the next-to-last paragraph just above). Moreover, as was alluded to in the introduction (i.e., Chapter 1), our general pointwise and distributional fractal tube formulas can be extended (under suitable hypotheses) to include the case where the associated fractal zeta functions have nonremovable singularities which are not poles. Several examples of such situations have been provided earlier in various contexts, but we plan to develop the corresponding systematic theory in a later work.

[74] The Koch curve is, of course, self-similar, but it is clearly not a self-similar spray or tiling. Furthermore, the Koch snowflake (on which the corresponding snowflake RFD is based) is obtained by putting together three isometric and abutting copies of the Koch curve. See, e.g., [Lap-vFr3, Figures 12.7 and 12.6], along with [Fal1] and [Man1].

Remark 5.5.27. We note that it follows from the results in [Lap-vFr3, Chapters 2 and 3] discussed elsewhere in this book that, under mild assumptions on their generators,[75] self-similar fractal sprays are fractal in dimension d for only a finite (but nonempty) set of values of d in the lattice case, whereas they are fractal in dimension d for an infinite countable and dense set of values of d in the nonlattice case. More specifically, the set of d's for which nonlattice (resp., generic nonlattice) self-similar sprays are fractal in dimension d is dense in finitely many nonempty compact intervals (resp., in a single interval of the form $[D_l, D]$, where $D_l \in \mathbb{R}$ and $D_l < D$). (This follows from the main result of [MorSepVi1] proving and extending a conjecture in [Lap-vFr2, Section 3.7], as well as more specifically, in reference [Lap-vF7] of [Lap-vFr2] or of [Lap-vFr3].)

More generally, we conjecture that under appropriate hypotheses, self-similar RFDs and sets satisfying the open set condition enjoy the same properties.

[75] For example, it suffices to assume that the generators are convex polytopes or, more generally, that they are "nonfractal" in our sense, so that they do not have any nonreal complex dimensions.

Chapter 6
Classification of Fractal Sets and Concluding Comments

If I were to awaken after having slept for a thousand years, my first question would be: Has the Riemann hypothesis been proven?

David Hilbert (1862–1943)

Abstract In this last chapter, we first introduce a refinement of the classification of bounded sets in $\mathbb{R}^N$ which had begun with the well-known distinction between Minkowski nondegenerate and Minkowski degenerate sets. Further distinction will be made by classifying fractals according to the properties of their tube functions and allowing, in particular, more general scaling laws than the standard power laws. We then provide a short historical survey concerning notions pertaining to Minkowski measurability and related topics which play an important role in this work. We conclude the book with a few remarks, a long list of open problems, and propose several directions for future research. The research problems and directions proposed here connect many different aspects of fractal geometry, number theory, complex analysis, functional analysis, harmonic analyis, complex dynamics and conformal dynamics, partial differential equations, mathematical physics, spectral theory and spectral geometry, as well as nonsmooth analysis and geometry.

Key words: classification of fractal sets, tube function, Minkowski degenerate set, Minkowski nondegenerate set, Minkowski measurability, gauge functions, historical survey, open problems, research directions.

In this last chapter, we introduce (in Section 6.1) a refinement of the classification of bounded sets in $\mathbb{R}^N$ which had begun with the well-known distinction between Minkowski nondegenerate and Minkowski degenerate sets. Further distinction will be made by classifying them according to the properties of their tube functions. See Subsection 6.1.1.

Towards the end of Section 6.1, we provide a short historical survey concerning notions pertaining to Minkowski measurability and related topics which play an important role in this work; see Subsection 6.1.2. We conclude the book with a few remarks (Subsection 6.2.1), a relatively long list of open problems (Subsection 6.2.2), and propose several directions for future research (Subsection 6.2.3).

The research problems and directions proposed here connect many different aspects of fractal geometry, complex analysis, functional analysis, harmonic analyis,

M.L. Lapidus et al., *Fractal Zeta Functions and Fractal Drums*,
Springer Monographs in Mathematics, DOI 10.1007/978-3-319-44706-3_6

complex dynamics and conformal dynamics, partial differential equations, mathematical physics, spectral theory and spectral geometry, as well as nonsmooth analysis and geometry.

6.1 Classification of Bounded Sets in Euclidean Spaces

We propose the following general classification of bounded sets A in $\mathbb{R}^N$. The roughest classification into Minkowski degenerate and Minkowski nondegenerate categories has already been introduced on page 32 in Section 1.3.

(a) A is *Minkowski nondegenerate* (or simply *nondegenerate*), if there exists $D \geq 0$ such that $0 < \mathscr{M}_*^D(A) \leq \mathscr{M}^{*D}(A) < \infty$. In particular, we then have $D = \dim_B A$.

(b) A is a *Minkowski degenerate set* (or simply *degenerate*) if

• either $D = \dim_B A$ exists and at least one of the corresponding D-dimensional Minkowski contents is degenerate (i.e., $\mathscr{M}_*^D(A) = 0$ or $\mathscr{M}^{*D}(A) = +\infty$)

• or else $\underline{\dim}_B A < \overline{\dim}_B A$.

Remark 6.1.1. Recall from Remark 1.3.4 that since the t-neighborhood of A is equal to that of its closure $\overline{A}$, the same is true of the tube function $t \mapsto |A_t|$ of A, as well as of the (upper, lower) Minkowski dimension and Minkowski content of A. Therefore, in what follows, we may as well asume that instead of being bounded, A is a compact subset of $\mathbb{R}^N$.

6.1.1 Classification of Compact Sets Based On the Properties of Their Tube Functions

We now introduce a finer classification of bounded sets in $\mathbb{R}^N$, based on the asymptotic behavior of their tube functions. Since, in light of Remark 6.1.1 just above, the tube function of a bounded subset of $\mathbb{R}^N$ coincides with that of its closure, which is a compact set, this classification amounts to a classification of the compact subsets of the N-dimensional Euclidean space $\mathbb{R}^N$.

First, we consider the case of Minkowski nondegenerate sets A. This is equivalent to saying that the tube function $t \mapsto |A_t|$ has the following form:

$$|A_t| = t^{N-D}(F(t) + o(1)) \quad \text{as} \quad t \to 0^+, \tag{6.1.1}$$

where the function $F : (0, \delta) \to \mathbb{R}$ is bounded away from zero and infinity, that is, $0 < \inf F \leq \sup F < \infty$. Clearly,

$$\liminf_{t \to 0^+} F(t) = \mathscr{M}_*^D(A), \quad \limsup_{t \to 0^+} F(t) = \mathscr{M}^{*D}(A).$$

The idea of this classification is to introduce function-theoretic notions for sets. More precisely, various properties of A will be expressed in terms of properties of an associated function F in the corresponding tube formula (6.1.1).

We shall need an auxiliary function $\rho = \rho(t)$, defined for $t > 0$ small enough, such that

$$\rho = \rho(t) \text{ is decreasing, positive, continuous, and } \lim_{t\to 0^+} \rho(t) = +\infty. \tag{6.1.2}$$

6.1.1.1 Classification of Minkowski Nondegenerate Sets

Let A be a Minkowski nondegenerate bounded (or, equivalently, compact)[1] set in $\mathbb{R}^N$. We say that

• A is a *constant set*, or a *Minkowski measurable set*, if there exists a finite and positive constant $\mathscr{M}$ such that (6.1.1) is satisfied with $F(t) \equiv \mathscr{M}$. It then follows that A is Minkowski measurable with Minkowski content $\mathscr{M}$.

• A is a *nonconstant set* if there is no positive constant function F satisfying (6.1.1).

We now classify Minkowski nondegenerate sets that are not constant, i.e., sets that are not Minkowski measurable. Let A be a nonconstant (i.e., Minkowski nonmeasurable) set in $\mathbb{R}^N$.

• A is a *periodic set* if (6.1.1) holds with F of the form $F(t) = G(\rho(t))$ for all t small enough, where G is a periodic function and ρ satisfies conditions (6.1.2). In the applications, we often have $\rho(t) = \log t^{-1}$, like in the case of the Cantor set or of the Sierpiński carpet. See Examples 2.3.31 and 2.3.36, Proposition 3.1.2, and also Example 2.3.33, dealing with general self-similar fractal strings. The value of the minimal period of G is called the *oscillatory period of the set* A, and is denoted by $\mathbf{p}$. At least in spirit, it is closely related to the definition of the oscillatory period of lattice self-similar sets studied in [Lap-vFr3]. We also introduce the notion of the *oscillatory amplitude of* A, denoted by $\mathbf{am} = \mathbf{am}(A)$, defined as the oscillation of the function G, $\mathbf{am}(A) = \operatorname{osc} G = \sup G - \inf G$; that is,[2]

$$\mathbf{am}(A) := \mathscr{M}^{*D}(A) - \mathscr{M}_*^D(A). \tag{6.1.3}$$

• A is a *nonperiodic set* if any function $F(t)$ appearing in (6.1.1) cannot be written in the form $F(t) = G(\rho(t))$ for all t small enough, where G is periodic and ρ satisfies conditions (6.1.2).

Nonperiodic sets can be further classified as follows. Let A be a nonperiodic set in $\mathbb{R}^N$.

---

[1] See Remark 6.1.1.

[2] The *oscillatory amplitude* of a set A, defined here, should not be mistaken for the 'amplitude' of the set A, introduced in [Žu4, Remark 2.4]. These two notions are different, though related. For example, the larger the amplitude of A, the larger its oscillatory amplitude; see [Žu4, Equation (23)].

• Let n be an integer ≥ 2 or else $n = \infty$. Then, A is a *transcendentally* (resp., *algebraically*) *n-quasiperiodic set* if $F(t) = G(\rho(t))$, where the function $G = G(\tau)$ is transcendentally (resp., algebraically) n-quasiperiodic (in the sense of Definition 3.1.9, for $n < \infty$, or of Definition 4.6.6, for $n = \infty$) and ρ satisfies condition (6.1.2). We say that A is an *n-quasiperiodic set* if it is either algebraically or transcendentally n-quasiperiodic. Several examples of such transcendentally n-quasiperiodic sets (along with associated RFDs and fractal strings) have been studied in Subsection 3.1.2 (for $n < \infty$), as well as in Subsections 4.6.1 and 4.6.2 (for $n = \infty$). Alternatively, one says that A is a *quasiperiodic set of finite* ($n < \infty$) *or infinite* ($n = \infty$) *order.* When no ambiguity may arise, we simply say that A is a quasiperiodic set.

• A is a *nonquasiperiodic* set if it is not a quasiperiodic set; that is, if any function $F(t)$ appearing in (6.1.1) cannot be written in the form $F(t) = G(\rho(t))$, with $G = G(\tau)$ being quasiperiodic (in the sense of Definition 3.1.9) and ρ satisfying condition (6.1.2).

Remark 6.1.2. The functions G and ρ in the decomposition $F = G \circ \rho$ are not uniquely determined. For example, let us fix any positive real number c. If we let $F_1(\tau) := F(c\tau)$ and $\rho_1(t) := c^{-1}\rho(t)$, then we have $G \circ \rho = G_1 \circ \rho_1$.

The value of the oscillatory period of A depends not only on A, but also on the choice of the function ρ; that is, $\mathbf{p} = \mathbf{p}(A, \rho)$. For example, since $G(\rho(t)) = G(c \cdot c^{-1}\rho(t))$ for any fixed positive real number c, then the oscillatory period of A with respect to the function ρ_1 defined by $\rho_1(t) = c^{-1}\rho(t)$ is equal to $\mathbf{p}(A, \rho_1) = \mathbf{p}(A, \rho)/c$. Indeed, this is precisely the value of the minimal period of the periodic function G_1 defined by $G_1(\tau) = G(c\tau)$.

Assuming that A is a bounded subset in $\mathbb{R}^N$ such that $D := \dim_B A$ exists and $\mathscr{M}^{*D}(A) < \infty$, then the oscillatory amplitude of A is *D-homogeneous*; that is, $\mathbf{am}(\lambda A) = \lambda^D \mathbf{a}(A)$ for any positive λ. This is a consequence of the *D-homogeneity* of the D-dimensional upper and lower Minkowski contents:

$$\mathscr{M}^{*D}(\lambda A) = \lambda^D \mathscr{M}^{*D}(A), \quad \mathscr{M}_*^D(\lambda A) = \lambda^D \mathscr{M}_*^D(A);$$

see [Žu4, Proposition 4.4], along with the discussion surrounding Equations (1.3.17)–(1.3.19).

Remark 6.1.3. The 'oscillatory amplitude' of A, am(A), as defined by Equation (6.1.3), is called by Mandelbrot in [Man2] the *lacunarity* of A. (See also [BedFi] for a related, but probably better, definition.) We refer to [Lap-vFr1–2] and [Lap-vFr3, Subsection 12.1.3] for a discussion of the possible connections between the heuristic notion and the definition of 'lacunarity' proposed in [Man1, esp., Chapter 35], [Man2] (as well as in [BedFi]) and the theory of complex dimensions developed in [Lap-vFr1–3]. In the abovementioned discussion from [Lap-vFr3, Subsection 12.1.3], not only the complex dimensions themselves (i.e., the poles of the given fractal zeta function, assumed to be simple), but also the values and the asymptotic behavior of the associated residues play a key role.

We summarize the above classification of bounded sets in $\mathbb{R}^N$, based on the properties of the associated tube function $t \mapsto |A_t|$, in Figure 6.1 on page 543.[3]

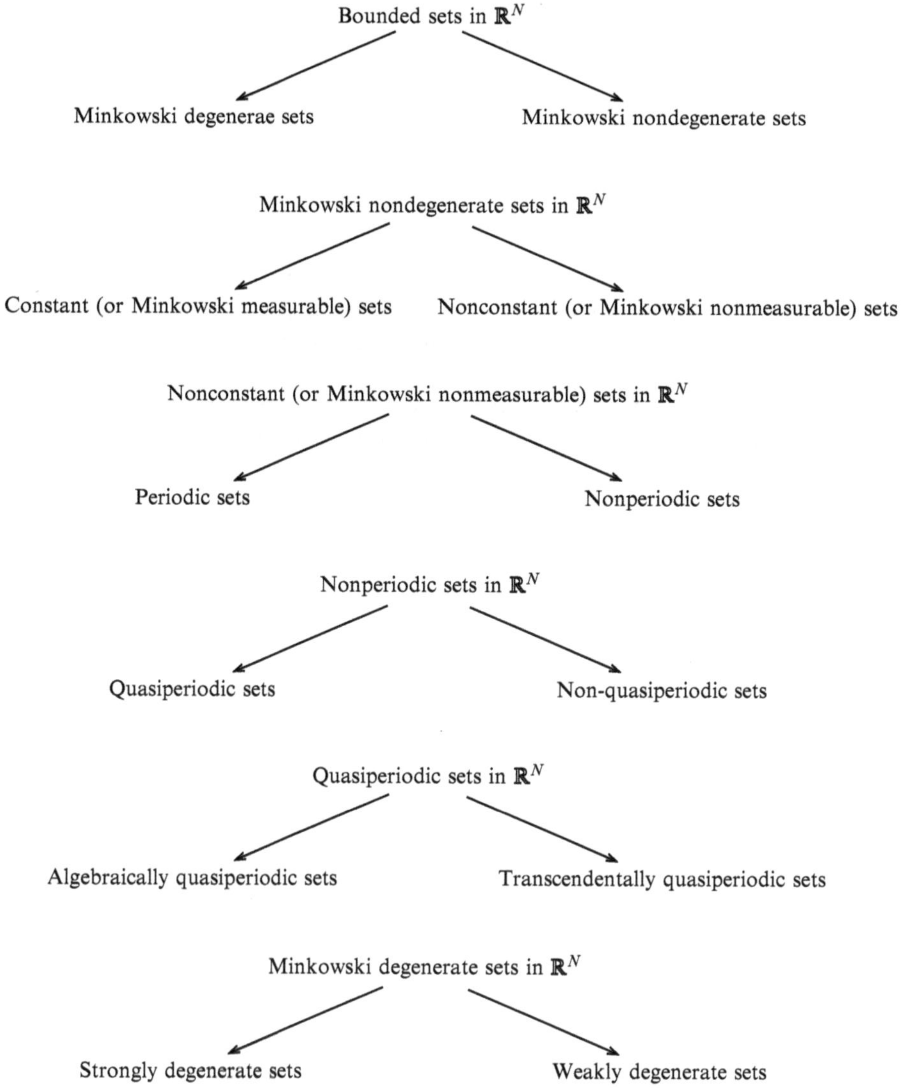

Fig. 6.1 Classification of bounded sets A in $\mathbb{R}^N$, depending on the asymptotic properties of the associated tube functions $t \mapsto |A_t|$ as $t \to 0^+$.

---

[3] Given an integer $n \geq 2$, we say that A is *algebraically n-quasiperiodic* if the periods $T_1, \ldots, T_n$ appearing in Definition 3.1.11 (see also Definition 3.1.9) are rationally independent, but not algebraically independent. For example, if we only have two periods T_1 and T_2, it suffices to assume that T_1/T_2 is an irrational algebraic number. (This definition can be extended without change to the case when $n = \infty$.)

6.1.1.2 Classification of Minkowski Degenerate Sets

If A is (Minkowski) degenerate and such that $D := \dim_B A$ exists, we assume that

$$|A_t| = t^{N-D}(F(t) + o(1)) \quad \text{as} \quad t \to 0^+, \tag{6.1.4}$$

where $F : (0, \varepsilon_0) \to (0, +\infty)$, for some sufficiently small $\varepsilon_0 > 0$.

Let A be a degenerate set in $\mathbb{R}^N$. Then:

• A is *weakly degenerate* if $D = \dim_B A$ exists and either $\mathcal{M}_*^D(A) = 0$ (i.e., $\liminf_{t\to 0^+} F(t) = 0$) or $\mathcal{M}^{*D}(A) = +\infty$ (i.e., $\limsup_{t\to 0^+} F(t) = +\infty$). See Equation (6.1.4).

• A is *strongly degenerate* if $\underline{\dim}_B A < \overline{\dim}_B A$. Note that here, (6.1.4) is impossible for any $D \geq 0$ (otherwise, D would be equal to the box dimension of A).

Weakly degenerate sets can be classified by their *gauge functions* h, if they exist (see Definition 6.1.4). We assume that the function $F(t)$ appearing in Equation (6.1.1) is of the form

$$F(t) = h(t) \quad \text{or} \quad F(t) = \tfrac{1}{h(t)}, \tag{6.1.5}$$

where $h : (0, \varepsilon_0) \to (0, +\infty)$, for some small $\varepsilon_0 > 0$, $h(t) \to +\infty$ as $t \to 0^+$ and

$$h(t) = O(t^{0)}) \quad \text{as} \quad t \to 0^+, \quad \text{with} \quad O(t^{0)}) := \bigcap_{\beta<0} O(t^\beta). \tag{6.1.6}$$

Note that we need to assume that $h(t) = O(t^{0)})$ as $t \to 0^+$ in order to fix the value $D = \dim_B A$; see Equation (6.1.4).

Definition 6.1.4. If a function $h : (0, \varepsilon_0) \to (0, +\infty)$ is of class $O(t^{0)})$ and converges to infinity as $t \to 0^+$, we then say that h is *of slow growth to infinity* as $t \to 0^+$. Analogously, a function $g : (0, \varepsilon_0) \to (0, +\infty)$ is said to be *of slow decay to zero* as $t \to 0^+$ if it is of the form $g(t) = 1/h(t)$, for some function h which is of slow growth to infinity as $t \to 0^+$. Such functions h and g are called *gauge functions*.

It is easy to see that a function $g : (0, \varepsilon_0) \to (0, +\infty)$ is of slow decay to zero as $t \to 0^+$ if and only if for every $\beta > 0$, $t^\beta = O(g(t))$ as $t \to 0^+$.

Example 6.1.5. If we define $h_1(t) = \log t^{-1}$, $h_2(t) = \log\log t^{-1}$, and more generally, $h_3(t) = (\log t^{-1})^a$, $h_4(t) = (\log^k t^{-1})^a$, for all $t \in (0,1)$ (here, $a > 0$, $k \in \mathbb{N}$, and $\log^k$ denotes the k-fold composition of logarithms), then all of these functions are of slow growth to infinity as $t \to 0^+$. Furthermore, their reciprocals are functions of slow decay to 0 as $t \to 0^+$.

Since for a weakly degenerate set A we have $\mathcal{M}^{*D}(A) = +\infty$ or $\mathcal{M}_*^D(A) = 0$, it will be convenient to define (as in [HeLap]) the *upper* and *lower* D-dimensional *Minkowski contents of A with respect to a given gauge function h*, as follows:

$$\mathscr{M}^{*D}(A,h) = \limsup_{t\to 0+} \frac{|A_t|}{t^{N-D}h(t)},$$
$$\mathscr{M}_*^D(A,h) = \liminf_{t\to 0+} \frac{|A_t|}{t^{N-D}g(t)}. \tag{6.1.7}$$

The aim is to find gauge functions h and g so that the upper and lower Minkowski contents of A with respect to h are *nondegenerate*, that is, belong to $(0,+\infty)$.

Definition 6.1.6. If $\mathscr{M}_*^D(A,h) = \mathscr{M}^{*D}(A,h) \in (0,+\infty)$, this common value is denoted by $\mathscr{M}^D(A,h)$ and called the *h-Minkowski content* of A. We then say that A is *h-Minkowski measurable*.

Definition 6.1.7. Assume that A is a bounded subset of $\mathbb{R}^N$ such that (6.1.4) holds under one of the conditions stated in (6.1.5) and that, in addition, (6.1.6) is satisfied. We then say that $h = h(t)$ or $g = 1/h(t)$ is a *gauge function* of A.[4] We also say that the set A is *weakly degenerate, of type h or* $1/h$, respectively.

Note that in the first case of (6.1.5), we have

$$\mathscr{M}^{*D}(A) = +\infty, \quad \mathscr{M}^{*D}(A,h) \in (0,+\infty),$$

while in the second case of (6.1.5), we have

$$\mathscr{M}_*^D(A) = 0, \quad \mathscr{M}_*^D(A,1/h) \in (0,+\infty).$$

Let A be a weakly degenerate set in $\mathbb{R}^N$ of type h, in the sense of Definition 6.1.7. We say that

- A is a *constant weakly degenerate set* of type h (or an *h-Minkowski measurable set*), if $\mathscr{M}^D(A,h)$ exists and belongs to $(0,+\infty)$. Then, $\mathscr{M}^D(A,h)$ is called the h-Minkowski content of A.

- A is a *nonconstant weakly degenerate set of type h* (or an *h-Minkowski nondegenerate set*) if
$$0 < \mathscr{M}_*^D(A,h) < \mathscr{M}^{*D}(A,h) < \infty.$$

We adopt a similar terminology in the case of the gauge function $1/h$ instead of h; see Definition 6.1.7.

At this stage, it would be of interest to develop general methods for finding gauge functions associated with various classes of weakly degenerate sets; see Problem 6.2.4. Some basic results in this direction can be found in [HeLap].

Concerning a function-theoretic terminology for bounded (or equivalently, compact) sets in $\mathbb{R}^N$, we can also, for example, propose the following notions.

[4] In the case when $F(t) = g(t)$, we also assume that the implied function $o(1)$ appearing in (6.1.4) satisfies $o(1)/g(t) \to 0$ as $t \to 0^+$.

Definition 6.1.8. We say that *a bounded set A has a pole at $s_1 \in \mathbb{C}$* if $\operatorname{Re} s_1 > D_{\mathrm{mer}}(\zeta_A)$ and its distance zeta function ζ_A admits a meromorphic extension which has a pole at s_1. The poles of A are also called *complex dimensions of* A, according to the terminology introduced by the first author and M. van Frankenhuijsen; see [Lap-vFr3].

In closing this subsection, we note that a classification entirely similar to the above one and all of the above definitions can be introduced for general RFDs (A, Ω) instead of merely for bounded (or compact) subsets A of $\mathbb{R}^N$; see, in particular, the discussion surrounding Equation (4.5.10) in Section 4.5 above.

6.1.2 A Short Historical Survey

From our perspective, the notions of lower and upper Minkowski contents are among some of the central objects in the study of the properties of fractal sets in Euclidean spaces, and their associated zeta functions. These important notions have been introduced and/or used (with different degrees of generality and precision) by various authors, including Bouligand [Bou], Hadwiger [Had], Kneser [Kne], Federer [Fed2] and Stachó [Sta], to only mention references ranging from the 1920s through the mid-1970s. The expression "Minkowski measurability" was perhaps used for the first time in [Sta] (and in a slightly weaker meaning in [Had, Definition 2], permitting the values 0 and $+\infty$) but the corresponding notion of "Minkowski content" was already used explicitly in [Had], [Kne], and [Fed2], and at least implicitly in [Bou].

Finally, we note that to our knowledge, the notion of 'Minkowski dimension' (now often called 'box dimension' in the literature on fractal geometry) was first used by Minkowski for integer values and then introduced and studied by Bouligand in the late 1920s in [Bou], in the general case of possibly noninteger values (but without making a clear distinction between the lower and upper limits). More recently, Tricot has used and studied various aspects of the Minkowski (or box) dimension; see, for example, [Tri1–3] and the relevant references therein. Furthermore, the notions of Minkowski dimension and Minkowski measurability have also played a key role in the first author and his collaborators' work on fractal drums [Lap1–3, HeLap, Lap7–8] and (as will be discussed next) fractal strings and sprays [LapPo2–3, LapMa1–2, Lap-vFr3, Lap6, HeLap, LapPeWi1–2].[5]

The notion of complex dimensions (poles) of sets, introduced in Definition 6.1.8, as well as much earlier in the book (for instance, in Chapters 2 and 4), is a continuation of the program of study of the complex dimensions of fractal strings and their generalizations, undertaken by the first author (M. L. Lapidus) and his collaborators in the early 1990s; see, for example, an extensive joint monograph by the first author and van Frankenhuijsen [Lap-vFr3], the earlier monographs [Lap-vFr1–2,

---

[5] See also, e.g., Section 4.3 and Remark 4.1.5 (resp., page 18 of Section 1.1) above for further relevant references about fractal drums (resp., about fractal strings and sprays).

Lap6] and the many references therein. The notions of nondegenerate and degenerate sets have been introduced by the third author (D. Žubrinić) in [Žu4]. A geometric and spectral characterization of nondegenerate fractal strings (or equivalently, of nondegenerate compact subsets of $\mathbb{R}$) appeared for the first time in a paper of the first author and Pomerance [LapPo2] (announced in [LapPo1]). Moreover, a geometric characterization of Minkowski measurable fractal strings (or, equivalently, of Minkowski measurable or 'constant' compact subsets of $\mathbb{R}$) was obtained by these same authors in [LapPo1–2]. (For more details, see Remark 6.1.10 at the end of this subsection.) Assuming some of the results of those same papers, the geometric characterization of Minkowski measurability obtained in [LapPo1–2] was then given a different proof in [Fal2] (by Falconer) and most concisely, in [RatWi2] (by Rataj and Winter).[6] In terms of complex dimensions (and under suitable assumptions on the associated geometric zeta function $\zeta_{\mathscr{L}}$), this characterization was further extended by the first author and van Frankenhuijsen in [Lap-vFr1–3] by showing that a fractal string $\mathscr{L}$ is Minkowski measurable if and only if the only complex dimension on the critical line $\{\mathrm{Re}\, s = D\}$ is D itself, and D is a simple pole of $\zeta_{\mathscr{L}}$. For self-similar strings (or equivalently, for self-similar sets in $\mathbb{R}$), this is equivalent to stating that the self-similar string (or equivalently, the self-similar set) is nonlattice (i.e., that the logarithms of its distinct scaling ratios are rationally independent), as was first shown in [Lap-vFr1] (thereby settling in the affirmative the geometric part of a conjecture formulated in [Lap3, §4.4.1a]). (See [Lap-vFr3, Chapter 8].)

The aforementioned characterization of (or criterion for) Minkowski measurability obtained in [LapPo2] was a key step in the proof of the (one-dimensional) Weyl–Berry conjecture for fractal drums (as formulated in [Lap1]) and also obtained in [LapPo1–2]. Accordingly, it was shown in [LapPo2] that if a fractal string $\mathscr{L}$ (or, equivalently, its boundary) is Minkowski measurable, then its spectral (or frequency) counting function $N_{\nu,\mathscr{L}}(\mu)$ admits a monotonic (i.e., nonoscillatory) asymptotic second term, of the form $-c_D \mathscr{M} \mu^{D/2}$, where $D \in (0,1)$ is the Minkowski (or box) dimension of $\mathscr{L}$, $\mathscr{M}$ is the Minkowski content of $\mathscr{L}$, and (for the present case of Dirichlet boundary conditions) $c_D > 0$. Moreover, the constant c_D depends only on D and is directly proportional to $-\zeta(D)$, where $\zeta = \zeta(s)$ $(= \zeta_R(s))$ denotes the classic Riemann zeta function.[7] The results of [LapPo1–2] have thereby established a direct connection between *Minkowski measurability, the direct spectral problem for fractal strings and the Riemann zeta function.*

Shortly afterwards, the first author and Maier obtained in [LapMa2] (announced in [LapMa1]) a natural *geometric and spectral reformulation of the Riemann hypothesis* stated in terms of the corresponding *inverse spectral problem for fractal strings*:

---

[6] The proof in [Fal2] is more of a dynamical systems nature while that in [RatWi2] is of a geometric measure-theoretic nature. Both proofs rely on a part of the original proof in [LapPo2], which is of a purely analytical nature.

[7] Since $D \in (0,1)$, we have that $\zeta(D) < 0$. Furthermore, recall that for a bounded fractal string, we always have $0 \leq D \leq 1$ (since $N = 1$); the cases where $D \in \{0,1\}$ are dealt with in [Lap-vFr1–3].

$(\mathrm{ISP})_D$ *Given that $\mathscr{L}$ is a fractal string for which the spectral counting function $N_{\nu,\mathscr{L}}(\mu)$ admits a monotonic asymptotic second term proportional to $\mu^{D/2}$ (as $\mu \to +\infty$), does it follow that $\mathscr{L}$ is Minkowski measurable?*[8]

It was shown in [LapMa1–2] that the above inverse spectral problem $(\mathrm{ISP})_D$ is intimately connected with the location of the critical zeros of $\zeta(s)$. More specifically, it was shown in [LapMa1–2] that for a given $D \in (0,1)$, the inverse spectral problem $(\mathrm{ISP})_D$ is true if and only if $\zeta(s)$ does not have any zeros on the vertical line $\{\mathrm{Re}\, s = D\}$. Consequently, $(\mathrm{ISP})_{1/2}$ is false (since $\zeta(s)$ is known to have zeros on the critical line $\{\mathrm{Re}\, s = 1/2\}$). *Moreover, the inverse spectral problem* $(\mathrm{ISP})_D$ *has an affirmative answer for all $D \in (0,1)$, with $D \neq 1/2$* (that is, except in the midfractal case when $D = 1/2$) *if and only if the Riemann hypothesis is true.*

Again, the above characterization of Minkowski measurability of fractal strings (obtained in [LapPo2]) played a significant role in one important step in the proof of the above results of [LapMa1–2]. Furthermore, the results of [LapPo1–2] (combined with the earlier works in [Lap1–3] and [HeLap], in particular) have provided an important motivation for the mathematical theory of complex dimensions (of fractal strings) developed by the first author and van Frankenhuijsen (see [Lap-vFr1–3]).

They have also been significantly extended in [Lap-vFr1–3], in order to obtain, in particular, a reformulation of the 'generalized Riemann hypothesis' by means of the generalized 'explicit formulas' (also obtained in the above books; see [Lap-vFr3, Chapter 5]). See, in particular, [Lap-vFr3, Sections 6.2, 6.3, as well as Chapters 8 and 9]; see also [Lap-vFr3, Chapter 11] where inverse spectral problems extending the ones considered in [LapMa1–2] are used to prove that the Riemann zeta function, along with many other Dirichlet series and integrals (including most arithmetic or number-theoretic zeta functions, with the exception of those associated with varieties over finite fields for which it is clearly not true), does not have infinitely many zeros in vertical arithmetic progression.

We refer the interested reader to [Lap8] for a survey of some of the main results obtained in [LapPo2] and [LapMa2], as well as for some of the later developments of fractal string theory and of the corresponding theory of complex dimensions.

Finally, we note that motivated in part by semi-heuristic suggestions made in [Lap-vFr2, Subsection 6.3.2] (and [Lap-vFr3, Subsection 6.3.2]) about a possible definition of the *spectral operator* on fractal strings (which sends the geometry onto the spectrum of fractal strings), a rigorous functional analytic definition of the spectral operator $\mathfrak{a}$ (acting on the weighted Hilbert space $\mathbb{H}_c := L^2(\mathbb{R}, \mathrm{e}^{-2ct}\mathrm{d}t)$, for any given $c \in \mathbb{R}$) was obtained by the first author and Herichi in [HerLap1–5]. In particular, the authors of [HerLap1–5] have shown in [HerLap1–3] that for a given $c \in \mathbb{R}$, the spectral operator $\mathfrak{a}_c$ is invertible (in a suitable sense)[9] if and only if $\zeta(s)$ does

---

[8] It then follows automatically that $\mathscr{L}$ has Minkowski (or box) dimension D and that (by the results of [LapPo2], see Remark 6.1.10) its Minkowski content can be explicitly computed.

[9] Specifically, the appropriate notion of invertibility used in [HerLap1–5] is the so called 'quasi-invertibility' of $\mathfrak{a}_c$, that is, the invertibility (in the usual sense, with the set-theoretic inverse being a bounded linear operator on $\mathbb{H}_c$) of each of the (suitably defined) truncated spectral operators $\{\mathfrak{a}_c^{(T)}\}_{T>0}$ of $\mathfrak{a}_c$.

not have any zero on the vertical line $\{\mathrm{Re}\, s = c\}$.[10] Consequently, $\mathfrak{a}_{1/2}$ is not invertible and the spectral operator $\mathfrak{a}_c$ is invertible (in the above sense) for all $c \in (0,1)$, $c \neq 1/2$ (or, equivalently, for all $c \in (0,1/2)$) if and only if the Riemann hypothesis is true. Accordingly, via an appropriate 'quantization' of the Riemann zeta function $\zeta = \zeta(s)$ (since $\mathfrak{a}_c = \zeta(\partial_c)$, where $\partial_c = \frac{\mathrm{d}}{\mathrm{d}t}$ is the suitably defined differentiation operator on $\mathbb{H}_c := L^2(\mathbb{R}, \mathrm{e}^{-2ct}\mathrm{d}t)$, also called the *infinitesimal shift* (of the real line), an operator-theoretic reformulation of the results of [LapMa1–2] on the Riemann hypothesis and inverse spectral problems for fractal strings has been obtained in [HerLap1–3]. Many other results concerning various aspects of 'quantized number theory' are obtained in [HerLap1–5], but they fall outside the scope of the present discussion.

An asymmetric reformulation of the Riemann hypothesis (expressed in terms of the standard notion of invertibility of the spectral operator $\mathfrak{a}_c$ for all $c \in (0,1/2)$) was obtained by the first author in [Lap7]; see also [Lap8].

We also refer to [Lap9] and [Lap10] for a later, related reformulation, as well as for further extensions, of these results (and beyond), expressed in terms of a different 'quantization' of $\zeta(s)$ via infinitesimal shifts and spectral operators acting on a suitable family of weighted Bergman spaces ([HedKoreZh]) of entire functions.

Among numerous contributions dealing with periodic and related sets, we mention a fundamental work by Hutchinson [Hut] on the definition and properties of self-similar sets, following the earlier seminal work of Moran [Mora] in the case when $N = 1$, then the papers by Lalley [Lall1–3] and Gatzouras [Gat], the papers [LapPo1–2, Lap3], an article by Falconer [Fal2], the detailed investigation of the geometry of self-similar fractal strings (and particularly, of lattice strings) conducted in [Lap-vFr1–3] (see, especially, [Lap-vFr3, Chapter 8]), a joint work by Pearse, the first author and Winter in [LapPeWi1] (building on the earlier work by the first author and Pearse in [LapPe2–3], as described in [Lap-vFr3, Section 13.1]), the work by Kesseböhmer and Kombrink [KeKom] on self-conformal sets (see also the survey article [Kom]), as well as the work of Kombrink, Pearse and Winter in [KomPeWi] providing a proof (for noninteger D and for any $N \geq 1$) of the geometric part of [Lap3, Conjecture 3] according to which (nontrivial) self-similar sets are Minkowski measurable if and only if they are nonlattice. The special case when $N = 1$ (i.e., the case of self-similar fractal strings) had been establlished earlier in [Lap-vFr1–3], by using the theory of complex dimensions and the associated explicit formulas; see [Lap-vFr3, Section 8.4].

The first attempt at carrying out a systematic study of weakly degenerate sets (in the Minkowski sense, as defined on page 544), and their respective gauge functions, has been undertaken by He and the first author in [HeLap], where, in particular, the main results of [Lap1] have been extended to nonstandard scaling laws. The notions of relative box dimensions and relative Minkowski contents, consid-

---

[10] The parameter c is closely related to the box (or Minkowski) dimensions D of the fractal strings heuristically represented by $\mathbb{H}_c$. Indeed, intuivitely, we have $D \leq c$ (modulo an infinitesimal). More precisely, c is the supremum of the Minkowski dimensions of all the possible (generalized) fractal strings with counting functions belonging to $\mathbb{H}_c$.

ered in Section 4.1.1, have been introduced by the third author in [Žu4]. Some general results about constant (i.e., Minkowski measurable) sets have been obtained by Stachó in [Sta], as well as by the first author and Pomerance in [LapPo2]. Furthermore, a class of constant weakly degenerate sets has been studied by Rataj and Winter in [RatWi2]. In particular, in [LapPo2], is obtained a useful characterization of Minkowski nondegenerate compact sets in $\mathbb{R}$ (or, equivalently, of fractal strings); see Remark 6.1.10 below for more details. Still when $N = 1$, the characterization of Minkowski nondegenerate sets, along with the aforementioned characterization of Minkowski measurable compact sets (or, equivalently, fractal strings) obtained in [LapPo2], was extended in [HeLap] to a broad class of gauge functions (obeying a non power scaling law).

Some classes of strongly degenerate sets have been constructed by the third author in [Žu4, Theorem 1.2], where one can find a class of *maximally degenerate sets* A in $\mathbb{R}^N$, in the sense that $\underline{\dim}_B A = 0$ and $\overline{\dim}_B A = N$. (See also [Tri1] for related examples.) A family of strongly degenerate sets within the class of inhomogeneous self-similar sets has been studied by Fraser in [Fra1]. While the upper box dimension is known to be *finitely stable* with respect to the union of any two bounded sets in $\mathbb{R}^N$, that is,

$$\overline{\dim}_B(A \cup B) = \max\{\overline{\dim}_B A, \overline{\dim}_B B\}, \tag{6.1.8}$$

this property does not hold for lower box dimensions; see [Fal1]. For example, it is even possible to construct two sets A and B in $\mathbb{R}^N$, such that $\underline{\dim}_B A = \underline{\dim}_B B = 0$, whereas $\underline{\dim}_B(A \cup B) = N$; [11] see [Žu4, Theorem 1.4]. A generalization of finite stability property (6.1.8) to the upper box dimensions of RFDs can be found in Proposition 4.1.26 in Subsection (4.1.1) of Chapter 4 above.

Remark 6.1.9. Both the lower and upper box dimensions are, in general, *unstable* with respect to Cartesian products of Minkowski degenerate sets. For example, there exist two bounded sets A and B in $\mathbb{R}^N$ such that $\underline{\dim}_B A = \underline{\dim}_B B = 0$, whereas $\underline{\dim}_B(A \times B) = N$ and $\overline{\dim}_B A = \overline{\dim}_B B = N$, $\overline{\dim}_B(A \times B) = N$; see [Žu4, Theorem 1.4]. Besides the known (and elementary) inequalities,

$$\underline{\dim}_B A + \underline{\dim}_B B \le \underline{\dim}_B(A \times B), \quad \overline{\dim}_B(A \times B) \le \overline{\dim}_B A + \overline{\dim}_B B,$$

which hold for any two bounded sets A and B in $\mathbb{R}^N$, Robinson and Sharples obtained in [RoSha] a new pair of inequalities

$$\underline{\dim}_B(A \times B) \le \underline{\dim}_B A + \overline{\dim}_B B \le \overline{\dim}_B(A \times B).$$

Since $\underline{\dim}_B(A \times B) = \underline{\dim}_B(B \times A)$ and similarly for the upper box dimension, it then follows that

$$\underline{\dim}_B(A \times B) \le \min\{C, D\} \le \max\{C, D\} \le \overline{\dim}_B(A \times B),$$

[11] Since we always have $\underline{\dim}_B(A \cup B) \le \overline{\dim}_B(A \cup B) \le N$, we conclude that $\dim_B(A \cup B)$ exists in this case and, moreover, $\dim_B(A \cup B) = N$.

where $C := \underline{\dim}_B A + \overline{\dim}_B B$ and $D := \overline{\dim}_B A + \underline{\dim}_B B$. These inequalities are sharp; see [RoSha] for more details.

A lot of additional information about the past and ongoing work, related to various aspects of the theory of fractal strings, can be found in a joint monograph of the first author and van Frankenhuijsen [Lap-vFr3], and in a monograph of the first author [Lap6], dedicated to the search for the Riemann zeros and a deeper understanding of why the Riemann hypothesis should be true.

We refer, in particular, the interested reader to [Lap-vFr3, Chapter 13] which contains a survey of some of the recent developments of the theory of complex dimensions (since the publication of the monographs [Lap-vFr1] and [Lap-vFr2]) in the higher-dimensional case [Lap-vFr3, Section 13.1] (based on [LapPe3, LapPeWi1, Pe, PeWi], which focuses on fractal tube formulas for self-similar tilings and fractal sprays), the p-adic case [Lap-vFr3, Section 13.2] (based on [LapLu1–3] and [LapLu-vFr1–2], which focuses on nonarchimedean fractal strings), the multifractal case [Lap-vFr3, Section 13.3] (based on [LapRo1, LapLéRo, ElLapMacRo], which focuses on various multifractal zeta functions and self-similar measures) and the random case [Lap-vFr3, Section 13.4] (based on [HamLap], which focuses on random fractal strings and the associated random zeta functions). See also [Lap-vFr3, Section 13.5], which briefly discusses a few aspects of the book [Lap6] (and of the paper [LapNes].)

Remark 6.1.10. More specifically, the geometric characterization of Minkowski nondegeneracy (resp., of Minkowski measurability) obtained in [LapPo2] can be stated as follows: A fractal string $\mathscr{L} = (\ell_j)_{j=1}^{\infty}$ is Minkowski nondegenerate (with Minkowski dimension $D \in (0,1)$) if and only if $\rho_* > 0$ and $\rho^* < \infty$, where ρ_* and ρ^* denote, respectively, the lower and upper limit of $\ell_j j^{1/D}$. (Clearly, we always have $0 \leq \rho_* \leq \rho^* \leq \infty$.)[12] Furthermore, it is Minkowski measurable (with Minkowski dimension $D \in (0,1)$) if and only if, in addition, $\rho_* = \rho^*$ (i.e., $\ell_j \sim \rho j^{-1/D}$ as $j \to \infty$, for some $\rho \in (0,+\infty)$). (Clearly, we then have $\rho = \rho_* = \rho^*$.) Moreover, it is shown in [LapPo2] that in that case, the Minkowski content of $\mathscr{L}$ is given by

$$\mathscr{M} = \rho^D \frac{2^{1-D}}{1-D}. \tag{6.1.9}$$

Note that the term ρ^D on the right-hand side of Equation (6.1.9) is due to the scaling property of the Minkowski content; see page 35.

Here, $(\ell_j)_{j\geq 1}$ denotes the sequence of lengths (arranged in nonincreasing order) of the connected components (bounded open intervals) of the fractal string (viewed as a bounded open subset of $\mathbb{R}$) or equivalently, the lengths of the connected components of the complement of a bounded set A (contained in the real line) with respect to its closed convex hull J (i.e., the lengths of the fractal string $J \setminus A$).

We mention in closing this remark that, under mild assumptions on h, both of these characterizations have been extended (by He and the first author) to a gen-

[12] It is easy to see that the condition $0 < \rho_* \leq \rho^* < \infty$ is equivalent to the existence of two positive constants a and b such that $a \leq \ell_j j^{1/D} \leq b$ for all $j \geq 1$; that is, with $\ell_j \asymp j^{-1/D}$ as $j \to \infty$.

eral class of gauge functions in [HeLap]. Furthermore, we mention that in the same memoir, [HeLap], the direct and inverse spectral problems for fractal strings (as well as their connections with the Riemann zeta function and the Riemann hypothesis, respectively) considered in [LapPo1–2] and [LapMa1–2], respectively (and discussed earlier in the present subsection), were extended and studied in the context of fractal strings with a general class of gauge functions.

6.2 Open Problems and Future Research Directions

We close this chapter by proposing a number of open problems (of varying difficulty and importance) along with several directions for future research closely connected with (or motivated by) the theory developed in this book.

6.2.1 Concluding Comments

We first make a few elementary remarks which could be further developed in more algebraic terms.

Assume that a sequence of RFDs $(\partial\Omega_j,\Omega_j)$ in $\mathbb{R}^N$ is given, such that $(\Omega_j)_{j\geq 1}$ is a disjoint family of open connected sets. We define the corresponding relative fractal drum

$$(A,\Omega) := \bigcup_{j\geq 1}(\partial\Omega_j,\Omega_j). \tag{6.2.1}$$

Let $(\Omega'_j)_{j\geq 1}$ be a disjoint family of open sets such that for each $j\geq 1$, Ω'_j is obtained from Ω_j by a rigid motion of $\mathbb{R}^N$. We can analogously define $(A',\Omega') = \cup_{j\geq 1}(\partial\Omega'_j,\Omega'_j)$. Then the distance zeta functions of the RFDs (A,Ω) and (A',Ω') coincide; that is,

$$\zeta_{A,\Omega}(s) = \zeta_{A',\Omega'}(s).$$

More precisely, $\overline{\dim}_B(A,\Omega) = \overline{\dim}_B(A',\Omega')$ and $\zeta_{A,\Omega}(s) = \zeta_{A',\Omega'}(s)$ for $\mathrm{Re}\, s > \overline{\dim}_B(A,\Omega)$ or, more generally, for all s in any given domain to which either of these zeta functions has a meromorphic extension. This remark follows immediately from Lemma 4.5.9; see, in particular, condition (4.5.44).

In light of this comment, we can generate a new subclass of RFDs, defined as above. Furthermore, it is natural to introduce a *geometric equivalence of relative fractal drums*,

$$(A,\Omega) \sim_g (A',\Omega'), \tag{6.2.2}$$

in order to identify the indicated RFDs of the form (6.2.1). As we have just explained, their distance relative zeta functions coincide, so that we can speak of the relative distance zeta function $\zeta_{[(A,\Omega)]}$ of the equivalence class $[(A,\Omega)]$ corresponding to (A,Ω):

$$\zeta_{[(A,\Omega)]}(s) := \zeta_{A,\Omega}(s). \tag{6.2.3}$$

Each bounded fractal string $\mathscr{L} = (\ell_j)_{j\geq 1}$ can be naturally identified with a relative fractal drum $(A_\mathscr{L}, \Omega_\mathscr{L})$ in $\mathbb{R}$, where $A_\mathscr{L} = \{a_k = \sum_{j\geq k} \ell_j : k \in \mathbb{N}\}$ and $\Omega_\mathscr{L} = \cup_{j\geq 1} \Omega_j$, $\Omega_j = (a_{j+1}, a_j)$. In view of the definition of geometric equivalence in (6.2.2), if a relative fractal drum (A', Ω') is geometrically equivalent to $(A_\mathscr{L}, \Omega_\mathscr{L})$, then (A', Ω') can also be viewed as a 'realization' of a given bounded fractal string $\mathscr{L}$. Note that Ω' can even be unbounded, but clearly, $|\Omega'| = |\Omega| = \sum_{j\geq 1} \ell_j < \infty$. In this way, the class of all bounded fractal strings is embedded into the class of all RFDs. This can be described more formally in terms of category theory, which may be the object of a later investigation.

Let us very briefly summarize a symbolic discussion of some of the basic objects encountered in this book. Starting with any bounded fractal string $\mathscr{L} = (\ell_j)_{j\geq 1}$, we have the associated geometric zeta function $\zeta_\mathscr{L}(s) := \sum_{j\geq 1} \ell_j^s$, which in turn generates the corresponding set of principal (geometric) complex dimensions $\dim_{PC} \mathscr{L}$; i.e., the set of poles of (the meromorphic extension of) $\zeta_\mathscr{L}$, located on the critical line. We have a similar sequence of constructions for bounded fractal subsets A of Euclidean spaces and for RFDs (A, Ω) (but this time, with the distance zeta function ζ_A and the relative zeta function $\zeta_{A,\Omega}$, respectively, instead of $\zeta_\mathscr{L}$):[13]

$$\begin{aligned} \mathscr{L} &\longrightarrow \zeta_\mathscr{L} \longrightarrow \dim_{PC} \mathscr{L}, \\ A &\longrightarrow \zeta_A \longrightarrow \dim_{PC} A, \\ (A,\Omega) &\longrightarrow \zeta_{A,\Omega} \longrightarrow \dim_{PC}(A,\Omega). \end{aligned}$$

Using the mapping $\mathscr{L} \longrightarrow A_\mathscr{L}$, where $A_\mathscr{L} := \{a_k := \sum_{j\geq k} \ell_j : k \in \mathbb{N}\}$, we see that the family of bounded fractal strings is naturally embedded into the family of bounded fractal subsets of $\mathbb{R}$. Also, we have the natural correspondence $\zeta_\mathscr{L} \longrightarrow \zeta_{A_\mathscr{L}}$ and $\dim_{PC} \mathscr{L} = \dim_{PC} A_\mathscr{L}$. Therefore, for each bounded fractal string, we obtain the following commutative diagram:

$$\begin{array}{ccccc} \mathscr{L} & \longrightarrow & \zeta_\mathscr{L} & \longrightarrow & \dim_{PC} \mathscr{L} \\ \downarrow & & \downarrow & & \| \\ A_\mathscr{L} & \longrightarrow & \zeta_{A_\mathscr{L}} & \longrightarrow & \dim_{PC} A_\mathscr{L} \end{array} \tag{6.2.4}$$

Starting with a bounded subset A of $\mathbb{R}^N$, we can assign to it a relative fractal drum (A, A_δ), where δ is a fixed positive number. In this way, we obtain the following commutative diagram:

[13] Instead of ζ_A, we could use the tube zeta functions $\tilde{\zeta}_A$ or $\tilde{\zeta}_{A,\Omega}$, respectively. Assuming that $\overline{\dim}_B A < N$ or $\overline{\dim}_B(A,\Omega) < N$, respectively, the resulting sets of principal complex dimensions $\dim_{PC} A$ or $\dim_{PC}(A,\Omega)$, respectively, would remain unchanged and an entirely parallel discusion could be provided.

$$\begin{array}{ccccc} A & \longrightarrow & \zeta_A & \longrightarrow & \dim_{PC} A \\ \downarrow & & \| & & \| \\ (A, A_\delta) & \longrightarrow & \zeta_{A,A_\delta} & \longrightarrow & \dim_{PC}(A, A_\delta) \end{array} \tag{6.2.5}$$

Note that here $\zeta_A = \zeta_{A,A_\delta}$ and $\dim_{PC} A = \dim_{PC}(A, A_\delta)$.

We can also consider the following self-explanatory sequences of constructions, dealing with spectral zeta functions, corresponding to bounded fractal strings $\mathscr{L}$, bounded open sets Ω in Euclidean spaces and RFDs (A, Ω), respectively:

$$\mathscr{L} \longrightarrow \zeta^*_{\mathscr{L}} \longrightarrow \dim^*_{PC} \mathscr{L},$$

$$\Omega \longrightarrow \zeta^*_{\Omega} \longrightarrow \dim^*_{PC} \Omega,$$

$$(A, \Omega) \longrightarrow \zeta^*_{A,\Omega} \longrightarrow \dim^*_{PC}(A, \Omega).$$

The set of *principal spectral complex dimensions of* $\mathscr{L}$, $\dim^*_{PC} \mathscr{L}$, is defined as the set of principal complex dimensions of the corresponding *spectral zeta function* $\zeta^*_{\mathscr{L}}$ (also denoted by $\zeta_{v,\mathscr{L}}$; see [Lap-vFr3, Section 1.3]), meromorphically extended to a neighborhood of the critical line $\{\mathrm{Re}\, s = 1\}$, and similarly for $\dim^*_{PC} A$ and $\dim^*_{PC}(A, \Omega)$ (see Definition 4.3.4). Much as in the case of geometric zeta functions and the corresponding sets of principal complex dimensions, we can sketch the following commutative diagram associated to spectral zeta functions:

$$\begin{array}{ccccc} \mathscr{L} & \longrightarrow & \zeta^*_{\mathscr{L}} & \longrightarrow & \dim^*_{PC} \mathscr{L} \\ \downarrow & & \| & & \| \\ \Omega_{\mathscr{L}} & \longrightarrow & \zeta^*_{\Omega_{\mathscr{L}}} & \longrightarrow & \dim^*_{PC} \Omega_{\mathscr{L}} \end{array}$$

Here, to any bounded fractal string $\mathscr{L} := (l_j)_{j \geq 1}$ we have assigned an open subset $\Omega_{\mathscr{L}} := \cup_{k \geq 1}(a_{k+1}, a_k)$ contained in the real line, where $a_k := \sum_{j \geq k} l_j$. Finally, to any bounded open subset Ω of $\mathbb{R}^N$ (or, more generally, to any open set Ω of finite N-dimensional Lebesgue measure, such that $\Omega \subset (\partial\Omega)_\delta$ for some $\delta > 0$) we can assign a relative fractal drum $(\partial\Omega, \Omega)$. Therefore, we obtain the following commutative diagram:

$$\begin{array}{ccccc} \Omega & \longrightarrow & \zeta^*_{\Omega} & \longrightarrow & \dim^*_{PC} \Omega \\ \downarrow & & \| & & \| \\ (\partial\Omega, \Omega) & \longrightarrow & \zeta^*_{\partial\Omega,\Omega} & \longrightarrow & \dim^*_{PC}(\partial\Omega, \Omega) \end{array}$$

6.2.2 Open Problems

Besides several open problems already mentioned in the text, one can formulate numerous open problems related to the classification of bounded sets in Euclidean spaces. We propose a few of them here to the attention of the reader.

Problem 6.2.1. Find a bounded set A in $\mathbb{R}^N$ (or a relative fractal drum (A,Ω) in $\mathbb{R}^N$) such that the corresponding distance zeta function ζ_A (relative distance zeta function $\zeta_{A,\Omega}$) possesses an analytic continuation which generates a nontrivial Riemann surface.

Problem 6.2.2. Describe explicitly as large as possible a set of

(a) periodic functions $G(\tau)$

and

(b) functions $\rho(t)$

that are associated with the family $\mathscr{S}_p$ of all possible periodic subsets A of Euclidean spaces. Periodic sets are defined in Section 6.1 on page 541. Furthermore, find necessary and sufficient conditions for a pair of functions (G,ρ) to be associated with the family $\mathscr{S}_p$.

Problem 6.2.3. A similar question can be asked for (algebraically or transcendentally) quasiperiodic sets. Describe explicitly as large as possible a set of

(a) n-quasiperiodic functions $G(\tau)$

and

(b) functions $\rho(t)$

associated with the family $\mathscr{S}_{\rm qp}(n)$ of all possible n-quasiperiodic sets A in Euclidean spaces, where $n \geq 2$ or $n = \infty$. (Quasiperiodic sets are defined in Subsection 6.1.1.1, on page 542.) Furthermore, find necessary and sufficient conditions for a pair of functions (G,ρ) to be associated with the family $\mathscr{S}_{\rm qp}(n)$.

Since the family of quasiperiodic sets in Euclidean spaces is the union of the family of algebraically n-quasiperiodic sets and the family of n-transcendentally quasiperiodic sets, that is,

$$\mathscr{S}_{\rm qp}(n) = \mathscr{S}_{\rm aqp}(n) \cup \mathscr{S}_{\rm tqp}(n),$$

one can ask the analogous questions for $\mathscr{S}_{\rm aqp}(n)$ and $\mathscr{S}_{\rm tqp}(n)$. Finally, show that the family $\mathscr{S}_{\rm aqp}(n)$ is nonempty, as we expect to be the case.

Problem 6.2.4. Let $\mathscr{S}_{wd}$ be the family of all weakly degenerate sets in Euclidean spaces (see page 544). Describe as large as possible a set of gauge functions h (and $1/h$) that are associated with the members of the family $\mathscr{S}_{wd}$.

Problem 6.2.5. Let A be a periodic set in $\mathbb{R}^N$. Let (G,ρ) be an associated ordered pair of functions (see page 541), where G is periodic, and $\rho(t) \to +\infty$ as $t \to 0^+$. Study the regularity of the functions G and h in terms of the regularity of A (and conversely). For example, characterize all of the periodic sets such that an associated periodic function G is of class C^m, and more generally, of class $C^{m,\theta}$ (for some $m \in \mathbb{N}$ and $\theta \in [0,1]$).[14] See Remark 2.3.32.

Problem 6.2.6. Find a *geometric* definition of the "relative box dimension" ("relative Minkowski dimension") of a relative fractal drum (A,Ω), as defined in Section 4.1 (see Remark 4.1.6 on page 250). The sought for geometric definition should involve suitable coverings of A and Ω by cubes (or balls).

Problem 6.2.7. Can we determine $\zeta_{A\times B}$, up to zeta function equivalence, in terms of ζ_A and ζ_B? What if ζ_A and ζ_B are only known up to equivalence? See also a related open problem stated in Remark 3.6.8.

Problem 6.2.8. Is there a set A satisfying the conditions of Theorem 2.3.25 and such that the corresponding set of principal complex dimensions $\dim_{PC} A = \{s_k = D + \frac{2\pi}{T}\mathrm{i}k : \hat{G}_0(\frac{k}{T}) \neq 0,\ k \in \mathbb{Z}\}$ is *nonarithmetic* (i.e., does not consist of an infinite arithmetic progression)? (According to the results of [Lap-vFr3, Subsection 8.4.2] and to Proposition 2.1.72, one would expect this to be the case if A is a (nontrivial) lattice self-similar subset of $\mathbb{R}$.) Furthermore, can one find rational conditions under which such a set A is arithmetic (i.e., consists of a full arithmetic progression) or more generally, consists of an infinite (but not necessarily full) arithmetic progression (i.e., for which $\hat{G}_0(\frac{k}{T}) \neq 0$ for infinitely many $k \in \mathbb{Z}$)? The answer to this question is probably "yes" and could possibly be found in the examples of lattice self-similar fractals studied in [Lap-vFr3].

Problem 6.2.9. Construct a subset A of $[0,1]$ such that $\dim_B A = D$ exists for some $D \in [0,1]$, while $\mathscr{M}_*^D(A) = 0$ and $\mathscr{M}^{*D}(A) = +\infty$.

Problem 6.2.10. Prove or disprove that the bound β in (2.3.7), appearing in Theorem 2.3.2, is optimal.

Problem 6.2.11. Assume that A is a bounded set in $\mathbb{R}^N$ which is strongly degenerate, that is, such that $\underline{\dim}_B A < \overline{\dim}_B A$. Does the abscissa of meromorphic continuation $D_{\mathrm{mer}}(\zeta_A)$ of the distance zeta function ζ_A depend only on the difference $\overline{\dim}_B A - \underline{\dim}_B A$?

Problem 6.2.12. Assume that $\zeta_{A,A_\delta}(s) = \zeta_{B,B_\delta}(s)$ for all s such that $\mathrm{Re}\, s > \sigma$, where $\sigma \in \mathbb{R} \cup \{-\infty\}$ and $\delta > 0$ are fixed (with $\sigma \geq \max(\overline{\dim}_B A, \overline{\dim}_B B)$). From this, we

[14] Here, given $m \in \mathbb{N}_0$ and $\theta \in [0,1]$, $C^{m,\theta}$ denotes the space of m-times continuously differentiable functions on $\mathbb{R}$ whose m-th derivative is Hölder continuous of order θ.

cannot infer that the fractal sets A and B are interrelated in any way. This can be seen in the special case where A and B are n-point sets in the plane, provided δ is sufficiently small, so that A_δ and B_δ are both equal to the disjoint union of n disks. However, the pairs (A, A_δ) and (B, B_δ) might be related somehow. For example, B_δ can be obtained from A_δ by the rigid motion of the connected components (in this case, the disks of radius δ) of A_δ. Compare with Problem 6.2.17 below. Recall that any bounded fractal string $\mathscr{L}$ is uniquely determined by its geometric zeta function; see Theorem 2.1.39.

In fact, from the results of Chapter 5, one can deduce that, under suitable hypotheses, the tube functions $t \mapsto |A_t|$ and $t \mapsto |B_t|$ will be asymptotically equivalent as $t \to 0^+$, up to an order of growth depending on the growth properties of the distance zeta function $\zeta_{A,A_\delta}(s) = \zeta_{B,B_\delta}(s)$.[15]

Problem 6.2.13. Here, we state a problem formulated in [LapRoŽu, Remark 2.21]. Let A and B be two constant (i.e., Minkowski measurable) sets on the real line. It is clear that if $\overline{A}$ and $\overline{B}$ are disjoint, then $A \cup B$ is constant as well. Prove or disprove that this property holds if $\overline{A} \cap \overline{B} \neq \emptyset$.

The following problem has also been stated in [LapRoŽu, Open Problem 2.20].

Problem 6.2.14. If A and B are two constant (i.e., Minkowski measurable) sets in $\mathbb{R}^N$, prove or disprove that $A \times B$ is constant.

Recall from the discussion immediately preceding Corollary 2.3.23 that if a bounded set $A \subset \mathbb{R}^N$ is constant (i.e., Minkowski measurable), then it is also constant in $\mathbb{R}^{N+1}$; that is, it is Minkowski measurable when viewed as a subset of $\mathbb{R}^{N+1}$; see [Kne, Satz 7] and [Res, Theorem 4]. In the next problem we will explore a closely related problem for periodic instead of constant sets.

Problem 6.2.15. Let A be a bounded periodic set in $\mathbb{R}^N$.[16] Does it follow that $A \subset \mathbb{R}^{N+1}$ is periodic in $\mathbb{R}^{N+1}$? Furthermore, does the converse hold? Namely, if $A \subset \mathbb{R}^N$ is periodic in $\mathbb{R}^{N+1}$, is it also periodic in $\mathbb{R}^N$? Again, this open problem can be (partially) answered by finding appropriate hypotheses on the set A and by using Theorem 4.7.3 about the invariance of complex dimensions on the dimension of the ambient space as well as the approximate functional equation (4.7.4) of Theorem 4.7.2. Of course, on should also use the results of Chapter 5 on fractal tube formulas.

The next problem complements Problem 6.2.15.

[15] Note that, by the principle of analytic continuation, this equality continues to hold on any connected open set $U \subseteq \mathbb{C}$ which contains the open half-plane $\{\mathrm{Re}\, s > \sigma\}$ and to which any (and hence, both) of the two distance zeta functions can be meromorphically continued.

[16] Recall that in the definition of a periodic set, the underlying periodic function $G = G_N$ appearing in Equation (2.3.30) (or, more generally, in its counterpart in Subsection 6.1.1.1) is associated with the asymptotic behavior as $t \to 0^+$ of the tube function $|A_t|_N$, the N-dimensional volume of A_t in $\mathbb{R}^N$.

Problem 6.2.16. Let $A \subset \mathbb{R}^N$ be a bounded periodic set satisfying the stronger assumptions of Theorem 2.3.25. Namely,

$$|A_t|_N = t^{N-D}(G_N(\log t^{-1}) + O(t^{\alpha})) \quad \text{as} \quad t \to 0^+,$$

for some $\alpha > 0$, where $G = G_N$ is periodic (and necessarily continuous, by Lemma 2.3.30).[17] Does it then follow that the *normalized lower and upper Minkowski contents* of A, defined respectively by

$$\frac{\min G_N}{\omega_{N-D}} \quad \text{and} \quad \frac{\max G_N}{\omega_{N-D}}$$

(where A is viewed as a subset of $\mathbb{R}^N$) are independent of N?

Problem 6.2.17. Assume that we have two domains (i.e., open connected sets) Ω_1 and Ω_2 in $\mathbb{R}^N$ such that $\zeta_{\partial\Omega_1,\Omega_1}(s) = \zeta_{\partial\Omega_2,\Omega_2}(s)$ on a set S of complex numbers which has an accumulation point; that is, in light of the principle of analytic continuation, we know that the zeta functions of the RFDs $(\partial\Omega_1, \Omega_1)$ and $(\partial\Omega_2, \Omega_2)$ coincide on $\{\mathrm{Re}\, s > \rho\}$, where

$$\rho := \max\left\{D_{\mathrm{mer}}(\zeta_{\partial\Omega_1,\Omega_1}), D_{\mathrm{mer}}(\zeta_{\partial\Omega_2,\Omega_2})\right\}.$$

Prove or disprove that the sets Ω_1 and Ω_2 are then congruent, that is, Ω_2 can be obtained from Ω_1 by a rigid motion. This problem is related to Proposition 2.1.39, which states that any fractal string $\mathscr{L} = (\ell_j)_{k\geq 1}$ is uniquely determined by its geometric zeta function $\zeta_{\mathscr{L}}(s) = \sum_{j\geq 1} \ell_j^s$. Compare with Problem 6.2.12 above.

Problem 6.2.18. It is well known that the function $f : B_1 = \{|s| < 1\} \to \mathbb{C}$, $f(s) := \sum_{k=0}^{\infty} s^{2^k}$, is holomorphic on B_1, and that the bounding circle $S^1 = \{|s| = 1\}$ is equal to the set of its nonisolated singularities (see, e.g., [Tit1, p. 163]). More precisely, S^1 is a (holomorphic) natural boundary of f, in the sense of Definition 1.3.6 of Subsection 1.3.2; equivalently, $\{|s| < 1\}$ is a domain of holomorphy for f. Using the Möbius transformation $T : \{\mathrm{Re}\, s > 0\} \to B_1$ defined by $T(s) = (1-s)/(1+s)$, we can introduce the function

$$g : \{\mathrm{Re}\, s > D\} \to \mathbb{C}, \quad g(s) = f(T(s-D)) = \sum_{k=0}^{\infty} \left(\frac{1-(s-D)}{1+(s-D)}\right)^{2^k},$$

where D is any given real number. This is a holomorphic function on $\{\mathrm{Re}\, s > D\}$, and the entire vertical line $\{\mathrm{Re}\, s = D\}$ is the set of nonisolated singularities of g. More precisely, $\{\mathrm{Re}\, s = D\}$ is a (holomorphic) natural boundary of g; equivalently, the half-plane $\{\mathrm{Re}\, s > D\}$ is a domain of holomorphy for g. Is g representable as

---

[17] Recall from Theorem 2.3.25, Equation (2.3.31), that it then follows that the lower and upper Minkowski contents of A in $\mathbb{R}^N$ are given respectively by

$$\mathscr{M}_* = \mathscr{M}_{*N}^D(A) = \min G_N \quad \text{and} \quad \mathscr{M}^* = \mathscr{M}_N^{*D}(A) = \max G_N.$$

a Dirichlet series, or, more generally, as a (generalized) Dirichlet integral, and with abscissa of (absolute) convergence equal to D? Is there a relative fractal drum (A,Ω) in $\mathbb{R}^N$, for some $N \geq D$, such that $\zeta_{A,\Omega}(s) = g(s)$ for $\mathrm{Re}\, s > D$ and $\dim_B(A,\Omega) = D$? Of course, as was shown in Section 4.6, there *are* RFDs (A,Ω) such that the whole critical line $\{\mathrm{Re}\, s = D\}$ of the corresponding zeta function $\zeta_{A,\Omega}$ consists of nonisolated singularities of $\zeta_{A,\Omega}$; see Theorem 4.6.9.

Problem 6.2.19. Prove or disprove that there is a *maximally degenerate* relative fractal drum (A,Ω) in $\mathbb{R}^N$, that is, such that $\underline{\dim}_B(A,\Omega) = -\infty$, whereas $\overline{\dim}_B(A,\Omega) = N$. (See Corollary 4.1.38.) Recall that there exists a bounded set A in $\mathbb{R}^N$ which is maximally degenerate, that is, such that $\underline{\dim}_B A = 0$ and $\overline{\dim}_B A = N$; see [Žu4, Theorem 1.2].

In the next problem, at least for Julia sets and the Mandelbrot set, one should possibly use an appropriate gauge function (in the sense of [HeLap] and Definition 6.1.4 above) before addressing the question.

Problem 6.2.20. Find $D_{\mathrm{mer}}(\zeta_A)$, the abscissa of meromorphic continuation of ζ_A, in the case where A is the von Koch curve,[18] the Menger sponge and its generalizations, Julia sets, the Mandelbrot set (see, e.g., [Man1], [Man2], [TanL]),[19] the limit set of a Fuchsian group or of a Kleinian group (see, e.g., [BedKS]), etc. Moreover, determine the complex dimensions of the meromorphic extension of ζ_A in $\{\mathrm{Re}\, s > D_{\mathrm{mer}}(\zeta_A)\}$. For each of these sets, find all the singularities on the critical lines of the corresponding distance zeta function ζ_A. For the case of the von Koch curve, compare with [LapPe1]; compare also with [LapPe3] and [LapPeWi1] (as described, for example, in [Lap-vFr3, Section 13.1]).

The following two problems complement Problem 6.2.20. (Again, one may wish to first find a suitable gauge function in order to address the questions.)

We refer, for example, to [Bea, Man1, Man3, TanL] for a discussion of the classic fractals arising in complex dynamics, such as Julia sets and the Mandelbrot set.

Problem 6.2.21. Is the Mandelbrot set maximally hyperfractal? Are there Julia sets with that same property? Is there an appropriate Riemann surface naturally associated with the fractal zeta functions (ζ_A or $\tilde{\zeta}_A$, say) of the Mandelbrot set or of those Julia sets?

---

[18] Recall from footnote 43 on page 492 that, at least in principle, this problem has now been resolved by the authors, although the corresponding result still needs to be fully explicited and compared with the earlier results of [LapPe1], as described in [Lap-vFr3, Subsection 12.2.1]. However, one can also ask these questions for a variety of (lattice and nonlattice) Koch-type curves as well as for other fractal (or multifractal) curves, such as the Weierstrass curve, the Riemann curve and the Takagi curve.

[19] Since, according to Shishikura's well known result [Shi], the boundary A of Mandelbrot's set satisfies $\dim_H A = \underline{\dim}_B A = \overline{\dim}_B A = 2$, it follows from Theorem 2.1.11 and Corollary 2.1.20(i) that $D(\zeta_A) = 2$.

Problem 6.2.22. Which Julia sets are hyperfractal, or even strongly hyperfractal, but not maximally hyperfractal? (See Definition 4.6.23.) Which ones are not hyperfractal but have infinitely many complex dimensions? If it is not maximally hyperfractal, is the Mandelbrot set a hyperfractal, or even a strong hyperfractal? What are its (visible) complex fractal dimensions?

Problem 6.2.23. Address questions similar to those raised in Problems 6.2.21 and 6.2.22, but now with the types of fractals naturally arising in conformal dynamics rather than in complex dynamics, such as the limit sets of Fuchsian groups or of Kleinan groups (see, e.g., [BedKS]).

Problem 6.2.24. Is the graph of the Weierstrass function hyperfractal? More generally, are the graphs of the Weierstrass–Mandelbrot functions and any of the other classic families of nowhere differentiable functions hyperfractal or even, strongly or maximally fractal? What are their (visible) complex dimensions? A simpler, but still interesting question, is to determine the complex dimensions of the graph of the Cantor function (i.e., of the devil's staircase), or of any natural relative fractal drum naturally associated to it, and, in particular, the half-plane of meromorphic continuation (possibly, the entire complex plane) of its distance and tube zeta functions. (See Example 5.5.14 for an answer to this question in the special case of a particular relative fractal drum generated by the Cantor graph.) This is of particular interest in view of the discussion of the notion of fractality given in [Lap-vFr3, Subsections 12.1.1 and 12.1.2, including Figures 12.1–12.3].

The next problem is motivated in part by the work of Ben Hambly and the first author in [HamLap], where a theory of random fractals, geometric zeta functions and of the associated complex dimensions was first developed (for random fractal strings, that is, in the one-dimensional case). (See also [Lap-vFr3, Section13.4] for an exposition of some of the main results of [HamLap].) Fully addressing it will require to extend to the random case the definitions of fractal zeta functions considered in this book, as well as possibly, choosing appropriate gauge functions, such as the *iterated logarithm* $h(x) = \log\log x^{-1}$, where $x \in (0, 1/e)$.

Problem 6.2.25. Is a typical Brownian motion path in $\mathbb{R}^N$ hyperfractal, strongly hyperfractal or maximally hyperfractal?[20] When applicable, determine its (visible) random complex dimensions (i.e., the poles of an appropriate meromorphic continuation of the associated pointwise random zeta function). Ask and answer analogous questions about the sample paths and the zero sets of other stochastic processes (such as Lévy and α-stable processes; see [HamLap] and the suitable references therein), as well as about other classes of random fractals, including stochastically self-similar fractals.

---

[20] By "typical" here, we mean that the corresponding property holds almost surely with respect to the underlying Wiener (probability) measure; see, e.g., [Sim] or [JohLap, Chapters 2–4] and the many relevant references therein.

Problem 6.2.26. Prove or disprove that the union of two (strong, maximal) hyperfractals is a (strong, maximal) hyperfractal. Similarly for Cartesian products of hyperfractals.

Problem 6.2.27. Prove or disprove that there is a hyperfractal set A in $\mathbb{R}^N$ which is constant (i.e., Minkowski measurable).

In [Lap-vFr1–3], it is shown, that under suitable hypotheses on the meromorphic continuation of $\zeta_{\mathscr{L}}$ to a connected open neighborhood of the critical line $\{\mathrm{Re}\, s = D\}$, a fractal string $\mathscr{L}$ is Minkowski measurable if and only if it does not have any nonreal complex dimension on $\{\mathrm{Re}\, s = D\}$ (and D is a simple pole of $\zeta_{\mathscr{L}}$); see [Lap-vFr3, Theorem 8.15].[21] This result (which has been extended to higher dimensions in Chapter 5 above and in [LapRaŽu4, 6]) would tend to suggest that the answer to Problem 6.2.27 should be that there does not exist a hyperfractal subset A of $\mathbb{R}^N$ which is Minkowski measurable. Caution is required, however, since clearly, the hypotheses of the above theorem are far from being satisfied in the case of a hyperfractal set $A \subset \mathbb{R}^N$ (even when $N = 1$). This fact makes Problem 6.2.27 all the more interesting.

Problem 6.2.28. Prove or disprove the following statement: for any bounded set A in $\mathbb{R}^N$, we have

$$\overline{\dim}_{av} A = \overline{\dim}_B A \quad \text{and} \quad \underline{\dim}_{av} A = \underline{\dim}_B A,$$

where the upper and lower average Minkowski dimensions are defined in Subsection 2.4.2. See, in particular, Definition 2.4.11 and Proposition 2.4.9.

Problem 6.2.29. Prove or disprove the following statement: there is a bounded set A in $\mathbb{R}^N$ such that there exists $D := \dim_{av} A$ and $\tilde{\mathscr{M}}_*^D(A) < \tilde{\mathscr{M}}^{*D}(A)$. Recall that for the Cantor ternary set A we have that $\tilde{\mathscr{M}}_*^D(A) = \tilde{\mathscr{M}}^{*D}(A)$; see Corollary 3.1.6. Average Minkowski contents are defined in Definition 2.4.1.

Problem 6.2.30. Construct a bounded set A in $\mathbb{R}^N$ such that $\underline{\dim}_{av} A < \overline{\dim}_{av} A$. Moreover, is it possible to find an example for which $\underline{\dim}_{av} A = 0$ and $\overline{\dim}_{av} A = N$? Study the properties of the lower and upper average Minkowski dimensions, introduced in Subsection 2.4.2, with respect to finite unions and Cartesian products of bounded subsets of Euclidean spaces.

Problem 6.2.31. Is the upper box dimension additive with respect to Cartesian products of RFDs? This is true in the case of Minkowski nondegenerate fractal drums; see Proposition 4.1.20(b). It is well known that the *lower* box dimension is not additive with respect to the Cartesian product of compact sets; see, e.g., [Fal1]; see also Remark 6.1.9 on page 550.

---

[21] Recall from [Lap-vFr3, Section 1.2] and from Subsection 2.1.4 above that provided $\mathscr{L}$ is nontrivial, the abscissa of convergence $D(\zeta_{\mathscr{L}})$ of $\zeta_{\mathscr{L}}$ necessarily coincides with the upper box dimension of (the boundary of) $\mathscr{L}$ as well as of $A_{\mathscr{L}}$: $D = \overline{\dim}_B \partial\Omega_{\mathscr{L}} = \overline{\dim}_B A_{\mathscr{L}}$. (See Corollary 2.1.57.)

Problem 6.2.32. Let (A,Ω) be a relative fractal drum and, in particular, let $(\partial\Omega,\Omega)$ be an ordinary fractal drum in $\mathbb{R}^N$. Assume, for simplicity, that we choose Dirichlet boundary conditions (in the variational sense) and that $|\Omega| < \infty$. For various concrete examples (e.g., the Koch snowflake drum or, more generally, self-similar drums, Julia sets, the Mandelbrot set, as, for instance, in [Lap1–3]), compare the set of geometric and spectral complex dimensions of the drum (that is, the complex dimensions of the distance or tube zeta function, on the one hand, and those of the normalized spectral zeta function of the Dirichlet Laplacian, on the other hand). Can one compare these two sets of complex dimensions (when they exist) for a general class of (relative) fractal drums? (Compare with related questions asked in [Lap3] and towards the end of [Lap5]. See also Problem 6.2.36 below for the case of the geometric complex dimensions of self-similar sets.)

In order to understand some of the statements, hypotheses and notation of parts of our last two problems (namely, Problem 6.2.35 and 6.2.36), the reader might first want to read (or review) Subsection 5.5.6, including Remark 5.5.26. In the sequel, we let (as in Subsection 5.5.6) σ_0 denote the *similarity dimension* of a self-similar set (or, equivalently, of the associated self-similar tiling or spray); that is, σ_0 is the unique real solution of the Moran equation $\sum_{j=1}^{J} r_j^s = 1$, where $J \geq 2$, $\sum_{j=1}^{J} r_j^N < 1$ and $\{r_j\}_{j=1}^{J}$ is the list of scaling ratios (counted according to their multiplicities) of the self-similar set (or, equivalently, of the associated self-similar tiling). The self-similar set is also assumed implicitly to satisfy the open set condition (in the sense of [Hut, Fal1]).[22] We have that $0 < \sigma_0 \leq D < N$, where D denotes the (upper) box dimension of the self-similar tiling (or spray), viewed as an RFD. (We exclude here the extreme case when $D = N$; see footnote 22 on page 562.) We have (see part (c) of Remark 5.5.26 in Subsection 5.5.6)

$$D := \dim_B(A,\Omega) = \max\{\sigma_0, D_G\}, \tag{6.2.6}$$

where G is the generator of the tiling and D_G denotes the (inner) box (or Minkowski) dimension of its boundary: $D_G := \dim_B(\partial G, G)$. Here, for simplicity, we assume that there is a single generator and that it is pluriphase, in the sense of [LapPe2–3, LapPeWi1–2] (as is the case, for example, of most polytopes, in light of [KoRati]); see Subsection 5.5.6 above. See also Remark 6.2.33 just below for the case of multiple generators.

Remark 6.2.33. In the case of multiple generators $\{G^{(q)}\}_{q=1}^{Q}$ and in light of the results of [Lap-vFr3, Chapter 3 and Section 8.4], we expect that analogous results should hold (see also part (b) of Remark 5.5.26), modulo suitable modifications. For example, D_G should be replaced by $\max\{D_{G^{(q)}}\}_{q=1}^{Q}$ and (when $Q \geq 2$) the equality should be replaced by the inequality $D \leq \max\{\sigma_0, D_G\}$ since there might be some cancellations between the zeros of $\zeta_{\partial G,G} := \sum_{q=1}^{Q} \zeta_{\partial G^{(q)},G^{(q)}}$ and the zeros of

[22] In fact, a little more is assumed since we assume that the associated self-similar tiling is nontrivial, which implies the open set condition of [Hut] and the strict inequality $D := \dim_B(A,\Omega) < N$; see [LapPe2–3, LapPeWi1–2, Pe, PeWi], along with [Lap-vFr3, Section 13.1].

$1-\sum_{j=1}^{J} r_j^s$.[23] Furthermore, in Problem 6.2.35, some of the potential scaling complex dimensions of the self-similar set (i.e., the poles of the geometric zeta function $\zeta_{\mathfrak{S}}$ of the fractal string associated with its self-similar tiling, in the sense of [Lap-vFr3, Chapter 3 and Section 13.1]) might be canceled by some of the zeros of the numerator of $\zeta_{\mathfrak{S}}$. However, according to the results of [Lap-vFr3, Section 8.4], at least infinitely many of these scaling complex dimensions should not be canceled on the vertical line $\{\mathrm{Re}\, s = \sigma_0\}$.) Moreover, under the above assumptions, and since (according to the results of Subsection 5.5.6; see, especially, Equations (5.5.105) and (5.5.172)), we have that

$$\zeta_{A,\Omega}(s) = \zeta_{\mathfrak{S}}(s) \cdot \zeta_{\partial G,G}(s), \tag{6.2.7}$$

where

$$\mathscr{P}(\zeta_{\partial G,G}) := \mathscr{P}(\zeta_{\partial G,G}, \mathbb{C}) \subseteq \{0,1,\ldots,N-1\}; \tag{6.2.8}$$

more specifically, $D_G = \dim_B(\partial G, G) \in \mathbb{N}_0$ and the set $\mathscr{P}(\zeta_{\partial G,G})$ of 'integer dimensions' is given by

$$\mathscr{P}(\zeta_{\partial G,G}) = \{0,1,\ldots,D_G\}, \tag{6.2.9}$$

or (due to the potential cancellations), a subset of $\{0,1,\ldots,D_G\}$, but with $D_G \in \mathscr{P}(\zeta_{\partial G,G})$. All of these integer dimensions are simple poles of $\zeta_{\partial G,G}$. Furthermore, $\zeta_{\mathfrak{S}}(s) = \left(1-\sum_{j=1}^{J} r_j^s\right)^{-1}$ for all $s \in \mathbb{C}$ and the multiset $\mathscr{P}(\zeta_{\mathfrak{S}})$ of 'scaling complex dimensions' is given (when $Q=1$) by

$$\mathscr{P}(\zeta_{\mathfrak{S}}) := \mathscr{P}(\zeta_{\mathfrak{S}}, \mathbb{C}) = \left\{ s \in \mathbb{C} : \sum_{j=1}^{J} r_j^s = 1 \right\}; \tag{6.2.10}$$

this multiset is described in [Lap-vFr3, Theorem 3.6], as well as throughout Chapter 3 of [Lap-vFr3].

Remark 6.2.34. In the case of truly multiple generators (i.e., $Q \geq 2$), due to the possible cancellations, the second equality of Equation (6.2.10) should be replaced by a containment, $\subseteq$. (Generally, however, it remains an equality.) Furthermore, we always have that $\sigma_0 \in \mathscr{P}(\zeta_{\mathfrak{S}})$. More specifically, according to the aforementioned results of [Lap-vFr3, Chapter 3 and Section 8.4], in the nonlattice case, σ_0 is the only principal pole (i.e., the only pole with real part σ_0) of $\zeta_{\mathfrak{S}}$, whereas in the lattice case, $\mathscr{P}(\zeta_{\mathfrak{S}})$ contains not only σ_0 but also an infinite arithmetic progression of principal poles of $\zeta_{\mathfrak{S}}$. Moreover, all of the principal poles of $\zeta_{\mathfrak{S}}$ (including σ_0) are simple. See [Lap-vFr3, Theorem 8.2.5 and Corollary 8.27].[24]

---

[23] Generally, however, such cancellations do not occur and hence, the exact counterpart of the identity (6.2.6) holds.

[24] The setting of [Lap-vFr3, Section 8.4] is that of general self-similar strings (for which we then have that $\sum_{j=1}^{J} r_j < 1$) but, in light, in particular, of the general framework considered in [Lap-vFr3, Chapter 3], where this condition is not assumed about the r_j's, the proofs and the statements of all of the results in that section can be immediately adjusted to our present situation (where $\sum_{j=1}^{J} r_j^N < 1$). This fact is also used in [LapPe2–3, LapPeWi1–2] and [Lap-vFr3, Section 13.1].

The content of Problem 6.2.36 will be in part to connect the complex dimensions of the (suitable) self-similar set F (or, more generally, self-similar RFD) and the complex dimensions of the associated self-similar tiling (or spray) (A,Ω), as described in Subsection 5.5.6, and taking into account the aforementioned results about the complex dimensions of (generalized) self-similar strings obtained in [Lap-vFr3, Chapter 3 and Section 8.4]. (See, in particular, footnote 24 on the previous page.) First, we consider Problem 6.2.35, which revisits some of the issues dealt with in Subsection 5.5.6, especially in parts (b) and (c) of Remark 5.5.26. In closing these introductory comments, we point out that, as will be clear to the reader, Problem 6.2.35 is closely related to Problem 6.2.36.

Problem 6.2.35. Prove that every (nontrivial) lattice self-similar subset F or, more generally, self-similar RFD (F,Ω_0) of $\mathbb{R}^N$ (satisfying the open set condition) and such that $D_G \neq \sigma_0$ is either periodic or strictly subcritically periodic; that is, more specifically, it is Minkowski nonmeasurable, when $D_G < \sigma_0$ (and hence, $D := \dim_B(F,\Omega_0) = \sigma_0$) or else (strictly subcritically) Minkowski measurable in dimension σ_0 (in the sense explained at the end of Subsection 5.5.6), when $\sigma_0 < D_G$ (and hence, $D := \dim_B(F,\Omega_0) = D_G \in \{0,1,\dots,N-1\}$). It would be natural to assume that the affine subspace of $\mathbb{R}^N$ generated by G is all of $\mathbb{R}^N$, in which case $D_G = N-1$.

More generally, show that provided $D_G \neq \sigma_0$, the self-similar set F (or, in the broader setting of self-similar RFDs, the self-similar RFD (F,Ω_0)) is Minkowski measurable in dimension σ_0 (in the sense of footnote 71 at the end of Chapter 5) if and only if it is nonlattice. [Equivalently, the self-similar set F (or, more generally the self-similar RFD (F,Ω_0)) is Minkowski nonmeasurable in dimension σ_0 if and only if it is lattice.] Exactly the same statement holds with the self-similar set F replaced by its associated self-similar tiling (or spray) (A,Ω). Also, under mild assumptions, we have that $D := \dim_B(A,\Omega) = \dim_B F$, whereas $\dim_B(F,\Omega_0) = \max\{\dim_B F, \dim_B(\partial G, G)\} = \max\{\sigma_0, D_G\}$, since $\dim_B F = \sigma_0$, and G is the generator of the self-similar RFD (F,Ω_0), viewed as an inhomogeneous self-similar set here (or as a suitable generalization thereof, in the spirit of Examples 4.2.33, 4.2.34 and 4.2.35). In the case of multiple generators $\{G^{(q)}\}_{q=1}^{Q}$, with $Q \geq 2$, we instead have the inequality $\dim_B(F,\Omega_0) \leq \max_{q=1,\dots,Q}\{\sigma_0, D_{G^{(q)}}\}$.

Finally, when $D_G = \sigma_0$, F (or, more generally, (F,Ω_0)) is not Minkowski measurable.[25] Show, however that, if we consider the gauge function $h(t) := \log t^{-1}$, for all $t \in (0,1)$, then the self-similar set F (or, more generally, the self-similar RFD (F,Ω_0)) is always h-Minkowski measurable, whether it is lattice or nonlattice (or, equivalently, whether the associated self-similar spray or tiling (A,Ω) is lattice or nonlattice).[26]

---

[25] Indeed, its distance zeta function $\zeta_F(s)$ should then have a multiple pole (of second order) at $s = D$ $(= D_G = \sigma_0)$ and hence, according to Theorem 5.4.10 and Lemma 5.4.11, F cannot be Minkowski measurable. On the other hand, as is stated here, it should follow from results of Chapter 5, that when $D_G = \sigma_0$, F is h-Minkowski measurable whether it is lattice or nonlattice, where $h(t) = \log t^{-1}$ for all $t \in (0,1)$.

[26] The intuitive reasoning behind this part of the conjecture (or open problem) is as follows. In both cases (i.e., the lattice and the nonlattice cases), we have that the only double pole is σ_0; so it

We note that the results of [Lap-vFr1–3] (see, especially, [Lap-vFr3, Section 8.4]) imply that the answer to Problem 6.2.35 is affirmative when $N = 1$ (except, of course, in the trivial case of an interval); see [Lap-vFr3, Theorem 8.25 and Corollary 8.27, along with Theorems 8.23 and 8.36], thereby proving the geometric part of [Lap3, Conjecture 3, pp. 163–164] in the special case of self-similar fractal strings. Part of these results have been extended in [LapPe2–3] and in [LapPeWi1–2] to certain lattice self-similar sprays (or tilings), as well as to a more restricted class of lattice self-similar sets in $\mathbb{R}^N$ ($N \geq 1$); see, in particular, [LapPeWi2].

Furthermore, for general self-similar sprays (or tilings) in $\mathbb{R}^N$ (with $N \geq 1$), we have essentially answered Problem 6.2.35 in the affirmative, in Subsection 5.5.6 (see, especially part (c) of Remark 5.5.26), by using the fractal tube formulas and Minkowski measurability criteria of Chapter 5. More specifically, when $D_G < \sigma_0$, as was discussed in case (i) of Remark 5.5.26(c), a combination of Theorem 5.4.2 and Theorem 5.4.20 enables us to obtain the required result, exactly as in the proof of Corollary 5.4.23 (which corresponds to the $N = 1$ case). Furthermore, when $D_G > \sigma_0$, as in case (iii) of Remark 5.5.26(c), a suitable adjustment (to the strictly subcritical case) of the statements and proofs of these two theorems should yield the discussed result. Moreover, for general self-similar sets in $\mathbb{R}^N$ (with $N \geq 1$) in the case when $D_G < \sigma_0$ (i.e., $D_G \neq \sigma_0$, although it is not formulated in this manner in that paper), the problem is essentially answered in the affirmative in [KomPeWi], thereby proving the geometric part of [Lap3, Conjecture 3] when D is not an integer, by using the renewal theorem and the main results of [Gat].

Recall from Subsection 5.5.6 that the case when $D_G > \sigma_0$ cannot occur for self-similar sets (satisfying the open set condition) since then, we have

$$\sigma_0 = D_F = D_{(A,\Omega)} =: D = \max\{D_G, \sigma_0\} \tag{6.2.11}$$

and hence, $D_G \leq \sigma_0$; see part (a) of Remark 5.5.26. (In the case of multiple generators, we have that $D \leq \max_{q=1,\dots,Q}\{D_{G^{(q)}}, \sigma_0\}$, also implying that $D_G := \max_{q=1,\dots,Q}\{D_{G^{(q)}}\} \leq \sigma_0$; see part ($b$) of Remark 5.5.26.) However, for a general self-similar RFD (or spray) we no longer have that $D_G = D_{(A,\Omega)} = \sigma_0$, allowing for D_G to exceed σ_0. (See, e.g., the case of the inhomogeneous Sierpiński N-gasket RFD discussed in Example 4.2.26, along with Examples 4.2.34 and 4.2.35.)

Strictly speaking, the case when $D_G > \sigma_0$ remains open for general self-similar RFDs (or sprays), although we expect the results and methods of Chapter 5 (combined with some of the results about the complex dimensions suggested in Problem 6.2.36 just below) should enable us to resolve it in the affirmative, as well as to

---

is clear that in the nonlattice case, F should be h-Minkowski measurable. But in the lattice case, all the other nonreal principal poles are simple; therefore, they generate powers of $t^{N-\sigma_0}$ times an oscillatory term in the tube formula, but this contribution is dominated by $t^{N-\sigma_0}\log t^{-1}$ when $t \to 0^+$. Hence, in the lattice case F should be also h-Minkowski measurable. On the other hand, the h-Minkowski nonmeasurable but h-Minkowski nondegenerate situation arises, for instance, for the second-order Cantor set discussed in Example 4.2.10, where we have an infinite sequence of double principal poles occurring in arithmetic progression along the critical line.

obtain a modified proof for all of the possible cases considered in Problem 6.2.36; namely, $D_G < \sigma_0$, $\sigma_0 < D_G$, and $D_G = \sigma_0$.

In the next problem, we propose to study the fractal zeta functions and the corresponding complex dimensions, as well as the Minkowski measurability, of self-similar sets in $\mathbb{R}^N$, much as is done when $N = 1$ for self-similar strings in [Lap-vFr3, Chapters 2, 3 and 8], and when $N \geq 1$ but for self-similar tilings (or for more general fractal sprays) and for special kinds of self-similar sets, in [LapPe2–3] and [LapPeWi1–2] (as is discussed in [Lap-vFr3, Section 13.1]).

Before stating this problem, we recall some terminology associated with self-similar sets (or, more generally, with self-similar RFDs or sprays) and the corresponding lattice/nonlattice dichotomy. Let F be a self-similar subset of $\mathbb{R}^N$ or, more generally, a self-similar RFD (with a single generator G for the associated self-similar tiling or spray), and consider the multiplicative group $\mathscr{G} = \prod_{i=1}^{n} (\rho_i)^{\mathbb{Z}}$ generated by its distinct scaling ratios $\rho_1, \ldots, \rho_n$ and viewed as a subgroup of $\mathbb{R}_*^+ := (0, +\infty)$. Then, recall that F is said to be *lattice* if $\mathscr{G}$ is of rank 1 (i.e., $\mathscr{G} = \rho^{\mathbb{Z}}$, for some ρ such that $0 < \rho < 1$), and *nonlattice* otherwise. Moreover, F is a *generic nonlattice* self-similar set if $\mathscr{G}$ has rank $n \geq 2$, where (as above) n is the number of distinct scaling ratios of F. See [Lap-vFr3, Chapters 2 and 3] and the relevant references therein.

In the lattice case, the *oscillatory period* (i.e., the common difference of the arithmetic progression of the imaginary parts of any two consecutive complex dimensions along the vertical lines $\{\mathrm{Re}\, s = w_u\}_{u=1}^{q}$), but ignoring the $\mathbb{i} := \sqrt{-1}$ factor, is given by $\mathbf{p} := 2\pi / \log \rho^{-1}$, where (as above) $\rho \in (0,1)$ is the positive generator of the (discrete) multiplicative group $\mathscr{G}$ generated by the distinct scaling ratios.

In the sequel, we will denote by F the self-similar set or, more generally, RFD under consideration and by (A, Ω) the associated self-similar tiling (or spray), viewed as an RFD. (See the discussion preceding Problem 6.2.35.) Implicit in the statement of Problem 6.2.36 is the fact that under suitable hypotheses, we can show (as was conjectured in part (a) of Remark 5.5.26 of Subsection 5.5.6) that

$$\zeta_F(s) = \zeta_{A,\Omega}(s) + \zeta_{O,\mathrm{out}}(s) + f(s), \tag{6.2.12}$$

where (as was explained towards the end of part (a) of Remark 5.5.26) $\zeta_{O,\mathrm{out}}$ represents the contributions of the outer neighborhoods of a feasible and admissible open set O relative to which F satisfies the open set condition. Furthermore, f is a holomorphic function in some open right half-plane $\{\mathrm{Re}\, s > \alpha\}$, with $-\infty \leq \alpha < \overline{\dim}_B(A, \Omega) =: D\ (< N)$. If $\alpha > -\infty$ (that is, if f is not entire), then clearly, the conclusion of Problem 6.2.36 concerning the structure of the complex dimensions of F should be suitably adjusted. Namely, the corresponding statements should be limited to the *visible* complex dimensions (i.e., the poles of ζ_F or of $\tilde{\zeta}_F$ with real part $> \alpha$).

Finally, recall that since $D := \overline{\dim}_B(A,\Omega) < N$,[27] the complex dimensions of F can be interpreted indifferently as the poles of the distance zeta function ζ_F or of the tube zeta function $\tilde{\zeta}_F$, and analogously, for the complex dimensions of the RFD (A,Ω), with ζ_F and $\tilde{\zeta}_F$ replaced by $\zeta_{A,\Omega}$ and $\tilde{\zeta}_{A,\Omega}$, respectively. This statement follows from Proposition 2.2.19 and its obvious counterpart for relative fractal zeta functions of RFDs.

Finally, we note that in light of the results of Subsection 5.5.6 and Section 5.4 (combined with those of [Lap-vFr3, Chapters 2–3 and Section 8.4]), most of the results expected to hold in Problem 6.2.36 have already been established in this book in the special case of a self-similar spray (with a single generator or, more generally, with finitely many generators).

Problem 6.2.36. (*Geometric complex dimensions of self-similar sets*). Calculate the distance and tube zeta functions of self-similar fractals F in $\mathbb{R}^N$, and, more generally, of self-similar RFDs (F,Ω_0) (or of self-similar sprays). Let (F,Ω_0) be such a self-similar RFD, with associated self-similar tiling (or spray) (A,Ω). If F is a self-similar set, we assume, in particular, that it satisfies the open set condition of [Hut] (see also [Fal1]). Can one show that these fractal zeta functions have a meromorphic continuation to all of $\mathbb{C}$?[28] Determine the poles of these zeta functions, that is, the complex dimensions of a self-similar set F or, more generally, of a self-similar RFD (F,Ω_0). Finally, compare the results with those obtained by means of the well-developed theory of complex dimensions of fractal strings. (See [Lap-vFr3, Chapters 2–3 and Section 8.4].) and, in the higher-dimensional case, with the results from [Lap-vFr3, Section 13.1] describing some of the work in [LapPe3] and [LapPeWi1] on tubular zeta functions and the complex dimensions of self-similar tilings and sets.) Compare also with the results of Subsection 5.5.6 above on self-similar sprays.

In particular, show that, under suitable hypotheses, we have the following inclusion between multisets (in the case of a single generator G), which would follow at once from Equation (6.2.12) above, combined with the results of Subsection 5.5.6 (see also Remark 6.2.33):[29]

$$\begin{aligned}\mathscr{P}(\zeta_F) &\subseteq \mathscr{P}(\zeta_{A,\Omega}) \cup \mathscr{P}(\zeta_{O,\mathrm{out}}) \\ &\subseteq \mathscr{P}(\zeta_{\mathfrak{S}}) \cup \mathscr{P}(\zeta_{\partial G,G}) \cup \mathscr{P}(\zeta_{O,\mathrm{out}}),\end{aligned} \tag{6.2.13}$$

---

[27] Recall, that in the case of multiple generators $\{G^{(q)}\}_{q=1}^{Q}$, we have, in general, $\overline{D} := \overline{\dim}_B(A,\Omega) \le \max\{D_{G^{(q)}}, \sigma_0 : q = 1,\dots,Q\}$; so that $D < N$ because σ_0 and each $D_{G^{(q)}}$ $(q = 1,\dots,Q)$ is strictly less than N. Note that the fact that $D_G \le N-1$ and the hypothesis according to which $\sum_{j=1}^{J} r_j^N < 1$, which guarantees that Ω has finite N-dimensional volume, imply that $\sigma_0 < N$.

[28] If not, determine $D_{\mathrm{mer}}(\zeta_F) = D_{\mathrm{mer}}(\tilde{\zeta}_F)$, the commmon abscissa of meromorphic continuation of ζ_F and $\tilde{\zeta}_F$.

[29] Equation (6.2.13) should still remain valid in the more general case when (A,Ω) has multiple generators. (See Subsection 5.5.6, including parts (a) and (b) of Remark 5.5.26.) Often, unless there are obvious cancellations (see, for example, Remark 6.2.33 for one of the sources of these possible cancellations, in the case of multiple generators), the inclusion signs can be raplaced by equalities in Equation (6.2.13).

where $\mathscr{P}(\zeta_{\mathfrak{S}})$ is the set of 'scaling complex dimensions' of the associated self-similar tiling (A,Ω) (as given by Equation (6.2.10)) and $\mathscr{P}(\zeta_{\partial G,G}) \subseteq \{0,1,\ldots,D_G\} \subseteq \{0,1,\ldots,N-1\}$ is the 'set of integer dimensions' of (A,Ω), as given by Equation (6.2.8). Accordingly, the elements of $\mathscr{P}(\zeta_{\mathfrak{S}})$ (respectively, $\mathscr{P}(\zeta_{\partial G,G})$) are also called the *scaling* (respectively, *integer*) *complex dimensions* of the self-similar set F. (Naturally, some integers in $\{0,1,\ldots,D_G\}$ might be common to $\mathscr{P}(\zeta_{\mathfrak{S}})$ and $\mathscr{P}(\zeta_{\partial G,G})$.)

Consequently, it follows from the results of [Lap-vFr3, Chapter 3] (see especially [Lap-vFr3, Chapter 3]) that if $D := \dim_B(A,\Omega)$ and $\sigma_0 \in (0,N)$ denotes the *similarity dimension* of F (or, equivalently, of (A,Ω)), as defined before Equation (6.2.6), that the scaling complex dimensions of nonlattice self-similar sets F have real parts strictly less than σ_0 (except for σ_0 itself, which is also a scaling complex dimension), but that there exists an infinite (and explicitly computable) sequence of such scaling dimensions approaching from the left the vertical line $\{\operatorname{Re} s = \sigma_0\}$, whereas the scaling complex dimensions of lattice self-similar sets F are distributed periodically along finitely many vertical lines $\{\operatorname{Re} s = w_u\}$, for $u = 1,\ldots,q$, with $w_q \le \cdots \le w_2 < w_1 = \sigma_0$. In all cases, the only scaling complex dimension which is real is σ_0 itself, and it is simple. (In the lattice case, all the scaling complex dimensions with real part σ_0 are simple as well.) Moreover, the scaling complex dimensions of a nonlattice string have a quasiperiodic structure, as shown and described in detail throughout [Lap-vFr3, Chapter 3]. Finally, according to [MorSepVi1] (proving and extending a conjecture in [Lap-vFr2–3]), the set of real parts of the scaling complex dimensions of a generic nonlattice (respectively, and more generally, of a suitable nonlattice) self-similar set is dense in a single compact interval, of the form $[D_*,D]$ (with $-\infty < D_* < D$), or in the union of finitely many disjoint compact intervals (see [Lap-vFr2, Subsection 3.7.1], [Lap-vFr3, Section 3.7], along with [MorSepVi1], [MorSep] and [DubSep]). Therefore, generic nonlattice (respectively, under mild assumptions, more general nonlattice) self-similar sets are fractal in dimension d, for d in an infinite countable and dense subset of a compact interval $[D_*,D]$, with $D_* < D$ (respectively, of a finite union of compact disjoint intervals).

Furthermore, assume for now that $D_G < \sigma_0$ (so that $D = \sigma_0$). Then, in the lattice case, F is Minkowski nondegenerate but is not Minkowski measurable. (See Problem 6.2.35 and the comments surrounding it; see also [KomPeWi], proving a conjecture of [Lap3] in this case.) In the nonlattice case, F is Minkowski measurable (see Problem 6.2.35, along with [Gat]).[30] Also, in the nonlattice case, the residue of the tube zeta function $\tilde{\zeta}_F$ at $s = D = \sigma_0$ is equal to $\mathscr{M}$, the Minkowski content of F, whereas in the lattice case, it is equal to $\tilde{\mathscr{M}}$, the average Minkowski content (which exists and belongs to $(0,+\infty)$). See Remark 6.2.37 below.

[30] Recall from Example 4.2.26 and Subsection 5.5.6 that for the inhomogeneous Sierpiński N-gasket (which is not a self-similar set unless $N = 2$), $D_G < \sigma_0$ for $N = 2$, $D_G = \sigma_0$ for $N = 3$, and $D_G > \sigma_0$ for every $N \ge 4$. See also the illustrative planar Examples 4.2.33, 4.2.34 and 4.2.35 (which too are not self-similar sets, but are self-similar RFDs and, in fact, are inhomogeneous self-similar sets, in a suitably generalized sense).

Moreover, for a general self-similar RFD (F,Ω_0) (but not for a self-similar set F, for which this case cannot arise),[31] if we assume instead that $D_G > \sigma_0$ (so that $D = D_G$),[32] then (F,Ω_0) is always Minkowski measurable. More interestingly, the self-similar RFD (F,Ω_0) is (strictly subcritically) Minkowski measurable in dimension σ_0, in the sense of footnote 71 at the end of Chapter 5, if and only if it is nonlattice, and in that case, its (strictly subcritical) σ_0-Minkowski content is equal to the residue of $\tilde{\zeta}_{F,\Omega_0}$ at $s = \sigma_0$. Also, in the lattice case, (F,Ω_0) is (strictly subcritically) σ_0-Minkowski nonmeasurable (and σ_0-Minkowski nondegenerate), with average (strictly subcritical) σ_0-Minkowski content equal to the residue of $\tilde{\zeta}_{F,\Omega_0}$ at $s = \sigma_0$.

In the case when $D_G = \sigma_0 = D$, then the self-similar set F is always Minkowski degenerate with Minkowski content equal to $+\infty$, due to the double pole of $\tilde{\zeta}_F$ at $s = D$, but it is also h-Minkowski measurable, where the *gauge function* h is given by $h(t) := \log t^{-1}$ for all $t \in (0,1)$. Furthermore, its h-Minkowski content is then equal to $\tilde{\zeta}_F[D]_{-2}$, the (-2)-nd coefficient in the Laurent series expansion of $\tilde{\zeta}_F$ around $s = D$. Still when $D_G = \sigma_0$ $(= D)$, but now for a general self-similar RFD (F,Ω_0) instead of a self-similar set F, exactly the same results are expected to hold, with F and $\tilde{\zeta}_F$ replaced by (F,Ω_0) and $\tilde{\zeta}_{F,\Omega_0}$, respectively.

Finally, completely analogous statements hold for the distance zeta function ζ_F (instead of the tube zeta function $\tilde{\zeta}_F$). More specifically, since $D < N$ (as was assumed throughout), its poles (i.e., the associated complex dimensions), are exactly the same as for the tube zeta function $\tilde{\zeta}_F$, whereas (when $D_G < \sigma_0$) its residue at $s = D$ is equal to $(N-D)\mathscr{M}$ and $(N-D)\tilde{\mathscr{M}}$, in the nonlattice and lattice case, respectively. (See Theorem 2.2.3, Equation (2.2.4).) And analogously, when $D_G > \sigma_0$ and when $D_G = \sigma_0$.[33]

Remark 6.2.37. Implicit in the statement of the latter part of Problem 6.2.36 concerning the (ordinary) Minkowski measurability statements is the fact that one could show that when $D_G < \sigma_0$, the hypotheses of Theorem 2.3.18 and Theorem 2.3.25 are satisfied in the nonlattice and lattice case, respectively. In particular, for a lattice self-similar set (or, more generally, RFD) F, it would then follow from Theorem 2.3.25 and Corollary 2.3.26 that $\mathscr{M}_* = \min G$ and $\mathscr{M}^* = \max G$ are the values of the lower and upper Minkowski contents of F, respectively, where G is the periodic function occurring in the counterpart of Equation (2.3.30). Just as in [Lap-vFr1–3] (and then in [LapPe3] and [LapPeWi1–2]), this periodic function is determined by the sequence of (simple) complex dimensions, with maximal real part $w_1 = D$, distributed in arithmetic progression (along with the associated residues).

[31] See the discussion following Problem 6.2.35 and preceding the present one, along with parts (a) and (b) of Remark 5.5.26.

[32] We assume here implicitly that in the case of multiple generators, there are no cancellations.

[33] Naturally, in the case of a general self-similar RFD (F,Ω_0) instead of a self-similar set F, one should replace ζ_F and $\tilde{\zeta}_F$ by ζ_{F,Ω_0} and $\tilde{\zeta}_{F,\Omega_0}$.

We note in closing this discussion that the results obtained about the principal complex dimensions of the Sierpiński gasket and the Sierpiński carpet in Subsection 4.2.3 (on pages 290-305) are consistent with the results conjectured to hold in Problem 6.2.36.

6.2.3 Future Research Directions

Since the definition of the upper relative box dimension $\overline{\dim}_B(A,\Omega)$ involves the function $t \mapsto |A_t \cap \Omega|$ (see Section 4.1), it is natural to study the problem of finding a tube formula for this function, which we call the *relative tube formula of A with respect to* Ω, that is, of the tube A_t intersected by Ω; see Chapter 5. It will be the subject of further investigation by the authors. (See Problem 6.2.38 below, along with the comments following it, for several concrete research directions in this area.) The tube formula for $t \mapsto |A_t|$ has been extensively studied in [Lap-vFr1–3], partly motivated by the present theory as well as, in particular, by the earlier work in [BesTay, Tri3, Lap1–3, LapPo1–3, LapMa1–2, HeLap], and in [LapPe1–3] and in [LapPeWi1–2] for various classes of fractal sets A.

We plan to apply the methods developed in Section 2.3 to study the meromorphic extensions of *box-counting zeta functions*, recently introduced in a paper of John Rock with the first and third authors [LapRoŽu], and partly motivated by the present higher-dimensional theory of complex dimensions and some of its predecessors (including [BesTay, Tri3, Lap1–3, LapPo1–3, LapMa1–2, HeLap, Lap-vFr1–3, LapPe1–3, LapPeWi1–2]). We expect that many of the results in this book will have a suitable counterpart in the context of box-counting zeta functions. For example, we expect that our results about meromorphic extension of geometric zeta functions, obtained in Section 2.3, could be applied to this new class of zeta functions. The residues of these zeta functions could also be closely related to an appropriate (but perhaps less geometric) notion of Minkowski content, as in the case of distance and tube zeta functions. Further related work of John Rock, joint with the first and third authors, is in preparation on this subject.

We also plan to study the behavior of relative box dimensions and relative zeta functions with respect to Lipschitz functions between pairs of RFDs.

As we have already mentioned in Subsection 6.2.1 on pages 553–554, it is possible to consider some parts of our monograph from the point of view of category theory. It is easy to recognize several related categories, like, for example, the category of RFDs, or its subcategory, the category of bounded fractal strings. Some of the results obtained here can be formulated in the language of category theory. This and other related questions will be the subject of a future work.

Given a relative fractal drum (A,Ω) in $\mathbb{R}^N$ and a suitable weight function $w : \Omega \to \mathbb{C}$, one can define the associated *weighted relative distance relative zeta function* $\zeta_{A,\Omega,w}$, much as in Section 3.4:

$$\zeta_{A,\Omega,w}(s) := \int_\Omega d(x,A)^{s-N} w(x)\,\mathrm{d}x, \tag{6.2.14}$$

for all $s \in \mathbb{C}$ with $\operatorname{Re} s$ sufficiently large. For example, the derivative of a given relative distance zeta function $\zeta_{A,\Omega}$, as given by Definition 4.1.1, coincides with the following weighted relative distance zeta function (see Theorem 4.1.7(a)):

$$\zeta_{A,\Omega}(s,w) := \int_\Omega d(x,A)^{s-N} \log d(x,A)\,\mathrm{d}x,$$

with the weight function defined by $w(x) := \log d(x,A)$. In our future work, we plan to study general properties of weighted relative distance zeta functions as well.

It should also be interesting and useful for the applications to extend the present theory of distance and tube zeta functions (and their complex dimensions) to *metric measure spaces*, in which the underlying measure is assumed to be a "doubling measure" or, better, to satisfy the Ahlfors condition. (Metric measure spaces and their applications are studied, for example, in [DaMcCS] and the many relevant references therein.) Working on general metric measure spaces (locally compact metric spaces equipped with a positive Borel measure satisfying a "doubling condition") rather than on Euclidean spaces would enable us, in particular, to extend aspects of the present theory (under appropriate hypotheses) and to establish contact with aspects of geometric analysis on "nonsmooth manifolds" (as, for example, in [DavSem], [Chee] and the survey article [Hei]), as well as with aspects of analysis on self-similar (or not necessarily self-similar) fractals (as, for example, in the monograph by Kigami [Ki1] and in [Lap3–5, KiLap1–2, Ki2–3, ChrIvLap, LapSar] and the relevant references therein). Some preliminary work in this direction is being carried out in [Wat, LapWat], where it is shown, in particular, that many of the general results obtained in this book (including the fractal tube formulas obtained in Chapter 5) extend to the setting of Ahlfors spaces and where several examples of nonsmooth metric measures are studied.

Metric measure spaces are also called *spaces of homogeneous type*, *d-sets*, or *Ahlfors regular spaces* in the literature. We caution the reader that the definitions of these various notions are not quite equivalent and that the terminology or the definitions themselves are not always used consistently in the literature. Part of the problem discussed here will consist in finding the proper definitions and conditions which will best fit the situation under consideration.

The following open problem (Problem 6.2.38) is motivated by [Lap-vFr3, Chapters 5 and 8], along with [LapPe2–3] and [LapPeWi1–2], as well as, of course, by the general higher-dimensional fractal tube formulas obtained in Chapter 5 (especially in Sections 5.1–5.3 and in Subsection 5.5.6). More precisely, recall that in [Lap-vFr1–2] and [Lap-vFr3, Chapter 8] are obtained (under suitable assumptions) and used distributional and pointwise fractal tube formulas (with or without error term) for fractal strings, and, in particular, (pointwise) exact fractal tube formulas for self-similar strings (both in the lattice and nonlattice cases). These fractal tube formulas are expressed in terms of the residues of the underlying geometric zeta functions. Moreover, in [LapPe2–3] and [LapPeWi1–2] (partially discussed in [Lap-vFr3, Section 13.1]), which build on and extend the aforementioned work in [Lap-vFr1–3], are obtained (under suitable assumptions) and used distributional and pointwise tube formulas (with or without error terms) for fractal sprays, and, in particular, (pointwise) exact tube formulas for self-similar sprays or 'self-similar tilings' (both in the lattice and nonlattice cases), where the embedding dimension $N \geq 1$ is arbitrary. (The results in [LapPeWi1–2] also yield fractal tube formulas for a restricted class of self-similar sets in $\mathbb{R}^N$; see [LapPeWi2].) These fractal tube formulas are expressed in terms of suitably defined 'tubular zeta functions' associated with the fractal sprays (or the self-similar tilings).[34] They involve, in particular, the residues of 'scaling zeta functions' which play, in this more general context, the rôle played by the geometric zeta functions of fractal strings. In addition to these scaling data, the tubular zeta functions, however, involve other geometric data about the corresponding fractal sprays, such as the 'curvatures' of the generators of the sprays.

The aforementioned results of [LapPe2–3] and [LapPeWi1–2] about fractal tube formulas for self-similar sprays (discussed in part in [Lap-vFr3, Section 13.1]) have ben significantly extended in Subsection 5.5.6 and [LapRaŽu5], where they have also been placed within the much more general framework of the (higher-dimensional) theory of fractal zeta functions and fractal tube formulas developed in this book. The corresponding general theory of complex dimensions and of (*global*) fractal tube formulas should serve as a key guide for addressing the corresponding problem.

We will state the announced open problem (Problem 6.2.38) in words and hence, mostly qualitatively (since only future research will tell us what is the precise form of the sought for *local* fractal tube formulas). The main point is that we can now hope to obtain local fractal tube formulas valid (under suitable assumptions) for a very large class of relative fractal drums in $\mathbb{R}^N$ and, in particular, of compact subsets of $\mathbb{R}^N$ (where the embedding dimension $N \geq 1$ is arbitrary), including self-similar sets in $\mathbb{R}^N$. Furthermore, in light of the results of Chapter 5 (especially, of Sections 5.1–5.3), it is natural to expect that they would now be expressed in terms of the residues of *local analogs* of the fractal zeta functions introduced in this book, namely, the distance and tube zeta functions associated with arbitrary RFDs and, in particular, with bounded subsets of $\mathbb{R}^N$.[35] At first, because in the present context, it is the most interesting case, geometrically, the reader may wish to focus on obtaining and interpreting local fractal tube formulas expressed in terms of suitably defined local distance zeta functions (in the spirit of Appendix B to the present book). It

[34] The '*tubular zeta functions*' of [LapPe2–3] and [LapPeWi1–2] are not to be mistaken with the tube zeta functions introduced in this book.

[35] For simplicity, we assume here implicitly that the (visible) complex dimensions are simple. In the general case of possible multiple complex dimensions, the local fractal tube formula should be expressed as a sum over the (visible) complex dimensions of the residues of suitable expressions involving the corresponding local fractal zeta functions, much as is the case for their global counterparts obtained in Sections 5.1–5.3.

should be easy to obtain corresponding fractal tube formulas expressed in terms of the other fractal zeta functions.[36]

Finally, we mention that in the process, we expect to establish contact with earlier tube formulas and their interpretation in terms of curvatures or curvature measures (in the sense of Federer [Fed1]) associated with integer dimensions. See, especially, in addition to the original and key reference [Fed1], which has unified into a single framework (that of 'sets of positive reach') the work of Steiner [Stein] and its successors (for compact convex sets in $\mathbb{R}^N$) and of Weyl [Wey3] (for smooth compact submanifolds of Euclidean spaces, as described in [BergGos], and in the more general context of Riemannian manifolds, in [Gra]), the informative book [Schn2] and the articles [Schn1, Zä1–3], along with [HugLasWeil] (for the case of compact sets in $\mathbb{R}^N$) and [Wi, WiZä, Zä4–5, KeKom] (for the case of certain self-similar sets and their self-conformal and random generalizations). Further relevant references are provided in the introduction of Chapter 5.

Problem 6.2.38. (*Local fractal tube formulas and curvatures*).

(*i*) Obtain local forms (in a sense akin to [Fed1]) of the fractal tube formulas established in Chapter 5 (especially, in Sections 5.1–5.3) for relative fractal drums (and, in particular, for bounded sets) in $\mathbb{R}^N$. Much as in Chapter 5, these local fractal tube formulas should be pointwise or distributional, as well as with error term or else exact (i.e., without error term), depending on the growth assumptions made on the underlying fractal zeta functions and (when applicable) on the screens used to formulate the results. Also by analogy with Chapter 5 (Sections 5.1–5.3), they should be expressed in terms of the residues (evaluated at the underlying visible complex dimensions) of expressions explicitly involving suitably defined local fractal zeta functions (for instance, local distance, tube or shell zeta functions).[37]

(*ii*) In the important special case of simple (visible) poles (for example),[38] interpret the coefficients of the (global) fractal tube formulas of Chapter 5 (Sections 5.1–5.3) as 'fractal curvatures' associated with each of the complex dimensions of the relative fractal drum (or bounded set) in $\mathbb{R}^N$. (See problem (*iii*) just below.)

(*iii*) Furthermore, under the same assumptions as in part (*ii*), interpret the coefficients of the local fractal tube formulas sought for in part (*i*) of this problem as the corresponding action (on Borel subsets of $\mathbb{R}^N$ or on a class of appropriately localized test functions) of suitably defined 'curvature measures (or distributions)', again

[36] It is likely, however, that exactly as in Chapter 5, it would be easier to first obtain local tube formulas expressed in terms of the local tube zeta functions (much as in Sections 5.1 and 5.2), and only then deduce their counterparts expressed in terms of local distance zeta functions (as in Section 5.3).

[37] The 'local zeta function' proposed in Appendix B may be helpful in this context, provided its definition is appropriately modified, for geometric reasons (and by analogy for example, with [Fed1, Schn2, HugLasWeil, Wi]).

[38] This hypothesis is not really necessary but assuming it makes more transparent the analogy with the classic literature on tube formulas (see, e.g., [Fed1, Schn2]).

in the spirit of [Fed1].[39] Consequently, interpret the 'fractal curvatures' of part (*ii*) as the total masses (or their smeared analogs) of the above curvature measures (or distributions).

(*iv*) Finally, whenever possible, obtain *concrete* realizations of the various fractal tube formulas obtained in Chapter 5 (especially, in Sections 5.1–5.3), as well as of the 'local fractal tube formulas' [and hence, of the associated fractal curvature measures (or distributions) and, in particular, of the fractal curvatures] sought for in part (*i*) (as well as in parts (*ii*) and (*iii*) just above). Do so, for example, in the case of self-similar sets (see Problem 6.2.36 and part (*a*) of Remark 5.5.26) and, in particular, of self-similar sprays or tilings (see, especially, Subsection 5.5.6) as well as in the case of Julia sets (see also, for instance, the geometric part of Problem 6.2.32, Problem 6.2.21 and 6.2.22), the Mandelbrot set, limit sets of Fuchsian and Kleinian groups, conformal fractals, nonlinear and 'approximately self-similar' fractals, along with random (or stochastically self-similar) fractals and other random fractals naturally occurring in mathematics and physics (see also Problem 6.2.25 and the brief discussion preceding it), without forgetting (for the question concerning fractal curvatures) the various classes of examples discussed in Section 5.5 from the point of view of the fractal tube formulas and the associated (geometric) complex dimensions.

We should caution the reader that Problem 6.2.38 is difficult, especially in the most interesting case of distance zeta functions. Its investigation and eventual resolution, however, should open new venues in a number of directions:

(*a*) The further development (and justification) of a theory of *complex dimensions* and of the corresponding *geometric oscillations*, valid (both globally and locally) for 'arbitrary' compact sets (or, more generally, RFDs) in $\mathbb{R}^N$, for any $N \geq 1$.

(*b*) The possible interpretation (sought for in parts (*ii*) and (*iii*) of Problem 6.2.38) of the coefficients of the (global or local) fractal tube formulas in terms of (global or local) 'fractal curvatures', along the lines suggested in a related context in [Lap-vFr3, Sections 8.2, 12.4, 12.5 and 13.1] as well as in [LapPe3] and [LapPeWi1], should eventually be connected with the notion of 'fractal cohomology' conjectured to exist in [Lap-vFr1–3] (see, especially, [Lap-vFr3, Sections 12.4 and 12.5]) and in [Lap6]. In the latter fractal (or complex) cohomology theory, a suitable finite-dimensional complex Hilbert space is associated with each (visible) complex dimension (and of dimension equal to the multiplicity of the complex dimension). Alternatively, or rather, in addition, to each $d \in \mathbb{R}$ for which the geometric object under consideration is 'fractal' (see the latter part of (*c*) just below), is associated a possibly infinite dimensional complex Hilbert space (equal to the direct sum of the finite-dimensional Hilbert spaces associated to each of the (visible) complex dimensions with real part d). (See also [Lap6] and the relevant references therein; see, especially, [Lap9, Lap10, CobLap][40] for the construction of a possible fractal

[39] Naturally, the curvature measure associated with a given visible complex dimension should be a complex measure (or else, a complex-valued distribution).

[40] The papers [Lap9, Lap10] may become joint with Tim Cobbler.

cohomology theory fulfilling these criteria in the context of a suitable class of meromorphic functions, including essentially all arithmetic zeta functions, as well as the scaling zeta functions of all self-similar strings and sprays.)

(c) A general definition of *fractality*, and especially its justification. More explicitly, as was discussed in various places in this book (see, e.g., Remark 4.6.24), a geometric object is said to be *fractal* if its associated fractal zeta function has at least one nonreal complex dimension. Furthermore, recall from the discussion in Remark 5.5.15 that given $d \in \mathbb{R}$, a geometric object is said to be *fractal in dimension d* if it has at least one nonreal complex dimension with real part d. These definitions could be extended to the local complex dimensions (i.e., the poles of the relevant local fractal zeta functions) and thereby justified in part by the general local fractal tube formulas sought for in part (i) of Problem 6.2.38, as well as by their concrete realizations sought for in a variety of situations in part (iv) of Problem 6.2.38; see also Problems 6.2.21, 6.2.22 and 6.2.32.

The *concrete* forms of the global (and local) fractal tube formulas sought for in part (iv) of Problem 6.2.38 would then further demonstrate the fact that the presence of geometric oscillations is intimately connected with the notion of fractality. Hence, for example, according to the answers conjectured in Problem 6.2.36, self-similar geometries are always fractal (except in the trivial case of N-dimensional cubes, say). (Compare with the discussion in [Lap-vFr3, Section 12.2].) More specifically, in the lattice case, they would be fractal for finitely many values of d (no less than one, but possibly just one), whereas in the nonlattice case, they would be fractal in dimension d for infinitely many countable values of d, dense in a single interval or in a finite union of intervals. Similarly, one would expect that Julia sets (except in the trivial case of circles, for example), the Mandelbrot set, the Cantor curve (i.e., the 'devil's staircase'; see [Lap-vFr3, Subsection 12.1.2]), the Weierstrass–Mandelbrot curve, the limit sets of Fuchsian and Kleinian groups, chaotic attractors (in the theory of dynamical systems), approximately self-similar attractors and their various nonlinear generalizations, stochastically self-similar sets, ..., to be fractal in the above sense.[41] A difficult challenge consists in actually calculating the visible poles of the associated fractal zeta functions (that is, the visible complex dimensions of the fractals under consideration). This is, in part, the object of several of the open problems stated in Subsection 6.2.2 above within the present significantly more general context of the theory of fractal zeta functions; see, especially, Problems 6.2.21–6.2.25 along with Problem 6.2.36.

We should add that, just as in [Lap-vFr3, Section 13.4.3], as well as in Subsections 4.6.2 and 4.6.3 above within the significantly more general context of the theory of fractal zeta functions, we must extend the definition of fractality as follows: A geometric object (say, an arbitrary bounded subset A of $\mathbb{R}^N$) is said to be *fractal* if its associated fractal zeta function (ζ_A or $\tilde{\zeta}_A$) has a (meromorphic) *partial natural boundary* along a screen or else has at least one nonreal (visible) complex

[41] We allow here the use of general gauge functions when trying to determine the complex dimensions and the nature of the corresponding oscillations.

dimension. We point out that according to this refined definition, even a maximal hyperfractal would be fractal (see Subsection 4.6.2, including Definition 4.6.23(iii)), and in a certain way, would be 'maximally fractal', since every point of the critical line $\{\mathrm{Re}\, s = D\}$ is a (nonisolated) singularity of the fractal zeta function.

We plan to further study the spectrum $\sigma(A,\Omega)$ of RFDs (A,Ω) in Euclidean spaces $\mathbb{R}^N$, as well as the associated spectral zeta functions $\zeta^*_{A,\Omega}$, introduced in Section 4.3.1. This theory is well understood for $N = 1$, that is, in the case of fractal strings; see [Lap1–3], [LapPo1–3], [LapMa1–2], and [Lap-vFr3, esp., Chapters 1, 6 and 9–11]. When $N \geq 2$, partial mathematical results are known in the cases where $A = \mathbb{R}^N$ or $A = \partial\Omega$ (corresponding to ordinary fractal drums with Neumann or Dirichlet boundary conditions, respectively). See, for example, [Lap1–3], [LapPo3], [HeLap], [Lap-vFr3, Section 12.5] and the many relevant references therein, including [BroCar, Cae, FlVa, Ger, GerSc, MolVai, vBGilk]. (See Section 4.3 above.)

A challenging problem in this context consists in determining the (visible) spectral complex dimensions (i.e., the visible poles of a nontrivial meromophic continuation, when it exists, of the spectral zeta function $\zeta^*_{A,\Omega}$) of a variety of interesting (relative) fractal drums (A,Ω) and to obtain spectral analogues of the (geometric) fractal tube formulas, sought for in Problem 6.2.38 above. The resulting spectral 'explicit formulas' would express the spectral counting function $N_\nu(\mu)$ as a (typically countably infinite) sum involving the residues of the spectral zeta function $\zeta^*_{A,\Omega}$. For practical or physical reasons (and because it is easier, mathematically), one may also wish to work with other spectral functions, such as the partition function (or trace of the heat semigroup), instead of the spectral counting function $N_\nu(\mu)$.

In the process of attempting to address this problem, one should naturally be led to investigate in depth the connections between the spectral zeta function $\zeta^*_{A,\Omega}$ of a relative fractal drum (A,Ω) and the corresponding (relative) fractal zeta functions (namely, the relative distance and tube zeta functions, $\zeta_{A,\Omega}$ and $\tilde{\zeta}_{A,\Omega}$) introduced in this book.

Appendix A
Tamed Dirichlet-Type Integrals

Abstract The goal of this appendix is to provide several properties and examples of Dirichlet-type integrals (DTIs) and their natural generalizations (extended DTIs) in our context, as well as to modify accordingly the relation $\sim$ (introduced and used in the book) so that it remains a true equivalence relation on the resulting space of extended DTIs. At the end of the appendix, we also introduce a notion of "asymptotic equivalence", which is no longer a true equivalence relation but allows more flexibility and, as a result, may potentially be more useful in certain practical situations. All of the fractal zeta functions studied in this book, along with the classic arithmetic zeta functions and Dirichlet series and integrals, are shown to be very special cases of the general DTIs introduced and studied in this appendix.

Key words: Dirichlet type integral (DTI), tamed DTI, extended DTIs, asymptotic equivalence.

The goal of this appendix is to provide several properties and examples of Dirichlet-type integrals (DTIs) and their natural generalizations (extended DTIs) in our context, as well as to modify accordingly the relation $\sim$ so that it remains a true equivalence relation on the resulting space of extended DTIs.

At the end of the appendix (see, especially, Definition A.6.6, along with the comment preceding it and Remark A.6.7), we will also introduce a notion of "asymptotic equivalence", which is no longer a true equivalence relation but allows more flexibility and, as a result, may potentially be more useful in certain practical situations.

This appendix shoud be read in conjunction with Subsection 2.1.3.2, which it complements in a variety of ways. Recall that in that subsection, the notion of DTI was introduced (albeit not as precisely as here).

All the fractal zeta functions studied in this book, along with the classic arithmetic zeta functions and Dirichlet series and integrals, are shown to be very special cases of the general DTIs introduced and studied in this appendix.

M.L. Lapidus et al., *Fractal Zeta Functions and Fractal Drums*,
Springer Monographs in Mathematics, DOI 10.1007/978-3-319-44706-3

A.1 Local Measures and DTIs

We begin by recalling the notion of local (positive or complex) measure on a given Hausdorff locally compact space, equipped with its Borel σ-algebra.[1] (See [DolFr], [JohLap], [JohLapNi], [HerLap1–4] and [Lap-vFr3, Chapter 4] for the important special case where E is an interval of $\mathbb{R}$.) In all of our applications to the various fractal zeta functions encountered in this book as well as to the modified equivalence relation to be discussed towards the end of this appendix, E is a subset of some Euclidean space. For example, in the applications of the theory, we can have $E = [0,+\infty)$, $[1,+\infty)$, $[0,\delta]$ for some $\delta > 0$ or else $E = A_\delta$ (or even $E = A_\delta \setminus \overline{A}$), where A_δ is the δ-neighborhood of a given bounded subset A of $\mathbb{R}^N$.

Definition A.1.1. A (positive or complex) *local measure* μ on E (or a locally bounded positive or complex Borel measure on E) is a set-function $\mu : \mathscr{B}(E) \to [0,+\infty]$ or $\mu : \mathscr{B}(E) \to \mathbb{C}$ whose restriction to $\mathscr{B}(K)$, the Borel σ-algebra of an arbitrary compact subset K of E, is either a bounded positive measure on K (case of a local positive measure) or is a complex (and hence, bounded) measure on K (case of a local complex measure). Thus, $|\mu|(K) < \infty$ for every compact subset K of E, where $|\mu|$ denotes the *total variation measure* of μ (see, e.g., [Coh], [Foll] and [Ru]) defined for each $B \in \mathscr{B}(E)$ by

$$|\mu|(B) = \sup\left\{\sum_{k=1}^{m} |\mu(B_k)|\right\}, \tag{A.1.1}$$

where $m \geq 1$ and $\{B_k\}_{k=1}^m$ ranges over all finite partitions of B into disjoint measurable subsets of E.

The total variation measure $|\mu|$ is a local positive measure. Furthermore, if μ itself is a (local) positive measure, then $|\mu| = \mu$. In addition, a positive local measure on E is nothing but a locally bounded (Borel) measure on E. Moreover, the reason why (in the case of complex local measures) we have to work with a set-function that is not a complex measure (in the usual sense) on E but only on its compact subsets is that, as is well known, an ordinary complex measure is always bounded (see, e.g., [Coh], [Foll] or [Ru]). We refer to [Coh, Foll, Ru] for the basic results on measure theory and, in particular, for the theory of standard positive or complex measures.

We next recall the definition of a Dirichlet-type integral associated with a given triple (E,φ,μ), where μ is a (positive or complex) measure on E and the Borel measurable function $\varphi : E \to \mathbb{R}$ satisfies $\varphi \geq 0$ $|\mu|$-a.e. on E (i.e., $|\mu|$-almost everywhere on E).[2]

---

[1] For our own purposes it would be sufficient to assume that E is also separable and metrizable, so that it has a countable basis of relatively compact open sets.

[2] In all but one of the applications of DTIs (or extended DTIs), we have $\varphi > 0$ $|\mu|$-a.e. an E (i.e., $|\mu|(\{\varphi = 0\}) = 0$). The one exception is the case of the distance or tube zeta function of a bounded set $A \subset \mathbb{R}^N$, for which we have explained what to do in the main text of the book. (An

Definition A.1.2. The *Dirichlet-type integral* (in short, DTI) $f = \zeta_{E,\varphi,\mu}$ associated with the triple (E, φ, μ) as above is given by

$$f(x) := \int_E \varphi(x)^s \mathrm{d}\mu(x), \tag{A.1.2}$$

for all $s \in \mathbb{C}$ with $\mathrm{Re}\, s$ sufficiently large, provided f is tamed in the sense of Definition A.1.3 below.

When $f = \zeta_{E,\varphi,\mu}$ can be meromorphically continued to some connected domain $U \subseteq \mathbb{C}$, we still denote its (unique) meromorphic extension to U in the same way, as usual. In the sequel, we will write, indifferently, $\mathrm{d}\mu(x)$ or $\mu(\mathrm{d}x)$. We will also write interchangeably $\zeta_{E,\varphi,\eta}$ or $\zeta_{(E,\varphi,\eta)}$.

The notion of Dirichlet-type integral is only truly useful for us when the DTI in question is tamed, in the following sense.

Definition A.1.3. A DTI $f = \zeta_{(E,\varphi,\mu)}$ (as in Definition A.1.2) is said to be *tamed* if there exists a positive and finite constant $C = C(f)$ such that $|\mu|(\{\varphi > C\}) = 0$; i.e., if

$$\varphi \leq C \quad |\mu|\text{-a.e. on } E. \tag{A.1.3}$$

We wish to stress the fact that a DTI $f = \zeta_{E,\varphi,\mu}$ is precisely defined only once the associated triple (E, φ, μ) has been specified. Hence, the choice of our notation, $\zeta_{E,\varphi,\mu}$.

We next recall from Definition 2.1.8 in Section 2.1 the definition of the abscissa of convergence $D(f)$ of a tamed DTI $f = \zeta_{E,\varphi,\mu}$ given by (A.1.2)[3]:

$$\begin{aligned} D(f) :&= \inf\left\{\alpha \in \mathbb{R} : \int_E \varphi(x)^\alpha |\mu|(\mathrm{d}x) < \infty\right\} \\ &= \inf\{\alpha \in \mathbb{R} : \varphi(x)^\alpha \text{ is } \mu\text{-integrable}\}. \end{aligned} \tag{A.1.4}$$

Occasionally, $D(f) = D(\zeta_{E,\varphi,\mu})$ is also referred to as the *abscissa of absolute convergence* of $f = \zeta_{E,\varphi,\mu}$, since a measurable function is Lebesgue μ-integrable if and only if it is absolutely $|\mu|$-integrable, where as before, $|\mu|$ is the total variation measure of μ (defined by (A.1.1) above).

Also as in Section 2.1, the *half-plane of convergence of a tamed DTI* $f = \zeta_{E,\varphi,\mu}$, denoted by $\Pi(f)$, is defined by

$$\Pi(f) := \{\mathrm{Re}\, s > D(f)\}, \tag{A.1.5}$$

where $D(f)$ is the abscissa of convergence of f given by (A.1.4) above. Sometimes, $\Pi(f)$ is called the *half-plane of absolute convergence* of f.

---

alternative would be to integrate $\varphi(x)^s$ on the complement in E of $\{\varphi = 0\}$ in Equation (A.1.2) of Definition A.1.2.)

[3] The second equality in (A.1.4) holds if (as will always be assumed) the DTI is tamed, in the sense of Definition A.1.3.

Using a classic theorem concerning the analyticity of the integrals depending on a parameter (see Theorem 2.1.47 in Subsection 2.1.3.2, along with the associated Remark 2.1.49 on page 83), it is easy to show that the tamed DTI $f = \zeta_{E,\varphi,\mu}$ is holomorphic on $\Pi(f) = \{\mathrm{Re}\, s > D(f)\}$. Hence, $D_{\mathrm{hol}}(f) \le D(f)$.

The point of the tameness condition introduced in Definition A.1.3 above is that under this condition, $D(f)$ and $\Pi(f)$ are always *well defined*, in the sense that $\Pi(f)$ is the *largest* open right half-plane $\{\mathrm{Re}\, s > \alpha\}$, with $\alpha \in \mathbb{R} \cup \{\pm\infty\}$, on which the Lebesgue integral $\int_E \varphi(x)^s \mathrm{d}\mu(x)$ is convergent (i.e., $\zeta_{E,\varphi,|\mu|}(\mathrm{Re}\, s) = \int_E \varphi(x)^{\mathrm{Re}\, s} \mathrm{d}|\mu|(x) < \infty$) and hence, $f(s) = \zeta_{E,\varphi,\mu}(s)$ is well defined by Equation (A.1.2) and furthermore, the second equality of Equation (A.1.4) holds.

Theorem A.1.4. *If the DTI $f = \zeta_{E,\varphi,\mu}$ is tamed, then $D(f)$ and $\Pi(f)$ are well defined (in the above sense). In particular, $f(s) = \zeta_{E,\varphi,\mu}(s)$ is given by* (A.1.2) *for* $\mathrm{Re}\, s > D(f)$ *and cannot be defined (by a Lebesgue integral) in this manner on any strictly larger open right half-plane.*

Proof. See the proof of Theorem 2.1.45(a) on page 83 in Subsection 2.1.3.2, which can easily be adapted to the present, more general, situation. □

A.2 Basic Properties of DTIs

In this section, we show that a (pointwise) product of tamed DTIs is still a tamed DTI, and examine when the tameness condition is preserved. We also show that a linear combination of tamed DTIs based on the same underlying pair (E, φ) is again a tamed DTI. In Section A.3, we will use the results of this section regarding products of tamed DTIs, while in the later sections, we will use the results concerning linear combinations of DTIs.

Theorem A.2.1. *The (pointwise) product $h = f \cdot g$ of two tamed DTIs f and g is again a tamed DTI. (See Corollary A.2.2 below for a more specific statement.)*

Proof. Assume that $f = \zeta_{E,\varphi,\mu}$ and $g = \zeta_{F,\psi,\eta}$ are the tamed DTIs. We claim that $h = \zeta_{E\times F,\varphi\otimes\psi,\mu\otimes\eta}$, where $\varphi \otimes \psi : E \times F \to \mathbb{C}$ is the tensor product of the functions φ and ψ, defined by $(\varphi \otimes \psi)(x,y) = \varphi(x)\,\psi(y)$ for $(x,y) \in E \times F$, and $\mu \otimes \eta$ is the usual tensor product of the measures μ and η (also called the product measure and denoted by $\mu \times \eta$; see, e.g., [Coh, Foll, Ru] for the latter notion as well as for the Fubini theorem used below). It can be easily checked that if (as in the case here) μ and η are local measures on E and F, respectively, then $\mu \otimes \eta$ is a local measure on $E \times F$. (Recall that $\mathscr{B}(E \times F) = \mathscr{B}(E) \otimes \mathscr{B}(F)$.)

We leave it as an exercise to the interested reader to verify (by using Equation (A.1.1), in particular) that

$$|\mu \otimes \eta| \le |\mu| \otimes |\eta|; \tag{A.2.1}$$

i.e., $|\mu \otimes \eta|(B) \le (|\mu| \otimes |\eta|)(B)$ for every $B \in \mathscr{B}(E \times F)$. Let us simply mention that since the rectangles generate the Borel σ-algebra $\mathscr{B}(E \times F) = \mathscr{B}(E) \otimes \mathscr{B}(F)$, it suffices to check the latter equality for B a rectangle of the form $B = C \times D$, where $C \in \mathscr{B}(E)$ and $D \in \mathscr{B}(F)$. Furthermore, for similar reasons, in the counterpart of (A.1.1) defining $|\mu \otimes \eta|(B)$, it suffices to take partitions of $B = C \times D$ of the form $C_j \times D_j$, with $C_j \in \mathscr{B}(E)$ and $D_j \in \mathscr{B}(F)$.

The claim follows from a relatively routine use of the Fubini theorem (for the Lebesgue integral). Nevertheless, let us give some of the steps involved:

Let $s \in \mathbb{C}$ be such that $\mathrm{Re}\, s > \max\{D(f), D(g)\}$. Then (since $|\mu \otimes \eta| \leq |\mu| \otimes |\eta|$, in light of Equation (A.2.1)), we can use Definition A.1.2 and the Fubini–Tonelli theorem (see [Coh, Foll, Ru]) in order to obtain successively:

$$\begin{aligned} \zeta_{E\times F,\varphi\otimes\psi,|\mu\otimes\eta|}(\mathrm{Re}\, s) &= \int_{E\times F} ((\varphi\otimes\psi)(x,y))^{\mathrm{Re}\, s}(|\mu\otimes\eta|)(\mathrm{d}x,\mathrm{d}y) \\ &\leq \int_{E\times F} ((\varphi(x)\psi(y))^{\mathrm{Re}\, s}(|\mu|\otimes|\eta|)(\mathrm{d}x,\mathrm{d}y) \\ &= \int_{E\times F} \varphi(x)^{\mathrm{Re}\, s}\psi(y)^{\mathrm{Re}\, s}(|\mu|\otimes|\eta|)(\mathrm{d}x,\mathrm{d}y) \\ &= \int_E \varphi(x)^{\mathrm{Re}\, s}|\mu|(\mathrm{d}x)\int_F \psi(y)^{\mathrm{Re}\, s}|\eta|(\mathrm{d}y) \\ &= \zeta_{E,\varphi,|\mu|}(\mathrm{Re}\, s)\,\zeta_{F,\psi,|\eta|}(\mathrm{Re}\, s) < \infty. \end{aligned} \tag{A.2.2}$$

(The last inequality follows from the fact that the DTI's f and g are tamed; see Theorem 2.1.45(a) on page 81.) We may therefore now apply the classic Fubini theorem to the integral

$$\int_{E\times F} ((\varphi\otimes\psi)(x,y))^{\mathrm{Re}\, s}(\mu\otimes\eta)(\mathrm{d}x,\mathrm{d}y)$$

to deduce the desired claim. More specifically, let $s \in \mathbb{C}$ be such that $\mathrm{Re}\, s > \max\{D(f), D(g)\}$. Then, we have successively (much as in (A.2.2) above, except with s instead of $\mathrm{Re}\, s$ and μ, η and $\mu\otimes\eta$ instead of $|\mu|$, $|\eta|$ and $|\mu\otimes\eta|$, respectively):

$$\begin{aligned} h(s) &= \zeta_{E\times F,\varphi\otimes\psi,\mu\otimes\eta}(s) \\ &= \int_{E\times F} ((\varphi\otimes\psi)(x,y))^{s}(\mu\otimes\eta)(\mathrm{d}x,\mathrm{d}y) \\ &= \int_{E\times F} \varphi(x)^{s}\psi(y)^{s}(\mu\otimes\eta)(\mathrm{d}x,\mathrm{d}y) \\ &= \int_E \varphi(x)^{s}\mu(\mathrm{d}x)\int_F \psi(y)^{s}\eta(\mathrm{d}y) \\ &= \zeta_{E,\varphi,\mu}(s)\,\zeta_{F,\psi,\eta}(s) = f(s)\,g(s), \end{aligned} \tag{A.2.3}$$

as desired. Note that in (A.2.3), we have used Definition A.1.2 in the second and fifith equalities, while we have used Fubini's theorem in the fourth equality.

Observe that the application of the Fubini theorem (see, e.g., [Coh, Ru]) is justified by the computation and the conclusion of Equation (A.2.2) above (namely, $\zeta_{E,\varphi,\mu}(\mathrm{Re}\, s)\,\zeta_{E,\varphi,\mu}(\mathrm{Re}\, s) < \infty$ for $\mathrm{Re}\, s > \max\{D(f), D(g)\}$). This proves that $f \cdot g$ is a DTI. Further, note that (A.1.4) implies that $D(h) \leq \max\{D(f), D(g)\}$.

To show that the DTI $f \cdot g$ is tamed, recall that, since f and g are tamed DTIs, then there exist two positive contants C_φ and C_ψ such that $0 \le \varphi(x) \le C_\varphi$ $|\mu|$-a.e. on E and $0 \le \psi(x) \le C_\psi$ $|\eta|$-a.e. on F. It follows immediately that

$$0 \le (\varphi \otimes \psi)(x,y) := \varphi(x)\,\psi(y) \le C_\varphi C_\psi \tag{A.2.4}$$

$(|\mu| \otimes |\eta|)$-a.e. on $E \times F$, and hence, also $(|\mu \otimes \eta|)$-a.e. on $E \times F$ (because the inequality (A.2.1), $|\mu \otimes \eta| \le |\mu| \otimes |\eta|$, implies that sets of $(|\mu| \otimes |\eta|)$-measure zero are also of $|\mu \otimes \eta|$-measure zero).

Finally, in light of (A.2.4), we see that the DTI $h = \zeta_{E\times F,\varphi\otimes\psi,\mu\otimes\eta}$ is tamed and that $C(h) \le C(f)C(g)$.

This completes the proof of the theorem. □

The following result is really a direct consequence of the proof of Theorem A.2.1.

Corollary A.2.2. *If $f = \zeta_{E,\varphi,\mu}$ and $g = \zeta_{F,\psi,\eta}$ are tamed DTIs, then their (pointwise) product $f \cdot g$ coincides with the DTI $\zeta_{E\times F,\varphi\otimes\psi,\mu\otimes\eta}$,*

$$f \cdot g = \zeta_{E\times F,\varphi\otimes\psi,\mu\otimes\eta}, \tag{A.2.5}$$

and it is also tamed, with $C(fg) \le C(f)C(g)$. Moreover,

$$D(f \cdot g) \le \max\{D(f), D(g)\}. \tag{A.2.6}$$

Next, we consider the stability of the class of tamed DTIs under linear combinations. The following simple theorem shows that the class of tamed DTIs associated with the same underlying pair (E,φ) is a vector space.

Theorem A.2.3. *The class of tamed DTIs attached to the same pair (E,φ) is stable under linear combinations; i.e., it is a complex vector space. More specifically, if $f := \zeta_{E,\varphi,\mu}$ and $g := \zeta_{E,\varphi,\eta}$ and if $\alpha, \beta \in \mathbb{C}$, then $h := \alpha f + \beta g$ coincides with the following DTI $\zeta_{E,\varphi,\nu}$, where $\nu := \alpha\mu + \beta\eta$.*[4] *Furthermore, $D(h) \le \max\{D(f), D(g)\}$ and h is tamed, with $C(h) \le \max\{C(f), C(g)\}$.*

Proof. First, note that clearly, since μ and η are local measures on E, then so is $\nu = \alpha\mu + \beta\eta$. Next, for $s \in \mathbb{C}$ such that $\operatorname{Re} s > \max\{D(f), D(g)\}$, it is immediate to verify that $h(s) = \zeta_{E,\varphi,\nu}(s)$, with ν defined as above (so that $|\nu| \le |\alpha|\,|\mu| + |\beta|\,|\eta|$). Also,

$$\zeta_{E,\varphi,|\nu|}(\operatorname{Re} s) \le |\alpha|\zeta_{E,\varphi,|\mu|}(\operatorname{Re} s) + |\beta|\zeta_{E,\varphi,|\eta|}(\operatorname{Re} s) < \infty.$$

Hence, $D(h) \le \max\{D(f), D(g)\}$, as claimed.

Let us next prove that the functions h is tamed, in the sense of Definition A.1.3 above. Since $f = \zeta_{E,\varphi,\mu}$ and $g = \zeta_{E,\varphi,\nu}$ are tamed, there exist two nonnegative constants constant $C(f)$ and $C(g)$ such that

[4] If $\beta = 0$, but with g arbitrary, then obviously, we have $D(\alpha f) = D(f)$ for $\alpha \ne 0$, and $D(\alpha f) = -\infty$ for $\alpha = 0$. Hence, the inequality $D(h) \le \max\{D(f), D(g)\}$ may be strict (because $D(g)$ can be arbitrary).

$$0 \leq \varphi \leq C(f) \quad |\mu|\text{-a.e. on } E \tag{A.2.7}$$

and

$$0 \leq \varphi \leq C(g) \quad |\mu|\text{-a.e. on } E. \tag{A.2.8}$$

Let us show that

$$0 \leq \varphi \leq \max\{C(f), C(g)\} \quad |\nu|\text{-a.e. on } E. \tag{A.2.9}$$

To this aim, let $T_f := \{x \in E : \varphi(x) > C(f)\}$ and $T_g := \{x \in E : \varphi(x) > C(g)\}$.

Now, in light of (A.2.7) and (A.2.8), we have $|\mu|(T_f) = 0$ and $|\eta|(T_g) = 0$. Therefore,

$$(|\alpha|\,|\mu| + |\beta|\,|\eta|)(T_f \cap T_g) = 0 \tag{A.2.10}$$

and hence, since $|\nu| = |\alpha\mu + \beta\eta| \leq |\alpha|\,|\mu| + |\beta|\,|\eta|$ (in the sense of positive measures), we also have $|\nu|(T_f \cap T_g) = 0$. Since $T_f \cap T_g = \{x \in E : \varphi(x) > T_h\}$, where $T_h := \max\{C(f), C(g)\}$, this implies the second inequality in (A.2.9); i.e., $\varphi \leq \max\{C(f), C(g)\}$ $|\nu|$-a.e. on E and hence, $C(h) \leq \max\{C(f), C(g)\}$.

We conclude the proof by establishing the first inequality in (A.2.9). It suffices to let $\mathscr{T} := \{x \in E : \varphi(x) < 0\}$; so that, due to (A.2.7) and (A.2.8), we have $|\mu|(\mathscr{T}) = |\eta|(\mathscr{T}) = 0$. Hence, since $0 \leq |\nu| = |\alpha\mu + \beta\eta| \leq |\alpha|\,|\mu| + |\beta|\,|\eta|$, we also have $|\nu|(\mathscr{T}) = 0$. This statement, in turn, precisely means that $0 \leq \varphi$ $|\nu|$-a.e. in E, which is the sought for first inequality in (A.2.9).

This completes the proof of the theorem. □

In light of Theorem A.2.3 just above, the following result is of interest. Note that it suplements large parts of Examples 2.1.40, 2.1.41, 2.1.43 and 2.1.44 in Subsection 2.1.3.2, as well as Lemma 2.2.9. (See also Corollary A.2.7 below.)

In view of part (1) just below, recall that, according to Definition 4.1.2, a relative fractal drum (A, Ω) is defined as an ordered pair of two subsets A and Ω of $\mathbb{R}^N$ such that Ω is open, $|\Omega|_N < \infty$ and $\Omega \subseteq A_\delta$ for some $\delta > 0$.

Furthermore, in view of part (2) below, recall from [Lap-vFr3, Section 4.1] that a generalized fractal string η is a local (positive or complex) measure on $(0, +\infty)$, which is supported on $[x_0, +\infty)$ for some $x_0 > 0$. Furthermore, by definition, we have $\zeta_\eta(s) := \int_0^{+\infty} x^{-s}\eta(\mathrm{d}x)$ for all $s \in \mathbb{C}$ with $\operatorname{Re} s$ sufficiently large.

Proposition A.2.4. *All of the fractal zeta functions used in this book are tamed DTIs.*

(1) *More specifically, the distance zeta function $\zeta_{A,\Omega}$ and the tube zeta function $\tilde{\zeta}_{A,\Omega}$ of a relative fractal drum (A, Ω) in $\mathbb{R}^N$ are tamed. In particular, for any $\delta > 0$, this is the case of $\zeta_A = \zeta_{A,A_\delta}$ and $\tilde{\zeta}_A = \tilde{\zeta}_{A,A_\delta}$, the distance and tube zeta functions of an arbitrary bounded subset A of $\mathbb{R}^N$.*

(2) *Moreover, if $\zeta_{\mathscr{L}}$ is the geometric zeta function of a bounded fractal string $\mathscr{L} = (\ell_j)_{j=1}^\infty$, with $\ell_j \downarrow 0^+$ as $j \to \infty$, then $\zeta_{\mathscr{L}}$ is a tamed DTI. More generally, the geometric zeta function ζ_η of a generalized fractal string η is a tamed DTI.*

Proof. (1) Since $\zeta_{A,\Omega}(s) := \int_\Omega d(x,A)^{s-N}\mathrm{d}x = \int_\Omega d(x,A)^s d(x,A)^{-N}\mathrm{d}x$ for all $s \in \mathbb{C}$ such that $\mathrm{Re}\, s > \overline{\dim}_B(A,\Omega)$, we can write $\zeta_{A,\Omega}(s)$ as a DTI:

$$\zeta_{A,\Omega}(s) = \zeta_{E,\varphi,\mu}(s),$$

where $E := \Omega$, $\varphi(x) := d(x,A)$ and $\mu(\mathrm{d}x) := d(x,A)^{-N}\mathrm{d}x$. This DTI is tamed, since $\varphi(x) \le \delta$ for all $x \in E$, where the constant $\delta > 0$ is from the definition of the RFD (A,Ω), that is, such that $\Omega \subseteq A_\delta$.

Similarly, the tube zeta function $\tilde{\zeta}_{A,\Omega}(s) := \int_0^\delta t^{s-N-1}|A_t \cap \Omega|\,\mathrm{d}t$ can be written as a DTI:

$$\tilde{\zeta}_{A,\Omega}(s) = \zeta_{E,\varphi,\mu}(s),$$

where $E := (0,\delta]$, $\varphi(t) := t$ and $\mu(\mathrm{d}t) := t^{-N}|A_t \cap \Omega|\,\mathrm{d}t/t$. Then, obviously, we have $\varphi(t) \le \delta$ for all $t \in E$ and therefore, $\tilde{\zeta}_{A,\Omega}(s)$ is a tamed DTI.

Note that one could also let $E := \Omega \setminus \overline{A}$, $\varphi(x) := d(x,A)$ for $x \in E$ and $\varphi(x) = 0$ for $x \in \overline{A}$, with N as above, in order to define $\zeta_{A,\Omega}$ and $\tilde{\zeta}_{A,\Omega}$.

(2) Since $\zeta_{\mathscr{L}}(s) := \sum_{j=1}^\infty \ell_j^s$ for all $s \in \mathbb{C}$ with $\mathrm{Re}\, s > D(\zeta_{\mathscr{L}})$, we can write the geometric zeta function of the fractal string $\mathscr{L}$ as a DTI:

$$\zeta_{\mathscr{L}}(s) = \zeta_{E,\varphi,\mu}(s),$$

where $E := [x_0,+\infty)$ with $x_0 := 1/\ell_1$, $\varphi(x) := 1/x$ for all $x \in E$, and $\mu(\mathrm{d}x) := \sum_{j=1}^\infty \delta_{1/\ell_j}$. The DTI is tamed since $\varphi(x) \le 1/x_0 = \ell_1$ for all $x \in E$. Finally, note that $D(\zeta_{\mathscr{L}}) = D(\zeta_{E,\varphi,\mu})$.

The reasoning is similar for ζ_η, the geometric zeta function of a generalized fractal string η. Indeed, it is a DTI, since

$$\zeta_\eta(s) := \int_{x_0}^{+\infty} x^{-s}\eta(\mathrm{d}x) = \zeta_{E,\varphi,\mu}(s),$$

where $E := [x_0,+\infty)$, $\varphi(x) := 1/x$ for all $x \in E$ and $\mu := \eta$. The DTI is tamed, because $\varphi(x) \le 1/x_0$ for all $x \in E$. Again, note that $D(\zeta_\eta) = D(\zeta_{E,\varphi,\mu})$.

This concludes the proof of the proposition. ☐

Remark A.2.5. It is useful to note that the zeta function $\zeta_{A,\Omega}$ of an RFD (A,Ω) in part (1) of Proposition A.2.4 can be related to another DTI $\zeta_{F,\psi,\eta}$ as follows:

$$\zeta_{A,\Omega}(s) := \int_\Omega d(x,A)^{s-N}\mathrm{d}x = \zeta_{F,\psi,\eta}(s-N), \tag{A.2.11}$$

where $F := \Omega$, $\psi(x) := d(x,A)$ and $\eta(\mathrm{d}x) := \mathrm{d}x$ is the usual Lebesgue measure on Ω.

Similarly, the tube zeta function $\tilde{\zeta}_{A,\Omega}$ of an RFD (A,Ω) in part (1) of Proposition A.2.4 can be related to the following DTI $\zeta_{F,\psi,\eta}$:

$$\tilde{\zeta}_{A,\Omega}(s) := \int_0^\delta t^{s-N-1}|A_t \cap \Omega|\,\mathrm{d}t = \zeta_{F,\psi,\eta}(s-N-1), \tag{A.2.12}$$

where $F := \Omega$, $\psi(x) := t$ and $\eta(\mathrm{d}x) := |A_t \cap \Omega|\,\mathrm{d}x$.

We close this section by stating the following result, which is a consequence of Theorem 2.1.47 (the classic theorem about the analyticity of an integral depending holomorphically on a parameter).

Theorem A.2.6. *Let $f = \zeta_{E,\varphi,\mu}$ be a tamed DTI, so that $D(f)$ and $\Pi(f)$ are well defined. Furthermore, assume, for simplicity, that $\varphi(x) > 0$ for $|\mu|$-a.e. $x \in E$.*[5] *Then, f is holomorphic on the half-plane of convergence $\Pi(f) := \{\mathrm{Re}\, s > D(f)\}$ and for all $s \in \mathbb{C}$ with $\mathrm{Re}\, s > D(f)$, its derivative is given by*

$$f'(s) = \int_E \varphi(x)^s \log \varphi(s)\, \mu(\mathrm{d}x) = \zeta_{E,\varphi,\eta}(s), \tag{A.2.13}$$

provided φ is μ-essentialy locally bounded away from zero (i.e., for every compact subset K of E there exists a positive constant c_K such that $c_K \leq \varphi(x)$ for $|\mu|$-a.e. $x \in K$),[6,7] *where η is the (positive, signed or complex) local measure given by $\eta(\mathrm{d}x) := \log \varphi(x)\, \mu(\mathrm{d}x)$.*

Furthermore, under the same assumptions, the DTI $f' = \zeta_{E,\varphi,\eta}$ is tamed becasue the DTI $f = \zeta_{E,\varphi,\mu}$ is tamed. It follows that

$$D_{\mathrm{hol}}(f) \leq D(f). \tag{A.2.14}$$

Note that in Theorem A.2.6, η would be a positive measure if μ were assumed to be positive and $\varphi(x) \geq 1$ μ-a.e. on E.

The following result is an immediate consequence of Theorem A.2.6, taking into account the last two footnotes to the statement of the theorem (specifically, footnotes 6 and 7 on this page).

Corollary A.2.7. *Let $f = \zeta_{E,\varphi,\mu}$ be a tamed DTI. Then f is holomorphic on its half-plane of convergence $\Pi(f) := \{\mathrm{Re}\, s > D(f)\}$ and in that half-plane, we have*

$$f'(s) = \int_E \varphi(x)^s \log \varphi(s)\, \mu(\mathrm{d}x) =: \zeta_{E,\varphi,\eta}(s). \tag{A.2.15}$$

Furthermore,

$$D_{\mathrm{hol}}(f) \leq D(f). \tag{A.2.16}$$

---

[5] We have seen in Chapter 2 that there are situations where this additional assumption does not have to be made. For example, in the case of the distance (or the tube) zeta function of a bounded subset of $\mathbb{R}^N$. Note that for the case of ζ_A, we have $\varphi(x) := d(x,A)$, so that $\varphi(x) > 0$ $|\mu|$-a.e. is equivalent to $|A| = 0$.

[6] This additional hypothesis on φ is only assumed to guarantee that η is a (that is, locally bounded) local signed measure, and hence, that $f' = \zeta_{E,\varphi,\eta}$ is a bona fide DTI (in the sense of Definition A.1.2). It is *not* needed to guarantee that f is holomorphic on $\{\mathrm{Re}\, s > D(f)\}$ and that f' is given by the first equality of (A.2.13); likewise, it is not needed to conclude that the inequality (A.2.14) holds, namely, that $D_{\mathrm{hol}}(f) \leq D(f)$.

[7] This condition on φ is satisfied if φ is a continuous function on E. Indeed, on a given compact subset K of E, it suffices to set $c_K := \min_K \varphi$. Then, clearly, we have $c_K > 0$.

Moreover, if f is a continuous function on E, then $f' = \zeta_{E,\varphi,\eta}$ is a local tamed DTI, associated with the (positive, signed or complex) local measure $\eta(\mathrm{d}x) := \log\varphi(x)\,\mu(\mathrm{d}x)$.

Remark A.2.8. We have seen in Chapter 2 that the inequality (A.2.14) of Theorem A.2.6 is sharp, in general. For example, under the assumptions of part (c) of Theorem 2.1.11 and when f is the distance zeta function of a suitable bounded subset of $\mathbb{R}^N$, we have $D_{\mathrm{hol}}(f) = D(f)$. On the other hand, there are many examples of tamed DTIs for which $D_{\mathrm{hol}}(f) < D(f)$. This is also the case if f is a Dirichlet L-function with a nontrivial primitive character, which is an ordinary Dirichlet series with complex coefficients and hence, is a tamed DTI, being the Mellin transform of a Dirac comb (a weighted countable sum of Dirac measures). Thus $f(s) = L(s,\chi) := \sum_{n=1}^{\infty}\chi(n)n^{-s}$ for $\mathrm{Re}\,s > 1$, where the "character" χ is a completely multiplicative function from $\mathbb{Z}$ to the unit circle in the complex plane and is periodic modulo some integer $m \geq 1$. (Furthermore, $\chi \not\equiv 1 \mod m$.) Then, $D(f) = 1$ and $D_{\mathrm{hol}}(f) \leq 0$; see, e.g., [HardWr] and [Ser, Sections VI.2 and VI.3]. If, in addition, the underlying character χ is primitive, then $f = L(\cdot\,,\chi)$ is an entire function, so that $D_{\mathrm{hol}}(f) = -\infty$ whereas $D(f) = 1$. (For instance, let $f(s) := \sum_{n=1}^{\infty}(-1)^{n-1}n^{-s}$ for $\mathrm{Re}\,s > 1$; see Equation (2.1.39) in Remark 2.1.29 on page 69.)

A.3 New Examples of DTIs

In Subsection 2.1.3.2, especially in Examples 2.1.40, 2.1.41, 2.1.43 and 2.1.44, we have given many examples of tamed DTIs, including all of the fractal zeta functions considered in this book. (See also Proposition A.2.4 above.) The goal of the present section is to provide new examples of DTIs, of a rather different nature but which, when combined with a suitable change of variable (see the next section, Section A.4), will play an important role for the modified equivalence relation defined in Section A.4 below.

The next result will provide the key step in the proof of the main result of this section, namely, Theorem A.3.2.

Lemma A.3.1. *Let $a \in \mathbb{C}$. Then $f(s) := 1/(s-a)$ is a tamed DTI.*

More specifically, we have $f = \zeta_{E,\varphi,\mu}$, where $E := [1,+\infty)$, $\varphi(x) := 1/x$ for all $x \in E$, and $\mu(\mathrm{d}x) := x^{a-1}\mathrm{d}x = x^a\mathrm{d}x/x$ (so that $|\mu|(\mathrm{d}x) = x^{\mathrm{Re}\,a-1}\mathrm{d}x = x^{\mathrm{Re}\,a}\mathrm{d}x/x$). Furthermore, we have $D_{\mathrm{mer}}(f) = -\infty$ and[8]

$$D_{\mathrm{hol}}(f) = D(f) = \mathrm{Re}\,a. \qquad \text{(A.3.1)}$$

[8] More specifically, the DTI $\zeta_{E,\varphi,\mu}$ can be meromorphically continued to all of $\mathbb{C}$ and $\zeta_{E,\varphi,\mu}(s) = f(s) := 1/(s-a)$ for all $s \in \mathbb{C}$.

Proof. A simple computation yields that for all $s \in \mathbb{C}$ with $\mathrm{Re}\, s > \mathrm{Re}\, a$,

$$\begin{aligned} f(s) = \zeta_{E,\varphi,\mu}(s) &= \int_1^{+\infty} x^{-s+a-1} \mathrm{d}x = \frac{x^{a-s}}{a-s}\Big|_{x=1}^{+\infty} \\ &= \frac{1}{a-s}(0-1) = \frac{1}{s-a}. \end{aligned} \tag{A.3.2}$$

Similarly,

$$f(\mathrm{Re}\, s) = \zeta_{E,\varphi,|\mu|}(\mathrm{Re}\, s) = \frac{1}{\mathrm{Re}\, s - \mathrm{Re}\, a} \quad \text{for} \quad \mathrm{Re}\, s > \mathrm{Re}\, a,$$

whereas

$$f(\mathrm{Re}\, s) = \zeta_{E,\varphi,|\mu|}(\mathrm{Re}\, s) = +\infty \quad \text{for} \quad \mathrm{Re}\, s \leq \mathrm{Re}\, a;$$

indeed, for $\mathrm{Re}\, s = \mathrm{Re}\, a$, $\zeta_{E,\varphi,\mu}(\mathrm{Re}\, a) = \int_1^{+\infty} \frac{\mathrm{d}x}{x} = \log x \,|_{x=1}^{+\infty} = +\infty$, while for $\mathrm{Re}\, s < \mathrm{Re}\, a$, $\zeta_{E,\varphi,\mu}(\mathrm{Re}\, s) = \frac{x^{\mathrm{Re}\, a - \mathrm{Re}\, s}}{\mathrm{Re}\, a - \mathrm{Re}\, s}\Big|_{x=1}^{+\infty} = +\infty$. From the definition of $D(f)$ given in (A.1.4), it then follows that $D(f) = D(\zeta_{E,\varphi,\mu}) = \mathrm{Re}\, a$.

Also, in light of (A.3.2), we see that $f(s) = 1/(s-a)$, first for $\mathrm{Re}\, s > \mathrm{Re}\, a$ and then, upon anlytic continuation, for all $s \in \mathbb{C}$. It follows that we also have $D_{\mathrm{hol}}(f) = \mathrm{Re}\, a$ since clearly, $f = f(s)$ has a pole at $s = a$. Therefore, (A.3.1) holds.

Finally, we check that the DTI $f = \zeta_{E,\varphi,\mu}$ is tamed, as claimed. But this is obvious since as we have seen just above, $E := [1, +\infty)$ and $\varphi(x) := 1/x$ for all $x \in E$, so that $\varphi(x) \leq 1$ on E. □

We can now state the main result of this section.

Theorem A.3.2. *Let $P \in \mathbb{C}[X]$ be an arbitrary polynomial with complex coefficients. Then $f(s) := 1/P(s)$ is a tamed DTI. (See Corollary A.3.3 below for a more specific statement.)*

Proof. Without loss of generality, we may assume (for notational simplicity) that P is a monic polynomial (i.e., its leading coefficient is equal to 1). Denote by n the degree of P.

Case 1: If $n = 0$, then $P(s) \equiv 1$, $f(s) \equiv 1$ and hence (since $1 = \int_{[1,+\infty)} 1^s \mathrm{d}\delta_1(x)$), we have $f = \zeta_{E,\varphi,\mu}$, where $E = [1, +\infty)$, $\varphi(x) \equiv 1$ and $\mu := \delta_1$, the Dirac measure concentrated at $x = 1$. Clearly, f is a tamed DTI, since $\varphi(x) \equiv 1$.

Case 2: Let us now assume that $n \geq 1$. Then, by the fundamental theorem of algebra (and since P is monic), we have that

$$P(s) = \prod_{m=1}^{n} (s - a_m),$$

where the complex numbers $a_1, \ldots a_n$ are the roots of P, repeated according to their multiplicities.

Therefore, in light of Theorem A.2.1 (according to which a product of tamed DTIs is again a tamed DTI) and its associated Corollary A.2.2, combined with Lemma A.3.1 just above (applied repeatedly, that is, n times) we deduce that $f(s) = 1/P(s)$ is a tamed DTI. More specifically, in light of Lemma A.3.1, we have that for each $m = 1,\dots,n$,

$$f_n(s) := \frac{1}{s - s_m} = \zeta_{E,\varphi,\mu_m},$$

where $E := [0,+\infty)$, $\varphi(x) := x^{-1}$ for $x \in E$ and $\mu_m(\mathrm{d}x) := x^{a_m - 1}\mathrm{d}x = x^{a_m}\mathrm{d}x/x$. Hence, according to Theorem A.2.1 and Corollary A.2.2 (applied n times), we have

$$f(s) = \frac{1}{P(s)} = \prod_{m=1}^{n} \frac{1}{s - a_m} = \zeta_{F,\Phi,\mu},$$

where $F := [1,+\infty)^n$, $\Phi : F \to \mathbb{R}$, $\Phi(x_1,\dots,x_n) = \varphi(x_1)\cdots\varphi(x_n) = (x_1\cdots x_n)^{-1}$, $\mu := \mu_1 \otimes \cdots \otimes \mu_n$ (so that $\mu(\mathrm{d}x_1,\dots,\mathrm{d}x_n) = x_1^{a_1-1}\mathrm{d}x_1 \dots x_n^{a_n-1}\mathrm{d}x_n$).

Since, obviously, $\Phi(x_1,\dots,x_n) = (x_1\cdots x_n)^{-1} \le 1$ for all $(x_1,\dots,x_n) \in F = [0,+\infty)^n$, the DTI $f = \zeta_{F,\Phi,\mu}$ is tamed.

This completes the proof of the theorem. □

The following result is really a direct consequence of the proof of Theorem A.3.2 combined (for the last part of Corollary A.3.3) with Theorem A.2.6 (as will be further explained just after the statement of Corollary A.3.3).

Corollary A.3.3. *Let P be a polynomial of degree $n \ge 1$. Then, for all $s \in \mathbb{C}$, we have*[9]

$$f(s) := \frac{1}{P(s)} = \zeta_{F,\Phi,\mu}(s), \tag{A.3.3}$$

where $F := [1,+\infty)^n$, $\Phi : F \to \mathbb{R}$, $\Phi(x_1,\dots,x_n) = \varphi(x_1)\cdots\varphi(x_n) := (x_1\cdots x_n)^{-1}$ for all $(x_1,\dots,x_n) \in F$, and

$$\mu(\mathrm{d}x_1,\dots,\mathrm{d}x_n) := c\cdot x_1^{a_1-1}\mathrm{d}x_1 \dots x_n^{a_n-1}\mathrm{d}x_n.^{10} \tag{A.3.4}$$

Here, $P(s) = c\Pi_{m=1}^{n}(s - a_m)$, $c \neq 0$, $c = \frac{1}{n!}P^{(n)}(0)$ and $a_1,\dots,a_n \in \mathbb{C}$ are the roots of P (repeated according to their multiplicities). Moreover,

$$D_{\mathrm{hol}}(f) = D(f) = \max\{\mathrm{Re}\,a_1,\dots,\mathrm{Re}\,a_n\}. \tag{A.3.5}$$

Finally, for all $s \in \mathbb{C}$ with $\mathrm{Re}\,s > D(f)$, f is holomorphic with derivative given by the tamed DTI:

$$f'(s) = \zeta_{E,\Phi,\eta}(s), \tag{A.3.6}$$

where $\eta(\mathrm{d}x) := \log\Phi(x)\,\mu(\mathrm{d}x)$ with $x := (x_1,\dots,x_n) \in F := [1,+\infty)^n$.

---

[9] More specifically, the DTI $\zeta_{F,\Phi,\mu}$ can be meromorphically continued to all of $\mathbb{C}$ and its analytic extension coincides with f on all of $\mathbb{C}$; namely, Equation (A.3.3) holds for all $s \in \mathbb{C}$.

[10] It follows that $|\mu|(\mathrm{d}x_1,\dots,\mathrm{d}x_n) := |c|\cdot x_1^{\mathrm{Re}\,a_1-1}\mathrm{d}x_1 \dots x_n^{\mathrm{Re}\,a_n-1}\mathrm{d}x_n$.

We note that the last part of Corollary A.3.3 follows from Corollary A.2.7, including the last part of that corollary (since $\Phi(x) = (x_1 \cdots x_n)^{-1}$ for all $x := (x_1, \ldots, x_n) \in F$ is a positive and continuous function on $F := [1, +\infty)^n$).

Remark A.3.4. We can use the fact that $\mathrm{d}x/x$ is the natural Haar measure on the multiplicative group $(0, +\infty)$ in order to rewrite $1/P(s)$ in many different, but completely equivalent ways, as a tamed DTI, assuming $n \geq 1$ (as in Corollary A.3.3). For example, if we make the change of variables $(y_1, \ldots, y_n) = (x_1^{-1}, \ldots, x_n^{-1})$, $(\mathrm{d}y_1/y_1, \ldots, \mathrm{d}y_n/y_n) = (-\mathrm{d}x_1/x_1, \ldots, -\mathrm{d}x_n/x_n)$, we obtain (using Corollary A.3.3):

$$f(s) := \frac{1}{P(s)} = \zeta_{F^*, \Phi^*, \mu^*}(s), \tag{A.3.7}$$

where $F^* := (0, 1]^n$, $\Phi^*(y_1, \ldots, y_n) := y_1 \cdots y_n$, and

$$\begin{aligned} \mu^*(\mathrm{d}y_1, \ldots, \mathrm{d}y_n) &:= c y_1^{-a_1 - 1} \mathrm{d}y_1 \cdots y_n^{-a_n - 1} \mathrm{d}y_n \\ &= c y_1^{-a_1} \frac{\mathrm{d}y_1}{y_1} \cdots y_n^{-a_n} \frac{\mathrm{d}y_n}{y_n}. \end{aligned} \tag{A.3.8}$$

Since $\Phi^*(y) \leq 1$ for all $y \in F^*$, the DTI $\zeta_{F^*, \Phi^*, \mu^*}$ is tamed as well.

Of course, many other equivalent ways of rewriting $f(s) := 1/P(s)$ as a tamed DTI are obtained by applying the change of variable $y_j = x_j^{-1}$ to some of the variables x_j ($j \in \{1, \ldots, n\}$) but not to others (as well as by permuting the variables in an arbitrary manner).

A.4 Extended Dirichlet-Type Integrals

In this section, we define the notion of an extended DTI (EDTI, in short) which will be the key to the modification of the equivalence relation $\sim$ to be defined in the next section.

Definition A.4.1. Given $r \in (0, 1)$, an *extended DTI of base* r is a function $g = g(s)$ of the form

$$g(s) := \zeta_{E, \varphi, \mu}(r^{-s}) \tag{A.4.1}$$

where $f(s) := \zeta_{E, \varphi, \mu}(s)$ is a standard DTI (in the sense of Definition A.1.2).

An extended DTI $g(s) := \zeta_{E, \varphi, \mu}(r^{-s})$ of base r is said to be *tamed* if the associated DTI $f(s) := \zeta_{E, \varphi, \mu}(s)$ enjoys the same property (in the sense of Definition A.1.3). However, contrary to intuition, the function $g(s)$ is not always holomorphic on an open right half-plane of the form $\{\mathrm{Re}\, s > \alpha\}$ with $\alpha \in \mathbb{R} \cup \{\pm\infty\}$, as shown by Proposition A.4.2 below. Moreover, there are examples of extended DTIs of base $r \in (0, 1)$ which are not holomorphic on *any* open right half-plane with $\alpha \in \mathbb{R} \cup \{-\infty\}$ (alternatively, we can write $D(g) = D_{\mathrm{hol}}(g) = +\infty$). In particular, the right open half-plane $\Pi(g) := \{\mathrm{Re}\, s > D(g)\}$ of (absolute) convergence of g is then equal to the empty set.

Proposition A.4.2. *Let $g = g(s)$ be an EDTI with base $r \in (0,1)$, generated by a tamed DTI $f := \zeta_{E,\varphi,\mu}$, as in Definition A.4.1. Then the following properties hold (see also Figure A.1):*

(a) If $D(f) > 0$, then g converges (absolutely) for all complex numbers $s := x + \mathrm{i}y$ in the open set $\Pi'(g) \subset \mathbb{C}$ defined by the following conditions

$$x > \frac{\log D(f) - \log\cos\left((\log r^{-1})\cdot y\right)}{\log r^{-1}}$$
$$y \in \bigcup_{k\in\mathbb{Z}} \Big\{\Big(-\frac{\pi}{2\log r^{-1}}, \frac{\pi}{2\log r^{-1}}\Big) + \frac{2\pi k}{\log r^{-1}}\Big\}. \tag{A.4.2}$$

The open set $\Pi'(g)$ of absolute convergence of g has countably many open connected components (corresponding to each $k \in \mathbb{Z}$) and is contained in the open right half-plane

$$\Big\{\mathrm{Re}\, s > \frac{\log D(f)}{\log r^{-1}}\Big\}. \tag{A.4.3}$$

In particular, $D(g) = +\infty$ and the boundary of the set $\Pi'(g)$ is described by the following equation:

$$x = \frac{\log D(f) - \log\cos\left((\log r^{-1})\cdot y\right)}{\log r^{-1}}, \tag{A.4.4}$$

where y takes the values indicated in (A.4.2).

(b) If $D(f) < 0$, then the convergence set $\Pi'(g)$ of the function $g = g(s)$ is described by

$$x < \frac{\log|D(f)| - \log\cos\left((\log r^{-1})\cdot y + \frac{\pi}{2}\right)}{\log r^{-1}}$$
$$y \in \bigcup_{k\in\mathbb{Z}} \Big\{\Big(-\frac{\pi}{\log r^{-1}}, 0\Big) + \frac{2\pi k}{\log r^{-1}}\Big\}. \tag{A.4.5}$$

In particular, $D(g) = +\infty$ and $\Pi'(g)$ is contained in the open left half-plane

$$\Big\{\mathrm{Re}\, s < \frac{\log|D(f)|}{\log r^{-1}}\Big\}. \tag{A.4.6}$$

The boundary of $\Pi'(g)$ can be described analogously as in case (a).

(c) If $D(f) = 0$, then $\Pi'(g)$ is defined by

$$x \in \mathbb{R}, \quad y \in \bigcup_{k\in\mathbb{Z}} \Big\{\Big(-\frac{\pi}{2\log r^{-1}}, \frac{\pi}{2\log r^{-1}}\Big) + \frac{2\pi k}{\log r^{-1}}\Big\}. \tag{A.4.7}$$

In particular, $D(g) = +\infty$ and $\Pi'(g)$ is equal to the disjoint union of countably many translation invariant equidistant open horizontal strips of the form $\mathbb{R} \times I_k$, with $k \in \mathbb{Z}$, where I_k is the (translated) open interval defined between curly brackets *in* (A.4.7).[11] *The boundary of* $\Pi'(g)$ *is equal to the family of equidistant horizontal lines; it contains the line* $y = \pi/(2\log r^{-1})$ *and the distance between any two consecutive lines in the family is* $\pi/(\log r^{-1})$.

The bounds defining the convergence set $\Pi'(g)$ *in cases* (a), (b) *and* (c) *are optimal for the class of tamed extended DTIs; i.e., they cannot be improved.*

Proof. The condition of absolute convergence of $g(s) := \int_E \varphi(x)^{r^{-s}} \mu(\mathrm{d}x)$ is

$$\int_E |\varphi(x)^{r^{-s}}|\,|\mu|(\mathrm{d}x) = \int_E \varphi(x)^{r^{-\mathrm{Re}\,s}\cos((\log r^{-1})\,\mathrm{Im}\,s)}|\mu|(\mathrm{d}x) < \infty, \tag{A.4.8}$$

and it is equivalent to

$$r^{-x}\cos\left((\log r^{-1})y\right) > D, \tag{A.4.9}$$

where we have let $D := D(f)$ and $s := x + \mathrm{i}y$, with $x := \mathrm{Re}\,s$ and $y := \mathrm{Im}\,s$. We consider the following cases:

(a) If $D > 0$, then $r^x < \frac{1}{D}\cos\left((\log r^{-1})y\right)$, that is,

$$x\log r < \log\left(\frac{1}{D}\cos\left((\log r^{-1})y\right)\right),$$

and the claim follows by dividing by $\log r < 0$.

(b) If $D < 0$, then multiplying (A.4.9) by -1 we deduce that

$$r^x|D| > -\cos\left((\log r^{-1})y\right) = \cos\left((\log r^{-1})\cdot y + \frac{\pi}{2}\right).$$

The claim then follows similarly as in case (a).

(c) If $D = 0$, then (A.4.9) reduces to $\cos\left((\log r^{-1})y\right) > 0$ and $x \in \mathbb{R}$.

The optimality of the bounds obtained in cases (a), (b) and (c) can be verified by considering an extended DTI $g(s) = f(r^{-s})$ generated by a DTI f of the form of the distance zeta function $f := \zeta_A$, where A is a maximal hyperfractal (in the sense of Definition 4.6.23(iii)). Recall that for maximal hyperfractal sets A, the associated critical line $\{\mathrm{Re}\,s = D(\zeta_A)\}$ consists of nonisolated singularities of ζ_A. In particular, ζ_A does not have a meromorphic extension to a connected open set containing any given compact subinterval of the critical line $\{\mathrm{Re}\,s = D(\zeta_A)\}$. The existence of such sets has been proved in Corollary 4.6.17. In the context of the present proposition, if the extended DTI g had a meromorphic extension containing a compact connected subset of the boundary of $\Pi'(g)$, then this would imply that f could be meromorphically extended to a connected open set containing the corresponding compact interval on the critical line $\{\mathrm{Re}\,s = D(f)\}$, which is impossible.

This completes the proof of the proposition. ☐

[11] In particular, the set $\Pi'(g)$ is not contained in any left or right open half-plane of the form $\{\mathrm{Re}\,s < \alpha\}$ or $\{\mathrm{Re}\,s > \alpha\}$, respectively, for some $\alpha \in \mathbb{R}$.

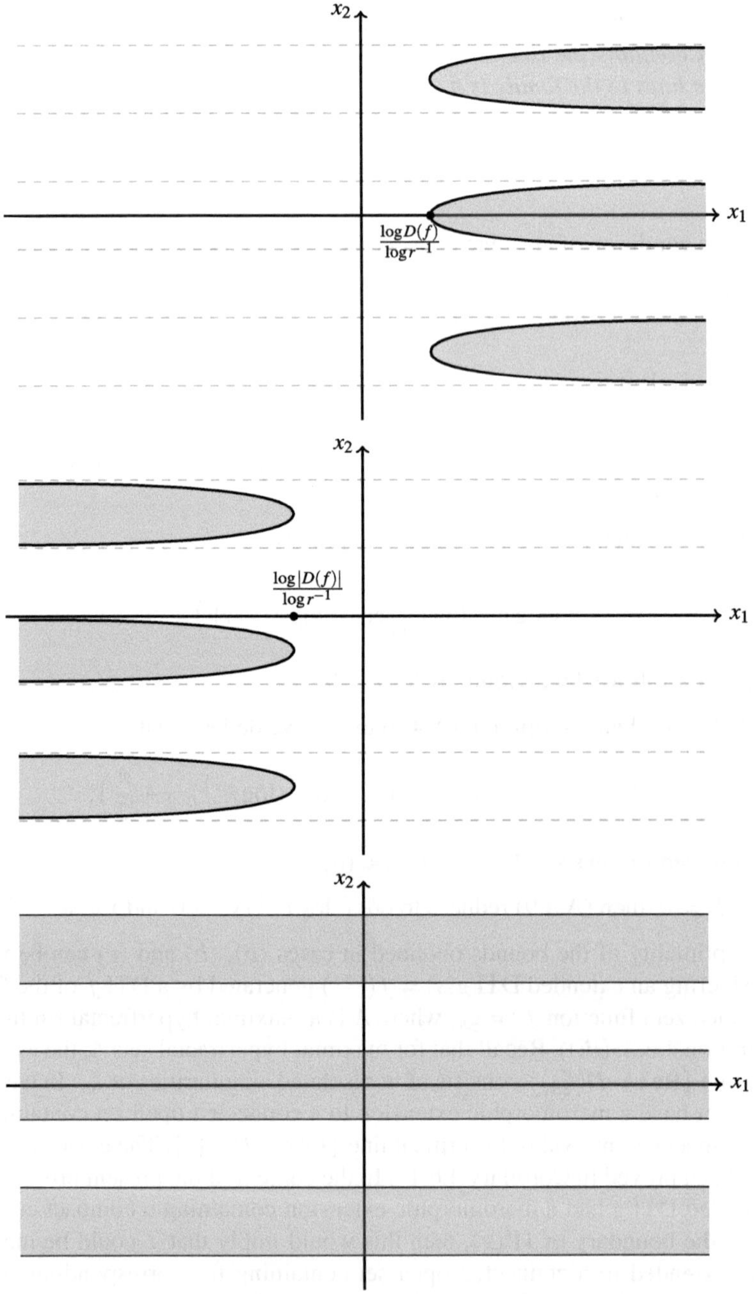

Fig. A.1 Typical domains of (absolute) convergence of an EDTI $g = g(s)$ of type II appearing in cases (a), (b) and (c) of Proposition A.4.2.

Remark A.4.3. It would be of interest to know if the optimality condition for tamed extended DTIs g appearing in Proposition A.4.2 is in some sense generic, i.e., satisfied for some "large" set of tamed extended DTIs g.

Remark A.4.4. As we have seen in Proposition A.4.2, for any EDTIs h of type II, the corresponding (absolute) convergence set $\Pi'(h)$ is never of the form of a nonempty open right half-plane. This class of EDTIs of type II includes, for example, the functions of the form

$$h(s) := \frac{\rho(s)}{a - b \cdot r^{-s}}, \tag{A.4.10}$$

where a and b are positive real numbers, $r \in (0,1)$, and ρ is a nowhere vanishing entire function. On the other hand, note that a function of the form given in (A.4.10) appears in the expression of the distance zeta functon ζ_A of a generalized Cantor set $C^{(m,a)}$; see Equation (3.1.5) on page 188 appearing in Proposition 3.1.1. In this case we view the function h defined by (A.4.10) as an EDTI of type I (i.e., as a geometric zeta function generated by the bounded fractal string $\mathscr{L}$ corresponding to $C^{(m,a)}$) and according to Proposition 3.1.1, the resulting open right half-plane of convergence is nontrivial and equal to $\{\operatorname{Re} s > \log_{1/a} m\}$. This example shows that the intersection of the class of EDTIs of type I and of the class of EDTIs of type II is not empty, and that one must always specify how to view a given EDTI in the intersection of these two classes.

We are now ready to give the general definition of an extended DTI (also abbreviated as an EDTI). As will be explained, we will distinguish between EDTI of type I (e.g., given by (A.4.11) below) or of type II (e.g., given by (A.4.12) below).

Definition A.4.5. An *extended Dirichlet-type integral* (in short, an *extended* DTI or simply, an EDTI) $h = h(s)$ is of the form

$$h(s) := \rho(s)\, \zeta_{E,\varphi,\mu}(s) \tag{A.4.11}$$

or of the form

$$h(s) := \rho(s)\, \zeta_{E,\varphi,\mu}(r^{-s}), \tag{A.4.12}$$

where $\rho = \rho(s)$ is a nowhere vanishing entire function and $f(s) := \zeta_{E,\varphi,\mu}$ is a DTI. More generally, ρ can be a holomorphic function which does not have any zeros in the given domain $U \subseteq \mathbb{C}$ under consideration, where U contains the closed half-plane $\{\operatorname{Re} s \geq D(\zeta_{E,\varphi,\mu})\}$.

Moreover, if the extended DTI is of the form (A.4.11), it is said to be *of type* I, and if it is of the form (A.4.12), it is said to be *of type* II (or of *type* II_r if one wants to keep track of the underlying base r).

Finally, if the DTI $f(s) := \zeta_{E,\varphi,\mu}(s)$ is tamed, then the extended DTI (or EDTI) $h(s)$ is said to be *tamed*.

The following comments supplement Definition A.4.5 just above:

Given any tamed EDTI $h = h(s)$ of type I (i.e., given by (A.4.11), with $f(s) := \zeta_{E,\varphi,\mu}(s)$), its *abscissa of convergence*, $D(h)$, is defined by $D(h) := D(f)$. As shown by Proposition A.4.2, in the case of a tamed EDTI of type II, the abscissa of convergence $D(h)$ does not exist as a real number.

Similarly, the *half-plane of convergence* of h is denoted by $\Pi(h)$ and defined by $\Pi(h) := \Pi(f)$ if h is of type I. Clearly, since the function ρ is entire and nonzero, the tamed EDTI h is then holomorphic on $\Pi(h)$. Hence, $D_{\rm hol}(h) \leq D(h)$.

The vertical line $\{\operatorname{Re} s = D(h)\}$, where h is a tamed EDTI of type I, is called the *critical line* of h. It coincides with the critical line of f.

Finally, still for a tamed EDTI $h = h(s)$ of type I, since $D(h)$ and $\Pi(h)$ are well defined, as explained above, we can define $\mathscr{P}_c(h)$ and $\mathscr{P}(h)$ in much the same way as in Definition 2.1.68 (where ζ_A is replaced by h). More specifically, if $h = h(s)$ admits a (necessarily unique) meromorphic continuation to a connected open subset $U \subseteq \mathbb{C}$ containing the closed half-plane $\{\operatorname{Re} s \geq D(h)\} = \overline{\Pi(f)}$,[12] we denote by $\mathscr{P}_c(h)$ the set of *principal complex dimensions* of h, that is, the set of poles of h located on the critical line $\{\operatorname{Re} s = D(h)\}$:

$$\mathscr{P}_c(h) := \{\omega \in U : \omega \text{ is a pole of } h \text{ and } \operatorname{Re}\omega = D(h)\}. \tag{A.4.13}$$

Clearly, $\mathscr{P}_c(h)$ does not depend on the choice of the domain U satisfying the above condition.

Under the same assumptions, we define similarly $\mathscr{P}(h) = \mathscr{P}(h,U)$, the *set of (visible) complex dimensions of h, relative to U*:

$$\mathscr{P}(h) := \{\omega \in U : \omega \text{ is a pole of } h\}. \tag{A.4.14}$$

Clearly, $\mathscr{P}(h) = \mathscr{P}(h,U)$ depends on the choice of U, in general. Also, since the function ρ in Definition A.4.5 does not have any pole or zero, for any tamed EDTI of type I (i.e., given as in (A.4.11), with the notation of the first part of Definition A.4.5), we have

$$\mathscr{P}_c(h) = \mathscr{P}_c(f) \quad \text{and} \quad \mathscr{P}(h) = \mathscr{P}(f), \tag{A.4.15}$$

where $f(s) := \zeta_{E,\varphi,\mu}(s)$,

Let us assume that h is a tamed EDTI of type II. As we have seen in Proposition A.4.2, we always have $D(h) = +\infty$. If there exists an analytic continuation of the function h such that it is holomorphic on an open right half-plane $\{\operatorname{Re} s > \alpha\}$[13] for some $\alpha \in \mathbb{R}$, we can define the *abscissa of holomorphic continuation* of h by

$$D_{\rm hol}(h) := \inf\{\alpha \in \mathbb{R} : h \text{ is holomorphic on } \{\operatorname{Re} s > \alpha\}\}; \tag{A.4.16}$$

or equivalently, $\mathscr{H}(h) := \{\operatorname{Re} s > D_{\rm hol}(h)\}$, the *half-plane of holomorphic continuation* of h, is the largest open right half-plane on which h is holomorphic. We can also define $\Pi(h) := \{\operatorname{Re} s > D(h)\}$ in the usual way, as the half-plane of (absolute) convergence of h.

---

[12] Since ρ is nowhere vanishing on U, if h is of type I, this is the case if and only if $f = f(s)$ admits a meromorphic continuation to U.

[13] If no such α exists, we set $D_{\rm hol}(h) = +\infty$ and hence, $\mathscr{H}(h) = \emptyset$, while if all α can be chosen, then $D_{\rm hol}(h) = -\infty$ and so, $\mathscr{H}(h) = \mathbb{C}$.

A.5 Modified Equivalence Relation and Tamed EDTIs

Recall from the discussion following Definition A.4.5 that if h is any tamed extended DTI of type I, then its abscissa of convergence $D(h)$, half-plane of convergence $\Pi(h) := \{\text{Re}\, s < D(h)\}$, critical line $\{\text{Re}\, s = D(h)\}$, as well as (under the hypotheses (ii) of Definition A.5.1 just below), its set of complex principal dimensions, $\mathscr{P}_c(h)$, are well defined.[14]

We can now modify as follows the definition of the equivalence relation $\sim$ provided in Definition 2.1.69 of Subsection 2.1.5.[15]

Definition A.5.1. Let h_1 and h_2 be two tamed extended DTIs of type I (or briefly, tamed EDTIs of type I), as in Definition A.4.5 of Section A.4 above. We say that h_1 and h_2 are *equivalent*, and write $h_1 \sim h_2$, if the following three conditions (i)–(iii) are satisfied:

(i) h_1 and h_2 have the same abscissa of convergence (assumed to be a real number): $D(h_1) = D(h_2)$ $(\in \mathbb{R})$; call D this common value.

(ii) The functions h_1 and h_2 admit a necessarily unique meromorphic continuation to a connected open neighborhood U of the closed half-plane $\{\text{Re}\, s \geq D\}$ (the closure of their common half-plane of convergence $\Pi := \Pi(h_1) = \Pi(h_2)$).

(iii) Finally, the sets of poles of h_1 and h_2 on their common critical line $\{\text{Re}\, s = D\}$ coincide (and have the same multiplicities):

$$\mathscr{P}_c(h_1) = \mathscr{P}_c(h_2), \tag{A.5.1}$$

where the equality holds between multi-sets (i.e., the multiplicities of the principal poles are taken into account).

If in addition to the above conditions (i), (ii) and (iii), the functions h_1 and h_2 are the EDTIs of the form $h = \zeta_{(E,\varphi,\mu_1)}$ and $h_2 = \zeta_{(E,\varphi,\mu_2)}$, for the same pair (E, φ), then we write

$$h_1 \overset{(E,\varphi)}{\sim} h_2, \tag{A.5.2}$$

and we say that h_1 *and* h_2 *are* (E, φ)*-equivalent.*

As was alluded to earlier, in practice, when we apply the (modified) definition of the equivalence relation (see Definition A.5.1 above)

$$h_1 \sim h_2, \tag{A.5.3}$$

the meromorphic function h_1 is a fractal zeta function (a tamed extended DTI of type I), while the function h_2 (which gives the "leading beahvior" of h_1, to mimick

[14] So is $\mathscr{P}(h) = \mathscr{P}(h, U)$, its set of (visible) complex dimensions, but this is not relevant to our present discussion.

[15] See also Definition A.6.6 in Section A.6 below (along with the text surrounding it, including Remark A.6.7), for a somewhat different, but potentially also useful, definition of 'asymptotic eqivalence', in the case when the function g is merely assumed to be meromorphic.

the terminology of the theory of asymptotic expansions) is another tamed extended DTI of type I but of a 'simpler' closed form. Hence, the importance of the Corollary A.5.1 in the theory developped in the present book, as well as in its companion research and survey articles [LapRaŽu1–8].

Remark A.5.2. The two definitions of the notion of equivalence $\sim$ provided in Definition 2.1.69 and Definition A.5.1 are clearly compatible. In fact, the first part of Definition A.5.1 merely extends Definition 2.1.69 (appearing on page 98) to the case where both f and g are allowed to be tamed extended DTIs (i.e., EDTIs) of type I.

Finally, it is possible, even likely, that in future applications of the theory of fractal zeta functions developed in this book and in [LapRaŽu1–8], we will need to deal with functions g which are no longer tamed EDTIs but are meromorphic functions of a suitable kind. In that case, we will have to suitably modify Definition 2.1.69 and Definition A.5.1 in order to deal with such a situation; see Definition A.6.6 and Remark A.6.7 below.

Proposition A.5.3. *Let us assume that a and b are nonzero complex numbers and $r \in (0,1)$. Then the function $f(s) = 1/(a - br^s)$ coincides (in all of $\mathbb{C}$) with (the meromorphic continuation of) the tamed DTI $\zeta_{E,\varphi,\mu}(s)$, where (for example) $E := (1/2, +\infty)$, $\varphi(x) := r^x$ for all $x \in E$ and*

$$\mu(\mathrm{d}x) := a^{-1} \sum_{k=0}^{\infty} b^k a^{-k} \delta_k, \tag{A.5.4}$$

with δ_k denoting the Dirac measure concentrated at $k \in \mathbb{N} \cup \{0\}$. Furthermore, the abscissa of (absolute) convergence $D(f)$ is given by

$$D(f) = \frac{\log(|b|/|a|)}{\log r^{-1}}. \tag{A.5.5}$$

Equivalently, the open right half-plane of convergence $\Pi(f)$ of f is given by

$$\Pi(f) = \left\{ \mathrm{Re}\, s > \frac{\log(|b|/|a|)}{\log r^{-1}} \right\}. \tag{A.5.6}$$

Proof. It suffices to represent the function f as a Dirichlet series, as follows:

$$f(s) = \frac{1}{a} \cdot \frac{1}{1 - ba^{-1} r^s} = a^{-1} \sum_{k=0}^{\infty} b^k a^{-k} r^{ks}, \tag{A.5.7}$$

Then, clearly, $f(s) = \zeta_{E,\varphi,\mu}(s)$. The largest open set of complex numbers s for which the series is absolutely convergent is defined by $\{s \in \mathbb{C} : |ba^{-1} r^s| < 1\}$, that is, by (A.5.6). Furthermore, the abscissa of convergence $D(f)$ of the Dirichlet series is therefore determined by $|b|\,|a|^{-1} r^{D(f)} = 1$, that is, by (A.5.5). □

Theorem A.5.4. *Assume that a_j and b_j are nonzero complex numbers and $r_j \in (0,1)$, where $j = 1, \ldots, m$. Then the function*

$$f(s) := \frac{1}{(a_1 - b_1 r_1^s)\ldots(a_m - b_m r_m^s)} \tag{A.5.8}$$

can be represented as a tamed DTI $\zeta_{F,\varphi,\mu}(s)$, for an explicit choice of (F,φ,μ) specified in the proof.[16] *Furthermore, the abscissa of convergence of the DTI f is given by*

$$D(f) = \max\left\{\frac{\log(|b_j|/|a_j|)}{\log r_j^{-1}} : j = 1,\ldots,m\right\}. \tag{A.5.9}$$

Proof. From Proposition A.5.3, applied to $f_j(s) := 1/(a_j - b_j r_j^s)$ for each $j = 1,\ldots,m$, we deduce that

$$f(s) = f_1(s)\cdots f_m(s) = \zeta_{(E_1,\varphi_1,\mu_1)}(s)\cdots\zeta_{(E_m,\varphi_m,\mu_m)}(s),$$

where (with the notation of (A.5.4) above) for $j = 1,\ldots,m$, we let

$$\mu_j(\mathrm{d}x) := a_j^{-1}\sum_{k=0}^{\infty} b_j^k a_j^{-k}\delta_k,$$

$E_j := E := [1/2,+\infty)$ and $\varphi_j(x) := r_j^x$ for all $x \in E_j$. Now, Theorem A.2.1 and Corollary A.2.2 imply that $f(s) = \zeta_{(F,\varphi,\mu)}(s)$, where $F := E_1 \times\cdots\times E_m = E^m$, $\varphi := \varphi_1 \otimes\cdots\otimes\varphi_m$ and $\mu := \mu_1 \otimes\cdots\otimes\mu_m$. This completes the proof of Theorem A.5.4 and specifies its statement. □

We illustrate Definition A.5.1 in the case of the distance zeta function of the ternary Cantor set.

Corollary A.5.5. *Let ζ_A be the distance zeta function of the ternary Cantor set $A := C^{(1/3)}$, contained in $[0,1]$:*

$$\zeta_A(s) := \int_{A_\delta} d(x,A)^{s-1}\mathrm{d}x, \tag{A.5.10}$$

where δ is a fixed positive real number. Then

$$\zeta_A(s) \overset{(E,\varphi)}{\sim} \frac{1}{1-2\cdot 3^{-s}}, \tag{A.5.11}$$

in the sense of the second part of Definition A.5.1, with respect to the connected open set $U := \{\mathrm{Re}\, s > 0\}$, where $E := A_\delta$ and $\varphi(x) := d(x,A)$ for all $x \in E$.

An analogous claim holds for the generalized Cantor sets $A := C^{(m,a)}$ introduced in Definition 3.1.1.

Proof. We assume without loss of generality that $\delta > 1/6$, in which case we obviously have that $A_\delta = (-\delta, 1+\delta)$. Then, the distance zeta function $h_1(s) := \zeta_A(s)$ is a tamed DTI which is given by

[16] More precisely, the DTI $\zeta_{F,\varphi,\mu}$ can be meromorphically continued to all of $\mathbb{C}$ and its analytic continuation satisfies $\zeta_{F,\varphi,\mu}(s) = f(s)$, for all $s \in \mathbb{C}$.

$$h_1(s) := \frac{2 \cdot 6^{-s}}{s(1 - 2 \cdot 3^{-s})} + 2\frac{\delta^s}{s}. \tag{A.5.12}$$

(See the second line of Equation (2.1.113) on page 105.) It can be meromorphically extended to the whole complex plane $\mathbb{C}$, so that (A.5.12) continues to hold for all $s \in \mathbb{C}$. (See the second line of Equation (2.1.113) on page 105.) Note that here, we have $E := A_\delta = (-\delta, 1 + \delta)$, $\varphi(x) := d(x, A)$ and μ is the standard Lebesgue measure on E. We next proceed to show that it is (E, φ)-equivalent in the sense of Definition A.5.1 to the function $h_2(s) := 1/(1 - 2 \cdot 3^{-s})$.

To this end, note that the function $p(s) := 2\delta^s/s$ is a DTI which is holomorphic on U, and that we can represent it as a DTI corresponding to the same underlying pair (E, φ) as above, but with a new measure ν defined by $\nu(dx) := 2\chi_{[0,1]}(x) \cdot dx$, where $\chi_{[0,1]} : E \to \mathbb{R}$ is the characteristic function of the interval $[0, 1]$. Therefore, by Theorem A.2.3, we conclude that the function

$$\frac{2 \cdot 6^{-s}}{s(1 - 2 \cdot 3^{-s})} = \zeta_A(s) - p(s) \tag{A.5.13}$$

is also a DTI corresponding to the same pair (E, φ). Defining the function $\rho(s) := 2^{-1} s 6^s$, and noting that $\rho(s) \neq 0$ on U, we obtain upon multiplying the left-hand side of Equation (A.5.13) by $\rho(s)$ (and noting that this product is equal to $h_1(s)$), that h_1 is (E, φ)-equivalent to $h_2(s)$ in the sense of the second part of Definition A.5.1, as desired. □

A.6 Further Generalizations: Stable Tamed DTIs and EDTIs

Recall that, according to Theorem A.2.1 and Corollary A.2.2, the class of tamed DTIs is stable under product (i.e., pointwise multiplication), while in light of Theorem A.2.3 the class of tamed DTIs $\zeta_{E,\varphi,\mu}$, associated with the same pair (E, φ) [but with variable μ], is stable under (finite, complex) linear combinations. Consequently, it is natural to introduce the following definitions, which extend the definitions of DTIs (Definition A.1.2).

Definition A.6.1. Given a fixed pair (E, φ), we denote by $\mathscr{E}_{(E,\varphi)}$ the class of all tamed DTIs $\zeta_{E,\varphi,\mu}$ (with variable μ) associated with (E, φ).

It follows from the discussion preceding Definition A.6.1 that $\mathscr{E}_{(E,\varphi)}$ is a (complex) vector space. More specifically, according to Theorem A.2.3, for any $\alpha, \beta \in \mathbb{C}$,

$$\alpha \zeta_{(E,\varphi,\mu)} + \beta \zeta_{(E,\varphi,\eta)} = \zeta_{(E,\varphi,\nu)}, \tag{A.6.1}$$

where $\nu := \alpha\mu + \beta\eta$. Moreover, in light of Theorem A.2.1 and its corollary (Corollary A.2.2), we have that

$$\mathscr{E}_{(E,\varphi)} \cdot \mathscr{E}_{(F,\psi)} \subseteq \mathscr{E}_{(E \times F, \varphi \otimes \psi)}. \tag{A.6.2}$$

More specifically, for $\zeta_{(E,\varphi,\mu)} \in \mathscr{E}(E,\varphi)$ and $\zeta_{(F,\psi,\eta)} \in \mathscr{E}(F,\psi)$, we have

$$\zeta_{(E,\varphi,\mu)} \cdot \zeta_{(F,\psi,\eta)} = \zeta_{(E\times F,\varphi\otimes\psi,\mu\otimes\eta)} \in \mathscr{E}_{(E\times F,\varphi\otimes\psi)}. \tag{A.6.3}$$

In light of (A.6.2) or (A.6.3), the vector space $\mathscr{E}(E,\varphi)$ is clearly not an algebra (i.e., is not stable under products). Therefore, we next imitate a well-known construction in order to obtain an associated algebra $\mathscr{A}(E,\varphi)$. Specifically, we let[17]

$$\mathscr{A}_{(E,\varphi)} := \bigoplus_{n=0}^{\infty} \mathscr{E}_{(E^n,\varphi^{\otimes n})}, \tag{A.6.4}$$

where in (A.6.4), the symbol $\oplus$ stands for the vector space direct sum, E^n denotes the Cartesian product of n copies of E, and $\varphi^{\otimes n}$ is the n-fold tensor product of φ by itself; so that

$$\varphi^{\otimes n}(x_1,\ldots,x_n) := \varphi(x_1)\cdots\varphi(x_n) \tag{A.6.5}$$

for all $(x_1,\ldots,x_n) \in E^n$.

Reading Equation (A.6.2) in reverse order (that is, from right to left), with $(F,\psi) := (E,\varphi)$, we see that for all $n \in \mathbb{N}\cup\{0\}$,

$$\mathscr{E}_{(E,\varphi)} = \mathscr{E}^{\cdot n}_{(E,\varphi)}, \tag{A.6.6}$$

where

$$\mathscr{E}^{\cdot n}_{(E,\varphi)} := \underbrace{\mathscr{E}_{(E,\varphi)}\cdots\mathscr{E}_{(E,\varphi)}}_{n \text{ times}} \tag{A.6.7}$$

stands for the space of products of n elements of $\mathscr{E}_{(E,\varphi)}$. The space $\mathscr{A}_{(E,\varphi)}$ is somewhat analogous to the classic *Fock space* (representing the interactions of n different particles, for every $n \geq 0$ and hence, of countable many particles), of broad use in quantum mechanics and quantum field theory. (See, e.g., [GliJaf, ItzZube, ReeSim1, Wein1–2]. Furthermore, essentially by construction, $\mathscr{A}(E,\varphi)$ is not only a vector space but also an algebra (over $\mathbb{C}$). In fact, it is a unital, abelian algebra (with unit 1, the constant DTI identically equal to 1).

We can now introduce the class of tamed, stable, extended DTIs (tamed SEDTIs, in short), as follows. Given a fixed pair (E,φ) as above and a fixed nowhere vanishing entire function ρ, we say that h is a *stable, extended DTI of type I* (associated with (E,φ) and ρ) if

$$h := \rho^n \cdot f, \tag{A.6.8}$$

for some $n \in \mathbb{N}\cup\{0\}$ and $f \in \mathscr{A}_{(E,\varphi)}$. We denote by $\mathscr{S}^{(I)}_{(E,\varphi),\rho}$ the resulting class of SEDTIs of type I. (Compare with Definition A.4.5 above.)

Similarly, $h := \rho^n \cdot g$ is called an SEDTI (a *stable* EDTI) *of type* II if $g(s) = f(r^{-s})$ for some $n \in \mathbb{N}\cup\{0\}$, $f \in \mathscr{F}_{(E,\varphi)}$ and $r \in (0,1)$. In that case, g is said to be a DTI *of type* II_r (as in Definition A.4.5 above) and h is called an SEDTI *of type* II_r for that value of $r \in (0,1)$. (Compare with Definition A.4.5.) We denote by $\mathscr{S}^{(II)}_{(E,\varphi),\rho}$ the

[17] By definition, 1 is the constant DTI equal to 1 and $\mathscr{E}(E^0,\varphi^{\otimes 0}) := \mathbb{C}\cdot 1 \simeq \mathbb{C}$.

resulting class of SEDTIs of class II. (Of course, much as in Definition A.4.5, an SEDTI is either an SEDTI of type I or an SEDTI of type II. It is implicitly assumed that we are only talking here about SEDTIs associated with a fixed pair (E,φ) and a function ρ.)

Theorem A.6.2. *Each of $\mathscr{S}^{(I)}_{(E,\varphi),\rho}$ and $\mathscr{S}^{(II)}_{(E,\varphi),\rho}$, for a given $r \in (0,1)$, is a unital algebra (over $\mathbb{C}$).*

Proof. This follows at once from the definitions and the fact that $\mathscr{A}_{(E,\varphi)}$, defined by Equation (A.6.4), is itself a unital algebra (over $\mathbb{C}$). □

Example A.6.3. As we have seen in Example A.5.5, for the function defined by $h(s) := 1/(1-2\cdot 3^{-s})$ on the open right half-plane $U := \{\operatorname{Re} s > 0\}$, we have that $h \in \mathscr{S}^{(I)}_{(E,\varphi),\rho}$, with $E := (-\delta, 1+\delta)$, where $\delta > 0$ is fixed and $\varphi(x) := d(x,A)$ for all $x \in E$, and where A is the ternary Cantor set. Note that $\varphi(x) \leq \max\{\delta, 1/6\}$ for all $x \in E$.

We still assume implicitly that the pair (E,φ) and the function ρ are fixed. We are now able to extend the definition of equivalence $\sim$ so that it becomes an equivalence relation on SEDTIs of type I. We do not formally state the definition since it suffices to substitute SEDTI of type I for EDTI of type I, in the statement of Definition A.5.1 (for the equivalence relation $\sim$ on the set of EDTIs).

This equivalence relation is probably sufficient for all the applications of interest in the present theory. In practice, when we write $h_1 \sim h_2$, the function h_1 (in the present counterpart of Definition A.5.1) is a (tamed) fractal zeta function. Recall from Proposition A.2.4 that essentially without loss of generality, all fractal zeta functions encountered in the theory can be assumed to be tamed.

Furthermore, in many cases, h_2 (also in the counterpart of Definition A.5.1) is an SEDTI of type II_r, for some $r \in (0,1)$ (also associated with (E,φ) and ρ). More specifically, it will be an SEDTI of the following form (for a given $r \in (0,1)$):

$$h_2 := \rho^n g, \tag{A.6.9}$$

for some $n \in \mathbb{N} \cup \{0\}$, $g(s) := f(r^{-s})$ and

$$f(s) := \sum_{j=1}^{J} \frac{\alpha_j}{P_j(s)},\text{[18]} \tag{A.6.10}$$

where P_j is an arbitrary polynomial (i.e., $P_j \in \mathbb{C}[X]$). Here, we let $E := [1,+\infty)$, $\varphi(x) := 1/x$ for all $x \in E$, and use the fact that according to Theorem A.3.2 and Corollary A.3.3,

$$F(s) := \frac{1}{P(s)}, \quad \text{where} \quad P \in \mathbb{C}[X], \tag{A.6.11}$$

[18] Clearly, the coefficients α_j can be replaced by 1, since they can be absorbed into the definition of the polynomials P_j.

coincides with the following tamed DTI:

$$F(s) = \zeta_{E^n,\Phi,\mu}(s), \tag{A.6.12}$$

where $E^n := E \times \cdots \times E$ and $\Phi := \varphi^{\otimes n}$ with $n := \deg P$, so that $f(s) \in \mathscr{E}_{(E^n,\varphi^{\otimes n})}$.

With the aforementioned choice of E and φ, the pair (E, φ) is *universal* for *all* polynomials P and furthermore, the integer n depends only on the degree n.

In closing, we note that the space of the SEDTIs h_2 of the form (A.6.9), with f as in (A.6.10), $n \in \mathbb{N} \cup \{0\}$, $J \in \mathbb{N}$, $\alpha_j \in \mathbb{C}$ and $P_j \in \mathbb{C}[X]$ arbitrary (but for a fixed ρ and with $E := [1, +\infty)$ and $\varphi(x) := 1/x$ for all $x \in E$, as above) is a subalgebra of the algebra $\mathscr{S}^{(II_r)}_{(E,\varphi),\rho}$ of stable EDTIs of type II$_r$.

Similary, the space of SEDTIs of the form $k_2 = \rho^n f$, where f is as in (A.6.10), $n \in \mathbb{N} \cup \{0\}$, $J \in \mathbb{N}$, $\alpha_j \in \mathbb{C}$ and $P_j \in \mathbb{C}[X]$ arbitrary (but for a fixed ρ and still with $E := [1, +\infty)$ and $\varphi(x) := 1/x$ for all $x \in E$), is a subalgebra of the algebra $\mathscr{S}^{(I)}_{(E,\varphi),\rho}$ of stable EDTIs of type I.

Remark A.6.4. Note that for each $j \in \mathbb{N}$,

$$\frac{1}{P_j} = \zeta_{(E^{n_j},\varphi^{\otimes n_j},\mu_j)}, \tag{A.6.13}$$

where $n_j := \deg P_j$ is as above, and the measure μ_j depends on j; more precisely, μ_j depends on the zeros of P_j counted according to their multiplicities (that is, viewed as a multi-set). (See Theorem A.3.2 and, especially, Corollary A.3.3 along with Equations (A.3.3) and (A.3.4).) It follows that $1/P_j$ belongs to $\mathscr{E}^{(I)}_{(E^{n_j},\varphi^{\otimes n})}$. Therefore, f, as given by (A.6.10), belongs to $\mathscr{A}^{(I)}_{(E,\varphi)}$ and so, g belongs to $\mathscr{A}^{(II)}_{(E,\varphi)}$. Consequently, h_2 belongs to $\mathscr{S}^{(II_r)}_{(E,\varphi),\rho}$, as claimed.

Remark A.6.5. Strictly speaking, when working with the above SEDTIs (i.e., the element of $\mathscr{S}^{II_r}_{(E,\varphi)\rho}$) of type II, one cannot use, in general, the above extension of Definition A.5.1. (See, however, the end of the second paragraph of this remark for a way of viewing a subclass of these SEDTIs as SEDTIs of type I.) The same is not true for the above SEDTIs of type I (i.e., the elements of $\mathscr{S}^{I}_{(E,\varphi),\rho}$).

Another class of SEDTIs of type I of interest in this context consists of functions h_2 of the form $h_2 = \rho^n f$, where $n \in \mathbb{N} \cup \{0\}$ and f is given by

$$f(s) = \sum_{j=1}^{J} \frac{\alpha_j}{(a_{1,j} - b_{1,j} r_{1,j}^s) \ldots (a_{m,j} - b_{m,j} r_{m,j}^s)}, \tag{A.6.14}$$

with $J \in \mathbb{N}$, and for $j = 1, \ldots, J$, $\alpha_j \in \mathbb{C}$, while for $k = 1, \ldots, m_j$, $a_{k,j}$, $b_{k,j}$ are nonzero complex numbers, and $r_{k,j} \in (0,1)$. (See Theorem A.5.4 above.) The class of all such SEDTIs of type I constitutes an algebra. If, furthermore, we assume that $r_{k,j} \equiv r$, where $r \in (0,1)$ is a fixed base, then we obtain a subalgbra of this aforementioned algebra. If, in addition, we allow some of the coefficients $a_{j,k}$ to be equal to zero (which can be easily accommodated), one then obtains the exact same algebra as $\mathscr{S}^{II_r}_{(E,\varphi)\rho}$ (with (E,φ) and ρ as in Remark A.6.4 and the text preceding it), which can therefore also be viewed as an algebra of SEDTIs of type I, but for different choices of (E,φ) (thanks to Proposition A.5.3 and Theorem A.5.4, combined with Theorem A.2.1 and Corollary A.2.2).

A possible alternative way of dealing with this latter issue and, more generally, with EDTIs and SEDTIs of type II$_r$, for a fixed $r \in (0,1)$, would be to make the change of variable $z := r^s$ much as in the theory of zeta functions of varieties over finite fields (see, e.g, [ParsSh1], [Lap-vFr3] or [Lap6, Appendix B] and the many relevant references therein) or of geometric zeta functions of lattice self-similar fractal strings (see [Lap-vFr3, Chapters 2 and 3]). In many cases of interest, we could work with power series in this new variable z and consider their radii of convergence (instead of the abscissae of convergence of the original EDTIs or SEDTIs of type II). By necessity of concision, however, we will not consider this possibility here.

In closing this appendix, we mention that in certain current and, likely, future applications of the theory of fractal zeta functions developed in this book (and in [LapRaŽu1–8]), we will need to deal with situations in which it is no longer the case that both of the functions f and g are DTIs (as in Definition 2.1.69 of Section 2.1.5), or even, EDTIs (as in Definition A.6.6 of Section A.5). Typically, f will be a tamed DTI or more generally, a tamed EDTI, with a suitable meromorphic extension, whereas g will only be a meromorphic function in an appropriate domain. In that more general situation, we propose to use the following definition (Definition A.6.6), which is a suitable modification of Definition A.5.1 (and, a fortiori, of Definition 2.1.69). Strictly speaking, it no longer gives rise to an equivalence relation (since f and g now belong to different classes of functions), but in this new sense, the statement that $f \overset{\text{asym}}{\sim} g$ still captures appropriately the idea according to which "*f is asymptotic to g*".

The situation is very analogous, in spirit, to the evaluation of the "leading part" ($g = g(s)$, in the present case) of a function ($f = f(s)$, here) in the theory of asymptotic expansions. In that situation, the "leading part" g belongs to a scale of typical functions (describing the possible asymptotic behaviors of the function f in the given asymptotic limit).

Definition A.6.6. (Asymptotic equivalence). Let f be a tamed EDTI of type I and let g be a meromorphic function, both defined and assumed to be meromorphic on a connected open subset U of $\mathbb{C}$ containing the closed right half-plane $\Pi(f) := \{\mathrm{Re}\, s > D(f)\}$. Then, the function f is said to be *asymptotically equivalent to* g and we write $f \overset{\text{asym}}{\sim} g$ if the following two conditions (i)–(ii) are satisfied:

(i) The abscissa of (absolute) convergence of f and the abscissa of holomorphic continutation of g coincide: $D(f) = D_{\text{hol}}(g)$ (and is a real number); call D this common value;[19]

and

[19] Note that this implies that $\Pi(f) = \mathscr{H}(g) = \{\mathrm{Re}\, s > D\}$; so that g is holomorphic on $\{\mathrm{Re}\, s > D\}$ and therefore, does not have any poles in this open right half-plane.

(*ii*) The poles of f and g located on the convergence critical line of f, which is also the holomorphy critical line of g, coincide and have the same multiplicities. More succinctly, and with the notation specified in Equation (A.6.16) below (compare with Equation (A.5.1) in Definition A.5.1 above), we have

$$f \overset{\text{asym}}{\sim} g \quad \overset{\text{def.}}{\Longleftrightarrow} \quad D(f) = D_{\text{hol}}(g)\ (\in \mathbb{R}) \quad \text{and} \quad \mathscr{P}_c(f) = \mathscr{P}_{c,\text{hol}}(g), \tag{A.6.15}$$

where the latter equality hold between multisets. Here, we let

$$\mathscr{P}_{c,\text{hol}}(g) := \{\omega \in U : \omega \text{ is a pole of } g \text{ and } \operatorname{Re}\omega = D_{\text{hol}}(g)\}. \tag{A.6.16}$$

Remark A.6.7. Observe that if g is assumed to be a tamed EDTI, Definition A.6.6 may differ, in general, from its counterpart stated in Definition A.5.1 (or in Definition 2.1.69 in the special case when g is a tamed DTI). (That is, in general $f \overset{\text{asym}}{\sim} g$ does not imply that $f \sim g$, and conversely.) Indeed, recall that there are many DTIs (and, a fortiori, EDTIs) g for which $D_{\text{hol}}(g) < D(g)$.[20] Therefore, strictly speaking, Definition A.6.6 does not extend Definition 2.1.69 or even Definition A.5.1. However, this should not cause any problem in practice and seems to provide us with additional flexibility in the actual and potential applications of the theory.

[20] There are also many fractal zeta functions g for which $D_{\text{hol}}(g) = D(g)$ $(\in \mathbb{R})$; see, for example, part (*c*) of Theorems 2.1.11, 2.2.11 and 4.1.7 (also, part (*ii*) of Corollary 4.1.10), along with Theorem 2.1.55 and Corollary 2.1.61.

Appendix B
Local Distance Zeta Functions

Abstract In this appendix, we briefly discuss a topic which should be much further expanded in future work, because of its potential connections with the fractal tube formulas of Chapter 5 and their yet to be established local versions, in the general setting of this monograph. More specifically, we propose a notion of local zeta function (and the associated notion of local complex dimensions) adapted to our work. We also illustrate these notions by means of a concrete example.

Key words: local distance zeta function.

In this appendix, we briefly discuss a topic which should be much further expanded in future work, because of its potential connections with the fractal tube formulas of Chapter 5 and their yet to be established local versions, in the general setting of this monograph. (The attentive reader will recognize some analogies with our earlier discussion of relative fractal sprays in Section 4.2.1; see especially, Theorem 4.1.44 and Remark 4.1.45.)

Definition B.0.1. (*Local distance zeta function*). Let A be an arbitrary bounded Borel subset of $\mathbb{R}^N$, where $\delta > 0$ is fixed. Then, its *local distance zeta function* Z_A is given by the family of relative distance zeta functions $\{\zeta_{A,\Omega} : \Omega \in \mathscr{B}\}$, where Ω runs through the class $\mathscr{B} = \mathscr{B}(A_\delta)$ of Borel subsets of A_δ. Note that up to now, relative distance zeta functions were only considered for open subsets Ω of $\mathbb{R}^N$, but there is no reason not to assume Ω to be an arbitrary Borel subset of A_δ.

Remark B.0.2. Later on (in Definition B.0.5), we will refine and revisit Definition B.0.1. Indeed, we will then define and view Z_A, the local distance zeta function of A, as a suitable zeta function-valued complex Borel measure. See Definition B.0.5, along with Corollary B.0.4 which fully justifies it.

Theorem B.0.3. *Let $A \subseteq \mathbb{R}^N$ be bounded. Let $\{\Omega_j\}_{j=1}^\infty$ be an arbitrary Borel partition of $\Omega \in \mathscr{B}$ (i.e., $\Omega_j \in \mathscr{B}$ for all $j \geq 1$, $\Omega_j \cap \Omega_k = \emptyset$ for $j \neq k$ and $\Omega = \cup_{j=1}^\infty \Omega_j$). Then, for all $s \in \mathbb{C}$ with $s > \overline{\dim}_B A$, we have that*

M.L. Lapidus et al., *Fractal Zeta Functions and Fractal Drums*,
Springer Monographs in Mathematics, DOI 10.1007/978-3-319-44706-3

$$\zeta_{A,\Omega}(s) = \sum_{j=1}^{\infty} \zeta_{A,\Omega_j}(s), \tag{B.0.1}$$

where the series is absolutely convergent (and hence, convergent) in $\mathbb{C}$, *and in the special case where* s *is real, is a series with nonnegative terms having a finite sum.*

Proof. This is an immediate consequence of Theorem 4.1.44 and Remark 4.1.45, in the special case where $A_j = A$ for all $j \geq 1$ (in the notation of that theorem).[1] The fact that now, the sets Ω_j are Borel (rather than open) subsets of A_δ (or, more precisely, of some δ-neighborhood A_δ of A) does not affect the proof in any way. Finally, note that clearly, $\overline{\dim}_B A \geq \overline{\dim}_B(A,\Omega)$, provided $\Omega \subseteq A_\delta$ for some $\delta > 0$. □

Corollary B.0.4. *Let* $A \subset \mathbb{R}^N$ *be an arbitrary bounded set, and let* P_A *denote the half-plane* $\{\operatorname{Re} s > \overline{\dim}_B A\}$ *of holomorphic convergence of* $\zeta_{A,\Omega}$. *Then, the map*

$$\Omega \in \mathscr{B} \mapsto \zeta_{A,\Omega} \in \operatorname{Hol}(P_A,)$$

is a $\operatorname{Hol}(P_A)$*-valued complex Borel measure, where* $\operatorname{Hol}(P_A)$ *denotes the space of holomorphic functions equipped with the topology of local uniform convergence on* P_A *(i.e., uniform convergence on all the compact subsets of* P_A*).*

Proof. It suffices to double-check that in the conclusion of Theorem B.0.3 (i.e., in Equation (B.0.1)), the convergence holds not only pointwise for $\operatorname{Re} s > \overline{\dim}_B A$ but also locally uniformly on P_A. To easily verify this, it also clearly suffices to show that the series in (B.0.1) converges uniformly on vertical strips of the form $\beta \geq \operatorname{Re} s \geq \alpha > \overline{\dim}_B A$, where $\alpha, \beta \in \mathbb{R}$ are otherwise arbitrary. If we assume, without loss of generality, that $\delta \leq 1$, this last claim follows from the following computation, valid for all $s \in \mathbb{C}$ in a strip of the above type (and for $p, q \in \mathbb{N}$, with $q \geq p$):

$$\begin{aligned}\sum_{j=p}^{q} \left|\zeta_{A,\Omega_j}(s)\right| &= \sum_{j=p}^{q} \left|\int_{\Omega_j} d(x,A)^{s-N}\,\mathrm{d}x\right| \\ &\leq \sum_{j=p}^{q} \int_{\Omega_j} d(x,A)^{\operatorname{Re} s-N}\,\mathrm{d}x \\ &\leq \sum_{j=p}^{q} \int_{\Omega_j} d(x,A)^{\alpha-N}\,\mathrm{d}x \to 0 \ \text{ as } p,q \to \infty,\end{aligned} \tag{B.0.2}$$

since $\alpha > \overline{\dim}_B A$ and by Theorem 4.1.44, the numerical series (of nonnegative terms)

$$\sum_{j=1}^{\infty} \int_{\Omega_j} d(x,A)^{\alpha-N}\,\mathrm{d}x = \sum_{j=1}^{\infty} \zeta_{A,\Omega_j}(\alpha,)$$

converges (to $\zeta_{A,\Omega}(\alpha)$) and therefore satisfies the Cauchy criterion. It follows that the series of holomorphic functions $\sum_{j=1}^{\infty} \zeta_{A,\Omega_j}$ satisfies the Cauchy criterion for

[1] Note that still in the notation of Theorem 4.1.44, condition (4.1.52) is then automatically satisfied.

uniform convergence in the strip $\{\alpha \le \operatorname{Re} s \le \beta\}$, as desired. Consequently, it is also convergent in that strip, and hence, locally uniformly convergent on P_A. □

As was suggested in Remark B.0.2, we can now use Corollary B.0.4 to give a more satisfactory definition of the notion of local distance zeta function in our context.

Definition B.0.5. (*Local distance zeta function, revisited*). Let A be a bounded subset of $\mathbb{R}^N$. Then the zeta function-valued complex Borel measure on A_δ, for some fixed $\delta > 0$, obtained in Corollary B.0.4, is called the *local distance zeta function* of A and is denoted by Z_A.

Let us now assume that Ω is a fixed Borel subset of $\mathbb{R}^N$ (not necessarily of finite volume) and $x \in \overline{\Omega}$. Our aim is to consider the fractal properties of Ω near x, using relative zeta functions. To this end, we consider the *local distance zeta function of Ω at x*, defined as the zeta function of the relative fractal drum $(A := B_r(x), B_{r+\delta}(x) \cap \Omega)$, where $\delta > 0$ is fixed:

$$\zeta_{x,\Omega}(s) := \int_{B_{r+\delta}(x)\cap\Omega} d(y, B_r(x))^{s-N}\,\mathrm{d}y. \tag{B.0.3}$$

Note that $B_{r+\delta}(x)$ is the δ-neighborhood of $A := B_r(x)$. Furthermore, we assume that $|\Omega| > 0$, since otherwise $\zeta_{x,\Omega}(s) \equiv 0$ for all $s \in \mathbb{C}$ such that $\operatorname{Re} s > N$.

The *local tube zeta function of the set Ω at $x \in \overline{\Omega}$* is defined by

$$\tilde{\zeta}_{x,\Omega}(x) := \int_0^\delta t^{s-N-1}|B_{r+t}(x)\cap\Omega|\,\mathrm{d}t. \tag{B.0.4}$$

Example B.0.6. Let us consider the case when $\Omega := \mathbb{R}^N$. Since $\mathbb{R}^N$ is homogeneous and translation invariant, it is clear that the local complex dimensions of $\mathbb{R}^N$ (generated by the local distance zeta function $\zeta_{x,\mathbb{R}^N}(s)$) do not depend on x (as well as on r and δ). In light of Equation (B.0.3), the local distance zeta function of $\mathbb{R}^N$ computed at $x = 0$ is then obtained as follows (for any fixed $r > 0$ and $\delta > 0$):

$$\begin{aligned}
\zeta_{0,\mathbb{R}^N}(s) &= \int_{\{r<|y|<r+\delta\}} (|y|-r)^{s-N}\mathrm{d}y \\
&= N\omega_N \int_r^{r+\delta} (\rho - r)^{s-N}\rho^{N-1}\mathrm{d}\rho \\
&= N\omega_N \int_0^\delta t^{s-N}(t+r)^{N-1}\mathrm{d}t \\
&= N\omega_N \int_0^\delta t^{s-N}\sum_{k=0}^{N}\binom{N-1}{k}t^{N-1-k}r^k\mathrm{d}t \\
&= N\omega_N \sum_{k=1}^{N-1}\binom{N-1}{k}\frac{r^k\delta^{s-k}}{s-k}.
\end{aligned} \tag{B.0.5}$$

By analytic continuation, we deduce that $\zeta_{0,\mathbb{R}^N}$ can be meromorphically continued to all of $\mathbb{C}$ and that, for all $s \in \mathbb{C}$,

$$\zeta_{0,\mathbb{R}^N}(s) = N\omega_N \sum_{k=1}^{N-1} \binom{N-1}{k} \frac{r^k \delta^{s-k}}{s-k}. \tag{B.0.6}$$

Therefore, the set of local complex dimensions of $\mathbb{R}^N$, generated by the local distance zeta function, is given by

$$\dim_{loc} \mathbb{R}^N = \{0,1,\ldots,N-1\}. \tag{B.0.7}$$

Furthermore, we have $\operatorname{res}(\zeta_{0,\mathbb{R}^N},k) = N\omega_N \binom{N-1}{k} r^k$ for each $k \in \{0,1,\ldots,N-1\}$.

Similarly, for $A := B_r(0)^c = \mathbb{R}^N \setminus B_R(0)$ and $\delta \in (0,r)$, we obtain that

$$\begin{aligned}\zeta_{0,\mathbb{R}^N}(s) := \zeta_{A,\mathbb{R}^N}(s) &= \int_{\{r-\delta<|y|<r\}} (r-|y|)^{s-N} \mathrm{d}y \\ &= N\omega_N \int_{r-\rho}^{r} (\rho - r)^{s-N} \rho^{N-1} \mathrm{d}\rho \\ &= \int_0^\delta t^{s-N}(r-t)^{N-1}\mathrm{d}t = \sum_{k=1}^{N-1} \binom{N-1}{k} r^{N-1-k} \frac{\delta^{s-N+k+1}}{s-N-k+1}.\end{aligned} \tag{B.0.8}$$

Upon analytic continuation, we conclude that $\zeta_{0,\mathbb{R}^N}(s)$ can be meromorphically continued to all of $\mathbb{C}$ and that

$$\zeta_{0,\mathbb{R}^N}(s) = \sum_{k=1}^{N-1} \binom{N-1}{k} r^{N-1-k} \frac{\delta^{s-N+k+1}}{s-N-k+1}, \tag{B.0.9}$$

for all $s \in \mathbb{C}$. It then follows, much as above, that the set of complex dimensions generated by the local distance zeta function (with respect to $A := B_r(0)^c$) is again given by $\dim_{loc} \mathbb{R}^N = \{0,1,\ldots,N-1\}$.

On the other hand, by using Equation (B.0.4), we deduce after a short computation that

$$\begin{aligned}\tilde{\zeta}_{x,\mathbb{R}^N}(s) &= \int_0^\delta t^{N-s-1} |B_{r+t}(x) \cap \Omega| \, \mathrm{d}t = \int_0^\delta t^{N-s-1} \omega_N (r+t)^N \, \mathrm{d}t \\ &= \omega_N \sum_{k=0}^{N} \binom{N}{k} r^k \int_0^\delta t^{s-k-1} \, \mathrm{d}t = \omega_N \sum_{k=0}^{N} \binom{N}{k} \frac{r^k \delta^{s-k}}{s-k},\end{aligned} \tag{B.0.10}$$

for all $s \in \mathbb{C}$ such that $\operatorname{Re} s > N$. In light of the principle of analytic continuation, we conclude that $\tilde{\zeta}_{x,\mathbb{R}^N}(s)$ can be meromorphically continued to all of $\mathbb{C}$ and that

$$\tilde{\zeta}_{x,\mathbb{R}^N}(s) = \omega_N \sum_{k=0}^{N} \binom{N}{k} \frac{r^k \delta^{s-k}}{s-k}, \tag{B.0.11}$$

for all $s \in \mathbb{C}$. Therefore, the set of complex dimensions generated by the local tube zeta function $\tilde{\zeta}_{x,\mathbb{R}^N}(s)$ is given by

$$\dim_{loc} \mathbb{R}^N = \{0, 1, \ldots, N\}. \tag{B.0.12}$$

(See also [LapRaŽu7, Exercise 5.22].) Remarkably, this result is, of course, in complete agreement with what one would expect, intuitively. It therefore seems to indicate that the local tube zeta function $\tilde{\zeta}_{x,\mathbb{R}^N}(s)$ is a more suitable tool for the computation of local complex dimensions than the local distance zeta function $\zeta_{x,\mathbb{R}^N}(s)$.

Appendix C
Distance Zeta Functions and Principal Complex Dimensions of RFDs

Abstract In this appendix, we review in a table form some of the basic relative fractal drums (A,Ω) in $\mathbb{R}^N$ appearing in this monograph, as well as the associated relative distance zeta functions:

$$\zeta_{A,\Omega}(s) := \int_{\Omega} d(x,A)^{s-N}\mathrm{d}x.$$

For a given relative fractal drum, we also indicate the corresponding set of principal complex dimensions $\dim_{PC}(A,\Omega)$. Recall that this set is defined as the set of poles of $\zeta_{A,\Omega}$ located on the critical line $\{\mathrm{Re}\, s = \overline{\dim}_B(A,\Omega)\}$, provided there exists a meromophic extension of $\zeta_{A,\Omega}$ to a connected open neighborhood of the critical line (this assumption is fulfilled for all relative fractal drums in the table):

$$\dim_{PC}(A,\Omega) = \mathscr{P}_c(A,\Omega).$$

Key words: relative fractal drum (RFD), relative distance zeta function, basic RFDs, complex dimensions of RFDs.

In this appendix, we review some of the basic relative fractal drums (A,Ω) in $\mathbb{R}^N$ appearing in this monograph (see Table C.1), as well as the associated relative distance zeta functions:

$$\zeta_{A,\Omega}(s) := \int_{\Omega} d(x,A)^{s-N}\mathrm{d}x.$$

For a given relative fractal drum, we also indicate the corresponding set of principal complex dimensions $\dim_{PC}(A,\Omega)$. Recall that this set is defined as the set of poles of $\zeta_{A,\Omega}$ located on the critical line $\{\mathrm{Re}\, s = \overline{\dim}_B(A,\Omega)\}$, provided there exists a meromophic extension of $\zeta_{A,\Omega}$ to a connected open neighborhood of the critical line (this assumption is fulfilled for all relative fractal drums in the table):

$$\dim_{PC}(A,\Omega) = \mathscr{P}_c(A,\Omega).$$

M.L. Lapidus et al., *Fractal Zeta Functions and Fractal Drums*,
Springer Monographs in Mathematics, DOI 10.1007/978-3-319-44706-3

For some of the relative fractal drums listed in Table C.1 we have to impose a number of natural conditions, but we do not mention them explicitly. For example, in the case of the generalized Cantor set $A = C^{(m,a)}$, viewed as the relative fractal drum with respect to the unit interval $\Omega = (0,1)$, the constant m is assumed to be an integer larger than 1 and a is a positive real number such that $a < 1/m$; see Definition 3.1.1 on page 186.

In order to save space, we have used the self-explanatory symbols ▲ for an equilateral triangle, ■ for the square (both viewed as subsets of $\mathbb{R}^2$), and △ and □ for their respective boundaries. Furthermore, any angle θ is expressed in radians and is assumed to be contained in the interval $[0, 2\pi]$.

By Sector$(0,\theta,\delta)$, we mean a planar sector with vertex at the origin $0 \in \mathbb{R}^2$, of radius $\delta > 0$ and opening angle θ:

$$\text{Sector}(0,\theta,\delta) = \{(r,\varphi) \in \mathbb{R}^2 : 0 \le r < \delta,\ 0 \le \varphi \le \theta\} \subset B_\delta(0).$$

Here, we have denoted by (r,φ) the polar coordinates of a point in the plane.

Remark C.0.1. In the case of the Sierpiński 3-gasket appearing in Table C.1 above (see also Example 4.2.26 on page 294–303), the relative Minkowski dimension $D = 2$ has multiplicity two (as a pole of $\zeta_{A,\Omega}$), whereas all the other complex dimensions are simple.

Remark C.0.2. The set of complex dimensions generated by the local tube zeta function $\widetilde{\zeta}_{x,\mathbb{R}^N}(s)$ is given by $\dim_{loc} \mathbb{R}^N = \{0,1,\ldots,N\}$; see Equation (B.0.12) at the end of Appendix B and the discussion preceding it.

$\mathbb{R}^N$	A	Ω	$\zeta_{A,\Omega}(s)$	$\dim_{PC}(A,\Omega)$
$\mathbb{R}$	$\{0\}$	$(0,\delta)$	$\frac{\delta^s}{s}$	0
$\mathbb{R}^2$	$\{(0,0)\}$	$B_\delta((0,0))$	$2\pi\frac{\delta^s}{s}$	0
$\mathbb{R}^2$	$\{(0,0)\}$	$\text{Sector}(0,\theta,\delta)$	$\theta\frac{\delta^s}{s}$	0
$\mathbb{R}^N$	$\{0\}$	$B_\delta(0)$	$N\omega_N\frac{\delta^s}{s}$	0
$\mathbb{R}^2$	vertices of ■	$■ = (0,1)^2$	$\frac{8}{2^s s}\int_0^{\pi/4}\cos^{-s}\varphi\,\mathrm{d}\varphi$	0
$\mathbb{R}^N$	$\partial B_\delta(0)$ p. 128	$B_\delta(0)$	$N\omega_N\delta^s\sum_{j=0}^{N-1}\binom{N-1}{j}\frac{(-1)^{N-j-1}}{s-j}$	$N-1$
$\mathbb{R}$	$C^{(1/3)}$ the ternary Cantor set, p. 105	$(0,1)$	$\frac{2\cdot 6^{-s}}{s(1-2\cdot 3^{-s})}$	$\log_3 2+\frac{2\pi}{\log 3}\mathrm{i}\mathbb{Z}$
$\mathbb{R}$	$C^{(m,a)}$ generalized Cantor set, p. 188 $m\geq 2,\ 0<a<1/m$	$(0,1)$	$\left(\frac{1-ma}{2(m-1)}\right)^{s-1}\frac{1-ma}{s(1-ma^s)}$	$\log_{1/a}m+\frac{2\pi}{\log(1/a)}\mathrm{i}\mathbb{Z}$
$\mathbb{R}$	$\{j^{-a}:j\geq 1\}$ a-string, p. 151	$(0,1)$	$\sum_{j=1}^{\infty}(j^{-a}-(j+1)^{-a})^s$	$\frac{1}{1+a}$
$\mathbb{R}^2$	$\{(0,0)\}$ p. 256	$\{(x,y):x\in(0,1),0<y<x^{-\alpha}\}$ $\alpha\in(0,1)$	$\frac{1}{s-\alpha}-\frac{1}{s-\alpha-1}$	$1+\alpha$
$\mathbb{R}^2$	$(0,a)\times\{0\}$	$(0,a)\times(0,\delta)$	$\frac{a\delta^{s-1}}{s-1}$	1
$\mathbb{R}^2$	$\triangle=\partial(▲)$ p. 292	equilateral triangle ▲ of side a	$6\frac{(\sqrt{3})^{1-s}}{s(s-1)}\left(\frac{a}{2}\right)^s$	1
$\mathbb{R}^2$	$\square=\partial(■)$ p. 206	$■ = (0,a)^2$	$\frac{8}{s(s-1)}\left(\frac{a}{2}\right)^s$	1

(The table is continued from the preceding page.)

$\mathbb{R}^N$	A	Ω	$\zeta_{A,\Omega}(s)$	$\dim_{PC}(A,\Omega)$
$\mathbb{R}^2$	fractal nest of center type: $\{(r,\varphi): r=k^{-\alpha},\, k\in\mathbb{N}\}$ $\alpha\in(0,1)$, p. 224	$B_1((0,0))$	$\sim\sum_{k=1}^{\infty}k^{1-(\alpha+1)s}$	$\frac{2}{1+\alpha}$
$\mathbb{R}^2$	fractal nest of outer type: $\{(r,\varphi): r=1-k^{-\alpha},\, k\in\mathbb{N}\}$ $\alpha>0$, p. 227	$B_1((0,0))$	$\sim\sum_{k=1}^{\infty}k^{(\alpha+1)(1-s)}$	$\frac{2+\alpha}{1+\alpha}$
$\mathbb{R}^2$	geometric (α,β)-chirp: $\bigcup_{k=1}^{\infty}\{k^{-1/\beta}\}\times(0,k^{-\alpha/\beta})$ $0<\alpha<\beta$, p. 229	$(0,1)^2$	$\sim\sum_{k=1}^{\infty}k^{\left(1+\frac{1}{\beta}\right)(1-s)-\frac{\alpha}{\beta}}$	$2-\frac{1+\alpha}{1+\beta}$
$\mathbb{R}^2$	Sierpiński carpet p. 204	unit square ■	$\sim\frac{1}{3^s-8}$	$\log_3 8+\frac{2\pi}{\log 3}\mathrm{i}\mathbb{Z}$
$\mathbb{R}^2$	Sierpiński gasket p. 208	equilateral triangle ▲ of side 1	$\sim\frac{1}{2^s-3}$	$\log_2 3+\frac{2\pi}{\log 2}\mathrm{i}\mathbb{Z}$
$\mathbb{R}^3$	inhomogeneous 3-gasket pp. 294–303	tetrahedron (3-simplex)	$\sim\frac{1}{(s-2)(2^s-4)}$	$2+\frac{2\pi}{\log 2}\mathrm{i}\mathbb{Z}$
$\mathbb{R}^N$	inhomogeneous N-gasket $N\geq 4$, pp. 294–303	N-simplex	$\sim\frac{1}{s-(N-1)}$	$N-1$
$\mathbb{R}^N$	Sierpiński N-carpet $N\geq 1$, pp. 294–303	N-cube $(0,1)^N$	$\sim\frac{1}{3^s-(3^N-1)}$	$\log_3(3^N-1)+\frac{2\pi}{\log 3}\mathrm{i}\mathbb{Z}$
$\mathbb{R}^2$	$\{(0,0)\}$ p. 262	$\{(x,y)\in(0,1)\times\mathbb{R}: 0<y<x^\alpha\}$ $\alpha>1$	$\int_\Omega(\sqrt{x^2+y^2})^{s-2}\mathrm{d}x\,\mathrm{d}y$	$1-\alpha<0$
$\mathbb{R}^2$	$\{(0,0)\}$ p. 265	$\{(x,y)\in(0,1)\times\mathbb{R}: 0<y<\mathrm{e}^{-1/x}\}$	$\int_\Omega(\sqrt{x^2+y^2})^{s-2}\mathrm{d}x\,\mathrm{d}y$	$-\infty$

Table C.1 Table of relative distance zeta functions of some relative fractal drums. All undetermined constants (for example, a, b, δ, etc.) appearing in the table are asumed to be positive.

Acknowledgements

We express our gratitude to Driss Essouabri, Machiel van Frankenhuijsen, Erin Pearse, John Rock, and Steffen Winter for useful discussions during our stay at the "First 2011 International Meeting of PISRS (Permanent International Session of Research Seminars)," Conference on "Analysis, Fractal Geometry, Dynamical Systems, and Economics," held in November 2011 at the University of Messina in Italy. We wish to thank Steffen Winter for having provided us with the two useful references [Had] and [Kne] in connection with the history of the notion of Minkowski content. We are also grateful to the members of the Seminar on "Nonlinear Analysis of Differential Equations and Dynamical Systems", at the University of Zagreb, Croatia, where parts of this monograph have been discussed.

Furthermore, we would like to thank the participants in the first author's seminars, "The Fractal Research Group" Seminar and "The Mathematical Physics and Dynamical Systems" Seminar (including the first author's many past and current Ph.D. students, postdoctoral fellows, mentees and visitors), held at the University of California, Riverside, and where many of these results were presented at various stages of this research program. In particular, the first author would like to thank his current Ph.D. students, Sean Watson and Alexander (Xander) Henderson, for several specific comments on an earlier version of the book.

Parts of the theory developed in this book were presented by the authors at many international conferences and meetings, including most recently by the first author in a research course entitled "Fractal Zeta Functions and Complex Dimensions" and given at the International Summer School and Conference on "Fractal Geometry and Complex Dimensions" held in June 2016 in San Luis Obispo, California, on the occasion of his birthday.

Moreover, the first author would like to thank Michael Barnsley and Martina Zähle for a useful query concerning a family of examples arising in Chapter 4. He would also like to acknowledge a stimulating conversation with Martina Zähle, many years ago, after one of his lectures on the theory of complex dimensions and the associated tube formulas for fractal strings and sprays, as well as on their possible geometric interpretations.

M.L. Lapidus et al., *Fractal Zeta Functions and Fractal Drums*,
Springer Monographs in Mathematics, DOI 10.1007/978-3-319-44706-3

In addition, the three authors of this monograph are very grateful to several anonymous and very conscientious expert referees who have made very helpful and constructive comments on the presentation and the content of two successive versions of this book. In particular, the suggestion to expand the introduction in order to make the book more accessible to nonexperts was most welcome, and we have done our best to follow it, along with a number of other specific and judicious comments. They would also like to thank their editor, Elizabeth Loew, executive director for Mathematics at Springer, for her constant encouragements and helpful suggestions.

The first author (Michel Lapidus) would like to thank the Institut des Hautes Études Scientifiques (IHÉS) in Bures-sur-Yvette (near Paris) for its hospitality in the Spring of 2012 while part of this monograph was being completed, as well as the US National Science Foundation (NSF) for its continued support under the research grants DMS-0707524 and DMS-1107750, along with many earlier NSF grants since the mid-1980s on research eventually leading to the present work. The second and the third authors (Goran Radunović and Darko Žubrinić) are grateful to the Ministry of Science of the Republic of Croatia for its support, as well as to the Croatian Science Foundation for its support of the project IP-2014-09-2285. A part of this monograph was completed during a visit of the third author (Darko Žubrinić) at the Okayama University of Science and at the Research Institute of Mathematical Sciences (RIMS) of Kyoto University, Japan, in the Autumn of 2012.

Finally, we would like to express our gratitude to Professor Francesco Nicolosi from the Univeristy of Catania, Italy, for having organized several international meetings which, among other things, enabled personal contacts between the first and third authors.

Bibliography

[Ba] A. Baker, *Transcendental Number Theory*, Cambridge Univ. Press, Cambridge, 1975.

[BakFraMa] S. Baker, J. M. Fraser and A. Máthé, Inhomogeneous self-similar sets with overlaps, preprint, 2015. (Also: e-print, `arXiv:1509.03589v1 [math.CA]`, 2015.)

[Bar] M. F. Barnsley, *Superfractals*, Cambridge Univ. Press, Cambridge, 2006.

[BarDemk] M. F. Barnsley and S. Demko, Iterated function systems and the global construction of fractals, *Proc. Roy. Soc. London Ser. A* **399** (1985), 243–275.

[Bea] A. F. Beardon, *Iteration of Rational Functions*, Springer-Verlag, Berlin, 1991.

[BedFi] T. Bedford and A. M. Fisher, Analogues of the Lebesgue density theorem for fractal sets of reals and integers, *Proc. London Math. Soc.* (3) **64** (1992), 95–124.

[BedKS] T. Bedford, M. Keane and C. Series (eds.), *Ergodic Theory, Symbolic Dynamics and Hyperbolic Spaces*, Oxford Univ. Press, Oxford, 1991.

[BerGay] C. A. Berenstein and R. Gay, *Complex Variables: An Introduction*, Springer-Verlag, New York, 1991.

[BergGos] M. Berger and B. Gostiaux, *Differential Geometry: Manifolds, Curves and Surfaces*, English translation, Springer-Verlag, Berlin, 1988.

[Berr1] M. V. Berry, Distribution of modes in fractal resonators, in: *Structural Stability in Physics* (W. Güttinger and H. Eikemeier, eds.), Graduate Texts in Mathematics 125, Springer-Verlag, Berlin, 1979, pp. 51–53.

[Berr2] M. V. Berry, Some geometric aspects of wave motion: Wavefront dislocations, diffraction catastrophes, diffractals, in: *Geometry of the Laplace Operator*, Proc. Sympos. Pure Math., vol. 36, Amer. Math. Soc., Providence, R. I., 1980, pp. 13–38.

[BesTay] A. S. Besicovitch and S. J. Taylor, On the complementary intervals of a linear closed set of zero Lebesgue measure, *J. London Math. Soc.* **29** (1954), 449–459.

[BiSo] M. Sh. Birman and M. Z. Solomyak, Spectral asymptotics of nonsmooth elliptic operators, I, *Trans. Moscow Math. Soc.* **27** (1972), 3–52; II, *ibid.* **28** (1973), 3–34.

[Bla] W. Blaschke, *Integralgeometrie*, Chelsea, New York, 1949.

[Boh] H. Bohr, *Almost Periodic Functions*, Chelsea, New York, 1951.

[Bon] M. Bonk, Quasiconformal geometry of fractals, in: *Proc. Internat. Congress Math.* (Madrid, Spain, 2006), vol. 2, European Math. Soc., Zürich, 2007, pp. 1349–1374.

[Bou] G. Bouligand, Ensembles impropres et nombre dimensionnel, *Bull. Sci. Math.* (2) **52** (1928), 320–344 and 361–376.

M.L. Lapidus et al., *Fractal Zeta Functions and Fractal Drums*,
Springer Monographs in Mathematics, DOI 10.1007/978-3-319-44706-3

[Bre] H. Brezis, *Analyse Fonctionnelle: Théorie et Applications*, Masson, Paris, 1983; expanded English version: *Functional Analysis, Sobolev Spaces and Partial Differential Equations*, Springer, New York, 2011.

[BroCar] J. Brossard and R. Carmona, Can one hear the dimension of a fractal?, *Commun. Math. Phys.* **104** (1986), 103–122.

[Cae] A. M. Caetano, On the search for the asymptotic behaviour of the eigenvalues of the Dirichlet Laplacian for bounded irregular domains, *Internat. J. Appl. Sci. Comput.* **2** (1995), 261–287.

[Cah] E. Cahen, Sur la fonction $\zeta(s)$ de Riemann et sur des fonctions analogues, Ph.D. Dissertation, *Ann. Sci. Ec. Norm. Sup.*, (1894), 75–164.

[Carl] B. C. Carlson, *Special Functions of Applied Mathematics*, Academic Press, New York, 1977.

[CarMi] R. D. Carmichael and D. Mitrović, *Distributions and Analytic Functions*, Pitman Research Notes in Mathematics Series, vol. 206, Longman Scientific and Technical, 1989.

[Chee] J. Cheeger, Differentiability of Lipschitz functions on metric measure spaces, *Geom. Functional Anal.* **9** (1999), 428–517.

[CheeMüSchr1] J. Cheeger, W. Müller and R. Schrader, On the curvature of piecewise flat manifolds, *Commun. Math. Phys.* **92** (1984), 405–454.

[CheeMüSchr2] J. Cheeger, W. Müller and R. Schrader, Kinematic and tube formulas for piecewise linear spaces, *Indiana Univ. Math. J.* **35** (1986), 737–754.

[ChrIvLap] E. Christensen, C. Ivan and M. L. Lapidus, Dirac operators and spectral triples for some fractal sets built on curves, *Adv. in Math.* No. 1, **217** (2008), 42–78. (Also: e-print, `arXiv:math.MG/0610222v2`, 2007.)

[CobLap] T. L. Cobler and M. L. Lapidus, Towards a fractal cohomology: Spectra of Polya–Hilbert operators, regularized determinants and Riemann zeros, preprint, 2016. [To appear in *Exploring the Riemann Zeta Function: 190 Years from Riemann's Birth* (H. Montgomery, A. Nikeghbali and M. Rassias, eds.), Springer, Basel, Berlin and New York, 2017.]

[Coh] D. L. Cohn, *Measure Theory*, Birkhäuser, Boston, 1980.

[Con] J. B. Conway, *Functions of One Complex Variable*, Springer-Verlag, New York, 1973.

[CouHil] R. Courant and D. Hilbert, *Methods of Mathematical Physics* I, Interscience Publ. Inc., New York, 1953.

[CranMH] A. Crannell, S. May and L. Hilbert, Shifts of finite type and Fibonacci harps, *Appl. Math. Lett.* No. 2, **20** (2007), 138–141.

[CriJo] T. Crilly (with the assistance of J. Johnson), The emergence of topological dimensions theory, in: *History of Topology* (I. M. James, ed.), Elsevier, Amsterdam, 1999, pp. 1–24.

[Cur] R. L. Curl, Fractal dimensions and geometries of caves, *Mathematical Geology*, No. 8, **18** (1986), 765–783.

[DaMcCS] G. Dafni, R. J. McCann and A. Stancu (eds.), *Analysis and Geometry of Metric Measure Spaces*, Lecture Notes of the 50th Séminaire de Mathématiques Supérieures (SMS), (Montréal, 2011), CRM Proceedings & Lecture Notes, vol. 56, Centre de Recherches Mathématiques (CRM), Montreal and Amer. Math. Soc., Providence, R. I., 2013.

[DavSem] G. David and S. Semmes, *Fractured Fractals and Broken Dreams: Self-Similar Geometry Through Metric and Measure*, Oxford Lecture Series in Mathematics and its Applications, vol. **7**, Oxford Univ. Press, Oxford and New York, 1997.

[DemDenKoÜ] B. Demir, A. Deniz, S. Koçak and A. E. Üreyen, Tube formulas for graph-directed fractals, *Fractals*, No. 3, **18** (2010), 349–361.

[DemKoÖÜ] B. Demir, Ş. Koçak, Y. Özdemir and A. E. Üreyen, Tube formulas for self-similar fractals with non-Steiner-like generators, in: *Proc. Gökova Geometry-Topology Conf.* (2012), Internat. Press, Somerville, MA, 2013, pp. 123–145. (Also: e-print, arXiv:0911.4966 [math.MG], 2009.)

[DenKoÖÜ] A. Deniz, Ş. Koçak, Y. Özdemir and A. E. Üreyen, Tube volumes via functional equations, *J. Geom.* **105** (2014), 1–10.

[deSLapRRo] R. de Santiago, M. L. Lapidus, C. A. Roby and J. A. Rock, Multifractal analysis via scaling zeta functions and recursive structure of lattice strings, in: *Fractal Geometry and Dynamical Systems in Pure and Applied Mathematics I: Fractals in Pure Mathematics* (D. Carfi, M. L. Lapidus, E. P. J. Pearse and M. van Frankenhuijsen, eds.), Contemporary Mathematics, vol. 600, Amer. Math. Soc., Providence, R. I., 2013, pp. 205–238. (dx.doi.org/10.1090/conm/600/11930.) (Also: e-print, arXiv:1027.6680v3 [math-ph], 2013; IHES preprint, IHES/M/12/19, 2013.)

[Dia] H. G. Diamond, On a Tauberian theorem of Wiener and Pitt, *Proc. Amer. Math. Soc.* **31** (1972), 152–158.

[DolFr] J. D. Dollard and C. N. Friedman, *Product Integration, with Application to Differential Equations*, Encyclopedia of Mathematics and Its Applications, vol. 10, Addison-Wesley, Reading, 1979.

[DubSep] E. Dubon and J. M. Sepulcre, On the complex dimensions of nonlattice fractal strings in connection with Dirichlet polynomials, *J. Experimental Math.* No. 1, **23** (2014), 13–24.

[DupMenTri] Y. Dupain, M. Mendès France and C. Tricot, Dimension de spirales, *Bull. Soc. Math. France* **111** (1983), 193–201.

[Ebe] W. Ebeling, *Functions of Several Complex Variables and Their Singularities*, Graduate Studies in Mathematics, vol. 83, Amer. Math. Soc., Providence, R. I., 2007.

[EdmEv] D. E. Edmunds and W. D. Evans, *Spectral Theory of Differential Operators*, Oxford Science Publications, Oxford Mathematical Monographs, Oxford University Press, Oxford, 1987.

[Edw] H. M. Edwards, *Riemann's Zeta Function*, Academic Press, New York, 1974.

[ElLapMacRo] K. E. Ellis, M. L. Lapidus, M. C. Mackenzie and J. A. Rock, Partition zeta functions, multifractal spectra, and tapestries of complex dimensions, in: *Benoît Mandelbrot: A Life in Many Dimensions* (M. Frame and N. Cohen, eds.), The Mandelbrot Memorial Volume, World Scientific, Singapore, 2015, pp. 267–322. (Also: e-print, arXiv:1007.1467v2 [math-ph], 2011; IHES preprint, IHES/M/12/15, 2012.)

[Es1] D. Essouabri, Singularités des séries de Dirichlet associées à des polynômes de plusieurs variables et applications en théorie analytique des nombres, *Ann. Inst. Fourier (Grenoble)*, **47** (1996.), 429–484.

[Es2] D. Essouabri, Zeta functions associated to Pascal's triangle mod p, *Japan J. Math.* (New Series) **31** (2005), 157–174.

[EsLapRRo] D. Essouabri, M. L. Lapidus, S. Roby and J. A. Rock, Analytic continuation of a class of multifractal zeta functions, in preparation, 2016.

[EsLi1] D. Essouabri and B. Lichtin, Zeta functions of discrete self-similar sets, *Adv. in Math.* 232 (2013), 142–187.

[EsLi2] D. Essouabri and B. Lichtin, k-point configurations of discrete self-similar sets, in: *Fractal Geometry and Dynamical Systems in Pure and Applied Mathematics I: Fractals in Pure Mathematics* (D. Carfi, M. L. Lapidus, E. P. J. Pearse and M. van Frankenhuijsen, eds.), Contemporary Mathematics, vol. 600, Amer. Math. Soc., Providence, R. I., 2013, pp. 21–50. (dx.doi.org/10.1090/conm/600/11947.)

[EstKa] R. Estrada and R. P. Kanwal, *A Distributional Approach to Asymptotics: Theory and Applications*, second edition, Birkhäuser, Boston, 2002.

[EvGa] L. C. Evans and R. F. Gariepy, *Measure Theory and Fine Properties of Functions*, CRC Press, Boca Raton, 1999.

[Fal1] K. J. Falconer, *Fractal Geometry: Mathematical Foundations and Applications*, third edition, John Wiley and Sons, Chichester, 2014. (First and second editions: 1990 and 2003.)

[Fal2] K. J. Falconer, On the Minkowski measurability of fractals, *Proc. Amer. Math. Soc.* **123** (1995), 1115–1124.

[FaZe] Y. Fang and Y. Zeng, Minkowski contents on two sets, *Journal of Convergence Information Technology* (*JCIT*) **7** (2012), 435–441.

[Fed1] H. Federer, Curvature measures, *Trans. Amer. Math. Soc.* **93** (1959), 418–491.

[Fed2] H. Federer, *Geometric Measure Theory*, Springer-Verlag, New York, 1969.

[FlVa] J. Fleckinger and D. Vassiliev, An example of a two-term asymptotics for the "counting function" of a fractal drum, *Trans. Amer. Math. Soc.* **337** (1993), 99–116.

[Foll] G. B. Folland, *Real Analysis: Modern Techniques and Their Applications*, second edition, John Wiley and Sons, New York, 1999.

[Fr] M. Frantz, Lacunarity, Minkowski content, and self-similar sets in $\mathbb{R}$, in: *Fractal Geometry and Applications: A Jubilee of Benoît Mandelbrot* (M. L. Lapidus and M. van Frankenhuijsen, eds.), Proc. Sympos. Pure Math., vol. 72, Part 1, Amer. Math. Soc., Providence, R. I., 2004, pp. 77–91.

[Fra1] J. M. Fraser, Inhomogeneous self-similar sets and box dimensions, *Studia Mathematica* No. 2, **213** (2012), 133–155.

[Fra2] J. M. Fraser, Inhomogeneous self-affine carpets, preprint, 2013.; (Also: e-print, `arXiv:1307.5474v2 [math.MG]`, 2013.)

[FreKom] U. Freiberg and S. Kombrink, Minkowski content and local Minkowski content for a class of self-conformal sets, *Geom. Dedicata* **159** (2012), 307–325.

[Fu1] J. H. G. Fu, Tubular neighborhoods in Euclidean spaces, *Duke Math. J.* **52** (1985), 1025–1046.

[Fu2] J. H. G. Fu, Curvature measures of subanalytic sets, *Amer. J. Math.* **116** (1994), 819–880.

[Gat] D. Gatzouras, Lacunarity of self-similar and stochastically self-similar sets, *Trans. Amer. Math. Soc.* **352** (2000), 1953–1983.

[Gau] Gauss, K. F., De nexu inter multitudinem classium, in quas formae binariae secundi gradus distribuntur, earumque determinantem, in *Werke*, **2**, 269–280. Georg Olms Verlag, Hildesheim, New York, 1981.

[Gel] A. O. Gel'fond, *Transcendental and Algebraic Numbers*, Dover Phoenix editions, Dover Publications, New York, 1960.

[Ger] J. Gerling, *Untersuchungen zur Theorie von Weyl–Berry–Lapidus*, Graduate Thesis (Diplomarbeit), Dept. of Physics, Universität Osnabrück, Germany, 1992.

[GerSc] J. Gerling and H.-J. Schmidt, Self-similar drums and generalized Weierstrass functions, *Phys. A.* Nos. 1–4, **191** (1992), 536–539.

[GilTru] D. Gilbarg and N. Trudinger, *Elliptic Partial Differential Equations of Second Order*, second edition, Springer-Verlag, Berlin, 1983.

[Gilk] P. B. Gilkey, *Invariance Theory, the Heat Equation, and the Atiyah–Singer Index Theorem*, second edition, Publish or Perish, Wilmington, 1984. (New revised and enlarged edition in *Studies in Advanced Mathematics*, CRC Press, Boca Raton, 1995.)

[GliJaf] J. Glimm and A. Jaffe, *Quantum Physics: A Functional Integral Point of View*, second edition, Springer-Verlag, Berlin and New York, 1987.

[Gra] A. Gray, *Tubes*, second and revised edition (of the 1990 edn.), Progress in Math., vol. 221, Birkhäuser, Boston, 2004.

[Had] H. Hadwiger, Zur Minkowskischen Dimensions- und Maßbestimmung beschränkter Punktmengen des euklidischen Raumes, *Math. Nachr.* **4** (1951), 202–212.

[HajKosTu1] P. Hajlasz, P. Koskela and H. Tuominen, Sobolev embeddings, extensions and measure density condition, *J. Funct. Anal.* No. 5, **254** (2008), 1217–1234.

[HajKosTu2] P. Hajlasz, P. Koskela and H. Tuominen, Measure density and extendability of functions, *Rev. Mat. Iberomaricana*, No. 2, **24** (2008), 645–669.

[HamLap] B. M. Hambly and M. L. Lapidus, Random fractal strings: their zeta functions, complex dimensions and spectral asymptotics, *Trans. Amer. Math. Soc.* No. 1, **358** (2006), 285–314.

[HardWr] G. H. Hardy and E. M. Wright, *An Introduction to the Theory of Numbers*, sixth edition, Oxford Univ. Press, Oxford, 2008.

[HarPol] R. Harvey and J. Polking, Removable singularities of solutions of linear partial differential equations, *Acta Math.* **125** (1970), 39–56.

[Haz] M. Hazewinkel (ed.), Essential singular point, *Encyclopedia of Mathematics*, European Math. Soc., Zürich, URL: `www.encyclopediaofmath.org`, last updated in 2014. [Adapted from the article by E. D. Solomentsev, *Encyclopedia of Mathematics*, Kluwer, Dordrecht, 2002. ISBN 1402006098; originally published in: I. M. Vinogradov (ed.), *Matematicheskaya entsiklopediya*, Sov. Entsiklopediya, Moscow, 1985, vol. 5, p. 282, in Russian.]

[HeLap] C. Q. He and M. L. Lapidus, *Generalized Minkowski Content, Spectrum of Fractal Drums, Fractal Strings and the Riemann Zeta-Function*, *Memoirs Amer. Math. Soc.* No. 608, **127** (1997), 1–97.

[HedKoreZh] H. Hedenmalm, B. Korenblum and K. Zhu, *Theory of Bergman Spaces*, Graduate Texts in Mathematics, Springer, New York, 2000.

[Hei] J. Heinonen, Nonsmooth calculus, *Bull. Amer. Math. Soc.* **44** (2007), 163–232.

[HerLap1] H. Herichi and M. L. Lapidus, *Quantized Number Theory, Fractal Strings and the Riemann Hypothesis: From Spectral Operators to Phase Transitions and Universality*, research monograph, World Scientific, Singapore, 2017, to appear, approx. 330 pages.

[HerLap2] H. Herichi and M. L. Lapidus, Riemann zeros and phase transitions via the spectral operator on fractal strings, *J. Phys. A: Math. Theor.* **45** (2012) 374005, 23 pp. (Also: e-print, `arXiv:1203.4828v2 [math-ph]`, 2012; IHES preprint, IHES/M/12/09, 2012.)

[HerLap3] H. Herichi and M. L. Lapidus, Fractal complex dimensions, Riemann hypothesis and invertibility of the spectral operator, in: *Fractal Geometry and Dynamical Systems in Pure and Applied Mathematics I: Fractals in Pure Mathematics* (D. Carfì, M. L. Lapidus, E. P. J. Pearse and M. van Frankenhuijsen, eds.), Contemporary Mathematics, vol. 600, Amer. Math. Soc., Providence, R. I., 2013, pp. 51–89. (dx.doi.org/10.1090/conm/600/11948.) (Also: e-print, `arXiv:1210.0882v3 [math.FA]`, 2013; IHES preprint, IHES/M/12/25, 2012.)

[HerLap4] H. Herichi and M. L. Lapidus, Truncated infinitesimal shifts, spectral operators and quantized universality of the Riemann zeta function, *Annales de la Faculté des Sciences de Toulouse*, No. 3, **23** (2014), 621–664. [Special issue dedicated to Christophe Soulé on the occasion of his 60th birthday.] (Also: e-print, `arXiv:1305.3933v1 [math-NT]`, 2013; IHES preprint, IHES/M/13/12, 2013.)

[HerLap5] H. Herichi and M. L. Lapidus, Quantized Riemann zeta functions: Its operator-valued Dirichlet series, Euler product and analytic continuation, in preparation, 2016.

[Hö1] L. Hörmander, The spectral function of an elliptic operator, *Acta Math.* **121** (1968), 193–218.

[Hö2] L. Hörmander, *The Analysis of Linear Partial Differential Operators*, vol. I, *Distribution Theory and Fourier Analysis*, second edition (of the 1983 edn.), Springer-Verlag, Berlin, 1990.

[Hö3] L. Hörmander, *The Analysis of Linear Partial Differential Operators*, vols. II–IV, Springer-Verlag, Berlin, 1983 & 1985.

[HorŽu] L. Horvat and D. Žubrinić, Maximally singular Sobolev functions, *J. Math. Anal. Appl.* No. 2, **304** (2005), 531–541.

[HugLasWeil] D. Hug, G. Last and W. Weil, A local Steiner-type formula for general closed sets and applications, *Math. Zeitschrift* **246** (2004), 237–272.

[Hut] J. Hutchinson, Fractals and self-similarity, *Indiana Univ. J. Math.* **30** (1981), 713–747.

[In] A. E. Ingham, *The Distribution of Prime Numbers*, second edition (reprinted from the 1932 edition), Cambridge Univ. Press, Cambridge, 1992.

[ItzZube] C. Itzykson and J.-B. Zuber, *Quantum Field Theory*, McGraw Hill, New York, 1980.

[IwSb] T. Iwaniec and C. Sbordone, On the integrability of the Jacobian under minimal hypotheses, *Arch. Rational Mech. Anal.* **119** (1992), 129–143.

[JaffMey] S. Jaffard and Y. Meyer, *Wavelet Methods for Pointwise Regularity and Local Oscillations of Functions*, *Memoirs Amer. Math. Soc.* No. 587, **123** (1996), 1–110.

[JohLap] G. W. Johnson and M. L. Lapidus, *The Feynman Integral and Feynman's Operational Calculus*, Oxford Science Publications, Oxford Mathematical Monographs, Oxford Univ. Press, Oxford, 2000. (Corrected reprinting and paperback edition, 2002.)

[JohLapNi] G. W. Johnson, M. L. Lapidus and L. Nielsen, *Feynman's Operational Calculus and Beyond: Noncommutativity and Time-Ordering*, Oxford Science Publications, Oxford Mathematical Monographs, Oxford Univ. Press, Oxford, 2015.

[Jon] P. W. Jones, Quasiconformal mappings and extendability of functions in Sobolev spaces, *Acta. Math.* **147** (1981), 1–61.

[Kac] M. Kac, Can one hear the shape of a drum?, *Amer. Math. Monthly* (Slaught Memorial Papers, No. 11) (4) **73** (1966), 1–23.

[Kat] Y. Katznelson, *An Introduction to Harmonic Analysis*, third edition, Cambridge Mathematical Library, Cambridge Univ. Press, Cambridge, 2009.

[KeKom] M. Kesseböhmer and S. Kombrink, Fractal curvature measures and Minkowski content for self-conformal subsets of the real line, *Adv. in Math.* **230** (2012), 2474–2512.

[Ki1] J. Kigami, *Analysis on Fractals*, Cambridge Univ. Press, Cambridge, 2001.

[Ki2] J. Kigami, Measurable Riemannian geometry on the Sierpinski gasket: The Kusuoka measure and the Gaussian heat kernel estimate, *Math. Ann.* No. 4, **340** (2008), 781–804.

[Ki3] J. Kigami, Volume doubling measures and heat kernel estimates on self-similar sets, *Memoirs Amer. Math. Soc.* No. 932, **199** (2009), 1–94.

[KiLap1] J. Kigami and M. L. Lapidus, Weyl's problem for the spectral distribution of Laplacians on p.c.f. self-similar fractals, *Commun. Math. Phys.* **158** (1993), 93–125.

[KiLap2] J. Kigami and M. L. Lapidus, Self-similarity of volume measures for Laplacians on p.c.f. self-similar fractals, *Commun. Math. Phys.* **217** (2001), 165–180.

[KlRot] D. A. Klain and G.-C. Rota, *Introduction to Geometric Probability*, Accademia Nazionale dei Lincei, Cambridge Univ. Press, Cambridge, 1999.

[Kne] M. Kneser, Einige Bemerkungen über das Minkowskische Flächenmaß, *Arch. Math.* (Basel) **6** (1955), 382–390.

[KoRati] S. Koçak and A. V. Ratiu, Inner tube formulas for polytopes, *Proc. Amer. Math. Soc.* No. 3, **140** (2012), 999–1010. (Also: e-print, `arXiv:1008.2040v1 [math.MG]`, 2010.)

[Kom] S. Kombrink, A survey on Minkowski measurability of self-similar sets and self-conformal fractals in $\mathbb{R}^d$, survey article, in: *Fractal Geometry and Dynamical Systems in Pure and Applied Mathematics I: Fractals in Pure Mathematics* (D. Carfi, M. L. Lapidus, E. P. J. Pearse and M. van Frankenhuijsen, eds.), Contemporary Mathematics, vol. 600, Amer. Math. Soc., Providence, R. I., 2013, pp. 135–159. (dx.doi.org/10.1090/conm/600/11931.)

[KomPeWi] S. Kombrink, E. P. J. Pearse and S. Winter, Lattice-type self-similar sets with pluriphase generators fail to be Minkowski measurable, *Math. Zeitschrift*, No. 3, **283** (2016), 1049–1070. (Also: e-print, `arXiv:1501.03764v1 [math.PR]`, 2015.)

[Kor] J. Korevaar, *Tauberian Theory: A Century of Developments*, Springer-Verlag, Heidelberg, 2004.

[Kow] O. Kowalski, Additive volume invariants of Riemannian manifolds, *Acta Math.* **145** (1980), 205–225.

[KraPa] S. G. Krantz and H. R. Parks, *The Geometry of Domains in Space*, Birkhäuser Advanced Texts, Birkhäuser, Boston, 1999.

[LalLap1] N. Lal and M. L. Lapidus, Hyperfunctions and spectral zeta functions of Laplacians on self-similar fractals, *J. Phys. A*: *Math. Theor.* **45** (2012), 365205, 14 pp. (dx.doi.org/10.1090/conm/601/11959.) (Also: e-print, `arXiv:1202.4126v2 [math-ph]`, 2012; IHES preprint, IHES/M/12/14, 2012.)

[LalLap2] N. Lal and M. L. Lapidus, The decimation method for Laplacians on fractals: Spectra and complex dynamics, in: *Fractal Geometry and Dynamical Systems in Pure and Applied Mathematics II*: *Fractals in Applied Mathematics* (D. Carfi, M. L. Lapidus, E. P. J. Pearse and M. van Flankenhuijsen, eds.), Contemporary Mathematics, vol. **601**, Amer. Math. Soc., Providence, R. I., 2013, pp. 227–249. (dx.doi.org/10.1090/conm/601/11959.) (Also: e-print, `arXiv:1302.4007v2 [math-ph]`, 2014; IHES preprint, IHES/M/12/31, 2012.)

[Lall1] S. P. Lalley, Packing and covering functions of some self-similar fractals, *Indiana Univ. Math. J.* **37** (1988), 699–709.

[Lall2] S. P. Lalley, Renewal theorems in symbolic dynamics, with applications to geodesic flows, noneuclidean tessellations and their fractal limits, *Acta Math.* **163** (1989), 1–55.

[Lall3] S. P. Lalley, Probabilistic counting methods in certain counting problems of ergodic theory, in [BedKS, pp. 223–258].

[Lan] E. Landau, Über einen Satz von Herrn Phragmén, *Acta. Math.* **30** (1905), 195–201.

[Lap1] M. L. Lapidus, Fractal drum, inverse spectral problems for elliptic operators and a partial resolution of the Weyl–Berry conjecture, *Trans. Amer. Math. Soc.* **325** (1991), 465–529.

[Lap2] M. L. Lapidus, Spectral and fractal geometry: From the Weyl–Berry conjecture for the vibrations of fractal drums to the Riemann zeta-function, in: *Differential Equations and Mathematical Physics* (C. Bennewitz, ed.), Proc. Fourth UAB Internat. Conf. (Birmingham, March 1990), Academic Press, New York, 1992, pp. 151–182.

[Lap3] M. L. Lapidus, Vibrations of fractal drums, the Riemann hypothesis, waves in fractal media, and the Weyl–Berry conjecture, in: *Ordinary and Partial Differential Equations* (B. D. Sleeman and R. J. Jarvis, eds.), vol. IV, Proc. Twelfth Internat. Conf. (Dundee, Scotland, UK, June 1992), Pitman Research Notes in Mathematics Series, vol. 289, Longman Scientific and Technical, London, 1993, pp. 126–209.

[Lap4] M. L. Lapidus, Analysis on fractals, Laplacians on self-similar sets, noncommutative geometry and spectral dimensions, *Topological Methods in Nonlinear Analysis* **4** (1994), 137–195. (Special issue dedicated to Jean Leray.)

[Lap5] M. L. Lapidus, Towards a noncommutative fractal geometry? Laplacians and volume measures on fractals, in: *Harmonic Analysis and Nonlinear Differential Equations* (*A Volume in Honor of Victor L. Shapiro*), Contemporary Mathematics, vol. 208, Amer. Math. Soc., Providence, R. I., 1997, pp. 211–252.

[Lap6] M. L. Lapidus, *In Search of the Riemann Zeros*: *Strings, Fractal Membranes and Noncommutative Spacetimes*, research monograph, Amer. Math. Soc., Providence, R. I., 2008.

[Lap7] M. L. Lapidus, Towards quantized number theory: Spectral operators and an asymmetric criterion for the Riemann hypothesis, *Philos. Trans. Royal Soc. Ser. A*, No. 2047, **373** (2015), 24 pp.; DOI:10.1098/rsta.2014.0240. (Special issue titled "*Geometric Concepts in the Foundations of Physics*"), 2015. (Also: e-print, `arXiv:1501.05362v2 [math-ph]`, 2015; IHES preprint, IHES/M/15/12, 2015.)

[Lap8] M. L. Lapidus, The sound of fractal strings and the Riemann hypothesis, in: *Analytic Number Theory: In Honor of Helmut Maier's 60th Birthday* (C. B. Pomerance and T. Rassias, eds.), Springer Internat. Publ. Switzerland, Cham, 2015, pp. 201–252; doi:10.1007/978-3-319-22240-0_14. (Also: e-print, arXiv: `1505.01548v1 [math-ph]`, 2015; IHES preprint, `IHES/M/15/11`, 2015.)

[Lap9] M. L. Lapidus, Quantized Weil conjectures, spectral operators and Polya–Hilbert operators (tentative title), in preparation, 2016.

[Lap10] M. L. Lapidus, Riemann hypothesis, weighted Bergman spaces and quantized Riemann zeta function (tentative title), in preparation, 2016.

[Lap11] M. L. Lapidus, Fractal geometry and applications—An introduction to this volume, in: *Fractal Geometry and Applications: A Jubilee of Benoît Mandelbrot* (M. L. Lapidus and M. van Frankenhuijsen, eds.), Proc. Symposia Pure Math., vol. 72, Parts 1 & 2, Amer. Math. Soc., Providence, R. I., 2008, pp. 1–25 (of Part 1).

[LapLéRo] M. L. Lapidus, J. Lévy-Véhel and J. A. Rock, Fractal strings and multifractal zeta functions, *Lett. Math. Phys.* No. 1, **88** (2009), 101–129 (special issue dedicated to the memory of Moshe Flato). (Springer Open Access: DOI 10.1007/s11005-009-0302-y.) (Also: e-print, `arXiv:math-ph/0610015v3`, 2009.)

[LapLu1] M. L. Lapidus and H. Lu, Nonarchimedean Cantor set and string, *J. Fixed Point Theory and Appl.* No. 2, **3** (2008), 181–190. (Special issue dedicated to the Jubilee of Vladimir I. Arnold, vol. I.)

[LapLu2] M. L. Lapidus and H. Lu, Self-similar p-adic fractal strings and their complex dimensions, *p-adic Numbers, Ultrametric Analysis and Applications* (Springer & Russian Academy of Sciences, Moscow), No. 2, **1** (2009), 167–180. (Also: IHES preprint, IHES/M/08/42, 2008.)

[LapLu3] M. L. Lapidus and H. Lu, The geometry of p-adic fractal strings: A comparative survey, in: *Advances in Non-Archimedean Analysis*, Proc. 11th Internat. Conf. on *"p-Adic Functional Analysis"* (Clermont-Ferrand, France, July 2010), (J. Araujo, B. Diarra and A. Escassut, eds.), Contemporary Mathematics, vol. 551, Amer. Math. Soc., Providence, R. I., 2011, pp. 163–206. (Also: e-print, `arXiv:1105.2966v1 [math.MG]`, 2011.)

[LapLu–vFr1] M. L. Lapidus, H. Lu and M. van Frankenhuijsen, Minkowski measurability and exact fractal tube formulas for p-adic self-similar strings, in: *Fractal Geometry and Dynamical Systems in Pure and Applied Mathematics I: Fractals in Pure Mathematics* (D. Carfi, M. L. Lapidus, E. P. J. Pearse and M. van Frankenhuijsen, eds.), Contemporary Mathematics, vol. 600, Amer. Math. Soc., Providence, R. I., 2013, pp. 161–184. (dx.doi.org/10.1090/conm/600/11949.) (Also: e-print, `arXiv:1209.6440v1 [math.MG]`, 2012; IHES preprint, IHES/M/12/23, 2012.)

[LapLu-vFr2] M. L. Lapidus, H. Lu and M. van Frankenhuijsen, Minkowski dimension and explicit tube formulas for p-adic fractal strings, preprint, 2016. [Invited paper for potential publication in *Reviews in Mathematical Physics*.] (Also: e-print, `arXiv:1603.09409v1 [math-ph]`, 2016.)

[LapMa1] M. L. Lapidus and H. Maier, Hypothèse de Riemann, cordes fractales vibrantes et conjecture de Weyl–Berry modifiée, *C. R. Acad. Sci. Paris Sér. I Math.* **313** (1991), 19–24.

[LapMa2] M. L. Lapidus and H. Maier, The Riemann hypothesis and inverse spectral problems for fractal strings, *J. London Math. Soc.* (2) **52** (1995), 15–34.

[LapNes] M. L. Lapidus and R. Nest, Fractal membranes as the second quantization of fractal strings, in preparation, 2016.

[LapNeuReGr] M. L. Lapidus, J. W. Neuberger, R. J. Renka and C. A. Griffith, Snowflake harmonics and computer graphics: Numerical computation of spectra on fractal domains, *Internat. J. Bifurcation & Chaos* **6** (1996), 1185–1210.

[LapPa] M. L. Lapidus and M. M. H. Pang, Eigenfunctions of the Koch snowflake drum, *Commun. Math. Phys.* **172** (1995), 359–376.

[LapPe1] M. L. Lapidus and E. P. J. Pearse, A tube formula for the Koch snowflake curve, with applications to complex dimensions, *J. London Math. Soc.* No. 2, **74** (2006), 397–414. (Also: e-print, `arXiv:math-ph/0412029v2`, 2005.)

[LapPe2] M. L. Lapidus and E. P. J. Pearse, Tube formulas for self-similar fractals, in: *Analysis on Graphs and its Applications* (P. Exner, *et al.*, eds.), Proc. Sympos. Pure Math., vol. 77, Amer. Math. Soc., Providence, R. I., 2008, pp. 211–230. (Also: e-print, `arXiv:math.DS/0711.0173`, 2007; IHES preprint, IHES/M/08/28, 2008.)

[LapPe3] M. L. Lapidus and E. P. J. Pearse, Tube formulas and complex dimensions of self-similar tilings, *Acta Applicandae Mathematicae* No. 1, **112** (2010), 91–137. (Springer Open Access: DOI 10.1007/S10440-010-9562-x.) (Also: e-print, `arXiv:math.DS/0605527v5`, 2010; IHES preprint, IHES/M/08/27, 2008.)

[LapPeWi1] M. L. Lapidus, E. P. J. Pearse and S. Winter, Pointwise tube formulas for fractal sprays and self-similar tilings with arbitrary generators, *Adv. in Math.* **227** (2011), 1349–1398. (Also: e-print, `arXiv:1006.3807v3` [math.MG], 2011.)

[LapPeWi2] M. L. Lapidus, E. P. J. Pearse and S. Winter, Minkowski measurability results for self-similar tilings and fractals with monophase generators, in: *Fractal Geometry and Dynamical Systems in Pure and Applied Mathematics I: Fractals in Pure Mathematics* (D. Carfì, M. L. Lapidus, E. P. J. Pearse and M. van Frankenhuijsen, eds.), Contemporary Mathematics, vol. 600, Amer. Math. Soc., Providence, R. I., 2013, pp. 185–203. (dx.doi.org/10.1090/conm/600/11951.) (Also: e-print, `arXiv:1104.1641v3 [math.MG]`, 2012; IHES preprint, IHES/M/12/33, 2012.)

[LapPo1] M. L. Lapidus and C. Pomerance, Fonction zêta de Riemann et conjecture de Weyl–Berry pour les tambours fractals, *C. R. Acad. Sci. Paris Sér. I Math.* **310** (1990), 343–348.

[LapPo2] M. L. Lapidus and C. Pomerance, The Riemann zeta-function and the one-dimensional Weyl–Berry conjecture for fractal drums, *Proc. London Math. Soc.* (3) **66** (1993), No. 1, 41–69.

[LapPo3] M. L. Lapidus and C. Pomerance, Counterexamples to the modified Weyl–Berry conjecture on fractal drums, *Math. Proc. Cambridge Philos. Soc.* **119** (1996), 167–178.

[LapRaŽu1] M. L. Lapidus, G. Radunović and D. Žubrinić, Distance and tube zeta functions of fractals and arbitrary compact sets, *Advances in Mathematics*, **307** (2017), 1215–1267. (dx.doi.org/10.1016/j.aim.2016.11.034.) (Also: e-print, `arXiv:1506.03525v3 [math-ph]`, 2016; IHES preprint, IHES/M/15/15, 2015.)

[LapRaŽu2] M. L. Lapidus, G. Radunović and D. Žubrinić, Complex dimensions of fractals and meromorphic extensions of fractal zeta functions, *Journal of Mathematical Analysis and Applications*, in press, 2017, 27 pp. (dx.doi.org/10.1016/j.jmaa.2017.03.059) (Also: e-print, `arXiv:1508.04784v3 [math-ph]`, 2016; IHES preprint, IHES/M/15/15, 2015.)

[LapRaŽu3] M. L. Lapidus, G. Radunović and D. Žubrinić, Zeta functions and complex dimensions of relative fractal drums: Theory, examples and applications, *Dissertationes Mathematicae*, in press, 2017, 101 pp. (Also: e-print, `arXiv:1603.00946v3 [math-ph]`, 2016.)

[LapRaŽu4] M. L. Lapidus, G. Radunović and D. Žubrinić, Fractal tube formulas and a Minkowski measurability criterion for compact subsets of Euclidean spaces, preprint, 2016. [Invited paper for potential publication in a special issue of *Discrete and Continuous Dynamical Systems, Ser. S*, 2017.] (Also: e-print, `arXiv:1411.5733.v4 [math-ph]`, 2016; IHES preprint, IHES/M/15/17, 2015.)

[LapRaŽu5] M. L. Lapidus, G. Radunović and D. Žubrinić, Fractal tube formulas for compact sets and relative fractal drums: Oscillations, complex dimensions and fractality, *Journal of Fractal Geometry*, in press, 2016, 104 pp. (Also: e-print, `arXiv:1604.08014v3 [math-ph]`, 2016.)

[LapRaŽu6] M. L. Lapidus, G. Radunović and D. Žubrinić, Minkowski measurability criteria for compact sets and relative fractal drums in Euclidean spaces, preprint, 2016. (Also: e-print, `arXiv:1609.04498v1 [math-ph]`, 2016.)

[LapRaŽu7] M. L. Lapidus, G. Radunović and D. Žubrinić, Fractal zeta functions and complex dimensions of relative fractal drums, *J. Fixed Point Theory and Appl.* No. 2, **15** (2014), 321–378. Festschrift issue in honor of Haim Brezis' 70th birthday. (DOI: 10.1007/s11784-014-0207-y.) (Also: e-print, `arXiv:1407.8094v3 [math-ph]`, 2014; IHES preprint, `IHES/M/15/14`, 2015.)

[LapRaŽu8] M. L. Lapidus, G. Radunović and D. Žubrinić, Fractal zeta functions and complex dimensions: A general higher-dimensional theory, survey article, in: *Fractal Geometry and Stochastics V* (C. Bandt, K. Falconer and M. Zähle, eds.), Proc. Fifth Internat. Conf. (Tabarz, Germany, March 2014), *Progress in Probability*, vol. 70, Birkhäuser/Springer Internat., Basel, Boston and Berlin, 2015, pp. 229–257; DOI: 10.1007/978-3-319-18660-3_13. (Based on a plenary lecture given by the first author at that conference.) (Also: e-print, `arXiv:1502.00878v3 [math.CV]`, 2015; IHES preprint, IHES/M/15/16, 2015.)

[LapRo1] M. L. Lapidus and J. A. Rock, Towards zeta functions and complex dimensions of multifractals, *Complex Variables and Elliptic Equations* No. 6, **54** (2009), 545–560. (Also: e-print, `arXiv:math-ph/0810.0789`, 2008; IHES preprint, IHES/M/15/16, 2015.)

[LapRo2] M. L. Lapidus and J. A. Rock, *An Invitation to Fractal Geometry: Dimension Theory, Zeta Functions and Applications I & II*, book in preparation, 2016. (Two-volume set for publication in the series *Student Mathematical Library*, Amer. Math. Soc., Providence, R. I.,)

[LapRoŽu] M. L. Lapidus, J. A. Rock and D. Žubrinić, Box-counting fractal strings, zeta functions, and equivalent forms of Minkowski dimension, in: *Fractal Geometry and Dynamical Systems in Pure and Applied Mathematics I: Fractals in Pure Mathematics* (D. Carfi, M. L. Lapidus, E. P. J. Pearse and M. van Frankenhuijsen, eds.), Contemporary Mathematics, vol. 600, Amer. Math. Soc., Providence, R. I., 2013, pp. 239–271. (Also: e-print, `arXiv:1207.6681v2 [math-ph]`, 2013; IHES preprint, IHES/M/12/22, 2012.)

[LapSar] M. L. Lapidus and J. J. Sarhad, Dirac operators and geodesic metric on the harmonic Sierpinski gasket and other fractal sets, *Journal of Noncommutative Geometry* No. 4, **8** (2014), 947–985. (DOI: 10.4171/JNCG/174.) (Also: e-print, `arXiv:1212:0878v3 [math.MG]`, 2015; IHES preprint, IHES/M/12/32, 2012.)

[Lap-vFr1] M. L. Lapidus and M. van Frankenhuijsen, *Fractal Geometry and Number Theory: Complex Dimensions of Fractal Strings and Zeros of Zeta Functions*, Birkhäuser, Boston, 2000.

[Lap-vFr2] M. L. Lapidus and M. van Frankenhuijsen, *Fractal Geometry, Complex Dimensions and Zeta Functions: Geometry and Spectra of Fractal Strings*, Springer Monographs in Mathematics, Springer, New York, 2006.

[Lap-vFr3] M. L. Lapidus and M. van Frankenhuijsen, *Fractal Geometry, Complex Dimensions and Zeta Functions: Geometry and Spectra of Fractal Strings*, second revised and enlarged edition (of the 2006 edn., [Lap-vFr2]), Springer Monographs in Mathematics, Springer, New York, 2013.

[LapWat] M. L. Lapidus and S. Watson, Ahlfors metric measure spaces, fractal zeta functions and complex dimensions (tentative title), in preparation, 2015.

[LéMen] J. Lévy-Véhel and F. Mendivil, Multifractal and higher-dimensional zeta functions, *Nonlinearity* No. 1, **24** (2011), 259–276.

[LioMag] J. L. Lions and E. Magenes, *Non-Homogeneous Boundary Value Problems and Applications*, vol. I, English translation, Springer-Verlag, Berlin, 1972.

[LlWi] M. Llorente and S. Winter, A notion of Euler characteristic for fractals, *Math. Nachr.* Nos. 1–2, **280** (2007), 152–170.

[Man1] B. B. Mandelbrot, *The Fractal Geometry of Nature*, English translation, revised and enlarged edition (of the 1977 edn.), W. H. Freeman, New York, 1983.

[Man2] B. B. Mandelbrot, Measures of fractal lacunarity: Minkowski content and alternatives, in: *Fractal Geometry and Stochastics* (C. Bandt, S. Graf and M. Zähle, eds.), Progress in Probability, vol. 37, Birkhäuser-Verlag, Basel, 1995, pp. 15–42.

[Man3] B. B. Mandelbrot, *Fractals and Chaos: The Mandelbrot Set and Beyond*, Springer, New York, 2004.

[Mat] P. Mattila, *Geometry of Sets and Measures in Euclidean Spaces: Fractals and Rectifiability*, Cambridge Univ. Press, Cambridge, 1995.

[Mattn] L. Mattner, Complex differentiation under the integral, *Nieuw Arch. Wiskd.* (5) No. 1, **2** (2001), 32–35.

[Maz] V. G. Maz'ja, *Sobolev Spaces*, Springer-Verlag, Berlin, 1985.

[Mét1] G. Métivier, Théorie spectrale d'opérateurs elliptiques sur des ouverts irréguliers, Séminaire Goulaic–Schwartz, No. 21, Ecole Polytechnique, Paris, 1973.

[Mét2] G. Métivier, *Etude asymptotique des valeurs propres et de la fonction spectrale de problèmes aux limites*, Thèse de Doctorat d'Etat, Mathématiques, Université de Nice, France, 1976.

[Mét3] G. Métivier, *Valeurs propres de problèmes aux limites elliptiques irréguliers*, *Bull. Soc. Math. France Mém.* **51–52** (1977), 125–219.

[Mil] J. Milnor, Euler characteristic and finitely additive Steiner measure, in: *John Milnor: Collected Papers*, vol. 1, *Geometry*, Publish or Perish, Houston, 1994, pp. 213–234. (Previously unpublished.)

[Mink] H. Minkowski, Theorie der konvexen Körper, insbesondere Begründung ihres Oberflächenbegriffs, in: *Gesammelte Abhandlungen von Hermann Minkowski* (part II, Chapter XXV), Chelsea, New York, 1967, pp. 131–229. (Originally reprinted in: Gesamm. Abh., vol. II, Leipzig, 1911.)

[MitŽu] D. Mitrović and D. Žubrinić, *Fundamentals of Applied Functional Analysis*, Pitman Monographs and Surveys in Pure and Applied Mathematics, Addison-Wesley-Longman, 1998.

[MolVai] S. Molchanov and B. Vainberg, On spectral asymptotics for domains with fractal boundaries, *Commun. Math. Phys.* **183** (1997), 85–117.

[MorSep] G. Mora and J. M. Sepulcre, Privileged regions in critical strips of non-lattice Dirichlet polynomials, *Complex Anal. Oper. Theory* No. 4, **7** (2013), 1417–1426.

[MorSepVi1] G. Mora, J. M. Sepulcre and T. Vidal, On the existence of exponential polynomials with prefixed gaps, *Bull. London Math. Soc.* No. 6, **45** (2013), 1148–1162.

[MorSepVi2] G. Mora, J. M. Sepulcre and T. Vidal, On the existence of fractal strings whose set of dimensions of fractality is not perfect, *Rev. R. Acad. Cienc. Exactas Fis. Nat. Ser. A Math. RACSAM* No. 1, **109** (2015), 11–14.

[Mora] P. A. P. Moran, Additive functions of intervals and Hausdorff measure, *Math. Proc. Cambridge Philos. Soc.* **42** (1946), 15–23.

[NaPaTaŽu] Y. Naito, M. Pašić, S. Tanaka and D. Žubrinić, Fractal oscillations near the domain boundary of radially symmetric solutions of p-Laplace equations, in: *Fractal Geometry and Dynamical Systems in Pure and Applied Mathematics II: Fractals in Applied Mathematics*, (D. Carfì, M. L. Lapidus, E. P. J. Pearse and M. van Frankenhuijsen, eds.), Contemporary Mathematics, vol. 601, Amer. Math. Soc., Providence, R. I., 2013, pp. 325–343. (dx.doi.org/10.1090/conm/601/11912.)

[Ol1] L. Olsen, Multifractal tubes: Multifractal zeta functions, multifractal Steiner tube formulas and explicit formulas, in: *Fractal Geometry and Dynamical Systems in Pure and Applied Mathematics I: Fractals in Pure Mathematics* (D. Carfì, M. L. Lapidus and M. van Frankenhuijsen, eds.), Contemporary Mathematics, vol. 600, Amer. Math. Soc., Providence, R. I., 2013, pp. 291–326. (dx.doi.org/10.1090/conm/600/11920.)

[Ol2] L. Olsen, Multifractal tubes, in: *Further Developments in Fractals and Related Fields*, Trends in Mathematics, Birkhäuser/Springer, New York, 2013, pp. 161–191.

[OlSni] L. Olsen and N. Snigireva, L^q spectra and Rényi dimensions of in-homogeneous self-similar measures, *Nonlinearity* **20**, (2007), 151–175.

[ParrPol1] W. Parry and M. Pollicott, An analogue of the prime number theorem and closed orbits of Axiom A flows, *Annals of Math.* **118** (1983), 573–591.

[ParrPol2] W. Parry and M. Pollicott, *Zeta Functions and the Periodic Orbit Structure of Hyperbolic Dynamics, Astérisque*, vols. 187–188, Soc. Math. France, Paris, 1990.

[ParsSh1] A. N. Parshin and I. R. Shafarevich (eds.), *Number Theory*, vol. I, *Introduction to Number Theory*, Encyclopedia of Mathematical Sciences, vol. 49, Springer-Verlag, Berlin, 1995. (Written by Yu. I. Manin and A. A. Panchishkin.)

[ParsSh2] A. N. Parshin and I. R. Shafarevich (eds.), *Number Theory*, vol. II, *Algebraic Number Fields*, Encyclopedia of Mathematical Sciences, vol. 62, Springer-Verlag, Berlin, 1992. (Written by H. Koch.)

[Pe] E. P. J. Pearse, Canonical self-affine tilings by iterated function systems, *Indiana Univ. Math. J.* No. 6, **56** (2007), 3151–3169. (Also: e-print, `arXiv:math.MG/0606111`, 2006.)

[PeWi] E. P. J. Pearse and S. Winter, Geometry of canonical self-similar tilings, *Rocky Mountain J. Math.* No. 4, **42** (2012), 1327–1357. (Also: e-print, `arXiv:0811.2187`, 2009.)

[PeitJüSa] H.-O. Peitgen, H. Jürgens and D. Saupe, *Chaos and Fractals: New Frontiers of Science*, second edition (of the 1992 edn.), Springer, New York, 2004.

[Per] O. Perron, Zur Theorie der Dirichletschen Reihen, *Crelle* **134** (1908), 95–143.

[Ph] Pham The Lai, Meilleures estimations asymptotiques des restes de la fonction spectrale et des valeurs propres relatifs au laplacien, *Math. Scand.* **48** (1981), 5–38.

[PiStVi] S. Pilipović, B. Stanković and J. Vindas, *Asymptotic Behavior of Generalized Functions*, Series on Analysis, Applications and Computations, vol. 5, World Scientific, Hackensack, NJ, 2012.

[Pit] H. R. Pitt, *Tauberian Theorems*, Oxford Univ. Press, London, 1958.

[PitWie] H. R. Pitt and N. Wiener, A generalization of Ikehara's theorem, *J. Math. and Phys. M.I.T.* **17** (1939), 247–258.

[Pom] Ch. Pommerenke, *Boundary Behavior of Conformal Maps*, Springer, New York, 1992.

[Pos] A. G. Postnikov, *Tauberian Theory and its Applications*, Proc. Steklov Institute of Mathematics, vol. 144, 1979 (English transl., issue 2, 1980), Amer. Math. Soc., Providence, R. I., 1980.

[Ra1] G. Radunović, *Fractal Analysis of Unbounded Sets in Euclidean Spaces and Lapidus Zeta Functions*, Ph. D. Thesis, University of Zagreb, Croatia, 2015.

[Ra2] G. Radunović, Fractality and Lapidus zeta functions at infinity, *Mathematical Communications* **21** (2016), 141–162. (Also: e-print, `arXiv:1510.06449v2 [math-ph]`, 2015.)

[RaŽuŽup] G. Radunović, D. Žubrinić and V. Županović, Fractal analysis of Hopf bifurcation at infinity, *Internat. J. Bifur. Chaos Appl. Sci. Engrg.* **22** (2012), 1230043-1-1230043-15.

[RatWi1] J. Rataj and S. Winter, On volume and surface area of parallel sets, *Indiana Univ. Math. J.* **59** (2010), 1661–1685.

[RatWi2] J. Rataj and S. Winter, Characterization of Minkowski measurability in terms of surface area, *J. Math. Anal. Appl.* **400** (2013), 120–132. (Also: e-print, `arXiv:1111.1825v2 [math.CA]`, 2012.)

[ReeSim1] M. Reed and B. Simon, *Methods of Modern Mathematical Physics*, vol. I, *Analysis of Operators*, Academic Press, New York, 1980.

[ReeSim2] M. Reed and B. Simon, *Methods of Modern Mathematical Physics*, vol. IV, *Analysis of Operators*, Academic Press, New York, 1979.

[Res] M. Resman, Invariance of the normalized Minkowski content with respect to the ambient space, *Chaos, Solitons & Fractals* **57** (2013), 123–128. (Also: e-print, `arXiv:1207.3279v1`, 2012.)

[RoSha] J. C. Robinson and N. Sharples, Strict inequality in the box-counting dimension product formulas, *Real Anal. Exchange*, **38** (2012), 95–120.

[Ru] W. Rudin, *Real and Complex Analysis*, third edition, McGraw-Hill, New York, 1987.

[Rue1] D. Ruelle, Generalized zeta-functions for Axiom A basic sets, *Bull. Amer. Math. Soc.* **82** (1976), 153–156.

[Rue2] D. Ruelle, Zeta functions for expanding maps and Anosov flows, *Invent. Math.* **34** (1978), 231–242.

[Rue3] D. Ruelle, *Thermodynamic Formalism*, Addison-Wesley, Reading, 1978.

[Rue4] D. Ruelle, *Dynamical Zeta Functions for Piecewise Monotone Maps of the Interval*, CRM Monographs Ser. (Centre de Recherches Mathématiques, Université de Montréal), vol. 4, Amer. Math. Soc., Providence, R. I., 1994.

[SapGoMar] B. Sapoval, Th. Gobron and A. Margolina, Vibrations of fractal drums, *Phys. Rev. Lett.* **67** (1991), 2974–2977.

[Schn1] R. Schneider, Curvature measures of convex bodies, *Ann. Mat. Pura Appl.* IV, **116** (1978), 101–134.

[Schn2] R. Schneider, *Convex Bodies: The Brunn–Minkowski Theory*, Encyclopedia of Mathematics and its Applications, vol. 44, Cambridge Univ. Press, Cambridge, 2003. (Reprinted from the 1993 edition.)

[Schw] L. Schwartz, *Théorie des Distributions*, revised and enlarged edition (of the 1951 edn.), Hermann, Paris, 1966.

[See1] R. T. Seeley, Complex powers of elliptic operators, in: *Proc. Sympos. Pure Math.*, vol. 10, Amer. Math. Soc., Providence, R. I., 1967, pp. 288–307.

[See2] R. T. Seeley, A sharp asymptotic remainder estimate for the eigenvalues of the Laplacian in a domain of $\mathbb{R}^3$, *Adv. in Math.* **29** (1978), 244–269.

[See3] R. T. Seeley, An estimate near the boundary for the spectral counting function of the Laplace operator, *Amer. J. Math.* **102** (1980), 869–902.

[Sen] M. Senechal, *Quasicrystals and Geometry*, Cambridge Univ. Press, Cambridge, 1995.

[Ser] J.-P. Serre, *A Course in Arithmetic*, English translation, Springer-Verlag, Berlin, 1973.

[Shi] M. Shishikura, The Hausdorff dimension of the boundary of the Mandelbrot set and Julia sets, *Ann. of Math.* **147** (1998), 225–267.

[Sim] B. Simon, *Functional Integration and Quantum Physics*, Academic Press, New York, 1979.

[Sma] S. Smale, Differentiable dynamical systems, *Bull. Amer. Math. Soc.* **73** (1967), 747–817.

[SriTod] H. M. Srivastava and P. G. Todorov, An explicit formula for the generalized Bernoulli polynomials, *J. Math. Anal. Appl.* **130** (1988), 509–513.

[Sta] L. L. Stachó, On the volume function of parallel sets, *Acta Sci. Math.* **38** (1976), 365–374.

[Stein] J. Steiner, Über parallele Flächen, *Monatsb. preuss. Akad. Wiss.*, Berlin, 1840, pp. 114–118. (Reprinted in: *Gesamm. Werke*, vol. II, pp. 173–176.)

[Steinh] B. Steinhurst, *Diffusions and Laplacians on Laakso, Barlow–Evans, and Other Fractals*, Ph.D. Dissertation, University of Connecticut, Storrs, USA, June 2010.

[TanL] Tan Lei (ed.), *The Mandelbrot Set: Theme and Variations*, London Mathematical Society Lecture Notes Series, No. 274, Cambridge Univ. Press, Cambridge, 2000.

[Tem] N. M. Temme, *Special Functions: An Introduction to the Classical Functions of Mathematical Physics*, John Wiley and Sons, New York, 1996.

[Tep1] A. Teplyaev, Spectral zeta functions of symmetric fractals, in: *Fractal Geometry and Stochastics III*, Progress in Probability, vol. 57, Birkhäuser-Verlag, Basel, 2004, pp. 245–262.

[Tep2] A. Teplyaev, Spectral zeta functions of fractals and the complex dynamics of polynomials, *Trans. Amer. Math. Soc.* **359** (2007), 4339–4358. (Also: e-print, `arXiv:math.SP/0505546`, 2005.)

[Tit1] E. C. Titchmarsh, *The Theory of Functions*, second edition, Oxford University Press, Oxford, 1939.

[Tit2] E. C. Titchmarsh, *Introduction to the Theory of Fourier Integrals*, second edition, Oxford University Press, Oxford, 1948.

[Tit3] E. C. Titchmarsh, *The Theory of the Riemann Zeta-Function*, second edition (revised by D. R. Heath-Brown), Oxford Univ. Press, Oxford, 1986.

[Tri1] C. Tricot, *Mesures et Dimensions*, Thèse de Doctorat d'Etat Es Sciences Mathématiques, Université Paris-Sud, Orsay, France, 1983.

[Tri2] C. Tricot, Dimensions aux bords d'un ouvert, *Ann. Sci. Math. Québec* **11** (1987), 205–235.

[Tri3] C. Tricot, *Curves and Fractal Dimension*, Springer-Verlag, Berlin, 1995.

[Tri4] C. Tricot, General Hausdorff functions, and the notion of one-sided measure and dimension, *Ark. Mat.* **48** (2010), 149–176.

[vBGilk] M. van den Berg and P. B. Gilkey, A comparison estimate for the heat equation with an application to the heat content of the s-adic von Koch snowflake, *Bull. London Math. Soc.* No. 4, **30** (1998), 404–412.

[Vin] I. M. Vinogradov (ed.), *Matematicheskaya entsiklopediya*, Sov. Entsiklopediya, Moscow, 1979, vol. 2, pp. 818–819. (Also: online, English translation, Quasi-periodic function. Yu. V. Komlenko and E. L. Tonkov (authors), *Encyclopedia of Mathematics*, public wiki monitored by an editorial board under the management of the European Mathematical Society, URL: `www.encyclopediaofmath.org`.)

[Wat] S. Watson, *Fractal Zeta Functions: Ahlfors Spaces and Beyond* (tentative title), Ph.D. Dissertation, University of California, Riverside, U.S.A., in preparation, June 2017.

[WaMLItz] M. Waldschmidt, P. Moussa, J.-M. Luck and C. Itzykson, *From Number Theory to Physics*, Springer-Verlag, Berlin, 1992.

[Wein1] S. Weinberg, *Quantum Theory of Fields*, vol. I, *Foundations*, Cambridge University Press, Cambridge, 1995.

[Wein2] S. Weinberg, *Quantum Theory of Fields*, vol. II, *Modern Applications*, Cambridge University Press, Cambridge, 1996.

[Weyl1] H. Weyl, Über die Abhängigkeit der Eigenschwingungen einer Membran von deren Begrenzung, *J. Reine Angew. Math.* **141** (1912), 1–11. (Reprinted in [Wey4, vol. I, pp. 431–441].)

[Wey2] H. Weyl, Das asymptotische Verteilungsgesetz der Eigenwerte linearer partieller Differentialgleichungen, *Math. Ann.* **71** (1912), 441–479. (Reprinted in [Wey4, vol. I, pp. 393–430].)

[Wey3] H. Weyl, On the volume of tubes, *Amer. J. Math.* **61** (1939), 461–472. (Reprinted in [Wey4, vol. III, pp. 658–669].)

[Wey4] H. Weyl, *Hermann Weyl: Gesammelte Abhandlungen* (Collected Works), vols. I and II, Springer-Verlag, Berlin and New York, 1968.

[Wi] S. Winter, Curvature measures and fractals, *Dissertationes Math.* (*Rozprawy Mat.*) **453** (2008), 1–66.

[WiZä] S. Winter and M. Zähle, Fractal curvature measures of self-similar sets, *Adv. Geom.* **13** (2013), 229–244.

[Zä1] M. Zähle, Integral and current representation of Federer's curvature measures, *Arch. Math.* **46** (1986), 557–567.

[Zä2] M. Zähle, Curvatures and currents for unions of sets with positive reach, *Geom. Dedicata* **23** (1987), 155–171.

[Zä3] M. Zähle, Approximation and characterization of generalized Lipschitz–Killing curvatures, *Ann. Global Anal. Geom.* No. 3, **8** (1990), 249–260.

[Zä4] M. Zähle, Lipschitz–Killing curvatures of self-similar random fractals, *Trans. Amer. Math. Soc.* No. 5, **363** (2011), 2663–2684. (DOI: 10.1090/S0002-9947-2010-05198-0.)

[Zä5] M. Zähle, Curvature measures of fractal sets, survey article, in: *Fractal Geometry and Dynamical Systems in Pure and Applied Mathematics I: Fractals in Pure Mathematics* (D. Carfì, M. L. Lapidus and M. van Frankenhuijsen, eds.), Contemporary Mathematics, vol. 600, Amer. Math. Soc., Providence, R. I., 2013, pp. 381–399. (dx.doi.org/10.1090/conm/600/11953.)

[Žu1] D. Žubrinić, Singular sets of Sobolev functions, *C. R. Acad. Sci. Paris Sér. I. Math., Analyse mathématique* **334** (2002), 539–544.

[Žu2] D. Žubrinić, Minkowski content and singular integrals, *Chaos, Solitons & Fractals* No. 1, **17** (2003), 169–177.

[Žu3] D. Žubrinić, Singular sets of Lebesgue integrable functions, *Chaos, Solitons & Fractals* **21** (2004), 1281–1287.

[Žu4] D. Žubrinić, Analysis of Minkowski contents of fractal sets and applications, *Real Anal. Exchange* No. 2, **31** (2005/2006), 315–354.

[Žu5] D. Žubrinić, Hausdorff dimension of singular sets of Sobolev functions and applications, in: *More Progress in Analysis*, Proc. of the 5th Internat. ISAAC Congress (H. G. W. Begher and F. Nicolosi, eds.), World Scientific, Singapore, 2009, pp. 793–802.

[ŽuŽup1] D. Žubrinić and V. Županović, Fractal analysis of spiral trajectories of some planar vector fields, *Bulletin des Sciences Mathématiques* No. 6, **129** (2005), 457–485.

[ŽuŽup2] D. Žubrinić and V. Županović, Box dimension of spiral trajectories of some vector fields in $\mathbb{R}^3$, *Qualitative Theory of Dynamical Systems* No. 2, **6** (2005), 251–272.

[ŽupŽu] V. Županović and D. Žubrinić, Fractal dimensions in dynamics, in: *Encyclopedia of Mathematical Physics* (J.-P. Françoise, G. L. Naber and S. T. Tsou, eds.), vol. 2, Elsevier, Oxford, 2006, pp. 394–402.

Author Index

M.L. Lapidus et al., *Fractal Zeta Functions and Fractal Drums*,
Springer Monographs in Mathematics, DOI 10.1007/978-3-319-44706-3

Subject Index

M.L. Lapidus et al., *Fractal Zeta Functions and Fractal Drums*,
Springer Monographs in Mathematics, DOI 10.1007/978-3-319-44706-3

G

MIX
Papier aus verantwortungsvollen Quellen
Paper from responsible sources
FSC® C105338

If you have any concerns about our products,
you can contact us on
ProductSafety@springernature.com

In case Publisher is established outside the EU,
the EU authorized representative is:
Springer Nature Customer Service Center GmbH
Europaplatz 3, 69115 Heidelberg, Germany

Printed by Libri Plureos GmbH
in Hamburg, Germany